OLEFIN METATHESIS

OLEFIN METATHESIS

Theory and Practice

Edited by

KAROL GRELA
Faculty of Chemistry, University of Warsaw, Warsaw, Poland

WILEY

Published by John Wiley & Sons, Inc., Hoboken, New Jersey.
Published simultaneously in Canada.

Library of Congress Cataloging-in-Publication Data:

Olefin metathesis : theory and practice / edited by Dr. Karol Grela, Warsaw University, Warsaw, Poland.
pages cm
"Zespol III."
Includes bibliographical references and index.
ISBN 978-1-118-20794-9 (cloth)
1. Metathesis (Chemistry) 2. Catalysis. 3. Alkenes. I. Grela, Karol, 1970- editor of compilation.
QD505.O44 2014
547′.215–dc23

2013041990

10 9 8 7 6 5 4 3 2 1

CONTENTS

PREFACE

Our goal is to create a comprehensive book that can be an everyday reference for synthetic chemists, with no prerequisite deep knowledge in inorganic and coordination chemistry, but at the same time providing the experts with a contemporary view on the theory and methods in the flourishing field of modern olefin metathesis.

The book comprises two major parts: the first one is devoted to the applications of metathesis (*Targets*), while the second one describes the metathesis *Tools*.

The most important types of the olefin metathesis reaction (cross metathesis (CM), ring-closing metathesis (RCM), enyne metathesis, ring-opening metathesis polymerization (ROMP), tandem processes, etc.) have been described in the first part of the book. This part also presents a short selection of the key applications of this methodology, for example, in the total synthesis of natural compounds, in the preparation of macromolecules and polymers, in medicinal chemistry, and in the conversion of renewable materials. The goal behind this part of the book is to present a detailed, yet clear description of all important flavors of the metathesis reaction.

The second part of the book describes the tools. A series of chapters introduce the most important classes of metal complexes that are active in metathesis, creating *the user's guide* to the galaxy of known olefin metathesis catalysts. The same attention is paid to optimization of the reaction conditions, including discussion on effects of the solvent and additives, methods of catalysts immobilization and recovery, purification of the products, computational methods, and so on. This part of the book is used exactly similarly to the famous "*The Hitch-Hiker's Guide to the Galaxy*," a fictional travel guide, invented by the writer Douglas Adams. With "the words *Don't Panic* inscribed in large friendly letters on its cover"(1), the second part of the book allows even the inexperienced end-user to select the most optimal catalyst and conditions for his or her important metathesis project easily and effectively.

I am delighted with the list of authors who have agreed to contribute, and I am honored to act as the editor. All chapters collected in this book come from the leading experts and practitioners in the area and nicely highlight the aspects mentioned above. I would like to thank all the authors for their excellent contributions. My personal wish is that the reader will savor the reading of this book as much as I personally enjoyed reading all chapters and editing the volume.

I want to give a special thanks to Polish artist, Ms. Katarzyna Felchnerowska (Effe.Fineart) who prepared the beautiful cover picture, presenting her personal idea of olefin metathesis—a change-your-partners dance (2). Last, but certainly not least, I want to thank the members of my research group who have reviewed the text at every stage of the editing process.

Altogether, I hope that this handy, one-volume book will take its common place on the desks and benches of researchers working in academic laboratories as well as in the industry.

Enjoy reading!

Karol Grela

REFERENCE

1. Adams D. *The Hitchhiker's Guide to the Galaxy*, 25th Anniversary Edition. Crown Publishing Group; New York, 2004. ISBN: 978-1400052929.
2. http://www.nobelprize.org/nobel_prizes/chemistry/laureates/2005/press.html.

CONTRIBUTORS

Didier Astruc, ISM, UMR CNRS N°5255, Univ. Bordeaux, Talence Cedex, France

Michał Barbasiewicz, Faculty of Chemistry, Warsaw University, Warsaw, Poland

Olivier Baslé, Ecole Nationale Supérieure de Chimie de Rennes, Rennes Cedex, France

Christian Bruneau, UMR 6226 Institut des Sciences Chimiques de Rennes, Organométalliques: Matériaux et Catalyse, Université de Rennes 1, Rennes Cedex, France

Michael R. Buchmeiser, Lehrstuhl für Makromolekulare Stoffe und Faserchemie, Institut für Polymerchemie, Universität Stuttgart, Stuttgart, Germany; Institut für Textilchemie und Chemiefasern (ITCF) Denkendorf, Denkendorf, Germany

Luigi Cavallo, Dipartimento di Chimica, Università di Salerno, Fisciano, SA, Italy; KAUST Catalyst Research Center, Physical Sciences and Engineering Division, King Abdullah University of Science and Technology, Thuwal, Kingdom of Saudi Arabia

Catherine S. J. Cazin, School of Chemistry, University of St. Andrews, St Andrews, UK

Shawn K. Collins, Department of Chemistry, Université de Montréal, Montréal, PQ, Canada

Janine Cossy, Laboratorie de Chimie Organique ESPCI ParisTech, Paris Cedex, France

Isabelle Dez, Ecole Nationale Supérieure de Chimie de Rennes, Rennes Cedex, France

Steven T. Diver, Department of Chemistry, University at Buffalo-SUNY Buffalo, NY

Laura Falivene, Dipartimento di Chimica, Università di Salerno, Fisciano, SA, Italy

Keith R. Fandrick, Chemical Development, Boehringer Ingelheim Pharmaceuticals Inc., Ridgefield, CT

Tobias Fiedler, Department of Chemistry, Texas A&M University, College Station, TX

Cédric Fischmeister, UMR 6226 Institut des Sciences Chimiques de Rennes, Organométalliques: Matériaux et Catalyse, Université de Rennes 1, Rennes Cedex, France

Deryn E. Fogg, Centre for Catalysis Research & Innovation, Department of Chemistry, University of Ottawa, Ottawa, Ontario, Canada

Annie-Claude Gaumont, Ecole Nationale Supérieure de Chimie de Rennes, Rennes Cedex, France

Régis M. Gauvin, Unité de Catalyse et de Chimie du Solide (UMR CNRS 8181), Axe "Catalyse et Chimie Moléculaire", Villeneuve d'Ascq Cedex, France

Subir Ghorai, Catalysis R&D, Sigma-Aldrich Chemical Co., Sheboygan Falls, WI

Yakov Ginzburg, Ben-Gurion University, Israel

John A. Gladysz, Department of Chemistry, Texas A&M University, College Station, TX

Karol Grela, Biological and Chemical Research Center, Department of Chemistry, Warsaw University, Warsaw, Poland; Institute of Organic Chemistry, Polish Academy of Sciences, Warsaw, Poland

Justin R. Griffiths, Department of Chemistry, University at Buffalo-SUNY Buffalo, NY

Stefano Guidone, School of Chemistry, University of St. Andrews, St Andrews, UK

Łukasz Gułajski, Apeiron Synthesis S.A., Wrocław, Poland

Anna Kajetanowicz, Institute of Organic Chemistry, Polish Academy of Sciences, Warsaw, Poland

Astrid-Caroline Knall, Institute for Chemistry and Technology of Materials, Graz University of Technology, Graz, Austria

Stefan Krehl, Institut für Chemie, Organische Synthesechemie, Universität Potsdam, Golm, Germany

N. Gabriel Lemcoff, Ben-Gurion University, Israel

Bianca J. van Lierop, Centre for Catalysis Research & Innovation, Department of Chemistry, University of Ottawa, Ottawa, Ontario, Canada

Bruce H. Lipshutz, Department of Chemistry, University of California, Santa Barbara, CA

Justin A. M. Lummiss, Centre for Catalysis Research & Innovation, Department of Chemistry, University of Ottawa, Ottawa, Ontario, Canada

Marc Mauduit, Ecole Nationale Supérieure de Chimie de Rennes, Rennes Cedex, France

Yohann Morin, Unité de Catalyse et de Chimie du Solide (UMR CNRS 8181), Axe "Catalyse et Chimie Moléculaire", Villeneuve d'Ascq Cedex, France

Andrew Nickel, Materia Inc., Pasadena, CA

Steven P. Nolan, School of Chemistry, University of St. Andrews, St Andrews, UK

Richard L. Pederson, Materia Inc., Pasadena, CA

Cezary Pietraszuk, Faculty of Chemistry, Adam Mickiewicz University in Poznań, Poznań, Poland

Albert Poater, Departament de Química, Institut de Química Computacional, University of Girona, Girona, Catalonia, Spain; Dipartimento di Chimica, Università di Salerno, Fisciano, SA, Italy

Pierre Queval, Ecole Nationale Supérieure de Chimie de Rennes, Rennes Cedex, France

Mathieu Rouen, Ecole Nationale Supérieure de Chimie de Rennes, Rennes Cedex, France

Cezary Samojłowicz, Institute of Organic Chemistry, Polish Academy of Sciences, Warsaw, Poland

Jolaine Savoie, Chemical Development, Boehringer Ingelheim Pharmaceuticals Inc., Ridgefield, CT

Bernd Schmidt, Institut für Chemie, Organische Synthesechemie Universität Potsdam, Golm, Germany

Chris H. Senanayake, Chemical Development, Boehringer Ingelheim Pharmaceuticals Inc., Ridgefield, CT

Krzysztof Skowerski, Apeiron Synthesis S.A., Wrocław, Poland

Christian Slugovc, Institute for Chemistry and Technology of Materials, Graz University of Technology, Graz, Austria

Jinhua J. Song, Chemical Development, Boehringer Ingelheim Pharmaceuticals Inc., Ridgefield, CT

Brice Stenne, Department of Chemistry, Université de Montréal, Montréal, PQ, Canada

César A. Urbina-Blanco, School of Chemistry, University of St. Andrews, St Andrews, UK

Georgios C. Vougioukalakis, Laboratory of Organic Chemistry, Department of Chemistry, University of Athens, Athens, Greece; Division of Physical Chemistry, IAMPPNM, NCSR Demokritos, Athens, Greece

Łukasz Woźniak, Institute of Organic Chemistry, Polish Academy of Sciences, Warsaw, Poland

Nathan Yee, Chemical Development, Boehringer Ingelheim Pharmaceuticals Inc., Ridgefield, CT

Grzegorz Zieliński, Institute of Organic Chemistry, Polish Academy of Sciences, Warsaw, Poland

Karolina Żukowska, Institute of Organic Chemistry, Polish Academy of Sciences; Biological and Chemical Research Center, Department of Chemistry, Warsaw University, Warsaw, Poland

PART I

APPLICATIONS OF OLEFIN METATHESIS REACTIONS

SECTION 1

INTRODUCTION

1

OLEFIN METATHESIS REACTIONS: FROM A HISTORICAL ACCOUNT TO RECENT TRENDS

Didier Astruc

ISM, UMR CNRS N°5255, Univ. Bordeaux, Talence Cedex, France

1.1 INTRODUCTION

Metathesis (1–5) occupies a central place in the synthesis of complex organic molecules and polymers, and the major problems concerning the catalysts have been solved, including the presence of various functional groups in the organic substrates. Unlike palladium catalysis of cross carbon–carbon bond formation that is the other breakthrough in the formation of organic skeleton architectures, the metathesis reactions do not consume stoichiometric amounts of base, producing stoichiometric amounts of salts as wastes. In this sense, the metathesis reactions belong to the field of green chemistry, saving a number of steps in total syntheses and avoiding the handling and production of inorganic wastes. The discovery of homogeneous catalysis by Osborn and Wilkinson (6), allowing the hydrogenation of olefins, and its efficient asymmetric version by Kagan (7) considerably enhanced the possibilities of bringing about high selectivity and approaching a perfect knowledge of the molecular mechanisms leading to improvements and optimization. Thus, after the pioneering research of American industrial chemists in the 1960s, a significant advance in metathesis chemistry has been the mechanistic insight of Yves Chauvin in the early 1970s in disentangling the "black box" and the intense academic research by organometallic chemists at the end of the twentieth century (8), in particular by the groups of Robert Grubbs at Caltech and Richard Schrock at MIT in their quest for transition-metal-alkylidene (or benzylidene) and alkylidyne metathesis catalysts. Therefore, in this historical chapter, we wish to underline the role of the development of ideas and research efforts that have led to a success story in the advancement of metathesis chemistry and its applications. This chapter also includes examples of the most recent and major developments and applications of the various metathesis reactions, with emphasis on catalyst design and sustainable chemistry.

1.2 THE METATHESIS REACTIONS: EXCHANGES OF CARBENES AND CARBYNES

The word metathesis comes from the Greek $\mu\epsilon\tau\alpha\theta\epsilon\sigma\iota\zeta$ that means transposition. Metathesis of ions between two ion pairs is a long-known example of metathesis in which two ion pairs exchange their partners (Eq. 1.1) (8).

$$A^+\,B^- + C^+\,D^- \rightleftharpoons A^+\,D^- + C^+\,B^- \qquad (1.1)$$

Likewise, the exchange of the two carbenes of an olefin with those of another olefin molecule (Eq. 1.2) was first called *olefin metathesis* by Calderon (9) in 1967, but this reaction requires a catalyst.

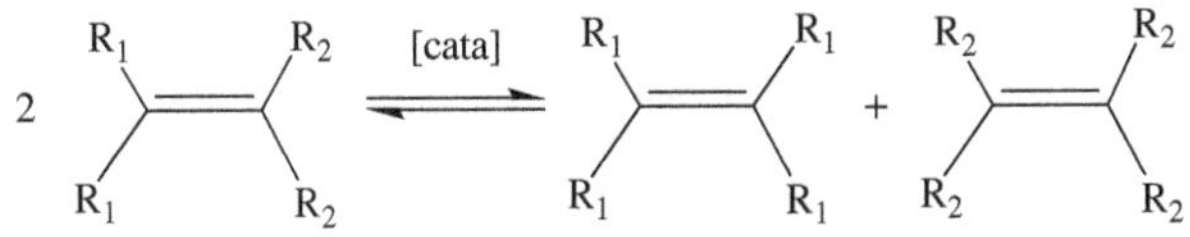

(1.2)

The principle is extended to the exchange of the two carbynes of alkyne molecules

$$2\ R^1C{\equiv}CR^2 \xrightleftharpoons{\text{cata}} R^1C{\equiv}CR^1 + R_2C{\equiv}CR^2 \qquad (1.3)$$

Olefin Metathesis: Theory and Practice, First Edition. Edited by Karol Grela.

The metathesis reactions are under thermodynamic control, which means that the reactions of Eqs 1.2 and 1.3 present the inconvenient of being equilibrated. The problem is usually solved by using terminal olefins that produce gaseous ethylene as one of the metathesis products, which displaces the reaction toward the metathesis products. For alkyne metathesis, terminal alkyne metathesis is possible (see Section 1.15) but of little use due to competitive alkyne polymerization, so methyl-terminated alkynes are used for metathesis, which produces 2-butyne that is also removed according to the same principle.

1.3 THE EARLY DAYS OF OLEFIN METATHESIS IN AMERICAN INDUSTRY

The uncatalyzed reaction of propene upon heating at 852 °C had been reported in 1931 by Schneider and Fröhlich (10) to give very low amounts of ethene and 2-butenes among other products, but the publication remained ignored for a long time. Ziegler and Natta's discoveries of ethylene and propylene polymerization in 1953 induced considerable research interest in olefin polymerization reactions and their mechanisms. The first catalyzed metathesis reactions were reported in the late 1950s when industrial chemists at Du Pont, Standard Oil and Phillips Petroleum (H. S. Eleuterio, E. F. Peters, B. L. Evering, R. L. Banks, and G. C. Bailey) reported that propene led to ethylene and 2-butenes when it was heated with molybdenum [in the form of the metal, oxide, or [$Mo(CO)_6$] on alumina (Fig. 1.1)] (11–16).

The first polymerization of norbornene by the system $WCl_6/AlEt_2Cl$ was independently reported in 1960 by Eleuterio (11,15) and by Truett et al. (13), but it was recognized only in 1967 by Calderon (9,17) at Goodyear that the polymerization of cyclic alkenes to polyalkenemers and the disproportionation of acyclic alkenes were the same type of reaction, the metathesis. The following year, labeling experiments by Calderon (9,17) at Goodyear and Mol (18) in Amsterdam confirmed this key finding.

1.4 UNSUCCESSFUL ATTEMPTS TO SOLVE THE MECHANISTIC PUZZLE

The first mechanistic suggestion to solve the mechanistic puzzle came in 1967 from Bradshaw et al. (19) who proposed a four-centered cyclobutane–metal intermediate. This hypothesis was supported the following year by Calderon (20). No other hypothesis appeared in the United States for three years, and this mechanism seemed to be adopted as being "conventional" by the metathesis community in the United States. Yet, cyclobutanes are not produced by metathesis, and they are not metathesis substrates either. Other mechanistic hypotheses by American chemists appeared only in the early 1970s. In 1971, the brilliant organometallic chemist Pettit (21,22), who had generated the first transition-metal

History of the olefin metathesis reactions

Ethylene and 2-butenes are found in small amounts upon heating propene at 852 °C: Schneider and Fröhlich (1931)

Heterogeneous metathesis catalysts: $M(CO)_6$ (M = Mo or W) on alumina or silica, or Re_2O_7 (150–500 °C): Eleuterio (1956) and Bank (1964)

Chauvin mechanism: metal–alkylidene + olefin and metallacyclobutane intermediate, 1971

First stoichiometric metathesis reaction between a metal–carbene and an olefin, Casey (1974)

First isolated metal–alkylidene complex, [$Ta{=}CHBu^t(CH_2Bu^t)_3$]: Schrock (1974)

First catalytic metathesis reaction using a unimolecular catalyst, [$LnTa{=}CHBu^t$]: Schrock (1980)

First unimolecular ruthenium–carbene metathesis catalyst, [$Ru{=}C{=}CHPh(PR_3)_2Cl_2$]: Grubbs (1992)

First ruthenium–benzylidene metathesis catalyst: [$Ru{=}CHPh(PCy_3)_2Cl_2$]: Grubbs (1995)

Chiral Mo–alkylidene catalyst for enantioselective olefin metathesis: Schrock-Hoveyda (1998)

Second-generation Grubbs-type catalysts with *N*-heterocyclic carbene ligands L [$Ru{=}CHPh(PCy_3)(L)(Cl)$: Grubbs, Nolan, Fürstner-Herrmann (1999) (and chiral version: Grubbs, 2001)

Robust isopropoxybenzylidene Ru catalysts: Hoveyda, Blechert (with NHC), Grela (with NO_2), 2000–2002

Figure 1.1 Dates of the history of olefin metathesis.

methylene species $[Fe(\eta^5\text{-}C_5H_5)(CO)_2(=CH_2)][BF_4]$, suggested the possibility of a tetra(methylene)metal intermediate in which the four methylene units were bonded to the transition metal (23–25). Double oxidative addition to a transition-metal center is not possible, however (24,25). Grubbs (26) proposed rearranging metallocyclopentane intermediates and cyclobutane complexed to a carbene (27). Neither of the mechanisms mentioned above (Fig. 1.2) fit the data, and the olefin metathesis mechanism remained rather mysterious in the United States until the mid 1970s.

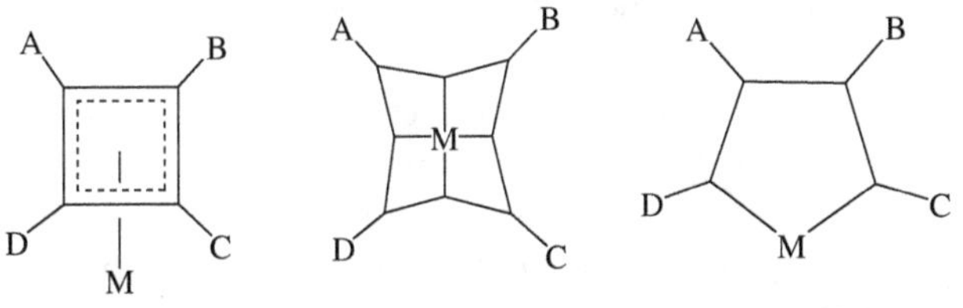

Figure 1.2 Erroneous intermediates proposed around 1970 for the olefin metathesis mechanism.

1.5 THE CHAUVIN MECHANISM: A METATHESIS DANCE

Chauvin from the Institut Français du Pétrole, had three key findings in mind when he envisaged the mechanism of olefin metathesis: the report of Fischer (28) on the synthesis of a tungsten–carbene complex, $[W(CO)_5\{C(CH_3)(OCH_3)\}]$, that of Natta (29) on the polymerization of cyclopentene by ring-opening catalyzed by a mixture of WCl_6 and $AlEt_3$, and that of Banks and Bailey (14) on the formation of ethylene and 2-butene from propene catalyzed by $[W(CO)_6]$ on alumina. Consequently, Chauvin and his student Hérisson published their proposition of metathesis mechanism in 1971 (Scheme 1.1) (30).

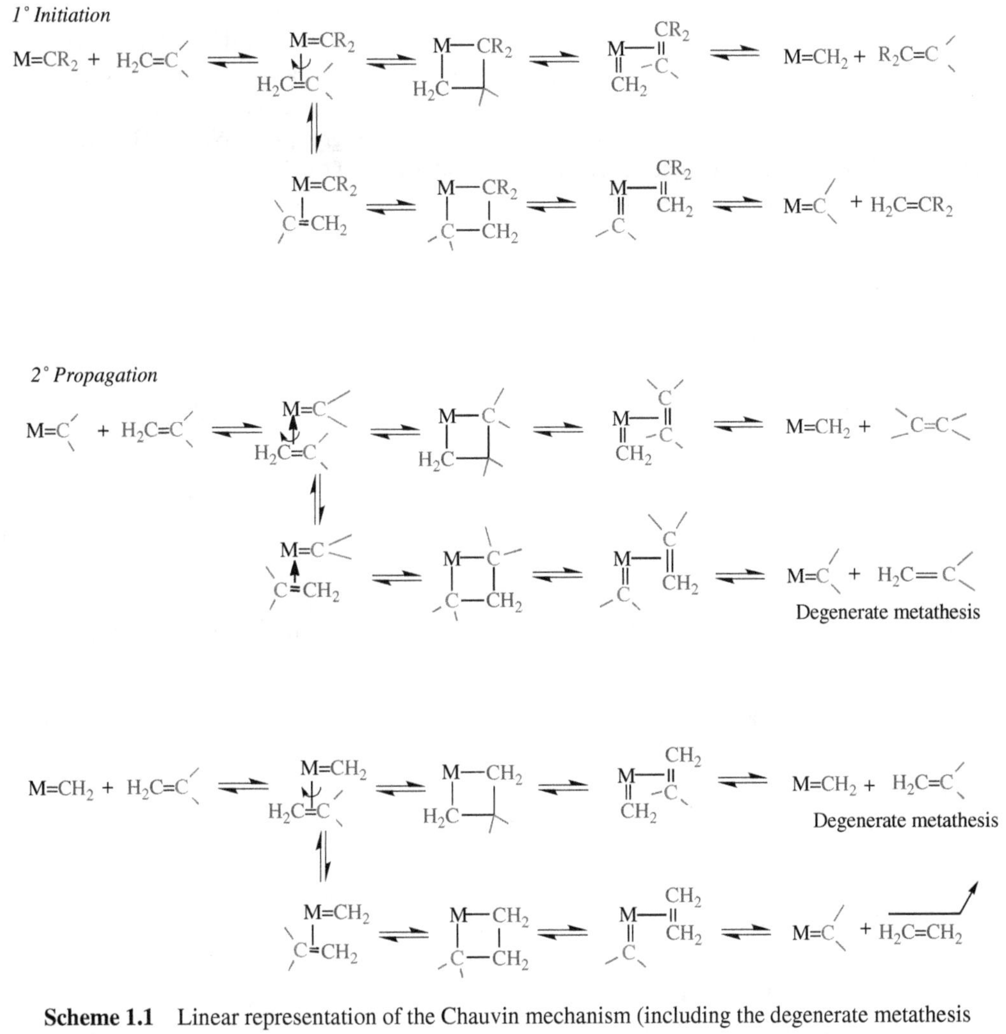

Scheme 1.1 Linear representation of the Chauvin mechanism (including the degenerate metathesis steps).

The Chauvin mechanism involves a metal–carbene species (or more precisely metal–alkylidene), the coordination of the olefin onto the metal atom of this species, followed by the shift of the coordinated olefin to form the metallocyclobutane intermediate, and finally the topologically identical shift of the new coordinated olefin in the metallocyclobutane in a direction perpendicular to the initial olefin shift. This forms a metal–alkylidene to which the new olefin is coordinated, then liberated. This new olefin contains a carbene from the catalyst and the other carbene from the starting olefin. The new metal–alkylidene contains one of the two carbenes of the starting olefin and it can re-enter a catalytic cycle of the same type as the first one (Schemes 1.1 and 1.2).

In fact, depending on the orientation of the coordinated olefin, the new catalytic cycle can give two different metallacyclobutenes, one leading to the symmetrical olefin and the other leading to the starting olefin. This latter cycle is said to be degenerate olefin metathesis. Thus, the catalytic cycles alternatively involve both metal–alkylidene species resulting from the combination of the metal with each of the two carbenes of the starting olefin (Scheme 1.1).

When the Chemistry Nobel prize was announced on October 5, 2005, Chauvin's metathesis mechanism was compared in a video to a dance in which couples exchange partners, which represents the two carbene fragments of the olefin. The dancers cannot exchange their partner directly, but they have to do the exchange by coupling with a master of ceremony that is the metal center. The master of ceremony also has a partner and, with the entering couple they form a circle so that the master of ceremony can exchange partners within the circle by taking a new partner from the couple. Then with his new partner, he can go to another couple for another exchange, and so on (31).

Chauvin and Hérisson not only suggested the metallacyclobutane mechanism, but also published several experiments to confirm it. For instance, they reported that reaction of a mixture of cyclopentene and 2-pentene led to C-9, C-10, and C-11 dienes in the ratio 1 : 2 : 1. Also, the reaction of a mixture of cyclooctene and 2-pentene led almost exclusively to the C-13 product. The latter reaction, but not the first one, was compatible with Calderon's mechanism. In 1973 and 1976, Chauvin (32,33) published other results showing that the mixture of $WCl_6 + MeLi$ catalyzes the formation of propene by reaction of 2-butene, which was proposed to proceed via methylation of tungsten, followed by the α-elimination in the tungsten–carbon bond of $W{-}CH_3$ to form a $W(=CH_2)(H)$ species, then metathesis. Here again, Chauvin's intuition was remarkable, because at that time, σ-bond metathesis in d^0 meta-alkyl complexes, that is the only available mechanism to activate such α-C–H bonds, was unknown and was disclosed only more than a decade later by the groups of Watson (Lu) (34), Bercaw (Sc) (35), and Marks (Th) (25,36,37).

The first recognition of Chauvin's valuable mechanism together with an elegant confirmation came from Casey and Burkhardt (38) when they reported that the carbene complex

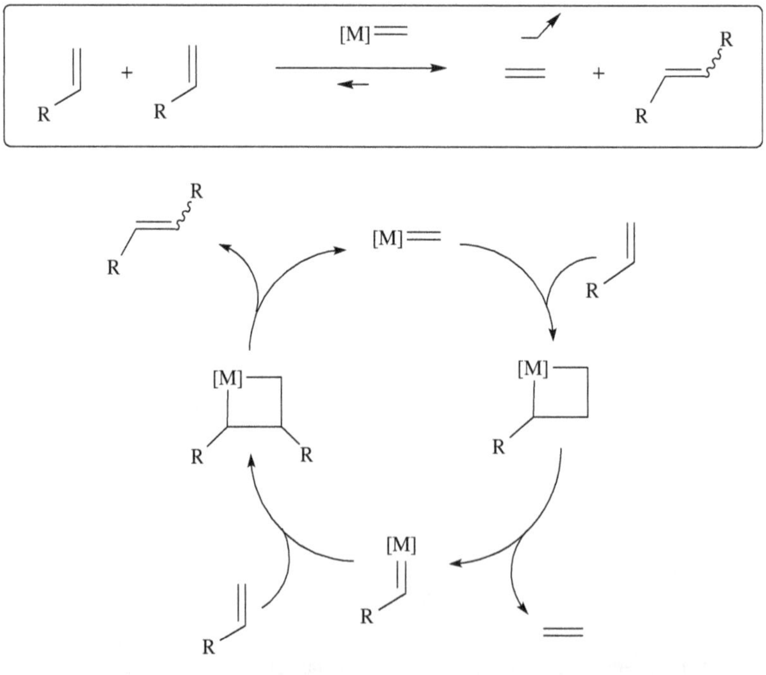

Scheme 1.2 Cyclic representation of the Chauvin metathesis mechanism.

$[W(CO)_5(=CPh_2)]$ reacted with isobutene to form a new olefin, 1,1-diphenylethene (Eq. 1.4), as the major product and that the same complex reacted with $H_2C=C(OCH_3)Ph$ to form 1,1-diphenylethene and the metal–carbene complex $[W(CO)_5\{=C(OCH_3)Ph\}]$ (Eq. 1.5), just as predicted in the Chauvin mechanism.

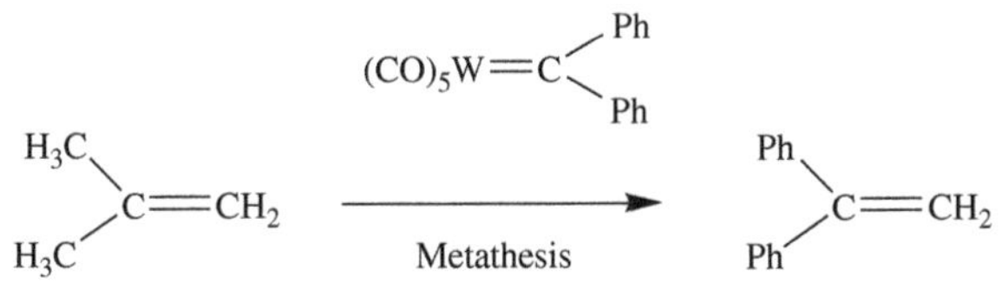

(1.4)

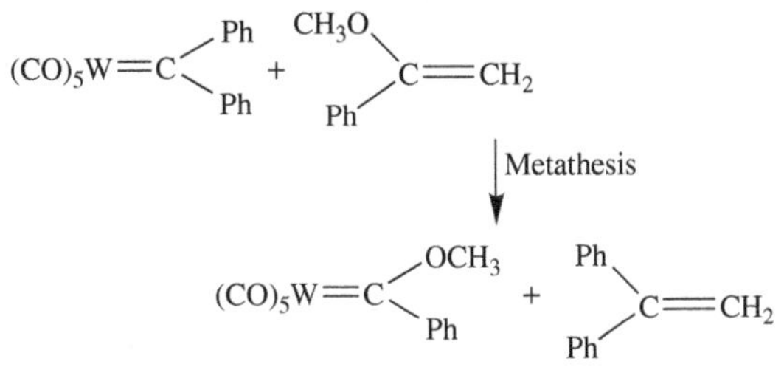

(1.5)

Later, labeling experiments by the groups of Grubbs and Katz (39–41) demonstrated that alkenes underwent non-pairwise exchange as required by the Chauvin mechanism. In particular, Grubbs (39) showed that a mixture of 1,7-octadiene and its analog that was deuterated on both methylene termini underwent metathesis to yield a statistical mixture of d_0-, d_2-, and d_4-ethylene, and that d_0- and d_4-ethylene were not scrambled after their formation. At that point, however, the actual catalytically active species was unknown, because the precatalysts used were 18-electron metal–carbene complexes such as $[W(CO)_5(=CPh_2)]$ or $[W(CO)_5\{C(OMe)Ph]$ or eventually metal precursors that did not contain a carbene ligand (38–41). Such complexes cannot bind olefins because the metal valence electron shell does not rise to 20 electrons; hence some decomposition of these pre-catalysts had to occur. Casey had nicely shown that the decomposition was partly limited to a single carbonyl ligand with the former complex, but also half of the starting material decomposed. The second complex that was sometimes used by Katz is more problematic, because carbenes bearing a heteroatom are, like CO itself, singlet carbenes. Thus, contrary to common usage, a double bond between this carbene and the metal should not be used in Fischer-type carbene complexes (28) such as this one (in the same way as the representation of the metal–carbonyl bond that is not written as M=C=O). Accordingly, we now know that the Fisher-type metal–carbene complexes are poor metathesis pre-catalysts, and that good tungsten olefin metathesis catalysts systematically have a high oxidation state (4).

1.6 THE IMPORTANCE OF THE CHAUVIN MECHANISM FOR OVERALL ORGANOMETALLIC CATALYSIS

Chauvin's mechanism introduced several new ideas. First, he proposed a metal–carbene complex to initiate the catalysis of the metathesis reaction. This idea first suggested that one could just synthesize unsaturated metal–alkylidene complexes (i.e., with 16 valence electrons on the metal or less) and let them react as catalysts or pre-catalysts with olefins to carry out the metathesis reaction. Of course, many authors later engaged in such research directions, first delineated by Chauvin. The induction time was long, however. Relatively few chemists became interested in such a route in the first half of the decade following Chauvin's proposal.

The second key point mentioned above was the explanation of the "black box" of the American industrial chemists: a d^0 metal–alkyl complex formed using such a mixture undergoes the formation of a metal–methylene or metal–alkylidene species that serves as a catalyst for alkene metathesis. For that purpose, Chauvin included in his metathesis mechanism the crucial proposition of an α-H elimination, a pioneering idea that was reactivated and rationalized by its σ-bond mechanism only much later.

Another very important aspect of the Chauvin mechanism concerns the intermediacy of the metallacyclobutane. Such metallocyclobutane complexes are sometimes stable, and some stable metallacyclobutenes have indeed been shown to be involved in metathesis. Elegant studies by Grubbs' group in 1980 showed that Tebbe's complex $[Cp_2Ti(CH_2)(ClAlMe_2)]$, reported in 1978 (42), reacted with olefins in the presence of dimethylaminopyridine to give titanacyclobutanes that slowly catalyze metathesis and could be used to identify all the intermediates in olefin metathesis (43–45).

Chauvin's mechanism applies to the whole range of olefin metathesis reactions, including cross metathesis (CM), ring-closing metathesis (RCM), acyclic diene metathesis polymerization (ADMEP), ring-opening metathesis polymerization (ROMP), enyne metathesis (EYM), and ring-opening cross metathesis (ROCM) (Fig. 1.3) (1–4).

Finally, it is possible to represent a generalization of the metallosquare formed by the shift of the olefin coordinated to the metal in the metal–alkylidene species (25). The metallosquare is not only involved as an intermediate or transition state in alkene metathesis, but also in many other catalytic organometallic mechanisms. Indeed, the metathesis of alkynes and the metathesis polymerization of cycloalkenes and alkynes formulated by Katz (41) are completely analogous. Moreover, it is possible to represent by a metallo-square scheme the mechanisms of σ-bond metathesis and β-elimination. Scheme 1.3 gathers together the different organometallic reactions involving a metallo-square (that can eventually have puckered shapes).

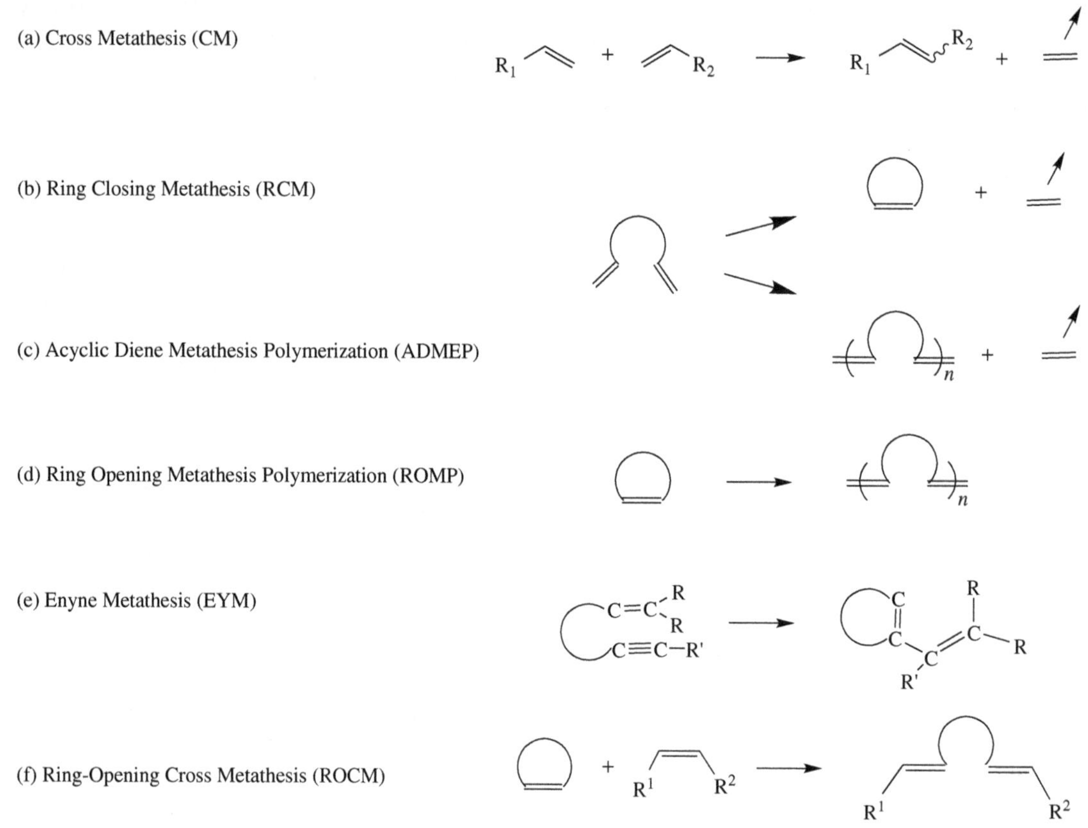

Figure 1.3 Various types of olefin metathesis reactions (all proceeding according to the Chauvin mechanism).

1.7 SCHROCK'S HIGH OXIDATION STATE ALKYLIDENE AND ALKYLIDYNE COMPLEXES

From the middle of the nineteenth century to the middle of the twentieth century, chemists believed that metal–alkyl compounds were intrinsically unstable, because of the supposedly too low energy of the metal–carbon bond. Wilkinson (46–48) then synthesized stable binary metal–alkyl complexes that did not contain β-hydrogen, showing that this instability was in fact kinetic, due to β-H elimination, because chemists had been trying to make binary metal–ethyl complexes. Organometallic chemists could then synthesize a whole series of thermally stable binary (and other) metal–alkyl complexes with alkyl groups lacking β-hydrogens, such as methyl, benzyl, neopentyl, trimethylsilylmethyl, and mesityl, even if the metal had less than 18 valence electrons in the valence shell (47,48). Such binary metal–poly(alkyl) complexes have indeed a low number of valence electrons in conflict with the 18-electron rule (49).

Richard Schrock was a PhD student at Harvard of John Osborn, who had been a PhD student of Geoffrey Wilkinson, who was at Imperial College, London, after Harvard had turned down his promotion for tenure. The influence of Wilkinson on his scientific grandson Schrock is seen clearly. Also inspired by Schmidbauer' synthesis of pentaalkyl phosphorous and arsenic derivatives, Schrock, then at Du Pont, synthesized $[TaMe_5]$, $[Ta(CH_2Ph)_5]$ (47) and tried to synthesize $[Ta(CH_2CMe_3)_5]$, which, analogously, would not contain β-hydrogens and thus, according to this principle, should have been stable. An α-elimination reaction occurred, however, upon attempting to coordinate the fifth neopentyl group, which produced one mole of neopentane and led to the isolation of the first stable metal–alkylidene complex, $[Ta(CH_2CMe_3)_3(=CHCMe_3)]$ that was reported in 1974 (Scheme 1.4) (50).

Schrock's group subsequently showed that the α-elimination reaction was quite general when the coordination sphere became crowded in these Ta and Nb complexes. This yielded a rich family of high oxidation state Ta and Nb alkylidene complexes in which the carbenic carbon is nucleophilic, somewhat resembling phosphorus ylids (51,52). At the time of this finding, the detailed mechanism of this reaction was unknown, but as stated above, the σ-bond metathesis that takes the α-elimination reaction into account was recognized 8 years later (34–37). Neither the 18-electron Fischer-type metal–carbene complexes (see above) nor the 10-electron Schrock-type Ta- or Nb-alkylidene complexes gave olefin metathesis upon

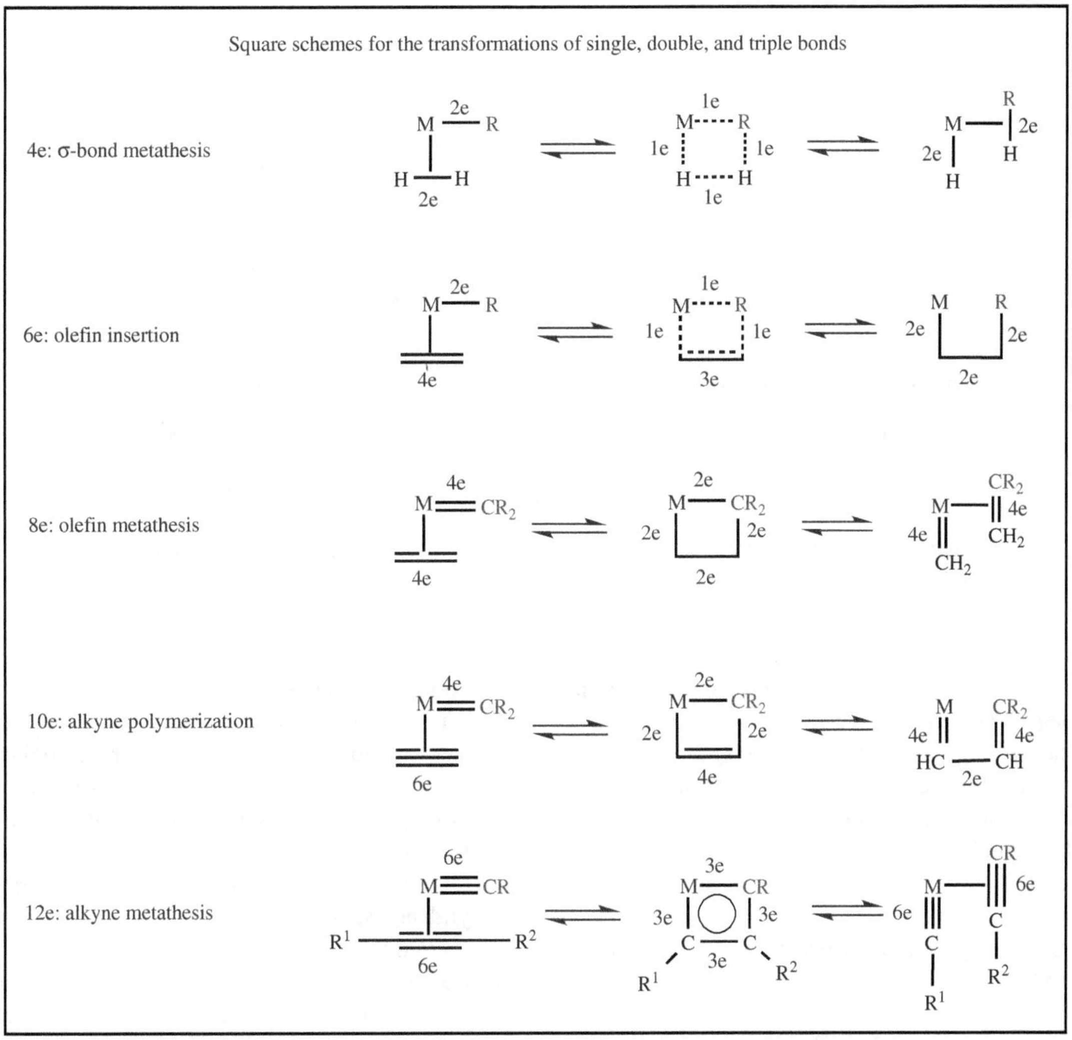

Scheme 1.3 Square schemes involved in the mechanisms of catalytic organometallic reactions (the metallosquares can eventually have puckered shapes).

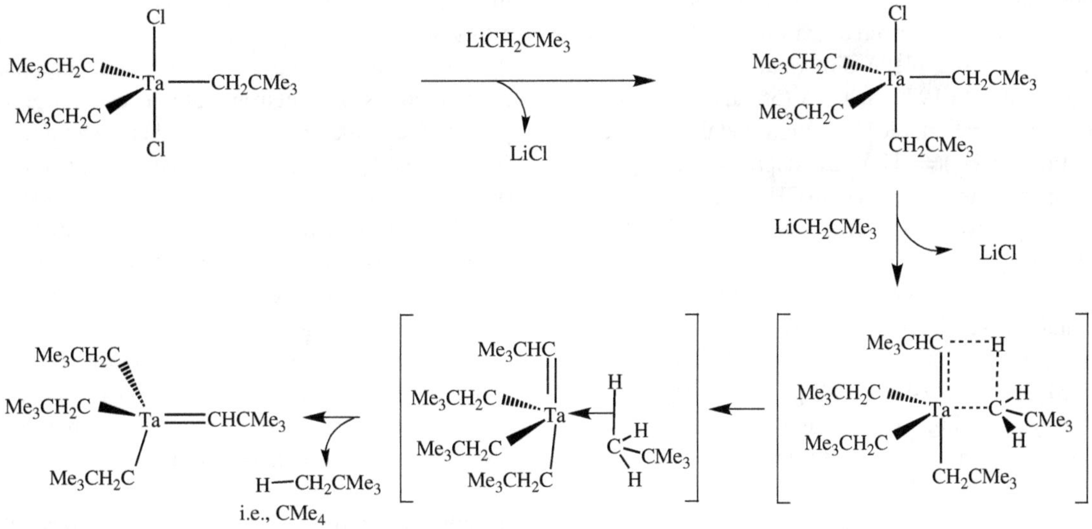

Scheme 1.4 σ-bond metathesis mechanism in the formation of Schrock's first neopentylidene complex.

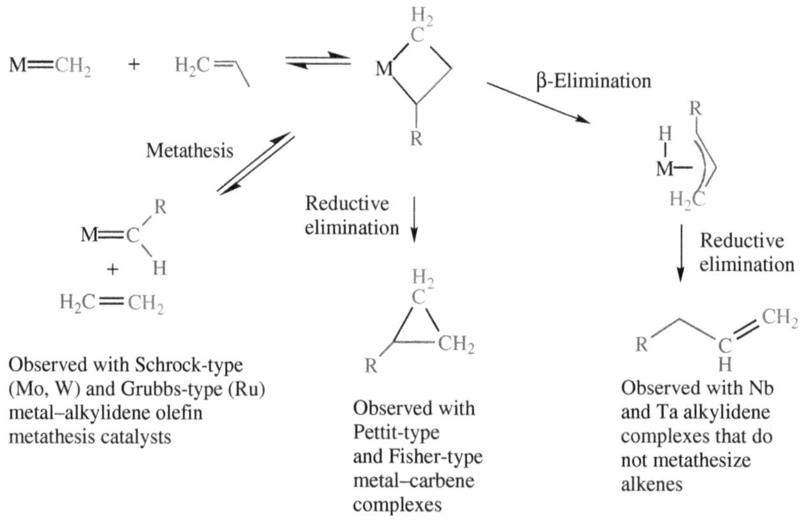

Scheme 1.5 The three modes of evolution of metallacyclobutanes formed by reaction between a metal–carbene complex and an olefin.

reaction with olefins, however, because the metallocyclobutane intermediates gave other products. Fischer-type metal–carbene complexes react with some olefins to give cyclopropanes eventually (but not always) by reductive elimination of intermediate metallacyclobutanes (53), whereas metallacyclobutanes resulting from the reaction of olefins with Schrock-type complexes gave β-H elimination, because they had less than 18 electrons in the Ta valence shell (Scheme 1.5).

The metal–alkenyl hydride species formed in the latter case gave reductive elimination. Then, the free coordination sites allowed reaction with 2 mol of olefins giving tantalacyclopentane intermediate that also underwent β-H elimination followed by reductive elimination to yield for instance 1-butene, when the olefin was ethylene. The metal species thus catalyzed olefin dimerization. Interestingly, Chauvin (54,55) also discovered extremely efficient and selective titanium-based olefin dimerization catalysts that are used industrially. In 1975, Schrock (56) also synthesized at Du Pont the first stable transition-metal–methylene complex, $[TaCp_2(CH_3)(=CH_2)]$ by deprotonation of the cationic Ta-methyl precursor $[TaCp_2(CH_3)_2][BF_4]$, and this methylene complex was characterized *inter alia* by its X-ray crystal structure.

It was only in 1980 that Schrock's group at MIT reported a tantalum–alkylidene complex, [Ta(=CH-*t*-Bu)(Cl)(PMe_3)(O-*t*-Bu)$_2$], **1** (Fig. 1.4, also Nb and W complexes), that catalyzed the metathesis of *cis*-2-pentene (57).

After Casey's finding in 1974 and stereochemical arguments from the Grubbs and Katz groups, this provided another evidence for Chauvin's mechanism of olefin metathesis with well-defined high oxidation state alkylidene complexes, almost a decade after Chauvin's proposal. The reason that these complexes catalyzed the metathesis reaction, whereas the other members of the family of niobium– and tantalum–alkylidene complexes failed to do so, was the presence of ancillary alkoxide ligands in the catalysts. Molybdenum and tungsten, however, were obviously the most active metals in alkene metathesis and, around 1980, Schrock and his group considerably increased their efforts in the search for stable molecular alkylidene and alkylidyne complexes of these metals including alkoxide ligands that would catalyze the metathesis of unsaturated hydrocarbons. This search was successful (58) and eventually produced a whole family of molybdenum– and tungsten–alkylidene complexes of the general formula $[M(=CHCMe_2Ph)(N-Ar)(OR_2]$, R and Ar being bulky groups. In these complexes, the imido ligand is supposed to be a four-electron ligand, because the lone pair of the nitrogen atom is engaged in a vacant tungsten orbital, so that these four-coordinate pseudo-tetrahedral complexes are best considered as 14-electron complexes. These compounds presently are the most active alkene metathesis catalysts (Fig. 1.4). Their metathesis mechanism involves direct (weak) coordination of the olefin to provide 16-electron M(=CH-*t*-Bu)(olefin) intermediates, presumably of trigonal bipyramidal structure that form 14-electron metallacyclobutanes according to the Chauvin (59–61) mechanism and further continue to give olefin metathesis.

Other chemists such as John Osborn in Strasbourg (26) and Jean-Marie Basset (27) in Lyon played an important role in the history of olefin metathesis by reporting tungsten complexes that were active as olefin metathesis catalysts in the 1980s. Osborn reported a well-defined W(VI) alkylidene metathesis catalysts, **2** (Fig. 1.4) and showed the living character of the polymerization system and the intermediacy

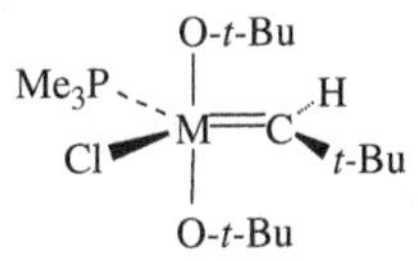

1

First Unimolecular olefin metathesis catalysts, M = Nb or Ta

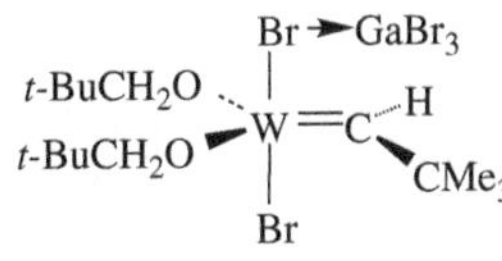

2

Kress–Osborn catalyst, 1983

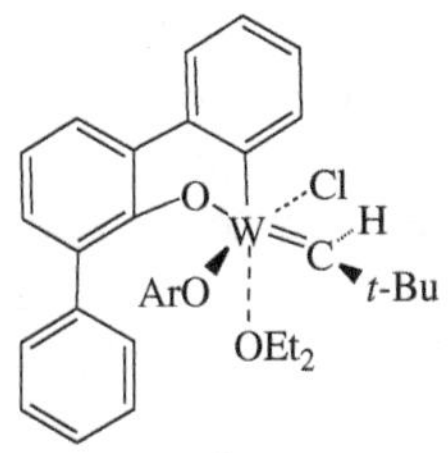

3

Basset's catalyst, 1985

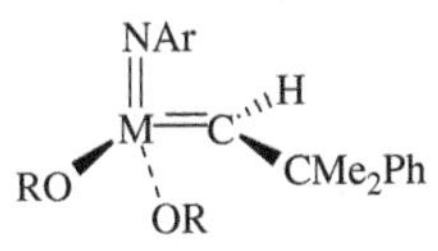

4

Family of Schrock's high oxidation-state catalysts (M = Mo or W; R and Ar bulky substituents), 1990

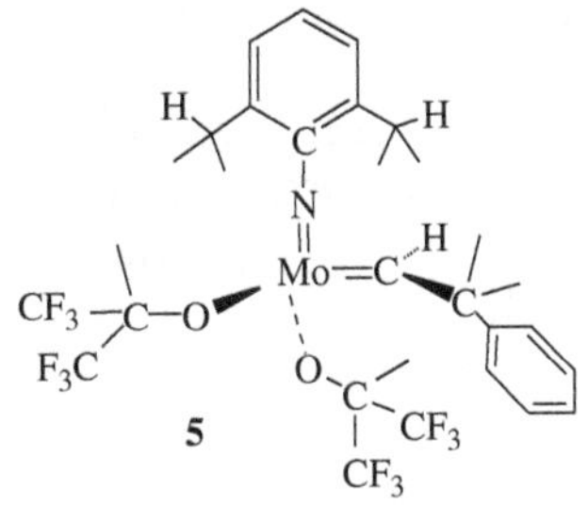

5

Prototype of Schrock's catalysts RCM of tri- and tetra substituted olefins, 1990

Figure 1.4 Early and optimized "unimolecular" Nb, Ta, Mo, and W metal–alkylidene catalysts of olefin metathesis.

of a tungstacyclobutane by ^{1}H NMR spectroscopy (62–65). Basset (66–68) reported aryloxoalkoxoalkylidene W(VI) catalysts and one of the first examples of Lewis-acid-free initiators, **3** (Fig. 1.4) that allowed the polymerization of substituted norbornenes following the ROMP mechanism.

The advantage of Schrock's (59–61) catalysts, whose most efficient members **4** and **5** (Fig. 1.3) were reported in 1990–1991, was that even though they are extremely active, they are molecular (without additives) and also provided a commercial catalyst and chiral versions for the first examples of asymmetric metathesis catalysis (69–76).

In 1975, Schrock (77) also reported the first high oxidation state alkylidyne complex, $[Ta(C\text{-}t\text{-}Bu)(CH_2t\text{-}Bu)_3]$. As its alkylidene analog and the Fischer-type carbyne complex $[W(CO)_4(Br)(C{-}Ph)]$, however, this Ta complex was inactive in metathesis. Schrock (78) applied the same principles as with olefin metathesis and thus turned to W complexes with alkoxide ancillary ligands. In 1982, his group reported a very active catalyst, $[W(C\text{-}t\text{-}Bu)(O\text{-}t\text{-}Bu)_3]$ for the metathesis of internal alkynes. Schrock's W-alkylidyne complexes do not react with olefins, but they selectively and efficiently metathesize alkynes without the need for a co-catalyst. For instance, the prototype $[W(CCMe_3)(O\text{-}t\text{-}Bu)_3]$ effects several hundred turnovers per minute of 2-heptyne metathesis under mild conditions. Some reactions even proceed at 25 °C. Here again, the alkoxide ligands are indispensible for the catalysis of the metathesis reactions.

1.8 GRUBBS' APPROACH AND THE RU OLEFIN METATHESIS CATALYSTS

Among the late transition metals, ruthenium is known as a *magic metal* by organometallic chemists. Despite a promising start with Pettit's (21,53) elusive methylene complex and its cyclopropanation chemistry upon reaction with olefins, an iron-based metathesis catalyst has not yet been discovered. The first ruthenium carbene complex, $[RuCp\{=C(Me)OMe\}(CO)(PCy_3)][PF_6]$, an 18-electron complex of Fischer-type, was synthesized by Malcolm Green's group in Oxford (78) but did not show any metathesis activity. The success of Grubbs' approach to stable benzylidene complexes containing the electrophilic benzylidene ligand may appear, by comparison, somewhat surprising, but it is due to the neutrality of the complexes, thus affording a considerably reduced electrophilicity of the carbene ligand compared to cationic complexes, and the great versatility of ruthenium that forms stable 16-electron complexes.

Grubbs (26,27) had been interested for a long time in the metathesis reaction, as indicated by his mechanistic proposals early on. He had eventually noticed Natta's 1965 publication on the catalysis by $RuCl_3$ of the polymerization of cyclobutene and 3-methylcyclobutene by ring opening (79). This process (in butanol) had been developed by Norsorex. In this context, the Ziegler–Natta polymerization of olefins under mild conditions obviously had a considerable impact on polymer chemistry (80). The delineation of a new polymerization mechanism, however, was not a simple task. Greatly inspired by this approach, Grubbs published in 1988 the polymerization of 7-oxanorbornene into a high molecular weight monodisperse polymer ($M_w = 1.3 \times 10^6$ g mol^{-1}; $M_w/M_n = 1.2$) by $RuCl_3 \cdot xH_2O$ or $[Ru(H_2O)_6](OTs)_2]$ (OTs = toluene sulfonate), 6 (Fig. 1.5).

This catalytic reaction was all the more remarkable as it was conducted in water (81). Shortly afterward, he could show, in the course of the same reaction, the formation of a Ru-alkylidene intermediate, then the polymerization of cyclooctene, an olefin with little constraints, when the alkylidene ligand source was ethyl diazoacetate added to the aqueous solution of $[Ru(H_2O)_6](OTs)_2$ (82). Consecutively and according to the same logic, a great step forward was accomplished by Grubbs in 1992. He reported the first molecularly well-defined ruthenium–carbene complex that promoted the ROMP of low strain olefins as well as the catalytic RCM of functionalized dienes without the need of a co-catalyst (83). Grubbs showed that these vinylidene complexes, $[RuCl_2(PR_3)(=CH-CH=CPh_2)]$ (R = Ph, **7**, or Cy), were efficient molecular catalysts for these polymerization reactions and other metathesis reactions such as those involving ring closing of terminal diolefins (83,84).

Interestingly, Noels' group (85) reported, also in 1992, the Ru-catalyzed ROMP of cycloolefins initiated by diazoesters. In 1995, this group showed that addition of such diazoesters to $[Ru(\eta^6\text{-cymene})PR_3]$ (R = Cy or *t*-Bu) produces very active arene-free ruthenium–carbene catalysts in which the carbene proton could be observed by ^1H NMR, shedding light onto the catalyst structure (86). In 1995, the new molecularly well-defined catalysts $[Ru(=CHPh)Cl_2(PR_3)_2]$, R = Ph or Cy, whose structures are closely related to the vinylidene ones published 3 years earlier, were reported by the Grubbs group and commercialized with R = Cy. The complex $[Ru(=CHPh)Cl_2(PCy_3)_2]$, **8**, is now known as the *first-generation Grubbs catalyst* and is even today the metathesis catalyst most used by organic chemists, because of its stability in air and compatibility with a large variety of functional groups (except for amines, nitriles, and basic media) (87). Subsequently, the best organometallic research groups also offered astute alternative syntheses to Grubbs' catalysts in the late 1990s (88–91).

Fine mechanistic studies with this catalyst led Grubbs' group to conclude that the mechanism first involved the dissociation of one phosphine to generate the reactive 14-electron ruthenium intermediate for coordination of the olefin. In order to supposedly favor this dissociative step, Grubbs introduced, in place of one phosphine, an *N*-heterocyclic (NHC) bis-amino carbene ligand that is relatively stable, even in the free form obtained by deprotonation of the corresponding imidazolium cation (92). These NHC ligands are excellent σ-donors without π-acceptor properties and have been known for several decades, but they have only recently become very popular in organometallic chemistry and catalysis (92–95). It was Herrmann's group (93) that first synthesized ruthenium complexes with two such NHC carbene ligands in the context of the catalysis of olefin metathesis, but their catalytic activity was shown to be modest. In Grubbs' second-generation catalysts **9** and **10** containing only one such ligand, the electron density at the ruthenium center is increased, which favors alkene coordination from the 14-electron intermediate. Thus, the second generation of Grubbs catalysts $[RuCl_2\{C(N(mesityl)CH_2)_2\}(PCy_3)(=CHPh)]$ containing a single NHC carbene ligand and its catalytic activity in metathesis were successively proposed within a few months by the groups of Grubbs (96,97), Nolan (98), and, with a variation of the NHC ligand, Fürstner and Herrmann (99) (Fig. 1.5). The complex **10**, and the third-generation of Grubbs' catalyst **15**, simply obtained upon addition of a pyridine to **10**, are presently the most used catalysts for efficient cross-metathesis reactions. These commercially available catalysts are even more active than Grubbs' first-generation catalyst **8**, although they are also more thermally stable than **8** (Fig. 1.5) (100–107).

In the late1990s, other ruthenium olefin metathesis catalysts were reported (108–113) including that of Peter Hofmann. The latter, also very active, was obtained by chloride abstraction providing a dicationic dimer from a ruthenium analog bearing a *cis*-diphosphine (108–110).

Hoveyda (114,115), Grela (116,117), Buchmeiser (118), Misutani (119), and Blechert (120–122) reported other related, very active, stable, and functional group-tolerant ruthenium metathesis catalysts of this family in the early 2000s. Some of the best catalysts are represented in Figure 1.4. The first Hoveyda metathesis catalyst **12** is derived from Grubbs' first-generation catalyst **8**, and the chelating benzylidene ligand improved the stability compared to **8**. It bears only one phosphine and a chelating carbene ligand. In 2000, Blechert and Hoveyda independently reported a more active and more stable complex **13**, also bearing a NHC ligand instead of the phosphine. Both catalysts are now commercially available, although expensive. Grela reported variations of the Hoveyda catalyst with increased efficiency (active even at 0 °C) when the aryl group of the benzylidene ligand bears a nitro group in the meta or para positions or two methoxy substituents (**14**, Fig. 1.5). Grela's successful idea was to destabilize the Ru–O(ether) bond in order to favor the ether decoordination that generates the catalytically active 14-electron species (116,117). This key finding was further confirmed by Grela's group upon reporting the poorer catalytic performances of this catalyst in which the nitro group was replaced by the electron-releasing NEt_2 group at the para position, whereas the introduction of the related electron-withdrawing diethyl methyl ammonium substituent brought about a remarkably excellent activity. This family of Grubbs–Hoveyda–Grela catalysts, whose

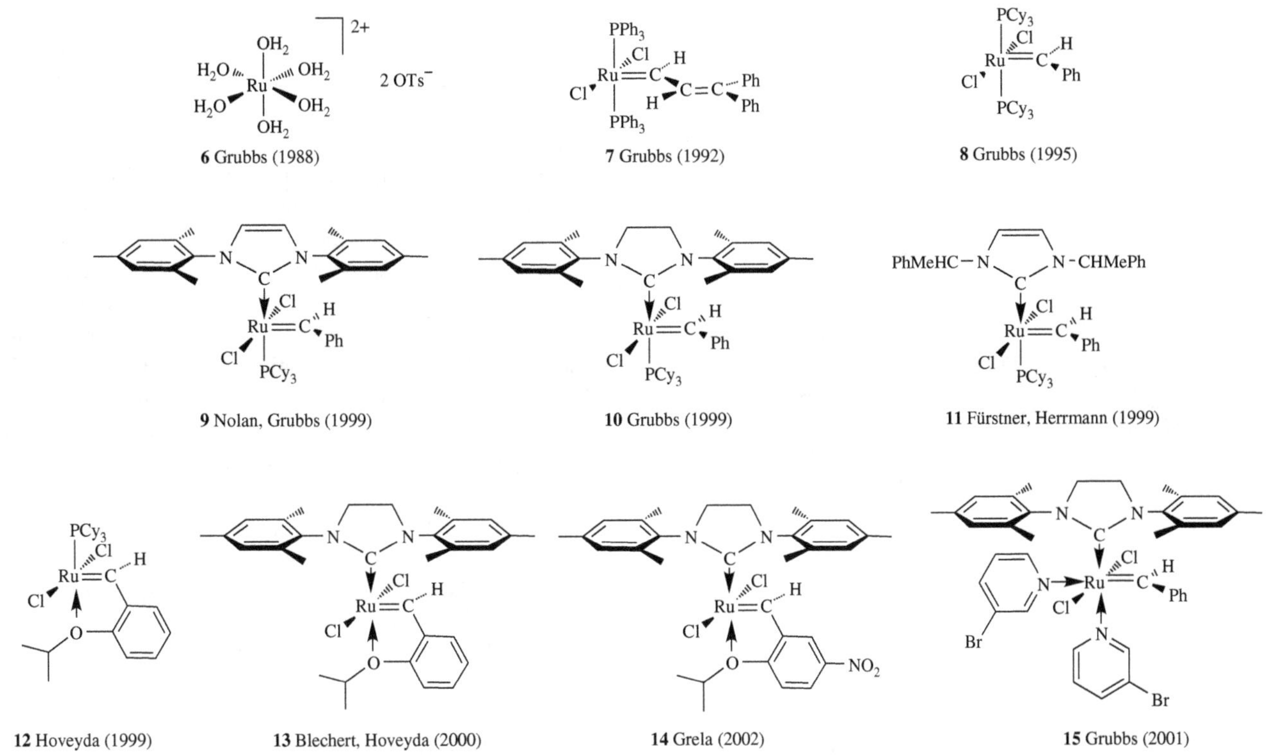

Figure 1.5 Key steps in the discovery and improvements or variations of ruthenium metathesis catalysts.

activity compares with that of the second-generation Grubbs catalyst, are especially useful for difficult cases of metathesis of polysubstituted olefins and selective CM in which homo-coupling needs to be avoided (114,115,123,124). A successful variation of these Ru-benzylidene catalysts so far has been reported by Blechert (120–122), whose strategy to sterically destabilize the Ru–O(ether) bond consisted of the introduction of an aryl (phenyl or naphthyl) substituent on the benzylidene aryl in the ortho position relative to the O(ether). The catalytic efficiency and stability of these Grela (116,117) and Blechert (120–122) complexes reported in the early 2000s surpasses those of many other Ru catalysts, although it has been shown several times that the catalytic efficiency depends on the type of metathesis reaction examined and the tolerance toward the required functional group.

Since the mid 2000s, the introduction of the NHC ligands and their chelate forms have led to myriads of structural variations of the ruthenium–NHC carbene–benzylidene complex structures toward the optimization of catalytic activities and recyclability in the various metathesis reactions. This very rich design research has been all the more attractive as these complexes are air- and water stable and of very practical use in organic chemistry, and this area has been the subject of excellent reviews (125–131). In-depth comparison of the catalytic efficiencies among various benzylidene and indenylidene ruthenium complexes conducted by Grela's group showed that different catalysts proved to be optimal for different applications, and that no single catalyst outperforms all others in all cases. A strong temperature effect was noted in all the reactions tested, and the authors suggested that difficult metathesis transformations should be conducted at 70 °C in toluene, rather than simply by increasing the catalyst loading at ambient temperature (127). For instance, under such conditions, Fürstner's and Nolan's (131–137) robust and readily accessible indenylidene complexes were shown to be in many cases fully equivalent to the second-generation Grubbs and Grubbs–Hoveyda complexes (although their activities under ambient conditions were lower, their practical and easy synthesis makes them interesting; Scheme 1.6) (126).

Probably more than 500 ruthenium metathesis catalysts have been reported to date, including all ligand variations, and all these complexes react with olefin eventually, after phosphine or ether (or other heteroatom) ligand decoordination, that is, at the 14-electron stage, to give metathesis via 14-electron ruthenacyclobutane intermediates (130).

Although the ruthenium metathesis catalysts are 16-electron complexes that require ligand decoordination before olefin coordination (Scheme 1.7), Piers' group (138–141) has isolated remarkable 14-electron ruthenium phosphonium–alkylidene complexes that are all better

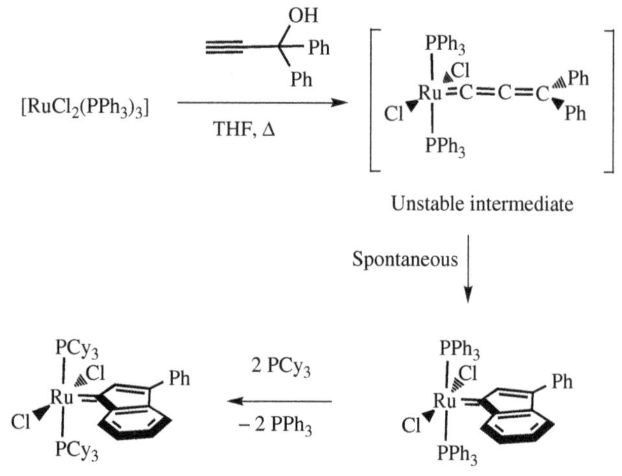

17 Fürstner (2001) **16** Fürstner (1999)

Scheme 1.6 Synthesis of the indenylidene Ru metathesis catalysts.

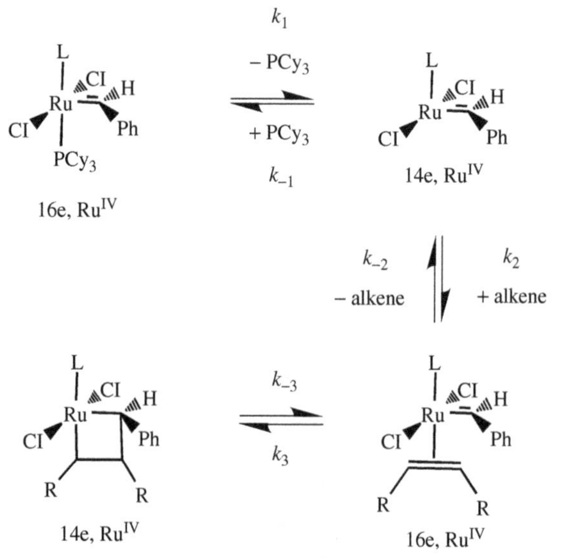

Scheme 1.7 General mechanism of the olefin metathesis reaction with the Ru catalysts. With L = PCy_3 (complex **8**), $k_1 = 10^2$; $k_2/k_{-1} = 10^{-4}$; with L = saturated NHC (complex **10**), $k_1 = 1$; $k_2/k_{-1} = 1$. It is the faster complexation of olefins by the 14e intermediate that makes the catalyst **10** more active than **8** (not the phosphine decoordination step) (107).

metathesis catalysts as they do not require ligand decoordination before olefin binding, which results in very low olefin binding energy, high catalytic activity in model RCM reactions, and direct relevant observation of ruthenacyclobutane intermediates resulting from olefin coordination (139–141).

An opposite strategy reported by the Lemcoff and Grela groups consisted in synthesizing ruthenium–NHC complexes that were thermo-switchable latent metathesis catalysts with a chelating benzylidene thioether, sulfoxide, or sulfone ligand. These robust complexes are not catalytically active under ambient conditions but show very good RCM and EYM catalytic activity upon heating (Fig. 1.6) (141–145).

Among the multiple ligand variations, Grubb's third-generation metathesis catalysts are especially remarkable five- and six-coordinate mono- and bis-pyridine complexes (see for instance the structure of the most active catalyst of this series, **15**, Fig. 1.4). The *cis*-bis(pyridine) complexes, simply prepared by adding an excess of the appropriate pyridine to Grubbs' second-generation catalyst **10** (146,147) are efficient in the challenging CM of acrylonitrile (147) and are among the fastest initiating ruthenium catalysts known, which, for instance, with 3-bromopyridine in **15** proved extremely useful for the synthesis of polymers with very narrow polydispersity and for the synthesis of block copolymers (148). The corresponding mono(pyridine) complexes are obtained from the bis(pyridine) complexes under vacuum and many structural variations are also known in this series of Grubbs' third-generation catalysts (129).

1.9 OLEFIN METATHESIS IN AQUEOUS SOLVENTS AND WATER

Water obviously is the greenest solvent, due to its abundance and lack of toxicity and risks. The ruthenium molecular metathesis catalysts withstand water, and often even benefit from its presence in the medium. Early studies by Grubbs (82) involved the ROMP of 7-oxanorbornene catalyzed by $RuCl_3$(hydrate) in which water acted favorably, and further studies conducted in water as the only solvent produced a larger polymer and with a better polydispersity index (PDI) than in organic solvents (82). Later, this catalyst and method was applied to neoglycopolymers (149). Since then, ruthenium benzylidene catalysts have been widely used as olefin metathesis catalysts in aqueous solvents or water only under homogeneous or heterogeneous conditions, water-soluble ruthenium catalysts have been synthesized and efficiently used, and the area of aqueous olefin metathesis has been the subject of an excellent review (150).

Classic examples include the use of ruthenium benzylidene catalysts such as **21–23** in aqueous solvents leading to homogeneous mixtures when water was added to a water-miscible organic solvent added in sufficient quantities to dissolve the substrates and catalyst (Fig. 1.7) (151,152).

Metathesis "on water" is a very interesting concept that was developed by Grela's group with standard ruthenium benzylidene catalysts that are, as substrates, water insoluble. Such reactions were conducted using ultrasonification that produces emulsions in which CM (with electron-poor olefins), RCM, and EYM took place smoothly in quantitative yields (153). Metathesis in aqueous emulsions containing a

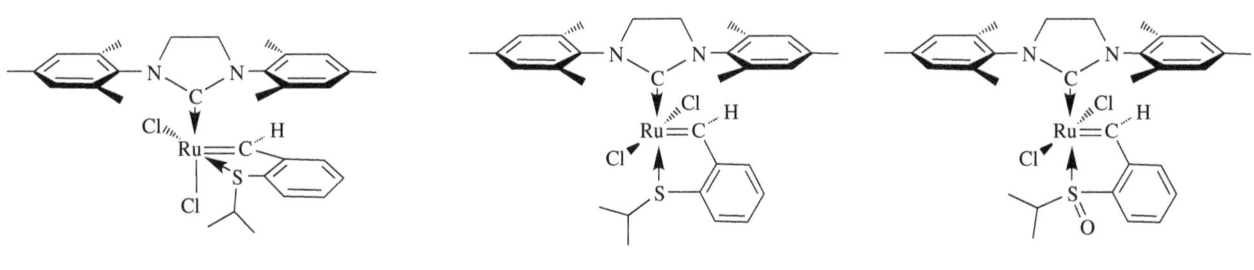

18 Lemcoff (2008) **19** Grela (2009) **20** Grela (2009)

isomers

Figure 1.6 Thermo-switchable latent ruthenium metathesis catalysts with chelating benzylidene–thioether and benzylidene–sulfoxide ligands. A whole family of Lemcoff's metathesis catalysts related to **18** was also synthesized with various sulfur substituents (129).

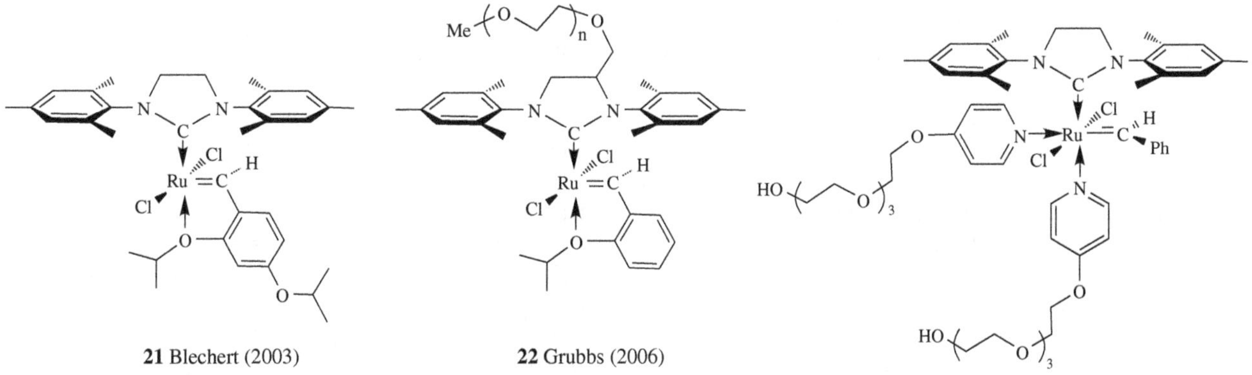

21 Blechert (2003) **22** Grubbs (2006) **23** Beitenkamp and Emrick (2005)

Figure 1.7 Examples of hydrophilic neutral ruthenium complexes that catalyze metathesis olefin reactions in aqueous solvents: **21** catalyzes RCM (but not CM) in CH_3OH/H_2O or DMF/H_2O mixtures. **22** and **23** catalyze RCM (22) and ROMP reactions of some water-soluble substrates (150).

surfactant has been broadly applied to polymer syntheses by ROMP metathesis reactions using hydrophilic or hydrophobic norbornene derivatives (150,154–158). Then this strategy was first applied by Sinou's group (159) to RCM and CM metathesis in water with and without a sodium dodecyl sulfate (SDS) surfactant using Grubbs' commercial catalysts **8** and **10**. Lipschulz's group (160,161) reported the use of a nonionic PTS amphiphile derived from vitamin E as an efficient surfactant for the metathesis of CM reactions in water with Grubbs catalyst **10**. Astruc's group reported efficient RCM, CM, and EYM of water-insoluble substrates with low amounts of commercial Grubbs catalysts **8** and **10** under ambient condition in water only and air, using a water-soluble dendrimer serving as a nanoreactor host with hydrophobic interior. The water-insoluble catalyst was recovered by filtration after the reaction, and the water-soluble dendrimer was used in very small amounts and re-used at least 10 times without loss (162). Nicolaou's group (163) reported the CM catalysis of vancomycin derivatives in a heterogeneous mixture of water and CH_2Cl_2 (>95 : 5) using $[C_{12}H_{25}NMe_3]^+Br^-$ as the phase-transfer catalyst. In another approach, the commercial Grubbs catalysts were occluded in a slab of polymethylsiloxane for CM and RCM metathesis reactions of substrates that are soluble in aqueous methanol (164). Another classic approach involves the modification of the ruthenium metathesis catalyst with a small (most often ionic) or hydrophilic polymer (polyethylene glycol (PEG) type) group on one of the ligands (phosphine, NHC carbene, or benzylidene) in order to render it water soluble. This method is applied to water-soluble substrates for metathesis in a homogeneous aqueous solution (Fig. 1.7) (150,165,166). It has also been applied in the outer aqueous phase of 1–4 μm liposomes for ROMP reactions by Maingotaud's group to heterogeneous mixtures of ruthenium catalysts that are in the same time-surfactants ("inisurf") via a polyalkyl or polyfluorinated alkyl tag (167,168). The use of ammonium substituents in ruthenium–benzylidene–isopropyl ether chelate catalysts such as **24–26** has proved very efficient, not only

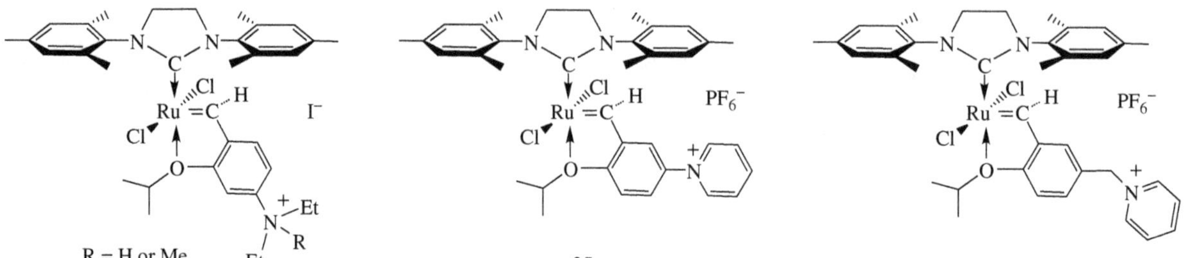

R = H or Me

24 Grela (2002)

25 Grela, Mauduit and Clavier (2006)

26 Grela, Mauduit and Clavier (2006)

Figure 1.8 Examples of Grela's cationic ruthenium benzylidene–isopropyl ether complexes **24–26** that catalyze RCM, CM, and EYM reactions in water. The ammonium substituent also serves as a non-covalent anchoring group for the fixation of anionic tags (150).

for metathesis catalysis in water, but also as a viable strategy to introduce tags noncovalently (Fig. 1.8).

In conclusion, metathesis in water is extremely attractive, using water- and oxygen-stable very active ruthenium catalysts, which, for instance, enables applications of metathesis in biological molecules (150,169).

1.10 OLEFIN METATHESIS IN OTHER "GREEN SOLVENTS": IONIC LIQUIDS AND sCO_2

The concept of the use of ionic liquids in catalysis (170–173) was raised by Chauvin and Olivier-Bourbigou (170,171) in 1995 when they efficiently used [bmin]Cl-$AlCl_3$-$EtAlCl_2$ ionic liquid as a solvent for the metathesis of pentene catalyzed by [$W(OAr)_2Cl_4$] (ArH = 2,6-diphenylphenol or 2,4,6-trimethylphenol). Many advantages are provided by their use, including high chemical and thermal stability, extremely low vapor pressure, insolubility, or immiscibility with aqueous and organic reaction media, and good ability to solvate both polar and nonpolar compounds. Since this first report, more than 50 publications have appeared on this specific subject that was reviewed in 2008 (172). The studies concerned, almost exclusively, ruthenium catalysts (173), the first example of which was reported by Buijsman (174), with the synthesis of hydantoin using Grubbs'catalysts. Indeed, in the early 2000s, commercial Grubbs's catalysts were used as such in ionic liquids without a tag. Our group reported several reactions in [bmim]PF_6 including the multiple RCM reaction of a nona-allyl dendritic core to a tetracyclic aromatic in 75% yield (175). Occasionally, microwave irradiations were efficiently used to accelerate reactions, but without catalyst recovery (176,177), and a biphasic catalyst composed of ionic liquid and $scCO_2$ was astutely applied to the macrocyclization of pharmaceutical precursors (178). Charged allenylidene ruthenium catalysts were also efficiently used in biphasic [bmim]BF_4–toluene at 80 °C (179). A step forward in the area was accomplished with the introduction of ruthenium catalysts incorporating the imidazolium motif of the ionic salt as a tag (180–185). In particular, the use of the biphasic system [bmim]PF_6–toluene at 25 °C allowed the formation of tri- and tetra- substituted olefins together with catalyst recycling and low ruthenium contamination (1–22 ppm, Fig. 1.9) (181). Alternatively, the introduction of pyridinium group in the tag instead of imidazolium brought about promising results (Fig. 1.8) (186,187).

Another "green" solution consists of conducting the metathesis reaction in supercritical carbon dioxide ($scCO_2$), rendering catalyst recovery easier than with organic solvents (126,188,189). Combined use of ionic liquid and $scCO_2$ has already been involved (see above) (178). This field was pioneered by DeSimone (190) with the ROMP of norbornene catalyzed by [$Ru(H_2O)_6$][OTs]$_2$ (Ts = *para*-toluenesulfonate), then by Fürstner and Leitner (112,191) with Schrock's and Grubbs' metal–carbene catalysts. Using the Grubbs–Hoveyda-type catalysts covalently immobilized on solid supports, Bannwarth et al. (192) obtained excellent recycling of the catalysts.

1.11 SOLID CATALYST SUPPORTS

Solid catalyst supports allow, in principle, to recover the solid catalyst simply by filtration and thus remove it from the substrate solution after the reaction. The solid support must be attached to one of the ligands. In the ruthenium catalysts, there is the choice between anionic ligands (by substitution of one or two Cl ligands in the Ru catalysts), phosphines (first-generation Grubbs catalysts), NHC carbenes (second and third-generation Grubbs catalysts), pyridines (third-generation Grubbs catalysts), arenes (Fürstner catalyst), or most commonly benzylidenes (Grubbs–Hoveyda catalysts). In the Schrock-type catalysts, the Basset group has replaced an alkoxide ligand by silica in Mo, W, and Re catalysts. The area has been reviewed several times, in particular by Buchmeiser (165,193), Basset (194–197), Weck (198), Toma (199), and in the review on sustainable concepts in olefin metathesis by Clavier et al. (126).

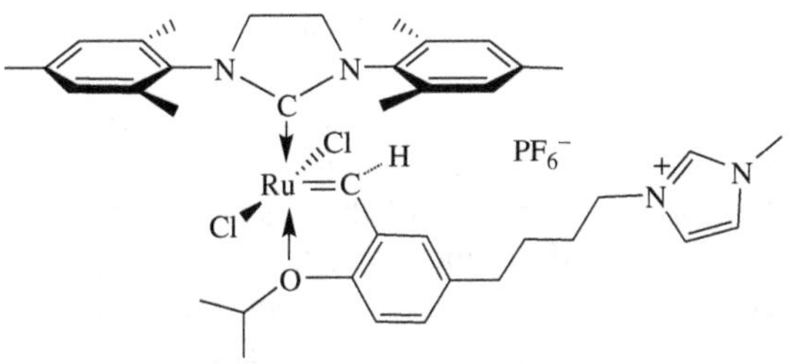

27 Mauduit and Clavier (2004)

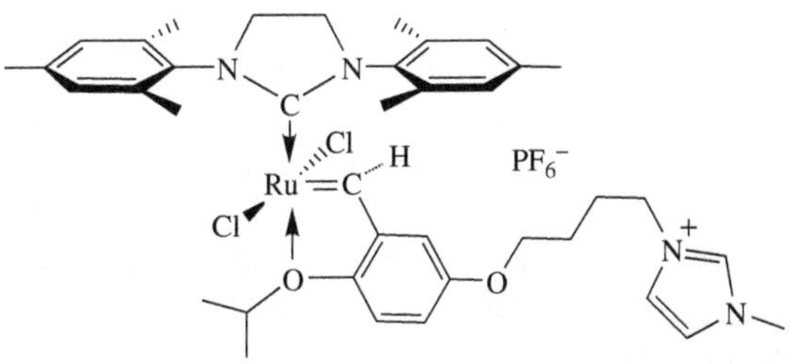

28 Yao (2005)

Figure 1.9 Examples of the ruthenium metathesis catalysts **27** and **28** derivatized with imidazolium tags used in biphasic medium dichloromethane/[BMI][PF_6] or toluene/[BMI][PF_6] and displaying high recyclability for the RCM of tri- and tetra-substituted olefins at room temperature with very low Ru contamination (1–22 ppm over several cycles) (181,182).

1.11.1 High Oxidation-State Mo, W, and Re Silica-Bonded Catalysts

Basset's group has provided well-defined heterogeneous catalysts for olefin metathesis upon coordinating active metal centers (Mo, W, and Re) to silica, with the metal-bearing ligands that have already proved useful in homogeneous catalysis and with silica as an additional ligand (60). Recall that Schrock had turned metathesis-inactive alkylidene complexes into active ones by the introduction of alkoxy groups. In Basset's catalysts, this beneficial role is played by a siloxide ligand from silica. Thus, the catalysts [(SiO)M(=CH-*t*-Bu)(CH_2-*t*-Bu)$_2$], M = Mo or W (200), and [(SiO)Mo(=NH)(=CH-*t*-Bu)(CH_2-*t*-Bu)] (200,201) are active at 25 °C, unlike previously reported ill-defined heterogeneous catalysts and the early Mo and W oxides on silica or alumina. The only oxide that had catalyzed olefin metathesis at 25 °C was Re_2O_7/Al_2O_3, but it suffers from a low number of active sites, side reactions caused by the acid support, and deactivation of the catalyst (193). On the other hand, Basset's silica-supported rhenium catalyst [(SiO)(Re(C-*t*-Bu)(=CH-*t*-Bu)(CH_2-*t*-Bu)], **29** (Fig. 1.10) catalyzes the metathesis of propene at 25 °C with an initial rate of 0.25 mol per mol Re per second.

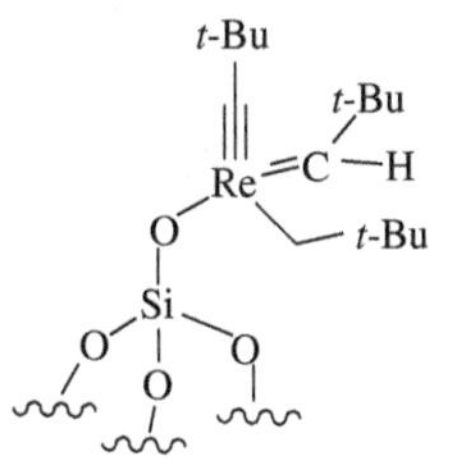

Silica

29 Basset (2001)

Figure 1.10 Example of Basset's highly active silica-supported alkene metathesis catalysts (196).

The formation of 3,3-dimethylbutene and 4,4-dimethylpentene in a 3 : 1 ratio results from cross metathesis between propene and the neopentylidene ligand, and the ratio of cross metathesis products matches the relative stability of the metallacyclobutane intermediates. Cross metathesis of propene and isobutene and self-metathesis of methyl oleate can also be achieved, and TON reaches 900 for the latter reaction, which is unprecedented for heterogeneous and most homogeneous catalysts (202,203). In his recent *Chemical Review* article on high oxidation state Mo and W imido alkylidene complexes, Schrock (204) has included a review on Mo and W metathesis catalysts supported on silica. In addition to the advantage of separating the solid catalyst from the products, Schrock has emphasized that another advantage of the support is to minimize bimolecular alkylidene coupling by retaining the metal centers far apart on the solid support. Finally, Basset's (205) work has produced silica-supported Schrock-type metathesis catalysts that are sometimes, as for [Mo(NAr)($CHCMe_2R$)(O-*t*-Bu)(OSi700)], dramatically more active in metathesis with their siloxide ligand than their soluble version with the *t*-BuO ligand, although it is difficult to deconvolute the various reasons for increased catalytic activity.

1.11.2 Ruthenium Benzylidene Olefin Metathesis Catalysts on Solid Supports

Anchoring catalysts on solid supports allows easy removal of catalysts and their re-use. As the subject has been reviewed (128–131,165,193,197–199), only some of the main representative examples are given here. Buchmeiser (197) has reported perfluoroalkylcarboxylate ligands that result in less than 0.015 ppm Ru contamination, but these supported catalysts have not been recycled. Air-stable supported pyridine complexes, active in CM and RCM, could be recycled five time with almost no yield decrease, but no Ru contamination was reported (206), as for polymer-supported

Ru–arene–allenylidenes complexes that catalyzed RCM upon recycling three times (207). An NHC ligand derivatized with a poly(norbornene) support provided a ROMP and RCM catalyst with only 70 ppm ruthenium leaching (208). However, silica-supported NHC ligand-containing catalysts were also efficiently recycled multiple times, and they did not leach ruthenium (<5 ppb by ICP-MS) (209). Anchoring the metathesis catalyst to a solid support via the benzylidene ligand is the most widely used method, because the benzylidene ligand with its support is released at the beginning of the metathesis mechanism, allowing catalysis to proceed homogeneously in solution, but is recovered at the Ru center according to the "boomerang" process of the chelate carbene ligand at the end of the reaction (126,165). The Grubbs–Hoveyda catalyst with the chelating benzylidene–ether ligand is particularly efficient for this strategy and has been developed with various supports (Scheme 1.8 and Fig. 1.11), but contamination usually reaches several hundred ppm, which is an inconvenience because of the re-anchoring of the active Ru species to the support at the end of the reaction (165). With a Wang resin attached to the benzylidene ligand, however, the ruthenium catalyst could be recycled up to five times (210). Likewise, the dendrimer technique (211) did not prove to be productive, as recycling failed (212).

1.12 METAL CONTAMINATION OF THE METATHESIS REACTION PRODUCTS

The preceding sections have summarized the efforts to recover and recycle the metathesis catalysts in the spirit

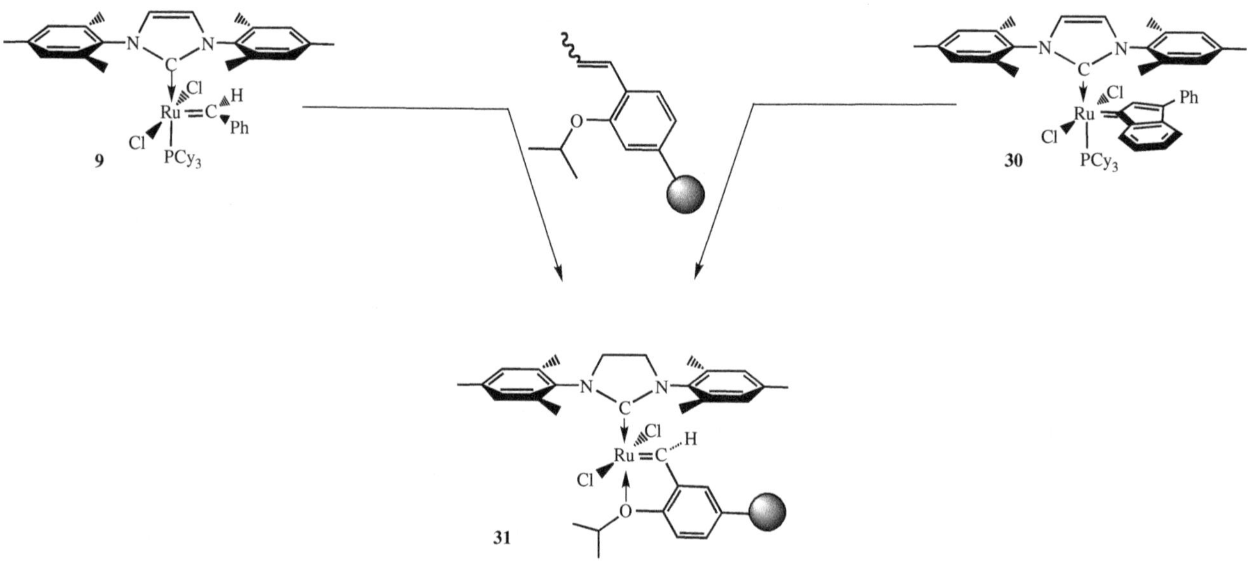

Scheme 1.8 Boomerang synthesis of Ru metathesis catalysts **31** derivatized on solid supports starting from the catalyst **9** or **30** (165).

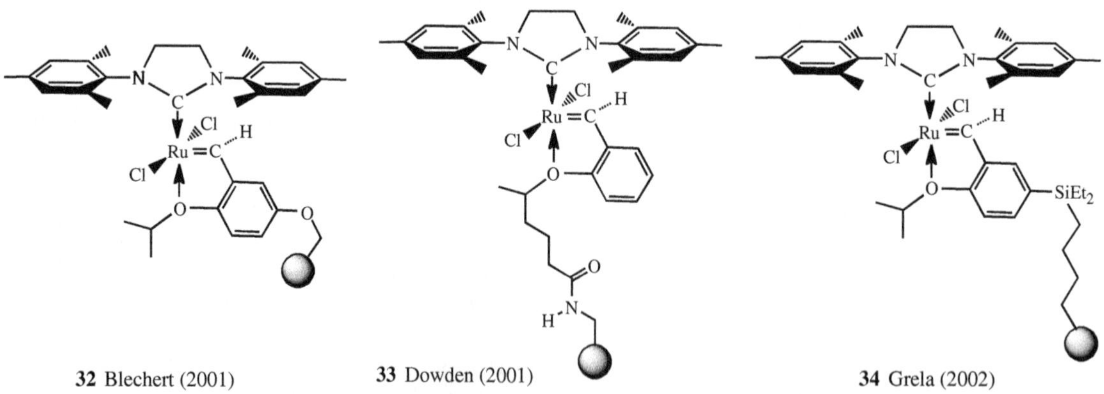

Figure 1.11 Examples **32–34** of ruthenium metathesis catalysts anchored on solid support and recycled several times with high yields. The isopropyl benzylidene-chelating ligand allows the so-called "boomerang" recycling that is allegedly efficient.

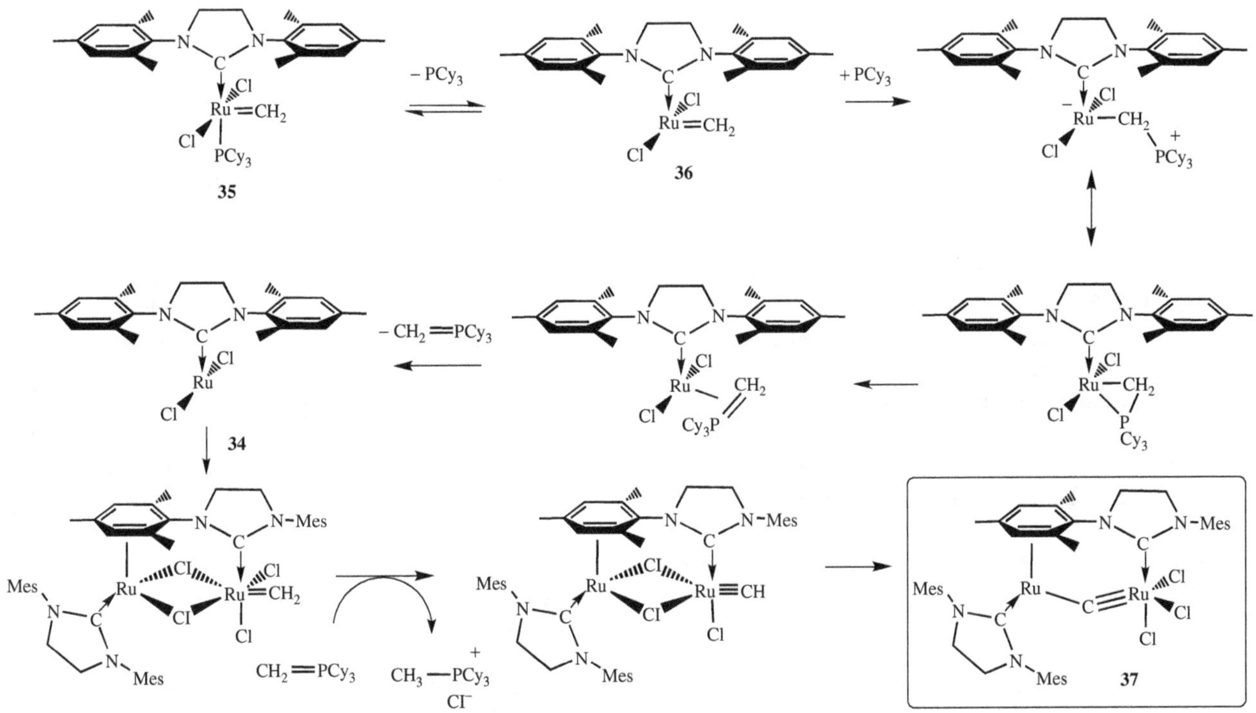

Scheme 1.9 Proposed mechanism for the decomposition of the second-generation Grubbs ruthenium–methylene complex **35** to the carbide-bridged diruthenium complex **37** (100).

of "green chemistry." Ruthenium catalysts have concentrated most studies on metal contamination of the products, although immobilization of Mo and W catalysts is also well known (165,193–196,200–204,213). Metathesis chemistry requires a rather large amount of catalyst due to the high reactivity of the metal–methylene species involved in the metathesis of terminal olefins. For instance, many metathesis steps in total synthesis use 20 mol% ruthenium (126,214,215). Yet, the acceptable limit for pharmaceutical industries concerning ruthenium contamination is 10 ppm. Thus, considerable efforts have been devoted to the removal of metal contamination either by chromatographic purification or by tagging the catalyst to separate it from the products. The most classic and efficient tag is a solid phase involving attachment of the catalyst to an insoluble organic or inorganic polymer as indicated in the above section, but water (Section 1.9), ionic liquids, or perfluorinated solvents (Section 1.10) are alternative media attracting tags that have specific affinities for these solvents and allow separation from organic products. A simple example is the derivatization of an NHC carbene ligand with PEG of the Grubbs–Hoveyda complex that catalyzes RCM, followed by extraction of the reaction mixture with water, leading to a contamination level of 41 ppm, which was decreased to 0.04 ppm (ICP-MS) by treatment with activated carbon (216). Other published strategies to decrease Ru contamination down to 60–200 ppm involve purification of the products on silica gel followed by treatment with activated carbon, use of DMSO, $Ph_3P{=}O$, or lead tetra-acetate as Ru scavengers in combination with chromatography and treatment with isocyanide $CNCH_2CO_2K$ (130). As the decomposition of Grubbs' dichlororuthenium catalyst **10** was shown to yield (via the methylene complex **35**) the neutral carbide-bridged organo-diruthenium complex **37** (Scheme 1.9) (100,217,218), an additional potential solution to remove them from the organic metathesis products might be to oxidize them.

A drawback of a large number of publications on metathesis catalysis is the lack of reports of the metal contamination of the reaction products, although inductively coupled plasma mass spectrometry (ICP-MS) is an ideal tool for this purpose. This issue is the key to the awaited industrial development of drugs involving olefin metathesis steps in their synthesis.

1.13 INDUSTRIAL APPLICATIONS

1.13.1 Simple Alkenes

Some simple alkenes are produced on an industrial scale. Although propene is mainly produced from naphtha steam crackers, strong demand has led to the development of the Phillips tri-olefin process from ethylene and 2-butene, using heterogeneous metathesis catalysts such as WO_3/SiO_2 at

>260 °C and 30–35 bar. 1-Butene can be used in such a process with MgO as a catalyst for isomerization to 2-butene. This process has been developed in Texas by Lyondell Petrochemical Co. and BASF Petrochemical, in Osaka by Mitsui Chemicals, in Singapore by Shell Chemicals and Sumitomo Chemical and in Caojing, China, by Shangai Secco Petrochemical. The Institut Français du Pétrole and the Chinese Petroleum Corporation at Kaoshiang, Taiwan, have also developed the Meta-4 process in liquid phase using Re_2O_7/Al_2O_3 at 35 °C and 60 bar (219).

Neohexene (3,3-dimethyl-1-butene), an intermediate in the synthesis of Tolalide, a synthetic musk perfume, and Terbinafine, an antifungal agent, is produced by metathesis of ethylene and the dimer of isobutene (a mixture of 2,4,4-trimethyl-2-pentene and 2,4,4-trimethyl-1-pentene).

In the Shell Higher Olefin Process (SHOP), lighter (<C6) and heavier (>C18) terminal alkenes produced by nickel-catalyzed oligomerization of ethylene are isomerized over a potassium metal catalyst, then metathesized over a MoO_3/Al_2O_3 catalyst at 100–125 °C and 10 bar to a statistical distribution of C_{11}–C_{14} linear internal alkenes sold by Shell Chemical as Neodene (>1,000,000 t per year) (220).

$$CH_3\text{-}CH = CH\text{-}CH_3 + CH_3(CH_2)_7CH = CH(CH_2)_9CH_3 \rightleftharpoons$$
$$CH_3\text{-}CH = CH(CH_2)_7CH_3 + CH_3\text{-}CH = CH(CH_2)_9CH_3$$

1.13.2 Polymers

The three main cycloolefins that have been transformed to polymers using the ROMP reaction in industrial processes are norbornene (the classic example), dicyclopentadiene, and cyclo-octene (221,222). Cyclopentene, however, was also polymerized to *trans*- and *cis*-cyclopentenamer for the rubber industry soon after the discovery of olefin metathesis (Scheme 1.10) (221–224).

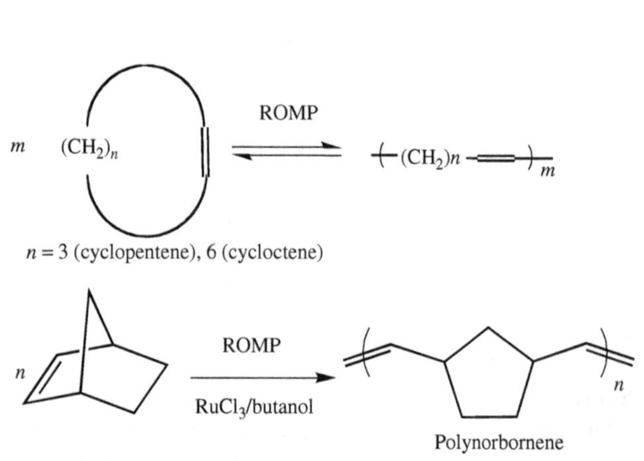

Scheme 1.10 Main cycloalkenes (together with dicyclopentadiene) utilized for ROMP reactions in industry.

In 1976, CdF Chimie and Elf Atochem in France started to produce norbornene polymers. The ROMP reaction of 2-norbornene catalyzed by $RuCl_3$/HCl in butanol operating in air gives a 90% trans polymer of molecular weight $>3 \times 10^6$ g/mol, Norsorex, produced by Atofina and developed by Nippon Zeon as Zeonex (225,226).

The ROMP reaction of endo-dicyclopentadiene (obtained from naphtha crackers) leads to opening of the strained norbornene ring, yielding linear polymers. Under certain conditions, however, the cyclopentene double bond also opens, giving cross linking. The catalyst can be WCl_6 + $WOCl_4$, with the $EtAlCl_2$ co-catalyst and additives, a tetrakis (tridodecylammonium)octa-molybdate activated with a mixture of Et_2AlCl, propanol and $SiCl_4$ or a Grubbs-type ruthenium catalyst. These polymers are largely used for heavy-vehicle applications (220).

Degussa has been producing Vestenamer 8012 by ROMP of cyclooctene since 1980. The reaction proceeds in hexane using WCl_6 as the catalyst. The trans double bond content is 80% and the molecular weight is 75,000. A minor fraction of the product results from cyclization of short oligomers. Vestenamer 8012 is hard at 20 °C and has an exceptionally high viscosity, then becomes fluid above 60 °C. It is useful in blends (225).

1.13.3 Fine Chemicals

The availability of highly chemo-selective catalysts is now developing applications in oleochemistry, agrochemicals (insect pheromones, etc.), fragrances, drugs, pharmacy, and so on (see Sections 1.14 and 1.15) (227).

1.14 APPLICATIONS TO ORGANIC SYNTHESIS

Metathesis reactions have a huge impact on the possibilities of transformation of organic molecules and are systematically thought of and involved in organic strategies, including those concerning complex structures and total syntheses (1,214,215). This area is well developed in the second volume of Grubbs' (1) metathesis handbook, and a book on metathesis in natural product synthesis edited by Janine Cossy (215) that has recently appeared. Several Grubbs-type and Schrock-type catalysts and derivatives are commercial and now of common use in organic synthesis. A study of the tolerance of Mo metathesis catalysts has indicated that Schrock's Mo catalysts, although air- and moisture-sensitive, are effective in the presence of phosphanes, thioethers, nitriles (whereas the Ru catalysts are decomposed by these substrates), sterically protected free alcohols, metal carbonyls and, in many cases, in the presence of amines (even giving unparalleled enantioselectivity with the latter groups) (4). Grubbs and the family of Grubbs-type catalysts benefit from the largest use in organic synthesis, because they are stable in air and moisture, and metathesis reactions can be conducted in water or aqueous solvents (Section 1.8). The main type of metathesis used in organic synthesis is olefin RCM, despite unpredictable and problematic *E*/*Z* selectivity (a problem that is avoided in alkyne RCM, see below Section 1.14), because complex rings are of great value in pharmaceuticals. CM and EYM are also largely used (228). Ethylene (or 2-butyne) is lost in RCM and CM reactions of terminal dienes (or diynes terminated by ethylidyne fragments), which displaces the thermodynamic equilibrium toward the formation of the metathesis products. On the other hand, EYM is "greener" in the sense that no carbon is lost in the metathesis reactions.

Remarkably, RCM metathesis allows the synthesis of small, medium, and large rings. The easiest RCM reactions are those affording synthesis of 5-membred rings (carbocycles and a large variety of heterocycles), and the entropic and enthalpic factors become less favorable as the ring size increases (229–231). Many eight-membered rings have been synthesized by metathesis, however. Large macrocycle formation (calixarenes, cyclodextrins, macrolactones, macrolactams, and catenanes) usually requires high dilution (to minimize oligomerization), high catalyst content, and higher reaction temperatures because of competition with catalyst decomposition (228). Tandem or domino reactions, that is, successive one-pot reactions involving RCM and another reaction such as ring-opening metathesis (ROM), CM, or a non-metathesis reaction are particularly elegant, ecologically advantageous, and useful, because they allow the rapid construction of complex structure from relatively simple precursors (228,232,233). Optically pure cyclic compounds are accessible from an optically pure precursor diene upon treatment with an achiral RCM catalyst or alternatively a racemic RCM product can be resolved using an enantiomerically pure catalyst, such as one of Schrock's or Grubbs' chiral catalysts (Eqs 1.6, 1.7) (3,4,234).

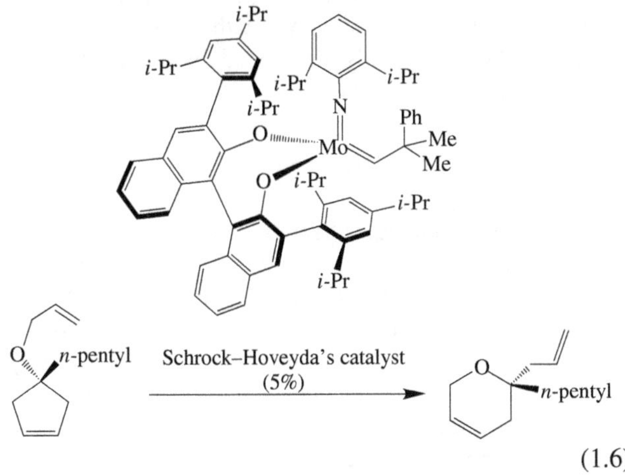

(1.6)

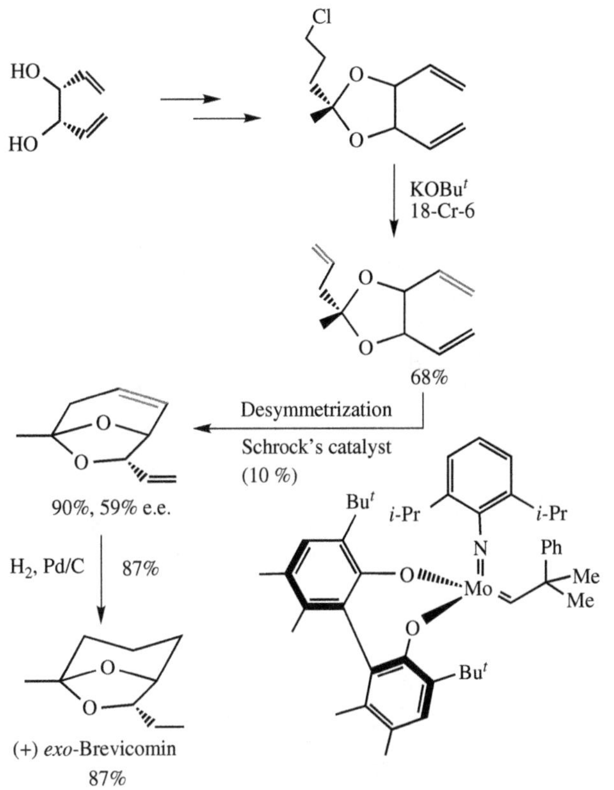

(1.7)

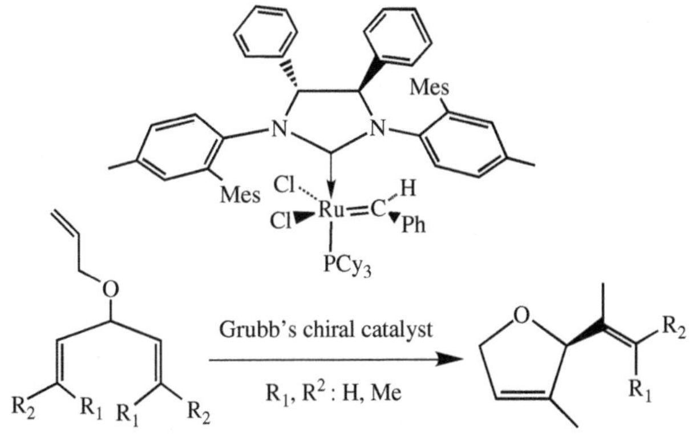

(1.8)

A plethora of RCM reactions have been reported for the synthesis of bioactive compounds, including total syntheses (233–236). As an example, the macrocyclic hepatitis C virus (HCV) S3 protease inhibitor-labeled BILN 2061 (Ciluprevir™), **39**, was synthesized via RCM (237). This RCM reaction gave the best results with Grela's (238) catalysts (in particular **14**), yielding the macrocycle **38** with the desired *Z* selectivity using an RCM reaction that has then been scaled-up to 400 kg (Scheme 1.11) (239).

Cross metathesis is another currently used metathesis reaction for the synthesis of a wide variety of biologically active compounds and organic materials, although it also suffers from the lack of *E*/*Z* selectivity with many olefins (Eq. 1.8) (239).

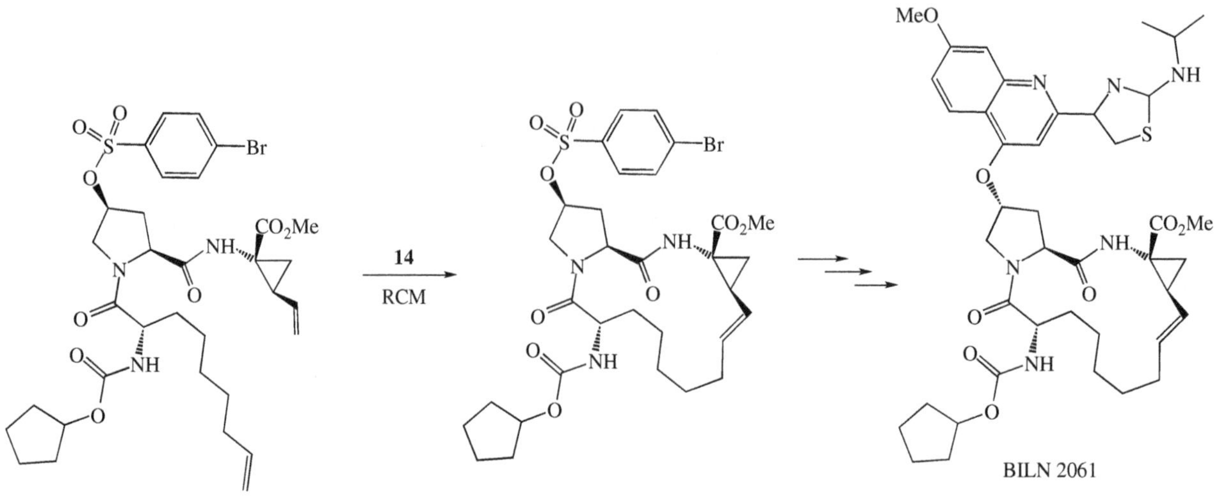

Scheme 1.11 Synthesis of BILN 2061 involving a stereospecific RCM step most efficiently catalyzed by Grela's complex **14** (238).

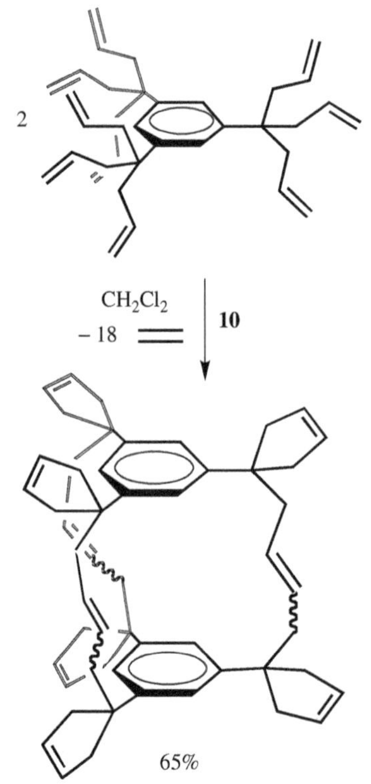

(1.9)

The problem of the lack of *Z* selectivity has been recently solved by Grubbs' and Schrock's groups (240) with new catalysts bringing about *Z* selectivity based on kinetic control in the metathesis products. When one of the olefin bears an electron-withdrawing or bulky substituent, however, excellent *E* selectivity can be obtained as a result of thermodynamic control (241–243). For instance, CM also is a key step in the synthesis of *E*-stilbenes that are precursors of anticancer compounds also useful for cardiovascular properties such as resveratrol, piceatanniol, and pinostilbene (244,245).

CM has been used for the total synthesis *inter alia* of plastensimycin (and carbaplatensimycin) and analogs that attract considerable interest due to their pharmaceutical properties related to strong activity against many drug-resistant bacteria. A CM step is a crucial common point in their total syntheses (236,246,247).

EYM is largely applied to the cyclization of enynes producing cyclic conjugated dienes, although intermolecular metathesis between an alkyne and an olefin (enyne cross metathesis) is almost limited to ethylene and thus much less applied in organic chemistry (particularly in total synthesis) to produce acyclic conjugated 2,3-disubstituted dienes due to selectivity problems. This metathesis reaction between internal alkynes and ethylene (1 atm) to produce such dienes, however, was introduced by Mori's group for the synthesis of anolignan A. High yields for cross EYM under 1 atm ethylene are usually obtained with the catalyst **10** (248). Earlier, in the mid 1990s, it was Mori who had pioneered intramolecular EYM using the first-generation Grubbs catalyst, and rapidly applied it to total synthesis, a field that has been considerably expanded (235,249).

Many applications of alkyne metathesis to total syntheses have also been elegantly developed by Fürstner's group (Section 1.15).

1.15 ALKYNE METATHESIS

Although the metathesis of internal (and sometimes terminal, see below) alkynes proceeds according to the same Chauvin-type mechanism (2) as olefin metathesis (250), via metallacyclobutadiene intermediates (that can be isolated and also serve as alkyne metathesis catalysts (78,251,252)), it is remarkable that olefin metathesis catalysts do not metathesize alkynes (except enynes), and alkyne metathesis catalysts do not metathesize olefins.

Besides the unpractical heterogeneous mixture of tungsten oxide and silica that was found in the late 1960s to metathesize alkynes at 200–450 °C (253), the first useful internal alkyne metathesis catalysts were reported by Mortreux and Blanchard (254) in 1972 and consisted of a mixture of $[Mo(CO)_6]$ and a simple phenol derivative in high boiling solvents. In 1998, Bunz's group (255,256) reported that phenols bearing electron-deficient substituents such as *p*-chlorophenol, a temperature of 130 °C, and purging 2-butyne with a N_2 stream gave optimized yields and rates, which proved to be synthetically valuable, and other groups indicated that microwave assistance was also helpful (257,258). Improvement of the $[Mo(CO)_6]$ catalyst with 2-flurorophenol was reported in 2002 by Grela and Ignatowska (117,259). In 2003, Lavigne and Chauvin's group (260) found that the temperature could be decreased to 50 °C in the presence of 1,2-diphenoxyethane and molecular sieves.

Schrock's high oxidation state Mo and W alkylidyne complexes bearing the key alkoxide ligands (as in olefin metathesis) are well defined, unimolecular catalysts that are more reactive than Mortreux' catalyst and had been disclosed in the early 1980s (78,252,261,262). For instance, the prototype of this family, [W(C-*t*-Bu)(O-*t*-Bu)$_3$], **42**, was obtained in a remarkable metathesis reaction between a triply bonded tungsten–tungsten dimer **40** (an alkyne metathesis catalyst (78,263)) and neoheptyne (or also conveniently with *t*-BuCN) via the W_2C_2 tetrahedrane intermediate **41** (Scheme 1.12) (264,265).

It is an efficient catalyst at ambient temperature for 3-heptyne that is equilibrated with the metathesized alkynes in 5 min at 20 °C (260). In the light of the structure and efficiency of Schrock's catalysts, it appears probable that in Mortreux's improved system the phenol molecules are turned into phenoxide ligands, diphenoxyethane eventually serves as a bulky stabilizing chelate ligand, and the carbyne ligand comes from the same metathesis reaction as that involved in the synthesis of Schrock's catalyst.

In the late 1990s, Fürstner discovered even more active alkyne metathesis pre-catalysts such as the binary compound [Mo{N-*t*-Bu(2,4-dimethylphenyl)}$_3$], **43** (Fig. 1.12) producing [MoX{N-*t*-Bu(2,4-dimethylphenyl)}$_3$], X = Cl or Br upon reaction with a halogen source, in particular CH_2Cl_2.

These complexes do not contain a carbyne ligand, and the X-ray crystal structure of the Mo–Cl derivative showed considerable shielding of the metal by the bulky ligands (266,267). These precatalysts could even initiate (presumably also through a high oxidation-state catalytically active

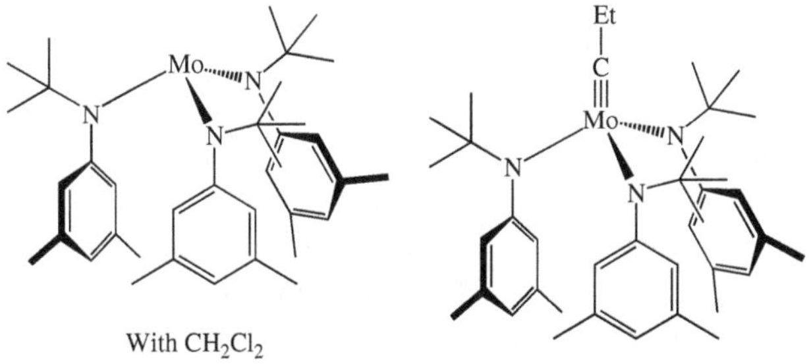

43 Fürstner (1999) **44** Moore (2003)

Figure 1.12 Fürstner's and Moore's efficient Mo alkyne metathesis initiators.

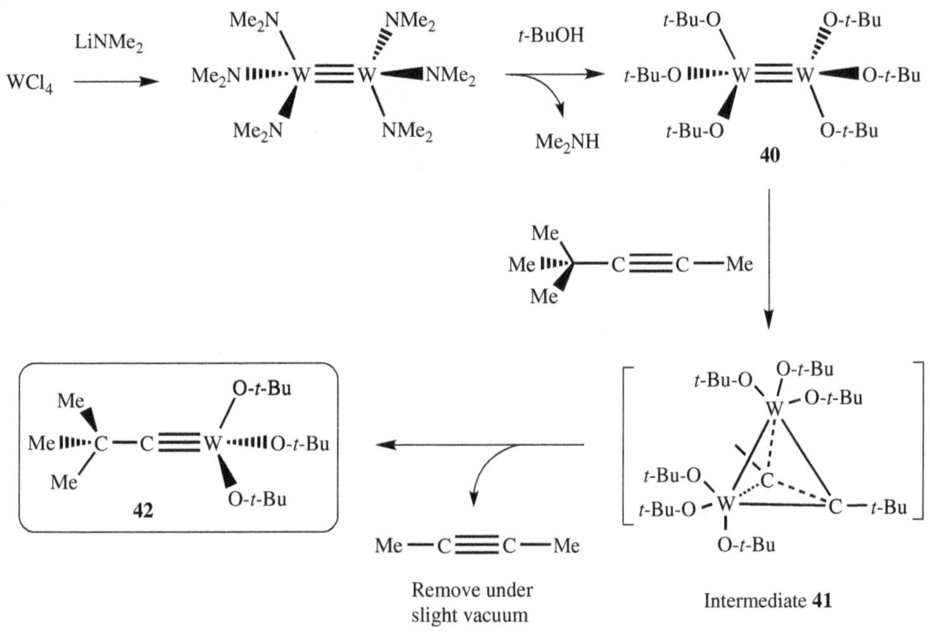

Scheme 1.12 Synthesis of Schrock's prototype alkyne metathesis catalyst **42** (264,265).

metal–carbyne species (268)) the metathesis of alkynes containing a thioether, amine, or crown ether fragment that could not be metathesized using Schrock's catalysts. The complex [Mo(Cl){N-*t*-Bu(2,4-dimethylphenyl)}$_3$], formed by reaction of **43** with CH_2Cl_2 is the pre-catalyst that was conveniently used in the metathesis reactions in this solvent.

Moore (268–270) has reported the use of [EtCMo {NAr(*t*-Bu)}$_3$], **44** (Fig. 1.12) for the remarkable synthesis of arylene ethynylene macrocycles by precipitation-driven alkyne metathesis. Discrete alkylidyne W and Mo complexes on silica prepared by Basset's group (271) also produce useful catalysts that are intermediate between homogeneous and heterogeneous, with the advantages of both types. Metathesis of 2-pentyne by the silica-supported alkylidyne–alkylidene complex [(SiO_xRe(C-*t*-Bu)(=CH-*t*-Bu)(CH_2-*t*-Bu)], **29** (Fig. 1.10) has been reported.

Applications of alkyne metathesis catalysis have been greatly developed using the improved Mortreux-type-, Schrock-type-, and Fürstner-type catalysts. The categories of metathesis reactions that have been exploited are the RCM, CM, and acyclic diyne metathesis (ADIMET). Although the metathesis of terminal alkynes is successful (272–275) using Schrock's catalyst [W(C-*t*-Bu)(O-*t*-Bu)$_3$] in the presence of quinuclidine, it has not been exploited in organic synthesis. However, 80% of the metathesis compounds were obtained from terminal alkynes at 80 °C within 1 min using 4 mol% catalyst. Even phenylacetylene that was earlier found to polymerize more easily than other terminal alkynes gave several turnovers of carbyne exchange metathesis (273,274). Besides, one can also find a literature report of an heterogeneous catalyst, $CoO{-}MoO_3{-}SiO_2$, for the carbyne exchange metathesis of terminal alkynes such as 1-heptyne (276). Terminal alkynes, however, polymerize in the presence of ordinary transition-metal initiators (such as [$W(CO)_6$]) to yield polymers according to the Katz mechanism that starts with the well-known isomerization of the terminal alkynes to a metal–vinylidene complex followed by coordination of another terminal alkyne molecule that forms a metallocyclobutene intermediate (41,250).

Thus, organic strategies involving alkyne metathesis use molecules terminated by a propynyl group for CM or two propynyl ends for diyne RCM. The byproduct of these alkyne metathesis reactions, 2-butyne, is volatile and removed from the reaction medium in the course of the reactions (a slight vacuum is necessary), which shifts the metathesis equilibrium toward the metathesis products. Owing to the ring strain of the cycloalkyne, no macrocycle smaller than 11 or 12 members can be obtained. A key point is the stereoselectivity that is obtained using alkyne metathesis, whereas the *E*/*Z* selectivity is a major drawback of many olefin RCM reactions. Lindlar reduction of alkynes gives *Z* olefins, whereas Birch reduction yields *E* olefins (Scheme 1.13).

This is exemplified here in the problem of the total synthesis of epothilone C. Several previous total syntheses of epothilone C using nonsteroselective olefin RCM had been less attractive than conventional methods. The total synthesis of epothilone C, **45**, was subsequently elegantly achieved by Fürstner's group via alkyne RCM followed by stereospecific Lindlar reduction to give the precursor yielding epothilone C in only one more step (Scheme 1.14). In this alkyne RCM metathesis reaction, Fürstner's pre-catalyst [MoCl{N-*t*-Bu(2,4-dimethylphenyl)}$_3$], **43**, in CH_2Cl_2/toluene was used whereas Schrock's catalyst [W(C-*t*-Bu)(O-*t*-Bu)$_3$] was not efficient due to the incompatible presence of sulfur and nitrogen atoms in the thiazole ring (277).

An impressive example of alkyne CM is the total synthesis of (+)dehydrohomoancepsenolide reported by Fürstner and Dierkes in 2000. It involves an alkene RCM catalyzed by the first-generation Grubbs' catalyst leaving the alkyne bond unreacted, followed by alkyne CM using Schrock's catalyst [W(C-*t*-Bu)(O-*t*-Bu)$_3$] (10%) at 100 °C leaving the olefinic bond unreacted, and finally stereoselective Lindlar reduction (277).

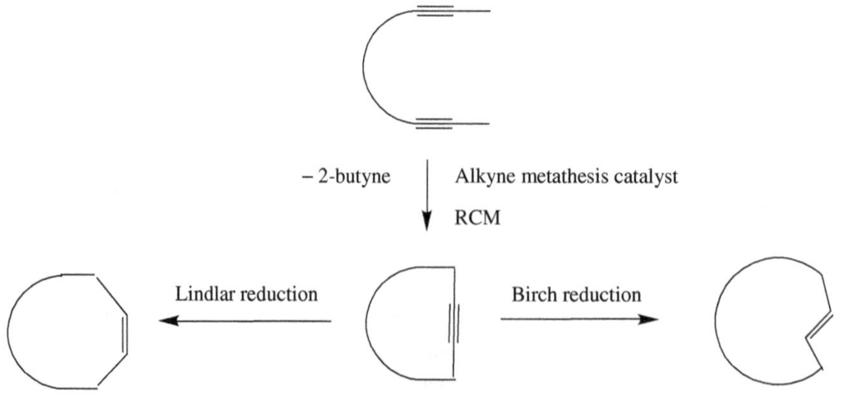

Scheme 1.13 Selective access to *E* and *Z* olefins initiated by Fürstner via alkyne RCM (131).

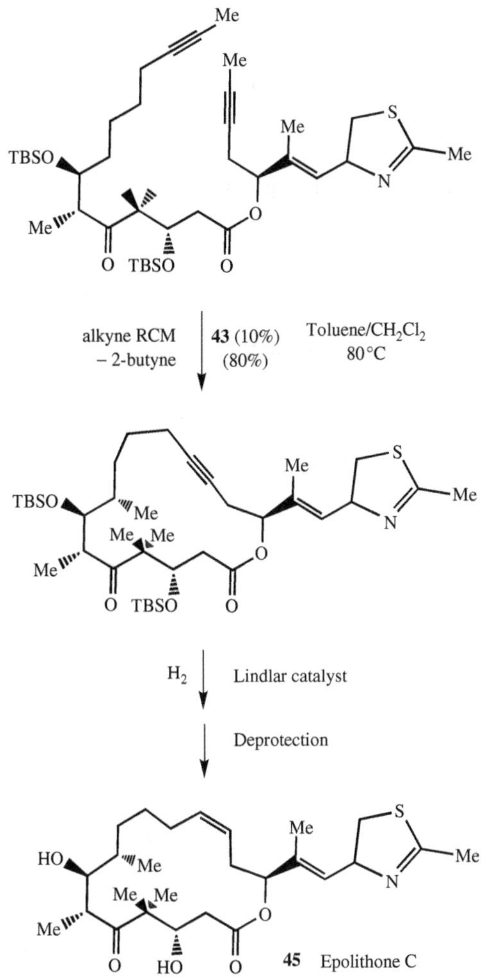

45 Epolithone C

Scheme 1.14 Synthesis of epothilone C: Fürstner's stereospecific synthesis via alkyne RCM followed by Lindlar semi-hydrogenation (277).

The alkyne CM reaction has also been largely developed by Bunz's group with the simple catalyst $[Mo(CO)_6]$ + p-ClC_6H_4OH for the synthesis of poly(dialkylparaphenyleneethynylene)s that are conjugated cyclic molecular nanostructures (256,279) and polymers with high molecular weights (with particularly interesting optical and liquid-crystalline properties) resulting from the metathesis of dipropynylated dialkylbenzenes (255,256).

1.16 ALKANE METATHESIS

The metathesis of alkanes was first reported by Burnett and Hugues (280) at the Chevron Company in 1973. When butane was passed at 400 °C on a heterogeneous Pt–Al_2O_3 mixture, a deshydrogenation catalyst, and WO_3–SiO_2 mixture, an olefin metathesis catalyst, the direct metathesis products propane and pentane were predominantly formed, together with lower and higher molecular weight alkanes (Scheme 1.15).

Basset's group reported a family of well-defined single-site heterogeneous Ta and W–alkylidene catalysts containing siloxy ligands that metathesize alkanes. These catalysts resulted from the reactions of silica with Schrock's high oxidation-state olefin metathesis catalysts. The siloxy ligand brought by silica played the role of alkoxy ligands in favoring metathesis activity with even an improved reactivity with the siloxy ligand resulting from the increased metal electrophilic properties compared to alkoxy complexes. The strategy was thus clearly to metathesize olefins resulting from alkane dehydrogenation (196,281–287). In particular, Basset et al. also noticed that propane and propene gave similar C_{n+1}/C_{n+2} ratio of cross-metathesis products on silica-supported Ta-neopentylidene catalysts at 150 °C. The complexes $[(SiO)_xTa(=CH\text{-}t\text{-}Bu)(CH_2\text{-}t\text{-}Bu)_{3-x}]$ ($x = 1$ or 2 catalyzed the metathesis of alkanes into a mixture of higher and lower alkanes at 150 °C, as did the hydride complex $[(SiO)_xTa–H]$. For instance, ethane reversibly yielded methane and propane. The mechanism was suggested to proceed by a composite series of σ-bond metathesis of C-H bonds and α and β-eliminations (rather than direct σ-bond metathesis of C–C bonds). The α-elimination from d^2 metal–methyl or metal–alkyl species formed, respectively $HTa{=}CH_2$ or HTa=CHR, and the mechanism was proposed to then follow an alkene metathesis pathway with olefins generated by β-elimination (including metallacyclobutane intermediates as in the Chauvin mechanism; Scheme 1.16) (194–196).

What is most elegant and original in this strategy is the multiple activation by a single site, and the in-depth characterization techniques of surface organometallic species that resulted in very efficient, well-controlled, and robust heterogeneous metathesis catalysts for alkanes, alkenes, and alkynes.

With this background in mind, Goldman and Brookhart also recently directly mimicked the system of the Chevron company (280), using well-defined homogeneous catalysts for alkane metathesis. The challenge was the compatibility between the alkane dehydrogenation catalyst and the olefin metathesis catalyst operating separately in solution. Successful "tandem" catalytic activation using homogeneous catalysts for both alkane dehydrogenation and olefin metathesis types was thus reported in 2006 (288,289). The dehydrogenation catalysts are Ir pincer complexes nicely designed and improved by Goldman (290–292) and further optimized by both research groups (293–297), and the olefin metathesis catalyst is a Schrock-type complex such as $[Mo(NAr)(=CHCMe_2Ph)(OR_{F6})_2]$ or a heterogeneous catalyst, Re_2O_7 on Al_2O_3. Reactions in neat octane or decane require heating for several days at more than 125 °C to approach alkane metathesis equilibrium, but the reaction is limited by the decomposition of the Mo–alkylidene catalyst.

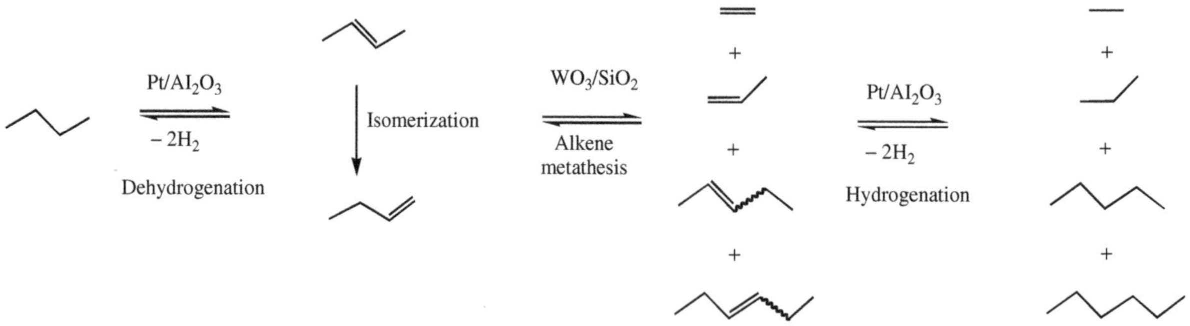

Scheme 1.15 Alkane metathesis: the Chevron process at 150 °C involving alkane dehydogenation followed by alkene metathesis and hydrogenation of the metathesized alkenes (280).

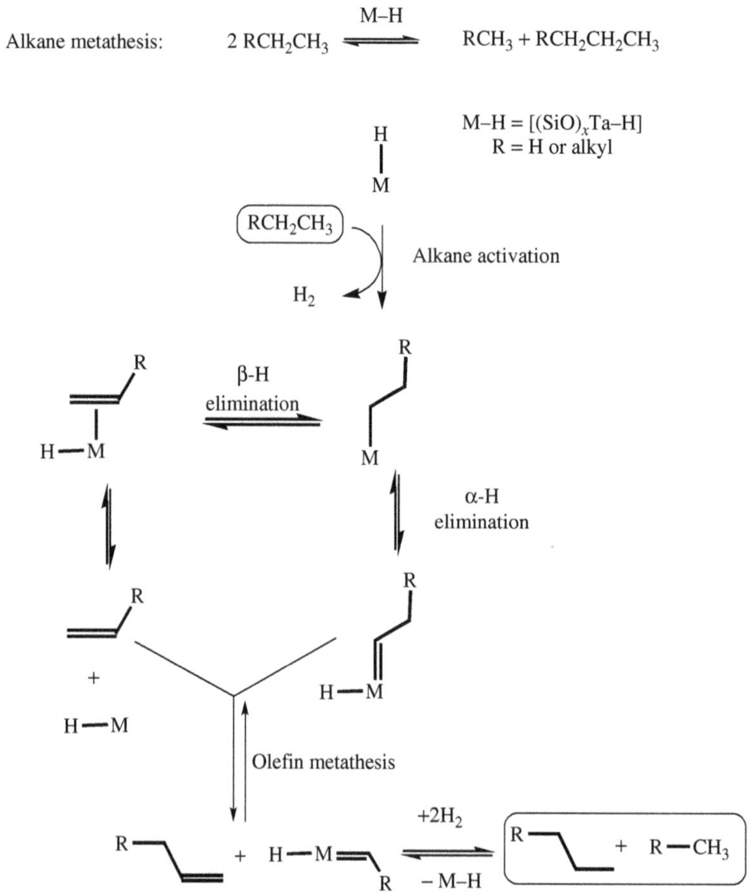

Scheme 1.16 Alkane metathesis at 150 °C on Basset' single-site early transition-metal catalysts containing siloxy ligands and its mechanism. The alkane activation step involves σ-bond metathesis between the M–H bond and an alkane C–H bond (196,281).

More than 40 Mo and W alkylidene catalysts were tested, W catalysts outperforming Mo ones, and the greatest activity was obtained using [W(NAr)(=CHCMe$_2$Ph)(OSiPh$_3$)$_2$] (297). Indeed, as in Basset's catalyst (see above), the siloxy ligand brings an advantage over the alkoxy ligand because it presumably reduces the donation of π electron density to the metal.

1.17 POLYMERIZATION METATHESIS AND POLYMER MATERIALS

Several types of metathesis reactions produce polymers and the polymer field occupies a major part of the overall activities in metathesis research. For instance, the entire third volume of Grubbs' (1) handbook on metathesis is devoted

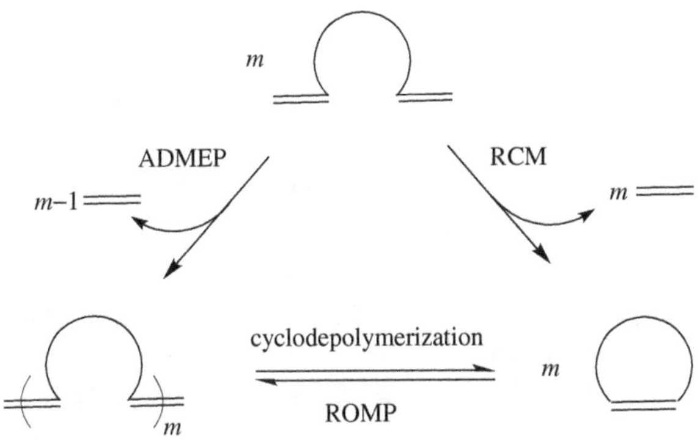

Scheme 1.17 Relationships between ring-closing, ring-opening, and polymerization/depolymerization processes in metathesis reactions initiated by metal–carbene complexes (301).

to polymers. The best known metathesis reaction leading to polymers is the ring opening metathesis polymerization (ROMP) that is mostly used for norbornene derivatives (the other cyclo-olefins undergoing ROMP are those that are also subjected to release of ring strain upon opening, which provides the driving force of the reaction, that is, in particular, cyclobutene, cyclopentene, cyclo-octene, and dicyclopentadience), including industrial developments (Section 1.8 and Scheme 1.10).

The acyclic diene metathesis (ADMET) polymerization is also well known to proceed when the RCM of the terminal diene is sterically inhibited (298). Its equivalent with aromatic methyl-terminated diynes (ADIMET) produces, as indicated in Section 1.15, poly(dialkylparaphenyleneethynylene)s in addition to sacrificial 2-butyne (256). Terminal alkynes polymerize according to the Chauvin–Katz metathesis mechanism subsequent to alkyne coordination and isomerization to metal–vinylidene complexes upon catalysis by tungsten carbonyl (41) and other simple organometallic compounds (299). This mechanism is different from the Ziegler–Natta-type polymerization of alkynes resulting from alkyne coordination followed by insertion into a metal alkyl bond (300). Finally, a unifying view should be highlighted by recalling that the ROMP and ADMET (or ADMEP) are also connected to the RCM, and that the reverse reaction of ROMP, cyclodepolymerization, is also known (Scheme 1.17). The equilibrium between RCM and its reverse leading to oligomerization, and the ring-chain equilibriums in ROMP have been emphasized and analyzed in depth in an excellent review by Monfette and Fogg (301) in 2009.

During the last decade, the stereo-control among the monomer units introduced by ROMP has been addressed, involving cis/trans configuration of the exo-cyclic double bond, configuration of the allylic bridgehead carbons, and linkage of unsymmetrically substituted monomers (302,303). For instance, a Schrock-type Mo initiator with hexa(isopropyl)terphenoxide and monopyrrolide ligand allowed cis selectivity in the ROMP of norbornadiene and cyclooctene derivatives, although these initiators suffered from high sensitivity toward moisture and oxygen (304). Another example is the alternative polymerization of cyclooctene and norbornene that was achieved with Ru initiators based on the different insertion rates for norbornene and cyclooctene that depend on the previous monomer inserted (305–310).

Most advances in polymer materials synthesized by ROMP have involved Grubbs' first-, second-, and third-generation catalysts and the indenylidene ruthenium catalysts (310,311). The ruthenium-based initiators are usually chosen for ROMP because of their functional-group tolerance and allowance to achieve copolymer syntheses (312,313). Highly functionalized polynorbornene homopolymers have been synthesized by ROMP, for instance, with radical moieties (TEMPO, 2,2,6,6-Tetramethylpiperidinyloxy) for applications as cathode-active materials in organic radical batteries (314). Random co-polymer synthesis allows to incorporate, for instance, optical sensors (309). Well-defined block copolymers with narrow size distributions have been reported (307). End-group functionalization can be implemented using a carbene-functionalized initiator, a chain-transfer agent during polymerization, or a terminating agent and is a valuable means for combining different polymerization techniques (reversible addition–fragmentation chain transfer (RAFT), atom transfer radical polymerization (ATRP), etc.) (315). The structural diversity that can be reached involves grafted, dendronized, and hyperbranched polymers with various material applications (resistant plastics, antifouling coating, thermoplastic elastomers, and emulsifiers) (316) and liquid crystals (317), porous polymers (318), and self-healing materials (310). Finally, implications of polymers in green chemistry often involve natural products (319) and water-soluble polymers (Section 1.7) (310).

1.18 CONCLUDING REMARKS

From the Chauvin mechanism, alkene, alkyne, and alkane metathesis reactions are rather well understood. All viable olefin metathesis catalysts proceed by olefin coordination of a 14-electron transition-metal species to produce 14-electron metallocyclobutane intermediates that can sometimes be isolated and shown to also be good metathesis catalysts. The various types of metathesis reactions lead to numerous applications in pharmacologically important drugs (alkene and alkyne RCM, and EYM) and polymer materials (ROMP, ADMET, and ADIMET) (1,320). The metathesis reactions are greener than many other bond forming reactions, because they do not require a base or product salt as in Pd-catalyzed cross carbon-carbon and carbon-element forming reactions. Yet there are a number of key challenges to address before metathesis catalysis can be extensively adopted in industrial processes. The present drawbacks are the metal contaminants related to the rather high loading of catalysts in reactions (from a few % to sometimes up to 50% in some steps of total syntheses) (321). Thus the robustness of the catalysts must still be improved, as well as the turnover numbers (322). Metathesis catalyst decomposition or aerobic sensitivity due to the intermediacy of fragile metal–methylene species intermediates in the catalytic cycles of common terminal olefin metathesis reactions remains a crucial problem. The definition and prevention of decomposition pathways is therefore a difficult challenge that remains to be faced. Olefin metathesis catalysts should be designed to reach (particularly) Z selectivity (323,324). The synthesis of large rings still requires the use of high dilution to minimize the formation of oligomers. This trend cannot be presently extended to large scales. This stereochemistry is especially challenging for the formation of di- and trisubstituted olefins. The design of chiral catalysts to produce highly enantioselective reactions is also called for (325–327). The progress that has been accomplished in the last thirty years is formidable and has brought the metathesis reactions to the forefront of organic synthesis and materials science. There is no doubt that the present exciting challenges to bring metathesis of current use in industry will encourage researches to make decisive new discoveries in the area (331–335).

ACKNOWLEDGMENTS

I am grateful to my students for their enthusiasm, hard work, and great ideas, to a number of colleagues for fruitful collaborations and discussions, and to the Université de Bordeaux 1, the Centre National de la Recherche Scientifique (CNRS), the Agence Nationale de la Recherche (ANR), the Ministère de la Recherche et de l'Enseignement Supérieur et (MRES), and the Institut Universitaire de France (IUF) for funding.

REFERENCES

1. Grubbs RH, editor. *Handbook of Metathesis*. Weinheim: Wiley-VCH; 2003. Volumes 1–3: *Catalyst Development* (Volume 1); *Applications in Organic Synthesis* (Volume 2); *Applications in Polymer Synthesis* (Volume 3).
2. Chauvin Y. *Angew Chem Int Ed* 2006;45:3740(Nobel Lecture).
3. Grubbs RH. *Angew Chem Int Ed* 2006;45:3760(Nobel Lecture).
4. Schrock RR. *Angew Chem Int Ed* 2006;45:3748(Nobel Lecture).
5. Recent review on metathesis reactions: Deraedt C, d'Halluin M, Astruc D. *Eur J Inorg Chem* 2013:4881.
6. Evans D, Osborn JA, Jardine FH, Wilkinson G. *Nature (London)* 1965;208:1203.
7. Kagan H, Dang TP. *J Chem Soc, Chem Commun* 1971:481 J Am Chem Soc 1972;94:6429.
8. Loupy A, Tchoubar B, Astruc D. *Chem Rev* 1992;92:1141.
9. Calderon N. *Tetrahedron Lett* 1967;34:3327.
10. Schneider V, Frölich PK. *Ind Eng Chem* 1931;23:1405.
11. Eleuterio HS. *German Pat* 1960:1072811.
12. Peters EF, Evering BLU.S. Patent. 19602,963,447.
13. Truett WL, Johnson DR, Robinson IM, Montague BP. *J Am Chem Soc* 1960;82:2337.
14. Banks RL, Bailey GC. *Int Eng Prod Dev* 1964;3:170.
15. For a detailed historical account of the early days, see Eleuterio HS. *J Mol Catal* 1991;65:55.
16. Rouhi AM. *Chem Eng News* 2002;80:29(December 23).
17. Calderon N. *Acc Chem Res* 1972;5:127.
18. Mol JC. *Chem Commun* 1968:633.
19. Bradshaw CPW, Howmen EJ, Turner L. *J Catal* 1967; 7:269.
20. Calderon N. *J Am Chem Soc* 1968;90:4133.
21. Jolly PW, Pettit R. *J Am Chem Soc* 1966;88:5044.
22. The synthesis of the C_5M_5 analogue of Pettit's complex later allowed the observation of the $Fe{=}CH_2$ bond and its rotation barrier by (1)H NMRGuerchais V, Astruc D. *J Chem Soc, Chem Commun* 1985:835.
23. Pettit R. *J Am Chem Soc* 1971;93:7087.
24. Note that contrary to the common double oxidative addition of O_2 to a metal center (typically group 4 to 6 transition metals) yielding a di(oxo)metal complex, double oxidative addition of an olefin to metal center to give a metal–bis-carbene species is still unknown to-date (although single oxidative addition to give a metallacyclopropane is known in the case of electron-poor olefins and electron-rich metal centers). See Ref. 25.
25. Astruc D. *Organometallic Chemistry and Catalysis*. Heidelberg: Springer; 2008.
26. Grubbs RH. *J Am Chem Soc* 1972;94:2538.
27. Grubbs RH. *Inorg Chem* 1973;12:2166.
28. Fischer EO. *Angew Chem Int Ed Engl* 1964;3:580.
29. Natta G. *Angew Chem Int Ed Engl* 1964;3:723.
30. Chauvin Y, Hérisson J-L. *Makromol Chem* 1971;141:161.
31. http://www.nobelprize.org/nobel_prizes/chemistry/laureates /2005/animation.html XXXX
32. Chauvin Y. *C R Seances Acad Sci, Sér C* 1973;276:169.
33. Chauvin Y, Commereuc D, Cruypelinck D. *Makromol Chem* 1976;177:2637.
34. Watson PL. *J Am Chem Soc* 1983;105:6491.
35. Thompson ME, Bercaw JE. *Pure Appl Chem* 1984;56:1.
36. Fendrick CM, Marks TJ. *J Am Chem Soc* 1984;108:425.
37. Brookhart M, Green MLH. *J Organomet Chem* 1983;250: 395.
38. Casey CP, Burkhardt TJ. *J Am Chem Soc* 1974;96:7808.
39. Grubbs RH, Carr DD, Hoppin C, Burk PL. *J Am Chem Soc* 1976;98:3478.

40. McGinnis J, Katz TJ, Hurwitz S. *J Am Chem Soc* 1976;98: 605.
41. Katz TJ. *Adv Organomet Chem* 1978;16:283.
42. Tebbe FN, Parshall GW, Reddy GS. *J Am Chem Soc* 1978;100:3611 Although Tebbe's compound was reported only in 1978, it was known several years earlier when Tebbe was at the du Pont de Nemours Company in the same time as Schrock. It is formed from Cp_2TiCl_2 and $AlMe_3$ according to an α-H elimination reaction yielding CH_4 and the bridging methylene complex that can generate a reactive Schrock-type $Ti{=}CH_2$ species upon reaction with a base. There is an analogy, in the formation and reactivity of Tebbe's reagent, with the formation of Schrock's Ta- and Nb alkylidene complexes also proceeding according to α-H elimination (see Section 1.7) (42–45).
43. Howard TR, Lee JB, Grubbs RH. *J Am Chem Soc* 1980;102:6876.
44. Anslyn EV, Grubbs RH. *J Am Chem Soc* 1987;109:4880.
45. Hawkins JM, Grubbs RH. *J Am Chem Soc* 1988;110:2821.
46. Wilkinson G. *Pure Appl Chem* 1959;71:627 and *Science* 1974;185:109.
47. Schrock RR, Parshall GW. *Chem Rev* 1976;76:243.
48. Lappert MF, Pierce R. *Chem Rev* 1976;76:219.
49. Langmuir I. *Science* 1921;54:59.
50. Schrock RR. *J Am Chem Soc* 1974;96:67796.
51. Schrock RR. *Science* 1983;219:13.
52. Feldman J, Schrock RR. *Prog Inorg Chem* 1991;39:1.
53. Brookhart M, Stubaker WB. *Chem Rev* 1987;87:432.
54. Bre A, Chauvin Y, Commereuc D. *Nouv J Chim* 1986;10: 535.
55. Chauvin Y, Olivier H. In: Cornils B, Herrmann WA, editors. *Applied Homogeneous Catalysis*. Weinheim: Wiley-VCH; 1996Volume 1, chaps 2.3.1 and 2.3.6.
56. Schrock RR. *J Am Chem Soc* 1975;97:6577.
57. Schrock RR, Rocklage SM, Wengrovius JH, Rupprecht G, Feldmann J. *J Mol Catal* 1980;8:73.
58. Wengrovius JH, Schrock RR, Churchill MR, Missert JR, Youngs WJ. *J Am Chem Soc* 1980;102:4515.
59. Schrock RR, Murdzek JS, Bazan GC, Robbins J, DiMare M, O'Regan M. *J Am Chem Soc* 1990;112:3875.
60. Bazan GC, Khosravi E, Schrock RR, Feast WJ, Gibson VC, O'Regan MB, Thomas JK, Davis WM. *J Am Chem Soc* 1990;112:8378.
61. Bazan GC, Oskam JH, Cho HN, Park LY, Schrock RR. *J Am Chem Soc* 1991;113:6899.
62. Kress J, Wesolek M, Osborn JA. *J Chem Soc, Chem Commun* 1982:514.
63. Kress J, Wesolek M, Osborn JA. *J Am Chem Soc* 1983;105:6346.
64. Kress J, Osborn JA, Green RME, Ivin KJ, Rooney JJ. *J Am Chem Soc* 1987;109:899.
65. Kress J, Aguero A, Osborn JA. *J Mol Catal* 1986;36:1.
66. Quignard F, Leconte M, Basset J-M. *J Chem Soc, Chem Commun* 1985:1816.
67. Leconte M, Basset J-M, Quignard F, Larroche C. In: Braterman PS, editor. *Reactions of Coordinated Ligands*. New York: Plenum; 1986 Volume 1. p 371–420.
68. Lefebvre F, Lecomte M, Pagano S, Mutch A, Basset J-M. *Polyhedron* 1995;14:3209.
69. Alexander JB, La DS, Cefalo DR, Hoveyda AH, Schrock RR. *J Am Chem Soc* 1998;120:4041.
70. Zu SS, Cefalo DR, La DS, Jamieson JY, Davis WM, Hoveyda AH, Schrock RR. *J Am Chem Soc* 1999;121:8251.
71. Aeilts SL, Cefalo DR, Bonitatebus PJ, Houser JH, Hoveyda AH, Schrock RR. *Angew Chem Int Ed* 2001;40:1452.
72. Teng X, Cefalo DR, Schrock RR, Hoveyda AH. *J Am Chem Soc* 2002;124:10779.
73. Hultzsch KH, Jernelius JA, Hoveyda AH, Schrock RR. *Angew Chem Int Ed* 2002;114:609.
74. Tsang WCP, Jernelius JA, Cortez GA, Weatherhead GS, Schrock RR, Hoveyda AH. *J Am Chem Soc* 2003;125: 2591.
75. Tsang WCP, Hultzsch KC, Alexander JB, Bonitatebus PJ, Schrock RR. *J Am Chem Soc* 2003;125:2652.
76. Schrock, R. R. In Ref. 1, Volume 1, chap. 1.2, XXXX, p 8.
77. Guggenberger LJ, Schrock RR. *J Am Chem Soc* 1975;97:2935.
78. Schrock RR, Listermann ML, Sturgeoff LG. *J Am Chem Soc* 1982;104:4291.
79. Green MLH, Mitchard LC, Swanwick MG. *J Chem Soc A* 1971:794.
80. Natta G, Dall' Asta G, Porri L. *Makromol Chem* 1965;81: 253.
81. Grubbs RH, Tumas WJ. The capture by Grubbs of the renaissance of polymerization chemistry is well represented by his review article. *Science* 1989;243:907.
82. Novak BM, Grubbs RH. *J Am Chem Soc* 1988;110:7542.
83. France MB, Grubbs RH, McGrath V, Paciello RA. *Macromolecules* 1993;26:4742.
84. Nguyen ST, Johnson LK, Grubbs RH, Ziller JW. *J Am Chem Soc* 1992;114:3974.
85. Demonceau A, Noels AF, Saive E, Hubert AJ. *J Mol Catal* 1992;76:123.
86. Stumpf AW, Saive E, Demonceau A, Noels AF. *J Chem Soc, Chem Commun* 1995:1127.
87. Schwab P, France MB, Ziller JW, Grubbs RH. *Angew Chem Int Ed Engl* 1998;37:1124.
88. Olivan M, Caulton K. *Chem Commun* 1997:1733.
89. Wolf W, Stüer C, Grünwald C, Werner H, Schwab P, Schulz M. *Angew Chem Int Ed Engl* 1998;37:1124.
90. Werner, H., Wolf, J. In Ref. 1, Grubbs, RH., editor. Wiley-VCH, Weinheim, 2003, Volume 1, XXXX, chap. 1.8.
91. Gandelman M, Rybtchinski B, Ashkenazi N, Gauvin RM, Milstein D. *J Am Chem Soc* 1997;119:3887.
92. Arduengo AJ. *Acc Chem Res* 1999;32:913.
93. Herrmann WA, Köcher C. *Angew Chem Int Ed Engl* 1997;109:2256.

94. Bourissou D, Guerret O, Gabbaï FP, Bertrand G. *Chem Rev* 2000;100:39.
95. Wetskamp T, Schattenmann WC, Herrmann WA. *Angew Chem Int Ed* 1998;37:2490.
96. Scholl M, Trnka TM, Morgan JP, Grubbs RH. *Tetrahedron Lett* 1999;40:2247.
97. Scholl M, Ding S, Lee CW, Grubbs RH. *Org Lett* 1999;1:953.
98. Huang J, Stevens ED, Nolan SP, Petersen JL. *J Am Chem Soc* 1999;121:2647.
99. Ackermann L, Fürstner A, Weskamp T, Kohl FJ, Herrmann WA. *Tetrahedron Lett* 1999;40:4787.
100. Trnka TM, Morgan JP, Sanford MS, Wilhelm TE, Scholl M, Choi TL, Ding S, Day MW, Grubbs RH. *J Am Chem Soc* 2003;125:2546.
101. Love A, Sanford MS, Day MW, Grubbs RH. *J Am Chem Soc* 2003;125:10103.
102. Chatterjee AK, Choi T-L, Sanders DP, Grubbs RH. *J Am Chem Soc* 2003;125:11360(this article is particularly useful for the *E* selectivity in cross metathesis).
103. Morril C, Grubbs RH. *J Org Chem* 2003;68:603.
104. Kilbinger FM, Candrill SJ, Waltman AW, Day MW, Grubbs RH. *Angew Chem Int Ed* 2003;42:3281.
105. Love JA, Morgan JP, Trnka TM, Grubbs RH. *Angew Chem Int Ed* 2003;42:4035.
106. The second-generation Grubbs' catalyst **10** thermally decomposes at 55 °C in benzene to give, after 72 h, a 46% isolated yield of a hydridic dinuclear carbyne (HRuCRu) complex that catalyzes olefin isomerizationHong H, Day MW, Grubbs RH. *J Am Chem Soc* 2004;126:7414.
107. Trnka TM, Grubbs RH. *Acc Chem Res* 2001;34:18.
108. Hansen SM, Volland MAO, Rominger F, Eisenträger F, Hofmann P. *Angew Chem Int Ed* 1999;38:1273.
109. Hansen SM, Rominger F, Metz M, Hofmann P. *Chem-Eur J* 1999;5:557.
110. Volland MAO, Ansen SM, Rominger F, Hofmann P. *Organometallics* 2004;23:800.
111. Demonceau A, Stumpf AW, Saive E, Noels AF. *Macromolecules* 1997;30:3127.
112. Fürstner A, Koch D, Langemann W, Leitner C. *Angew Chem Int Ed* 1997;36:2466.
113. Fürstner A, Picquet C, Bruneau C, Dixneuf PH. *Chem Commun* 1998:1315.
114. Kingsbury JS, Harrity JPA, Bonitatebus PJ, Hoveyda AH. *J Am Chem Soc* 1999;121:791.
115. Garber SB, Kingsbury JS, Gray BL, Hoveyda AH. *J Am Chem Soc* 2000;122:8168.
116. Grela K, Harutyunyan S, Michrowska A. *Angew Chem Int Ed* 2002;41:4038.
117. Michrowska A, Bujok R, Harutyunyan S, Grela K. *J Am Chem Soc* 2004;126:9318.
118. Krause O, Zarka MT, Anders U, Weberskirch R, Nuyken O, Buchmeiser MR. *Angew Chem Int Ed* 2003;42:5965.
119. Honda T, Namiki H, Kaneda K, Misutani H. *Org Lett* 2004;6:87.
120. Connon SJ, Dunne A, Blechert S. *Angew Chem Int Ed* 2002;41:3835.
121. Wakamatsu H, Blechert S. *Angew Chem Int Ed* 2002;41:2403.
122. Dunne AM, Mix S, Blechert S. *Tetrahedron Lett* 2003;44:2733.
123. Cossy J, Bargiggia F, Bouzbouz S. *Org Lett* 2003;5:459.
124. Bouzbouz S, De Lemos E, Cossy J. *Adv Syn Catal* 2002;344:627.
125. Connon SJ, Blechert S. *Angew Chem Int Ed* 2003;42:1900.
126. Clavier H, Grela K, Kirschning A, Mauduit M, Nolan SP. *Angew Chem Int Ed* 2007;46:6786.
127. Bieniek M, Michowska A, Usanov DL, Grela K. *Chem Eur J* 2008;14:806.
128. Samojlowicz C, Bieniek M, Grela K. *Chem Rev* 2009;109:3708.
129. Disendruck CE, Tzur E, Lemcoff NG. *Eur J Inorg Chem* 2009:4185.
130. Vougioukalakis GC, Grubbs RH. *Chem Rev* 2010;110:1746.
131. Fürstner A. *Angew Chem Int Ed* 2000;39:3012.
132. Fürstner A, Hill AF, Liebl M, Wilton-Ely JDET. *Chem Commun* 1999:601.
133. Fürstner A, Radkowski K, Wittz C, Goddard R. *J Am Chem Soc* 2002;124:7061.
134. Fürstner A. *Proc Natl Acad Sci U S A* 2004;101:11960.
135. Jafarpour L, Schane H-J, Stevens ED, Nolan SP. *J Organomet Chem* 1999;18:5416.
136. Clavier H, Petersen JL, Nolan SP. *J Organomet Chem* 2006;691:5444.
137. For a review of indenylidene ruthenium metathesis catalyst, see Dragutan V, Dragutan I, Verpoort F. *Platinum Met Rev* 2005;49:33.
138. Romero PE, Piers WE, Mc Donald R. *Angew Chem Int Ed* 2004;43:6161.
139. Romero PE, Piers WE. *J Am Chem Soc* 2005;127:5032.
140. Romero PE, Piers WE. *J Am Chem Soc* 2007;129:1698.
141. Van der Eide EF, Romero PE, Piers WE. *J Am Chem Soc* 2008;130:4485.
142. Ben-Asuly A, Tzur E, Diesendrück CE, Sigalov M, Goldberg I, Lemcoff NG. *Organometallics* 2008;27:811.
143. Kost T, Sigalov M, Goldberg I, Ben Asuly A, Lemcoff NG. *J Organomet Chem* 2008;693:2200.
144. Szadkowska A, Makal A, Wozniak K, Kadyrov R, Grela K. *Organometallics* 2009;28:2693.
145. Szadkowska A, Samojlowicz C, Grela K. *Pure Appl Chem* 2011;83:553.
146. Sanford MS, Love JA, Grubbs RH. *Organometallics* 2001;20:5314.
147. Love JA, Morgan JP, Trnka TM, Grubbs RH. *Angew Chem Int Ed* 2002;41:4035.
148. Choi TL, Grubbs RH. *Angew Chem Int Ed* 2003;42:1743.
149. Mortell KH, Wertherman RV, Kiesling LL. *J Am Chem Soc* 1996;118:2297.
150. Burtscher D, Grela K. *Angew Chem Int Ed* 2009;48:442.

151. Connon SJ, Rivard M, Zaja M, Blechert S. *Adv Syn Catal* 2003;345:572.
152. Binder JB, Blank JJ, Raines RT. *Org Lett* 2007;9:4885.
153. Gulajski L, Sledz P, Lupa A, Grela K. *Green Chem* 2008;10:279.
154. Lynn DM, Kanaoka S, Grubbs RH. *J Am Chem Soc* 1996;118:784.
155. Gordon EJ, Sanders WJ, Kiessling LL. *Nature* 1998;392:30.
156. Claverie JP, Viala S, Maurel V, Novat C. *Macromolecules* 2001;34:382.
157. Quemeneur D, Heroguez V, Gnanou Y. *Macromolecules* 2005;38:7977.
158. Monteil V, Wehrmann P, Mecking S. *J Am Chem Soc* 2005;127:14568.
159. Davis KJ, Sinou D. *J Mol Catal* 2002;177:173.
160. Lipschulz BH, Aguinaldo GT, Ghorai S, Voigtritter K. *Org Lett* 2008;10:1325.
161. Lipschulz BH, Ghorai S, Aguinaldo GT. *Adv Synth Catal* 2008;7–8:953.
162. Diallo AK, Boisselier E, Ruiz J, Astruc D. *Chem Eur J* 2010;132:2729.
163. Nicolaou KC, Hughes R, Cho SY, Wissinger H, Labischinki H, Endermann R. *Chem Eur J* 2001;7:3824.
164. Mwangi LT, Runge MB, Bowden NB. *J Am Chem Soc* 2006;128:14434.
165. Buchmeiser MR. *New J Chem* 2004;28:549.
166. Michrowska A, Grela K. *Pure Appl Chem* 2008;80:31.
167. Jarroux N, Keller P, Mingotaud AF, Sykes C. *J Am Chem Soc* 2004;126:15958.
168. Mingotaud AF, Krämer M, Mingotaud C. *J Mol Catal A* 2007;263:39.
169. Zamman S, Curnow OJ, Abell AD. *Aust J Chem* 2009;62:91.
170. Chauvin Y, Mussmann L, Olivier-Bourbigou H. *Angew Chem Int Ed* 1995;34:2698.
171. Chauvin Y, Olivier-Bourbigou H. *CHEMTECH* 1995:26.
172. Sledz P, Mauduit M, Grela K. *Chem Soc Rev* 2008;37:2433.
173. Dupont J, de Souza RF, Suarez PA. *Chem Rev* 2002;102:3667.
174. Buijsman RC, van Vuuren E, Sterrenburg JG. *Org Lett* 2001;3:3785.
175. Martinez V, Blais JC, Astruc D. *Org Lett* 2002;4:651.
176. Mayo KG, Nearhoof, Kiddle JJ. *Org Lett* 2002;4:1567.
177. Garbacia S, Desai B, Lavastre O, Kappe CO. *J Org Chem* 2003;68:9136.
178. Boehringer Ingelheim Pharma GmbH § Co. KG, Patent Appl. WO2006/075021 A1, 2006.
179. Csihony S, Fishmeister C, Bruneau C, Horvath I, Dixneuf PH. *New J Chem* 2002;26:1667.
180. Audic N, Clavier H, Mauduit M, Guillemin J-C. *J Am Chem Soc* 2003;125:9248.
181. Chavier H, Audic N, Mauduit M, Guillemin J-C. *Chem Commun* 2004:2282.
182. Yao Q, Sheets M. *J Organomet Chem* 2005;690:3577.
183. Consorti CC, Aydos GLP, Ebeling G, Dupont J. *Org Lett* 2008;65:3397.
184. Consorti CC, Aydos GLP, Ebeling G, Dupont J. *Organometallics* 2009;28:4527.
185. Chen SW, Kim JH, Ryu KY, Lee W-W, Hong J, Lee S-g. *Tetrahedron* 2009;65:3397.
186. Rix D, Clavier H, Gulajski L, Grela K, Mauduit MJ. *Organomet Chem* 2006;691:5397.
187. Rix D, Caïjo F, Laurent I, Grela K, Mauduit M. *Chem Commun* 2007:3771.
188. Jessop PG, Ikariya T, Noyori R. *Chem Rev* 1999;99:475.
189. Prajapati D, Gohain R. *Tetrahedron* 2005;61:815.
190. Mistele CD, Thorp HH, DeSimone JM. *J Macromol Sci Pure Appl Chem* 1996;A33:953.
191. Fürstner A, Ackermann L, Beck K, Hiro H, Koch D, Lagemann K, Liebl M, Six C, Leitner W. *J Am Chem Soc* 2001;123:9000.
192. Michalek F, Mädge D, Rühe J, Bannwarth W. *Eur J Org Chem* 2006:577.
193. Copéret C, Lefebvre F, Basset J.-M. In Ref. 1, Volume 1, XXXX, chap. 1.12, p 190.
194. Copéret C, Chabanas M, Saint-Arroman RP, Basset J-M. *Angew Chem Int Ed* 2003;42:156.
195. Copéret C, Basset J-M. *Adv Syn Catal* 2007;349:78.
196. Basset J-M, Copéret C, Soulivong D, Taoufik M, Cazat JT. *Acc Chem Res* 2010;43:323.
197. Buchmeiser MR. *Chem Rev* 2009;109:303.
198. Sommer WJ, Weck M. *Coord Chem Rev* 2007;251:860.
199. Sebesta R, Kmentova I, Toma S. *Green Chem* 2008;37:2433.
200. Chabanas M, Baudouin A, Copéret C, Basset J-M. *J Am Chem Soc* 2003;125:492.
201. Basset J-M, Copéret C, Lefort L, Maunders BL, Maury O, Le Roux E, Saggio G, Soigner S, Soulovong D, Sunley GJ, Taoufik M, Thivole-Cazat J. *J Am Chem Soc* 2005;127:864.
202. Herrmann WA, Stumpf AW, Priermeier T, Bogdanovic S, Dufaud V, Basset J-M. *Angew Chem Int Ed Engl* 1996;35:2803.
203. Saint-Aroman PR, Chabanas M, Copéret C, Basset J-M, Emsley L. *J Am Chem Soc* 2001;123:3820.
204. Schrock RR. *Chem Rev* 2009;109:3211.
205. Blanc F, Rendon N, Berthoud R, Basset J-M, Copéret C, Tonzetich ZJ, Schrock RR. *J Chem Soc Dalton Trans* 2008:3156.
206. Mennecke K, Grela K, Kunz U, Kirschning A. *Synlett* 2005:2948.
207. Akyiama R, Kobayashi S. *Angew Chem Int Ed* 2002;41:2602.
208. Mayr M, Mayr B, Buchmeiser MR. *Angew Chem Int Ed* 2001;40:3839.
209. Allen DP, van Wingerden MM, Grubbs RH. *Org Lett* 2009;11:1261.
210. Randl S, Buschman N, Connon SJ, Blechert S. *Synlett* 2001;10:1547.
211. Astruc D, Boisselier E, Ornelas C. *Chem Rev* 2010;110:1857.

212. Garber SB, Kingsbury JS, Gray BL, Hoveyda AH. *J Am Chem Soc* 2000;41:9973.
213. Dolman SJ, Hultzsch KC, Pezet F, Teng X, Hoveyda AH, Schrock RR. *J Am Chem Soc* 2004;126:10945.
214. Nicolaou KC, Buger PG, Sarlah D. *Angew Chem Int Ed* 2006;118:4564.
215. (a) Cossy J, Arseniyadis S, Meyer C, editors. *Metathesis in Natural Product Synthesis*. Weinheim: Wiley-VCH; 2010. (b) Cossy J. In: Grela K, editor. *Metathesis Reactions*. Wiley; 2013chap. 3.
216. Hong SH, Grubbs RH. *Org Lett* 2007;9:1955.
217. Dinger MB, Mol JC. *Eur J Inorg Chem* 2003:2827.
218. Banti D, Mol JC. *J Organomet Chem* 2004;689:3113.
219. Parkinson G. *Chem Eng* 2001;108:27.
220. Mol JC. *J Mol Cat A: Chem* 2004;213:39.
221. Dragutan V, Streck R. *Catalytic Polymerization of Cycloolefins*. Amsterdam: Elsevier Science B. V.; 2000.
222. Schuster S, Blechert S. *Angew Chem Int Ed* 1997;36:2036.
223. Graulich W, Swodenk W, Theisen D. *Hydrocarbon Process* 1972;51:71.
224. Dall's Asta G. *Rubber Technol* 1974;47:511.
225. Singh OM. *J Sci Ind Res* 2006:957.
226. Schrock RR, Krouse SA, Knoll K, Felmann J. *J Mol Catal* 1988;46:243.
227. Pedersen, R.L. In Ref. 1, Volume 2, chap. 2.14, XXXX, p 491.
228. Han S-Y., Chang S. In Ref. 1, Volume 2, chap. 2.2, XXXX, p 5.
229. For the RCM synthesis of heterocycles, see Deithers A, Martin SF. *Chem Rev* 2004;104:2199.
230. For applications of ring-closing metathesis to alkaloid synthesis, see: Martin, S.F. In Ref. 1, Volume 2, XXXX, p 338.
231. For olefin metathesis strategies in synthesis of biologically relevant molecules, see Love, J. A. In Ref. 1, Volume 2, chap. 9, XXXX, p 295.
232. Alexakis A, Croset K. *Org Lett* 2002;4:4147.
233. Randl S., Blechert S. In Ref. 1, Volume 2, chap. 2.4, XXXX, p 151.
234. Hoveyda, AH. In Ref. 1, Volume 2, chap. 2.3, XXXX, p 128.
235. Nicolaou KC, Bulger PG, Sarlah. *Angew Chem Int Ed* 2005;44:4490.
236. Bicchielli D, Borguet Y, Delaude L, Demonceau A, Dragutan I, Dragutan V, Jossifov C, Kalinova R, Nicks F, Sauvage X. In: Dragutan V, Demonceau A, Dragutan I, Finkelshtein ES, editors. *Green Metathesis Chemistry*. Dordrecht: Springer; 2010. p 207.
237. Tsantrizos YS, Ferland JM, McClory A, Poirier M, Frina V, Yee NK, Wang X-J, Haddad N, Wei X, Zhang L. *J Organomet Chem* 2006;691:5163.
238. Bienek M, Bujok R, Stepowska H, Jacobi A, Hagenkötter R, Arlt D, Jarzembska K, Makal A, Wozniak K, Grela K. *J Organomet Chem* 2006;691:5289.
239. Nicola T, Brenner M, Donsbach K, Kreye P. *Org Process Res Dev* 2005;9:513.
240. (a) Keitz BK, Endo K, Patel PR, Herbert MB, Grubbs RH. *J Am Chem Soc* 2012;134:693. This new family of Ru metathesis catalysts features a C–H-activated ligand and a chelating nitrato ligand. It appears from DFT calculations that it is the steric repulsions between the substituents on the olefin and the N-substituent on the NHC carbene that leads to kinetic control of the highly selective formation of the Z product. See Liu P, Xu XF, Dong FX, Keitz BK, Herbert MB, Grubbs RH, Houk KN. J Am Chem Soc 2012;134:1464. For an example of application, see Herbert MB, Marx VM, Pederson RL, Grubbs RH. Angew Chem Int Ed 2013;52:310. (b) Kinetic control for *Z* selectivity in the stereoselective Schrock' Mo and W catalysts is provided by a bulky pyrrolidine ligandJiang AJ, Zhao Y, Schrock RR, Hoveyda AH. *J Am Chem Soc* 2009;131:16630 Example of application: Meek SJ, O'Brien RV, Llaveria J, Schrock RR, Hoveyda AH. Nature 2011;471:461.
241. For an example of cross metathesis with complete E selectivity in polymer and dendrimer syntheses, see Ornelas C, Méry D, Cloutet E, Ruiz J, Astruc D. *J Am Chem Soc* 2008;130:1495.
242. Ornelas C, Méry D, Blais J-C, Ruiz J, Astruc D. *Angew Chem Int Ed* 2005;44:7399.
243. Brümmer O, Rückert S, Blechert S. *Chem Eur J* 1997;3:441.
244. Ferré-Filmon K, Delaude L, Demonceau A, Noels F. *Eur J Org Chem* 2005:3319.
245. Velder J, Ritter S, Lex J, Schmalz H-G. *Synthesis* 2006;2:273.
246. Nicolaou KC, Pappo D, Tsang KY, Gibe R, Chen DY-K. *Angew Chem Int Ed* 2008;47:944.
247. Tiefenbacher K, Mulzer J. *Angew Chem Int Ed* 2008;47:2548.
248. Mori M, Tonogaki K, Nishigushi N. *J Org Chem* 2002;43:2235.
249. Mori, M. In Ref. 1, Volume 2, chap. 2.5, XXXX, p 176.
250. Katz TJ, McGinnis J. *J Am Chem Soc* 1975;97:1592.
251. Pedersen SF, Schrock RR, Chruchill MR, Wassermann HJ. *J Am Chem Soc* 1982;104:6808.
252. Schrock RR. *Acc Chem Res* 2001;19:342.
253. Pennella F, Bank RL, Bailey GC. *Chem Commun* 1968:1548.
254. Mortreux A, Blanchard M. *Bull Soc Chim Fr* 1972:1641.
255. Kloppenburg L, Song D, Bunz UHF. *J Am Chem Soc* 1998;79:7973.
256. Bunz, U. H. F. In: *Modern Arene Chemistry*; Astruc, D., editor; Wiley-VCH, Weinheim, 2002, chap. 7, p 217.
257. Villemin D, Héroux M, Blot V. *Tetrahedron Lett* 2001;42:3701.
258. Fürstner A, Stelzer F, Rumbo A, Krause H. *Chem Eur J* 2002;2:1856.
259. Grela K, Ignatowska J. *Org Lett* 2002;4:3747.
260. Huc V, Weihofen R, Martin-Jimenez I, Oulie P, Lepetit C, Lavigne G, Chauvin R. *New J Chem* 2003;27:1412.
261. Wengrovius JH, Sancho J, Schrock RR. *J Am Chem Soc* 1981;103:3932.
262. Sancho J, Schrock RR. *J Mol Catal* 1982;15:75.
263. Schrock RR, Listemann ML, Sturgeoff LG. *J Am Chem Soc* 1982;104:6729.

264. Listermann ML, Schrock RR. *Organometallics* 1985;4:74.
265. Schrock RR. *J Chem Soc Dalton Trans* 2001:2541.
266. Fürstner A, Mathes C, Lehmann CW. *J Am Chem Soc* 1999;121:9453.
267. Fürstner A, Mathes C, Lehmann CW. *Chem Eur J* 2001;7:5299.
268. Zhang W, Kraft S, Moore JS. *Chem Commun* 2003:832.
269. Zhang W, Kraft S, Moore J. *Am Chem Soc* 2004;126:329.
270. Zhang W, Moore JS. *J Am Chem Soc* 2004;126:12796.
271. Le Roux E, Taoufik M, Copéret C, de Mallmann A, Thivole-Cazat J, Basset J-M, Maunders BM, Sunley GJ. *Angew Chem Int Ed* 2005;44:6755.
272. Bray A, Mortreux A, Petit F, Petit M, Szymanska-Buzar T. *Chem Commun* 1993:197.
273. Mortreux A, Petit F, Petit M, Szymanska-Buzar T. *J Mol Catal A: Chem* 1995;96:95.
274. Mortreux A, Coutelier O. *J Mol Cat A: Chem* 2006;254:96.
275. Astruc D. *New J Chem* 2006;30:1848.
276. Mushegyan AV, Ksipterides VK, Dzhylakyan RK, Gevorkyan NA, Chukhadzhyan GA. *Armyan Khim Zh* 1975:672.
277. Fürstner A, Dierkes T. *Org Lett* 2000;2:2463.
278. Fürstner A, Mathes C, Grela K. *Chem Commun* 2001:1057.
279. Ge PH, Fu W, Campana C, Hertweck E, Herman WA, Adams RD, Bunz UHF. *Angew Chem Int Ed* 2000;39:3607.
280. Burnett RL, Hughes TR. *J Catal* 1973;31:55.
281. Vidal V, Theolier A, Thivole-Cazat J, Basset J-M. *Science* 1997;276:99.
282. Copéret C, Maury O, Thivolle-Cazat J, Basset J-M. *Angew Chem Int Ed* 2001;40:2331.
283. Basset J-M, Copéret C, Soulivong D, Taoufik M, Thivole-Cazat J. *Angew Chem Int Ed* 2006;45:6082.
284. Blanc F, Copéret C, Thivole-Cazat J, Basset J-M. *Angew Chem Int Ed* 2006;45:6201.
285. Schinzel S, Chermette H, Copéret C, Basset J-M. *J Am Chem Soc* 2008;130:7984.
286. Blanc F, Thivole-Cazat J, Basset J-M, Copéret C. *Chem Eur J* 2008;14:9030.
287. Maury O, Lefort L, Vidal V, Thivolle-Cazat J, Basset J-M. *Organometallics* 2010;29:6612.
288. Goldman AS, Roy AH, Huang Z, Ahuja R, Schinski W, Brookhart M. *Science* 2006;312:257.
289. Ahuja R, Kundu S, Goldman AS, Brookhart M, Vincente BC, Scott SL. *Chem Commun* 2008:253.
290. Xu W, Rosini GP, Krogh-Jespersen K, Goldman AS, Gupta M, Jensen CM, Kaska WC. *Chem Commun* 1997:2273.
291. Liu F, Pak EB, Singh B, Jensen CM, Goldman AS. *J Am Chem Soc* 1999;121:4086.
292. Zhu K, Achord PD, Zhang X, Krogh-Jespersen K, Goldman AS. *J Am Chem Soc* 2004;126:13044.
293. Göttker-Schnetmann I, White P, Brookhart M. *J Am Chem Soc* 2004;126:1804.
294. Göttker-Schnetmann I, Brookhart M. *J Am Chem Soc* 2004;126:9330.
295. Ahuja R, Punji B, Findlater M, Supplee C, Schinski W, Brookhart M, Goldman AS. *Nat Chem* 2011;3:167–171.
296. Coi J, MacArthur AHR, Brookhart M. *Chem Rev* 2011;111:1761.
297. Bailey BC, Schrock RR, Kundu S, Goldman AS, Huang Z, Brookhart M. *Organometallics* 2009;28:355.
298. Lehman S.E., Wagener K.B. In Ref. 1, Volume 3, chap. 3.9, XXXX, p 283.
299. This alkyne polymerization mechanism (41) can be considerably accelerated by coupling with electron-transfer-chain catalysisDesbois M-H, Astruc D. *New J Chem* 1989;13:595.
300. Masuda T, Sanda F. In Ref. 1, Volume 3, chap. 3.11, XXXX, p 375.
301. Monfette S, Fogg DE. *Chem Rev* 2009;109:3783.
302. Delaude L, Demonceau A, Noels AF. *Macromolecules* 2003;36:1446.
303. Hayano, Takeyama Y, Tsunogae Y, Igarashi I. *Macromolecules* 2006;39:4663.
304. Flook MM, Jiang AJ, Schrock RR, Müller P, Hoveyda AH. *J Am Chem Soc* 2009;131:7962.
305. Bornand M, Chen P. *Angew Chem Int Ed* 2005;44:7909.
306. Bornand M, Tocker S, Chen P. *Organometallics* 2007;26:3585.
307. Vehlow K, Wang D, Buchmeiser MR, Blechert S. *Angew Chem Int Ed* 2008;47:2615.
308. Lichtenheldt M, Wang D, Vehlow K, Reinhardt I, Kühnel C, Decker U, Buchmeiser MR, Blechert S. *Chem Eur J* 2009;15:9451.
309. Sutthasupa S, Shiotsuki M, Masuda T, Sanda F. *J Am Chem Soc* 2009;131:10546.
310. Leitgeb A, Wappel J, Slugovc C. *Polymer* 2010;51:2927.
311. Bielawski CW, Grubbs RH. *Prog Polym Sci* 2007;32:1.
312. Bielawski CW, Grubbs RH. *Controlled and Living Polymerizations*. Weinheim: Wiley-VCH; 2009. p 297.
313. Qu J, Katsumata T, Satoh M, Wada J, Matsuda T. *Polymer* 2009;50:391.
314. Noormofidi N, Slugovc C. *Eur J Chem* 2010;46:694.
315. Hilf S, Kilbinger AFM. *Nat Chem* 2009;1:537.
316. Hadjichristidis N, Pispas S, Pitsikalis M, Iatrou H, Lohse DJ. *Graft Copolymers, Encyclopedia of Polymer Science and Technology*. 3rd ed. Hoboken, NJ: Wiley; 2004.
317. Trimmel G, Riegler S, Fuchs G, Slugovc C, Stelzer F. *Adv Polym Sci* 2005;176:43.
318. Buchmeiser MR. *J Sep Sci* 2008;31:1907.
319. Meir MAR. *Macromol Chem Phys* 2009;210:1073.
320. Ivin KJ, Mol JC. *Olefin Metathesis and Metathesis Polymerization*. San Diego, CA: Academic Press; 1997.
321. Vougioukalakis GC. *Chem Eur J* 2012;18:8868.
322. Peeck LH, Savka RD, Plenio H. *Chem Eur J* 2012;18:12845.
323. Khan RKM, Zhugralin AR, Torker S, O'Brien RV, Lombardi PJ, Hoveyda AH. *J Am Chem Soc* 2012;134:12438.
324. Wegner HA. *Nachricht Chem* 2012;60:738.

325. Lujan C, Nolan SP. *Catal Sci Technol* 2012;2:1027.
326. Kress S, Blechert S. *Chem Soc Rev* 2012;41:4389.
327. Bicchielli D, Borguet Y, Delaude L, Demonceau A, Dragutan I, Dragutan V, Hans M, Jossifov C, Nicks F, Willem Q. *Curr Org Chem* 2012;9:397.
328. Schrock RR, McLain S, Sancho J. *J Am Chem Soc* 1979;101:4558.
329. Huang Z, Rolfe E, Carson EC, Brookhart M, Goldman AS, El-Khalafi SH, MacArthur AHR. *Adv Syn Catal* 2010;352:125.
330. Halbach MC, Kundu S, Brookhart M, Goldman AS. *Acc Chem Res* 2012;45:947.
331. Catalytic synthesis of *n*-alkyl arenes through alkyl group cross metathesis: Dobereiner GE, Yuan J, Schrock RR, Goldman AS, Hackenberg JD. *J Am Chem Soc* 2013;135:12572.
332. Recent review on alkyne metathesis by Mo and W alkylidyne complexes: Schrock RR, *Chem Commun* 2013;49:5529.
333. Olefin metathesis under continuous fow mode: Skoerski K, Wierzbicka C, Grela K. *Curr Org Chem* 2013;17:2740.
334. Silica-OWMe$_5$ efficient alkane metathesis catalyst with observed methyl/methylidyne species: Samantaray MK, Callens E, Abou-Hamad E, Rossini AJ, Widdifield CM, Dey R, Emsley L, Basset JM. *J Am Chem Soc* 2014;136:1054.
335. Ru-catalyzed Z-selective cross metathesis of allylic-substituted olefins: Quigley BL, Grubbs RH. *Chem Sci* 2014;5:501.

SECTION 2

TYPES OF OLEFIN METATHESIS REACTIONS

2

CROSS METATHESIS

KAROLINA ŻUKOWSKA

Institute of Organic Chemistry, Polish Academy of Sciences, Warsaw, Poland; Biological and Chemical Research Center, Department of Chemistry, Warsaw University, Warsaw, Poland

KAROL GRELA

Biological and Chemical Research Center, Department of Chemistry, Warsaw University, Warsaw, Poland

2.1 INTRODUCTION

Over the past decades, olefin metathesis (1) has become one of most widely utilized techniques of C–C double bond formation. This is reflected in awarding the 2005 Nobel Prize to Grubbs, Schrock, and Chauvin for their pioneering work in this area (2). Within the reaction, double bonds of two olefins are broken and rearranged by interchange of alkylidene species, resulting in the formation of two olefinic products. Undoubtedly, the greatest advantage of the transformation described is its universality. In fact, the term *olefin metathesis* includes several types of reactions that have been classified on the basis of the structure of both substrates and products. This chapter is devoted to olefin cross metathesis (CM), which is presented in the first equation in Scheme 2.1 (3).

The most common definition of CM describes it as a transformation of two alkenes, but this is not the only option. Depending on the number of multiple bonds present in both partners, we may have alkene, enyne, or alkyne CM variants (Scheme 2.1). This chapter focuses only on the alkene variant of CM, whereas information on the reactivity of enynes may be found in Chapter 3. As stated in the *Outlook*, alkyne metathesis is not covered in this book.

2.1.1 Principles

During CM, the double bonds of olefinic substrates are broken and subsequently rearranged. The mechanism of this transformation was described by Chauvin and Hérisson (4) and is now widely accepted (Scheme 2.2).

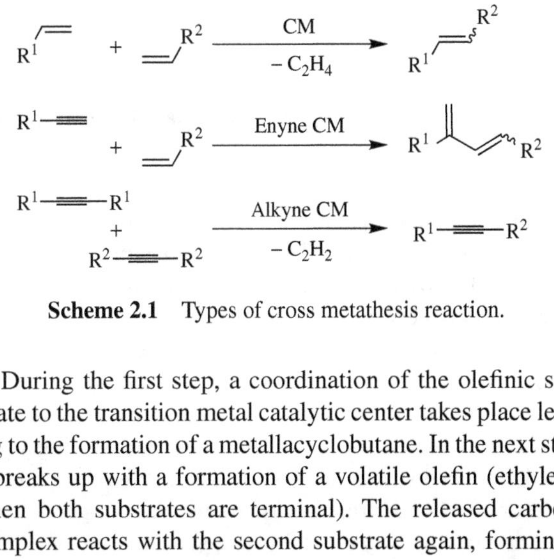

Scheme 2.1 Types of cross metathesis reaction.

During the first step, a coordination of the olefinic substrate to the transition metal catalytic center takes place leading to the formation of a metallacyclobutane. In the next step, it breaks up with a formation of a volatile olefin (ethylene, when both substrates are terminal). The released carbene complex reacts with the second substrate again, forming a cyclic intermediate. Decomposition of this species produces the final product, simultaneously releasing the carbene metal complex to the next step of the catalytic cycle. One of the most important characteristics of metathetic transformations stem from the fact that they are reversible. The fact that the release of the volatile product allows to drive the reaction to completion has a very important ramifications on the possible applications. Particular difficulties occur when two olefins *of similar reactivity* are subjected to CM (Scheme 2.3).

In case of full substrate conversion, the desired product will be statistically obtained with a maximum yield of 50%. At the same time, two undesired by-products originating

Olefin Metathesis: Theory and Practice, First Edition. Edited by Karol Grela.

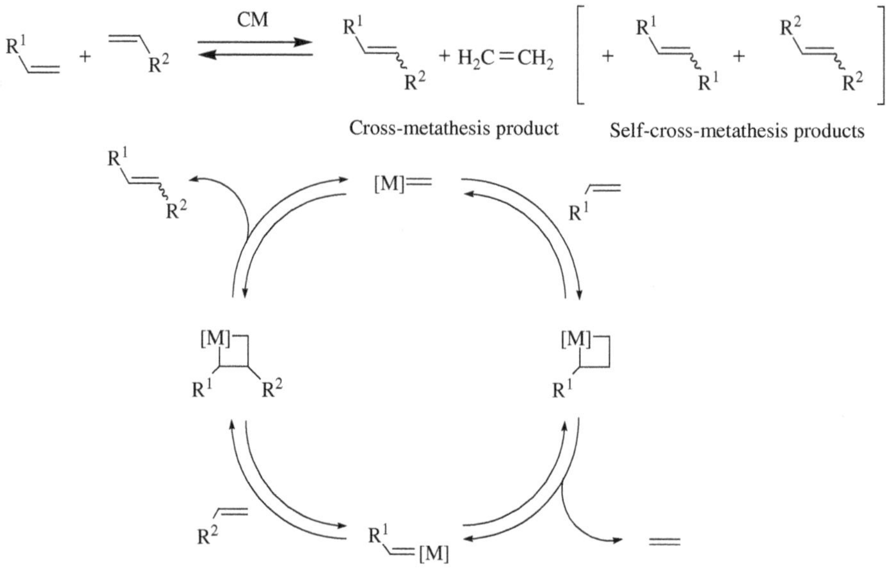

Scheme 2.2 Mechanism of cross metathesis.

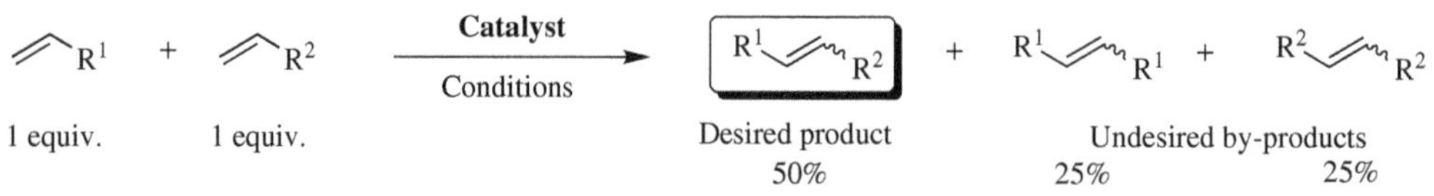

Scheme 2.3 Statistical distribution of cross metathesis products.

from the self-cross metathesis (S-CM) process (in literature terms "homodimerization" or "homometathesis" are also used) will be observed with 25% yield. There are a few solutions to this issue. One of these is to modify the reactivity of one of the partners by either electronic or steric factors. If this is not possible, applying an excess of one of the partners usually leads to a more synthetically suitable reaction outcome.

The described issue is not the only one leading to the broad product distribution. Another issue is the configuration of the double bond formed. Formation of *E*-configured bonds is thermodynamically more favored, yet a mixture of isomers is usually obtained. This is due to the small energy difference between those compounds. Obtaining a *Z*-configured product is a challenging task that requires a special approach; this is presented in Section 2.3.

Unfortunately, owing to the possible alkene C–C double bond isomerization (migration) during the metathesis reaction, an even more complicated mixture of products can be formed. Because the double bond isomerization products can also undergo further CM and isomerization, a really large number of by-products can be formed (Scheme 2.4).

A number of disparate pathways that emerge during such reactions prevents potential applications, therefore efforts were made to decrease the unwanted isomerization processes. A few solutions were proposed. Utilization of first-generation ruthenium complexes was found to be beneficial as they do not cause a noticeable amount of isomerization. The more active second-generation catalysts are known to shift the double bonds more easily, but this can be effectively suppressed by using a variety of additives such as benzoquinones (5), phenols (6), or tin halides (7).

The possibility of multiple metathesis pathways combined with reversibility issues makes it difficult to make any predictions on the reaction outcome. An additional difficulty is the catalyst choice, which influences not only the yield but also the selectivity of the reaction. For this reason, a wide selection of ruthenium, molybdenum, and tungsten complexes have been synthesized and are commercially available (Fig. 2.1) (8).

Their great functional group (FG) tolerance has made ruthenium complexes more renowned for their synthetic applications. Molybdenum and tungsten complexes are very active and often allow to obtain products with high selectivity, yet their applications are limited by lower tolerance toward air, moisture, and polar FGs. However, recent development of new, stable *Z*-selective catalysts should change this situation soon (see also Chapter 13).

The reactivity of olefins is a combination of steric and electronic factors. Focusing only on steric factors, one can note that differently substituted olefins have various reaction

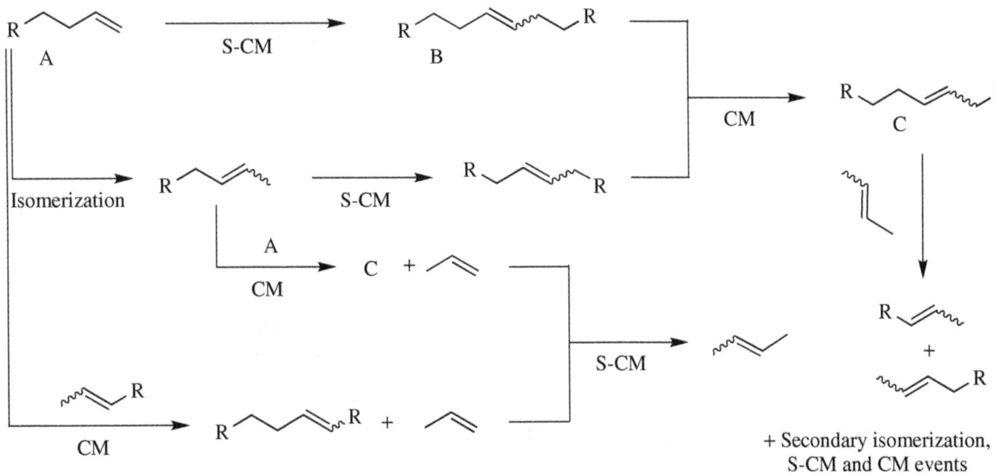

Scheme 2.4 Isomerization problem in cross metathesis.

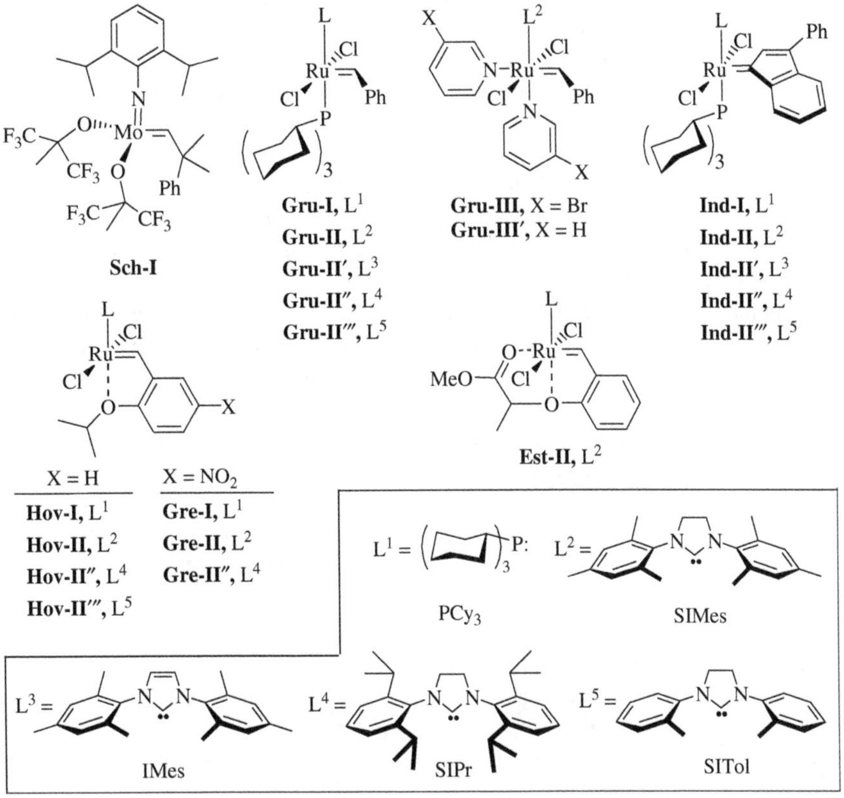

Figure 2.1 Selected olefin metathesis catalysts.

rates toward a metathesis catalyst. This can be illustrated using isomers of hexene as an example (Fig. 2.2) (9).

α-Olefins (alkenes with monosubstituted terminal C–C double bond, such as **1**) are typically the easiest substrates for CM reaction. Metathesis is also fast for substrates where the source of the steric hindrance is located relatively far from the reaction center (**2**). Internal olefins are less reactive then the terminal ones. The central part of Figure 2.2 presents *Z*- and *E*-configured hex-3-enes. Comparison of their reaction rates shows that (*Z*)-hex-3-ene (**3**) reacts faster than (*E*)-hex-3-ene (**4**). Although substrate **5** is a terminal olefin, because of the increased steric hindrance, its CM reaction is slower than that of **4**. The right part of Figure 2.2 is devoted to substrates with considerably increased steric hindrance. Irrespective of whether the double bond is geminally disubstituted or the source of steric hindrance is a bulky group located close to

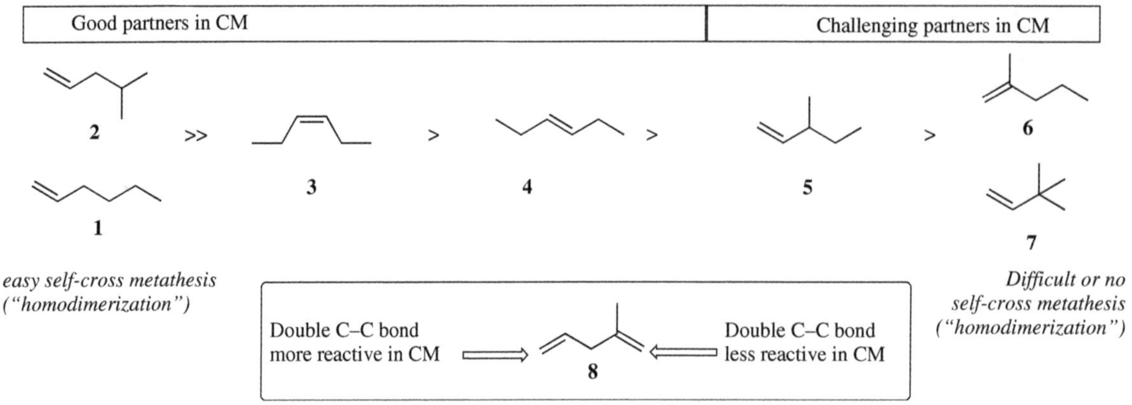

Figure 2.2 Relative reaction rates of variously substituted alkenes and dienes.

the reaction center, the reactivities of **6** and **7** are strongly diminished.

Because there is a significant difference in reaction rates between terminal mono- and disubstituted double bonds, in the case of substrates such as **8**, it is possible to selectively functionalize the more reactive C–C double bond in the presence of another olefinic fragment (for practical examples, see Schemes 2.13 and 2.50).

Some olefins can easily undergo S-CM. When such olefins are used as cross partners, their S-CM products are present in the reaction mixture and can be also converted into the expected CM product **9**. Alternative to using olefins **10** or **11**, one can consider using their "dimers" **12** or **13** as CM partners (Scheme 2.5). Such popular replacements for some terminal olefins are 1,4-butendiol (as an alternative to allylic alcohol) 1,4-diacetoxybutene (alternative to allyl acetate), 1,4-dihalobutene (replacement for allyl halides), etc. Many examples of this strategy are presented in this chapter.

2.1.2 Attempts in Olefin Classification

As CM has been developing rapidly for a long time, and the number of possible partners is on the increase, the need for introducing a classification is obvious. It is not so obvious, however, how to categorize the olefinic partners and what criteria to use. The most widely known system was established by Grubbs et al. (10) The authors selected three model catalyst types and divided the known CM partners into four categories depending on their rate of S-CM with these three catalysts: (i) rapid "homodimerization," (ii) slow "homodimerization," (iii) no "homodimerization," and (iv) spectators to CM. Then the authors generalized the behavior of each class in a reaction with another. This classification is without doubt very useful, but a complex issue such as this one generates exceptions and requires constant updates. Olefins previously classified as inert in CM (e.g., vinyl chlorides) are now successfully employed with the help of modern catalysts. Additionally, some reactions presumed to be selective in the light of this model have been shown to yield many products under different conditions. The main drawback of the described classification is putting all subtle *catalyst–substrate interactions* into one category, which sometimes leads to an oversimplification of the issue.

In this chapter, we attempt to introduce another classification scheme for CM substrates that can serve as a first-aid guide, useful in everyday laboratory practice.

2.2 FUNCTIONAL GROUP INFLUENCE ON THE METATHESIS OUTCOME

Taking into consideration the wealth of current knowledge on CM, making any categorizations is an overwhelming task. However, such classification is crucial to encourage potential users by guiding them through the entanglements of the CM methodology. Typically, CM is applied to "glue together" two different alkene fragments or to "decorate" a double bond with a fragment bearing some required

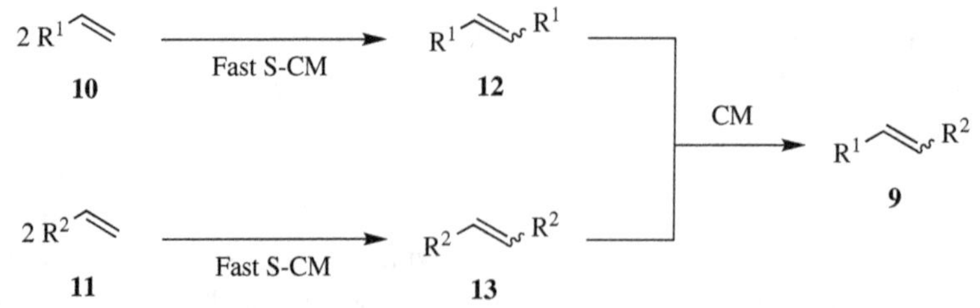

Scheme 2.5 Application of self-cross metathesis (S-CM) products in CM.

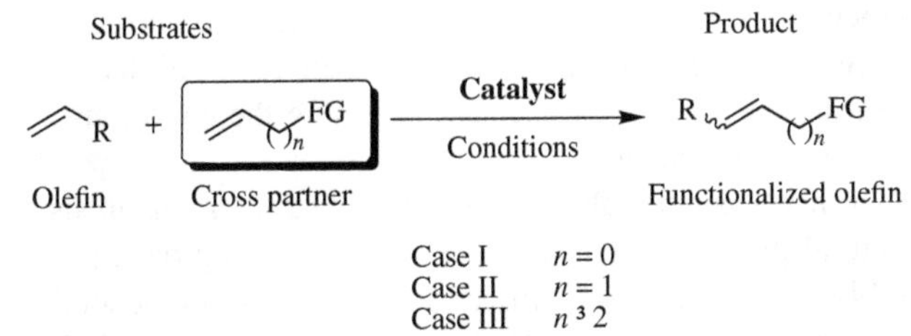

Scheme 2.6 General concept of olefin categorization used in this chapter.

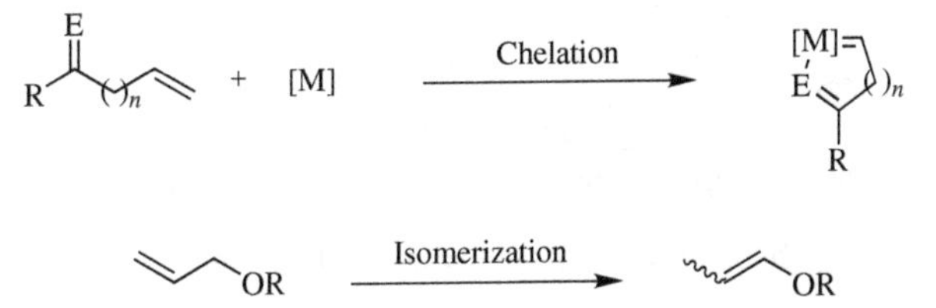

Scheme 2.7 Selected processes competing with cross metathesis.

functionality. The idea behind this chapter is simply to order all possible *CM partners* according to the FGs that they may contain. In fact, we decided to use the approach similar to the one used in *Beilstein's Handbook of Organic Chemistry*, brought out by Friedrich Konrad Beilstein in 1881. The character of a FG present in the CM partner molecules combined with the information on the distance between the FG and the reacting double bond will be the basis for our categorization. Let us discuss the following hypothetical CM reaction where an olefin is coupled with a functionalized CM partner—a process widely used in total synthesis (Scheme 2.6). In our classification, we are assuming that the reacting alkene bears in the residue R "innocent" FGs that do not influence the metathesis. The model for such an "olefin," presented in Scheme 2.6 can be, for example, *n*-hexene, or vinylferrocene, or a chiral homoallylic alcohol obtained in a stereoselective manner from an aldehyde and allyl metal reagent. In other words, this is our starting material, which we wish to functionalize in a CM reaction. The carrier for this functionality is the "CM partner" shown in Scheme 2.6 inside the box. Such a CM partner can be, for example, acrolein (FG = CHO, $n = 0$), styrene (FG = Ph, $n = 0$, allyl bromide (FG = Br, $n = 1$), or even *n*-hexene again (FG = C_4H_9).

This approach seems to be beneficial for some reasons. The most important one is that the user, interested in a given CM reaction, can quickly look under the FG(s) he or she needs to introduce, and is not obliged to have a broad literature knowledge of the entire metathesis methodology to benefit from this classification system. When both CM partners bear some non-innocent FGs, the user shall refer to both of them and observe if their nature and their location relative to the reacting double bonds, will lead to any problems.

The classification used in this chapter is therefore based on the FG(s) present in one or both CM partners. The nature of the FG is an obvious criterion for categorization, as it influences the behavior of the substrate molecule. The length of the spacer between the double bond and the FG, although less intuitive, is also of crucial importance. FG influences the CM reaction through both steric and electronic factors. It can even stop the CM reaction by forming a strong complex with the catalyst or arresting it in the form of a cyclic chelate (Scheme 2.7). It is also possible that FG can act as a trigger for other processes, such as double bond isomerization (Scheme 2.7) or catalyst decomposition.

In case of the model CM reaction shown in Scheme 2.6, three scenarios can apply. The first one ($n = 0$, α,β-unsaturated substrates) involves the FG directly attached to the reacting C=C bond and influencing the reaction by tuning the electronic character of the olefin. If the functionality does not have any detrimental effect, the yields and selectivities are usually good.

The second scenario is with the FG situated in the allylic position (Scheme 2.6, $n = 1$, β,γ-unsaturated substrates). In such cases, the direct electronic influence of the FG is diminished. The drawback is that molecules bearing chelating groups (such as carbonyls or amines) are able to arrest the olefin metathesis catalyst in the form of five-membered cyclic chelates of relative stability (Scheme 2.7).

The last scenario involves longer spacers (Scheme 2.6, $n \geq 2$). In such cases, an FG rarely has a strong influence on the reacting C–C double bond. These substrates are usually metathesized with no problem; however, the formation of cyclic chelates is still possible, which sometimes can shut down the catalyst's activity.

The all-cases exception are FGs known to be not compatible with a given metathesis catalyst (e.g., basic amines, isocyanides, some phosphines), that can cease olefin metathesis independently of their relative distance from the reacting C–C double bond.

The choice of FGs in this chapter was subjective. We tried to include the most important and widely utilized FGs in organic synthesis for which literature reports exist. Our motivation was to make this chapter useful in everyday laboratory practice, so the FGs are ordered according to the increasing atomic number of the atom directly connected to the carbon chain, similarly to the Cahn–Ingold–Prelog priority system.

2.2.1 Ethenolysis—Hydrogen as a Functional Group (FG = H)

The special and simplest type of CM involves ethylene—a cross partner bearing formally a hydrogen atom as an FG. This type of reaction is called *ethenolysis* and is presented in Scheme 2.8.

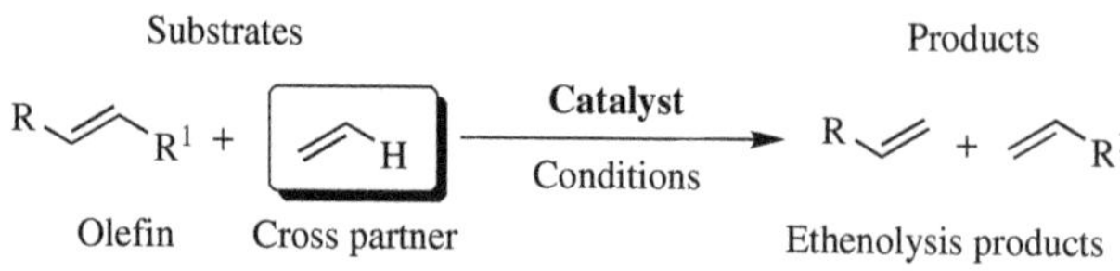

Scheme 2.8 General scheme of ethenolysis.

Owing to the simplicity of the process, ethenolysis is widely utilized in academia and in the industry. There are examples of both molybdenum- (11) and ruthenium-catalyzed (12) transformations. Probably the most widespread applications of ethenolysis are the ones connected to selective conversion of renewable feedstock, such as methyl oleate or other biomass-derived internally unsaturated compounds to terminal olefins (13). Such applications, especially on a large scale, are discussed in more detail in Chapter 11. Recent research has been directed also into exploring the potential of ethenolysis as a recycling method that would solve the problems such as those connected with storing large quantities of end-of-life tires (14).

Ethenolysis can be effectively applied in total synthesis. A very elegant example of *allyl to vinyl transformation* is presented in Scheme 2.9 (15). First, the allyl double bond isomerization utilizing **Gru-II** (similar processes are described in Chapter 5) led to product **15**, and then the subsequent shortening of the carbon chain length by ethenolysis promoted by **Hov-II** was made. Further hydrogenation of the resulted vinyl derivative **16** led to a (−)-tuberostemonine molecule, revealing the enormous potential of this method.

Ethenolysis has also significance in structural analysis of natural products. Scheme 2.10 presents an approach where CM with ethylene was utilized to establish a stereochemical composition of (+)-falcarindiol—a natural product derived from *Tetraplasandra hawaiiensis* (16). Comparison of the simplified compound **19** with an independently synthesized sample led to the determination of stereochemistry in the studied compound (Scheme 2.10). The presence of a diyne fragment that survived ethenolysis is noticeable.

Other numerous applications of CM with ethylene involve ethenolysis of silicon-containing cycloolefins (17), β-carotene (18) and many more.

2.2.2 Boron-Containing Alkenes (FG = $B(OR)_2$)

Boron compounds are of utmost importance largely due to the fact that a large number of FGs can be introduced via boranes or boronates. Some examples of these are hydroxyl, amine, aldehyde, ketone or halide functionalities. Last, but not the least, boron compounds are utilized in the Suzuki–Miyaura-type cross couplings, making them invaluable for organic synthesis. Thus CM of boron compounds allows many beneficial approaches in synthesis (Scheme 2.11).

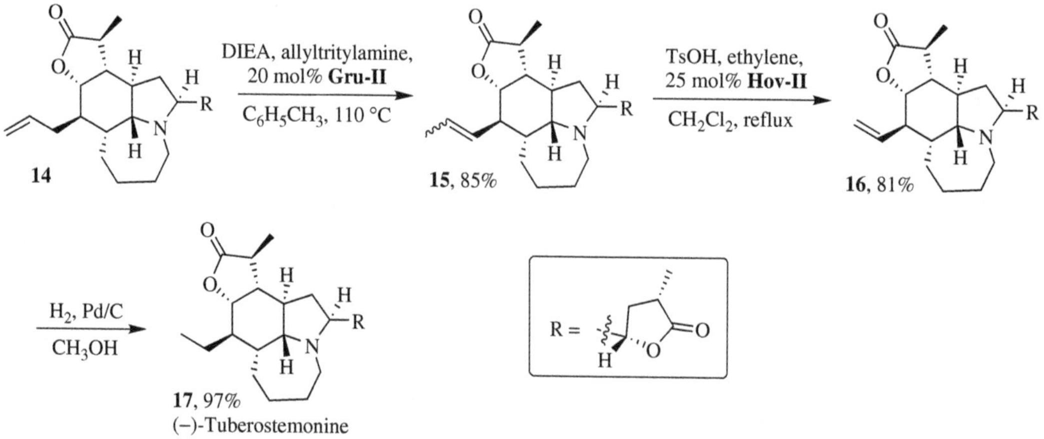

Scheme 2.9 Ethenolysis in total synthesis of (−)-tuberostemonine.

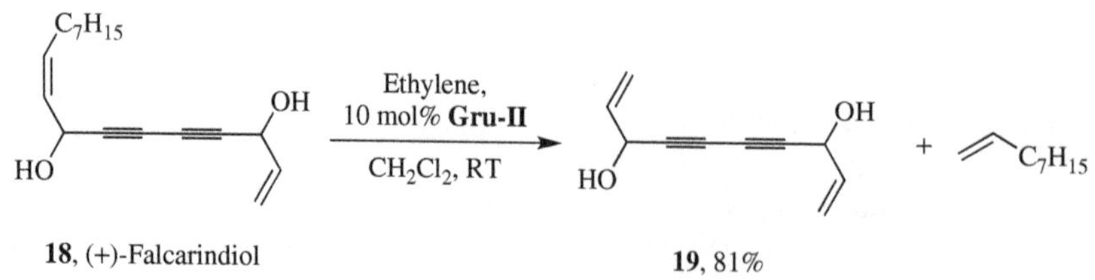

Scheme 2.10 Ethenolysis as a tool in natural product characterization.

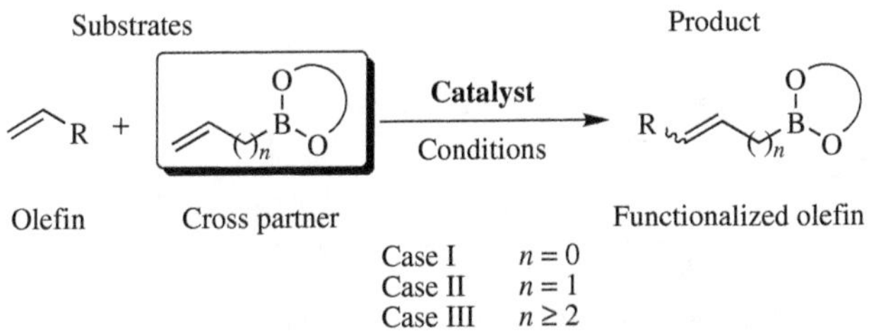

Scheme 2.11 General scheme of CM with boron compounds.

Scheme 2.12 presents a set of examples concerning functionalization of vinyl boronates via CM. It is a very convenient and high yielding method of incorporating boron-derived substituents into various molecules. One of the most popular boron-containing cross partners is the commercially available pinacol ester of vinylboronic acid (**23**) (19). However, this is not the only option. Propenyl boronates (**24**) (20) or those with increased steric hindrance (**25**) (21) have also been employed as cross partners. Other esters of vinylboronic acid (**26**), although not so commonly used, are also compatible with CM methodology (22).

This approach is often adapted in total synthesis. CM leading to compound **28** and its subsequent palladium-catalyzed coupling was utilized in synthesis of aureothin, which possesses antitumor, antifungal, and pesticidal activities (20). CM with cross partner **23** followed by a stereospecifical conversion to a *cis*-vinyl bromide was used in resorcylic acid lactone formation (23).

Unfortunately, not all boron-containing groups seem to be compatible with this method. During studies on synthesis of the C10-C26 fragment of amphidinolide E, CM of potassium vinyltrifluoroborate **34** ($CH_2{=}CHBF_3^-K^+$) with substrate **8** was approached (24). The reaction failed to proceed, so the authors conducted the metathesis step with pinacol derivative **23**, leading to product **32**, which was later converted to the required trifluoroborate **33** (Scheme 2.13).

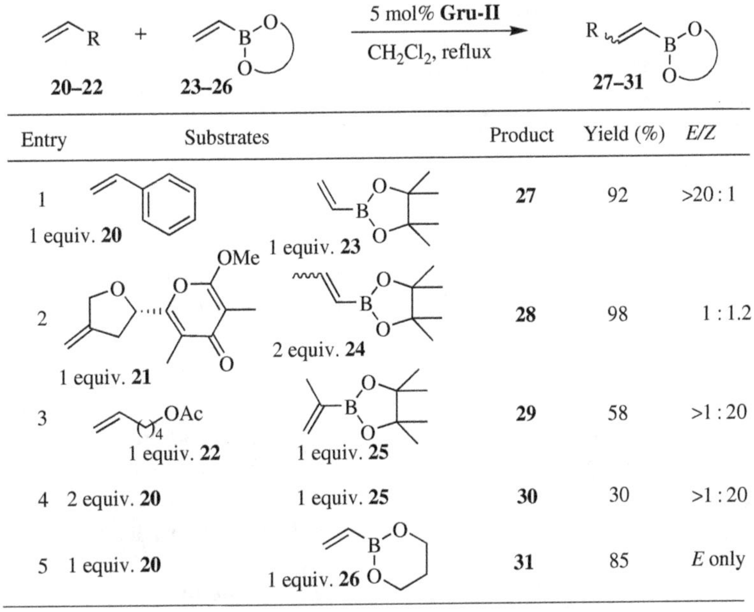

Entry	Substrates		Product	Yield (%)	E/Z
1	1 equiv. **20**	1 equiv. **23**	**27**	92	>20 : 1
2	1 equiv. **21**	2 equiv. **24**	**28**	98	1 : 1.2
3	1 equiv. **22**	1 equiv. **25**	**29**	58	>1 : 20
4	2 equiv. **20**	1 equiv. **25**	**30**	30	>1 : 20
5	1 equiv. **20**	1 equiv. **26**	**31**	85	*E* only

Scheme 2.12 CM of vinyl boronates.

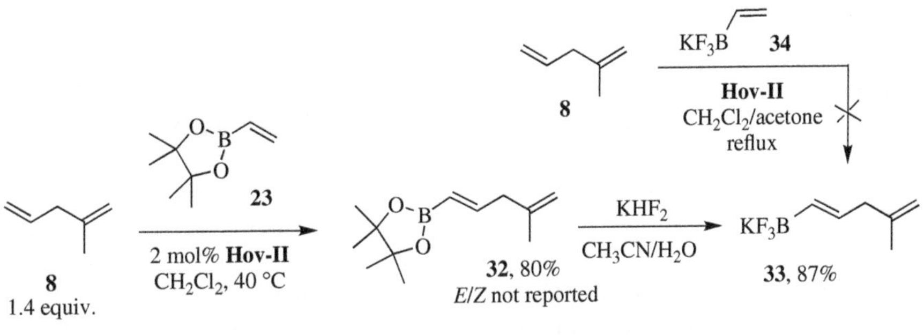

Scheme 2.13 Unsuccessful CM with potassium vinyltrifluoroborate.

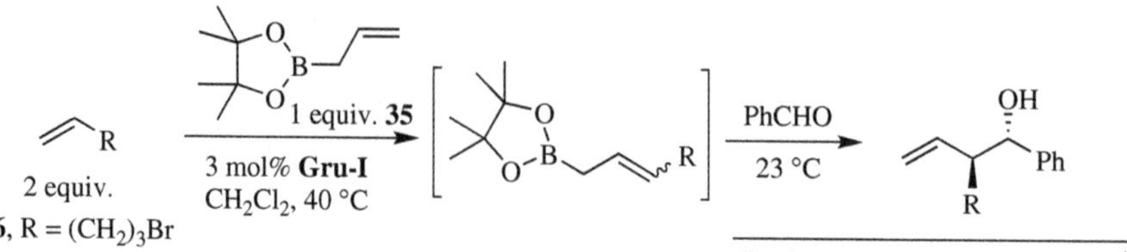

36, R = $(CH_2)_3Br$
37, R = CH_2COOMe

Product	R	Yield (%)	*anti:syn*
38	$(CH_2)_3Br$	71	69 : 31
39	CH_2COOMe	94	96 : 4

Scheme 2.14 One-pot CM of allylboronates and allylation.

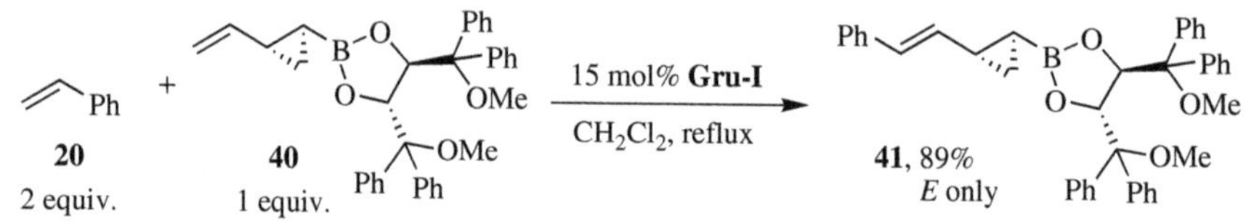

Scheme 2.15 CM with vinylcyclopropylboronic ester.

> **Representative Procedure of Cross Metathesis with Alkenyl Pinacol Boronate (Scheme 2.12) (20)**
>
> Substrate **21** (869 mg, 3.68 mmol) and the alkenyl pinacol boronate **24** (1.24 g, 7.36 mmol) were dissolved in CH_2Cl_2 (5 ml). Catalyst **Gru-II** (156 mg, 0.184 mmol) was added and the resulting mixture was refluxed for 3 h. After cooling to room temperature (RT), the solvent was evaporated *in vacuo* and the residue was purified by means of column chromatography. Product **28** (1.31 g, 98%) was obtained as a mixture of isomers (E/Z = 1:1.2).

CM with allyl boronates (n = 1; e.g., **35**) proceeds smoothly. It is often utilized in a one-pot sequence involving allylation of aldehydes with the boron reagents formed during the CM process (Scheme 2.14) (25,26). It is worth noting that a wide variety of FGs is tolerated within this process. For an example of S-CM of **35**, catalyzed by a tungsten–alkylidene complex, see Scheme 2.96.

Cases where the double bond is located more away from the boron functionality ($n \geq 2$) are rare, yet such substrates are easy transformed into CM products. An interesting example is that of CM transformation of vinylcyclopropylboronic ester **40** presented in Scheme 2.15 (27).

In general, boron compounds are tolerated within the metathesis processes and usually give good-to-excellent yields when other limiting factors are not present.

2.2.3 Olefins Bearing Carbon-Based Functional Groups

Carbon units are the basic building blocks in organic chemistry and their compatibility with any synthetic method is of crucial importance. CM proceeds readily with simple

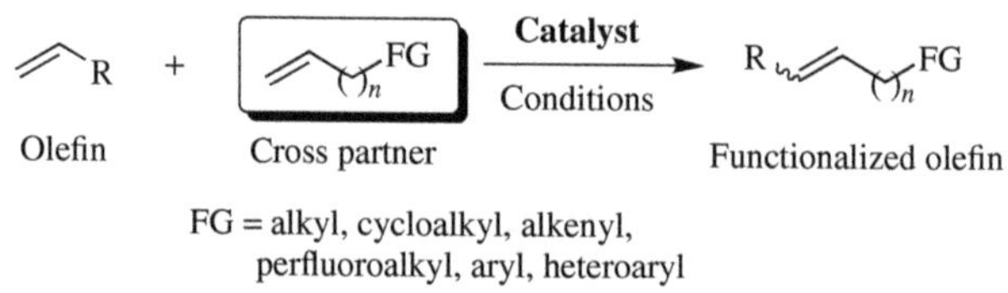

Scheme 2.16 CM partners with carbon-based functional groups.

α-olefins. In addition to these substrates, other CM partners, such as dienes, polyenes, and styrenes, are discussed in this section (Scheme 2.16).

As far as CM partners bearing hydrocarbon-type functionality are concerned, defining the length of the spacer between the double bond and the FG leads to confusion, so in this case, substrates were simply divided into classes containing alkyl, cycloalkyl, alkenyl (dienes and polyenes), aryl, and heteroaryl groups, etc.

2.2.3.1 *Olefins Containing Alkyl and Cycloalkyl Groups (FG = Alkene or* cyclo*-Alkene)*

A certain drawback of metathesizing simple olefins is their well-known ability to undergo S-CM (Scheme 2.5). To improve the yield of the desired cross product, usually an excess of one of the olefins is required. Practically, 2–10 equivalents of one partner may be required to obtain a CM product in a reasonable yield (28,29). Usually, the partner that is cheaper, more accessible, or simply easier to remove after the reaction is used in excess. Another approach is based on the fact that compounds originating from dimerization that have an internal double bond are also active in CM. Often, utilization of such dimers leads to obtaining better yields of the CM products (cf. Scheme 2.5). If the CM product can be easily separated from the substrates and further recycled, such transformation can be of significant practical value. This approach is utilized in the production of a pheromone of the peach twig borer moth, which is used as an environmentally friendly alternative to classical pesticides (30). On the other hand, the S-CM of volatile Fischer–Tropsch fractions or other *n*-olefins into heavier internal olefins (also known as *CM dimers*) can be used for conversion of low value feedstocks into useful chemicals. Some of such processes are possible with relatively low loading of modern Ru catalysts (50–20 ppm) (31).

Linear Alkenes A CM approach involving terminal linear alkenes as cross partners was adapted in the synthesis of sphingolipids, compounds that are involved in signal transduction and molecular recognition processes in cell membranes. As already described, utilization of excess of the *n*-alkene reduced the formation of homodimer of the more expensive starting materials, **42**, **45**, and **47**, in the synthesis of D-*erythro*-sphingosine (32), D-*erythro*-ceramide (33), and sphingomyelin (34) precursors (Scheme 2.17, note the influence of protecting groups in CM of **45** and **47**).

In all the above cases, good yields and selectivity were obtained. It is worth highlighting that "homodimers" resulting from S-CM of **43** that were produced as by-products (not shown on Scheme 2.17) should be separated from the desired CM products. Fortunately, this is usually is not a big problem owing to the low polarity of these by-products. More examples will be presented in this chapter.

Sterically Hindered Alkenes As described in the introduction, steric hindrance is an obstacle, when localized too close to the reacting C–C double bond. When the reacting fragments of both substrates have hindered surroundings, CM may be impossible, but if only one is sterically hindered, and the other olefin is reactive, the reaction proceeds with various degree of success (Scheme 2.18). A major problem arises with metathesizing molecules bearing a sterically hindered group in α position (10) (**5** (35), **7**, **52**, **55** (36)) or their being geminally substituted such as in **59**. In such cases, only an olefin of significant reactivity can be successfully coupled with such a cross partner. Metathesis of *S*-(+)-carvone **59** proved to be an especially difficult example (37). Having tested the usual set of conditions and proving the results negative, the authors decided to employ fluorinated aromatic hydrocarbons (FAHs) as solvents along with microwave irradiation. This beneficial combination led to increased formation of product **60**. More information about unconventional conditions in metathesis reactions is available in Chapter 23. It should be also noted that CM of sterically hindered olefins (e.g., **5**) is sometimes promoted more effectively by specialized catalysts with reduced NHC ligands (35, 38).

A particularly interesting example is the application of 2-methylpropene (isobutylene, **61**, boiling point −6.9 °C) and 2-methyl-2-butene (**65**, boiling point 39 °C) as cross partners (Scheme 2.19).

These olefins are usually used in excess, or neat, which assures high conversions. Owing to higher boiling point, **65** can be handled in a more user-friendly manner, similarly to dichloromethane. What makes this compound an interesting cross partner is the fact that it can lead to the formation of two differently substituted products depending on the character of the olefin that it is coupled with. If the CM partner bears an electron-withdrawing group (e.g., **64**) the product of the reaction is almost exclusively the methyl-substituted one (e.g., **63**, Scheme 2.19) (39). However, this selectivity is not always easy to achieve, as it can be concluded from CM with 2-nitrostyrene (mostly the prenylated product is formed) (39) or from CM of some flavonoid derivatives (40). The selectivity can be to some extent controlled by reaction conditions (41). In this context, the CM with isobutylene (**61**) seems to be more selective, and has been used in a number

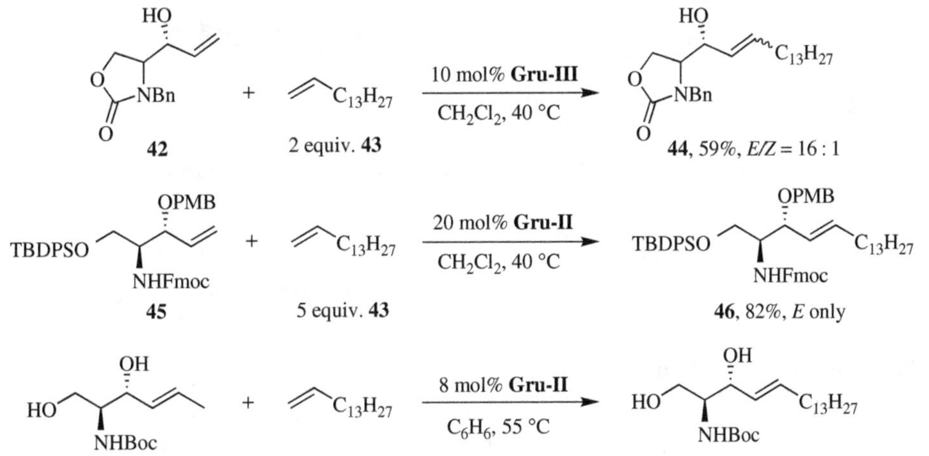

Scheme 2.17 CM in syntheses of bioactive products.

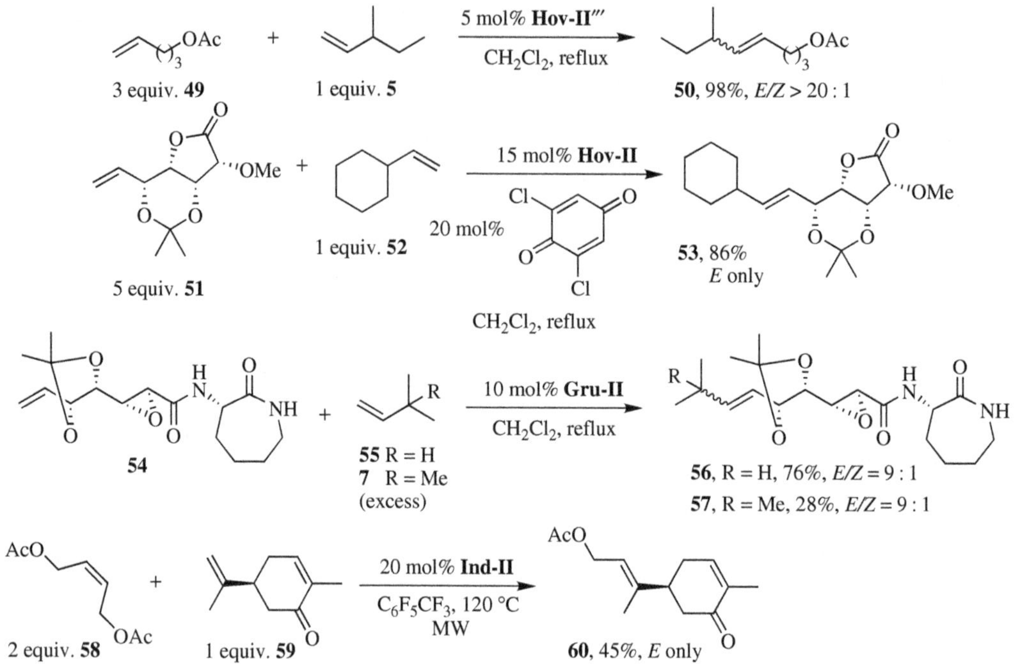

Scheme 2.18 CM of sterically hindered alkenes.

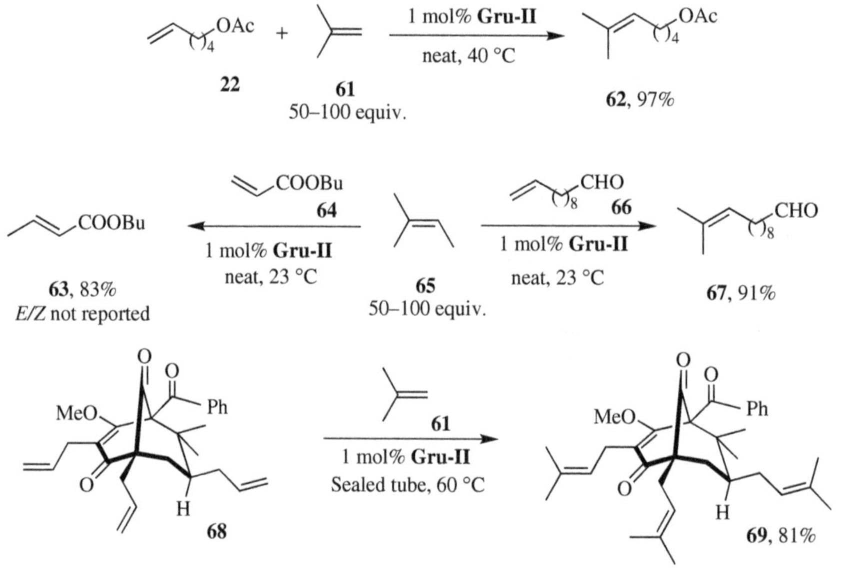

Scheme 2.19 CM of 2-methylpropene and 2-methyl-2-butene.

of stereoselective total syntheses, such as in preparation of nemorosone precursor **69** (Scheme 2.19) (42), clusianone (43), and others.

Representative Procedure of CM with a Sterically Hindered Alkane (Scheme 2.18) (37)

A 10 ml vial was charged with *cis*-1,4-diacetoxy-2-butene **58** (172.2 mg, 1 mmol) and *S*-(+)-carvone **59** (150.2 mg, 0.5 mmol), which were dissolved in octafluorotoluene (2.5 ml). Next, solid **Ind-II** was added (23.7 mg, 5 mol%), and the vial was sealed under nitrogen atmosphere. The reaction was performed in a microwave reactor at 120 °C for 10 min. The next portion of the catalyst was then added and the manipulation was repeated. The procedure was repeated four times. After that the solvent was evaporated and the crude product was purified by flash chromatography to yield product **60** (50 mg, 45%).

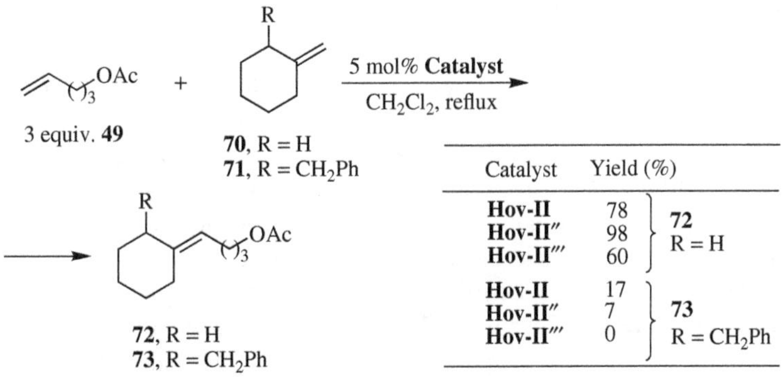

Catalyst	Yield (%)	
Hov-II	78	**72** R = H
Hov-II″	98	
Hov-II‴	60	
Hov-II	17	**73** R = CH_2Ph
Hov-II″	7	
Hov-II‴	0	

Scheme 2.20 CM with cycloolefins bearing exocyclic double bond.

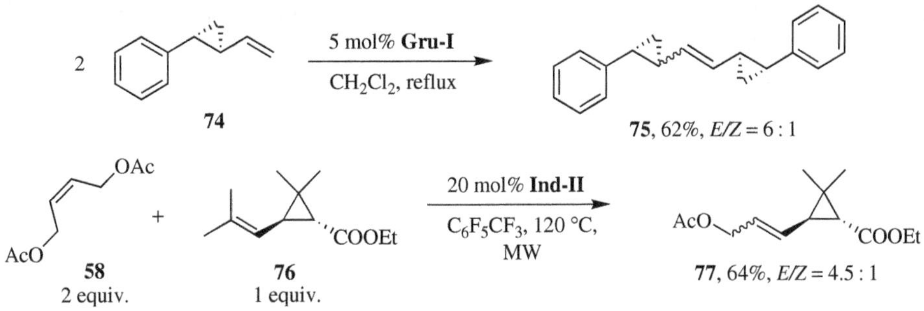

Scheme 2.21 CM of substrates containing cyclopropropane unit.

Another group of sterically hindered cross partners worth noting are carbo- (37) and heterocycles (44) bearing an exocyclic double bond. Unlike isobutene and 2-methyl-2-butene, these compounds undergo CM in a more sluggish manner, often requiring elaborated conditions (Scheme 2.20).

One of the most reluctant substrates in CM is compound **71**, which bears a substituent in the neighboring position of the reacting exocyclic double bond. Comparing with **70**, yields of CM transformations are dramatically lower in the case of **71**. Interestingly, according to the study by Grubbs et al. (35). the formation of trisubstituted olefins by CM can be sometimes more efficient using catalysts bearing bulkier NHC ligands (e.g., SIPr), likely due to the selectivity of productive versus unproductive pathways (Scheme 2.20).

Olefins Containing a Cyclopropane Fragment Unstained cycloalkyl groups such as cyclopentane and cyclohexane are well tolerated in the structure of the CM partner and there are numerous examples of such transformations (45). Smaller and more strained carbocycle—cyclopropane—seems to be also compatible with metathesis catalysts. For example, both **74** (46) and **76** (37) led to respective CM products in good yields (Scheme 2.21). However, it was once observed during a study on an anti-HIV agent, BILN 2061, that ring-closing metathesis (RCM) catalyzed by **Gru-I** of a diene, containing a chiral cyclopropane moiety in their structure, resulted in its racemization (46). The authors hypothesize that this was caused by a ring-opening ring-closing event of the cyclopropane ring, but no such observation was made when the second-generation complex **Gru-II** was utilized instead of **Gru-I**. However, studies focusing on CM of compounds bearing a cyclopropane moiety do not confirm the lability of the discussed moiety.

2.2.3.2 Alkenes Bearing Additional C=C Bonds (Allenes, Dienes, and Polyenes) When an additional double bond is present in the substrate molecule, such substrates generally come from two classes—cumulenes and polyenes. Only derivatives of the simplest compound from the cumulene group were tested in CM, and those were *substituted allenes* (47).

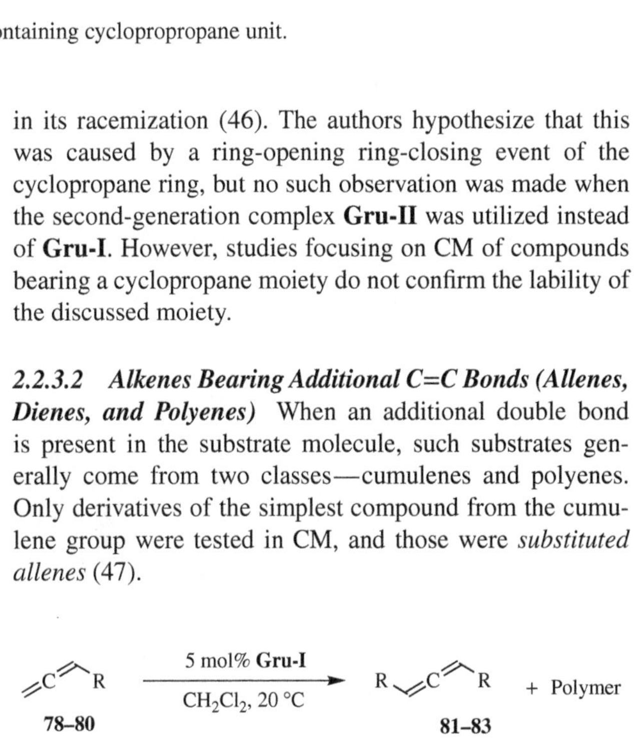

Substrate	Product	Yield (%)
78, R = $(CH_2)_5CH_3$	**81**	70
79, R = $(CH_2)_3OTHP$	**82**	20
80, R = Ph	**83**	0

Scheme 2.22 CM of allene derivatives.

As presented in Scheme 2.22, the CM of allene derivatives is difficult, yet possible. It is important to note, that this transformation seems strictly dependent on the electronic character of the substrate. Allenes bearing aliphatic functionalities (e.g., **78–79**) are consumed within the reaction leading to the expected product and undefined polymeric material in various proportions. Allenes bearing an aromatic substituent (e.g., **80**) rarely participate in the CM reaction, yielding only the polymeric products. Allene motifs in distant positions usually survive the CM reaction untouched (see Scheme 2.27 as an example).

Nonconjugated dienes undergo metathesis smoothly. The least hindered bond usually reacts first, leading to formation of a single product with good selectivity (Fig. 2.2). Such compounds are often utilized in organic synthesis and examples of their CM are presented in other sections of this chapter, particularly in Scheme 2.13 and Scheme 2.42.

Conjugated dienes are widely utilized in CM processes, yet more care must be applied when planning the reaction. In most cases, the less hindered C=C bond participates in the reaction, but caution must be taken as product mixtures can be obtained.

Conjugated dienes having an electron-withdrawing group close to the C–C double bonds are useful building blocks and therefore are of interest in organic chemistry. For that reason, multiple CM reactions of dienes with EWG-containing olefins (48) or EWG-functionalized dienes (**85** and **89**) (49) are being conducted. Such transformations are not always extremely effective owing to low yields and selectivities, but they represent an important tool in synthesis (Scheme 2.23).

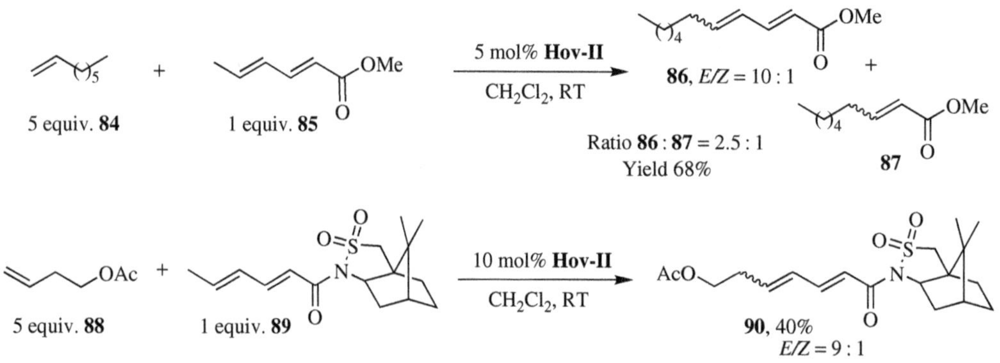

Scheme 2.23 CM of EWG-functionalized dienes.

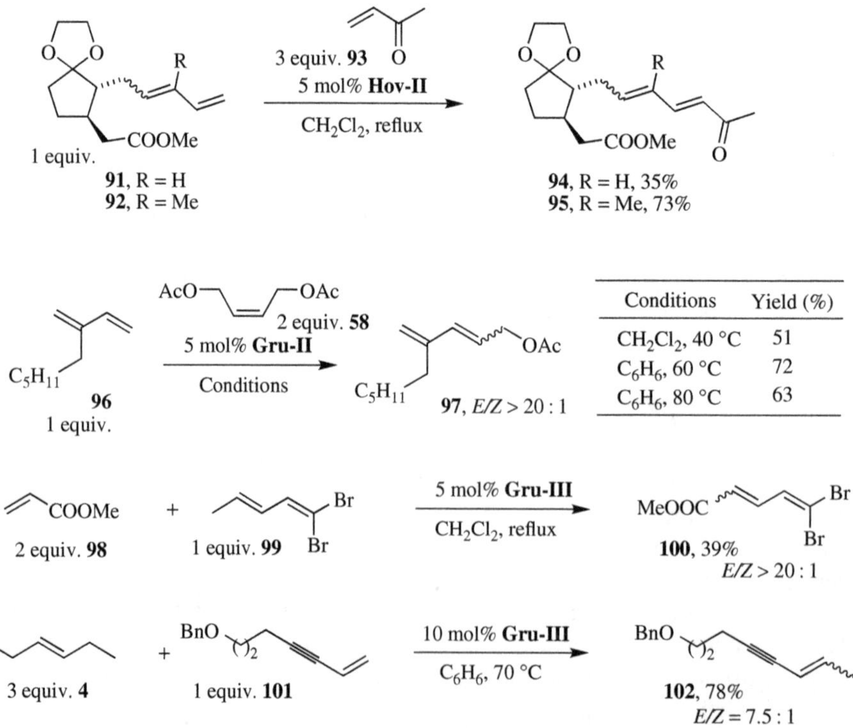

Conditions	Yield (%)
CH_2Cl_2, 40 °C	51
C_6H_6, 60 °C	72
C_6H_6, 80 °C	63

Scheme 2.24 Cross metathesis of other conjugated dienes and enynes.

Next, Scheme 2.24 presents a selection of other conjugated dienes (50) and enyne (51) substrates in CM. It should be noted that the CM transformation of these substrates leads to valuable functionalized products (e.g., **95**, **97**, and **102**) in just one step, with good selectivity in general.

CM of *conjugated polyenes* is especially spectacular because of the potential complexity of this process. If the reaction conditions are not chosen properly, complex mixtures of products resulting from statistical cleavage of conjugated double bonds can be formed.

A good example is CM of two polyene substrates: β-carotene **103** and ethyl 3-methylhexa-2,4-dienoate **104** (Scheme 2.25). In spite of the fact that C11–C12 and C15–C15′ double bonds in **103** are preferably cleaved during CM, the product mixture was rather complex. However, the CM reaction course could be controlled to some extent by a proper choice of conditions (52). A number of carotenoids was obtained using the CM approach (53). Other polyenes, for example, derivative **110**, react in a selective manner (Scheme 2.26).

Self-metathesis of polyne **112** also proceeded selectively, leading to a natural product—mimulaxanthin (54). One should note that the allene fragment survives the CM event unengaged (Scheme 2.27).

CM with polyene substrates was recently reviewed (55).

2.2.3.3 *Alkenes Containing Perfluoroalkyl Groups (FG = C_nF_{2n+1})*

The presence of fluorine or a perfluorinated alkane fragment within a molecule often gives it unique properties. Some of the most significant properties are increased metabolic stability and lipophilicity. CM is one of few methods for incorporating perfluorinated fragments into organic compounds (Scheme 2.28).

Cross partners having the perfluoroalkyl fragment attached to the reacting C–C double bond (n = 0, e.g., **115–116**, **120**) are not readily metathesized under typical conditions owing to the low electron density of the olefinic center. CM leading to compound **121** is a good example of such a situation (Scheme 2.29) (56). A significant improvement in yields is obtained when utilizing FAHs as solvents (for more on this, see Chapter 23). In such cases, good-to-excellent yields can be obtained (57), sometimes accompanied by the selective formation of the product (58).

Representative CM Reaction of (Perfluorohexyl)ethylene (Scheme 2.29) (58)

Compounds **119** (132.7 mg, 0.32 mmol) and **120** (1.1 g, 3.2 mmol) were dissolved in hexafluorobenzene (6.5 ml). A solid catalyst, **Ind-II**, was added (0.016 mmol, 5 mol%) and the mixture was heated at 60 °C for 20 h. The solvent was removed using reduced pressure and the residue was purified by column chromatography to give product **121** as a single E isomer (210.6 mg, 0.278 mmol) in 90% yield.

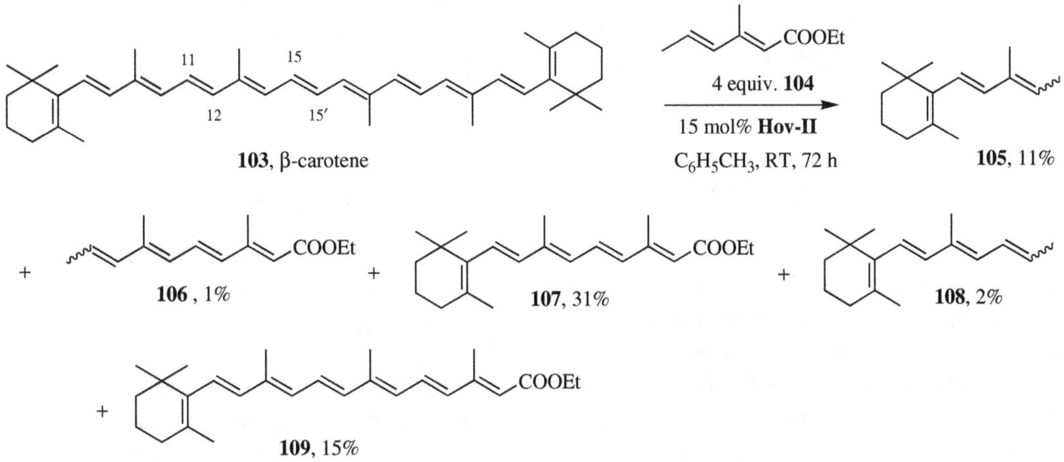

Scheme 2.25 CM of β-carotene and 3-methylhexa-2,4-dienoate.

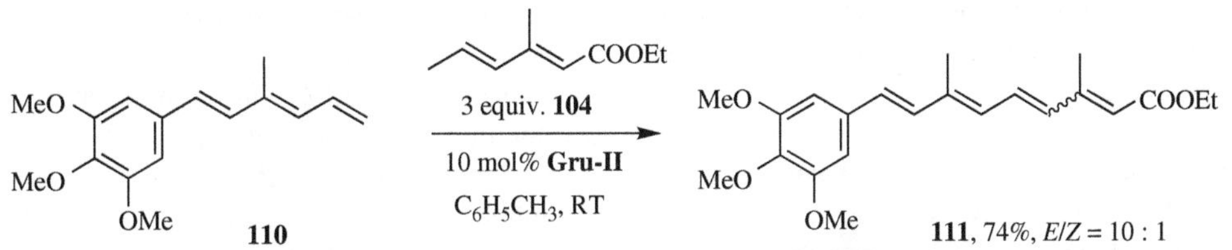

Scheme 2.26 CM of polyene partners leading to etretinate analog.

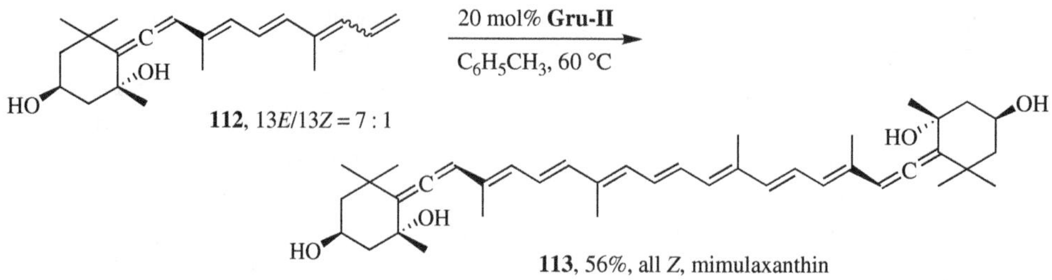

Scheme 2.27 Application of CM in mimulaxanthin synthesis.

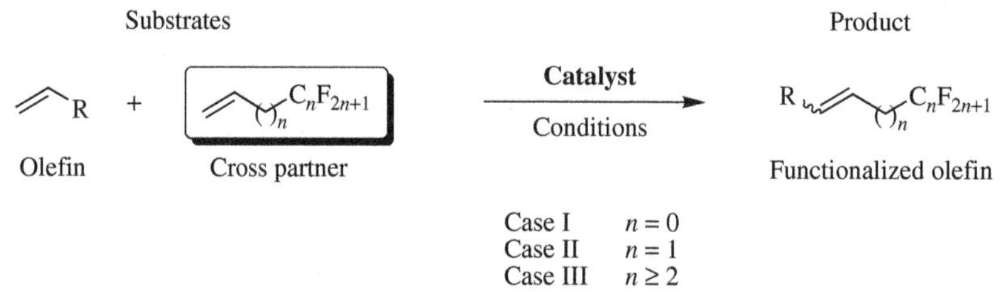

Scheme 2.28 General scheme of CM with cross partner bearing perfluoroalkyl group.

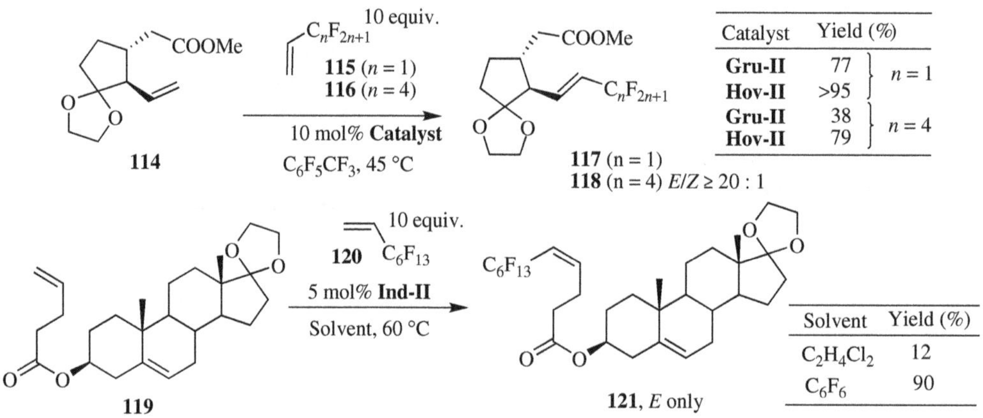

Scheme 2.29 CM with partners bearing perfluoroalkyl groups.

Once the electron-withdrawing perfluorinated alkane moiety is in a one carbon more distant position ($n = 1$), CM reactions proceed more efficiently (Scheme 2.30) (59). In compounds **123** and **126**, the electron deficiency in the double bond region is less pronounced leading usually to a better CM outcome. For more difficult cases, such as, for example, vinylferrocene **122**, further optimization may probably be required as the yields remain unsatisfactory.

Once the fluorine atoms are present at a significant distance from the metathesis active site, CM transformation is no longer an issue (60).

*2.2.3.4 **Alkenes Bearing Aryl and Heteroaryl Substituents (FG = Ar)*** Metathesis is widely applied to functionalize various olefins with cross partners bearing aromatic substituents. Irrespective of whether they are simple benzene derivatives, heteroaryl- or metallocene-bearing compounds can be metathesized quite easily.

Styrenes undergo CM and S-CM reaction readily, unless a steric hindrance is present in the ortho position (10). Self-metathesis of styrenes has found utility in the synthesis of stilbenes, which often have significant biological activity (61). Sometimes, the affinity of reactive styrenes for S-CM can lead to lower yields in CM reactions with other partners, as the resulting stilbenes are much more reluctant substrates in CM.

Examples of CM with styrene derivatives promoted by ruthenium- (10) and molybdenum-based (62) complexes are presented in Scheme 2.31. The same scheme summarizes application of azulenes (e.g., **136**) in CM (63). In such reactions, the products are formed with moderate yields and no self metathesis of vinylazulenes was reported (63).

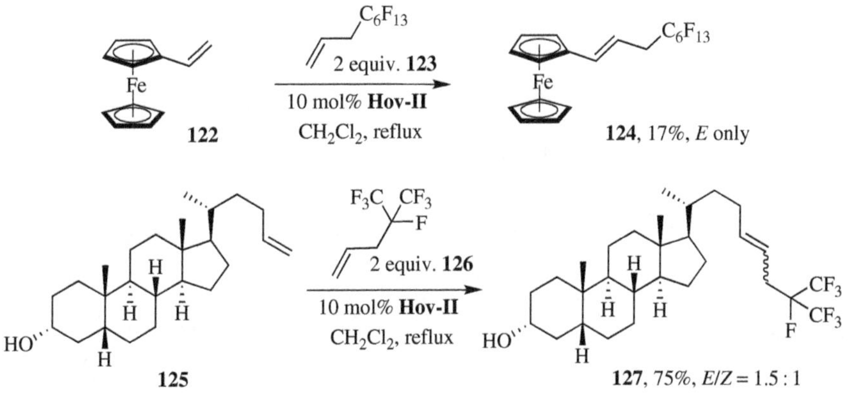

Scheme 2.30 CM of alkenes bearing perfluorinated fragment in allylic position.

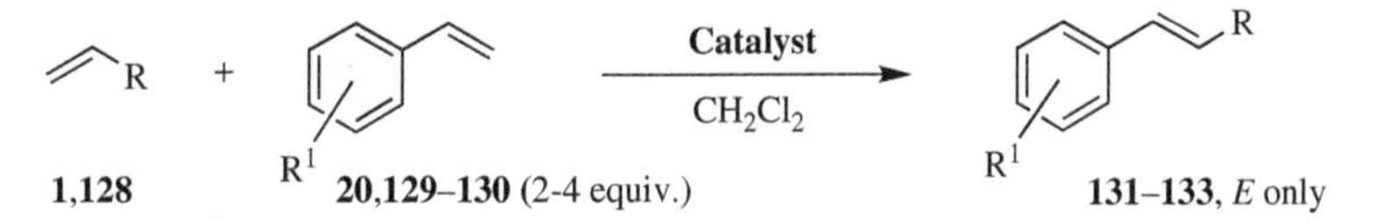

Substrates		Product	Catalyst	Temperature	Yield (%)
128, R = $(CH_2)_4$OTHP	**20**, R[1] = H	**131**	5 mol% **Gru-II**	40 °C	71
1, R = $(CH_2)_3CH_3$	**129**, R[1] = *p*-OMe	**132**	1 mol% **Sch-I**	RT	88
1, R = $(CH_2)_3CH_3$	**130**, R[1] = *m*-NO_2	**133**	1 mol% **Sch-I**	RT	48

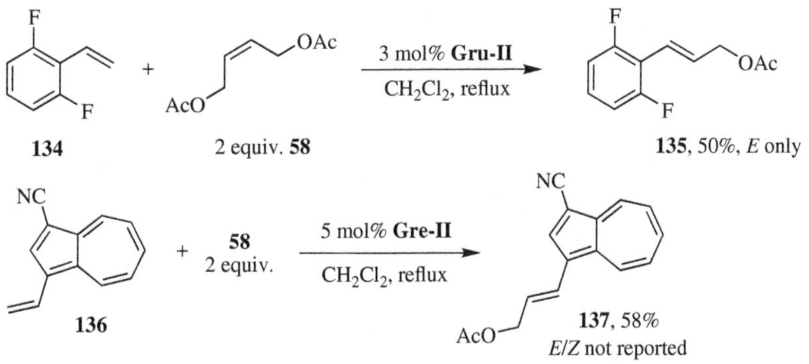

Scheme 2.31 Vinyl aromatic compounds as partners in CM reactions.

Wojtkielewicz investigated the CM reactions of styrene (**20**) and cinnamaldehyde (**138**) with ethyl 3-methylhexa-2,4-dienoate (**104**). Surprisingly, the desired CM product **139** was obtained in a better yield when cinnamaldehyde was used as the benzylidene fragment source. Probably, competitive self-metathesis of styrene is faster than that of cinnamaldehyde, and for this reason a significant lowering of the CM product yield was observed in the former case. Both second-generation catalysts (Grubbs and Grubbs–Hoveyda complexes) efficiently promoted the reaction of cinnamaldehyde, affording the desired CM product in yields above 90% (Scheme 2.32).

Furthermore, vinyl-substituted aromatic heterocycles (e.g., **140**, **142**, **144**, and **148**) were widely explored in CM transformations (Scheme 2.33). Virtually all heteroatoms and ring sizes are tolerated and lead to product formation in good-to-excellent yields (64–66). In case of substrates that may chelate the ruthenium center slightly, lower yields may result, but application of additives to overcome this unproductive pathway should solve the problem. When appropriate catalyst and conditions are applied, a variety of densely functionalized compounds may be obtained. A good example may be the CM transformation leading to the porphyrin derivative **146** (67).

Another interesting group is that of aromatic cross partners bearing vinyl-substituted metallocenes, such as ferrocene derivatives **122** and **150**. These compounds are compatible with metathetic transformations and undergo both CM (68) and S-CM ("homodimerization") (69). Depending

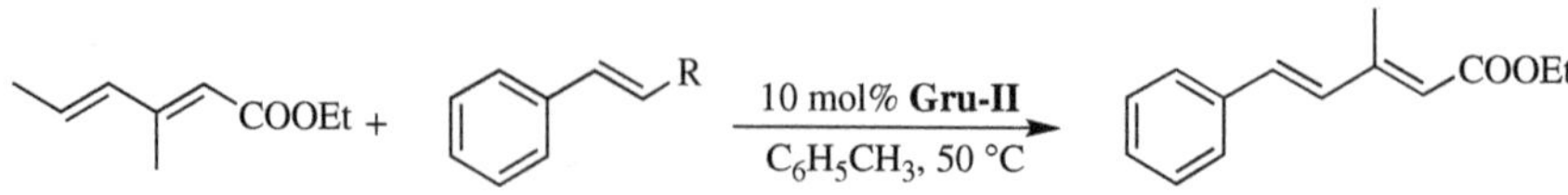

3 equiv. **104** — **20**, R = H; **138**, R = CHO — **139**, all *E*

56%, from **20** (R = Ph)
95%, from **138** (R = CHO)

Scheme 2.32 Cinnamaldehyde as an alternative to styrene in CM reaction.

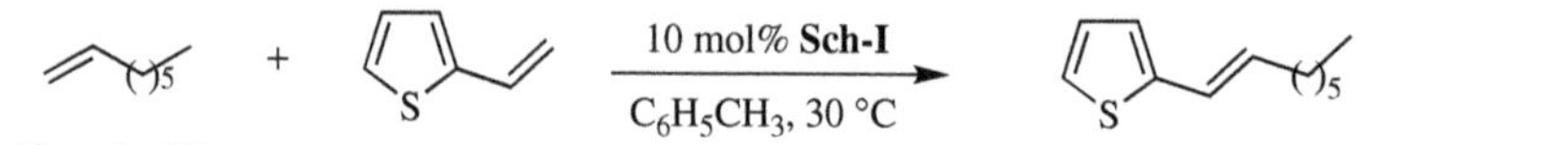

13 equiv. **84** — **140** — **141**, 98%, *E*/*Z* not reported

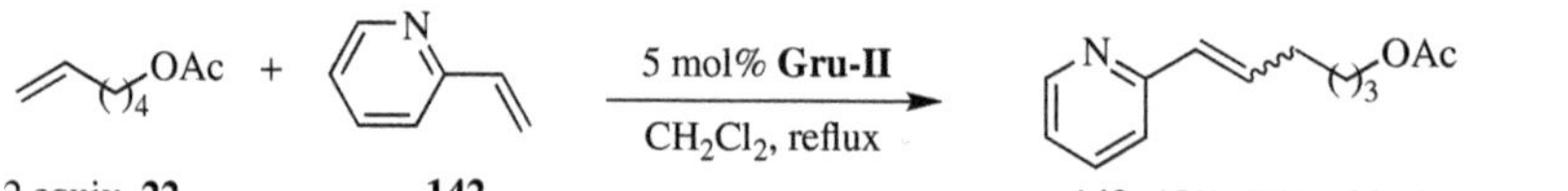

2 equiv. **22** — **142** — **143**, 45%, *E*/*Z* > 20 : 1

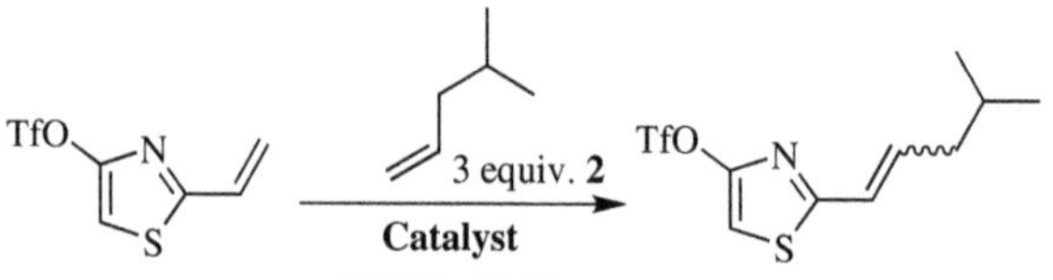

144 — **145**, *E*/*Z* > 20 : 1

Catalyst	Yield (%)
10 mol% **Gru-I**	67
10 mol% **Gru-II**	91
5 mol% **Hov-II**	84

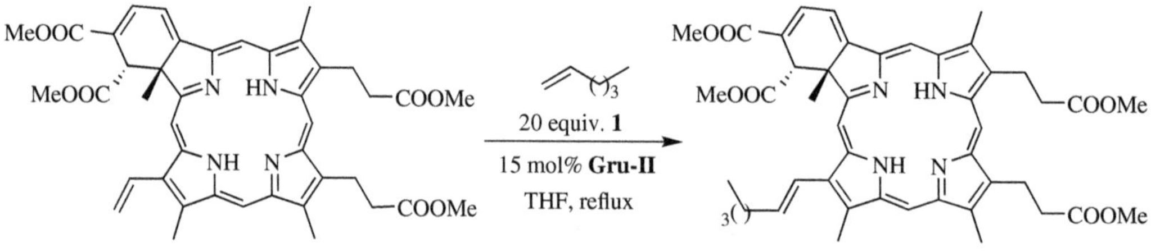

146 — **147**, 70%, *E* only

Scheme 2.33 CM of vinyl-substituted heteroaryl compounds.

on the exact structure of the substrate, moderate-to-excellent yields are obtained (Scheme 2.34).

Aromatic partners with a longer alkenyl chain are easy to metathesize. Examples involving CM of allylbenzene (70) and titanocenes (71) ($n = 1$) are presented in Scheme 2.35.

Aromatic and heteroaromatic rings in more distant positions are motifs often present in target-oriented syntheses utilising the CM as a key step (72). CM with aromatic cross partners is a valuable method of molecule functionalization. Numerous examples of applications in target-oriented synthesis can be found in the literature and some of these are presented in this chapter.

2.2.3.5 Unsaturated Nitriles (FG = CN) Compounds bearing a cyano group are very important in organic synthesis as they give access to a multitude of other functionalities, such as acids, amides, amines and others. This versatility makes unsaturated nitriles a significant group of cross partners (Scheme 2.36).

The simplest unsaturated nitrile—acrylonitrile (**153**) is a CM partner commonly utilized in reactions. It should be noted that it has a strongly electron-deficient double bond and due to this fact it is one of the most difficult cross partners (Scheme 2.36). The CM reactions involving **153** were successful with Schrock's Mo-based catalyst **Sch-I** (73), with **Hov-II**, and with its activated analog, **Gre-II** (44). Similar to other α,β-unsaturated EWG-bearing compounds, acrylonitrile does not undergo S-CM ("homodimerization") under typical conditions.

Attempts to optimize the conditions by lowering the loading, or utilizing **Gru-II** met with limited success and even in reactions with reactive olefins, the yields were moderate (product **159**) (74). In the case of acrylonitrile, CM with more demanding olefins, forcing conditions need to be applied.

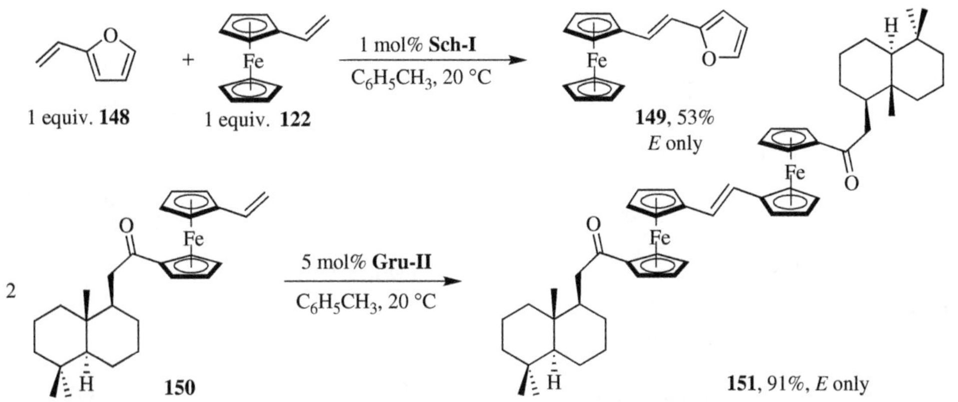

Scheme 2.34 Vinyl ferrocenes as partners in cross metathesis.

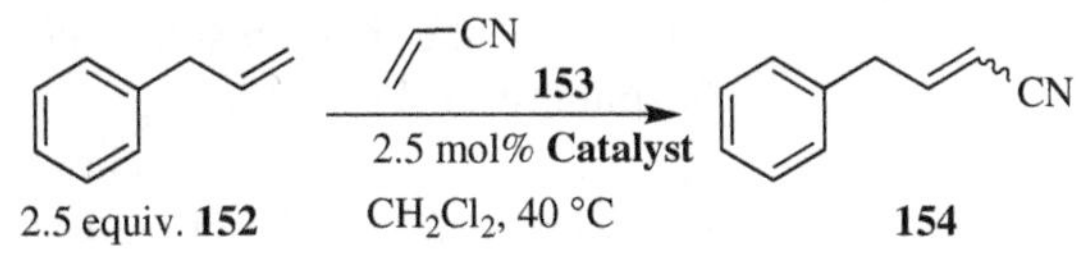

Catalyst	Yield (%)	E/Z
Hov-II	68	1 : 1.9
Gru-II	21	1 : 1.7
Gru-III	67	1 : 1.8

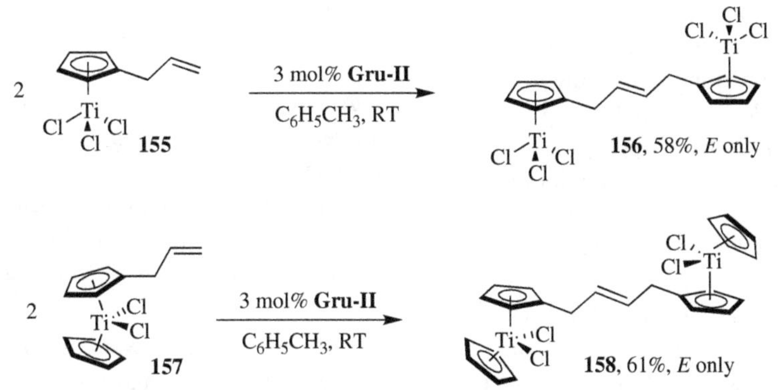

Scheme 2.35 CM of aromatic compounds bearing allyl fragment.

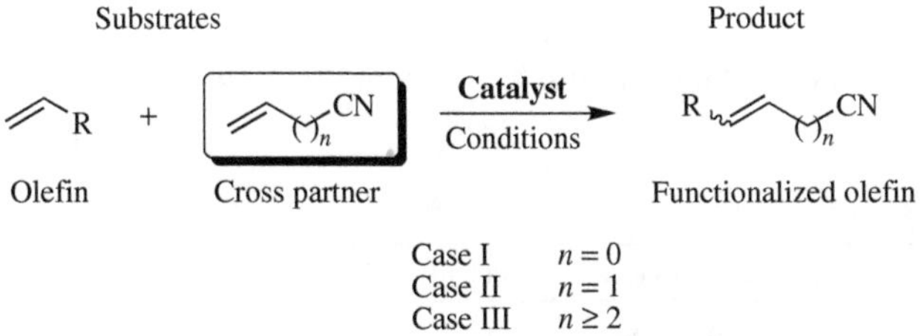

Scheme 2.36 General scheme of CM with cyano-substituted partners.

Some of those may include microwave irradiation (products **160** and **161**) (75), introducing third-generation Ru catalysts (76) or application of inorganic additives (Scheme 2.37) (77). There are also literature reports on successful utilization of fumaronitrile in CM (78).

Utilization of allylcyanide ($n = 1$, **164**) merely changes the issue. The strong electron-withdrawing character of the cyano group and its ability to act as a ligand to Ru makes this compound a rather demanding CM partner (Scheme 2.38) (79).

Further elongation of the carbon chain between the cyano group and the reacting C–C double bond leads to significant improvement in metathesis results (Scheme 2.39). The yield of the cross product generated from unsaturated nitriles **153**, **164**, **166** of varying chain lengths clearly demonstrates the diminishing influence of the cyano group (79).

The last example presents the utility of long-chain unsaturated nitrile CM. The high yielding CM of **171** led to a functionalized product **172**, which can serve after reduction as a monomer for the production of polyamides (Scheme 2.39) (80).

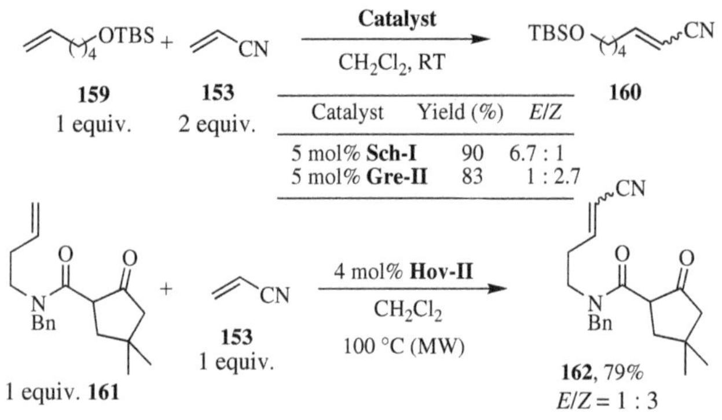

Catalyst	Yield (%)	*E/Z*
5 mol% **Sch-I**	90	6.7 : 1
5 mol% **Gre-II**	83	1 : 2.7

Scheme 2.37 Acrylonitrile as cross-metathesis partner.

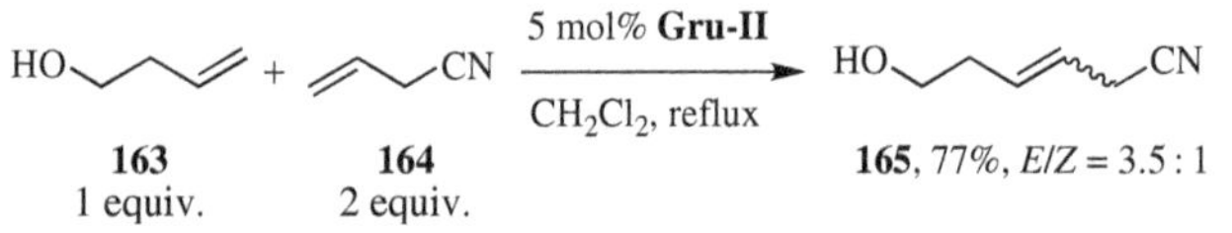

Scheme 2.38 CM of allylcyanide.

2.2.3.6 Unsaturated Aldehydes (FG = CHO) Owing to significant synthetic value of the aldehyde moiety, compounds bearing such functionality are of constant interest to organic chemists. Their applications in metathesis is not straightforward because of the electron-withdrawing character of the formyl group (Scheme 2.40). If appropriate conditions are applied, however, the reactions proceed with various degrees of success.

There are numerous examples of utilization of the simplest unsaturated aldehyde—acrolein (**174**) as a cross partner in target-oriented syntheses. One of these is presented in Scheme 2.41 (81). A consecutive series of CM transformations was utilized in the elegant synthesis of the C1–C14 fragment of amphidinol 3 (**180**) (82). Three CM reactions, two with acrolein and one with ethyl acrylate, were performed in the course of the synthesis. Importantly, in each case, the *E/Z* ratio and the chemoselectivity of the transformations led to obtaining the final product in a good overall yield.

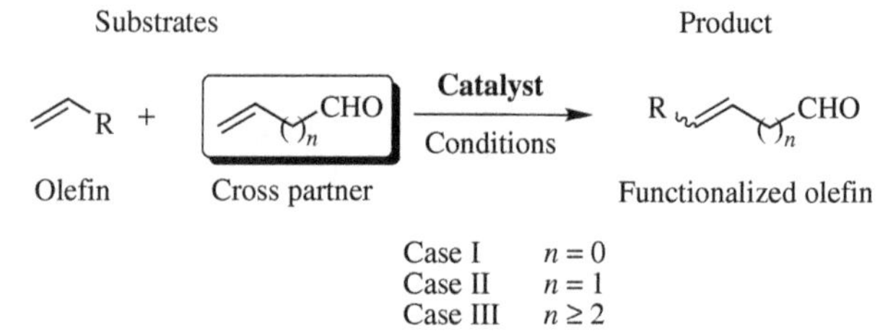

Scheme 2.40 General scheme of CM with unsaturated aldehydes.

It should be noted that acrolein is a rather unpleasant reagent, strongly odored and unstable, so it can be replaced by a more user-friendly analog—crotonaldehyde (**182**). This approach was presented in ricciocarpin A (**184**) synthesis by Michrowska and List (83) (Scheme 2.42).

Usually, the best results should be obtained with a phosphine-free **Hov-II** complex or its more active analogs (e.g., **Gre-II**); however, other second-generation catalysts were used in the literature as well. As with other α,β-unsaturated EWG-bearing compounds, acrolein and crotyl aldehyde hardly undergo S-CM under typical conditions.

Alkenes bearing a formyl group at distant positions are utilized in CM transformations with no difficulty (e.g., **185**, Scheme 2.43) (84).

Application of CM partners bearing other EWG functionalities is also an important issue, as these compounds are extremely valuable from the industrial point of view.

2.2.3.7 Unsaturated Ketones (FG = COR) Unsaturated ketones are valuable and relatively easily reacting partners in CM transformations (Scheme 2.44). They are commonly utilized in target-oriented synthesis of complex biologically active molecules.

Methyl vinyl ketone (**93**) is one of the most popular cross partners in this class. Usually, the best results should be obtained with a phosphine-free **Hov-II** complex;

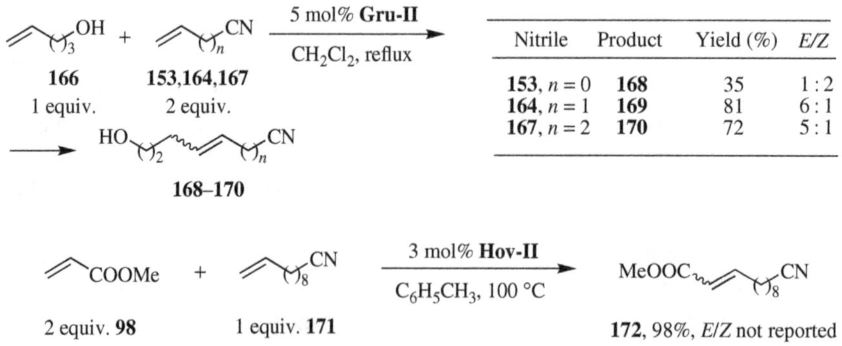

Nitrile	Product	Yield (%)	*E/Z*
153, *n* = 0	**168**	35	1 : 2
164, *n* = 1	**169**	81	6 : 1
167, *n* = 2	**170**	72	5 : 1

Scheme 2.39 CM with unsaturated nitriles of varying chain length.

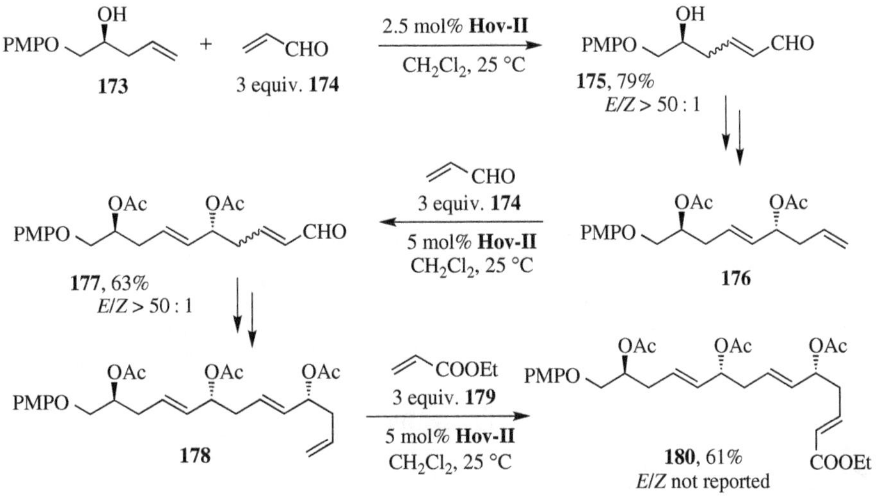

Scheme 2.41 Sequential CM with acrolein in total synthesis amphidinol 3.

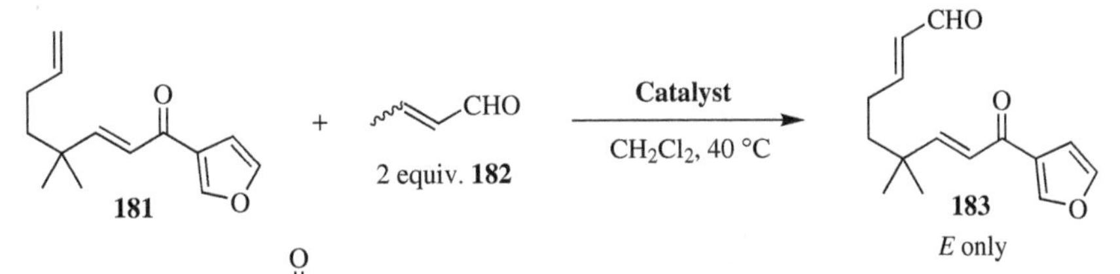

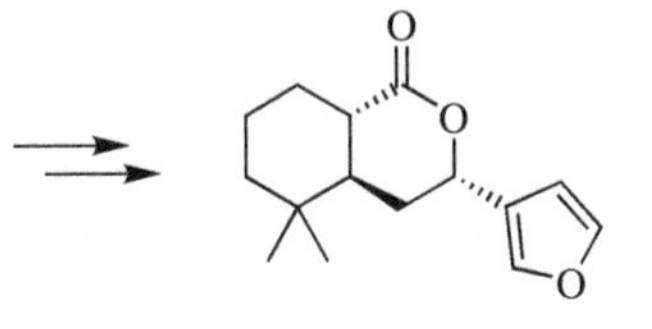

Catalyst	Yield (%)
10 mol% **Gru-II**	84
2 mol% **Gre-II**	90

Scheme 2.42 CM with crotyl aldehyde in synthesis of ricioccarpin A.

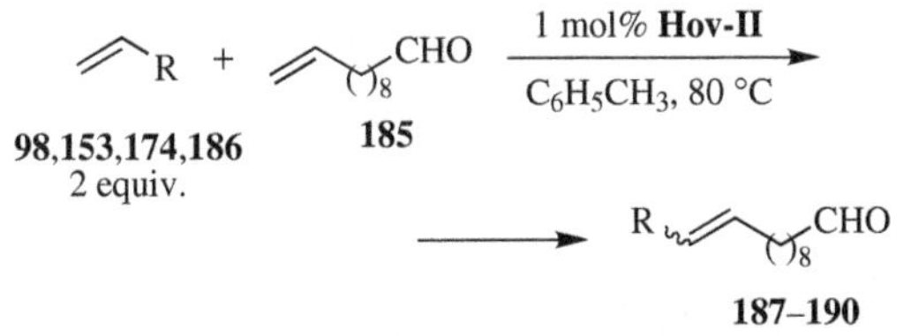

Partner	Product	Yield (%)	E/Z
153, R = CN	**187**	94	1 : 4
174, R = CHO	**188**	78	*E* only
186, R = COOH	**189**	84	*E* only
98, R = COOMe	**190**	92	*E* only

Scheme 2.43 CM of long-chain unsaturated aldehydes.

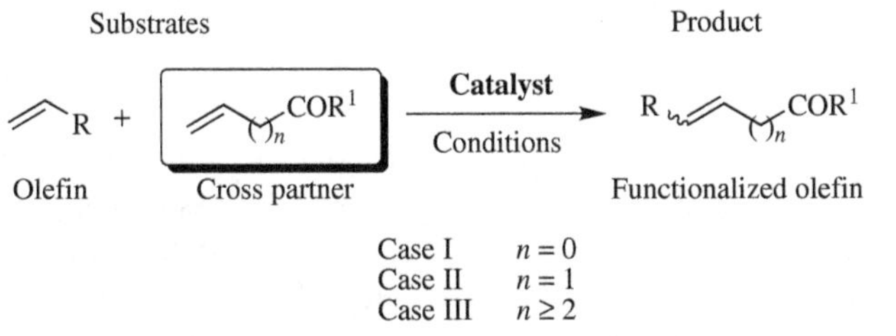

Scheme 2.44 General scheme of CM with ketones.

however, other second-generation catalysts were used as well. Similarly to other α,β-unsaturated EWG-bearing compounds, **73** hardly undergoes S-CM under typical conditions. Scheme 2.45 presents its utilization in synthesis, as well as CM reactions with other α,β-unsaturated ketones.

CM of methyl vinyl ketone presented in Scheme 2.45 was successfully utilized as a last step in (−)-dihydroanthatin (**192**) synthesis (85). Compounds obtained by CM functionalization of olefins **191** and **194** with vinyl ketones **193**

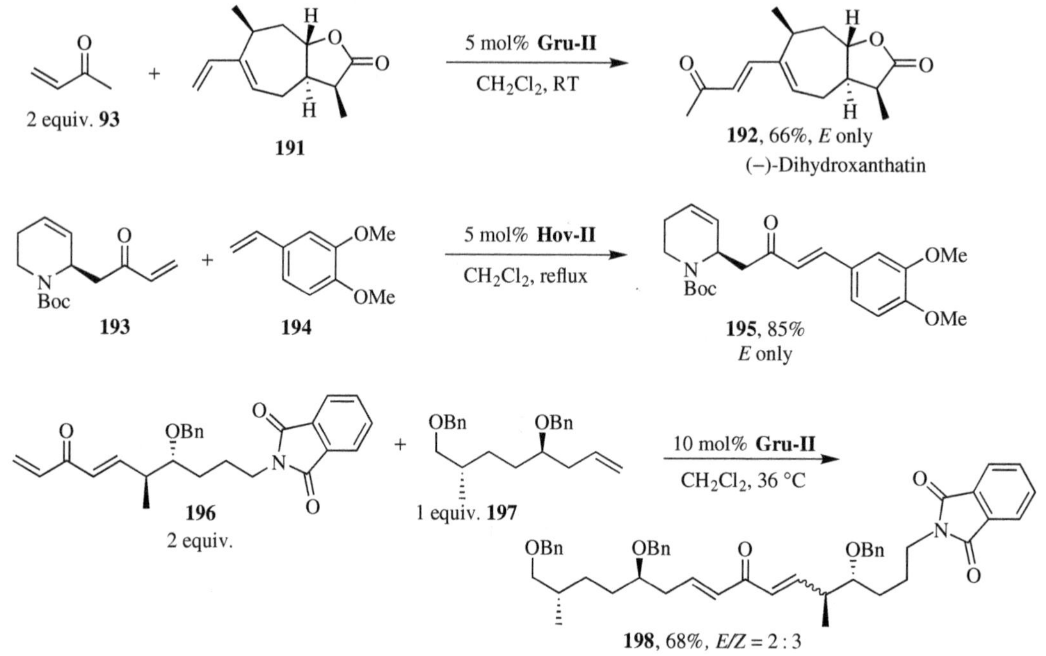

Scheme 2.45 Vinyl ketones CM in total synthesis.

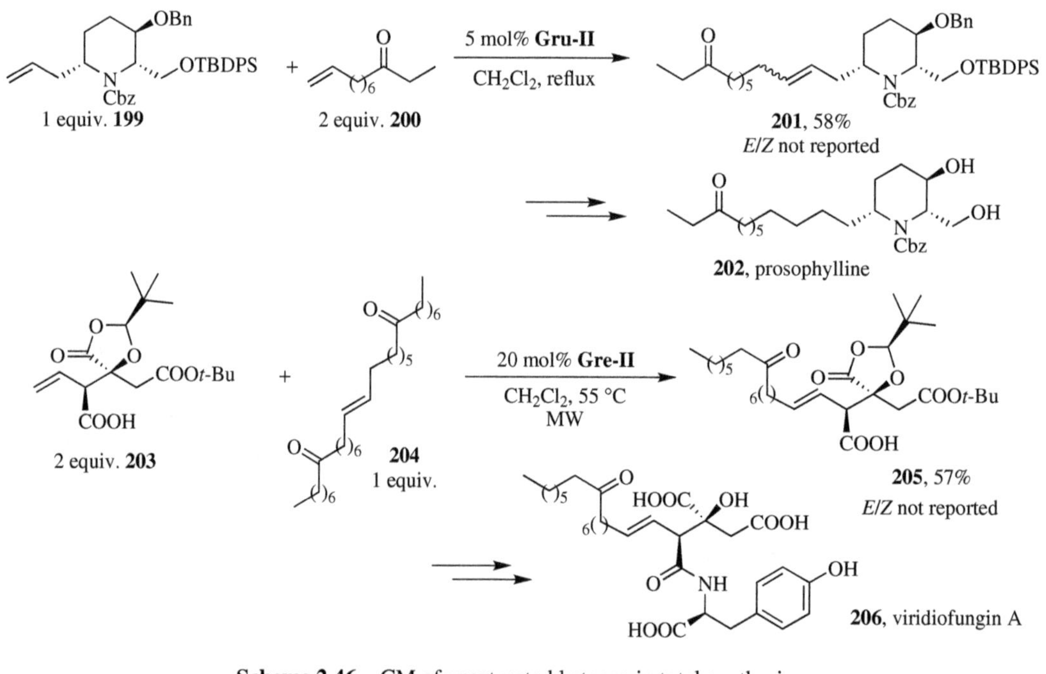

Scheme 2.46 CM of unsaturated ketones in total synthesis.

and **196** were utilized in synthetic projects leading to (−)-lasubine II (86) and bistramide A, respectively (87).

Alkenes containing keto function in more distant positions are not problematic in CM either. Scheme 2.46 presents applications of such substrates in synthesis of prosophylline (**202**) (88) and viridiofungin A (**206**) (89). It should be noted that a "dimeric" metathesis partner **204** was used in the latter case (refer to Scheme 2.5, for related discussion). In both cases, excellent conversions were achieved in the CM step. Unfortunately, owing to the presence of polar functionalities, the purification steps resulted in a noticeable decrease in the yield.

2.2.3.8 Unsaturated Carboxylic acids and Their Derivatives Unsaturated carboxylic acids and their derivatives are of utmost importance in synthesis. Utilization of unsaturated amides, acids, or esters as partners in CM is very common (Scheme 2.47). Molecules obtained in this manner can be easily coupled with other building blocks. Importantly, the reactions between terminal alkenes and unsaturated amides, acids, and esters are usually characterized by good selectivity. Transformations of particular acid derivatives are briefly described in the following sections.

Unsaturated Amides (FG = CONRR′) A detailed study on CM reactions of α,β-unsaturated amides with terminal olefins is presented in Scheme 2.48 (90). It should be noted that α,β-unsaturated amides are considered to be rather challenging substrates for Ru catalysts, and usually require more forcing conditions (e.g., 5 mol% of the second-generation catalyst, reflux). Variously substituted

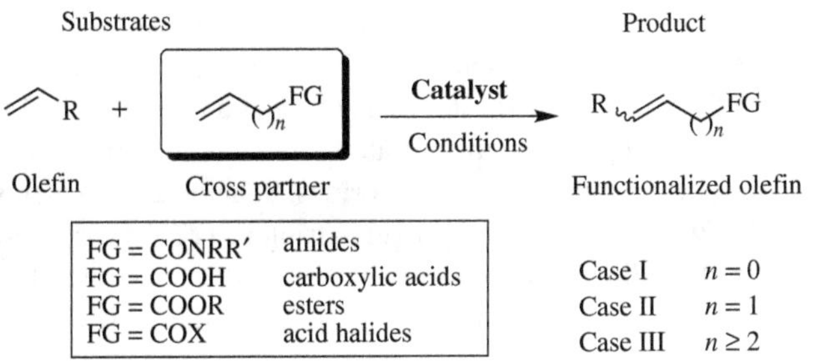

Scheme 2.47 General scheme for CM of unsaturated acid derivatives.

α,β-unsaturated amides, such as **207**, **210**, and **212**, can be applied in CM. The CM results vary from acceptable to excellent. More electron-rich amides give lower yields owing to lower metathesis activity resulting from carbonyl chelation to the Ru center.

Furthermore, cyclic amides bearing an exocyclic C–C double bond were utilized in CM. Studies in this area mainly focus on utilization of CM of alkenes with the β-lactams, such as **214**, because of potential biological activity of the resulting products. Examples of such transformations are presented in Scheme 2.49. Considering the previously described difficulty associated with CM of compounds with exocyclic C–C double bonds the obtained yields of lactams (e.g., **216**) bearing a tetrasubstituted C–C double bond are surprisingly high (91). During the course of the study by Liang et al., a number of substituted olefins were tested, revealing total prevention of the reaction by any α-branching in the olefin structure (e.g., **217**).

Although amides can form strong complexes with transition metal complexes, it is important to note that to the best of our knowledge there are no reports on distant amide FGs being detrimental to the metathetic transformations.

Unsaturated Carboxylic Acids (FG = COOH) Carboxylic acids are fully compatible with ruthenium-based complexes. CM of free carboxylic acids is possible, although difficult in the case of α,β-unsaturated acids; this is due to the electron-deficient character of the double bond. As presented in Scheme 2.50, the reaction proceeds equally well with acrylic acid (**186**) (77) and its higher β,γ-unsaturated homolog (**221**) (92). Another example is given on Scheme 2.43.

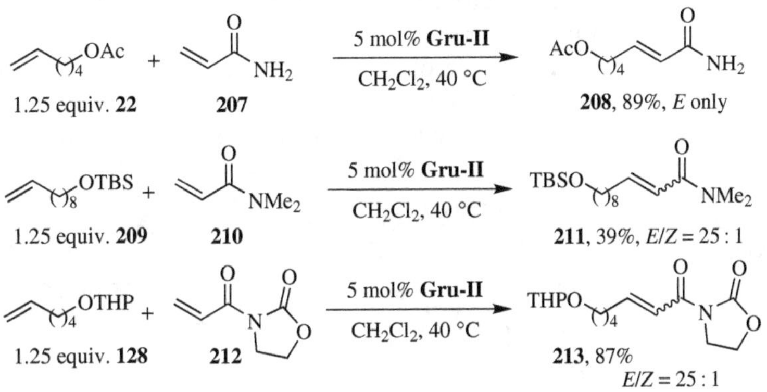

Scheme 2.48 CM of α,β-unsaturated amides.

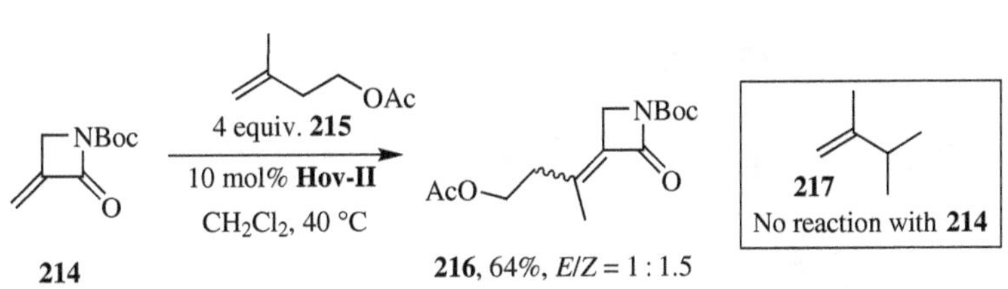

Scheme 2.49 CM reactions of β-lactams bearing an exocyclic C–C double bond.

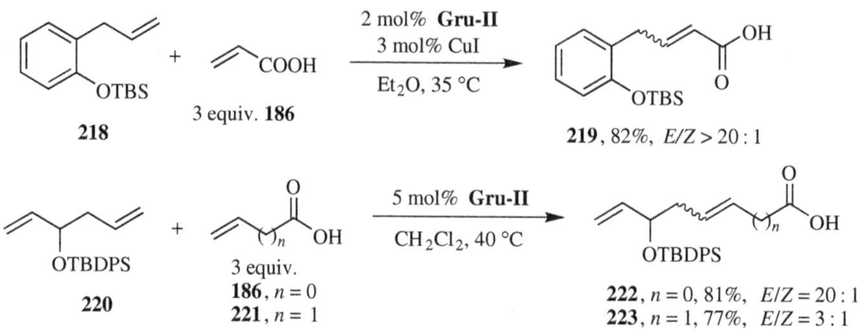

Scheme 2.50 CM of unsaturated carboxylic acids.

The common difficulty related to the use of unsaturated acids as CM partners is the removal of decomposed catalyst impurities from the polar products; this is problematic and often leads to decrease in the yield.

Unsaturated Esters and Lactams (FG = COOR) Unsaturated carboxylic acid esters are commonly used as partners in CM transformation. This group of substrates is very diverse as it contains simple derivatives of linear carboxylic acids, naturally occurring triglycerides, and cyclic esters—lactams. Metathesis of such compounds leads to the formation of functionalized molecules important in target-oriented synthesis.

The most popular CM partners from this class are acrylic acid esters ($n = 0$). Despite being a α,β-unsaturated compounds, they are rather easy targets for active modern metathesis catalysts. However, S-CM of **98** and other acrylic esters is not trivial and requires precisely optimized conditions. Numerous cases of CM between various alkenes and acrylic acid esters have been reported in the literature and various examples are provided in this chapter, one of them in this section (Scheme 2.51). CM with methyl acrylate was utilized *en route* to the synthesis of cochleamycin A (**226**) (93). Notably, α,β-unsaturated thioesters (e.g., **227**) were also tested as cross partners. A study was conducted utilizing the sulfur analog of ethyl acrylate, which is metathesized efficiently if second-generation Ru catalysts are applied (Scheme 2.51) (94).

There are numerous substrates where an ester group is placed at more distant location in the alkene chain ($n \geq 1$); some of these are presented in this book. One example of CM of β,γ-unsaturated ester **37** is presented in Scheme 2.14. An emerging group of unsaturated esters was obtained from renewable resources, the triglycerides. Scheme 2.52 presents a CM with a model compound—triolein (**229**) (95). Low catalyst loading was used to metathesize this substrate, thus proving that this methodology is efficient in the utilization of renewables.

Exocyclic C–C double bond in lactones of various ring sizes can also be functionalized by CM. A significant number of studies have been published on the utilization of such lactones, which, similarly to previously discussed cycloalkenes with exocyclic double bonds, are demanding substrates for metathesis transformations (Scheme 2.53).

In the case of four-membered lactones (e.g., **233**), the obtained results are surprisingly good (96). Irrespective of

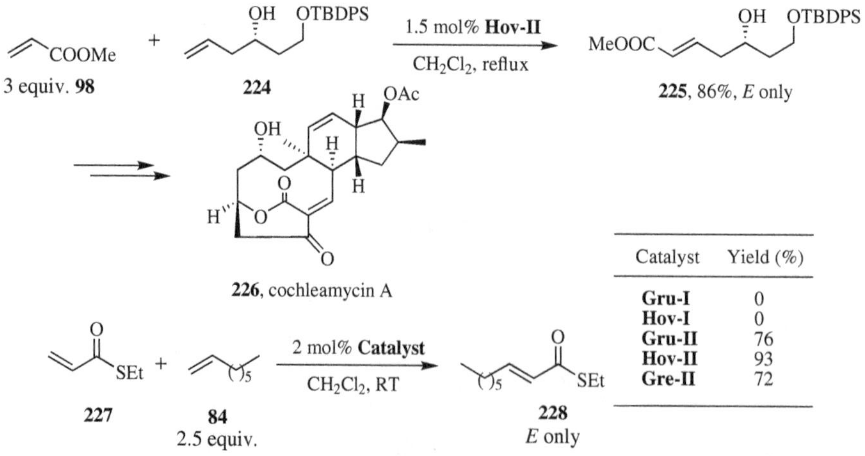

Catalyst	Yield (%)
Gru-I	0
Hov-I	0
Gru-II	76
Hov-II	93
Gre-II	72

Scheme 2.51 CM of α,β-unsaturated esters and thioesters.

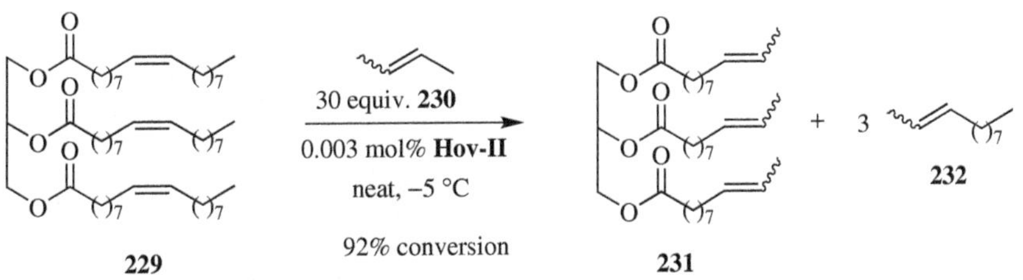

Scheme 2.52 CM of triolein.

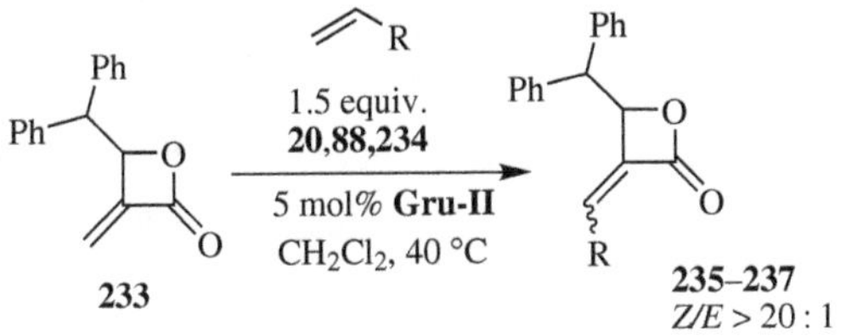

Olefin	Product	Yield (%)
88, R = CH_2OAc	235	84
20, R = Ph	236	55
234, R = CH_2Br	237	93

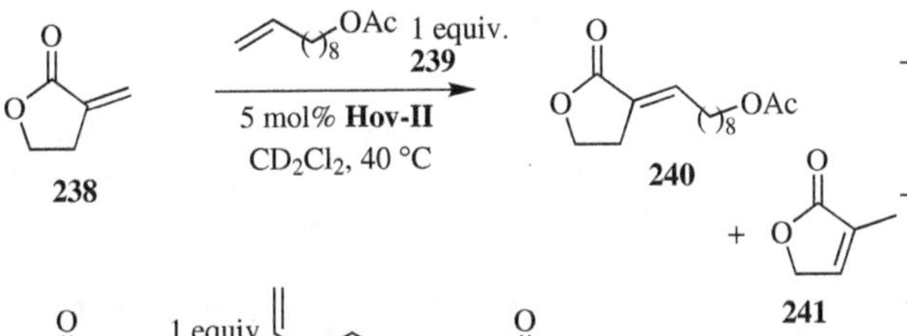

Additive	Product (%) 240	241
None	0	87
2,6-Dichloro benzoquinone	67	0

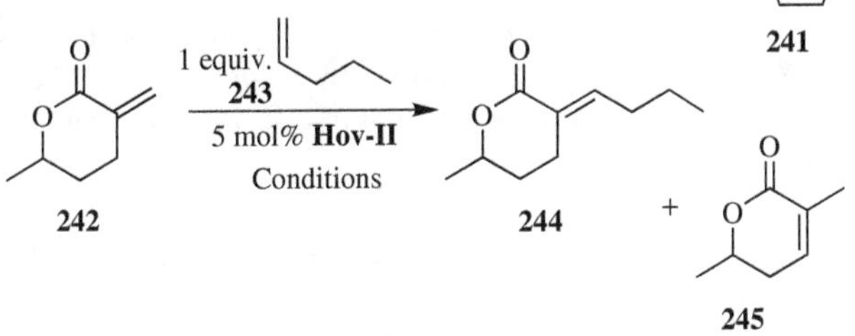

Conditions	Additive	Product (%) 244	245
CD_2Cl_2, reflux	none	19	0
CD_2Cl_2, reflux	AcOH	54	trace
$C_6D_5CD_3$, reflux	none	0	66

Scheme 2.53 CM of lactones bearing exocyclic C–C double bond.

high ring strain of the tested systems the transformation usually proceeds with high yield and excellent *Z*-selectivity. Less strained lactones (five- **238** and six-membered **242**) are slightly more difficult to metathesize (97,98). In these cases, the C–C double bond isomerization promoted by the ruthenium complexes was observed as a competing process, leading to products such as **241** and **245**. If no additives are added, this was the major process taking place during the reaction. Application of additives leads to formation of the desired products albeit in rather moderate yields (Scheme 2.53).

Unsaturated Acid Chlorides (FG = COCl) Although it may seem surprising, acid chlorides are compatible with metathetically active Ru complexes. Scheme 2.54 presents results of a study on CM with acryloil chloride (**246**) (99).

Products of the CM transformation were utilized in a subsequent reaction without isolation. This one-pot sequence containing addition of a nucleophile to the formed CM product leads to a variety of α,β-unsaturated compounds in good yields.

2.2.4 Alkenes Bearing Nitrogen-Based Substituents

CM with nitrogen-containing compounds is a complex issue. Generally, groups with donor properties tend to chelate the metal center, slowing down the reaction. This is the case when ruthenium-catalyzed complexes are discussed. In such cases, stable Fischer carbenes may be formed, which are not active in metathetic processes. A solution for this is avoidance of basic nitrogen functionalities, such as free amines. Application of protective groups, usually amide derivatives, leads to delocalization of the electron density, making the metathesis process more efficient.

2.2.4.1 Unsaturated Amines and Unsaturated Protected Amines (FG = NRR′) Metathesis of unsaturated amines used as free bases is very problematic owing to chelation issues (Scheme 2.55). The reaction can be conducted with no protection only if basic character of the amine is relatively weak, for example, in the case of *N*-allylaniline derivative **251** (101) (Scheme 2.56).

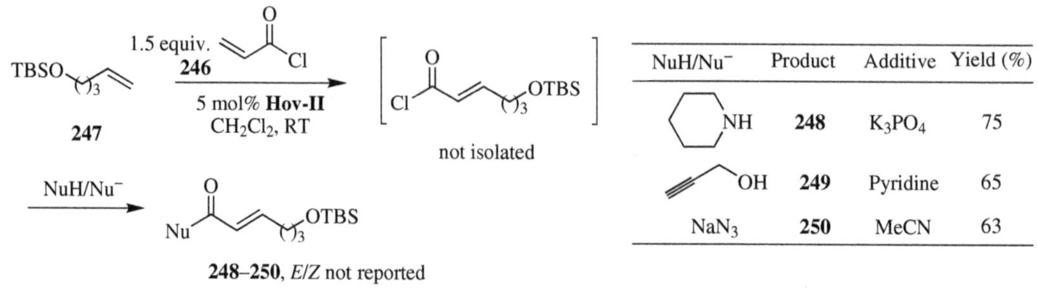

Scheme 2.54 CM with acryloil chloride.

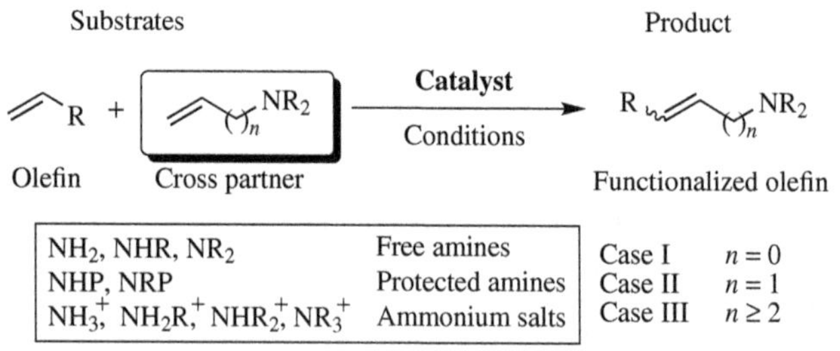

Scheme 2.55 General scheme for CM of unsaturated amines.

In some cases, Lewis acid additives may be necessary to overcome the catalyst chelation and obtain a good yield (e.g., in CM of **255**) (100). In the most cases, however, protection of the basic functionality is essential. Usually, amines are converted to the respective amide derivatives (Boc, Cbz, Ts) for the CM step and the yields of the obtained products are very good (101,102). The last example, presenting the functionalization of macrocycle **257**, is a great demonstration of how successful this approach can be in synthesis (103).

A method of temporary amine masking by formation of stable ammonium salts was introduced (104). The only issue while adapting this approach is the solubility of the amine salts **260–262** in organic solvents where metathesis is conducted. Apart from that, the desired products **263–265** can be often obtained in good yields (Scheme 2.57). Deprotection is very efficient and requires mild bases such as triethylamine or potassium carbonate.

2.2.4.2 Alkenes Containing Other Nitrogen-Based Groups (Guanidine, Diazoester, Azide) Compounds containing nitrogen in a form other than amine or protected amine are rarely metathesized. Nonetheless, there are some reports of such transformations. A very interesting case, where a polyfunctional substrate **266** that contains a guanidine group and a number of peptidic amido groups is presented in Scheme 2.58 (105). Despite moderate yield,

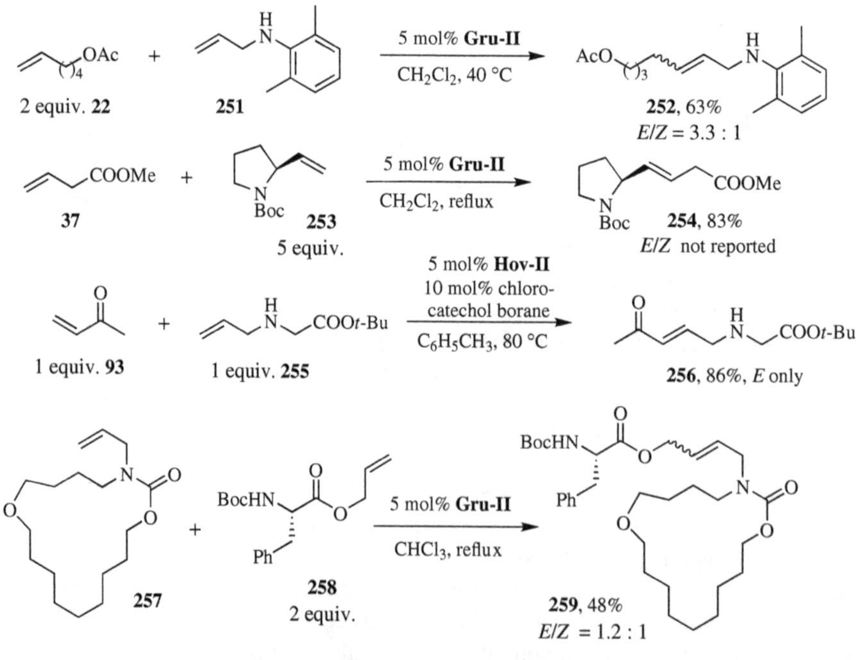

Scheme 2.56 Allyl amines in cross metathesis.

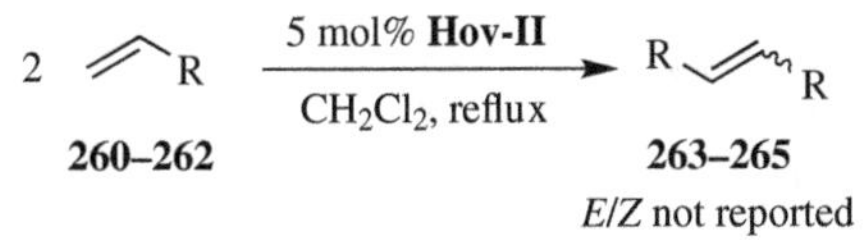

Substrate	Product	Yield (%)
260, R = $CH_2NH_3^+TsO^-$	**263**	74
261, R = $CH_2N^+H_2CH_2Ph$ TsO^-	**264**	46
262, R = $(CH_2)_9NH_3^+ Cl^-$	**265**	82

Scheme 2.57 CM of alkenes bearing ammonium salt fragment.

the formation of such densely functionalized product in CM should be considered as a success, especially because the cross partner—acryloamide **267** can also be described as very demanding.

It was found that in the presence of the Grubbs second-generation catalyst, ethyl diazoacetate is converted into diethyl maleate in a highly stereoselective manner (Scheme 2.59) (106). However, unsaturated α-diazocarbonyl compounds can be utilized in CM (107,108). For example, Ru-catalyzed CM of **269** led to formation of **270** with very good yield. The CM event was followed by one-pot carbonyl ylide formation reaction promoted by Rh, eventually leading to the intermolecular cycloaddition product **271** (108). The same strategy was used to access complex oxapolycycles, as well as hyperolactone C, an anti-HIV agent (108).

Azide moiety was also tested in CM (Scheme 2.60). Taking advantage of mild conditions typical for the CM transformation, azide functionality can be present in the reacting alkene (Scheme 2.60). The compatibility of this FG was proven for second- (33) and third-generation complexes (109). Unfortunately, there are also reports of unsuccessful

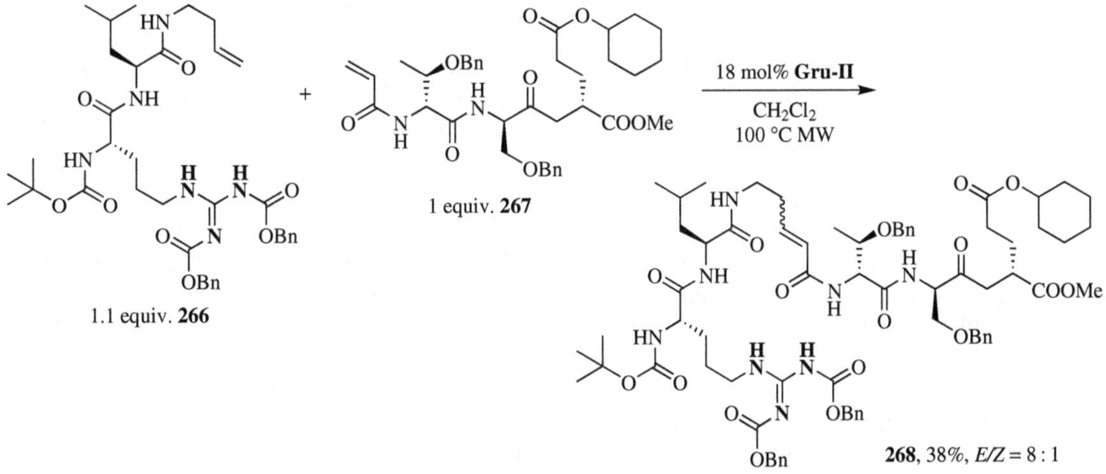

Scheme 2.58 CM reaction with peptide bearing guanidine functionality.

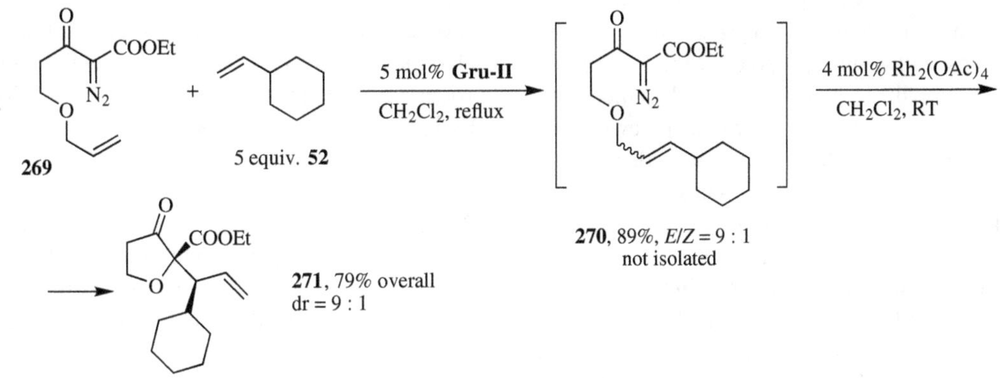

Scheme 2.59 CM of compounds bearing diazoester moiety.

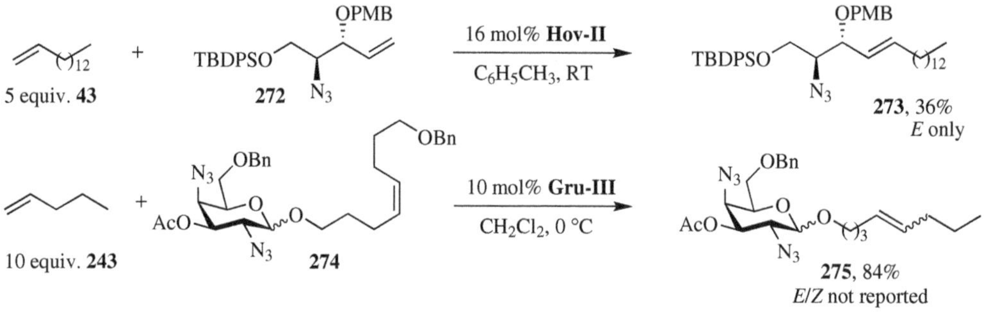

Scheme 2.60 CM of azide-containing substrates.

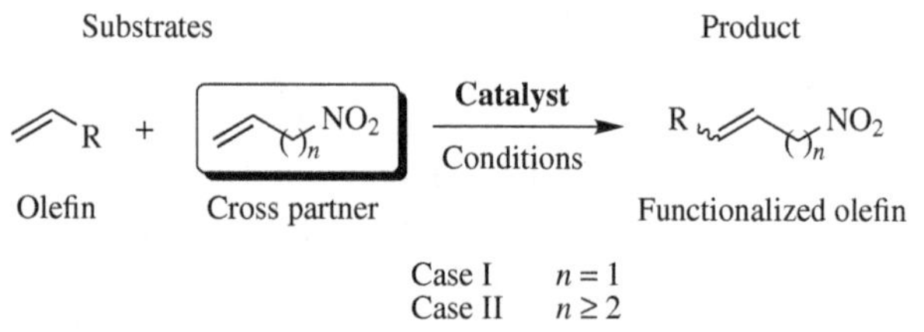

Scheme 2.61 CM of unsaturated nitro compounds—a general scheme.

CM with azide-containing compounds (110) so this matter has to be approached cautiously.

2.2.4.3 Unsaturated Nitro Compounds (FG = NO_2) The significance of the nitro group in organic synthesis is obvious. Reduction to hydroxyloamine or amine moiety, dehydrogenation leading to nitriles or nitrile oxides, or the Nef reaction leading to formation of the carbonyl group are just a small fraction of the affluent reactivity of this functionality. It is also possible to apply some unsaturated nitro compounds in CM (Scheme 2.61).

To the best of our knowledge, no examples of CM with nitroethylene or its derivatives ($n = 0$) were described. This makes α,β-unsaturated nitroalkanes one of the few classes of compounds insusceptible to CM.

The situation improves when β,γ-unsaturated nitrocompounds ($n = 1$) are considered. There is a report on CM of terminal alkenes with 3-nitropropene (**267**), leading to corresponding functionalized products in good-to-moderate yields. Such transformations proceed when a Lewis acid additive is utilized (Scheme 2.62). Importantly, during this study, S-CM ("homodimerization") of 3-nitropropene was not observed (111). It is worth mentioning that if no Lewis acid is applied, the reaction of **276** with **Gru-II** leads to isolation of a chelated ruthenium complex bearing a cyclic nitronate skeleton (112). The complex has been fully analyzed and proved to have low activity in metathesis. This explains why no productive CM of **276** with olefin is observed in the absence of Brønsted or Lewis acids.

Representative CM Reaction of 3-Nitropropene (Scheme 2.62) (111)

Triphenyl boronate (79.9 mg, 0.28 mmol) was added to a solution of 3-nitropropene **276** (193.0 mg, 2.22 mmol) and 10-bromo-1-decene **278** (245.5 mg, 1.12 mmol) in anhydrous dichloromethane (4.5 ml). Addition of **Gre-II** followed in one portion (37.0 mg, 0.06 mmol) and the mixture was stirred for 5 h in the protective atmosphere of argon. The solvent was then removed utilizing reduced pressure and the residue was purified by flash chromatography. Combined organic phases were evaporated to yield product **281** as a yellow oil (248.1 mg, 80%) that is a mixture of *E/Z* isomers (7.5:1, respectively).

Substrates containing the nitro groups situated further from the double bond ($n \geq 2$) undergo CM readily (113). In CM reactions with nitro partners **283–285**, good-to-excellent yields of the cross product can be obtained. Selected examples are presented in Scheme 2.63.

The presence of a longer chain between the nitro group and the olefinic center makes these substrates prone to S-CM process, so olefins **283–285** may need to be used in excess.

2.2.5 Alkenes Bearing Oxygen-Based Functionalities

Oxygen-based functionalities such as free and protected alcohols, phenols, diols, ethers, acetals, and epoxides are widespread and present in myriads of reported CM reactions (Scheme 2.64). Several examples can be found in this chapter and within this book.

2.2.5.1 Unsaturated Alcohols In general, the free OH group is fully compatible with modern, Ru-based catalysts. Allylic (10, 35) ($n = 1$) and homoallylic alcohols (114) ($n = 2$) react well, the same is true, of course, for higher homologs (e.g., **302**) (44) (Scheme 2.65). Chiral triol **299** bearing one trityl-protected and two free hydroxy groups can be very effectively functionalized via CM and therefore

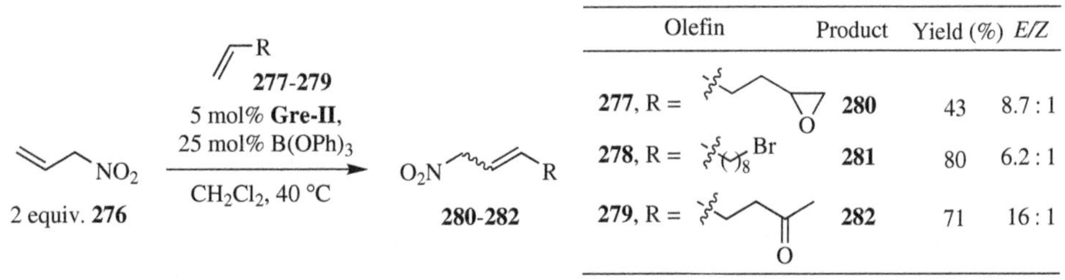

Scheme 2.62 Cross metathesis with 3-nitropropene.

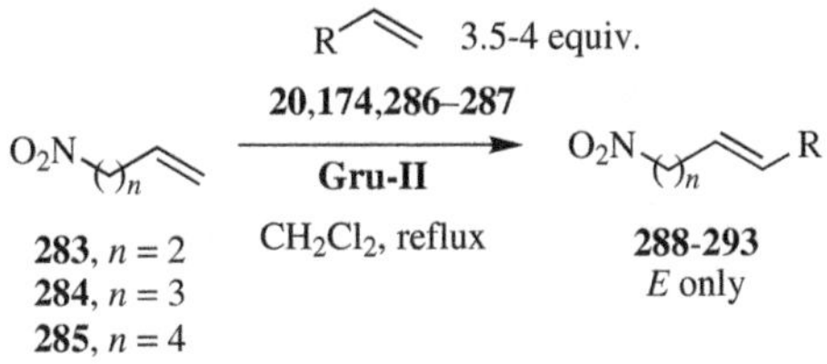

Substrates		Product	Yield (%)
20, R = Ph	**283**	**288**	68
20, R = Ph	**284**	**289**	63
20, R = Ph	**285**	**290**	69
286, R = *p*-C_6H_5Br	**285**	**291**	65
174, R = CHO	**285**	**292**	51
287, R = COO*t*-Bu	**285**	**293**	85

Scheme 2.63 CM of long-chain unsaturated nitro compounds.

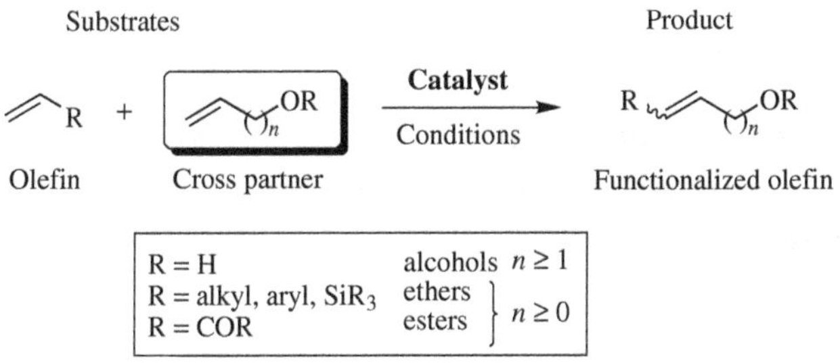

Scheme 2.64 CM of alkenes bearing oxygen-based functionalities.

is a very useful building block in target-oriented syntheses (115). 2-Buten-1,4-diol is often used as a more user-friendly replacement for allyl alcohol (3-propen-1-ol) (116).

Sometimes, the CM of *unprotected* allylic alcohols is accompanied by the formation of ketones, resulting from undesired C–C double bond isomerization (117). By conducting the reaction in the presence of some additives, reducing the CM reaction time, etc., the isomerization can be suppressed and the expected product (e.g., **312**) can be obtained (Scheme 2.66) (118). On the other hand, consecutive isomerization can be deliberately promoted after completing CM by adding a base (NaOH) or polymethylhydrosiloxane (PMHS) as chemical triggers leading to saturated ketone (e.g., **311**) (118), or simply by heating to 200 °C (119). This allows for a single-flask process that provides functionalized ketone-containing products from simple allylic alcohols (e.g., **310**).

Many examples presented in this chapter and in the others show that the free hydroxyl function is very well tolerated by Ru-based catalysts; however, sometimes is worth considering the use of protected alcohols as CM substrates, in order to obtain better yields. This is illustrated in Scheme 2.67, when a CM step was used en route to a natural product (−)-*iso*-cladospolide B, **318** (120).

2.2.5.2 *Ethers and Epoxides*

Vinyl ethers (n = 0) are very problematic substrates for Ru-based catalysts (although there are known fully successful RCM reactions of dienes bearing a vinyl ether fragment, see Chapter 3 for more details). Fortunately, the newly developed molybdenum and tungsten-based catalysts are suitable for CM involving vinyl ethers (see Scheme 2.84).

Allyl ethers (n = 1) are much easier substrates, even for first-generation Ru catalysts, and they were frequently utilized in CM reactions. S-CM ("homodimerization") of allylic ether derivatives is commonly utilized in synthesis of biologically active molecules, such as antibiotics (121), amino acid, peptide (28) steroids (122), and many others. In some cases, the self-metathesis was performed *in vacuo*, allowing for more efficient removal of ethylene, thus shifting the reaction equilibrium to the left (Scheme 2.68). The preparation of linked oligo-aminodeoxysugars—potential binders to polynucleotides—was studied by Kirschning (123). Saccharide dimers such as **322** were assembled by the self-metathesis reaction. The first-generation Grubbs catalyst **Gru-I** turned out to be the catalyst of choice for this reaction. In the case of allyl glycoside **323**, however, in addition to the expected product **324**, a by-product **325** was formed that contained a shifted C–C double bond, despite using a more selective catalyst (Scheme 2.68). Such side reactions (described in detail in Chapter 5) are sometimes problematic and require rigorous optimization of conditions (usually, using first-generation catalysts, lowering temperature, shortening reaction time, or using some additives, such as quinones of phenols, is of help).

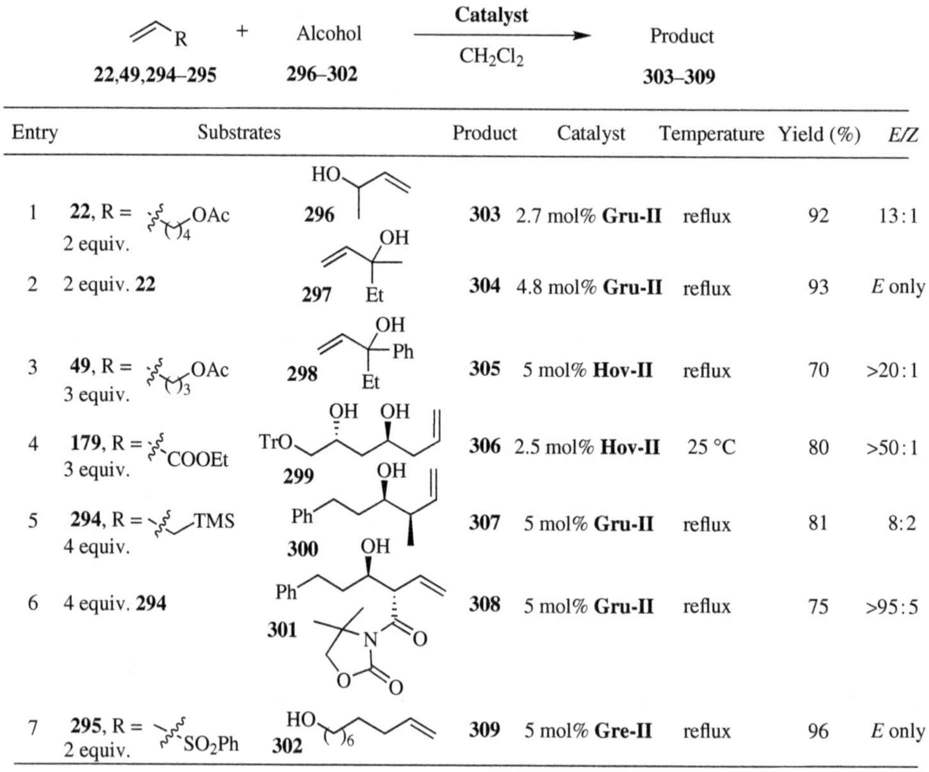

Scheme 2.65 CM of unsaturated alcohols and diols.

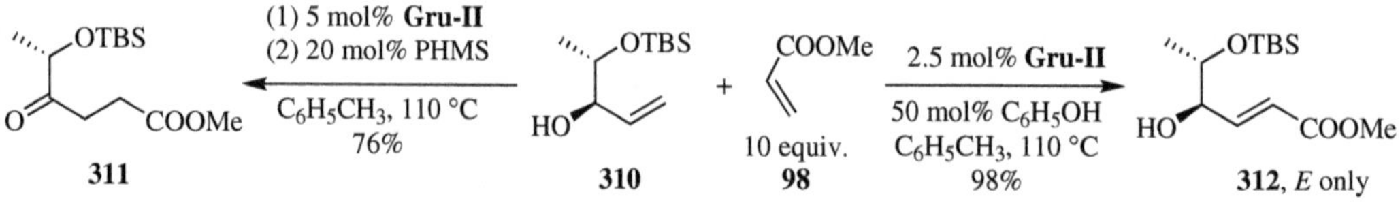

Scheme 2.66 CM of allylic alcohol and product isomerization.

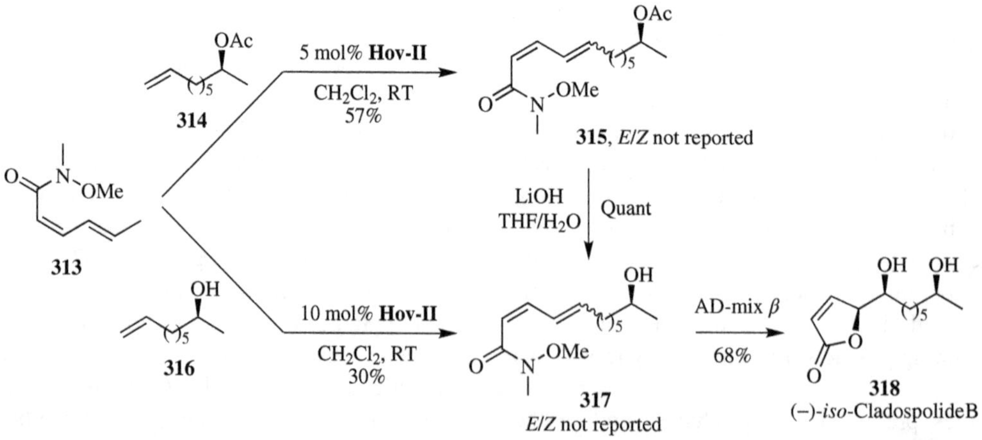

Scheme 2.67 CM reaction in the synthesis of (−)-*iso*-cladospolide B.

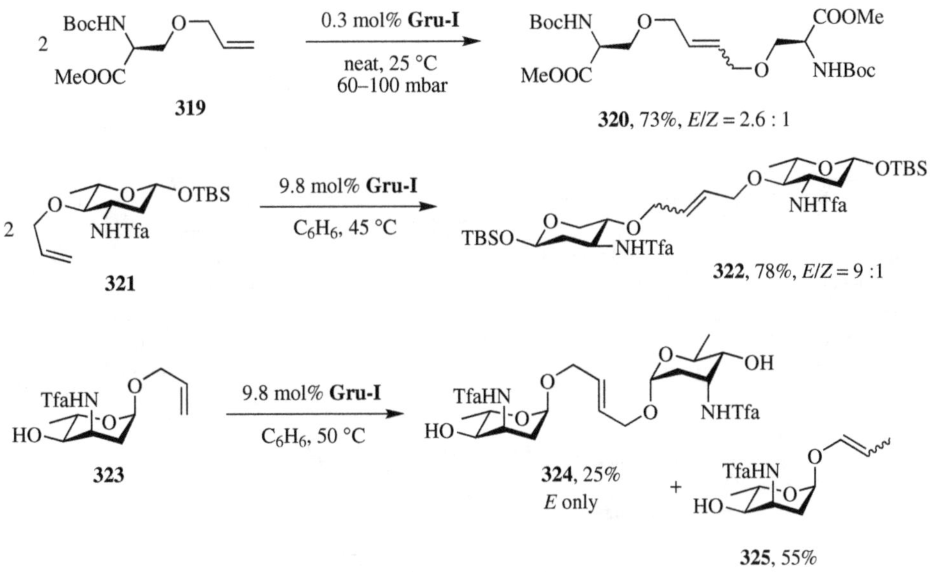

Scheme 2.68 S-CM reactions of *O*-allyl derivatives of biologically active compounds.

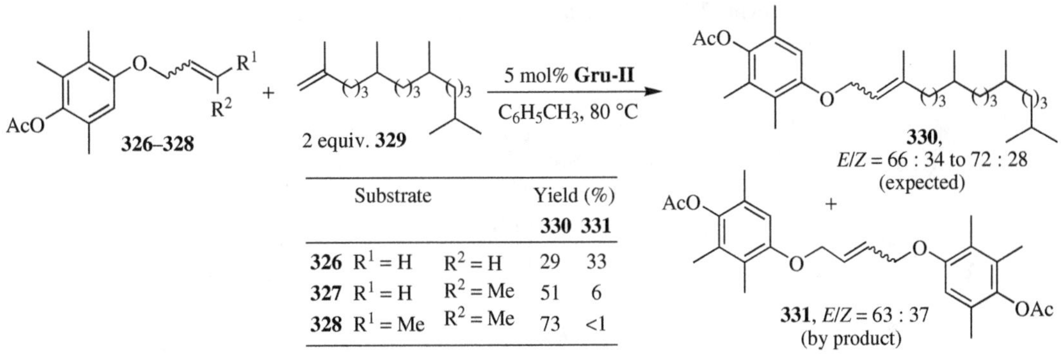

Substrate		Yield (%) 330	Yield (%) 331
326 $R^1 = H$	$R^2 = H$	29	33
327 $R^1 = H$	$R^2 = Me$	51	6
328 $R^1 = Me$	$R^2 = Me$	73	<1

Scheme 2.69 Application of allyl ethers in CM.

Susceptibility of allyl ethers to undergoing S-CM can be sometimes a problem if these substrates are applied in CM with other olefins. Interestingly, this problem can be controlled by substitution of the reacting double bond (Scheme 2.69) (124).

Epoxide is a synthetically important FG. Fortunately, this FG is very well tolerated by modern olefin metathesis catalysts, and survives untouched. There are examples of CM of epoxides placed at various distances to the reacting C–C double bond (one is shown in Scheme 2.62). Scheme 2.70 shows application of vinyloxirane **333** in modification of tacrolimus (FK506, **332**) (125) and two examples where the epoxide is located at longer distance from the reacting C–C double bond (126).

Not only simple benzyl ethers (28) but also a wide collection of other oxygen-based FGs are tolerated as well. Scheme 2.71 shows two representative examples (please note also the presence of spiroketals and 1,3-dithiane fragments in the azaspitracid-1 precursor **338** (127) and reactive peroxide bridge in artemisinin-derived dimer **343** (128)). Such excellent compatibility is of great importance, allowing CM to be used in synthesis of numerous polyoxygenated biologically active compounds.

2.2.6 Haloalkenes in Cross Metathesis

CM transformations with haloalkenes as CM partners are extremely beneficial. These reactions lead to the formation of compounds that can participate as building blocks in the following stages such as alkylation or cross coupling (Scheme 2.72). To avoid confusion, the reader is advised, that in this section CM reactions with substrates such as vinyl chloride or allyl bromide are described, while alkenes substituted with perfluoroalkyl groups have been reviewed in Section 2.2.3.3.

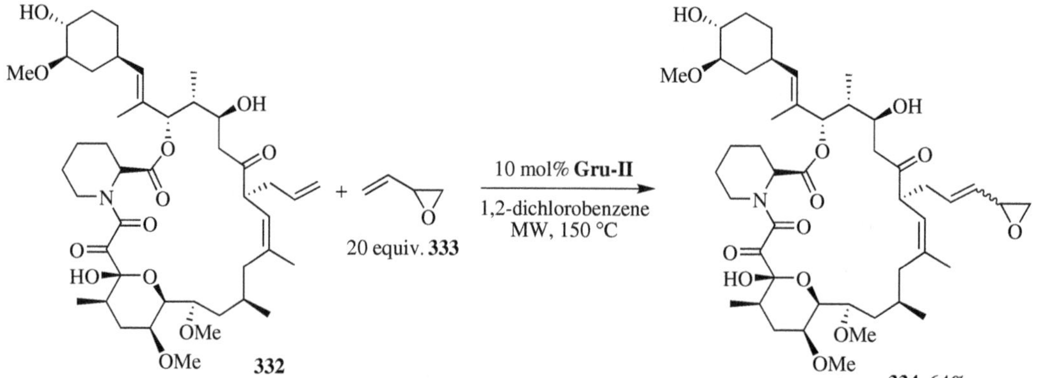

332 + 20 equiv. 333 → (10 mol% **Gru-II**, 1,2-dichlorobenzene, MW, 150 °C) → **334**, 64% *E*/*Z* not reported

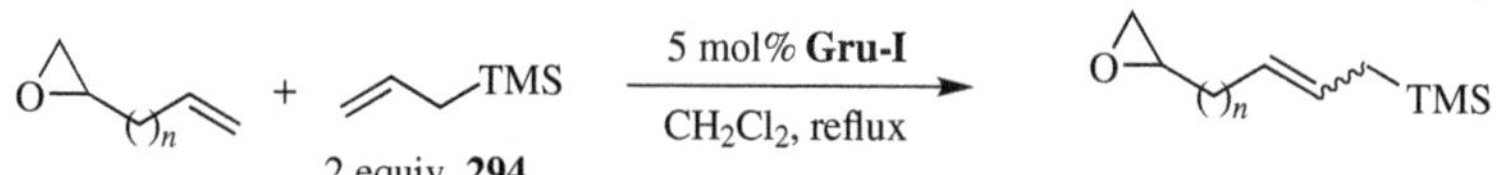

277,335 + 2 equiv. **294** → (5 mol% **Gru-I**, CH_2Cl_2, reflux) → **336–337**

Substrate	Product	Yield (%)	*E*/*Z*
277, *n* = 2	**336**	45	1 : 1
335, *n* = 6	**337**	41	1 : 1

Scheme 2.70 Unsaturated epoxides as partners in CM.

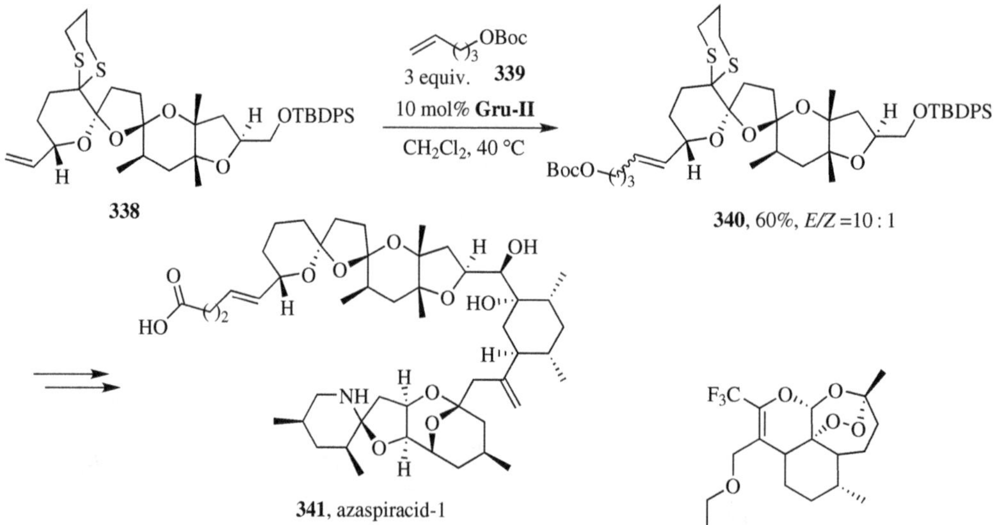

338 → (3 equiv. **339**, 10 mol% **Gru-II**, CH_2Cl_2, 40 °C) → **340**, 60%, *E*/*Z* =10 : 1 ⟶⟶ **341**, azaspiracid-1

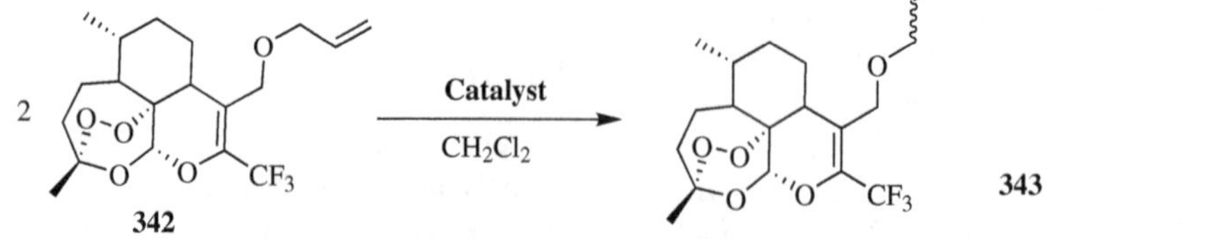

2 **342** → (**Catalyst**, CH_2Cl_2) → **343**

Catalyst	Temperature	Yield (%)	*E*/*Z*
5 mol% **Gru-I**	reflux	71	9 : 1
10 mol% **Gru-II**	RT	69	*E* only

Scheme 2.71 CM in synthesis of polyoxygenated biologically active compounds.

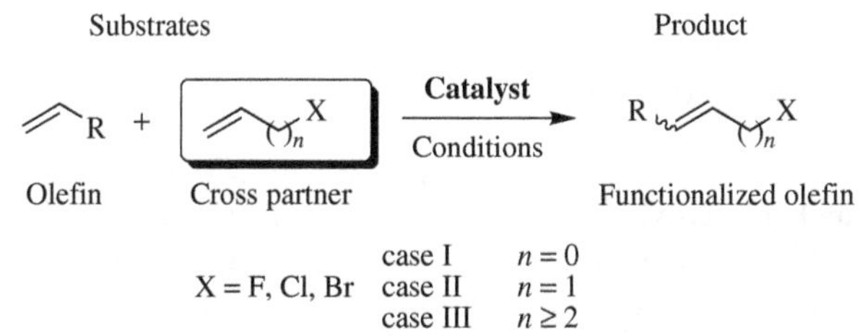

Scheme 2.72 Haloalkenes in CM—a general scheme.

There are a few examples involving metathesis of vinyl halides (n = 0, Scheme 2.73). Both vinyl halides and 1,2-dihaloethylene can be utilized. It seems that phosphine-free ruthenium complexes must be employed in transformation, as Grubbs-type complexes are deactivated (129). The CM reaction with 1,2-dichloroethene **347** is very interesting from a practical point of view; however, this transformation varies in the degree of selectivity and the overall yield. The best results were obtained with Hoveyda catalyst (130) and its activated derivative, **Gre-II** (131). In the latter case, **347** was used as a solvent, and was effectively recovered after the reaction. Unfortunately, under the same conditions, 1,2-dibromoethene gives much lower yields in CM with terminal olefins (131). In general, vinyl halides shall be considered as very challenging CM partners.

Allyl halides (n = 1) are very useful CM partners that allow to introduce the 3-halopropene fragment into alkene of interest. Scheme 2.74 shows some representative examples of terminal (132) and internal (133) olefin CM functionalization with allyl halides. One should note that (2*Z*)-1,4-dichloro-2-butene (**353**) can serve as a convenient alternative to allyl chloride (134). Alkenes bearing halogen atoms at more distant positions also react smoothly in CM (135).

2.2.7 Silicon-Containing Unsaturated Compounds

In the presence of transition metal complexes, vinylsilanes (n = 0) undergoes catalytic transformation with olefins leading to the synthesis of the respective functionalized vinyl-silicon compounds (Scheme 2.75). Depending on the initiator and conditions, two different mechanisms are operative—the one based on carbene intermediates is of interest in this chapter. The second one, so-called silylative coupling, was developed and studied extensively by the group of Pietraszuk and Marciniec, and has been reviewed elsewhere (136–138). In the presence of ruthenium alkylidene complexes, vinylsilanes, such as **362** participates in CM with a range of alkenes. In the molecule of vinylsilane various substituents are tolerated (e.g., **364–366**),

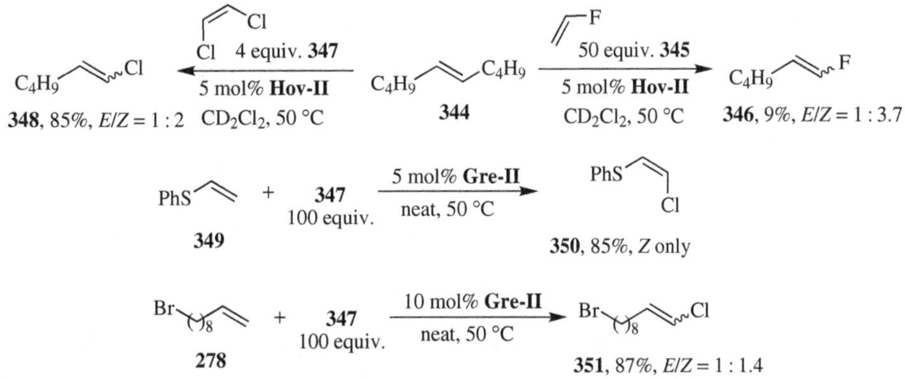

Scheme 2.73 CM of vinyl halides.

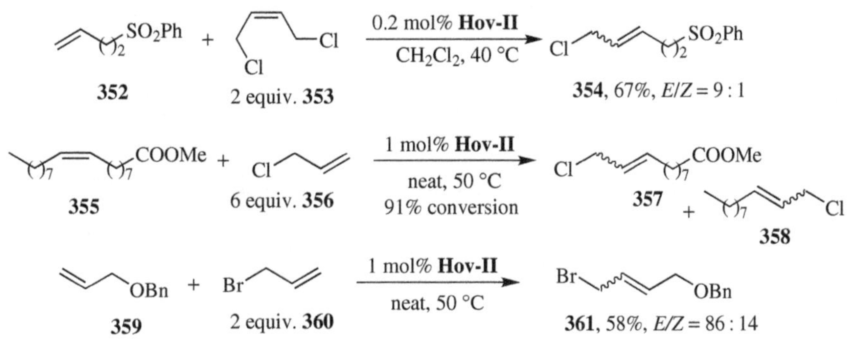

Scheme 2.74 CM of allyl halides.

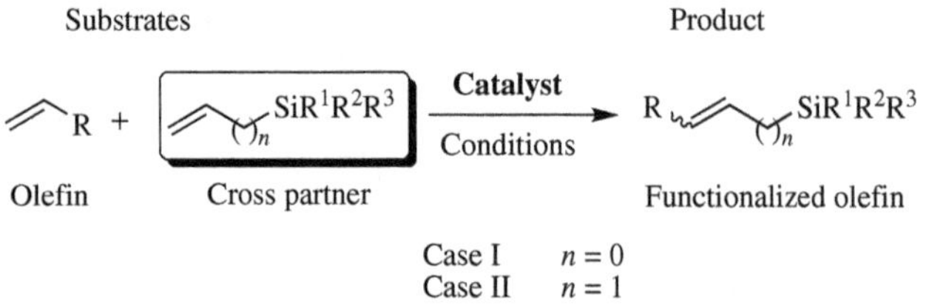

Scheme 2.75 CM of silicon-containing unsaturated compounds—a general scheme.

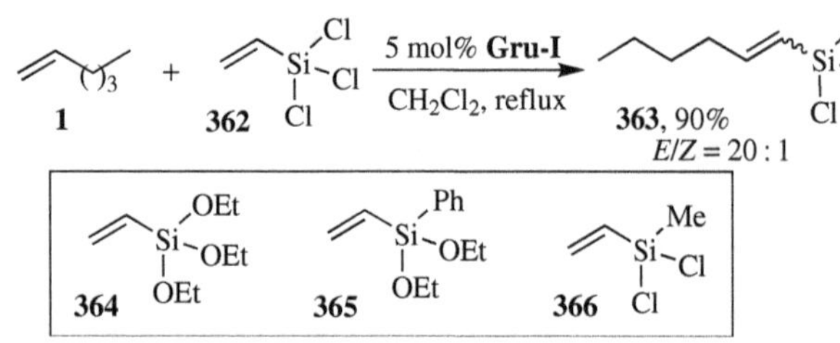

Scheme 2.76 Substrate scope in CM of vinylsilanes.

as it is exemplified in Scheme 2.76 (139). In the case of vinyl(trialkoxy)silanes, the more active **Gru-II** catalyst was required as **Gru-I** led to a diminished yield (56).

Another interesting example is related to CM of vinyl-substituted polyhedral oligosilsesquioxanes (POSS). These interesting materials of the general formula $(RSiO_{1.5})_n$ constitute a class of organosilicon compounds that currently have a strong impact on both catalytic research and materials science. The eight vinyl groups attached to the silsesquioxane core are particularly interesting for producing hybrid materials as they can be easily functionalized by a variety of transformations. Scheme 2.77 presents a representative example (140–143). It should be also noted that CM of vinylsilsesquioxane **368** in the presence of Schrock's molybdenum catalyst **Sch-I**, also gave very good isolated yields of the desired products (144). Efficient functionalization of a monovinyl silsesquioxane via Ru-catalyzed CM was reported by Pietraszuk and Marciniec (145).

Representative Procedure of POSS-Containing Compounds(Scheme 2.77) (142)

Octavinylsilsesquioxane **368** (20 mg, 0.032 mmol) and dimethyl 4′-vinylbiphenyl-3,5-dicarboxylate **367** (164 mg, 0.55 mg) were put in a tube under argon and dissolved in anhydrous CH_2Cl_2 (10 ml). Next, the solution of **Gru-II** in CH_2Cl_2 was added (1 mg, 0.001 mmol in 3 ml). The flask was equipped with a condenser and the mixture was heated to 55 °C. After 90 h the reaction was cooled to room temperature and filtered. The residue was concentrated and precipitated by addition of a solution of ethyl acetate and petroleum ether. The residue was purified by a silica gel column to afford product **369** as a white powder (54 mg, 62%).

Allylsilanes (n = 1) are also very useful CM partners in organic synthesis. Scheme 2.78 presents representative examples of CM with allyltrimethylsilane (**294**), a popular CM partner, leading to a facile introduction of silicon moiety to the structure of functionalized olefinic substrate under mild conditions (146–148). Interestingly, allyltrimethylsilane self-metathesis ("dimerization") is not causing problems during a CM reaction, rendering it as a very useful partner in CM (36). Allyltriphenylsilane reacts in CM similarly (149). The functionalized allylsilanes obtained through the CM reaction provide useful synthetic intermediates.

2.2.8 Unsaturated Phosphorus-Containing Compounds (FG = $P(E)R_2$)

Phosphines are privileged ligands in numerous catalytic transformations involving transition metals. At the same time, phosphorus-containing FGs such as phosphonates are present as pharmacophores in various classes of bio-

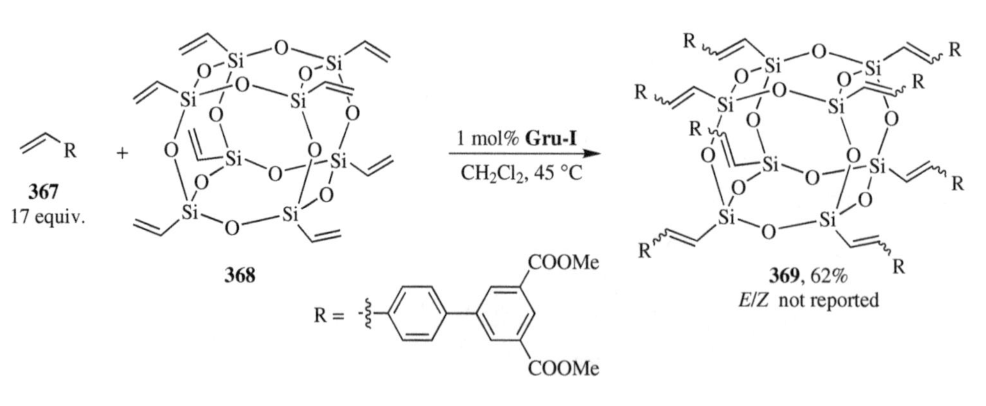

Scheme 2.77 CM of vinyl-substituted polyhedral oligosilsesquioxanes.

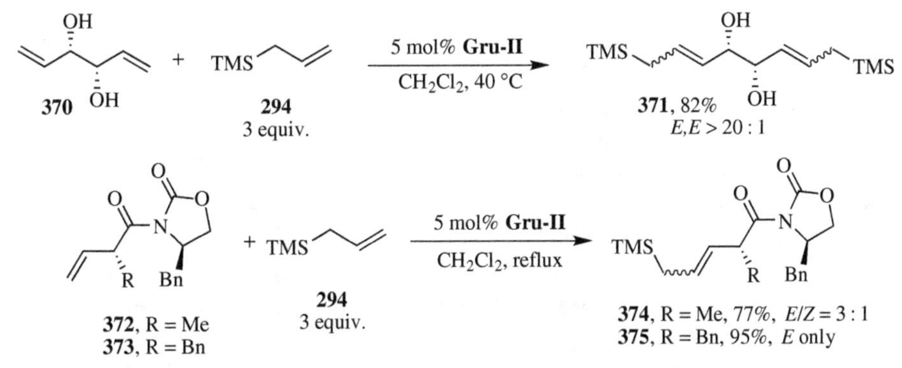

Scheme 2.78 CM of allyl silanes.

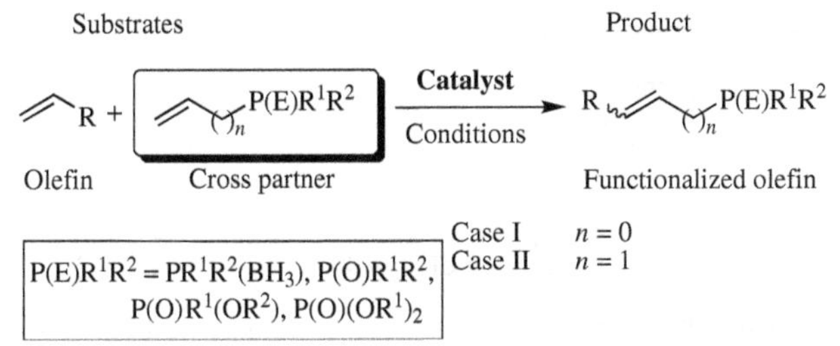

Scheme 2.79 CM of phosphorus-containing alkenes—a general scheme.

logically important molecules. Therefore, new ways of synthesizing of phosphorus-containing molecules are of importance (Scheme 2.79).

2.2.8.1 Alkenes Containing Phosphine–Borane Fragments (FG = $PR_2(BH_3)$) Because free phosphines usually make strong bonds to transition metals, it can be expected that protection of the phosphor center is required before the metathesis reaction. Indeed, it was found that unprotected diallylphenylphosphine did not react in the presence of Ru-based catalysts **Gru-I**, **II**, and **II′**. Interestingly, the same substrate underwent ring-closing metathesis in the presence of Schrock's Mo-based catalyst **Sch-I** giving the expected cyclic product with 95% conversion (150).

Even if CM was possible with unprotected phosphines, it is preferable and more practical to use protected phosphines, such as phosphine oxides or phosphine boranes, as these substrates are easier to handle and give products less prone to decomposition upon purification. A detailed study on CM of unsaturated phosphine–borane complexes (e.g., **377–380**) was made by Gouverneur et al. (151). In the presence of **Gru-II** catalyst, vinyl-substituted phosphine boranes ($n = 0$) underwent smooth CM with various olefinic partners providing products with exclusive *E*-selectivity (Scheme 2.80). Although the conversions were consistently very good to excellent, the yields of purified products were sometimes

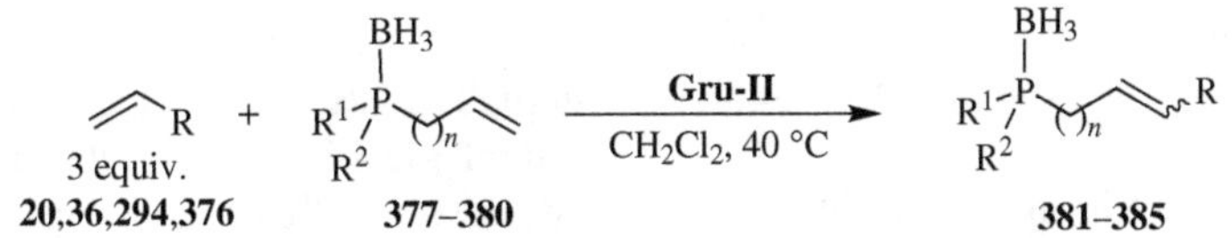

Substrates		Product	Catalyst	Yield (%)	E/Z
376, R = C_9H_{19}	**377**, $R^1 = R^2 = Ph$ $n = 0$	**381**	8 mol%	35	*E* only
36, R = $(CH_2)_3Br$	**378**, $R^1 = Ph$, $R^2 = OMe$ $n = 0$	**382**	6 mol%	55	*E* only
20, R = Ph	**378**, $R^1 = Ph$, $R^2 = OMe$ $n = 0$	**383**	8 mol%	66	*E* only
294, R = TMS	**379**, $R^1 = R^2 = Ph$ $n = 1$	**384**	4 mol%	85	5 : 1
20, R = Ph	**380**, $R^1 = Ph$, $R^2 = OMe$ $n = 1$	**385**	2 mol%	43	10 : 1

Scheme 2.80 CM of vinyl and allyl phosphine boranes.

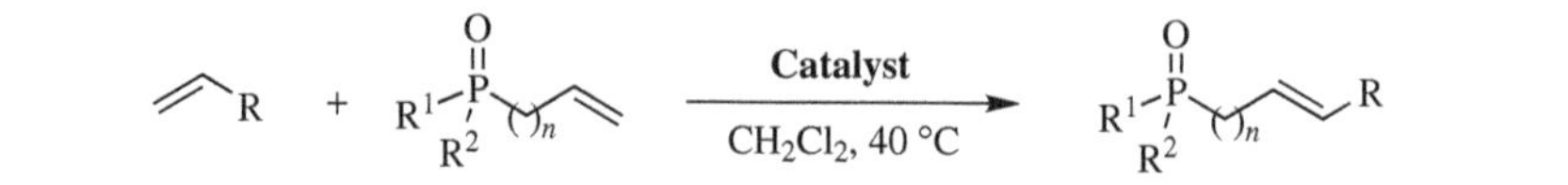

22,93,98,386 387–389 44–58, *E* only

Substrates			Product	Catalyst	Yield (%)
22, R = ; 3 equiv.	387, $R^1 = R^2 = Ph$	$n = 0$	390	2–6 mol% **Gru-II**	94
	388, $R^1 = R^2 = Ph$	$n = 1$	391	2–6 mol% **Gru-II**	70
93, R = ; 3 equiv.	387, $R^1 = R^2 = Ph$	$n = 0$	392	2–6 mol% **Gru-II**	0
	388, $R^1 = R^2 = Ph$	$n = 1$	393	2–6 mol% **Gru-II**	46
386, R = ; 1 equiv.	387, $R^1 = R^2 = Ph$; 2 equiv. 387	$n = 0$	394	5 mol% **Gre-II**	76
98, R = ; 30 equiv.	389, $R^1 = Ph$, $R^2 = Bn$	$n = 0$	395	5 mol% **Gre-II**	34

Scheme 2.81 CM of vinyl and allyl phosphine oxides.

significantly lower, probably because of troublesome purification. Allyl-substituted phosphine–borane complexes ($n = 1$) also react smoothly in CM to give the *E*–products predominantly. S-CM ("homodimerization") has been investigated and proved successful only for allyl-substituted, borane-protected phosphines (151).

2.2.8.2 *Alkenes Containing Phosphine-Oxide Fragments (FG = P(O)R_2)*

Gouverneur (152) reported on the preparation of substituted vinyl ($n = 0$) and allyl ($n = 1$) phosphine oxides by Ru-catalyzed CM providing a number of products (e.g., **390–395**) with exclusive *E*-olefin stereochemistry (Scheme 2.81). Interestingly, the CM reaction fails when vinyl phosphine oxide was coupled with electron-poor CM partner (**93**). A similar study was published in parallel by Pietrusiewicz and Grela (41). By using the nitroactivated catalyst, **Gre-II**, it was possible to obtain good yields even in case of sterically demanding partners such as **386**. Also CM with electron-poor partners **98** was possible, albeit in moderate yield (Scheme 2.81) (153).

CM functionalization of di- and tri(vinyl) phosphine oxides with various olefinic partners was studied by Gouverneur. This reaction allows for direct access to novel racemic *P*-stereogenic products featuring two or three different alkenyl groups (154).

Chiral phosphines are key compounds for the discovery of new catalytic asymmetric transformations. Unfortunately, preparation of *P*-stereogenic phosphines is often complicated, so innovative synthetic routes are of importance. Unsymmetrically substituted vinylphosphine oxides have been prepared in good yield and exclusive *E*-olefin selectivity using activated Hoveyda-type ruthenium catalyst **Gre-II**. Importantly, metathesis of chiral vinylphosphine oxides (e.g., (−)-**397**) proceeds without racemization of the phosphorus

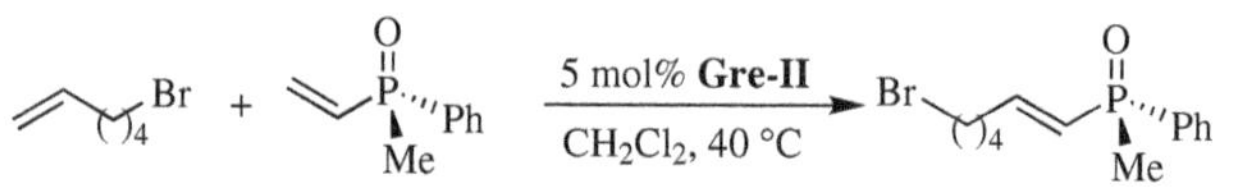

2.5 equiv. **396** (−)-**397**, 98% ee **398**, 86%, *E* only, 98% ee

Scheme 2.82 CM with *P*-chiral vinyl phosphine oxide **37**.

chirality center, providing easy access to functionalized chiral non-racemic *E*-alkenylphosphine oxides (such as **398**, Scheme 2.82) (41).

Utility of bisphosphines as ligands in transition metal catalysis was the reason to investigate the scope and limitations of S-CM of unsaturated phosphine oxides. Transformation of vinyl-, allyl-, and 4-penten-1-yl(diphenyl)phosphine oxides **287**, **388**, and **399** promoted by **Gru-II** catalyst was studied by Gouverneur, who observed no "dimerization" of the vinylphosphine oxide. In contrast, S-CM of the nonconjugated allyl- and 4-penten-1-yldiphenyl phosphine oxides was feasible and afforded the homodimeric products in high yields and *E*-selectivity (Scheme 2.83) (152). Interestingly, less bulky vinylphosphine oxides, such as **403** and **387**, were successfully coupled in the presence of **Gre-II** (41) as well as other activated analogs of Hoveyda–Grubbs catalysts, designed specifically for this reaction (155). It should be noted that chiral vinylphosphine oxides also undergo catalytic olefin S-CM leading to *P*-stereogenic diphosphine dioxides (e.g., **404**) with exclusive *E*-selectivity and without racemization of the phosphorus center of chirality (Scheme 2.83) (153). This finding provides a potentially useful method for preparing chiral bidentate phosphine ligands.

2.2.8.3 *Unsaturated Phosphonates (R = P(O)$(OR)_2$)*

Functionalized phosphonates are very interesting compounds not only because of their biological activity, but also

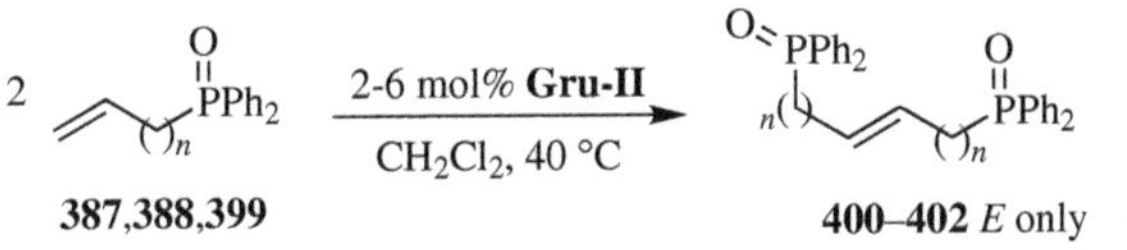

Substrate	mol%	Product	Yield %
387, $n = 0$	6	**400**	0
388, $n = 1$	2	**401**	96
399, $n = 3$	4	**402**	96

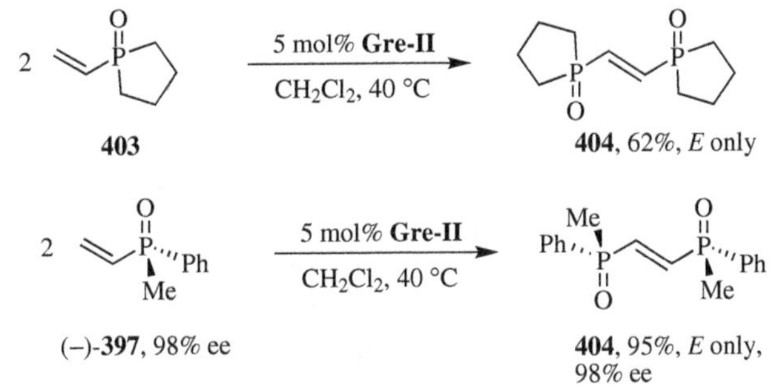

Scheme 2.83 Self-cross metathesis of phosphine oxides.

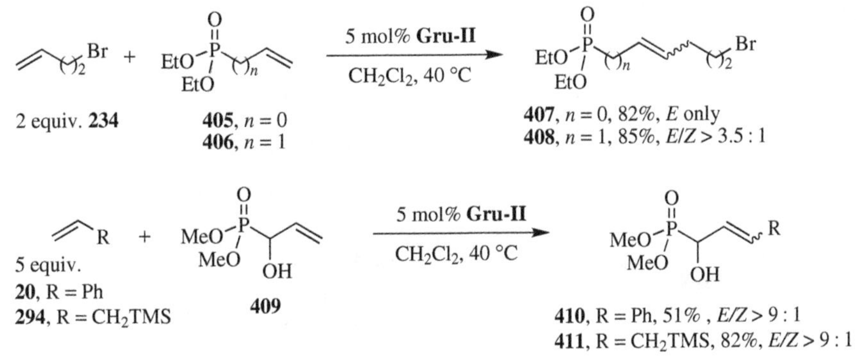

Scheme 2.84 CM of unsaturated phosphonates.

because they are attractive precursors in synthesis. Examples of CM involving vinyl ($n = 0$) and allyl phosphonates ($n = 1$) **405–406** and related compounds have been reported (156). Similarly, a number of allylic hydroxy phosphonates (e.g., **409**) and their derivatives were used as partners in CM with second-generation Grubbs catalyst (Scheme 2.84) (157). Agrofoglio et al. (158) studied CM of (acyloxymethyl) or (hexadecyloxypropyl)allylphosphonate building blocks as a short synthetic route to phosphonate prodrugs. Similarly, vinylphosphonate-linked nucleic acids (159) and other substrates were successfully applied in Ru-promoted CM. Allyl phosphonates undergo S-CM (157).

2.2.9 Alkenes Bearing Sulfur-Based Moieties

Sulfur-containing unsaturated compounds are very useful building blocks in organic chemistry. Unfortunately, some sulfur-containing substrates, such as vinyl sulfones were initially claimed to be nonreactive in CM (56). Examples of CM reactions with partners containing sulfur in various oxidation state are discussed systematically in a mini review-type paper by Samojłowicz et al. (Scheme 2.85) (160).

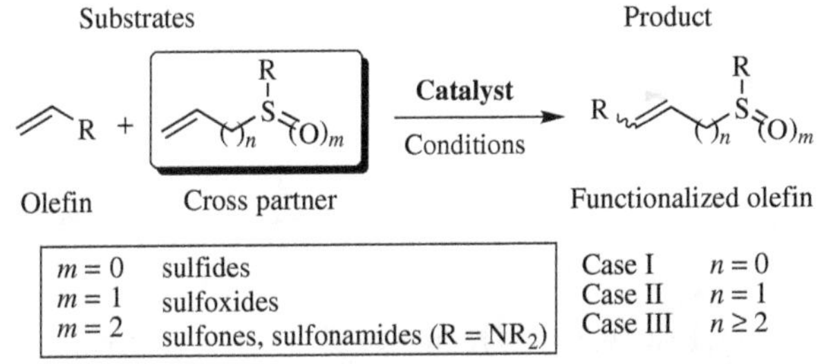

Scheme 2.85 General scheme of CM with sulfur-containing alkenes.

2.2.9.1 Unsaturated Thiols and Sulfides (FG = SR) Vinyl sulfides ($n = 0$) are electron-rich alkenes that usually do not participate in olefin metathesis reactions owing to the formation of stable Fischer-type carbenes. Grela et al. (160) have shown that the CM reaction of phenyl vinyl sulfide (**349**) with 5-hexenyl tertbutyldimethylsilyl ether (**154**) in the presence of **Gre-II** in DCM gives no CM product. However, it should be noted that vinyl sulfides (e.g., **349**) can react in CM reactions with vinyl halides (Scheme 2.73) (131) and

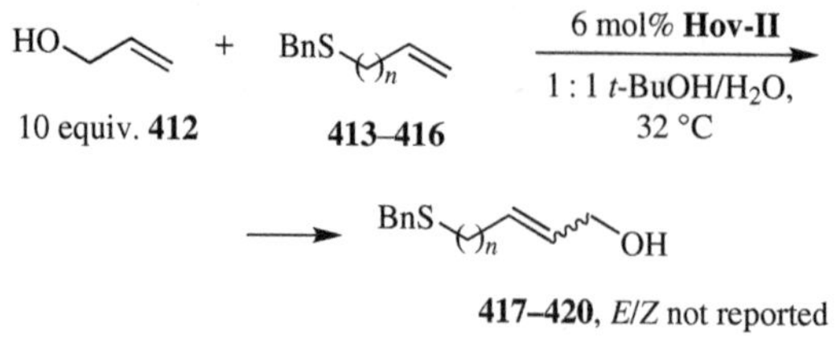

Sulfide	Product	Yield (%)
413, $n = 0$	**417**	0
414, $n = 1$	**418**	52
415, $n = 2$	**419**	19
415, $n = 3$	**420**	8

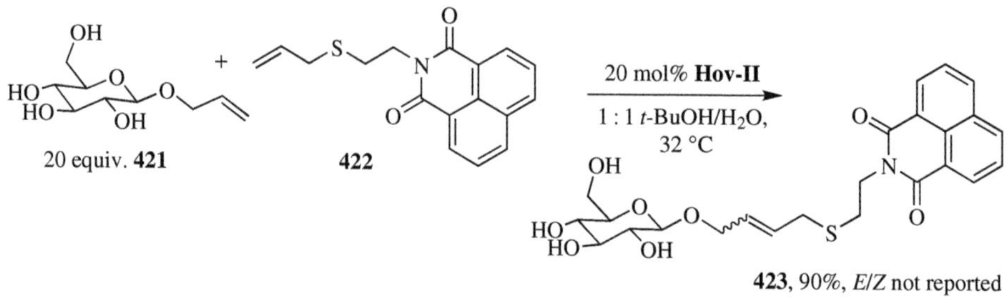

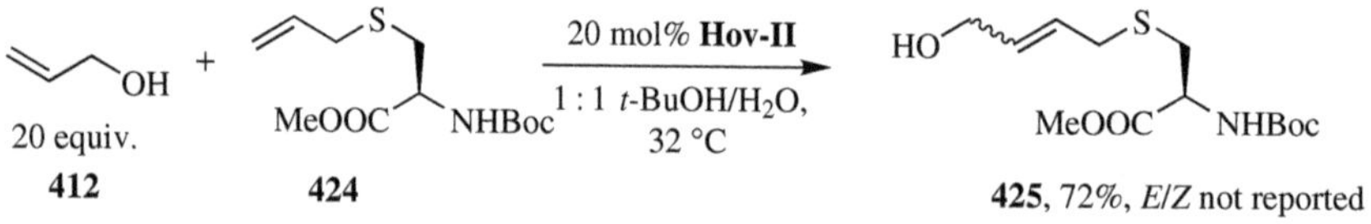

Scheme 2.86 CM of unsaturated sulfides in aqueous media.

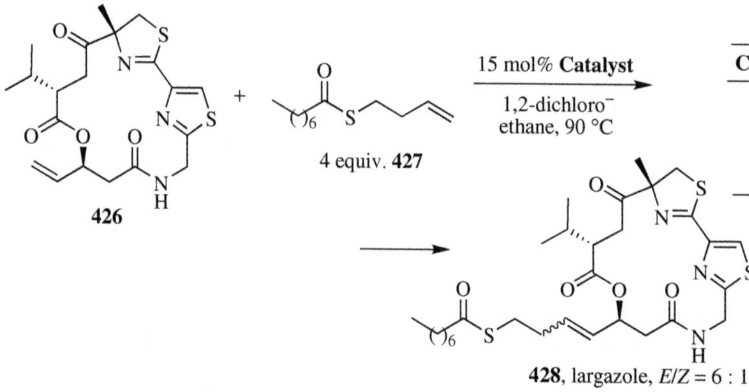

Catalyst	Yield (%)
Gru-II	11
Hov-II	44
Hov-II‴	13
Gre-II	75

Scheme 2.87 CM of unsaturated thioester in preparation of largazole.

that vinyl sulfides (161) and selenides (162) participate in some ring-opening/CM sequences.

Nolan and Mioskowski et al. (163) have published an elegant study to show that allyl ($n = 1$) sulfides and even allyl mercaptan can be successfully employed in S-CM. On the other hand, allyl sulfides have been used as privileged partners in CM with various olefins (164). CM of allyl sulfides with olefinic partners in aqueous media have been studied by Davies et al. who showed that even more promising results can be obtained for allyl selenides. This reaction can be used for protein modification (165,166). Interestingly, lengthening of the carbon chain between the thiol and the double bond does not improve the reactivity of the resulting substrate. On the contrary, the reactivity of the pentenyl and hexenyl sulfides is noticeably diminished (Scheme 2.86) (167). Probably the formation of stable cyclic Ru chelates disrupts the catalytic cycle.

Thioesters are the products of esterification between a carboxylic acid and a thiol, therefore the thioesters derived from *unsaturated thiols* are also described in this section (for CM of S-ethyl 2-propenethioate—an ester of acrylic acid and *a saturated thiol*, see Scheme 2.51). Thioesters are widespread in biochemistry, so their behavior in CM is of interest. Cramer (168) described a very elegant synthesis of largazole, a cytotoxic compound isolated from cyanobacteria. Modular synthesis, using CM of **426** and a short library of thioesters (incl. **427**) as cross partners, enabled the preparation of largazole **428** and its several analogs (Scheme 2.87). Other examples of CM with unsaturated thioesters are also known (169,170).

2.2.9.2 Sulfoxides Grela et al. (172) reported that neither vinyl ($n = 0$) nor allyl ($n = 1$) sulfoxides (**429–430**) are reactive in CM with terminal olefins (Scheme 2.88). This is intriguing, because there are known examples of successful RCM reactions of sulfoxide-containing dienes (see Chapter 3). Even when more promoting conditions (e.g., adding a Lewis acid, or using perfluorobenzene as a solvent) were used, the maximum yield obtained in CM with allyl benzyl sulfoxide was in the range 17–31% (160). However, when a sulfoxide function is located at a more distant position in the alkene chain (**431**), good-to-excellent reactivity in CM and S-CM is observed (Scheme 2.88, see also Scheme 2.92 for another example (171)). Interestingly, a ruthenium-based catalyst bearing a sulfoxide fragment in one of the ligands is known to possess very good application profiles (173). Therefore, the lack of reactivity in CM of vinyl and allyl sulfoxides is surprising.

2.2.9.3 Unsaturated Sulfones Substituted α,β-unsaturated sulfones ($n = 0$) are generally well accepted as useful intermediates in organic synthesis. Thus, vinyl sulfones serve efficiently, for example, as Michael acceptors and as 2π partners in cycloaddition reactions. In addition, the stability and ease of further transformation of the sulfonyl group renders further advantages to vinyl sulfones as synthetic intermediates. In addition, this FG has also recently been shown to potently inhibit a variety of many types of cysteine proteases, providing unique properties for drug design and medicinal chemistry (174).

Interestingly, vinyl sulfones were initially claimed to be nonreactive or in CM.[3d,57] Using standard second-generation Grubbs' catalyst **Gru-II**, Grela and Bieniek (175) have successfully conducted CM reactions between phenyl vinyl sulfone **295** and a variety of terminal olefins (e.g., **154**) in moderate-to-high yields with remarkable stereoselectivity

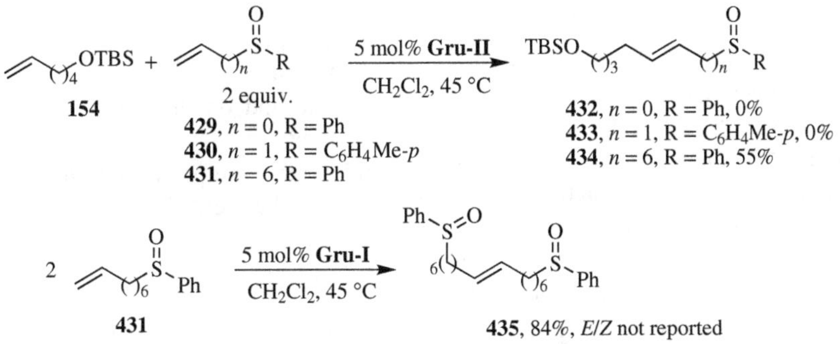

Scheme 2.88 CM of unsaturated sulfoxides.

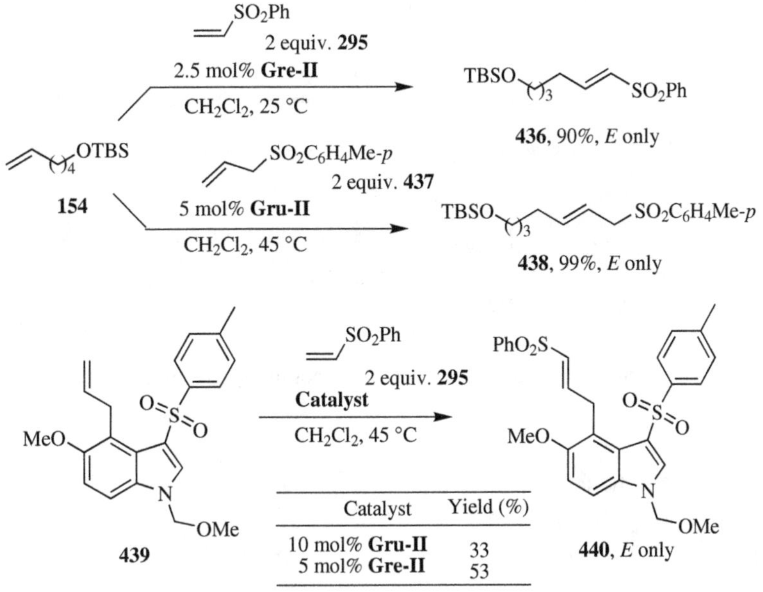

Catalyst	Yield (%)
10 mol% **Gru-II**	33
5 mol% **Gre-II**	53

Scheme 2.89 CM of vinyl and allyl sulfones.

(only the *E*-isomer was detected, Scheme 2.89). It was later found that **Hov-II** and especially **Gre-II** catalysts would also promote metathesis, even of sterically more demanding substrates (e.g., **439**)(172). Subsequently, the CM reaction of vinyl sulfones have been used in a number of synthetic projects. Ettari et al. synthesized a new class of peptidomimetic cysteine protease inhibitors bearing a vinyl sulfone moiety. The introduction of the vinyl sulfone "warhead" has been accomplished by CM using methyl, ethyl, and aryl vinyl sulfones as partners, and the products have been proved to be potent and selective inhibitors of falcipain-2, displaying interesting antiplasmodial activity (176).

Asymmetric dihydroxylation of vinyl sulfones has recently been used in the formation of enantioenriched α-hydroxyaldehydes and furan-2(5*H*)-ones (Scheme 2.90). After the CM step, the resulted functionalized aryl vinyl sulfone (e.g. **442**) was reacted with one of the AD-mix reagents leading—after the spontaneous 1,2-elimination of aryl sulfinic acid—to the formation of α-hydroxyaldehydes (e.g., **443**). This strategy was used by Evans and Leffray (177) in the synthesis of (*R*)-(−)-maritolide (**444**). In the route of this synthesis, aldehyde **443** was converted into the conjugated ester via the *cis*-selective Stille–Gennari variant of the Horner–Emmons–Wadsworth (HEW) reaction that cyclized to the furan-2(5*H*)-one natural product **444**. The same strategy was used for quercus lactone (177), swansonine (178), myrioxazine A (179), and other natural products. Vinyl sulfone CM was recently utilized in total synthesis of (±)-lepadiformine A (180). In summary, vinyl sulfones are demanding yet very useful partners in CM.

Interestingly, in the case of divinyl sulfone **445**, it is possible to selectively functionalize only one of the two vinyl groups present in this substrate (Scheme 2.91) (181). Next, by repeating the CM event, the second vinyl group in **446** can be functionalized by CM, leading to asymmetrically substituted products, such as **448**, with very little or no scrambling. This result revealed that CM technology gives easy access to synthetically useful desymmetrized mono- or disubstituted divinyl sulfones with excellent *E*-selectivity under relatively mild conditions.

The CM of phenyl allyl sulfone ($n = 1$) has been reported by Grubbs (28, 182). In addition, other allyl sulfones (e.g., **437**, Scheme 2.89) react in CM with ease (160, 172). For an example of CM of alkenes bearing a sulfone group in a more distant position ($n \geq 1$) see Scheme 2.74.

*2.2.9.4 **Unsaturated Sulfonamides*** Vinyl sulfonamides ($n = 0$), such as **449** and **451** undergo CM with a number of terminal olefins (171) (Scheme 2.92). These substrates seem to be similarly "difficult" as vinyl sulfones, so the second-generation Ru catalysts should be used, preferably the Hoveyda-type ones. The reaction is however highly stereoselective, leading to *E*-configured products. Liskamp et al. described CM of allyl sulfonamides ($n = 1$, e.g., **454**) with functionalized alkenes, as a method for the preparation of new peptidomimetics (182) (Scheme 2.92). The CM of allyl sulfonamides was successfully conducted both in solution and on the solid phase, in the presence of up to 20 mol% of **Gru-II** or **Hov-II**. It was also noted by the authors that the presence of a Lewis acid (e.g., Cy_2BCl) can increase the yields in some cases (183).

2.2.10 Unsaturated Stannanes (FG = SnR_3)

Blechert (184) shown that allylstannanes successfully participate in CM with a variety of alkenes using the molybdenum catalyst (Scheme 2.93). Moderate-to-good yields were obtained with a number of highly functionalized alkenes, containing ester, cyano, acetal, and other FGs. Interestingly, no

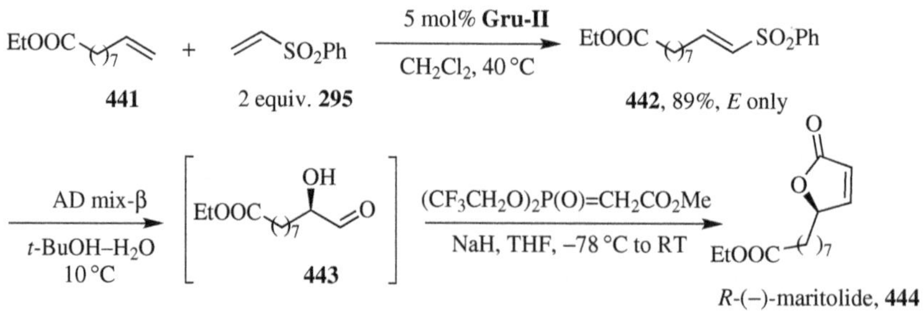

Scheme 2.90 CM of vinyl sulfones in total synthesis.

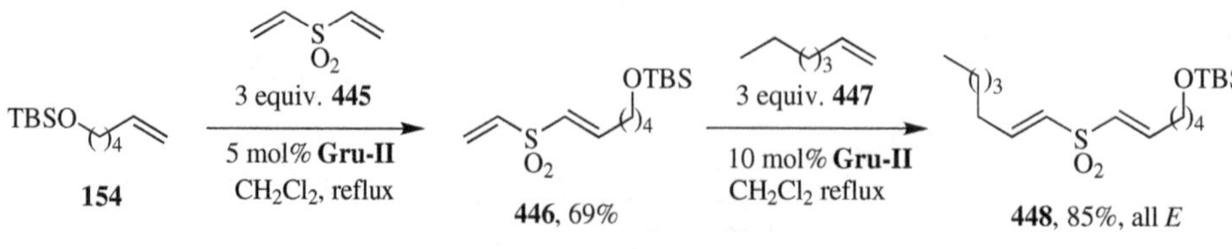

Scheme 2.91 CM of divinyl sulfone.

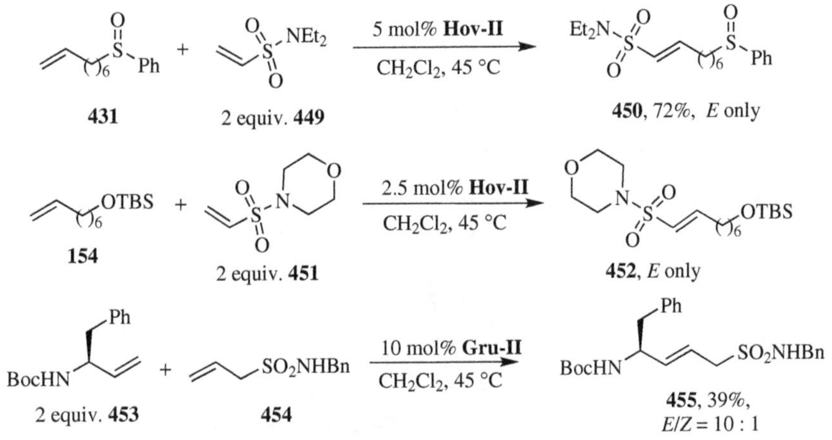

Scheme 2.92 CM of vinyl and allyl sulfonamides.

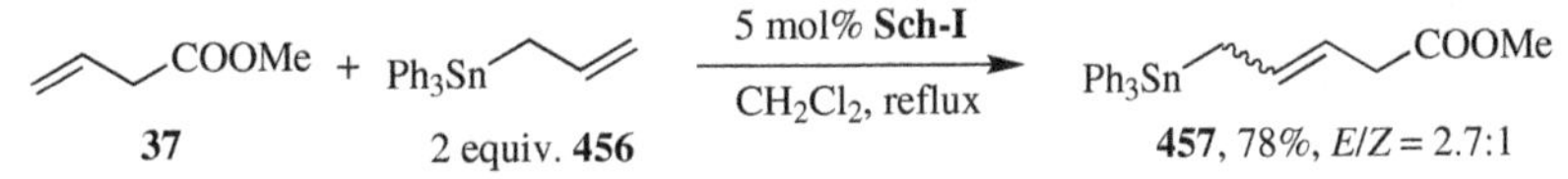

Scheme 2.93 CM of allylstannane.

significant quantities of allyl stannane self-metathesis ("homodimerization") products were detected. Allyltributyl stannane can be used, as well, although the yields of the CM products were lower in this case.

2.3 *E/Z* SELECTIVITY CONTROL

One of the remaining issues of CM is lack of selectivity in the resulting products. Throughout the chapter examples of *E*- or *Z*-selective transformations were reported, but these are the minority. In most cases, a mixture of products with various double bond configurations is obtained. A common way to obtain pure *E*-configured product is utilizing substrates with increased steric hindrance, but even this approach leads sometimes to limited success. If a Z-olefin is desired, no such manipulations will provide formation of the desired product. Observing this drawback in possible applications such as target-oriented synthesis, the scientific community focused on the development of a methodology necessary to overcome this difficulty. This effort led to establishment of novel series of catalysts for this exact purpose, a narrow selection of which is presented in Figure 2.3.

Depending on the metal in the core of the complex, different approaches were used, so that the promoted transformation would lead to the *Z*-configured product predominantly. During the Mo-catalyzed metathetic cycle, the ligands do not dissociate, thus constantly influencing the metal center. This rendered a possibility to install chirality within the complex that would be translated onto the metal center. A combination of a partially substituted alkoxide ligand and a pyrrolide moiety (**Mo*** and **Mo****) leads to the establishment of a stereocenter at the molybdenum atom, having a profound effect on the enantioselectivity of the resulting initiator (185). Such complexes are described in the literature as stereogenic-at-Mo or MAP—mono-alkoxide-pyrrolide complexes. Although *Z*-selective CM reactions are a fairly new addition to the metathesis toolkit, studies showing that this process can be mediated by metals other than Mo have already emerged. Another metal leading to significantly *Z*-enhanced olefinic products is tungsten. Expanding the idea of MAP complexes, a W-based compound was prepared (186). Owing to increased activity and stability, a cyclobutane derivative **W*** was recommended.

Also certain ruthenium-based complexes were examined in *Z*-selective homocoupling of olefins. Compared to the well-established classes of Mo- and W-containing MAP complexes, several ruthenium initiators that show enhanced *Z*-selectivity have been reported, but selectivity is still not satisfactory for precisely stereocontrolled syntheses (187).

The best results were obtained by modification of the anionic ligands, which do not dissociate from the metal center throughout the catalytic cycle. The more successful studies have been focusing on application of bidentate anionic ligands (e.g., bulky carboxylates, **Ru***) (188), which lead to the best metathesis activity and selectivity, while monodentate ones were usually less optimal. The application of Chen's nitrato ligands (189) in place of carboxylates provided an improvement in activity and selectivity of the resulting catalyst **Ru**** (190). The great advantage of Ru-based systems is that in contrast to Mo- and W-based

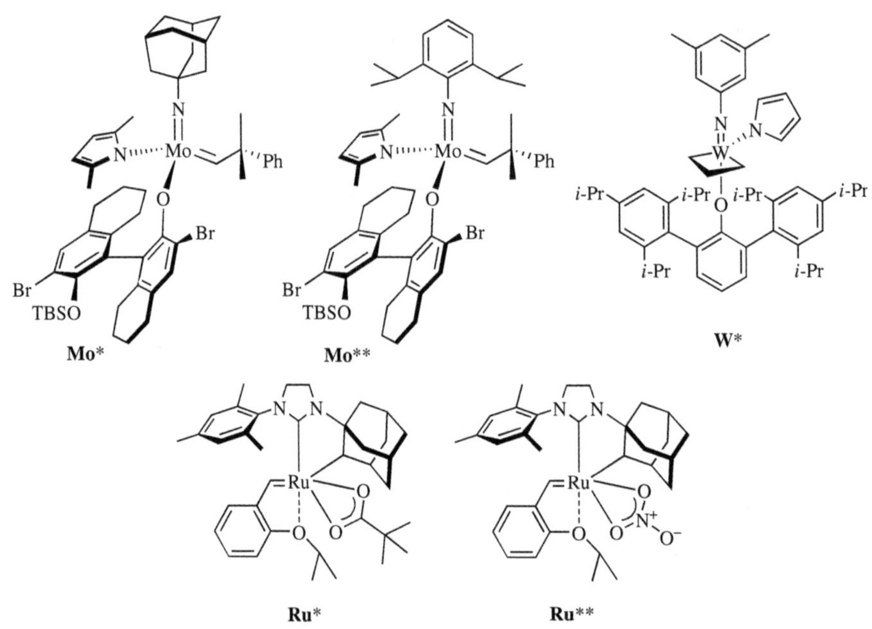

Figure 2.3 A selection of complexes successfully utilized in Z-selective CM.

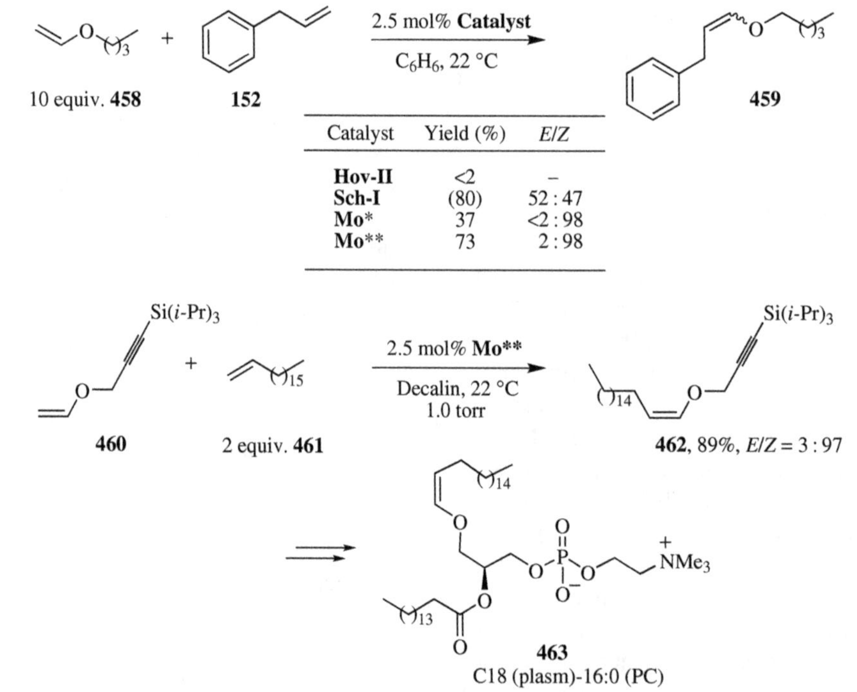

Catalyst	Yield (%)	*E*/*Z*
Hov-II	<2	–
Sch-I	(80)	52 : 47
Mo*	37	<2 : 98
Mo**	73	2 : 98

Scheme 2.94 *Z*-selective CM of vinyl ethers.

catalysts, the former are compatible with polar groups and protic solvents.

Complexes prepared for *Z*-selective catalysis were tested in a few CM transformations only, but the results have been very encouraging. The novel class of MAP Mo complexes in a reaction of utmost importance that has not been reported earlier—CM of enol ethers (185). To allow comparison, other metathetically active complexes were examined as well. Results of this study can be observed in Scheme 2.94. As expected, the ruthenium-based complex **Hov-II** failed to promote the reaction leading to **459**. It had already been demonstrated that in such cases formation of metathetically

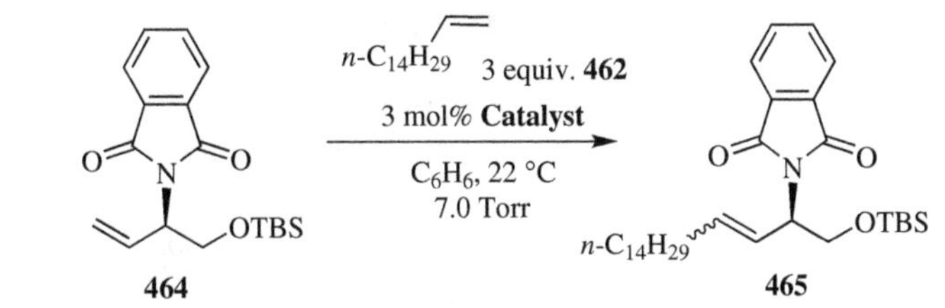

Catalyst	Yield (%)	*E/Z*
Hov-II	64	89 : 11
Sch-I	68	88 : 12
Mo*	88	3 : 97

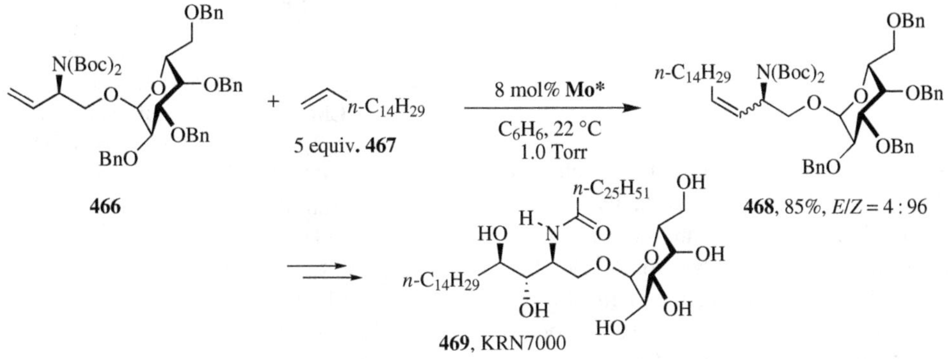

Scheme 2.95 Allylic amides in *Z*-selective CM transformation.

inactive Fischer carbenes takes place. The molybdenum complex **Sch-I** catalyzed the reaction albeit with no selectivity. The novel stereogenic-at-Mo complexes **Mo-I** and **Mo-II** led to the formation of the desired product in a fully stereoselective manner. Unfortunately, reaction catalyzed by the former did not lead to a satisfactory yield.

The same study utilized *Z*-selective CM of an enol ether **460** in a diastereoselective synthesis of an antioxidant plasmalogen phospholipid, C18 (plasm)-16:0 (PC) **463**. This transformation required an application of a vacuum, as ethylene that originated from self-metathesis of **461** increased the concentration of the Mo-methylidene promoting isomerization. Under such conditions, the CM product was obtained efficiently and converted in a consecutive elegant sequence to C18 (plasm)-16:0 (PC) (Scheme 2.94).

In the same pioneering work (185), a selective CM of allylic amides was presented. The transformation is of enormous potential, as in many biologically active molecules the presence of nitrogen in this position is crucial. CM with olefin **464** was conducted in previously optimized conditions (Scheme 2.95). Mild vacuum was again indicated to assure good *Z*-selectivity of this process, which proceeded with excellent yield. For comparison purposes, other complexes were tested in the aforementioned transformation, but their application led to the product in moderate yield only and as the *E*-isomer predominantly.

This approach was also utilized in synthesis of KRN7000 (469)—a potent antitumor agent. The key step—the metathetic transformation of a carbohydrate derivative **466** containing an allylic amide fragment with the terminal alkene **467**—proceeded with excellent yield as well as diastereoselectivity. Further transformations, such as dihydroxylation, esterification, and carbohydrate deprotection, completed the stereoselective synthetic pathway (Scheme 2.95).

Molybdenum is not the only metal proven to promote highly *Z*-selective metathesis transformations. Previously presented complex **W*** was also confirmed to provide excellent homometathesis reaction outcomes (Scheme 2.96). Transformation of substrate **470** is especially noteworthy as it leads to a product bearing three double bonds, which is nearly exclusively an *E,Z,E*-isomer obtained with excellent selectivity. This is the first example to date that reports formation of a high percentage of a *Z*-homocoupled product in the presence of *E*-olefinic bonds.

The outcome of ruthenium-mediated CM highly depends on type of the bidentate anionic ligand (191). A number of terminal olefins were tested in CM and S-CM promoted by such complexes. Complex **Ru**** was proved to be superior in cases of the latter, simultaneously exhibiting excellent *Z*-selectivity (Scheme 2.97).

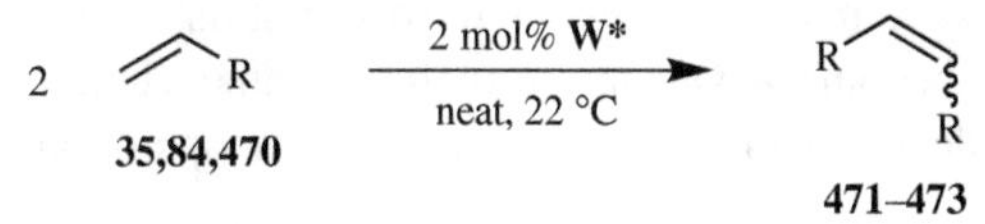

Substrate	Product	Yield (%)	*E/Z*
84, R = 	**471**	89	1 : 99
35, R = 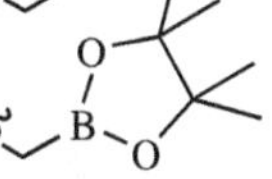	**472**	61	8 : 92
470, R = 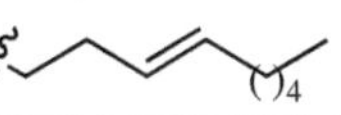	**473**	60	9 : 91

Scheme 2.96 Tungsten-catalyzed *Z*-selective cross metathesis.

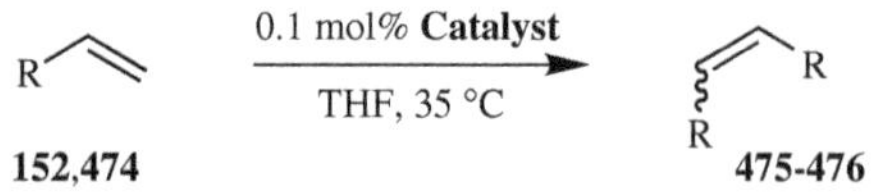

Substrate	Product	Catalyst	Time (h)	Yield (%)	Z (%)
152, R = Ph	475	**Ru***	3	73	86
		Ru**	3	89	95
474, R = ( )$_8$COOMe	476	**Ru***	12	13	90
		Ru**	12	75	94

Scheme 2.97 *Z*-selective CM catalyzed by Ru complexes.

Owing to the great potential of the presented *Z*-selective CM, further development of Ru, Mo, and W catalysts is expected. This would largely contribute toward many synthetic pathways where the stereoselective construction of *Z*-olefins is required and is now achieved by nonmetathetic transformations. Once complexes of increased activity and stability are examined in large-scale applications, the methodology has a chance of being rapidly adapted by the industry.

2.4 OUTLOOK AND PERSPECTIVES

Functionalized olefins are important building blocks in organic synthesis. Catalytic olefin CM can be a direct and convenient route to obtain functionalized olefins from simple precursors. One of the most appealing facets of this transformation is that a carbon-carbon double bond produced in CM can be substituted by a huge selection of FGs. In addition, advanced architectures such as polyfunctional natural and biologically active products can be relatively easily assembled using the CM reaction, often in the last step of a multistep synthesis.

A lot of efforts have been made toward understanding of the mechanism and the origin of selectivity in CM transformations. Development of new efficient catalysts and conditions is another key factor that allows the use of previously incompatible substrates. While some problems related to controlling of chemo- and stereoselectivities during CM reactions remain unsolved, this reaction has already become a valuable tool for organic chemists and further development of and increase in its scope is expected.

ACKNOWLEDGMENTS

KG is grateful for financial support from the Foundation for Polish Science for TEAM Program co-financed by the European Regional Development Fund, Operational Program Innovative Economy 2007–2013. KŻ thanks for “Diamond Grant” research project financed from the governmental funds for science 2012–2015.

REFERENCES

1. Grubbs RH. *Handbook of Metathesis*. 1st ed. Weinheim: Wiley-VCH Verlag GmbH & Co. KGaA; 2003.
2. http://nobelprize.org/nobel_prizes/chemistry/laureates/2005/advanced-chemistryprize2005.pdf (Accessed: 17 Feb 2014).
3. (a) Selected reviews on types of olefin metathesis **RCM** Monfette S, Fogg DE. *Chem Rev* 2009;109:3783–3816. (b) **ADMET** Mutlu H, Espinosa LM, Meier MAR. *Chem Soc Rev* 2011;40:1404–1445. (c) **ROMP** Bielawski CW, Grubbs RH. *Prog Polym Sci* 2007;32:1–29. (d) **CM** Connon SJ, Blechert S. *Angew Chem Int Ed* 2003;42:1900–1923. (e) **ene-yne** Diver ST, Giessert AJ. *Chem Rev* 2004;104:1317–1382.
4. Hérisson PJ, Chauvin Y. *Die Makromolekulare Chemie* 1971;141:161–176.
5. Hong SH, Sanders DP, Lee CW, Grubbs RH. *J Am Chem Soc* 2005;127:17160–17161.
6. Forman GS, McConnell AE, Tooze RP, van Rensburg WJ, Meyer WH, Kirk MM, Dwyer CL, Serfontein DW. *Organometallics* 2005;24:4528–4552.
7. Meyer WH, McConnell AE, Forman GS, Dwyer CL, Kirk MM, Ngidi EL, Blignaut A, Saku D, Slawin AMZ. *Inorg Chim Acta* 2006;359:2910–2917.
8. Commercial suppliers offer nowadays a huge selection of various olefin metathesis catalysts. A selection of them is listed in Chapter 6 of this book.
9. Ulman M, Grubbs RH. *Organometallics* 1998;17:2484-2489.
10. Chatterjee AK, Choi T-L, Sanders DP, Grubbs RH. *J Am Chem Soc* 2003;125:11360–11370.
11. Marinescu SC, Schrock RR, Müller P, Hoveyda AH. *J Am Chem Soc* 2009;131:10840–10841.
12. Schrodi Y, Ung T, Vargas A, Mkrtumyan G, Lee CW, Champagne TM, Pederson RL, Hong SH. *Clean Soil Air Water* 2008;36:669–673.
13. (a) Thomas RM, Keitz BK, Champagne TM, Grubbs RH. *J Am Chem Soc* 2011;133:7490–7496. (b) van der Klis F, Le Nôtre J, Blaauw R, van Haveren J, van Es DS. *Eur J Lipid Sci Technol* 2012;114:911–918.
14. Wolf S, Plenio H. *Green Chem* 2013;15:315–319.
15. Wipf P, Rector SR, Takahashi H. *J Am Chem Soc* 2002;124:14848–14849.
16. Ratnayake AS, Hemscheidt T. *Org Lett* 2002;4:4667–4669.
17. Karlou-Eyrisch K, Müller BKM, Herzig C, Nuyken O. *J Organomet Chem* 2000;606:3–7.
18. Jermacz I, Maj J, Morzycki JW, Wojtkielewicz A. *Toxicol Mech Methods* 2008;6:469–471.
19. Morrill C, Grubbs RH. *J Org Chem* 2003;68:6031–6034.
20. Jacobsen MF, Moses JE, Adlington RM, Baldwin JE. *Org Lett* 2005;7:641–644.
21. Morrill C, Funk TW, Grubbs RH. *Tetrahedron Lett* 2004;45:7733–7736.
22. Marciniec B, Jankowska M, Pietraszuk C. *Chem Commun* 2005:663–665.

23. Dakas P-Y, Jogireddy R, Valot G, Barluenga S, Winssinger N. *Chem Eur J* 2009;15:11490–11497.
24. Esteban J, Costa AM, Vilarasa J. *Org Lett* 2008; 10:4843–4846.
25. Goldberg SD, Grubbs RH. *Angew Chem Int Ed* 2002; 41:807–810.
26. Yamamoto Y, Takahashi M, Miyaura N. *Synlett* 2002: 128–130.
27. Garcia Garcia P, Hohn E, Pietruszka J. *J Organomet Chem* 2003;680:281–285.
28. Blackwell HE, O'Leary DJ, Chatterjee AK, Washenfelder RA, Bussmann DA, Grubbs RH. *J Am Chem Soc* 2000;122:58–71.
29. Brümmer O, Rückert A, Blechert S. *Chem Eur J* 1997; 3:441–446.
30. (a) Pederson RL, Fellows IM, Ung TA, Ishihara H, Hajela SP. *Adv Synth Catal* 2002;344:728–735. (b) Grubbs RH. *Angew Chem Int Ed* 2006;45:3760–3765. (c) Pederson, R. L., Grubbs, R. H., U.S. Patent No. 6,696,597. 2004.
31. Czaban J, Schertzer BM, Grela K. *Adv Synth Catal* 2013;355:1997–2006.
32. Torsell S, Somfai P. *Org Biomol Chem* 2004;2:1643–1646.
33. Rai AN, Basu A. *Org Lett* 2004;6:2861–2863.
34. Hasegawa H, Yamamoto T, Hatano S, Hakogi T, Katsumara S. *Chem Lett* 2004;33:1592–1593.
35. Stewart IC, Douglas CJ, Grubbs RH. *Org Lett* 2008; 10:441–444.
36. Sarabia F, Sánchez-Ruiz A. *Tetrahedron Lett* 2005;46: 1131–1135.
37. Samojłowicz C, Borré E, Mauduit M, Grela K. *Adv Synth Catal* 2011;353:1993–2002.
38. Torborg C, Szczepaniak G, Zieliński A, Malińska M, Woźniak K, Grela K. *Chem Commun* 2013;49:3188–3190.
39. Chatterjee AK, Sanders DP, Grubbs RH. *Org Lett* 2002;4:1939–1942.
40. Tischer S, Metz P. *Adv Synth Catal* 2007;349:147–151.
41. Demchuk OM, Pietrusiewicz KM, Michrowska A, Grela K. *Org Lett* 2003;5:3217–3220.
42. Uwamori M, Nakada M. *J Org Chem* 2012;77:5098–5107.
43. Uwamori M, Nakada M. *Nat Prod Commun* 2013;8:955–959.
44. Michrowska A, Bujok R, Harutyunyan S, Sashuk V, Dolgonos G, Grela K. *J Am Chem Soc* 2004;126:9318–9325.
45. Alam S, Dhimane H. *Synlett* 2010;19:2923–2927.
46. Yee NK, Farina V, Houpis IN, Haddad N, Frutos RP, Gallou F, Wang X-J, Wei X, Simpson RD, Feng X, Fuchs V, Xu Y, Tan J, Zhang L, Xu J, Smith-Keenan LL, Vitous J, Ridges MD, Spinelli EM, Johnson M, Donsbach K, Nicola T, Brenner M, Winter E, Kreye P, Samstag W. *J Org Chem* 2006;71:7133–7145.
47. Ahmed M, Arnauld T, Barrett AGM, Braddock DC, Flack K, Procopiou PA. *Org Lett* 2000;2:551–553.
48. Dewi P, Randl S, Blechert S. *Tetrahedron Lett* 2005;46:577–580.
49. Ferrié L, Amans D, Reymond S, Bellosta V, Capdevielle P, Cossy J. *J Organomet Chem* 2006;691:5456–5465.
50. Funk TW, Efskind J, Grubbs RH. *Org Lett* 2005;7:187–190.
51. Kang B, Lee J-M, Kwak J, Lee YS, Chang S. *J Org Chem* 2004;69:7661–7664.
52. Wojtkielewicz A, Maj J, Morzycki JW. *Tetrahedron Lett* 2009;50:4734–4737.
53. Fontán N, Domínguez M, Álvarez R, de Lera ÁR. *Eur J Org Chem* 2011:6704–6712.
54. Kajikawa T, Iguchi N, Katsumura S. *Org Biomol Chem* 2009;7:4586–4589.
55. Wojtkielewicz A. *Curr Org Synth* 2013;10:43–66.
56. Chatterjee AK, Morgan JP, Scholl M, Grubbs RH. *J Am Chem Soc* 2000;122:3783–3784.
57. Imhof S, Randl S, Blechert S. *Chem Commun* 2001: 1692–1693.
58. Samojłowicz C, Bieniek M, Zarecki A, Kadyrov R, Grela K. *Chem Commun* 2008:6282–6284.
59. Eignerová B, Dračínský M, Kotora M. *Eur J Org Chem* 2008:4493–4499.
60. Fustero S, Sánchez-Roselló M, Sanz-Cervera JF, Aceña JL, Carlos del Pozo C, Fernández B, Bartolomé A, Asensio A. *Org Lett* 2006;8:4633–4636.
61. Velder J, Ritter S, Lex J, Schmalz H-G. *Synthesis* 2006;2:273–278.
62. Crowe WE, Zhang ZJ. *J Am Chem Soc* 1993;115: 10998–10999.
63. Mikus A, Sashuk V, Kędziorek M, Samojłowicz C, Ostrowski S, Grela K. *Synlett* 2005:1142–1146.
64. Kawai T, Komaki M, Iyoda T. *J Mol Catal A* 2002; 190:45–53.
65. Chatterjee AK, Toste FD, Choi T-L, Grubbs RH. *Adv Synth Catal* 2002;344:634–637.
66. Dash J, Arseniyadis S, Cossy J. *Adv Synth Catal* 2007;349:152–156.
67. Liu X, Sternberg E, Dolphin D. *J Org Chem* 2008; 73:6542–6550.
68. Yasuda T, Abe J, Yoshida H, Iyoda T, Kawai T. *Adv Synth Catal* 2002;344:705–711.
69. de la Torre M, Deometrio AM, Álvaro E, Garcia I, Sierra MA. *Org Lett* 2006;8:593–596.
70. Love JA, Morgan JP, Trnka TM, Grubbs RH. *Angew Chem Int Ed* 2002;41:4035–4037.
71. Sierra JC, Hüerländer D, Hill M, Kehr G, Erker G, Fröhlich R. *Chem Eur J* 2003;9:3618–3622.
72. Plettenburg O, Mui C, Bodmer-Narkevitch V, Wong C-H. *Adv Synth Catal* 2002;344:622–626.
73. Crowe WE, Goldberg DR. *J Am Chem Soc* 1995;117: 5162–5163.
74. Rivard M, Blechert S. *Eur J Org Chem* 2003:2225–2228.
75. Boddaert T, Coquerel Y, Rodriguez J. *C R Chimie* 2009; 12:872–875.
76. Bai C-X, Lu X-B, He R, Zhang W-Z, Feng X-J. *Org Biomol Chem* 2005;3:4139–4142.

77. Voigtritter K, Ghorai S, Lipshutz BH. *J Org Chem* 2011; 76:4697–4702.
78. Malacea R, Fischmeister C, Bruneau C, Dubois J-L, Couturierb J-L, Dixneuf PH. *Green Chem* 2009;11:152–155.
79. Hoveyda HH, Vezina M. *Org Lett* 2005;7:2113–2116.
80. Miao X, Fischmeister C, Bruneau C, Dixneuf PH, Dubois J-L, Couturier J-L. *ChemSusChem* 2012;5:1410–1415.
81. Gessler S, Randl S, Blechert S. *Tetrahedron Lett* 2000; 41:9973–9976.
82. BouzBouz S, Cossy J. *Org Lett* 2001;3:1451–1454.
83. Michrowska A, List B. *Nat Chem* 2009;1:225–228.
84. Miao X, Fischermeister C, Bruneau C, Dixneuf PH. *ChemSusChem* 2009;2:542–545.
85. Evans MA, Morken JP. *Org Lett* 2005;7:3371–3373.
86. Zaja M, Blechert S. *Tetrahedron* 2004;60:9629–9634.
87. Statsuk AV, Liu D, Kozmin SA. *J Am Chem Soc* 2004;126:9546–9547.
88. (a) Cossy J, Willis C, Bellosta V. *Synlett* 2001:1578–1580. (b) Cossy J, Willis C, Bellosta V, BouzBouz S. *J Org Chem* 2002;67:1982–1990.
89. Goldup SM, Pilkington CJ, White AJP, Burton A, Barrett AGM. *J Org Chem* 2006;71:6185–6191.
90. Choi T-L, Chatterjee AK, Grubbs RH. *Angew Chem Int Ed* 2001;40:1277–1279.
91. Liang Y, Raju R, Le T, Taylor CD, Howell AR. *Tetrahedron Lett* 2009;50:1020–1022.
92. BouzBouz S, Simmons R, Cossy J. *Org Lett* 2004;6:3465–3467.
93. Dineen TA, Roush WR. *Org Lett* 2004;6:2043–2046.
94. van Zjil AW, Minnaard AJ, Feringa BL. *J Org Chem* 2008;73:5651–5653.
95. Patel J, Mujcinovic S, Jackson WR, Robinson AJ, Serelis AK, Such C. *Green Chem* 2006;8:450–454.
96. Raju R, Howell AR. *Org Lett* 2006;8:2139–2141.
97. Raju R, Allen LJ, Le T, Taylor CD, Howell AR. *Org Lett* 2007;9:1699–1701.
98. Moïse J, Arseniyadis S, Cossy J. *Org Lett* 2007;9:1695–1698.
99. Ferrié L, Bouzbouz S, Cossy J. *Org Lett* 2009;11:5446–5448.
100. Vedrenne E, Dupont H, Oualef S, Elkaïm L, Grimaud L. *Synlett* 2005;4:0670–0672.
101. Donohoe TJ, Race NJ, Bower JF, Callens CKA. *Org Lett* 2010;12:4094–4097.
102. Vasbinder MM, Miller SJ. *J Org Chem* 2002;67:6240–6242.
103. Enholm E, Low T. *J Org Chem* 2006;71:2272–2276.
104. Woodward CP, Spiccia ND, Jackson WR, Robinson AJ. *Chem Commun* 2011:779–781.
105. Morris T, Sandham D, Caddick S. *Org Biomol Chem* 2007;5:1025–1027.
106. Hodgson DM, Angrish D. *Chem Eur J* 2007;13:3470–3479.
107. Hodgson DM, Angrish D. *Adv Synth Catal* 2006;348: 2509–2514.
108. Hodgson DM, Angrish D, Erickson SP, Kloesges J, Lee CH. *Org Lett* 2008;10:5553–5556.
109. Kanemitsu T, Seeberger PH. *Org Lett* 2003;5:4541–4544.
110. (a) Barrett AG, Beall JC, Braddock DC, Flack K, Gibson VC, Salter MM. *J Org Chem* 2000;65:6508–6514. (b) Randl S, Blechert S. *J Org Chem* 2003;68:8879–8882.
111. Wdowik T, Samojłowicz C, Jawiczuk M, Zarecki A, Grela K. *Synlett* 2010:2931–2935.
112. Wdowik T, Samojłowicz C, Jawiczuk M, Malińska M, Woźniak K, Grela K. *Chem Commun* 2013:674–676.
113. Marsh GP, Parsons PJ, McCarthy C, Corniquet XG. *Org Lett* 2007;9:2613–2616.
114. Engelhardt FC, Schmitt MJ, Taylor RE. *Org Lett* 2001;3: 2209–2212.
115. Cossy J, BouzBouz S, Hoveyda AH. *J Organomet Chem* 2001;624:327–332.
116. Barluenga S, Dakas P-Y, Ferandin Y, Meijer L, Winssinger N. *Angew Chem Int Ed* 2006;45:3951–3954.
117. Trost BM, Gutierrez AC, Livingston RC. *Org Lett* 2009;11:2539–2542.
118. Schmidt B, Haukea S. *Org Biomol Chem* 2013;11:4194–4206.
119. Finnegan D, Seigal BA, Snapper ML. *Org Lett* 2006;8: 2603–2606.
120. Ferrié L, Reymond S, Capdevielle P, Cossy J. *Synlett* 2007;18:2891–2893.
121. Nicolaou KC, Hughes R, Cho SY, Winssinger N, Labischinski H, Endermann R. *Chem. Eur. J.* 2001;7:3824–3843.
122. Rega M, Jiménez C, Rodríguez J. *Steroids* 2007;72:729–735.
123. Chen G-W, Kirschning A. *Chem. Eur. J.* 2002;8:2717–2729.
124. Netscher T, Malaisé G, Bonrath W, Breuninger M. *Catalysis Today* 2007;121:71–75.
125. Marinec PS, Evans CG, Gibbons GS, Tarnowski MA, Overbeek DL, Gestwicki JE. *Bioorg Med Chem* 2009; 17:5763–5768.
126. Langer P, Holtz E. *Synlett* 2002;1:110–112.
127. Nicolaou KC, Koftis TV, Vyskocil S, Petrovic G, Ling T, Yamada YMA, Tang W, Frederick MO. *Angew Chem Int Ed* 2004;43:4318–4324.
128. Grellepois F, Crousse B, Bonnet-Delpon D, Bégué J-P. *Org Lett* 2005;7:5219–5222.
129. Macnaughtan ML, Johnson MJA, Kampf JW. *J Am Chem Soc* 2007;129:7708–7709.
130. Macnaughtan ML, Gary JB, Gerlach DL, Johnson MJA, Kampf JW. *Organometallics* 2009;28:2880–2887.
131. Sashuk V, Samojłowicz C, Szadkowska A, Grela K. *Chem Commun* 2008:2468–2470.
132. Bandini M, Cozzi PG, Licciulli S, Umani-Ronchi A. *Synthesis* 2004;3:409–414.
133. Jacobs T, Rybak A, Meier MAR. *Appl Catal A* 2009;353: 32–35.
134. Bieniek M, Samojłowicz C, Sashuk V, Bujok R, Śledź P, Lugan N, Lavigne G, Arlt D, Grela K. *Organometallics* 2011;30:4144–4158.
135. Baughman TW, Sworen JC, Wagener KB. *Tetrahedron* 2004; 60:10943–10948.

136. Marciniec B, Pietraszuk C. In: Grubbs RH, editor. *Handbook of Metathesis*. Vol. 2. Weinheim: Wiley-VCH; 2003. p 463–490.
137. Marciniec B, Pietraszuk C. *Curr Org Chem* 2000;7:691–743.
138. Marciniec B. *Coord Chem Rev* 2005;249:2374–2390.
139. Pietraszuk C, Fischer H, Rogalski S, Marciniec B. *J Organomet Chem* 2005;690:5912–5921.
140. Itami Y, Marciniec B, Kubicki M. *Chem Eur J* 2004; 10:1239–1248.
141. Cheng G, Vautravers NR, Morris RE, Cole-Hamilton DJ. *Org Biomol Chem* 2008;6:4662–4667.
142. Vautravers NR, André P, Slawin AMZ, Cole-Hamilton DJ. *Org Biomol Chem* 2009;7:717–724.
143. Żak P, Marciniec B, Majchrzak M, Pietraszuk C. *J Organomet Chem* 2011;696:887–891.
144. Feher FJ, Soulivong D, Eklund AG, Wyndham KD. *J Chem Soc Chem Commun* 1997:1185–1186.
145. Żak P, Pietraszuk C, Marciniec B, Spólnik G, Danikiewicz W. *Adv Synth Catal* 2009;351:2675–2682.
146. Purser S, Claridge TDW, Odell B, Moore PR, Gouverneur V. *Org Lett* 2008;10:4263–4266.
147. Thibaudeau S, Gouverneur V. *Org Lett* 2003;5:4891–4893.
148. Tredwell M, Tenza K, Pacheco MC, Gouverneur V. *Org Lett* 2005;7:4495–4497.
149. BouzBouz S, De Lemos E, Cossy J. *Adv Synth Catal* 2002;344:627–630.
150. Slinn CA, Redgrave AJ, Hind SL, Edlin C, Nolan SP, Gouverneur V. *Org Biomol Chem* 2003;1:3820–3825.
151. Dunne KS, Lee SE, Gouverneur V. *J Organometal Chem* 2006;691:5246–5259.
152. Bisaro F, Gouverneur V. *Tetrahedron Lett* 2003;44: 7133–7135.
153. Vinokurov N, Michrowska A, Szmigielska A, Drzazga Z, Wójciuk G, Demchuk OM, Grela K, Pietrusiewicz KM, Butenschön H. *Adv Synth Catal* 2006;348:931–938.
154. Bisaro F, Gouverneur V. *Tetrahedron* 2005;61:2395–2400.
155. Vinokurov N, Garabatos-Perera J-R, Zhao-Karger , Wiebcke M, Butenschön H. *Organometallics* 2008;27:1878–1886.
156. Chatterjee AK, Choi T-L, Grubbs RH. *Synlett* 2010: 1034–1037.
157. He A, Yan B, Thanavaro A, Spilling CD, Rath NP. *J Org Chem* 2004;69:8643–8651.
158. Pradere U, Clavier H, Roy V, Nolan SP, Agrofoglio LA. *Eur J Org Chem* 2011:7324–7330.
159. Lera M, Hayes CJ. *Org Lett* 2001;3:2765–2768.
160. Samojłowicz C, Grela K. *Arkivoc* 2011:71–81.
161. Rainier JD, Liu Z. *Org Lett* 2004;7:131–133.
162. Katayama H, Urushima H, Nishioka T, Wada C, Nagao M, Ozawa F. *Angew Chem Int Ed* 2000;39:4513–4515.
163. Spagnol G, Heck M-P, Nolan SP, Mioskowski C. *Org Lett* 2002;4:1767–1770.
164. For a review, seeLin YA, Davis BG. *Beilstein J Org Chem* 2010;6:1219–1228.
165. Lin YA, Chalker JM, Floyd N, Bernardes GJL, Davis BG. *J Am Chem Soc* 2008;130:9642–9643.
166. Lin YA, Chalker JM, Davis BG. *J Am Chem Soc* 2010; 132:16805–16811.
167. Hunter L, Condie GC, Harding MM. *Tetrahedron Lett* 2010;51:5064–5067.
168. Seiser T, Kamena F, Cramer N. *Angew Chem Int Ed* 2008; 47:6483–6485.
169. Bieniek M, Michrowska A, Usanov DL, Grela K. *Chem Eur J* 2008;14:806–818.
170. Barbasiewicz M, Michalak M, Grela K. *Chem Eur J* 2012;45:14237–14241.
171. Woźniak Ł, Rajkiewicz A, Grela K. unpublished results.
172. Michrowska A, Bieniek M, Kim M, Klajn R, Grela K. *Tetrahedron* 2003;59:4525–4531.
173. Żukowska K, Szadkowska A, Trzaskowski B, Pazio A, Pączek Ł, Woźniak K, Grela K. *Organometallics* 2013; 32:2192–2198.
174. Meadows DC, Gervay-Hague J. *Med Res Rev* 2006; 26:793–814.
175. Grela K, Bieniek M. *Tetrahedron Lett* 2001;42:6425–6428.
176. Ettari R, Nizi E, di Francesca ME, Dude MA, Pradel G, Vicík R, Schirmeister T, Micale N, Grasso S, Zappalà M. *J Med Chem* 2008;51:988–996.
177. Evans P, Leffray M. *Tetrahedron* 2003;59:7973–7981.
178. Au CWG, Pyne SGJ. *Org Chem* 2006;71:7097–7099.
179. Burrell AJM, Coldham I, Oram N. *Org Lett* 2009;11: 1515–1518.
180. Fujitani M, Tsuchiya M, Okano K, Takasu K, Ihara M, Tokuyama H. *Synlett* 2010;5:822–826.
181. Bieniek M, Kołoda D, Grela K. *Org Lett* 2006;8:5689–5692.
182. Chatterjee AK, Grubbs RH. *Org Lett* 1999;1:1751–1753.
183. Brouwer AJ, Elgersma RC, Jagodzinska M, Rijkers DTS, Liskamp RMJ. *Bioorg. Med. Chem. Lett.* 2008;18:78–84.
184. Feng J, Schuster M, Blechert S. *Synlett* 1997:129–130.
185. (a) **Mo*** Malcolmson SJ, Meek SJ, Sattely ES, Schrock RR, Hoveyda AH. *Nature* 2008;456:933–937. (b) Sattely ES, Meek SJ, Malcolmson SJ, Schrock RR, Hoveyda AH. *J Am Chem Soc* 2008;131:943–953. (c) **Mo**** Meek SJ, O'Brien RV, Llaveria J, Schrock RR, Hoveyda AH. *Nature* 2011;471:461–466.
186. Marinescu SC, Schrock RR, Müller P, Takase MK, Hoveyda AH. *Organometallics* 2011;30:1780–1782.
187. (a) Rosen EL, Sung DH, Chen Z, Lynch VM, Bielawski CW. *Organometallics* 2010;29:250–256. (b) Teo P, Grubbs RH. *Organometallics* 2010;29:6045–6050.
188. Fournier PA, Collins SK. *Organometallics* 2007;26: 2945–2949.
189. Jović M, Torker S, Chen P. *Organometallics* 2011;30: 3971–3980.
190. Kannenberg A, Rost D, Eibauer S, Tiede S, Blechert S. *Angew Chem Int Ed* 2011;50:3299–3302.
191. Keitz BK, Endo K, Herbert MB, Grubbs RH. *J Am Chem Soc* 2011;133:9686–9688.

3

RING-CLOSING METATHESIS

BIANCA J. VAN LIEROP‡, JUSTIN A. M. LUMMISS‡, AND DERYN E. FOGG

Centre for Catalysis Research & Innovation, Department of Chemistry, University of Ottawa, Ottawa, Ontario, Canada

‡*Equal contributions*

3.1 INTRODUCTION

Ring-closing metathesis (RCM) has transformed synthetic approaches to the construction of cyclic organic compounds, and the literature is now enormous. Excellent and still instructive general reviews (1–6) are to be found in the earlier literature: in recent years, the trend has been toward focused coverage of specific areas of advance. Citations to leading specialized reviews, many of which present a wealth of examples, are provided in the relevant sections below. This chapter provides a practical guide to RCM. It describes opportunities and challenges associated with different ring sizes and functional groups, and experimental parameters that can be modulated to improve yields. Examples are selected for illustrative, rather than comprehensive, purposes, and while an effort has been made to highlight recent examples (up to December 2012), case studies are drawn from the older literature where these are most illuminating.

This is an area in which synthetic applications have far outstripped mechanistic study, however, and some caveats must be given at the outset. First, catalyst loadings are rarely optimized. Significant improvements have resulted where such efforts were made (as illustrated in the final section dealing with industrial implementation), and the catalyst loadings cited throughout should thus be regarded as a rule-of-thumb guide to relative synthetic difficulty in standard practice, rather than a lower limit.

Second, target-oriented syntheses provide the most diverse and relevant examples, but rarely the most systematic data sets. Extraction of trends is complicated by the common omission of data for less successful experiments, and the fact that multiple experimental parameters influence yields (especially isolated yields). Nevertheless, broad patterns can be discerned. In the discussion below, we seek wherever possible to integrate established synthetic behavior with the emerging mechanistic picture, in the belief that this dual perspective will provide a more robust guide to the novice user, while uncovering potentially fruitful perspectives for the experienced practitioner.

3.2 GENERAL FACTORS IN CYCLIZATION EFFICIENCY

Metathesis is generally regarded as involving a series of fully reversible cycloaddition–cycloreversion equilibria, as shown in Figure 3.1a. However, the majority of RCM reactions liberate ethylene as a coproduct. The limited solubility of ethylene in organic solvents, and the inverse dependence of its solubility on temperature (7), result in relatively efficient volatilization under standard synthetic conditions: that is, stirred vessels open to an atmosphere of Ar or N_2. In consequence, the initial metathesis event is irreversible, and the reaction pathways simplify to those of Figure 3.1b. This is advantageous in minimizing nonproductive (degenerate) metathesis with ethylene. Conversely, however, reaction conditions that inhibit efficient volatilization of these by-product gases (e.g., use of limited headspace; mass-transfer-limited conditions in NMR-tube reactions or on scale-up) will reduce catalyst productivity.

Similar behavior is found when coupling an α-olefin with an α-methyl olefin. Here, intermolecular metathesis results in initial dimerization, rather than oligomerization

Olefin Metathesis: Theory and Practice, First Edition. Edited by Karol Grela.

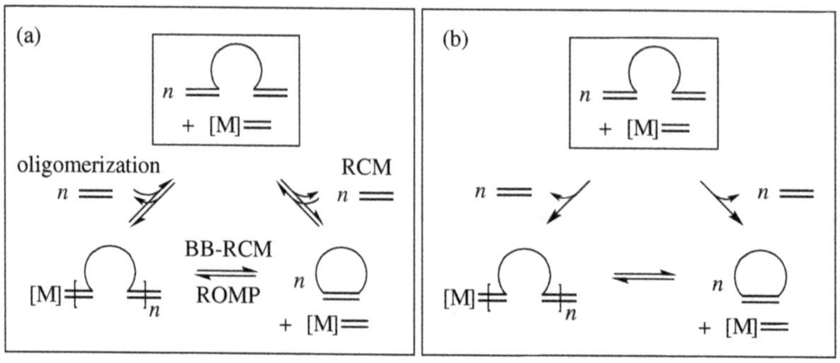

Figure 3.1 (a) Metathesis manifolds relating α,ω-divinyl species, RCM target, and oligomers (only linear, metallated oligomers shown). (b) Simplified pathways resulting from irreversible loss of ethylene. ROMP = ring-opening metathesis polymerization, BB = backbiting.

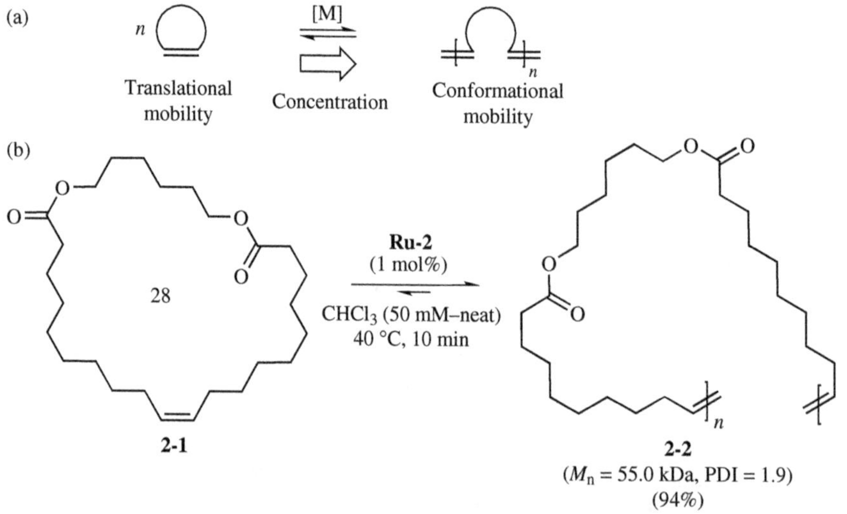

Figure 3.2 (a) General entropic factors that influence the concentration-dependence of RCM yields. (b) Example illustrating entropically driven ROMP of a macrolactone at high concentrations (9). **Ru-2** denotes the second-generation Grubbs catalyst $RuCl_2(H_2IMes)(PCy_3)(=CHPh)$; see Section 3.2.2.

(with ethylene as a coproduct). Formation of the RCM product is accompanied by the release of propylene. Again, as temperatures increase, the solubility of the by-product gas drops (8).

The entropic gain associated with volatilization of ethylene (propylene) provides a strong driving force for reaction. Frequently overlooked, however, is the fact that this driving force is unselective: it promotes consumption of diene, but does not discriminate between the RCM target and oligomers/dimers. This balance is ultimately controlled by the concentration dependence of the equilibria between these species (Fig. 3.2a). Increasing dilutions favor the smaller molecules owing to their higher translational entropy. Conversely, high concentrations favor oligomers, the conformational mobility of which is little affected by increases in viscosity that curtail translational motion. Figure 3.2b exemplifies the latter point: ring-opening metathesis polymerization (ROMP) of the 28-membered macrolactone **2-1** proceeds efficiently at concentrations above 50 mM, and the polymer can be isolated in 94% yield (9). At 5 mM, however, studies of related lactones show a comparable preference for the macrolactones; see Sections 3.2.1 and 3.2.2.3 below (10). It will be noted that enthalpy has played no part in the discussion so far. We now turn to this important point.

3.2.1 Ring Size and Strain

For the majority of RCM targets, ring strain is the most important contributor to the enthalpic cost of reaction, and hence the probability of cyclization. The entropic gains that can potentially compensate for this cost are concentration-dependent, as noted above. The relationship between starting diene concentration and the probability of cyclization is sometimes expressed as the effective molarity (EM). The thermodynamic EM value is defined as the concentration at which the equilibrium constants for the intra- and intermolecular reactions are equal: higher EM_K values thus correspond to higher yields of cyclic species at a given concentration (Equations 3.1, 3.2). Kinetic EM values (EM_k) are similarly defined as the concentration at which the rate constants for the intra- and intermolecular reactions are equal (Equations 3.3, 3.4).

$$EM_K = \frac{K_{intra}}{K_{inter}} \tag{3.1}$$

$$= \exp\left[\frac{\Delta G_{inter} - \Delta G_{intra}}{RT}\right] \tag{3.2}$$

$$EM_k = \frac{k_{intra}}{k_{inter}} \tag{3.3}$$

$$= \exp\left[\frac{\Delta G_{inter}{}^{\dagger} - \Delta G_{intra}{}^{\dagger}}{RT}\right] \tag{3.4}$$

Relatively few RCM studies have sought to measure EM values (11,12). One common difficulty lies in the frequent spectroscopic similarity of oligomeric and "monocyclic" RCM products, which hampers assessment of product ratios by NMR methods (13). Care must also be taken to ensure that these ratios are not perturbed by catalyst decomposition, especially during sampling. Nevertheless, Percy and coworkers (11) recently reported that by applying a range of 2D NMR techniques, including $^1H-^{13}C$ heteronuclear single quantum coherence-total correlation spectroscopy (HSQC-TOCSY) and ge-1D ^1H TOCSY, in conjunction with T_1 relaxation measurements, they were able to distinguish and quantify endgroups arising from unreacted diene, dimers, and higher oligomers for symmetrical C7–C12 α,ω-dienes. In other cases, particularly (but not invariably) (13) associated with more complex substitution patterns, the key signals can be readily distinguished (12). More general, straightforward access to EM data would clearly be beneficial. Efforts to develop computational methods to accurately predict EM values are under way, but have not yet been perfected (14).

For simple carbocycles, strain energy is at a minimum for five- to seven-membered rings. It increases sharply in the medium-ring regime (Fig. 3.3), then declines again (15). Strain can be attenuated by the presence of trigonal centers (e.g., sp^2 carbons, nitrogen atoms), or heteroatoms with large, polarizable orbitals within the ring (16). Medium-ring azacycles (17), silaketals (18), and silyl ethers (19) are hence more readily accessible than medium-ring carbocycles at comparable concentrations (20,21).

The ester functionality is less privileged, and dilutions of 0.5 mM are required for simple lactones of 7–10 members (Fig. 3.4) (10). (Not shown is the five-membered butenolide, which requires high dilutions for other reasons: this special case is treated under directly-functionalized olefins in Section 3.3.3). In contrast, RCM of *N*-Boc-diallylamine is essentially quantitative at 570 mM (22). Consistent with the general trend in ring size and strain seen in Figure 3.3, caprolactone **4-1** gave higher yields than eight-membered **4-2** at comparable concentrations: indeed, **4-2** was not observed

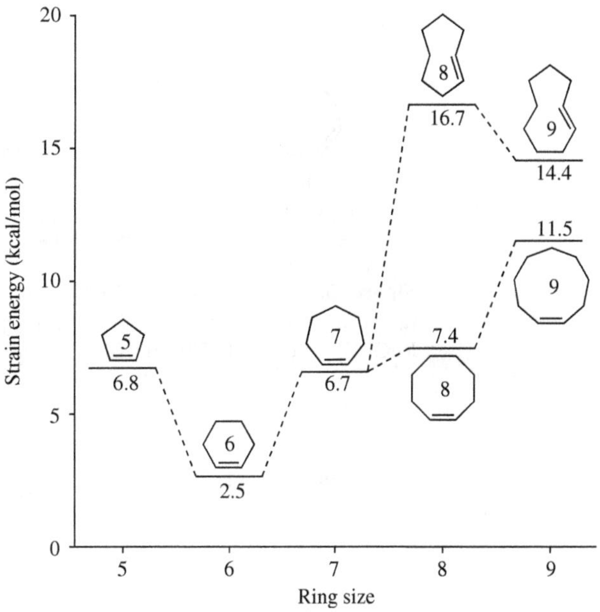

Figure 3.3 Calculated strain energies for unsubstituted, carbocyclic olefins of five to nine members (15).

even at 0.5 mM. Ten-membered lactone **4-3** was accessible in 95% yield at the same concentration (cf. 78% for **4-1**, implying slightly lower strain in the larger ring) (10). Strain decreases further for the larger macrolactones, and dilutions of 5 mM enabled near-complete selectivity for 14- and 16-membered **4-4** and **4-5**.

Challenging ring sizes can be attained even at relatively high substrate concentrations, where conformational constraints reduce the entropic penalty of cyclization (20,23,24). Fused rings (especially aromatic rings, which introduce sp^2 carbon centers) are common structural elements used to promote cyclization. The conformational influence of ring fusion can be further reinforced by additional substituents near the rigid structural element that increase the probability of encounter (3). Acyclic conformational constraints, often associated with vicinal substitution (gearing effects), can likewise aid cyclization (3). Thorpe–Ingold effects have long been regarded as a powerful driver for RCM of diethyl diallyl malonate (DeDAM) and are invoked to explain the complete selectivity for five-membered **5-1** (Fig. 3.5), even in neat diene (25,26) (conditions under which cyclopentene is polymerized) (27). That the beneficial effect of allylic substitution is solely thermodynamic is implied by a remarkable recent study in which Percy and coworkers (28) described much faster formation of cyclopentene, versus **5-1**, at diene concentrations of 10 mM (RCM in CH_2Cl_2 using 1 mol% $RuCl_2(H_2IMes)(PCy_3)(=CHPh)$, **Ru-2**).

Early evidence for the Thorpe–Ingold effect in RCM was reported by Forbes, Wagener, and coworkers. Exemplary behavior is that shown for seven-membered ketones **5-2** and **5-3** (Fig. 3.5): in neat diene, unsubstituted **5-2** was not formed, where yields of substituted **5-3** were near-quantitative (29).

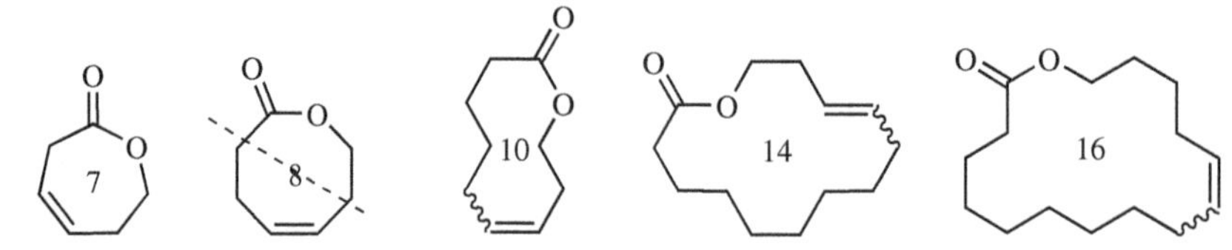

4-1	4-2	4-3	4-4	4-5
5 mM (29%)	5 mM (0%)	5 mM (41%)	5 mM	100 mM (10%)
0.5 mM (78%)	0.5 mM (0%)	0.5 mM (95%)	(94%)	5 mM (99%)

Figure 3.4 Dependence of RCM equilibrium yields on initial diene concentration for a range of lactone ring sizes. Oligomers represent the balance of material. Conditions: 5 mol% $RuCl_2(IMes)(PCy_3)(=CHPh)$ (**Ru-2′**); CH_2Cl_2, 40 °C (10).

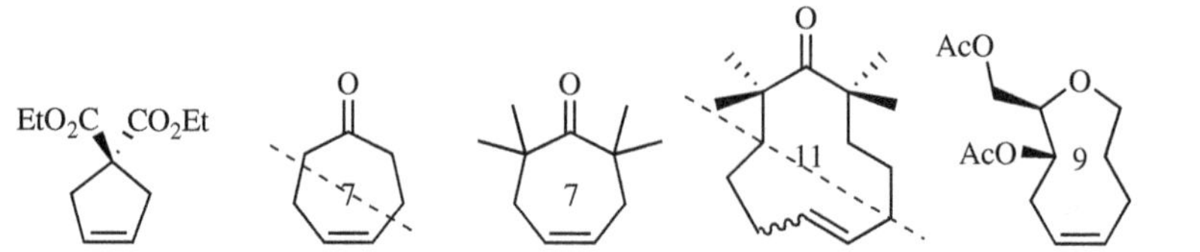

5-1	5-2	5-3	5-4	5-5
neat, 25 °C, 1 min	neat, 25 °C, 2 h	neat, 25 °C, 1 h	neat, 25 °C, 2 h	CH_2Cl_2 (3 mM)
(99%)	(0%)	(95%)	(0%)	40 °C, 2 h
				(89%; 10% dimer)

Figure 3.5 The operation and limits of the Thorpe–Ingold (gem-dialkyl) effect. Various catalysts used; loadings of 7 mol% or lower.

With increasing ring size and greater separation between the gearing element and the cyclization site, the effectiveness of these constraints diminishes (see 11-membered **5-4**) (29), although they remain useful where transmission is efficient. A now-classic example reported by Crimmins and coworkers (30) described the success of such effects in the synthesis of a challenging nine-membered cyclic ether (see **5-5**) at 3 mM. The gauche disposition of the substituents was proposed to bring the olefins into proximity.

The predictive utility of acyclic conformational constraints diminishes with increasing ring size and complexity. The impact of functionalization on conformational mobility is rarely evident a priori for large rings with multiple substituents and/or functional groups. Even remote substituents can have an unexpected impact on RCM yields (31–35), while the positions of the double bonds relative to functional groups has a profound influence (36–38), in part due to the potential for Lewis base chelation (see Section 3.3.4.2). "Protectable" centers near the site of cyclization can offer a superb opportunity for trial-and-error tuning of effective molarities. A high-profile example appears in the synthesis of BILN 2061, as discussed in Section 3.6. A recent book presents useful overviews of these and related strategies for RCM synthesis of macrocyclic and medium-sized carbocycles (21) and heterocycles (17,39), within the challenging context of natural products synthesis.

A more specialized methodology is shown in Figure 3.6. Collins and coworkers (40) explored the use of non-covalent π-stacking interactions as control elements for the assembly of paracyclophanes (see **6-1**, Fig. 3.6a). These eliminate the need to install and remove the gearing unit, but do not affect the need for very high dilutions to form these strained rings. Total RCM yields are then limited by catalyst decomposition, which competes with that of RCM below diene concentrations of about 0.5 mM.

3.2.1.1 Ring Size, Reaction Rates, and Metathesis Dimerization RCM rates are normally considerably slower for macrocycles and medium-sized rings than the corresponding five- or six-membered rings. This is due not merely to lower encounter probabilities and the steric barrier to back-biting within a pendant oligomer chain, but also the retarding effect of high dilutions on reaction kinetics. Slow metathesis is problematic because it increases the opportunity for competing catalyst decomposition, which reduces yields (as noted above), and can also contribute to side-reactions.

For dienes in which one olefin is deactivated, additional complications can arise. RCM is still observed for favored ring sizes, following initial installation of the metal at the unperturbed olefin. The first-generation Grubbs catalyst **Ru-1**, for example, was competent to effect RCM of α-methyl-DeDAM (**7-1a**, Fig. 3.7a) (41), despite the feeble reactivity of this catalyst in CM of 2,2-disubstituted olefins (see Section 3.3.2). Cyclization was much slower for isopropyl analog **7-1b** (48 h vs 24 h) but the product **7-2b** was still obtained in essentially quantitative yield (41).

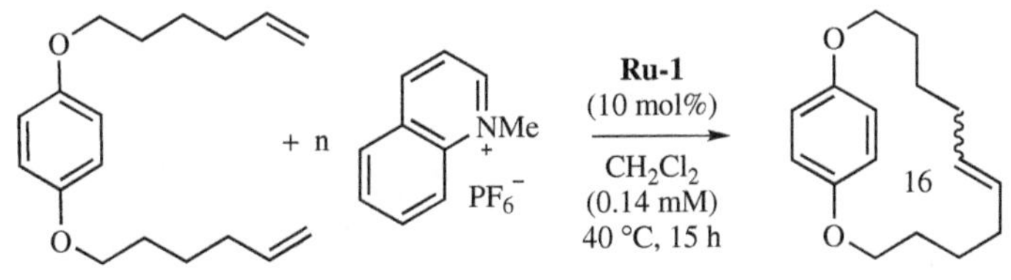

6-1

n	yield (%)
0	0
20	50

Figure 3.6 Use of an intermolecular, noncovalent tether to promote RCM (40).

(a)

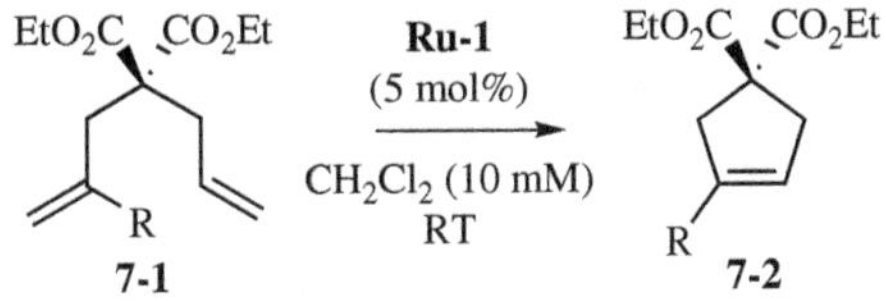

7-1 7-2

	R	t (h)	yield (%)
a	Me	24	91
b	^{i}Pr	48	98

(b)

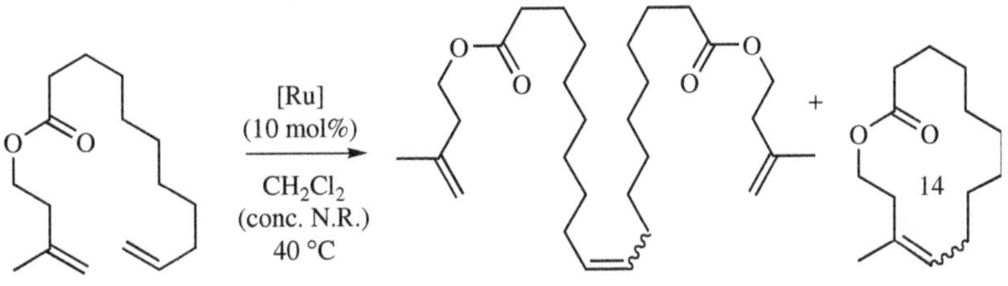

7-3 7-4 7-5

catalyst	*t* (h)	**7-4** (%)	**7-5** (%)
Ru-1	17	79	0
Ru-2'	40	trace	65

Figure 3.7 Examples of attempted RCM of a diene in which one olefin is sterically deactivated. (a) Successful RCM for a five-membered ring (41). (b) Terminal, head-to-head dimerization of **7-3** with **Ru-1**, versus successful macrocyclization with **Ru-2′** (42). N.R. = not reported.

For bulkier substituents, or for more demanding ring sizes, reaction may terminate at acyclic dimers. In the example of Figure 3.7b, metathesis of **7-3** by **Ru-1** occurred only at the accessible olefinic site, and the head-to-head linear dimer **7-4** was isolated in about 80% yield (42). Synthesis of the desired 14-membered **7-5** required use of **Ru-2′**; again, however, reaction was slow, and yields were limited to 65%. Nevertheless, addition of a more reactive catalyst offers a general strategy for inducing RCM of linear dimers formed from unsymmetrically deactivated dienes, meaning that such dimers should not be regarded as a dead end (given appropriate dilutions).

Dienes bearing one electron-deficient olefin exhibit similar behavior. For example, Fürstner (43) showed that the conjugated styrenyl olefin in **8-1** resisted RCM with **Ru-1**, but underwent smooth cyclization with **Ru-2′** (Fig. 3.8a). Many examples of such behavior appear in RCM of directly-functionalized olefins: Section 3.3.3.

A more fundamental challenge is illustrated in Figure 3.8b (44). Here the desired RCM target, a nine-membered ring, would require coupling of the acrylate and unperturbed olefin groups in **8-3**. Instead, the dominant product is cyclodimer **8-5**, presumably formed by RCM of linear **8-4**. A head-to-tail orientation within such cyclodimers is a clear marker for difficulties of ring strain rather than catalyst activity. The catalyst is evidently competent to couple the two olefin types present, but cannot overcome ring strain to convert the linear dimer into a nine-membered "cyclomonomer" ring.

In the examples above, cyclization is possible—where ring strain is not prohibitive—because the unperturbed olefin provides a site for initiation. Where both olefins are deactivated, RCM becomes much more difficult to achieve. A work-around strategy, relay RCM (RRCM), is discussed in Section 3.3.6.

3.2.2 Catalyst Choice: General Considerations

This section surveys general catalyst activity and stability, and catalyst selectivity for RCM products over oligomers, where known. Catalyst choice as it relates to the capacity to cope with bulk, specific functional groups, and *E*/*Z*-selectivity is dealt with in the sections focusing on those topics.

Classic, stoichiometric cyclization methodologies involve irreversible reactions, and kinetic selectivities are therefore critical to the yield of the cyclic targets. This is rarely the

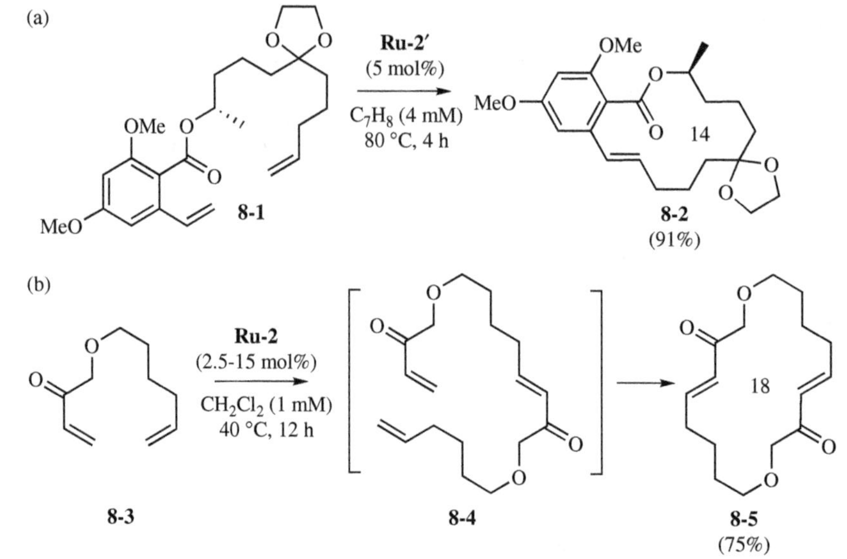

Figure 3.8 Examples of attempted RCM of dienes bearing an electron-deficient olefin. (a) Successful RCM (43). (b) Failed synthesis of a nine-membered ring: successive formation of linear and low-strain cyclic dimers (44).

case in RCM. Observation of oligomers during an attempted RCM reaction does not imply failure of the reaction (provided catalyst activity and lifetime are sufficient), but is simply a reflection of the kinetic bias of the catalyst. Long recognized in ROMP chemistry is the kinetic bias exhibited by some metathesis catalysts in favor of oligomerization, where others are biased toward cyclization (45). Given appropriate dilutions, catalyst activity, and sufficient time, however, the ring-chain equilibrium will "recycle" oligomers into cyclic species, as shown below.

*3.2.2.1 **Group 6 Catalysts*** The modular nature of the Schrock catalysts (Fig. 3.9) facilitates catalyst–substrate matching. This has led to remarkable advances, particularly in collaborative work by Schrock and Hoveyda on asymmetric metathesis (Chapter 6). The standard group 6 catalyst used in RCM is **Mo-1**, the reactivity of which is significantly greater than that of the Ru catalysts. While this is advantageous in many contexts, lower reactivity can be desirable where issues of selectivity arise. A case in point is *Z*-selective macrocyclization (Section 3.3.1.2), in which the high selectivity of recently developed tungsten alkylidene **W-1** (46) is attributed to its lower reactivity, which limits its ability to participate in ring-opening–ring-closing equilibria.

The functional-group tolerance of the group 6 catalysts is limited chiefly by their reactivity toward protic and aldehydic groups. Greater difficulties arise from the sensitivity of these invaluable catalysts to trace air or water (47), and their thermal instability. Best practices in handling are described in Section 3.4.1. Where high reactivity or high, tunable selectivity is required, however, the group 6 catalysts remain the systems of choice. A recent advance by Heppekausen and Fürstner described phenanthroline- and bipyridine-stabilized adducts of **Mo-1** and related catalysts, which exhibit convenient air-stability. *In situ* treatment with $ZnCl_2$ was reported to release the catalyst with uncompromised reactivity, albeit at only high temperatures: see Figure 3.10 (48). The advent of tungsten complex **W-1**, which is sufficiently stable to weigh in air, may also expand uptake of the group 6 complexes (46).

A final advantage of the group 6 catalysts, somewhat paradoxically, lies in their air-sensitivity. This enables rapid, efficient quenching once RCM is complete, eliminating the potential for secondary metathesis (see Section 3.4.6). In contrast to some of the quenched Ru products, the resulting metal-oxo species do not promote additional, unwanted catalysis such as isomerization, and their polarity greatly facilitates separation from organic products.

*3.2.2.2 **Ruthenium Catalysts*** The enormous success of ruthenium metathesis catalysts stems from their greater robustness and ease of handling. Following the discovery of the first- and second-generation Grubbs catalysts (**Ru-1** (49), **Ru-2** (50,51); Fig. 3.11), a very large number of alternatives emerged, and even when considering only commercially available examples, the choices are many. It is not our intention here to formulate trends in advance of the evidence: most of these catalysts have not been explored within a sufficiently broad context to provide a clear picture of their advantages and limitations, or even to identify any real performance differences between these and the dominant catalyst **Ru-2**. (Exceptions are the catalysts of Figure 3.11b, which have shown good performance in specialized contexts, particularly synthesis of *Z*- substituted or tetrasubstituted olefins (Sections 3.3.1.2 and 3.3.2.)

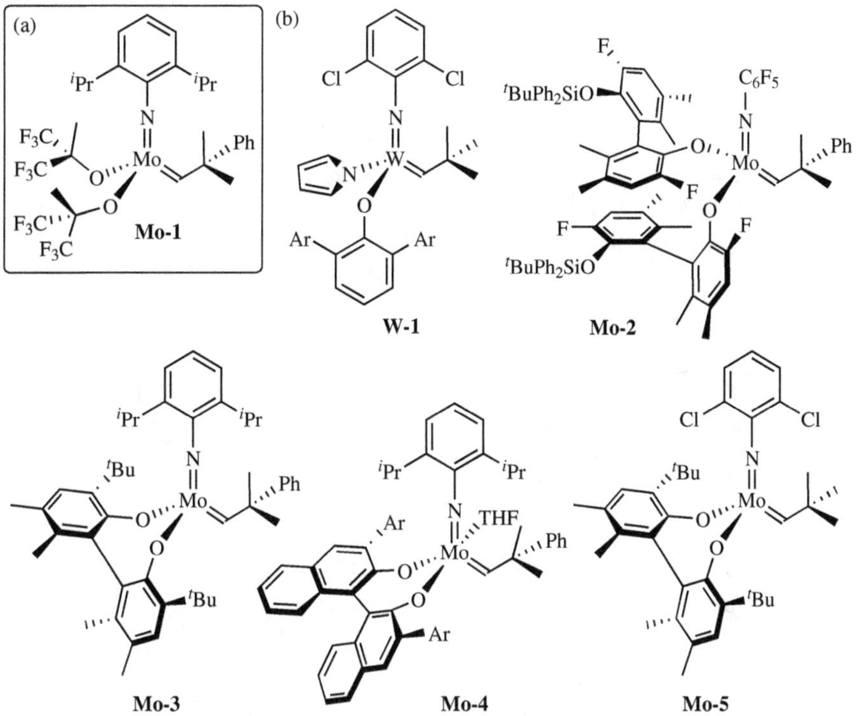

Figure 3.9 Key group 6 metathesis catalysts. (a) Catalyst in dominant use. (b) Catalysts used for more specialized purposes discussed in the text (Ar = 2,4,6-^{i}Pr$_3$-C$_6$H$_2$).

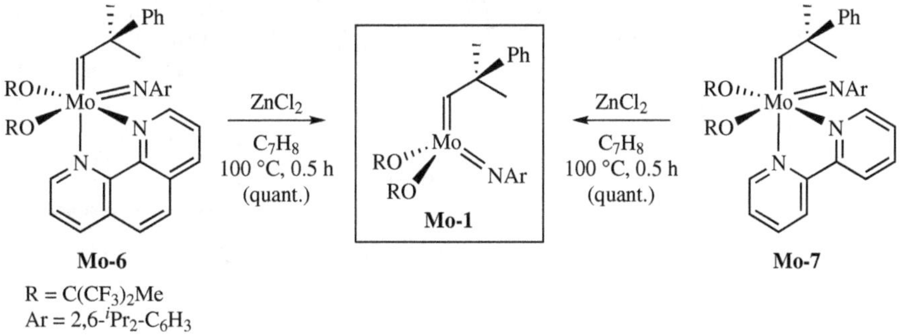

Figure 3.10 Reversibly-protected forms of **Mo-1** (48).

As a general statement, differences in initiation efficiency certainly exist, and a correlation has long been proposed between rapid turn-on and rapid deactivation, which tends to suggest that (outside ROMP) productivity is not best served by delivering the entirety of the catalyst charge in its active form at once. Excessively slow turn-on, on the other hand, means that a high proportion of the catalyst charge may not be used at all but simply adds to operating costs and purification requirements.

By far the most widely used RCM catalyst is **Ru-2**. Ruthenium–NHC complexes of this general type exhibit much lower turn-on efficiency than **Ru-1** (52) (a function of the reduced lability of the PCy$_3$ ligand when trans to an *N*-heterocyclic carbene), and elevated temperatures are often necessary to achieve significant levels of active catalyst. Their "commitment" to reaction with olefin, versus reuptake of PCy$_3$, is very high compared to **Ru-1**, however (53). In consequence, they can participate in many more cycles of metathesis before being trapped by recoordination of PCy$_3$ to form five-coordinate methylidene complex **Ru-12 b/b′** (Fig. 3.11c). This is critical to performance, as the drastically reduced lability of the phosphine ligand in **Ru-12** inhibits re-entry into the catalytic cycle.

The high stability of the five-membered chelate ring in the Hoveyda catalyst **Ru-3** confers a remarkable level of resistance to degradation by air and water, as evidenced by its tolerance toward purification by column chromatography in air (54). This catalyst is generally viewed (55,56) as being rather slow to turn on: indeed, its first-generation analog was reportedly 30 times slower to initiate than **Ru-1** (57).

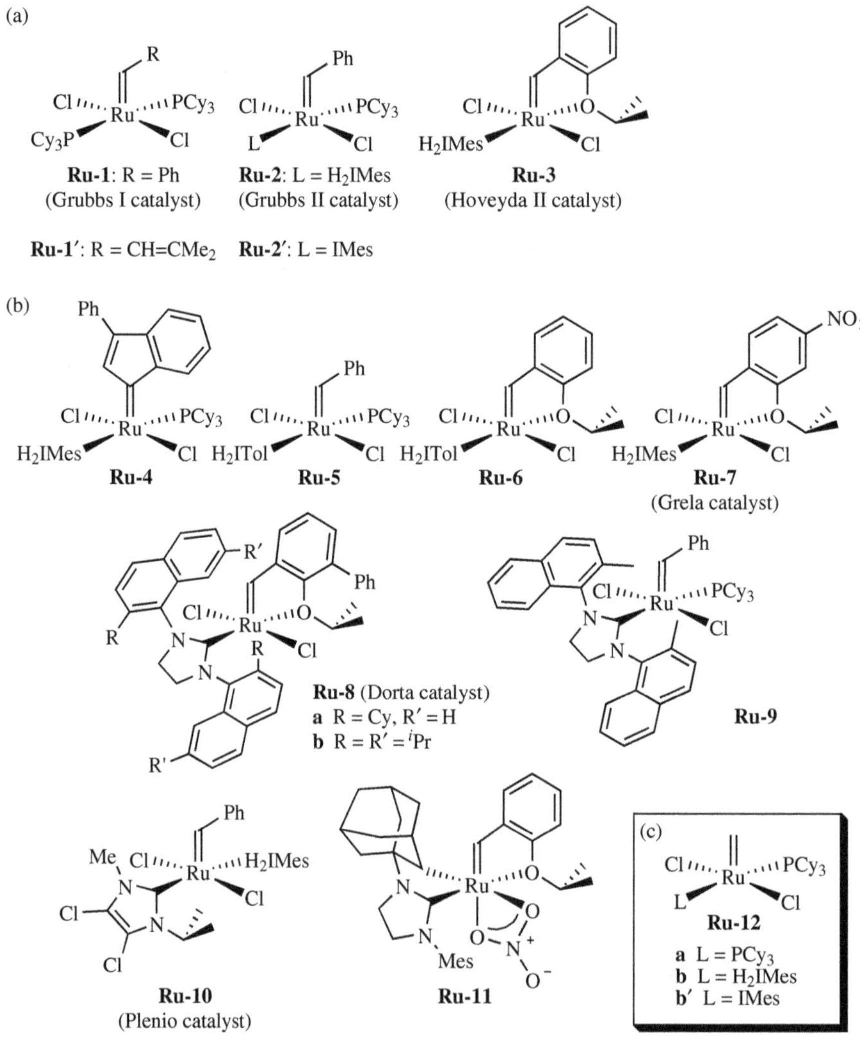

Figure 3.11 Key ruthenium catalysts for olefin metathesis. (a) Catalysts in dominant use. (b) Additional catalysts discussed in the text. (c) Resting-state methylidene species derived from the Grubbs catalysts.

Recent studies, however, suggest that turn-on of the NHC catalyst **Ru-3** can be rapid in the presence of sterically accessible olefins. Plenio and coworkers (58) showed that initiation rates increased significantly with higher concentrations of butyl vinyl ether or 1-hexene, suggesting that **Ru-3** reacts with small olefins via an interchange-associative mechanism, rather than the dissociative pathway common to the Grubbs catalysts (52). This undoubtedly contributes to the very impressive performance of **Ru-3** in cross-metathesis (see Chapter 2), where high olefin concentrations are practical. The relevance to RCM is less clear-cut. While much faster RCM in neat DeDAM has been reported for an analogue of **Ru-3** (26), the Plenio study found that the dissociative pathway was most important in RCM of DeDAM by **Ru-3** itself (58). In any case, the potentially beneficial impact of higher concentrations on metathesis rates is limited where high dilutions are required to effect cyclization (Section 3.2). Perhaps more relevant to total turnover numbers is the much-debated boomerang mechanism (54). The Hoveyda catalyst is surprisingly little used in RCM, however, despite impressive performance in a number of cases illustrated below.

3.2.2.3 Substrate–Catalyst Matching RCM synthesis of common, low-strain rings can be achieved with any of the catalysts discussed above. Vinylic or allylic bulk (see Section 3.3.2) necessitates the use of either **Mo-1** or a Ru–NHC catalyst. RCM of directly-functionalized olefins (Section 3.3.3) presents a challenge to both reactivity and productivity. In general, the Ru catalysts are most broadly tolerated, but vinyl ethers and silyl ethers exhibit higher compatibility with **Mo-1**. For substrates with polar allylic substituents, or allylic aromatic groups, competing isomerization is commonly observed, especially for Ru–NHC catalysts (Section 3.3.1). All of these complications occur independent of ring size, but

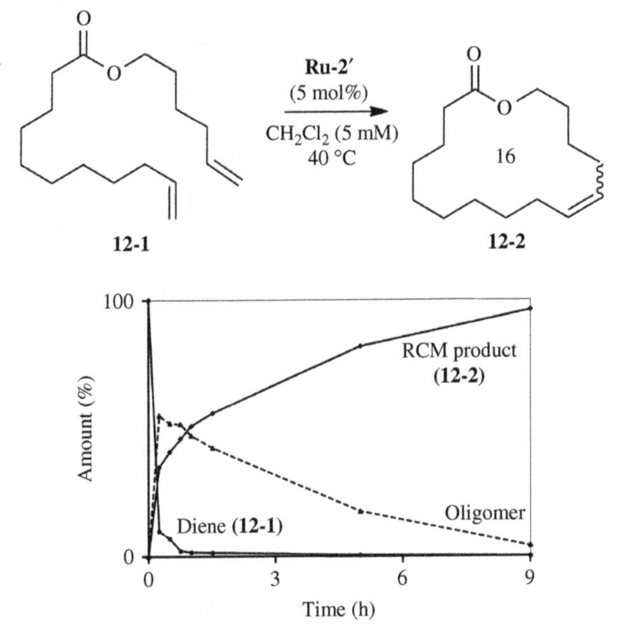

Figure 3.12 Rate plot showing rapid consumption of diene **12-1** and slow transformation of oligomers into RCM target **12-2** (10).

their impact increases as ring size or strain increases, because slower rates of cyclization allow non-metathetical processes to compete.

For medium and large rings, additional issues emerge relating to the kinetic bias of the catalyst toward inter- versus intramolecular reaction, the capacity of the catalyst to recycle oligomers into RCM products via backbiting at internal olefinic sites (see Fig. 3.1), and catalyst longevity, the latter being required to cope with the unfavorable kinetic effect of the high dilutions normally required.

Use of high dilutions is commonly invoked as a means of (kinetically) privileging RCM over oligomerization. In fact, this has no impact on kinetic selectivity. For the Grubbs-class Ru–NHC catalysts (and for the resting-state methylidene species **Ru-12**), phosphine dissociation is rate-determining,

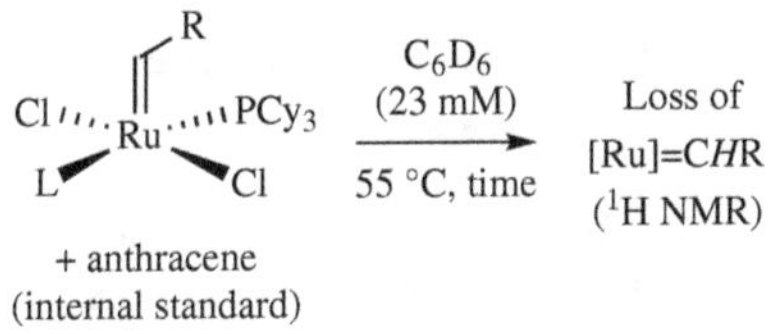

Catalyst	L	R	$t_{1/2}$ (h)
Ru-1	PCy_3	Ph	192
Ru-13	PCy_3	Et	8
Ru-12a	PCy_3	H	0.7
Ru-2	H_2IMes	Ph	>744
Ru-12b	H_2IMes	H	5.7

Figure 3.13 Half-lives reported for the Grubbs catalysts **Ru-1** and **Ru-2** and their alkylidene or methylidene derivatives at 55 °C in benzene (59,60).

and rates of metathesis are therefore independent of olefin concentration (52). The second-generation catalyst **Ru-2′** was shown by Fogg and coworkers (10) to exhibit a strong kinetic bias toward oligomerization even in dilute solution (5 mM diene; Fig. 3.12), for a wide range of α,ω-dienes (precursors to 7–20-membered lactones, catechol ethers, malonates, and cyclic polyethers). In all cases, oligomers were formed as the dominant products within 15 min in refluxing CH_2Cl_2. Continued reaction yielded the desired RCM products via backbiting. Similar behavior is presumed for the corresponding H_2IMes catalyst **Ru-2**, and was reported by the Boehringer–Ingelheim group for the Grela catalyst **Ru-7** (12), although no comprehensive catalyst study has yet been undertaken.

The relevance of this behavior to catalyst choice lies in the capacity of the catalyst to efficiently recycle oligomers into RCM products. Essential for this is a level of reactivity sufficient to tackle 1,2-disubstituted olefins containing a high percentage of *trans*-olefins. The low activity of the first-generation Grubbs catalyst **Ru-1** toward such olefins (Section 3.3.2) (41,42) translates into very slow backbiting (9), and the utility of the first-generation complexes in macrocyclization is thus determined by the extent of their kinetic bias toward intramolecular reaction.

Where oligomers form, and catalyst activity supports backbiting, RCM yields depend on whether the catalyst is sufficiently long-lived to reach equilibrium under the conditions of catalysis. Premature catalyst deactivation will result in isolation of oligomers even where reactions are carried out in a concentration regime suitable for backbiting. Catalyst decomposition almost certainly accounts for the surprisingly poor performance of the Schrock catalyst **Mo-1** in some macrocyclization reactions.

Metathesis of terminal olefins necessarily generates methylidene complexes, as does ensuing CM with ethylene or propylene by-products. In the case of the Grubbs catalysts, the resting-state methylidene complexes **Ru-12** are not only less thermally stable than their benzylidene/alkylidene analogs (see Fig. 3.13) (59,60), but are also slower to initiate. Metathesis of internal olefins, in comparison, is sterically impeded, but generates a more reactive and less vulnerable Ru–alkylidene complex. Methyl-substituted olefins, specifically, have the advantage of combining relatively low bulk with expulsion of a volatile by-product, thus preserving the entropic driving force for metathesis.

3.3 SCOPE, CHALLENGES, AND OPPORTUNITIES

3.3.1 Olefin Isomerization

3.3.1.1 Regiochemical Control: Double-Bond Migration

Among the wide range of non-metathesis pathways promoted by the Grubbs catalysts (61), olefin isomerization is particularly common. Such reactions can be beneficial or detrimental. Chapter 5 discusses the deliberate coupling of metathesis and isomerization manifolds via tandem catalysis, which can enable synthesis of targets that are otherwise more challenging, or indeed inaccessible. Here we focus on isomerization as an unintended side-reaction encountered during attempted RCM: that is, as a problem to circumvent.

We divide such reactions into three classes (Fig. 3.14). In **Class A** behavior, RCM is faster than isomerization. Because double-bond migration occurs only after cyclization, the expected ring sizes are obtained, but the position of unsaturation varies (62–64). An example is the formation of **14-1** (62). In **Class B** reactions, isomerization is faster than metathesis, and RCM of the isomerized (internal) olefin leads to ring-contracted (**Class B1**) products (18,65–68). An example is the formation of **14-2** (68). Where cyclization would yield a strained product, however, or where the catalyst is not competent to react with the internal olefin, dimers are obtained (**Class B2**). An example is the formation of **14-3** (69). In **Class C** systems, isomerization occurs without metathesis. Typically, the isomerized olefin itself is obtained (66,69) (**Class C1**), as exemplified by formation of **14-4** (66). This reflects a level of catalyst activity inadequate to cope with the increased steric bulk of the internal olefin. A more specialized example (**Class C2**) is illustrated by **14-5**, in which Fürstner (70) observed sequential C=C migration and isomerization of the secondary allylic alcohol, yielding the ethyl ketone. Similar behavior was subsequently reported by Gurjar et al. (71).

Double-bond migration is especially common for allylic ethers or amines and allylbenzenes, but is also observed with broader substrate classes (72–75). Work from the Wagener group (72,76,77) points toward the possibility of ring expansion, as well as ring contraction, resulting from isomerization along the carbon backbone of metathesis oligomers. While instances of isomerization have been reported for **Mo-1** (65,67,72,78,79) and for the first-generation Grubbs catalyst **Ru-1** (18,78,80,81), such side-reactions appear more common with the use of Ru–NHC catalysts (9,66,72,75,77,80). Several comparative studies indicate that isomerization becomes successively more problematic in the order **Mo-1** < **Ru-1** < **Ru-2** (9,72,77,78).

Metal hydride species are generally accepted as the culprits in these unwanted side-reactions. Ruthenium hydrides, in particular, can be highly effective isomerization catalysts (82). Contaminants generated during catalyst synthesis were recognized in early work as a potential source of hydrides. Snapper (83) reported that side-reactions due to isomerization dropped sharply when commercial samples of the Ru–H_2IMes catalyst **Ru-2** were purified by chromatography prior to use. This finding was subsequently corroborated by Rutjes and coworkers (84). Formation of Ru hydrides is indeed readily envisaged in the standard syntheses of **Ru-2**, which involves a combination of a

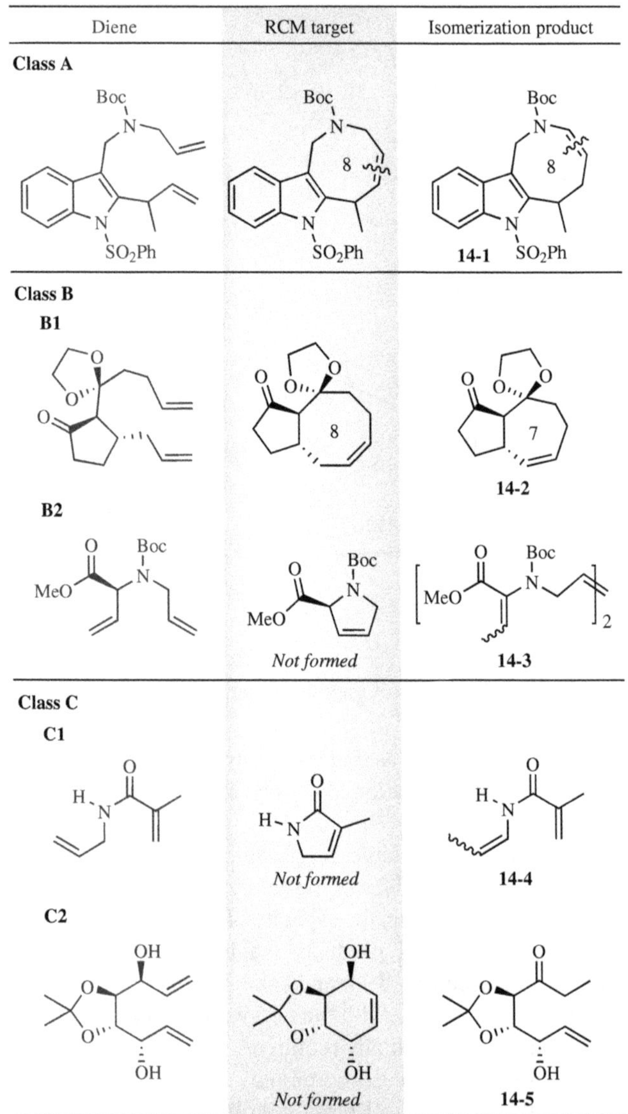

Figure 3.14 Influence of isomerization in RCM. **Class A** behavior: RCM is faster than isomerization; the position of unsaturation is affected. **Class B**: isomerization is faster than RCM, resulting in (**B1**) ring-contracted products; (**B2**) CM products, reflecting catalyst activity too low for ensuing RCM. **Class C**: isomerization without metathesis. **C1**: C=C migration. **C2**: Sequential C=C and C=O isomerization. Note: shading indicates the intended RCM product.

strong base (used to generate the free carbene *in situ*) and methanol washing to extract the inorganic salts (Fig. 3.15a) (85,86). Methanol itself is relatively innocuous toward the Grubbs catalysts at room temperature (RT), especially over short exposure times (more forcing conditions are another matter; see below in this section). Methoxide ion, however, is a very aggressive agent for decomposition of Ru–alkylidene complexes into carbonyl hydrides, as shown in Figure 3.15b and c (87). The recent development of a high-purity route to **Ru-2** via ligand exchange of **Ru-1** with *isolated* free H_2IMes should alleviate problems of

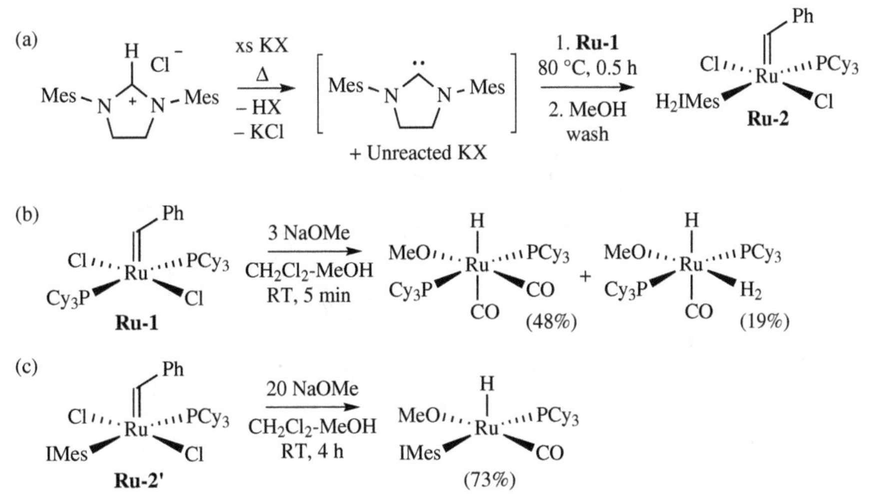

Figure 3.15 (a) Standard route to **Ru-2**, involving *in situ* generation of free H_2IMes by treatment with strong base (KO^{t}Bu, KOCMe$_2$Et, etc.) (85,86) (b) Rapid, room-temperature (RT) annihilation of **Ru-1** by methoxide ion (87); assessed by NMR integration against internal standard. (c) RT decomposition of **Ru-2′** by methoxide (87); assessed as in (b). Approximately 25% paramagnetic material is formed in both cases.

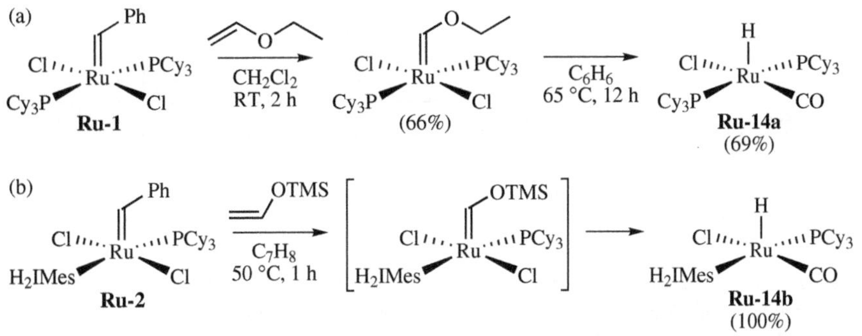

Figure 3.16 Formation and decomposition of Fischer carbenes from **Ru-1** and **Ru-2** (isolated yields). (a) By reaction with ethyl vinyl ether (90). (b) By reaction with a vinyloxysilane (91).

isomerization arising from catalyst contamination during synthesis (88).

In other cases, contaminants present in the reaction medium have been implicated. In the large-scale synthesis of BILN 2061, for example, problems of isomerization were traced to morpholine, present at parts per million levels in the toluene solvent (89). Generally, however, isomerization involves catalyst decomposition under conditions of metathesis, especially where high temperatures, long reaction times, or high catalyst loadings are required. Vinyl ethers (90) and (especially) vinyloxysilanes (91) are expected to be particularly susceptible to problems, owing to the readiness with which their Fischer carbene derivatives decompose into the long-lived hydride complexes **Ru-14** (Fig. 3.16). Work from the Sasol group supports the long-suspected accessibility of β-elimination pathways for the ruthenacyclobutane intermediate, implying that isomerization may be inextricable from metathesis (92).

Solvent choice can be critical. The Fürstner group (66) reported in early work that isomerization/ring contraction of **17-1** was more aggressive in toluene than in chlorinated solvents (Fig. 3.17a). Likewise, Prunet and coworkers (75) (Fig. 3.17b) described 30–50% isomerization of **17-4** in benzene, versus only 10% in dichloroethane. Prolonged exposure to higher temperatures (110 °C) in toluene was catastrophic, resulting in 80% isomerization to **17-6**. Strikingly, 100% isomerization was observed in dimethoxyethane at 85 °C, although this is a solvent rarely used for RCM. The long reaction times relative to those in Figure 3.17a should be noted. Prompt interception of the product once RCM is complete is advisable: for methods for monitoring RCM, see Section 3.4.5.

Alcohol solvents (93,94) are also reported to trigger isomerization during RCM, as are polyethylene glycols (95) now being explored as "green" solvents. (The heightened tendency toward isomerization in dimethoxyethane (DME)

(a)

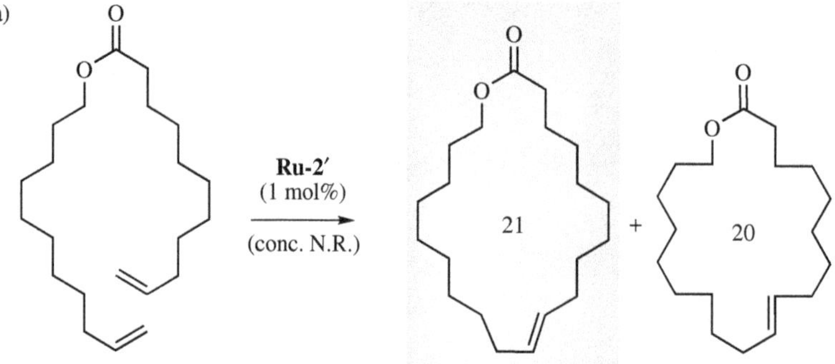

17-1 **17-2** **17-3**

solvent	T (°C)	t (h)	**17-2** (%)	**17-3** (%)
toluene	80	0.8	67	12
toluene	40	6.5	65	10
CH_2Cl_2*	40	23	71	2
$[CH_2Cl]_2$ (DCE)	80	12	23	2

(b)

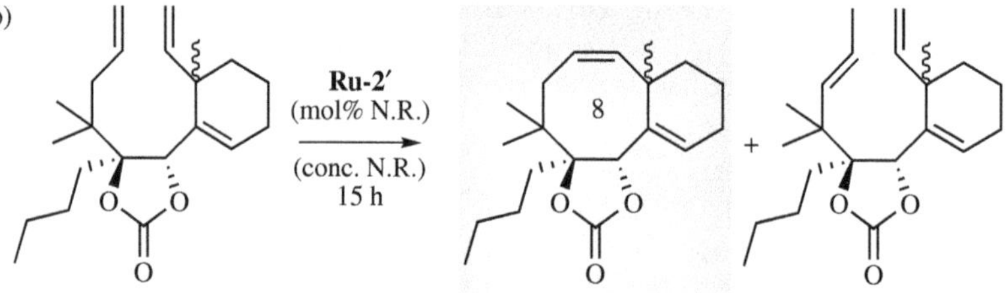

17-4 **17-5** **17-6**

solvent	T (°C)	**17-5** (%)	**17-6** (%)
benzene	80	70-50	30-50
DCE	83	90	10
toluene	110	20	80
$[CH_2(OMe)]_2$ (DME)	85	0	100

Figure 3.17 Effect of solvent and temperature on RCM yields. (a) Competing **Class B** isomerization (ring contraction; see **17-3**) (66). (b) Competing **Class C** isomerization (solely C=C migration; see **17-6**) (75). N.R. = not reported. The label (*) denotes 4 mol% **Ru-2′**. Note: shading indicates the intended RCM product.

may be a warning factor for use of the latter.) Both first- and second-generation (IMes, H_2IMes) Grubbs catalysts have been shown to decompose into Ru hydrides in the presence of primary (or secondary) alcohols and amine base (96–98). **Ru-1** and **Ru-2′** were completely degraded in 20% methanol–CH_2Cl_2 within 4–8 h at 60 °C (Fig. 3.18) (98). The second-generation catalyst decomposed at double the rate of **Ru-1**. Methoxide effects complete the destruction of both catalysts much more rapidly, even at ambient temperatures, as noted above (87).

Despite the efficiency with which the Grubbs catalysts can be transformed into Ru hydrides by these and other methods, it should be recognized that where the goal of catalysis is solely isomerization, commercially available hydrides (particularly $RuHCl(CO)(PCy_3)_2$, **Ru-14a**) are significantly cheaper than **Ru-1** or **Ru-2** and more convenient, as they require no activation to generate the hydride species.

Various measures have been proposed to inhibit isomerization. Grubbs and coworkers (99) reported that added

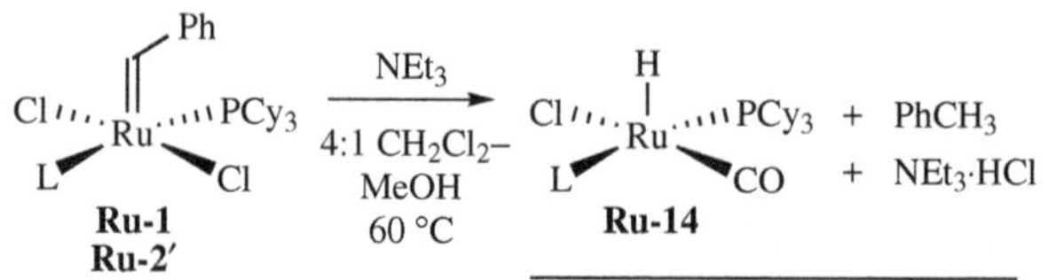

	L	*t* (h)	Yield (%)
a	PCy_3	8	58
b	IMes	4	83

Figure 3.18 Degradation of Grubbs catalysts into hydrides **Ru-14** by treatment with methanol and base. Times are those for complete consumption of the benzylidene precursor (98).

1,4-benzoquinone is effective (BQ; 2–10 equiv vs Ru). Quinones are now the accepted solution of first resort in attempting to suppress such side-reactions. Less widely adopted inhibitors include acetic acid (99), phenylphosphoric acid (100), and tricyclohexylphosphine oxide (75) (though the latter may impair conversions conversions;

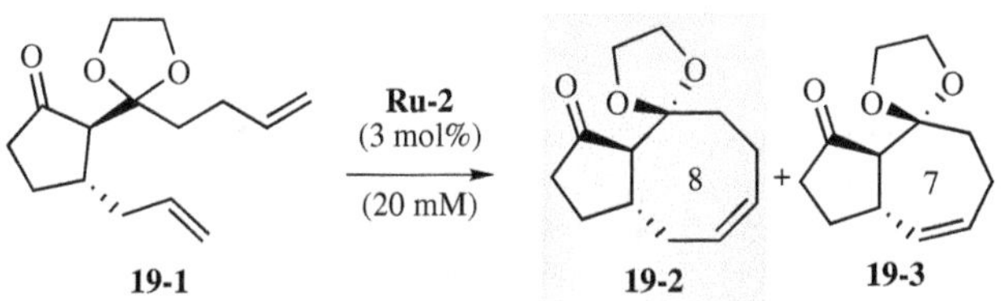

solvent	additive[a]	T (°C)	t (h)	19-2 (%)	19-3 (%)
benzene	–	81	4	31	31
DCE	–	83	24	29	20
DCE	O=PCy$_3$	83	24	18	12
DCE	BQ	83	24	54	3

[a]5 mol%.

Figure 3.19 Inhibiting isomerization during RCM. Combined effect of benzoquinone (BQ) and chlorinated solvent on RCM of **19-1** (68). Note: shading indicates the intended RCM product.

see Fig. 3.19); molecular sieves were apparently also effective in one case (101). Numerous studies have reported that chlorinated solvents diminish isomerization (63,66,68,102,103) (presumably via chlorination of the metal–hydride bond), although an anomalous report described more extensive isomerization in dichloromethane than diethyl ether (81). In a recent example, Marko reported that RCM of **19-1** to give eight-membered **19-2** (Fig. 3.19) could be increased to 54% yield by the combined use of BQ and a chlorinated solvent. Isomerization was not completely suppressed, and minor amounts of ring-contracted **19-3** were also formed. Added O=PCy$_3$ was less effective (68).

3.3.1.2 Stereochemical Control: E/Z Selectivity Ring strain results in selective formation of *cis*-olefins for rings of seven members or less, and strongly favors the cis-isomer for cyclooctene and cyclononene (see Fig. 3.3). For macrocyclic rings, an *E*/*Z* mixture is typically obtained: indeed, control over olefin stereochemistry has long constituted one of the chief limitations of olefin metathesis as a methodology for RCM assembly of macrocycles. The state of the art in this important area is briefly summarized below. Wherever possible, the selectivities for the crude products are provided, as these give direct insight into catalyst selectivity.

E-Selective RCM Ring-opening–ring-closing equilibria favor the thermodynamically more stable *trans*-olefins. The proportion of *E*-olefin products therefore increases over time, and as diene conversions approach synthetically relevant levels. For slow reactions, double-bond isomerization resulting from catalyst decomposition (see Section 3.3.1.1) also contributes to the *E* content of the products. Isomerization was implicated in an early Grubbs study using **Ru-2**, in which 92% *E*-lactone **20-2** was observed at full conversions in RCM of **20-1** (Fig. 3.20a) (104). The bar for *E*-selectivity is still set, however, by the figure of greater than 98% attained by Fürstner's (105) two-step protocol, involving alkyne metathesis, followed by Birch reduction or hydrosilylation–desilylation protocols (Fig. 3.20b).

Z-Selective RCM As the greatest erosion of *Z*-selectivity occurs in the last 20% conversion to the product, high *Z*-selectivities are often attained at the price of yields. High selectivities at intermediate diene conversions are of mechanistic interest from the perspective of catalyst development, but are synthetically unsatisfactory.

The combination of high selectivities and high macrocycle yields has only very recently been attained in RCM. In a spectacular series of advances, Schrock and Hoveyda (46,106) demonstrated that the monoaryloxide–pyrrolide (MAP) catalyst **W-1** (Fig. 3.9) enabled up to 97% *Z*-selectivity at 97–98% conversion, for the disubstituted olefins present in macrocycles **21-1** and **21-2** (Fig. 3.21). The products represent key intermediates in the total syntheses of nakadomarin A and epothilone C, respectively. Tempered catalyst activity was thought to be critical to limiting re-opening of the newly formed ring; also relevant

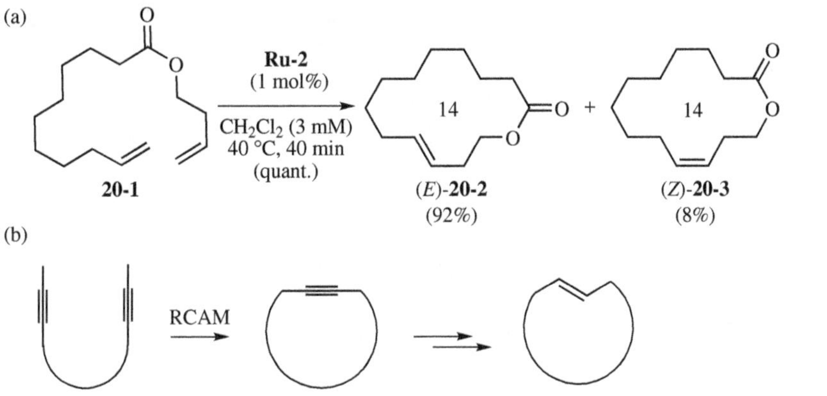

Figure 3.20 Strategies for attaining high *E*-selectivities in RCM. (a) Via thermodynamic control; exemplified by RCM synthesis of 14-membered **20-2** (104). (b) Via ring-closing alkyne metathesis (RCAM), followed by stereoselective reduction (105).

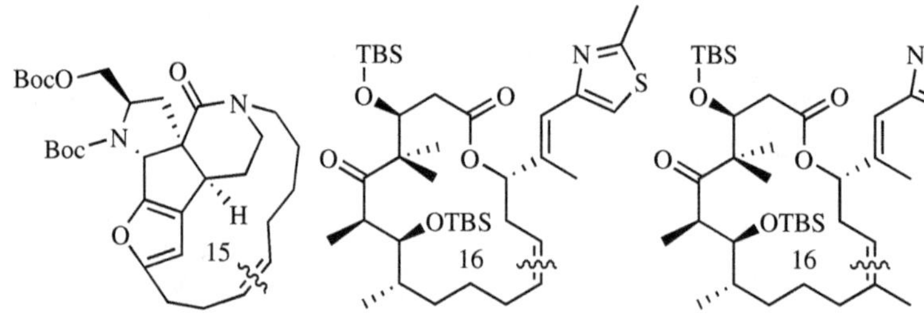

	21-1	21-2	21-3
	W-1 (5 mol%) C_7H_8 (5 mM) 7 torr, 22 °C, 2 h	**W-1** (10 mol%) C_6H_6 (1 mM) 1 torr, 22 °C, 2.5 h	**Mo-2** (10 mol%) C_6H_6 (2 mM) 1 torr, 22 °C, 2.5 h
diene conversion:	(98%)	(97%)	(88%)
isolated yield:	(90%)	(85%)	(82%)
selectivity:	(97% *Z*)	(96% *Z*)	(91% *Z*)

Figure 3.21 Leading results in *Z*-selective macrocyclization mediated by group 6 metathesis catalysts (46,106). For catalyst structures, see Figure 3.9.

is the use of room-temperature conditions and short reaction times (<3 h) (46). The kinetic selectivity of the catalyst with respect to oligomerization has not yet been examined.

High cis-content was also achieved in the synthesis of trisubstituted macrocycles. Thus, **21-3** (the cyclic core of epothilones B and D) was delivered with 91% *Z*-selectivity after 2.5 h reaction with **Mo-2** (106). The diene conversion at this stage was 88%.

Efficient Ru-catalyzed routes have been slower to develop. Alkyne metathesis followed by Lindlar reduction was for long the only reliable route to *Z*-configured macrocycles via Ru catalysis (107). Much recent attention focuses on developing ruthenium catalysis for *Z*-selective olefin metathesis. The state of the art, current as of the date of writing (December 2012), is shown in Figure 3.22a. Grubbs and coworkers reported that adamantyl catalyst **Ru-11** enabled up to 94% selectivity for 20-membered (*Z*)-**22-2**, after chromatographic purification (75% isolated yield) (108). Related 16- and 17-membered lactones were obtained with 84% or 89% *Z*-selectivity, respectively, at about 70% isolated yield.

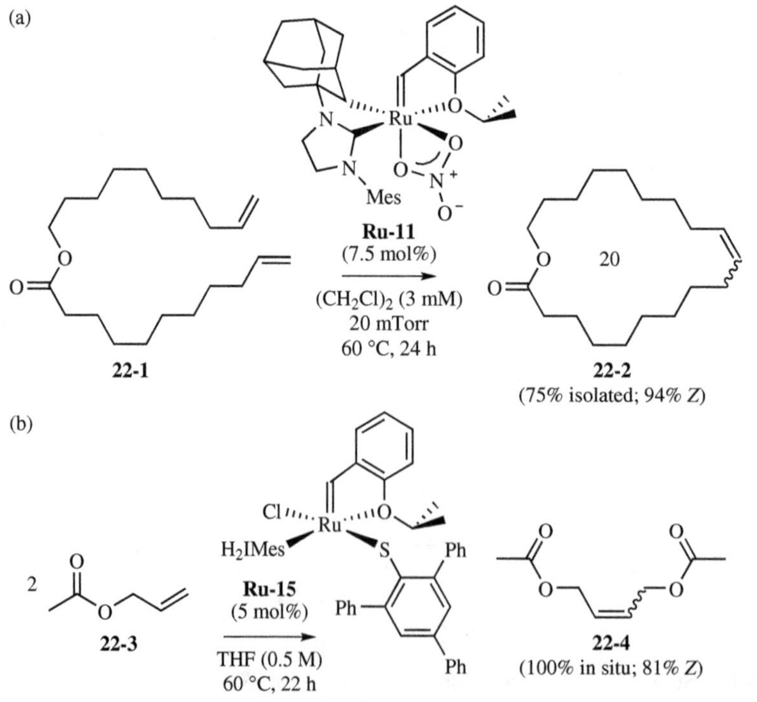

Figure 3.22 State-of-the-art Z-selectivities in Ru-catalyzed olefin metathesis, as of December 2012. (a) Macrocyclization via **Ru-11** (108). (b) CM using **Ru-15** (109).

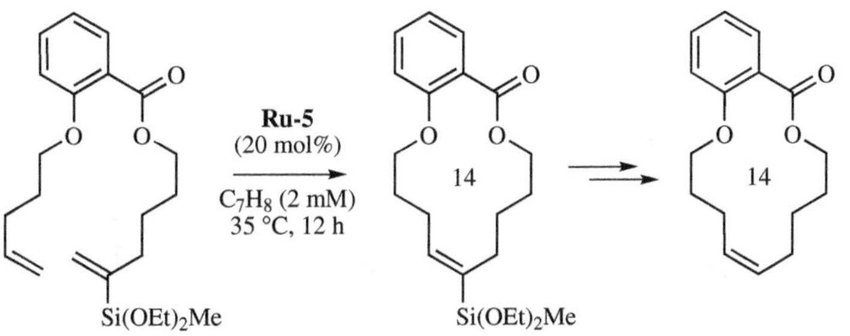

23-1 **23-2** (95%; >98% *E*) (64%; >98% Z)

Figure 3.23 Use of silylated olefins to promote "cis-selective" RCM macrocyclization, following desilylation (110).

A very recent report by Jensen, Occhipinti, and coworkers (109) highlights new directions in the design of stereoselective ruthenium catalysts. Thioaryloxide complex **Ru-15** demonstrated up to 81% *Z*-selectivity at 100% yield in the CM homocoupling of allyl acetate **22-3** (Figure 3.22b). Applications of this interesting system to macrocyclization have not yet been reported.

A work-around utilizes H_2ITol catalyst **Ru-5**, in conjunction with a sterically demanding silyl substituent near the site of ring closure. Thus, the Young group reported that RCM of siloxy-functionalized **23-1** (Fig. 3.23) enabled greater than 98% cis-selectivity in the formation of **23-2** (95% yield) (110). The bulk of the silicon substituent on the olefin protects it from re-opening to form an equilibrium distribution of *E*/*Z* products.

3.3.2 Olefin Substitution Patterns: Steric Impact

Retrosynthetic analysis is a powerful design tool in target-driven synthesis. Within the context of RCM, it has led to a broadly accepted practice of classifying reactions according to the olefin substitution pattern in the target compound—that is, formation of a 1,2-disubstituted, a trisubstituted, or a tetrasubstituted olefin. The implicit assumption is that the simplest precursor will be adopted.

The simplest precursor is not always the most relevant, however, and a larger set of possibilities exists. Disubstituted **24-1**, for example, can be assembled not merely from diene **A**, containing two terminal olefins, but—in principle—from any of the six dienes **A** to **F** (Fig. 3.24a). All six are equivalent from a product-focused perspective. A product-focused analysis would similarly equate the diene precursors to, respectively, trisubstituted **24-2** and tetrasubstituted **24-3** (Fig. 3.24b–c). In practice, profound differences in reactivity are present within each set (in the extreme of **F**, the present generation of Ru catalysts is probably inadequate to the task), and in the probability of reversible metathesis. The reduced volatility and higher solubility of the olefin by-product generated by metathesis of an internal olefin (vs ethylene) (111) increases the reversibility of diene consumption (see Fig. 3.1a). Intermolecular metathesis, on the other hand, may be limited to dimerization, while backbiting is likely to be sharply curtailed.

Increases in catalyst activity expand the range of dienes that can be efficiently cyclized. The risk of unwanted secondary metathesis (and, commonly, catalyst susceptibility to decomposition) increases in parallel. It is therefore essential to match catalyst activity to the synthetic challenge presented by a given substrate/target pair. The discussion immediately below centers on the competence of the standard metathesis catalysts to react with olefins bearing different levels of substitution.

The order of reactivity shown in Figure 3.25 is widely inferred from an early study by Ulman and Grubbs (112), which assessed the rate of initiation of **Ru-1** by various olefins at 35 °C in benzene. Under these conditions, reaction with the terminal olefin **25-1** occurred about 50% faster than with the cis-disubstituted olefin **25-2**, which in turn reacted about two times faster than its trans isomer **25-3**. Even for terminal olefins, allylic bulk is important. Thus, *sec*-butylethylene **25-4**, bearing a single allylic methyl group, reacted at about one-fourth the rate of 1-hexene **25-1**; *tert*-butylethylene (not shown) evinced negligible reaction over 4 days. No reaction was observed with 2,2-disubstituted **25-5** under these mild conditions. More heavily substituted substrates such as **25-6** and **25-7** were therefore not examined.

In RCM of unsymmetrically substituted dienes, initial metathesis will occur preferentially at the more accessible olefin, to an extent determined by the steric congestion at the metal. In Figure 3.26, for example, initiation at the 2,2-disubstituted olefin (i.e., reaction via Path (a)) is higher in energy, owing to the placement of a quaternary carbon α to the Ru center. Reaction via Path (b) is therefore favored. The high probability of encounter between the metal center and the pendant olefin aids subsequent metathesis at the 2,2-disubstituted olefin. The overall process can be viewed as converting one high-energy step into two steps, each of lower energy.

Reviews of early advances by Armstrong (1), Majumdar (113), and others demonstrate that both **Mo-1** and **Ru-1** give access to trisubstituted alkenes via RCM (see **27-1**, **27-2**,

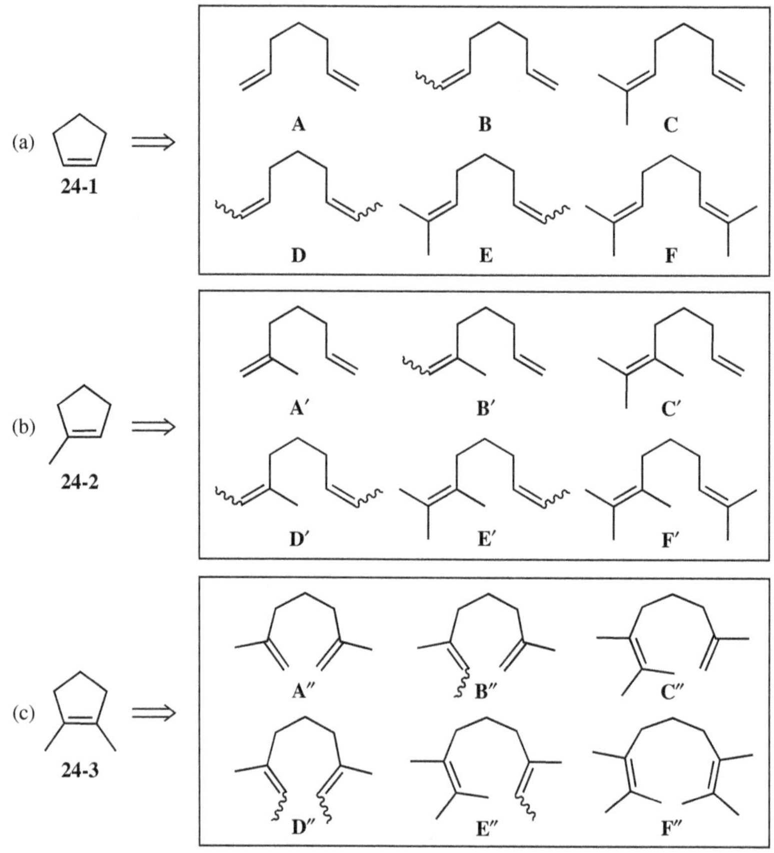

Figure 3.24 Potential precursors to RCM targets bearing (a) disubstituted, (b) trisubstituted, and (c) tetrasubstituted olefins. Six distinct diene substitution patterns funnel down to each target.

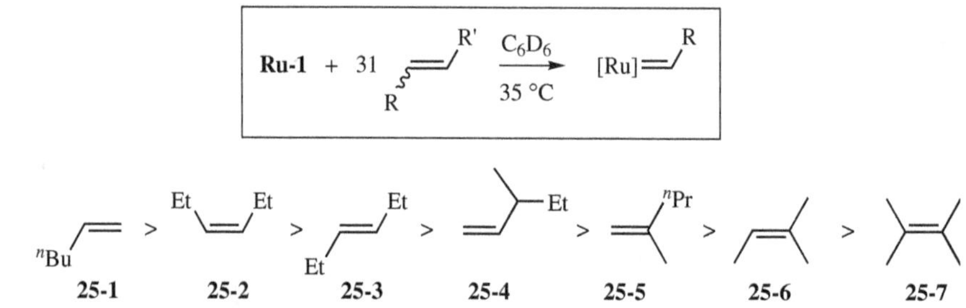

k_{rel}	100	76	30	25	0	(0)	(0)

Figure 3.25 Trend in olefin reactivity inferred from relative rates of initiation of **Ru-1** (112). That for **25-6** and **25-7** is inferred.

Fig. 3.27a) (41,114). Unsurprisingly, reactions using **Ru-1** are considerably slower, even for five- to seven-membered rings. This catalyst fails in attempted synthesis of macrocycles containing a trisubstituted olefin. Whereas such targets are accessible in excellent yields using **Mo-1** (see **27-3**) (115), attempted RCM of **27-4** using **Ru-1** terminated at the head-to-head dimer **27-5** (Fig. 3.27b) (42).

Between the extremes of **Ru-1** and **Mo-1** lie the Ru–NHC catalysts. The higher reactivity of **Ru-2′**, relative to **Ru-1**, toward internal olefins has been discussed above (Section 3.2.2). Five- and six-membered rings containing trisubstituted olefins are efficiently formed using **Ru-2** (see **28-1** to **28-3**, Fig. 3.28) (116–118), while medium or macrocyclic rings are obtained more slowly, as expected (see **28-4**, **28-5**) (42,119). The moderate yield reported for 14-membered **28-5** (65%) should be compared with complete failure of RCM using **Ru-1**, which yielded only dimer **27-5**, as noted above.

Tetrasubstituted olefin targets are considerably more challenging, as their synthesis constrains placement of

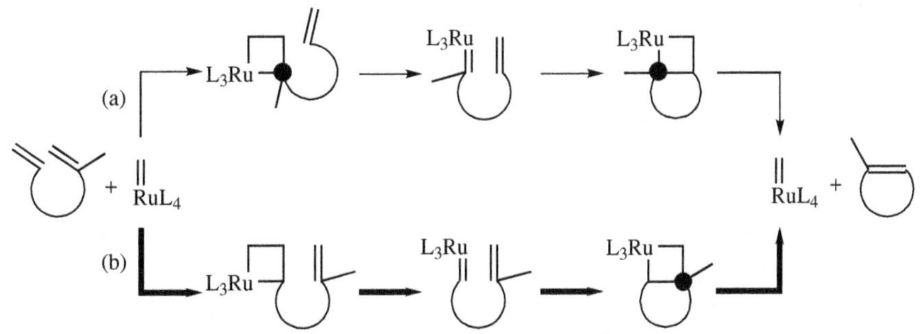

Figure 3.26 General schematic for RCM where one olefin is 2,2-disubstituted. (a) Initiation at the disubstituted olefin. (b) Initiation at the vinylic olefin. Preferred pathway indicated with bold arrows; (•) denotes tetrasubstituted carbon centers.

(a)

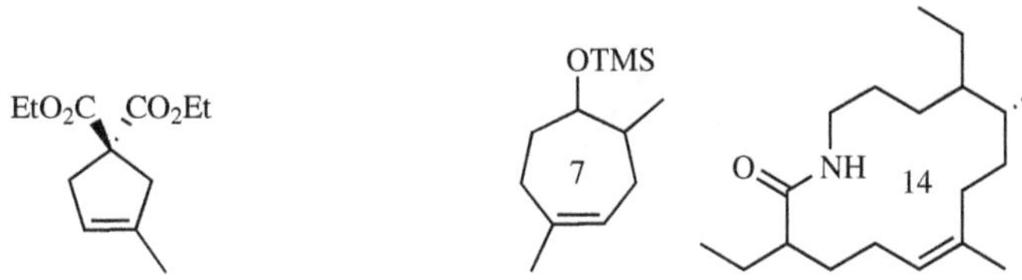

27-1 **27-2** **27-3**

catalyst (mol%)	solvent	conc. (mM)	T (°C)	t (h)	yield (%)
Mo-1 (5)	C_6H_6	100	65	0.2	100
Ru-1 (5)	CH_2Cl_2	10	RT	0.2	20
				24	93

27-2: **Ru-1** (10 mol%), CH_2Cl_2 (5 mM), 40 °C, 24 h (100%)

27-3: **Mo-1** (20 mol%), C_6H_6 (6 mM), 50 °C, 14 h (90%)

(b)

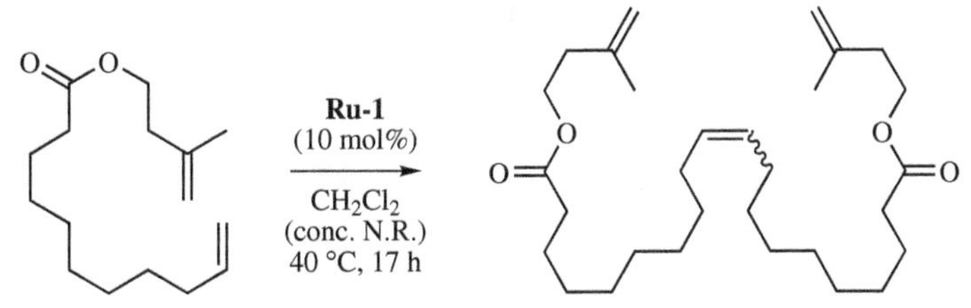

27-4 **27-5** (79%)

Figure 3.27 RCM of terminal and 2,2-disubstituted olefins using **Ru-1** and **Mo-1**. (a) Successful synthesis of trisubstituted olefins. (b) Termination at acyclic dimer, due to insufficient catalyst activity. N.R. = not reported.

quaternary carbons α to the metal center in the metallacyclobutane intermediate. Advances in this area were recently reviewed: the scope is limited to ring sizes of five to seven members (120). **Mo-1** is the sole example of the common metathesis catalysts capable of coupling two 2,2-disubstituted olefins with high conversions. In RCM assembly of **29-1**, for example (Fig. 3.29a), the Grubbs group reported 93% yield for **Mo-1**, no reaction for **Ru-1**, and maximum yields of 50% with **Ru-2** (41,121). The Hoveyda catalyst **Ru-3** was even less productive, albeit under different conditions of solvent and temperature (122). Some of the best results to date using Ru catalysts have been achieved using the H_2ITol catalysts: **Ru-5** enabled up to 86% yield of **29-1** in 1 h (121), while styrenyl ether derivative **Ru-6** gave 100% yield after 4 h in benzene (vs 70% yield after 24 h in CH_2Cl_2) (122). **Ru-6** has also delivered standout results in other cases. The Stoltz group described quantitative yields of spirocyclic **29-2** on the use of this catalyst (Fig. 3.29b): notable is the presence of a challenging chlorinated double bond (123,124). General approaches to halide-functionalized olefins are discussed in the next section, 3.3.3.1.

The success of the H_2ITol catalysts is attributed to the reduced bulk of the NHC ligand, and consequently improved access of the metal to the sterically congested olefin. Naphthyl-functionalized **Ru-9** also performed well under suitable reaction conditions (125). While the Dorta group (126) had earlier reported modest yields of **29-1** using this catalyst in CD_2Cl_2, Blechert and coworkers reported significantly improved performance in hexafluorobenzene. Synthesis of seven-membered **29-3**, for example, proceeded in

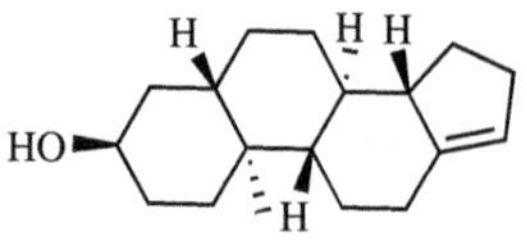

28-1

Ru-2 (5 mol%)
CH_2Cl_2 (45 mM)
40 °C, 1 h
(98%)

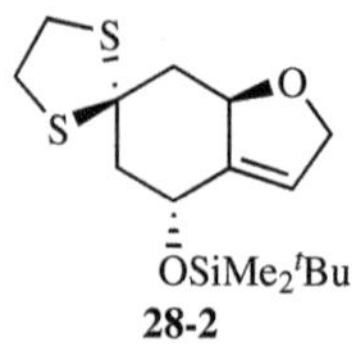

28-2

Ru-2 (9 mol%)
C_7H_8 (4 mM)
80 °C, 40 min
(81%)

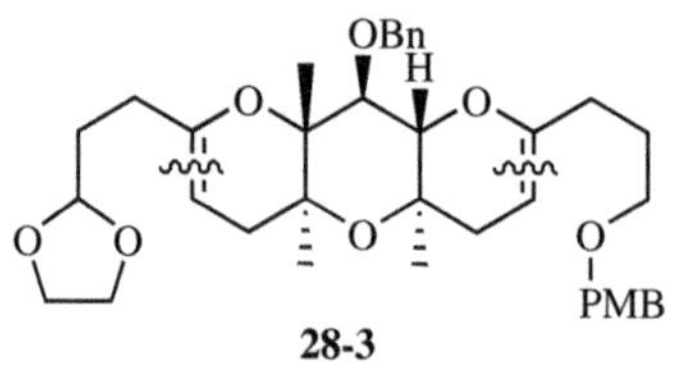

28-3

Ru-2 (10 mol%)
C_7H_8 (conc. N.R.)
70 °C, time N.R.
(89%)

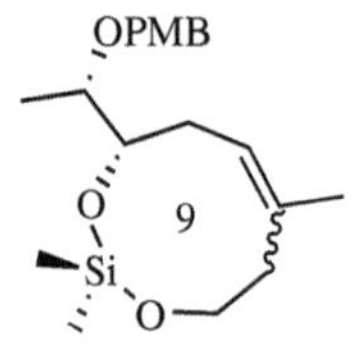

28-4

Ru-2 (15 mol%)*
CH_2Cl_2 (2.6 mM)
40 °C, 16 h
(98%)

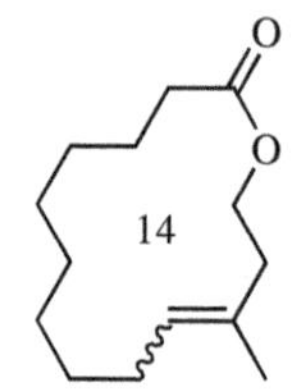

28-5

Ru-2′ (10 mol%)
CH_2Cl_2 (conc. N.R.)
40 °C, 40 h
(65%)

Figure 3.28 RCM synthesis of trisubstituted cyclic olefins using Ru-NHC catalysts. PMB = *p*-methoxybenzyl. N.R. = not reported; the label (*) for **28-4** denotes slow addition of catalyst.

(a)

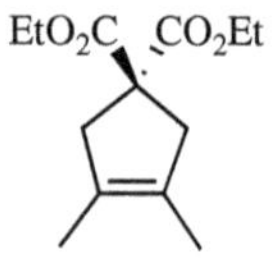

29-1

[M] (5 mol%)
100 mM

catalyst	solvent	T (°C)	t (h)	yield (%)
Mo-1	C_6H_6	65	24	93
Ru-1	CH_2Cl_2	RT	24	0
Ru-2	CH_2Cl_2	40	24	50
Ru-3	C_6H_6	60	24	30
Ru-5	CH_2Cl_2	40	1	86
Ru-6	C_6D_6	60	4	100

(b)

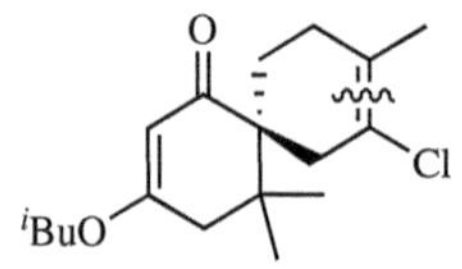

29-2

Ru-6 (5 mol%)
C_6H_6 (107 mM)
60 °C, 18 h
(97%)

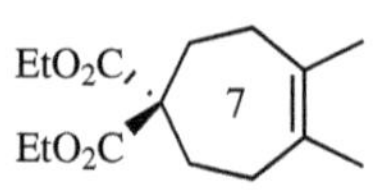

29-3

Ru-9 (3 mol%)
C_6F_6 (100 mM)
80 °C, 12 h
(99%)

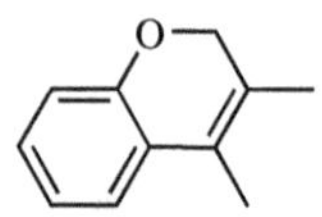

29-4

Ru-10 (0.5 mol%)
C_7H_8 (20 mM)
80 °C, 20 h
(99%)

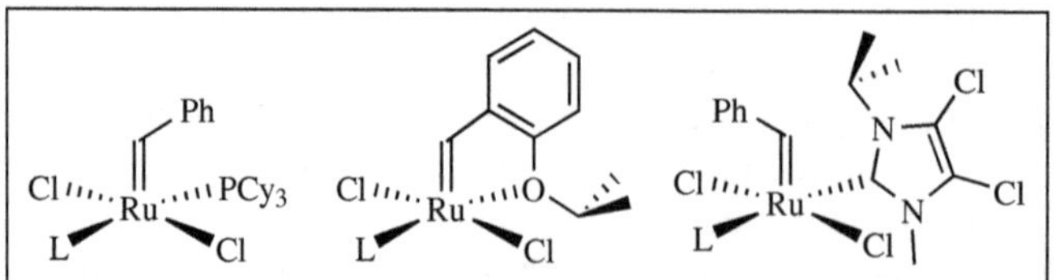

Ru-5: L = H_2ITol
Ru-9: L = H_2INap-Me

Ru-6: L = H_2ITol

Ru-10: L = H_2IMes

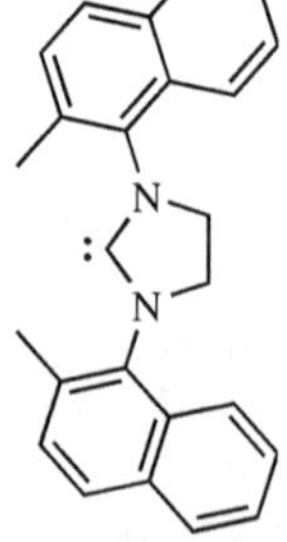

H_2INap-Me

Figure 3.29 Exemplary syntheses of tetrasubstituted cyclic olefins by RCM. (a) Benchmarked against a standard model substrate. (b) In synthesis of six-membered rings. Structures of the more specialized catalysts are reproduced here for convenience.

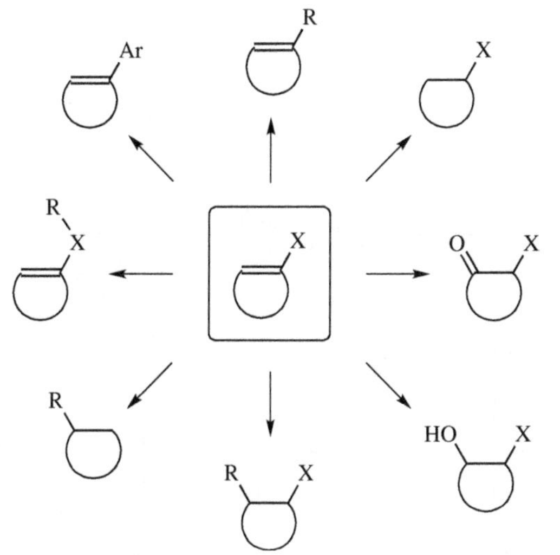

Figure 3.30 Synthetic opportunities arising from post-RCM reactions of directly-functionalized cyclic olefins.

quantitative yields within 12 h at 80 °C (125). The extended π system of the NHC ligand was proposed to stabilize the active catalyst, and to increase initiation efficiency via π-stacking interactions with the solvent.

The origin of the high TONs attainable with the Plenio bis-NHC catalysts (e.g., **Ru-10**) is less clear, if these deliver the same active species as **Ru-2** or **Ru-3**. Nevertheless, their high performance is evident from the quantitative formation of **29-4** at a catalyst loading of 0.5 mol% (127).

3.3.3 Directly-Functionalized Olefins

RCM of directly-functionalized olefins remains a frontier for discovery. While these reactions are frequently challenging, the products have unique synthetic potential (128). Such heteroatom-functionalized cycloolefins represent potentially highly versatile platforms for elaboration via a rich set of post-metathesis reaction manifolds, including regioselective functionalization of the unsymmetrical double bond (Fig. 3.30). Highlighted below is progress toward the coupling of vinylic and 2,2-disubstituted olefins, to yield RCM products in which the olefin substituent X may be exocyclic (as illustrated) or endocyclic.

3.3.3.1 RCM of Electron-Deficient Olefins

Vinyl Halides RCM of vinyl halides is of keen interest, given the potential for further elaboration of the products via Pd-catalyzed cross-coupling. To date, vinyl iodides are spectator substituents (even toward highly reactive catalysts such as **Mo-1**) (129,130), but advances have been made with other haloalkenes in Ru-catalyzed RCM. These substrates present multiple challenges: they are electronically and sterically deactivated toward metathesis, and they can promote catalyst decomposition. Johnson and coworkers (131) reported that **Ru-2** was decomposed into metathesis-inactive carbide **Ru-16** and phosphiniomethylidene **Ru-17** by vinyl chloride even at ambient temperature (Fig. 3.31). Reaction with vinyl bromide was broadly similar (131); **Ru-3** was also decomposed (132). In a computational study, Fomine, Tlenkopatchev, and coworkers (133) suggested that the energy barrier to metathesis is raised by the steric impact of the halogen substituent, and by the relatively low energy of the chlorinated alkylidene.

The high catalyst loadings often required for RCM of haloalkenes (see Fig. 3.34 below) have been widely presumed to originate in rapid decomposition via pathways such as those delineated by Johnson. It is unclear, however, whether haloalkylidenes are in fact involved in RCM. The electron-deficiency of the halide-functionalized olefin, compounded by its 2,2-disubstitution, is expected to be a powerful driver promoting initial metathesis at the unperturbed olefin. An important recent study by the Dorta group (134) showed that **Ru-2** was inert to 2-bromoheptene (Fig. 3.32a), but rapidly deactivated by 2-bromohepta-1,6-diene (Fig. 3.32b). This implies that the deleterious effect of vinyl bromides is mediated by the bromine substituent on the *pendant* olefin. Decomposition was impeded by installing a phenyl group cis to bromine to block the incipient Ru–Br interaction (Fig. 3.32c). The choice of a 1-phenyl substituent results in the release of **Ru-2** on cyclization.

This strategy afforded unprecedentedly efficient access to carbocyclic and heterocyclic bromoalkenes of five to

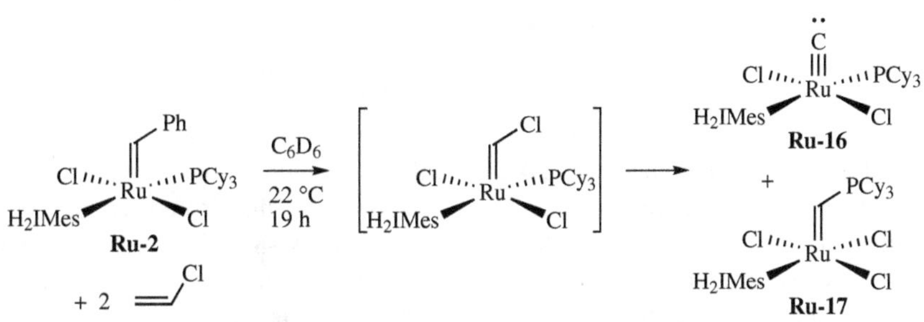

Figure 3.31 CM reaction of **Ru-2** with vinyl chloride (131).

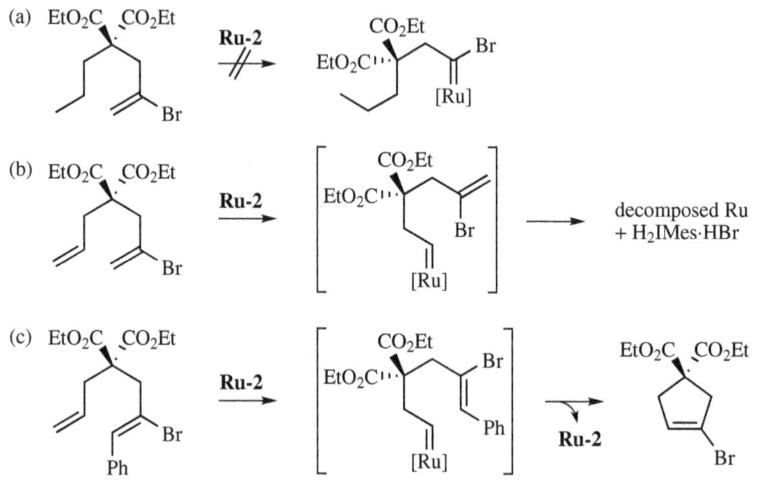

Figure 3.32 Influence of alkene substitution on metathesis of vinyl bromides by **Ru-2**. (a) Failed CM of 2-bromoheptene. (b) Decomposition by 2-bromohepta-1,6-diene. (c) Successful RCM. Note: RCM fails if the olefinic phenyl and bromide substituents in (c) are mutually trans (134).

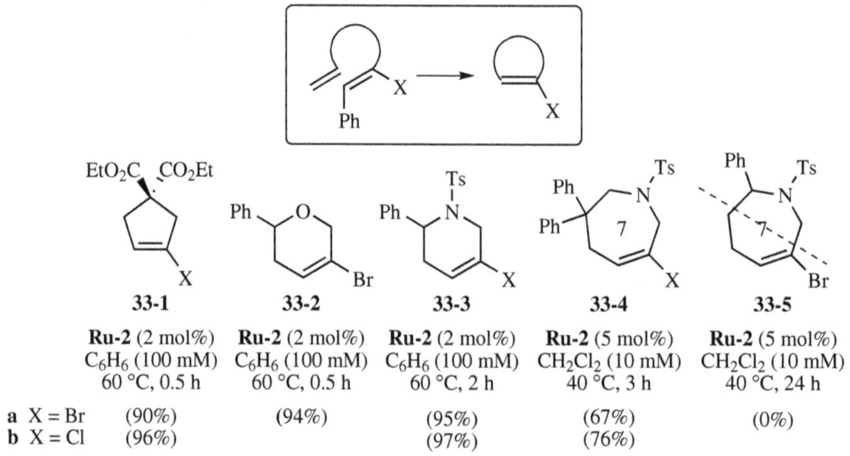

Figure 3.33 Strategy for synthesis of halogenated cyclic olefins (134).

seven members (see **33-1a** to **33-4a**, Fig. 3.33). Formation of seven-membered **33-4** required Thorpe–Ingold assistance from two phenyl groups at the allylic position, as well as dilutions of 10 mM (vs 100 mM for the smaller ring sizes: see discussion in Section 3.2.1 above). Where a substituent was present only on the homoallylic carbon (**33-5**), RCM failed even at 10 mM. Cyclic chloroalkenes were likewise accessible by this approach (see **33-1b**, **33-3b**, **33-4b**) (134).

Prior work by Weinreb, Rutjes, and their respective coworkers (124,135,136) demonstrated that RCM of vinyl chlorides and vinyl fluorides can work very well without such assistance, for rings of up to seven members (see **34-1** to **34-4**, Fig. 3.34). Catalyst loadings of 5 mol% were typically required for complete diene consumption. Also notable in this context is the Stoltz synthesis of a tetrasubstituted chloroalkene described in the previous section (see **29-2**, Fig. 3.29) (123,124). Larger rings are more challenging. Eight-membered azacycle **34-5** was formed in modest yields (ca. 40%), while 12-membered **34-6** proved inaccessible, presumably reflecting the relative kinetics of RCM versus decomposition (136). A complex mixture of unidentified products was reported, despite use of potentially appropriate dilutions (10 mM; recall the higher concentrations tolerated by medium-ring azacycles vs carbocycles discussed in Section 3.2.1).

More unexpected is the failed synthesis of five-membered **34-7** (136), which may indicate that the deactivation pathway of Figure 3.32b is relevant even for fluorinated olefins, where the run length favors approach of fluorine to the metal center. Finally, fluorinated cyclic hydrazines have also been prepared (**34-8**) (137). Complete conversions required high catalyst loadings (20 mol%) with serial addition: depending on the protecting groups, competing isomerization was observed.

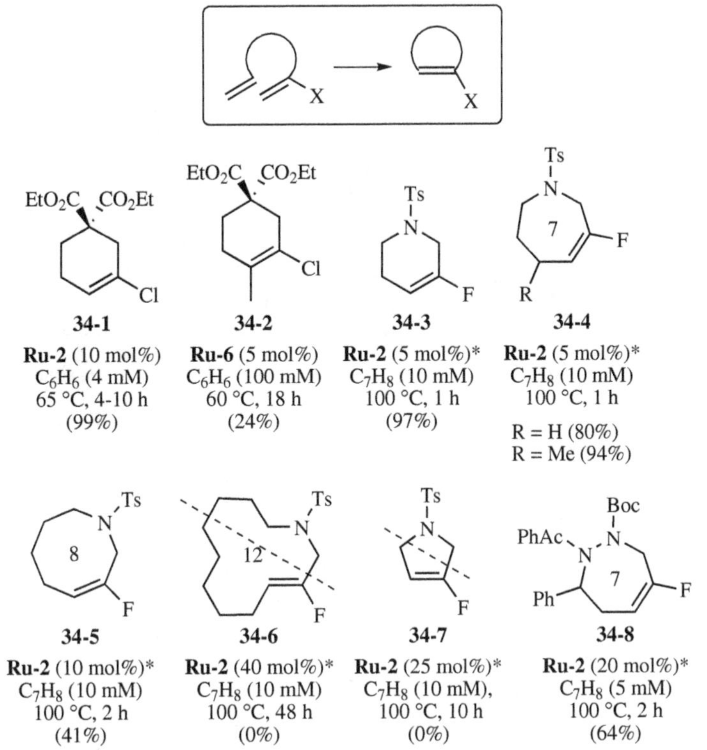

Figure 3.34 Representative examples of the RCM of 2-chloro- and 2-fluoroolefins. The label (*) denotes slow addition of catalyst.

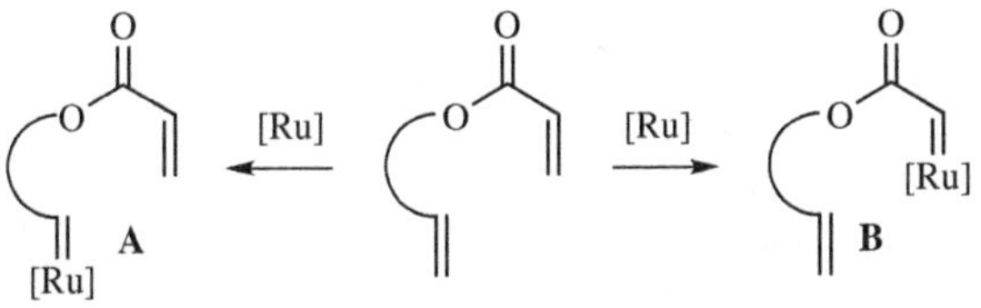

Figure 3.35 Potential ruthenium alkylidene (**A**) and enoic alkylidene (**B**) intermediates in RCM of alkenyl acrylates.

α,β-Unsaturated Carbonyl Compounds Acrylate, acrylamide, and enone derivatives appear to be incompatible with the Mo catalysts, based on early CM studies by the Crowe and Schrock groups (138,139), in which such substrates reportedly inhibited catalysis. Similarly, an RCM study by Fu and Grubbs (140) reported that an *N*-allylacrylamide derivative resisted RCM with **Mo-1**. Notable in this context is a report from the Schrock group describing carbonyl-chelated metallacyclobutanes on reaction of W and Mo analogues of **Mo-1** with acrylamide (141). The discussion below therefore focuses primarily on the ruthenium catalysts, beginning with acrylates, which afford the broadest set of examples.

Acrylate metathesis is hampered by the instability of ruthenium enoic alkylidenes (e.g., **B**, Fig. 3.35), which are reported to decompose much more rapidly than conventional alkylidenes of type **A** (142,143). In an early CM study, Chatterjee and Grubbs (142) suggested that these electron-deficient alkylidenes function as vectors for decomposition, not metathesis. Improved CM performance for **Ru-2**, versus **Ru-1**, was subsequently attributed to stabilization by strong donation by the NHC ligand (144). Subsequent reports demonstrated that the Hoveyda catalyst **Ru-3** exhibits considerably higher CM productivity than **Ru-2** (145–148), perhaps reflecting sustained retention of a reservoir of viable catalyst. Undoubtedly also relevant is the beneficial impact of higher substrate concentration on metathesis rates for **Ru-3**; see Section 3.2.2.2.

In RCM synthesis of targets bearing an exocyclic acrylate, the electron-deficiency of the acrylate olefin is exacerbated by 2,2-disubstitution. The impediments to reaction at the acrylate site are thus both steric and electronic, as seen with the haloolefins above. In a comparative study, Rutjes and coworkers (84) reported that **Mo-1** and **Ru-1** exhibited low reactivity toward the acrylate olefin in **36-1** (Fig. 3.36). Rather than the desired **36-2**, the dominant product was linear dimer **36-3**, indicating preferential intermolecular coupling at the unperturbed olefin. On raising the temperature of the **Mo-1**-catalyzed reaction from 50 to 85 °C, a small proportion of RCM was observed (14%, including ring-contracted **36-4**), but diene conversions dropped to about 60%, indicating catalyst decomposition. The greater robustness of **Ru-2** improved yields of the RCM target **36-2** to 51%. Conversions were again incomplete, however, and isomerization resulted in nearly 30% **36-4**. Again, these observations point toward catalyst decomposition.

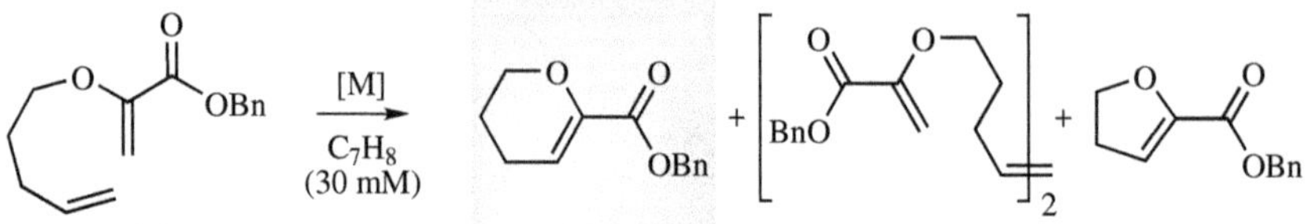

36-1 36-2 36-3 36-4

catalyst (mol%)	T (°C)	t (h)	**36-2** (%)	**36-3** (%)	**36-4** (%)
Mo-1 (15)	50	3	0	59	0
Mo-1 (15)	85	2	12	47	2
Ru-1 (20)	70	40	0	56	0
Ru-2 (10)	70	2	51	0	28

Figure 3.36 Comparison of catalyst efficiency in acrylate RCM (84). Note: shading indicates the intended RCM product.

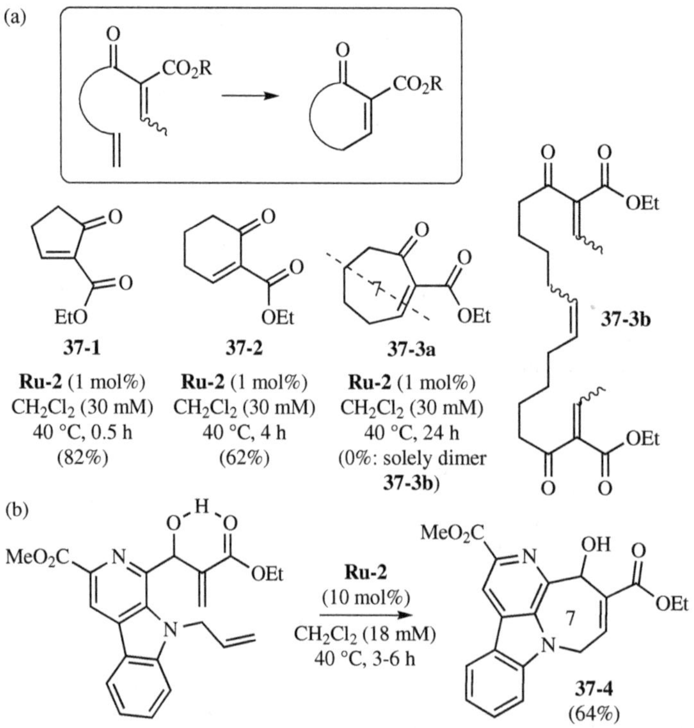

Figure 3.37 (a) RCM synthesis of five- to seven-membered rings bearing an exocyclic acrylate functionality. (b) RCM synthesis of a seven-membered example, with conformational constraints and potential H-bonding.

Yields of about 80% and 60%, respectively, were reported by Toueg and Prunet for five- and six-membered targets (see **37-1**, **37-2**, Fig. 3.37a) (149). A possible contributor to success was their use of a trisubstituted acrylate olefin, which would greatly reinforce the bias toward initiation at the distal olefin. The corresponding seven-membered ring **37-3a** was not formed, this reaction yielding linear dimer **37-3b** instead. Notable here are the 30 mM concentrations employed: cf. the dilutions of 0.5 mM required for synthesis of caprolactone **4-1** (see Fig. 3.4). In contrast, seven-membered **37-4** (Fig. 3.37b) was obtained in about 65% yield by Hutait and Batra (150). Here RCM was aided by the conformational constraint of the fused pyridoindole unit (and almost certainly amplified by hydrogen bonding between the ester and hydroxyl groups), as well as the nearly twofold higher dilution.

Selected endocyclic targets are depicted in Figure 3.38. Conjugated δ-lactone **38-1** was accessible in moderate yields (65%) on use of **Ru-1** (151). Compound **38-2** presents a greater challenge because it requires coupling of an electron-deficient acrylate with a 2,2-disubstituted olefin. Poor results were found using **Ru-1**, with less than 25% RCM even on dropwise addition of the catalyst, and about 75% unreacted diene (151). The resistance of **Ru-1** to

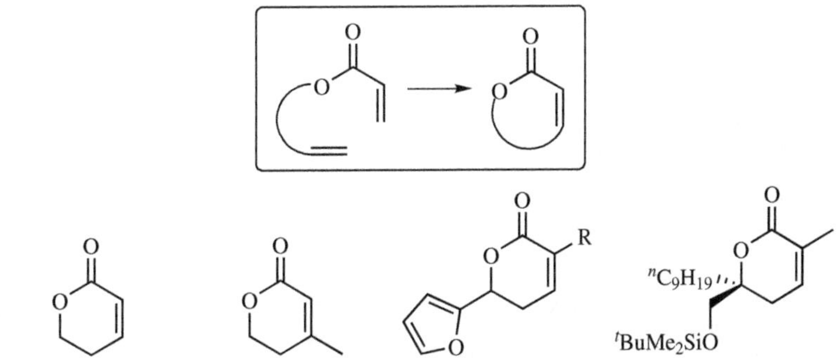

38-1
Ru-1 (10 mol%)*
CH_2Cl_2 (10 mM)
40 °C, 6 h
(65%)

38-2
Ru-1 (10 mol%)*
CH_2Cl_2 (10 mM)
40 °C, 18 h (20%)
Ru-2 (5 mol%)*
C_7H_8 (100 mM)
80 °C, 20 h (73%)

38-3
Ru-2 (5 mol%)
CH_2Cl_2 (125 mM)
40 °C
R = H: 80 min (55%)
R = Me: 4 h (23%)

38-4
[M] (5 mol%)
C_6H_6 (5 mM)
70 °C
Ru-2: 15 h (34%)
Ru-3: 7 h (88%)

Figure 3.38 Effect of substitution on access to pentenolides via RCM of acrylates. The label (*) denotes slow addition of catalyst.

reaction with 2,2-substituted olefins (Section 3.3.2) directs the metal to the acrylate site, hence promoting formation of the unstable enoic alkylidene (144,151). The penalty is clear from the catastrophic drop in RCM productivity, versus **38-1**.

Use of **Ru-2** at higher temperatures in toluene increased RCM yields to 73% (66). This catalyst is better able to cope with 2,2-substituted olefins (a function of its higher activity relative to **Ru-1**), and is less rapidly decomposed by the acrylate functionality, as noted above. The reaction remains challenging, however, as indicated by the rather poor yields reported for **38-3** (where R = H) (152) and **38-4** (153) (55% and 34%, respectively, using **Ru-2**). Honda and coworkers (153) found that **Ru-3** offered a significant performance advantage, delivering **38-4** in nearly 90% yield.

Synthesis of the corresponding seven-membered γ-lactones is hampered by competing dimerization (Fig. 3.39a). Independent reports by the Fürstner and Grubbs groups (66,154) reported that reaction of **39-1a** with **Ru-2′** or **Ru-2**, respectively, yielded none of the desired lactone **39-2a**, instead affording solely head-to-tail cyclic dimer **39-3** even at concentrations below 5 mM. Use of **Ru-1** gave only linear dimer **39-4** (154). (An earlier erroneous report of the successful synthesis of **39-2a** at 20 mM concentrations attests to the deceptive similarity between the NMR spectra for the cyclic monomers and dimers: this issue is discussed in Section 3.4.5.) Less surprising is the inaccessibility of the corresponding eight-membered ring **39-2b**, attempted preparation of which yielded only cyclic dimers and trimers (see **39-5**, **39-6**) at diene concentrations down to 2 mM (154). Use of a fused aromatic ring to constrain the mobility of the diene **39-7** aids in cyclization, but Dudley and coworkers (155) reported that seven-membered **39-8** (Fig. 3.39b) was formed in only 55% yield even with this assistance.

We turn last to synthesis of butenolides, as these present several puzzling features. One expected complication is carbonyl chelation. Initiation at the unfunctionalized olefin is favored for alkenyl acrylates, as discussed at the beginning of this Section, and Ru intermediates of type **A** (Fig. 3.40) have been proposed (142) to be the key vectors for RCM. For butenolides, cycloaddition of the pendant acrylate olefin in **A** is impeded by the capacity of the carbonyl group to compete for binding to the metal (see **B**, Fig. 3.40). The stability of the six-membered oxygen chelate, and the donor strength of the oxygen atom relative to the electron-deficient acrylate, combine to hamper RCM. Lewis acids such as $Ti(O^iPr)_4$ have been used in RCM to sequester the potentially chelating carbonyl group by binding to the oxygen site (38). This strategy is especially common in butenolide synthesis using **Ru-1**.

Intermolecular interactions may also be important. D'Annibale and coworkers (156) demonstrated that reasonable butenolide yields (ca. 70%) could be achieved without additives, at dilutions of 10 mM (see **40-1**, Fig. 3.40a). This should be contrasted with the concentrations of 100 mM and higher tolerated in the assembly of other five-membered carbocycles, as discussed in prior sections (in particular, Section 3.2.1), and as also reported in the synthesis of pentenolide **38-2** above (66). Schmidt and coworkers (157) made the same observation with **Ru-2**, and suggested that the chelate complex itself reacts with a further equivalent of substrate, to give a metathesis-inactive product. While the mechanistic details are not yet clear, several important points emerged from the Schmidt study. First, catalyst decomposition was rapid at acrylate concentrations of 50 mM or higher (for the butenolides studied, 20 mM proved sufficiently dilute for successful RCM; see **40-2**, **40-3**; Fig. 3.40b). Second, increasing catalyst loadings merely

(a)

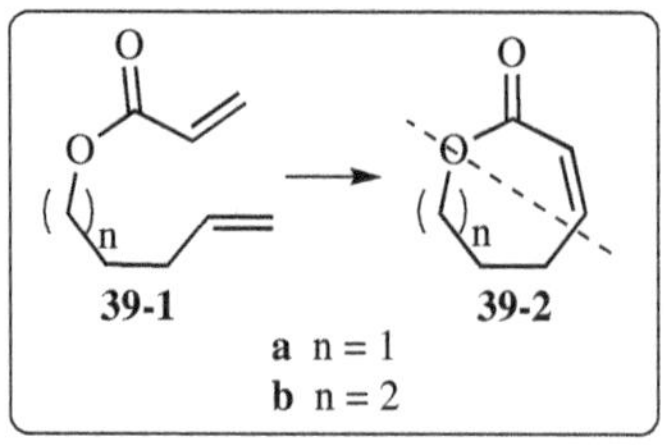

n = 1

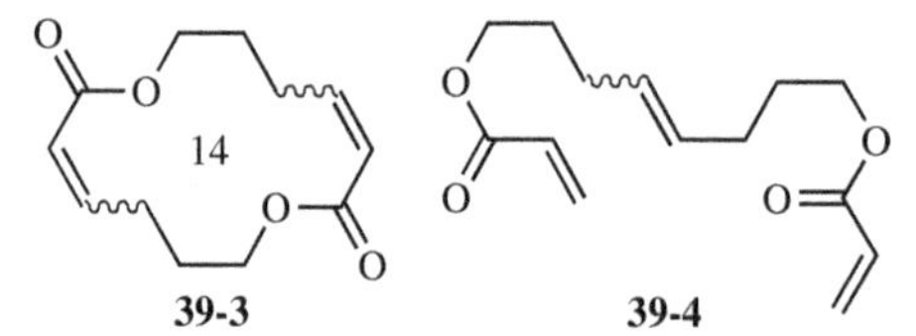

catalyst (mol%)	conditions	**39-3** (%)	**39-4** (%)
Ru-2' (5)	C_7H_8, 4 mM 80 °C, 1 h	62	0
Ru-2 (5)	CH_2Cl_2, 3 mM 40 °C, time N.R.	59	0
Ru-1 (5)	CH_2Cl_2, 30 mM 40 °C, time N.R.	0	62

n = 2

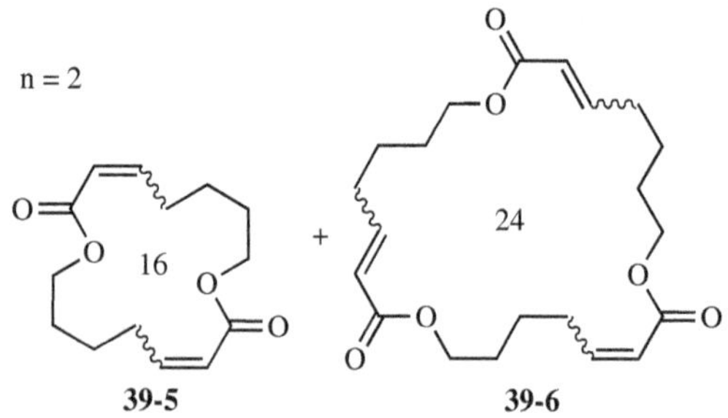

catalyst (mol%)	conditions	**39-5** (%)	**39-6** (%)
Ru-2 (8)	CH_2Cl_2, 2 mM 40 °C, 11 h	38	38

(b)

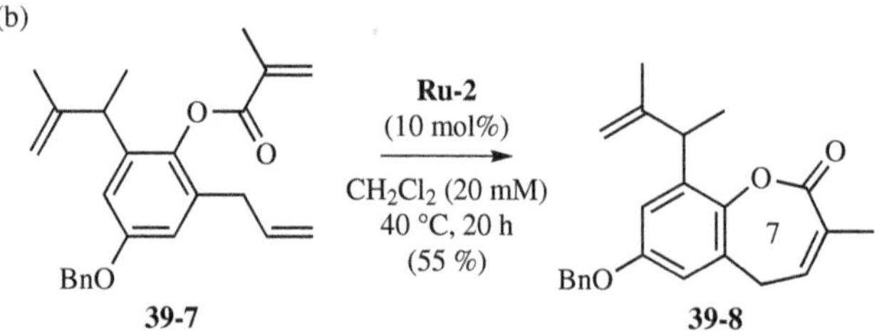

Figure 3.39 (a) Attempted synthesis of seven- and eight-membered conjugated lactones by RCM. (b) Successful RCM of a seven-membered example, aided by conformational constraints. N.R. = not reported.

increased rates of decomposition (a conclusion also drawn by D'Annibale) (156). Third, intriguingly, the acrylate substrate appeared to inhibit isomerization in experiments in which diallyl ether was also present, suggesting either quenching or inhibition of the putative Ru–hydride isomerization catalyst. Ultimately, butenolides were attainable in yields of up to 86% using 2.5 mol% **Ru-2**.

A counterpoint to these data is Feringa's finding (158) that compound **40-4**, a precursor to the whiskey lactone, was attainable in higher yields with much greater efficiency on increasing diene concentrations to 200 mM from 5 mM (Fig. 3.40c). Noteworthy here is the use of a 1,2-disubstituted cinnamyl precursor for the RCM reaction: the internal olefin may suppress bimolecular decomposition. The excellent performance of the Hoveyda catalyst **Ru-3** should also be noted.

Acrylamides Generally similar trends emerge in the synthesis of lactams by RCM of acrylamides. The area was recently reviewed (159). Again, Ru–NHC catalysts typically deliver the most reliable performance, although impressive results have been reported for **Ru-1** in some instances (see **41-1**, Fig. 3.41) (160). Sturino and coworkers (161) reported that secondary amides are tolerated (see **41-2**), although 10% higher yields were obtained for tertiary **41-3**.

Best results are found where initiation at the acrylamide olefin is suppressed. The Donohoe group (82) reported that substitution of the non-acrylamide olefin (which increases the likelihood of forming the enoic alkylidene) disrupted RCM almost completely for **41-4a**: in contrast, unsubstituted **41-4b** could be obtained in quantitative yields. Similarly, steric factors may limit initiation at the unperturbed olefin for tricyclic **41-5** and pivalate-protected **41-6**, accounting for the high catalyst loadings (20 mol%) reported by Arya and coworkers for the former (162), and the modest yields (54%) reported by Pritchard and Wilden (163) for the latter. A further likely complication in the synthesis of **41-6** is chelation of the pivalate carbonyl: this will further inhibit

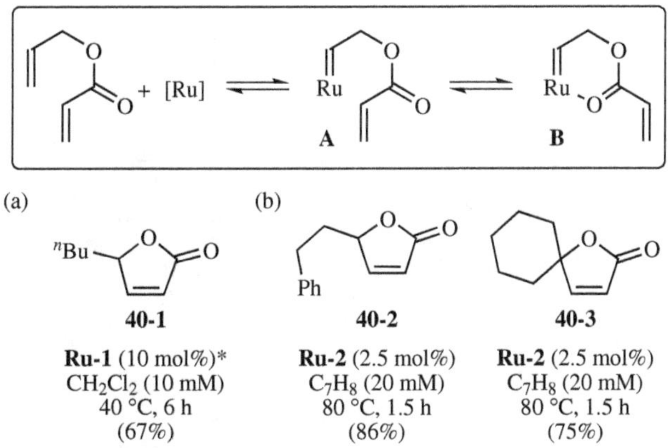

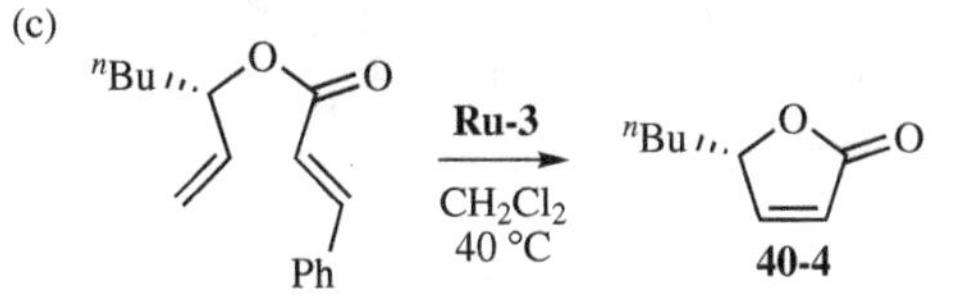

Ru-3 (mol%)	conc. (mM)	t (d)	yield (%)
6	5	7	74
3	200	1	83

Figure 3.40 RCM of representative five-membered conjugated lactones (butenolides), showing conditions required for (a) **Ru-1**; (b) **Ru-2**; (c) **Ru-3**. The label (*) denotes slow catalyst addition.

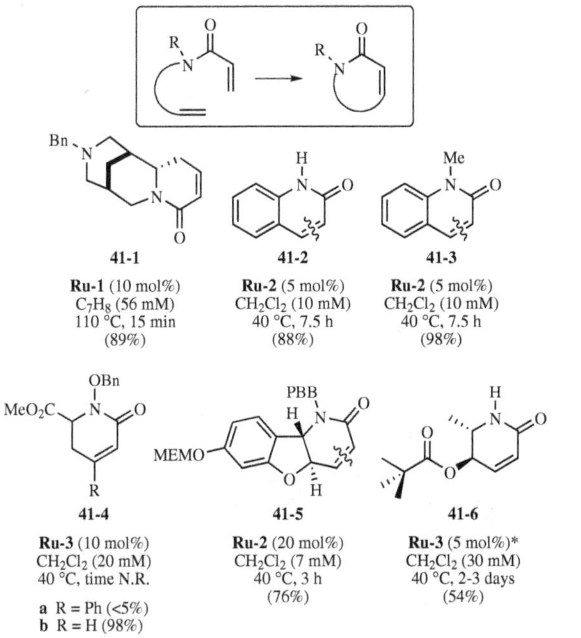

Figure 3.41 Examples of the synthesis of six-membered lactams by acrylamide RCM. N.R. = not reported; (*) denotes slow catalyst addition; PBB = *p*-bromobenzyl, MEM = methoxyethoxymethyl.

productivity, following initiation at the "desired" olefinic site, as discussed in Section 3.3.4.2.

More challenging examples of acrylamide RCM are shown in Figure 3.42a. RCM of five-membered conjugated lactams has seen less discussion than that of butenolides, but similar issues of deactivation could contribute to the null conversions reported by Chen et al. (164) for secondary amide **42-1a**. More clearly relevant is the inhibiting effect of carbonyl chelation, following initiation at the vinylic olefin (which can again result in a six-membered chelate ring). Performance was dramatically improved by the use of a bulky amide substituent to inhibit chelation: 76% yield was reported for **42-1b** within just 2 h at RT.

Seven-membered and medium-sized rings can be efficiently obtained given suitable conformational constraints, as illustrated by the high yield syntheses of seven-membered **42-2** (165) and eight-membered **42-4** (162). Dihydroazepinone **42-3** was obtained in modest yields (53%) despite such constraints, however (166), perhaps owing to chelation by the nearer ether oxygen.

RCM of unsaturated secondary amides may be affected by intramolecular hydrogen bonding. Such interactions probably contribute to success in the synthesis of **42-5**, reported by the Georg group (167). They can also play a deleterious role, however. In RCM of **42-6a**, H-bonding of the secondary amide (see dashed line, Fig. 3.42b) is expected to stabilize a conformation that will promote carbonyl chelation if initiation occurs at the acrylamide olefin (see inset). Consistent with this, Nan and coworkers (164) reported low yields of caprolactam **42-7a** (35%) even after 36 h in refluxing CH_2Cl_2. Yields of **42-7** were raised to 81% by installing an α-methyl group at the acrylamide olefin (see **42-6b**) to direct initiation to the distal olefin. This strategy failed for the corresponding eight-membered ring **42-7c**, for which the limited conformational constraints present appear insufficient to overcome ring strain.

Vinyl Ketones RCM of vinyl ketones follows the same trends. Again, cyclization of five-membered rings is hampered by carbonyl chelation (Fig. 3.43; in the case of **43-1**, the bulk of the quaternary allylic carbon also contributes) (168). Six- and seven-membered cyclic enones

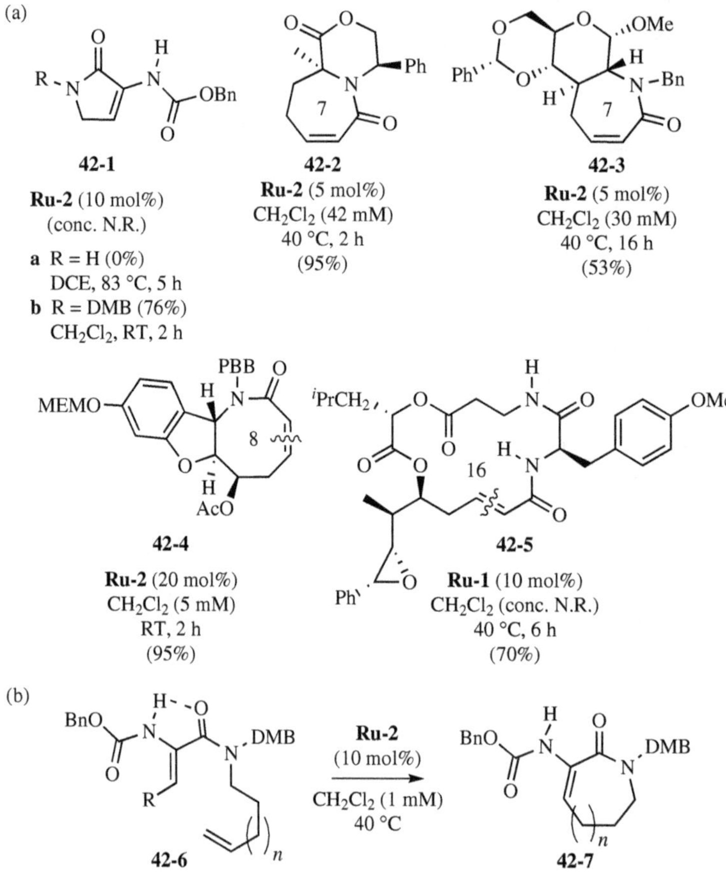

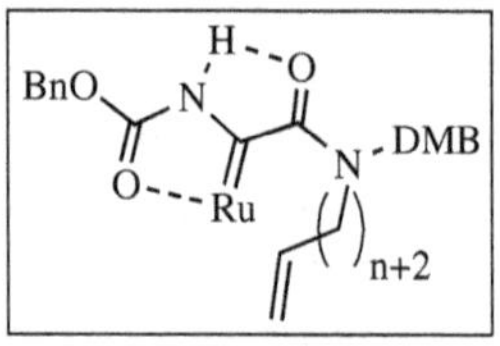

	n	R	T (°C)	t (h)	Yield (%)
a	1	H	RT	36	35
b	1	Me	40	20	81
c	2	Me	RT	36	0

Figure 3.42 (a) More challenging lactams obtained by acrylamide RCM. (b) Potential for inhibition by carbonyl chelation. Inset depicts putative chelate formed from **42-6**, following initiation at the 2,2-disubstituted olefin. Note, however, that reaction at the α-olefin site will be kinetically preferred. N.R. = not reported; DMB = 2,4-dimethoxybenzyl, PBB = p-bromobenzyl, MEM = methoxyethoxymethyl.

are most readily accessible (see **43-2**, **43-3**) (169,170), and comparative studies demonstrate higher yields with **Ru-2**, versus **Ru-1**. In the synthesis of cyclooctenone **43-4**, the fused furan ring provides an effective conformational constraint, and Paquette and coworkers (171) reported 81% yield using **Ru-2**. (With **Ru-1**, in comparison, yields reached only 35%, even at catalyst loadings of 35 mol% and a reaction time of 3.5 days) (172). The trans-fused ring in polycyclic ether **43-5**, in contrast, was obtained by Clark and coworkers (173) in 50% yield; whether the balance of material was dimeric was not indicated. Somewhat unexpected is the rapidity with which macrocyclic **43-6** was formed (30 min) (174). Couladouros and coworkers noted that significant amounts of the cyclic dimer (29%) were observed if the benzyl protecting groups were omitted, and yields of the 12-membered target dropped to 40%.

Vinyl Aldehydes A rare example of RCM of a vinylic aldehyde is shown in Figure 3.44. Chavez and Jacobsen (175) reported 80% yield in the synthesis of cyclopentene **44-2** using **Ru-2**, despite the anticipated challenge associated with coupling a trisubstituted and a 2,2′-disubstituted, carbonyl-conjugated olefin. The group 6 catalysts are incompatible with aldehydic substrates, from which they rapidly abstract oxygen via a Wittig-like reaction (176). The utility of this reaction in quenching catalysis via **Mo-1** is discussed in Section 3.4.6.

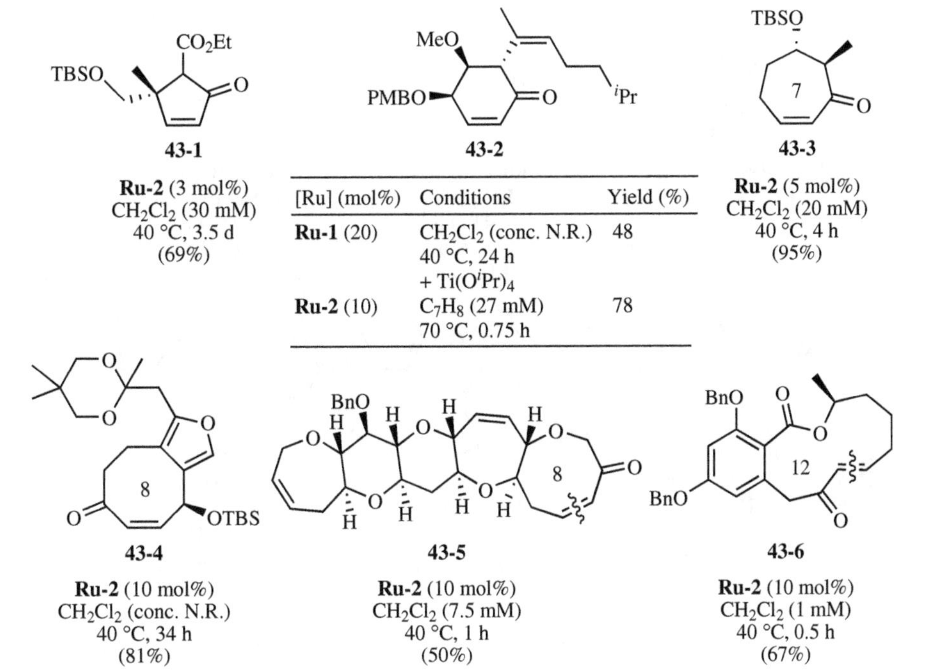

[Ru] (mol%)	Conditions	Yield (%)
Ru-1 (20)	CH_2Cl_2 (conc. N.R.) 40 °C, 24 h + $Ti(O^iPr)_4$	48
Ru-2 (10)	C_7H_8 (27 mM) 70 °C, 0.75 h	78

Figure 3.43 Representative examples of the RCM of vinyl ketones. PMB = *p*-methoxybenzyl, TBS = *tert*-butyldimethylsilyl. N.R. = not reported.

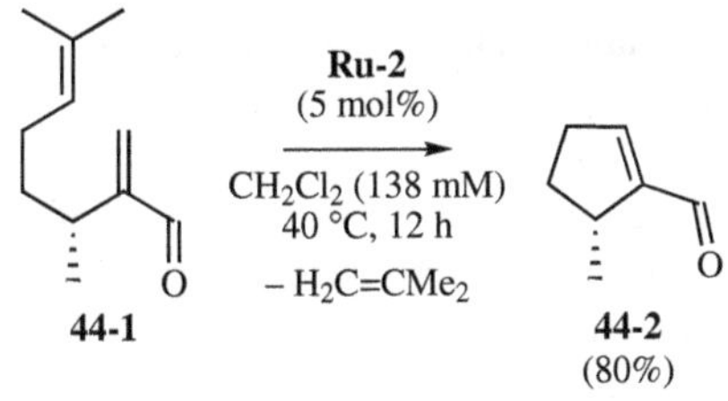

Figure 3.44 RCM of a vinylic aldehyde (175).

Conjugated Dienes Considered in this section are, first, olefins conjugated to an aromatic ring, and second, 1,3-dienes. An olefin conjugated to an aromatic ring is slightly deactivated toward metathesis (177). RCM has been reported for five- and six-membered rings using **Ru-1** even at RT, however (see, e.g., **45-1**, **45-2**; Fig. 3.45) (178,179). For more challenging targets, reaction rates often benefit from the higher activity of **Ru-2**. Example **45-3** illustrates a successful cyclization despite the potential for oxygen chelation (180), while **45-4** demonstrates the high yields attainable despite coupling of two electron-deficient olefins (181). In a recent comparative study, Grisi and coworkers (182) reported 99% yields of 14-membered macrocycle **45-5** within 1 h using **Ru-2** (cf. <30% after 24 h using the less electron-rich catalyst **Ru-1**). Synthesis of the corresponding lactam **45-6** showed a yet more dramatic difference, with just 3% yield for **Ru-1**, versus 99% for **Ru-2**. (182) A contributing factor is presumably chelation of the unprotected secondary amide, which renders cyclization slow even for **Ru-2**. Other instances of inhibition by Lewis-base chelation are discussed in Section 3.3.4.2: this example stands out because the chelate ring size suggests binding of the amide nitrogen (rather than the carbonyl), notwithstanding the rather weak basicity of *N*-arylamides.

In RCM of conjugated 1,3-dienes, the challenge lies in discriminating between the terminal and the internal olefinic sites. Here the higher reactivity of the Ru–NHC catalysts can be problematic, leading to competing formation of ring-contracted products. An example of this behavior is seen in Menche's attempted RCM of **46-1**, which yields 20-membered **46-2** as well as the desired 22-membered **46-3** (Fig. 3.46a) (183). Improved selectivity was attained with **Ru-1**, the lower reactivity of which promotes reaction at the terminal olefin. Very high catalyst loadings were required, however (50 mol%).

Alternatively, ring contraction can be limited by introducing a substituent on the internal olefin. This maximizes the steric distinction between the internal and terminal olefins, effectively directing metathesis to the more accessible terminal site. RCM of **46-4** (Fig. 3.46b) illustrates this point: Fürstner and coworkers (184) reported 92% yield of silyl-protected **46-5a**, versus 82% for unprotected analog **46-5b**. The silyl substituent also promotes trans-stereoselective cyclization. The more complex target **46-6** was obtained in 76% yield under more forcing conditions (refluxing toluene) (184). These experiments employed the Grela and Dorta catalysts, respectively (**Ru-7** and **Ru-8a**), in conjunction with tricyclohexylphosphine oxide or tetrafluoro-*p*-quinone to inhibit isomerization.

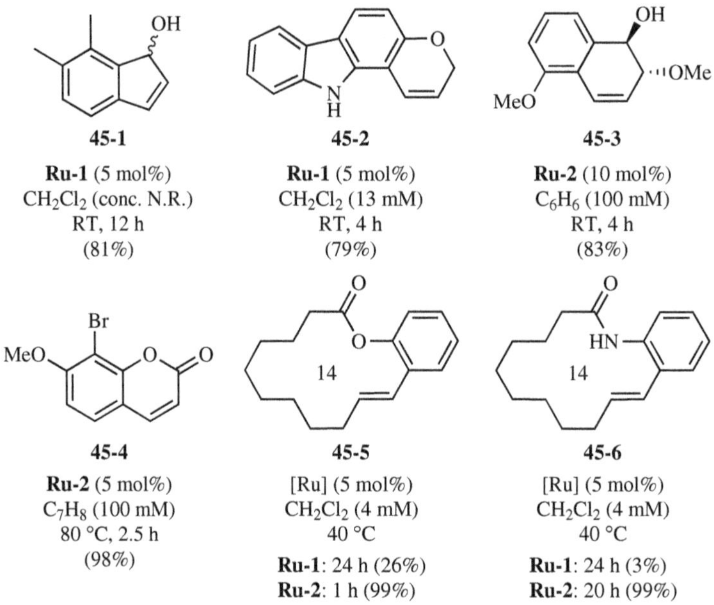

Figure 3.45 RCM of dienes in which one olefin is conjugated with an aromatic group. N.R. = not reported.

Vinyl Boronates RCM products in which the double bond bears a boronate group are of keen interest for elaboration via palladium-catalyzed coupling reactions. Pioneering work by Renaud and Ouellet (185) demonstrated that five- to seven-membered trisubstituted cyclic vinylboronates could be produced in high yield using **Ru-1**. Surprisingly little work has been undertaken in this area since the handful of examples discussed by Eustache (128) and Brown (186) in their 2006 reviews. To date, examples of post-metathesis modification of vinylboronates by cross-coupling are limited to enyne metathesis or CM reactions.

3.3.3.2 Electron-Rich Olefins Metathesis of heteroatom-functionalized vinyl olefins yields Fischer carbene complexes. While this does not appear to impede the activity of the Mo derivatives (see below in this section), [Ru]=CHER species suffer from diminished stability, as well as reduced metathesis reactivity. Louie and Grubbs (90) reported that both stability and activity decrease in the order N > S > O, for Fischer carbenes derived from **Ru-1** and **Ru-2**. Enol ethers thus yield the least active and most unstable Ru intermediates. This is offset, however, by the potential synthetic significance of their RCM reactions, and they are therefore examined in greatest detail.

Enol Ethers The thermal sensitivity of ruthenium Fischer carbenes has been established in both catalytic and stoichiometric studies (90,91). In a study of enol ether metathesis using **Ru-2**, Arisawa, Nishida, and coworkers reported rapid, quantitative transformation of the ethoxy or silyoxy derivatives **Ru-18** into the ruthenium hydride **Ru-14b** at elevated temperatures (see Fig. 3.47) (91). This results not only in loss of the catalyst charge, but also formation of a known olefin isomerization catalyst. (For the deleterious effect of isomerization on regioselectivity and control over ring size, see Section 3.3.1.1.) This transformation probably accounts for the fact that Ru-catalyzed RCM of vinyl ethers typically works well only for five- to seven-membered rings, or when initiation is directed to the distal olefinic site (128). For larger rings, RCM yields are often low, even at high catalyst loadings.

Where 2,2-disubstitution of the vinyl ether constrains initiation at the ω-olefin (that is, the Fischer carbene is not formed), **Ru-2** has been used with good success to prepare five- to seven-membered enol ethers. High-profile examples shown in Figure 3.48 are intermediates in natural product synthesis. These include the six-membered G and I rings of gambieric acid (**48-1**; the ca. 90% yield is particularly notable, given that this target required a double-RCM reaction) (118), and the bicyclic frameworks of the sesquiterpene lactone thapsigarguin (**48-2**) and (+)-1-deoxynojirimycin (**48-3**) (187,188). Notable in the last example is the improved performance found on use of the Hoveyda catalyst **Ru-3**, as well as the need for equimolar benzoquinone to suppress isomerization (188).

In other examples, the Ru Fischer carbene cannot be avoided. A number of such cases have been encountered in the RCM synthesis of aromatic compounds with 1,2-heteroatom substituents (189). Synthesis of **48-4**, for example, required RCM coupling of two vinyloxy functionalities (190). The 80% yield of this reaction is notable given the need for forcing conditions (refluxing toluene),

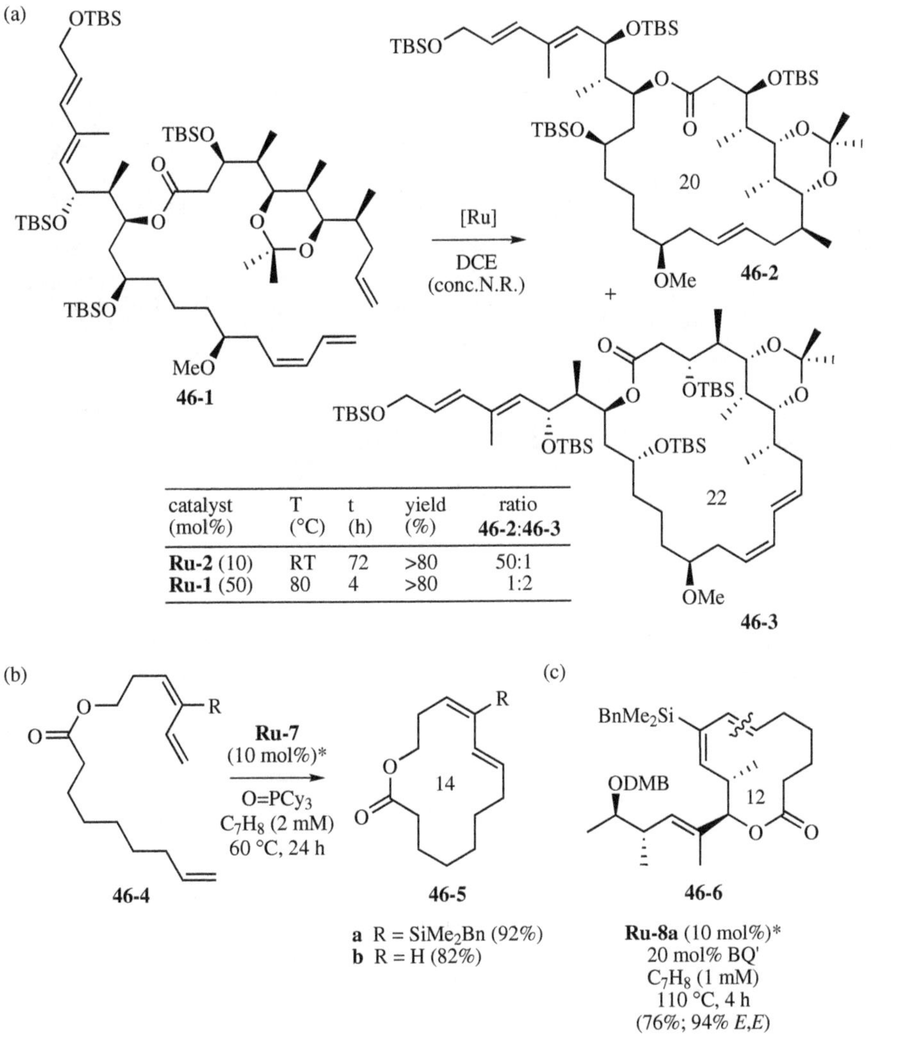

catalyst (mol%)	T (°C)	t (h)	yield (%)	ratio **46-2**:**46-3**
Ru-2 (10)	RT	72	>80	50:1
Ru-1 (50)	80	4	>80	1:2

Figure 3.46 Regio- and stereoselectivities in RCM macrocyclization of conjugated dienes. (a) Competing ring contraction. (b, c) Use of a silyl substituent to enhance selectivity. The label (*) denotes slow catalyst addition. DMB = dimethoxybenzyl; BQ′ = tetrafluoro-1,4-benzoquinone; N.R. = not reported.

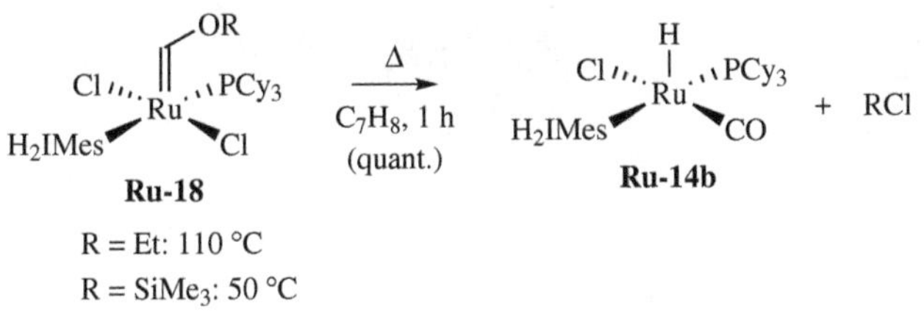

Figure 3.47 Decomposition of Fischer carbenes into Ru hydrides (91).

which would normally trigger rapid catalyst decomposition, as judged from the conditions seen in Figure 3.47. The stability of the ruthenium intermediate may reflect a stabilizing influence of the thiophene unit. Higher yields (90%) under milder conditions were reported by van Otterlo and coworkers (191) for the mixed enamine-enol **48-5**, consistent with the trend in reactivity for Fischer carbene derivatives of **Ru-2** (E = N > O) noted above. A further asset in this case was reaction via an internal olefin bearing an α-methyl group, rather than the vinyl precursor: this may contribute to improved yields by circumventing the formation of the methylidene species **Ru-12b**.

In RCM synthesis of enol ether **49-1** (Fig. 3.49), 2,2-disubstitution at the distal olefin constrains formation of the Fischer carbene. This does not impair the reactivity of **Mo-1**, which effected quantitative RCM within 3 h at RT (192). The Grubbs catalyst **Ru-2** required prolonged heating (14 h at 65 °C) to achieve comparable yields. Much faster reaction was seen for the Hoveyda catalyst **Ru-3**, but high temperatures were still required.

(a)

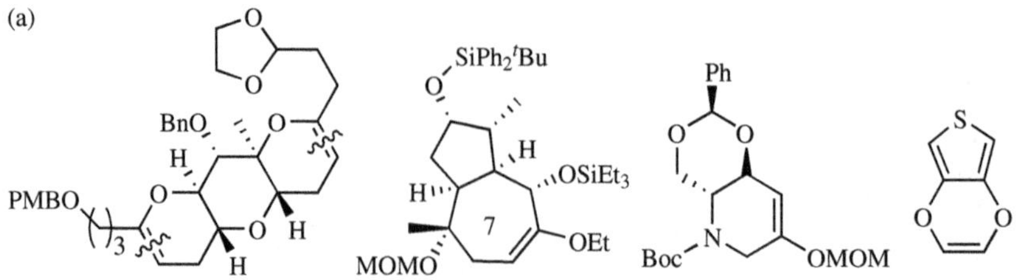

48-1	48-2	48-3	48-4
Ru-2 (10 mol%)	**Ru-2** (2.5 mol%)	[Ru] (10 mol%)	**Ru-2** (2.5 mol%)
C_7H_8 (conc. N.R.)	CH_2Cl_2 (32 mM)	BQ (10 mol%)	C_7H_8 (conc. N.R.)
70 °C, time N.R.	40 °C, 21 h	C_7H_8 (86 mM)	110 °C, time N.R.
(89%)	(88%)	110 °C, 4.5 h	(80%)
		Ru-2: 70%	
		Ru-3: 87%	

(b)

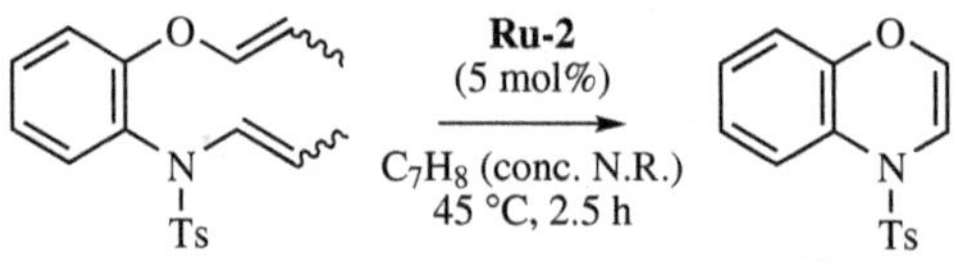

48-5
(90%)

Figure 3.48 Synthesis of cyclic enol ethers by Ru-catalyzed RCM. (a) Via coupling of vinylic and 2,2-disubstituted enol ethers. (b) Via coupling of 1,2-disubstituted olefins. PMB = *p*-methoxybenzyl, MOM = methoxymethyl. N.R. = not reported.

49-1

[M] (5 mol%), C_6H_6 (conc. N.R.)

Catalyst	T (°C)	t (h)	Yield (%)
Mo-1	22	3	>98
Ru-2	65	1	19
		14	94
Ru-3	65	1	88

49-2

C_6H_6 (100 mM), 22 °C
Mo-1: 5 mol%, 1 h
(>98%)
Mo-4: 15 mol%, 20 h
(96%)

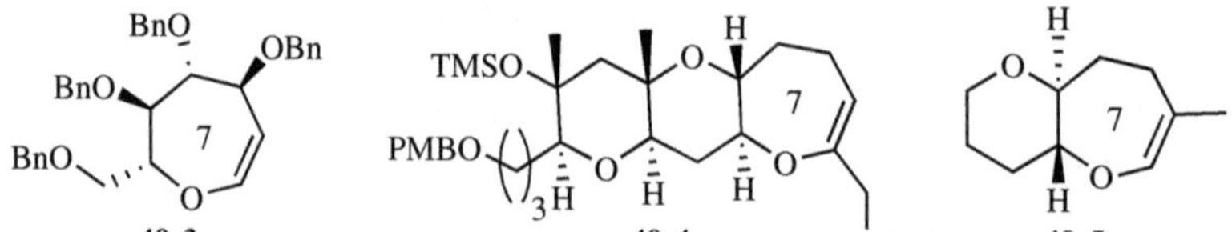

49-3

[M] (20 mol%)

Catalyst	Solvent	Conc. (mM)	T (°C)	t (h)	Yield (%)
Mo-1	C_7H_8	2	60	4	92
Ru-2	CH_2Cl_2	5	40	16	0

49-4

Mo-1 (11 mol%)
hexanes (conc. N.R.)
65 °C, time N.R.
(88%)

49-5

Mo-1 (12 mol%)
pentane (21 mM)
25 °C, 4-5 h
(94%)

Figure 3.49 High performance of **Mo-1** in RCM synthesis of five- to seven-membered enol ethers. N.R. = not reported.

Hoveyda, Schrock, and coworkers (192) reported that even less electrophilic 1,1′-bi-2-naphthol (BINOL) catalysts can cyclize such enol ethers at ambient temperatures (see **49-2**), although **Mo-1** acts much faster at a third of the catalyst loading. For more challenging seven-membered rings such as **49-3**, yields above 90% could be obtained using **Mo-1**, where **Ru-2** failed (193). Excellent yields were also reported by the Rainier group on use of **Mo-1** for assembly of the seven-membered ring in polycyclic ether **49-4** (194).

Mo-1 yields impressive performance even where steric constraints favor a Fischer carbene intermediate (21,195). In

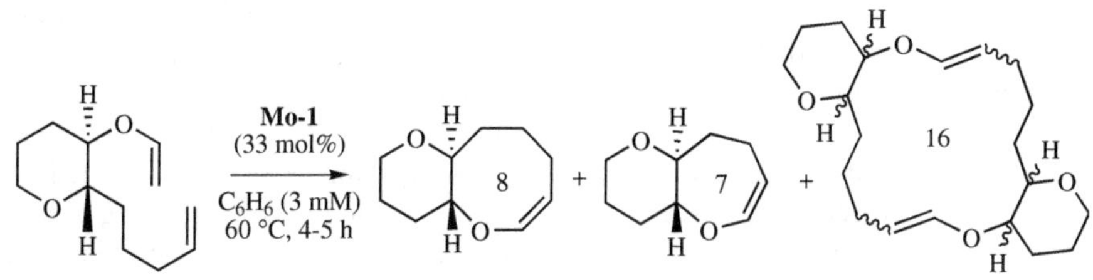

50-1 → **50-2** (27%) + **50-3** (13%) + **50-4** (dominant product; % N.R.)

Figure 3.50 Attempted synthesis of an eight-membered enol ether. Competing ring contraction (**Class B** isomerization) and dimerization in Mo-catalyzed RCM of a vinyl ether (67). N.R. = not reported.

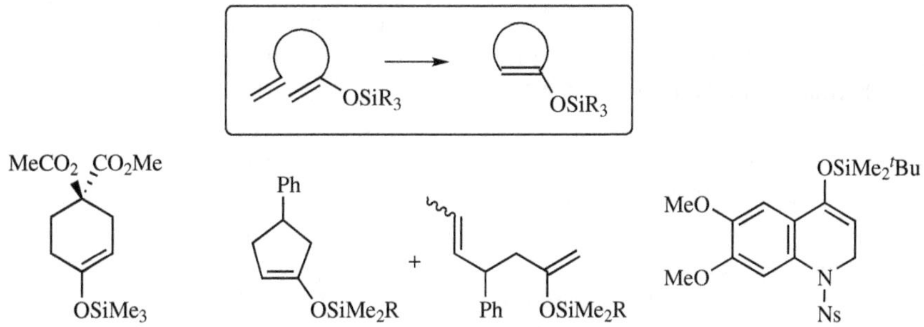

51-1: **Ru-2** (7 mol%), C_6H_6 (5 mM), 65 °C, 1 h (99%)

51-2 + **51-3**: **Ru-2**, C_6H_6, 80 °C

	R	[Ru] (mol%)	Conc. (mM)	*t* (h)	51-2 (%)	51-3 (%)
a	Me	10	10	0.5	0	31
b	^{t}Bu	20	5	N.R.	60	15

51-4: **Ru-2** (5 mol%), CH_2Cl_2 (24 mM), 40 °C, 2 h (92%)

Figure 3.51 RCM synthesis of carbocycles and heterocycles containing exocyclic silyl enol ethers. Ns = 2-nitrophenylsulfonyl. N.R. = not reported.

addition to the example of **49-1** cited above, near-quantitative yields of six- or seven-membered rings containing trisubstituted olefins were obtained at RT (see **49-2** (192) and **49-5** (77)).

Access to medium-ring enol ethers is capricious and often low-yielding, even on use of highly reactive **Mo-1**. An excellent overview of advances and failures in such reactions appears in a recent review (39). Long reaction times often result in competing isomerization: as noted in Section 3.3.1, the Mo catalysts are not immune to isomerization, although they are less prone to this side-reaction than the Ru systems. In the attempted RCM of **50-1**, for example (Fig. 3.50), the target **50-2** was accompanied by smaller amounts of the seven-membered, ring-contracted product **50-3** (27% vs 13%) (67). The dominant product was cyclodimer **50-4**, even at dilutions of 3 mM. The conformational constraints associated with the antifused ring in **50-1** are evidently inadequate to favor the eight-membered ring. The fact that the dimer is head-to-tail indicates that the problem lies in ring strain, not catalyst reactivity.

Success has been reported in RCM of silyl enol ethers using **Ru-2**, typically for five- and six-membered rings bearing an exocyclic silyl ether (Fig. 3.51). These are expected to favor initiation at the sterically more accessible vinylic olefin, circumventing formation of the Fischer carbene. Shibasaki and coworkers (196) reported quantitative yields for six-membered **51-1**, although catalyst loadings were relatively high (7 mol%). The bulk of the silyl substituents can be critical. In the RCM synthesis of five-membered **51-2a**, Aggarwal and coworkers (101) found that a trimethylsilyl group resulted in a **Class C** isomerization, rather than metathesis, affording acyclic **51-3a** in 31% yield. Replacing one methyl group with a *tert*-butyl substituent gave access to the RCM product **51-2b** in 60% yield, although a small proportion of **51-3b** (15%) was still observed. This reduced isomerization was attributed to a greater bias toward initiation at the sterically more accessible vinylic site, and hence reduced Ru-hydride formation. In comparison, Inoue and coworkers (197) reported greater than 90% yields of TBS-protected **51-4**: the use of a chlorinated solvent, lower catalyst loadings, and lower temperatures should be noted.

Endocyclic silyl enol ethers represent a special subclass of vinyl ethers, owing to the rapidity with which [Ru]=CHOSiR$_3$ derivatives decompose into hydrides (see

Fig. 3.47) (91). Competing isomerization can be anticipated as a major issue, unless steric constraints dictate initiation at the ω-olefin (195). This may account for the rarity of this class of compounds in RCM.

Clark and others (195) have demonstrated that cyclic enol ethers are often more fruitfully accessed via the RCM of allylic ethers, and this is now a generally endorsed strategy (64,198). RCM of allyl ethers, followed by isomerization, sidesteps the challenges of vinyl ether metathesis (although premature isomerization can be problematic, as discussed above). Efficient tandem catalysis methodologies that accomplish this transformation are described in Chapter 5.

Vinyl Amines and Amides As expected from the higher reactivity and improved thermal stability of Ru=CHNRR species noted above (NRR = carbazolide), vinyl amines are less intractable. Issues of poisoning by nitrogen binding, often observed for sterically unencumbered amines (Section 3.3.4.2), are mitigated by the reduced basicity of enamines, and even more so for enamides: see later in this section. Simple five- and six-membered azacycles were prepared in 80–85% yield by Rutjes and coworkers (199) using **Ru-1** (see **52-1**, **52-2**, Fig. 3.52). Where the functionalized olefin was 2,2-disubstituted (see **52-3–52-5**), the more reactive catalyst **Ru-2** was required to achieve similar yields (199,200). Arisawa and coworkers (91,201) reported that RCM of **52-6a** (in which the Fischer carbene is disfavored by the reactivity difference between the two olefins), afforded **52-6b** quantitatively, using **Ru-2**. For electron-deficient **52-7**, yields above 80% were obtained using **Ru-3** (202).

While applications of ene-enamide RCM to macrocyclization remain few, a highly interesting example was reported by the Evano group. Thus, RCM of enamide **53-1a** was promoted by **Ru-2**, affording 13-membered **53-2** in nearly 50% yield (Figure 3.53) (203). Use of the α-methyl enamide was essential: with the terminal olefin **53-1b**, deallylation resulted in the acyclic primary amide **53-2** as the major product (204). (A competing sigmatropic rearrangement was proposed, but the resemblance to the deallylation of *N*-allyl amines described in Section 3.3.4.2 below is striking). The forcing conditions required are unsurprising, given the deactivated nature of both olefins, and the probability of ether or carbonyl chelation following initiation at either ene or enamide sites, respectively.

3.3.3.3 *Special-Case Olefins: Vinylsilanes*

Vinylsilanes As the electronic impact of a silicon group on the C=C double bond is largely a function of the silicon substituents, vinylsilanes are treated separately from directly-functionalized olefins. Alkenylsilanes are useful synthetic intermediates, which can be converted to a range of products via cross-coupling, oxidative cleavage (this enabling transformation of endocyclic alkenylsilanes to acyclic dihydroxyolefins), or silyl-halide exchange. Protodesilylation of exocyclic silyl groups affords the cycloolefin, a strategy discussed above in the context of stereoselective RCM (Section 3.3.1.2). Cleavage of endocyclic silanes following cyclization forms the basis of "temporary silicon-tethered" strategies, in which intermolecular CM targets are achieved via more favorable, intramolecular RCM reactions (205,206).

Studies by the Barrett, Gibson, and Grubbs groups (19,207) described the Mo-catalyzed RCM of 2, 2-disubstituted vinylsilanes to afford cycloolefins containing an endocyclic silyl group (see **54-1**, **54-2**; Fig. 3.54). The steric bulk of the silyl substituent limits the efficacy of **Ru-1** in RCM of vinylsilanes. However, where a methylene spacer was introduced between the silicon group and the olefin, **Ru-1** proved competent for synthesis of eight-membered rings (see **54-3**) (19). Denmark has employed vinylsilane metathesis with great success in

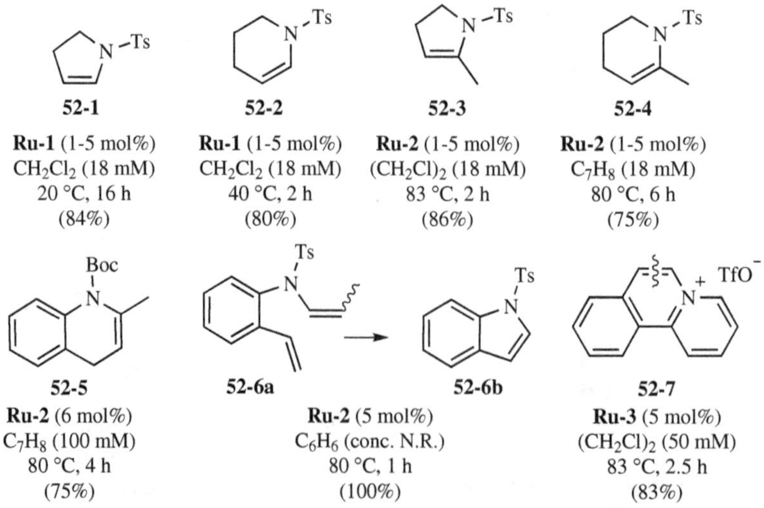

Figure 3.52 RCM of five- and six-membered, *N*-protected vinyl amines.

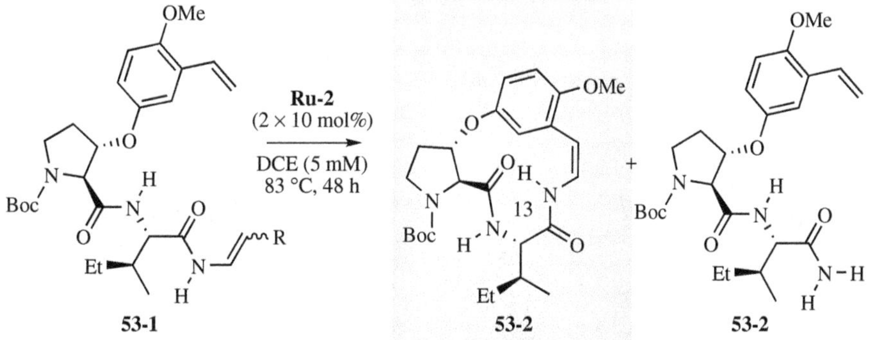

	R	53-2 (%)	53-3 (%)
a	Me	49	13
b	H	8	78

Figure 3.53 RCM synthesis of a 13-membered enamide. Note: shading indicates the intended RCM product.

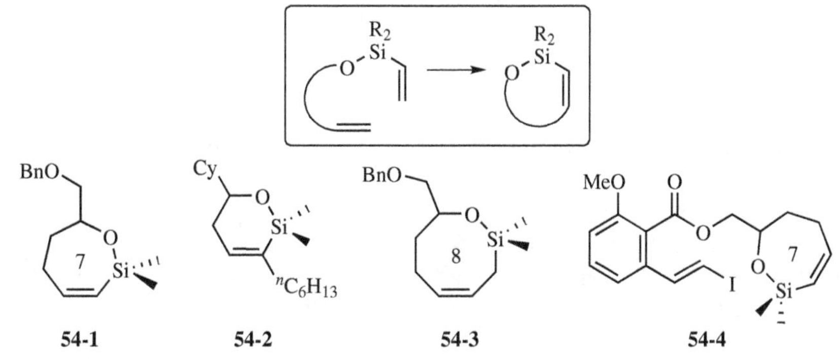

54-1	54-2	54-3	54-4
Mo-1 (5 mol%) C_6H_6 (50 mM) RT, 5 h (91%)	**Mo-1** (2 mol%) $CDCl_3$ (20 mM) RT, 2 h (>95%)	**Ru-1** (5 mol%) CH_2Cl_2 (50 mM) 25 °C, 3 h (96%)	**Mo-1** (10 mol%) C_6H_6 (85 mM) RT, 15 h (83%)

Figure 3.54 Exemplary syntheses of silacycles by RCM of vinylsilanes.

the synthesis of a number of natural products, a recent example being seven-membered **54-4**, a precursor to oximidine II. The iodide-functionalized olefin does not react with **Mo-1**, as noted above (129). Notable are the high concentrations at which the medium rings were obtained (**54-3**: 50 mM; **54-4**, 85 mM). The larger size and more diffuse orbitals of silicon, relative to carbon, significantly reduce ring strain in the silacycle, versus the corresponding carbocycle, as noted in Section 3.2.1. The mild conditions under which these reactions proceed should also be noted: this is a key factor, given the susceptibility of the siloxy group to nucleophilic attack under thermally forcing conditions (206).

Gouverneur and coworkers demonstrated that **Ru-2** effects RCM of 2,2-disubstituted vinylsilanes to afford five- and six-membered cycloolefins with exocyclic silyl groups (**55-1**, **55-2**, Fig. 3.55). Where the distal olefin was electronically or sterically deactivated (i.e., an acrylate or an internal olefin), yields of greater than 80% were attainable, but required catalyst loadings of about 10 mol% and long reaction times (>24 h) (208). RCM again failed with **Ru-1** and, more surprisingly, **Ru-2′**.

Work by Fischer, Marciniec, and coworkers (209,210) established that electron-withdrawing silicon substituents improve CM performance by retarding β-silyl elimination from the metallacyclobutane intermediate. Consistent with this, Young and coworkers observed significantly higher yields for RCM of vinyl groups bearing alkoxysilane, versus alkylsilane substituents (110). The improved catalyst lifetimes afforded access to 14- and 15-membered macrocycles in good to excellent yields (see **55-3**, **55-4**). Siloxy derivatives also benefit from the reduced steric bulk at silicon, which is presumed to reduce the energy barrier to cycloaddition.

3.3.4 Functional-Group Tolerance: Non-olefinic Substituents

Issues of functional-group tolerance specific to vinylic functionalization have been described in the previous section. This section considers the impact of functional groups at non-olefinic sites. The majority of effects, for both group 6

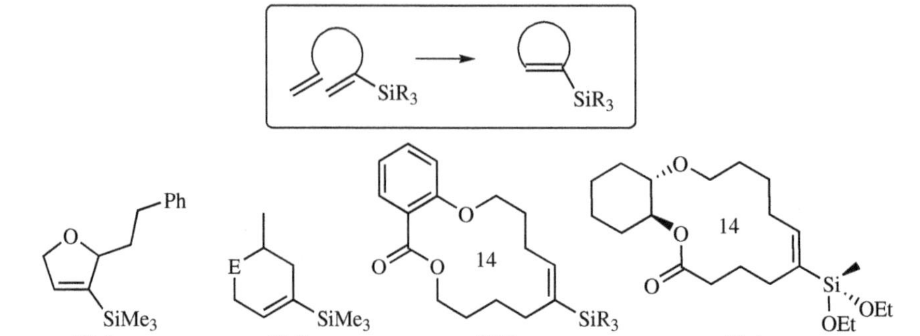

55-1: **Ru-2** (2 mol%), CH_2Cl_2 (conc. N.R.), 40 °C, 1 h (90%)

55-2: **Ru-2** (3 mol%), CH_2Cl_2 (conc. N.R.), 40 °C; **a** E = O, 3 h (93%); **b** E = $C(CO_2Et)_2$ 8 h (98%)

55-3: **Ru-5** (20 mol%), C_7H_8 (2 mM), 35 °C, 12 h; **a** R = Et (10%); **b** R = OEt (92%)

55-4: **Ru-5** (20 mol%), C_7H_8 (2 mM), 35 °C, 12 h (76%)

Figure 3.55 Syntheses of exocyclic silyl-functionalized olefins by RCM of 2,2-disubstituted olefins.

and Ru catalysts, arise from the presence of Bronsted acids or Lewis base donors. Oxygen-atom donors (e.g., aldehydes) are an added problem for Mo and W catalysts; these will be treated last.

Intra- and intermolecular interactions with non-innocent functionalities are expected to have qualitatively similar effects. Higher encounter probabilities result in more dramatic effects for intramolecular interactions, as does positioning of the functional group near the metal alkylidene. Donor atoms present as substituents at the allylic position (i.e., β to the alkylidene: see **56-1**, Fig. 3.56) pose the greatest likelihood for interaction with the metal center, owing to the stability of five-membered chelate rings. Substituents on the homoallylic site are also a risk factor for formation of six-membered chelates (see **56-2**). Seven-membered chelate rings (**56-3**), while relevant for larger donor atoms such as phosphines, are typically more labile. Functional groups positioned β or γ (and, to a lesser extent, δ) to unhindered, reactive olefins are therefore of greatest potential concern.

3.3.4.1 ***Bronsted Acids: E–H Bond Donors*** Protic functionalities (e.g., alcohols and phenols, carboxylic acids, and thiols) are inimical toward the group 6 systems, which readily undergo ligand protonolysis. In the ruthenium systems, this susceptibility is reduced and a number of beneficial effects are possible. For catalysts such as **Ru-1** and **Ru-2**, which are activated by loss of PCy_3, Bronsted acids improve "commitment" to metathesis, by sequestering the free phosphine (Fig. 3.57a). This inhibits back-reactions that remove the catalyst from the active cycle, and also inhibits deactivation pathways involving attack of phosphine on the methylidene (60) (Fig. 3.57b). The Sasol group pointed out that phenols exert a stabilizing effect on **Ru-1**, possibly via reversible coordination to the metal center (211). Hydrogen-bonding interactions with the chloride ligand (212) can also increase the probability of encounter between a pendant olefin and the metal center, thus favoring metathesis.

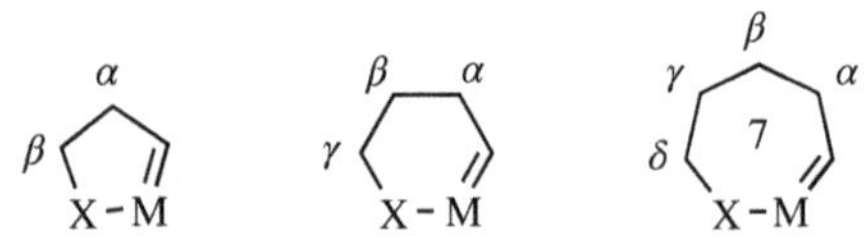

56-1 **56-2** **56-3**

Figure 3.56 Proximity effects for functional groups relative to the metal alkylidene.

Alcohols The reactions of the Grubbs catalysts with primary and secondary alcohols described in Section 3.3.1 point toward the potential for catalyst degradation where metathesis is slow or conditions are forcing (96–98). Decomposition is mediated by a ruthenium alkoxide species, which undergoes facile β-elimination once formed. The slow rate of these reactions is due to the sluggishness with which the alkoxide intermediate is formed (even, in the reactions described, in the presence of NEt_3 as base). Alkali alkoxides such as KOMe, in contrast, undergo rapid salt metathesis with Ru–Cl bonds, triggering aggressive decomposition of Ru-alkylidenes. *In situ* formation of alkoxide salts should thus be strenuously avoided.

Allylic alcohols—in which the metathesis reaction itself places the alcohol functionality in proximity to the metal—have diverse, conflicting effects, depending on whether they are primary, secondary, or tertiary. The Werner group reported that **Ru-1** reacts with allylic alcohol **58-1** at RT to form propionaldehyde and $RuCl_2(CO)(PCy_3)_2$ (**Ru-19**, Fig. 3.58a) (213). Tertiary allylic alcohols such as **58-2**, which cannot undergo β-elimination, are stabilized against this deactivation pathway. They also undergo accelerated RCM with **Ru-1**. In a competition experiment between

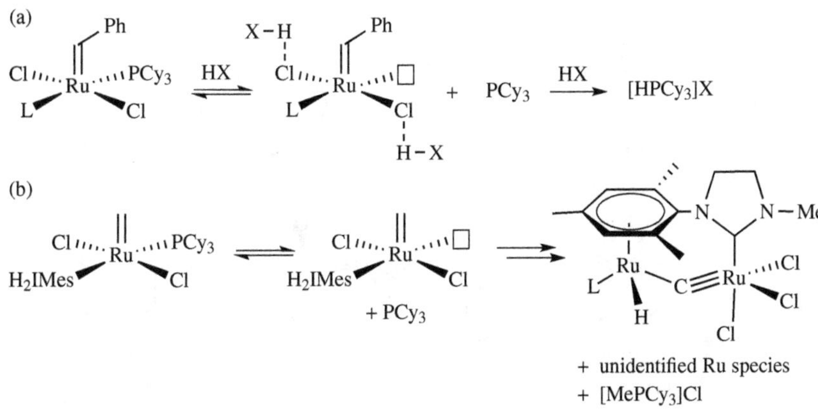

Figure 3.57 (a) Sequestration of phosphine by acid inhibits reuptake of PCy_3. (b) Catalyst deactivation mediated by attack of PCy_3 on the methylidene group.

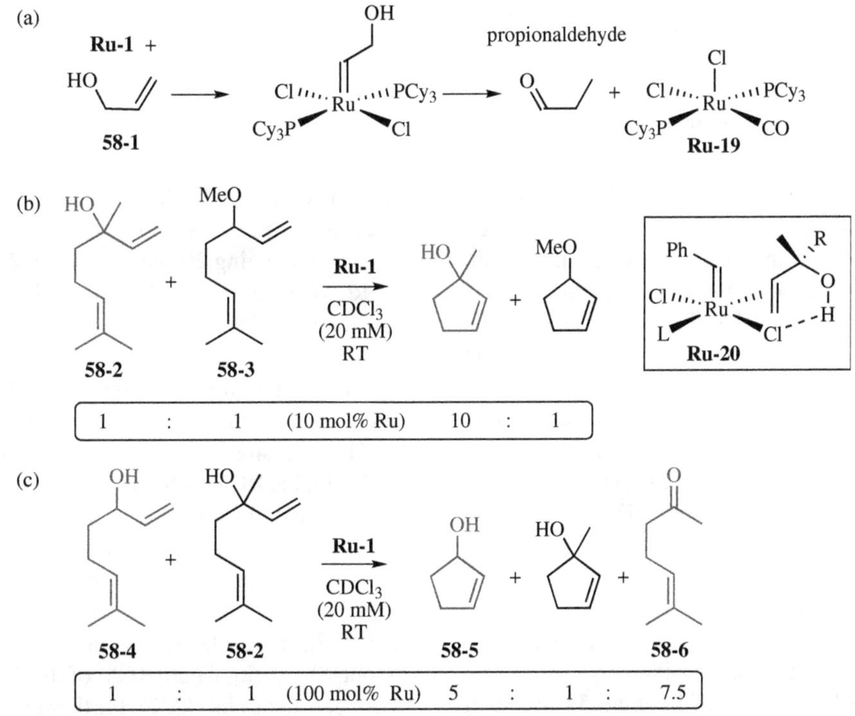

Figure 3.58 Reactions of **Ru-1** with allylic alcohols. (a) Decomposition by a primary allylic alcohol. (b) Preferential RCM of tertiary allylic alcohol **58-2**, versus its methyl ether analog **58-3**. Inset depicts suspected H-bonding. (c) Preferential RCM of secondary allylic alcohol **58-4**, versus tertiary **58-2**, and competing truncation of **58-4**. Structures in gray indicate fastest reactions.

58-2 and the corresponding allylic ether **58-3**, Hoye and coworkers (214) showed a 10 : 1 preference for RCM of the former (Fig. 3.58b). Preferential reaction of the allylic alcohol is most plausibly due to hydrogen bonding to the chloride ligand in the pre-assembly phase, as shown for **Ru-20** in the inset. This increases the probability of encounter between the olefin and the metal center, as discussed earlier (cf. the anchor effects proposed for carbonyl groups by Fürstner and Langemann; see Section 3.2.1.1) (37). H-Bonding to the PCy_3 or styrenyl ether ligands may also improve initiation efficiency by weakening their donation to Ru.

RCM of the secondary alcohol **58-4** was in turn faster than RCM of tertiary **58-2**, probably because of its reduced allylic bulk. Importantly, however, yields of the cyclic secondary alcohol **58-5** were impaired by competing truncation to yield ketone **58-6** (Fig. 3.58c) (214). This process also decomposes the catalyst: 100 mol% **Ru-1** was required for 80% conversion of **58-5**.

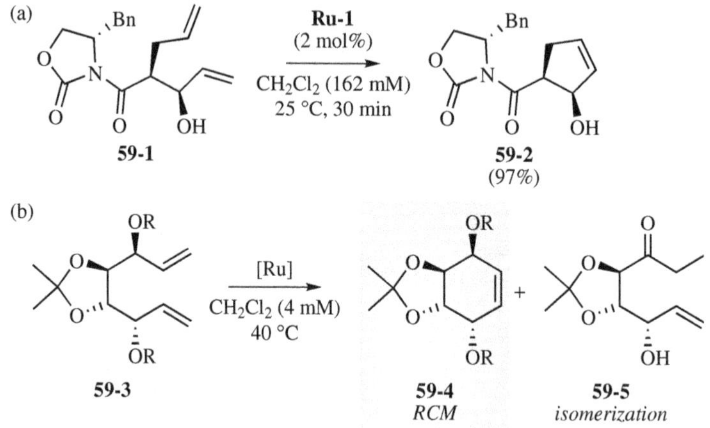

R	Catalyst (mol%)	t (h)	**59-4** (%)	**59-5** (%)
H	**Ru-1** (20)	20	0	29
Ac	**Ru-1** (20)	120	77	0
H	**Ru-2'** (1.5)	2	69	0
Ac	**Ru-2'** (1)	5	71	0

Figure 3.59 RCM of secondary allylic alcohols using **Ru-1**. (a) Successful RCM. (b) RCM and competing isomerization. Note: shading indicates the intended RCM product.

This side-reaction did not impede cyclization of the sterically more accessible diene **59-1** with **Ru-1**, which led to five-membered **59-2** in excellent yields under mild conditions (RT, 0.5 h; Fig. 3.59a) (215). That this success is due to initiation at the distal olefin is implied by the poor results found with **59-3**, in which both olefins bear an allylic alcohol (70). None of the expected RCM products **59-4** was observed, but acyclic **59-5** was isolated in about 30% yield (indicating a **Class C2** isomerization; Fig. 3.59b). Protection of the alcohol increased RCM yields to 77%, but reaction then took 120 h in refluxing CH_2Cl_2, even at a catalyst loading of 20 mol%. **Ru-2′** appears less susceptible to decomposition by secondary allylic alcohols. RCM of unprotected **59-3** using this catalyst (1.5 mol%) gave **59-4** in about 70% yield within 2 h. Acyl protection offered no improvement, and slowed reaction considerably.

As indicated in the Hoye study of Figure 3.58, RCM of secondary allylic alcohols by **Ru-1** can be subverted by truncation to the acyclic methyl ketone, as well as isomerization. Tori and coworkers (216) observed both processes during the attempted RCM of **60-1**; the 10-membered RCM product **60-2** was not formed (Fig. 3.60a). In the attempted macrocyclization of **60-3** (Fig. 3.60b), Maleczka and coworkers (217) reported formation of solely truncated ketone **60-4** when **Ru-1** was used as catalyst. On use of **Ru-2**, only traces of **60-4** were observed, but yields of the desired 20-membered RCM product reached only 35%. It was unclear whether isomerized material forms part of the mass balance.

Use of the more reactive catalyst **Ru-2** for RCM of secondary allylic alcohols can dramatically improve yields, as indicated above. Further examples appear in Fig. 3.61. Six-membered **61-1**, a precursor to ovalicine, was obtained in 94% yield by Takahashi and coworkers (218), while the Nicolaou group (219) reported nearly 80% yield in synthesis of the challenging 20-membered **61-2**.

RCM of unprotected homoallylic alcohols can be more challenging. The complications rise from Lewis base behavior (see next section; 3.3.4.2), rather than Bronsted acidity; that is, the impact of oxygen chelation on reaction rates and yields. Briefly, chelation is particularly problematic for **Ru-1**, as illustrated by the low yields (40%) of **62-2** obtained in the double RCM of trisallylcarveol **62-1** (Fig. 3.62a) (220). The five-membered chelate ring is shown in the inset. Alcohol protection is generally advantageous, even for **Ru-2**, as illustrated by the greatly improved RCM yields reported for acyl derivative **62-3a**, versus unprotected **62-3b** (ca. 90% vs 24%) (221). Similarly, silyl-protected **62-4a** is accessible in about 90% yield, where RCM of the alcohol derivative **62-4b** was reported to fail (222). Further examples of this behavior will be seen in RCM of homoallylic ethers below.

Thiols Thiols have been relatively little studied for either Ru or group 6 catalysts, and we are aware of no reported examples of thiol RCM. The acidity of the S–H bond is anticipated as an issue for the group 6 catalysts, which are readily decomposed by proton donors. With the ruthenium catalysts, Lewis base behavior is anticipated as a more significant problem. CM experiments (though not always a reliable guide to RCM) revealed that allyl mercaptan **63-1a** undergoes no reaction with **Ru-1**; see Figure 3.63. Metathesis was observed with **Ru-2′** at 80 °C, but ceased at 45% yield (223). Similar behavior was observed for the α-methyl olefin **63-1b**, indicating that the problem is sulfur poisoning, rather than catalyst decomposition via, for

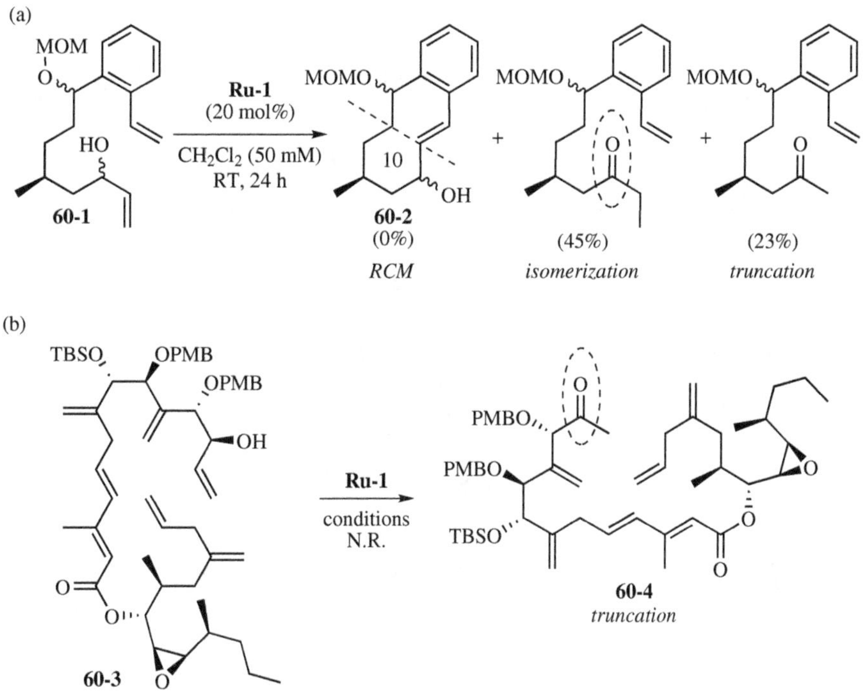

Figure 3.60 Attempted RCM of secondary allylic alcohols using **Ru-1**. (a) Competing **Class C2** isomerization and truncation. (b) Competing truncation. MOM = methoxymethyl, PMB = *p*-methoxybenzyl, TBS = *tert*-butyldimethylsilyl.

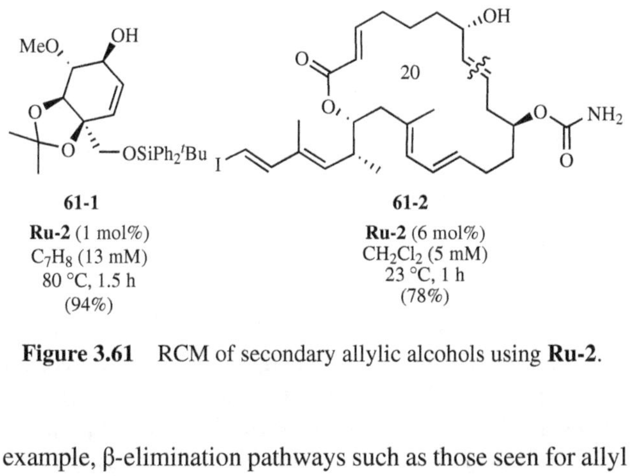

61-1
Ru-2 (1 mol%)
C_7H_8 (13 mM)
80 °C, 1.5 h
(94%)

61-2
Ru-2 (6 mol%)
CH_2Cl_2 (5 mM)
23 °C, 1 h
(78%)

Figure 3.61 RCM of secondary allylic alcohols using **Ru-2**.

example, β-elimination pathways such as those seen for allyl alcohol above.

3.3.4.2 ***Lewis Base Donors: Metal Binding*** This section examines the potentially inhibiting effect of Lewis donors on catalyst activity. A beneficial effect was recently proposed by Lin and Davis (224), who suggested that transiently bound sulfur and selenium donors can function as relays that promote CM (cf. Fürstner's (37) early suggestion that weak metal–carbonyl interactions promote RCM macrocyclization). The generality of such effects has not yet been systematically examined, but will clearly depend on the strength of the donor interaction. In many cases, and perhaps counter-intuitively, the Mo catalysts are more resistant to Lewis base poisoning than their ruthenium counterparts. Within the Ru systems, the more electron-rich Ru–NHC donors tend to perform better; also relevant is their greater stability at the higher temperatures used to improve Ru–L lability.

Nitrogen Donors: Amines The donor capacity of the amine functionality is an important consideration in designing RCM routes to nitrogen heterocycles. In an important 2007 review, Compain (225) analyzed the factors responsible for success and failure. In general, RCM efficiency increased with greater steric hindrance at the amino group: that is, the presence of substituents on or adjacent to nitrogen that create a steric impediment to binding. Hindered tertiary amines have indeed been used with outstanding success in RCM, including in natural product synthesis (17,64). Electron-withdrawing substituents that reduce N-basicity are advantageous, though not always sufficient to limit poisoning in the absence of accompanying steric inhibition. Aromatic heterocycles, in which the donor ability of the nitrogen lone pair is drastically reduced, have been regarded as typically well tolerated (225). Pyridine has been reported to rapidly decompose the methylidene species **Ru-12b**, however (60). Best results can therefore be expected where nitrogen coordination is sterically inhibited.

As this suggests, primary aliphatic amines are recognized poisons for both **Mo-1** (226) and the Ru catalysts (227). The basis for the latter incompatibility was examined by Moore

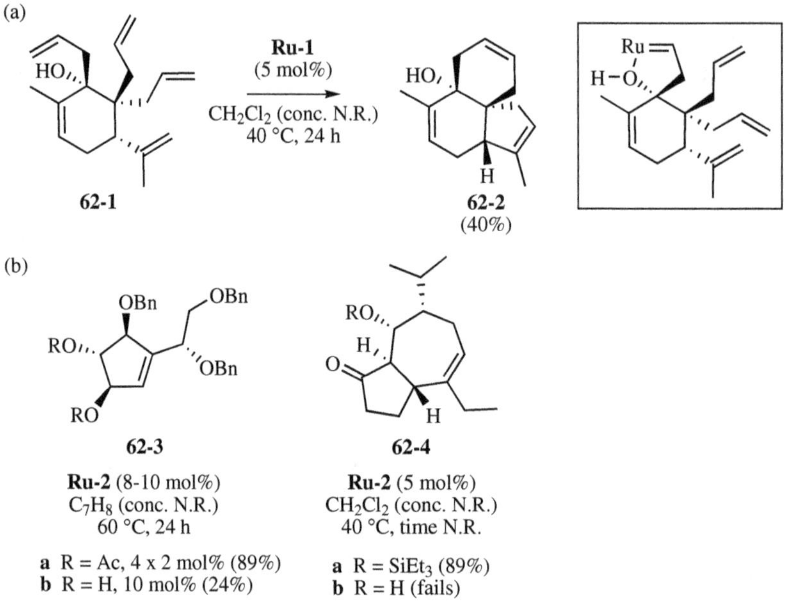

Figure 3.62 RCM of homoallylic alcohols. (a) Complications associated with oxygen chelation. (b) Beneficial effect of alcohol protection.

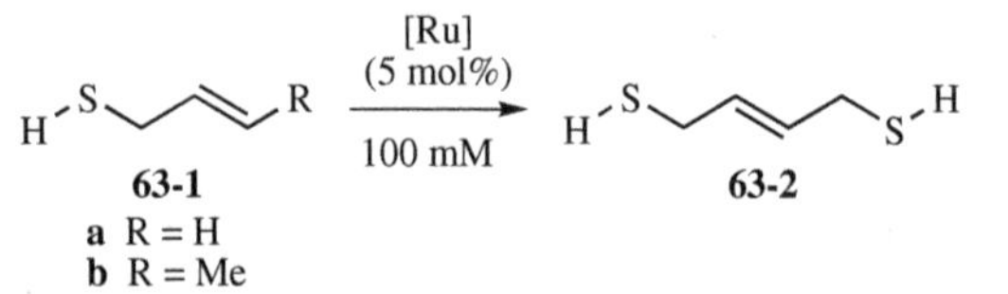

R	Catalyst	Solvent	T (°C)	t (h)	Yield (%)
H	**Ru-1**	C_6D_6	RT	20	0
H	**Ru-2'**	C_7D_8	80	2	45
Me	**Ru-1**	C_6D_6	RT	20	0
Me	**Ru-2'**	C_7D_8	80	2	38

Figure 3.63 Poor performance of Ru catalysts in metathesis (CM) of allyl mercaptans (223).

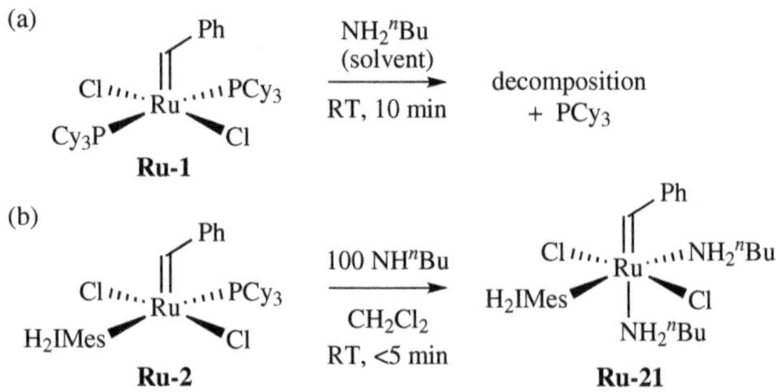

Figure 3.64 Poisoning of (a) **Ru-1** and (b) **Ru-2** by n-butylamine. The Ru products show zero or low RCM activity (227).

and coworkers (227) in a study involving reaction of **Ru-1** and **Ru-2** with excess n-butylamine. Complete loss of PCy_3 from **Ru-1** resulted in an unidentified, metathesis-inactive product (Fig. 3.64a). In the corresponding reaction of **Ru-2**, the H_2IMes ligand was retained: the ruthenium product was identified as six-coordinate **Ru-21** (see Fig. 3.64b) on the basis of X-ray crystallographic analysis. This species exhibited very low RCM activity even toward the readily cyclized diethyl diallylmalonate, DeDAM, with a plateau at 10% conversion after 15 min at 23 °C (2 mol% Ru; cf. conversions of greater than 90% for **Ru-2** under the same conditions).

RCM of tertiary amines by **Mo-1** was first reported more than 20 years ago by Fu and Grubbs (140). The unexpected tolerance of this catalyst for unprotected secondary amines was subsequently uncovered by Shon and Lee (228), and applied to the assembly of pyrroline and tetrahydropyridine derivatives (see **65-1a**, Fig. 3.65). Benzyl protection of the otherwise very accessible allylic nitrogen in the 3-pyrroline derivative **65-1b** gave higher yields at much shorter reaction times, however. The improved performance reported by Hoveyda, Schrock, and coworkers (229) for unprotected **65-2** probably reflects the presence of a tertiary carbon center adjacent to nitrogen, which impedes N-binding. In studies directed toward synthesis of chiral nitrogen heterocycles, the less reactive biphenolate catalyst **Mo-5** was found to be unreactive toward **65-2**, but introduction of a phenyl group to generate a quaternary center α to the amine delivered piperidine derivative **65-3** in 95% yield (229).

Even with **Mo-1**, it will be noted that unencumbered secondary amines generally require rather high catalyst loadings and/or relatively long reaction times (228). Sterically encumbered secondary amines or tertiary amines are readily cyclized, however (1,225). An impressive example is Fürstner's (230) high-yield synthesis of azepine **65-4**, which leaves the pendant azide functionality intact.

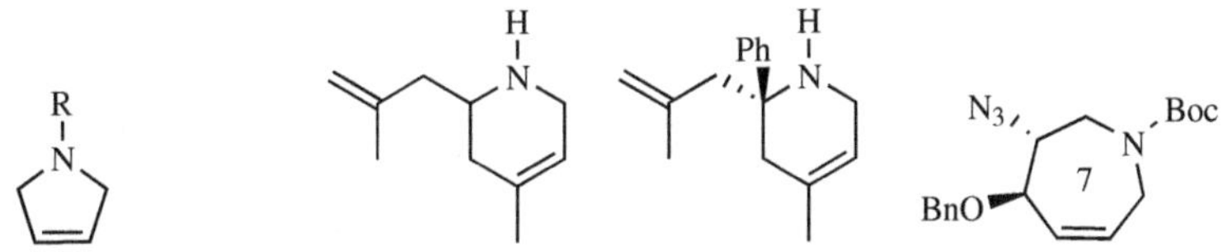

65-1

Mo-1, C_6H_6 (40 mM), 20 °C

	R	[Mo] (mol%)	t (h)	yield (%)
a	H	10	2	58
			24	82
b	Bn	5	3	92

65-2

[Mo] (5 mol%)
C_6H_6 (10 mM)
22 °C, 12 h

Mo-1: 75%
Mo-5: <2%

65-3

Mo-5 (5 mol%)
C_6H_6 (10 mM)
22 °C, 24 h
(95%)

65-4

Mo-1 (5 mol%)
CH_2Cl_2 (10 mM)
22 °C, 0.5 h
(94%)

Figure 3.65 Representative examples of Mo-catalyzed RCM of allylic secondary amines.

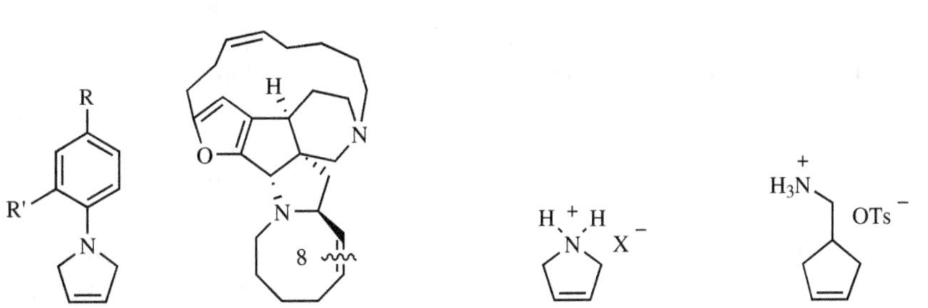

66-1

Ru-1 (2 mol%)
CH_2Cl_2 (50 mM)
RT, 12-24 h

a R = R' = OMe (29%)
b R = CO_2Me, R' = H (83%)

66-2

Ru-1 (29 + 8 mol%)
(+)-CSA (3 equiv.)
CH_2Cl_2 (0.3 mM)
40 °C, 5 + 3 h
(70%)

66-3

CH_2Cl_2 (100 mM), 40 °C, 24 h

	X	Catalyst (mol%)	Yield (%)
a	OTs	**Ru-3** (5)	>95
b	Cl	**Ru-2'** (5)	<20

66-4

Ru-2 (5 mol%)
CH_2Cl_2 (100 mM)
40 °C, 24 h
(>95%)

Figure 3.66 Ru-catalyzed RCM of amines. CSA = camphorsulfonic acid.

For **Ru-1**, attempted RCM of secondary amines is often unsatisfactory, even on use of high loadings and long reaction times (23,225,231). **Ru-2** performs better, but again, performance is generally improved by N-protection or increased bulk adjacent to nitrogen. Even tertiary arylamines can inhibit metathesis by **Ru-1**, despite the attenuated basicity of aromatic amines. In reactions of diallylanilines with **Ru-1**, Grigg and coworkers (232) established an inverse correlation between amine basicity and RCM activity (see **66-1**, Fig. 3.66). Yields improved from 29% for methoxy-functionalized **66-1a** to 83% for the less basic **66-1b**. It is unclear whether this reflects intermolecular N-binding, or chelation via a five-membered ring in the metallacyclobutane intermediate.

Both Lewis acids (233) and Bronsted acids (e.g., camphorsulfonic acid; see **66-2** (46,234), and phenols, as noted above) have been used to improve performance. Protection as the ammonium salts can likewise be useful, presuming the absence of acid-sensitive entities elsewhere in the substrate (66,235–237). With **Ru-1**, however, at least one report indicated unsatisfactory performance, attributed to catalyst inhibition and degradation (238). The Robinson group (239) recently pointed out that yields can be greatly improved by choosing a counter-ion that maximizes substrate solubility. Thus tosylate **66-3a** was obtained in greater than 95% yield (cf. <20% for chloride **66-3b** under comparable conditions) (66). In RCM of primary amines by **Ru-2** (see **66-4**), such ammonium masking strategies were reportedly superior in both yield and convenience to acyl protection (240). Other protection strategies involve conversion of the amine group into an amide, carbamate, or sulfonamide derivative. A key consideration for each of these is the potential for chelation, as discussed below.

A final note of caution concerns side-reactions that have been observed during Ru-catalyzed RCM of allylic amines, particularly under forcing conditions. In reactions with **Ru-1**, cleavage of the N–C(allyl) bond has been reported (Fig. 3.67). Both aromatic and aliphatic amines are susceptible to loss of the allyl unit. Alcaide and coworkers (240,241) have exploited the efficiency of such pathways to develop amine deprotection strategies for secondary and tertiary nitrogen centers. Deallylation appears to be less problematic with **Ru-2**, although it has been reported for a 1,4-diene (222). A common drawback of the Ru–NHC catalysts, however, is competing double-bond isomerization, as discussed in Section 3.3.1.1 and elsewhere (62,66,70,72,242,243). This

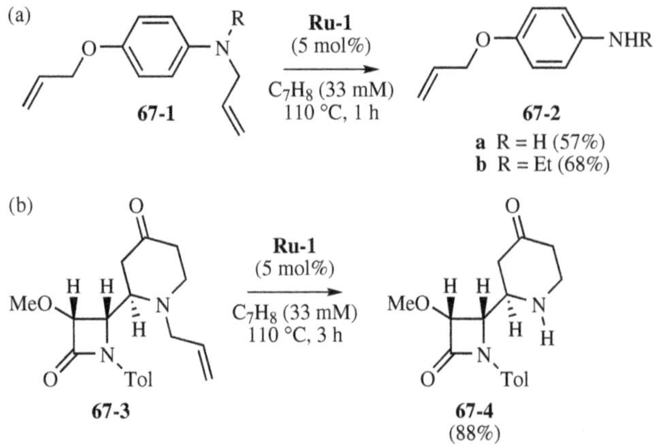

Figure 3.67 Side-reactions in the attempted RCM of *N*-allyl amines. Cleavage of the N–C(allyl) bond by **Ru-1**. (a) Observed for secondary and tertiary aromatic amines. (b) Observed for an aliphatic amine.

susceptibility has attracted much attention within the context of tandem metathesis. Such methodologies are treated in detail in Chapter 5.

Nitrogen Donors: Amides The greater bulk and lower N-basicity of amide derivatives, relative to their amine parents, reduce their tendency to poison metathesis catalysts by nitrogen binding. Indeed, it is for this reason that secondary amines are frequently protected as their amide derivatives prior to RCM. (Carbonyl binding can be a significant, often unrecognized impediment, however, as discussed in the next section.)

The resistance of **Mo-1** to poisoning by secondary amides was demonstrated by a series of reports from the Hoveyda group (115,244,245) in the mid-1990s. A representative RCM target, 14-membered **68-1** (Fig. 3.68a), was obtained in 41% yield, with about 30% dimers (the product distribution reflects the 10 mM concentrations of diene adopted in this early study (115); cf. required dilutions of 5 mM for closely related macrolides discussed in Section 3.2.1 above). A related target, trisubstituted **68-2**, was obtained in 92% yield at a 2 mM concentration of diene.

An unexpectedly high tolerance of **Ru-1** for primary amides is suggested by the greater than 80% yields reported by Barrett, Gibson, and coworkers (246) in the synthesis of dioxepine **68-3** (Fig. 3.68b). Where a five-membered chelate is accessible, however, secondary amides can strongly inhibit RCM by **Ru-1**. Thus, Grisi and coworkers reported 28% yield of 14-membered **68-4a**, versus 97% for lactone **68-4b** (182). In comparison, **Ru-2** delivered the lactam in 84% yield, illustrating the greater resistance of this catalyst to poisoning by nitrogen donors (84%; yields of the lactone were also superior) (182). A further example from the same study was depicted in Figure 3.45 above; in that case, the lactam yield reached 99% for **Ru-2**, reflecting the lower basicity of the aromatic amide.

Six-membered chelates are also detrimental, as evidenced by the poor performance of **Ru-2** in the attempted synthesis of **68-5a** (247). The sole observed product was the acyclic dimer **68-5b** (formed despite dilutions of <1 mM). Conversions were limited to about 40%, implying strong chelation even in refluxing toluene. Much improved performance was seen on use of the Hoveyda catalyst **Ru-3**, with 75% yield of the desired **68-5a**, accompanied by 8% dimer.

Finally, the combination of a remote, non-chelating amide and steric bulk α to nitrogen enables good yields even using **Ru-1** (see **68-6**) (248), as seen from the discussion above, although performance does not reach the levels seen for **Mo-1**. More generally, spectacular advances in the **Ru-2**-catalyzed RCM of amino acids and peptidomimetics point toward excellent compatibility of this catalyst with non-chelating secondary amides. It should be noted, however, that these reactions are often most successful where microwave irradiation is used (see Section 3.3.5).

Tertiary amides are often well tolerated by both **Mo-1** and the Ru catalysts. Many such entities in RCM are masked secondary amines, and the capacity to tune protecting groups to maximize EM values can be invaluable, as illustrated most prominently in the BILN 2061 campaign (Section 3.6). However, amide and carbamate protecting groups have the potential to chelate to the alkylidene intermediate via the carbonyl donor (Fig. 3.69). Carbonyl chelation is discussed in the next section: particularly for challenging RCM targets, its consequences are inefficient metathesis, poor turnover numbers, and competing isomerization. In metathesis of olefins bearing an allylic nitrogen, Compain advocated the use of non-chelating protecting groups (e.g., α-methylbenzyl, diphenylmethyl, and triphenylmethyl) (225). Phenylamine protecting groups (e.g., *p*-methoxyphenyl, PMP) have the added advantage of reducing nitrogen basicity.

Oxygen Donors: Carbonyls The carbonyl oxygen is a relatively weak Lewis donor, and does not poison either Mo or Ru metathesis catalysts via intermolecular coordination. More problematic, as expected, is intramolecular chelation via a five- or six-membered ring (i.e., allylic or homoallylic positioning of the carbonyl group) (249,250). Chelation is thus a common complication in metathesis of β,γ- and γ,δ-unsaturated carbonyl compounds. Many examples have been reported, a number of which are alluded to in the sections above.

Even in norbornene ROMP, where the release of ring strain provides a strong driving force for reaction, carbonyl chelation has been shown to impede metathesis. The Fürstner and Khosravi groups (250–253) demonstrated that homoallylic carbonyl groups retard propagation for **Ru-1**. Slugovc and coworkers (254) reported that ketone derivative **70-1** was consumed much more slowly than ester **70-2**, implying that

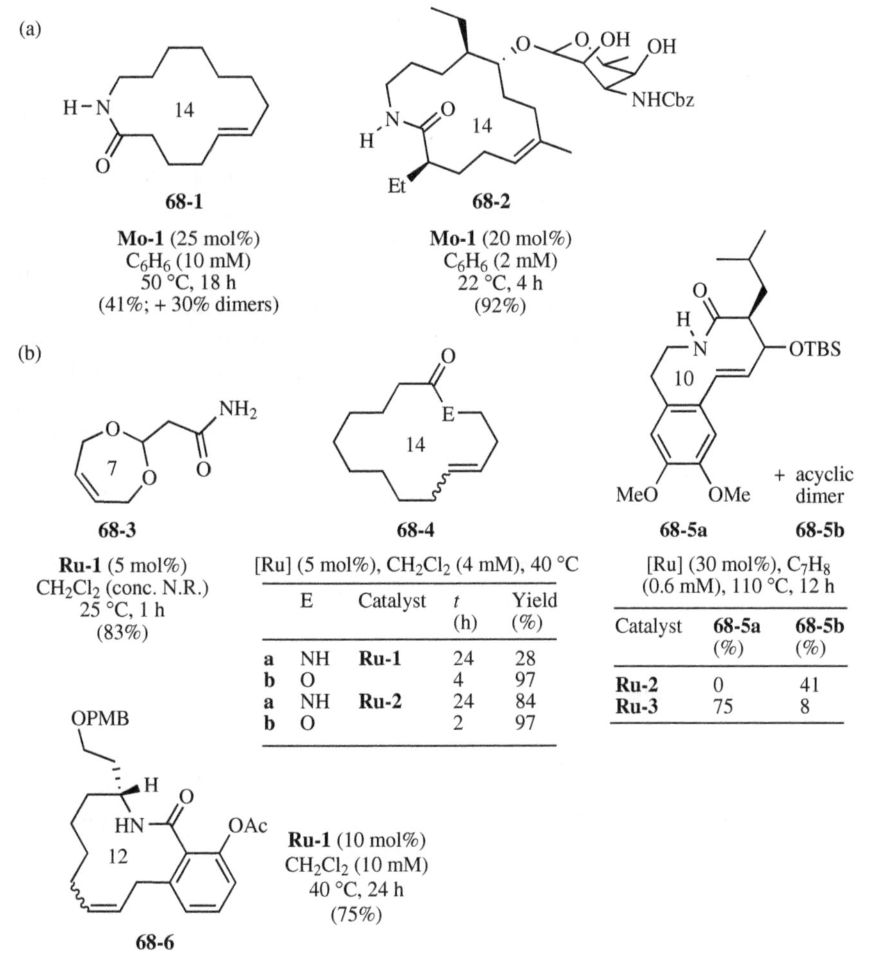

	E	Catalyst	t (h)	Yield (%)
a	NH	**Ru-1**	24	28
b	O		4	97
a	NH	**Ru-2**	24	84
b	O		2	97

Catalyst	**68-5a** (%)	**68-5b** (%)
Ru-2	0	41
Ru-3	75	8

Figure 3.68 RCM of unprotected amides. (a) Examples of Mo-catalyzed macrolactamization. (b) Ru-catalyzed RCM of a primary amide, and various secondary amides.

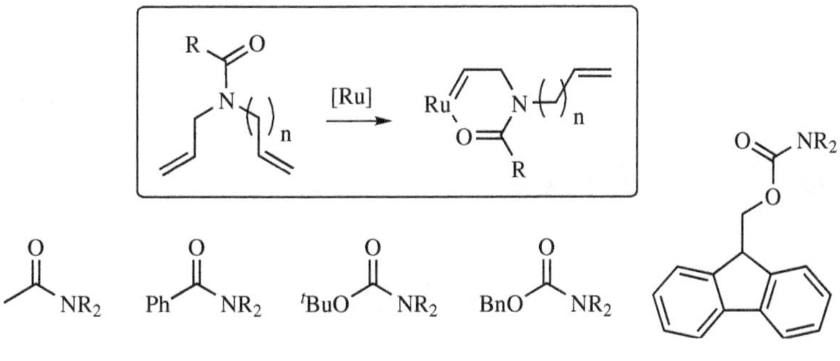

Figure 3.69 Potentially chelating nitrogen protecting groups: acetyl (Ac), benzoyl (Bz), *tert*-butyloxycarbonyl (Boc), carboxybenzyl (CBz), and 9-fluorenylmethoxycarbonyl (Fmoc).

ketones are more strongly inhibiting than esters (Fig. 3.70). Much faster reaction was observed for **70-3**, for which the large chelate ring size (eight members) disfavors chelation.

Even highly reactive **Mo-1** can show reduced RCM activity where a five-membered chelate can be formed (140,229). For readily cyclized dienes (see **71-1**, Fig. 3.71a), the effect is minimal (140), but for more demanding substrates such as **71-2**—in which 2,2-disubstitution of the distal olefin directs initiation to the site where chelation is favored—the effects are more extreme (229). Similarly, an early study by the Weiler group (36) using **Ru-1** highlighted five-membered chelate rings as most detrimental to RCM synthesis of lactones and lactams. Similar effects are seen with **Ru-2** (see **71-3**, **71-4**) (164,255), despite long reaction

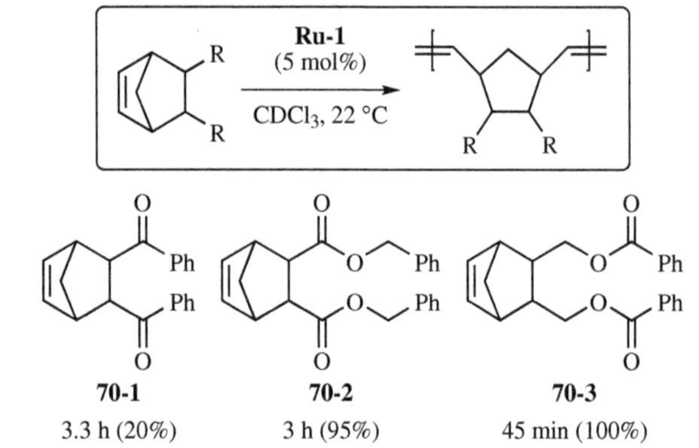

Figure 3.70 Stronger inhibiting effect of ketone versus ester functionalities, and impact of positioning of the carbonyl group.

times or high temperatures. It should be noted that the incidence of initiation at the 2,2-disubstituted olefin in **71-3** should normally be low. The observed resistance to RCM could reflect either sequestration of the catalyst when this does occur, or potentially synergic positioning effects of H-bonding and carbonyl chelation. A further example of the latter behavior will be seen below.

Six-membered carbonyl chelates likewise inhibit RCM, as shown explicitly by Fürstner and coworkers (250) for isolated chelate complexes of Ru styrenyl esters. The effect is less pronounced than for five-membered chelates, however, and drops off further as chelate ring size increases (36). Where the RCM targets are "easy" five- or six-membered rings, olefin binding is competitive and RCM is not significantly impeded (Fig. 3.71b). Thus, **Mo-1** cyclized **71-5** within 15 min at RT in 89% yield (140), while even **Ru-1** readily effects RCM of N-Boc-protected diallylamine (**71-6**) (256). Six- and seven-membered lactams are also accessible, but longer reaction or more aggressive heating is required where the probability of chelation is high (see **71-7**, **71-8**) (257,258). The competition for binding between a CO ligand and an electron-deficient olefin can be especially problematic, as discussed above within the context of butenolide synthesis, and depicted in Figure 3.40.

Synthesis of larger rings using **Ru-1** is more readily disrupted by chelation, as illustrated by the Weiler study just discussed. Similarly, RCM of **72-1**, for which a six-membered ketone chelate is anticipated, failed entirely (Fig. 3.72) (259), while less than 25% RCM of **72-2** was observed even after prolonged reaction (38). Addition of $Ti(O^iPr)_4$ improved yields of the 14-membered lactone to 55% (38). Use of this mild, sterically protected Lewis acid additive (sometimes in large amounts: see **72-3**) has evolved into a standard strategy for **Ru-1**, the relatively low thermal stability of which limits recourse to thermolytic methods of breaking the Ru–O bond. While good success was reported in a number of examples, $Ti(O^iPr)_4$ can also catalyze transesterification, isomerization, and nucleophilic addition to carbonyl compounds (260), and

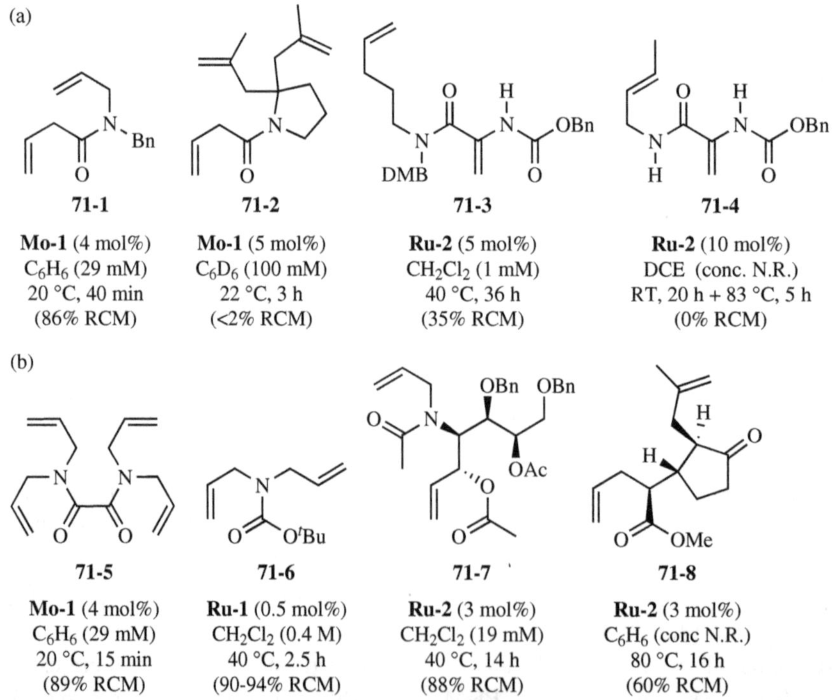

Figure 3.71 Impact of metal–carbonyl chelation on RCM synthesis of five- and six-membered targets. Dienes are depicted in the conformation expected to promote chelation. (a) Deleterious impact of five-membered chelates. (b) Reduced impact of six-membered chelates.

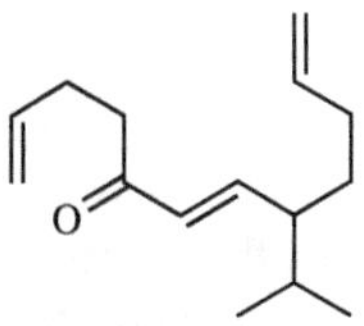

72-1

Ru-1 (10 mol%)
CH_2Cl_2 (0.6 mM)
40 °C, 48 h
(0% RCM)

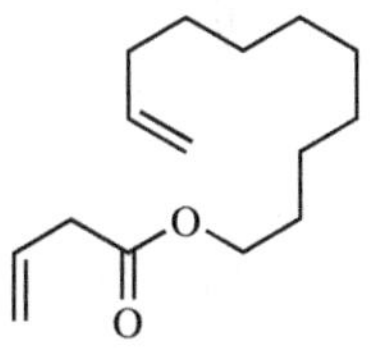

72-2

Ru-1 (5 mol%)
CH_2Cl_2 (conc N.R.)
3 d, 25 °C (22% RCM)
+ Ti(O^iPr)$_4$ (5 mol%)
40 °C (55% RCM)

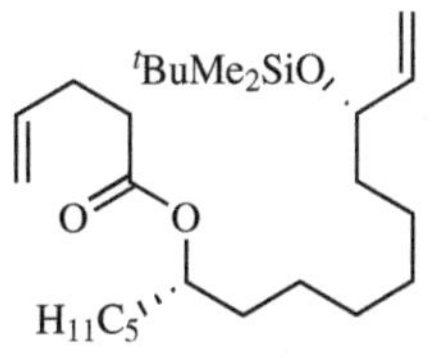

72-3

Ru-1 (4 mol%)
Ti(O^iPr)$_4$ (30 mol%)
CH_2Cl_2 (2.5 mM)
40 °C, 2 x 24 h
(80% RCM)

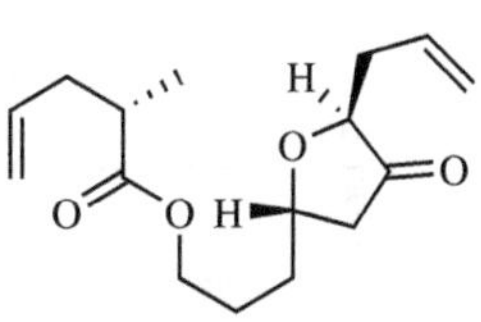

72-4

Ru-2 (9 mol%)
DCE (3 mM)
50 °C, 15 min
(72% RCM)

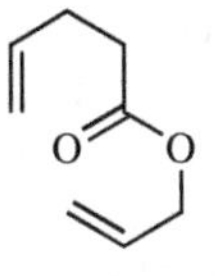

72-5

Ru-2 (10 mol%), CH_2Cl_2 (100 mM), 40 °C, 5 h

Additive	Cyclo-monomer :	Cyclo-dimer (%)
–	0	100
1.05 ATPH	90	10

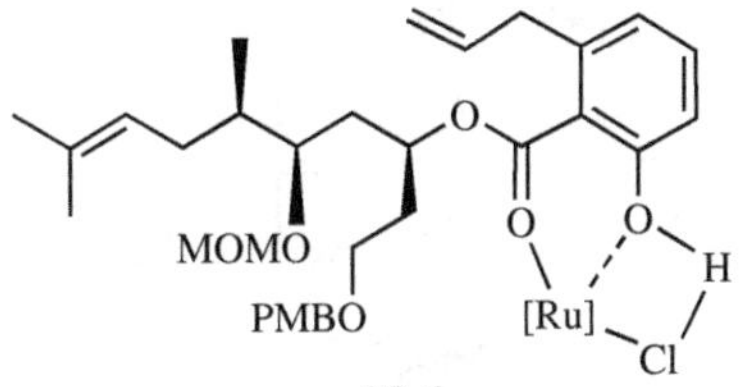

72-6

Ru-2′ (10 mol%)
C_7H_8 (5 mM)
80 °C, 20 h
(69% RCM)

Figure 3.72 Deleterious impact of five- and six-membered carbonyl chelates on RCM synthesis of medium or macrocyclic rings. Dienes are depicted in the conformation expected to promote chelation. ATPH = aluminum tris(2,6-diphenylphenoxide).

has been reported to trigger isomerization during intended RCM reactions (261). The higher reactivity and reduced Lewis acidity of second-generation ruthenium catalysts typically give better results, as shown for **72-4** (262). Nguyen and coworkers (263) have reported that use of a bulky aluminum Lewis acid to sequester the carbonyl donor also improves selectivity for seven-membered lactone "cyclomonomer" at concentrations of 100 mM (see **72-5**).

Finally, a special case of carbonyl chelation is shown in **72-6** (35). The unexpected poor performance of **Ru-2′** is unlikely to be due to CO chelation to the alkylidene, which would yield a seven-membered ring. Instead, hydrogen bonding between the phenol and a chloride ligand on ruthenium may promote chelation of the *o*-carbonyl functionality. The methoxy-protected derivative showed much improved yields, reaching 93% after 1.5 h.

Oxygen Donors: Ethers The excellent tolerance of metathesis catalysts for ether functionalities is illustrated by the successful RCM of ether-protected alcohols (e.g., Bn, MOM, PMB, and Tr), and the accessibility of crown ethers via RCM (264). Again, however, oxygen chelation can retard metathesis, particularly for homoallylic ethers. The stability of five-membered ether chelates is exemplified by the impressive robustness of the Hoveyda catalyst **Ru-3**. Although the high productivity of this catalyst is itself evidence for successful dechelation, this is not an entirely valid comparison, as the styrenyl ether is removed from the coordination sphere of the metal following the initial metathesis event, while the potential for substrate chelation exists in every turnover.

Synthesis of crown ethers that are not susceptible to chelation is fast and efficient using **Ru-2′**, given appropriate dilutions. Thus, 14-membered **73-1** (Fig. 3.73), the precursor to which would yield a four-membered chelate ring, was obtained in greater than 90% yield after just 0.5 h, at a diene concentration of 0.5 mM (10). (At 5 mM, metathesis yielded almost wholly oligomeric products.) In comparison, Wu and coworkers (265) reported very slow reaction (60 h) in RCM of **73-2**. For this substrate, installation of the metal at either olefinic site can form an ether chelate, of five or six members.

Chelation by the oxygen atom in homoallylic alcohols is also suspected in some cases (221,222,266), as noted at the end of Section 3.3.4.1. Binding is favored by the reduced steric encumbrance at oxygen for alcohols, versus ethers.

Sulfur Donors: Sulfides Relatively few examples of RCM of sulfur-containing substrates appear in the recent literature,

(a)

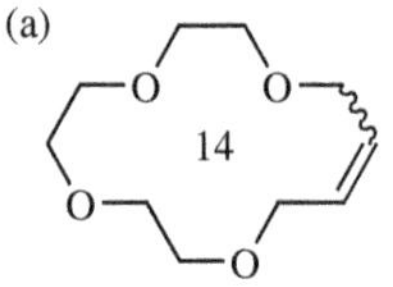

Conc. (mM)	RCM at equil. (%)	Oligomers at 15 min (%)
5	4	100
0.5	92	24

73-1

Ru-2' (5 mol%)
CH_2Cl_2, 40 °C, 0.5 h

(b)

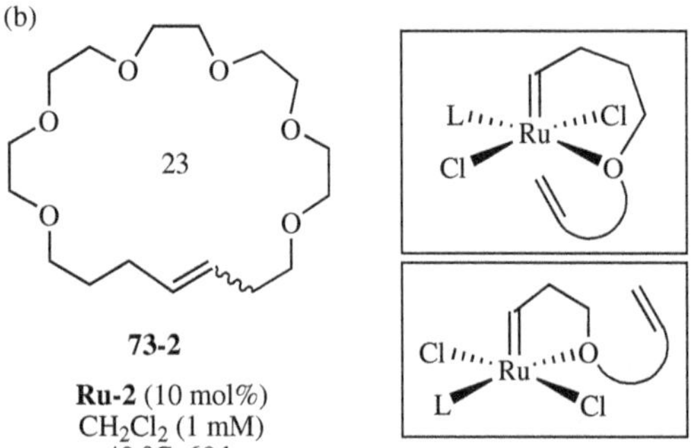

73-2

Ru-2 (10 mol%)
CH_2Cl_2 (1 mM)
40 °C, 60 h
(75%)

Figure 3.73 RCM of polyether derivatives. (a) Allylic: ether chelation is unfavorable. (b) Chelation is highly likely following initiation at either end. Inset shows the potential five- or six-membered chelate complexes.

but developments from the early 1990s to the present have been recently reviewed (267). Two earlier reviews remain useful for their detailed overviews of the performance of both molybdenum and ruthenium systems (1,268).

Divalent sulfur (sulfides or thioethers, disulfides, and dithioacetals) is remarkably well tolerated by **Mo-1**, as evidenced most strikingly by the rapid, efficient RCM of diallyl sulfide **74-1** (full conversions within an hour at 20 °C, at a catalyst loading of 5 mol%; Fig. 3.74) (228). In comparison, yields of only about 10% were achieved with **Ru-1** even after 20 h (223). Quantitative RCM was also observed at high temperatures (80 °C) using **Ru-2′**: thermolysis presumably aids in breaking the Ru–S bond. A rare example of the successful RCM of a sulfide by **Ru-1** was reported by the Hanson group (269). While yields of less than 20% were obtained with **74-2a** even at a catalyst loading of 15 mol%, RCM was near-quantitative for **74-2b**, in which an ester substituent was installed α to sulfur. As already discussed in the context of nitrogen binding, the joint steric and inductive effects of α-substitution prove decisive in inhibiting sulfur coordination (in this specific case, they may also inhibit chelation of the phosphine oxide).

The greater resistance of the Ru–NHC systems to poisoning is highlighted by the successful RCM of both diallyl disulfide **74-1** and dithiine **74-3** using **Ru-2′** (223). The success of the latter reaction is particularly notable, given that a five-membered sulfur chelate (see **74-4**) can be formed, irrespective of the site of initiation. A sulfur-chelated analog of the Hoveyda catalyst **Ru-3** was found to require aggressive thermal conditions (100 °C) for reasonable rates in RCM of DeDAM (270).

Unsurprisingly, given the susceptibility of **Ru-1** to sulfur chelation, attempted macrocyclization of dithioacetal **75-1** returned only unreacted diene (Fig. 3.75a) (271). In contrast, Danishefsky and coworkers (272) reported that **Ru-2** successfully cyclized the challenging conjugated macrocycle **75-3** (Fig. 3.75b). In this example, sulfur chelation is limited by the electron-deficiency and substitution of the conjugated diene, which favor initiation at the distal site.

The results above suggest that **Ru-2** may be more useful than has generally been considered for RCM of substrates containing divalent sulfur. In some cases, however, the nucleophilic properties of the sulfur center can give access to additional side-reactions. In a recent example reported by van Otterlo and coworkers (273), attempted RCM synthesis of eight-membered **76-1a** afforded comparable amounts of the unexpected rearrangement product **76-2** (Fig. 3.76a). Very high catalyst loadings were required (ca. 50% **Ru-2**): a deactivation pathway involving intramolecular nucleophilic attack on the alkylidene was proposed. The problem was curbed by oxidation of the sulfur atom, as indicated by the 95% yield of **76-1b** (Fig. 3.76b). Oxidation is indeed a general strategy for successful metathesis of sulfur-containing substrates, and we turn to such examples next.

Curbing Sulfur Donation: Sulfur Oxides The tolerance of both Mo and Ru catalysts toward sulfonates and sulfonamides is suggested by the ubiquity of tosylate-protected alcohols and amines in RCM. Many examples have been reported of the RCM of dienes containing sulfonates (267,268,274). Of note, the Hanson group (275) established in early work that protection of the NH group in secondary sulfonamides was not required for effective RCM, even using **Ru-1**.

Several studies suggest that the potential formation of five- and six-membered S=O chelates does not impede RCM by Ru–NHC catalysts. Good initiation efficiency was reported for a sulfonate analog of the Hoveyda catalyst containing a six-membered chelate ring (276) (implying tolerance for γ,δ-unsaturated sulfonates), while tolerance for β,γ-unsaturated sulfoxides is indicated by the van Otterlo (276) study seen in Figure 3.76, and by the fact that diallylsulfoxide itself is quantitatively cyclized within 1 h using 0.5 mol% **Ru-2**.

Examples **77-1a–c** (Fig. 3.77) illustrate RCM of medium-ring allylic sulfonates with minimal conformational constraints (277,278). Thus, Metz and coworkers reported 76–82% yield of seven- and eight-membered **77-1a** and **77-1b** at loadings of about 2 mol% **Ru-2**, although higher dilutions were required for the latter, as

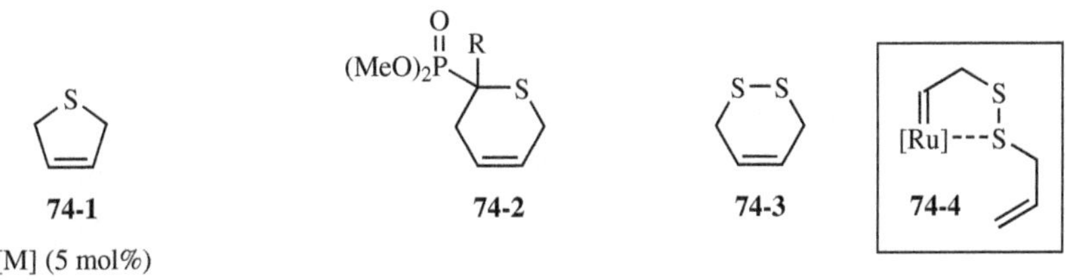

[M] (5 mol%)

[Ru] (mol%)	Conditions	Yield (%)
Mo-1	C_6H_6 (37 mM) 20 °C, 1 h	99
Ru-1	C_6D_6 (100 mM) RT, 20 h	12
Ru-2'	C_7D_6 (100 mM) 80 °C, 1 h	100

Ru-1
CH_2Cl_2 (50 mM)
40 °C
a R = H
15 mol%, 3 d (19%)
b R = $CO_2{}^tBu$
5 mol%, 20 h (99%)

[Ru] (5 mol%)
CD_2Cl_2 (100 mM)
40 °C
Ru-1: 12 h (15%)
Ru-2': 4 h (100%)

Figure 3.74 Synthesis of five- and six-membered targets by RCM of dienes bearing divalent sulfur groups. Inset: formation of a stable five-membered chelate ring in RCM of substrates bearing a homoallylic sulfide.

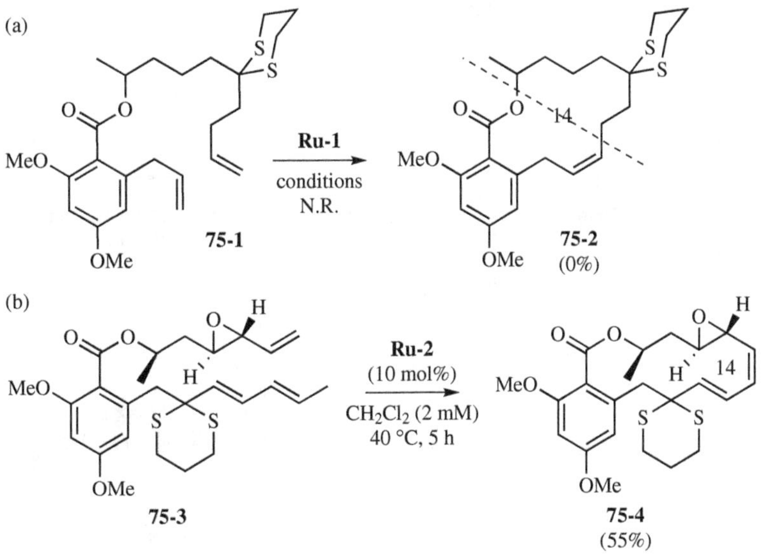

Figure 3.75 RCM macrocyclization of dienes bearing divalent sulfur groups. (a) Failed RCM by **Ru-1**. (b) Successful RCM by **Ru-2**.

expected (5 mM vs 15 mM). Even nine-membered **77-1c** was isolated in 53% yield at 5 mM concentrations. The sulfonate unit, in conjunction with β,γ-unsaturation, is clearly an effective gearing element. A γ,δ-unsaturated sulfone likewise appears effective (even using **Ru-1**), as judged by Yao's (278) 85–98% yields of seven- and eight-membered **77-2a** and **77-2b**. The exocyclic phenyl group may aid in cyclization. Notably, however, the α,β-unsaturated sulfonate rings **77-3b/c** were not formed. Chelation is not the issue here, given the small size of the chelate ring (four members). The position of the sulfonate appears critical to the conformational bias in this system (cf. the successful synthesis of **77-1b/c**). The dependence of RCM yields on the position of unsaturation has been mentioned previously (279).

A related report from the Bates group (280), describing 89% yield of eight-membered sulfone **78-1** (Fig. 3.78), versus 40% for the corresponding sulfoxide **78-2**, suggests that two S=O bonds may be essential. This is not unreasonable, but the 13 mM higher concentration for the sulfoxide may also have limited RCM. A detailed study by Gais and coworkers (266) indicates that sulfoximine-functionalized medium rings (see **78-3**) are likewise accessible. Finally, exocyclic sulfones were successfully employed by the Fürstner group (249) in macrocyclization (see 13-membered **78-4**).

3.3.4.3 Oxygen-Atom Donors: Aldehydes Few examples have been reported of RCM of aldehyde-containing dienes.

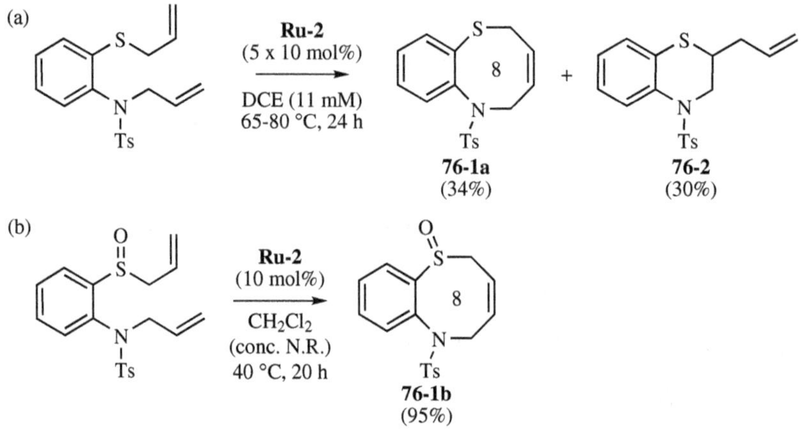

Figure 3.76 (a) Observation of a sulfide-mediated side-reaction during RCM synthesis of **76-1a**. (b) High-yield route to RCM product **76-1b** enabled by oxidation of sulfur.

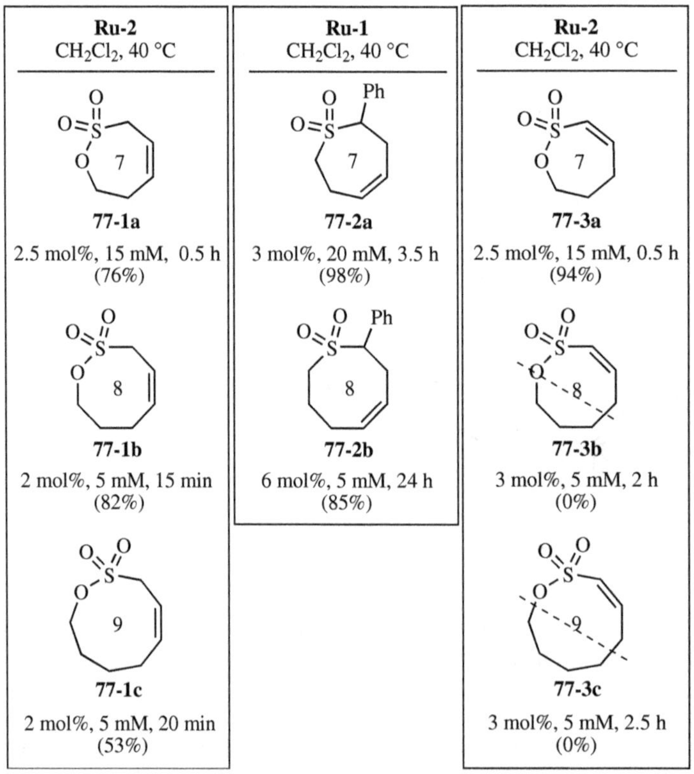

Figure 3.77 RCM synthesis of medium-ring sulfonates (**77-1**, **77-3**) and sulfones (**77-2**).

For the group 6 catalysts, this is unsurprising: the molybdenum and tungsten catalysts rapidly abstract oxygen from aldehydes. Indeed, this forms the basis for the very efficient methodologies developed to quench such catalysts (Section 3.4.6) (281). The situation for ruthenium catalysts is less clear-cut. Grubbs (235) demonstrated that first-generation Ru systems can tolerate an allylic aldehyde, albeit within a highly favored RCM reaction (Fig. 3.79). While the tendency of Ru(II) species to decarbonylate aldehydes might be expected to complicate slower, more challenging RCM reactions, no evidence for such behavior has been reported.

3.3.5 Challenging Substrates: Microwave RCM

Microwave methodologies are particularly useful for RCM of sterically and electronically deactivated olefins, or RCM of densely functionalized substrates. In some cases, microwave heating has enabled productive RCM where conventional thermal reactions failed (283). An important study by

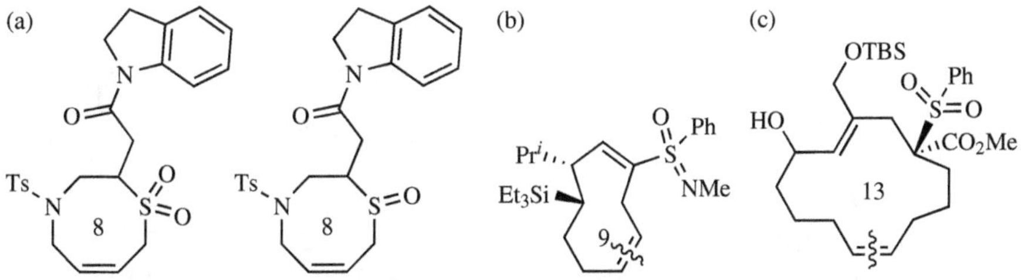

Figure 3.78 RCM synthesis of medium sized or macrocyclic rings containing sulfur oxides. (a) Relative yields in RCM cyclization of an eight-membered sulfone (**78-1**) or sulfoxide (**78-2**). (b) Exemplary synthesis of a medium ring sulfoximine. (c) Synthesis of 13-membered **78-3**, bearing an exocyclic sulfone. TBS = $SiMe_2{}^tBu$. The label (*) denotes slow catalyst addition.

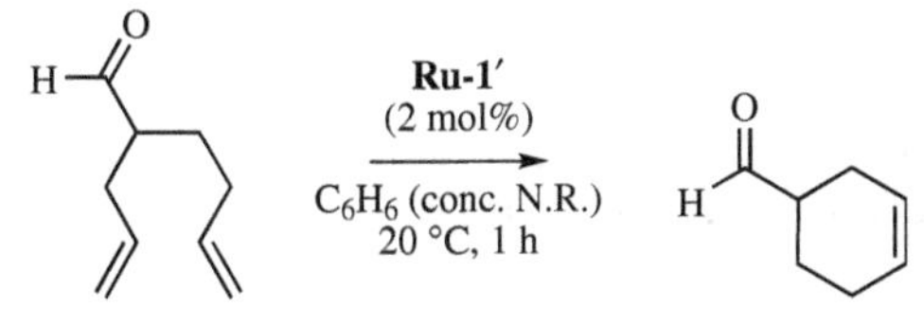

Figure 3.79 RCM of an aldehyde-containing substrate.

Kappe and coworkers (283) strongly suggests that the beneficial effect is simply much faster heating. When oil-bath temperatures were raised sufficiently to match the microwave solution heating profile (measured with an immersed fiber-optic probe, rather than an external IR sensor), the extent of catalyst decomposition was identical. Under standard heating methods, in which the bath temperature was simply set at the intended operating temperature, the solvent thermal ramp was much slower, and considerably more catalyst degradation occurred before operating temperatures were reached. The chief advantage of microwave methodologies therefore lies in the convenience and rapidity with which they reach operating temperatures.

Selected, representative studies that compare the effects of microwave versus standard thermal heating are shown in Figure 3.80a. In RCM of an electron-deficient acrylate, Wilson and coworkers (285) reported 97% yield of **80-1** within 5 min at 150 °C (microwave; 60–80 W), as compared to 75% after 20 min at 50 °C. The latter figure did not increase even after heating for 5 h. Much improved rates were also observed for the μW-RCM process in synthesis of eight- and nine-membered **80-2** and **80-3** (286,287). It will be noted that the putative temperatures are well above the boiling point of the solvents. In the case of **80-1**, reaction was carried out in a pressure vessel; it is unclear whether this is also the case for **80-2** and **80-3**. The difficulties of accurately assessing temperatures under microwave conditions are apparent from the discussion above.

Reiser (288) has pointed out that ethylene must be removed from the reaction to realize the full benefits of microwave methodologies for demanding RCM reactions. Highly substituted **80-4** was quantitatively formed on use of a small Teflon tube to introduce a flow of argon (thus effecting mixing as well as ethylene removal). In a comparative study designed to assess the impact of ethylene under optimized microwave conditions (Fig. 3.80b), Kappe and coworkers (284) reported 85% yield of eight-membered **80-5** in just 5 min in argon-sparged vessels, using 1.0 mol% **Ru-1** (DCE, 83 °C). Yields dropped to 60% in a sealed vessel, even at 1.5 mol% of the catalyst. Longer reaction did not improve performance, indicating catalyst decomposition.

RCM under microwave conditions has had a major impact on access to stapled and backbone-stabilized peptidomimetics. Even for small peptidomimetics (see **81-1**, Fig. 3.81) (288), microwave heating can significantly improve rates and yields. For larger targets with a greater density of polar functional groups, conventional thermal RCM often fails, even at high catalyst loadings. Microwave irradiation is thought to aid reaction by disrupting secondary structures that inhibit access of the catalyst to the double bonds (and, for off-resin reactions, reducing solubility in metathesis-compatible organic solvents). A breakthrough report from the Robinson group (289) described the quantitative synthesis of carbocyclic analogs of α-conotoxin ImI (see, for example, **81-2**) within 2 h by microwave heating with 5 mol% **Ru-2**, in conjunction with solid supports to isolate substrate sites. No intermolecular CM products were observed, even at a relatively high resin substitution level (>0.5 mmol/g). Microwave methodologies using these or closely related conditions have since become the default (288,290–292).

(a)

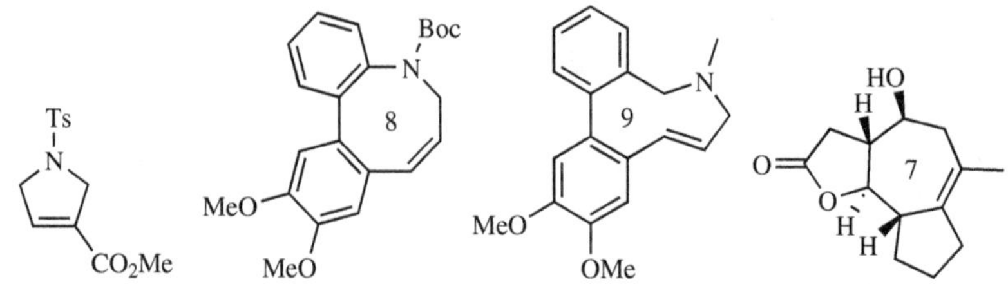

80-1	80-2	80-3	80-4
Ru-2 (3 mol%) DCE (25 mM)	**Ru-2** (3 mol%) C_7H_8 (50 mM)	**Ru-2** (3 mol%) C_7H_8 (67 mM)	**Ru-2** (3 × 5 mol%) C_7H_8 (20-30 mM)*
μW: 150 °C, 5 min 60-80 W (97%)	150 °C, 5 min 150 W (69%)	150 °C, 15 min 150 W (55%)	110 °C, 1.5 h 300 W (98%)
Δ: 50 °C, 20 min (75%)	110 °C, 3 h (58%)	110 °C, 12 h (15%)	110 °C, 2 h (57%)

(b)

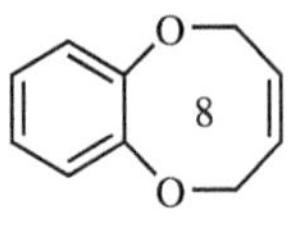

80-5

DCE (20 mM)
83 °C, 5 min
μW (150 W)

	Ru-1 (mol%)	Yield (%)
Ar-sparged*	1	85
sealed	1.5	60

Figure 3.80 (a) Comparison of microwave (μW) versus thermal (Δ) heating in RCM. (b) Assessing the impact of ethylene retention or removal. The label (*) denotes sparging with inert gas.

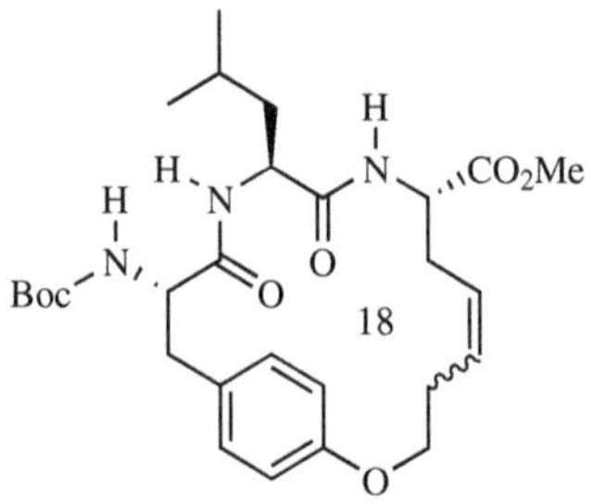

81-1

Ru-2 (3 × 10 mol%), TCE (10 mM); + Cy_2BCl (10 mol%)

	T (°C)	t (h)	Yield (%)
μW[a]	115	1	100
Δ	115	18	74

[a]1200 W

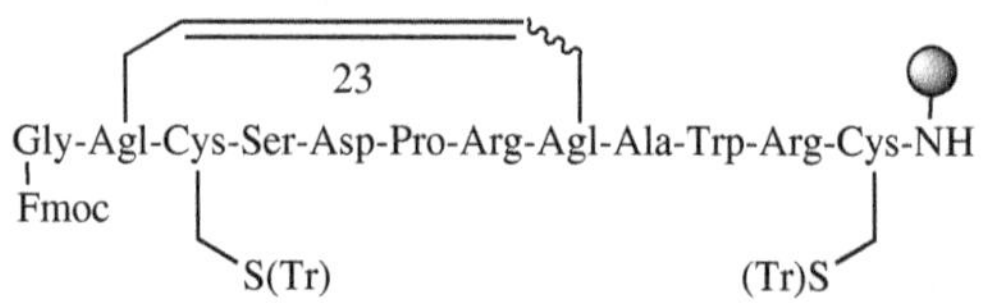

81-2

Ru-2, CH_2Cl_2 (27 mM); + LiCl/DMF (10 mol%)

	Ru-2 (mol%)	T (°C)	t (h)	Yield (%)
μW[a]	5	100	2	100
Δ	50	40	12	70

[a]100 W

Figure 3.81 Peptidomimetics via μW-enabled RCM. Sealed-tube μW reaction for **81-2**; the high temperatures sharply reduce the solubility of ethylene. Letter codes denote amino-acid residues; Agl = allylglycine. TCE = trichloroethane; Tr = trityl; Fmoc = 9-fluorenylmethoxycarbonyl.

3.3.6 Challenging Substrates: Relay RCM (RRCM)

RRCM, a strategy originally developed by the Hoye group, enables specification of the site of initiation. This can address problems of chelation or preferential dimerization, and is particularly useful in promoting RCM of substrates that suffer from deactivation at both olefinic sites (293). As discussed in Section 3.3, steric or electronic deactivation of one olefin can be compensated for by initiation at an unperturbed site. When both ends are strongly deactivated, RCM commonly fails. In RRCM, an easily initiating ω-olefin chain is tethered to one olefin (Site **A**, Fig. 3.82). Initiation at the accessible ω-olefin, followed by 2+2cycloaddition,

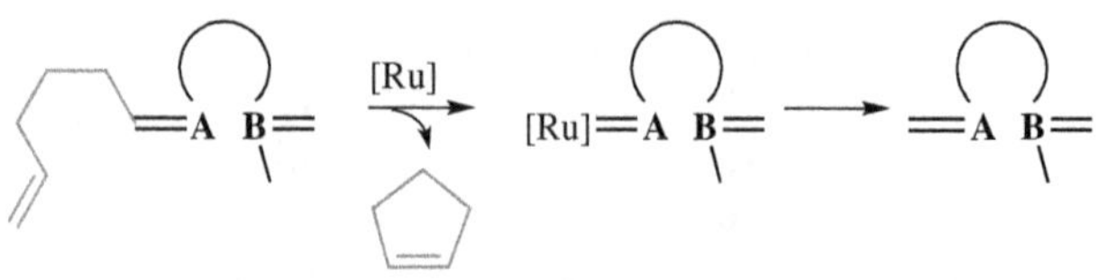

Figure 3.82 Schematic illustrating relay RCM. Bold letters indicate Sites **A** and **B**.

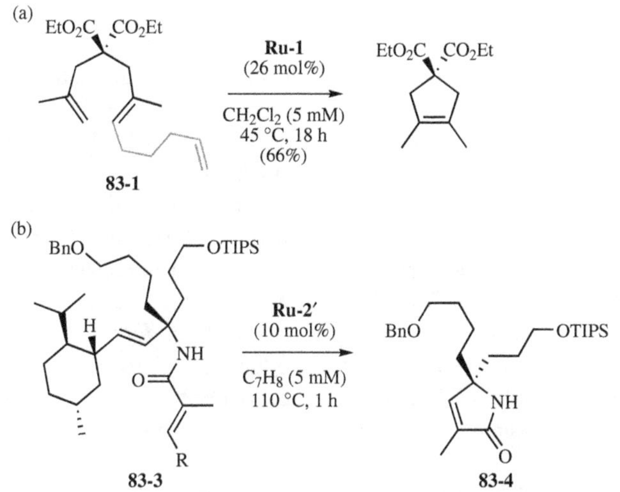

	R	83-4 (%)
a	H	0
b		63
c	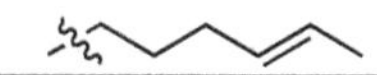	91

Figure 3.83 Use of RRCM to overcome constraints on cyclization efficiency. (a) Steric constraints. (b) Steric and electronic constraints.

effects installation of the alkylidene at Site **A**, accompanied by ejection of a stable cycloolefin, which is typically a five-membered ring (293,294). The target RCM reaction can then be consummated, if deactivation at Site **B** is not excessive. While RRCM can improve RCM yields, the problems they are designed to address are inherently challenging, and yields are often modest even using high catalyst loadings and long reaction times.

RRCM offers a "back-door" means of amplifying the reactivity scope of metathesis catalysts, enabling cyclization of more demanding substrates than would otherwise be possible. This principle is illustrated in Figure 3.83a. Hoye and coworkers (293) reported cyclization of **83-1** by **Ru-1** under moderate conditions to afford nearly 70% of the tetrasubstituted cyclopentene (see also **29-1**). This is a remarkable result, given the failure of the corresponding "non-relay" reaction shown in Figure 3.29, and the low reactivity of **Ru-1** toward 2,2-disubstituted dienes in general (Section 3.3.2). In Figure 3.84, RRCM is used to overcome the combined steric and electronic deactivation of the parent diene **83-3a**. Spino and coworkers (295) observed much improved yields of **83-4** for the *n*-pentene derivative **83-3b**. Significant further improvements accompanied installation of a terminal methyl group (see **83-3c**). This inhibits re-formation of tether-free **83-3a** by CM with the ethylene by-product that accompanies installation of the metal endgroup onto **83-3b**.

RRCM has also been used to good effect in cyclization of unsymmetrically deactivated dienes, which are otherwise susceptible to formation of linear, head-to-head dimers, as discussed above (Section 3.2.1). As illustrated in Figure 3.84a, metathesis of **84-1** yielded principally the acyclic dimer **84-2** after 15 h, even at very high dilutions (0.4 mM). Installation of a relay at the sterically more encumbered site (see **84-3**) gave access to the desired cyclophane **84-4** in nearly 70% yield (Fig. 3.84b) (296).

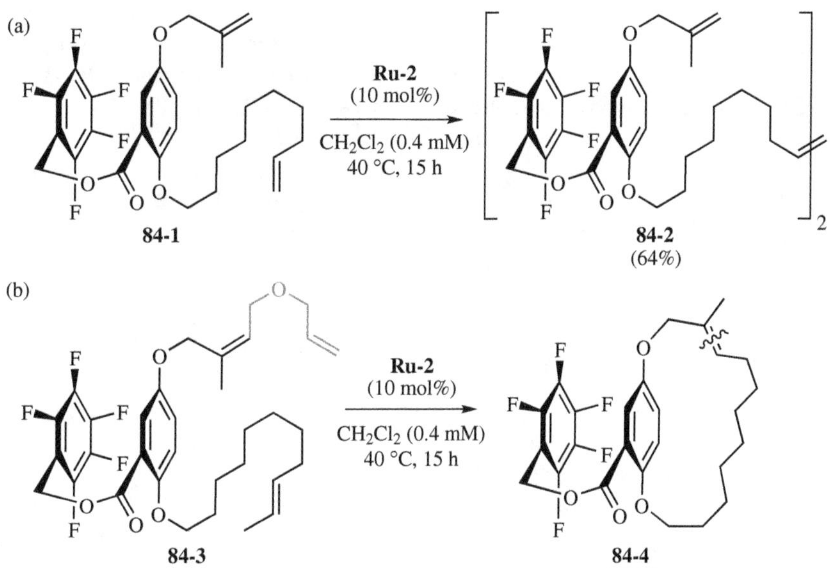

Figure 3.84 Use of RRCM to circumvent dimerization of unsymmetrically deactivated dienes (297). (a) Preferential dimerization via RCM. (b) Preferential cyclization via RRCM.

A final example illustrates the point that RRCM strategies are effective only where the bottleneck to reaction lies in installation of the alkylidene at a suitable site. The Arndt group examined the utility of RRCM in the attempted synthesis of the 19-membered ring in Jasplakinolide. Yields of **85-2** were essentially identical to those obtained in RCM (24% vs 18%; Fig. 3.85) (298). The failure of RRCM to improve reaction outcomes was attributed to difficulties in the final cyclization step.

A useful recent review presents an analysis of success and failure in RRCM reactions (294). Particular problems lie in regeneration of the tether-free, RCM substrate, as noted above, and in competing oligomerization. Hoye and coworkers (299) recently demonstrated that a diethyl allylmalonate unit serves as a highly effective "relay activator," which benefits from Thorpe–Ingold assistance and from faster decomplexation of the cyclic alkene.

3.4 PRACTICAL ASPECTS

This section provides a step-by-step guide to setting up RCM syntheses. Key operational parameters that must be considered before, during, and after the reaction are summarized.

3.4.1 Catalyst Purity and Loadings

Catalyst purity—often taken for granted—is one of the most fundamental considerations in RCM. The colors of the solid catalysts offer an immediate visual indicator of purity. The Schrock catalyst **Mo-1** is a bright yellow microcrystalline powder (176). As it decomposes in the solid state (even at RT), the powder darkens to brown. The catalyst should be stored at −30 °C in a glovebox, and recrystallized from rigorously dry, degassed pentane if discolored.

The ruthenium catalysts vary in color from purple or pink (for the Grubbs catalysts **Ru-1** and **Ru-2**), to vivid dark green for the Hoveyda catalyst **Ru-3**, and deep red for the indenylidene catalyst **Ru-4**. Any of the H_2IMes catalysts prepared via *in situ* carbene deprotection may be contaminated by excess base, metal salts, or (at worst) ruthenium hydrides formed during methanol washing (Section 3.3.1.1). While hydrides can be detected from their upfield NMR signals (Table 3.1), other contaminants can be more difficult to observe. Standard practice in our own laboratories involves taking up newly-prepared or obtained catalyst in benzene, and filtering through Celite in the glovebox to remove these polar impurities. The clean catalysts can be stored for years at RT in the solid state under N_2: while they are relatively air- and water-stable solids, storage in the glovebox is recommended.

As indicated by the examples cited throughout this chapter, typical catalyst loadings are on the order of 1–10 mol%, albeit extensive optimization has not been pursued in most cases. For the Ru catalysts, in particular, loadings should be kept to the minimum required to effect RCM. This is important not only to limit purification problems (see Section 3.4.6 below), but to minimize the potential for secondary reactions that erode yields and generate by-products.

3.4.2 Solvent Choice and Concentration

Solvent Choice Standard considerations in choosing the solvent for RCM syntheses include diene and product solubility and stability, the desired reaction temperature, and the potential for isomerization. The solvent medium affects the rates of catalyst initiation, productive metathesis, and catalyst deactivation, all of which contribute to the overall success of an RCM reaction. Caution should be used in extrapolating from initiation rates alone, unless this step is known to be rate-determining. (This is a particular concern for studies carried out with both **Ru-1** (52) and—depending on substrate—**Ru-3**) (58).

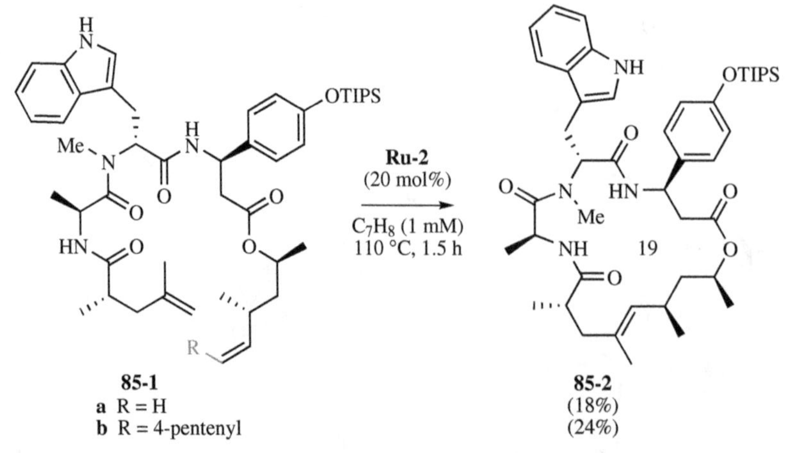

85-1
a R = H
b R = 4-pentenyl

85-2
(18%)
(24%)

Figure 3.85 Example of failed RRCM, attributed to difficulties in the final cyclization step (298).

TABLE 3.1 Colors and Key NMR Chemical Shifts for Common Catalysts; NMR Values for Common Hydride Contaminants

Catalysts (Color)	^{1}H NMR[a]	^{31}P{^{1}H} NMR	Solvent	References
Mo-1 (yellow)	12.12 (s)	–	C_6D_6	176
Ru-1 (purple)	20.02 (s)	36.61	CD_2Cl_2	49
Ru-2 (pink)	19.16 (s)	31.41	CD_2Cl	85
Ru-2′ (pink)	19.91 (s)	34.9	C_6D_6	300
Ru-3 (green)	16.56 (s)	–	$CDCl_3$	54
Decomposition Products				
RuHCl(CO)(L)(PCy_3)				
L = PCy_3 (**Ru-14a**)	−24.21 (t, J = 18 Hz)	46.9	C_6D_6	87
L = H_2IMes (**Ru-14b**)	−24.9 (d, J = 21 Hz)	46.8	C_6D_6	301
L = IMes (**Ru-14b′**)	−24.82 (d, J = 21 Hz)	47.8	C_6D_6	87
RuHCl($CO)_2$(L)(PCy_3)				
L = PCy_3	−4.9 (t, J = 20 Hz)	49.8	C_6D_6	87
L = IMes	−4.77 (d, J = 23 Hz)	48.2	C_6D_6	87
Grubbs hydride (Fig. 3.57b)	−8.61 (s)	–	CD_2Cl_2	303
$MePCy_3Cl$	2.42 (CH_3)	34.5	C_6D_6	302

[a]Alkylidene or hydride signal, except for $MePCy_3Cl$.

Importantly, however, solvent effects can also be strongly substrate-dependent. Early solvent screens of diallylanilines and diallylmalonates indicated faster, more efficient RCM by **Ru-1** in CH_2Cl_2 than aromatic solvents (41,232). Likewise, acrylate CM reactions promoted by **Ru-2** were found to afford maximum yields in CH_2Cl_2; use of C_6H_6, THF, $CHCl_3$, or CCl_4 reportedly gave much poorer results (144). The opposite trend was described for RCM of malonates (303) and sulfoximine-functionalized dienes (266) by **Ru-2**, with maximum yields being attained in toluene. Reports are conflicting in RCM macrocyclizations: CH_2Cl_2 was found to result in higher conversions for macrolactams (304), whereas toluene gave higher rates for mactolactones (66). However, significant isomerization in toluene undermined (66) the benefit of faster metathesis.

Chlorinated solvents can reduce isomerization and improve RCM yields, as discussed in Section 3.3.1.1. Chloroform, however, should be used with caution (especially at elevated temperatures), as it can be more aggressive in triggering catalyst decomposition than CH_2Cl_2. While Ru-catalyzed RCM is typically carried out in chlorinated or aromatic solvents, THF is occasionally used for reasons of substrate solubility (though competing dehydrogenation in refluxing THF has been described) (232). More strongly coordinating solvents (e.g., acetonitrile) are strongly inhibiting. The higher solubility of **Mo-1** enables its use in pentane or hexanes, as well as the solvents indicated above. THF slows reaction, although this may be advantageous where selectivity is an issue. Fluorinated aromatics such as octafluorotoluene or hexafluorobenzene, though not widely used, have been shown to significantly improve RCM rates in some cases (125,305).

Among other less widely used solvents, Adjiman and coworkers (303) reported that acetic acid improved reaction rates in RCM of DeDAM by **Ru-2**. Solubility and substrate compatibility are likely issues for broader use, as is faster catalyst deactivation in this solvent. Schulz and Wagener (306) recently reported that acetic acid leads to substantially lower molecular weights in metathesis polymerization of dienes, a marker for catalyst deactivation. In these studies, highest molecular weights were attained in CH_2Cl_2. Metathesis in alcohol and water is discussed in Chapter 21 and unconventional "green" solvents in Chapter 22. Risk factors for metathesis in primary alcohols have been discussed in Section 3.3.1.1. Adventitious alkoxides are a particular hazard for catalyst decomposition (see Fig. 3.15).

Concentrations For ring sizes of seven or larger, RCM yields are commonly concentration-limited (Section 3.2.1). Where concentrations are too high, oligomers or dimers will dominate. Reaction concentration is therefore one of the major tools available to manipulate product distributions. Dilutions must be sufficient to shift the ring-chain equilibrium completely in favor of RCM: at intermediate concentrations, the reaction will "stall" at the equilibrium proportion of the oligomer. Rate curves are important to identify a stalled reaction, in which the product distribution can normally be corrected by diluting further. A 10-fold increase is recommended: it is also essential to allow sufficient time (normally hours) for backbiting to occur. An additional catalyst charge should be added only if the product yield shows no increase over time, following dilution.

Within the limits imposed by selectivity, concentrations should be maximized, to promote RCM rates and catalyst productivity, and to minimize the Sheldon E-factor. Maximum concentrations must be assessed on an individual basis, but good starting points for five- or six-membered, seven-membered, medium and macrocyclic rings are given

TABLE 3.2 Recommended Starting Concentrations of Diene in Exploratory RCM Reactions

	Recommended Concentration	
Ring Size	Carbocycles, Lactones (10), mM	Azacycles (307)
5	100	Neat
6	100	1 M
7	5	50–200 mM
8–13	0.5	N.R.
14+	5	N.R.

in Table 3.2. The dilutions required are highly sensitive to the presence of heteroatoms such as nitrogen or silicon in the ring, and to conformational constraints, as discussed in the preceding sections. In general, dilutions beyond about 0.5 mM are impractical with the catalysts currently available, as metathesis is then too slow to compete with catalyst deactivation.

From the industrial perspective, access to macrocyclic and medium rings (see Section 3.6) is bedeviled by major concerns surrounding the environmental and economic costs of purchasing, operating, removing, and disposing of large volumes of solvent. Collins and James (308) have proposed an "Emac" index to take into account the dual impact of yields and dilutions on the efficiency of macrocyclization (Eq. 3.5).

$$\text{Emac} = \log[(\%\text{yield})^3 \times (\text{concentration in millimolar})] \quad (3.5)$$

These authors suggest that the high dilutions at which many RCM macrocyclizations are carried out are not in fact required. While this is clearly inaccurate for many of the cases outlined above, it presents an important challenge to both the research and the R&D communities. Remarkable successes have resulted from determined efforts to raise effective molarities in some cases (see Section 3.6: Industrial Applications of RCM), although the conformational constraint present in the tripeptide substrates should also be recognized.

3.4.3 Reaction Setup and Ethylene Removal

Total turnover numbers in RCM are adversely affected by unproductive exchange with ethylene (309,310). This is due in part to the extended time spent by the catalyst in its most vulnerable metallacyclobutane state: the unsubstituted metallacyclobutane is highly susceptible to decomposition, as is the methylidene resting state **Ru-12** (59). Deliberately sparging with an inert gas to accelerate loss of ethylene has been shown to increase RCM yields (288,311), though care must be taken to limit solvent evaporation. Use of sealed vessels with minimal headspace (e.g., NMR tubes, small vials in high-throughput wellplates, plug-flow reactors, etc.) inhibits volatilization of ethylene, with adverse effects on reaction rates and RCM yields, the extent of which will vary according to the vulnerability of the catalyst. For best performance, RCM should be carried out in a vessel open to an atmosphere of N_2 or Ar, with good stirring to ensure efficient mass transfer. Slightly reduced pressures have also been used to aid in removing higher-boiling olefins (propylene, butylene); see Section 3.3.1.3.

3.4.4 Temperature

High temperatures are generally required to achieve acceptable rates for slowly-initiating catalysts such as **Ru-2**. Where oligomers can be formed, high temperatures add to the driving force for cyclization. Not only do they weight the entropic term in the Gibbs–Helmholtz equation, but (by reducing viscosity and increasing thermal motion) they also maximize translational and rotational entropy. An early study by Danishefsky and coworkers (312) illustrated the improved yields of "cyclomonomer" **86-2a**, relative to the cyclic dimer **86-2b**, at 80 °C versus 40 °C (Fig. 3.86a).

Where the cyclic dimer is thermodynamically favored, however, higher temperatures are of no avail. Thus, Brynaert and coworkers (313) found that attempted RCM of **86-3** gave none of the desired **86-4** at temperatures ranging from ambient to 83 °C: instead, cyclic dimers **86-5** were obtained as the dominant products (Fig. 3.86b).

Heating accelerates the decomposition of both group 6 and Ru catalysts, and rate curves are essential to identify premature termination. The presence of unreacted diene is a key marker, distinguishing problems of catalyst deactivation from those of excessive concentrations (see Section 3.2.2 above). The thermal sensitivity of **Mo-1** is well established: as pointed out by Armstrong (1) this catalyst appears to react extremely rapidly or not at all, and decomposition in solution appears correspondingly rapid. Of interest in this context is the greater robustness of tungsten catalyst **W-1** (46).

The Grubbs precatalysts are thermally robust in the absence of substrate, with half-lives of 8 days (**Ru-1**) (59) and greater than 1 month (**Ru-2**) (60), respectively, at 55 °C in C_6D_6. Their methylidene resting states are much more fragile. The half-life for **Ru-12a** is reported to be only about 40 min under the same conditions, versus about 6 h for H_2IMes analog **Ru-12b** (59,60,309). Furthermore, the methylidene complexes are believed to decompose via unimolecular pathways (59,60), meaning that although RCM is slower at higher dilutions, rates of catalyst decomposition are unaffected. Unimolecular decomposition has obvious additional, adverse implications for the efficacy of catalyst anchoring strategies for recyclable RCM.

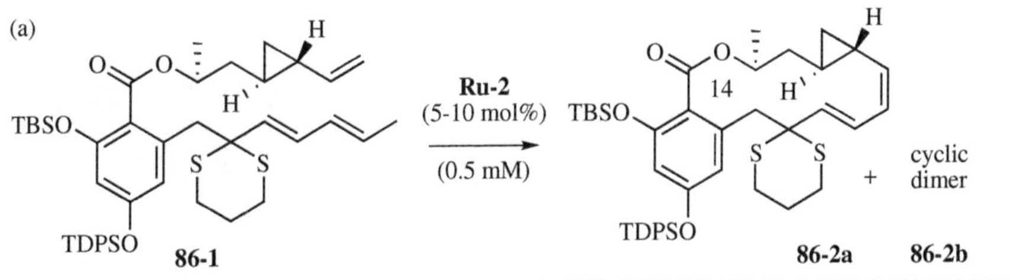

solvent	T (°C)	time	**86-2a** (%)	**86-2b** (%)
toluene	42	19 h	27	48
benzene	80	35 min	33	36

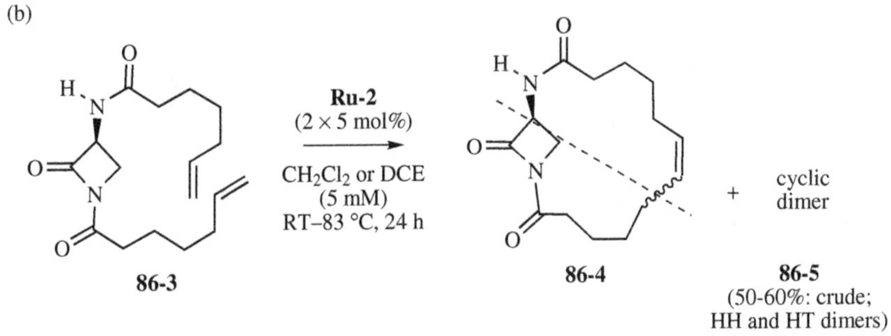

Figure 3.86 Impact of reaction temperature on RCM yields. (a) Increased yields at higher temperature (312). (b) Failure of high temperatures to promote a thermodynamically unfavorable cyclization. TBS = Si^tBuMe_2, TDPS = Si^tBuPh_2.

3.4.5 Reaction Time and Monitoring RCM

RCM synthesis of macrocyclic and medium rings typically involves rapid consumption of diene, followed by slower transformation of oligomers into products (see Fig. 3.12) (10). To maximize RCM yields, sufficient time must be allowed to attain equilibrium. Many apparently failed RCM reactions are due to premature termination: that is, workup after diene consumption is complete, but before oligomers have been recycled into RCM products. (Use of appropriate concentrations is also essential: see guidelines above; Section 3.4.2.)

In monitoring RCM reactions, it is essential to be able to gauge not only consumption of diene, but formation of the RCM target. ^1H NMR analysis is widely used, but is misleading, because spectra of oligomers are often indistinguishable from the composite spectrum of starting diene and RCM product (Fig. 3.87) (10,13). Where unreacted diene is present, more time or more catalyst is required: a plateau in the rate curve indicates the latter. Where oligomers (including dimers, trimers, etc.) are present, dilution is required. If this effects no change over a timescale of hours, more catalyst should be added.

Discrepancies in the reported accessibility of medium-ring targets arise from the difficulty in distinguishing the NMR spectra of cyclic dimers from the intended RCM target. As a case in point, greater than 90% yield of a conjugated caprolactone was initially reported at a concentration of 20 mM (142), based on NMR analysis. Subsequent work indicated that the major product, even at less than 5 mM, was the head-to-tail cyclic dimer, as confirmed by mass spectrometry (MS) (66,154). This underscores the point that MS analysis is essential to confirm product identity, particularly for targets in the medium-ring regime.

Thin layer chromatography (TLC) is a convenient qualitative means of monitoring the progress of RCM reactions. A particular strength of this technique lies in its ability to reliably report on the presence and disappearance of oligomeric species (Fig. 3.88). Caution must be taken to minimize introduction of oxygen, but this is easily achieved by using a syringe to sample through a septum-capped Schlenk sidearm.

Gas chromatography (GC) is often used as a quantitative means of assessing RCM rates. Given an internal standard and GC calibration curves (to correct for differences in the response factors of the constituents), GC analysis can accurately report on the amount of diene and RCM product present. Oligomers are normally too involatile to observe by GC methods, and are therefore assessed by difference. Without calibration curves for the product, the presence of oligomers at equilibrium (i.e., use of excessively high concentrations for the reaction at hand) is readily overlooked. Liquid chromatography with mass spectrometric (LC-MS) detection is a complementary method in this respect: the combination of ultrahigh performance liquid chromatography (UHPLC) and a self-calibrating time-of-flight (TOF) detector appears particularly attractive from the perspective of efficiency.

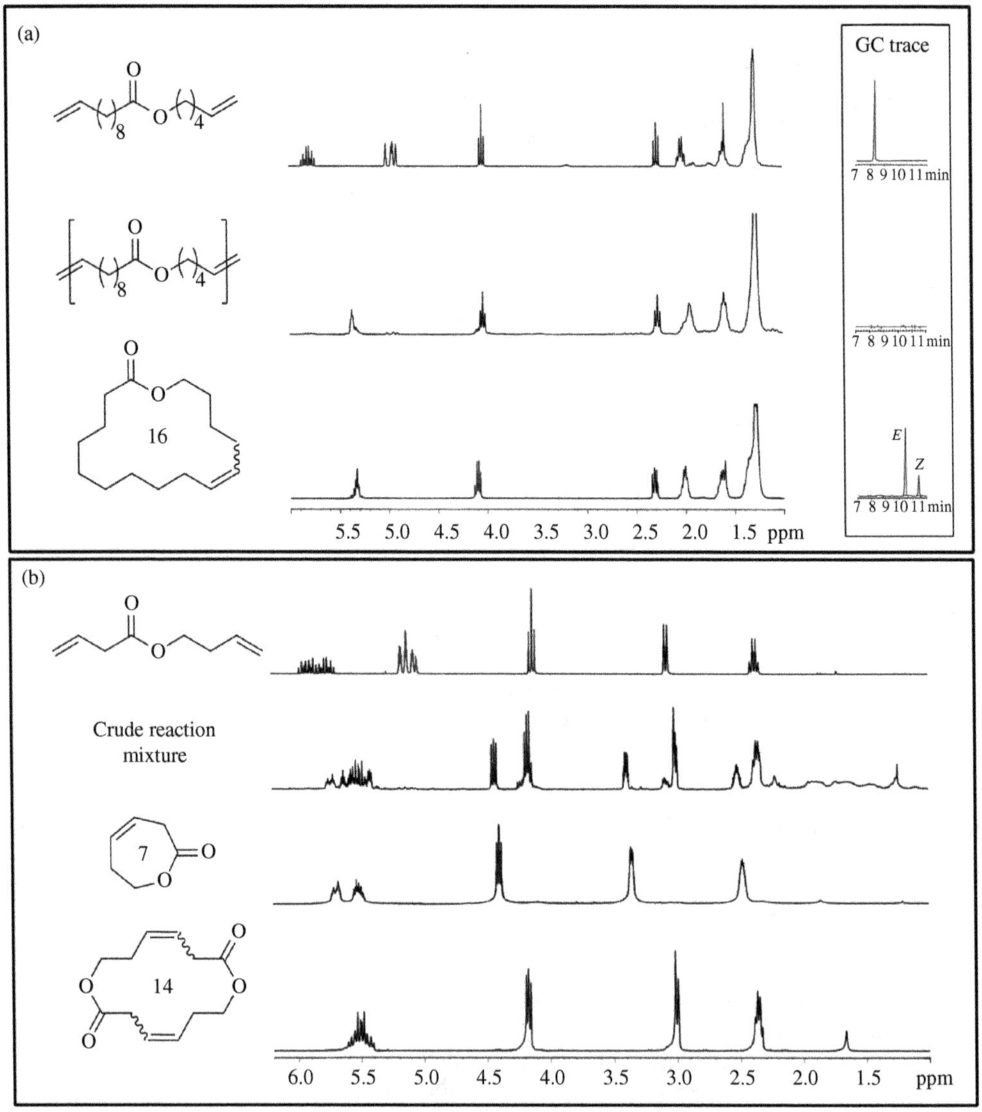

Figure 3.87 Difficulties in monitoring the progress of RCM reactions by ^{1}H NMR analysis, arising from the spectroscopic similarities of the species involved. (a) A representative macrolactonization. (b) A representative synthesis of a medium-ring lactone (13).

3.4.6 Catalyst Quenching

Catalyst quenching is essential to prevent "run-on" metathesis in the interval between sampling and queued analysis, and to prevent degradation of RCM products by oligomerization during workup (315).

The group 6 catalysts are rapidly and efficiently quenched by exposure to air or addition of benzaldehyde (Fig. 3.89) (176), eliminating the potential for secondary metathesis. Ligand protonolysis almost certainly occurs on subsequent exposure to water, whether atmospheric or via wet solvents. The polarity of the resulting metal-oxo species facilitates separation of transition-metal residues from organic products during chromatographic workup.

Given the lower oxophilicity of the Ru catalysts, and the extraordinary catalytic versatility of this metal, complete annihilation of activity requires greater care. Oligomerization during solvent removal has been reported even at ambient temperatures for macrocyclizations promoted by the Hoveyda catalyst **Ru-3** (315). Particular risk factors for degradation of metathesis products, in addition to ring strain, include high catalyst loadings, large-scale reactions (316,317), and high temperatures (see Section 3.3.1 above).

For the Ru catalysts, in particular, quenching has the added benefit of facilitating removal of heavy-metal residues. Isolation of RCM products by silica-gel chromatography is hampered by the tendency of the catalyst residues to

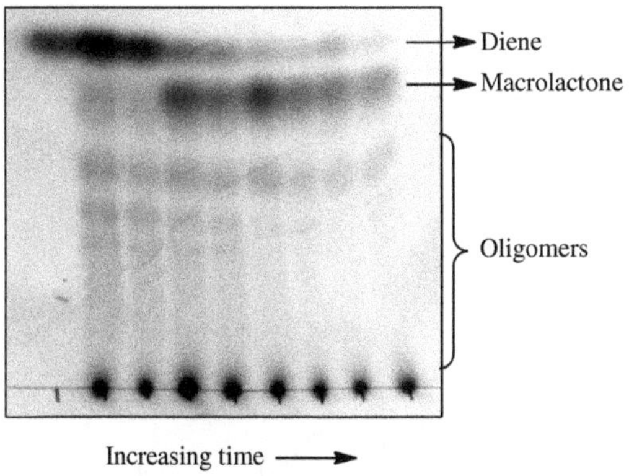

Figure 3.88 Visualizing rates of oligomer formation and backbiting using TLC analysis: monitoring the RCM reaction of Figure 3.87a (see also Figure 3.12) (314).

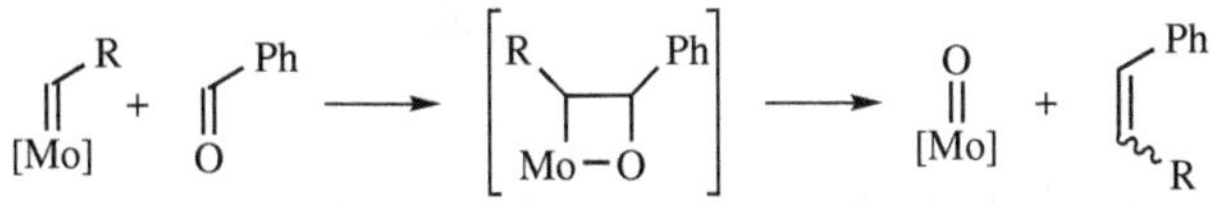

Figure 3.89 Quenching of **Mo-1** by added benzaldehyde (176).

streak on the column, which results in contamination of the products by trace ruthenium. Stable, polar ruthenium adducts formed by quenching typically streak less, resulting in easier purification of RCM products.

Most common of the quenching agents used for Ru catalysis, but far from the most efficient, is ethyl vinyl ether, EVE. Among the disadvantages of EVE are its slow quenching rates, the residual activity of the Fischer carbene products (90), the low polarity of these species, which impedes their separation from organic products, and their potential for thermal decomposition into isomerization-active Ru hydrides (90,318). A related strategy utilizes di(ethylene glycol) vinyl ethers bearing alcohol or amine endgroups (319). While these suffer from many of the same limitations as EVE, their higher polarity aids in ruthenium removal. Post-workup ruthenium levels were below 1 ppm for a Hoveyda-class catalyst.

Fast quenching methodologies were developed by the Diver and Fogg groups. The Diver approach is based on polar isocyanides such as the commercially available ethyl 2-cyanoacetate (**90-1**). Addition of about 10 equiv of **90-1** effected complete loss of the ruthenium alkylidene in **Ru-1** or **Ru-2** within 5 min at RT (Fig. 3.90a) (320). A major asset with this quenching agent is the polarity of the Ru products, which facilitates their removal by chromatography. Residual ruthenium levels were less than 2 ppm.

In work directed at establishing accurate rate plots in high-throughput catalysis, Blacquiere and Fogg demonstrated that potassium tris(pyrazolyl)borate **90-2** effects immediate, room-temperature knockdown of all three of the dominant metathesis catalysts following RCM (Fig. 3.90b). Samples analyzed by gas chromatography flame ionization detector (GC-FID) indicated complete annihilation of RCM activity within 5 min. Excellent agreement was seen between samples analyzed immediately, versus samples that spent up to 24 h queued for GC analysis (321).

In reports describing the large-scale synthesis of BILN 2061 (see Section 3.6), process chemists at Boehringer–Ingelheim highlighted the need to completely quench the ruthenium catalyst prior to workup by distillation (316). Decomposition of the macrocylic target occurred during evaporation, consistent with the concentration-dependence of the ring-chain equilibrium. Stirring with excess 2-mercaptionicotinic acid (MNA; 2-mercapto-3-carboxypyridine) effectively quenched catalysis (6 h, 30 °C); subsequent extraction with aqueous bicarbonate

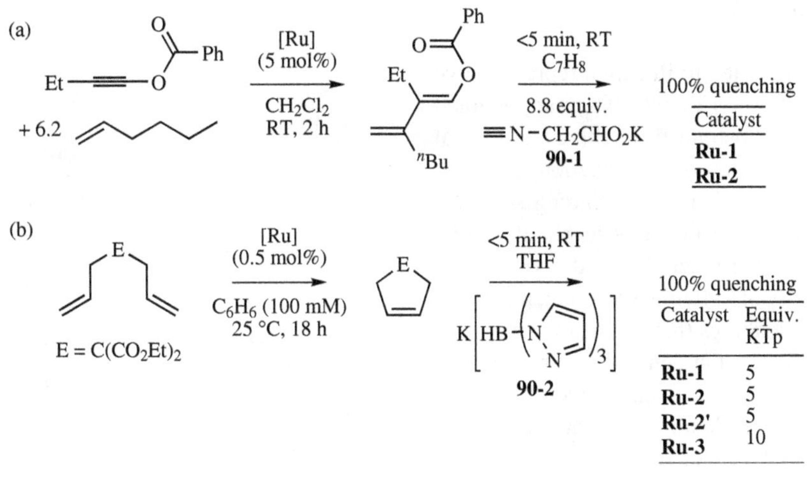

Figure 3.90 Rapid quenching of Ru metathesis catalysts. (a) With potassium 2-isocyanoacetate (**90-1**). (b) With potassium tris(pyrazolyl)borate (KTp, **90-2**)).

TABLE 3.3 Methods for Removing Ruthenium Residues Following RCM

Treatment[a]	Catalyst loading, mol%	Residual Ru, ppm	References
Polymer-bound $ArPh_2PO$ (50 equiv), RT, 6 h; silica filtration	5	<3	323
MNA (1 equiv), NEt_3 (1 equiv); 30 °C, 0.5 h, then 55 °C, 6 h; extraction with aqueous bicarbonate; activated charcoal, RT, 12 h; filtration	5	<30	316
Supercritical CO_2 extraction	5	56	324
$Me_2N(CH_2)_3NMe(CH_2)_2OCH{=}CH_2$ (4 equiv); 50 °C, 2 h; extraction with aqueous HCl	2	<15	319
MNA (5 equiv); 60 °C, 2 h; extraction with aqueous bicarbonate	0.2–0.3	50	12
For ROMP: MNA or $PhN{=}NC(S)NMe_2$ (50 equiv); RT, 20 h; MeOH precipitation	0.2	30	325
Membrane nanofiltration; STARMEM™ 228	0.15	<9	326

[a]Equivalents indicated for scavengers are relative to RCM substrate.

also reduced the proportion of residual ruthenium. Large amounts of MNA (2 kg for 1 kg diene) were required in the first scale-up campaign, which used catalyst loadings of 3–5 mol%. The second-generation process used an order of magnitude less ruthenium, and MNA levels of 50-fold less enabled efficient quenching, as well as ruthenium removal to 50 ppm levels.

Removal of spent ruthenium following metathesis is essential in industrial applications, to minimize heavy-metal contamination of the products. The state of the art was recently reviewed (322). Table 3.3 summarizes methods that enable effective removal of ruthenium residues without chromatography, a decisive parameter within the industrial context.

3.5 PRACTICAL EXAMPLES

The examples below are selected to illustrate representative procedures for a range of RCM targets. They are adapted from literature reports to illustrate different experimental approaches, and incorporate additional precautions aimed at addressing challenges discussed in the foregoing sections. The level of care taken to exclude air and water should not be interpreted as specific to the product classes described: this will depend on the difficulty of individual reaction, the yields required, and the catalyst loadings that can be tolerated, and must be modified according to the demands of the reaction at hand. More challenging, slower reactions require greater stringency, especially if catalyst loadings are low. A constant slow flow of nitrogen is advisable. Both air-exclusion and ethylene loss benefit from use of Schlenk adapters rather than needle-pierced septa.

3.5.1 A Representative, Straightforward RCM Reaction

An N_2-filled flask equipped with a stirbar and reflux condenser was charged with neat diene **91-1a** (100 mg, 0.295 mmol) and dry CH_2Cl_2 (3 ml), to give a final diene concentration of 3 mM. Solid **Ru-1** (9.7 mg, 0.012 mmol; 4 mol%) was added against a flow of inert gas. The solution was heated to reflux. Once reaction was complete (2 h, as judged by TLC analysis), the catalyst was quenched with KO_2CCH_2NC (Aldrich 705438). The solvent was stripped off by rotary evaporation and the product was purified by flash chromatography to give 82 mg (90%) of **91-2a**. The homologous seven- and eight-membered rings **91-2a** and **91-2b** were obtained similarly, at higher dilutions (Fig. 3.91). Product identity was confirmed by NMR and MS analyses.

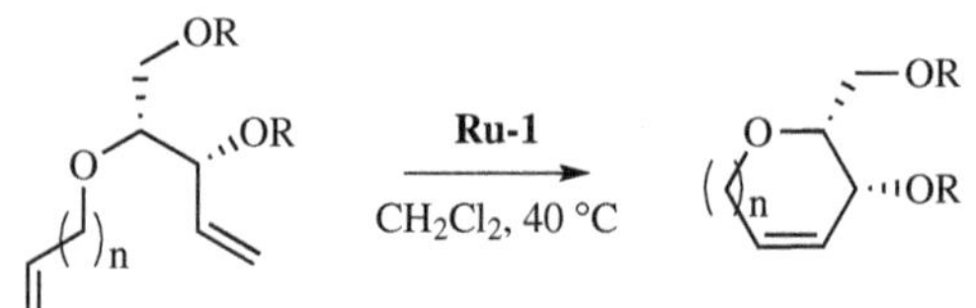

91-1 **91-2**

	R	n (ring size)	**Ru-1** (mol%)	conc. (mM)	t (h)	Yield (%)
a	Bz	1 (6)	4	100	1	90
b	Ac	2 (7)	7	3	2	95
c	Ac	3 (8)	7	3	2	73 (+ 17% dimer)
				100	2	chiefly dimer

Figure 3.91 RCM synthesis of cyclic ethers: adapted from Ref. 327.

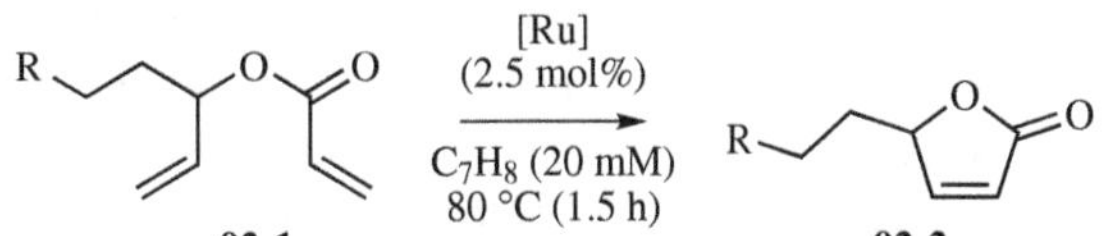

92-1 **92-2**

	R	[Ru]	Yield (%)
a	Ph	**Ru-2**	86
		Ru-4	90
b	^{n}Pr	**Ru-2**	90
		Ru-4	85

Figure 3.92 Butenolide synthesis: adapted from Ref. 157.

3.5.2 RCM of Acrylates to Yield Butenolides

To an N_2-filled Schlenk flask containing a stirbar and **Ru-2** (21.2 mg, 0.025 mmol), connected via a condenser to a Schlenk line, was added by cannula dry, degassed toluene (25 ml), followed by a solution of degassed acrylate **92-1a** (216 mg, 1.0 mmol; Fig. 3.92) in 25 ml toluene. The reaction was heated at 80 °C for 1.5 h. After this time, an aliquot was removed by syringe via the stopcock on the septum-sealed Schlenk sidearm, and analyzed by GC-FID (to assess conversion), and TLC (to detect any oligomers). The reaction was allowed to cool to RT and quenched with KTp (63 mg, 0.25 mmol, 10 equiv vs Ru) in 5 ml THF. The solvent was stripped off (rotavap), and the residue purified by chromatography on silica to give 162 mg (86%) of lactone **92-2a**. With **Ru-4**, yields increased to 90%. Product identity was confirmed by NMR and MS analyses. For the corresponding *n*-butyl product **92-2b**, **Ru-2** slightly outperformed **Ru-4**.

3.5.3 RCM Macrolactonization

An argon-filled flask equipped with a stirbar and a reflux condenser was charged with diene **93-1a** (450 mg, 0.811 mmol; Fig. 3.93) and 600 ml of dry toluene to give a final diene concentration of 1 mM. Solid **Ru-2′** (68.0 mg, 0.0803 mmol, 10 mol%) was added against a flow of argon, after which the solution was heated to 80 °C. The reaction was monitored by TLC and terminated after 90 min by addition of KTp (202 mg, 0.803 mmol, 10 equiv vs Ru). The solvent was stripped off (rotavap), and the product was purified by flash chromatography (hexanes/Et_2O 4 : 1) to give a final isolated yield of 376 mg (93%; 66% *E*). Product identity was confirmed by NMR and HR-MS analyses. The *t*-butyldimethylsilyl (TBS)- and methoxymethyl (MOM)-protected analogs **93-1b** and **93-1c** gave similar yields. Note: the unprotected phenol derivative was obtained in 69% yield using **Ru-3** (69%; see **72-5**, Fig. 3.72).

3.5.4 Microwave RCM

A 25 ml oven-dried Pyrex round-bottom flask equipped with a stirbar and reflux condenser was purged with argon for 5 min. To the flask was added diene **94-1a** (29 mg, 0.15 mmol; Fig. 3.94), 14.4 ml of DCE purified by distillation from P_2O_5 under N_2, and 600 μl from a 2 mM stock solution of **Ru-1** in DCE (1.2 mg, 1.5 μmol, 1 mol%). The final diene concentration was 10 mM. A fiber-optic (FO) temperature probe and glass capillary for Ar sparging were immersed in the solution via the condenser. The apparatus was transferred to a microwave chamber and irradiated for 5 min at 150 W of constant power (FO temperature 83 °C) while sparging. An aliquot was removed to confirm complete reaction (GC-FID). The solvent was stripped off (rotavap), and the product was purified by flash chromatography on silica gel (10 : 1 petroleum ether–ethyl acetate) to give 28 mg (85%) of **94-2a** as a light yellow oil. Product identity was confirmed by MS and NMR analyses. The corresponding tetrasubstituted olefin **94-2b** was obtained in 71% yield on use of **Ru-2** (2× 10 mol%).

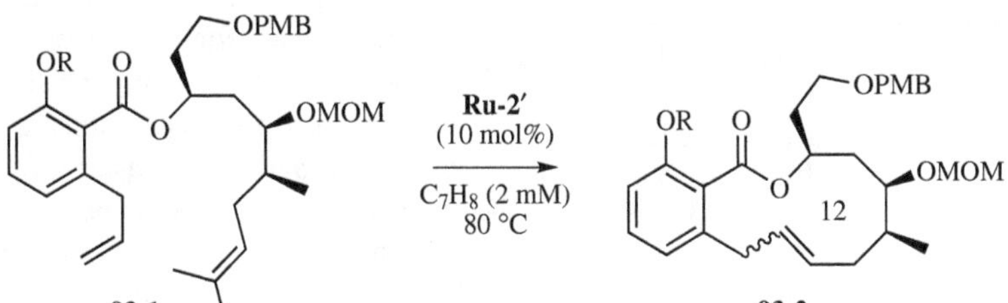

93-1 **93-2**

	R	*t* (h)	Yield (%)
a	Me	1.5	93
b	TBS*	1	91
c	MOM	3	91

Figure 3.93 Macrolactone synthesis: adapted from Ref. 35. The label (*) denotes use of 5 mol% **Ru-2′**.

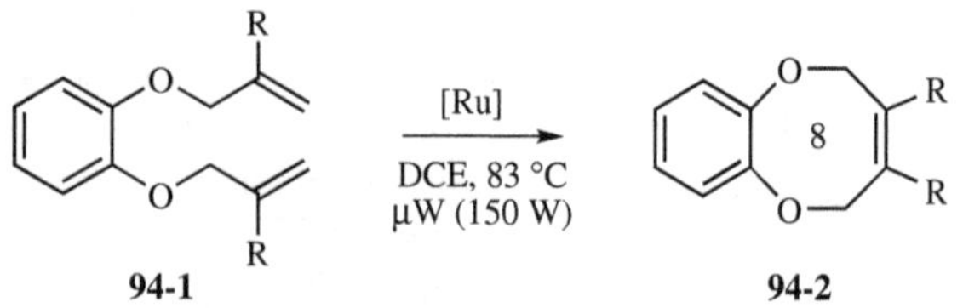

	R	[Ru] (mol%)	Conc. (mM)	t (h)	Yield (%)
a	H	**Ru-1** (1)	10	0.1	85
b	Me	**Ru-2** (2 x 10)	2	1	71

Figure 3.94 RCM under microwave conditions. Source: Adapted from Ref. 293.

3.6 INDUSTRIAL APPLICATIONS OF RCM

An increasing number of publications and patent filings from pharmaceutical R&D programs testify to keen interest in deploying RCM for the synthesis of drugs and drug candidates (Table 3.4). Macrocyclic targets dominate these reports, as indicated in Table 3.4, and in two 2011 reviews of catalysis in pharma (328,329). While no industrial processes utilizing RCM are yet on stream, several reports describe implementation of RCM within large-scale syntheses, including pilot-scale processes. Cost is a primary, overriding consideration, even given the value-added nature of these targets, and these campaigns bring into sharp focus the practical consequences resulting from many of the challenges discussed above.

Specific issues highlighted are the large volumes of solvent required; catalyst costs; side-reactions and product selectivity; and purification requirements associated with removal of by-products (routinely dimers, oligomers, and isomerized material), as well as catalyst residues. Ruthenium contamination is a particular concern, given the stringent limits for total platinoid metals in active pharmaceutical ingredients (APIs). Other aspects cited were the commercial availability of the required catalysts in bulk and high purity, environmental issues associated with use of chlorinated solvents, and potential safety hazards associated with carrying forward large amounts of ruthenium into downstream chemistry. Three recently disclosed, high profile cases are presented below, with a focus on key challenges and insights extracted during scale-up. Discussed first and in greatest detail is work from Boehringer–Ingelheim on the first example of large-scale RCM.

TABLE 3.4 Recently Disclosed Examples of RCM in Pharmaceutical Research

Company	Ring Size	Catalyst	References
Aileron Therapeutics	>12	**Ru-1**	330
Boehringer–Ingelheim	15	**Ru-7**	12
GlaxoSmithKline	7	**Ru-3**	331, 332
Johnson & Johnson	8–9	**Ru-2**	333
Merck	15, 19	Zhan 1B[a]	334, 335
Merck	20	**Ru-3**	336, 337
Novartis	14–17	**Ru-2**	338
Pfizer	28	**Mo-1**	339
		Ru-1	339
Pfizer	5,7 (bicyclic)	**Ru-2**	340
Roche	15	N.R.	341
Tibotec Pharmaceuticals	12	Zhan 1B[a]	341

[a]Variant on **Ru-3** commercialized by Zannan SciTech: the styrenyl ether bears a p-SO_2NMe_2 group.

Ciluprevir (BILN 2061) In 2005 and 2006, Boehringer–Ingelheim reported the first implementation of RCM on production scale (89,316). The target was the 15-membered macrocycle ciluprevir, a protease inhibitor used in the treatment of hepatitis C, reported in 2004 by the Quebec R&D team (343). While Phase 1b clinical trials were halted in 2003, owing to cardiac toxicity in primate models (344,345), the very promising clinical performance of this first-in-class hepatitis C virus (HCV) protease inhibitor stimulated development of new potentially therapeutic macrocycles based on this peptidomimetic scaffold (342,346–348).

Bench-scale experiments indicated that 5–7 mol% of **Ru-1** was required to effect complete cyclization of the tripeptide diene **95-1** in refluxing CH_2Cl_2. The desired Z-olefin was formed in 83% yield, but significant epimerization (see *epi*-**95-2**, Fig. 3.95) occurred at the vinylcyclopropane site prior to cyclization (349). This side-reaction was correlated with the presence of free PCy_3, as well as pyrrolidine residues from the previous synthetic step. Use of the NHC catalysts **Ru-2** and **Ru-3** increased reaction rates and reduced epimerization, but resulted in up to 10% cyclic dimers. The first-generation Hoveyda catalyst **Ru-22** afforded minimal dimerization or epimerization, and was therefore selected for initial scale-up.

Production of 400 kg to support clinical studies was carried out in batches of 20 kg for the critical RCM step (316). To put this in perspective, for each batch about 2,400 l of toluene was required to reach the dilution of about 10 mM needed to minimize oligomerization. Toluene was chosen in preference to CH_2Cl_2 for the sake of higher temperature operation, and hence faster throughput. Space–time yields were a key consideration because the scale of these reactions was incompatible with available pilot-plant reactor volumes, necessitating direct transfer of operations to a commercial production facility, using 3,000 l stirred-tank reactors.

New complications that emerged on this scale included inefficient mass transfer, which inhibited ethylene volatilization (despite sparging with N_2 at 1,000 l/h), resulting in competing regeneration of the starting diene. Also, 10–15% epimerization was observed in an early batch and traced to morpholine present at about 20 ppm levels

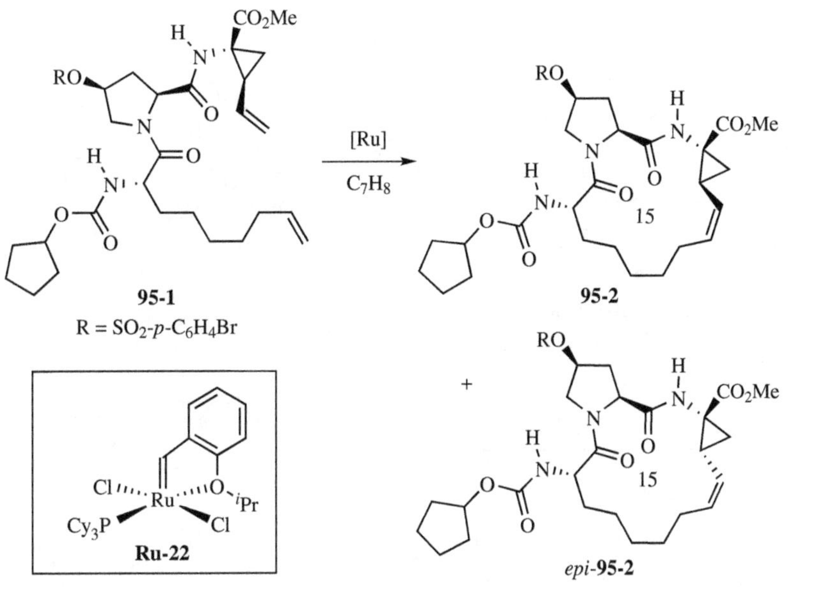

Figure 3.95 Epimerization encountered in early-stage RCM studies toward ciluprevir.

in the technical-grade toluene used. (A recent mechanistic study describes the role of morpholine, pyrrolidine, and other coordinating nitrogen bases in deactivating related Ru catalysts) (351). Given the large volume of toluene used, this translates into a 1 : 2 molar ratio of the secondary amine versus **Ru-22**. Morpholine contamination was addressed by pre-washing the toluene with aqueous HCl and drying by azeotropic distillation prior to RCM. An additional problem involved ring-opening on distillation of the RCM product from the unquenched reaction: this issue, as well as the nicotinic acid treatment developed to address it, has been discussed in Section 3.4.6. Finally, and re-emphasizing the enormous challenges associated with high-dilution chemistry on the industrial scale, solvent evaporation proved the bottleneck in the process as a whole.

Exhaustive efforts in subsequent process development led to a second-generation process, reported in 2009 (Fig. 3.96) (12). Explicit performance goals were higher concentrations, reduced catalyst loadings, and improved space–time yields, while maintaining product selectivity. The problem of dimerization by the more active and productive Ru–NHC catalysts was addressed by protecting the secondary amine as an NBoc derivative to increase the EM values. The increased bulk near the vinylcyclopropane site also favored initiation at the distal olefin: this suppressed epimerization even in the presence of amine nucleophiles (350), and also increased overall reaction rates by preventing carbonyl chelation (see Section 3.3.4.2). As ultimately implemented, the process utilized the Grela catalyst **Ru-7** at loadings of 0.1 mol% or lower in refluxing toluene. RCM proceeded in 93% yield with minimal oligomerization at the remarkably high concentration of 200 mM (12). The total volume of solvent per ton of diene was reduced from about 150,000 to 7,500 l, enabling the use of standard process equipment and resulting in a much more economical, more environmentally benign process.

Vaniprevir (MK-7009) Merck research laboratories recently reported the 200 g scale synthesis of vaniprevir, an HCV protease inhibitor containing a 20-membered macrocycle (337). A Hoveyda-class catalyst (**Ru-3**) was used to cyclize **97-1**, at an initial loading of 5 mol% and a diene concentration of 40 mM (toluene, 60 °C; Fig. 3.97). Class B isomerization led to contamination by up to 9% of ring-contracted **97-3**. Isomerization was reduced to less than 1% by addition of 2,6-dichloroquinone and use of higher concentrations (ca. 100 mM); the latter presumably reduces the proportion of catalyst decomposition. Only minor amounts of oligomer (5%) were observed even at this concentration, again suggesting conformational constraint in the tripeptide–isoindoline diene. Lower catalyst loadings and still higher concentrations were attained by the simultaneous slow addition of catalyst and diene: the optimized procedures (0.2 mol% **Ru-3**, 130 mM) delivered the target **97-2** in 91% yield after 1 h.

Relacatib (SB-462795) In 2009, the process chemistry group at GSK reported the large-scale RCM production of relacatib, a cathepsin K inhibitor designed to treat osteoporosis and osteoarthritis (332). At the discovery stage, catalyst **Ru-2** was effective, but the process was found to suffer from irreproducibility and diminished yields on scale-up. Better results were obtained with **Ru-3**, although contaminating DBU and other amine bases (secondary amines, ureas, etc.) from prior synthetic steps limited conversions (331). These issues, as well as problems of isomerization, were addressed

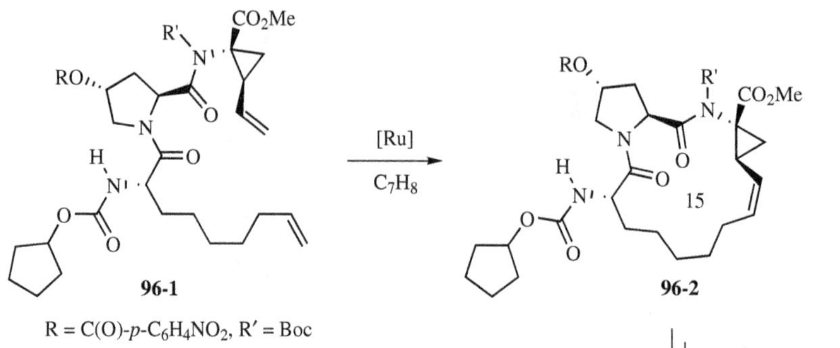

	2006 **95-1**	2009 **96-1**
max. scale	20 × 20 kg	not disclosed
[Ru]	**Ru-22** (3 mol%)	**Ru-7** (0.1 mol%)
T; time	80 °C, 4-9 h	110 °C, 0.5 h
conc.	10 mM	200 mM
yield	83%	93%

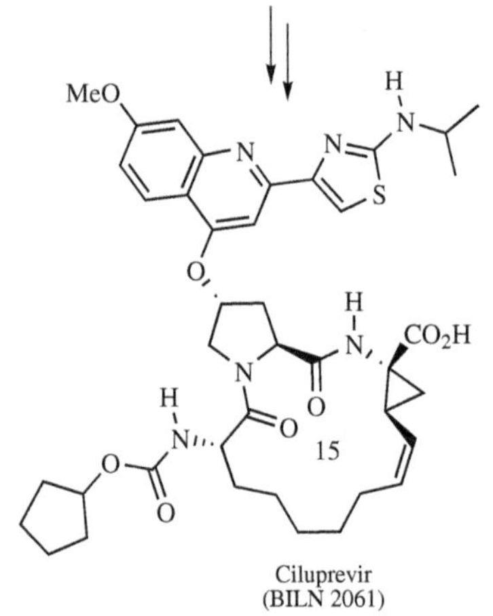

Figure 3.96 Synthesis of ciluprevir (BILN 2061) using RCM as a key step (12).

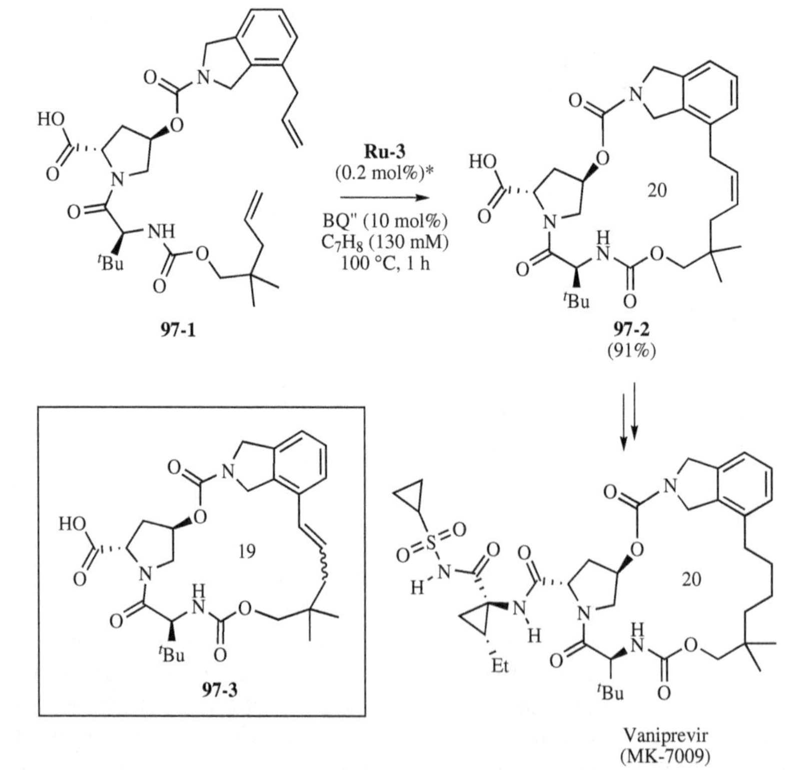

Figure 3.97 Synthesis of vaniprevir (MK-7009), using RCM as a key step (337). The label (*) denotes simultaneous slow addition of catalyst and substrate. BQ″ = 2,6-dichloro-1,4-benzoquinone.

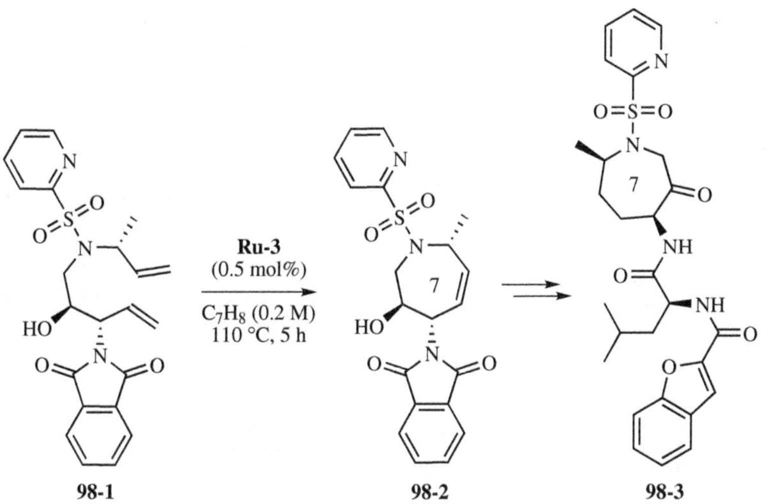

98-1 98-2 (96%) 98-3 Relacatib (SB-462795)

Figure 3.98 Large-scale RCM synthesis employed en route to relacatib (SB-462795) (332).

by identifying a readily purified RCM precursor, and increasing the steric bulk of the tetrahydroazepine.

Cyclization of **98-1** (Fig. 3.98) was readily amenable to operation at high concentrations, consistent with the low strain of medium-sized azacycles discussed in Section 3.2.1 and elsewhere. The pilot process generated a total of 200 kg of the therapeutic candidate for clinical studies (332). The RCM step was carried out on 80 kg batches (880 l toluene; 0.2 M) at a catalyst loading of 0.5 mol%, and delivered the tetrahydroazepine **98-2** in 96% yield. With highly pure diene, catalyst loadings could be reduced to 0.1–0.2 mol%. The low-solubility product crystallized as it formed, facilitating separation from the ruthenium by-product.

3.7 CONCLUSIONS

RCM is an extraordinarily powerful synthetic tool, which over the past 20 years has transformed approaches to the assembly of cyclic molecules. Uptake into the industrial context, particularly in pharma, is well under way and implementation into manufacturing processes is now on the near horizon.

This chapter offers a practical guide to the challenges and opportunities associated with different ring sizes and functional groups, as well as insight into experimental parameters that can be modulated to improve RCM yields. Throughout, we have sought to tie synthetic outcomes with the developing mechanistic picture. It is our hope that this approach will provide an informed guide of greatest benefit to the novice, while uncovering details that enrich the perspective of the experienced practitioner.

ACKNOWLEDGMENTS

We thank the many scientists involved in metathesis, who have provided the perspective and examples on which this work was based. Financial support from NSERC of Canada and the Government of Ontario (Ontario Graduate Scholarship to JAML) is gratefully acknowledged.

REFERENCES

1. Armstrong SK. *J Chem Soc, Perkin Trans* 1998:371–388.
2. Fürstner A. *Top Catal* 1998;4:285–299.
3. Maier ME. *Angew Chem Int Ed* 2000;39:2073–2077.
4. Gradillas A, Perez-Castells J. *Angew Chem Int Ed* 2006;45:6086–6101.
5. Hoveyda AH, Zhugralin AR. *Nature* 2007;450:243–251.
6. Grubbs RH, editor. *Handbook of Metathesis*. Weinheim: Wiley-VCH; 2003.
7. Lee L-S, Ou H-J, Hsu H-L. *Fluid Phase Equilib* 2005;231:221–230.
8. Dariva C, Lovisi H, Mariac LCS, Coutinho FMB, Oliveira JV, Pinto JC. *Can J Chem Eng* 2003;81:147–152.
9. Kamau SD, Hodge P, Hall AJ, Dad S, Ben-Haida A. *Polymer* 2007;48:6808–6822.
10. Conrad JC, Eelman MD, Duarte Silva JA, Monfette S, Parnas HH, Snelgrove JL, Fogg DE. *J Am Chem Soc* 2007;129:1024–1025.
11. Nelson DJ, Ashworth IW, Hillier IH, Kyne SH, Pandian S, Parkinson JA, Percy JM, Rinaudo G, Vincent MA. *Chem Eur J* 2011;17:13087–13094.
12. Farina V, Shu C, Zeng X, Wei X, Han Z, Yee NK, Senanayake CH. *Org Process Res Dev* 2009;13:250–254.

13. Monfette S, Crane AK, Duarte Silva JA, Facey GA, dos Santos EN, Araujo MH, Fogg DE. *Inorg Chim Acta* 2010;363:481–486.
14. Ashworth IW, Hillier IH, Nelson DJ, Percy JM, Vincent MA. *Chem Comm* 2011;47:5428–5430.
15. Schleyer PVR, Williams JE, Blanchard KR. *J Am Chem Soc* 1970;92:2378–2386.
16. Ivin KJ. *J Polym Sci, Part A* 2000;38:2137–2146.
17. Rutjes FPJT. Natural products containing medium-sized nitrogen heterocycles synthesized by RCM. In: Cossy J, Arseniyadis S, Meyer C, editors. *Metathesis in Natural Product Synthesis: Strategies, Substrates and Catalysts.* Weinheim: Wiley-VCH; 2010. p 45–86.
18. Hoye TR, Promo MA. *Tetrahedron Lett* 1999;40:1429–1432.
19. Chang S, Grubbs RH. *Tetrahedron Lett* 1997;38:4757–4760.
20. Michaut A, Rodriguez J. *Angew Chem Int Ed* 2006;45: 5740–5750.
21. Blanchard N, Eustache J. Synthesis of natural products containing medium-size carbocycles by ring-closing alkene metathesis. In: Cossy J, Arseniyadis S, Meyer C, editors. *Metathesis in Natural Product Synthesis: Strategies, Substrates and Catalysts.* Weinheim: Wiley-VCH; 2010. p 1–43.
22. Sturmer R, Schafer B, Wolfart V, Stahr HKU, Helmchen G. *Synthesis* 2001:46–48.
23. Deiters A, Martin SF. *Chem Rev* 2004;104:2199–2238.
24. Blankenstein J, Zhu JP. *Eur J Org Chem* 2005:1949–1964.
25. Dinger MB, Mol JC. *Adv Synth Catal* 2002;344:671–677.
26. Gatti M, Vieille-Petit L, Luan X, Mariz R, Drinkel E, Linden A, Dorta R. *J Am Chem Soc* 2009;131:9498–9499.
27. Ivin KJ, Mol JC. *Olefin Metathesis and Metathesis Polymerization.* New York: Academic Press; 1997.
28. Nelson DJ, Carboni D, Ashworth IW, Percy JM. *J Org Chem* 2011;76:8386–8393.
29. Forbes MDE, Patton JT, Myers TL, Maynard HD, Smith DW Jr, Schulz GR, Wagener KB. *J Am Chem Soc* 1992; 114:10978–10980.
30. Crimmins MT, Choy AL. *J Am Chem Soc* 1999;121: 5653–5660.
31. Vassilikogiannakis G, Margaros I, Tofi M. *Org Lett* 2004;6: 205–208.
32. Fürstner A, Thiel OR, Blanda G. *Org Lett* 2000;2:3731–3734.
33. Yang K, Blackman B, Diederich W, Flaherty PT, Mossman CJ, Roy S, Ahn YM, Georg GI. *J Org Chem* 2003;68: 10030–10039.
34. Aissa C, Riveiros R, Ragot J, Fürstner A. *J Am Chem Soc* 2003;125:15512–15520.
35. Fürstner A, Dierkes T, Thiel OR, Blanda G. *Chem Eur J* 2001;7:5286–5298.
36. Goldring WPD, Hodder AS, Weiler L. *Tetrahedron Lett* 1998;39:4955–4958.
37. Fürstner A, Langemann K. *Synthesis* 1997:792–803.
38. Fürstner A, Langemann K. *J Am Chem Soc* 1997;119: 9130–9136.
39. Rainier JD. General RCM approaches to medium rings; synthesis of natural products containing medium-size oxygen heterocycles. In: Cossy J, Arseniyadis S, Meyer C, editors. *Metathesis in Natural Product Synthesis: Strategies, Substrates and Catalysts.* Weinheim: Wiley-VCH; 2010. p 87–128.
40. Bolduc P, Jacques A, Collins SK. *J Am Chem Soc* 2010; 132:12790–12791.
41. Kirkland TA, Grubbs RH. *J Org Chem* 1997;62:7310–7318.
42. Fürstner A, Thiel OR, Ackermann L. *Org Lett* 2001;3: 449–451.
43. Fürstner A, Thiel OR, Kindler N, Bartkowska B. *J Org Chem* 2000;65:7990–7995.
44. Taillier C, Hameury T, Bellosta V, Cossy J. *Tetrahedron* 2007;63:4472–4490.
45. Höcker H. *J Mol Catal* 1991;65:95–99.
46. Yu M, Wang C, Kyle AF, Jakubec P, Dixon DJ, Schrock RR, Hoveyda AH. *Nature* 2011;479:88–93.
47. Moore JS. Ring-opening metathesis polymerization. In: Hegedus LS, editor. *Comprehensive Organometallic Chemistry II.* Vol. 12. Elsevier; 1994. p 1209–1232.
48. Heppekausen J, Fürstner A. *Angew Chem Int Ed* 2011;50: 7829–7832.
49. Schwab P, Grubbs RH, Ziller JW. *J Am Chem Soc* 1996;118: 100–110.
50. Scholl M, Trnka TM, Morgan JP, Grubbs RH. *Tetrahedron Lett* 1999;40:2247–2250.
51. Huang J, Stevens ED, Nolan SP, Petersen JL. *J Am Chem Soc* 1999;121:2674–2678.
52. Sanford MS, Love JA, Grubbs RH. *J Am Chem Soc* 2001;123:6543–6554.
53. Adlhart C, Chen P. *Helv Chim Acta* 2003;86:941–949.
54. Garber SB, Kingsbury JS, Gray BL, Hoveyda AH. *J Am Chem Soc* 2000;122:8168–8179.
55. Vougioukalakis GC, Grubbs RH. *Chem Rev* 2010;110: 1746–1787.
56. Michrowska A, Grela K. *Pure Appl Chem* 2008;80:31–43.
57. Kingsbury JS, Harrity JPA, Bonitatebus PJ, Hoveyda AH. *J Am Chem Soc* 1999;121:791–799.
58. Thiel V, Hendann M, Wannowius K-J r, Plenio H. *J Am Chem Soc* 2012;134:1104–1114.
59. Ulman M, Grubbs RH. *J Org Chem* 1999;64:7202–7207.
60. Hong SH, Wenzel AG, Salguero TT, Day MW, Grubbs RH. *J Am Chem Soc* 2007;129:7961–7968.
61. Alcaide B, Almendros P, Luna A. *Chem Rev* 2009;109:3817–3858.
62. Bennasar ML, Zulaica E, Sole D, Alonso S. *Chem Commun* 2009:3372–3374.
63. Fustero S, Sanchez-Rosello M, Jimenez D, Sanz-Cervera JF, Del Pozo C, Acena JL. *J Org Chem* 2006;71:2706–2714.
64. Prunet J. *Eur J Org Chem* 2011:3634–3647.
65. Joe D, Overman LE. *Tetrahedron Lett* 1997;38:8635–8638.
66. Fürstner A, Thiel OR, Ackermann L, Schanz H-J, Nolan SP. *J Org Chem* 2000;65:2204–2207.

67. Clark JS, Kettle JG. *Tetrahedron* 1999;55:8231–8248.
68. Bo GD, Markó IE. *Eur J Org Chem* 2011:1859–1869.
69. Miller SJ, Blackwell HE, Grubbs RH. *J Am Chem Soc* 1996;118:9606–9614.
70. Ackermann L, El Tom D, Fürstner A. *Tetrahedron* 2000;56: 2195–2202.
71. Gurjar MK, Yakambram P. *Tetrahedron Lett* 2001;42: 3633–3636.
72. Lehman SE, Schwendeman JE, O'Donnell PM, Wagener KB. *Inorg Chim Acta* 2003;345:190–198.
73. Kuźnik N, Krompiec S. *Coord Chem Rev* 2007;251:222–233.
74. Krompiec S, Krompiec M, Penczek R, Ignasiak H. *Coord Chem Rev* 2008;252:1819–1841.
75. Bourgeois D, Pancrazi A, Nolan SP, Prunet J. *J Organomet Chem* 2002;643–644:247–252.
76. Petkovska VI, Hopkins TE, Powell DH, Wagener KB. *Macromolecules* 2005;38:5878–5885.
77. Courchay FC, Sworen JC, Wagener KB. *Macromolecules* 2003;36:8231–8239.
78. Chuchuryukin AV, Chase PA, Dijkstra HP, Suijkerbuijk B, Mills AM, Spek AL, van Klink GPM, van Koten G. *Adv Synth Catal* 2005;347:447–462.
79. Ahuja R, Kundu S, Goldman AS, Brookhart M, Vicente BC, Scott SL. *Chem Commun* 2008:253–255.
80. van Gerven PCM, Elemans JAAW, Gerritsen JW, Speller S, Nolte RJM, Rowan AE. *Chem Commun* 2005:3535–3537.
81. Edwards SD, Lewis T, Taylor RJK. *Tetrahedron Lett* 1999;40:4267–4270.
82. Donohoe TJ, Bower JF, Basutto JA, Fishlock LP, Procopiou PA, Callens CKA. *Tetrahedron* 2009;65:8969–8980.
83. Sutton AE, Seigal BA, Finnegan DF, Snapper ML. *J Am Chem Soc* 2002;124:13390–13391.
84. Hekking KFW, van Delft FL, Rutjes F. *Tetrahedron* 2003;59:6751–6758.
85. Scholl M, Ding S, Lee CW, Grubbs RH. *Org Lett* 1999;1:953–956.
86. Jafarpour L, Hillier AC, Nolan SP. *Organometallics* 2002;21:442–444.
87. Beach NJ, Lummiss JAM, Bates JM, Fogg DE. *Organometallics* 2012;31:2349–2356.
88. van Lierop BJ, Reckling AM, Lummiss JAM, Fogg DE. *ChemCatChem* 2012;4:2020–2025.
89. Nicola T, Brenner M, Donsbach K, Kreye P. *Org Process Res Dev* 2005;9:513–515.
90. Louie J, Grubbs RH. *Organometallics* 2002;21:2153–2164.
91. Arisawa M, Terada Y, Takahashi K, Nakagawa M, Nishida A. *J Org Chem* 2006;71:4255–4261.
92. van Rensburg WJ, Steynberg PJ, Meyer WH, Kirk MM, Forman GS. *J Am Chem Soc* 2004;126:14332–14333.
93. Yang Z, Zhang B, Zhao G, Yang J, Xie X, She X. *Org Lett* 2011;13:5916–5919.
94. Wdowik T, Samojłowicz C, Jawiczuk M, Malińska M, Woźniak K, Grela K. *Chem Comm* 2013;49:674–676.
95. Bantreil X, Sidi-Ykhlef M, Aringhieri L, Colacino E, Martinez J, Lamaty F. *J Catal* 2012;294:113–118.
96. Dinger MB, Mol JC. *Organometallics* 2003;22:1089–1095.
97. Dinger MB, Mol JC. *Eur J Inorg Chem* 2003:2827–2833.
98. Beach NJ, Camm KD, Fogg DE. *Organometallics* 2010;29:5450–5455.
99. Hong SH, Sanders DP, Lee CW, Grubbs RH. *J Am Chem Soc* 2005;127:17160–17161.
100. Gimeno N, Formentin P, Steinke JHG, Vilar R. *Eur J Org Chem* 2007:918–924.
101. Aggarwal VK, Daly AM. *Chem Commun* 2002:2490–2491.
102. Campbell MJ, Johnson JS. *J Am Chem Soc* 2009;131: 10370–10371.
103. Garbacia S, Desai B, Lavastre O, Kappe CO. *J Org Chem* 2003;68:9136–9139.
104. Lee CW, Grubbs RH. *Org Lett* 2000;2:2145–2147.
105. Fürstner A, Radkowski K. *Chem Commun* 2002:2182–2183.
106. Wang C, Haeffner F, Schrock RR, Hoveyda AH. *Angew Chem Int Ed* 2013;52:1939–1943.
107. Fürstner A, Davies PW. *Chem Commun* 2005:2307–2320.
108. Marx VM, Herbert MB, Keitz BK, Grubbs RH. *J Am Chem Soc* 2013;135:94–97.
109. Occhipinti G, Hansen FR, Törnroos KW, Jensen VR. *J Am Chem Soc* 2013;135:3331–3334.
110. Wang Y, Jimenez M, Hansen AS, Raiber E-A, Schreiber SL, Young DW. *J Am Chem Soc* 2011;133:9196–9199.
111. Stephen H, Stephen T. *Solubilities of Inorganic and Organic Compounds*. Vol. 1. New York: Pergamon; 1963. p 1177–1183.
112. Ulman M, Grubbs RH. *Organometallics* 1998;17:2484–2489.
113. Majumdar KC, Muhuri S, Islam RU, Chattopadhyay B. *Heterocycles* 2009;78:1109–1169.
114. Nakashima K, Inoue K, Sono M, Tori M. *J Org Chem* 2002;67:6034–6040.
115. Xu Z, Johannes CW, Houri AF, La DS, Cogan DA, Hofilena GE, Hoveyda AH. *J Am Chem Soc* 1997;119:10302–10316.
116. Wang C, Rath NP, Covey DF. *Tetrahedron Lett* 2006;47:7837–7839.
117. Glebocka A, Sokolowska K, Sicinski RR, Plum LA, DeLuca HF. *J Med Chem* 2009;52:3496–3504.
118. Clark JS, Kimber MC, Robertson J, McErlean CSP, Wilson C. *Angew Chem Int Ed* 2005;44:6157–6162.
119. Gaich T, Mulzer J. *Org Lett* 2005;7:1311–1313.
120. Paek S-M. *Molecules* 2012;17:3348–3358.
121. Stewart IC, Ung T, Pletnev AA, Berlin JM, Grubbs RH, Schrodi Y. *Org Lett* 2007;9:1589–1592.
122. Berlin JM, Campbell K, Ritter T, Funk TW, Chlenov A, Grubbs RH. *Org Lett* 2007;9:1339–1342.
123. White DE, Stewart IC, Grubbs RH, Stoltz BM. *J Am Chem Soc* 2008;130:810–811.
124. White DE, Stewart IC, Seashore-Ludlow BA, Grubbs RH, Stoltz BM. *Tetrahedron* 2010;66:4668–4686.

125. Rost D, Porta M, Gessler S, Blechert S. *Tetrahedron Lett* 2008;49:5968–5971.
126. Luan X, Mariz R, Gatti M, Costabile C, Poater A, Cavallo L, Linden A, Dorta R. *J Am Chem Soc* 2008;130:6848–6858.
127. Sashuk V, Peeck LH, Plenio H. *Chem Eur J* 2010;16: 3983–3993.
128. Van de Weghe P, Bisseret P, Blanchard N, Eustache J. *J Organomet Chem* 2006:5078–5108.
129. Denmark SE, Muhuhi JM. *J Am Chem Soc* 2010;132: 11768–11778.
130. Hodgson DM, Foley AM, Lovell PJ. *Synlett* 1999;6:744–746.
131. MacNaughtan ML, Johnson MJA, Kampf JW. *J Am Chem Soc* 2007;129:7708–7709.
132. MacNaughtan ML, Gary JB, Gerlach DL, Johnson MJA, Kampf JW. *Organometallics* 2009;28:2880–2887.
133. Fomine S, Ortega JV, Tlenkopatchev MA. *J Mol Catal A* 2007;263:121–127.
134. Gatti M, Drinkel E, Wu L, Pusterla I, Gaggia F, Dorta R. *J Am Chem Soc* 2010;132:15179–15181.
135. Chao W, Weinreb SM. *Org Lett* 2003;5:2505–2507.
136. De Matteis V, Dufay O, Waalboer DCJ, van Delft FL, Tiebes J, Rutjes FPJT. *Eur J Org Chem* 2007:2667–2675.
137. De Matteis V, van Delft FL, Tiebes J, Rutjes FPJT. *Synlett* 2008:351–354.
138. Crowe WE, Goldberg DR. *J Am Chem Soc* 1995;117: 5162–5163.
139. Fox HH, Schrock RR, O'Dell R. *Organometallics* 1994; 13:635–639.
140. Fu GC, Grubbs RH. *J Am Chem Soc* 1992;114:7324–7325.
141. Feldman J, Murdzek JS, Davis WM, Schrock RR. *Organometallics* 1989;8:2260–2265.
142. Chatterjee AK, Morgan JP, Scholl M, Grubbs RH. *J Am Chem Soc* 2000;122:3783–3784.
143. Claverie JP, Viala S, Maurel V, Novat C. *Macromolecules* 2001;34:382–388.
144. Choi T-L, Lee CW, Chatterjee AK, Grubbs RH. *J Am Chem Soc* 2001;123:10417–10418.
145. Rybak A, Meier MAR. *Green Chem.* 2007;9:1356–1361.
146. Bilel H, Hamdi N, Zagrouba F, Fischmeister C, Bruneau C. *Green Chem* 2011;13:1448–1452.
147. Miao X, Malacea R, Fischmeister C, Bruneau C, Dixneuf PH. *Green Chem* 2011;13:2911–2919.
148. Lummiss JAM, Oliveira KC, Pranckevicius A, Santos A, dos Santos EN, Fogg DE. *J Am Chem Soc* 2012;134:18889–18891.
149. Toueg J, Prunet J. *Synlett* 2006:2807–2811.
150. Hutait S, Batra S. *Tetrahedron Lett* 2010;51:5781–5783.
151. D'Annibale A, Ciaralli L, Bassetti M, Pasquini C. *J Org Chem* 2007;72:6067–6074.
152. Sardan M, Sezer S, Günel A, Akkaya M, Tanyeli C. *Bioorg Med Chem Lett* 2012;22:5814–5818.
153. Mizutani H, Watanabe M, Honda T. *Tetrahedron* 2002;58:8929–8936.
154. Lee CW, Grubbs RH. *J Org Chem* 2001;66:7155–7158.
155. Briggs TF, Dudley GB. *Tetrahedron Lett* 2005;46: 7793–7796.
156. Bassetti M, D'Annibale A, Fanfoni A, Minissi F. *Org Lett* 2005;7:1805–1808.
157. Schmidt B, Geissler D. *ChemCatChem* 2010;2:423–429.
158. Mao B, Geurts K, Fananas-Mastral M, Zijl v A W, Fletcher SP, Minnaard AJ, Feringa BL. *Org Lett* 2011;13:948–951.
159. Hassan HMA. *Chem Commun* 2010;46:9100–9106.
160. Stead D, O'Brien P, Sanderson AJ. *Org Lett* 2005;7: 4459–4462.
161. Minville J, Poulin J, Dufresne C, Sturino CF. *Tetrahedron Lett* 2008;49:3677–3681.
162. Nandy JP, Rakic B, Sarma BVNB, Babu N, Lefrance M, Enright GD, Leek DM, Daniel K, Sabourin LA, Arya P. *Org Lett* 2008;10:1143–1146.
163. Pritchard DR, Wilden JD. *Tetrahedron Lett* 2010;51: 1819–1821.
164. Chen YH, Zhang HJ, Nan FJ. *J Combinat Chem* 2004; 6:684–687.
165. Fustero S, Mateu N, Albert L, Acena JL. *J Org Chem* 2009;74:4429–4432.
166. Laventine DL, Cullis PM, García MD, Jenkins PR. *Tetrahedron Lett* 2009;50:3657–3660.
167. Tripathy NK, Georg GI. *Tetrahedron Lett* 2004;45: 5309–5311.
168. Trost BM, Jiang C. *Org Lett* 2003;5:1563–1565.
169. Rodeschini V, Boiteau J-G, Van de Weghe P, Tarnus C, Eustache J. *J Org Chem* 2004;69:357–373.
170. Yokoe H, Sasaki H, Yoshimura T, Shindo M, Yoshida M, Shishido K. *Org Lett* 2007;9:969–971.
171. Paquette LA, Efremov I. *J Am Chem Soc* 2001;123: 4492–4501.
172. Efremov I, Paquette LA. *J Am Chem Soc* 2000;122: 9324–9325.
173. Clark JS, Conroy J, Blake AJ. *Org Lett* 2007;9:2091–2094.
174. Couladouros EA, Vidali VP. *Chem Eur J* 2004;10: 3822–3835.
175. Chavez DE, Jacobsen EN. *Org Lett* 2003;5:2563–2565.
176. Fox HH, Lee JK, Park LY, Schrock RR. *Organometallics* 1993;12:759–768.
177. Chatterjee AK, Choi T-L, Sanders DP, Grubbs RH. *J Am Chem Soc* 2003;125:11360–11370.
178. Chen VX, Boyer F-D, Rameau C, Retailleau P, Vors J-P, Beau J-M. *Chem–Eur J* 2010;16:13941–13945.
179. Chattopadhyay SK, Ghosh D, Mondal P, Ghosh SK. *Synthesis* 2012;44:2448–2454.
180. Mukherjee P, Sarkar TK. *Org Biomol Chem* 2012;10: 3060–3065.
181. Schmidt B, Krehl S, Kelling A, Schilde U. *J Org Chem* 2012;77:2360–2367.
182. Grisi F, Costabile C, Grimaldi A, Viscardi C, Saturnino C, Longo P. *Eur J Org Chem.* 2012:5928–5934.
183. Li P, Li J, Arikan F, Ahlbrecht W, Dieckmann M, Menche D. *J Org Chem* 2010;75:2429–2444.

184. Gallenkamp D, Furstner A. *J Am Chem Soc* 2011;133: 9232–9235.

185. Renaud J, Ouellet SG. *J Am Chem Soc* 1998;120:7995–7996.

186. Brown RCD, Satcharoen V. *Heterocycles* 2006;70:705–736.

187. Andrews SP, Ball M, Wierschem F, Cleator E, Oliver S, Hogenauer K, Simic O, Antonello A, Hunger U, Smith MD, Ley SV. *Chem Eur J* 2007;13:5688–5712.

188. Danoun G, Ceccon J, Greene AE, Poisson J-F. *Eur J Org Chem* 2009:4221–4224.

189. van Otterlo WAL, de Koning CB. *Chem Rev* 2009;109: 3743–3782.

190. Leriche P, Blanchard P, Frere P, Levillain E, Mabon G, Roncali J. *Chem Commun* 2006:275–277.

191. van Otterlo WAL, Morgans GL, Khanye SD, Aderibigbe BAA, Michael JP, Billing DG. *Tetrahedron Lett* 2004;45: 9171–9175.

192. Lee A-L, Malcolmson S, J, Puglisi A, Schrock R, R, Hoveyda A, H. *J Am Chem Soc* 2006;128:5153–5157.

193. Peczuh MW, Snyder NL. *Tetrahedron Lett* 2003;44: 4057–4061.

194. Majumder U, Cox JM, Johnson HWB, Rainier JD. *Chem Eur J* 2006;12:1736–1746.

195. Clark JS. *Chem Commun* 2006:3571–3581.

196. Okada A, Ohshima T, Shibasaki M. *Tetrahedron Lett* 2001;42:8023–8027.

197. Theeraladanon C, Takahashi N, Shiina M, Hamada K, Takada Y, Endo H, Tateishi U, Oka T, Ogata K, Inoue T. *Cancer Biother Radiopharm* 2010;25:479–485.

198. Chattopadhyay SK, Karmakar S, Biswas T, Majumdar KC, Rahaman H, Roy B. *Tetrahedron* 2007;63:3919–3952.

199. Kinderman SS, van Maarseveen JH, Schoemaker HE, Hiemstra H, Rutjes FPJT. *Org Lett* 2001;3:2045–2048.

200. Bennasar M-L, Roca T, Monerris M, Garcia-Diaz D. *Tetrahedron Lett* 2005;46:4035–4038.

201. Kasaya Y, Hoshi K, Terada Y, Nishida A, Shuto S, Arisawa M. *Eur J Org Chem* 2009:4606–4613.

202. Nunez A, Cuadro AM, Alvarez-Builla J, Vaquero JJ. *Org Lett* 2007;9:2977–2980.

203. Toumi M, Rincheval V, Young A, Gergeres D, Turos E, Couty F, Mignotte B, Evano G. *Eur J Org Chem* 2009:3368–3386.

204. Toumi M, Couty F, Evano G. *J Org Chem* 2008;73: 1270–1281.

205. Čusak A. *Chemistry* 2012;18:5800–5824.

206. Bracegirdle S, Anderson EA. *Chem Soc Rev* 2010;39: 4114–4129.

207. Ahmed M, Barrett AGM, Beall JC, Braddock DC, Flack K, Gibson VC, Procopiou PA, Salter MM. *Tetrahedron* 1999;55:3219–3232.

208. Schuman M, Gouverneur V. *Tetrahedron Lett* 2002;43: 3513–3516.

209. Pietraszuk C, Fischer H, Rogalski S, Marciniec B. *J Organomet Chem* 2005;690:5912–5921.

210. Pietraszuk C, Fischer H. *Organometallics* 2001;20: 4641–4646.

211. Forman GS, McConnell AE, Tooze RP, Van Rensburg WJ, Meyer WH, Kirk MM, Dwyer CL, Serfontein DW. *Organometallics* 2005;24:4528–4542.

212. Hoveyda AH, Lombardi PJ, O'Brien RV, Zhugralin AR. *J Am Chem Soc* 2009;131:8378–8379.

213. Werner H, Grunwald C, Stuer W, Wolf J. *Organometallics* 2003;22:1558–1560.

214. Hoye TR, Zhao H. *Org Lett* 1999;1:1123–1125.

215. Crimmins MT, King BW. *J Org Chem* 1996;61:4192–4193.

216. Nakashima K, Okamoto S, Sono M, Tori M. *Molecules* 2004;9:541–549.

217. Maleczka RE Jr, Terrell LR, Geng F, Ward JS III,. *Org Lett* 2002;4:2841–2844.

218. Takahashi S, Hishinuma N, Koshino H, Nakata T. *J Org Chem* 2005;70:10162–10165.

219. Nicolaou KC, Leung GYC, Dethe DH, Guduru R, Sun Y-P, Lim CS, Chen DYK. *J Am Chem Soc* 2008;130:10019–10023.

220. Srikrishna A, Pardeshi VH, Satyanarayana G. *Tetrahedron Lett* 2007;48:4087–4090.

221. Frigell J, Cumpstey I. *Tetrahedron Lett* 2007;48:9073–9076.

222. Dowling MS, Vanderwal CD. *J Org Chem* 2010;75: 6908–6922.

223. Spagnol G, Heck M-P, Nolan SP, Mioskowski C. *Org Lett* 2002;4:1767–1770.

224. Lin YA, Davis BG. *Beilstein J Org Chem* 2010;6:1219–1228.

225. Compain P. *Adv Synth Catal* 2007;349:1829–1846.

226. Schrock RR, Hoveyda AH. *Angew Chem Int Ed* 2003;42:4592–4633.

227. Wilson GO, Porter KA, Weissman H, White SR, Sottos NR, Moore JS. *Adv Synth Catal* 2009;351:1817–1825.

228. Shon Y-S, Lee TR. *Tetrahedron Lett* 1997;38:1283–1286.

229. Sattely ES, Cortez GA, Moebius DC, Schrock RR, Hoveyda AH. *J Am Chem Soc* 2005;127:8526–8533.

230. Fürstner A, Thiel OR. *J Org Chem* 2000;65:1738–1742.

231. Perlmutter P, Rose M, Vounatsos F. *Eur J Org Chem* 2003:756–760.

232. Evans P, Grigg R, Monteith M. *Tetrahedron Lett* 1999; 40:5247–5250.

233. Yang Q, Xiao W-J, Yu Z. *Org Lett* 2005;7:871–874.

234. Kyle AF, Jakubec P, Cockfield DM, Cleator E, Skidmore J, Dixon DJ. *Chem Commun* 2011;47:10037–10039.

235. Fu GC, Nguyen ST, Grubbs RH. *J Am Chem Soc* 1993;115:9856–9857.

236. Fürstner A, Leitner A. *Angew Chem Int Ed* 2003;42:308.

237. Fürstner A, Grabowski J, Lehmann CW. *J Org Chem* 1999;64:8275–8280.

238. Wright DL, Schulte JP, Page MA. *Org Lett* 2000;2:1847–1850.

239. Woodward CP, Spiccia ND, Jackson WR, Robinson AJ. *Chem Commun* 2011;47:779–781.

240. Alcaide B, Almendros P, Alonso JM, Luna A. *Synthesis* 2005:668–672.

241. Alcaide B, Almendros P. *Chem Eur J* 2003;9:1259–1262.
242. Arisawa M, Nishida A, Nakagawa M. *J Organomet Chem* 2006;691:5109–5121.
243. Higman CS, Plais L, Fogg DE. *ChemCatChem* 2013;5:3548–3551.
244. Houri AF, Xu Z, Cogan DA, Hoveyda AH. *J Am Chem Soc* 1995;117:2943–2944.
245. Xu Z, Johannes CW, Salman SS, Hoveyda AH. *J Am Chem Soc* 1996;118:10926–10927.
246. Barrett AGM, Baugh SPD, Gibson VC, Giles MR, Marshall EL, Procopiou PA. *Chem Commun* 1996:2231–2232.
247. Johannes M, Altmann KH. *Org Lett* 2012;14:3752–3755.
248. Balan D, Burns CJ, Fisk NG, Hugel H, Huang DCS, Segal D, White C, Wagler J, Rizzacasa MA. *Org Biomol Chem* 2012;10:8147–8153.
249. Fürstner A, Gastner T, Weintritt H. *J Org Chem* 1999;64: 2361–2366.
250. Fürstner A, Thiel OR, Lehmann CW. *Organometallics* 2002;21:331–335.
251. Czelusniak I, Heywood JD, Kenwright AM, Khosravi E. *J. Mol. Catal. A* 2008;280:29–34.
252. Haigh DM, Kenwright AM, Khosravi E. *Macromolecules* 2005;38:7571–7579.
253. Haigh DM, Kenwright AM, Khosravi E. *Tetrahedron* 2004;60:7217–7224.
254. Slugovc C, Demel S, Riegler S, Hobisch J, Stelzer F. *Macromol Rapid Commun* 2004;25:475–480.
255. Liu G, Tai W-Y, Li Y-L, Nan F-J. *Tetrahedron Lett* 2006;47:3295–3298.
256. Ferguson ML, O'Leary DJ, Grubbs RH. *Org Synth* 2003;80:85–92.
257. Kumar A, Alam MA, Rani S, Vankar YD. *Carbohydr Res* 2010;345:1142–1148.
258. Michalak K, Michalak M, Wicha J. *Tetrahedron Lett* 2005;46:1149–1153.
259. Nevalainen M, Koskinen AMP. *Angew Chem Int Ed* 2001;40:4060–4062.
260. Banwell MG, Garcia-Garcia P, Aguilar E. *e-EROS Encycl Reagents Org Synth* 2001No pp given.
261. Miller B, Mao S, Rosenker KMG, Pierce JG, Wipf P. *Beilstein J Org Chem* 2012;8:1091–1097.
262. Clark JS, Labre F, Thomas LH. *Org Biomol Chem* 2011;9:4823–4830.
263. Pentzer EB, Gadzikwa T, Nguyen ST. *Org Lett* 2008;10: 5613–5615.
264. Ibrahim YA. *J Mol Catal A* 2006;254:43–52.
265. Dasgupta S, Wu J. *Chem Sci* 2012;3:425–432.
266. Lejkowski M, Banerjee P, Schller S, Mnch A, Runsink J, Vermeeren C, Gais H-J. *Chem Eur J* 2012;18:3529–3548.
267. Samojlowicz C, Grela K. *Arkivoc* 2011:71–81.
268. McReynolds MD, Dougherty JM, Hanson PR. *Chem Rev* 2004;104:2239–2258.
269. Moore JD, Sprott KT, Hanson PR. *Synlett* 2001:605–608.
270. Kost T, Sigalov M, Goldberg I, Ben-Asuly A, Lemcoff NG. *J Organomet Chem* 2008;693:2200–2203.
271. Furstner A, Seidel G, Kindler N. *Tetrahedron* 1999;55: 8215–8230.
272. Garbaccio RM, Danishefsky SJ. *Org Lett* 2000;2:3127–3129.
273. Morgans GL, Yadav DB, Fernandes MA, de Koning CB, Michael JP, van Otterlo WAL. *Tetrahedron Lett* 2012;53:2384–2387.
274. Thomas CD, Hanson PR. Phosphorus and sulfur heterocycles via ring-closing metathesis. In: Cossy J, Arseniyadis S, Meyer C, editors. *Metathesis in Natural Product Synthesis: Strategies, Substrates and Catalysts*. Weinheim: Wiley-VCH; 2010. p 129–144.
275. Hanson PR, Probst DA, Robinson RE, Yau M. *Tetrahedron Lett* 1999;40:4761–4764.
276. Szadkowska A, Zukowska K, Pazio AE, Wozniak K, Kadyrov R, Grela K. *Organometallics* 2011;30:1130–1138.
277. Karsch S, Freitag D, Schwab P, Metz P. *Synthesis* 2004:1696–1712.
278. Yao Q. *Org Lett* 2002;4:427–430.
279. Fürstner A. *Angew Chem Int Ed* 2000;39:3012–3043.
280. Bates DK, Li XF, Jog PV. *J Org Chem* 2004;69:2750–2754.
281. Schrock RR, Murdzek JS, Bazan GC, Robbins J, DiMare M, O'Regan M. *J Am Chem Soc* 1990;112:3875–3886.
282. Coquerel Y, Rodriguez J. *Eur J Org Chem* 2008:1125–1132.
283. Dallinger D, Irfan M, Suljanovic A, Kappe CO. *J Org Chem* 2010;75:5278–5288.
284. Yang C, Murray WV, Wilson LJ. *Tetrahedron Lett* 2003;44:1783–1786.
285. Appukkuttan P, Dehaen W, Van der Eycken E. *Org Lett* 2005;7:2723–2726.
286. Appukkuttan P, Dehaen W, Van der Eycken E. *Chem Eur J* 2007;13:6452–6460.
287. Nosse B, Schall A, Jeong WB, Reiser O. *Adv Synth Catal* 2005;347:1869–1874.
288. Abell AD, Alexander NA, Aitken SG, Chen H, Coxon JM, Jones MA, McNabb SB, Muscroft-Taylor A. *J Org Chem* 2009;74:4354–4356.
289. Robinson AJ, Elaridi J, van Lierop BJ, Mujcinovic S, Jacksonnobreak WR. *J Pept Sci* 2007;13:280–285.
290. Andersson H, Demaegdt H, Johnsson A, Vauquelin G, Lindebergnobreak G, Hallberg M, Erdélyi M, Karlén A, Hallberg A. *J Med Chem* 2011;54:3779–3792.
291. Barragan F, Moreno V, Marchan V. *Chem Commun* 2009:4705–4707.
292. Robinson F, van Lierop BJ, Garland RD, Teoh E, Elaridi J, Illesinghe JP, Jackson WR. *Chem Commun* 2009:4293–4295.
293. Hoye TR, Jeffrey CS, Tennakoon MA, Wang J, Zhao H. *J Am Chem Soc* 2004;126:10210–10211.
294. Hoye TR, Jeon J. Metathesis involving a relay and applications in natural product synthesis. In: Cossy J, Arseniyadis S, Meyer C, editors. *Metathesis in Natural Product Synthesis: Strategies, Substrates and Catalysts*. Weinheim: Wiley-VCH; 2010. p 261–285.

295. Spino C, Boisvert L, Douville J, Roy S, Lauzon S, Minville J, Gagnon D, Beaumier F, Chabot C. *J Organomet Chem* 2006;691:5336–5355.
296. El-azizi Y, Schmitzer A, Collins SK. *Angew Chem Int Ed* 2006;45:968–973.
297. Collins SK. *J Organomet Chem* 2006;691:5122–5128.
298. Tannert R, Hu T-S, Arndt H-D, Waldmann H. *Chem Commun* 2009;0:1493–1495.
299. Hoye TR, Jeon J, Tennakoon MA. *Angew Chem Int Ed* 2011;50:2141–2143.
300. Bantreil X, Nolan SP. *Nature Protoc* 2011;6:69–77.
301. Beach NJ, Dharmasena UL, Drouin SD, Fogg DE. *Adv Synth Catal* 2008;350:773–777.
302. Hong SH, Day MW, Grubbs RH. *J Am Chem Soc* 2004;126:7414–7415.
303. Adjiman CS, Clarke AJ, Cooper G, Taylor PC. *Chem Commun* 2008:2806–2808.
304. Dandapani S, Lowe JT, Comer E, Marcaurelle LA. *J Org Chem* 2011;76:8042–8048.
305. Samojlowicz C, Bieniek M, Zarecki A, Kadyrov R, Grela K. *Chem Commun* 2008:6282–6284.
306. Schulz MD, Wagener KB. *ACS Macro Lett* 2012;1:449–451.
307. Kuhn KM, Champagne TM, Hong SH, Wei W-H, Nickel A, Lee CW, Virgil SC, Grubbs RH, Pederson RL. *Org Lett* 2010;12:984–987.
308. Collins JC, James K. *Med Chem Commun* 2012;3: 1489–1495.
309. Lysenko Z, Maughon BR, Mokhtar-Zadeh T, Tulchinsky ML. *J Organomet Chem* 2006;691:5197–5203.
310. van der Eide EF, Romero PE, Piers WE. *J Am Chem Soc* 2008;130:4485–4491.
311. Sieck SR, McReynolds MD, Schroeder CE, Hanson PR. *J Organomet Chem* 2006;691:5307–5311.
312. Yamamoto K, Biswas K, Gaul C, Danishefsky SJ. *Tetrahedron Lett* 2003;44:3297–3299.
313. Sliwa A, Dive G, Jiwan J-LH, Marchand-Brynaert J. *Tetrahedron* 2010;66:9519–9527.
314. Monfette, S. From analysis of ligands effects to reactor design: a rational approach to efficient macrocyclization via ring-closing metathesis. PhD, University of Ottawa: Ottawa, 2010.
315. Smith AB III, Basu K, Bosanac T. *J Am Chem Soc* 2007;129:14872–14874.
316. Yee NK, Farina V, Houpis IN, Haddad N, Frutos RP, Gallou F, Wang X-J, Wei X, Simpson RD, Feng X, Fuchs V, Xu Y, Tan J, Zhang L, Xu J, Smith-Keenan LL, Vitous J, Ridges MD, Spinelli EM, Johnson M, Donsbach K, Nicola T, Brenner M, Winter E, Kreye P, Samstag W. *J Org Chem* 2006;71:7133–7145.
317. Shu C, Zeng X, Hao M-H, Wei X, Yee NK, Busacca CA, Han Z, Farina V, Senanayake CH. *Org Lett* 2008;10:1303–1306.
318. Wu Z, Nguyen ST, Grubbs RH, Ziller JW. *J Am Chem Soc* 1995;117:5503–5511.
319. Liu W, Nichols PJ, Smith N. *Tetrahedron Lett* 2009;50: 6103–6105.
320. Galan BR, Kalbarczyk KP, Szczepankiewicz S, Keister JB, Diver ST. *Org Lett* 2007;9:1203–1206.
321. Blacquiere JM, Jurca T, Weiss J, Fogg DE. *Adv Synth Catal* 2008;350:2849–2855.
322. Vougioukalakis GC. *Chem Eur J* 2012;18:8868–8880.
323. Haack KL, Ahn YM, Georg GI. *Mol Diversity* 2005;9: 301–303.
324. Gallou F, Saim S, Koenig KJ, Bochniak D, Horhota ST, Yee NK, Senanayake CH. *Org Process Res Dev* 2006;10: 937–940.
325. Lambeth RH, Pederson SJ, Baranoski M, Rawlett AM. *J Polym Sci A* 2010;48:5752–5757.
326. van der Gryp P, Barnard A, Cronje JP, de Vlieger D, Marx S, Vosloo HCM. *J Membr Sci* 2010;353:70–77.
327. Crimmins MT, Choy AL. *J Org Chem* 1997;62:7548–7549.
328. Magano J, Dunetz JR. *Chem Rev* 2011;111:2177–2250.
329. Busacca CA, Fandrick DR, Song JJ, Senanayake CH. *Adv Synth Catal* 2011;353:1825–1864.
330. Sawyer TK. *Chem Biol Drug Des* 2009;73:3–6.
331. Wang H, Goodman SN, Dai Q, Stockdale GW, Clark WM. *Org Process Res Dev* 2008;12:226–234.
332. Wang H, Matsuhashi H, Doan BD, Goodman SN, Ouyang X, Clark WM. *Tetrahedron* 2009;65:6291–6303.
333. Wilson LJ, Malaviya R, Yang C, Argentieri R, Wang B, Chen X, Murray WV, Cavender D. *Bioorg Med Chem Lett* 2009;19:3333–3338.
334. Liverton NJ, Holloway MK, McCauley JA, Rudd MT, Butcher JW, Carroll SS, DiMuzio J, Fandozzi C, Gilbert KF, Mao S-S, McIntyre CJ, Nguyen KT, Romano JJ, Stahlhut M, Wan B-L, Olsen DB, Vacca JP. *J Am Chem Soc* 2008;130: 4607–4609.
335. McCauley JA, Rudd MT, Nguyen KT, McIntyre CJ, Romano JJ, Bush KJ, Varga SL, Ross CW III, Carroll SS, DiMuzio J, Stahlhut MW, Olsen DB, Lyle TA, Vacca JP, Liverton NJ. *Angew Chem Int Ed* 2008;47:9104–9107.
336. Song ZJ, Tellers DM, Journet M, Kuethe JT, Lieberman D, Humphrey G, Zhang F, Peng Z, Waters MS, Zewge D, Nolting A, Zhao D, Reamer RA, Dormer PG, Belyk KM, Davies IW, Devine PN, Tschaen DM. *J Org Chem* 2011;76:7804–7815.
337. Kong J, Chen C-Y, Balsells-Padros J, Cao Y, Dunn RF, Dolman SJ, Janey J, Li H, Zacuto MJ. *J Org Chem* 2012;77:3820–3828.
338. Machauer R, Veenstra S, Rondeau J-M, Tintelnot-Blomley M, Betschart C, Neumann U, Paganetti P. *Bioorg Med Chem Lett* 2009;19:1361–1365.
339. Content S, Dutton CJ, Roberts L. *Bioorg Med Chem Lett* 2003;13:321–325.
340. Brawn RA, Guimaraes CRW, McClure KF, Liras S. *Org Lett* 2012;14:4802–4805.
341. Kazmierski WM. *Antiviral Drugs*. Hoboken, NJ: John Wiley & Sons; 2011.
342. Kazmierski WM, Hamatake R, Duan M, Wright LL, Smith GK, Jarvest RL, Ji J-J, Cooper JP, Tallant MD, Crosby RM, Creech K, Wang A, Li X, Zhang S, Zhang Y-K, Liu Y,

Ding CZ, Zhou Y, Plattner JJ, Baker SJ, Bu W, Liu L. *J Med Chem* 2012;55:3021–3026.

343. Llinàs-Brunet M, Bailey MD, Bolger G, Brochu C, Faucher A-M, Ferland JM, Garneau M, Ghiro E, Gorys V, Grand-Maître C, Halmos T, Lapeyre-Paquette N, Liard F, Poirier M, Rhéaume M, Tsantrizos YS, Lamarre D. *J Med Chem* 2004;47:1605–1608.

344. Hinrichsen H, Benhamou Y, Wedemeyer H, Reiser M, Sentjens Roel E, Calleja Jose L, Forns X, Erhardt A, Cronlein J, Chaves Ricardo L, Yong C-L, Nehmiz G, Steinmann Gerhard G. *Gastroenterology* 2004;127:1347–1355.

345. Georgopapadakou N. *Expert Opin Investig Drugs* 2007; 16:1–10.

346. Chatel-Chaix L, Baril M, Lamarre D. Pharmacology and Mechanisms of Action of Antiviral Drugs: Protease Inhibitors. In: McCaughan GW, McHutchison J, Pawlotsky J-M, editors. *Advanced Therapy for Hepatitis C*. Toronto: Wiley-Blackwell; 2011.

347. Tsantrizos YS. *Acc Chem Res* 2008;41:1252–1263.

348. Chatel-Chaix L, Baril M, Lamarre D. *Viruses* 2010;2:1752–1765.

349. Poirier M, Aubry N, Boucher C, Ferland J-M, LaPlante S, Tsantrizos YS. *J Org Chem* 2005;70:10765–10773.

350. Lummiss JAM, Ireland BJ, Fogg DE. *Chem Cat Chem* 2012;6:459–463.

4

ENE-YNE METATHESIS

STEVEN T. DIVER AND JUSTIN R. GRIFFITHS
Department of Chemistry, University at Buffalo-SUNY, Buffalo, NY, USA

4.1 INTRODUCTION

Ene-yne metathesis provides a catalytic method for the synthesis of 1,3-dienes. Conjugated dienes are valued building blocks in organic synthesis. For example, conjugated 1,3-dienes are widely used in Diels–Alder reactions for the synthesis of six-membered rings. With catalytic ene-yne metathesis, the 1,3-diene is synthesized in a single reaction step from alkene and alkyne reactants. The chemical reaction is triggered by the Grubbs ruthenium carbenes. The ruthenium carbenes are uniquely suited to react with π bonds, which is the origin of their remarkable chemoselectivity. Functional group tolerance allows the method to be used in complex molecules with dense arrays of organic functional groups. The ene-yne metathesis typically employs the mild, chemoselective, and readily available Grubbs carbene complexes. Related ene-yne bond reorganization promoted by the metal salts such as $PtCl_2$ will not be covered in this chapter. For these, the reader is directed to other reviews (1).

Ene-yne metathesis is the carbon–carbon bond coupling that occurs between an alkene and an alkyne with pi bond reorganization. By far the most common use of ene-yne metathesis is the ring-closing enyne metathesis (RCEYM, Eq. 4.1) (2). This variation is discussed in Section 4.2.1. When the reaction occurs between two separate molecules in an intermolecular reaction, it is known as a *cross* ene-yne metathesis (CEYM, Eq. 4.2) (3). For the reaction to occur, the alkene must be reactive with the Grubbs carbene being used. This variation is discussed in Section 4.2.2. Enyne metathesis is also valuable as a part of cascading metathesis reactions that form two or more rings (Eq. 4.3) (4). The cascade or tandem processes are discussed in Section 4.3.3. Ring rearrangement is discussed in Section 4.3.4 (Scheme 4.1).

Ene-yne metathesis has been widely used in organic synthesis. A better mechanistic understanding of the reaction, the timing of the steps, relative reactivity of unsaturated components, and selectivity—though by no means complete—has helped move applications of the reaction into more complex systems to form multiple rings. Recent topics of interest to synthetic chemists include complexity-building potential, which makes cascade processes attractive, and diversity-oriented synthesis, which uses that principle for the assembly of polycyclic scaffolds that can be used in drug discovery. A number of excellent review articles have appeared on tandem reactions conjoined with ene-yne metathesis (5–8). Tandem processes using both traditional reactions and tandem metathesis cascades are discussed in Section 4.3.

4.1.1 Catalysts Used

The development of ene-yne metathesis has paralleled improvements in the well-defined ruthenium carbene catalysts. The most important catalyst in ene-yne metathesis is Grubbs' second-generation carbene complex **Ru2** (Scheme 4.2), which was reported in 1999 (9) and became commercially available in 2003. Although the Grubbs complexes were not available in 1985, Katz and Sivavec (10) managed to show the catalytic effectiveness of RCEYM using a tungsten Fischer carbene complex. The early contribution of Katz was important as it identified a catalytic reaction distinct from that of alkene metathesis, and furthermore laid the mechanistic groundwork for the reaction. Alkene metathesis was

Olefin Metathesis: Theory and Practice, First Edition. Edited by Karol Grela.

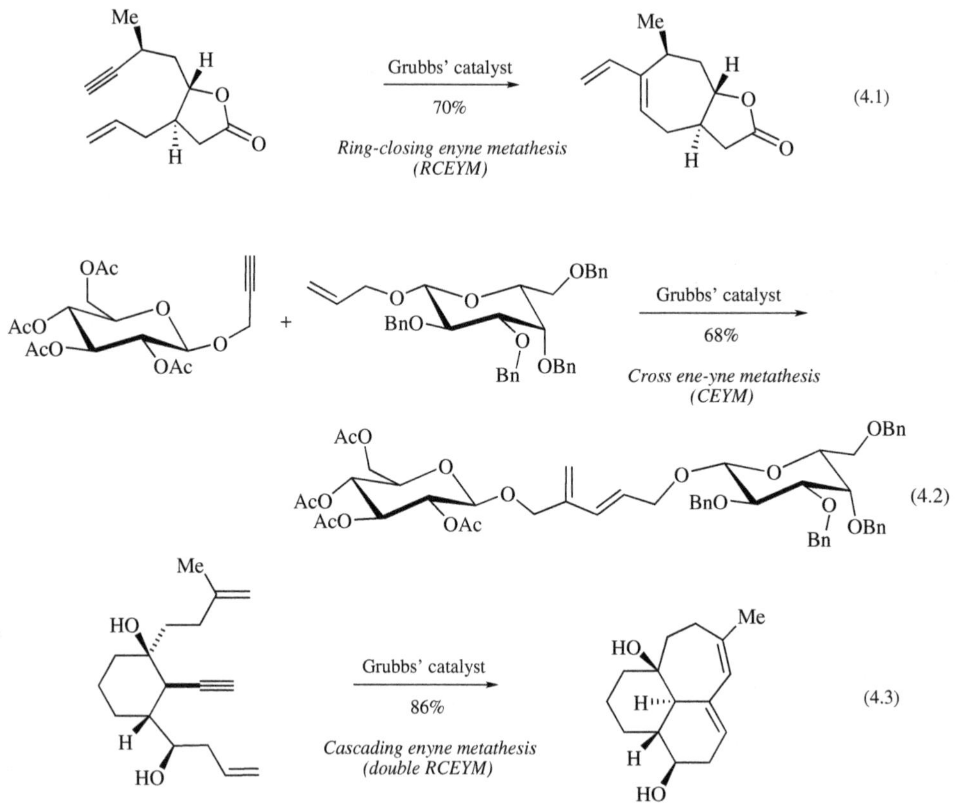

Scheme 4.1 Three types of en(e)-yne metathesis (2–4).

being studied contemporaneously by Grubbs, Schrock, and Chauvin. However, the use of early transition metal carbene complexes in enyne metathesis was limited to isolated cases of ring-closing metatheses (RCMs) and suffered from harsh conditions. This and the perception that early metal carbenes were harsh limited adoption of this chemistry by the synthetic community. After a decade, Mori's group applied the metathesis to a challenging nitrogen heterocycle synthesis, dispelling the perception that enyne metathesis could not be used in complex molecule synthesis (11–13). However, the real breakthrough came from catalyst development. The Grubbs group discovered a well-defined, group VIII ruthenium carbene complex that promoted alkene metathesis (14). This led to a number of applications by Grubbs and Fu, which captured the interest of the synthetic community (ruthenium (15) and molybdenum (16)). Having already investigated catalytic enyne metathesis, the Mori group revisited the problem using the Grubbs ruthenium carbene complex and found excellent reactivity in enyne metathesis (17). Mori, Kinoshita, and coworkers followed this work with additional studies using ethylene, which secured enyne metathesis a place in organic synthesis (18–20). Blechert and his group then showed that the ene-yne metathesis could be performed between two separate molecules of alkene and alkyne. This was a breakthrough because at the time it was unclear how well the catalytic reaction would work without the alkene and alkyne connected together. This showed that the scope could be even more general and not just be used to form rings. Last, as the Grubbs' second-generation catalyst became more widely available, additional applications capitalized on the greater reactivity of this complex, which had been extracted from alkene metathesis.

The second-generation Grubbs carbene complex **Ru2** has dominated ene-yne metathesis applications in the past several years (Scheme 4.2). Previously, the remarkable chemoselectivity of the first-generation Grubbs complex commanded the attention of the synthetic community because the reactivity profile was so unique to this organometallic complex. Orthogonal reactivity, which is selective reactivity of unactivated groups under neutral reaction conditions, is a valuable tool in the synthetic chemist's workshop. It requires poise for the ruthenium carbene to be able to split carbon–carbon bonds but yet not be destroyed by functional groups. In addition, the Grubbs catalysts are extremely easy to use and commercially available. Due to ease-of-use and catalyst availability, the use of metathesis in a synthetic plan has a low activation barrier that has facilitated adventurous synthetic pursuits. Moreover, it can partly

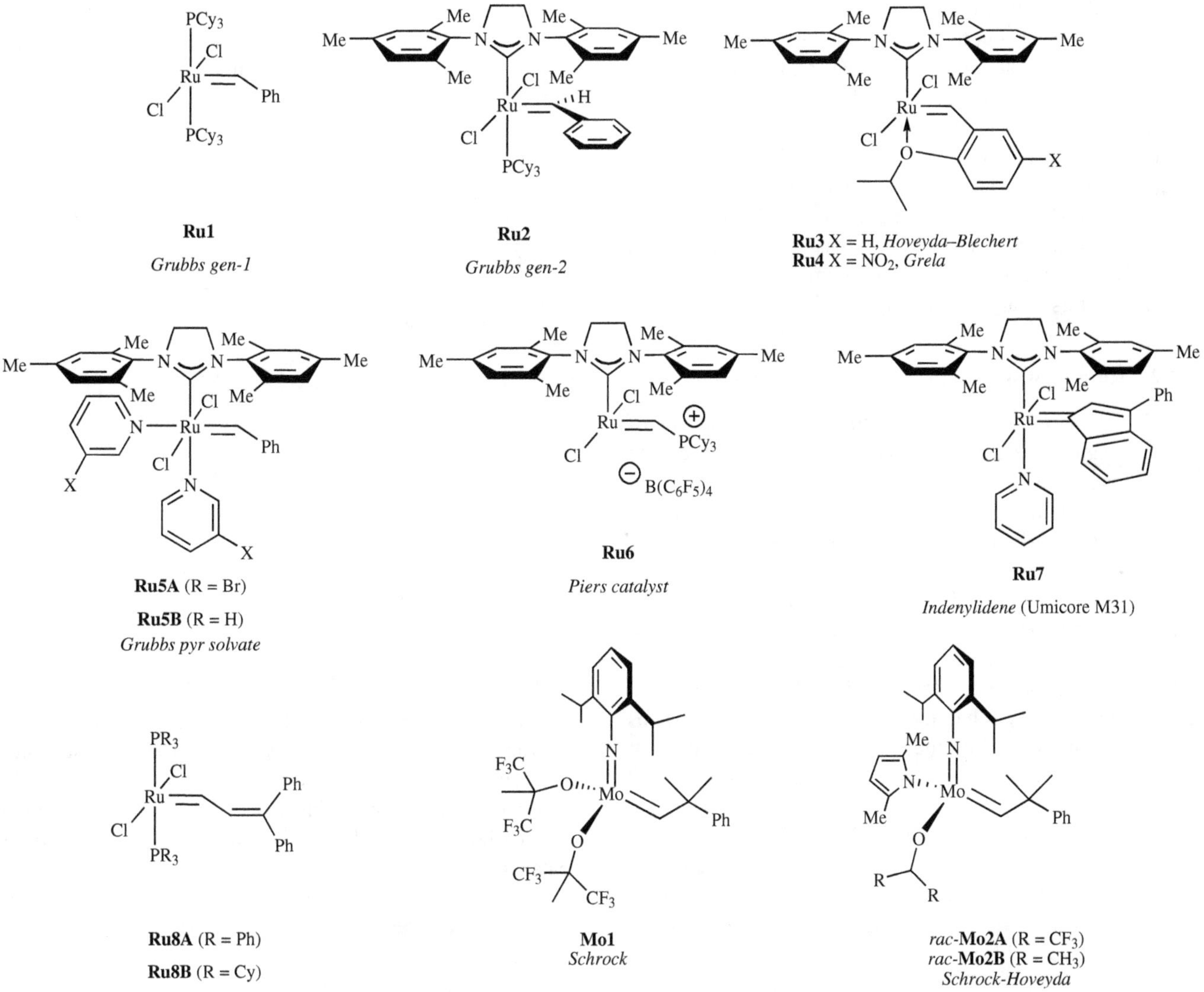

Scheme 4.2 Metal carbene complexes used to promote ene-yne metathesis.

explain the explosion in metathesis applications and the widespread use of the Grubbs catalyst. The widely accepted utility of this catalytic carbon–carbon bond formation led to the recognition of the groundbreaking work and leadership from Grubbs (21), Schrock (22), and Chauvin (23), who were awarded the Nobel Prize in Chemistry in 2005.

The Grubbs catalysts **Ru1** and **Ru2** are primarily used in ene-yne metathesis, but other ruthenium carbenes can also be used. The bisphosphine complex **Ru1** is called the *first-generation Grubbs complex* which is now used less frequently because it is less reactive than the second-generation Grubbs complex **Ru2** (24,25). Carbene complex **Ru2** bears a single *N*-heterocyclic carbene (NHC) ligand and retains a dissociable Cy_3P ligand, reported by the Grubbs group in a highly cited 1999 paper (9). Because its higher reactivity in alkene metathesis derives from higher binding affinity of unsaturated functionality, **Ru2** is also found to be superior in ene-yne metathesis (26). New carbenes are continually being developed for challenging applications in alkene metathesis. Phosphine-free catalysts are also used in ene-yne metathesis. With the saturated NHC ligand, dihydroIMes, the chelated complex **Ru3** is known as the *Hoveyda–Blechert complex*, as it was independently developed by the Hoveyda (27,28) and Blechert research groups (29) (the catalyst is commercially available as the Hoveyda–Grubbs complex). The electron-deficient complex **Ru4** is known as the *Grela* (30) *complex*, which has a more weakly coordinated ether that increases catalyst initiation rate. Phosphine-free catalysts **Ru3–Ru7** are useful in cases where higher temperatures are needed. Other new catalysts are based on the indenylidene such as **Ru7**; some of these have been used to promote ene-yne metathesis (31–33).

By far, the ruthenium carbenes are used most to promote ene-yne metathesis, but there are a few examples of molybdenum carbenes that can promote the intramolecular reaction. The parent Schrock carbene **Mo1** is not used in enyne metathesis because it tends to polymerize alkynes. However, the catalyst can be tuned to give endo-selective RCEYM (section 4.2.1.3). In addition, the Schrock–Hoveyda catalysts such as **Mo2** are stereogenic at the metal, which could be exploited in asymmetric RCEYM applications. Though the molybdenum carbenes are not commonly used in enyne metathesis, this catalyst subfamily is unique.

4.1.2 Ring-Closing Enyne Metathesis (RCEYM)

RCEYM is a well-established method for the synthesis of rings. A variety of carbocyclic and heterocyclic rings can be formed using this protocol. Good results have been obtained in molecules with a high density of functional groups such as those found in carbohydrates and natural products. RCEYM has been used reliably in natural products synthesis. The synthesis of small rings is relatively straightforward using metathesis methods (7,21). Larger rings can also be formed under typical macrocyclization conditions involving high dilution of the enyne reactant. In ring-closing metathesis, the mode selectivity is based on whether the alkyne carbons become incorporated into the newly formed ring (endo) or not (exo). In smaller rings such as five- and six-membered rings, RCEYM is controlled by geometric constraint that dictates the exo ring-closure mode. Macro-RCEYM follows endo mode selectivity. The diene products can be subjected to subsequent reactions, such as cycloaddition or a second cross metathesis (CM) reaction, useful for the rapid assembly of molecular scaffolds.

Historically, the RCEYM had pushed into synthetic frontiers as far as catalyst technology would allow. Organic chemists were interested in Mori's early application of Fischer carbenes to nitrogen heterocycle synthesis, though the catalysts were not particularly reactive or functional group tolerant. Currently, the second-generation Grubbs carbene is widely employed for RCEYM, though in earlier examples and for the simplest cases, the first-generation Grubbs complex **Ru1** was successfully used. The most difficult ring closures to make small rings or those containing highly substituted alkenes require longer lived and thermally robust catalysts such as the Hoveyda–Blechert catalyst **Ru3**. Finally, RCEYM with the first-generation Grubbs carbene benefits from the presence of ethylene.

Mori and Kinoshita (34) used RCEYM as a key step in their synthesis of the *Stemona* alkaloid stemoamide (Eq. 4.4). This was a landmark contribution because it demonstrated that the Grubbs carbenes could be used catalytically in complex molecule settings. The Grubbs catalyst **Ru8B** was new at that time and offered a tremendous advance over early transition metal Fischer carbenes, which lacked the activity, selectivity, and functional group tolerance of the well-defined ruthenium carbene complex. This ushered in the modern era of catalytic diene synthesis using the Grubbs ruthenium carbenes.

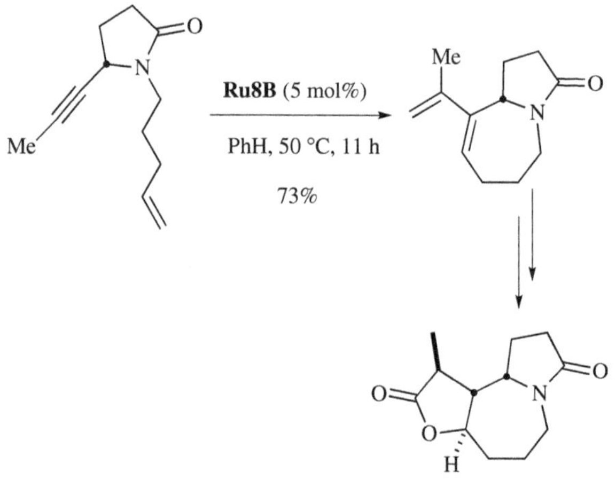

(–)-*stemoamide*

(4.4)

4.1.3 Cross (Intermolecular) Ene-Yne Metathesis

Intermolecular metathesis between an alkene and an alkyne is called *cross* ene-yne metathesis. The intermolecular reaction was developed after the intramolecular ring-closing application, but has become a reliable method for carbon–carbon bond formation using the Grubbs catalysts. As such, the CM represents a metathesis cross coupling. It is cross selective, resulting in the selective joining of the alkene with the alkyne. This selectivity tends to supersede other combinations of the unsaturated reactants. An excess of alkene is common.

A critical experimental parameter is that of alkene concentration. The alkene plays two distinct roles in CM. First, it must react with the metal precatalyst to form a metal carbene. The metal carbene is needed to carry out catalysis. This is commonly called *the ene-then-yne* pathway. Second, higher alkene concentrations, usually expressed as higher alkene equivalents (3–6 equiv), are beneficial because they tend to accelerate the kinetically slow step of catalysis—the turnover step. The alkene is important in sustaining ene-yne metathesis catalysis, but in so doing plays another important role—reducing competing side reactions. We term this as *manifold* selectivity—where the alkene helps maintain ene-yne metathesis manifold reactivity and thereby suppresses other metal carbene-promoted pathways such as alkyne oligomerization. The use of several equivalents of alkene is quite common. For more precious alkenes, lower equivalents can be used, but side reactions may occur depending on the alkyne reactivity.

As powerful as intermolecular ene-yne metathesis (EYM) is for carbon–carbon bond formation, it is not kinetically

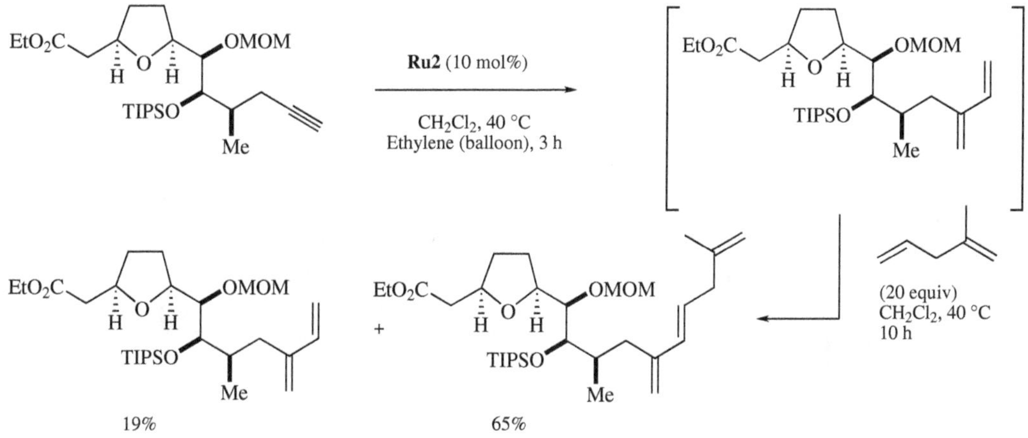

Scheme 4.3 Use of ethylene for *E*-selective cross ene-yne metathesis (39).

stereoselective: *E/Z* mixtures typically result. Though the *E*-isomer tends to be favored when using **Ru2**, the *Z*-isomer is also present and can be difficult to separate by column chromatography. Presently, there are no ruthenium carbene catalysts that exert *Z*-selectivity in CEYM. There are isolated cases where *E*-selectivity can be obtained under thermodynamic control using large excess of a reactive alkene (35,36). However, achieving *E*-selectivity is not general. Use of ethylene can provide an *E*-selective diene synthesis (37). In this system, ethylene first reacts with the alkyne to give an intermediate 2-substituted 1,3-diene. The *E*-selectivity comes in the CM of the 1,3-diene with excess 1-alkene (38), a more selective alkene metathesis. In the synthetic example shown in Scheme 4.3 (39), excess ethylene is removed after the first step, and then excess of the CM partner is introduced. In this case, the 1,4-diene reacts site selectively at the more highly reactive alkene site (40). The success of 1,3-diene equilibration by secondary metathesis depends on the reactivity of the alkene and the sustained activity of the ruthenium carbene (35).

4.1.4 Mechanism of Ene-Yne Metathesis

Mechanistic studies have developed to provide a generally accepted mechanism. The reaction between alkene and alkyne is an alkene-first reaction, whereby the alkene generates a reactive metal carbene. This has been termed an *alkylidene-first* mechanism and, more recently, called an *ene-then-yne* mechanism. With certain unsaturated reactants (i.e., with poorly reactive alkenes), the sequence may be altered and using different catalysts may result in a change of the rate-determining step (41,42).

An accepted mechanism has emerged based on mechanistic studies, DFT calculations, and kinetic investigation of the ene-yne metathesis reaction mechanism (Scheme 4.4). Kinetic evidence of the cross ene-yne metathesis promoted by **Ru2** indicates an ene-then-yne sequence of reactivity. Supporting evidence for an alkylidene-first mechanism can be found in early independent contributions by Hoye (43) and Kozmin (44). Zero-order dependence on the alkyne concentration indicates that the alkyne is not involved in the rate-determining step. First-order dependence was found for both the ruthenium complex and the terminal alkene. This kinetic evidence suggests that last steps of the catalytic cycle (either step V or step VI) are the rate-determining steps (alkene binding or metallacycle formation/collapse). On the basis of DFT calculations using a simplified ruthenium catalyst, Lippstreu and Straub (45) suggested that the cycloreversion step VI is rate limiting, which is consistent with the experimental kinetic data. The mechanistic hypothesis that vinyl carbene turnover should be susceptible to alkene reactivity was tested in a Hammett study on vinyl carbene reactions with styrenes (46). In light of the kinetic dependence of the rate on alkene concentration, higher concentrations or higher equivalents of alkene are beneficial to keep the catalytic cycle turning. Typically, the alkene is used in molar excess, but not always (47).

4.1.5 Role of Ethylene in Ring-Closing Enyne Metathesis (Mori's Conditions)

The helpful effect of ethylene on ring-closing metathesis of a 1,ω-enyne is known as *Mori's conditions.* (19) For example, ring-closing metathesis of a simple enyne gave low yield under inert atmosphere, but under an atmosphere of ethylene (an ethylene balloon) high yield was obtained (Eq. 4.5). Ethylene benefits the reaction in several ways: by stabilizing reactive intermediates, facilitating catalyst turnover, and preventing alkyne oligomerization pathways. The importance of alkene concentration had been noted in earlier studies by Hoye (43) and Lee (48), and a mechanistic corollary can be made with ene-yne metathesis, studied kinetically by Diver and Keister (42).

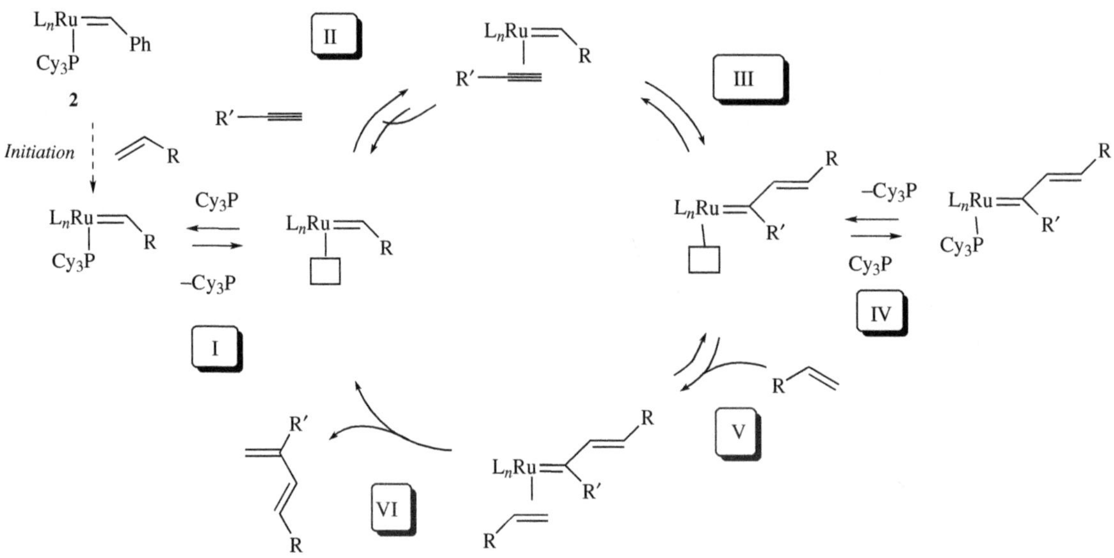

Scheme 4.4 Mechanism for intermolecular ene-yne metathesis based on kinetic studies (42).

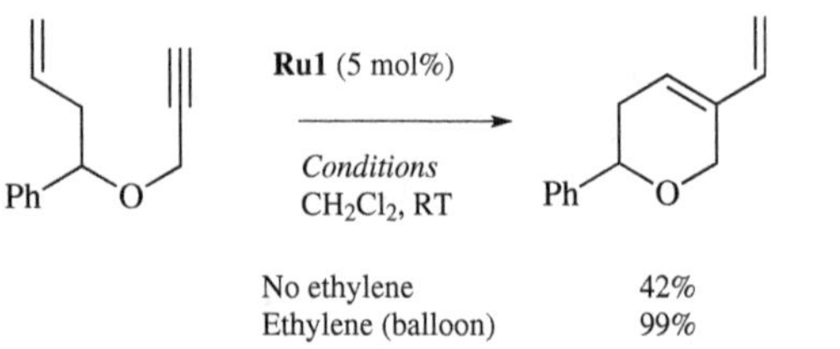 (4.5)

The ethylene effect is primarily due to the need for higher concentrations of alkene to maintain catalysis. Hoye (43) showed that enyne concentration affects the *rate* of RCEYM. In an elegant study, Lloyd-Jones (49) showed that ^{13}C-labeled ethylene produced vinyl carbene turnover. Because this occurred by another pathway that did not involve enyne reactant, the ethylene turnover occurred by a "second cycle." This is consistent with the earlier concentration study by Hoye and Vos (43) and by the mechanistic proposal by Diver et al (50). that ethylene speeds up the slow step of cross ene-yne metathesis. Interestingly, both intramolecular enyne metathesis and intermolecular ene-yne metathesis benefit from higher alkene concentration that helps propel the vinyl carbene forward in catalysis, bypassing competing pathways ("manifold" selectivity).

Ethylene helps suppress alkyne polymerization. Insight into the ethylene effect has been provided by observations made using matrix-assisted laser desorption/ionization (MALDI) mass spectral analysis under inert atmosphere conditions (51). Fogg and coworkers had developed mass spectral analysis as a tool for studying intermediates and decomposition products of ruthenium carbenes relevant to metathesis. A ring-closing metathesis was investigated under nitrogen or ethylene. Under an ethylene atmosphere, an Ru=CH_2 resting state was detected by MALDI-MS analysis. Under nitrogen, a decomposition species was found that had incorporated two molecules of enyne substrate. At longer reaction times, additional alkyne oligomerization had occurred. The dimer and oligomer observed under nitrogen atmosphere (in the absence of ethylene) are suggestive of alkyne oligomerization through a vinyl carbene intermediate.

For some ring-closing metathesis, the use of ethylene can lead to intermolecular ene-yne metathesis. For example, if ring closure is slow, the competitive ethylene–alkyne CM can be expected. Formation of eight-membered rings is particularly difficult. Ring closing under standard inert atmosphere conditions gave a low yield of the RCM product (Eq. 4.6). Using an ethylene atmosphere resulted in only the

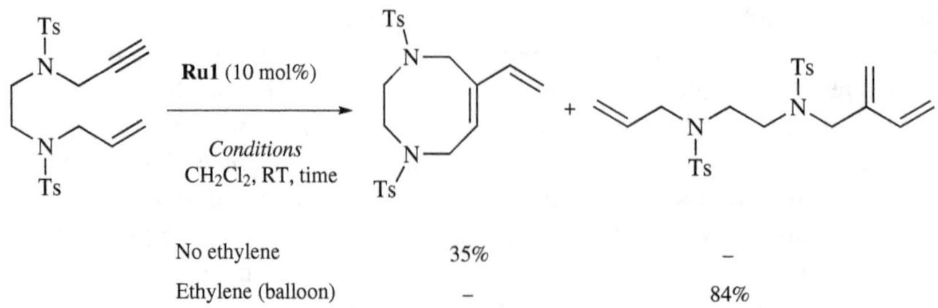 (4.6)

ethylene CM product, a 2-substituted butadiene (Eq. 4.6) (52). In other cases, the acyclic 1,3-diene may undergo a subsequent RCM with a pendant alkene, but this requires the use of **Ru2**.

4.2 ENE-YNE METATHESIS NOT INVOLVING CASCADES

Enyne metathesis is a powerful means of carbon–carbon bond coupling. Ring-closing reactions are highly successful because the reactive unsaturated elements are held in close proximity. The atoms joining the reactive units may be quite diverse, allowing a variety of heterocycles to be formed. RCEYM is reliable, although coordination by functionality can become problematic if it resides in close proximity to the alkene or alkyne. Ene-yne metathesis is an effective cross coupling that has wide alkene scope when coupled with the increased reactivity of **Ru2**. In this section, reaction involving a single metathesis event is overviewed. This provides the reader with a snapshot of reactivity and hopefully stimulates new uses of this reaction.

4.2.1 Synthesis of Rings

The RCEYM has been used to furnish a variety of carbocyclic and heterocyclic products. RCEYM provides unique access to functionalized heterocyclic ring systems from linear unsaturated elements joined by heteroatom-containing linkers. The ring-closure method is effective in building bicyclic and macrocyclic rings. In this application, there are many synthetic examples; a few representative examples are provided.

Formation of strained four-membered rings is possible by RCEYM. In general, formation of four-membered rings by ring-closing metathesis is difficult due to ring strain. However, the π bond reorganization has a favorable enthalpy of reaction that can compensate for the strain found in the product. To overcome the kinetic barrier, microwave heating with complex **Ru3** promoted RCEYM to form strained cyclobutene rings (Eq. 4.7) (53). In all cases, the alkyne was internal which helped suppress competing oligomerization. Goess et al (54). uses this method to synthesize grandisol (Table 4.1, entry 3). In a separate study, Grela et al. (54). found that a combination of microwave irradiation with fluorinated aromatic solvents promoted this reaction in high yield. The presence of the quaternary center at the propargylic position is rare in enyne metathesis, yet gives the best chemical yield. This can be understood in terms of Fogg's study discussed above.

RCEYM to form carbocyclic rings is generally straightforward. RCEYM of 1,7-enynes accesses vinyl cyclohexenes that are amenable to intramolecular Diels–Alder cycloaddition (Scheme 4.5) (56). In this example, the 1,3-diene product

TABLE 4.1 Microwave-Assisted RCEYM

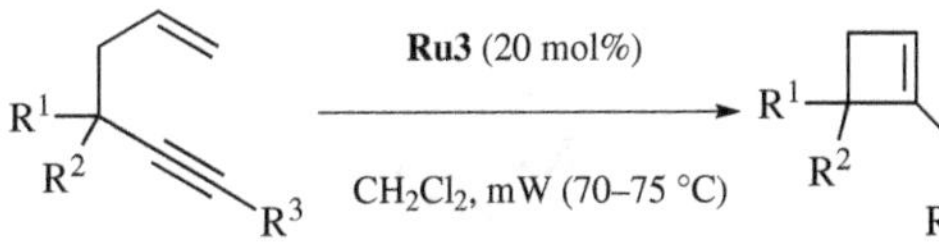

(4.7)

Entry	R^1	R^2	R^3	Yield, %
1	Ph	–	Bu	58
2	Ph	–	CH_2CH_2Ph	39
3	Me	CH_2CH_2OTBS	Me	83

Refs 53, 54.

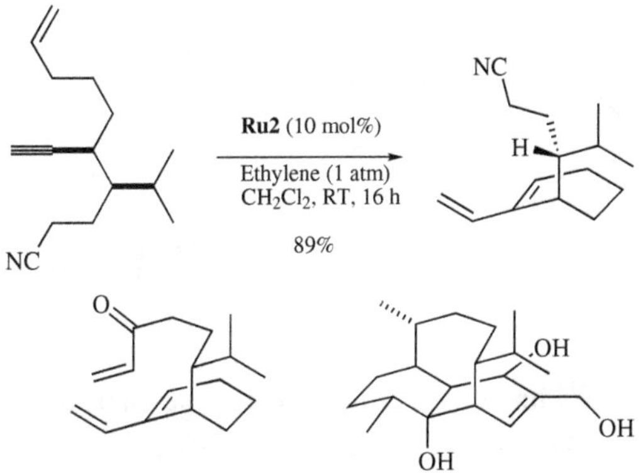

(4.8)

1 Vinigrol

Scheme 4.5 Ring-closing enyne metathesis used to set up intramolecular cycloaddition (56).

was transformed into the enone **1**, which underwent a Lewis acid-catalyzed Diels–Alder reaction to provide the carbon framework of the tricyclic core of the diterpenoid vinigrol.

Highly functionalized cyclic ethers are accessible from carbohydrate-based enynes by RCEYM. Spiroketals were formed from protected carbohydrates bearing unsaturated elements at the anomeric position (Eq. 4.9) (57). When a vinyl substrate **2** was used, the ring closure failed due to the steric hindrance associated with the highly substituted anomeric center.

RCEYM has been used to access sensitive heterocyclic ring systems such as the carbpenams and their structural analogs. The ring-closing metathesis conditions are sufficiently mild such that the delicate β-lactam ring system remains intact during annulation (Eq. 4.10) (58,59). Larger B-rings were made with selenium present in the nascent ring using **Ru2** (Eq. 4.11) (60). This reaction proved quite challenging due to the ring size and the presence of the selenium atom.

The functional group tolerance of the Grubbs carbenes and the efficiency of RCEYM are highly effective for rapid assembly of diverse molecular structures, a requirement for diversity-oriented synthesis. These two examples illustrate that the enynes can be adorned with a rich array of

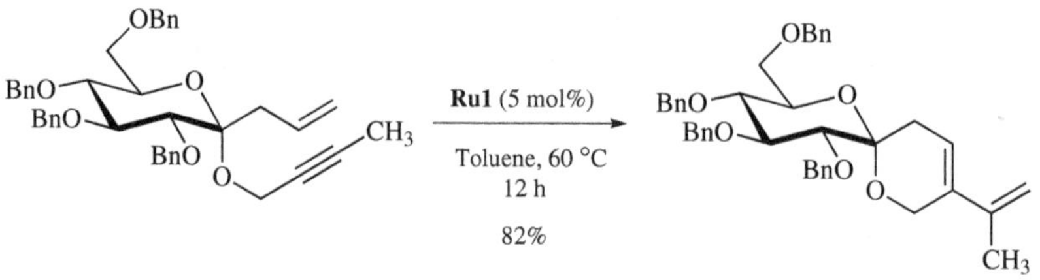

(4.9)

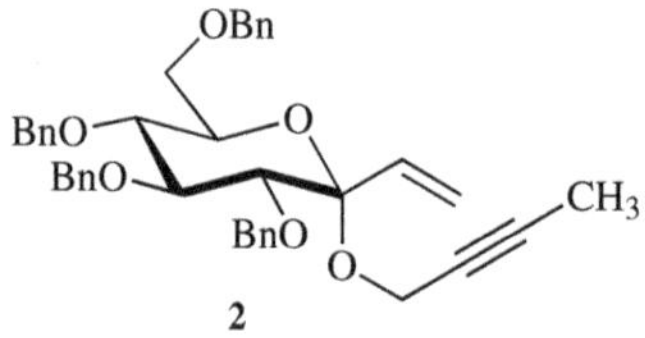

2

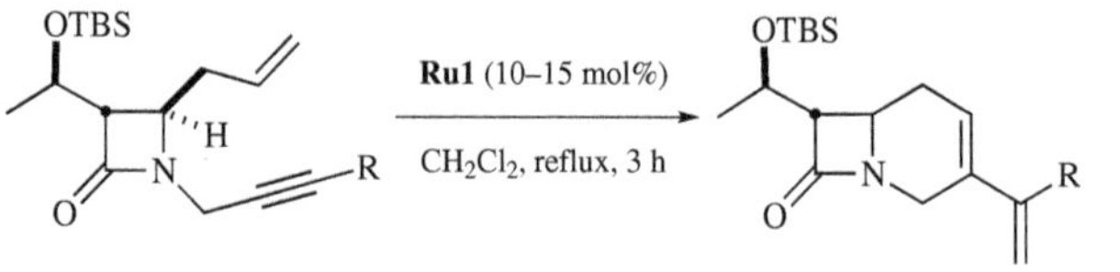

R = CH_3, quant.
R = $CH_2OTBDMS$, 72%

(4.10)

functionality without deleterious effect on the catalytic enyne metathesis (Eq. 4.12 and 4.13) (61).

A wide variety of heteroatoms are tolerated in the linker. One recent example produced an unusual heterocycle, the sulfoximine (Eq. 4.14) (62). This RCEYM benefitted from smaller ring sizes used to drive the reaction; attempts to form larger rings were not successful. The resulting 1,3-dienes underwent further transformation such as a Diels–Alder/aromatization with excess benzoquinone, allowing access to benzofused sulfoximines.

Five-membered ring-containing sultams can be accessed by RCEYM (Eq. 4.15) (63). Heterocyclic ring systems

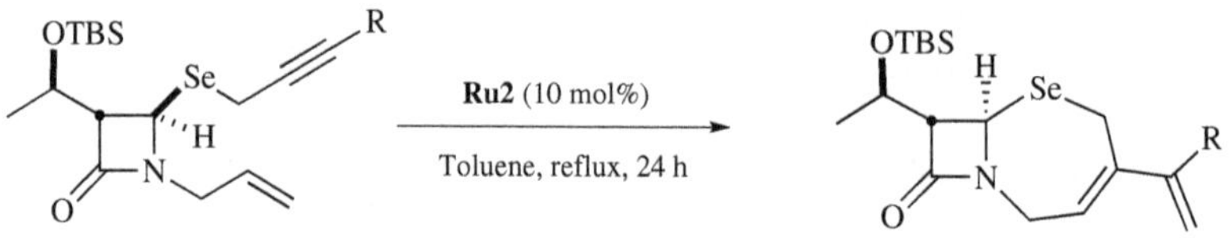

(4.11)

R = CH_3, 44%
R = Et, 46%
R = Ph, 59%

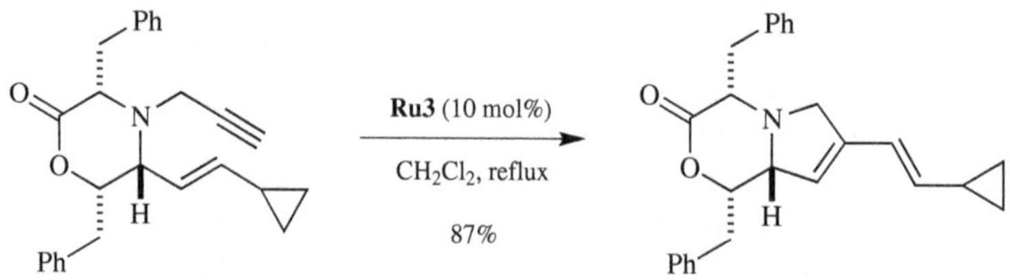

(4.12)

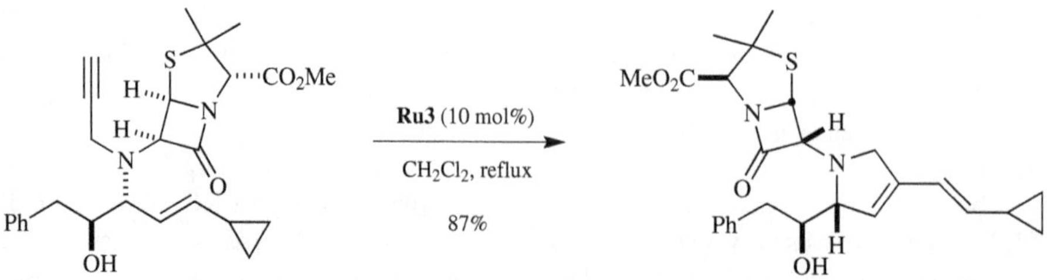

(4.13)

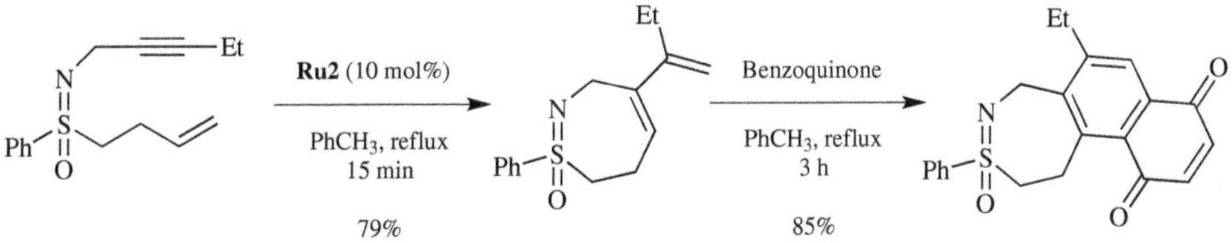

(4.14)

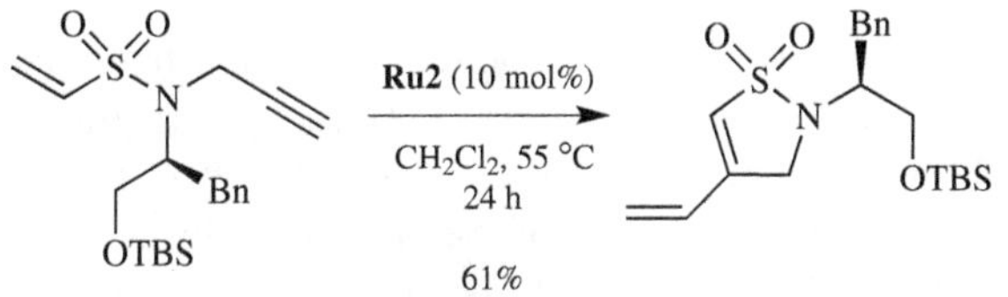

(4.15)

appear in many pharmaceutical agents and play an important role in drug development. Heterocycle synthesis has become a major focus for chemical library synthesis where robust ring synthesis methods are needed. Initiating metathesis at an electron-deficient alkene such as a vinyl sulfonamide is challenging. For an intermolecular reaction, this would not be a good partner, but with smaller ring-closing reactions, higher effective molarity of the alkyne favors the reaction (Eq. 4.15).

Procedure for RCEYM (63)

Vinylsulfonamide (202 mg, 0.51 mmol) was dissolved into 50 mL CH_2Cl_2. After it was degassed by argon for 5 min, **Ru2** (10 mol%) was added; the mixture was refluxed at 55 °C (oil bath) overnight; and then cooled to room temperature (RT), concentrated under reduced pressure, and purified by flash chromatography (10 : 1 hexane/EtOAc) to afford the γ-sultam (123 mg, 0.31 mmol) in 61% yield.

RCEYM can also be used to synthesize bicyclic structures in the presence of nitrogen heterocycles such as pyrimidines (Eq. 4.16) (64).

Bicyclic nitrogen heterocycles have been synthesized by RCEYM. Retrosynthetic analysis envisioned that the bicyclic enone would be derived from a 1,3-diene by an oxidative transformation. The groups of Martin (65) and Mori (66) had independently pursued two closely related strategies toward anatoxin-a. For example, Martin found that the bridged azabicycle was formed under nominal conditions of RCEYM (Eq. 4.17). The exocyclic geminal alkene of the diene product was elaborated by position-selective osmylation and oxidative cleavage of the 1,2-diol. After protecting group removal, the natural enantiomer of (+)-anatoxin was formed.

Procedure for Bicyclization RCEYM (67)

To a degassed solution of **Ru2** (212 mg, 0.250 mmol) in CH_2Cl_2 (500 mL), a solution of enyne (744 mg, 2.50 mmol) in degassed CH_2Cl_2 (30 mL) was added. The mixture was stirred under a blanket of argon for 16 h, and then dimethyl sulfoxide (DMSO) (0.89 mL) was added to decompose the catalyst. The mixture was stirred for an additional 23 h, whereupon the solvent was removed under reduced pressure and the resulting residue was purified by flash chromatography (silica gel) eluting with Et_2O/pentane (1 : 2) to afford 623 mg (84%) of product as a yellow oil.

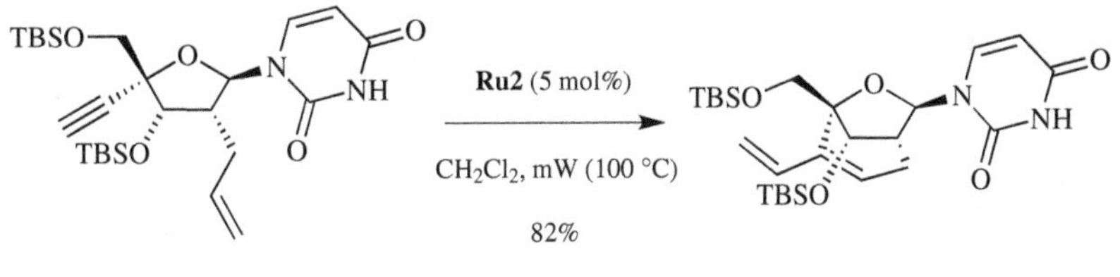

(4.16)

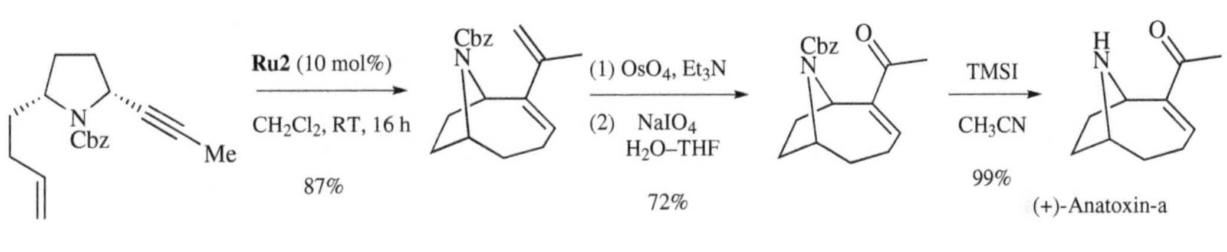

(4.17)

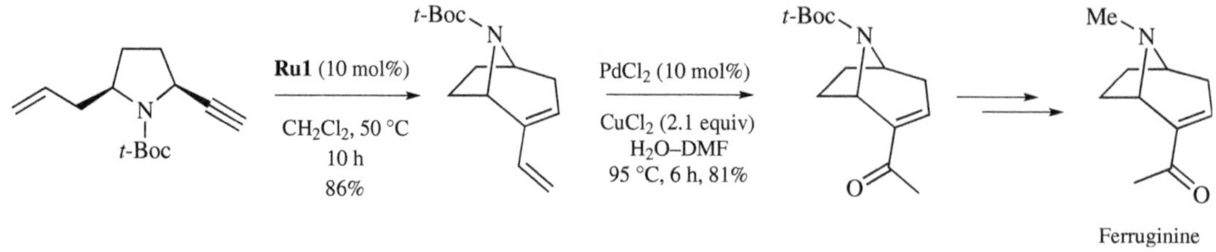

(4.18)

The bicyclic ring system of the alkaloid (+)-ferruginine was assembled by RCEYM (Eq. 4.18). The use of Mori's conditions proved unsuccessful and instead led to ethylene–alkyne CM as the major reaction pathway (not shown) (68). In the event, the best catalyst was found to be **Ru1** because **Ru2** and **Ru5A** caused unwanted dimerization of the product subsequent to the bicyclization. A Wacker-type oxidation sequence was used to transform the exocyclic geminal alkene to a methyl ketone.

Diastereoselective ene-diyne metathesis can access phosphorus-containing heterocycles (Scheme 4.6) (69). It can be seen that electron-deficient alkynes participate in the RCEYM. The desymmetrization affords one major diastereomer shown. These unique phosphorus heterocycles provide two sites for subsequent functionalization, leading to interesting bicyclic structures. The alkyne underwent click chemistry (Huisgen 3+2 cycloaddition) with BnN_3. The 1,3-diene could be engaged in thermal 4+2 cycloaddition with DMAD to furnish the cycloadduct.

4.2.1.1 Macrocyclic Ring Closure In larger RCEYM, there are two orientations that can be obtained, which affects the ring size formed. If the ring is smaller than 10 members, then ring constraint forces the enyne to come together with the 1,2-disubstitution pattern (Scheme 4.7). This is called *exo-selectivity* because the intermediate vinyl carbene leading to *exo*-**3** features the ruthenium exocyclic to the newly formed ring. However, if the ring is very large, the ene and yne react with the regiochemistry observed for the intermolecular reaction (as in **4**). In this case, the vinyl carbene forms endocyclic to the macroring formed. Another way of describing exo- or endo-selectivity is by considering where the alkyne atoms end up in the newly formed ring. If both are contained in the ring, then they are endocyclic. If one atom of the alkyne ends up exocyclic to the ring as in *exo*-3 (atom b), then an exo-closure has occurred. Lee and Hansen (70) studied this in a systematic way and rationalized the data in terms of reactive intermediate stabilities. With slightly greater flexibility in the enyne tether, the ring closure may result in the larger ring size by switching regiochemical preference. In general, rings smaller than 10 members will form with exo-selectivity where alkyne carbon b is exocyclic with respect to the formed ring. Rings in the 10–11 range are borderline cases, and larger rings result in the endo-selectivity, where both alkyne atoms a and b are contained in the newly formed ring. The endo-selectivity matches the 1,3-disubstitution that results from cross ene-yne metathesis, where there is no tether joining the reactants. A study of macrocyclization in a related system has also appeared, with an emphasis on diversity-oriented synthesis (71).

A key macro-RCEYM was used to assemble the monomeric fragments used to synthesize longithorone A (72). The synthesis of one cyclophane monomer (right half) is illustrated in Scheme 4.8.

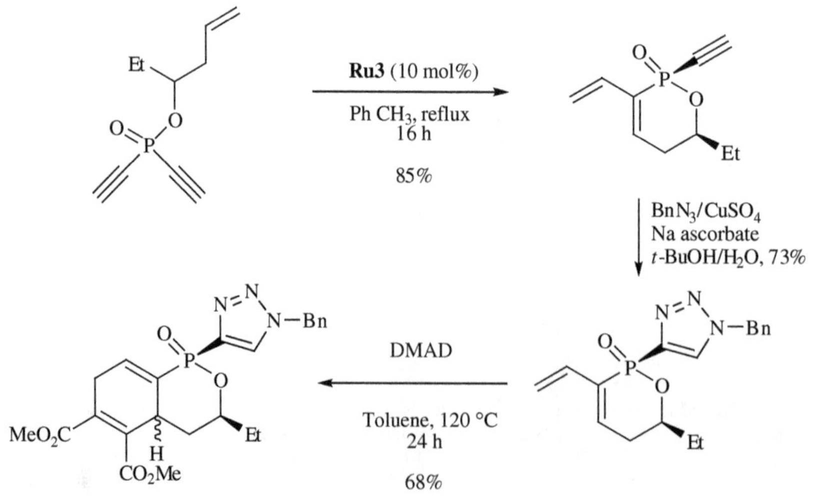

Scheme 4.6 Desymmetrization ring-closing metathesis on diyne (69).

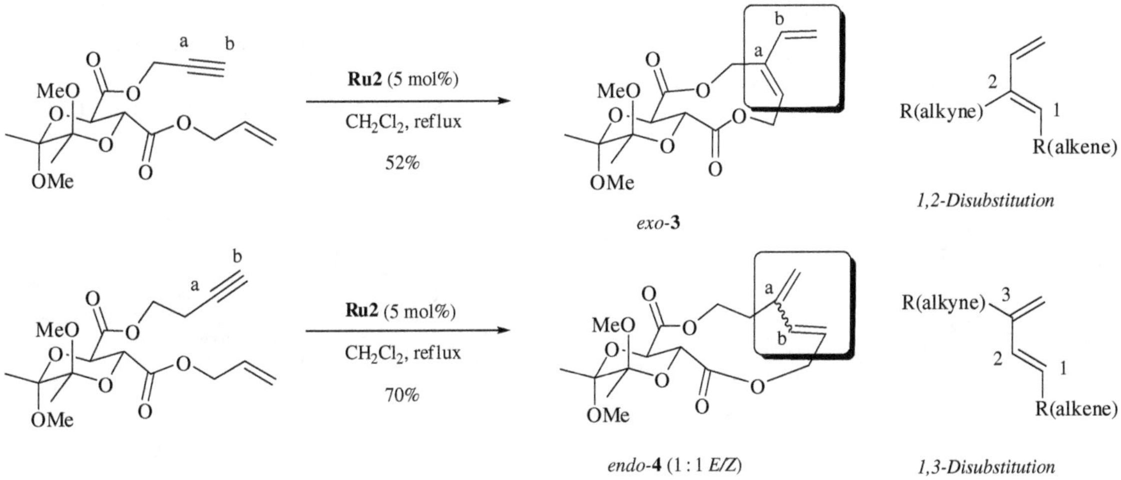

Scheme 4.7 Ring-closure mode selectivity (70).

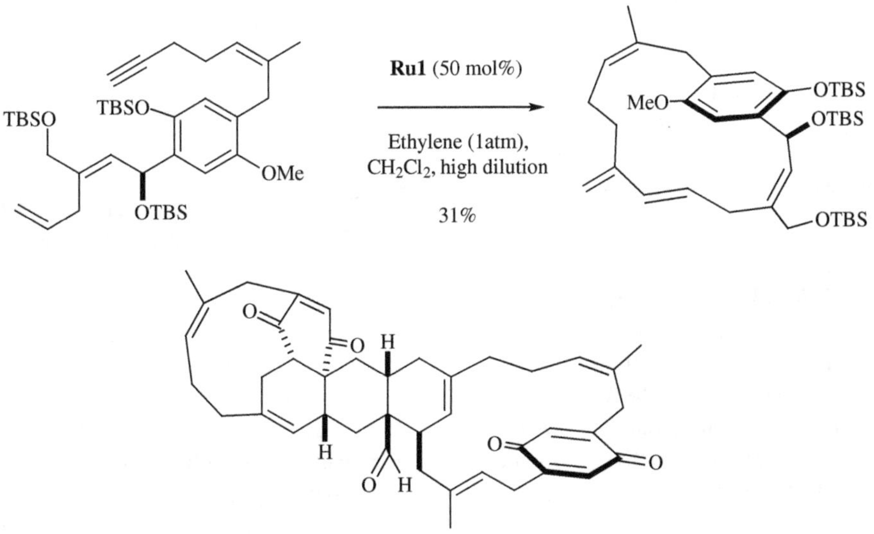

Longithorone A

Scheme 4.8 Macroring closure *en route* to longithorone A (72).

Greater steric bulk near the alkyne helps increase the yield of RCEYM. Higher propargylic substitution reduces competing processes such as alkyne oligomerization that gives intractable material and results in poor mass recoveries. The loss of mass can be seen in entries 1 and 2 (Table 4.2) (51). Though some alkyne conversion is seen, very little enyne is converted to the diene product. Alkyne oligomerization has been observed in ene-yne metathesis using **Ru3** (73), and alkyne trimerization has been reported using **Ru1** (74,75). Monosubstitution at the alkyne propargylic position enhanced conversion but did not appreciably increase the yield (entries 3 and 4). Disubstitution gives significantly enhanced conversions and yields (entries 5 and 6). Increased degree of propargylic substitution limits alkyne oligomerization and gives higher yield of the desired 1,3-diene without the use of ethylene.

4.2.1.2 Mode Selectivity A tandem metathesis sequence allowed endo ring closure to form a six-membered ring. In an RCEYM of 1,6-enynes, it is typical for the exo ring closure to occur, giving a five-membered ring product. If a pendant alkene could be made to close onto the distal carbon of the alkyne, a larger ring would result. Endo-selectivity was achieved in a tandem ethylene CM/ring-closing metathesis sequence as a means to access 4-methylene-2-cyclohexanols (76). Conducting the reaction under ethylene results in intermolecular ethylene–alkyne metathesis, providing the triene intermediate shown (Eq. 4.19, Table 4.3). The intermediate engages in a slower alkene-1,3-diene ring-closing metathesis. Site-selectivity originates from the preferred site of reactivity in the 1,3-diene. This is an interesting example that uses "Mori's conditions," which results in ethylene metathesis instead of direct RCEYM,

TABLE 4.2 RCEYM Yields Versus Propargylic Substitution

Entry		Catalyst	Time, h	Alkyne conversion, %	Diene yield, %
1	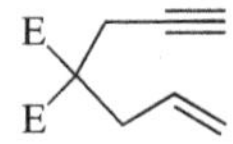	**Ru1**	24	33	18
2		**Ru2**	24	17	3
3	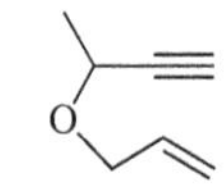	**Ru1**	–	–	–
4		**Ru2**	24	42	3
5	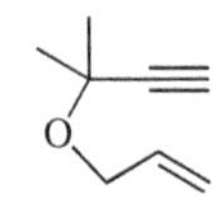	**Ru1**	–	–	–
6		**Ru2**	20	>99	67

Ref. 51.

probably due to the decreased reactivity of the allylic alcohol.

Ring-closing mode selectivity can be used to introduce an element of diversity in library synthesis. Depending on the presence of a propargylic protecting group, either the six- or seven-membered nitrogen heterocycle was formed by endo or exo ring closures, respectively (Eqs 4.20 and 4.21) (77).

4.2.1.3 Using the Schrock Molybdenum Carbene Catalysts The Schrock molybdenum carbene complexes have not been widely used in enyne metathesis. This can be attributed to the high reactivity of the catalysts with terminal alkynes: alkyne polymerization outcompetes enyne metathesis (i.e., inverted "manifold" reactivity). In recent years, the Schrock catalyst has proven to be a versatile platform for asymmetric catalyst design, resulting in successful asymmetric alkene metathesis applications and in highly *Z*-selective alkene CM. Catalyst development efforts resulted in a collateral benefit for enyne metathesis: improved cross selectivity. Recent stereogenic-at-molybdenum catalysts give endo-selective RCEYM. In some cases, the mode selectivity can be tuned based on the catalyst structure.

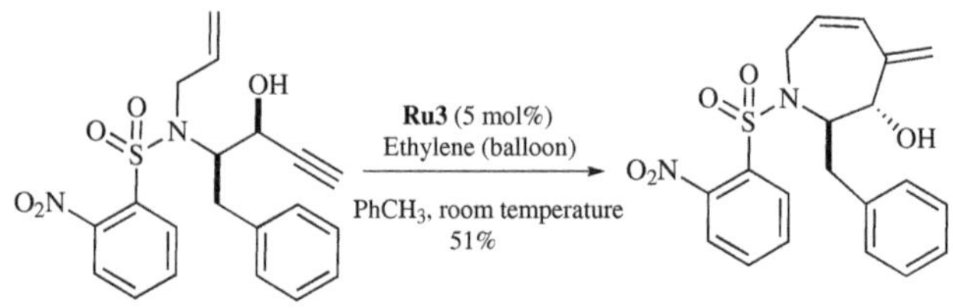

(4.20)

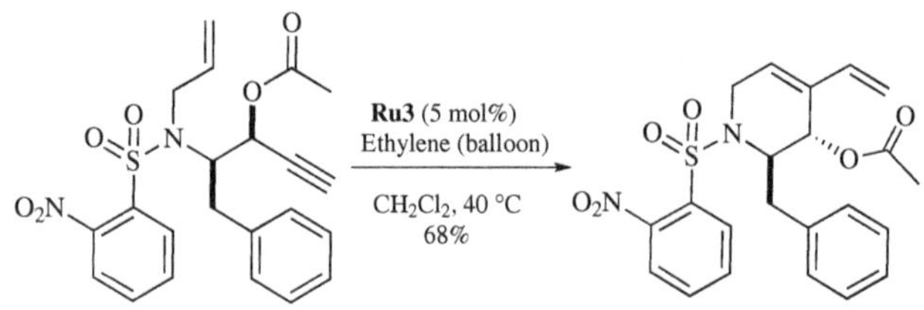

(4.21)

The parent Schrock catalyst was found to promote the RCEYM of an allenyne (78). Due to the higher reactivity of the allene versus alkenes, this isolated example is a well-balanced case of intramolecular reactivity over alkyne polymerization (Eq. 4.22, Table 4.4). Based on the initiation studies, a mechanistic sequence was proposed where the molybdenum allenylidene first reacted with the alkyne and then underwent ring closure. Though the ether linkage failed due to chelation with the oxophilic metal, a basic amine gave successful enyne metathesis.

By changing the ligands in the Schrock complex, ring-closing metathesis of 1,ω-enynes is possible (79). The Schrock complex **Mo1** has two alkoxy groups bound

TABLE 4.3 An Unusual Ring-Closure Mode for Enyne Metathesis

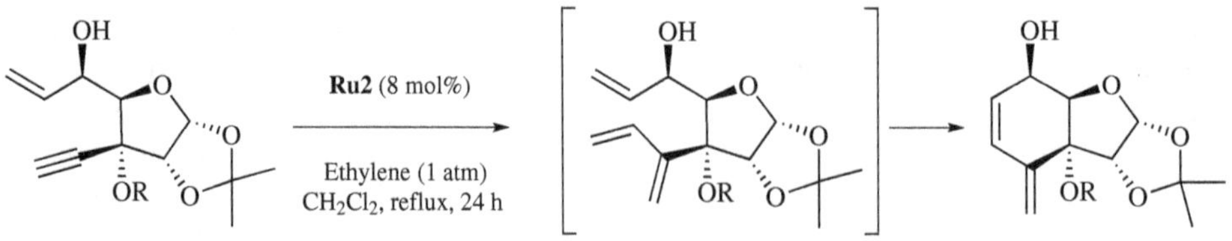

(4.19)

Entry	R	Yield, %
1	Me	83
2	Et	89
3	Bn	87
4	MOM	88

Ref. 76.

TABLE 4.4 Allene–Alkyne Ring-Closing Metathesis

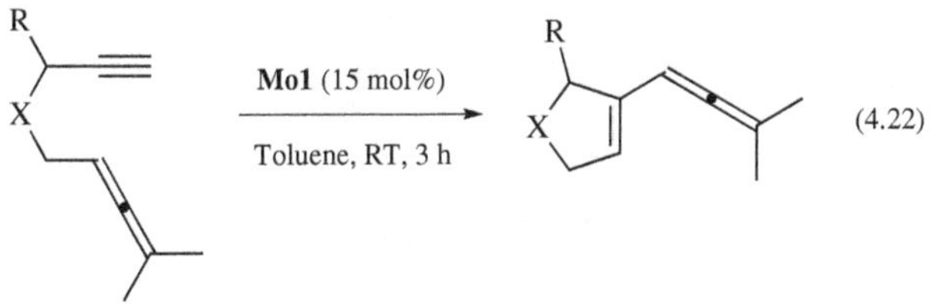

Entry	X	R	Yield, %
1	O	H	0
2	NTs	H	71
3	NTs	Me	68
4	NBn	H	73
5	$C(CO_2Et)_2$	H	95

Ref. 78.

to the molybdenum center. The chiral-at-molybdenum carbene complex **Mo2A** bears one alkoxy group and one pyrrole ligand directly bound to the metal (79,80). This slight alteration in ligand environment resulted in enyne metathesis reactivity. Monopyrrole catalyst **Mo2A** promotes endo ring-closing metathesis (Eq. 4.23) (79). Interestingly, this ring-closure "mode selectivity" is different from that observed with the Grubbs ruthenium carbene. The mode selectivity observed in Eq. 4.23 is endo-selectivity. This complementary mode selectivity suggests a niche for the molybdenum (and tungsten catalysts) (81). The ligand environment of the molybdenum catalyst has a profound effect on the ring-closure mode selectivity (Table 4.5). For instance, **Mo2A** gives primarily the endo ring-closure mode in a 1,7-enyne (entry 2), but changing to catalyst **Mo2B** results in mostly exo ring closure (entry 3).

TABLE 4.5 Schrock–Hoveyda Molybdenum Carbene-Promoted Ring-Closing Enyne Metathesis

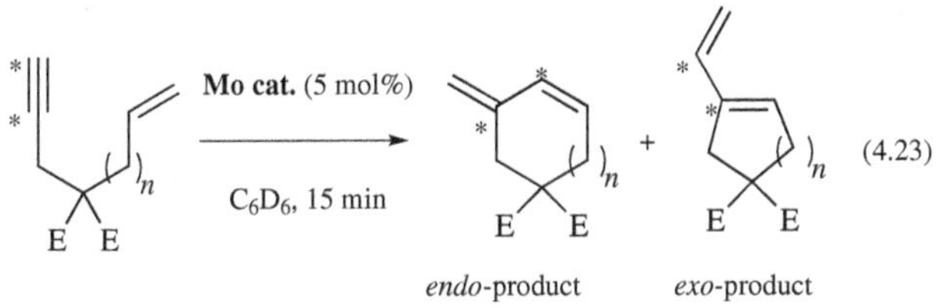

Entry	Mo cat.	*n*	*endo* : *exo*	Yield, %
1	**Mo2A**	1	100:0	68
2	**Mo2A**	2	100:0	79
3	**Mo2B**	2	20:80	70

Ref. 79.

4.2.2 Selectivity and Reactivity in Cross Ene-Yne Metathesis

The simple alkene ethylene performs well in cross ene-yne metathesis (20,50,82,83). For example, ethylene–alkyne metathesis involving an unsaturated cycloalkyne furnished the macrocyclic framework of amphidinolide V (Scheme 4.9) (84). Ethylene metathesis has also been used to form anolignan (85).

Simple aliphatic 1-alkenes, used in molar excess, perform best in cross enyne metathesis with either generation Grubbs complex. Blechert et al (3). reported the first cross enyne metathesis using **Ru1** (Eq. 4.24). Simple 1-alkenes were found to participate in the reaction, giving the conjugated 1,3-diene products as *E/Z* mixtures (Table 4.6). With **Ru1**, neither the alkene nor alkyne can be substituted at the allylic or propargylic positions, respectively (Table 4.6, entries 1–3). Follow-up studies using **Ru2** showed substantial improvements in reactivity and allowed propargylic substitution (entries 4,5) (86). Alkenyl-substituted sugars also undergo CM effectively to form glycoconjugates (87,88).

Ene-yne CM can be performed with free hydroxyl functionality in the alkene reactant. Because alcohols can cause decomposition of **Ru2** (89), the CM was run at higher temperatures with excess alkenol reactant (Eq. 4.25, Table 4.7). This allowed the desired ene-yne metathesis to outpace decomposition. The reactions were quenched using the polar isocyanide (90) that simplifies the purification of organic products from ruthenium catalyst residue.

Typical Procedure for Cross Enyne Metathesis (91)

Into an oven-dried, 50-mL Schlenk tube equipped with magnetic stir bar and cold finger condenser was added alkyne (0.3 mmol), unsaturated alcohol (1.8 mmol, 6 equiv), and 1,2-dichloroethane (5 mL), followed by Grubbs catalyst **Ru2** (0.015 mmol, 5 mol%). This solution was heated at 85 °C until the reaction was judged to be complete by thin layer chromatography (TLC) analysis (typical reaction times: 5–45 min). Reactions were then quenched with isocyanide KO_2CCH_2NC (0.075 mmol, 0.25 equiv) in 1 mL of methanol. Reactions were subsequently filtered (silica gel plug eluted with dichloromethane), concentrated *in vacuo* (rotary evaporator), and purified by flash chromatography on silica gel using ethyl acetate–petroleum ether to provide the corresponding dienes as colorless oils.

4.2.2.1 Enol Ethers The enthalpic advantage of ene-yne metathesis over alkene metathesis permits the use of enol ethers as alkene reactants (92). Enol ethers such as ethyl vinyl ether are used to quench alkene metatheses because the resulting ruthenium Fischer carbenes are unreactive with respect to further alkene metathesis. For unreactive

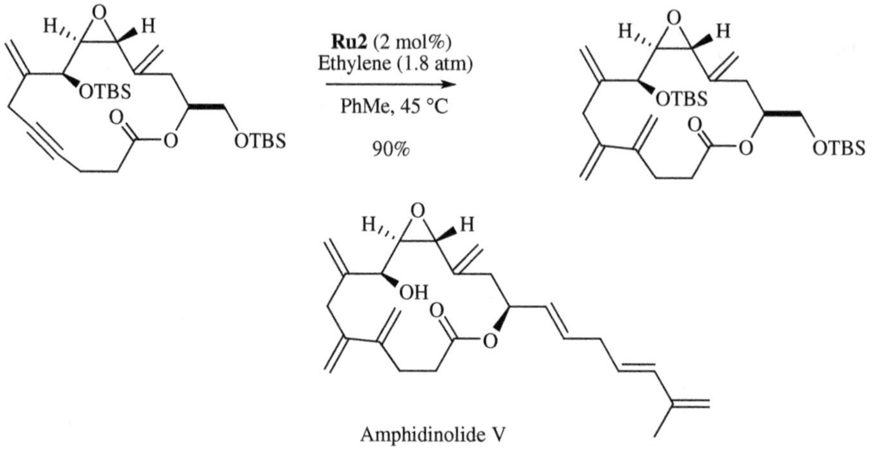

Amphidinolide V

Scheme 4.9 Fürstner's synthesis of amphidinolide V by ethylene–alkyne cross metathesis (84).

TABLE 4.6 Cross Ene-Yne Metathesis

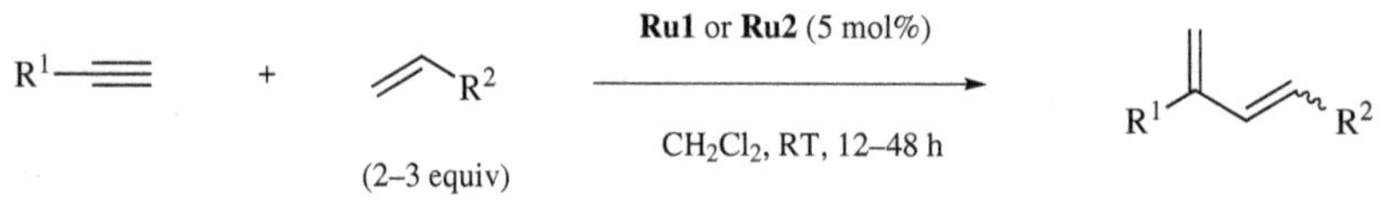

(4.24)

Entry	R^1	R^2	Catalyst	Yield, %
1	$BnOCH_2CH_2CH$-	$-CH_2SiMe_3$	**1**	86
2	$BnOCH_2CH_2CH_2$	$-CH_2CH_2CH(CO_2Me)_2$	**1**	76
3	$BnOCH_2CH_2CH_2$	$-CH_2OTBS$	**1**	83
4	cyclohexyl-	$-CH_2SiMe_3$	**2**	69
5	Me_3Si-	$-CH_2SiMe_3$	**2**	77

Ref. 3, 86.

TABLE 4.7 Allylic Alcohols and Homologs in Cross Ene-Yne Metathesis

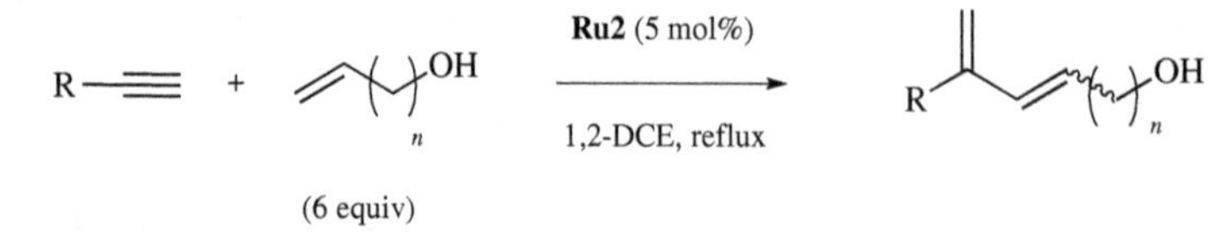

(4.25)

Entry	R^1	n	Yield, %
1	CH_2OBz	1	74
2	CH_2OBz	2	91
3	CH_2OBz	3	70
4	CH_2OTHP	1	58
5	CH_2CH_2OTBS	2	72
6	$CH_2CH_2N(Ts)Ph$	1	45

Ref. 91.

TABLE 4.8 Enol Ether–Alkyne Cross Metathesis Producing Electron-Rich 1,3-Dienes

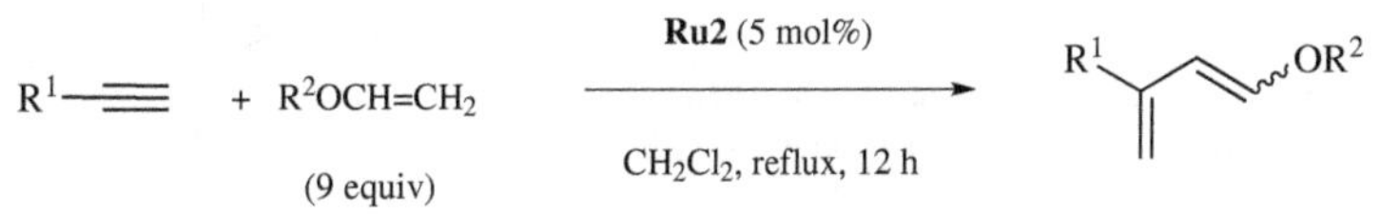

(4.26)

Entry	R^1	R^2	Yield, %	E/Z
1	$CH(OAc)CH_2Ph$	Et	88	1.9 : 1.0
2	$CH_2OTBDPS$	Et	97	2.1 : 1.0
3	$CH(OBz)CH_3$	Et	99	1.7 : 1.0
4	$CH_2N(Ts)(n\text{-}Bu)$	n-Bu	92	1.7 : 1.0

Ref. 92.

TABLE 4.9 Strain-Assisted Ene-Yne Metathesis

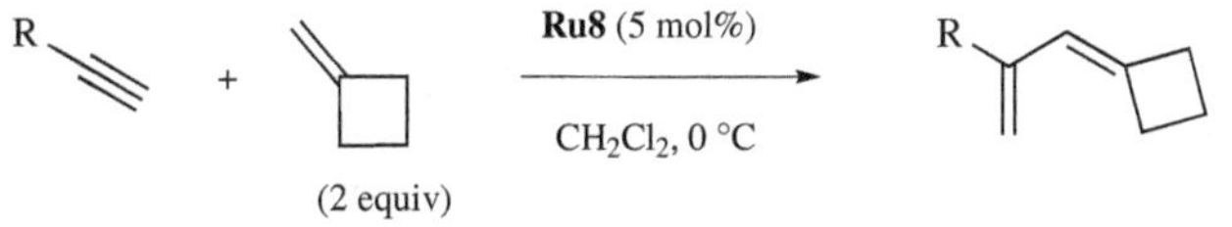

(4.27)

Entry	R^1	Yield, %
1	$CH_2CH_2N(Ts)Ph$	95
2	CH_2CH_2OBn	99
3	$CH(OAc)CH_2Ph$	96
4	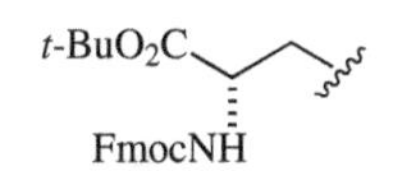	99

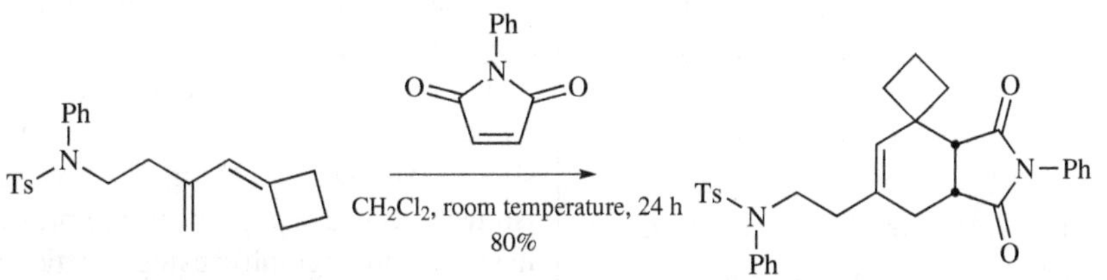

(4.28)

Ref. 93.

alkynes, excess alkene is commonly employed. For an unreactive Fischer carbene, excess alkene is similarly effective in coaxing catalytic reactivity. Use of excess enol ether gives ene-yne metathesis, which is complete in a matter of minutes (Eq. 4.26, Table 4.8). The excess enol ether is thought to accelerate catalysis in the turnover step (50). Ethylene coadditive can be used with enol ethers (proceeding through ruthenium Fischer carbene intermediates) to similarly accelerate the slow turnover step (50).

4.2.2.2 Strain-Assisted Strain can be used to promote reaction of geminally substituted alkenes with 1-alkynes (93). Normally, geminal (1,1-disubstituted) alkenes are unreactive in ene-yne metathesis because they do not initiate with **Ru2** and cannot participate in the ene-then-yne mechanism. To overcome this lack of reactivity, ring strain in the alkene can be used to drive the initiation (94) (Eq. 4.27, Table 4.9). Interestingly, these reactions could be conducted at low reaction temperatures, whereas most ene-yne metatheses are run at or above rt. Cycloaddition proceeded at ambient temperature due to angle strain in the 1,3-diene (Eq. 4.28). More recently, a relay strategy has been developed to obtain reactivity from geminally substituted alkenes (95,96).

It is possible to conduct ene-yne metathesis as a cross coupling using equimolar amounts of unsaturated reactants. As previously noted, cross ene-yne metathesis is typically

TABLE 4.10 Atom Economy in Ene-Yne Metathesis Cross Coupling

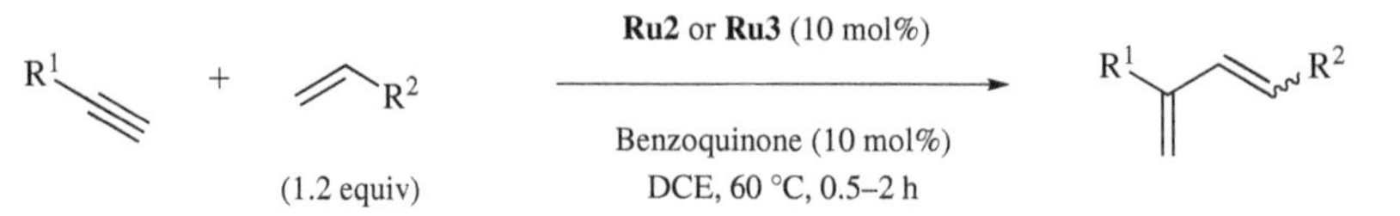

(4.29)

Entry	R^1	R^2	*E/Z*	Yield, %
1	$-CH(OBz)CH_3$	$-C_4H_9$	*E* only	79
2	$-CH(OBz)CH_3$	$-CH_2OBz$	10 : 1.0	62
3	$-CH(OBz)CH_3$	$CH_2CH_2CH_2OAc$	*E* only	72
4	CH(OAc)Ph	$CH_2CH_2CH_2OH$	1.5 : 1.0	79
5	$CH_2CH_2CH_2OTBS$	$CH_2CH_2CH_2OH$	3 : 1.0	68
6	$CH(OTBS)CH_2CH_2Ph$	$CH_2CH_2CH_2OH$	1.5 : 1.0	63

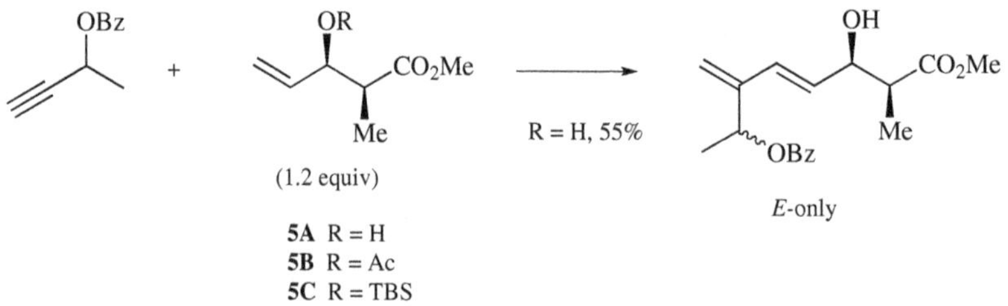

(4.30)

Ref. 47.

done with an excess of the alkene reactant. To make cross ene-yne metathesis truly "atom economical," (97,98) the alkyne is paired with an alkene in equimolar amounts (Eq. 4.29, Table 4.10) (47). In addition, enantiopure *syn*-allylic alcohols were shown to perform well in the CM with branched alkynes (Eq. 4.30). In contrast, allylic alcohols with protecting groups such as tert-butyldimethyl silyl (TBS) or acetate failed to give CM under conditions of atom economy. Benzoquinone was used to suppress the formation of unwanted by-products.

Typical Procedure for Atom Economical Cross Ene-Yne Metathesis (Table 4, entry 4)

Into an oven-dried, 50-mL Schlenk tube equipped with a magnetic stir bar and purged with argon was added a solution of 80 mg of alkyne (0.46 mmol, 1 equiv) dissolved in 1,2-dichloroethane, a solution of 47 mg of alkene (0.55 mmol, 1.2 equiv) dissolved in 1,2-dichloroethane, benzoquinone (0.092 mmol, 20 mol%), an additional 4 mL of 1,2-dichloroethane to achieve a reaction concentration of 0.08 M, followed by 28.8 mg of **Ru3** (0.046 mmol, 10 mol%). The solution was immersed in a preheated 60 °C oil bath for 30 min and the reaction was judged to be complete by TLC analysis at this time. The reaction was quenched by addition of a methanolic solution of commercially available KO_2CCH_2NC (90) (0.115 mmol). The reaction was concentrated under reduced pressure, and 1,3-diene was obtained as a slightly yellow oil (94 mg, 79% yield, 1.5 : 1 *E:Z*). The isomers proved separable during purification by silica gel flash chromatography using gradient elution (10–20% ethyl acetate in petroleum ether).

*4.2.2.3 **Internal Alkynes*** Internal alkynes have good reactivity in CM, but generally suffer from poor regiocontrol (17). The orientation preference of the metal carbene is determined by the alkyl substituents on the alkyne. With internal alkynes bearing two different alkyl groups, there is poor distinction between the two ends resulting in a mixture of regioisomers (Scheme 4.10).

Silylalkynes give regioselective cross ene-yne metathesis. The bulkier silyl group directs the orientation of ruthenium carbene cycloaddition step, giving good yields with an excess of alkene reactant (Eq. 4.31, Table 4.11) (99).

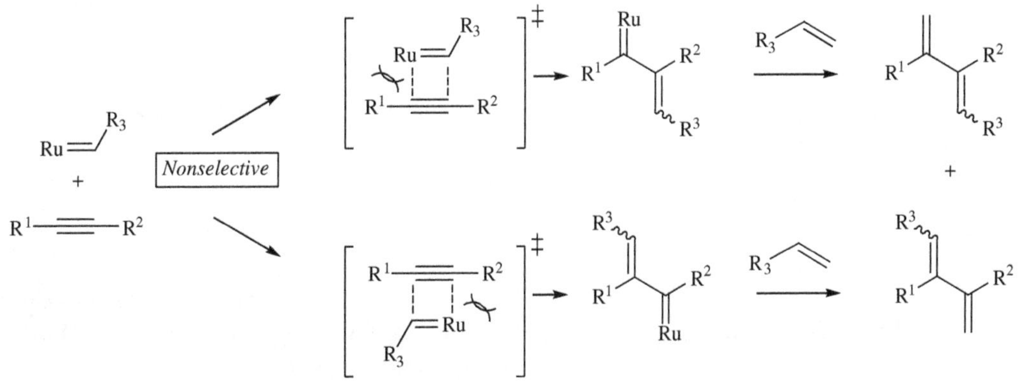

Scheme 4.10 The regioselectivity problem in ene-yne cross metathesis.

TABLE 4.11 Regioselective Cross Ene-Yne Metathesis of Internal Silylalkynes

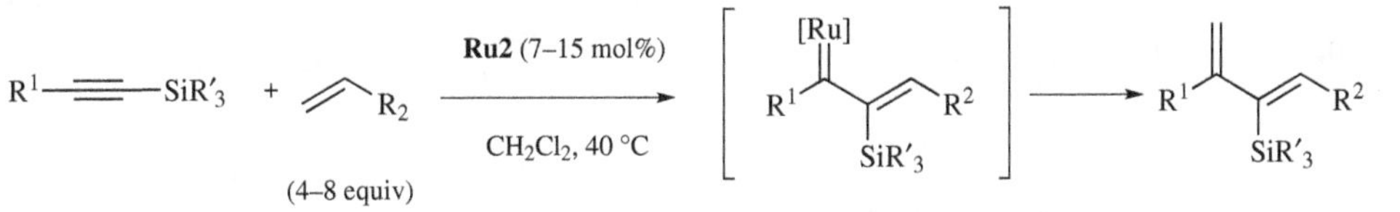

(4.31)

Entry	R^1	R^2	R'_3	Yield, %
1	$AcOCH_2$	CH_2OAc	Me_3	63
2	$AcOCH_2$	$CH_2OC_6H_4CH_3$	Me_3	43
3	$AcOCH_2$	$CH_2CH_2CH_2Br$	Me_3	53
4	$MeOCH_2$	CH_2CH_2OAc	$Ph_2(MeO)$	64

Ref. 99.

Typical Procedure for Cross Metathesis of Silylated Internal Alkynes (Table 4.6, entry 1) (99)

Silyated alkyne (19.9 mg, 0.117 mmol, 1.00 equiv) and 1-alkene (82 mg, 0.819 mmol, 7 equiv) were dissolved in 3.9 mL of freshly distilled CH_2Cl_2 (0.03 M). Grubbs catalyst **2** (6.8 mg, 0.008 mmol, 7 mol%) was added to the reaction mixture. After stirring under reflux for 40 h, the reaction mixture was concentrated under reduced pressure. The residue was purified by flash column chromatography using ether/hexanes (gradient elution from 1 : 10 to 1 : 8 ether–hexanes) to yield 19.9 mg of the 1,3-diene product (63%).

Symmetry in either the alkyne or alkene is a useful strategy to limit regioisomer formation. Symmetrical internal alkynes can be successfully reacted with alkenes to give a single product in high yield (Eq. 4.32) (46). Alternatively, a symmetrical alkene such as ethylene may be employed to form a single product without any possible regioisomers (Eq. 4.33) (20). In the future, a more general solution to this difficult problem may be achieved through catalyst control.

4.3 TANDEM PROCESSES INVOLVING EN(E)YNE METATHESIS

In recent years, enyne metathesis has been used as a step in tandem bond-forming reactions. A nice recent review has appeared (8). Since the inception of enyne metathesis, the 1,3-diene products have been used for subsequent reactions such as cycloaddition. Section 4.3.1 provides an overview of some useful tandem reactions. Section 4.3.2 describes sequential metathesis reactions that form a single ring. Section 4.3.3 describes cascade ring-closing applications that form two rings or more. Ring-closing cascade reactions are very popular because cascade processes can form multiple rings in one metathesis step. In these cases, end differentiation and relay metathesis offer means to control ring size. Last, tandem metathesis reactions that result in ring rearrangement are detailed in Section 4.3.4.

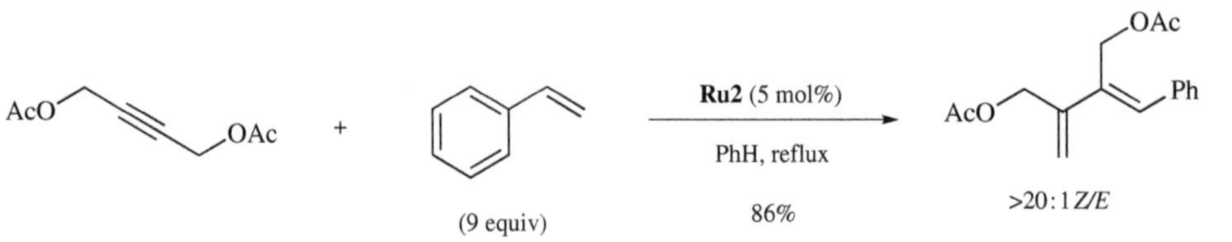

(4.32)

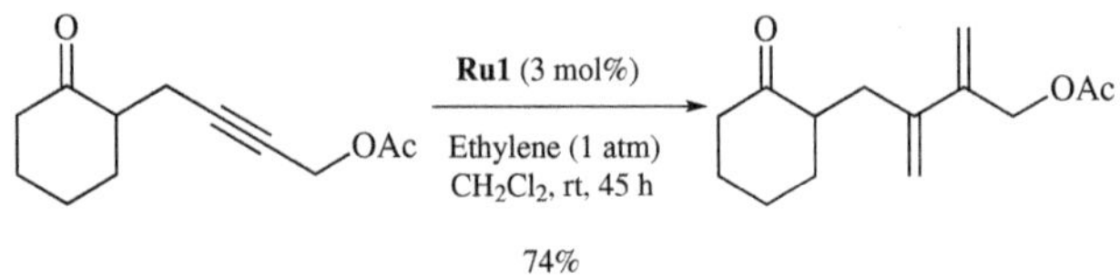

(4.33)

4.3.1 Metathesis with Consecutive Non-Metathesis Reaction

The allure of the enyne metathesis is the synthetic utility of the conjugated diene products. Cycloaddition is one of the most familiar combinations and is widely used. Allylic transposition through Claisen rearrangement offers access to 1,3-dienes with unique substitution patterns. When the alkyne bears unique element functionality (Si, B), synthetic access to enones is possible. Boronate-substituted dienes are good substrates for Suzuki cross coupling.

4.3.1.1 With Diels–Alder Cycloaddition RCEYM can be combined with cycloaddition; the latter serves as an annulation method forging a ring onto that created in the RCEYM step. The RCEYM forms the first ring, which may be a heterocycle, and cycloaddition is used to engage the diene in order to form the second ring. This is by now a classical tandem application, with dozens of examples using both thermal and Lewis-acid-catalyzed conditions. The diene intermediate need not be isolated (Eq. 4.34) (100). The tandem sequence can be followed by an oxidative aromatization as well (Eq. 4.35) (101,102).

The conditions of ene-yne metathesis can be used to promote an *in situ* intramolecular 4+2 cycloaddition. This tandem approach was used to access taxol-like structures (103). In the metathesis event, an ethylene CM was employed. The elevated temperature allowed the intermediate 1,3-diene to undergo a type II intramolecular Diels–Alder reaction (Eq. 4.36).

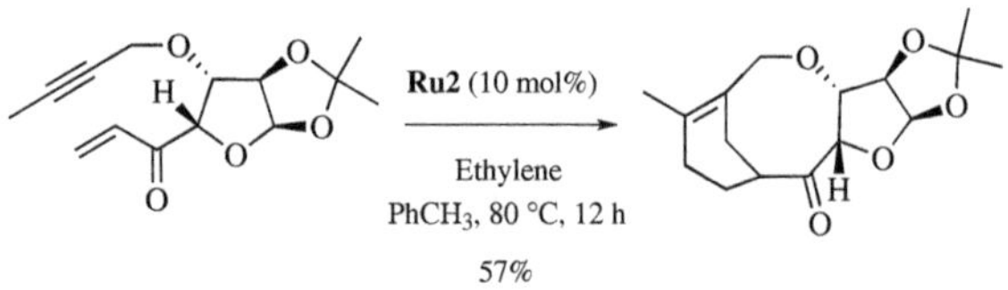

(4.36)

Additional examples of Diels–Alder cycloaddition can be found in the cascade applications for diversity-oriented synthesis (104) and in ring rearrangement applications (105).

4.3.1.2 With Claisen Rearrangement A tandem metathesis/Claisen rearrangement provides synthetic access to highly substituted 1,3-dienes (Scheme 4.11) (36). Since ene-yne CM can only access certain conjugated diene substitution patterns, an allylic rearrangement was sought to increase the degree of substitution on the 1,3-diene. After an *E*-selective cross ene-yne metathesis, the allylic acetate was converted to the silyl ketene acetal and subjected to Ireland ester enolate Claisen rearrangement, giving a 1,2-disubstituted-1,3-diene as a single stereoisomer (Eq. 4.37). Similarly, cyclic dienes gave an *s*-trans diene following the Claisen rearrangement, which then underwent

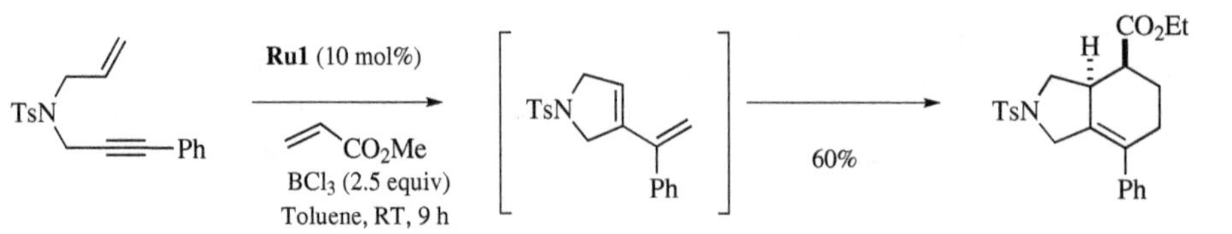

(4.34)

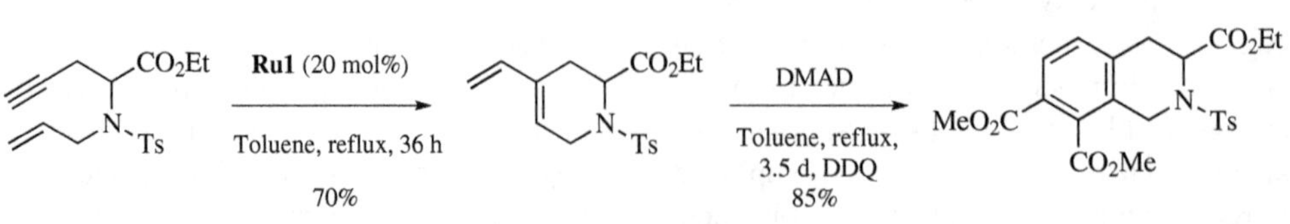

(4.35)

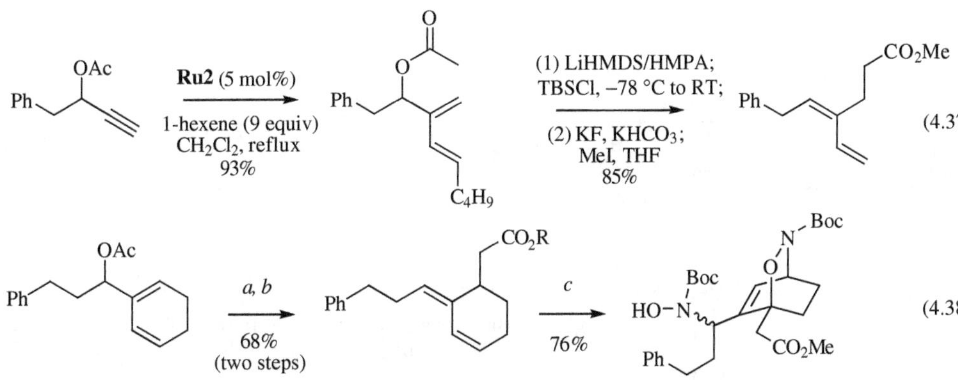

(4.37)

(4.38)

(a) LiHMDS/HMPA,–78 °C; TBSCl, RT; (b) $TMSCHN_2$, CH_2Cl_2/MeOH;
(c) BocNHOH (4 equiv), Pr_4NIO_4 (4 equiv), CH_2Cl_2, 0 °C.

Scheme 4.11 Tandem ene-yne metathesis and Claisen rearrangement (36).

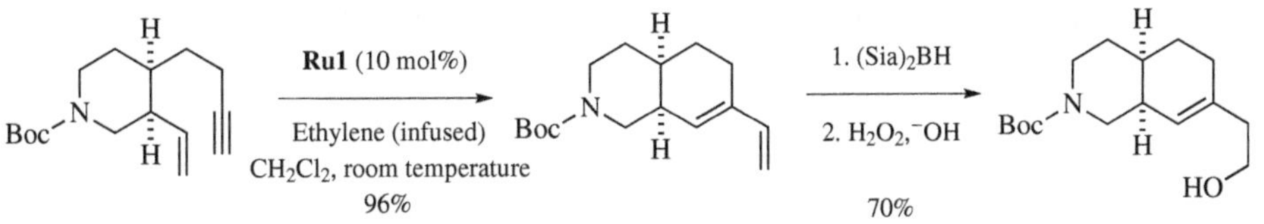

(4.39)

TABLE 4.12 Methyl Ketones by Tandem RCEYM/Silyl Enol Ether Hydrolysis

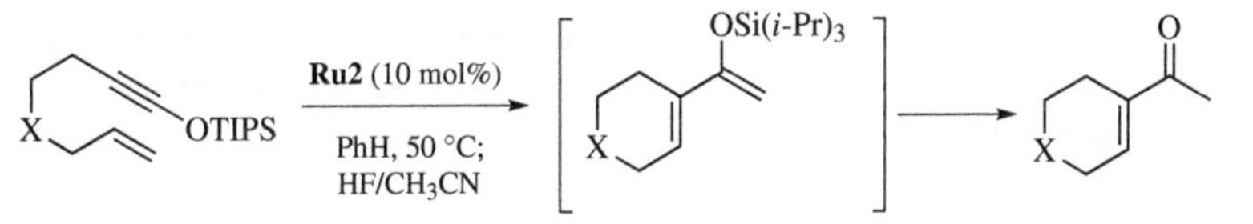

(4.40)

Entry	X	*n*	Yield, %
1	CH_2	1	81
2	O	1	88
3	$N(CO_2Me)$	0	91
4	$N(CO_2Me)$	1	95

Ref. 44.

a selective ene reaction and an *in situ* [4+2] cycloaddition (Eq. 4.38).

4.3.1.3 With Site-Selective Hydroboration When the 1,3-diene produced contains a terminal vinyl group, site-selective hydroboration is possible, as demonstrated in a synthetic study directed toward the manzamines (Eq. 4.39) (106). An additional example of site-selective hydrogenation, which is also quite rare, is provided below (107).

4.3.1.4 Of Alkynyl Silyl Ethers: Access to Methyl Ketones Through a simple tandem reaction, the RCEYM can access methyl ketone functionality (44). Silyl alkynyl ethers undergo RCEYM to give an intermediate silyl enol ether that was desilylated *in situ* to obtain a methyl ketone (Eq. 4.40, Table 4.12). Formally, this is the product that would result if one could affect a ring-closing metathesis of an aldehyde across a terminal alkyne. A variety of ring sizes can be formed using this tandem sequence.

A temporary ring has been used to assist the formation of acyclic enones (Eq. 4.41) (108,109). Under the conditions of metathesis with Grubbs catalyst **Ru2**, a boronate ester exchange takes place, thereby collecting the reactants for a RCEYM. Having served its function as a connector, the resulting dienylboronate was hydrolyzed to provide the enone product.

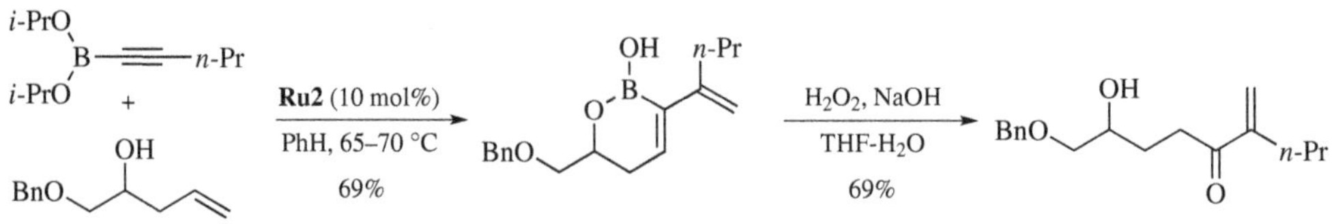

(4.41)

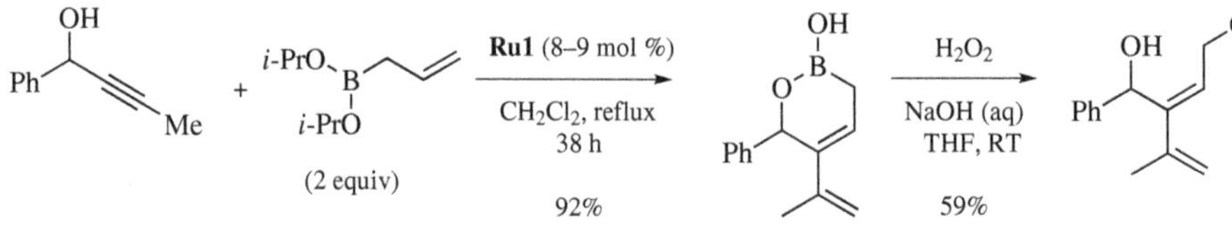

(4.42)

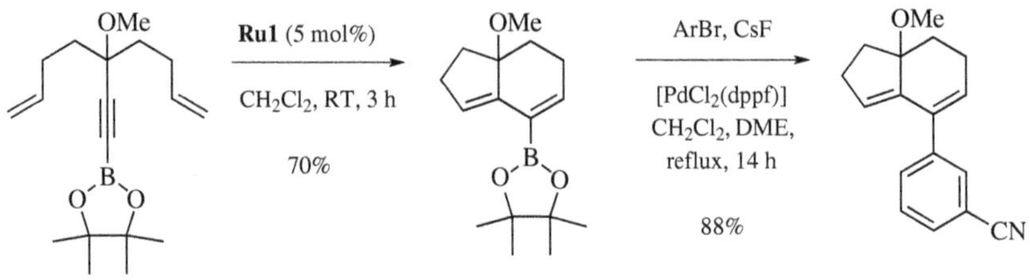

(4.43)

The boronate approach can be used in alternative ways to access highly functionalized acyclic 1,3-dienes. The boronate linker strategy offers a means to control the regiochemistry of alkene addition in reaction with internal alkynes. The cyclic boronates could also be subsequently hydrolyzed to provide acyclic dienes (Eq. 4.42). The geometric constraint enforced by the boronate linker in the ring-closing step affords a single stereoisomer, which addresses a limitation of the direct ene-yne metathesis.

4.3.1.5 Followed by Suzuki Coupling Dienes with boronate functionality can be synthesized by RCEYM, setting the stage for tandem Suzuki coupling (Eq. 4.43) (110).

Ene-yne CM can be used to access highly substituted 1,3-dienes as demonstrated in the total synthesis of amphidinolide K (Scheme 4.12) (111). The macrocyclic ring of the natural product features a highly substituted 1,3-diene. The alkynyl boronate guided the intermolecular reaction to form a single regioisomer in the CM step. The boronate was converted to the trisubstituted alkene through Suzuki coupling using MeI. In addition to providing an elegant solution to carbon–carbon bond coupling in this total synthesis, the tandem approach should prove generally useful to access more highly substituted 1,3-dienes.

4.3.2 Tandem Metathesis Applications Resulting in a Single Ring Formed

An enyne metathesis can be followed by another metathesis step occurring *in situ*. For example, the 1,3-diene can be produced by ring closing and directly react with an added alkene by CM. Another useful application is a ring synthesis from an alkyne and 1,5 diene reactants. Last, an existing ring can undergo a ring expansion process to give a larger ring. These three types of reactions are briefly described in this section. Those tandem processes involving cascades or ring rearrangements are discussed in Sections 4.3.3 and 4.3.4.

4.3.2.1 With Tandem Cross Metathesis Tandem metathesis offers a method for both ring synthesis and cross coupling metathesis, in one pot. Use of a less reactive alkene delays the CM event until the more rapid intramolecular RCEYM has already taken place. In this case, the Hoveyda–Blechert complex **Ru3** is well suited to performing the second CM reaction because it is reactive with electron-deficient alkenes (Eq. 4.44) (112).

Martin (113) used a similar tandem metathesis-based approach to make the diterpene 8-*epi*-xanthatin using the Hoveyda–Blechert complex **Ru3** (Eq. 4.45).

Tandem RCEYM/CM can be conducted with aminophosphonates present. In general, free amines are considered problematic and are usually protected as amides during the metathesis step. In this tandem reaction, the intermediate 1,3-diene did not fully convert to the CM product (Eq. 4.46, Table 4.13) (114). This reflects the attenuated reactivity of 1,3-dienes in cross alkene metathesis, though they can be successfully converted with additional alkene (38,112). By adding 5 equiv of the alkene after the first 2 h of reaction, high conversion was obtained. At the end of the reaction,

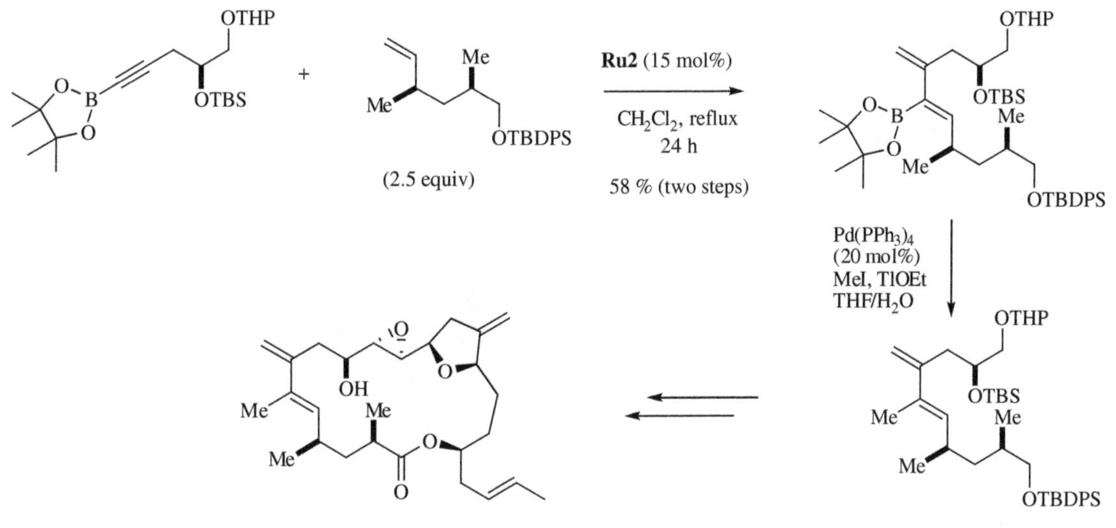

(−)-amphidinolide K

Scheme 4.12 Cross ene-yne metathesis in synthesis of amphidinolide K (111).

about 10% of the intermediate butadiene remained unreacted but proved separable by flash chromatography. Benzo-fused seven-membered structures are considered privileged structures due to their appearance in a variety of bioactive natural products and their structural similarity to the benzodiazepines.

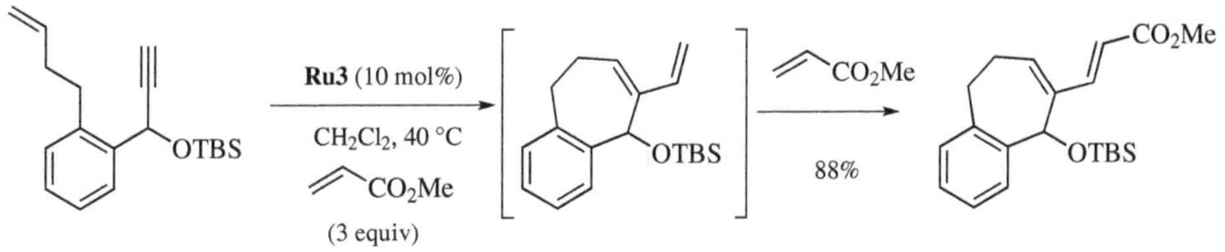

(4.44)

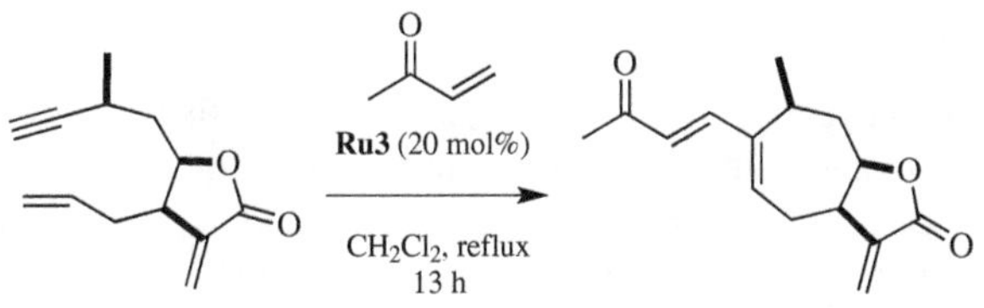

(+) 8-*epi*-xanthatin

(4.45)

4.3.2.2 *Ring Synthesis*

Rings can be formed by ene-yne metathesis between 1,ω-dienes and alkynes (115). To accomplish this ring synthesis, both a cross ene-yne metathesis and an *in situ* ring-closing metathesis occur in the same reaction. The use of a diene as a four-carbon donor to an alkyne results in the synthesis of 1,3-cyclohexadienes. Larger rings can be formed by CM/ring closing performed on enyne substrates.

The methylene-free tandem metathesis approach offers a synthetically useful ring synthesis (Eq. 4.47, Table 4.14). The reaction is preparatively useful and can be conducted on a larger scale at reduced catalyst loadings (116). Either 1,5-cyclooctadiene or polybutadiene can be used as the alkene reactant. Because there are no terminal methylene

TABLE 4.13 RCEYM to Access Phosphono-Substituted Azepines

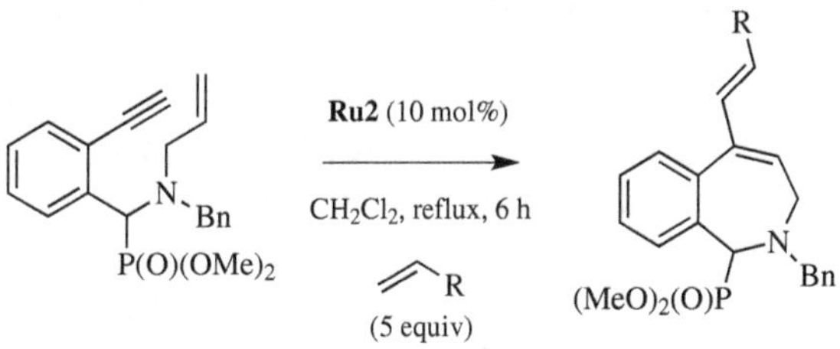

(4.46)

Entry	R	Yield, %	*E/Z* ratio
1	Ph	78	*E* only
2	CH_2SiMe_3	74	66 : 34
3	$CH_2CH_2C(O)CH_3$	68	67 : 33

Ref. 114.

TABLE 4.14 Methylene-Free Enyne Metathesis: Formation of 1,3-Cyclohexadienes

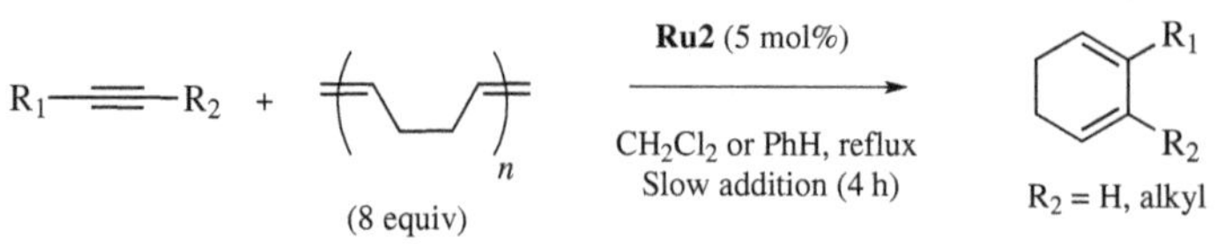

(4.47)

Entry	R^1	R^2	Yield, %
1	$(EtO_2C)_2CHCH_2$	H	82
2	$BnOCH_2CH_2$	H	70
3	$BocN(Ts)CH_2CH_2$	CH_3	72
4	$TsNHCH_2CH$	H	73
5	$AcOCH_2$	C_2H_5	65
6	MeO_2C	Me	53

Refs 117, 118.

groups, this is termed *methylene-free conditions*. These conditions are thought to permit equilibration between *E*- and *Z*-vinyl carbene intermediates resulting in higher chemical yields of 1,3-cyclohexadienes (Table 4.14).

Typical Methylene-Free Ring Synthesis Procedure (117)

Into an oven-dried, argon-cooled, 100-mL Schlenk tube is added all-*cis* polybutadiene (*Note: 1,5-cyclooctadiene can also be used.*) (116,118) (8.66 g, 160 mmol, 8 equiv) followed by 40 mL CH_2Cl_2 and Grubbs carbene **Ru2** (271 mg, 0.32 mmol, 1.6 mol%) heated to reflux. Propargyl benzoate (3.2 g, 20 mmol) was added in about 1 mL of CH_2Cl_2 over a period of 2 h by means of a syringe pump. After a further hour of heating, ethyl vinyl ether (612 μl, 0.64 mmol, 2 equiv based on ruthenium catalyst) is added and heating is continued for an additional 30 min. The reaction mixture is concentrated, passed through silica gel plug, and further purified by bulb-to-bulb distillation (124–128 °C at 1–2 mmHg) to afford the cyclohexadiene as a colorless oil (2.64 g, 62% yield).

Ene-yne metathesis has been used as a key step in a diversity-oriented approach to macrocyclic structures through a CM/ring-closing ring synthesis. A diverse collection of macroheterocyclic derivatives were produced by cross ene-yne metathesis with 1,5-hexadiene (Eq. 4.48). The *Z*-isomer underwent macroring-closing metathesis, but the *E*-isomer did not. The isolated *E*-tetraene was resubjected to Grubbs catalyst **Ru2**, but failed to undergo ring-closing metathesis due to geometric isolation of the unsaturated elements.

4.3.2.3 Ring Expansion: Synthesis of 1,3-Cycloheptadienes Like the ring rearrangement, an existing ring can be converted into a larger ring by net alkyne insertion. This is a ring expansion ene-yne metathesis that occurs as a result of a cascade process. An example of the ring expansion with cyclopentene is shown in Eq. 4.49 (119).

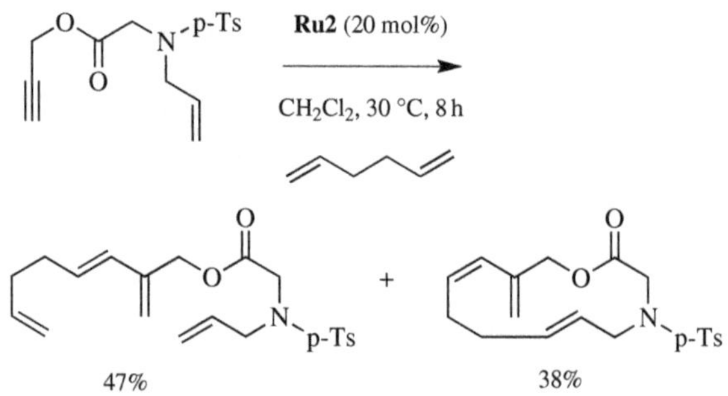

(4.48)

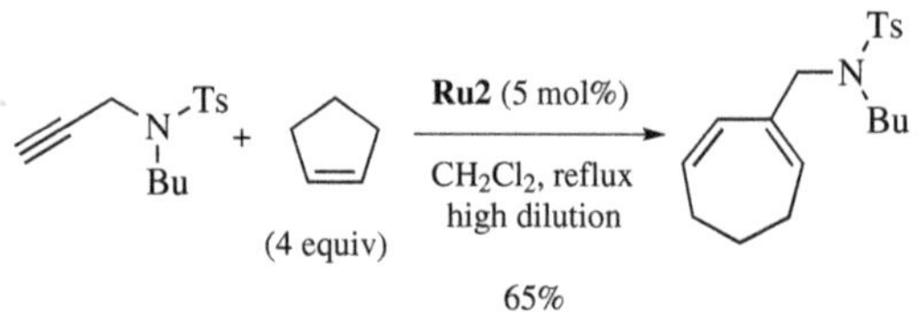

(4.49)

In bicyclic systems, the double bond in the five-membered ring reacts selectively. The ring opening of the bicyclic diene occurred with high site-selectivity and regioselectivity, producing a single product (Eq. 4.50) (120). To obtain the desired product, slow addition of the reactive components to the Grubbs catalyst was needed via syringe pump addition, which was thought to allow *E/Z* isomerization of vinyl carbenes *in situ*.

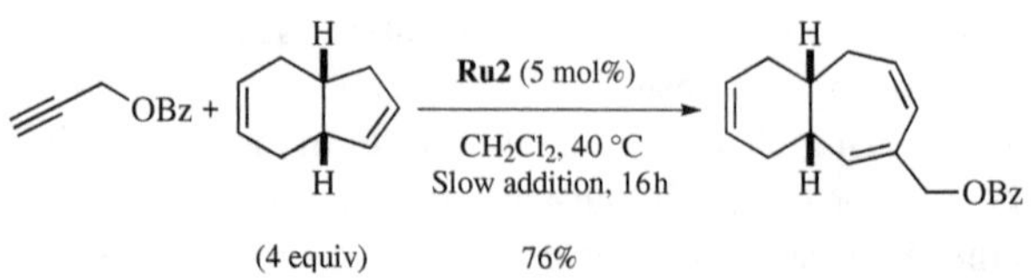

(4.50)

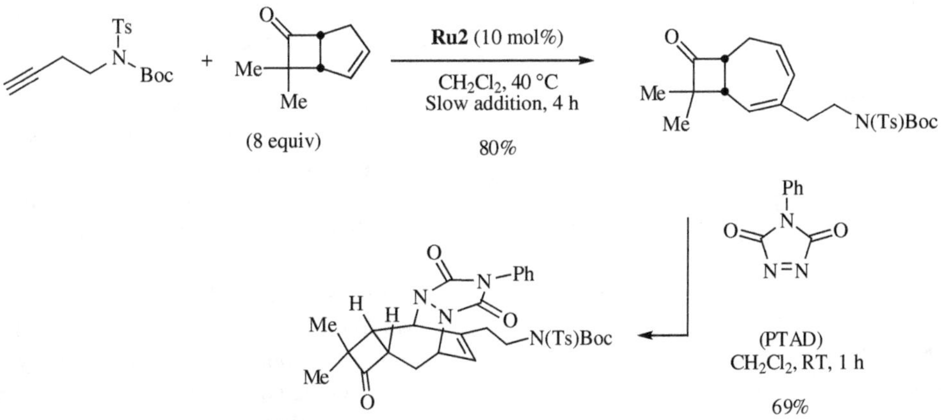

Scheme 4.13 Regioselectivity in the ring expansion of a bicyclic cyclopentene (120).

To guide regioselectivity in bicyclic systems, a bulky group at the allylic position is required. In the bicyclic ketone shown, the geminal dimethyl substituents guide the ring opening of the cyclopentene and give a single cycloheptadiene product (Scheme 4.13). Facile [4+2] cycloaddition with 4-phenyl-4H-1,2,4-triazole-3,5-dione (PTAD) gives a crystalline tetracyclic product (120).

4.3.3 Cascading Double Ring-Closing Metathesis

Cascade metathesis involves setting up two or more ring closures in a single catalytic reaction. With alkynes bearing pendant alkenes, many different ring sizes and heterocycles can be formed. In most cases, two different products are possible depending on which end initiates the cascade. Control of end differentiation uses the basic reactivity principles that are the basis of the Grubbs model for cross alkene metathesis. In cases where reactivity is similar, relay metathesis can be used to initiate the cascade. Because of the rapid gain in molecular complexity and good guidelines for end differentiation, the cascade process has been applied to total synthesis and in diversity-oriented synthesis.

4.3.3.1 End Differentiation in Cascade Dienyne Metathesis Cascade dienyne metathesis is a powerful ring-building method that furnishes multiple rings from acyclic precursors. To control ring sizes formed, the metathesis cascade must begin at one of the two pendant alkenes. This is called *end differentiation*, which results in selective formation of bicyclic products. The more reactive terminal alkenes are commonly used to initiate cascade enyne metathesis. In cases where alkene reactivity is similar, a directing element can be used to guide reaction at the desired end of the dienyne. This latter process is called *relay ring-closing metathesis* (RRCM) and has emerged as a valuable tactic to control end differentiation in cascade enyne metathesis.

The differential substitution of the pendant alkenes controls the ring size of the bicyclic system formed. An early example by Grubbs is instructive (Eq. 4.51) (121,122). This process begins by the reaction of the more reactive 1-alkene, which guides the formation of the first ring. End differentiation is achieved by relative reactivity of the alkene. In a synthesis application, Honda employed end differentiation for the synthesis of the alkaloid securinine (123).

4.3.3.2 **Relay Metathesis** Relay metathesis was developed to influence site-selectivity in a substrate containing multiple sites of unsaturation possessing similar reactivity. The RRCM strategy involves the installation of an alkene tether (relay element) that is used to direct catalyst initiation. This alkene tether directs the reaction to one alkene over another, allowing for selective ruthenium carbene formation. Site-selective placement of carbene reactivity directs the next metathesis reaction. This allows for "choreographed movement" of ruthenium carbenes. The relay metathesis is useful in controlling the directionality of a metathesis cascade and inducing the activation of an otherwise unreactive alkene. The relay concept was introduced by Hoye and coworkers (124). This concept was embraced quickly after it was discovered in the arena of natural product synthesis (125,126) where localization of reactivity is *sine qua non*.

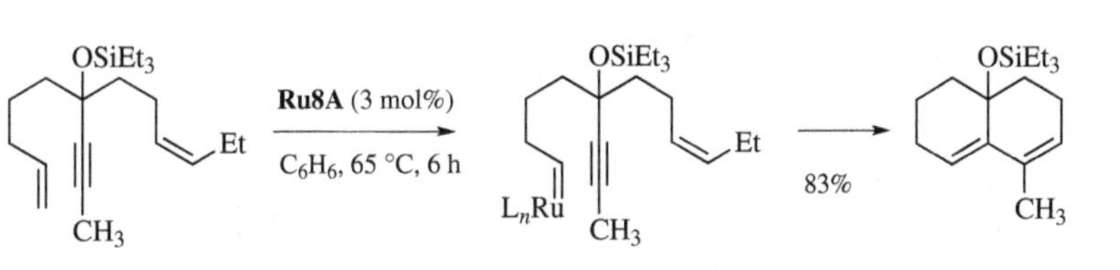

(4.51)

TABLE 4.15 End Differentiation by Relay Metathesis

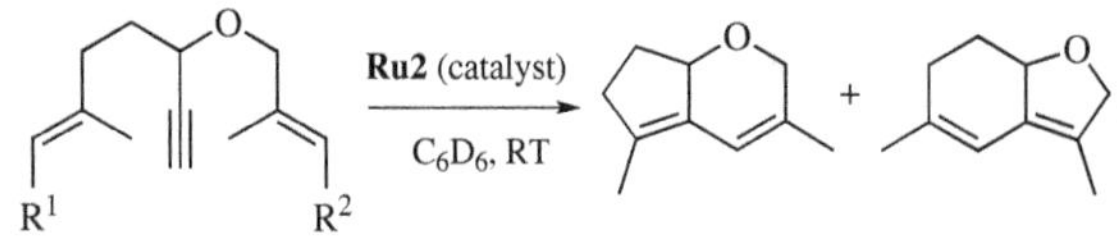

(4.52)

Entry	R^1	R^2	Product ratio	End differentiation
1	H	H	1.0 : 2.0	*random*
2	**R**	H	4.7 : 1.0	*le ft-to-right*
3	H	**R**	1.0 : 7.0	*right-to-le ft*

R = 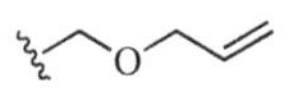

Ref. 124.

RRCM was developed to address site-selective initiation of ruthenium carbene reactivity. With no directing relay element, a 1 : 2 mixture results (Eq. 4.52, Table 4.15) (127). Installing the tether and directing the metathesis sequence from left to right afforded a 4.7 : 1 mixture (entry 2) and right to left afforded a 1 : 7 mixture (entry 3). RRCM was used to selectively direct catalyst initiation at a specific alkene, thereby directing the metathesis sequence to preferentially form one product.

A cascading RRCM sequence was also used to construct the bicyclic core of (−)-englerin A (Eq. 4.53) (128). The mixture of carbonate diastereomers was subjected to the highly active Grubbs–Stewart catalyst, both diastereomers reacting to afford the desired bicyclic ring system. The major *cis*-diastereomer was carried forward in the synthesis.

Relay metathesis was used to initiate a cascade of RCEYM with a 1,3-metallotropic shift to synthesize (−)-tricholomenyn A (129). In Eq. 4.54, the relay element was used to start the cascade near the allylic acetate. After the initial ring-closing metathesis, an RCEYM forms the six-membered ring and is followed by the 1,3-metallotropic shift. This strategy was also used for the total synthesis of (+)-asperpentyn and (−)-harveyone.

The relay metathesis strategy has been expanded to include cross ene-yne metathesis (95), where the relay is used to gain reactivity from unreactive alkene substrates (96). Because it is an intermolecular process, it is not detailed here.

4.3.3.3 Selected Examples of Cascade Dienyne Metathesis The cascade metathesis has also been used as an entry point to bicyclic ring systems that contain seven-membered rings. Seven-membered rings can be found in a variety of natural products. The polycyclization sequence provides the guanacastepene carbotricyclic framework (130,131) (Eq. 4.55, Table 4.16). A variety of substituents on the alkyne were tolerated in the metathesis, with the exception of a trimethylsilyl (TMS) group which is too bulky to permit the alkyne insertion step. Boyer et al (132). used a similar cascade to form colchicine.

The strained bicyclic [5.3.1] ring system found in taxol can be formed by a metathesis cascade (133,134). Taxosteroids, hybrids of taxol and steroids, were sought due to their taxol-like ability to stabilize microtubules. The metathesis cascade was initiated at the homoallylic alcohol to form the eight-membered ring in the initial enyne metathesis ring closure (Eq. 4.56). The second ring-closing step was optimal when an isopropyl-substituted pendant alkene was used.

Tandem dienyne ring-closing metathesis can be used to access bicyclic nitrogen heterocycles. The bicyclic skeleton of the lepadins was produced through tandem dienyne metathesis (Eq. 4.57, Scheme 4.14) (135).

A temporary siloxane linker can be used to assemble the unsaturated components and then removed to furnish acyclic diene, such as that found in cochleamycin

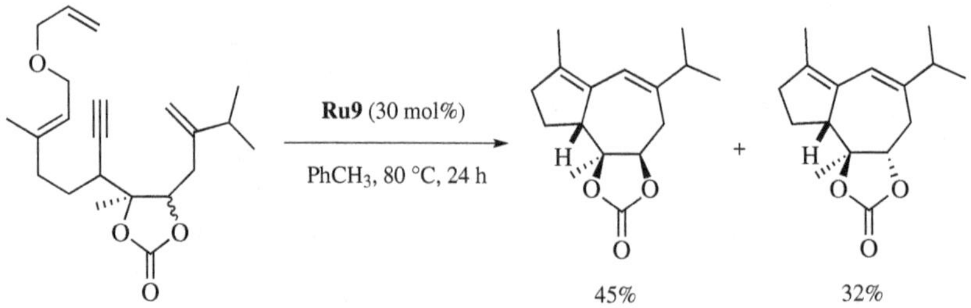

(4.53)

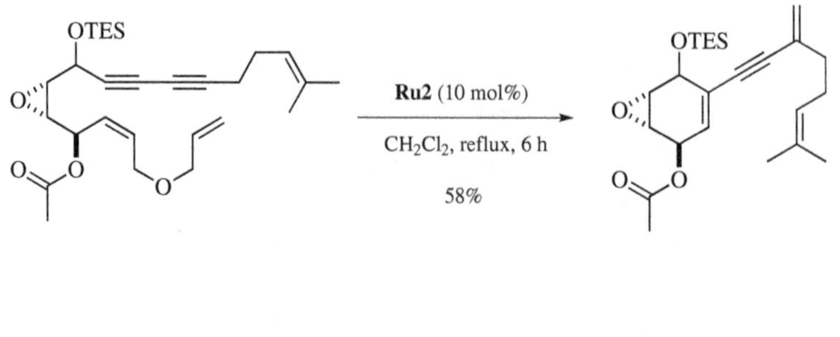

(4.54)

TABLE 4.16 Cascade Metathesis Approach to Guanacastepene

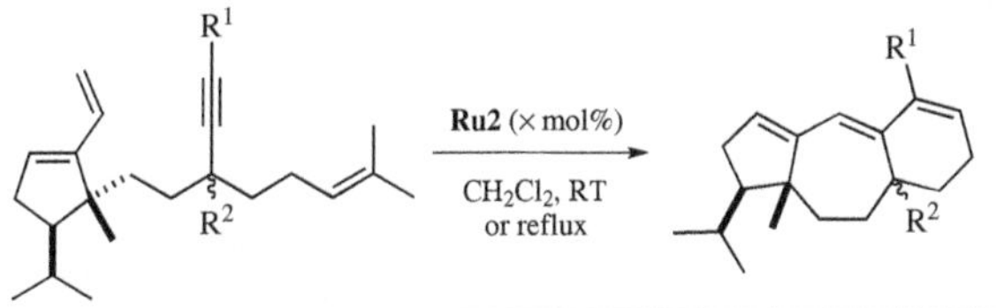

(4.55)

Entry	R^1	R^2	Loading	Yield, %
1	H	OTES	7.5	80
2	Me	OTES	7.5	70
3	$SiMe_3$	OTES	7.5	N.R.
4	CO_2Me	OTES	10	93
5	CO_2Me	Me	12	82

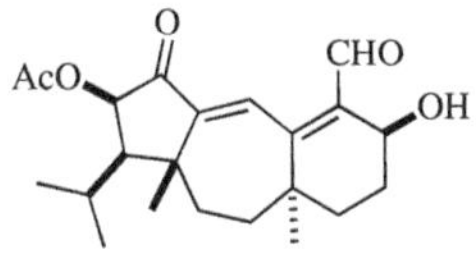

Guanacastepene A

Ref. 130.

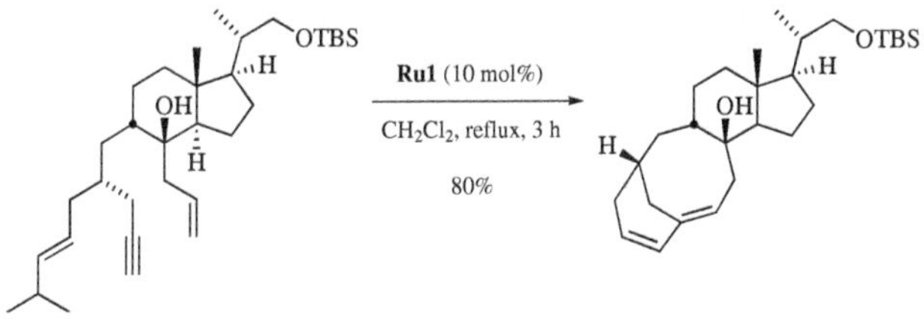

(4.56)

(Scheme 4.15) (136). The mixed silaketal sets the stage for the cascade ene-yne metathesis.

The polycyclic core of cortistatin A was accessible by cascade dieneyne ring-closing metathesis. Though using a mixture of diastereomers, only one diastereomer underwent the cascade, giving a 37% yield of the desired product (Scheme 4.16) (137).

A cascade ring-closing metathesis was used to assemble the core ring system of the spirotricyclic natural product, acylfulvene (138). The metathesis cascade (which had been previously studied in model systems) (139) was initiated at the allyl ether and terminated by the less reactive geminal alkene (Eq. 4.58). After the tandem reaction, the 1,3-diene **6** was reductively transposed using the Myers protocol for the *in situ* generation of the diazene, giving **7**. This synthetic tactic converts the conjugated 1,3-diene into a 1,4-diene. A ring-closing alkene metathesis of **7** provided the natural product (−)-acylfulvene (Eq. 4.59).

A concise tandem ring-closing dienyne metathesis was used to assemble the core polycyclic framework of the alkaloid lycoflexine (107). The ring-closing event was initiated at the more reactive terminal alkene to first form the five-membered ring (Scheme 4.17). The second ring closure generated the nine-membered azacyclic ring. The 1,3-diene was semi-hydrogenated in a site-selective reduction step where the residual Grubbs catalyst was transformed into a ruthenium hydride, giving a ruthenium hydride catalyst that selectively reduced the diene fragment. Such use of tandem catalysis under hydrogenation (H_2) or transfer hydrogenation conditions has been previously employed in ring-closing alkene metathesis/hydrogenation tandem reactions (140).

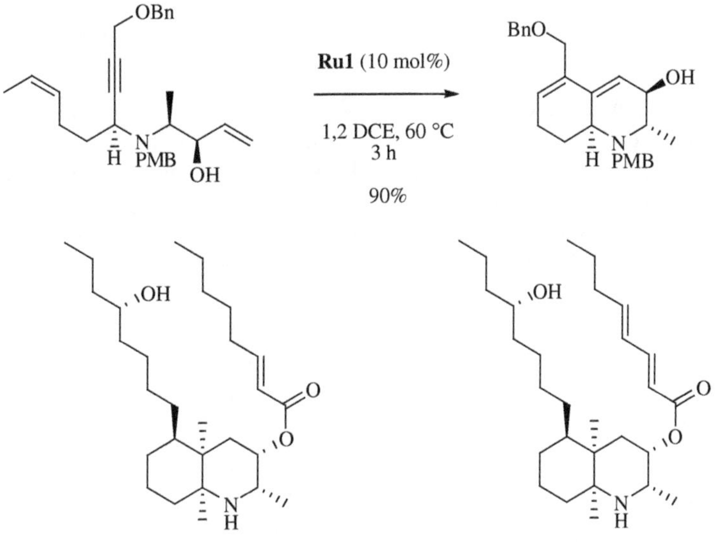

(4.57)

ent-Lepadin F

ent-Lepadin G

Scheme 4.14 Dienyne metathesis cascade approach to the lepadins (135).

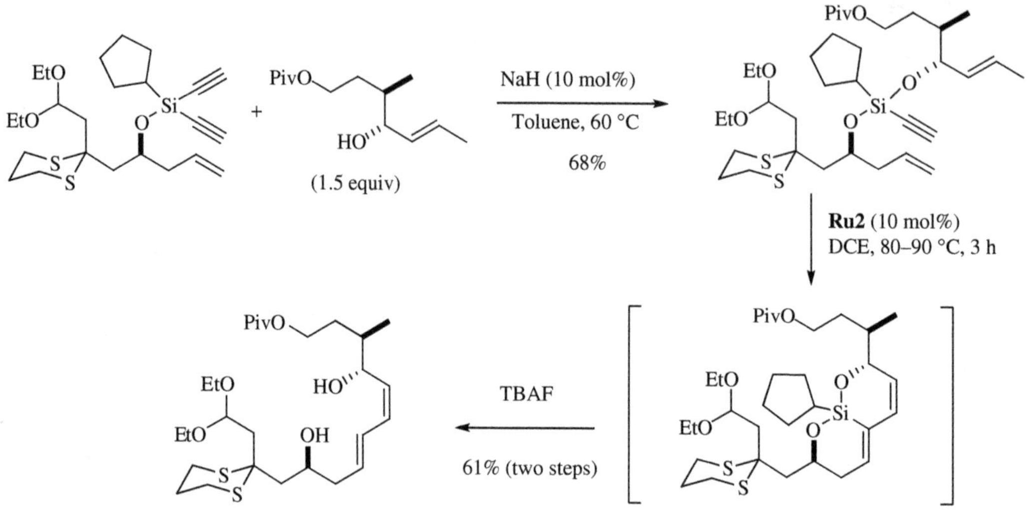

Scheme 4.15 Siloxane-mediated cascade ene-yne metathesis for stereocontrolled diene synthesis (136).

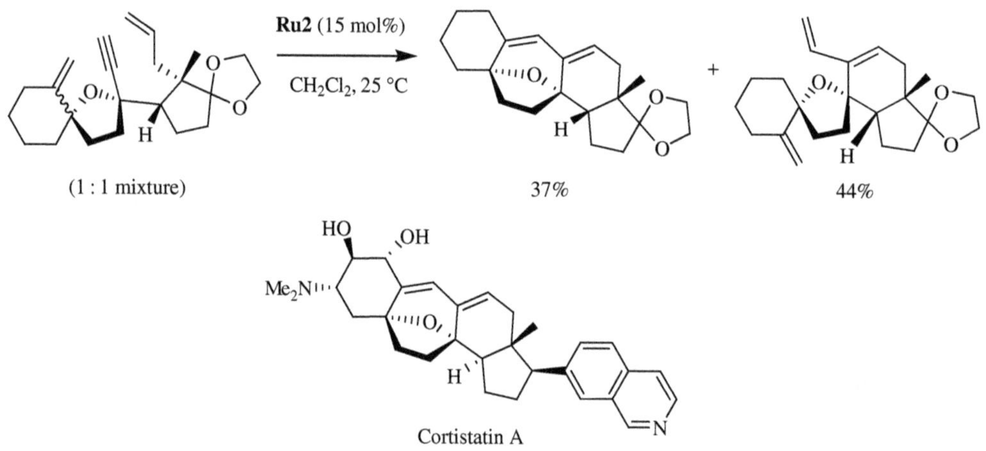

Scheme 4.16 Cascade metathesis approach to the core structure of cortistatin A (137).

Cascade RCEYM can be used to form diverse heterocycles in polycyclic frameworks. Several examples were detailed above. In one key step, the Grubbs catalyst **Ru1** performs chemoselective ring closure with the unsaturated reactants (Scheme 4.18). Mori's conditions were used to assist ring closure. Using this approach coupled with the additional versatility of a Diels–Alder cycloaddition, Schreiber and coworkers (141) constructed a library of structurally diverse heterocycles.

Cascade metathesis has been used to form more than two rings in one cascade process (142).

4.3.4 Ring Rearrangement Metathesis

RCEYM onto a pendant cycloalkene shuffles the ring. In the process, the existing cycloalkene (YZ linker atoms) is converted into a new ring incorporating the atoms that separated the cycloalkene and alkyne in the first place (linker atoms W–X, Eq. 4.60). Often the linker W–X includes a heteroatom, which distinguishes this metathesis as a useful conversion of carbocyclic rings to heterocyclic rings. To the best of our knowledge, the reaction was developed independently by Blechert (143) and Mori (105). Blechert dubbed the process to be a "ring rearrangement;" (143–147) this terminology will be used throughout this section. The ring rearrangement is driven forward to products by π bond reorganization, the same enthalpic driving force of enyne metathesis.

Banti and North (148) nicely demonstrated ring rearrangement with a doubly armed norbornene scaffold (Eq. 4.61). Under an atmosphere of ethylene, a two-directional synthesis occurred with a double-ring rearrangement metathesis. The resulting bis(1,3-diene) was trapped with maleic anhydride in a Diels–Alder reaction. In this simple

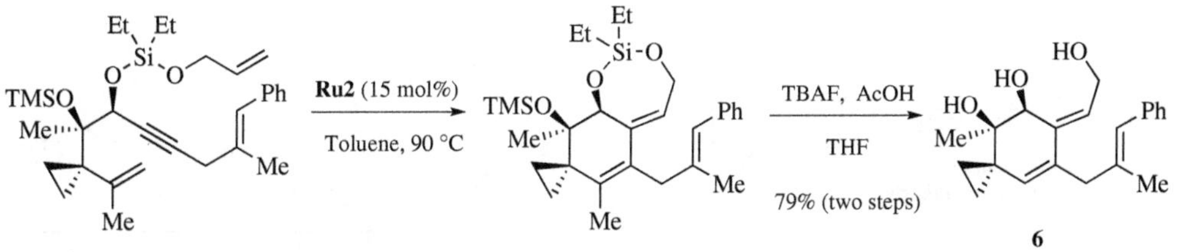

6 (4.58)

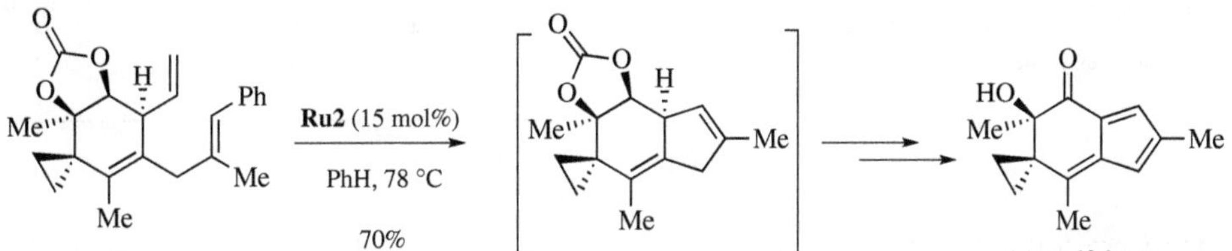

7 (–)-acylfulvene (4.59)

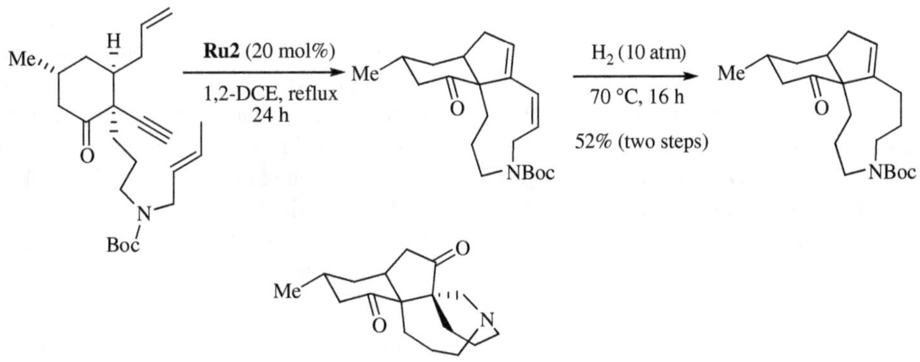

(+)-lycoflexine

Scheme 4.17 Cascade metathesis approach to the alkaloid lycoflexine (107).

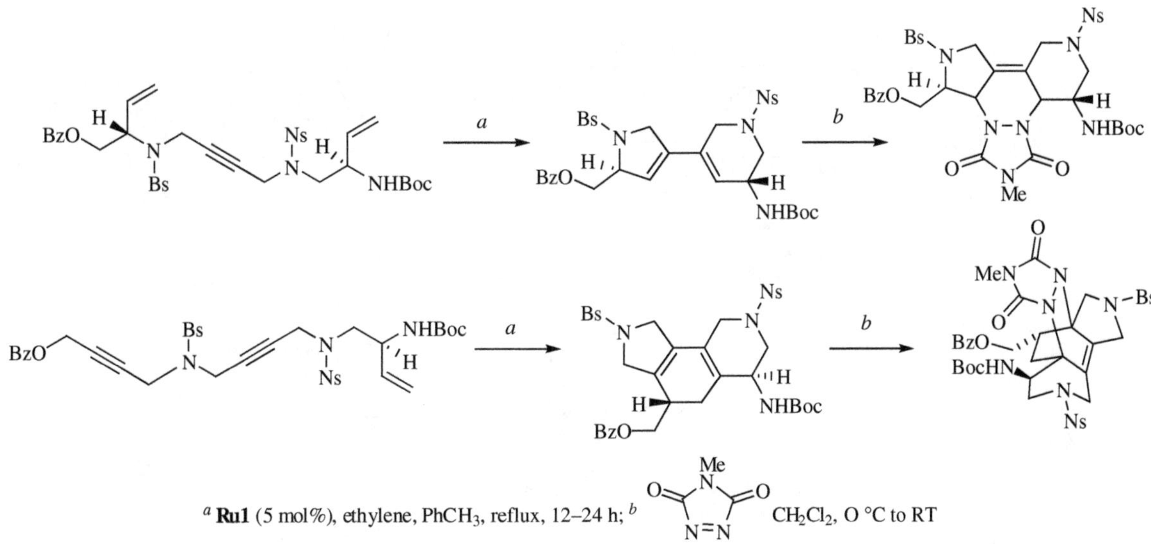

[a] **Ru1** (5 mol%), ethylene, $PhCH_3$, reflux, 12–24 h; [b] 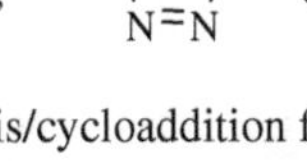

CH_2Cl_2, O °C to RT

Scheme 4.18 Tandem ring-closing enyne metathesis/cycloaddition for diversity synthesis (141).

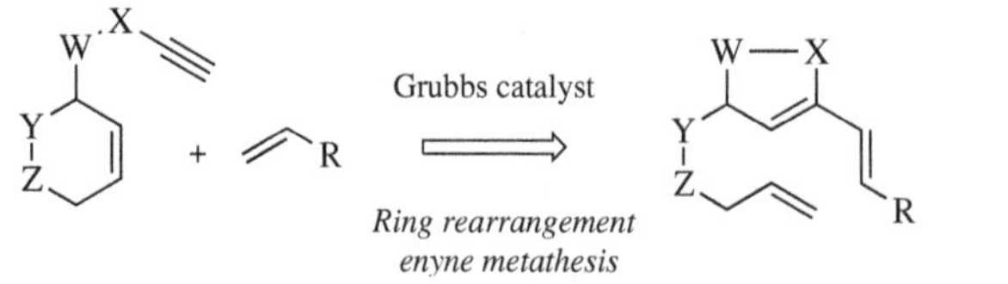

(4.60)

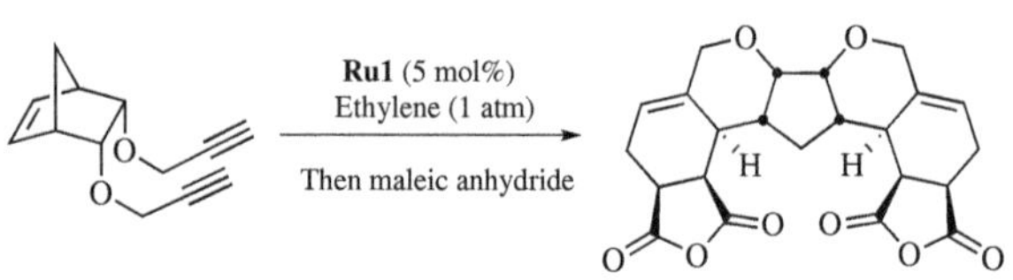

(4.61)

sequence, seven rings were formed from a single bicyclic molecule.

Ring rearrangement is very efficient for rapidly generating skeletal diversity. In Scheme 4.19, metathesis events are controlled in a stepwise process that capitalizes on individual catalyst attributes (149). The overall process most likely begins with ethylene CM joining the strained norbornene with the alkyne. The greater reactivity of **Ru2** is needed to form the seven-membered ring. To facilitate purification of the intermediate triene from the ruthenium catalysts, a polar isocyanide (90) was used to quench the catalysts and facilitate removal of the organometallic impurities. Microwave irradiation was used to accelerate the metathesis and cycloaddition steps.

The ring rearrangement enyne metathesis occurs in the presence of added ethylene or higher alkenes (Eq. 4.62, Table 4.17). With added alkenes, ring rearrangement and CM occur (143). "Low strain" cycloalkenes, such as seven-

TABLE 4.17 Ring Rearrangement with Added Alkenes

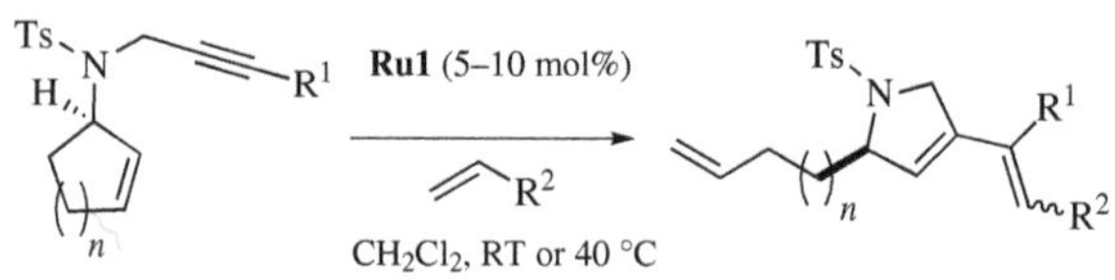

(4.62)

Entry	R^1	n	R^2	Yield, %
1	H	1	CH_2OTBS	78
2	H	2	CH_2OTBS	85
3	H	3	CH_2OTBS	81
4	Me	3	H	56
5	Me	4	H	75

Refs 105, 143.

and eight-membered cycloalkenes, perform well in the ring rearrangement enyne metathesis using ethylene (105).

Diastereodivergent enyne metathesis pathways were observed in cyclohexenes bearing an endocyclic chiral center (Scheme 4.20) (105). In the *cis*-diastereomer (Eq. 4.63), the intermediate vinyl carbene bumps into the pseudoaxial TBS ether, forbidding an RCM and leading to ethylene metathesis instead of ring rearrangement. In the *trans*-diastereomer, the absence of this steric interaction allows the ring rearrangement to proceed uneventfully (Eq. 4.64).

More difficult ring rearrangements may result in alkene isomerization. Under more forcing conditions possible with **Ru2**, an eight-membered ring rearrangement product was expected. In the event, a seven-membered ring carbocycle was obtained along with a trace dimeric product (Scheme 4.21) (150). The higher reaction temperatures caused alkene isomerization via the intermediate triene **8**. In reactions involving $L_nRu{=}CH_2$, Grubbs identified a ruthenium hydride that

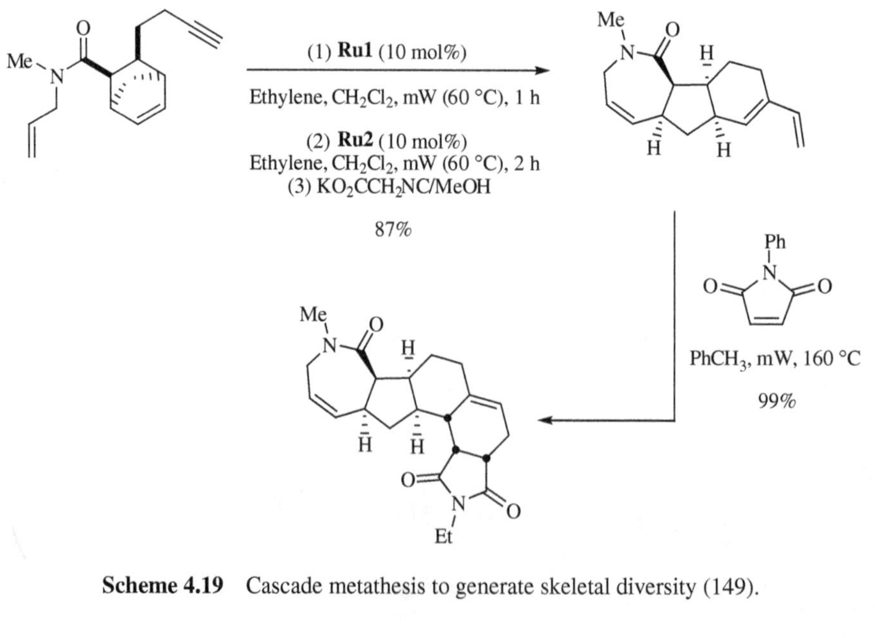

Scheme 4.19 Cascade metathesis to generate skeletal diversity (149).

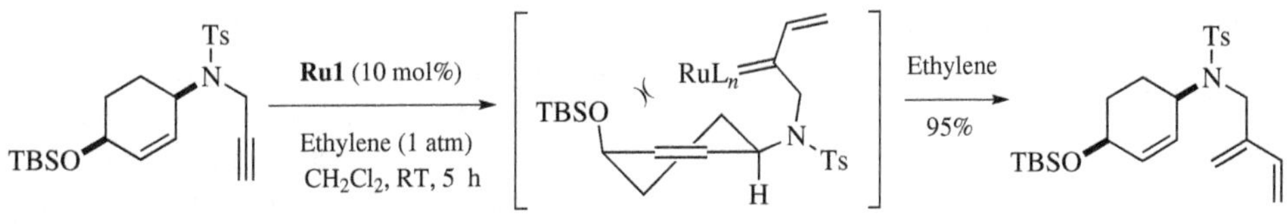

(4.63)

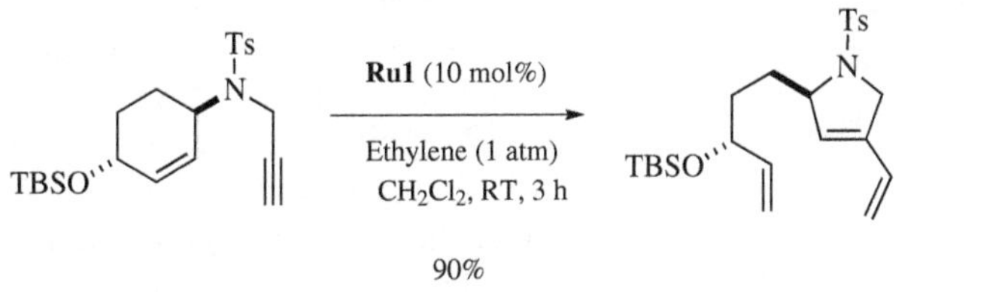

(4.64)

Scheme 4.20 Diastereospecific ring rearrangement enyne metathesis (105).

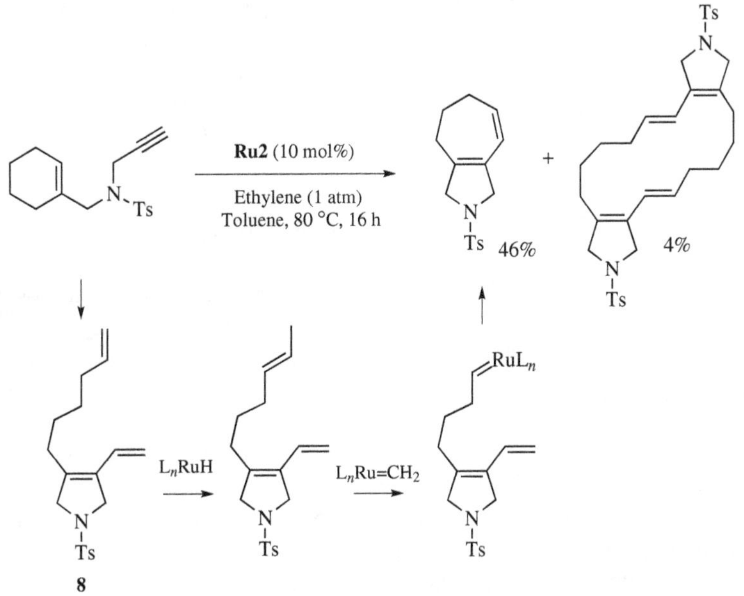

Scheme 4.21 Unexpected truncation product due to alkene isomerization (150).

promotes alkene isomerization (151). This has been limited in some cases through the use of benzoquinone (47,152).

Ring rearrangement metathesis has been used to generate a dienophile–diene duo leading to an *in situ* intramolecular Diels–Alder reaction (Eq. 4.65) (153). The metathesis sequence generated an intermediate butenolide that serves as dienophile in cycloaddition with the 1,3-diene previously formed by RCEYM. It is notable that this compact pentacycle is formed in two steps from a monocyclic precursor.

Not surprisingly, strained cycloalkenes also give the ring rearrangement (Eq. 4.66, Table 4.18) (154). This net process is similar to the intermolecular synthesis of 1,3-cyclohexadienes using methylene-free conditions (117).

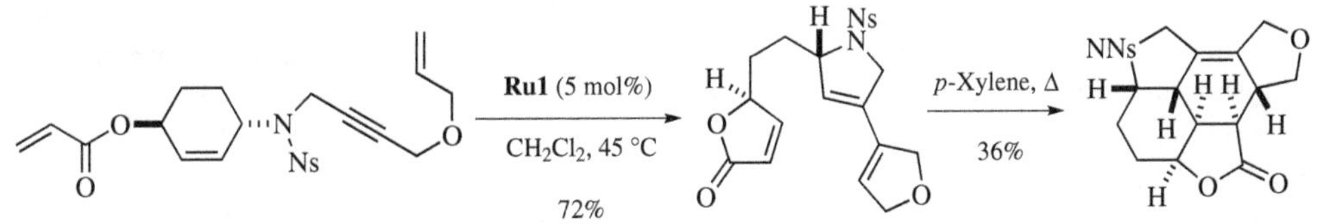

(4.65)

TABLE 4.18 Ring Rearrangement to Form Isoquinoline Derivatives

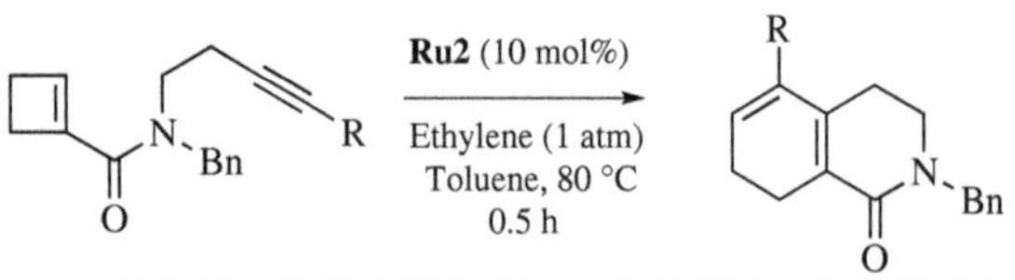

(4.66)

Entry	R	Yield, %
1	H	61
2	Me	56
3	Ph	82

Ref. 154.

TABLE 4.19 Ring Rearrangement to Form Skipped Trienes

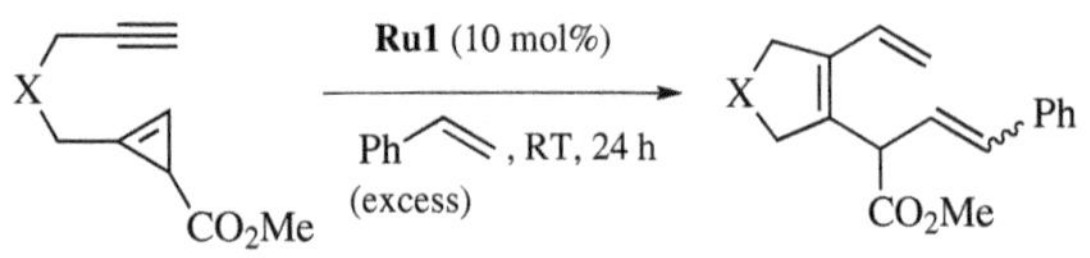

(4.67)

Entry	X	Yield, %	E/Z ratio
1	NTs	60	1:1.5
2	O	55	1:1.4

Ref. 155.

Representative Procedure for Ring Rearrangement (154)

A crude 1,3-cyclohexadiene was prepared from cyclobutenylamide (54.0 mg, 0.23 mmol) and **Ru2** (19.2 mg, 22.56 μmol, 10 mol%) in toluene (8 mL) by warming at 80 °C for 0.5 h under ethylene, and was purified by column chromatography on silica gel (hexane/ethyl acetate, 3 : 2) to give the bicyclic amide (33.0 mg, 61%) as a pale yellow liquid.

Ring rearrangement with cyclopropene-containing reactants provides skipped triene products (Eq. 4.67, Table 4.19). The added alkene was used in molar excess to help prevent homopolymerization of the cyclopropene (155).

4.4 CONCLUSION AND OUTLOOK

Ene-yne metathesis provides an efficient catalytic method for diene synthesis using simple unsaturated reactants. The reaction shows excellent functional group tolerance making the catalytic reaction suitable for diverse synthetic applications. Ene-yne metathesis delivers conjugated diene products from simple reactants. The well-known reactivity of the conjugated diene in cycloaddition reactions makes further ring building a natural extension demonstrating synthetic utility of the ene-yne metathesis. This tandem sequence results in a rapid gain of structural complexity. Ene-yne metathesis is successfully used in tandem with other bond-forming processes, an area that has seen tremendous growth in the past few years. These successes, combined with continuing improvements in substrate scope and a refined understanding of reaction mechanism, have made ene-yne metathesis a useful synthetic method for C–C bond construction.

Ring-closing metathesis is an efficient ring synthesis. Some recent topics that use the economy of the enyne metathesis and recognize the utility of the 1,3-diene product include diversity-oriented synthesis or the synthesis of platforms that array functional groups in a stereodefined way. It was also seen that a number of tandem reactions are conjoined with the enyne metathesis, some distinct bond-forming processes such as cycloaddition or metallotropic shift, and others involving additional metathesis steps. The potential chemistry here is expansive and is open to the imagination and limited perhaps only by the level of difficulty in obtaining the acyclic polyeneyne precursors. Control of the formed ring in polypendant dienynes is controlled through alkene reactivity or through use of the relay strategy.

In cross ene-yne metathesis applications, a molar excess of the alkene is typically used to drive the ene-yne metathesis forward. However, in the best cases, nearly equimolar amounts of the alkene and alkyne can be used. In these cases, the ene-yne metathesis is a carbon–carbon cross coupling, with a reorganization of the π bonds. Some of the other exciting applications include the ring rearrangements to form heterocyclic rings, ring construction and ring expansion methods.

There are still areas that need further research. Some of the unresolved difficulties include poor control of alkene geometry, high catalyst loadings, and oftentimes the need for a large excess of alkene reactant. Though the mechanism has been studied, the rate profile with a wide assortment of alkynes and alkenes has not been fully evaluated, so some mechanistic nuances limit reaction design. A more complete understanding of mechanism will guide future synthetic efforts with less reactive unsaturated reactants. Strategies to increase reactivity in CM applications and fragment coupling, an important activity in total synthesis, would make the reaction more attractive and more efficient. Significant growth is expected in cascade metathesis and applications that functionalize the resulting 1,3-dienes in new ways, further increasing the synthetic potential of ene-yne metathesis.

REFERENCES

1. Enyne cycloisomerization reviews: (a) Herrero-Gomez E, Echavarren AM. *Handbook of Cyclization Reactions*

2010;2:625–686. (b) Gandon V, Thorimbert S, Malacria M. *Science of Synthesis* 2009;46:173–238. (c) Michelet V, Toullec PY, Genet J-P. *Angew. Chem., Int. Ed.* 2008;47:4268–4315. (d) Jimenez-Nunez E, Echavarren AM. *Chem. Rev.* 2008;108:3326–3350. (e) Zhang L, Sun J, Kozmin SA. *Adv. Synth. Catal.* 2006;348:2271–2296. (f) Nieto-Oberhuber C, Lopez S, Jimenez-Nunez E, Echavarren AM. *Chem. - Eur. J.* 2006;12:5916–5923. (g) Ma S, Yu S, Gu Z. *Angew. Chem., Int. Ed.* 2006;45:200–203. (h) Anorbe L, Dominguez G, Perez-Castells J. *Chem. - Eur. J.* 2004;10:4938–4943. (i) Aubert C, Buisine O, Malacria M. *Chem. Rev.* 2002;102:813–834.

2. Evans MA, Morken JP. *Org Lett* 2005;7:3371–3373.
3. Stragies R, Schuster M, Blechert S. *Angew Chem Int Ed* 1997; 36:2518–2520.
4. Oguri H, Hiruma T, Yamagishi Y, Oikawa H, Ishiyama A, Otoguro K, Yamada H, Omura S. *J Am Chem Soc* 2011; 133:7096–7105.
5. Mori M. *Ene-Yne Metathesis*, in Handbook of Metathesis: Catalyst Development (ed R. H. Grubbs), Weinheim: Wiley-VCH; 2008 Vol. 2.
6. Poulsen CS, Madsen R. *Synthesis* 2003:1–18.
7. Diver ST, Giessert AJ. *Chem Rev* 2004;104:1317–1382.
8. Li J, Lee D. *Eur J Org Chem* 2011;23:4269–4287.
9. Scholl M, Ding S, Lee CW, Grubbs RH. *Org Lett* 1999; 1:953–956.
10. Katz TJ, Sivavec TM. *J Am Chem Soc* 1985;107:737–738.
11. Watanuki S, Ochifuji N, Mori M. *Organometallics* 1995;14: 5062–5067.
12. Watanuki S, Mori M. *Organometallics* 1995;14:5054–5061.
13. Watanuki S, Ochifuji N, Mori M. *Organometallics* 1994;13: 4129–4130.
14. Nguyen ST, Grubbs RH, Ziller JW. *J Am Chem Soc* 1993;115:9858–9859.
15. Fu GC, Nguyen ST, Grubbs RH. *J Am Chem Soc* 1993;115: 9856–9857.
16. Fu GC, Grubbs RH. *J Am Chem Soc* 1992;114:5426–5427.
17. Kinoshita A, Mori M. *Synlett* 1994:1020–1022.
18. Kinoshita A, Sakakibara N, Mori M. *Tetrahedron* 1999;55: 8155–8167.
19. Mori M, Sakakibara N, Kinoshita A. *J Org Chem* 1998;63: 6082–6083.
20. Kinoshita A, Sakakibara N, Mori M. *J Am Chem Soc* 1997;119:12388–12389.
21. Grubbs RH. *Angew Chem Int Ed* 2006;45:3760–3765.
22. Schrock RR. *Angew Chem Int Ed* 2006;45:3748–3759.
23. Chauvin Y. *Angew Chem Int Ed* 2006;45:3741–3747.
24. Schwab P, Grubbs RH, Ziller JW. *J Am Chem Soc* 1996; 118:100–110.
25. Schwab P, France MB, Ziller JW, Grubbs RH. *Angew Chem Int Ed* 1995;34:2039–2041.
26. Smulik JA, Diver ST. *Org Lett* 2000;2:2271–2274.
27. Garber SB, Kingsbury JS, Gray BL, Hoveyda AH. *J Am Chem Soc* 2000;122:8168–8179.
28. Kingsbury JS, Harrity JPA, Bonitatebus Jr PJ, Hoveyda AH. *J Am Chem Soc* 1999;121:791–799.
29. Gessler S, Randl S, Blechert S. *Tetrahedron Lett* 2000; 41:9973–9976.
30. Grela K, Harutyunyan S, Michrowska A. *Angew Chem Int Ed* 2002;41:4038–4040.
31. Clavier H, Correa A, Escudero-Adán EC, Benet-Buchholz J, Cavallo L, Nolan SP. *Chem- Eur J* 2009;15: 10244–10254.
32. Clavier H, Nolan SP. *Chem- Eur J* 2007;13:8029–8036.
33. Fürstner A, Guth O, Düffels A, Seidel G, Liebl M, Gabor B, Mynott R. *Chem- Eur J* 2001;7:4811–4820.
34. Kinoshita A, Mori M. *J Org Chem* 1996;61:8356–8357.
35. Giessert AJ, Diver ST. *J Org Chem* 2005;70:1046–1049.
36. Clark DA, Kulkarni AA, Kalbarczyk K, Schertzer B, Diver ST. *J Am Chem Soc* 2006;128:15632–15636.
37. Lee H-Y, Kim BG, Snapper ML. *Org Lett* 2003;5:1855–1858.
38. Funk TW, Efskind J, Grubbs RH. *Org Lett* 2005;7:187–190.
39. Kim CH, An HJ, Shin WK, Yu W, Woo SK, Jung SK, Lee E. *Angew Chem Int Ed* 2006;45:8019–8021.
40. Chatterjee AK, Choi T-L, Sanders DP, Grubbs RH. *J Am Chem Soc* 2003;125:11360–11370.
41. Marshall JE, Keister JB, Diver ST. *Organometallics* 2011;30: 1319–1321.
42. Galan BR, Giessert AJ, Keister JB, Diver ST. *J Am Chem Soc* 2005;127:5762–5763.
43. Hoye TR, Donaldson SM, Vos TJ. *Org Lett* 1999;1:277–279.
44. Schramm MP, Reddy DS, Kozmin SA. *Angew Chem Int Ed* 2001;40:4274–4277.
45. Lippstreu JJ, Straub BF. *J Am Chem Soc* 2005;127: 7444–7457.
46. Giessert AJ, Diver ST. *Org Lett* 2005;7:351–354.
47. Clark JR, Diver ST. *Org Lett* 2011;13:2896–2899.
48. Maifeld SV, Miller RL, Lee DJ. *Am ChemSoc* 2004;126: 12228–12229.
49. Lloyd-Jones GC, Margue RG, de Vries JG. *Angew Chem Int Ed* 2005;44:7442–7447.
50. Giessert AJ, Brazis NJ, Diver ST. *Org Lett* 2003;5: 3819–3822.
51. Grotevendt AGD, Lummiss JAM, Mastronardi ML, Fogg DE. *J Am Chem Soc* 2011;133:15918–15921.
52. Mori M, Kitamura T, Sato Y. *Synthesis* 2001:654–664.
53. Debleds O, Campagne J-M. *J Am Chem Soc* 2008;130: 1562–1563.
54. Graham TJA, Gray EE, Burgess JM, Goess BC. *J Org Chem* 2010;75:226–228.
55. Samojłowicz C, Borré E, Mauduit M, Grela K. *Adv Synth Cat* 2011;353:1993–2002.
56. Grise CM, Tessier G, Barriault L. *Org Lett* 2007;9:1545–1548.
57. Leeuwenburgh MA, Appeldoorn CCM, van Hooft PAV, Overkleeft HS, van der Marel GA, van Boom JH. *Eur J Org Chem* 2000;6:873–877.

58. Barrett AGM, Baugh SPD, Braddock DC, Flack K, Gibson VC, Procopiou PA. *Chem Commun* 1997:1375–1376.
59. Barrett AGM, Baugh SPD, Braddock DC, Flack K, Gibson VC, Giles MR, Marshall EL, Procopiou PA, White AJP, Williams DJ. *J Org Chem* 1998;63:7893–7907.
60. Bankar DB, Koketsu M. *Eur J Org Chem* 2010;14:2742–2745.
61. Kumagai N, Muncipinto G, Schreiber SL. *Angew Chem Int Ed* 2006;45:3635–3638.
62. Füger B, Bolm C. *Synlett* 2009:1601–1604.
63. Zhou A, Rayabarapu D, Hanson PR. *Org Lett* 2009;11: 531–534.
64. Kumar S, Hansen MH, Albæk N, Steffansen SI, Petersen M, Nielsen P. *J Org Chem* 2009;74:6756–6769.
65. Brenneman JB, Martin SF. *Org Lett* 2004;6:1329–1331.
66. Mori M, Tomita T, Kita Y, Kitamura T. *Tetrahedron Lett* 2004;45:4397–4399.
67. Brenneman JB, Machauer R, Martin SF. *Tetrahedron* 2004;60:7301–7314.
68. Aggarwal VK, Astle CJ, Rogers-Evans M. *Org Lett* 2004;6: 1469–1471.
69. Harvey JS, Giuffredi GT, Gouverneur V. *Org Lett* 2010;12: 1236–1239.
70. Hansen EC, Lee D. *J Am Chem Soc* 2003;125:9582–9583.
71. Grimwood ME, Hansen HC. *Tetrahedron* 2009;65: 8132–8138.
72. Layton ME, Morales CA, Shair MD. *J Am Chem Soc* 2002;124:773–775.
73. Diver ST, Kulkarni AA, Clark DA, Peppers BP. *J Am Chem Soc* 2007;129:5832–5833.
74. Peters J-U. *Chem Commun* 1997:1983–1984.
75. Das SK, Roy R. *Tetrahedron Lett* 1999;40:4015–4018.
76. Betkekar VV, Panda S, Kaliappan KP. *Org Lett* 2012;14: 198–201.
77. Pizzirani D, Kaya T, Clemons PA, Schreiber SL. *Org Lett* 2010;12:2822–2825.
78. Murakami M, Kadowaki S, Matsuda T. *Org Lett* 2005; 7:3953–3956.
79. Singh R, Schrock RR, Müller P, Hoveyda AH. *J Am Chem Soc* 2007;129:12654–12655.
80. Lee Y-J, Schrock RR, Hoveyda AH. *J Am Chem Soc* 2009; 131:10652–10661.
81. Zhao Y, Hoveyda AH, Schrock RR. *Org Lett* 2011; 13:784–787.
82. Smulik JA, Giessert AJ, Diver ST. *Tetrahedron Lett* 2002;43:209–211.
83. Smulik JA, Diver ST. *J Org Chem* 2000;65:1788–1792.
84. Fürstner A, Larionov O, Flügge S. *Angew Chem Int Ed* 2007;46:5545–5548.
85. Mori M, Tonogaki K, Nishiguchi N. *J Org Chem* 2002;67:224–226.
86. Stragies R, Voigtmann U, Blechert S. *Tetrahedron Lett* 2000;41:5465–5468.
87. Schürer SC, Blechert S. *Chem Commun (Cambridge, United Kingdom)* 1999:1203–1204.
88. Zheng G, Graham A, Shibata M, Missert JR, Oseroff AR, Dougherty TJ, Pandey RKJ. *Org Chem* 2001;66:8709–8716.
89. Dinger MB, Mol JC. *Organometallics* 2003;22:1089–1095.
90. Galan BR, Kalbarczyk KP, Szczepankiewicz S, Keister JB, Diver ST. *Org Lett* 2007;9:1203–1206.
91. Clark DA, Clark JR, Diver ST. *Org Lett* 2008;10:2055–2058.
92. Giessert AJ, Snyder L, Markham J, Diver ST. *Org Lett* 2003;5:1793–1796.
93. Clark DA, Basile BS, Karnofel WS, Diver ST. *Org Lett* 2008;10:4927–4929.
94. Wu Z, Nguyen ST, Grubbs RH, Ziller JW. *J Am Chem Soc* 1995;117:5503–5511.
95. Clark JR, French JM, Diver ST. *J Org Chem* 2012;77: 1599–1604.
96. Clark JR, French JM, Jecs E, Diver ST. *Org Lett* 2012; 14:4178–4181.
97. Trost BM. *Angew Chem Int Ed Engl* 1995;34:259–281.
98. Trost BM. *Science* 1991;254:1471–1477.
99. Kim M, Park S, Maifeld SV, Lee D. *J Am Chem Soc* 2004; 126:10242–10243.
100. Bentz D, Laschat S. *Synthesis* 2000:1766–1773.
101. Kotha S, Sreenivasachary N. *Eur J Org Chem* 2001: 3375–3383.
102. Kotha S, Sreenivasachary N. *Chem Commun* 2000:503–504.
103. Nandurdikar RS, Subrahmanyam AV, Kaliappan KP. *Eur J Org Chem* 2010;14:2788–2799.
104. Comer E, Rohan E, Deng L, Porco Jr JA. *Org Lett* 2007; 9:2123–2126.
105. Kitamura T, Mori M. *Org Lett* 2001;3:1161–1163.
106. Clark JS, Townsend RJ, Blake AJ, Teat SJ, Johns A. *Tetrahedron Lett* 2001;42:3235–3238.
107. Ramharter J, Weinstabl H, Mulzer J. *J Am Chem Soc* 2010; 132:14338–14339.
108. Micalizio GC, Schreiber SL. *Angew Chem Int Ed* 2002; 41:3272–3276.
109. Micalizio GC, Schreiber SL. *Angew Chem Int Ed* 2002; 41:152–154.
110. Renaud J, Graf C-D, Oberer L. *Angew Chem Int Ed* 2000; 39:3101–3104.
111. Ko HM, Lee CW, Kwon HK, Chung HS, Choi SY, Chung YK, Lee E. *Angew Chem Int Ed* 2009;48:2364–2366.
112. Royer F, Vilain C, Elkaïm L, Grimaud L. *Org Lett* 2003; 5:2007–2009.
113. Kummer DA, Brenneman JB, Martin SF. *Org Lett* 2005; 7:4621–4623.
114. Dieltiens N, Stevens CV. *Synlett* 2006:2771–2776.
115. Smulik JA, Diver ST. *Tetrahedron Lett* 2001;42:171–174.
116. Kulkarni AA, Diver ST. *Org Synth* 2006;83:200–208.
117. Kulkarni AA, Diver ST. *J Am Chem Soc* 2004;126: 8110–8111.

118. Peppers BP, Kulkarni AA, Diver ST. *Org Lett* 2006;8: 2539–2542.
119. Kulkarni AA, Diver ST. *Org Lett* 2003;5:3463–3466.
120. Diver ST, Clark DA, Kulkarni AA. *Tetrahedron* 2008;64:6909–6919.
121. Kim S-H, Zuercher WJ, Bowden NB, Grubbs RH. *J Org Chem* 1996;61:1073–1081.
122. Kim S-H, Bowden N, Grubbs RH. *J Am Chem Soc* 1994;116:10801–10802.
123. Honda T, Namiki H, Kaneda K, Mizutani H. *Org Lett* 2004;6:87–89.
124. Hoye TR, Jeffrey CS, Tennakoon MA, Wang J, Zhao H. *J Am Chem Soc* 2004;126:10210–10211.
125. Wallace DJ. *Angew Chem Int Ed* 2005;44:1912–1915.
126. Hoye TR, Jeon J. *Metathesis Involving a Relay and Applications in Natural Product Synthesis, in Metathesis in Natural Product Synthesis: Strategies, Substrates and Catalysts* (eds J. Cossy, S. Arseniyadis and C. Meyer), Weinheim: Wiley-VCH; 2010. pp. 261–285.
127. Hoye TR, Jeffrey CS, Tennakoon MA, Wang J, Zhao H. *J Am Chem Soc* 2004;126:10210–10211.
128. Lee J, Parker KA. *Org Lett* 2012;14:2682–2685.
129. Li J, Park S, Miller RL, Lee D. *Org Lett* 2009;11:571–574.
130. Boyer F-D, Hanna I. *Tetrahedron Lett* 2002;43:7469–7472.
131. Boyer F-D, Hanna I, Ricard L. *Org Lett* 2004;6:1817–1820.
132. Boyer F-D, Hanna I. *Org Lett* 2007;9:2293–2295.
133. Aldegunde MJ, Garcia-Fandino R, Castedo L, Granja JR. *Chem- Eur J* 2007;13:5135–5150.
134. Garcia-Fandino R, Codesido EM, Sobarzo-Sanchez E, Castedo L, Granja JR. *Org Lett* 2004;6:193–196.
135. Niethe A, Fischer D, Blechert S. *J Org Chem* 2008;73: 3088–3093.
136. Mukherjee S, Lee D. *Org Lett* 2009;11:2916–2919.
137. Baumgartner C, Ma S, Liu Q, Stoltz BM. *Org Biomol Chem* 2010;8:2915–2917.
138. Siegel DS, Piizzi G, Piersanti G, Movassaghi M. *J Org Chem* 2009;74:9292–9304.
139. Movassaghi M, Piizzi G, Siegel DS, Piersanti G. *Tetrahedron Lett* 2009;50:5489–5492.
140. Louie J, Bielawski CW, Grubbs RHJ. *Am Chem Soc* 2001;123:11312–11313.
141. Spiegel DA, Schroeder FC, Duvall JR, Schreiber SLJ. *Am Chem Soc* 2006;128:14766–14767.
142. Zuercher WJ, Scholl M, Grubbs RHJ. *Org Chem* 1998;63: 4291–4298.
143. Randl S, Lucas N, Connon SJ, Blechert S. *Adv Synth Cat* 2002;344:631–633.
144. Holub N, Blechert S. *Ring Rearrangement Metathesis (RRM) in Alkaloid Synthesis Nicole Holub and Siegfried Blechert in Asymmetric Synthesis - The Essentials* (eds M. Christmann and S. Bräse), Weinheim: Wiley-VCH; 2007:254–259.
145. Böhrsch V, Neidhöfer J, Blechert S. *Angew Chem Int Ed* 2006;45:1302–1305.
146. Imhof S, Blechert S. *Synlett* 2003:609–614.
147. Voigtmann U, Blechert S. *Synthesis* 2000:893–898.
148. Banti D, North M. *Tetrahedron Lett* 2002;43:1561–1564.
149. Spandl RJ, Rudyk H, Spring DR. *Chem Commun* 2008: 3001–3003.
150. Mori M, Kuzuba Y, Kitamura T, Sato Y. *Org Lett* 2002;4: 3855–3858.
151. Hong SH, Day MW, Grubbs RH. *J Am Chem Soc* 2004;126:7414–7415.
152. Hong SH, Sanders DP, Lee CW, Grubbs RH. *J Am Chem Soc* 2005;127:17160–17161.
153. O'Leary-Steele C, Pedersen PJ, James T, Lanyon-Hogg T, Leach S, Hayes J, Nelson A. *Chem Eur J* 2010;16:9563–9571.
154. Mori M, Wakamatsu H, Tonogaki K, Fujita R, Kitamura T, Sato Y. *J Org Chem* 2005;70:1066–1069.
155. Zhu Z-B, Shi M. *Org Lett* 2010;12:4462–4465.

5

DOMINO AND OTHER OLEFIN METATHESIS REACTION SEQUENCES

Bernd Schmidt and Stefan Krehl

Institut für Chemie, Organische Synthesechemie, Universität Potsdam, Golm, Germany

5.1 INTRODUCTION

For quite some time, the terms "domino" (1) and "tandem" reactions were used synonymously, following Tietze's definition, for processes "involving two or more consecutive reactions in which subsequent reactions result as a consequence of the functionality formed by bond formation or fragmentation in the previous step" (2). Unfortunately, the increasing popularity of the concept, in particular in the field of homogeneous catalysis, lead to an inflationary use of terms such as "sequential catalysis," "multicatalysis," "cascade catalysis," or "tandem catalysis." Recently, Ambrosini and Lambert criticized that "the nomenclature describing concepts encompassed by the term "multicatalysis" have not been used uniformly and sometimes lead to confusion," (3) and Chapman and Frost stated even more pointedly that "indeed, the term tandem catalysis is often misrepresented in chemical literature" (4). Thus, occasionally typical one-flask reactions that proceed by subsequent addition of reagents to the reaction mixture in due course or even multistep syntheses proceeding with isolation of each intermediate have sometimes been classified as tandem reactions.

Some years ago, Fogg and dos Santos proposed a taxonomy that is especially designed for the classification of catalytic reaction sequences (5). They distinguish between the following conditons:

a. Catalytic one-pot reactions: These are conducted in one reaction vessel without isolation of the intermediates. Typically, the required precatalysts or reagents are added one after another.
b. Domino catalytic reactions: These reactions are characterized by the use of only one catalyst that promotes different steps in a defined order via just one catalytic mechanism.
c. Tandem catalytic reactions: In Fogg's and dos Santos' taxonomy, a tandem reaction sequence proceeds via more than one catalytic mechanism. This can be accomplished by using multiple precatalysts ("orthogonal tandem catalysis"), which need to be compatible under the reaction conditions, or by using just one precatalyst, which is first activated to catalyze one reaction, and then undergoes a transformation to a different catalytic species, which promotes a second reaction via a different mechanism. This catalyst interconversion may occur spontaneously ("auto-tandem catalysis") or by a change of reaction conditions, for example, addition of a "chemical trigger" ("assisted tandem catalysis").

An obvious advantage of all catalytic reaction sequences outlined above is their workup-step economy (6). The isolation of reaction products in multistep syntheses is generally considered as unproductive, time consuming, and labor consuming. If chromatographic purification is required in one or more steps, considerable expenses for solvents and silica are required, which might render a certain synthesis unattractive on industrial scale. However, the development of catalytic reaction sequences of all types highlighted above normally requires high yields and selectivities for each step, otherwise the product isolation after the final step might be hampered by the accumulation of unreacted starting materials,

Olefin Metathesis: Theory and Practice, First Edition. Edited by Karol Grela.

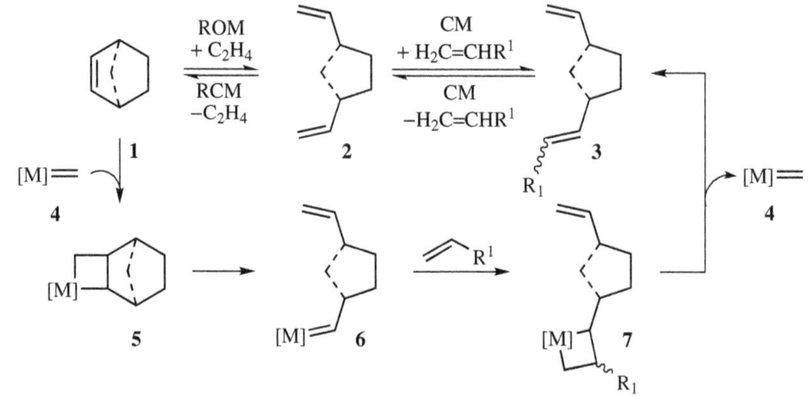

Scheme 5.1 ROM/CM sequence as an example for a domino metathesis transformation.

intermediates, and by-products. In addition to workup-step economy, over the past few years "catalyst economy" (7,8) has been recognized as an additional major advantage of catalytic domino-, auto-, or assisted tandem transformations.

Shortly after the discovery of the first stable and defined precatalysts for olefin metathesis by Schrock (9) and Grubbs (10) and parallel to the development of improved and more active variants, it was discovered that the olefin metathesis reaction is particularly well suited for implementation in domino reaction sequences because in this transformation two alkenes react to two different alkenes. Thus, no functional group interconversion occurs, and the reaction of the alkenes may proceed as a cross metathesis (CM, due to the reversibility in principle in both directions), ring-opening metathesis (ROM), or ring-closing metathesis (RCM). All these modes of olefin metathesis are catalyzed by the same catalytically active species, that is, a metal carbene complex via essentially the same mechanism, that is, Chauvin's mechanism (11), and can be connected in a defined order of events. The principle is outlined in Scheme 5.1 for the example of an ROM/CM sequence. Via this sequence, a cycloalkene **1** is converted into a diene **3** in the presence of a catalyst **4**. In a stepwise manner, this conversion would require ethenolysis (ROM) to the diene **2**, which would then in a separate metathesis step undergo a CM with a terminal alkene to yield the product **3**. In a domino process, the cycloalkene **1** would directly react to the desired product **3** in the presence of the CM partner and the catalyst via the carbene complex **6** (Scheme 5.1).

In the late 1990s, a conceptually different class of single-precatalyst olefin metathesis reaction sequences emerged. In these sequences, a metathesis step is followed by a subsequent catalytic non-metathesis transformation of the previously formed C–C double bond or the allylic position. Notably, both steps are catalyzed by different metal complexes; however, only one precatalyst is required. This means that after completion of the first step, an organometallic transformation of the metal carbene into another species, which is capable of catalyzing a non-metathesis-type functional group interconversion, takes place. Normally, this organometallic transformation of the catalytically active species will be accomplished by the addition of a reagent, a so-called "chemical trigger." Based on the terminology introduced by Fogg and dos Santos (5), these sequences can be categorized as "assisted tandem catalyses," although it should be noted that the use of the term "tandem" for such reactions is not indisputable. For instance, Ambrosini and Lambert restrict the term "tandem catalysis" to "a multicatalytic reaction whereby the product of one catalytic transformation immediately undergoes a subsequent catalytic transformation (by the same or a different catalyst), without any change in reaction conditions" (3). In Scheme 5.2, the concept is illustrated for the general example of an RCM/non-metathesis sequence: a metathesis precursor **8** is first converted into the corresponding RCM product **9**; upon completion of the metathesis reaction, the metal carbene is converted into a different catalytically active species [L_nM], which will then promote a metal-catalyzed functionalization of the C–C double bond or the allylic position, leading to the cyclic product **10**. Essentially, via this approach formerly independent catalytic cycles are connected by organometallic transformations *in situ* (12).

This chapter will cover domino metathesis reactions, as exemplified in Scheme 5.1, and single-precatalyst-induced metathesis/non-metathesis transformations ("assisted tandem catalytic reactions") as outlined in Scheme 5.2. For the readers' convenience, the structural formulae of all precatalysts mentioned in this chapter are compiled in Table 5.1, along with the numbering system used and the selected references for these precatalysts.

5.2 DOMINO REACTIONS

This section covers combinations of the various metathesis modes. It is subdivided according to the structural patterns

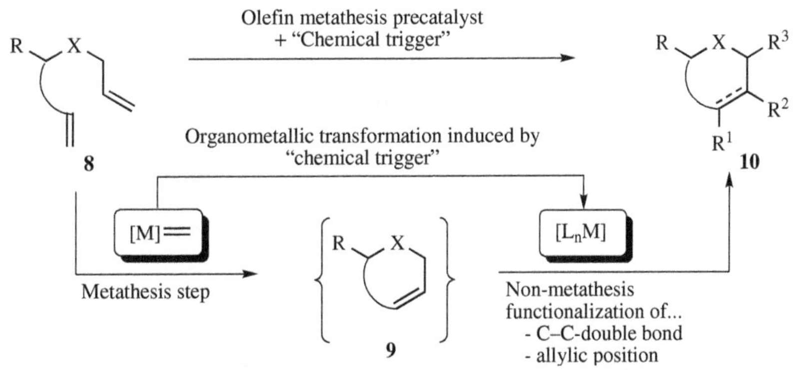

Scheme 5.2 General scheme for RCM/non-metathesis transformation sequences.

of the products rather than the type of domino sequence. Reaction sequences comprising enyne metathesis steps, that is, those starting from precursors with C–C triple bonds, are covered only briefly in this section, because they are discussed in more detail in Chapter 4.

5.2.1 Acyclic 1,n-Dienes

Acyclic 1,*n*-dienes are accessible via ROM/CM sequences starting from strained cycloolefins, normally cyclopropenes and cyclobutenes (27). In the absence of a suitable CM partner, ring-opening metathesis polymerization (ROMP) will occur (28), a reaction which is covered in Chapter 19. Examples for synthetically useful ROM/CM sequences of larger rings are rare, because release of ring strain is an important driving force for these reactions. In this context, one example for an ROM/CM sequence starting from a strained medium sized ring will be presented at the end of this section.

A simple example for an ROM/CM sequence is the ethenolysis of a cyclopropene, yielding a 3,3-disubstituted 1,4-pentadiene. Hoveyda et al. demonstrated in the initial steps of their quebrachamine synthesis that such a transformation can be synthetically useful and provide a sufficient amount of starting material for a multistep total synthesis. The required cyclopropene **13** was synthesized from the diazo compound **11**, which is in turn available from a 1,3-dicarbonyl compound, via Rh-catalyzed [2 + 1] cycloaddition to TMS acetylene. Base-catalyzed removal of the TMS group in **12** furnished the cyclopropene **13**, which reacts with ethylene at 1 bar in the presence of second-generation Hoveyda–Grubbs' catalyst **C1** (Scheme 5.3) (29).

Earlier, the spirocyclic cyclopropenone acetal **15** had been discovered as a useful precursor for protected unsymmetrical 1,4-pentadiene-3-ones. Thus, treatment of **15** with a slight excess of a CM partner such as allyl trimethylsilane in the presence of first-generation Grubbs' catalyst **C2** gives **16** in high yield and *E*/*Z*-selectivity (Scheme 5.4). Lower stereoselectivities were observed with this catalyst for other CM partners, in particular linear α-olefins (30).

Synthesis of (*E*)-trimethyl(3-(2-vinyl-1, 3-dioxan-2-yl)allyl)silane (16) (30).

Acetal **15** (112 mg, 1.0 mmol) and allyltrimethylsilane (137 mg, 1.2 mmol) were dissolved in dry and degassed benzene (8 ml) under an atmosphere of argon. To this solution was added Ru-catalyst **C1** (41 mg, 5 mol%) and the mixture was stirred at ambient temperature until full consumption of **15**. The solvent was evaporated, and the residue was purified by chromatography on silica to furnish the title compound (195 mg, 86%).

Later, the ROM/CM sequence of cyclopropenone acetals was used by Kozmin et al. in the synthesis of complex natural products, such as bistramide A (31), spirofungin A (32), and routiennocin (33). Notably, these authors could demonstrate that the terminal C–C double bond generated via ring opening/cross metathesis can be efficiently exploited in another CM step, which effectively sets the stage for a bidirectional synthesis. Because of the C_{2v} symmetry of **15**, its application in ROM/CM sequences does not result in the formation of stereoisomers. If, however, the cyclopropene is prochiral and the CM partner is chiral, diastereomers might result. Based on theoretical considerations, Hoveyda et al. proposed hydrogen bonding between the allylic OH group of the CM partner and one chloro ligand of the Ru-carbene complex as the decisive stereocontrol element in these reactions, as highlighted for the ROM/CM of **17** and *R*-1-butene-2-ol (**18**). In the presence of just 0.5 mol% of catalyst **C1**, product **19** could be obtained in moderate *E*/*Z*-selectivity, but with good yield and diastereoselectivity (Scheme 5.5) (34).

With prochiral cyclopropenes and achiral CM partners, enantiocontrol must be achieved. This approach was pursued by Giudici and Hoveyda, who used a chiral Ru-based metathesis catalyst **C3** for this purpose. For example, cyclopropene **20** and allyl ester **21** react in the presence of

TABLE 5.1 Olefin Metathesis Precatalysts Mentioned in This Chapter.

No.	Structure	References	No.	Structure	References
C1	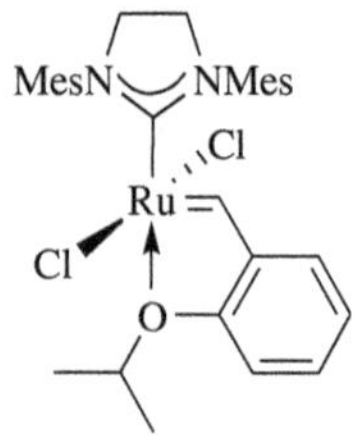	(13,14)	**C2**	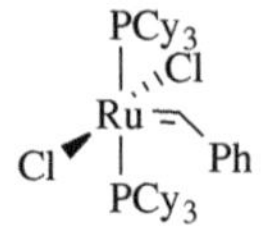	(15)
C3	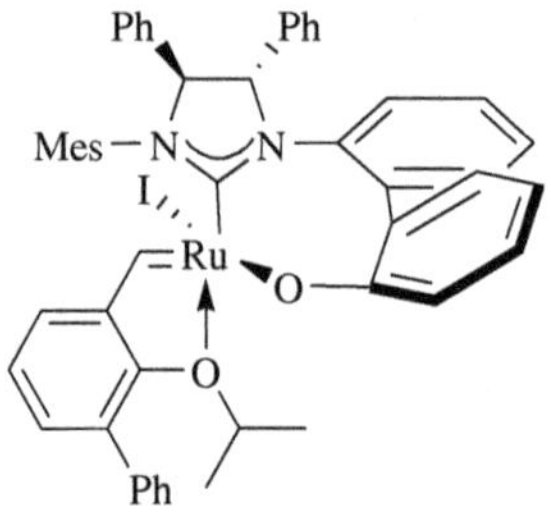	(16)	**C4**	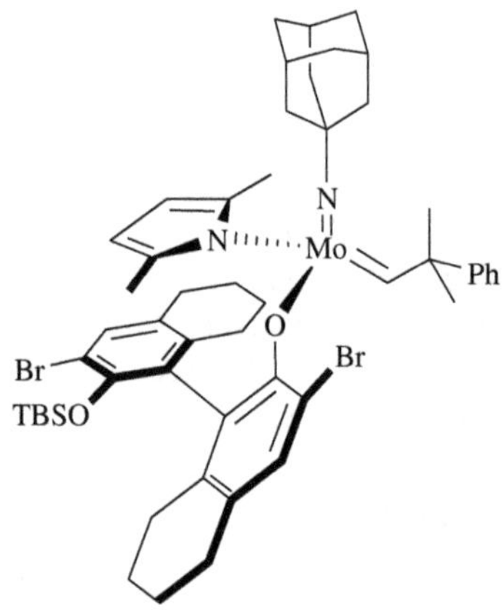	(17)
C5	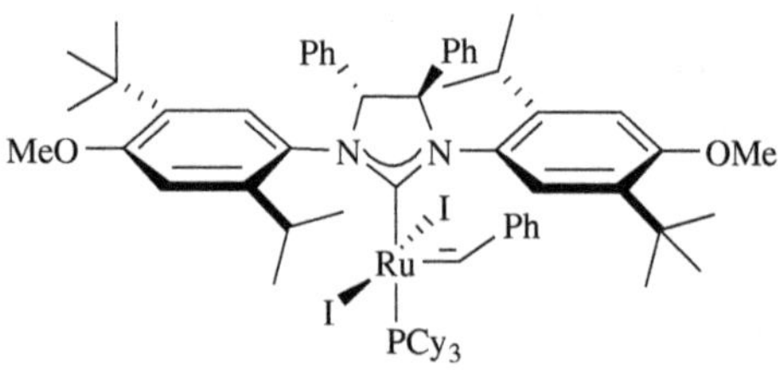	(18)	**C6**	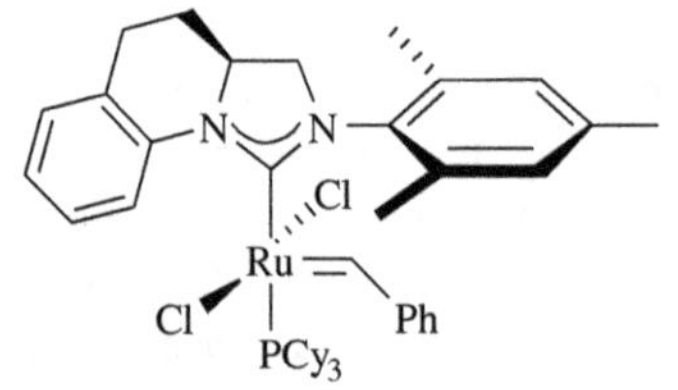	(19)
C7	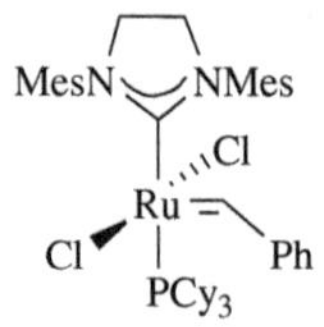	(20)	**C8**	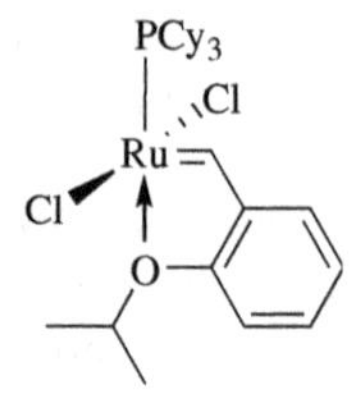	(21)
C9	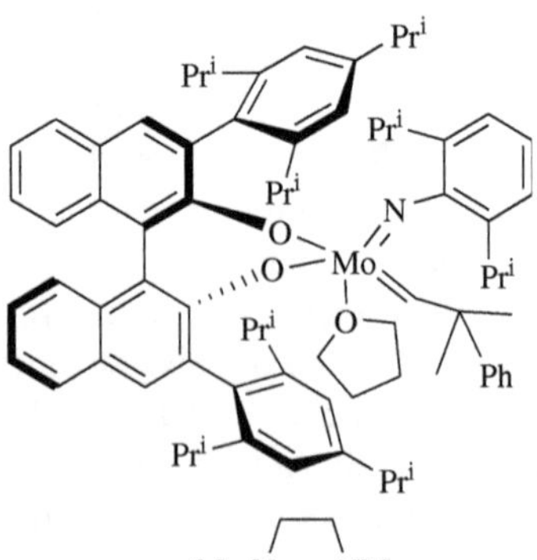	(22)	**C10**	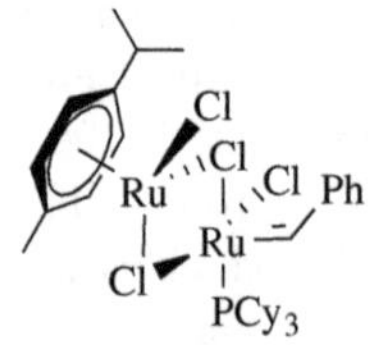	(23)
C11	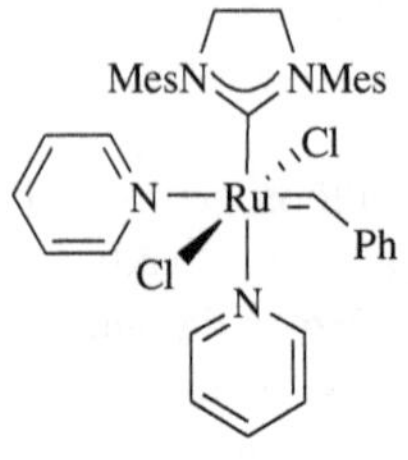	(24)	**C12**	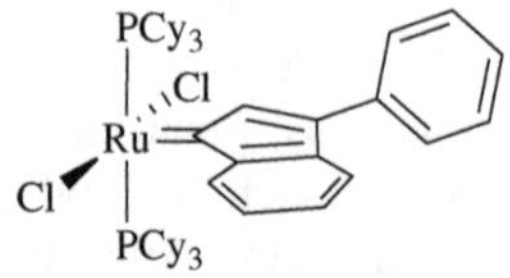	(25,26)

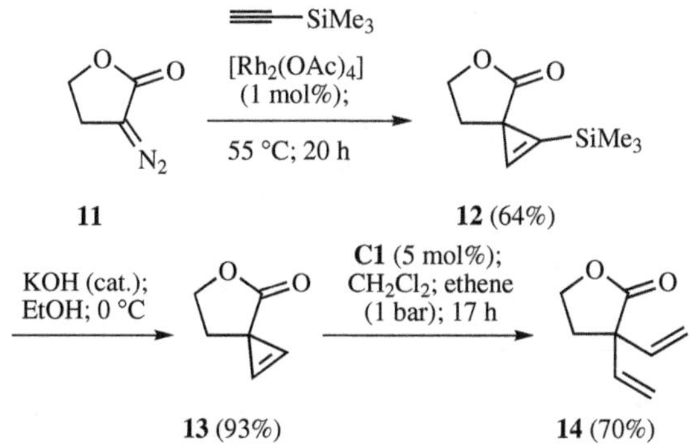

Scheme 5.3 ROM/CM for the synthesis of quebrachamine precursor **14**.

precatalyst **C3** in THF (a solvent which is rather unusual for olefin metathesis reactions) to the triene **22**. The authors discovered that allyl enoates such as **21** give significantly better enantioselectivities than other, sterically similarly demanding allyl esters, which is attributed to the formation of Ru-enoate chelate complexes in the transition state (Scheme 5.6) (35).

Similarly, substituted 1,5-hexadienes can be synthesized from cyclobutenes and olefins via ROM/CM. This reaction has been used with ethene as a CM partner in natural product total syntheses, for example, of *rac*-sporochnol A (36) and (−)-asteriscanolide (37). With a view toward isoprostanes, Schrader and Snapper investigated the ring opening/cross metathesis of TBS-protected allyl alcohol (**24**) and cyclobutene **23**, which is synthesized through photochemically induced electrocyclic ring closure of 1,3-cycloheptadiene. In the presence of first-generation catalyst **C2**, the desired diene **25** was obtained as a mixture of *Z*- and *E*-alkenes in a 2.3 : 1 ratio (Scheme 5.7) (38).

Occasionally, the ROM/CM strategy is tainted by competing ROMP. This problem has been reported for the attempted synthesis of **28** from **26** and oct-1-en-3-one (**27**). In this particular case, a two-step reaction turned out to be advantageous: cyclobutene **26** was first subjected to ethenolysis in the presence of **C2** under high dilution conditions, followed by CM of the intermediate **29** and enone **27** catalyzed by **C1**. Via this detour, the isoprostane precursor **28** could be obtained in yields varying between 70 and 90% (Scheme 5.8) (39).

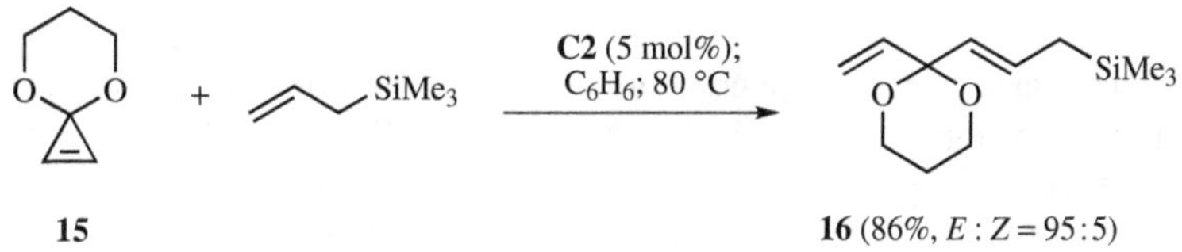

Scheme 5.4 ROM/CM with cyclopropene acetal **15**.

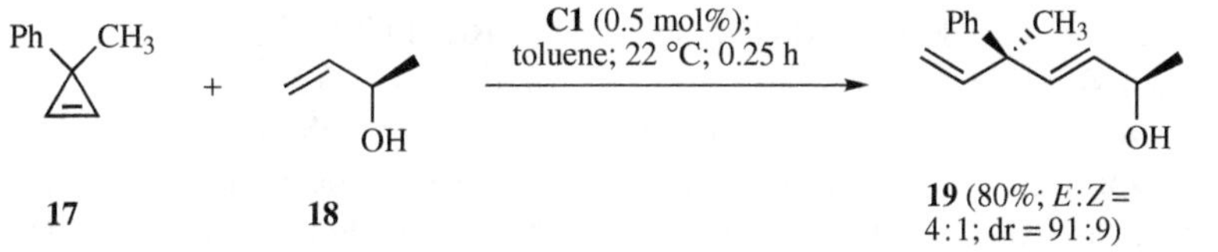

Scheme 5.5 H-bonding as stereochemical control element for ROM/CM sequences.

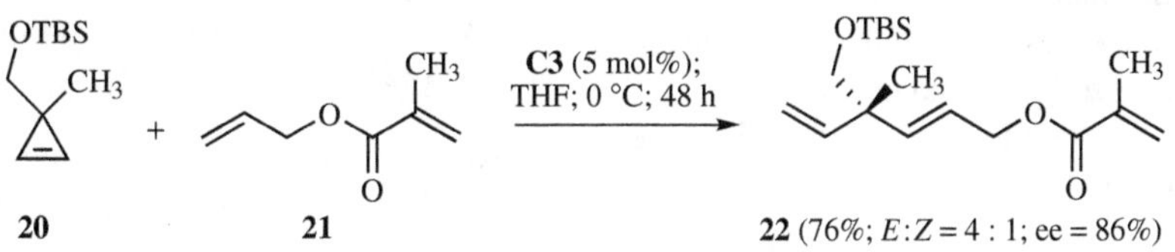

Scheme 5.6 Enantioselective ROM/CM.

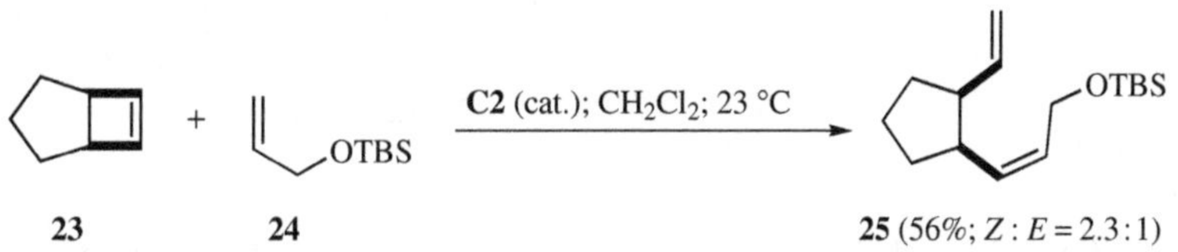

Scheme 5.7 Toward isoprostane derivatives via ROM/CM.

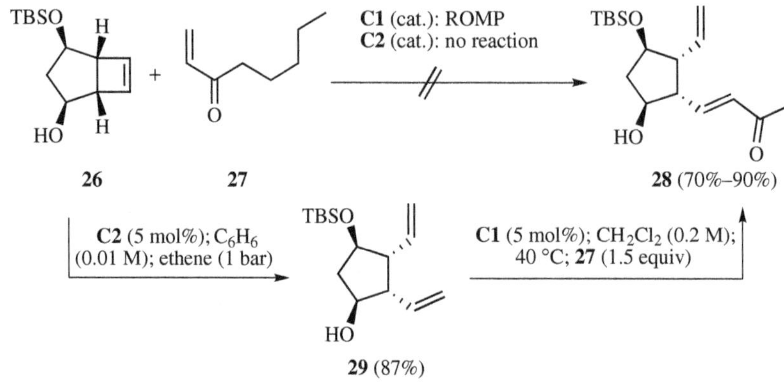

Scheme 5.8 Two-step sequence to isoprostane derivatives.

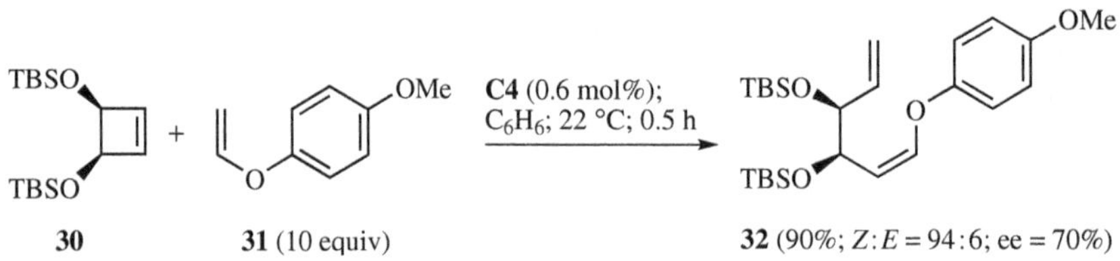

Scheme 5.9 Desymmetrization of meso-cyclobutenes with "stereogenic-at-metal"-Mo catalysts.

Asymmetric ROM/CM has also been applied to meso-cyclobutenes by using the "stereogenic-at-metal"-Mo catalysts introduced by Schrock and Hoveyda. A representative example is the reaction of **30** and an excess of vinyl ether **31** in the presence of Mo-catalyst **C4**. The formation of diene **32** proceeds with high *Z*-selectivity and 70% ee (Scheme 5.9) (17).

An example for the application of an ROM/CM–CM sequence to a strained medium sized ring has recently been published by Wulff et al., who used cyclononenone **33** in a bidirectional synthesis of spiroacetals. In their approach, **33** (obtained inter alia through an ROM/RCM sequence, see Section 5.2.4 for details) was treated with excess methyl vinyl ketone in the presence of **C1**, resulting not only in a ROM/CM but in an additional CM. Desymmetrization of the resulting *meso*-product **34** was accomplished via Sharpless' dihydroxylation, followed by spontaneous hemiacetalization/conjugate addition of the intermediate diol **35** to the spiroacetal **36** (Scheme 5.10) (40).

5.2.2 Cyclic Products with Pendant ω-Unsaturated Side Chains

As outlined above, release of ring strain is an important driving force for synthetically useful ROM/CM sequences. Although small and medium sized rings normally have a considerable ring strain due to their ring size, five- and six-membered rings are strained if they are part of a bicyclic system. This is, for instance, the case with norbornenes or 7-heteronorbornenes that have been extensively investigated in ROMP and ROM/CM sequences. This chemistry has been thoroughly reviewed by Plumet et al. (41), and we will therefore only discuss a few examples to illustrate the principle and some more recent contributions, in particular those dealing with enantioselective ROM/CM of strained bicyclic compounds.

ROM/CM of norbornene derivatives results in the formation of 1,3-*cis*-disubstituted five-membered rings, as illustrated for the reaction of **37** (the Diels–Alder adduct of cyclopentadiene and maleic anhydride) and styrene **38** to racemic cyclopentane **39** (42) (Scheme 5.11).

Synthesis of (3a*R*′,4*S*,6*R*,6a*S*,*E*)-6-(4-methoxystyryl)-4-vinyl-tetrahydro-3a*H*-cyclopenta[c]furan-1,3-dione (39) (42)

To a solution of **37** (295 mg, 1.8 mmol) and styrene **38** (482 mg, 3.6 mmol) in CH_2Cl_2 (25 ml) was added first-generation catalyst **C2** (75 mg, 5 mol%). The solution was stirred at ambient temperature for 24 h, all volatiles were evaporated, and the residue was purified by chromatography on silica using hexane/ethyl acetate (3 : 1) as an eluent to furnish the title compound **39** (327 mg, 61%) as a colorless solid.

An enantioselective version of this transformation was published shortly afterwards by Hoveyda et al., who used

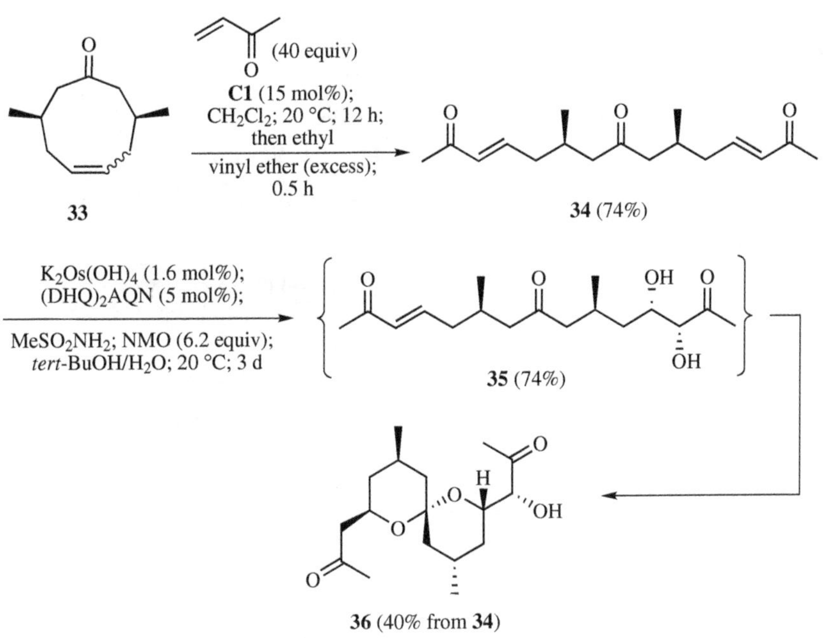

Scheme 5.10 Spiroacetals via ROM/CM–CM sequence and Sharpless' dihydroxylation.

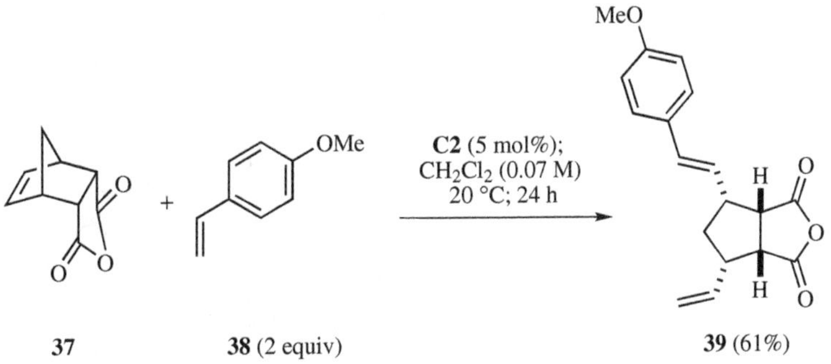

Scheme 5.11 ROM/CM of a 7-oxanorbornene derivative.

Mo-catalysts that are modified with chiral biphenolate ligands (43). By using this methodology, 1,2,3-trisubstituted cyclopentanes were synthesized in perfect diastereoselectivity and ee's up to 98% (44). For Ru-based metathesis catalysts, chiral information is mostly implemented via the NHC backbone and transferred to the metal center by the *N*-aryl groups attached to the NHC ring in 1- and 3-position. For quite some time, these catalysts were inferior to the chiral Mo-catalysts with respect to enantioselectivity, but recently significant progress was achieved. For example, Grubbs et al. reported the enantioselective ROM/CM of **40** with styrene. In the presence of a catalytic amount of **C5**, **41** was obtained in short reaction times and quantitative yield but as a mixture of *E*/*Z*-isomers. Interestingly, these isomers are formed in significantly different enantioselectivities (18). Recently, Blechert and coworkers introduced a novel type of Ru-precatalysts such as **C6**, in which the NHC ligand is annulated to a tetrahydroquinoline system. For the same reaction, the ROM/CM of **40** and styrene, **C6** appears to be less reactive, but both *E*/*Z*-selectivity and enantioselectivity are significantly better, which has been attributed to a restricted rotation around the *N*-aryl bond (Scheme 5.12).

Oxabicycles derived from a [4+3] addition to furans are also useful substrates for ROM/CM sequences. These reactions open up a rapid diastereoselective and—if chiral metathesis catalysts are used—enantioselective access to 2,6-disubstituted dihydropyrans, a common structural pattern in many natural products. In the following example, the "stereogenic-at-metal"-Mo catalyst **C4** was used to synthesize **43** from the meso-oxabicyclic compound **42** and vinyl ether **31** in high yield, high *E*/*Z*-selectivity, and high enantioselectivity (Scheme 5.13) (17).

5.2.3 Unsaturated Monocyclic Products

Unsaturated monocyclic products with a pendant ω-unsaturated side chain may result from domino metathesis

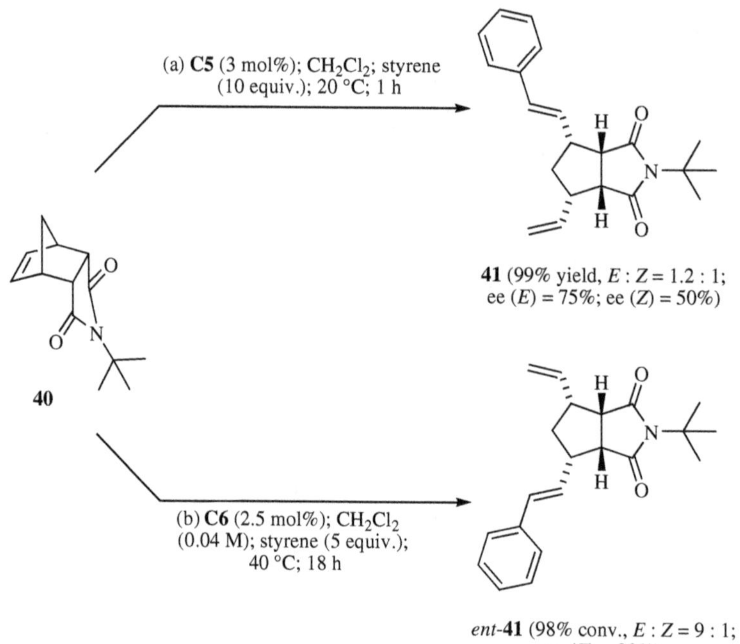

40

41 (99% yield, $E:Z = 1.2:1$; ee (E) = 75%; ee (Z) = 50%)

ent-**41** (98% conv., $E:Z = 9:1$; ee (E) = 92%)

Scheme 5.12 Enantioselective ROM/CM of norbornene derivatives with chiral Ru-catalysts.

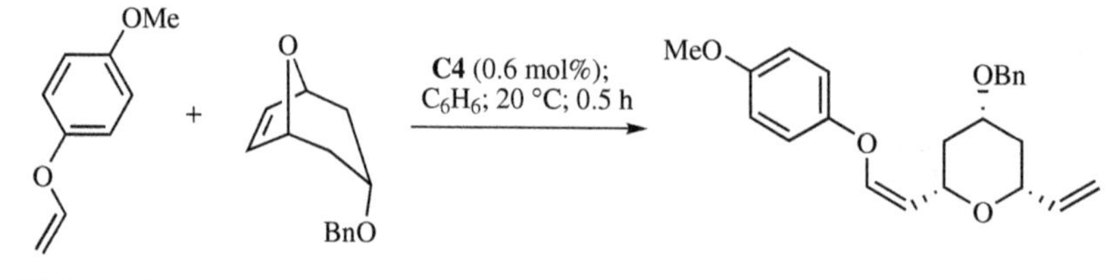

31 (1.1 equiv) **42** **43** (80%; $Z:E > 98:2$; ee = 94%)

Scheme 5.13 ROM/CM of oxabicyclic precursors.

sequences if another cycloolefin with an attached alkenyl substituent is treated with a suitable metathesis catalyst. The reaction resembles skeleton rearrangement reactions, and hence the term ring rearrangement metathesis (RRM) has been coined to appropriately describe this type of transformations (45). Generally to proceed efficiently, an RRM sequence requires a driving force that may come from one of the following sources:

a. Release of ring strain. This is typically the case for the reactions discussed in this section, if a small ring (cyclopropene or cyclobutene) undergoes a rearrangement to a five- or six-membered cycle.
b. Release of strain caused by steric interactions of substituents. Several examples have been reported where a five-membered ring undergoes rearrangement to a different five-membered ring efficiently. In these cases, release of ring strain seems to be a less important contributor to the driving force of the reaction. Instead, unfavorable interactions of proximal substituents in the starting cycloalkene may be reduced if these substituents are located at a side chain after the rearrangement.
c. Release of ethylene or another volatile metathesis product. An RRM may be terminated by a CM reaction with an α-olefin added to the reaction mixture. In these cases, a volatile metathesis product is formed that can shift the equilibrium in the desired direction.

In a synthesis of the natural product deoxypukalide, Pattenden et al. exploited an ROM/RCM/CM sequence of a 1-cyclobutene carboxylate **44**, which underwent RRM and finally CM with 2-methyl-2-propene-1-ol (**45**) in the presence of second-generation Grubbs' catalyst **C7** (Scheme 5.14) (46). Presumably, the sequence proceeds via initiation at the sterically least hindered double bond to give the carbene complex **44a**, which then reacts with the cyclobutene ring in

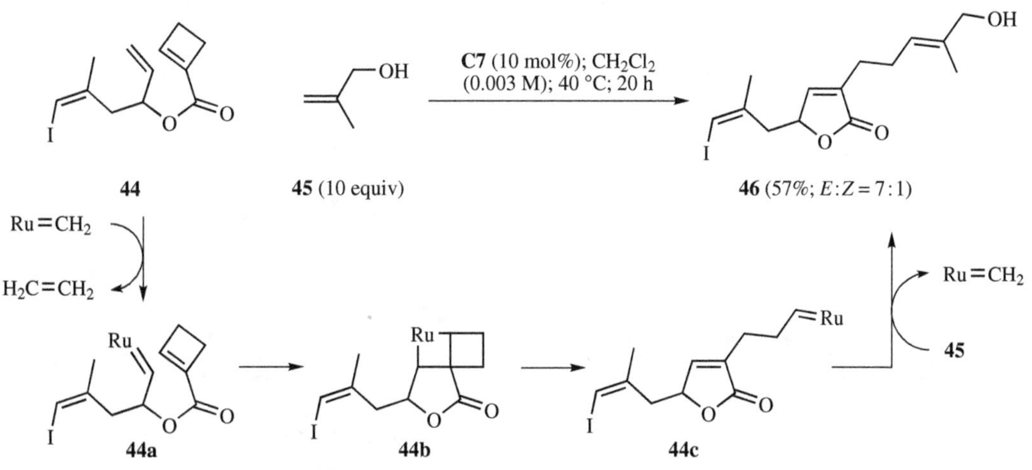

Scheme 5.14 Toward deoxypukalide via ROM/RCM/CM of a cyclobutene carboxylate.

an ROM/RCM sequence to give—via the ruthenacyclobutane intermediate **44b**—the Ru-carbene complex **44c**. Termination of the sequence occurs by CM of **49** and alkene **45**, with regeneration of the Ru-methylidene catalyst and liberation of the product **46**.

A related approach has been pursued by Tadano et al. in synthetic studies toward clavilactone A (47), and White and Snapper combined an ROM/RCM sequence of cyclobutenes with an anion-induced oxy-Cope rearrangement to access medium sized rings (48). Cyclopropenes with a pendant unsaturated side chain have also been used in ROM/RCM sequences (Scheme 5.15) (49).

At least for some of these substrates it appears much more likely that initiation of the metathesis sequence occurs by ROM of the cyclopropene, because high conversion and synthetically useful yields have also been obtained in those cases where the pendant olefinic side chain is less reactive. For example, cyclopropene **47** with the unsaturated side chain being located at the 3-position undergoes RRM to the lactone **48** presumably via Ru-carbene **49a**, rather than **49b** (49).

An ROM/RCM rearrangement of a cyclopentenone **50** has been combined with a terminating CM using a styrene **51** en route to the piperidine alkaloid (−)-lasubine II (50). The first-generation catalyst **C2** turned out to be ineffective in this reaction, but both second-generation catalysts **C1** and **C7** promoted the formation of piperidine derivative **52** in comparable yield. Acid-mediated cleavage of the Boc group in **52** induced a conjugate addition, resulting in the formation of **53** and its epimer. Under basic conditions, epimerization to the desired product **53** was accomplished, which was converted to (−)-lasubine II via hydrogenation of the C–C double bond and diastereoselective reduction of the carbonyl group (Scheme 5.16). Several natural products, in particular alkaloids, have been synthesized along these lines and a representative collection of other examples has been included in reviews (51).

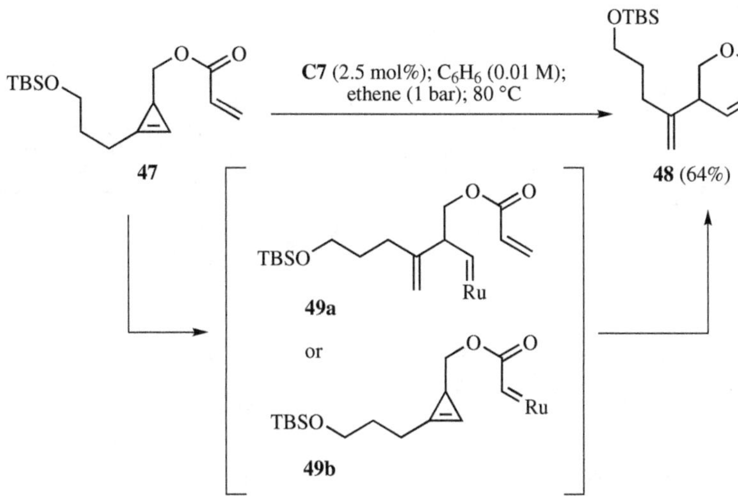

Scheme 5.15 ROM/RCM of cyclopropenes.

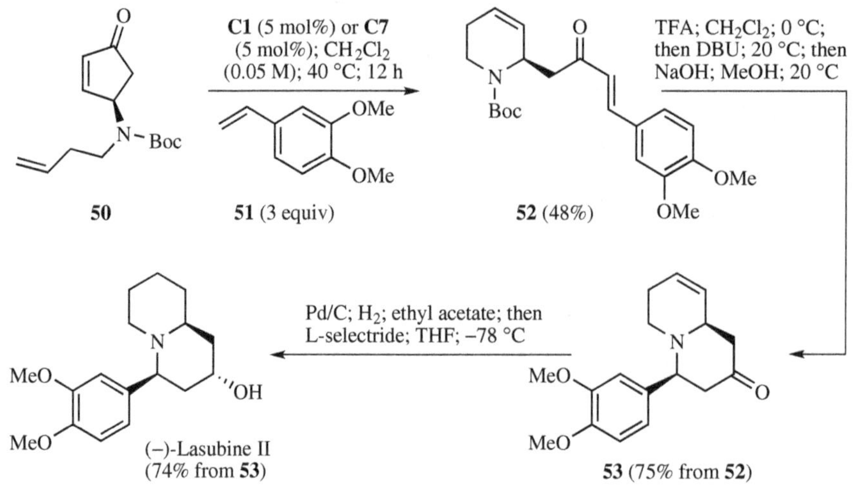

Scheme 5.16 ROM/RCM/CM route to (−)-lasubine II.

An RRM sequence that proceeds without a change of the ring size has been used for the total synthesis of the alkaloid (+)-*trans*-195A (52). Enantiomerically pure homoallyl amine **54** and cyclohexenol **55** were coupled via Mitsunobu inversion to yield the RRM precursor **56**, which underwent ROM/RCM in the presence of first-generation catalyst **C2** under an atmosphere of ethene. The *cis*-configured dehydropiperidine **57** was obtained in high yield and high optical purity as a single diastereomer, which was then converted in few steps, including a diastereoselective Zr-promoted Negishi cyclization, to the target molecule (+)-*trans*-195A (Scheme 5.17).

In the previous example, the *cis*-configuration of the piperidine derivative obtained through RRM results from previous enantioselective steps and the highly diastereoselective Mitsunobu inversion. During the actual RRM sequence, no stereocenter is generated or inverted. An attractive extension to this strategy is stereoselective RRM reactions, in which either diastereotopic or enantiotopic halves of a cycloalkene are differentiated by the metathesis event. The rearrangement of cyclopentene **58** is an illustrative example: upon treatment with first-generation Hoveyda's catalyst **C8** under an atmosphere of ethene, the bicyclic piperidine derivative **59** is formed in high yield and as a 5 : 1 ratio of *trans*-isomer and *cis*-isomer (Scheme 5.18).

Synthesis of 5-allyl-5,6-dihydro-1*H*-oxazolo[3,4-a] pyridin-3(8a*H*)-one (59) (53)

A solution of **58** (180 mg, 1.0 mmol) in CH_2Cl_2 (50 ml) was saturated with ethene. Under an atmosphere of ethene, catalyst **C8** (12 mg, 2 mol%) was added, and the mixture was stirred at an ambient temperature until the starting material was fully consumed. The reaction was quenched by addition of ethyl vinyl ether (1 ml), and the mixture was stirred for an additional 0.5 h. After evaporation of all volatiles, the residue was purified by

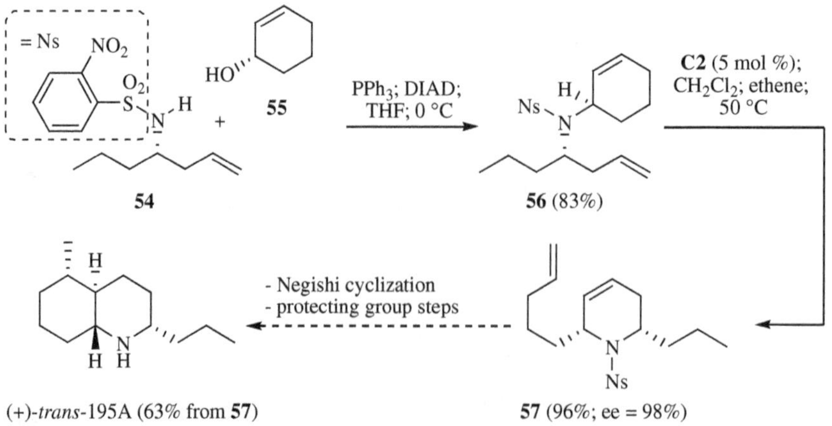

Scheme 5.17 ROM/RCM sequence toward (+)-*trans*-195A.

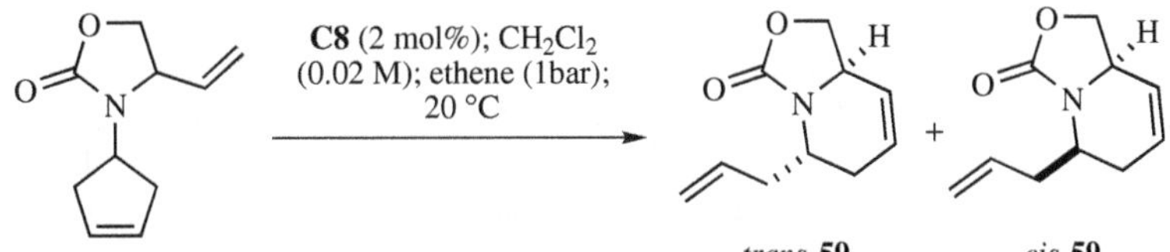

58

trans-**59** *cis*-**59** (89%; *trans* : *cis* = 5 : 1)

Scheme 5.18 Diastereoselective RRM of a cyclopentene.

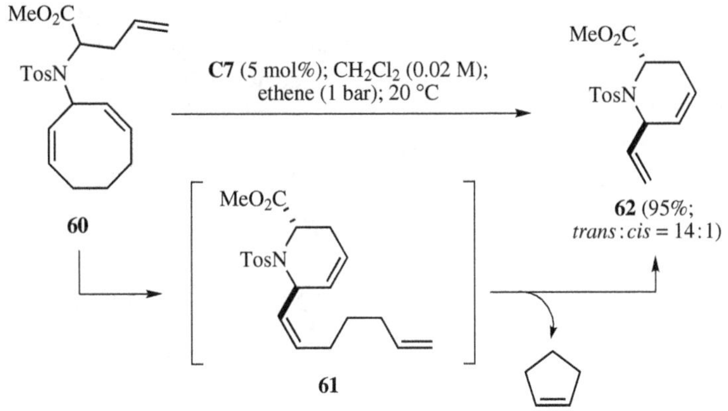

Scheme 5.19 Diastereoselective RRM and subsequent backbite metathesis.

column chromatography on silica, using hexane/ether (2 : 1) mixture as an eluent, to give the title compound **59** (160 mg, 89%) as a colorless oil in a 5 : 1 ratio of diastereomers.

The diastereoselective RRM of cyclooctadiene **60** proceeds with a consecutive backbite metathesis reaction when the reaction is run under an atmosphere of ethene and in the presence of second-generation catalyst **C7**. This results in a truncation of the side chain with liberation of cyclopentene and formation of the 6-vinyl piperidine derivative **62**, rather than the expected product **61**. When **C7** was replaced by the less active first-generation catalyst **C2**, mixtures of **61** and **62** were obtained (Scheme 5.19).

An illustrative example for enantioselective RRM is the synthesis of dihydropyran **64** via desymmetrization of *meso*-cyclopentene **63**, which was catalyzed by a chiral Mo-catalyst **C9** (Scheme 5.20) (54). The enantiomerically enriched dihydropyran **64** was converted in few steps to the lactone fragment **65** of the antiviral agent tipranavir.

5.2.4 Spirobicyclic Products

If an unsaturated pendant side chain is located at the bridgehead of a bicycloalkene or heteroalkene, RRM leads to

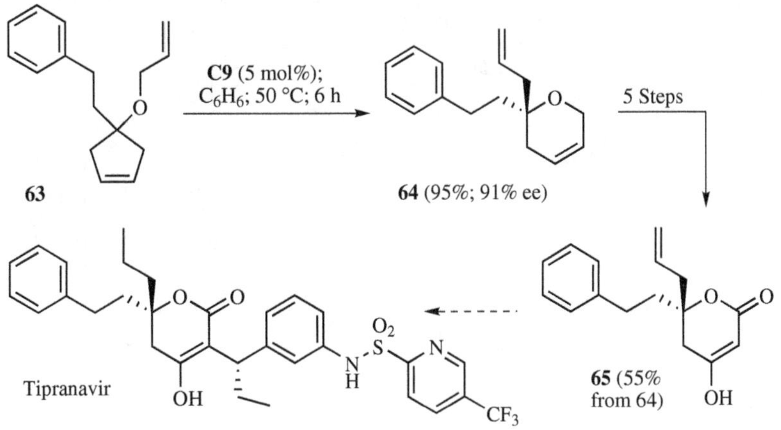

Scheme 5.20 Toward tipranavir via enantioselective RRM.

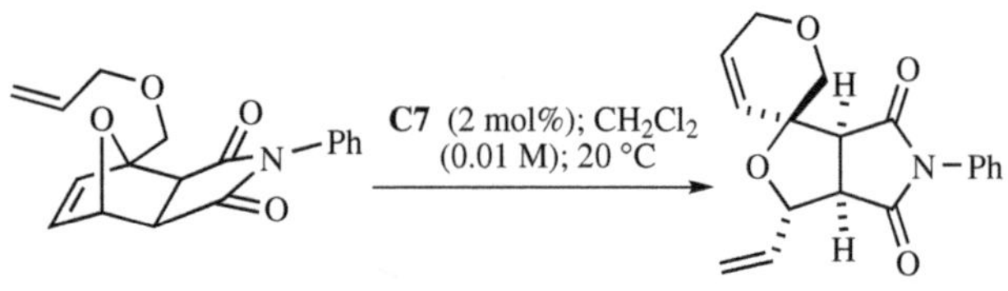

66 **67** (95%)

Scheme 5.21 Formation of spirocycles from furan [4 + 2] cycloadducts.

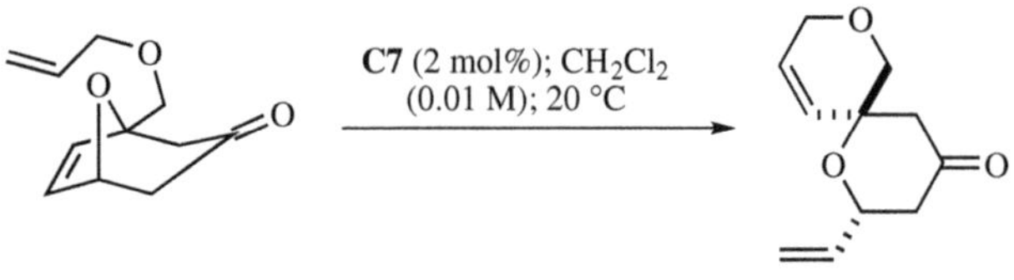

68 **69** (80%)

Scheme 5.22 Formation of spirocycles from furan [4 + 3] cycloadducts.

spirocycles. This type of transformation is rare compared to those cases where the ω-unsaturated side chain is attached to one of the bridges of the bicyclic RRM precursor; in these cases, which are discussed in Section 5.2.5, annulated bicyclic products result from RRM.

Wright et al. investigated the formation of various spirocycles from appropriately functionalized oxabicyclic starting materials, in particular [4 + 2] and [4 + 3] cycloaddition products of furans. An example for the application of Diels–Alder adducts of furans is the rearrangement of **66** to product **67** in the presence of second-generation catalyst **C7**. Apparently, no ethene atmosphere is required for this reaction (Scheme 5.21). Identical conditions were used by the same authors for the RRM of a [4 + 3] cycloadduct **68**, derived from a 2-substituted furan, to the oxaspirocyclic product **69** (Scheme 5.22) (55).

More recently, Wulff et al. nicely combined an ROM/RCM with a tethered olefin metathesis reaction and an anionic oxy-Cope rearrangement to access a cyclononenone required for a bidirectional ROM/CM sequence. The latter step has been discussed Section 5.2.1, Scheme 5.10. Starting from **70**, the spirocyclic siloxane **71** was synthesized using a remarkably small amount of second-generation catalyst **C7**. Cleavage of the silyl tether was accomplished with TBAF and *Z*/*E*-isomerization of the electron-deficient double bond by treatment with DBU, leading to cyclopentanol **72**. Treatment of this intermediate with DBU at more forcing conditions resulted in a base-induced oxy-Cope rearrangement with concomitant ring enlargement. Decarboxylation and photochemical *E*/*Z*-isomerization furnished cyclononenone **33** (Scheme 5.23) (40).

5.2.5 Annulated Bicyclic Products

Bicycloalkenes or heterosubstituted derivatives with an ω-unsaturated side chain located at the position adjacent to the bridgehead are probably the most common, and with respect to target molecule applications, they are the most useful starting materials for RRM sequences. These reactions lead to the formation of annulated bicyclic or polycyclic products and numerous examples have been covered in comprehensive reviews (41,45,51).

The total synthesis of the alkaloid (+)-lepadin B is an illustrative example (Scheme 5.24). The required 2-azabicyclo[2.2.2]octene **76** was obtained from pyridine through auxiliary controlled stereoselective addition of a methyl Grignard reagent, followed by endo-selective Diels–Alder addition of the resulting diene **74** and methyl acrylate. The bicyclic product **75** was converted into the ROM/RCM precursor **76** in four steps, which underwent RRM to product **77**. From **77**, (+)-lepadin B, the enantiomer

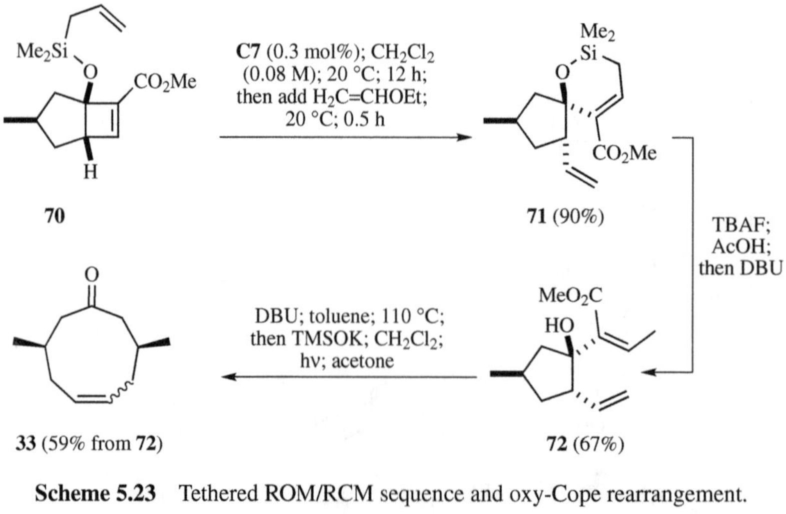

Scheme 5.23 Tethered ROM/RCM sequence and oxy-Cope rearrangement.

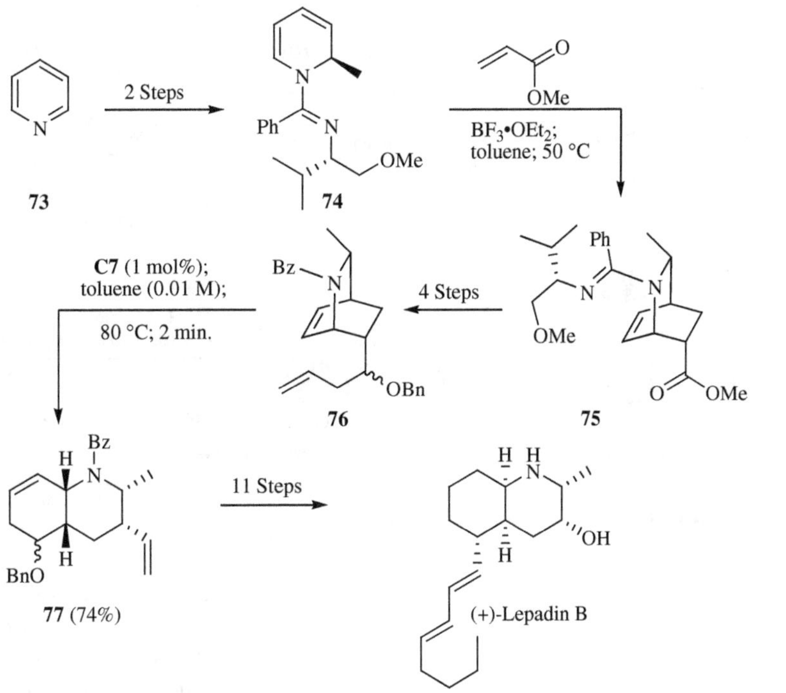

Scheme 5.24 Total synthesis of (+)-lepadin B via RRM of a pyridine-derived cycloadduct.

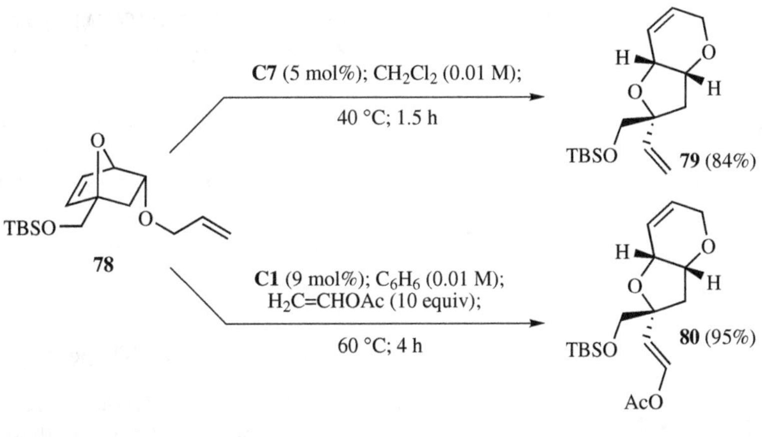

Scheme 5.25 Toward dysiherbaine via RRM of a furan-derived Diels–Alder adduct.

of the naturally occurring alkaloid, was accessible in 11 steps (56).

The following example for the RRM of bicyclo[2.2.1] derivatives to the annulated bicyclic products has been reported in the course of a formal total synthesis of the excitatory amino acid dysiherbaine (Scheme 5.25) (57).

An ROM/RCM sequence of 7-oxanorbornene derivative **78** to rearrangement product **79** was accomplished in good yield using second-generation catalyst **C7** in the absence of an ethene atmosphere. Extension of the sequence by a CM step, using vinyl acetate as a CM partner, is possible and yields the masked aldehyde **80**, but requires a higher catalyst loading of **C1**, elevated temperatures, and longer reaction times.

A recent study directed at the synthesis of the sesquiterpene anthecularin revealed that RRM reactions do not necessarily proceed in the intended way, but that additional steric constraints and the formation of unfavorable ring sizes can prevent the formation of the desired product and lead to ring-contracted products. For example, the bicyclo[2.2.2]octene **81** reacts in the presence of second-generation catalyst **C7** to the decaline derivative **83** rather than the expected product **84**. This result can be explained with a double-bond isomerization, which is often

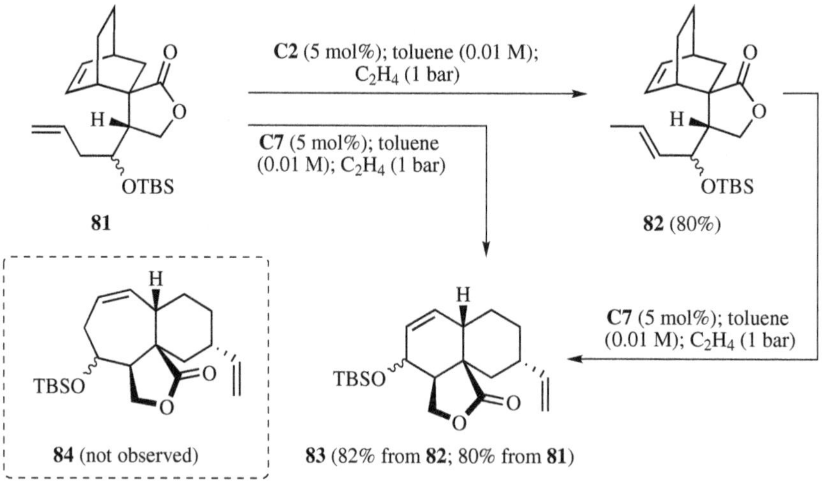

Scheme 5.26 Unexpected formation of a decaline product rather than a 6/7 fused product.

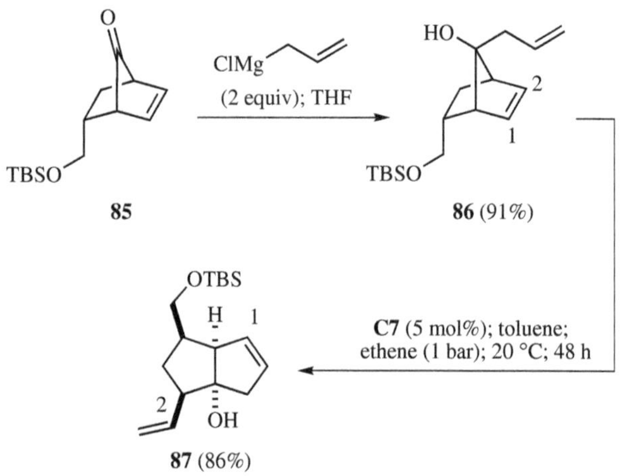

Scheme 5.27 RRM of norbornene derivatives substituted at the 7-position.

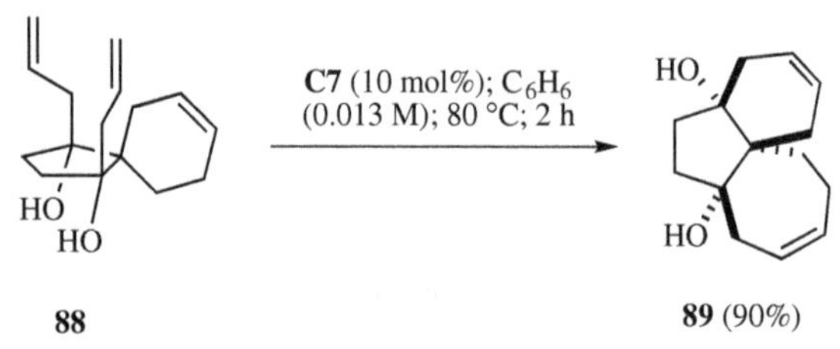

Scheme 5.28 RRM of a spirocyclic precursor.

observed as a side reaction if the intended olefin metathesis step is slow (58). RRM of the resulting internal alkene **82** is obviously much more facile, presumably due to reduced steric constraints, and gives **83**. Indeed, with first-generation catalyst **C2**, only isomerized product **82** was observed that underwent RRM to **83** upon treatment with **C7** (Scheme 5.26) (59).

In most cases where RRM is applied to norbornene derivatives, the unsaturated side chain is located at the ethano bridge. An interesting example for the RRM of a derivative with the unsaturated side chain at the methano bridge, that is, at position 7, has recently been published (Scheme 5.27) (60a). The RRM precursor **86** was obtained as a single diastereomer by the addition of allyl magnesium chloride to ketone **85**. ROM/RCM of **86** proceeded with very high regioselectivity to **87**. The carbon atom 1 is preferentially engaged in the formation of the annulated cyclopentene.

Apart from bicyclo[2.2.1]- and bicyclo[2.2.2]-alkenes, spirocyclic alkenes with ω-unsaturated side chains located in the α-position to the spiro atom may also undergo RRM to annulated bicyclic or polycyclic compounds, as illustrated for the reaction of **88** to **89** in the presence of **C7** (Scheme 5.28) (60b).

5.2.6 "Dumbbell"-Type Bicyclic or Polycyclic Products

"Dumbbell"-type products are characterized by two rings formed through olefin metathesis, which are connected by a tether. The tether can be a single bond, an alkyl chain, or an aromatic ring. Typical starting materials for dumbbell-type products are cycloalkenes with two ω-unsaturated side chains located in both allylic positions of the endocyclic double bond participating in the ROM step of the sequence. Blechert and coworkers have applied this type of RRM to many total syntheses of natural products, in particular alkaloids, and instructive examples have been included in reviews (61,62).

An instructive example to illustrate the principle is the reaction of cyclobutene **92** to the tetra-oxacyclic system **93**, a structural pattern commonly found in marine polyether toxins (Scheme 5.29) (63). The synthesis of the precursor was accomplished via a highly diastereoselective S_N2' reaction of tetrahydropyran-3-ol **90** and *cis*-dichlorocyclobutene (**91**). In the presence of second-generation catalyst **C7**, **92**

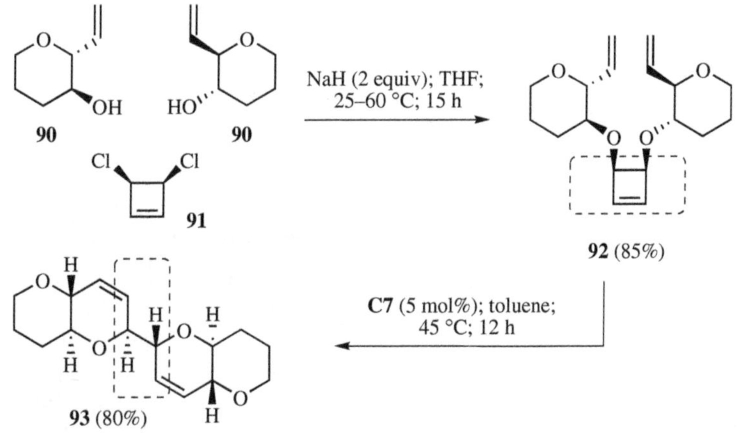

Scheme 5.29 RCM/ROM/RCM sequence applied to disubstituted cyclobutenes.

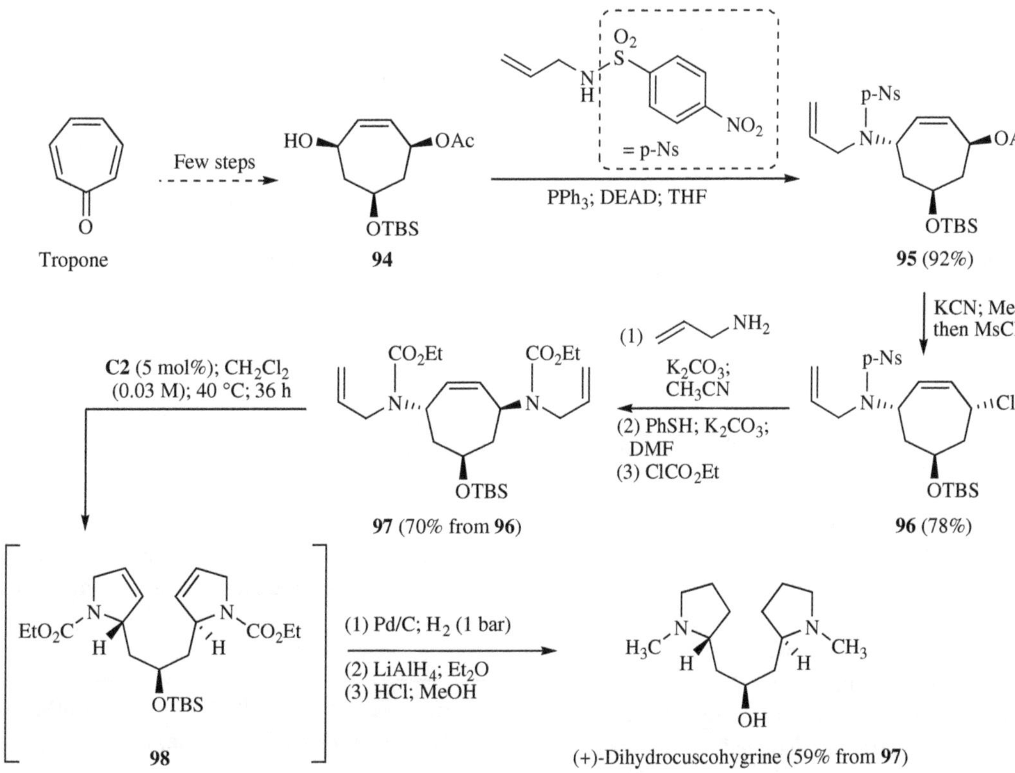

Scheme 5.30 (+)-Dihydrocuscohygrine and cuscohygrine via RCM/ROM/RCM.

undergoes rearrangement to **93** through an RCM/ROM/RCM sequence.

In the example described in Scheme 5.29, a dumbbell-type product is formed in which the two ring systems resulting from the metathesis reaction are tethered by a single bond. To illustrate how an alkyl chain can be incorporated as a tether, we discuss the total synthesis of the *Solanaceae* alkaloid (+)-dihydrocuscohygrine (Scheme 5.30) (64). In this example, the central cycloalkene to be used for the ROM step is a cycloheptene **94** derived from tropone. The two pendant ω-unsaturated side chains are attached via Mitsunobu inversion and S_N2 reaction, respectively, and the actual RRM precursor **97** was eventually obtained after protecting group operations. RCM/ROM/RCM of **97** yields **98**, in which two dihydropyrroles are tethered by a three-carbon chain. This product is converted, without isolation, in three steps to the target compound (+)-dihydrocuscohygrine.

A tether does not necessarily need to be an all-carbon chain, and the RRM sequences discussed in this section are not restricted to the formation of just two rings by olefin metathesis, although the number of examples describing RRM cascades leading to the formation of

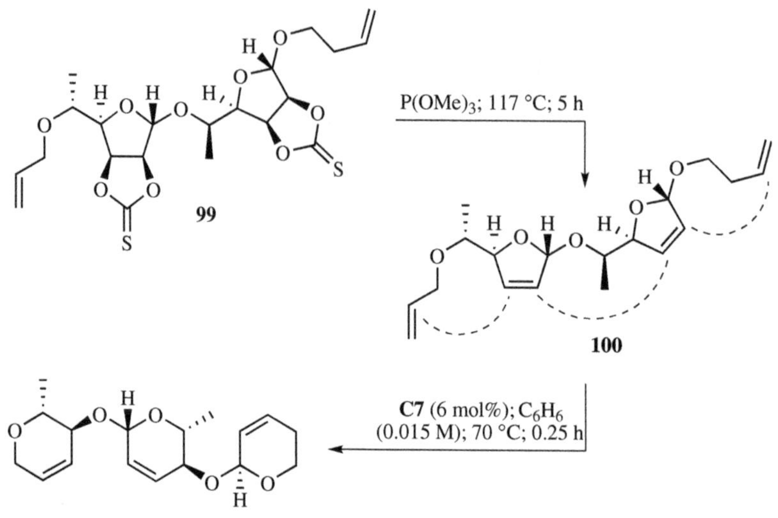

99

100

101 (42% from 99)

Scheme 5.31 A trisaccharide derivative via RCM/ROM/RCM/ROM/RCM sequence.

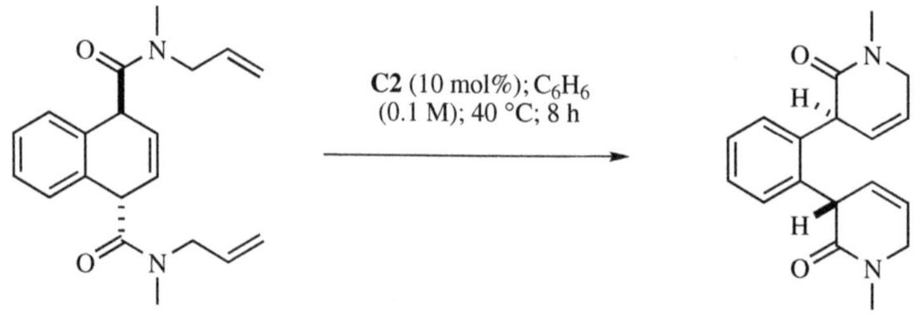

102

103 (95%)

Scheme 5.32 Dumbbell-type product with an *o*-disubstituted benzene ring as a tether.

three or more rings is rare. One of these examples is an interesting synthesis of a trisaccharide derivative via an RCM/ROM/RCM/ROM/RCM sequence (Scheme 5.31) (65). The required RRM precursor **100** was obtained from thiocarbonate **99** (which, in turn, was synthesized from D-mannose in several steps) via Corey–Winter fragmentation and subjected to second-generation catalyst **C7** without purification. The product **101** was isolated in 42% overall yield, and the olefinic carbons to be connected during the RRM sequence are indicated by dashed lines in the molecular formula of **100**.

To conclude this section on the formation of dumbbell-type products through RCM/ROM/RCM sequences, an example for an *ortho*-disubstituted benzene as a tether is given in Scheme 5.32 (66). The required RRM precursor is a *trans*-configured dihydronaphthalene **102**, which rearranges in the presence of a comparatively high loading of first-generation catalyst **C2** to the bis-lactam **103** in nearly quantitative yield. This is one of the earliest examples for domino ROM/RCM sequences, which was published by Grubbs et al. several years before the discovery of the more active second-generation catalyst **C7**. One might assume, although this has not been investigated, that the more reactive NHC-ligated precatalysts available today would allow a significant reduction of the amount of catalyst required.

Synthesis of *trans*-3,3′-(1,2-phenylene)bis(1-methyl-1,6-dihydropyridin-2(3*H*)-one) (103): (66)

To a solution of **102** (130 mg, 0.4 mmol) in dry and degassed benzene (4 ml) was added first-generation catalyst **C2** (33 mg, 10 mol%). The solution was heated to 40 °C for 8 h, cooled to ambient temperature, and then evaporated. The residue was purified by column chromatography on silica using hexane/ether mixture as an eluent, to furnish the title compound **103** (110 mg, 95%) as an oil.

5.2.7 Sequences Involving Triple Bonds

Metathesis reactions of enynes yield conjugated dienes. These reactions are covered in detail in Chapter 4, and

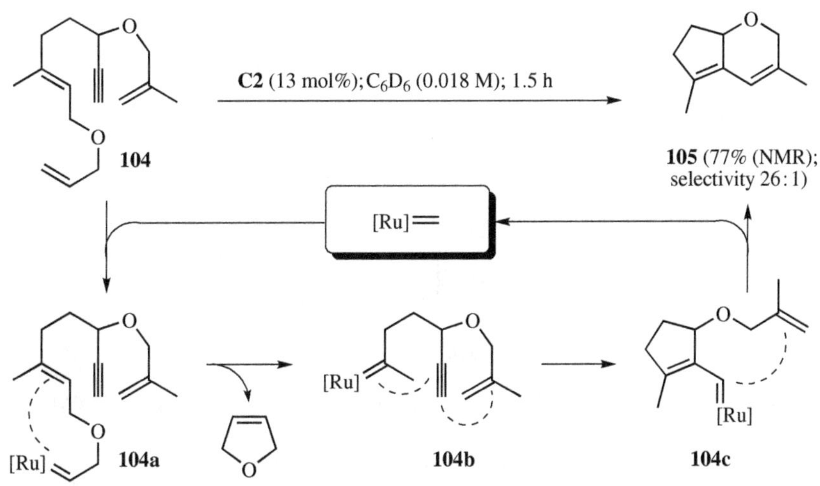

Scheme 5.33 Domino ring-closing enyne metathesis sequence with a relay.

we will therefore restrict the discussion to two selected examples, to illustrate how an enyne metathesis step can be combined with other metathesis steps to a domino sequence. Several instructive examples have been included in specialized reviews (51,61,67).

The precursor **104** has been equipped with a pendant allyloxymethyl side chain to direct the initiation to a preferred site (Scheme 5.33) (68). This concept has been introduced as a relay RCM by Hoye et al. and facilitates the formation of tetrasubstituted double bonds or directs, as in the present case, the metathesis catalyst to a certain position (69). In this case, **104** cyclizes in high selectivity to the bicyclic product **105** via initiation at the least hindered position, giving carbene complex **104a**, which then reacts via "backbite" metathesis with liberation of 2,5-dihydrofuran to **104b**. This carbene complex attacks the central triple bond with concomitant cyclization of the cyclopentene ring and the formation of a new Ru-carbene **104c**, which finally undergoes RCM to yield **105**.

If the allyloxymethyl substituent is attached to the other C–C double bond, as in precursor **106**, initiation is directed to this position and the isomeric product **107** is formed in even higher selectivity through an analogous sequence

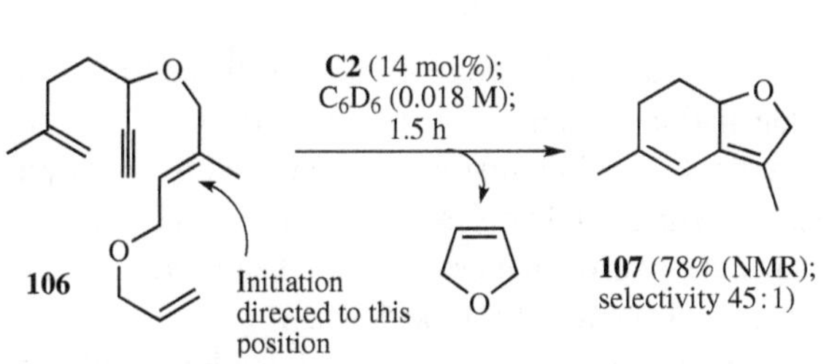

Scheme 5.34 Alternative cyclization mode using the relay strategy.

of steps (Scheme 5.34). In the absence of the directing allyloxymethyl substituent, first-generation catalyst **C2** was found to be ineffective and second-generation catalyst **C7** resulted in the formation of both isomers **105** and **107** with virtually no selectivity (68).

The reaction of norbornene derivative **108** to the tricyclic product **109** illustrates the combination of an ROM step with a ring closing enyne metathesis (RCEYM) step (Scheme 5.35) (70). This transformation requires the presence of ethene, in the absence of ethene, unreacted starting material was recovered or—at higher substrate concentrations—ring opening metathesis polymerization occurred. A plausible mechanism involves initial ROM/CM of the norbornene system to compound **110**, with ethene being the CM partner. Cyclopentane **110** may then undergo double RCEYM to the tricyclic product **109** (70). It should be noted that the question whether initiation of enyne metathesis reactions occurs preferentially at the alkyne or at the alkene has been discussed controversially, and an alternative mechanism starting with an intermolecular enyne metathesis step might also be possible.

5.3 METATHESIS/NON-METATHESIS REACTION SEQUENCES

In this section, sequences consisting of metathesis steps and subsequent non-metathesis transformations are discussed, in which the non-metathesis reaction is catalyzed by a metal complex resulting from a triggered degradation of the original metathesis catalyst. In terms of Fogg's and dos Santos' taxonomy, the reaction sequences discussed in this section can be described as "assisted tandem catalytic" sequences. The section is structured according to the type of non-metathesis transformation.

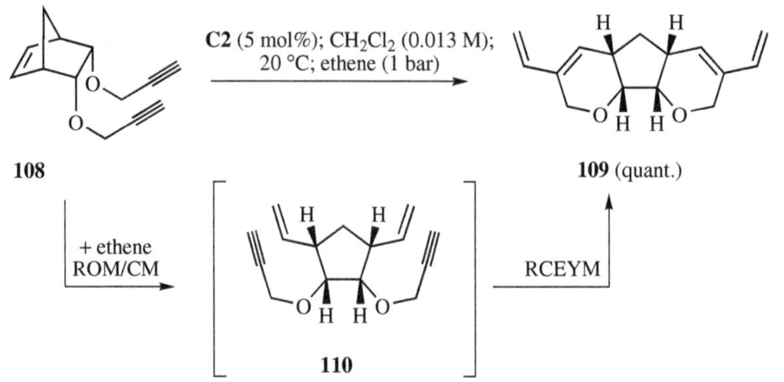

Scheme 5.35 ROM/CM followed by ring-closing enyne metathesis (RCEYM).

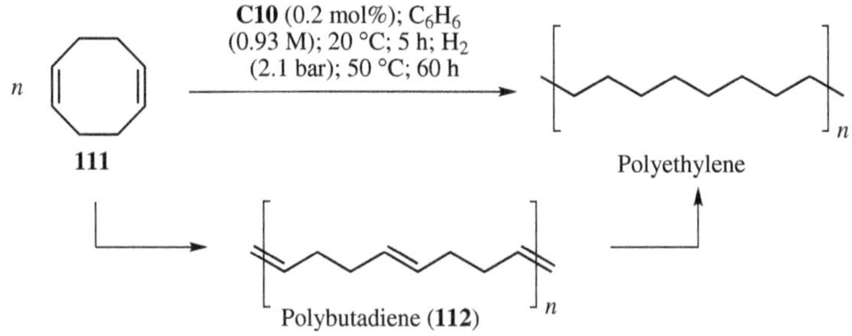

Scheme 5.36 Assisted tandem ROMP/hydrogenation of COD.

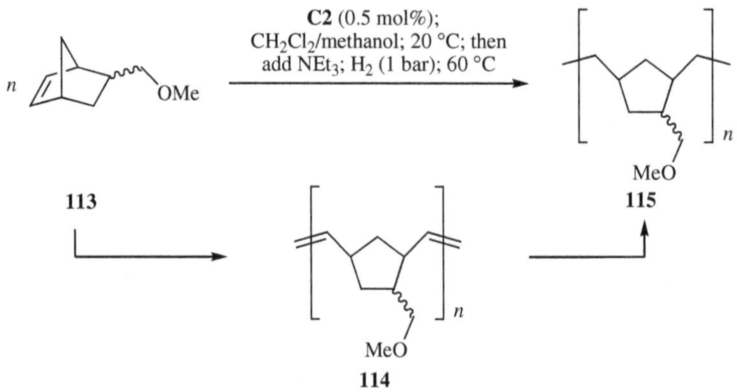

Scheme 5.37 Assisted tandem ROMP/hydrogenation of norbornene derivative **113**.

5.3.1 Metathesis/Hydrogenation

Historically, the olefin metathesis/hydrogenation sequence was the first assisted tandem catalytic metathesis /non-metathesis sequence to be discovered. In the first examples, polyolefins (inter alia some, which are not accessible through Ziegler–Natta-type olefin polymerization) were obtained in a polymer analogous reaction by combining ROMP or acyclic diene metathesis polymerization (ADMET) with hydrogenation. For instance, Dias and Grubbs reported that polyethylene with less than 5% of C–C double bonds results from ROMP of cyclooctadiene (**111**) in the presence of the bimetallic Ru catalyst **C10**, if—after completion of the polymerization—the reaction vessel is pressurized with about 2.1 bar of hydrogen (Scheme 5.36) (23).

Shortly afterwards, Fogg et al. investigated the ROMP/hydrogenation sequence for cyclooctene and a norbornene derivative **113** using first-generation Grubbs' catalyst **C2** (71). By using methanol as a co-solvent and NEt_3 as an additive, the hydrogen pressure could be lowered to 1 bar for complete conversion (Scheme 5.37).

The individual steps of assisted tandem catalytic sequences can proceed with different efficiency, depending on the precatalyst chosen. For instance, sterically demanding

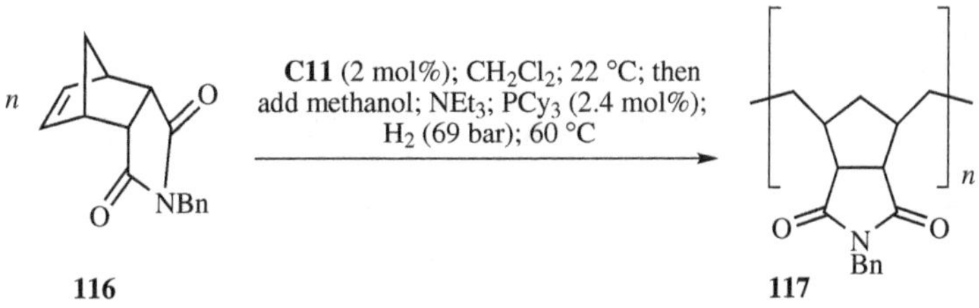

Scheme 5.38 Assisted tandem ROMP/hydrogenation of sterically demanding norbornenes.

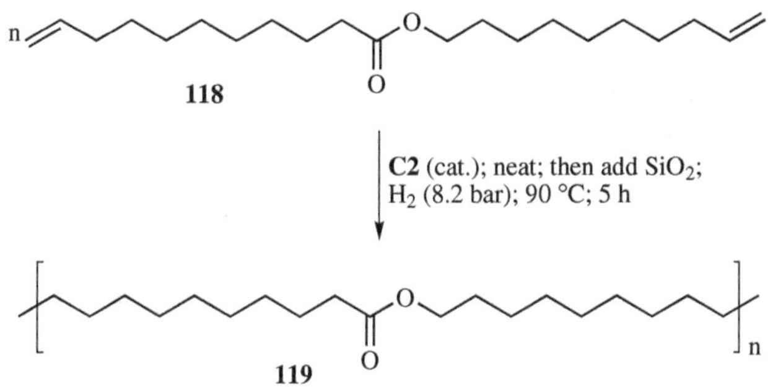

Scheme 5.39 Assisted tandem ADMET/hydrogenation.

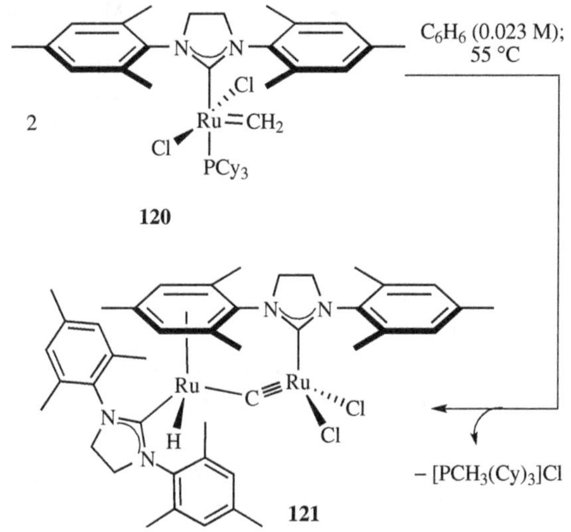

Scheme 5.40 Thermal bimolecular decomposition of Ru-NHC-methylidene complex **120**.

norbornene derivatives such as **116** undergo ROMP only slowly with first- or second-generation catalyst **C2** or **C7**, respectively, but the efficiency of the hydrogenation catalysts generated *in situ* from these precatalysts is very high. On the other hand, the phosphine-free, third-generation catalyst **C11** promotes ROMP of **116** very efficiently, but hydrogenation ceases at 40% after 24 h. This problem has been attributed to insufficient stability of the hydrogenation catalyst, and addition of PCy_3 as a stabilizing ligand prior to the hydrogenation was found to improve the activity in hydrogenation to the level observed with precatalysts **C2** or **C7** (Scheme 5.38) (72).

ADMET has also been combined with hydrogenation. In the following example, the catalyst was adsorbed on silica after completed metathesis polymerization of **118**, followed by addition of hydrogen to yield the polyester **119** (Scheme 5.39) (73).

Shortly after the discovery of the first metathesis polymerization/hydrogenation sequences, it had been suspected that these transformations rely on the *in situ* conversion of metathesis-active alkylidene complexes to hydrogenation-active Ru-hydride complexes. Indeed, several Ru-hydrides could be identified as products resulting from hydrogenolysis (74,75) and also from methanolysis (76,77), or thermal degradation of Fischer-type carbene complexes (78,79). Interestingly, the conversion of Ru-carbene complexes into Ru-hydrides is not one way: (80,81) by adding alkynes to Ru-hydrides previously generated through hydrogenolysis of first-generation catalyst **C2**, metathesis-active Ru-carbenes could be regenerated, allowing a cycling between metathesis and hydrogenation reactions (71). A more detailed discussion of these important contributions can be found in the review on catalysis at the Ru-carbene and Ru-hydride interface (82), and so the present discussion is restricted to two more recent investigations into Ru-carbene to Ru-hydride pathways. Shortly after publication of our review, Grubbs et al. reported that the propagating species resulting from second-generation catalyst **C7**, the methylidene complex **120**, dimerizes upon heating to a binuclear Ru-hydride complex **121** with a carbyne bridge between the two Ru centers (Scheme 5.40) (83a). The structure assigned to **121** was confirmed by single-crystal X-ray structure analysis, and a plausible mechanism for its formation was presented. In addition, it was demonstrated that **121** efficiently promotes double-bond isomerization (83b), which is often an undesired side reaction in olefin metathesis reaction, but may also become useful if combined with metathesis in an assisted tandem catalytic sequence (see Section 5.3.2 for details).

Very recently, Fogg et al. isolated and characterized hitherto undetected Ru-methoxyhydride species resulting from **C2** or **C7**, respectively, by methanolysis under basic conditions (84). The proposed mechanism is outlined for second-generation catalyst **C7** in Scheme 5.41.

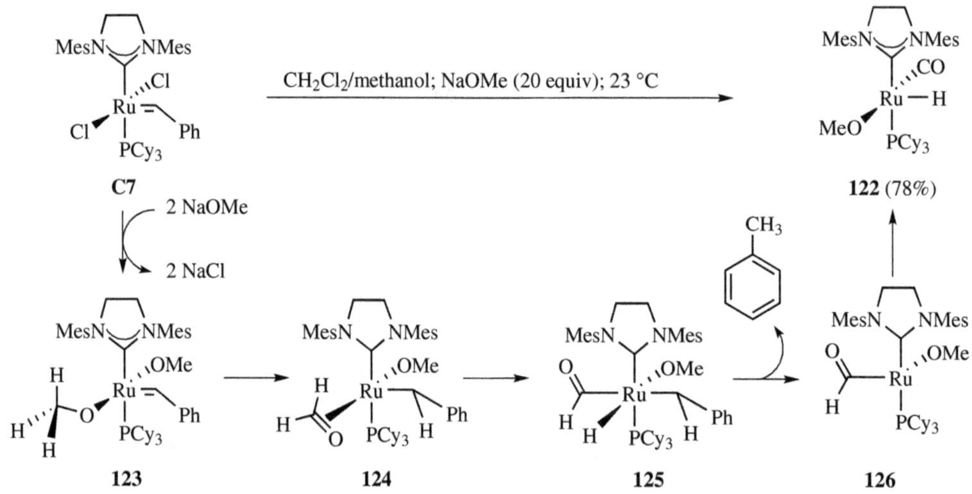

Scheme 5.41 Formation of a methoxy-hydride complex from catalyst **C7**.

First, **C7** undergoes dual chloride-methoxide exchange to **123**, which then reacts via H-transfer from one methoxide to the alkylidene group, giving benzyl-σ-complex **124** with a π-bound formaldehyde. The following steps proceed via formyl-hydride complex **125**, reductive elimination of toluene, and extrusion of CO from the formyl complex **126**, which is bound in the isolable Ru-hydride **122** as a carbonyl ligand.

The major product obtained from hydrogenation of the first-generation catalyst **C2** is the Ru-hydride $[RuHCl(H_2)(PCy_3)_2]$ (80,85). The mechanism of alkene hydrogenation reactions catalyzed by this and other Ru-hydrides has been investigated by kinetic studies (86,87) and DFT studies (88), and it is therefore possible to present a comprehensive mechanistic scenario for the assisted Ru-catalyzed tandem olefin metathesis/hydrogenation sequence. In this context, we discuss the mechanism for a low molecular weight example, the RCM/hydrogenation of diallyl malonate **127** to cyclopentane **135** (Scheme 5.42) (85). First, substrate **127** undergoes the metathesis cycle to produce the intermediate RCM product **131**; the metathesis cycle is depicted in a simplified manner (e.g., omitting the necessary phosphine dissociation to generate the catalytically active 14-electron species). After completed RCM, hydrogen is added and the metathesis catalyst reacts to the complex $[RuHCl(H_2)(PCy_3)_2]$. To enter the catalytic hydrogenation cycle, dissociation of one phosphine ligand and presumably the dihydrogen ligand is required (omitted in Scheme 5.42), and in the first step coordination of the olefin occurs, giving **132**, followed by formation of a dihydrogen

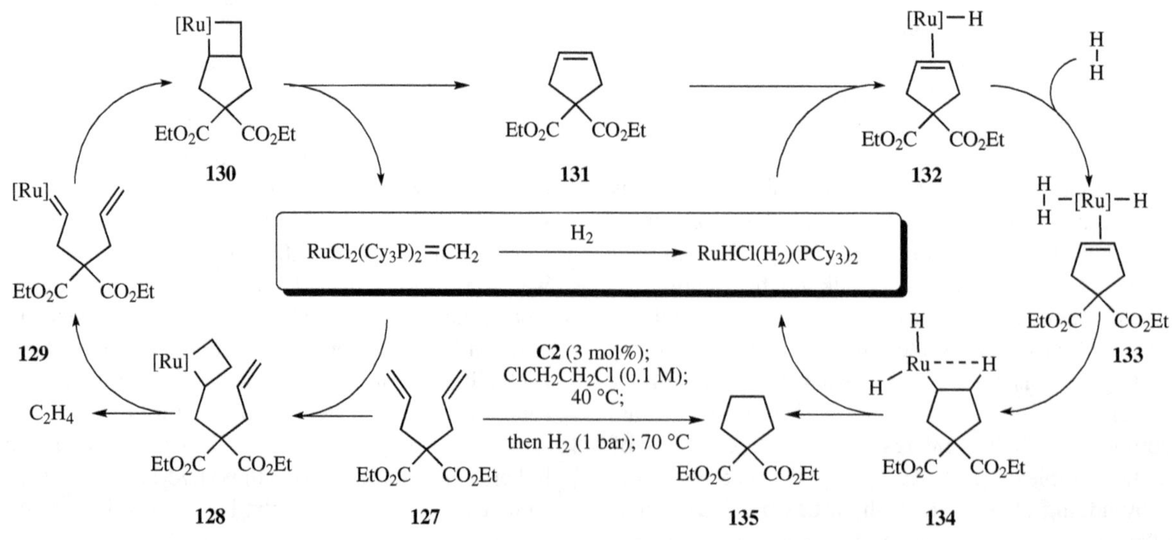

Scheme 5.42 Connecting catalytic cycles: RCM/hydrogenation of diallyl malonate **127**.

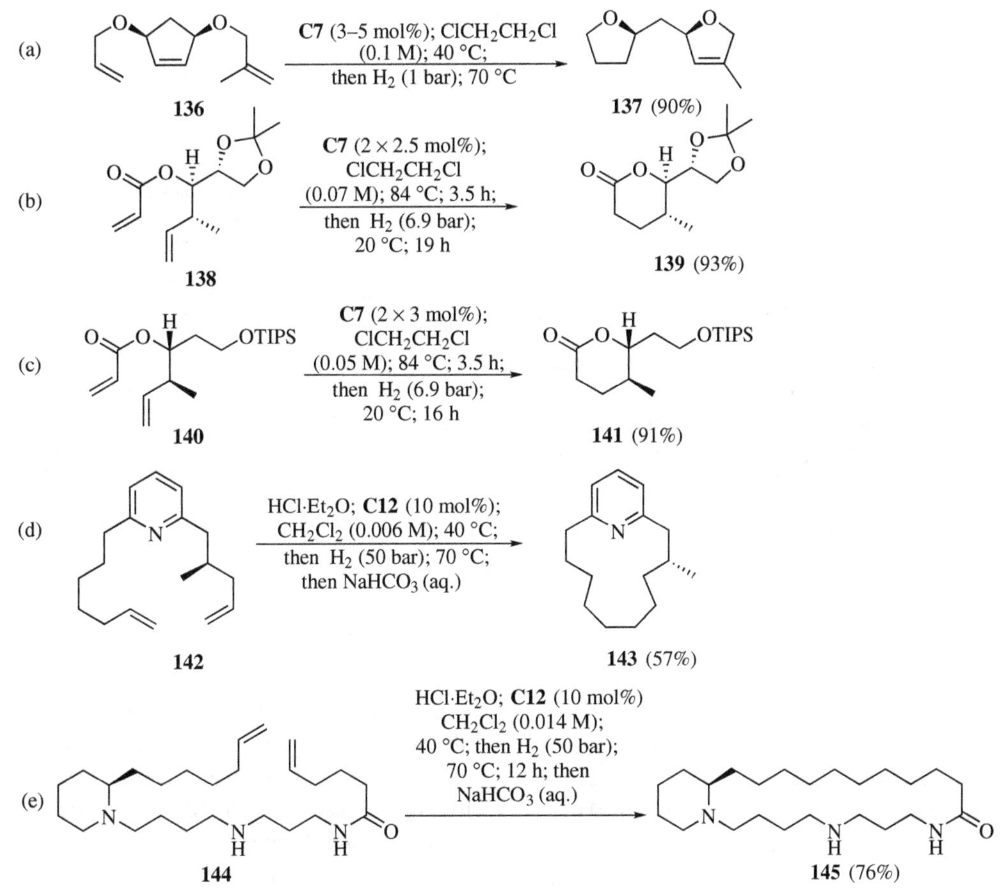

Scheme 5.43 **a–e**. RCM/hydrogenation sequences with H_2 as trigger and reagent.

complex **133**. The order of events shown in Scheme 5.42 is supported by a DFT study, while an investigation of the kinetics suggests that hydrogen coordinates first, followed by the alkene. The tentative catalytic cycle continues with an insertion of the alkene into the Ru–H bond and simultaneous oxidative addition of hydrogen, resulting in the σ-complex **134**. Evidence for the involvement of **134** in the catalytic cycle comes from the DFT study, but the authors also note that the formation of such Ru(IV) intermediates might be questionable due to the generally assumed instability of Ru(IV) dihydrides (88). Eventually, the product **135** is liberated by reductive elimination and regeneration of the catalytically active species, which may enter its resting state (depicted in Scheme 5.42) by coordinating one phosphine ligand and dihydrogen.

Other examples for assisted tandem catalytic RCM/hydrogenation reactions triggered by molecular hydrogen are compiled in Scheme 5.43, along with the reaction conditions. In the first example (Equation a), the RRM of **136** is combined with a selective hydrogenation of the less substituted double bond to furnish **137**. It can be assumed that at elevated hydrogen pressures the other double bond would also undergo hydrogenation, although this has not been investigated (85).

The second and third examples are the syntheses of saturated δ-valerolactones **139** and **141** from homoallylic acrylates **138** and **140**, respectively. These steps have been used by Rychnosky et al. in their total syntheses of the natural products attenol A (Equation b) (89) and leucascandrolide (Equation c) (90). The examples given in Equations d and e are steps in Fürstner's total syntheses of (+)-muscopyridine (91) and isooncinotine (92), respectively. In both cases, the first-generation Ru-indenylidene catalyst **C12** has been used, but successful RCM required conversion of the amine functionalities into the corresponding HCl adducts. Significantly higher hydrogen pressures are required than for the examples discussed previously.

The conversion of **146** to **147** is the final step in a synthesis of the C-1–C-12 fragment of amphidinolide recently published by Clark et al. Scheme 5.44 (equation a) (93). Notably, in this example, two double bonds are hydrogenated: the double bond formed by RCM and the exo-methylene group located at the tetrahydrofuran ring. Hydrogenation of this double bond results in the formation of a new stereocenter, hence the product **147** is obtained as a 4 : 1 mixture of

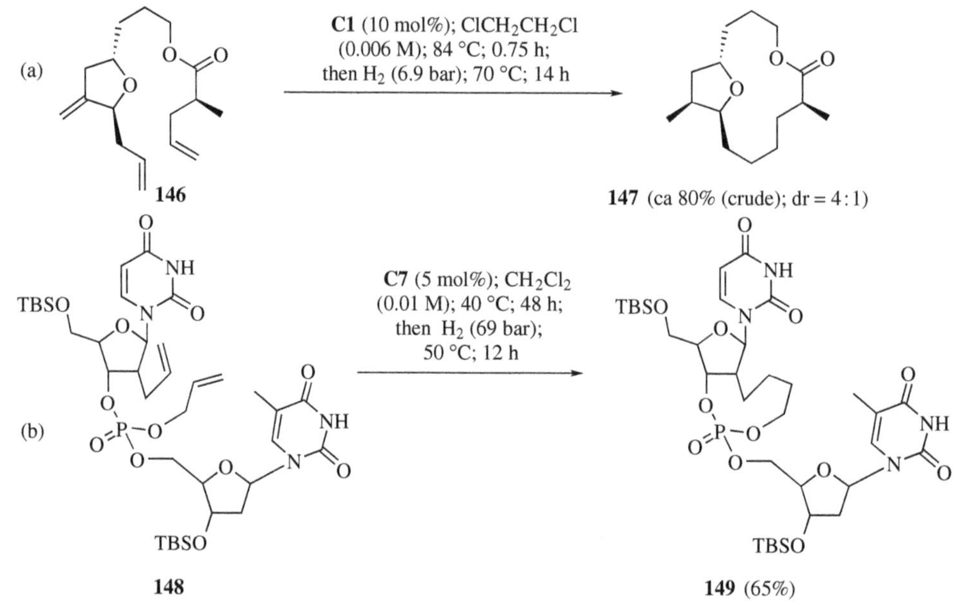

Scheme 5.44 **a, b**. RCM/hydrogenation sequences with H_2 as trigger and reagent.

epimers, with the major epimer shown in the scheme. The final example in Scheme 5.44 (Equation b) describes the installation of a saturated butylene linker group between the 2′–C and 3′–O-phosphate groups of a dinucleotide **149** from the RCM precursor **148** (94). Similar strategies have been applied by the same authors in other syntheses of oligonucleotide derivatives (95,96).

In their seminal contribution on the application of tandem metathesis/hydrogenation to low molecular weight compounds, Grubbs and coworkers also described the combination of CM and hydrogenation. These reactions proceeded with moderately elevated hydrogen pressures of 6.9 bar. An example is the reaction of **150** and methyl vinyl ketone (**151**) in the presence of second-generation catalyst **C7**, followed by hydrogenation, to furnish 8-acetoxyoctan-2-one (**152**) (Scheme 5.45). Notably, the carbonyl group is not affected under the hydrogenation conditions.

A domino enyne metathesis/hydrogenation sequence has been used by Ramharter et al. in a total synthesis of the lycopodium alkaloid (+)-lycoflexine (Scheme 5.46) (97). The enyne metathesis intermediate, diene **154**, is selectively hydrogenated at the disubstituted double bond with 10 bars of hydrogen.

As can be seen from the examples discussed so far, in almost all cases, the hydrogenation step requires elevated temperatures and hydrogen pressures, with particularly high hydrogen pressures of up to 70 bar being necessary if trisubstituted double bonds need to be hydrogenated. The only examples where rapid hydrogenation is observed at 1 bar of hydrogen are five-membered rings with no additional substituents located at the endocyclic double bond of the intermediate RCM product (see, e.g., the conversions of **127** to **135** (Scheme 5.42) or **136** to **137** (Scheme 5.43a). Nevertheless, elevated temperatures, presumably to trigger the conversion of the metathesis catalyst to the hydrogenation catalyst, are still necessary. A notable exception was discovered by preactivating the catalyst with inorganic hydrides, in particular NaH, prior to the addition of hydrogen. A representative example is the conversion of RCM precursor **156** to cyclopentanol **157** (Scheme 5.47a). After complete conversion of **156** to the intermediate RCM product, a catalytic amount of NaH was added as a dispersion in mineral oil. Stirring the reaction mixture under a hydrogen atmosphere induces quantitative hydrogenation of the double bond and the cyclopentanol **157** was isolated in high yield. Alternatively, hydrogen can be generated *in situ* if a more acidic primary or secondary alcohol is present and if NaH is used

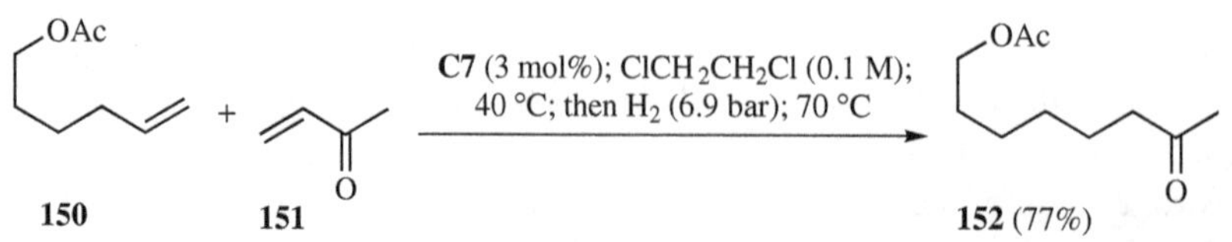

Scheme 5.45 Example for a CM/hydrogenation sequence with H_2 as trigger and reagent.

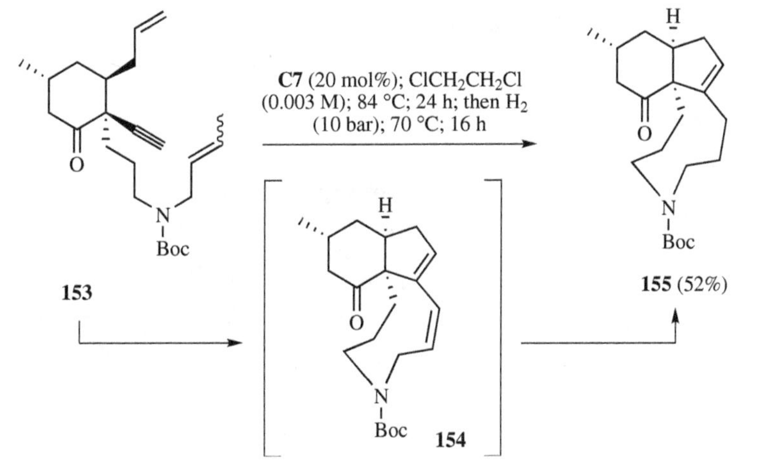

Scheme 5.46 Enyne metathesis/hydrogenation sequence with H_2 as trigger and reagent.

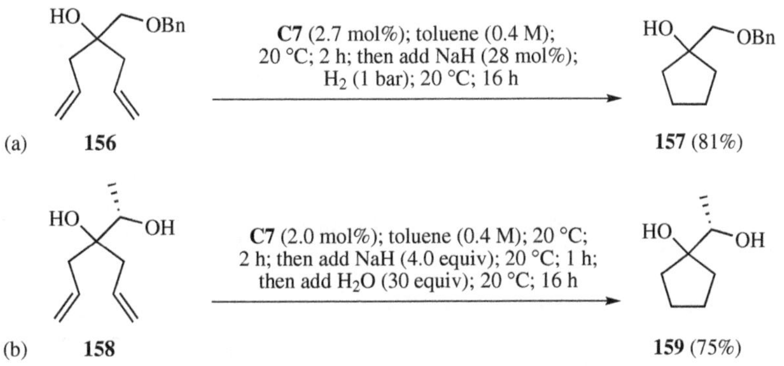

Scheme 5.47 RCM/hydrogenation triggered by NaH.

in excess, as illustrated for the conversion of precursor **158** to cyclopentanol **159**. To ensure that a sufficient quantity of hydrogen is generated *in situ*, an additional amount of water can be added (Scheme 5.47b) (98).

Synthesis of 1-(benzyloxymethyl)-cyclopentanol (157)

To a solution of **156** (1160 mg, 5.0 mmol) in dry and degassed benzene (10 ml) was added second-generation catalyst **C7** (114 mg, 2.7 mol%) under an atmosphere of argon. The solution was stirred at 20 °C until the starting material was fully consumed (approx. 2.5 h), followed by the addition of NaH (56 mg of a 60% dispersion in mineral oil, 0.28 equiv). Hydrogen gas was bubbled through the solution for 5 min, and the argon atmosphere was replaced by a hydrogen atmosphere. The reaction vessel was closed with a balloon filled with hydrogen, and the reaction mixture was stirred at 20 °C for 16 h. The reaction mixture was diluted with ether (20 ml) and washed with water (two times, 10 ml). The organic layer was separated, dried with $MgSO_4$, filtered, and all volatiles were evaporated. The residue was purified by flash chromatography on silica (hexane/methyl tert-butyl ether (MTBE) mixtures of increasing polarity) to furnish the title compound **157** (830 mg, 81%) as a colorless liquid.

It is generally desirable to replace molecular hydrogen by more conveniently to handle hydrogen sources, such as formic acid, formiates, secondary alcohols, or silanes. Unfortunately, as for hydrogenation reactions in general, the scope of these hydrogen substitutes also appears to be limited for metathesis/hydrogenation sequences. A recently published procedure uses 2-propanol as a chemical trigger and as a hydrogen source in an RCM/hydrogenation sequence to chromanes (Scheme 5.48) (99).

Interestingly, this protocol is specific for the synthesis of chromanes from styrenyl allyl ethers and is not applicable to the synthesis of other non-benzo-fused tetrahydropyrans, because without an annulated aromatic ring RCM/isomerization occurs under these conditions (see Section 5.3.2 for details). Presumably, the Ru-hydride results from the formation of a Ru-alkoxide in the presence

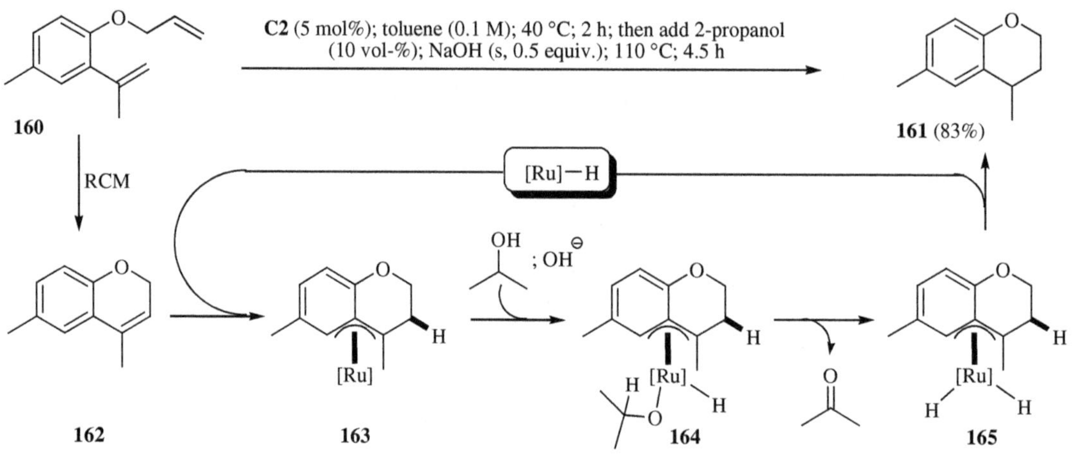

Scheme 5.48 Synthesis of chromanes via RCM/transfer hydrogenation.

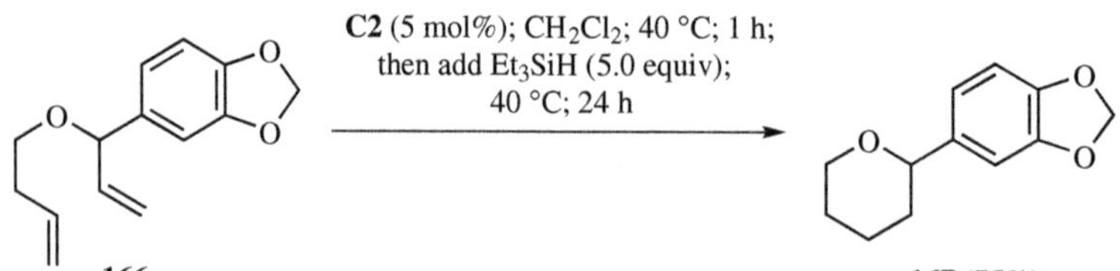

Scheme 5.49 RCM/hydrogenation with Et_3SiH as hydrogenating agent.

of a base, which undergoes β-hydride elimination with formation of acetone. Insertion of the endocyclic double bond of the intermediate RCM product **162** into the Ru−H bond might give a η^3-π-allyl-Ru complex **163**, which presumably undergoes a further transformation to dihydride **165** via alkoxide **164**. Reductive elimination regenerates the catalytically active Ru-hydride and liberates the chromane **161**.

An alternative hydrogen substitute which has been explored in tandem metathesis/hydrogenation sequences is triethylsilane (TES). A representative example is the conversion of allyl ether **166** to tetrahydropyran **167**, which was accomplished using first-generation catalyst **C2** and 5 equiv of TES (Scheme 5.49). The authors propose that the reaction proceeds via oxidative addition of the silane to the Ru catalyst and that hexaethyldisilane is formed as a by-product (100).

For the Ru-catalyzed hydrogenation of electron-deficient C−C double bonds, lower excess of silane and shorter reaction times are often sufficient. An example, which was recently described in the course of a synthesis of the natural product curvulalic acid, is the reaction of homoallylic alcohol **168** and methyl acrylate (**169**) to TES-protected ester **170** in a one-flask sequence comprising three Ru-catalyzed steps (Scheme 5.50) (101). First, **168** and methyl acrylate react in a CM to enoate **171**. Excess methyl acrylate is then removed *in vacuo*, and the residue is redissolved in toluene. Upon addition of TES and heating the reaction mixture to 80 °C two Ru-hydride-catalyzed steps occur: first, the secondary alcohol is protected as a TES ether via dehydrogenative silylation (102) to give **172**. In the next step, it can be assumed that TES undergoes a Ru-catalyzed conjugative addition to the electron-deficient double bond, presumably giving the silyl ketene acetal **173**, which might undergo hydrolysis to **170** upon chromatography on silica.

5.3.2 Metathesis/Isomerization

Olefin isomerization has for quite some time solely been considered as a highly undesirable complication of "difficult" olefin metathesis reactions, resulting in a complete failure of an intended metathesis reaction or in the formation of mixtures of products. Different reasons were originally discussed for the occurrence of isomerization reactions: the Ru-carbene complex itself might show isomerization activity, or Ru-hydrides might be responsible for this reaction. Sources of Ru−H species can be the impurities present in the precatalyst, resulting from washing the crude precatalyst with methanol or other alcohol solvents or the degradation products of the metathesis catalyst (58). Considering these diverse reasons for the occurrence of isomerization reactions, the main difficulty in developing an assisted tandem metathesis/isomerization sequence is obviously to induce isomerization at a defined point in a sequence after the metathesis step is completed, and not competing to the metathesis reaction.

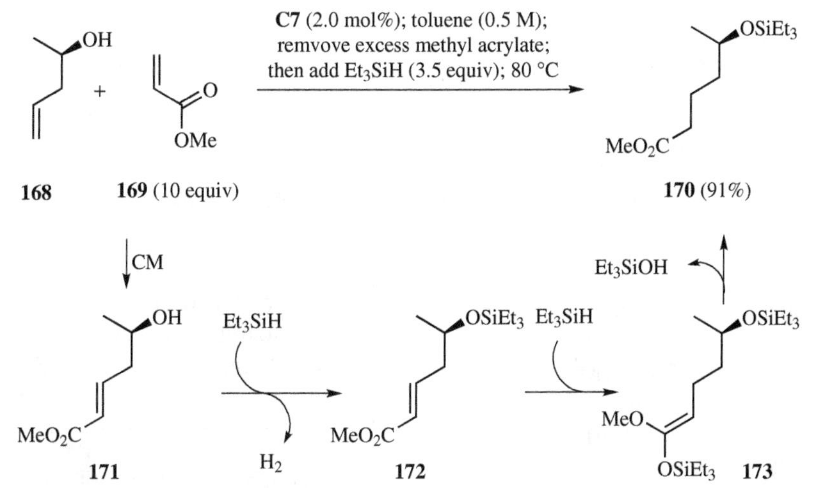

Scheme 5.50 CM/hydrogenation/dehydrogenative silylation sequence.

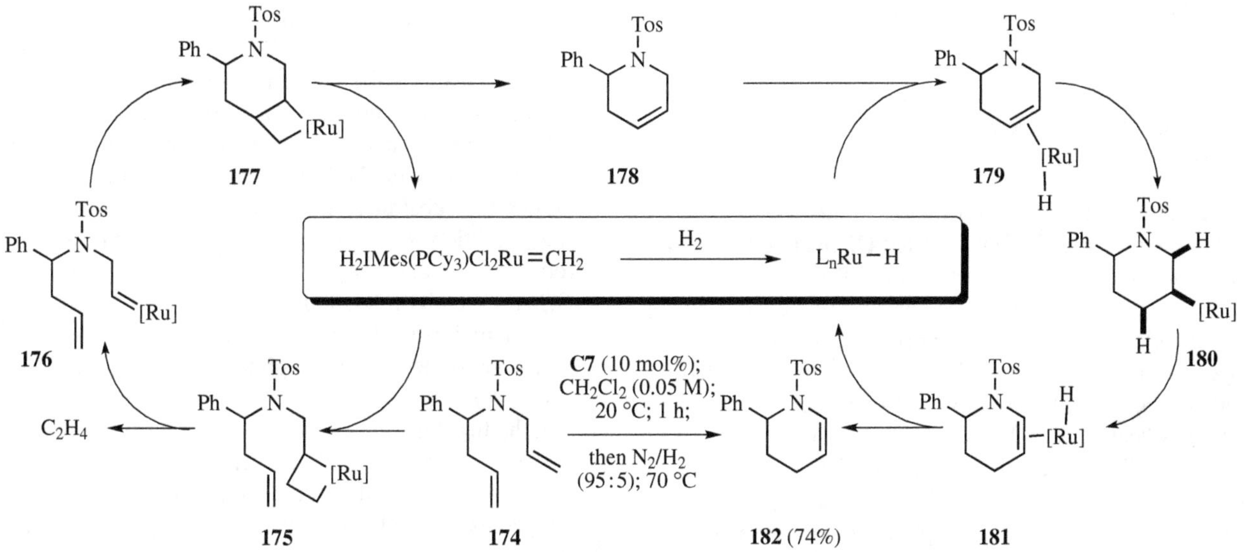

Scheme 5.51 Connecting catalytic cycles: RCM/isomerization of allyl amide **174**.

The first solution for this problem was proposed by Snapper et al., who used molecular hydrogen to trigger the conversion of the metathesis catalyst to an isomerization active Ru-hydride (103). The underlying organometallic transformation is identical to that discussed in the previous section for metathesis/hydrogenation sequences, but an additional difficulty that had to be overcome was competing hydrogenation. This was accomplished by diluting the hydrogen atmosphere required to trigger the conversion of the metathesis catalyst to the isomerization catalyst with 95% of nitrogen. In Scheme 5.51, the sequence is illustrated for the conversion of homoallyl-allyl amide **174** to cyclic enamide **182** (103). At the beginning of the sequence, the substrate enters the olefin metathesis catalytic cycle and reacts via the sequence of [2 + 2] cycloaddition and cycloreversion steps to the intermediate RCM product **178**, using second-generation catalyst **C7**. In Snapper's protocol, the inert gas atmosphere is replaced by a 95 : 5 N_2/H_2 mixture after completion of the RCM, and the reaction solution is heated to 70 °C. Under these conditions, an Ru-hydride complex $L_nRu\text{-}H$ is formed, which catalyzes a double-bond isomerization reaction to the thermodynamically more stable isomer **182**, through a sequence of olefin coordination, giving π-complex **179**, migratory insertion to σ-complex **180**, and productive β-hydride elimination to π-complex **181**, which eventually dissociates into the catalytically active Ru-hydride and the enamide **182**. For clarity, the reverse reactions have been omitted in both catalytic cycles in Scheme 5.51, and only the direction leading to the thermodynamically preferred product **182** with the more electron-rich double bond is shown. Similarly, a variety

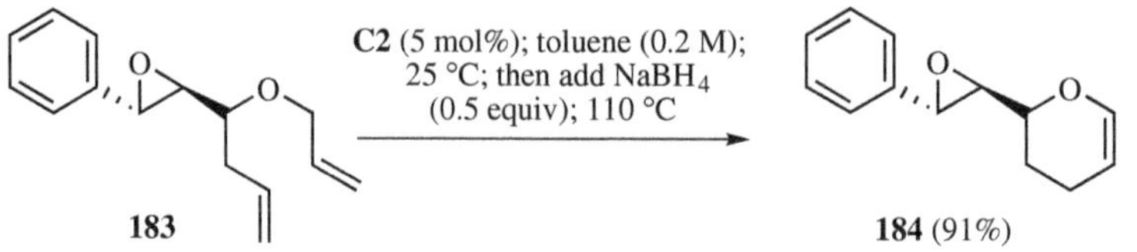

183 184 (91%)

Scheme 5.52 RCM/isomerization triggered by $NaBH_4$.

of allyl ethers were converted in one flask to cyclic enol ethers, which are regioisomers of the primary RCM products (103).

Parallel to and independent from Snapper's work, one of us developed an alternative protocol that uses first-generation catalyst **C2** and the inorganic hydrides $NaBH_4$ or NaH to induce the Ru-carbene to Ru-hydride interconversion (104). A representative example is the transformation of **183** to the six-membered cyclic enol ether **184**. Notably, the sensitive styrene oxide moiety tolerates the reducing agent $NaBH_4$, the isomerization proceeds with very high regioselectivity, and no epimerization was observed (Scheme 5.52).

Synthesis of 3-phenyloxiran-2-yl-3,4-dihydro-2*H*-pyran (184) (104)

To a solution of **183** (460 mg, 2.0 mmol) in dry and degassed toluene (10 ml) was added **C2** (82 mg, 5 mol%) under an atmosphere of dry argon. The solution was stirred at 20 °C until complete consumption of the starting material (approx. 1 h; monitoring the reaction by thin layer chromatography (TLC) is strongly recommended), and $NaBH_4$ (20 mg, 0.3 equiv) was then added. The mixture was heated to 110 °C until the primary RCM product was fully converted to the enol ether **184** (approx. 4–5 h; monitoring the reaction by TLC is strongly recommended; developing the TLC plates in an iodine chamber is the best method to detect the enol ether, on the other hand, RCM precursors and the intermediate primary RCM products are not detected by staining with iodine, therefore the TLC plates should also be developed using molybdophosphoric acid in ethanol). After cooling the reaction mixture to ambient temperature, it was washed with water, the aqueous layer was separated, extracted twice with ether, and the combined organic layers were dried with $MgSO_4$. After filtration, all volatiles were evaporated and the residue was purified by flash chromatography on silica using hexane/MTBE mixtures of increasing polarity to give the title compound **184** (360 mg, 1.8 mmol, 90%) as a colorless oil.

Although this protocol is applicable to numerous examples, occasionally unsatisfactory results were obtained if additional C–C double bonds are present in a side chain, or if the substrate carries unprotected hydroxy groups or sterically demanding substituents. For example, the RCM/isomerization of **185** *to* **186** does not proceed efficiently using $NaBH_4$ as an additive; however, by using 2-propanol in the presence of a base, a highly efficient isomerization catalyst is formed that allows complete conversion to the enol ether within 3 h (Scheme 5.53) (105). For certain substrates, in particular styrenyl allyl ethers, these conditions do not result in the formation of isomerized products, but transfer hydrogenation is induced (see Section 5.3.1 for details). It should, however, be noted that for the vast majority of examples, these conditions lead to a rapid chemoselective and epimerization-free double-bond isomerization.

In the few cases where competing transfer hydrogenation with the 2-propanol/NaOH procedure cannot be suppressed by reducing the reaction temperature or reaction time, TES emerged as a useful alternative. An example is the transformation of naphthyl-substituted derivative **187** to dihydropyran **188** (Scheme 5.54) (107).

Although $NaBH_4$ was ineffective in this case, both 2-propanol/NaOH and Et_3SiH were found to be the suitable chemical triggers to induce the isomerization step, and the isomerized dihydropyran was obtained in comparable yields under both conditions. While it could be established by NMR methods that in all these metathesis/isomerization protocols Ru-hydrides are involved, the exact structure of the isomerization catalysts is unknown. A notable exception in this

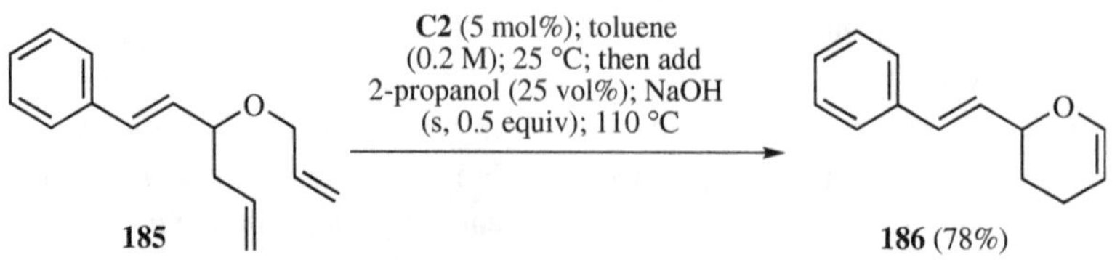

185 186 (78%)

Scheme 5.53 RCM/isomerization triggered by 2-propanol/NaOH.

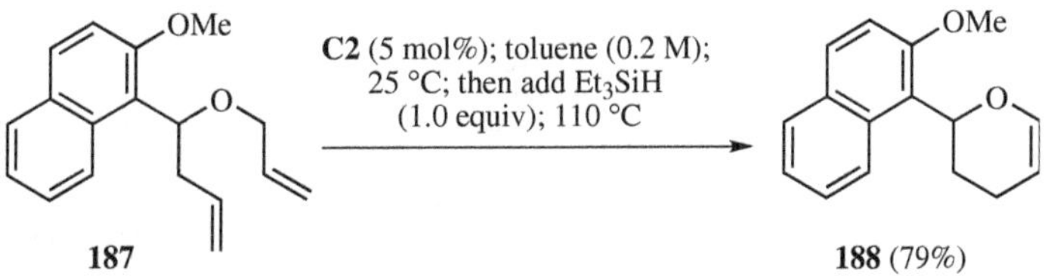

187 188 (79%)

Scheme 5.54 RCM/isomerization triggered by TES.

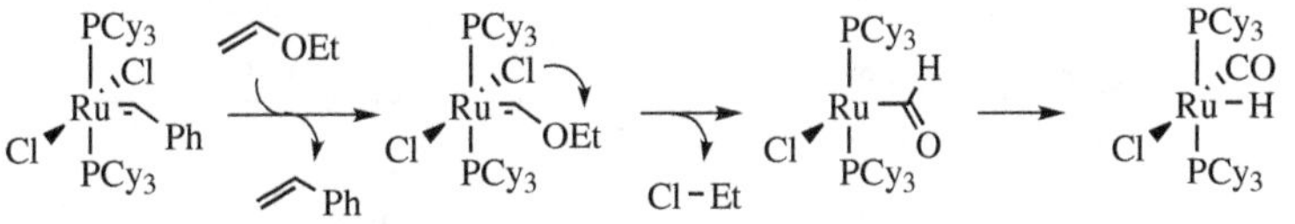

C2 189 190 191 ($\delta(^{31}P) = 47$ ppm)

Scheme 5.55 Degradation of **C2** to $[RuH(CO)(PCy_3)_2Cl]$ (**190**).

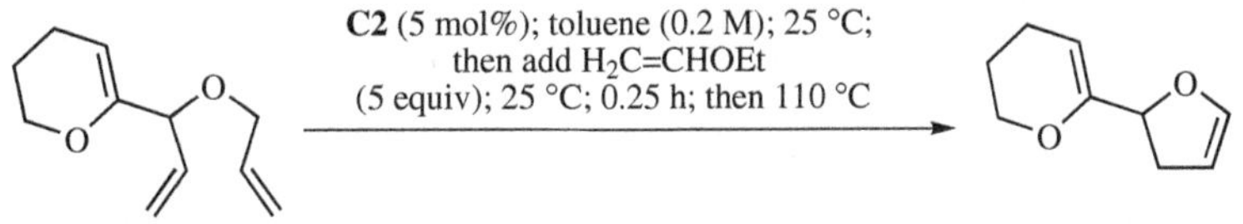

192 193 (69%)

Scheme 5.56 RCM/isomerization triggered by ethylvinyl ether.

regard is the next protocol, which uses ethyl vinyl ether as an additive and was found to be particularly useful for the synthesis of five-membered cyclic enol ethers from diallyl ethers. The underlying organometallic transformations were mechanistically investigated by Grubbs et al. (78,106), who found that the Fischer-type carbene complex **189**, which is rapidly formed from **C2** and ethylvinyl ether, decomposes to the Ru-hydride complex **191** upon heating to 80 °C in a benzene solution, presumably via elimination of chloroethane to give the 14-electron formyl complex **190**, which then undergoes CO extrusion and recoordination (Scheme 5.55).

A representative example for the exploitation of this organometallic transformation in a tandem RCM/isomerization is the conversion of **192** into spirocyclic bis-enol ether **193** (Scheme 5.56) (107). The formation of **191** under these conditions was proven by monitoring the reaction with ^{31}P NMR spectroscopy (108).

Fustero et al. discovered that in certain cases isomerization can be induced simply by heating an olefin metathesis reaction without adding a chemical trigger (109). However, this procedure is limited to second-generation catalysts such as **C1** or **C7** and is not applicable to first-generation catalysts, which require the addition of a chemical trigger to induce conversion to the Ru-hydride species. Thermally induced RCM/isomerization with second-generation catalysts relies on the bimolecular decomposition pathway (83a) established for these catalysts by Grubbs et al. (see Section 5.3.1, Scheme 5.40, for details) (83b). An example is the formation of cyclic enamide **195** from the allylamide precursor **194** (Scheme 5.57) (109).

A major benefit of RCM/isomerization sequences is that notoriously difficult metathesis reactions can be circumvented without additional synthetic steps. For a direct synthesis of cyclic enol ethers or enamides based on RCM, enolethers or enamides are required as precursors. These are more difficult to obtain than the corresponding allyl compounds, and the subsequent RCM of these electron-rich alkenes is normally slow and requires the more reactive second-generation catalysts (sometimes in comparatively high amounts) and high dilution (110,111). Two examples for the implementation of a tandem RCM/isomerization reaction in the synthesis of diarylheptanoid natural products are discussed in the following. The synthesis of (3*S*,7*R*)-5,6-dehydro-de-*O*-methyl-centrolobine (**199**), a natural product isolated from the plant *Alpinia*

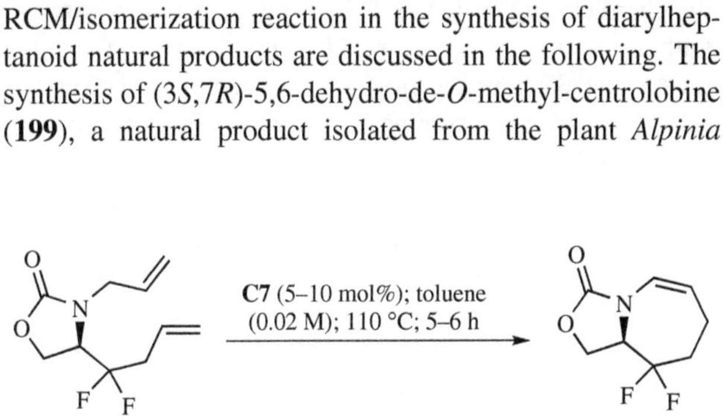

194 195 (85%)

Scheme 5.57 RCM/isomerization triggered by bimolecular thermal decomposition of second-generation catalysts.

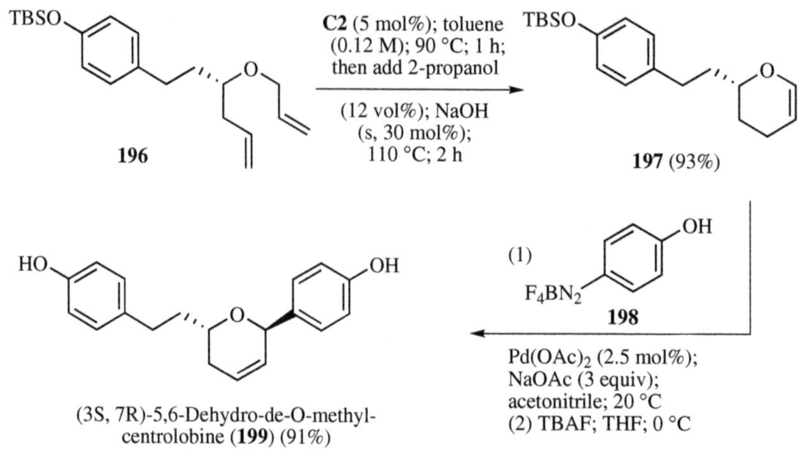

Scheme 5.58 RCM/isomerization in the synthesis of natural product **199**.

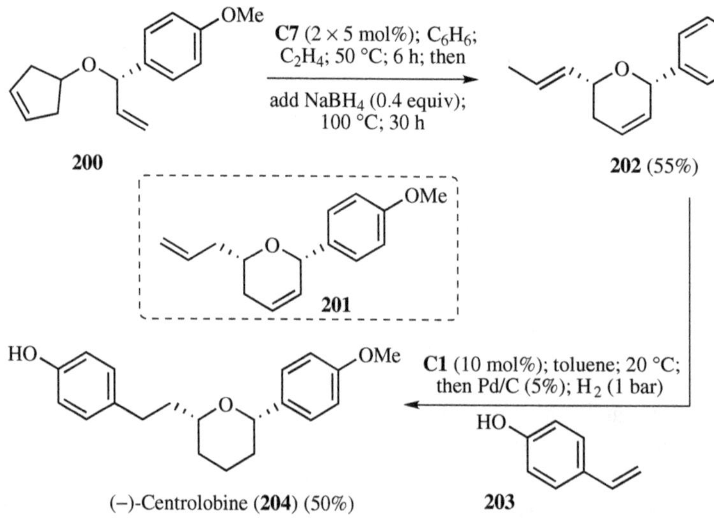

Scheme 5.59 Diastereoselective RRM/isomerization in the synthesis of (−)-centrolobine.

blepharocalyx, was accomplished starting from allyl ether **196**, which was converted to the cyclic enol ether **197** using the RCM/isomerization sequence with 2-propanol/NaOH as the chemical trigger (112). Dihydropyran **197** was obtained without racemization, and the TBS ether was not affected by the presence of base and alcohol. In the next step, **197** was reacted with phenol diazonium salt **198** in a Heck reaction, and then the TBS ether was cleaved to furnish the natural product in high *trans*-diastereoselectivity (Scheme 5.58) (113).

The structurally closely related natural product (−)-centrolobine was synthesized similarly (112), and also via an alternative application of a tandem metathesis/isomerization sequence (Scheme 5.59): (114) Böhrsch and Blechert have combined a diastereoselective RRM reaction of cyclopentene **200** with a $NaBH_4$-induced isomerization in an assisted tandem sequence. Under these conditions, the sterically more hindered endocyclic double bond of the initially formed RRM product **201** was not affected, but the pendant allyl substituent was isomerized to furnish **202**. In the following step, CM with styrene **203**, catalyzed by **C1**, and Pd/C-catalyzed hydrogenation of both double bonds gave (−)-centrolobine (**204**).

Further examples for assisted tandem metathesis/isomerization reactions leading to cyclic enamides or enol ethers are listed in Schemes 5.60 and 5.61. The synthesis of **206** from allyl ether **205** illustrates that the various protocols presented above are complementary: while 2-propanol and NaOH result in extensive transfer hydrogenation, clean isomerization was observed by using NaH as isomerization-inducing reagent (Scheme 5.60a) (115). A synthesis of a highly deoxygenated glycal **208** from precursor **207** is outlined in Scheme 5.60b (116). Formation of glycal **210** from precursor **209** contains, apart from RCM and isomerization, an additional dehydrogenative silylation, resulting in the TES protection of the primary alcohol (Scheme 5.60c). It was shown that dehydrogenative

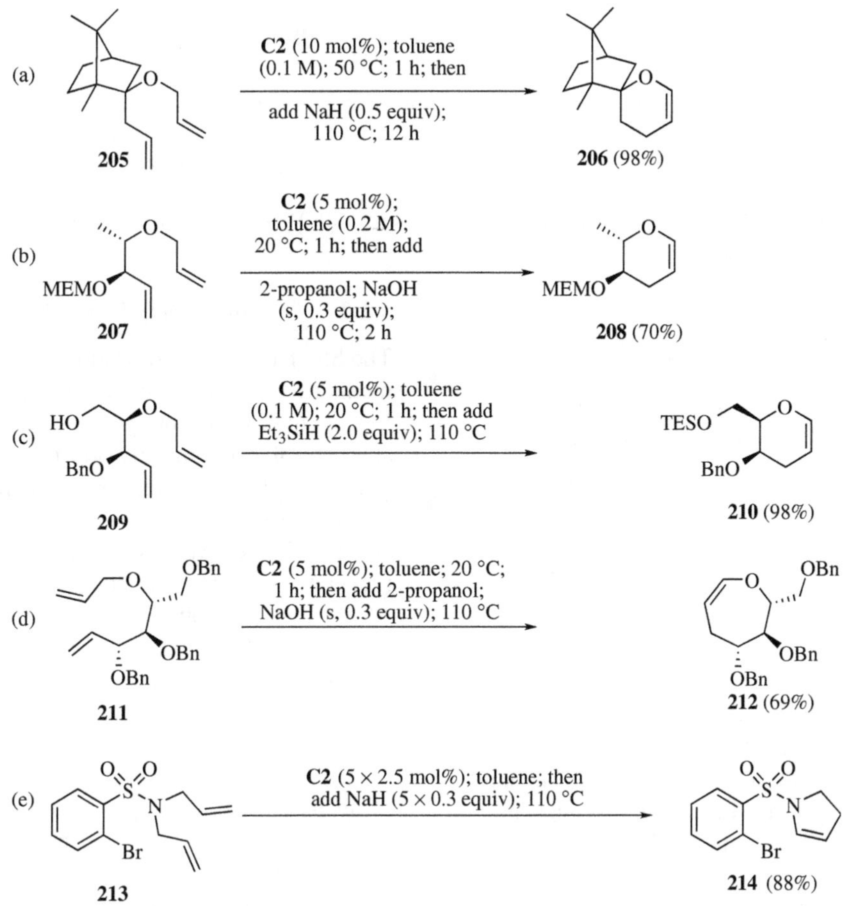

Scheme 5.60 **(a–e)**. Syntheses of enamides and enol ethers via RCM/isomerization.

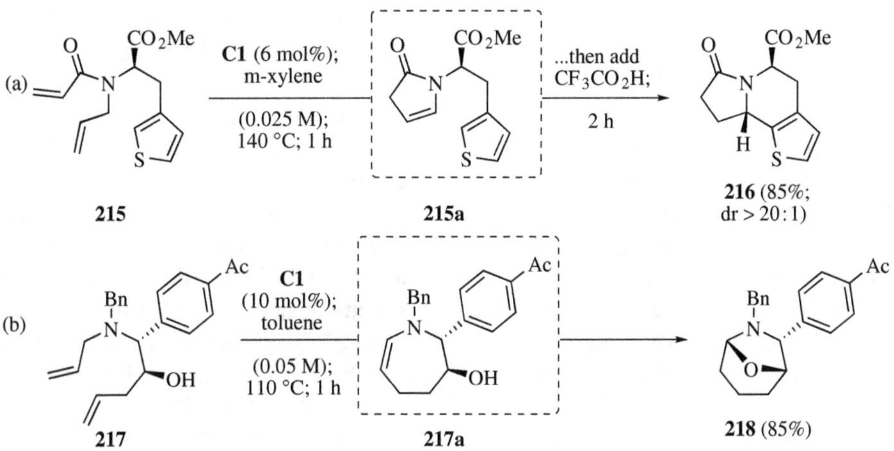

Scheme 5.61 **(a, b)**. Syntheses of enamides and enol ethers via RCM/isomerization.

silylation occurs faster and at lower temperatures than isomerization (117). A synthesis of a ring-expanded glycal **212** from **211** based on RCM/isomerization is presented in the following equation (Scheme 5.60d) (118). Application of RCM/isomerization, catalyzed by first-generation catalyst **C2** and induced by NaH, lead to the formation of a cyclic enamide **214** from arenesulfonamide **213**. In a subsequent step, **214** was subjected to an intramolecular Heck reaction, giving a tricyclic product (Scheme 5.60e) (119).

The two final examples, shown in Scheme 5.61a (120) and Scheme 5.61b (121), illustrate thermally induced Ru-hydride

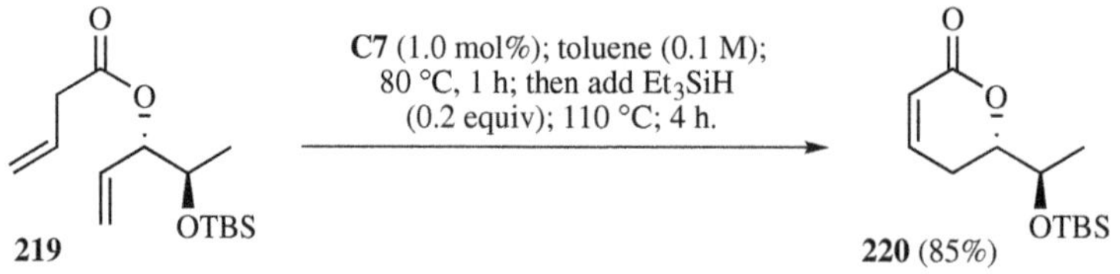

Scheme 5.62 α,β-Unsaturated γ-valerolactones via RCM/isomerization of butenoates.

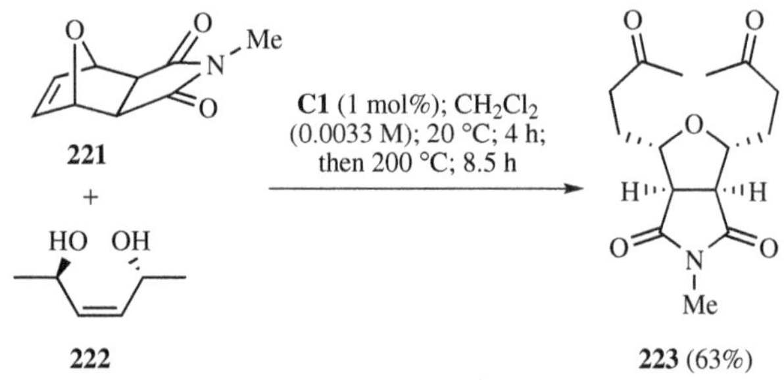

Scheme 5.63 ROM/CM sequence combined with isomerization.

formation, starting from second-generation catalyst **C1**. In both cases, the products resulting from RCM/isomerization, the heterocycles **215a** and **217a**, were not isolated but underwent iminium ion formation and intramolecular nucleophilic trapping, giving the tricyclic products **216** and **218**, respectively. Notably, in both cases, the use of second-generation catalyst **C7** resulted predominantly in the formation of RCM products, with subsequent isomerization occurring only to a minor extent, if at all.

Similar to RCM reactions involving electron-rich double bonds, those of electron-deficient alkenes are also often rather slow and require elevated loadings of precatalyst and high dilution. In these cases, RCM/isomerization can lead to a significant improvement in the reaction conditions. For example, RCM/isomerization of **219** catalyzed by just 1.0 mol% of second-generation catalyst **C7** and triggered by Et_3SiH yields the lactone **220** in high yield, even at an initial substrate concentration of 0.1 M (122). The direct synthesis of δ-lactones such as **220** via RCM of the corresponding acrylates would normally require higher amounts of catalyst and significantly higher dilution (123) (Scheme 5.62).

Metathesis/isomerization reactions involving allylic alcohols are a special case, because in these reactions the C–C double bond is no longer present in the molecule. Isomerization leads to enols, which rapidly and irreversibly tautomerize to carbonyl compounds (124–126), as illustrated for the ROM/CM/isomerization sequence of norbornene derivative **221** and allylic alcohol **222** to diketone **223**, shown in Scheme 5.63 (127). Occasionally, stoichiometric reactions of Grubbs' catalysts and allylic alcohols were observed, resulting in their irreversible deactivation (124,128).

5.3.3 Metathesis/Dihydroxylation

The Sharpless' dihydroxylation is the most commonly used method for the conversion of alkenes to *vic*-diols, because it gives high yields and proceeds with comparatively small amounts of osmium tetroxide in most cases. In addition, efficient and user friendly protocols for the enantioselective version exist (129,130). Although it is supposed to be considerably less volatile and less toxic, ruthenium tetroxide (generated *in situ* by oxidation of $RuCl_3$) (131) has attracted considerably less attention as an oxidation catalyst. This is often attributed to the significantly higher redox potential of Ru(VIII)/Ru(VI) compared to Os(VIII)/Os(VI), resulting in overoxidation and cleavage of the C–C double bond with formation of carboxylic acids (132). In spite of the high reactivity of Ru-(VIII)-oxide, this oxidative cleavage of alkenes was found to be slow and sluggish, which was attributed to an inhibition of the catalyst by complexation of the carboxylates (131). A significant improvement was achieved in this regard, when Sharpless et al. introduced acetonitrile as a co-solvent in RuO_4-catalyzed oxidation reactions: for example, 5-decene could be converted into pentanoic acid in 88% yield by using 2.2 mol% of $RuCl_3$ and 4 equiv of $NaIO_4$ as an oxidant in a solvent mixture of CCl_4/CH_3CN and water. In the absence of acetonitrile, 80% of the starting alkene was recovered and only 17% of pentanal was formed (133). Shing et al. were the first to describe conditions for a less rigorous Ru-catalyzed oxidation that allows the synthesis of diols from alkenes in useful yields and selectivities. Key elements of their protocol are the use of an ethyl acetate/acetonitrile/water solvent mixture in an exact 3 : 3 : 1 ratio, low temperatures (between 0 and 5 °C), comparatively high catalyst loadings of 7 mol% and short, controlled reaction times of 0.5–3 min (134). A representative example for a dihydroxylation using Shing's conditions is the highly diastereoselective conversion of cyclohexene **224** to *cis*-diol **225** (Scheme 5.64) (135).

Later, Plietker et al. discovered that a significant rate acceleration and thereby a significant reduction of the catalyst loading could be achieved by conducting the reaction in the presence of Brønsted (136) or Lewis acids, such as $CeCl_3$ (137). An example for the application of their protocol, extended by a dehydrogenation of a primary alcohol to an aldehyde, is the two-step synthesis of the carbohydrate derivative **230** (Scheme 5.65) (138). Starting from alkenes **226** and

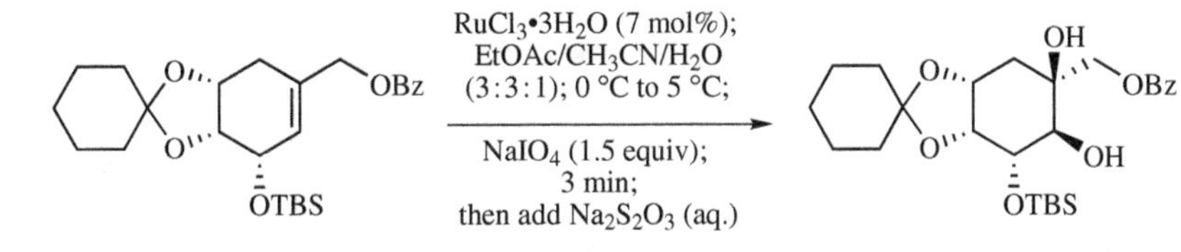

224 **225** (75%; dr > 95:5)

Scheme 5.64 RuO_4-catalyzed dihydroxylation using Shing's protocol.

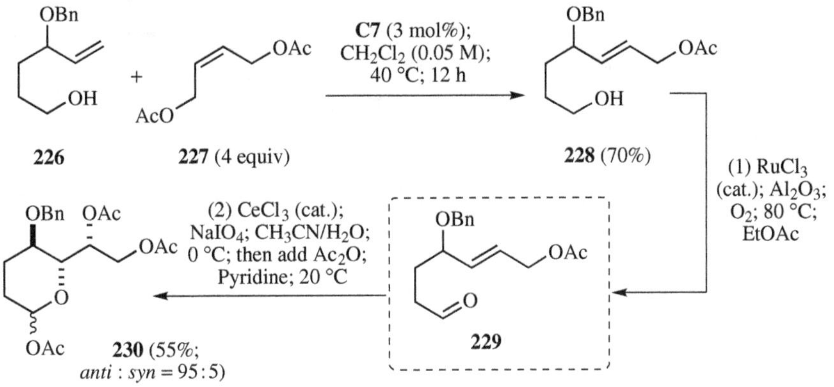

Scheme 5.65 Two-step CM and dehydrogenation/dihydroxylation sequence.

227, the CM product **228** was obtained as a single *E*-isomer in 70% yield, using second-generation catalyst **C7**. Alkene **228** was then exposed to a Ru-Al_2O_3 catalyst (obtained from $RuCl_3$ and Al_2O_3) in the presence of oxygen, which induced the dehydrogenation to aldehyde **229**, followed by addition of the Lewis-acidic co-catalyst $CeCl_3$ and the oxidant $NaIO_4$. Finally, all hydroxy groups were acetylated to yield **230** as a mixture of anomers but in high *syn*/*anti* diastereoselectivity.

A mechanistically related RuO_4-catalyzed reaction is the ketohydroxylation, which yields α-hydroxyketones directly from alkenes (132). A representative example is the formation of the 1,3-dicarbonyl compound **232** from cyclohexene carboxylate **231**: (139) in contrast to the Ru-catalyzed dihydroxylation, the triple salt "oxone" is used as an oxidant, with $KHSO_5$ being the active ingredient. A mechanistic rationale for the formation of α-hydroxy carbonyl compounds and a comparison with the Ru-catalyzed dihydroxylation have been published by Plietker et al. (132,139) (Scheme 5.66).

The first procedure for a one-flask/single-precatalyst metathesis/dihydroxylation sequence was published by Blechert et al. (140). A major difficulty in developing this sequence was the incompatibility of the metathesis reaction with the optimum solvent system for the Lewis-acid-accelerated dihydroxylation. Blechert et al. solved this problem by evaporating the solvent after completion of the metathesis step, and re-dissolving the residue in an acetonitrile/ethyl acetate/water mixture, which was then cooled to 0 °C and treated with the oxidant and the Lewis-acidic co-catalyst $YbCl_3{\cdot}6H_2O$. A representative example, together with a mechanistic rationale for this catalytic sequence, is the conversion of RCM precursor **233** to piperidinediol **241** shown in Scheme 5.67 (140).

Allyl homoallyl amide **233** first enters the metathesis cycle and is converted to the RCM product **237** via Ru-carbene- and ruthenacyclobutane intermediates **234** *to* **236**. The oxidant $NaIO_4$ then serves as the chemical trigger and most likely induces the oxidation of the propagating species of the RCM cycle, the Ru-methylidene complex, to RuO_4. The second catalytic cycle is based on Plietker's rationale for the Brønsted-acid-accelerated RuO_4-catalyzed dihydroxylation: (136) alkene **237** and RuO_4 react to the ruthenate **238**, in which the oxidation state of Ru is + VI. Ruthenate **238** is then

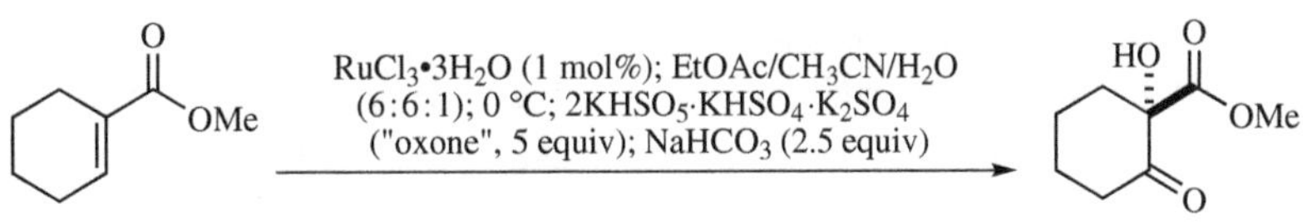

231 **232** (78%)

Scheme 5.66 RuO_4-catalyzed ketohydroxylation.

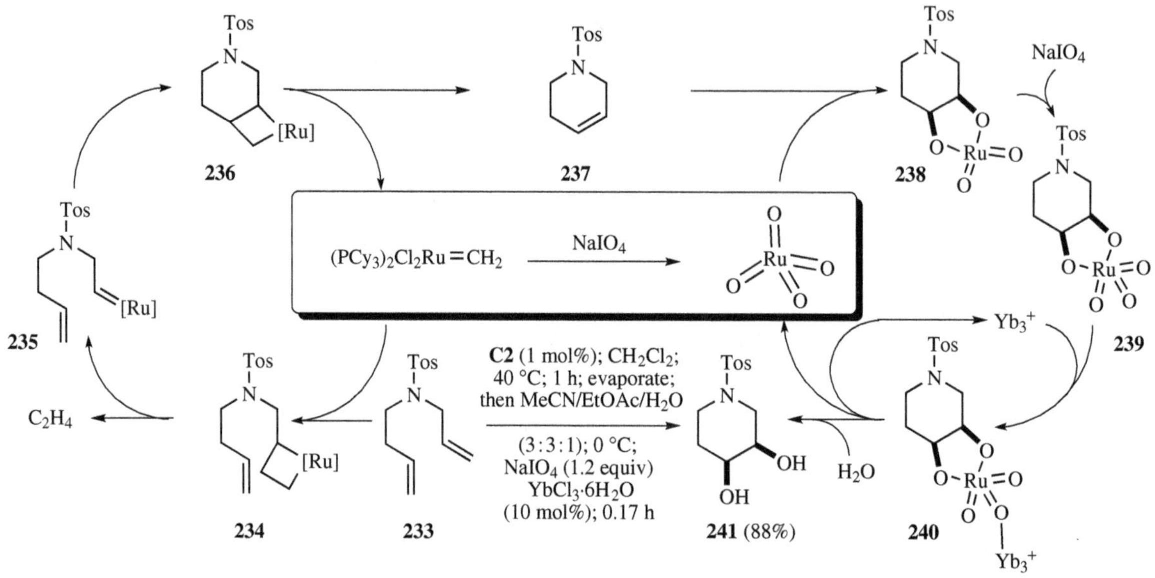

Scheme 5.67 Connecting catalytic cycles: RCM/dihydroxylation for allyl amide **233**.

oxidized by periodate to **239** (oxidation state + VIII), which is subsequently hydrolyzed to liberate the diol **241** and regenerate the catalyst RuO_4. At this stage, the Lewis-acidic co-catalyst comes into play: the hydrolysis of perruthenate **239** would normally be a slow process, which means that an electrocyclic fragmentation with cleavage of the C–C bond might compete. This scission is efficiently suppressed by accelerating the hydrolysis of perruthenate **239** with a Lewis acid, which presumably forms a complex **240** with enhanced electrophilicity at the ruthenium center. Shortly after publication of Blechert's sequential metathesis/dihydroxylation protocol, Snapper et al. proposed an alternative solution to the solvent compatibility problem. Although there is only a very limited flexibility in the choice of solvents for the Ru-catalyzed dihydroxylation, the Ru-catalyzed olefin metathesis often tolerates rather uncommon and surprisingly polar solvents, such as esters (141) or acetic acid (142,143). For instance, Snapper et al. found that many dienes undergo complete conversion to the corresponding RCM products in ethyl acetate at synthetically useful substrate concentrations within 1 h at ambient temperature, if 5 mol% of the second-generation catalyst **C7** is used (144). This observation was exploited to develop a protocol that avoids solvent exchange between the individual Ru-catalyzed steps: after completion of the metathesis reaction, a preformed mixture of the Lewis-acidic co-catalyst $CeCl_3{\bullet}7H_2O$ and the oxidant $NaIO_4$ in an acetonitrile/water solvent mixture was added at 0 °C, resulting in the oxidation of the metathesis catalyst and subsequent rapid dihydroxylation. Similarly, addition of a solution of oxone and $NaHCO_3$ in acetonitrile/water to a completed metathesis reaction in ethyl acetate induced formation of the analogous ketohydroxylation products. A representative example for both tandem sequences is the reaction of

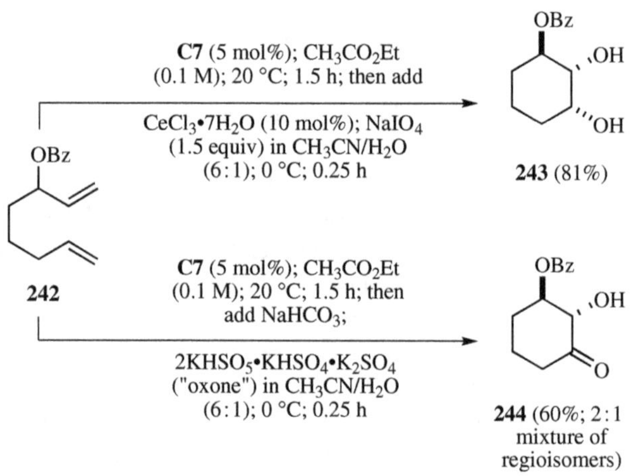

Scheme 5.68 Assisted tandem RCM/dihydroxylation and RCM/ketohydroxylation.

benzoate **242**, which, depending on the choice of oxidant, undergoes RCM/dihydroxylation to **243** or ketohydroxylation to **244**, respectively (Scheme 5.68) (144).

Synthesis of (1*RS*, 2*RS*, 3*RS*)-2,3-dihydroxycyclohexyl benzoate (243): (144)

To a solution of **242** (59 mg, 0.25 mmol) in dry and degassed ethyl acetate (3 ml) was added second-generation catalyst **C7** (11 mg, 5 mol%). The solution was stirred at 20 °C for 1.5 h and then cooled to 0 °C. Then $CeCl_3{\bullet}7H_2O$ (10 mg, 10 mol%) and $NaIO_4$ (81 mg, 0.38 mmol) were suspended in water (0.3 ml) and heated

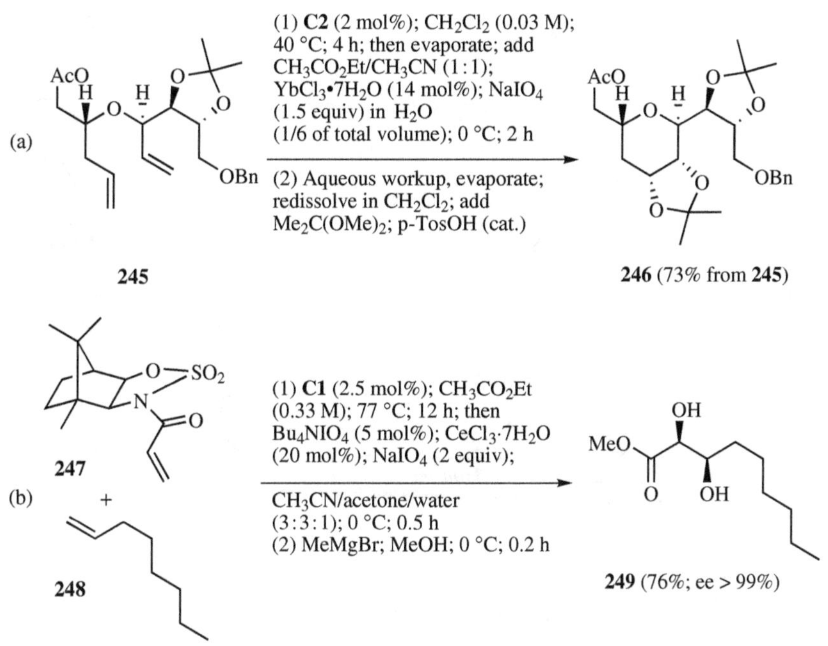

Scheme 5.69 Examples for assisted tandem metathesis/dihydroxylation reactions.

until the suspension turned yellow. Acetonitrile (3 ml) was then added, and this mixture was cooled to 0 °C followed by the addition of the completed metathesis reaction at 0 °C. After completion of the dihydroxylation step, Na_2SO_3 (s) was added, the mixture was stirred for another 0.2 h and then filtered through celite. After evaporation of all volatiles, the residue was purified by flash chromatography on silica using hexanes/ethyl acetate (1 : 4) as an eluent to furnish the title compound **243** (48 mg, 81%) as an oil.

Two further examples for tandem metathesis/dihydroxylation sequences are described in Scheme 5.69.

Crimmins et al. used Blechert's protocol in the total synthesis of the natural product amphidinol 3. The RCM precursor **245** was subjected to RCM/dihydroxylation conditions using first-generation catalyst **C2**, and the resulting diol was immediately converted to the acetal **246**. Notably, the reaction was run on a 7 g scale and the acetonide was furnished in a good overall yield (Scheme 5.69a) (145). In contrast, Kocovsky et al. reported that these conditions were less successful in their hands for the synthesis of aryl-*C*-ribonucleosides, and therefore a one-flask dual-catalyst approach, using $RuCl_3{\bullet}3H_2O$ as additional dihydroxylation catalyst, was preferred (146). Neisius and Plietker reported an auxiliary controlled enantioselective CM/dihydroxylation sequence, which is illustrated for the synthesis of diol **249** from sulfamidate **247** and 1-octene (**248**). Cleavage of the auxiliary was accomplished using magnesium methoxide, giving the diol **249** in virtually perfect enantioselectivity and good yield (Scheme 5.69b) (147).

5.3.4 Metathesis/Dehydrogenation and Metathesis/Allylic Oxidation

In this section, we summarize mechanistically distinct single-precatalyst reaction sequences that combine an olefin metathesis step and a subsequent formation of a carbonyl group. This carbonyl group can be located in the allylic position of the C–C double bond generated through metathesis or in a more remote position.

An example for the combination of metathesis, dehydrogenation, and alkene hydrogenation was published by Grubbs et al., who used the diene **250** as a precursor for a one-flask synthesis of the natural product (−)-muscone (Scheme 5.70) (85). After completion of the RCM step, NaOH and the co-solvent 3-pentanone were added to induce the transfer dehydrogenation of the secondary alcohol **251**. 3-Pentanone served as a hydrogen acceptor and was converted to 3-pentanol with concomitant formation of the ketone **252**. This ketone was not isolated, but subjected immediately to hydrogenation conditions (see Section 5.3.1 for details) to give (−)-muscone (**253**) in 56% yield based on diene **250**. Similarly, the CM of allylic alcohols and other olefins was combined with transfer dehydrogenation, using again 3-pentanone as a hydrogen acceptor and resulting in the formation of enones (85). In the same publication, the authors also described a tandem RCM/transfer hydrogenation,

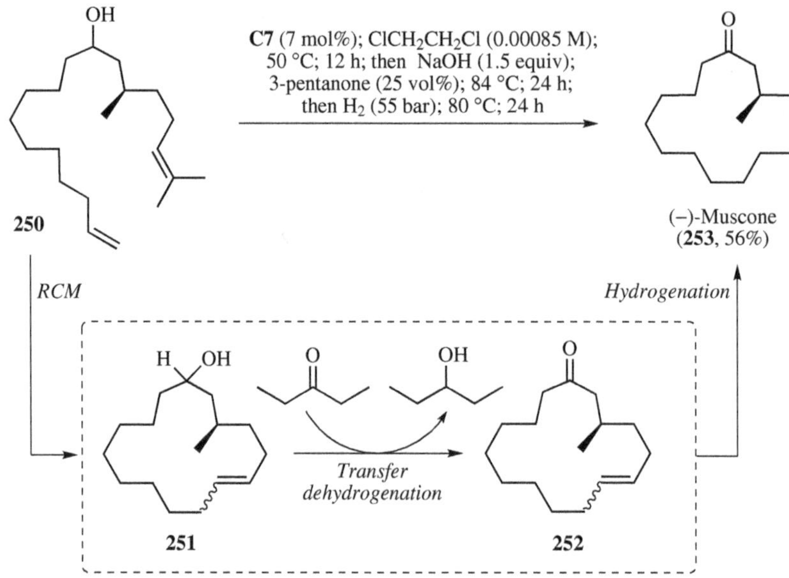

Scheme 5.70 Synthesis of muscone via RCM/dehydrogenation/hydrogenation sequence.

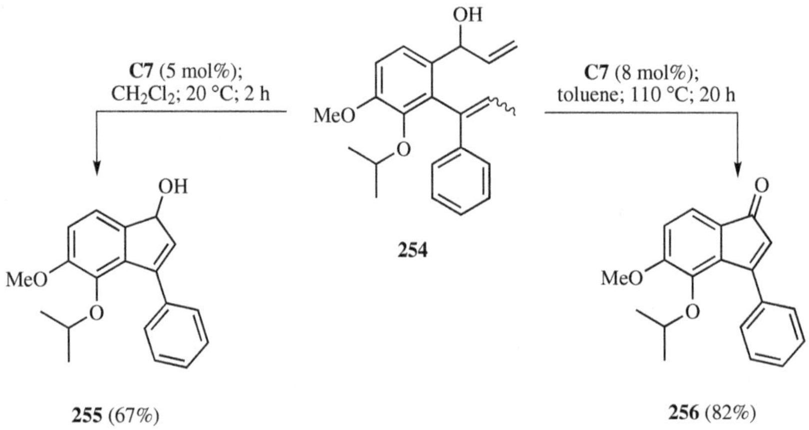

Scheme 5.71 RCM/dehydrogenation in the absence of hydrogen acceptors.

where the RCM product is a cyclic ketone that is reduced to the alcohol by treatment with 2-propanol as a co-solvent and a hydrogen donor in the presence of NaOH (85).

RCM/dehydrogenation reactions that do not require the addition of hydrogen acceptors have been reported by van Otterlo et al. (Scheme 5.71) (148). They discovered that RCM precursors for indene-3-ols such as **254** are converted to inden-3-ones (e.g., **256**) under more forcing conditions, even if oxygen is excluded. It is not fully clear whether molecular hydrogen is the co-product, but in the absence of any obvious hydrogen acceptor this appears to be a possible explanation. Interestingly, lower reaction temperatures and shorter reaction times lead to the formation of the RCM products (**255** from **254**) without subsequent dehydrogenation.

An example for an RCM/dehydrogenation that proceeds more efficiently in the presence of air was reported in the course of a synthesis of the carbazole alkaloid murrayaquinone A (Scheme 5.72) (149). Subjecting **257** to metathesis conditions did not result in the formation of the expected tautomer **258** of the RCM product, but lead directly to the MOM-protected murrayaquinone A **259**. It is, however, not clear whether the oxidation step leading from **258** to **259** is actually Ru catalyzed or occurs spontaneously.

Other substrates which occasionally undergo RCM with subsequent dehydrogenation in the absence of an obvious oxidant are diallyl amides (150–155) or diallyl ethers (156). In these cases, pyrroles or furans, respectively, are observed as by-products or instead of the expected 2,5-dihydropyrroles or 2,5-dihydrofurans. As for the example shown in Scheme 5.72, it is not always clear if the Ru-catalyst plays a role in these aromatization reactions, or if the dehydrogenation is an uncatalyzed oxidation caused by residual oxygen. A notable exception in this regard is the RCM/aromatization protocol

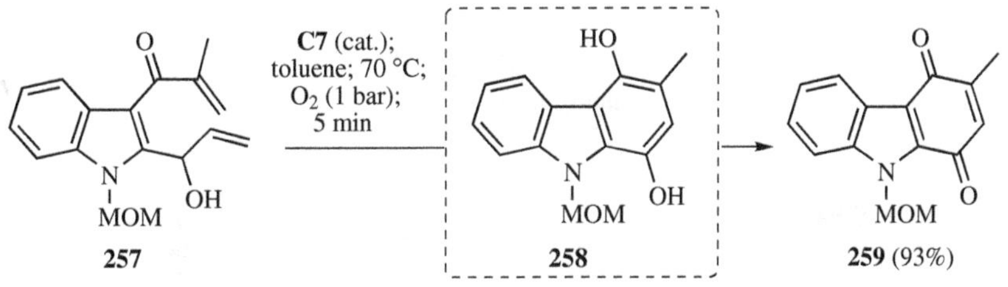

Scheme 5.72 RCM/dehydrogenation promoted by oxygen.

developed by Stevens et al., who used a dual-catalyst system consisting of first-generation Grubbs' catalyst (**C2**) to promote the RCM of diallyl amides and $RuCl_3$ to catalyze the aromatization of the RCM products to pyrroles (157). Reaction sequences using more than one precatalyst are beyond the scope of this chapter and will therefore not be discussed in detail.

The second group of tandem reaction sequences that can be used for introducing carbonyl groups do not rely on the dehydrogenation of alcohols but on Ru-catalyzed allylic oxidation. Similar to the RCM/isomerization sequence (see Section 5.3.2), the combination of RCM and allylic oxidation is an opportunity to circumvent a "difficult" metathesis step, that is, the metathesis of electron-deficient double bonds, such as enones or acrylates. These require often elevated catalyst loadings in combination with Lewis-acidic co-catalysts or the more active second-generation catalysts, and almost always high dilution conditions (123). Pioneering work on the field of Ru-catalyzed oxidation of sp^3-hybridized carbon centers was done by Murahashi and coworkers, who described numerous syntheses using Ru-precatalysts, in particular $[RuCl_2(PPh_3)_3]$ and *tert*-butyl hydroperoxide as an oxidant (131,158). An example is the highly selective benzylic oxidation of fluorene (**262**) to fluorenone (**266**) (Scheme 5.73) (159).

Based on kinetic studies, Murahashi et al. proposed a mechanism for this and similar oxidative transformations, which involves initial oxidation of the Ru-(II)-precatalyst **260** to a Ru-(IV)-oxo complex **261**. Complex **261** abstracts a H-radical from the substrate **262**, giving a caged Ru-(III)-hydroxy complex/fluorenyl radical pair **263**, which reacts further by transfer of the hydroxy group from the Ru to the substrate radical, giving alcohol **265**. At this stage, the Ru-catalyst **264** has a formal oxidation state of + II and is re-oxidized by *tert*-butyl hydroperoxide to the Ru-oxo complex **261**. The alcohol **265** is then oxidized with a second

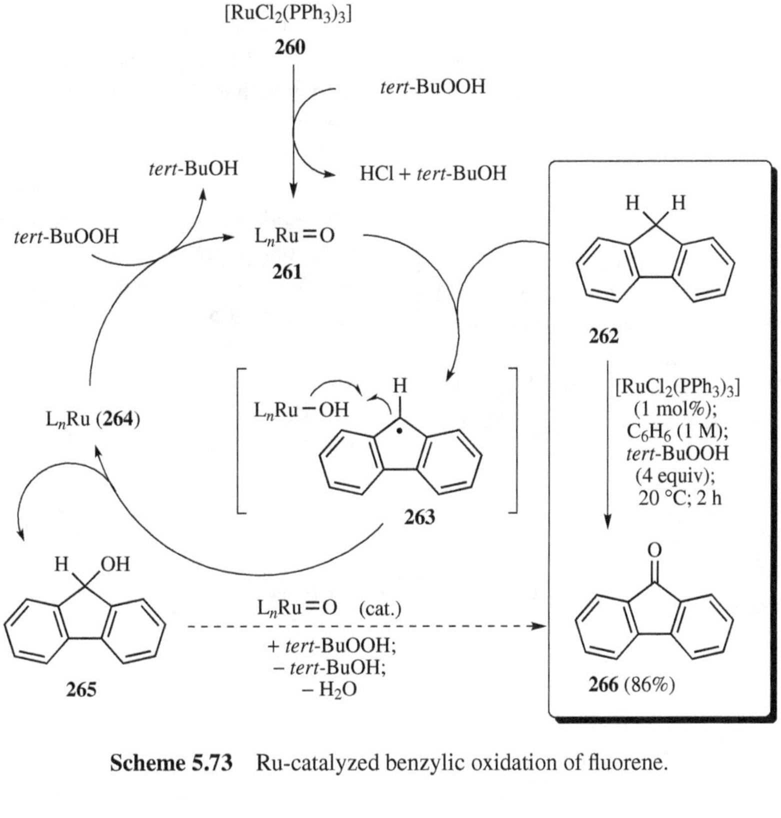

Scheme 5.73 Ru-catalyzed benzylic oxidation of fluorene.

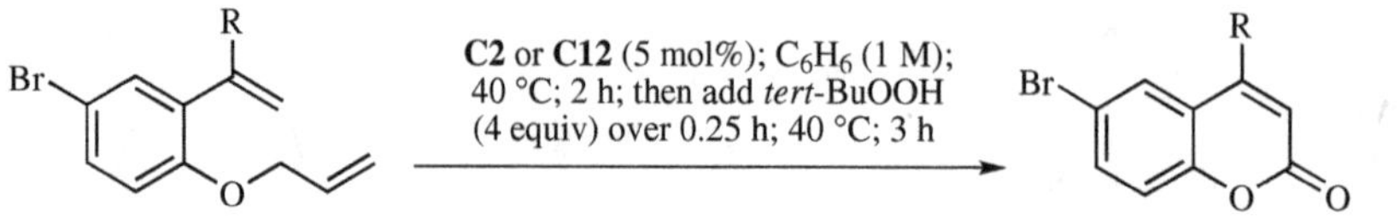

267a (R = –H)

267b (R = –CH_3)

268a (R = –H; 61% for **C2**; 62% for **C12**)
268b (R = –CH_3; 50% for **C2**; 45% for **C12**)

Scheme 5.74 Coumarins via RCM/allylic oxidation triggered by *tert*-BuOOH.

equivalent of *tert*-BuOOH to yield the ketone **266**. As can be seen from the example in Scheme 5.73, benzene is a suitable solvent for this type of Ru-catalyzed oxidations. Gratifyingly, most olefin metathesis reactions can be conducted in this solvent, and therefore, in contrast to the tandem sequences discussed in the previous section, no solvent compatibility problems need to be considered for a metathesis/allylic oxidation sequence. Such a sequence has recently been developed and applied to the synthesis of coumarins by us, as illustrated for the conversion of styrenyl allyl ethers **267** to coumarins **268**. For all examples, first-generation Grubbs' catalyst **C2** or the Umicore M1 catalyst **C12** could be used and gave very similar results. Remarkably high initial substrate concentrations of 1 M were tolerated and a commercially available solution of *tert*-BuOOH in decane could be used as purchased from the supplier (Scheme 5.74) (160a).

Synthesis of 6-bromocoumarin (268a) (160a)

To a solution of **267a** (239 mg, 1.0 mmol) in dry and degassed benzene (1 ml) was added **C12** (46 mg, 5 mol%). The solution was stirred at 40 °C for 2 h. After this time, a commercially available solution of *tert*-BuOOH in decane (5.5 M, 0.72 ml; 4.0 mmol) was added via syringe pump over a period of 15 min to the metathesis reaction mixture. After addition of this solution was complete, stirring at 40 °C was continued for 3 h. The solution was then cooled to ambient temperature, diluted with MTBE (50 ml), and $MgSO_4$ was added to absorb water and Ru-residues. The mixture was filtered through a pad of celite, and all volatiles were evaporated. The residue was chromatographed on silica (hexane/MTBE mixtures of increasing polarity) to furnish the title compound **268a** (140 mg, 62%) as a yellowish crystalline solid, mp 160–163 °C.

Although the fate of the metathesis catalyst on treatment with the chemical trigger *tert*-BuOOH could not be elucidated, it can be assumed from the results and some peroxide side products isolated from the reaction mixture that the Ru-(II)-methylidene species is oxidized to a Ru-(IV)-oxo complex, which then promotes the allylic oxidation analogous to the pathway proposed by Murahashi et al. (see Scheme 5.73). Parallel to our work, Arisawa et al. published a similar RCM/allylic oxidation sequence to access 2-quinolones (e.g., **270**) from benzylprotected allylamines such as **269** (Scheme 5.75) (161). In contrast to our conditions, these authors used the second-generation catalyst **C7**, most likely because the RCM step is more challenging as it involves the formation of a trisubstituted double bond, and the substrate contains a tertiary amine. Arisawa et al. used an aqueous solution of *tert*-BuOOH as the oxidant, which is also commercially available and could be used directly.

Ruthenium precatalysts in combination with hydroperoxides as oxidant do not necessarily result in the formation of allylic oxidation products. Instead, aromatization can be strongly preferred, in particular, for diallylethers or diallylamines, which are converted to furans or pyrroles, respectively. For example, the *N*-aryldiallyl amine **271** was converted to the corresponding *N*-aryl pyrrole **272** in high yield, using second-generation catalyst **C7** and aqueous *tert*-BuOOH as oxidizing agent (Scheme 5.76) (160b). For substrates without geminally disubstituted double bonds,

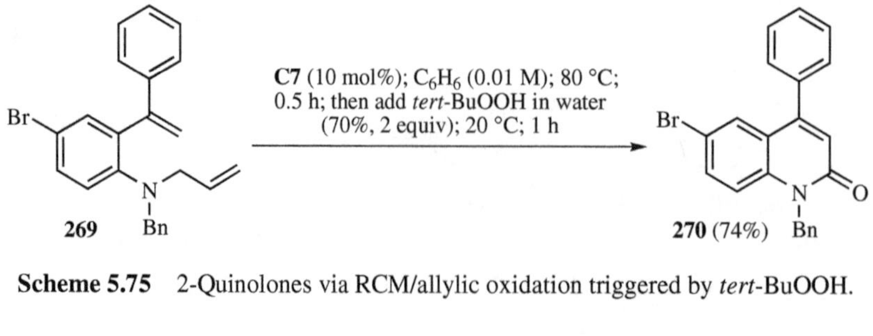

269

270 (74%)

Scheme 5.75 2-Quinolones via RCM/allylic oxidation triggered by *tert*-BuOOH.

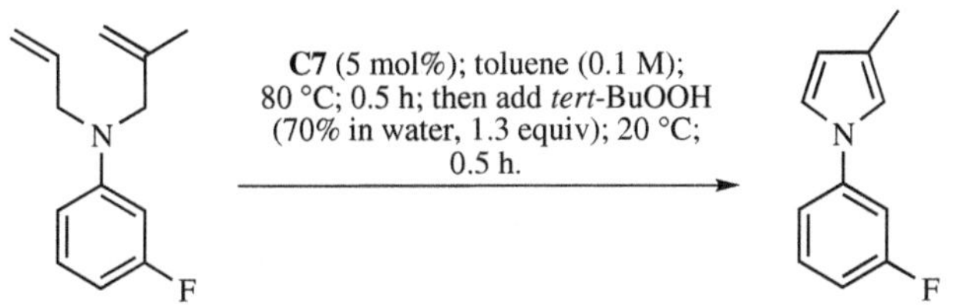

271 **272** (92%)

Scheme 5.76 *N*-Aryl pyrroles via RCM/aromatization triggered by *tert*-BuOOH.

first-generation catalyst **C2** was sufficiently reactive, and the RCM step could be conducted at ambient temperature with an initial substrate concentration of 1 M.

5.3.5 Metathesis/Atom Transfer Radical Reactions

Many ruthenium (162) and other transition metal (163–165) complexes were found to catalyze atom transfer radical reactions, often referred to as Kharasch additions (166). In the transition-metal-catalyzed version, a halogen radical is transferred from a polyhalogenated reactant, for example, a tetrahalomethane CX_4, to a metal complex that is thereby formally oxidized. The remaining carbon-centered radical $\bullet CX_3$ then adds to an alkene, and the new radical abstracts $\bullet X$ from the oxidized metal complex, thereby regenerating the catalyst. An early example for this type of reaction is the highly stereoselective addition of CCl_4 to cyclohexene, resulting in the formation of **273** (Scheme 5.77) (167).

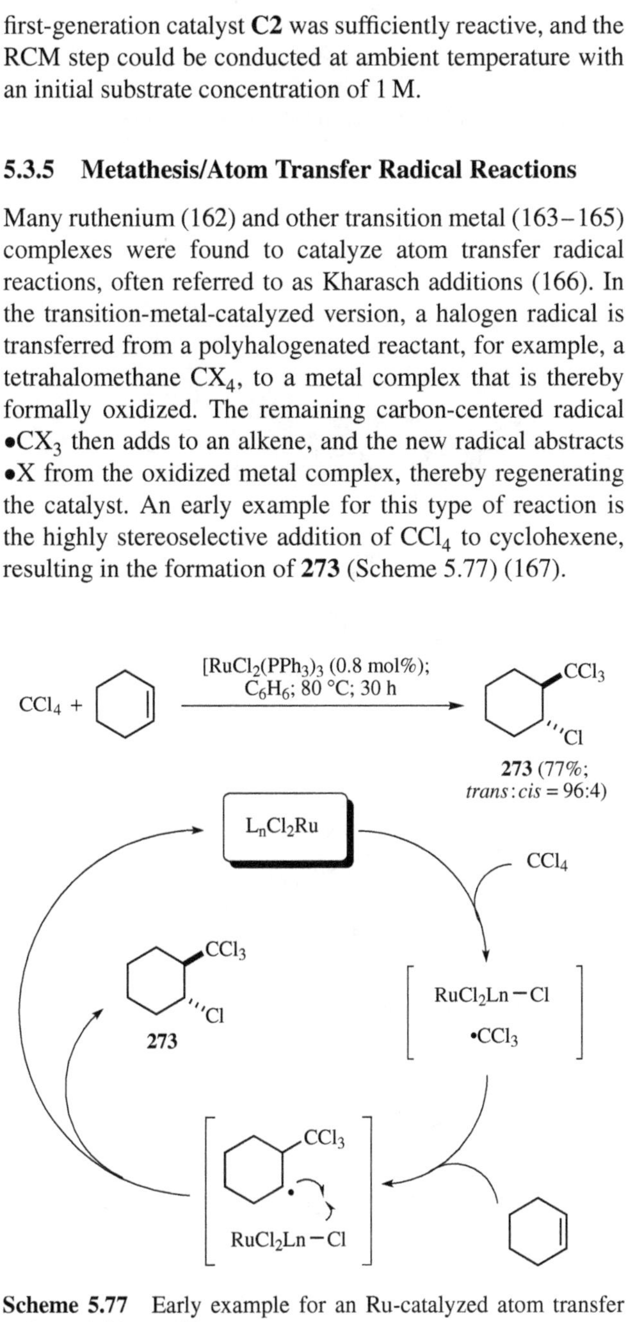

Scheme 5.77 Early example for an Ru-catalyzed atom transfer radical addition (ATRA).

A few years after the introduction of the first-generation Grubbs' catalyst **C2**, Snapper et al. discovered that this Ru-benzylidene complex is capable of promoting the atom transfer radical addition (ATRA) of chloroform to alkenes even more efficiently than the standard Ru complex $[RuCl_2(PPh_3)_3]$ (168).

An example is the reaction of methyl methacrylate (**274**) and chloroform to ester **275** (Scheme 5.78a) (168,169). A similar observation was reported by De Clercq and Verpoort for their Schiff-base ligated Ru-alkylidene complexes, which showed activity in metathesis and ATRA reactions (170). Shortly after Snapper's report, Demonceau et al. used **C2** and its PPh_3 analog in combination with CCl_4 to initiate the atom transfer radical polymerisation (ATRP) of styrene (**276**) to a polystyrene (**277**) with a comparatively narrow molecular weight distribution (Scheme 5.78b) (171). If a polyhalogenated substituent and a C–C double bond are present in the same molecule, an intramolecular reaction may occur. In these cases, the term atom transfer radical cyclization (ATRC) is commonly used. A typical example is the cyclization of trichloroacetamide **278** to lactam **279**, which proceeds in 70% yield with catalyst **C2** (Scheme 5.78c) (172), whereas 35% yield was reported with $[RuCl_2(PPh_3)_3]$ under otherwise similar conditions (173). A fourth type of atom transfer radical reaction is the cyclizative ATRA of 1,6-heptadienes and tetrahalomethanes, which had been accomplished using $[RuCl_2(PPh_3)_3]$ as a precatalyst (174). As illustrated for the cyclization of diallylcarbinol **280** to cyclopentanol **281** (Scheme 5.78d), the metathesis catalyst **C2** can also catalyze this transformation with very high chemoselectivity and diastereoselectivity (175). Notably, no cyclopentenol was observed, because the RCM of unprotected diallylcarbinols such as **280** is very slow with the less active first-generation catalyst **C2**. If this catalyst is replaced by the second-generation catalyst **C7**, only RCM products are observed under otherwise identical conditions (175).

The first assisted catalytic tandem sequences comprising metathesis and Ru-promoted radical reactions were RCM/ATRC sequences (176,177). A typical starting material is **282**, which contains the radical precursor trichloroacetate in an allylic position (176). In RCM/ATRC sequences, no additional reagent is required to trigger the non-metathesis reaction. Instead, the radical reaction is initiated by heating, and consequently synthetically useful selectivities can only be achieved if the temperature gap between metathesis and ATRC step is sufficiently large. This is indeed the case for **282**, which readily undergoes RCM to cyclopentene **283** at an ambient temperature in the presence of first-generation catalyst **C2**, whereas the radical reaction is initiated upon heating to reflux. For this particular example, the expected product **284** could not be isolated, because it underwent spontaneous elimination of HCl to give **285**, which cleanly isomerized to the bicyclic lactone **286** (Scheme 5.79) (176).

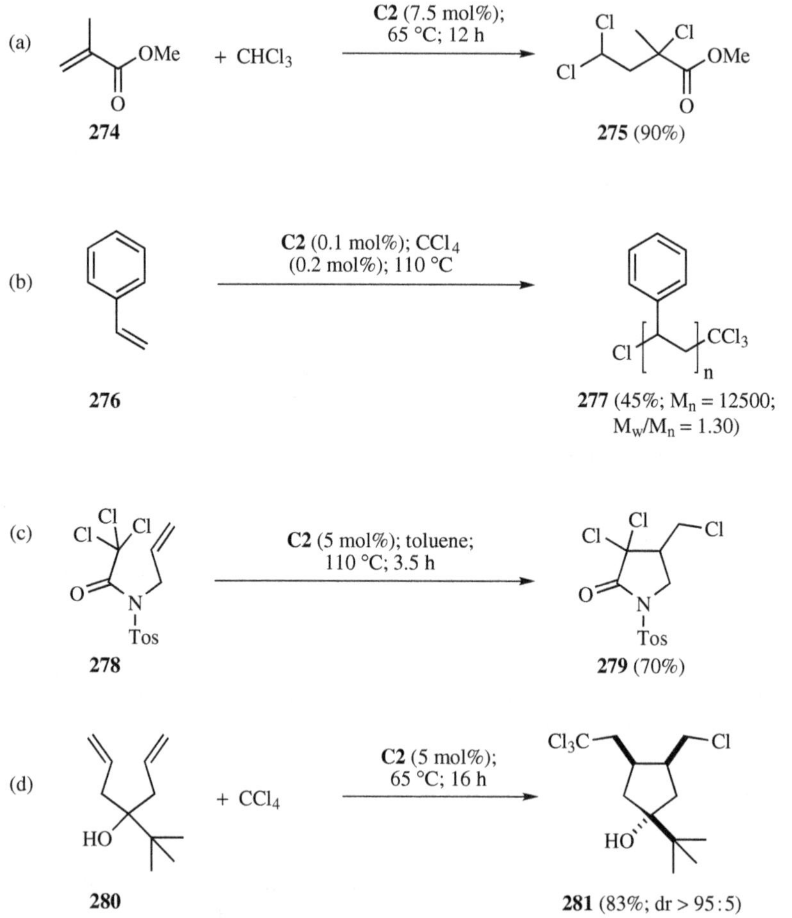

Scheme 5.78 Application of first-generation Grubbs' catalyst in ATR reactions.

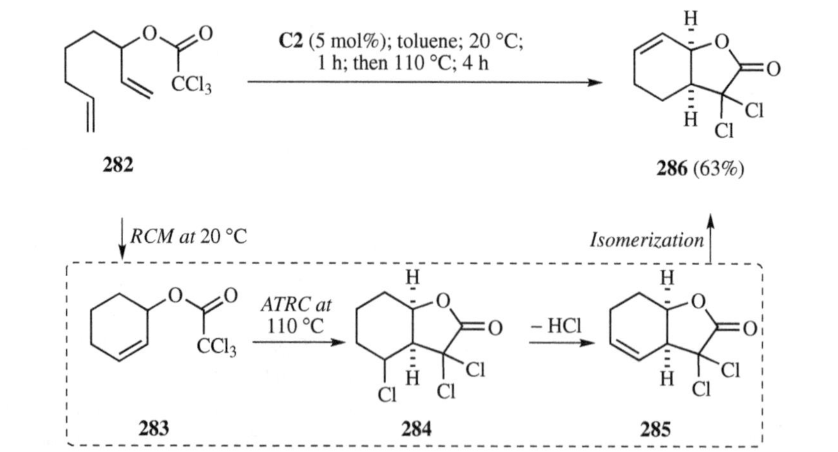

Scheme 5.79 Unsaturated bicyclic lactones through RCM/ATRC/elimination sequence.

Synthesis of *rac*-3,3-dichloro-3,3a,4,5-tetrahydro-benzofuran-2(7a*H*)-one (286)

To a solution of **282** (270 mg, 1.0 mmol) in dry and degassed toluene (10 ml) was added **C2** (41 mg, 5 mol%). The solution was stirred at 20 °C until complete consumption of the starting material as indicated by TLC (approx. 2 h). The mixture was subsequently heated to reflux until the intermediate RCM product was completely consumed (approx. 4 h). All volatiles were evaporated, and the residue was chromatographed on silica (cyclohexane/MTBE 20 : 1 as an eluent) to give the title compound **286** (130 mg, 63%) as a colorless oil.

Further examples for assisted tandem RCM/ATRC sequences are listed in Scheme 5.80. In contrast to **282**, the precursors **287** and **289** did not undergo RCM efficiently in the presence of first-generation catalyst **C2**. Therefore, the second-generation catalyst **C7** was used, and the five-membered lactones **288** and **290** were isolated in comparable yields as single diastereomers (Scheme 5.80a and Scheme 5.80b) (176). These results were somewhat unexpected, because second-generation catalysts had previously been reported to be significantly less active in promoting atom transfer radical reactions than first-generation catalysts (169).

All other examples given in Scheme 5.80 were reported by Snapper et al., who used only the first-generation catalyst **C2** to promote RCM/ATRC sequences (177). The conversion of **291** to **292** (Scheme 5.80c) illustrates that occasionally, in particular, for substrates containing amide bonds, even higher temperatures of up to 155 °C are necessary to induce the radical reaction. The presence of a carbonyl group adjacent to the radical precursor is not mandatory, as illustrated for the reaction of **293** to **294** that proceeds in high yield at 110 °C (Scheme 5.80d). The RCM/ATRC sequence can even be extended by an additional ATRA step, which is shown for the reaction of **295** and styrene to compound **296** (Scheme 5.80e). The final example (Scheme 5.80f) illustrates the problems caused by a small temperature gap between metathesis and radical reaction, because the ATRC step is already initiated at an ambient temperature and might compete with the RCM of **297**, resulting in the partial consumption of the starting material and hence a reduced yield of **298** (177). More recently, Sutherland et al. extended this type of reaction sequence by combination with a precursor synthesis through one-flask formation of a trichloroacetimidate **300** from allylic alcohol **299** and its enantioselective Pd-catalyzed Overman rearrangement to the RCM/ATRC precursor **301**. Diene **301** then underwent the RCM/ATRC sequence to the bicyclic lactam **302** in 70% yield based on the allylic alcohol **299** and 89% see (Scheme 5.81) (178).

Examples for a CM/ATRC sequence were reported by Quayle et al., who used the trichloroacetamide **303** in combination with 4-fluorostyrene (**304**) and other styrenes. These CM/ATRC sequences were accomplished with second-generation catalyst **C7** but required slightly elevated temperatures to promote the formation of the CM product (**305**) and a long reaction time of 3 days for the ATRC step. The expected γ-lactams, for example, **306**, were isolated in moderate yields between 23% and 32%, depending on the CM partner (Scheme 5.82) (179).

Very recently, intermolecular enyne metathesis and atom transfer radical addition were combined to a reaction sequence by Severin et al. (Scheme 5.83) (180). Intermolecular enyne metathesis has emerged as a useful method for the synthesis of conjugated dienes over the past decade (67,181,182), and the Ru-catalyzed ATRA to these substrates had previously been investigated and was found to proceed normally in a 1,4-fashion (162). In a typical example, Severin et al. reacted an arylalkyne such as **307** under 1 bar of ethene in the presence of second-generation catalyst **C7**. Upon completion of the enyne metathesis reaction, an organohalogen compound, for example, ethyl dichloroacetate, and activated magnesium were added. Magnesium had earlier been established as a rate-accelerating additive in Ru-catalyzed atom transfer radical reactions by Severin et al (183,184). It is believed to serve as a reducing agent and facilitate the regeneration of the catalytically active Ru-(II) species (185). Under these conditions, enyne metathesis/ATRA products such as **309** were obtained with good-to-fair regioselectivities, but in most cases with low levels of *E*/*Z*-selectivity.

5.3.6 Metathesis/Other C–C Bond Forming Reactions

Over recent years, there have been scattered reports describing other metathesis/non-metathesis reaction sequences in which the second step is a non-metathesis Ru-catalyzed C–C bond formation.

In 2005, Hodgson and Angrish described a hitherto unknown non-metathesis activity of Ru-carbene complex **C7**, the *Z*-selective dimerization of ethyl diazoacetate **310** to diethyl maleate **311** (Scheme 5.84a). Interestingly, after promoting the diazoacetate dimerization the catalytic activity in olefin metathesis is retained, and therefore both reactions can be coupled. This is illustrated for the dimerization of **312** to macrocycle **313**, which contains a *Z*-configured C–C double bond formed through olefin metathesis and a *Z*-configured C–C double bond, resulting from diazoacetate dimerization (Scheme 5.84b) (186,187). 2-Diazo-1,3-dicarbonyl compounds do not undergo this dimerization, but are tolerated in CM reactions and can be used in further Rh-catalyzed transformations via a one-flask dual-catalyst approach (188,189).

As diazoalkanes are useful reagents for the cyclopropanation of alkenes, a tandem sequence comprising a metathesis

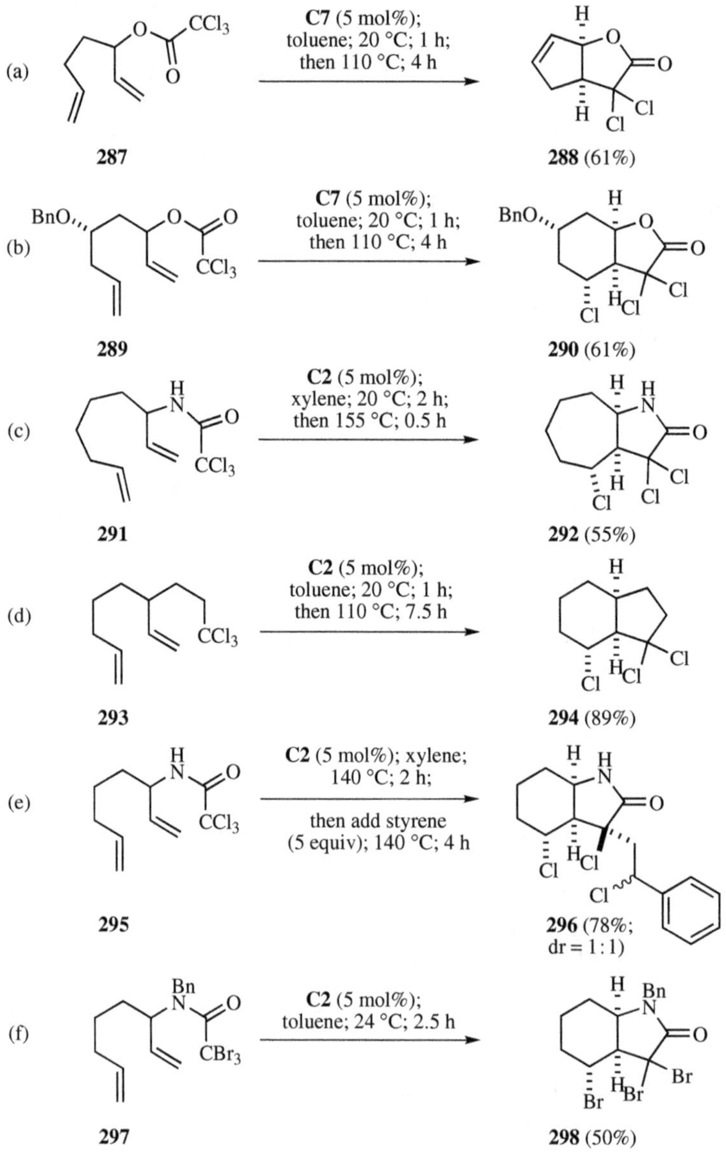

Scheme 5.80 Examples for RCM/ATRC sequences.

step and a cyclopropanation has also been investigated. An example is the reaction of amide **314** and ethyl diazoacetate to bicyclopropane **315** (Scheme 5.85) (190).

Seminal investigations into the Ru-hydride catalyzed hydrovinylation of conjugated dienes by Yi et al (191). inspired the development of an enyne metathesis/hydrovinylation sequence by Snapper et al. (Scheme 5.86) (192). A typical example is the transformation of enyne **316** to the diene **318**. The metathesis reaction was run under an atmosphere of ethene using first-generation catalyst **C2** to give initially the conjugated diene **317**. Addition of methanol in the presence of base induced the conversion of the Ru-carbene species into a Ru-hydride, which then catalyzes the hydrovinylation. Essentially, hydrovinylation results in a formal 1,4-addition of ethene to the conjugated diene. It proceeds via hydroruthenation of the conjugated diene **317** and π-coordination of ethene to give the π-allyl-Ru complex **319**, which reacts further in a migratory insertion of ethene into the Ru-C bond to give the Ru-σ complex **320**. The product **318** is liberated and the catalytically active Ru-hydride is regenerated by β-hydride elimination.

A very recently published cyclizative hydrovinylation of tri- or tetraynes promoted by typical metathesis catalysts is strictly speaking beyond the scope of this chapter, because no olefin or enyne metathesis reaction is involved. Nevertheless, we present one example for this transformation, because it is mechanistically closely related to the chemistry outlined above. The dipropargyl ether **321** undergoes cyclization

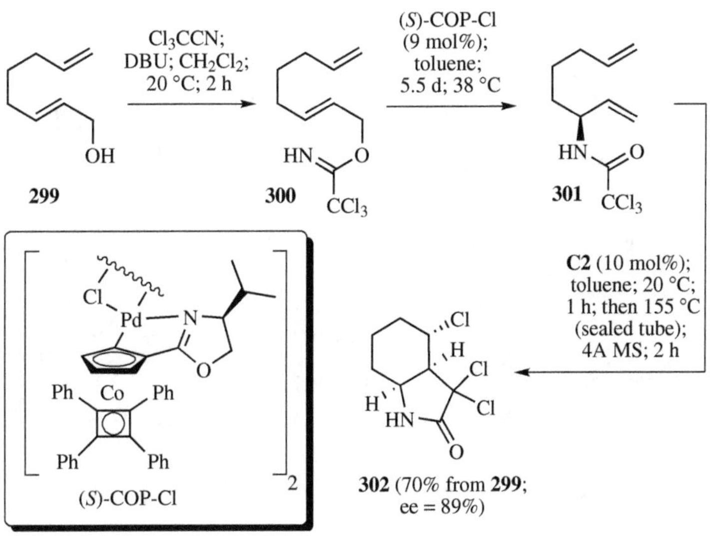

Scheme 5.81 One-flask synthesis of bicyclic lactams from allylic alcohols.

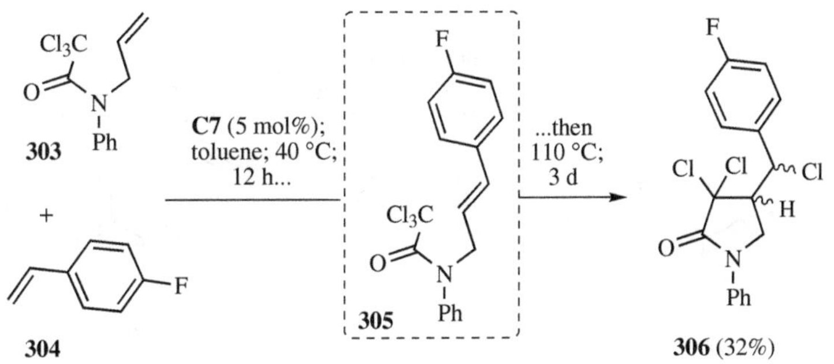

Scheme 5.82 Example for a CM/ATRC sequence.

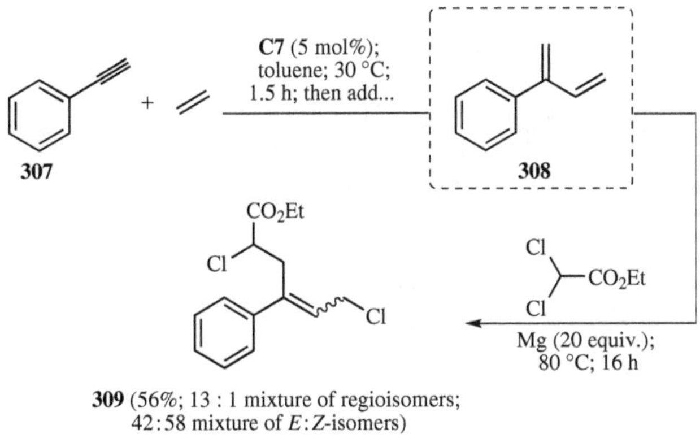

Scheme 5.83 Example for an enyne metathesis/ATRA sequence.

to the 3,4-bis-exo-alkylidene tetrahydrofuran **322** in the presence of second-generation Grubbs' catalyst **C7** and under an atmosphere of ethylene, with incorporation of 1 equiv of ethylene (location of ethene is highlighted by a dotted box in structure **322**) (193) (Scheme 5.87).

Ring closing metathesis has been combined with a Ru-catalyzed formal [2 + 2 + 1] carbonylative cycloaddition reaction by Snapper et al. (Scheme 5.88) (194). This transformation, sometimes characterized as hetero-Pauson–Khand (HPK) reaction, can be catalyzed by $Ru_3(CO)_{12}$ and other metal complexes. The Ru-catalyzed version has been thoroughly studied by Chatani et al. (195), who found that ketones, alkenes, and carbon monoxide react to γ-butyrolactones, and that a 2-pyridyl substituent adjacent to the ketone can serve as a coordinating group, leading to better results.

A representative example for this reaction sequence is the conversion of 2-pyridyl ketone **323**, which undergoes RCM to cycloheptene **324**. Transformation of the metathesis catalyst into the carbonylation catalyst was accomplished with NaOMe and carbon monoxide at elevated pressure. The resulting tricyclic lactone **325** was isolated as a single diastereomer in 71% yield (194). Taking the mechanistic rationale provided by Chatani into account (195), a Ru-chelate complex **326** is formed in which the C–C double bond, the carbonyl oxygen, the CO, and the directing pyridyl substituent

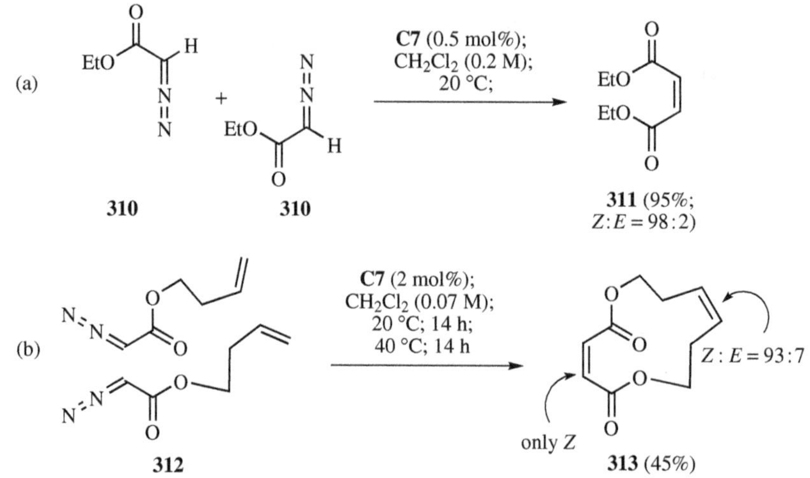

Scheme 5.84 Ru-carbene-catalyzed Z-selective diazoacetate dimerization and metathesis.

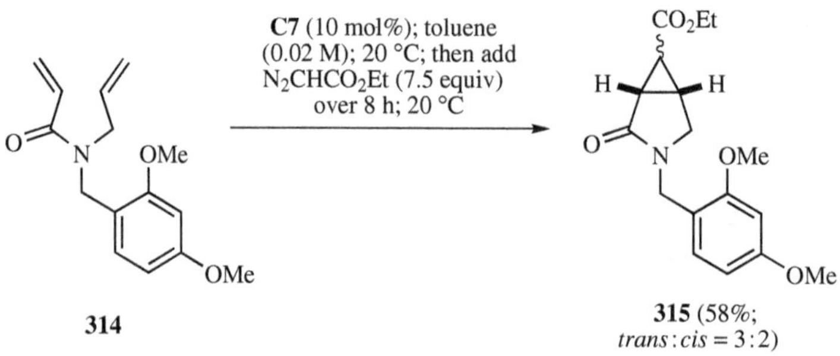

Scheme 5.85 Ru-catalyzed RCM/cyclopropanation with diazoacetate.

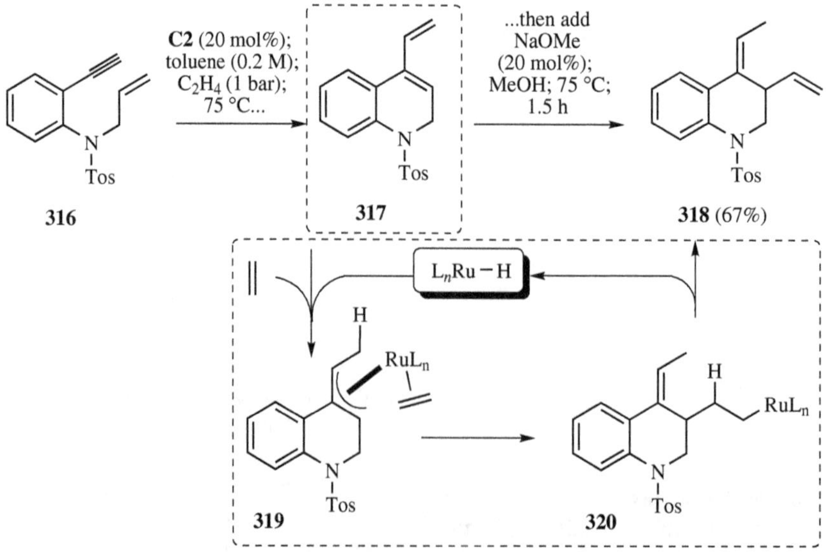

Scheme 5.86 Enyne metathesis/hydrovinylation sequence.

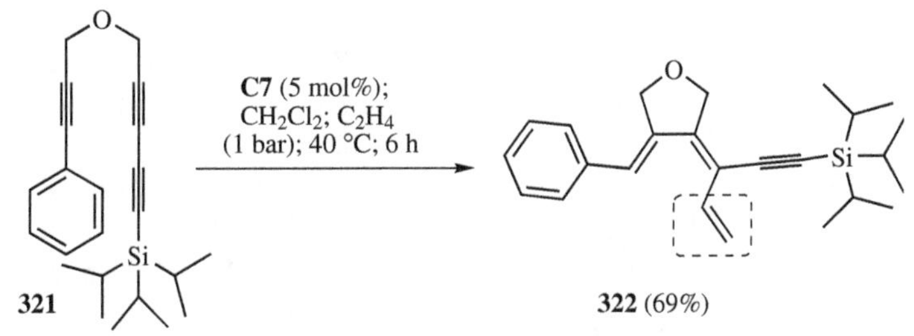

Scheme 5.87 Cyclizative hydrovinylation of a triyne catalyzed by Ru-carbenes.

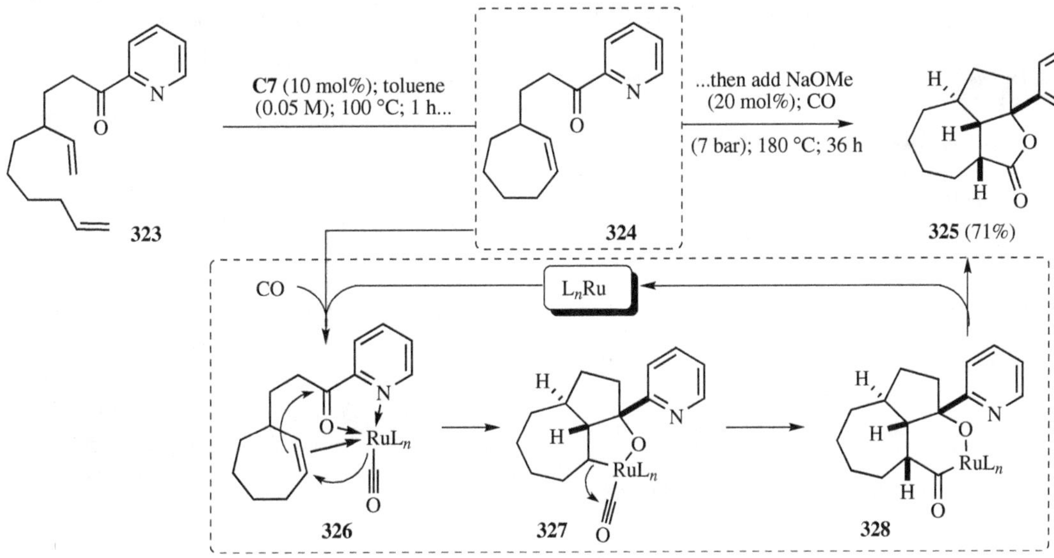

Scheme 5.88 RCM/hetero-Pauson–Khand sequence.

are bound to the Ru. Oxidative cyclization leads to **327**, followed by migratory insertion of the CO ligand into either the Ru-C-σ-bond as shown in Scheme 5.88 for intermediate **328**, or into the Ru-O bond (not shown). Reductive elimination of L_nRu finally furnishes the product, lactone **325**.

REFERENCES

1. Tietze LF. *Chem Rev* 1996;96:115–136.
2. Tietze LF, Schneider C, Grote A. *Chem Eur J* 1996;2: 139–148.
3. Ambrosini LM, Lambert TH. *ChemCatChem* 2010; 2:1373–1380.
4. Chapman CJ, Frost CG. *Synthesis* 2007:1–21.
5. Fogg DE, dos Santos EN. *Coord Chem Rev* 2004;248: 2365–2379.
6. Behr A. *Angewandte homogene Katalyse*. Weinheim: Wiley-VCH; 2008.
7. Faulkner J, Edlin CD, Fengas D, Preece I, Quayle P, Richards SN. *Tetrahedron Lett* 2005;46:2381–2385.
8. Ajamian A, Gleason JL. *Angew Chem Int Ed* 2004;43: 3754–3760.
9. Schrock RR, Murdzek JS, Bazan GC, Robbins J, DiMare M, O'Regan M. *J Am Chem Soc* 1990;112:3875–3886.
10. Fu GC, Nguyen ST, Grubbs RH. *J Am Chem Soc* 1993;115: 9856–9857.
11. Hérisson J-L, Chauvin Y. *Makromol Chem* 1971;141: 161–176.
12. Schmidt B. *Pure Appl Chem* 2006;78:469–476.
13. Garber SB, Kingsbury JS, Gray BL, Hoveyda AH. *J Am Chem Soc* 2000;122:8168–8179.
14. Gessler S, Randl S, Blechert S. *Tetrahedron Lett* 2000; 41:9973–9976.
15. Schwab P, Grubbs RH, Ziller JW. *J Am Chem Soc* 1996;118:100–110.
16. Van Veldhuizen JJ, Campbell JE, Giudici RE, Hoveyda AH. *J Am Chem Soc* 2005;127:6877–6882.
17. Yu M, Ibrahem I, Hasegawa M, Schrock RR, Hoveyda AH. *J Am Chem Soc* 2012;134:2788–2799.
18. Berlin JM, Goldberg SD, Grubbs RH. *Angew Chem Int Ed* 2006;45:7591–7595.

19. Kannenberg A, Rost D, Eibauer S, Tiede S, Blechert S. *Angew Chem Int Ed* 2011;50:3299–3302.
20. Scholl M, Ding S, Lee CW, Grubbs RH. *Org Lett* 1999; 1:953–956.
21. Kingsbury JS, Harrity JPA, Bonitatebus PJ Jr,, Hoveyda AH. *J Am Chem Soc* 1999;121:791–799.
22. Zhu SS, Cefalo DR, La DS, Jamieson JY, Davis WM, Hoveyda AH, Schrock RR. *J Am Chem Soc* 1999;121:8251–8259.
23. Dias EL, Grubbs RH. *Organometallics* 1998;17:2758–2767.
24. Sanford MS, Love JA, Grubbs RH. *Organometallics* 2001;20:5314–5318.
25. Fürstner A, Hill AF, Liebl M, Wilton-Ely JDET. *Chem Commun* 1999:601–602.
26. Harlow KJ, Hill AF, Wilton-Ely JDET. *J Chem Soc Dalton Trans* 1999:285–291.
27. Grela K. *Angew Chem Int Ed* 2008;47:5504–5507.
28. Song A, Parker KA, Sampson NS. *J Am Chem Soc* 2009;131:3444–3445.
29. Sattely ES, Meek SJ, Malcolmson SJ, Schrock RR, Hoveyda AH. *J Am Chem Soc* 2009;131:943–953.
30. Michaut M, Parrain J-L, Santelli M. *Chem Commun* 1998:2567–2568.
31. Statsuk AV, Liu D, Kozmin SA. *J Am Chem Soc* 2004; 126:9546–9547.
32. Marjanovic J, Kozmin SA. *Angew Chem Int Ed* 2007; 46:8854–8857.
33. Matsumoto K, Kozmin SA. *Adv Synth Catal* 2008; 350:557–560.
34. Hoveyda AH, Lombardi PJ, O'Brien RV, Zhugralin AR. *J Am Chem Soc* 2009;131:8378–8379.
35. Giudici RE, Hoveyda AH. *J Am Chem Soc* 2007; 129:3824–3825.
36. Bassindale MJ, Hamley P, Harrity JPA. *Tetrahedron Lett* 2001;42:9055–9057.
37. Limanto J, Snapper ML. *J Am Chem Soc* 2000; 122:8071–8072.
38. Schrader TO, Snapper ML. *Tetrahedron Lett* 2000; 41:9685–9689.
39. Shizuka M, Snapper ML. *Synthesis* 2007:2397–2403.
40. Davy JA, Moreau B, Wulff JE. *Synthesis* 2012;44:1854–1862.
41. Arjona O, Csaky AG, Plumet J. *Eur J Org Chem* 2003: 611–622.
42. Cuny GD, Cao J, Sidhu A, Hauske JR. *Tetrahedron* 1999;55:8169–8178.
43. Hoveyda AH, Schrock RR. *Chem Eur J* 2001;7:945–950.
44. La DS, Sattely ES, Ford JG, Schrock RR, Hoveyda AH. *J Am Chem Soc* 2001;123:7767–7778.
45. Holub N, Blechert S. *Chem Asian J* 2007;2:1064–1082.
46. Tang B, Bray CD, Pattenden G, Rogers J. *Tetrahedron* 2010;66:2492–2500.
47. Yasui H, Yamamoto S, Takao K-i, Tadano K-i. *Heterocycles* 2006;70:135–141.
48. White BH, Snapper ML. *J Am Chem Soc* 2003; 125:14901–14904.
49. Miege F, Meyer C, Cossy J. *Org Lett* 2010;12:248–251.
50. Zaja M, Blechert S. *Tetrahedron* 2004;60:9629–9634.
51. Schmidt B, Hermanns J. *Curr Org Chem* 2006;10:1363–1396.
52. Holub N, Neidhöfer J, Blechert S. *Org Lett* 2005;7:1227–1229.
53. Böhrsch V, Neidhöfer J, Blechert S. *Angew Chem Int Ed* 2006;45:1302–1305.
54. Cefalo DR, Kiely AF, Wuchrer M, Jamieson JY, Schrock RR, Hoveyda AH. *J Am Chem Soc* 2001;123:3139–3140.
55. Usher LC, Estrella-Jimenez M, Ghiviriga I, Wright DL. *Angew Chem Int Ed* 2002;41:4560–4562.
56. Barbe G, Charette AB. *J Am Chem Soc* 2008; 130:13873–13875.
57. Lee H-Y, Lee S-S, Kim HS, Lee KM. *Eur J Org Chem* 2012:4192–4199.
58. Schmidt B. *J Mol Catal A* 2006;254:53–57.
59. Bose S, Ghosh M, Ghosh S. *J Org Chem* 2012;77:6345–6350.
60. (a) Standen PE, Dodia D, Elsegood MRJ, Teat SJ, Kimber MC. *Org Biomol Chem* 2012;10:8669–8676. (b) Gao F, Stamp CTM, Thornton PD, Cameron TS, Doyle LE, Miller DO, Burnell DJ. *Chem. Commun.* 2012;48:233–235.
61. Randl S, Blechert S. In: Grubbs RH, editor. Handbook of Metathesis. Vol. 2. Weinheim: Wiley-VCH; 2003. p 151–175.
62. Porta M, Blechert S. In: Cossy J, Arseniyadis S, Meyer C, editors. *Metathesis in Natural Product Synthesis – Strategies, Substrates and Catalysts*. Weinheim: Wiley-VCH; 2010. p 313–341.
63. Nicolaou KC, Vega JA, Vassilikogiannakis G. *Angew Chem Int Ed* 2001;40:4441–4445.
64. Stapper C, Blechert S. *J Org Chem* 2002;67:6456–6460.
65. Donnard M, Tschamber T, Le Nouën D, Desrat S, Hinsinger K, Eustache J. *Tetrahedron* 2011;67:339–357.
66. Zuercher WJ, Hashimoto M, Grubbs RH. *J Am Chem Soc* 1996;118:6634–6640.
67. Diver ST, Giessert AJ. *Chem Rev* 2004;104:1317–1382.
68. Hoye TR, Jeffrey CS, Tennakoon MA, Wang J, Zhao H. *J Am Chem Soc* 2004;126:10210–10211.
69. Wallace DJ. *Angew Chem Int Ed* 2005;44:1912–1915.
70. Banti D, North M. *Adv Synth Catal* 2002;344:694–704.
71. Drouin SD, Zamanian F, Fogg DE. *Organometallics* 2001;20:5495–5497.
72. Camm KD, Castro NM, Liu Y, Czechura P, Snelgrove JL, Fogg DE. *J Am Chem Soc* 2007;129:4168–4169.
73. Watson MD, Wagener KB. *Macromolecules* 2000;33: 3196–3201.
74. Drouin SD, Yap GPA, Fogg DE. *Inorg Chem* 2000;39: 5412–5414.
75. Beach NJ, Camm KD, Fogg DE. *Organometallics* 2010;29:5450–5455.
76. Dinger MB, Mol JC. *Organometallics* 2003;22:1089–1095.
77. Dinger MB, Mol JC. *Eur J Inorg Chem* 2003:2827–2833.
78. Louie J, Grubbs RH. *Organometallics* 2002;21:2153–2164.

79. Arisawa M, Terada Y, Nakagawa M, Nishida A. *Angew Chem Int Ed* 2002;41:4732–4734.
80. Wilhelm TE, Belderrain TR, Brown SN, Grubbs RH. *Organometallics* 1997;16:3867–3869.
81. Oliván M, Caulton KG. *Inorg Chem* 1999;38:566–570.
82. Schmidt B. *Eur J Org Chem* 2004:1865–1880.
83. (a) Hong SH, Day MW, Grubbs RH. *J Am Chem Soc* 2004;126:7414–7415. (b) During typesetting of this chapter an investigation into the isomerization activiy of complex 121 was published by Fogg et al., who provide strong evidence that this particular Ru-hydride is not responsible for double bond isomerization. See, Higman CS, Plais L, Fogg DE. *Chem Cat Chem* 2013;5:3548–3551.
84. Beach NJ, Lummiss JAM, Bates JM, Fogg DE. *Organometallics* 2012;31:2349–2356.
85. Louie J, Bielawski CW, Grubbs RH. *J Am Chem Soc* 2001;123:11312–11313.
86. Yi CS, Lee DW. *Organometallics* 1999;18:5152–5156.
87. Yi CS, Lee DW, He Z, Rheingold AL, Lam K–C, Concolino TE. *Organometallics* 2000;19:2909–2915.
88. Rowley CN, Foucault HM, Woo TK, Fogg DE. *Organometallics* 2008;27:1661–1663.
89. LaCruz TE, Rychnovsky SD. *J Org Chem* 2007; 72:2602–2611.
90. VanOrden LJ, Patterson BD, Rychnovsky SD. *J Org Chem* 2007;72:5784–5793.
91. Fürstner A, Leitner A. *Angew Chem Int Ed* 2003;42:308–311.
92. Scheiper B, Glorius F, Leitner A, Fürstner A. *Proc Natl Acad Sci U S A* 2004;101:11960–11965.
93. Clark JS, Labre F, Thomas LH. *Org Biomol Chem* 2011;9:4823–4830.
94. Børsting P, Nielsen KE, Nielsen P. *Org Biomol Chem* 2005;3:2183–2190.
95. Børsting P, Nielsen P. *Chem Commun* 2002:2140–2141.
96. Børsting P, Freitag M, Nielsen P. *Tetrahedron* 2004; 60:10955–10966.
97. Ramharter J, Weinstabl H, Mulzer J. *J Am Chem Soc* 2010;132:14338–14339.
98. Schmidt B, Pohler M. *Org Biomol Chem* 2003;1:2512–2517.
99. Schmidt B, Krehl S, Sotelo-Meza V. *Synthesis* 2012; 44:1603–1613.
100. Menozzi C, Dalko PI, Cossy J. *Synlett* 2005:2449–2452.
101. Schmidt B, Kunz O. *Eur J Org Chem* 2012:1008–1018.
102. Lukevics E, Dzintara M. *J Organomet Chem* 1985; 295:265–315.
103. Sutton AE, Seigal BA, Finnegan DF, Snapper ML. *J Am Chem Soc* 2002;124:13390–13391.
104. Schmidt B. *Eur J Org Chem* 2003:816–819.
105. Schmidt B. *Chem Commun* 2004:742–743.
106. Wu Z, Nguyen ST, Grubbs RH, Ziller JW. *J Am Chem Soc* 1995;117:5503–5511.
107. Schmidt B. *J Org Chem* 2004;69:7672–7687.
108. Schmidt B. *Synlett* 2004:1541–1544.
109. Fustero S, Sánchez-Roselló M, Jiménez D, Sanz-Cervera JF, del Pozo C, Aceña JL. *J Org Chem* 2006;71:2706–2714.
110. Sturino CF, Wong JCY. *Tetrahedron Lett* 1998; 39:9623–9626.
111. Peczuh MW, Snyder NL. *Tetrahedron Lett* 2003; 44:4057–4061.
112. Schmidt B, Hölter F. *Chem Eur J* 2009;15:11948–11953.
113. Schmidt B, Hölter F, Kelling A, Schilde U. *J Org Chem* 2011;76:3357–3365.
114. Böhrsch V, Blechert S. *Chem Commun* 2006:1968–1970.
115. Schmidt B, Staude L. *J Organomet Chem* 2006; 691:5218–5221.
116. Schmidt B, Biernat A. *Chem Eur J* 2008;14:6135–6141.
117. Schmidt B, Biernat A. *Eur J Org Chem* 2008:5764–5769.
118. Schmidt B, Biernat A. *Org Lett* 2008;10:105–108.
119. Bressy C, Menant C, Piva O. *Synlett* 2005:577–582.
120. Ascic E, Jensen JF, Nielsen TE. *Angew Chem Int Ed* 2011;50:5188–5191.
121. Ascic E, Le Quement ST, Ishoey M, Daugaard M, Nielsen TE. *ACS Comb Sci* 2012;14:253–257.
122. Schmidt B, Kunz O. *Synlett* 2012;23:851–854.
123. Schmidt B, Geißler D. *ChemCatChem* 2010;2:423–429.
124. Hoye TR, Zhao H. *Org Lett* 1999;1:169–171.
125. Gurjar MK, Yakambram P. *Tetrahedron Lett* 2001; 42:3633–3636.
126. Greenwood ES, Parsons PJ, Young MJ. *Synth Commun* 2003;33:223–228.
127. Finnegan D, Seigal BA, Snapper ML. *Org Lett* 2006;8:2603–2606.
128. Werner H, Grunwald C, Stuer W, Wolf J. *Organometallics* 2003;22:1558–1560.
129. Sharpless KB, Amberg W, Bennani YL, Crispino GA, Hartung J, Jeong K-S, Kwong H-L, Morikawa K, Wang Z-M, Xu D, Zhang X-L. *J Org Chem* 1992;57:2768–2771.
130. Kolb HC, VanNieuwenhze MS, Sharpless KB. *Chem Rev* 1994;94:2483–2547.
131. Murahashi S-I, Komiya N. In: Murahashi S-I, editor. *Ruthenium in Organic Synthesis.* Weinheim: Wiley-VCH; 2004. p 53–93.
132. Plietker B, Niggemann M. *Org Biomol Chem* 2004; 2:2403–2407.
133. Carlsen PHJ, Katsuki T, Martin VS, Sharpless KB. *J Org Chem* 1981;46:3936–3938.
134. Shing TKM, Tai VW-F, Tam EKW. *Angew Chem Int Ed* 1994;33:2312–2313.
135. Shing TKM, Tam EKW, Tai VW-F, Chung IHF, Jiang Q. *Chem Eur J* 1996;2:50–57.
136. Plietker B, Niggemann M, Pollrich A. *Org Biomol Chem* 2004;2:1116–1124.
137. Plietker B, Niggemann M. *J Org Chem* 2005;70:2402–2405.
138. Niggemann M, Jelonek A, Biber N, Wuchrer M, Plietker B. *J Org Chem* 2008;73:7028–7036.
139. Plietker B. *J Org Chem* 2004;69:8287–8296.

140. Beligny S, Eibauer S, Maechling S, Blechert S. *Angew Chem Int Ed* 2006;45:1900–1903.
141. Miao X, Fischmeister C, Bruneau C, Dixneuf PH. *ChemSusChem* 2008;1:813–816.
142. Adjiman CS, Clarke AJ, Cooper G, Taylor PC. *Chem Commun* 2008:2806–2808.
143. Krehl S, Geißler D, Hauke S, Kunz O, Staude L, Schmidt B. *Beilstein J Org Chem* 2010;6:1188–1198.
144. Scholte AA, An MH, Snapper ML. *Org Lett* 2006; 8:4759–4762.
145. Crimmins MT, Martin TJ, Martinot TA. *Org Lett* 2010; 12:3890–3893.
146. Stambasky J, Kapras V, Stefko M, Kysilka O, Hocek M, Malkov AV, Kocovsky P. *J Org Chem* 2011;76:7781–7803.
147. Neisius NM, Plietker B. *J Org Chem* 2008;73:3218–3227.
148. van Otterlo WAL, Coyanis EM, Panayides J-L, de Koning CB, Fernandes MA. *Synlett* 2005:501–505.
149. Nishiyama T, Choshi T, Kitano K, Hibino S. *Tetrahedron Lett* 2011;52:3876–3878.
150. Evans P, Grigg R, Monteith M. *Tetrahedron Lett* 1999; 40:5247–5250.
151. Bujard M, Briot A, Gouverneur V, Mioskowski C. *Tetrahedron Lett* 1999;40:8785–8788.
152. Sletten EM, Liotta LJ. *J Org Chem* 2006;71:1335–1343.
153. Yang C, Murray WV, Wilson LJ. *Tetrahedron Lett* 2003;44:1783–1786.
154. Yang Q, Li X-Y, Wu H, Xiao W-J. *Tetrahedron Lett* 2006;47:3893–3896.
155. Ahmed-Omer B, Barrow DA, Wirth T. *Arkivoc* 2011;iv:26–36.
156. Evanno L, Nay B, Bodo B. *Synth Commun* 2005; 35:1559–1565.
157. Dieltiens N, Stevens CV, De Vos D, Allaert B, Drozdzak R, Verpoort F. *Tetrahedron Lett* 2004;45:8995–8998.
158. Murahashi S-I. *Angew Chem Int Ed* 1995;34:2443–2465.
159. Murahashi S-I, Komiya N, Oda Y, Kuwabara T, Naota T. *J Org Chem* 2000;65:9186–9193.
160. (a) Schmidt B, Krehl S. *Chem Commun* 2011;47:5879–5881. (b) Schmidt B, Krehl S, Jablowski E. *Org Biomol Chem* 2012;10:5119–5130.
161. Kato H, Ishigame T, Oshima N, Hoshiya N, Shimawaki K, Arisawa M, Shuto S. *Adv Synth Catal* 2011;353:2676–2680.
162. Nagashima H. In: Murahashi S-I, editor. *Ruthenium in Organic Synthesis*. Weinheim: Wiley-VCH; 2004. p 333–343.
163. Clark AJ. *Chem Soc Rev* 2002;31:1–11.
164. Iqbal J, Bhatia B, Nayyar NK. *Chem Rev* 1994;94:519–564.
165. Giese B, Kopping B, Göbel T, Dickhaut J, Thoma G, Kulicke KJ, Trach F. *Org React* 1996;48:301–856.
166. Kharasch MS, Jensen EV, Urry WH. *J Am Chem Soc* 1947;69:1100–1105.
167. Matsumoto H, Nikaido T, Nagai Y. *Tetrahedron Lett* 1975;16:899–902.
168. Tallarico JA, Malnick LM, Snapper ML. *J Org Chem* 1999;64:344–345.
169. Lee BT, Schrader TO, Martín-Matute B, Kauffman CR, Zhang P, Snapper ML. *Tetrahedron* 2004;60:7391–7396.
170. De Clercq B, Verpoort F. *Tetrahedron Lett* 2001; 42:8959–8963.
171. Simal F, Demonceau A, Noels AF. *Tetrahedron Lett* 1999;40:5689–5693.
172. Quayle P, Fengas D, Richards S. *Synlett* 2003:1797–1800.
173. Nagashima H, Wakamatsu H, Ozaki N, Ishii T, Watanabe M, Tajima T, Itoh K. *J Org Chem* 1992;57:1682–1689.
174. Grigg R, Devlin J, Ramasubbu A, Scott RM, Stevenson P. *J Chem Soc Perkin 1* 1987:1515–1520.
175. Schmidt B, Pohler M, Costisella B. *J Org Chem* 2004;69:1421–1424.
176. Schmidt B, Pohler M. *J Organomet Chem* 2005; 690:5552–5555.
177. Seigal BA, Fajardo C, Snapper ML. *J Am Chem Soc* 2005;127:16329–16332.
178. McGonagle FI, Brown L, Cooke A, Sutherland A. *Org Biomol Chem* 2010;8:3418–3425.
179. Edlin CD, Faulkner J, Quayle P. *Tetrahedron Lett* 2006; 47:1145–1151.
180. Kiefer G, Ruiz J, Solari E, Hilt G, Severin K. *Eur J Org Chem* 2012:93–98.
181. Diver ST, Giessert AJ. *Synthesis* 2004:466–471.
182. Diver ST. *Coord Chem Rev* 2007;251:671–701.
183. Thommes K, Içli B, Scopelliti R, Severin K. *Chem Eur J* 2007;13:6899–6907.
184. Fernández-Zúmel MA, Thommes K, Kiefer G, Sienkiewicz A, Pierzchala K, Severin K. *Chem Eur J* 2009;15: 11601–11607.
185. Pintauer T. *Eur J Inorg Chem* 2010:2449–2460.
186. Hodgson DM, Angrish D. *Chem Commun* 2005:4902–4904.
187. Hodgson DM, Angrish D. *Chem Eur J* 2007;13:3470–3479.
188. Hodgson DM, Angrish D, Labande AH. *Chem Commun* 2006:627–628.
189. Hodgson DM, Angrish D. *Adv Synth Catal* 2006; 348:2509–2514.
190. Mallagaray A, Dominguez G, Gradillas A, Perez-Castells J. *Org Lett* 2008;10:597–600.
191. Yi CS, Lee DW, Chen Y. *Organometallics* 1999; 18:2043–2045.
192. Gavenonis J, Arroyo RV, Snapper ML. *Chem Commun* 2010;46:5692–5694.
193. Yun SY, Wang K-P, Kim M, Lee D. *J Am Chem Soc* 2012;134:10783–10786.
194. Finnegan DF, Snapper ML. *J Org Chem* 2011;76:3644–3653.
195. Chatani N. *Chem Rec* 2008;8:201–212.

6

ENANTIOSELECTIVE OLEFIN METATHESIS

Brice Stenne and Shawn K. Collins
Department of Chemistry, Université de Montréal, Montréal, PQ, Canada

6.1 GENERAL INTRODUCTION

As chemical methodologies advance in their development and utility, very often enantioselective variants are targeted as key areas of research interest. As olefin metathesis has become one of the "go-to" methods for alkene synthesis in organic chemistry, it stands to reason that there has been considerable interest in adding enantioselective methods to the standard "toolbox" of organic chemists. The olefin metathesis reaction is characterized by a transition-metal-catalyzed reorganization of two olefins. In theory, the process would afford a thermodynamic mixture of four different olefins from two nonsymmetrical disubstituted starting alkenes, but in practice the process has advanced to the point where organic chemists can use retrosynthetic analysis and strategically plan how to transform existing olefins into new highly substituted alkenes, often with control of the olefin geometry (1). The relatively recent development of the olefin metathesis method can be traced to the discovery of well-defined Mo- and Ru-based catalysts (2).

Conceptually, an enantioselective olefin metathesis reaction must necessarily involve formation of a new carbon–carbon double bond. However, chemists have creatively developed reaction schemes in which a metathesis event can define one or more prochiral centers or even define an element of planar or helical chirality within a molecule, all through the use of enantioselective metathesis. In the past dozen years or so, enantioselective alkene metathesis has emerged as a powerful method for stereocontrolled synthesis in that it allows the formation of the aforementioned chirality elements in conjunction with the synthesis of carbon–carbon double bonds under relatively mild reaction conditions.

Enantioselective processes are now possible in all metathesis-related transformations including ring-closing metathesis, ring-opening metathesis, cross metathesis, ene–yne metathesis, polymerization reactions, and tandem processes involving one or more of the aforementioned transformations. These advances have been due to intensive efforts in the area of achiral and chiral catalyst design for Mo-, W-, and Ru-based catalysts (3–6). In general, the design of catalysts based upon these metals for enantioselective metathesis involves modification of existing achiral ligands into chiral versions that would effectively create a chiral environment about the metal center. This was first achieved with Mo-based catalysts (7) and the design principles (types of ligands and their effect on catalyst reactivity and selectivity) between Mo- and W-based are very similar. Given the Lewis acidity of the Mo- and W-based complexes, these catalysts are inherently reactive toward water and oxygen. Careful design of the ligand sphere about these metals has afforded catalysts that can induce high levels of asymmetry in the products of metathesis reactions with equally high levels of efficiency (Scheme 6.1). This is routinely achieved through judicious selection of the ligands, each of which remains coordinated to the Mo- or W-center throughout the metathesis cycle. Apart from the alkenylidene that acts as the key functionality in the metathesis cycle, the imido and alkoxy ligands play a large role in the development of chiral catalysts. For example, the imido ligand acts to shield the metal center from degradation, but can also affect the reactivity of the catalyst. The alkoxy ligands in achiral catalysts are typically alkyl alkoxides, but in chiral variations, the alkoxy ligands are typically bidentate ligands derived from BINOL (2,2′-binaphthol). The catalyst's activity and the enantioselectivity in different processes

Olefin Metathesis: Theory and Practice, First Edition. Edited by Karol Grela.

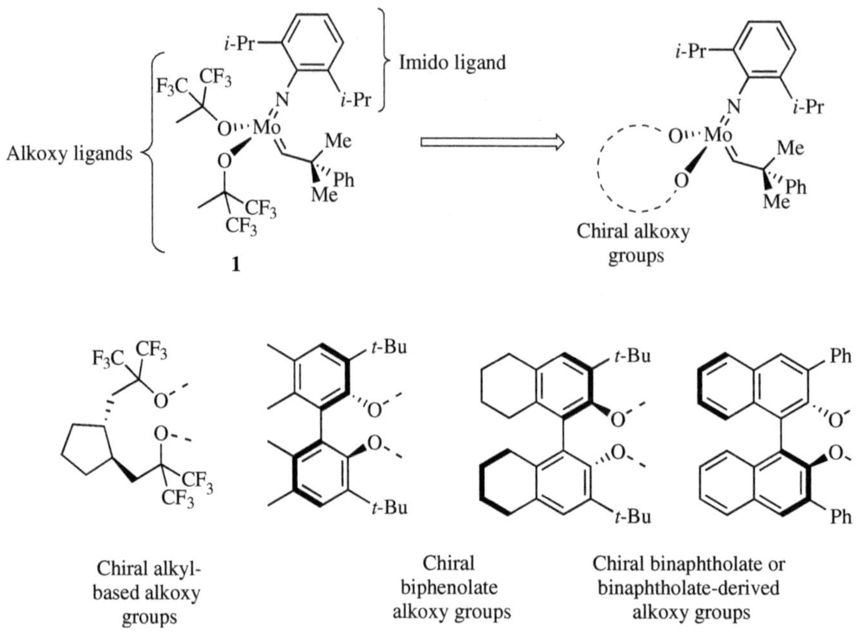

Scheme 6.1 Typical chiral motifs/ligands used in the design of Mo-based catalysts.

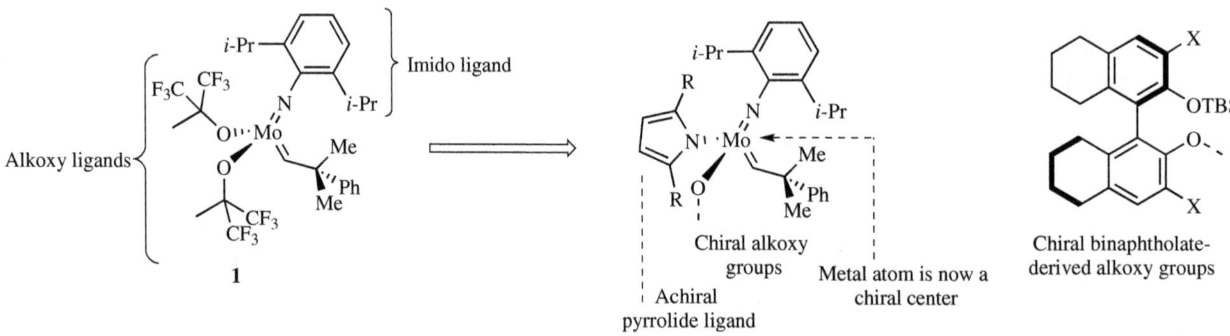

Scheme 6.2 Motifs/ligands used in the design of chiral-at-Mo-based catalysts.

are controlled by substituents on the BINOL backbone. Some examples of catalysts having biphenolate ligands do exist, but the BINOL-derived catalysts have become increasingly popular because of their ease of functionalization and commercial availability in an enantiomerically pure form.

More recently, catalysts having one monodentate binaphtholate-derived ligand and one pyrrolide ligand have emerged as a new and powerful class of Mo- and W-based catalysts (Scheme 6.2) that have significantly higher reactivity profiles in a number of reactions and can afford products with high levels of diastereo- and enantioselectivities in the formed olefin. These catalysts, having four different ligands attached to the metal center, create catalysts where the metal itself is a chiral center. These so-called chiral-at-Mo or chiral-at-W catalysts are also moisture and oxygen sensitive; however, new procedures for the *in situ* generation of all types of the above catalysts have been developed and are described in detail in the following sections.

Although achiral Ru-based catalysts for olefin metathesis have been prepared with a variety of different ligand types, the development of chiral Ru-based catalysts has focused on the design of new chiral *N*-heterocyclic carbene (NHC) ligands. It should be noted, however, that the halide and benzylidene ligands also play a role in determining the activity and selectivity of the Ru-based catalyst. The halide ligands are typically chloride ligands as in most achiral Ru-based catalysts. However, these halides can be exchanged for others, such as iodides, which modulate the reactivity and selectivity of the resulting catalyst (see below). Also of note is the nature of the benzylidene. Chiral Ru-based catalysts can be prepared bearing phosphine ligands, or with a chelated benzylidene (referred to as the Hoveyda versions). In many cases, the Hoveyda versions of most Ru-based

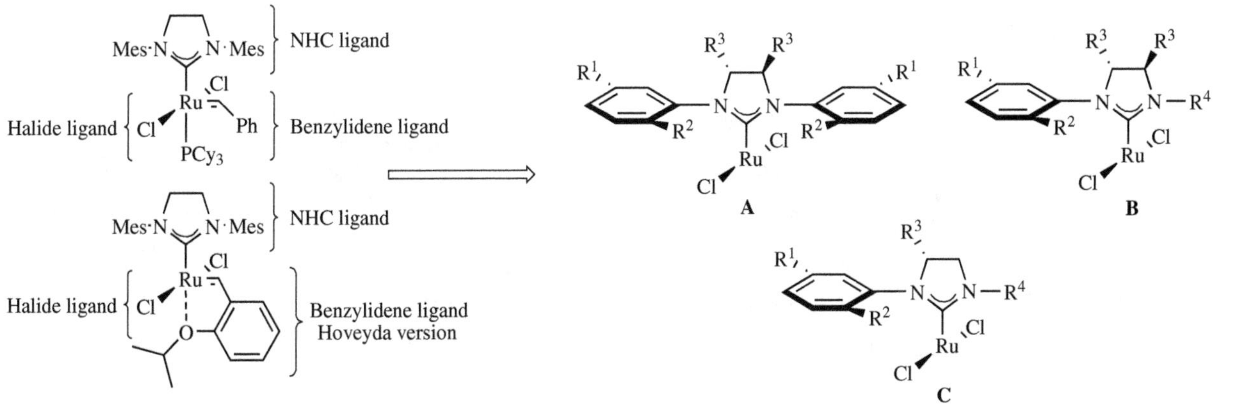

Scheme 6.3 Monodentate NHC ligands used in the design of chiral Ru-based catalysts.

catalysts show improved stability and activity. In general, the chiral Ru-based catalysts contain NHC ligands that display low reactivity to moisture and oxygen and are tolerant of a wide range of functional groups. In most cases, two different classes of chiral NHC designs have emerged: monodentate- and bidentate-type ligands. Chiral monodentate NHC ligands were first developed for Ru-based chiral catalysts based upon a chiral relay design. Chiral substituents are placed along the backbone of the NHC skeleton (R^3, Scheme 6.3). The steric encumbrance of the chiral groups influences and restricts the rotation of the N-substituents of the NHC. These substituents are typically aryl with large groups in the ortho- and meta-positions, which can be oriented away and toward the metal's coordination sphere (R^2 and R^1 respectively, Scheme 6.3). Chiral NHC ligands of this type can be prepared from easily available chiral diamines and the *N*-aryl groups can be installed through Buchwald–Hartwig type cross-coupling strategies. The C_2-symmetric design of these catalysts is noteworthy, as it eliminates complications that could arise from rotation about the NHC–Ru bond. Chiral catalysts utilizing this ligand design typically have good reactivities in asymmetric olefin metathesis reactions, but do not always provide high enantioselectivities across a wide substrate range. To circumvent these issues, halide additives (typically NaI) can be added, which results in halide exchange. The substitution of the chlorides for iodide ligands imparts increased steric bulk about the Ru center, which often increases the enantioselectivity observed in the process and also decreases the overall reactivity and lifetime of the catalyst.

Chiral Ru-based catalysts having monodentate NHCs can also be designed with ligands having C_1-symmetry (Scheme 6.3). In some cases, C_1-symmetry is achieved through the NHC having only a single chiral substituent along its skeleton (**C**, Scheme 6.3). However, in most instances, the C_1-symmetry is achieved through two different *N*-substituents on the NHC ligand. The NHC ligands can effectively use one of the *N*-substituents to create a chiral environment about the Ru atom and the other *N*-substituent can be used to help control the catalyst's stability and reactivity (**B**, Scheme 6.3). In this regard, if small alkyl substituents are used (R^4, Scheme 6.3), the resulting catalysts have high reactivities, but their solution stability is significantly decreased when compared to catalysts having C_2-symmetric NHC ligands. If an *N*-aryl substituent is used in place of a small *N*-alkyl group, the respective catalysts have increased stability and can be used in much smaller catalyst loadings and at higher temperatures. Interestingly, these catalysts (**B** and **C**, Scheme 6.3) can potentially have different rotamers, and the rate of rotation of the NHC at different steps in the catalytic cycle as well as the efficiency of each of the rotamers remains only vaguely studied.

An alternative ligand design for preparing chiral catalysts is to use a bidentate C_1-symmetric NHC ligand. One of the *N*-substituents is transformed from a simple aryl group into either a binaphtholate or a biphenolate, in which the alkoxy substituent can diastereoselectively displace one of the chloride ligands (**A** and **B** respectively, Scheme 6.4). The resulting Ru-based catalyst becomes chiral-at-Ru. The chirality of the complex can be created in two ways. Chiral binaphtholates derived from 2-amino-2′-hydroxy-1,1′-binaphthyl (NOBIN) can also impart an asymmetric environment about the Ru center. A second possibility has been to use an NHC ligand in which chiral substituents are located along the backbone. A chiral relay effect as described previously then controls the asymmetry in a biphenolate *N*-substituent.

The introduction of the chiral bidentate NHC ligand has dramatic consequences on the geometry of the catalyst and its reactivity. Firstly, the substitution of the chloride ligand for an alkoxy group forces the *N*-mesityl group away from the Ru atom, slightly decreasing the steric encumbrance about the metal. Unfortunately, an alkoxy ligand has the same effects as an iodide ligand. A catalyst having an alkoxy ligand will have its reactivity decreased because of

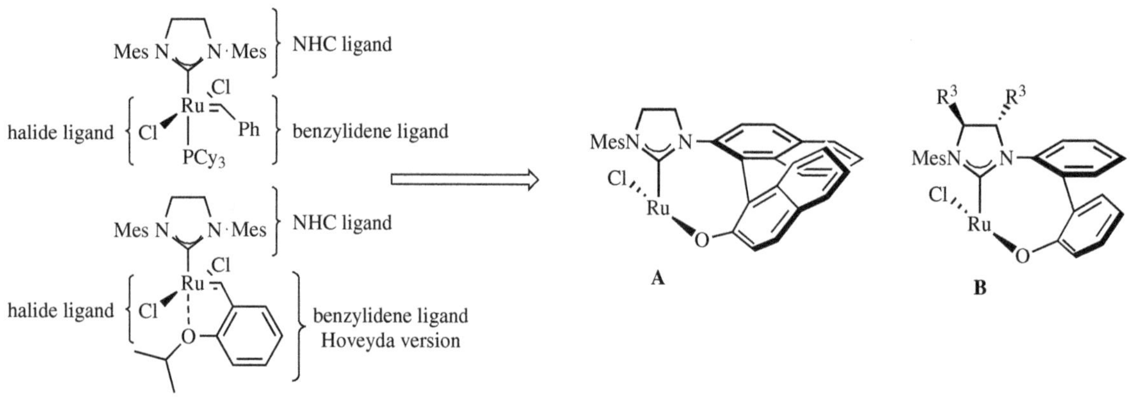

Scheme 6.4 Bidentate NHC ligands used in the design of chiral Ru-based catalysts.

both electronic and steric factors, but the alkoxy ligand does impart impressive stability to this catalyst class. The chiral Ru-based catalysts having the bidentate NHCs can typically promote asymmetric olefin metathesis reactions in undistilled solvents and/or in air and can often be recovered by column chromatography. In some cases, substitution of the binaphtholate unit can help improve catalytic activity. The biphenolate complexes have better reactivity than the binaphtholate-derived catalysts, and have an added advantage in that they can be prepared *in situ*. A detailed preparation of the Ru-based biphenolate catalysts is found in later sections.

The novel chiral metathesis catalysts developed to date have not only been effective in inducing high levels of enantioselectivity, efficiency, and olefin stereoselectivity, but have also served as the seed for advancements made in the field of achiral catalyst design (8). Despite a variety of new catalyst ligand designs, the field remains in its infancy, and general and highly enantioselective methods for each of the transformations that are described in the following sections have not been achieved. This is perhaps best exemplified by the fact that achiral metathesis reactions are routinely exploited in the synthesis of natural products and pharmaceuticals (9), but enantioselective metathesis reactions have been relatively underutilized in the synthesis of complex molecules.

The focus of this chapter is to describe the advances in enantioselective olefin methods so that the reader is given a full description of the current state of the art in the practice of enantioselective olefin metathesis. As such, the following text is organized in terms of the type of enantioselective process. The following sections also aim to provide a review of how the current methods have evolved so that the reader can put them in proper context. Consequently, each section on a specific enantioselective metathesis process is in general organized chronologically. It is hoped that in this way, the reader can also realize the immense future potential of enantioselective metathesis that remains to be explored.

6.2 ASYMMETRIC RING-CLOSING METATHESIS (ARCM)

6.2.1 General Introduction

The first examples of enantioselective metathesis processes were developed using ring-closing olefin metathesis, and the synthetic strategy has since evolved to provide access to a wide variety of structural motifs in good to excellent enantioselectivities. In most instances, an asymmetric ring-closing event is used to prepare enantioenriched carbocycles as well as oxa- and aza-heterocycles, which could all have medicinal applications or be used as key building blocks in natural product synthesis. In a general sense, the advantages of the method include the formation of a stereocenter in high enantiopurity, the formation of a cycle, and synthesis of a carbon–carbon bond as part of a stereodefined olefin, all from relatively simple acyclic starting materials. Two examples of asymmetric ring-closing metathesis (ARCM) strategies are shown in Scheme 6.5.

The simplest asymmetric transformation involves the ring closure of two olefins to form a ring in which a racemic stereocenter is already present somewhere along the acyclic precursor diene (Scheme 6.5, (1)). Treatment of the acyclic diene with a chiral olefin metathesis catalyst would promote a kinetic resolution. In order to achieve high enantioinduction and control of the ring-closing process, one of the olefins is usually terminal while the second bears some substitution. Through judicious control of the substitution patterns of the olefins, the chiral olefin metathesis catalyst will preferentially react faster at the least hindered olefin but proceed with a slower cyclization with the most hindered olefin. Given that the final cyclization event is normally slower and the first metathesis event is reversible, the catalyst can effectively select which enantiomer it would preferentially cyclize. If the cyclization is not favored (or slow), a simple nonproductive olefin metathesis event with another molecule of starting material could occur to free the catalyst for metathesis with

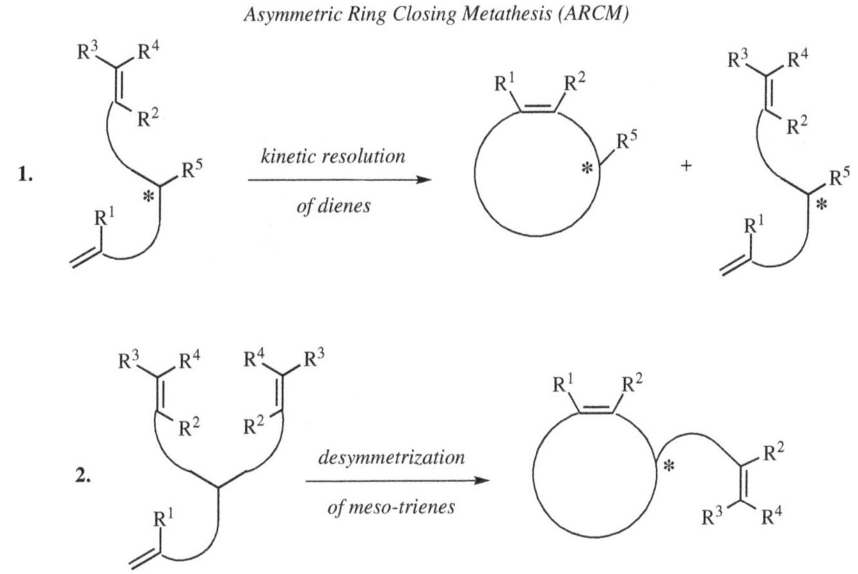

Scheme 6.5 Asymmetric ring-closing metathesis processes.

another molecule. In these kinetic resolution processes, the maximum yield of the product would be 50% and the unreacted starting diene could be re-isolated, also in 50% yield. In an ideal reaction, each compound would be isolated in high enantiopurity. The recovered acyclic diene could then be cyclized using an achiral olefin metathesis catalyst to provide the opposite enantiomer of the cyclized product. As 50% of the starting diene is usually an unwanted enantiomer, many consider the desymmetrization of trienes to be a more efficient ARCM process.

The desymmetrization of a *meso*-triene differs from the kinetic resolution shown in Scheme 6.5 only in the substitution of R^5 for an identical olefin side chain. Again, the substitution pattern of the olefins is normally chosen to help the catalyst begin the metathesis sequence at the desired olefin. Once a catalyst has reacted with the olefin that is the least substituted, it would have the choice to cyclize with either of the other two olefins. An advantage of desymmetrization versus kinetic resolution as a strategy to form enantioenriched compounds is that the product could be theoretically formed in 100% yield. Consequently, desymmetrization has emerged as the strategy of choice for the synthesis of a variety of challenging motifs such as quaternary carbon centers, sterically congested bicycles in natural product synthesis, as well as in the preparation of nontraditional forms of chirality, such as molecules possessing *P*-chiral centers, helical chirality, or planar chirality.

6.2.2 Kinetic Resolution of Dienes

The kinetic resolution of racemic substituted 1,6- and 1,7-dienes via cyclization to form all-carbon or heterocyclic ring systems containing a chiral carbon atom was first studied using chiral molybdenum-based catalysts. In 1996, Grubbs and coworkers (10,11) employed molybdenum-based catalysts adorned with enantiopure chiral bidentate alkoxide ligands (Scheme 6.6). These catalysts afforded a variety of five- and six-membered carbocycles, but the enantioselectivities (19–84% ee)

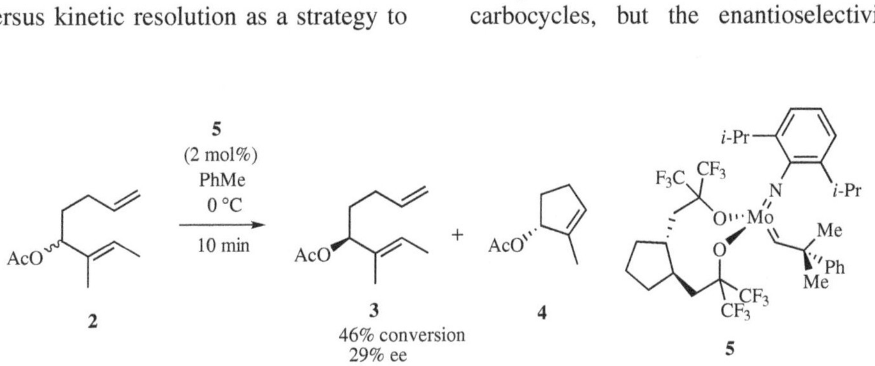

Scheme 6.6 First molybdenum-catalyzed kinetic resolution of dienes.

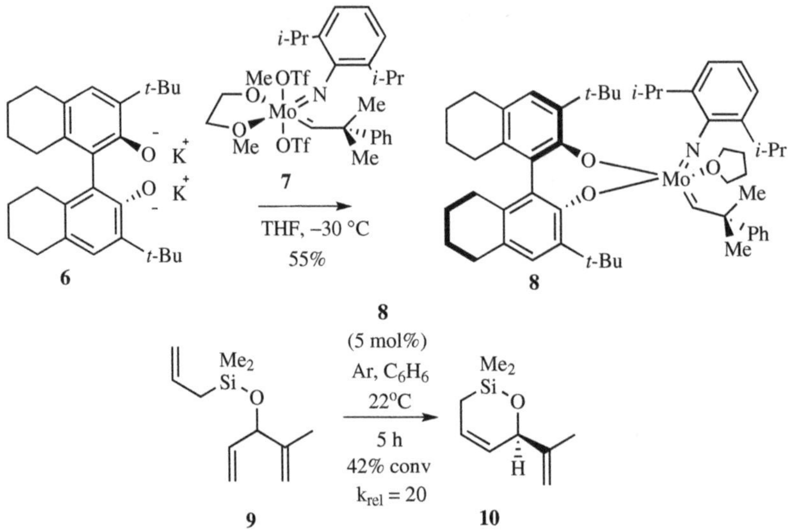

Scheme 6.7 Molybdenum-catalyzed kinetic resolution of dienes (11).

depended heavily upon the structure of the substrate, and high enantioselectivities were usually only obtained at very high conversions.

Schrock and Hoveyda (12) were able to obtain much higher enantioselectivities in kinetic resolution processes by employing molybdenum-based catalysts substituted with biphen- or BINOL-derived ligands (Scheme 6.7). Their report described several advances, one of which was that the active catalyst **8** could be formed *in situ* from two commercially available components, simplifying the synthesis and usage of the highly reactive Mo-based catalysts (13). Secondly, the studies demonstrated that more complex heterocycles, such as **10**, possessing different functionalities, could be prepared and exploited in synthesis.

The Mo-catalyzed kinetic resolution of dienes can also be used for the cyclization of some challenging substrates (Scheme 6.8). For example, a kinetic resolution of tertiary amines (whose basicity can be problematic with Mo-based catalysts) was possible to form six-, seven-, and even eight-membered ring systems (eight-membered ring systems undergo very slow ring closing and are regarded as some of the most challenging substrates in ARCM) (14). Kinetic resolution of dienes containing 1,2-disubstituted olefins or of dienes containing tertiary ethers that cyclize to form quaternary chiral carbon centers is possible (15).

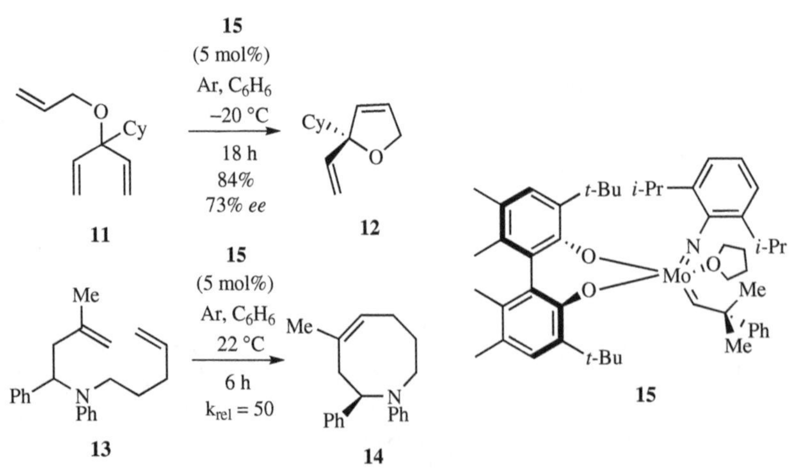

Scheme 6.8 Molybdenum-catalyzed kinetic resolution of challenging substrates.

General Procedure for *In Situ* Synthesis of Mo-Based Catalyst 8 for ARCM (Scheme 6.7) (13)

A solution of (*R*)-**6** (48.3 mg, 0.100 mmol) and **7** (79.2 mg, 0.100 mmol) was each prepared in tetrahydrofuran (THF) (volume of each solution 1 ml). Under an N_2 atmosphere, each solution was cooled to −30 °C. In another flask previously cooled to −30 °C was added 100 μl (0.01 mmol) of the solution of **8** and **9**. The reaction mixture was then warmed to 22 °C for 1 h, after which the catalyst solution was ready for use in ARCM reactions.

Kinetic resolution of dienes has also been used to form both helical and planar chiral molecules. The asymmetric synthesis of [7] helicene was accomplished in good ee (80%) by kinetic resolution via an asymmetric ring-closing olefin metathesis and was the first reported application of asymmetric olefin metathesis toward helically chiral molecules (Scheme 6.9) (16). The kinetic resolution of diene **16** utilized Ru-based olefin metathesis catalysts bearing C_1-symmetric NHC ligands and benefitted from two key additions to the reaction conditions. First simple olefins were investigated as additives to control the nature of the propagating alkylidene, with vinylcyclohexane being determined as optimal for the synthesis of **18**. The concept suggested that control of the propagating alkylidene can be an effective method for improving enantioselection in ARCM processes and can at times be more effective than the use of halide additives that can lower catalyst reactivity. Second, hexafluorobenzene was used as solvent, denoting the importance of solvent in olefin metathesis reactions.

Planar chiral ferrocenes can also be formed via kinetic resolution of dienes. Planar chiral ferrocenes are important chiral scaffolds in organic and organometallic chemistry (17). The synthesis of these compounds in enantioenriched form remains challenging, and examples of catalytic asymmetric syntheses of planar chiral ferrocene derivatives are rare (18). Ogasawara and coworkers (19) published a report that demonstrated that ferrocene derivatives bearing different olefin side chains on each of the cyclopentadienyl rings could be used to promote efficient asymmetric ring closure of one enantiomer to form bridged ferrocenes **19→21** (Scheme 6.10). The remaining uncyclized enantiomer could be cyclized using an achiral olefin metathesis catalyst (Grubbs second-generation catalyst) and be obtained in good enantiopurity (>80% ee). Ferrocenes containing bulky groups such as **19**, **20**, and **21** were obtained in good to high enantiomeric excess (Scheme 6.10).

Ogasawara and coworkers (20) extended the use of kinetic resolution of planar chiral organometallic complexes to (η^6-arene)chromium complexes. The planar chiral Cr complexes have found widespread application as chiral ligands for asymmetric catalysis and as chiral building blocks for natural product syntheses (21). A variety of methods exist for the preparation of enantiomerically enriched planar chiral (arene)chromium species but are typically based

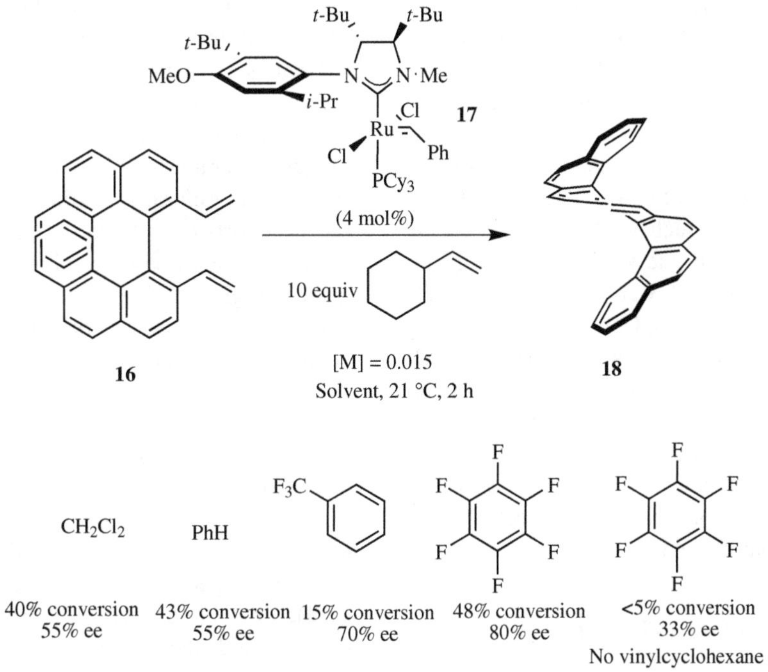

Scheme 6.9 Kinetic resolution to form enantioenriched [7]helicene.

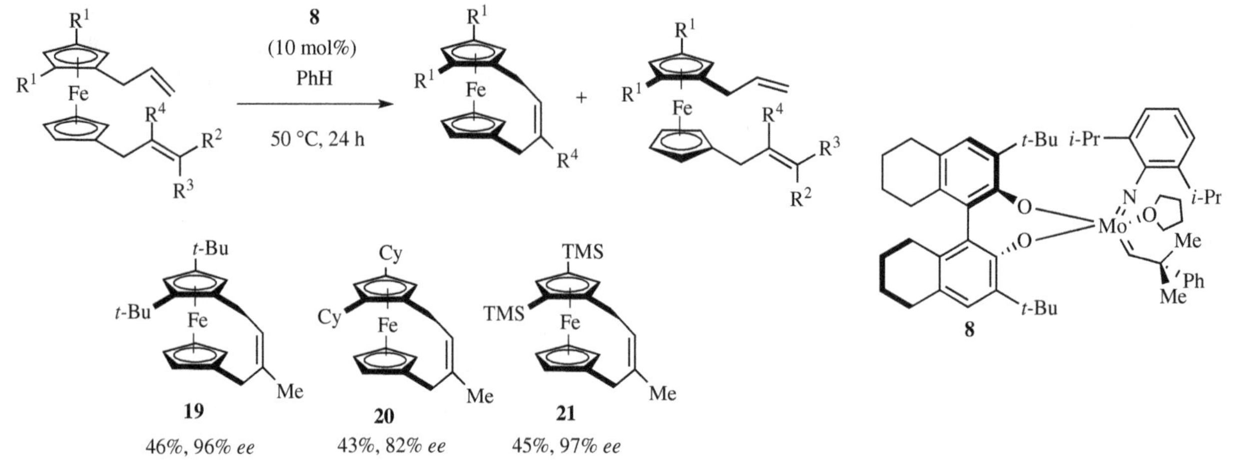

Scheme 6.10 Kinetic resolution to form enantioenriched ferrocenes.

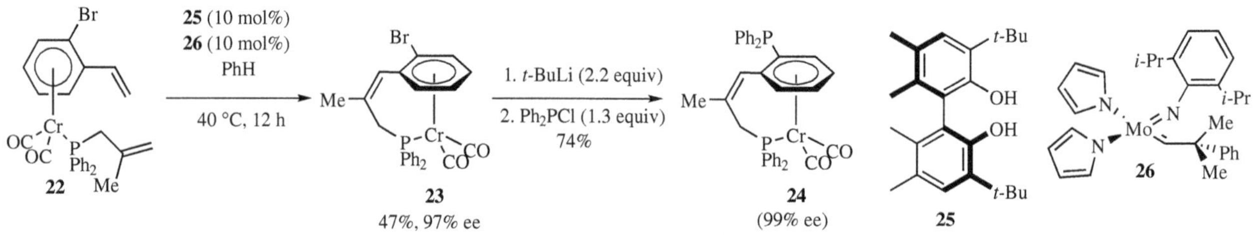

Scheme 6.11 Kinetic resolution of planar chiral (η^6-Arene)chromium complexes by molybdenum-catalyzed asymmetric ring-closing metathesis.

on either resolution of racemates (22) or methods that require a stoichiometric amount of chiral reagents or auxiliaries (23). Asymmetric catalytic methods for the preparation of optically active (η^6-arene)chromium complexes are rare (24).

A representative example of the kinetic resolution of Cr arene phosphine complexes studied by Ogasawara and coworkers involved complexes such as **22** (Scheme 6.11). Treatment of **22** with a Mo-based catalyst formed from the *in situ* reaction of the pyrrolide complex **26** and the axially chiral biphenolate ligand **25** afforded the Cr complex **23** in high enantiopurity. Furthermore, the authors described how **23** could be further transformed into a planar chiral phosphine ligand **24** on gram scale. The Cr complex **24** formed from an ARCM protocol could then be used to promote asymmetric addition of phenylboronic acid to cyclohexenone and the addition of phenyl boroxine to an aromatic imine, each in high enantioselectivity and isolated yield.

6.2.3 Desymmetrization of *meso*-Trienes

The desymmetrization of *meso*-trienes is one of the most well-studied asymmetric transformations involving chiral olefin metathesis catalysts. The reaction typically involves an achiral triene in which one of the olefins is terminal and serves as the most reactive olefin toward olefin metathesis, while the two other olefins are normally di- or trisubstituted. The reaction was initially studied with chiral molybdenum-based catalysts but has also been investigated using chiral ruthenium-based catalysts. Despite the recent reports of different types of air- and moisture-stable chiral ruthenium-based catalysts, these catalysts still cannot achieve high enantioselectivities with a wide range of substituted olefins that can be found with molybdenum-based catalysts. The synthesis of oxygen-containing heterocycles is the most common asymmetric transformation performed using both families of catalysts. For chiral molybdenum-based catalysts, a wide range of catalysts bearing different ligands have been explored for various substrate classes. For example, the Mo-based complex **15** can be used to cyclize the *meso*-triene **27** to afford the seven-membered ring **28** in good yield and enantiomeric excess (Scheme 6.12). The dihydropyran **30** bearing a quaternary carbon center could also be formed via desymmetrization of **29** at room temperature, also in good yield and enantiomeric excess (25,12).

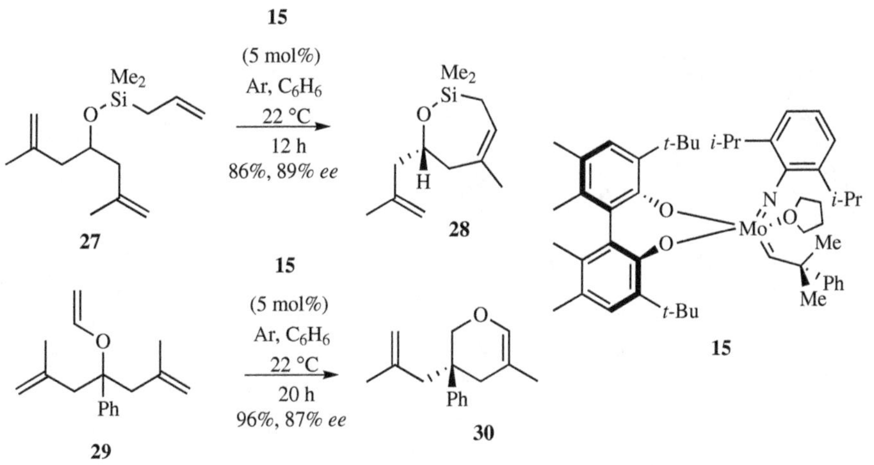

Scheme 6.12 Molybdenum-catalyzed desymmetrization of meso-trienes to form oxygen-containing heterocycles.

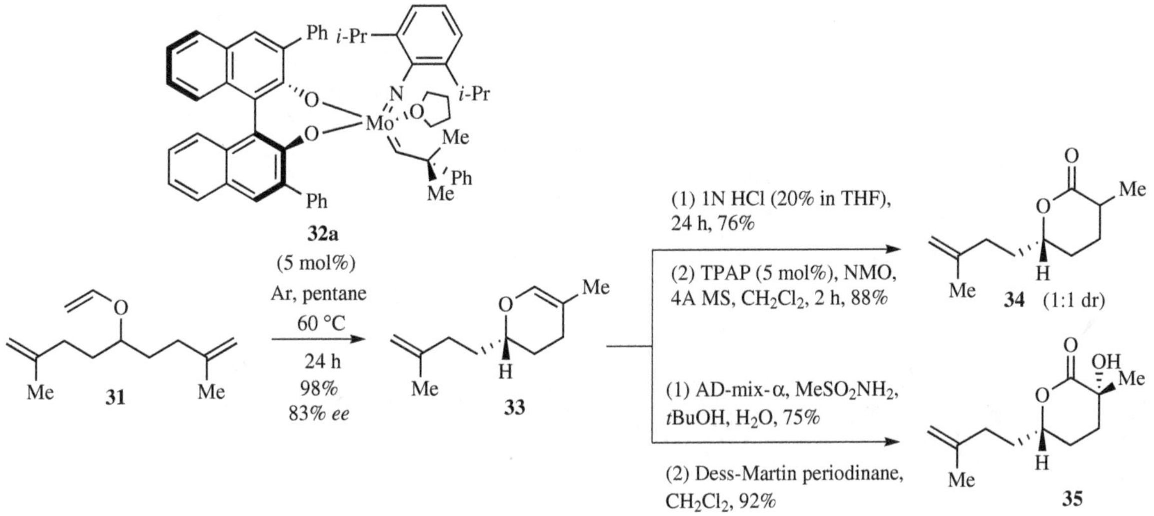

Scheme 6.13 Molybdenum-catalyzed synthesis of cyclic enol ethers.

General Procedure for Desymmetrization Using Mo-Based Catalysts: Synthesis of Enol Ether 30 (Scheme 6.12) (26)

To a one-dram vial in a N_2-filled glove box was added Mo catalyst **15** (12.8 mg, 50.0 μmol). To the vial *meso*-triene **29** was added as a solution in PhH (7.5 mmol, 0.5 ml of solution). The vial was then covered with a Teflon-lined cap. The reaction mixture was stirred at 22 °C for 20 h, at which point it was removed from the glove box. Exposure of the reaction mixture to air quenches the reaction and the reaction mixture was then purged with a stream of N_2 to concentrate and afford a brown residue. Purification by silica gel column chromatography (49 : 1 pentane:Et_2O) afforded the cyclic enol ether **30** (48.0 μmol, 96% yield).

The preparation of enantioenriched cyclic enol ethers can also be achieved with high levels of enantioselectivity (>90% ee); however, much greater catalyst loadings are often necessary to achieve reasonable levels of conversion (Scheme 6.13) (26). The enol ether **31** could be treated with the catalyst **32a** and cyclized to afford the cyclic enol ether **33** in nearly quantitative yield and good enantiomeric excess (83%). The versatility of the enol ether **33** was then demonstrated by oxidation to the corresponding lactone **34** and the α-hydroxy ester **35** in good yields. The desymmetrization to form enol ethers effectively demonstrates the better reactivity achieved with Mo-based catalysts, as metathesis reactions involving enol ethers are often problematic with Ru-based catalysts (27).

It should be noted that as an alternative to forming Mo-based catalysts such as **15** *in situ* for desymmetrizations,

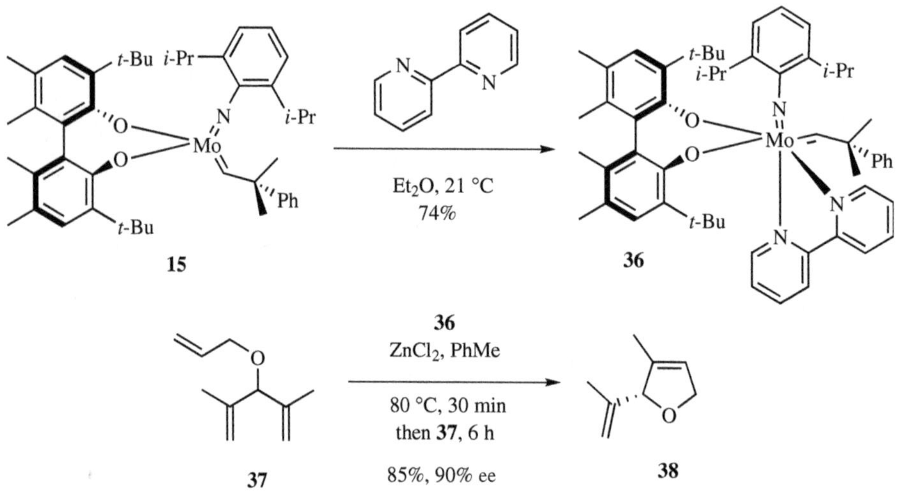

Scheme 6.14 Air-stable Schrock-type molybdenum alkylidene pre-catalysts.

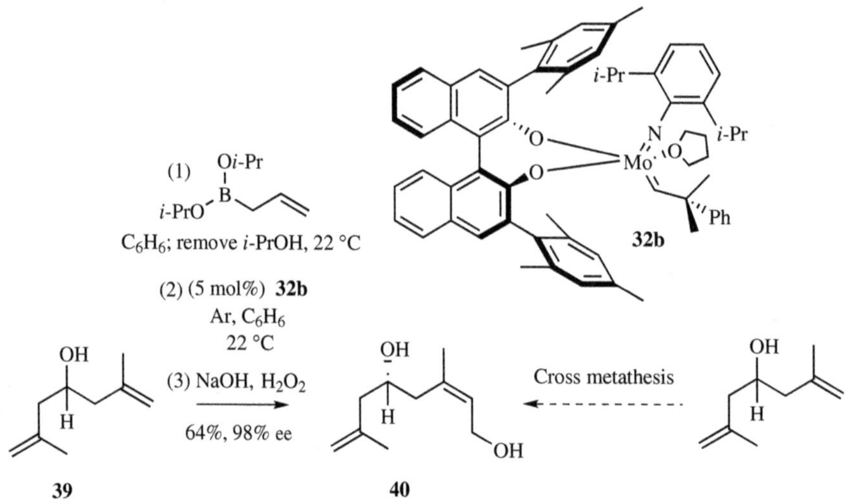

Scheme 6.15 Molybdenum-catalyzed synthesis of cyclic allyl boronates and subsequent oxidation to enantioenriched secondary alcohols.

one could use more stable pre-catalysts developed by Fürstner and Heppekausen (28) (Scheme 6.14). The Mo-based catalyst **15** could be reacted with 2,2′-bipyridine to form a bench-stable complex. The structure of the resulting complex **36** in the solid state depicts the coordination geometry about the Mo(VI) center as a distorted octahedron. The robust complex **36** could be activated prior to use in a metathesis reaction through complexation with a weak Lewis acid such as $ZnCl_2$. Through the Lewis-acid-mediated activation, complex **15** was unmasked and resulted in the desymmetrization of *meso*-triene **37** in 85% yield and 90% ee. The fact that this strategy of catalyst activation combines ease of handling with a crystalline and bench-stable compound has made it a method of choice for handling Mo-based catalysts in alkyne metathesis (29).

Hoveyda and coworkers (30) also demonstrated that desymmetrizations of *meso*-trienes could take place on substrates having temporary tethers (Scheme 6.15). A temporary boronate-derived tether could be employed to link together an alkene and a *meso*-alcohol **39** through a simple transesterification-like procedure. Once the new boron-derived triene was prepared, treatment with Mo-based catalyst **32** caused the asymmetric desymmetrization reaction. The cyclic boronates could then be oxidized *in situ* to afford linear diene products. It is worth noting that using the temporary tether strategy the authors could prepare enantioenriched acyclic products in good yields and enantioselectivities, which would not easily be prepared via an asymmetric cross metathesis (ACM) process. ACM between the *meso*-alcohol **39** and allyl alcohol to provide a

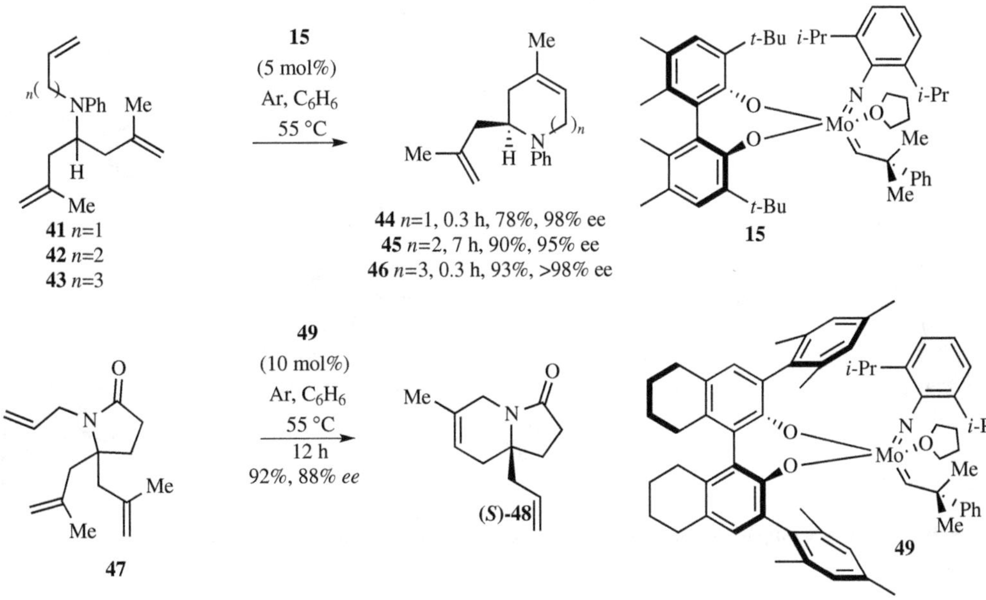

Scheme 6.16 Molybdenum-catalyzed desymmetrization of *meso*-trienes to form secondary and tertiary amines and cyclic amides.

stereodefined trisubstituted olefin is still not possible using even state-of-the-art catalysts.

Mo-based catalysts could also be used to prepare nitrogen-containing heterocycles. This substrate class is normally regarded as challenging, but Hoveyda and coworkers published a desymmetrization protocol using the less basic *N*-phenyl substituted trienes and amide-derived trienes (Scheme 6.16) (31,32). The desymmetrization of a series of tertiary amines (**41**, **42**, and **43**) was studied in which six-, seven-, and eight-membered rings were all prepared in good yields and in high enantioselectivities. Once again, the preparation of an eight-membered ring **46** is noteworthy as desymmetrization reactions using Ru-based catalysts are normally incapable of promoting cyclizations to form medium-sized rings. In subsequent examples, the Hoveyda group also demonstrated that other Mo-based catalysts were effective in promoting desymmetrization reactions to bicyclic amides such as **48** (Scheme 6.16).

With the recent development of new chiral-at-Mo catalysts bearing a large monodentate aryloxide ligand such as the chloro derivative **50**, the asymmetric desymmetrization of *meso*-trienes was reinvestigated to compare the new class of Mo-based catalysts with the existing Mo-based catalysts having bidentate aryloxide ligands (Scheme 6.17) (33). When the cyclization of triene **42** was compared with catalysts **15** and **50**, the six-membered ring **45** could be obtained in almost identical yield and enantiomeric excess, in decreased reaction time and temperature with a much lower catalyst loading. Similarly, the synthesis of the bicyclic amide (**R**)-**48** was also much more efficient when employing the chiral-at-Mo catalyst **50**. The desymmetrization of triene **47** once again occurred in almost identical yield and enantiomeric excess, with lower reaction time, temperature, and catalyst loading.

Mo-based chiral olefin metathesis catalysts have also been exploited for the preparation of *P*-chiral heterocycles (Scheme 6.18) (34). Despite the importance of *P*-chiral compounds (35), particularly in the transition metal catalysis, the report by Gouverneur and Hoveyda remains the only example of the application of asymmetric olefin metathesis to this class of compounds. In addition, the study by Gouverneur and Hoveyda was the first application of ARCM to the synthesis of compounds having a stereogenic heteroatom. In general, *P*-chiral phosphinates and phosphine oxides could be prepared in up to 98% ee. For example, the acyclic phosphinate **51** could be cyclized to afford the seven-membered ring product **52** in 96% ee and 79% yield when employing catalyst **55**. The same Mo-based catalyst **55** could also promote the asymmetric desymmetrization of the phosphine oxide **53** in good yield and enantiomeric excess.

Some Ru-based catalysts have been developed for the asymmetric desymmetrization of *meso*-trienes. Although Mo-based catalysts have demonstrated high enantiomeric excesses and yields at low catalyst loadings, the ease of handling and functional group tolerance of Ru-based catalysts make their development and application attractive to synthetic chemists. Grubbs and coworkers (36) reported the first chiral Ru-based catalysts for asymmetric olefin metathesis. The chiral Ru-based catalysts **55**, **56**, and **57** are also stable to air and moisture similarly to their achiral analogs (Scheme 6.19). The catalysts employed monodentate chiral NHC ligands in which chirality was directed toward the Ru center via a chiral relay effect originating from chiral substituents

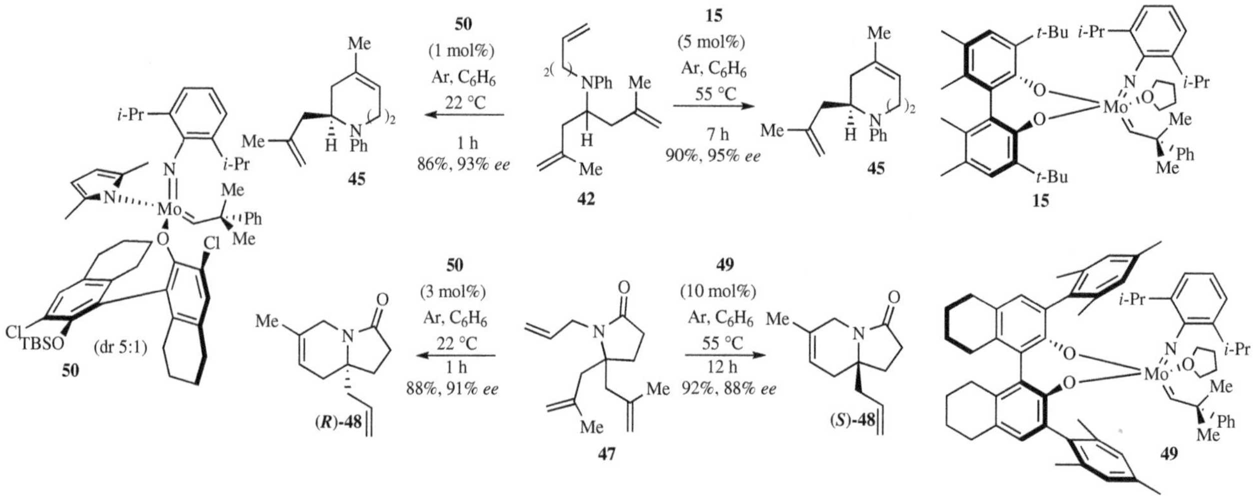

Scheme 6.17 Comparison of desymmetrization of *meso*-trienes to form secondary and tertiary amines and cyclic amides using molybdenum and chiral-at-Mo-based catalysts.

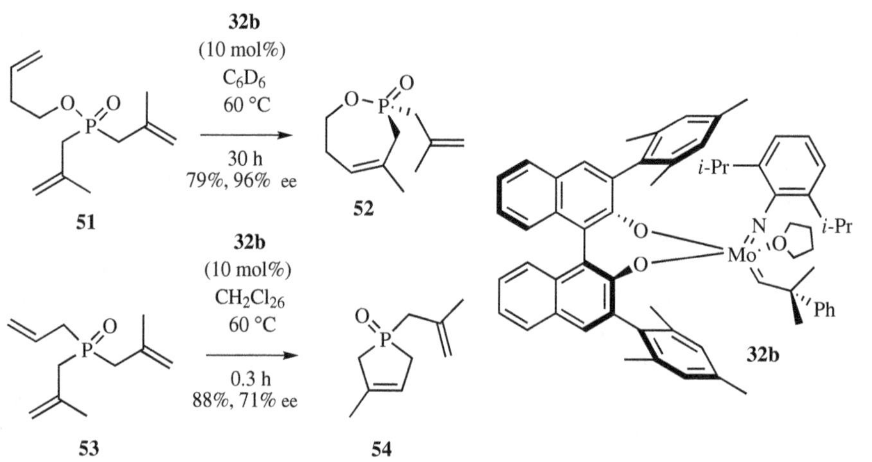

Scheme 6.18 Molybdenum-catalyzed desymmetrization of *meso*-trienes to form phosphinates and phosphine oxides.

along the NHC backbone. This strategy for developing chiral NHC ligands has since been heavily exploited in asymmetric synthesis (37). The desymmetrization of *meso*-trienes to afford enantioenriched five-, six-, and seven-membered rings was described. Within the series of heterocycles investigated, different catalysts were used to optimize the enantiomeric excesses for each substrate. In some instances, to improve enantioselectivities the catalysts were pre-treated with NaI to promote exchange of the chloride ligands to iodides. The increased steric bulk about the Ru center resulted in much greater enantioselectivities for heterocycles **58** and **59**.

An alternative class of Ru-based catalysts for ARCM reactions of *meso*-trienes are catalysts in which the NHC ligand is C_1-symmetric and monodentate (Scheme 6.20). Modification of the size of one of the *N*-substituents has been shown to be an effective strategy to improve the reactivity of Ru-based catalysts (38). Collins and coworkers prepared catalysts **64**, **17**, and **65** each having one *N*-Me group and an *N*-aryl group inspired form the previous Ru catalysts developed by Grubbs and coworkers (39). These catalysts typically afforded high enantioselectivities without the need for NaI additives. In addition, the catalysts were found to be highly reactive and could promote ring closure in a matter of minutes.

General Procedure for the Desymmetrization of *meso*-Trienes using Ru-Based Catalyst (Scheme 6.20) (39)

Catalyst **64** (2.5 mol%) was added to CH_2Cl_2 (0.055 M) and the appropriate *meso*-triene was added. The reaction mixture was stirred at 40 °C for 2 h, after which point ethylvinylether (approximately 0.1 ml) was added to

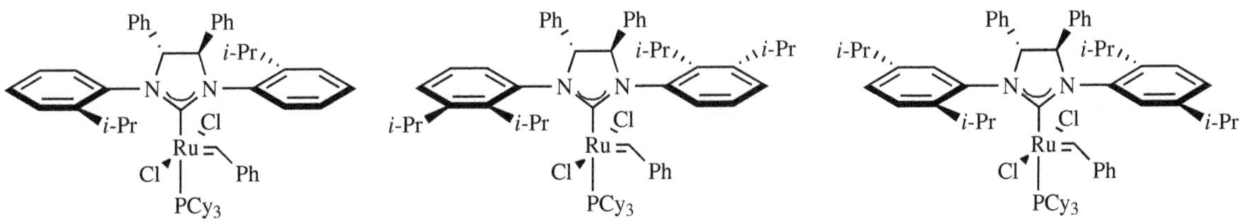

55 56 57

***meso*-Triene** → **Product**

Method A: catalyst (2 mol%), CH_2Cl_2, reflux, 2 h

Method B: catalyst (2 mol%), NaI (50 mol%), THF, 40°C, 2 h

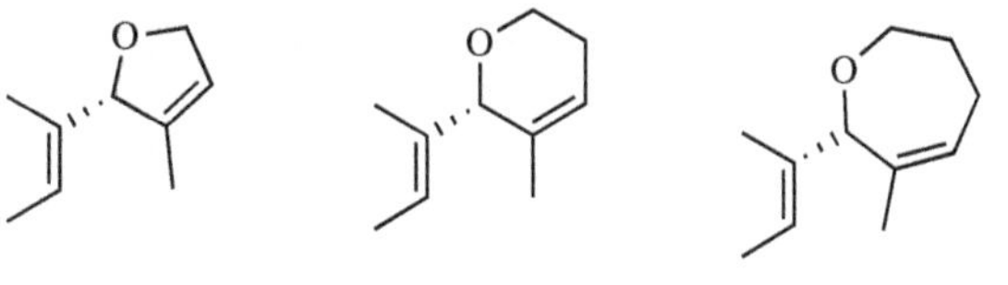

90% ee, >98% conv. **58** (Method B) — 92% *ee*, >98% conv. **59** (Method B) — 76% *ee*, 93% conv. **60** (Method A)

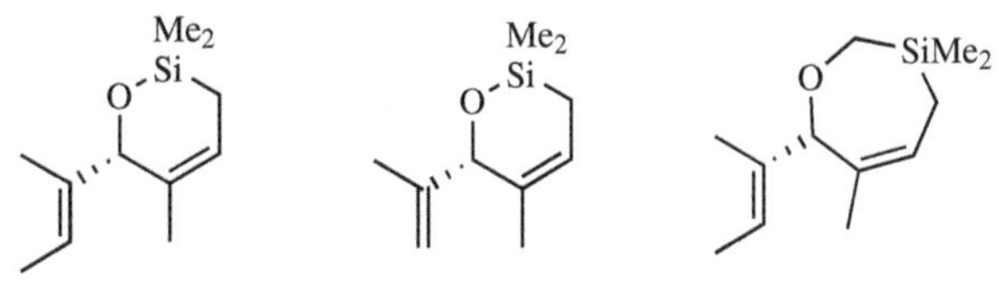

92% *ee*, >98% conv. **61** (Method A) — 45% *ee*, 92% conv. **62** (Method A) — 92% *ee*, 65% conv. **63** (Method A)

Scheme 6.19 Heterocycles prepared via desymmetrization of *meso*-trienes using chiral Ru-based olefin metathesis catalysts.

quench the reaction mixture. The reaction mixture was cooled to room temperature, filtered on neutral alumina and concentrated. Purification by flash chromatography (1–5% Et_2O in pentane) afforded the desired product.

Unfortunately, catalysts **64**, **17**, and **65** bearing *N*-Me substituents on the NHC ligand were found to be unstable in solution at prolonged reaction times. However, replacing the *N*-Me group with other bulkier alkyl groups, such as a Bn group in catalyst **66**, could improve stability (Scheme 6.21) (40). Catalyst **66** was subsequently used to promote desymmetrization of *meso*-trienes to afford cycles containing tetrasubstituted olefins (41). The cyclization to form tetrasubstituted olefins typically is studied with substrates in which the olefin formed has two Me groups as substituents, but catalyst **66** was used to promote the formation of heterocycle **68** in which the tetrasubstituted olefin possessed both a Me and an Et group. Reaction with catalyst **66** in C_6D_6 at 100 °C in a microwave afforded the desired product **68** in 49% conversion and 49% ee after 15 min. The desymmetrization to form the six-membered ring **69** using catalyst **66** was achieved in 47% conversion and 78% ee. Preliminary computational studies to suggest a model for the origin of enantioinduction were also reported.

The desymmetrization of *meso*-trienes has also been utilized in the formation of new ligand structures as well as natural product synthesis. Ogasawara and coworkers (42) reported that a catalytic enantioselective desymmetrization of planar-prochiral phosphaferrocenes was possible, even though the substrates for desymmetrization possessed a Lewis basic *P*-substituent (Scheme 6.22). The reactions proceeded efficiently in the presence of an appropriate chiral Mo catalyst, typically in high yields and up to 99% ee. The phosphaferrocene **71** was prepared in 99% ee using catalyst **72** and was capable of coordinating to a transition metal as a two-electron donor. The resulting complexation created an effective chiral environment around the metal center and demonstrated application as a potential chiral ligand in transition metal catalysis.

One of the first examples of the use of desymmetrization of *meso*-trienes in natural product synthesis was reported by Burke and coworkers (43) in the synthesis of the simple insect pheromones *exo*-brevicomin **75** (Scheme 6.23). Routes toward the diene **75** were first investigated with

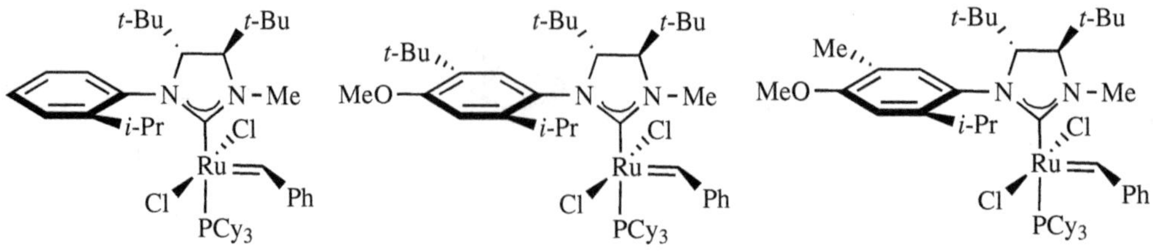

64 **17** **65**

catalyst (2-2.5 mol%)

***meso*-Triene** ⟶ **Product**

CH_2Cl_2, 40 °C, 5–10 min

64 82% *ee*, >98% conv.
17 81% *ee*, >98% conv.
65 82% *ee*, >98% conv.

64 28% *ee*, >98% conv.
17 94% *ee*, >98% conv.

64 60% *ee*, >98% conv.
17 86% *ee*, >98% conv.
65 71% *ee*, >98% conv.

64 >95% *ee*, <5% conv.

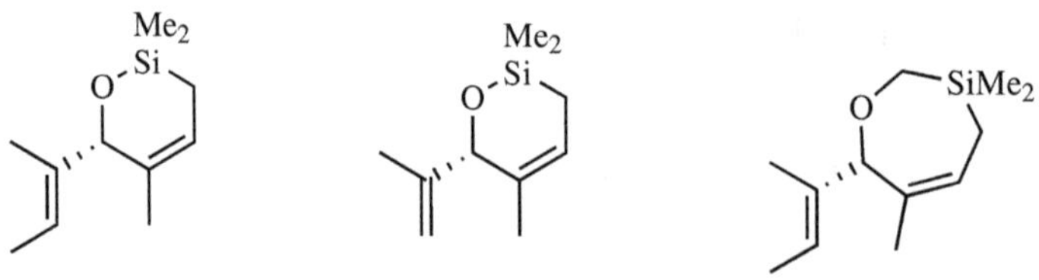

64 6% *ee*, >98% conv.
17 94% *ee*, >98% conv.
65 71% *ee*, >98% conv.

64 15% *ee*, >98% conv.
17 36% *ee*, >98% conv.

17 94% *ee*, >98% conv.

Scheme 6.20 Heterocycles prepared via desymmetrization of *meso*-trienes using chiral olefin metathesis catalysts bearing C_1-symmetric monodentate NHC ligands.

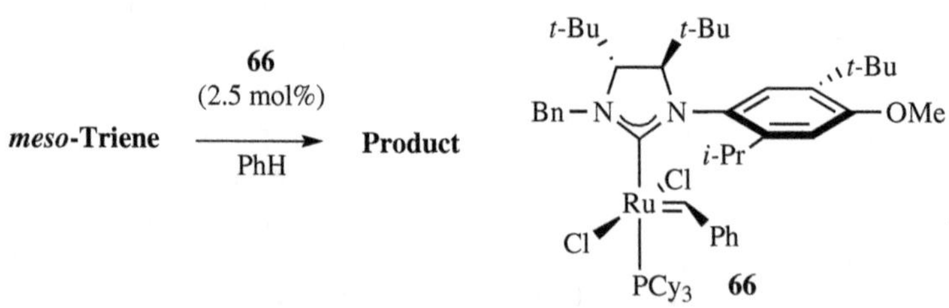

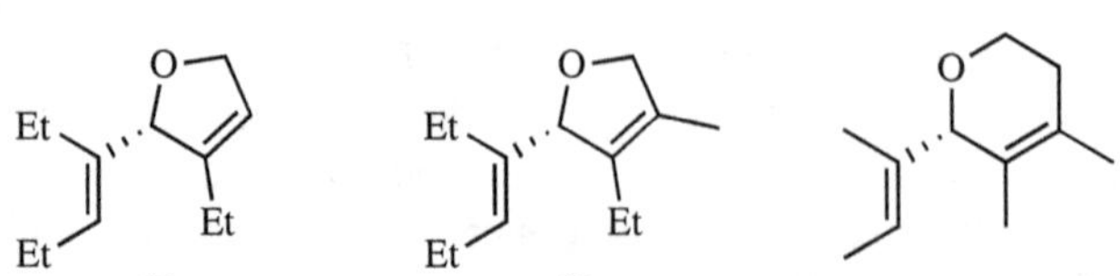

67 **68** **69**

67	**68**	**69**
60 °C, 6 h	100 °C (microwave), 15 min	60 °C, 6 h
12% *ee*, 46% conv.	49% *ee*, 49% conv.	78% ee, 47% conv.

Scheme 6.21 Heterocycles having tetra-substituted olefins prepared via desymmetrization of *meso*-trienes.

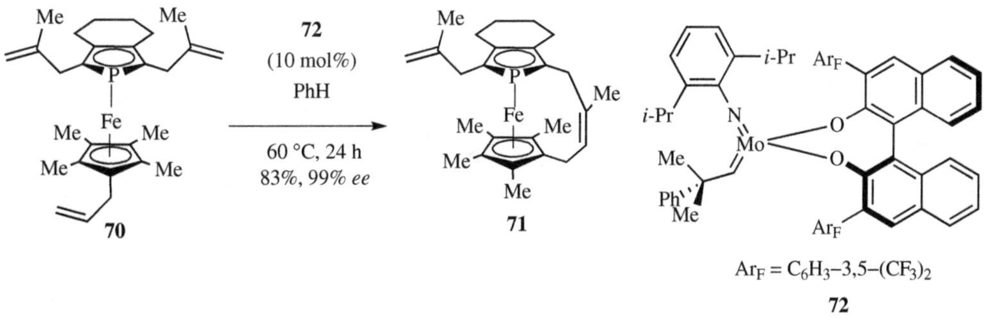

Scheme 6.22 Enantioselective synthesis of planar chiral phosphaferrocenes.

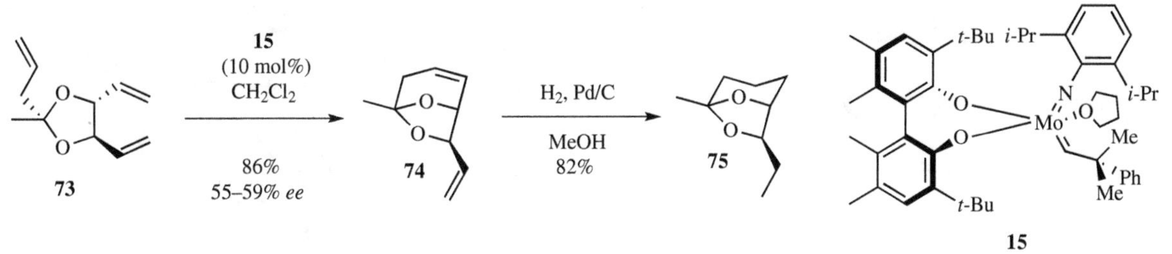

Scheme 6.23 Desymmetrization by ring-closing metathesis leading to 6,8-dioxabicyclo[3.2.1] octanes: a new route for the synthesis of (+)-*exo*-brevicomin.

achiral catalysts, but the desymmetrization of **73** using the Mo catalyst **15** (10 mol%) afforded **74** in 55–59% ee (44).

Perhaps one of the most impressive examples of ARCM involved in natural product synthesis is in the preparation of the polycyclic alkaloid quebrachamine **80** (Scheme 6.25) (33). Hoveyda and coworkers reported that successful desymmetrization of precursor **78** was only possible using a chiral-at-Mo-based catalyst such as **50**. The more common Mo-based chelate-complexes bearing diolate ligands were believed to be inactive in the desymmetrization of **78** because of strains within the transition states of the catalytic cycle caused by the rigid BINOL-like ligands attached to the Mo-center. The chiral-at-Mo systems bearing monodentate ligands were conceived to be more flexible and thus open to greater possible conformational change. In addition, the use of a chiral monodentate ligand such as **77** creates a stereocenter at the Mo-center, which could help promote more efficient enantioselective transformations. Catalyst **50** could be prepared *in situ* from the pyrrolide precursor **76** and ligand **77** to afford the resulting Mo complex **50** as a 5 : 1 mixture of diastereomers (Scheme 6.24).

The synthesis of quebrachamine as reported by Hoveyda and coworkers involved a desymmetrization process in the second-to-last step of the total synthesis (Scheme 6.25). Not only did **78** possess two sterically hindered vinyl groups at a congested all-carbon quaternary center but also a nearby basic tertiary amine, which made the transformation very challenging. Chiral BINOL-based Mo-complexes such as **82** had

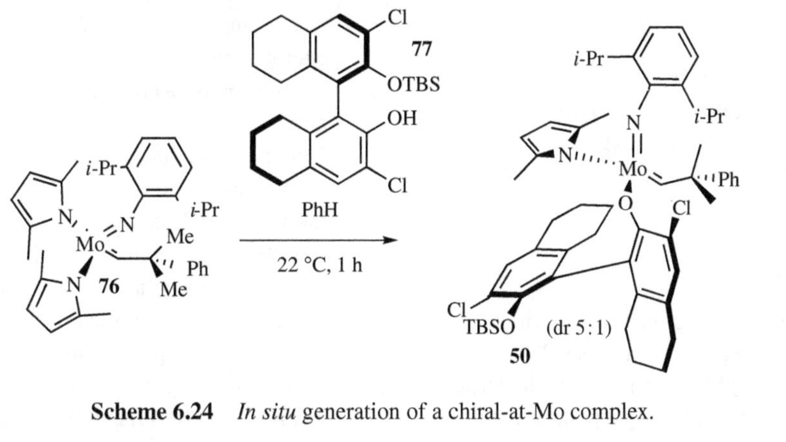

Scheme 6.24 *In situ* generation of a chiral-at-Mo complex.

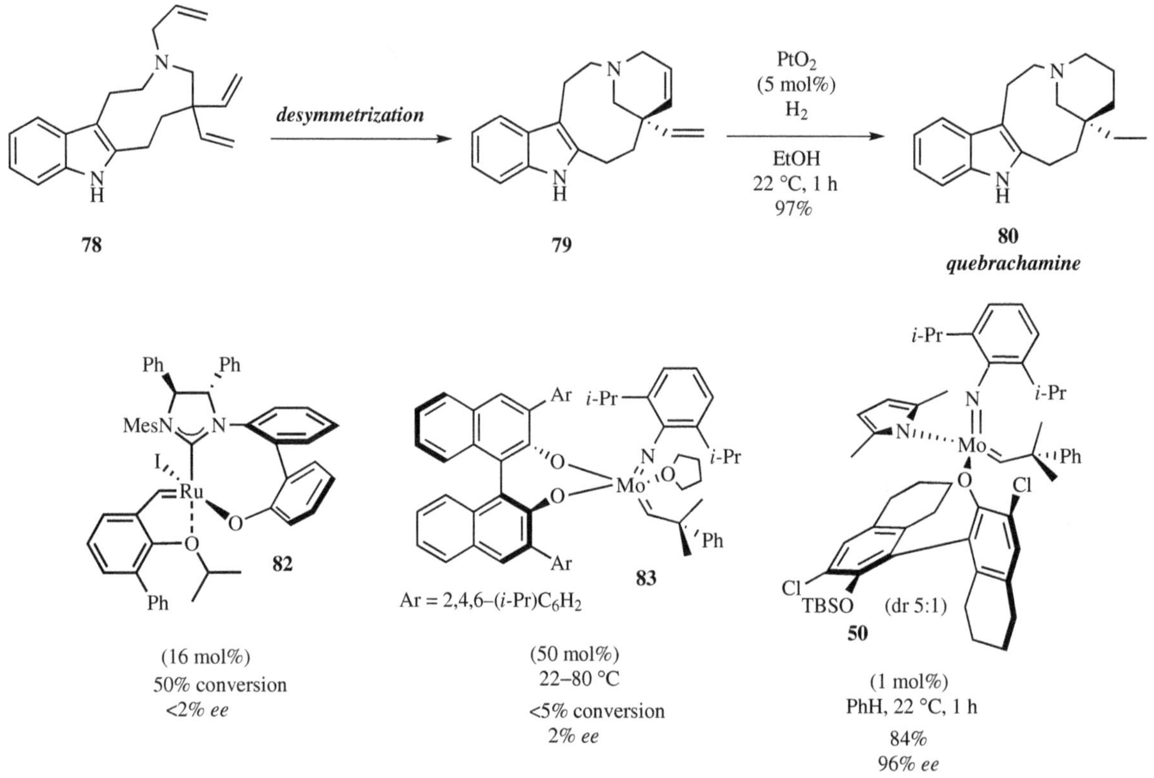

Scheme 6.25 Desymmetrization as a route toward the alkaloid quebrachamine.

great difficulty in promoting ring closure (less than 5% conversion), even when using high catalyst loadings (50 mol% and high temperature (80 °C). Chiral Ru-based catalyst **83** was successful at effecting the cyclization (conversion up to 98%, catalyst loading 15 mol%) but with low levels of enantioinduction (~10% ee). Chiral-at-Mo catalyst **50**, however, not only afforded the desired product in 84% yield at low catalyst loadings and at room temperature, but the polycyclic product **79** was also obtained in 96% ee. Interestingly, both diastereomers of catalyst **50** were investigated separately and although they catalyze the reaction with different levels of efficiency, they both afford the same enantiomer of **79**.

6.3 TANDEM ASYMMETRIC RING-OPENING/HBCROSS METATHESIS (AROM/CM) OR TANDEM ASYMMETRIC RING-OPENING/CROSS-CLOSING METATHESIS (AROM/RCM)

6.3.1 General Introduction

A number of both strained and unstrained rings can be ring-opened using olefin metathesis catalysts. When the ring opening is performed on highly symmetric cyclic structures having a plane of symmetry, the ring opening results in a desymmetrization of the molecule and can define multiple stereocenters in a single transformation. Traditionally, when ring opening happens in such a compound, an additional olefin is present within the system so that a second metathesis event can take place, resulting in a tandem process. One such case is when the olefin is connected to the molecule containing the strained or unstrained ring to be ring-opened, which usually results in a ring-closing metathesis process occurring during the reaction (tandem AROM/RCM, Scheme 6.26). Ring opening could occur via two possible mechanisms. The propagating metallocarbene species could react with the strained (or unstrained) cyclic olefin, resulting in the opening of the ring (Path A, Scheme 6.26). The newly generated metallocarbene then undergoes ring closing with the other olefin as the second metathesis event to afford one or more new rings with either bicyclic or spiro skeletons. More evidence has appeared for an alternative mechanistic pathway in which the catalyst reacts first with the exocyclic olefin (Path B, Scheme 6.26). The metallocarbene could then promote ring opening through an intramolecular reaction to afford the new cyclic product. These ring-rearrangement processes (45) have become powerful synthetic strategies for the synthesis of complex natural products (46).

Ring-opening processes could also be followed by an intermolecular cross metathesis reaction with another olefin in

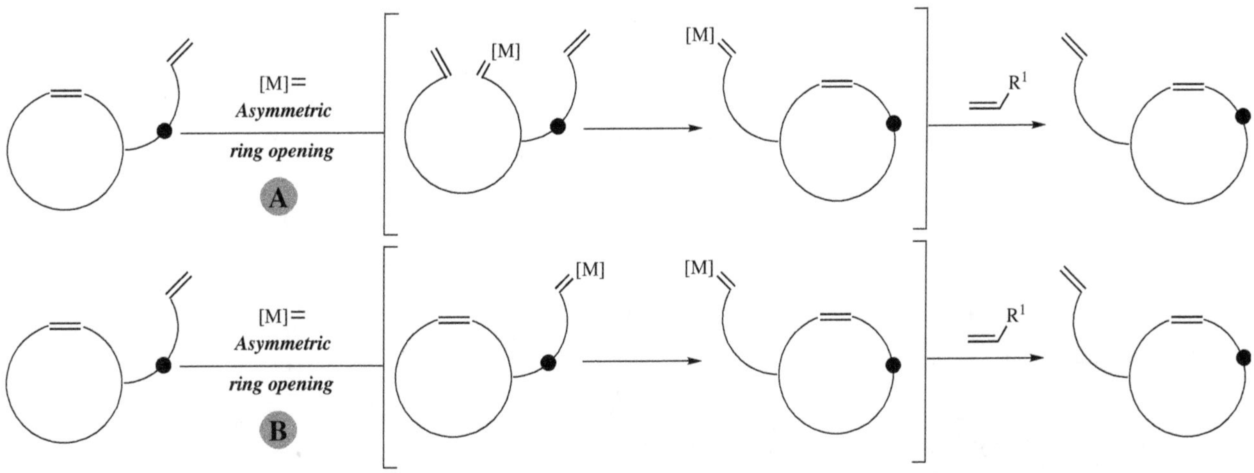

Scheme 6.26 Asymmetric ring-opening metathesis/ring-closing metathesis (AROM/RCM) processes.

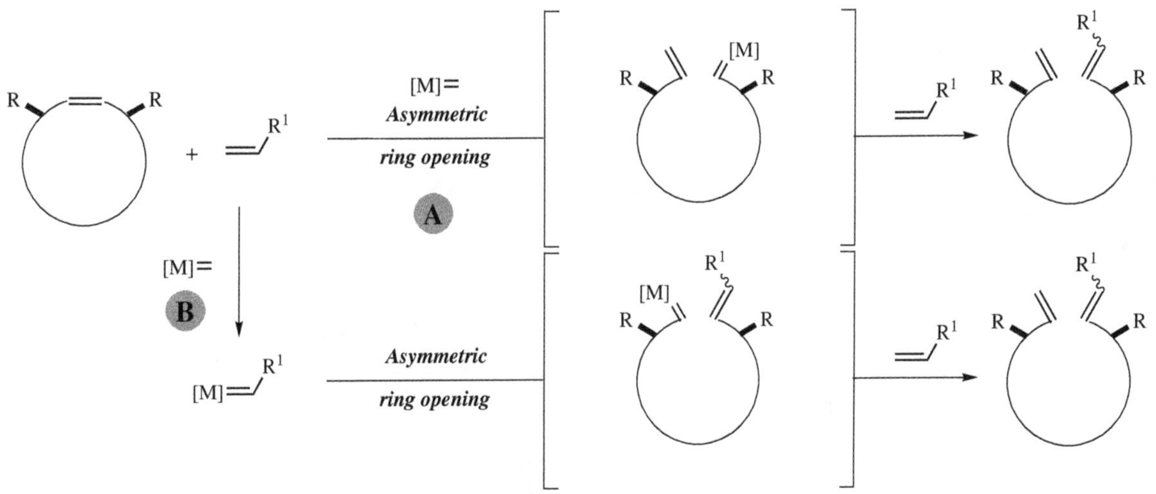

Scheme 6.27 Asymmetric ring opening metathesis/cross metathesis (AROM/CM) processes.

solution (Scheme 6.27). In this process, the metallocarbene propagating species could react with the cyclic olefin first, resulting in opening and desymmetrization of the ring (Path A, Scheme 6.27). The acyclic compound would undergo a cross metathesis process to generate the final product containing a stereodefined olefin and regenerate the propagating metallocarbene. Alternatively, the metallocarbene catalyst could react with the cross metathesis partner in solution to form a new metallocarbene, which could then promote ring opening (Path B, Scheme 6.27). Reaction of the resulting metallocarbene with another molecule of olefin would regenerate the propagating species and provide the desired product. The tandem AROM/CM process allows for greater complexity to be created in the final product as both the inclusion of new functionality and the stereochemistry of the resulting disubstituted olefin can be controlled.

6.3.2 Ring Opening of Strained Rings

The ring opening/cross metathesis of cyclopropenes have recently become increasingly studied using ruthenium-based olefin metathesis catalysts (47–49). Hoveyda and coworkers have shown that symmetric cyclopropenes bearing a quaternary carbon center (Scheme 6.28) can undergo efficient tandem AROM/CM with specific olefins. When **83** was reacted with an allyl ester and the Ru-based catalyst **81**, the ring opening occurred to afford an acyclic compound **84** bearing a skipped diene and a stereodefined quaternary carbon center. The resulting olefins formed in the reaction were obtained in good to excellent yields and enantioselectivities with useful levels of E:Z selectivity.

The enantioselectivity of the tandem AROM/CM of cyclopropenes was highly dependent on the nature of the cross

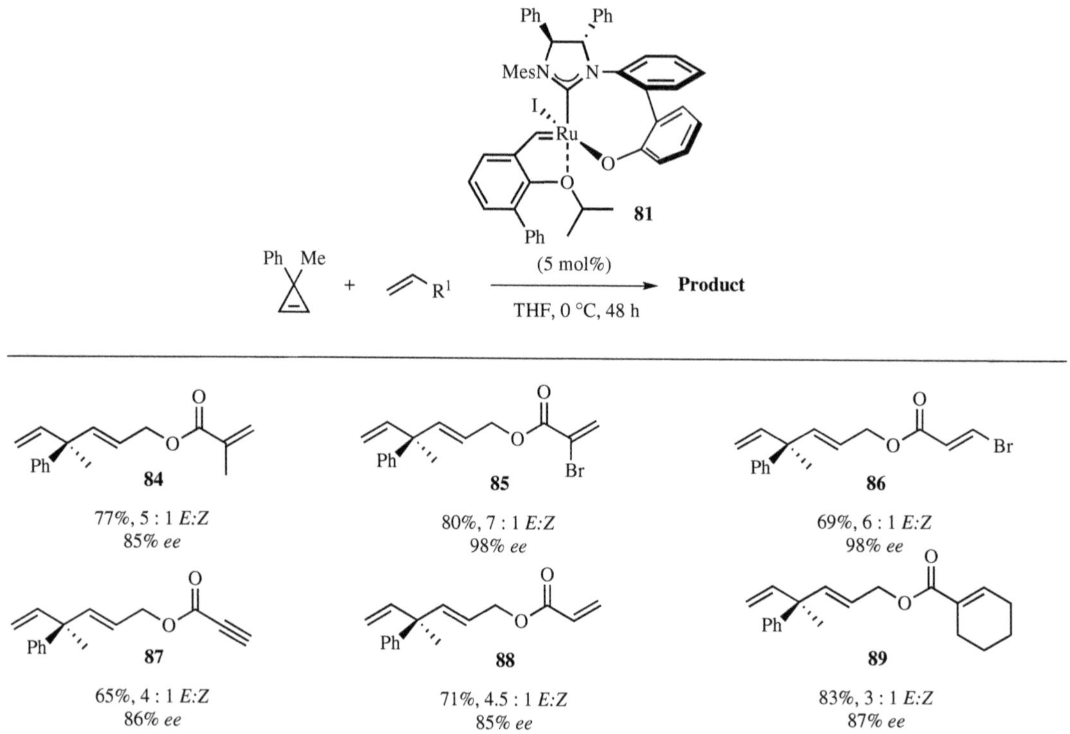

Scheme 6.28 Ruthenium-catalyzed tandem ring-opening metathesis/cross metathesis of cyclopropenes.

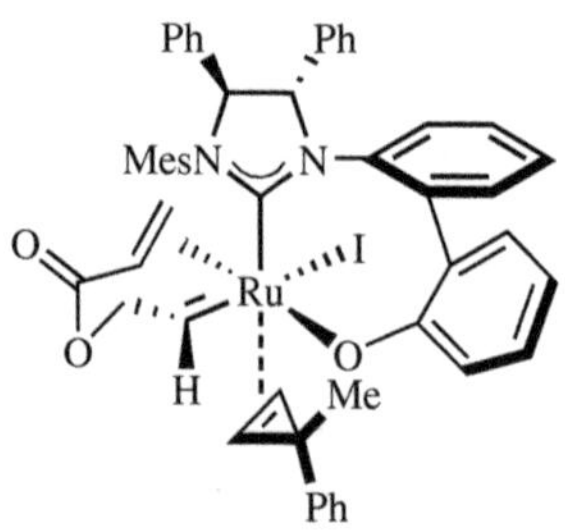

Scheme 6.29 Proposed model for the high selectivities observed in the ruthenium-catalyzed tandem ring-opening metathesis/cross metathesis of cyclopropenes.

metathesis partner. All of the highly selective examples possessed an additional olefin, typically in the form of an acrylate, which was proposed to theoretically chelate to the Ru catalyst (Scheme 6.29). In a proposed model, Hoveyda and coworkers suggest that reaction with the cross metathesis partner would allow for chelation of an olefin and formation of an octahedral Ru complex. Complexation of the *meso*-cyclopropene derivative to the Ru atom and subsequent ring opening would provide the desired products.

The ring-opening metathesis/cross metathesis of norbornenes or norbornene-like bicycles is perhaps the most well-studied AROM/CM process. Both Mo- and Ru-based catalysts have been developed for these reactions and in some cases have demonstrated interesting complementarity (50). Given that norbornene structures have more sites for functionalization than do cyclopropenes, the subsequent AROM/CM reactions of these norbornene-like compounds have the potential to afford heavily functionalized cyclic products. One such example is the AROM/CM of norbornenes to afford functionalized cyclopentanes with three to four stereodefined contiguous stereocenters (Scheme 6.30). The AROM/CM reaction of the simple bicycle **90** with catalyst **15** and *p*-methoxystyrene afforded the cyclopentane **91** in 92% yield and >98% ee. The same process could be performed on the more heavily substituted norbornenes **92** and **94**, and the Mo-based catalyst **15** could promote the formation of the products **93** and **95** with high enantioselectivities (51).

Ruthenium-based catalysts have also been extensively studied in AROM/CM processes, in part because they have been shown to be effective for a wide range of substrate classes (52,53). The synthesis of functionalized cyclopentanes could also be performed with Ru catalyst **102** (Scheme 6.31). The AROM/CM of the anhydrides **96** and **98** could be performed to afford the corresponding [3.3.0] ring systems **97** and **99** in good yield and selectivity. Heavily

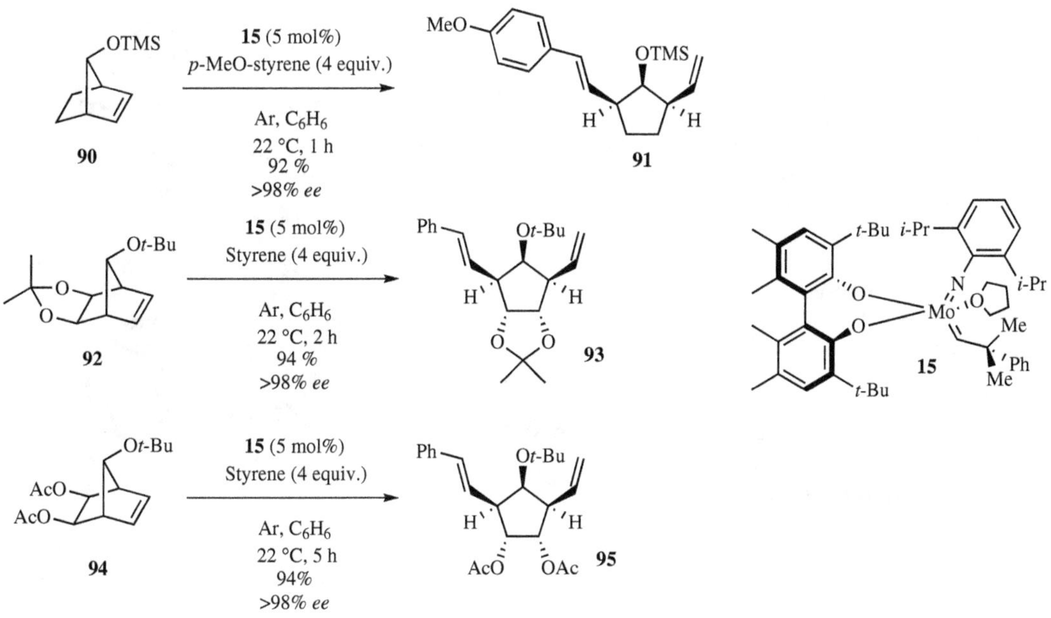

Scheme 6.30 Molybdenum-catalyzed tandem ring-opening metathesis/cross metathesis of norbornyl ethers.

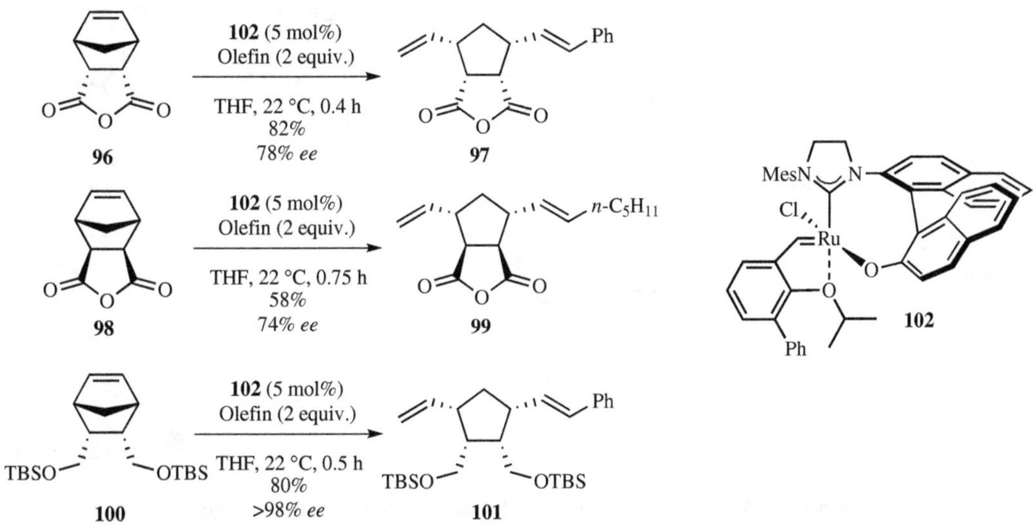

Scheme 6.31 Ruthenium-catalyzed tandem ring-opening metathesis/cross metathesis of norbornene derivatives.

substituted cyclopentanes such **101** could be formed in good yield (80%) and high ee (>98%).

The enantioselective AROM/CM process using Ru-based catalysts has also been exploited in the synthesis of pyrans through the ring opening of oxabicyclic olefins (Schemes 6.32). The [3.2.1]bicyclic ring systems may be less strained than the previously mentioned cyclopropenes and norbornenes, but the processes can be conducted in good yields using either of two different ruthenium-based catalysts (**103** and **81**). When the AROM/CM of oxabicyclic alcohol **104** was examined, very high enantioselectivities could be obtained using either catalyst. The pyran alcohol **106** was obtained in 97% ee and 81% yield using Ru catalyst **81**, and in 89% yield and >98% *ee* with catalyst **103**. Disappointingly, only catalyst **81** could promote the AROM/CM of the corresponding pyran ketone **106** (87%, 83% ee) while the catalyst **103** afforded only traces of the desired ring-opened product **107** (54,55).

The greater reactivity and functional group tolerance of Ru-based catalyst **81** presents other advantages to catalyst

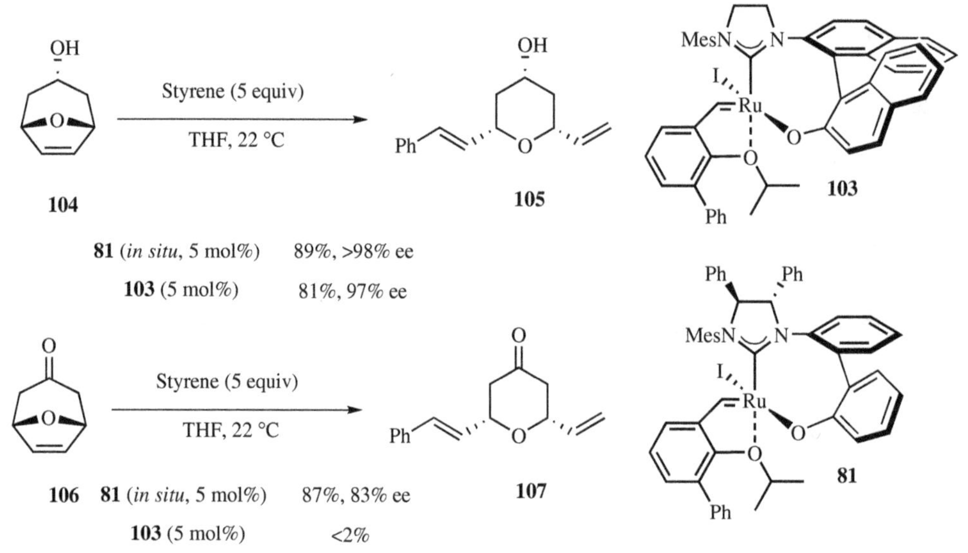

Scheme 6.32 Ruthenium-catalyzed tandem ring-opening metathesis/cross metathesis to form tetrahydropyrans.

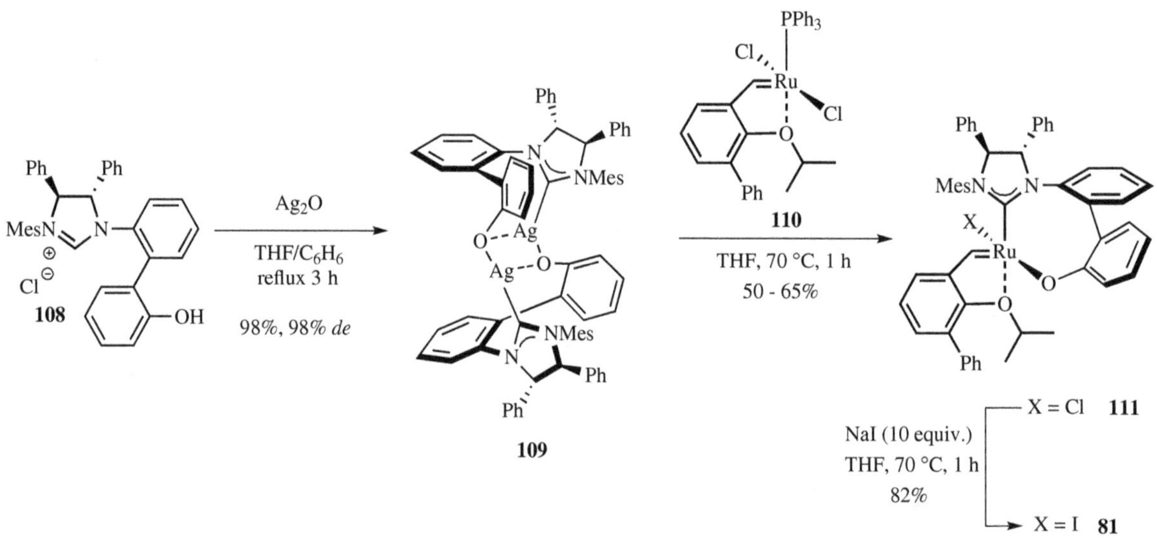

Scheme 6.33 Synthesis of Ru-based catalysts using an *in situ* protocol.

103. The Ru-based catalyst **81** can be prepared *in situ* from the corresponding silver salt **109**, a Ru-based fragment **110**, and NaI (Scheme 6.33). In a first step, the imidazolinium salt **108** can be treated with Ag_2O to form the bis-Ag salt **109**. Treatment of the salt **109** with the Ru complex **111** in THF results in transmetallation and NHC transfer, affording the Ru catalyst **111** in 50–65%. Exchange of the pendant chloride anion for iodide is possible in the same pot or as a second transformation.

Tandem AROM/CM reactions using the *in situ* derived catalyst **81** were found to be highly sensitive to the nature of the cross metathesis partner (Scheme 6.34). Hoveyda and coworkers observed a dramatic change in the enantioselectivity of the process between nonaromatic and aromatic cross metathesis partners. Mechanistic studies revealed that the diminished enantioselectivities obtained with vinylcyclohexane could be induced by a non-olefin metathesis-based isomerization of the carbene through polytopal rearrangements (56). During the catalytic cycle, two different propagating metallocarbenes are possible, relating to their orientation with regard to the substituents of the NHC ligand (exo and endo, Scheme 6.34). The authors theorized that high enantioselectivity likely arises from ring-opening reaction of the *exo*-carbene with **104** as the

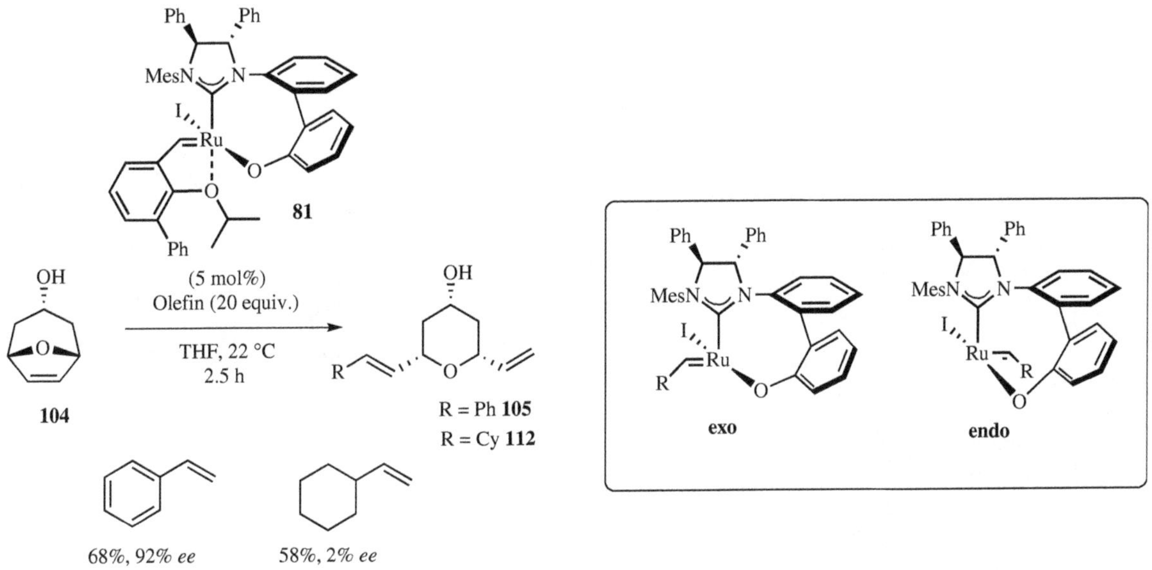

Scheme 6.34 Influence of the cross metathesis partner in ruthenium-catalyzed tandem ring-opening metathesis/cross metathesis to form tetrahydropyrans.

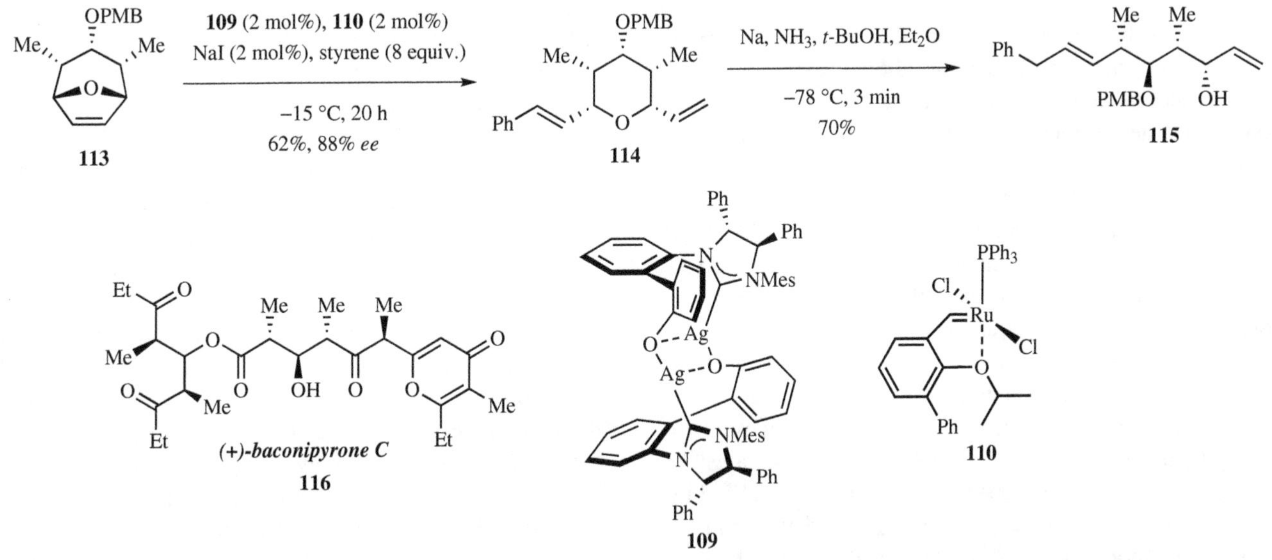

Scheme 6.35 Tandem ring-opening metathesis/cross metathesis forms a key subunit in the total synthesis of (+)-baconipyrone.

stereochemistry-determining step (Scheme 6.34). Consequently, the higher energy *endo*-carbene must be involved in product-releasing cross metathesis. Any deviation from these roles results in diminished enantioselectivities. These considerations provide a rationale for the enantioselectivity in tandem AROM/CM reactions of oxabicyclic alkenes with aryl- versus alkyl-substituted terminal olefins.

The ability of the tandem AROM/CM strategy to form heavily substituted tetrahydropyrans with up to five stereodefined contiguous stereocenters was exploited by Hoveyda and coworkers (57) in a synthesis of (+)-baconipyrone (Scheme 6.35) A key acyclic fragment of (+)-baconipyrone was to be prepared via ring opening of the pyran **114**, itself obtained from a Ru-catalyzed AROM/CM of oxabicycle **113**. When the oxabicycle **113** was treated with catalyst **81** (formed *in situ* from **109**, **110**, and NaI), the resulting pyran **114** was obtained in an efficient and selective process. The catalyst loading could eventually be reduced to 2 mol% and when the AROM/CM was performed at −15 °C, the pyran **114** was isolated in 62% yield and

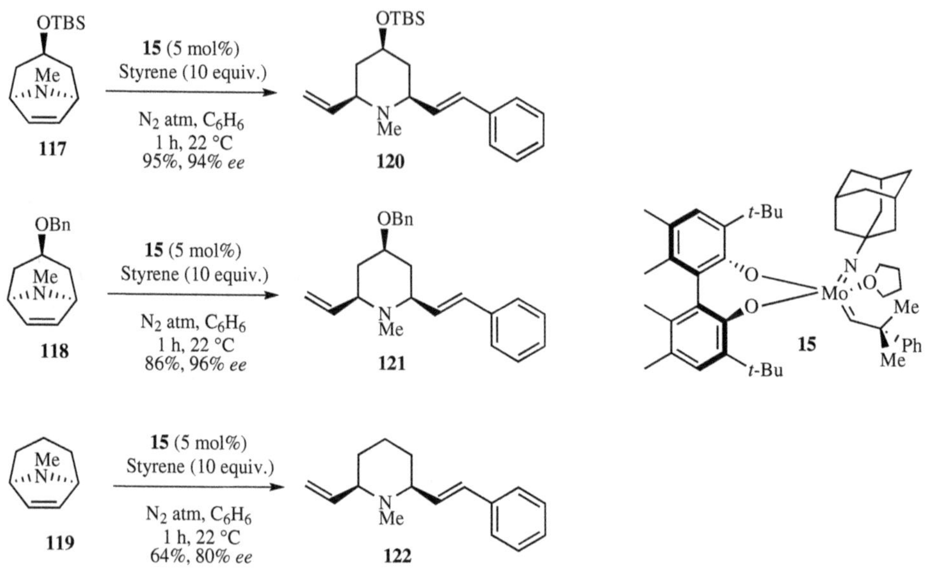

Scheme 6.36 Molybdenum-catalyzed tandem ring-opening metathesis/cross metathesis to form piperidines.

88% *ee*. The Ru-catalyzed AROM/CM was easily carried out on a multigram scale and pyran **114** could be reductively ring-opened to form the acyclic diene **115** in which four contiguous stereocenters were established, three of which were present in (+)-baconipyrone.

The tandem AROM/CM of bicycles could also be extended to nitrogen-containing derivatives that could be used for the synthesis of piperidines. When examining the tandem AROM/CM of azabicycles, the optimal catalyst was found to depend on the nature of the *N*-substituent, and the molybdenum- and ruthenium-based catalysts were found to be effective for different substrates (58,50). Early examples of the AROM/CM to form piperidines was reported by Hoveyda and coworkers and utilized Mo-based catalyst **15** (Scheme 6.36). The ring opening of the bicycles **117**, **118**, and **119** could be promoted by Mo catalyst **15** at room temperature, in short reaction times (1 h) and in typically good yield (>80%) and high enantioselectivity (>90% ee). Substrate bicycles without substituents such as the N-Me substituted **119** were found to be problematic, affording the piperidine **122** in lower yield and enantiomeric excess.

The complexity of AROM/CM of azabicycles is best demonstrated in the ring opening of structures having different nitrogen-protecting groups (Scheme 6.37) (50). When the *N*-Me azabicycle **117** was treated with Mo-based catalyst **15** under standard AROM/CM conditions, the piperidine **120** was isolated in high yield (95%) and enantiomeric excess (94%). Unfortunately, the authors reported that the same AROM/CM of **117** with Ru catalyst **102** under neat conditions afforded the corresponding piperidine in only 34% ee. The opposite effect was observed when examining the AROM/CM of the *N*-Bn azabicycle **123**. During the reaction of **123** in the absence of solvent, with Ru catalyst, the product **124** was obtained efficiently in 80% yield and >98% ee while the Mo-based catalyst only afforded a 21% ee.

Given the development of AROM/CM reactions to date, it is no surprise that it has become the method of choice for examining the efficiency and selectivity of new catalyst designs. One such example has been provided by Blechert and coworkers who reported new chiral ruthenium-based catalysts bearing a C_1-symmetric monodentate NHC ligand (59). The authors proposed that the presence of one bulky *N*-substituent on the NHC ligand (mesityl) and the use of the Hoveyda chelating benzylidene were structural features that afforded the new catalyst improved solution stability (Scheme 6.38) (60). Indeed, no sign of decomposition of the catalyst **125** was observed even after 12 days at 40 °C in CH_2Cl_2, allowing for AROM/CM reactions with extended reactions times or low catalyst loadings to be investigated if needed.

Standard testing of catalyst **125** in AROM/CM reactions of bicyclic anhydrides was investigated with only 1 mol% catalyst loading (Scheme 6.38). The catalyst provided average to good enantioselectivities with a wide range of substrates, but excellent *E* selectivity was observed in the cyclopentane products (up to >30 : 1 *E*:*Z*). The authors noted that changing the electron density of the coupling partner did not improve enantioselectivities, although in some cases, tuning the temperature could be beneficial.

Blechert and coworkers also observed that catalyst **125** could promote challenging AROM/CM reactions (Scheme 6.39). The reaction of anhydride **129** with Ru-based catalyst

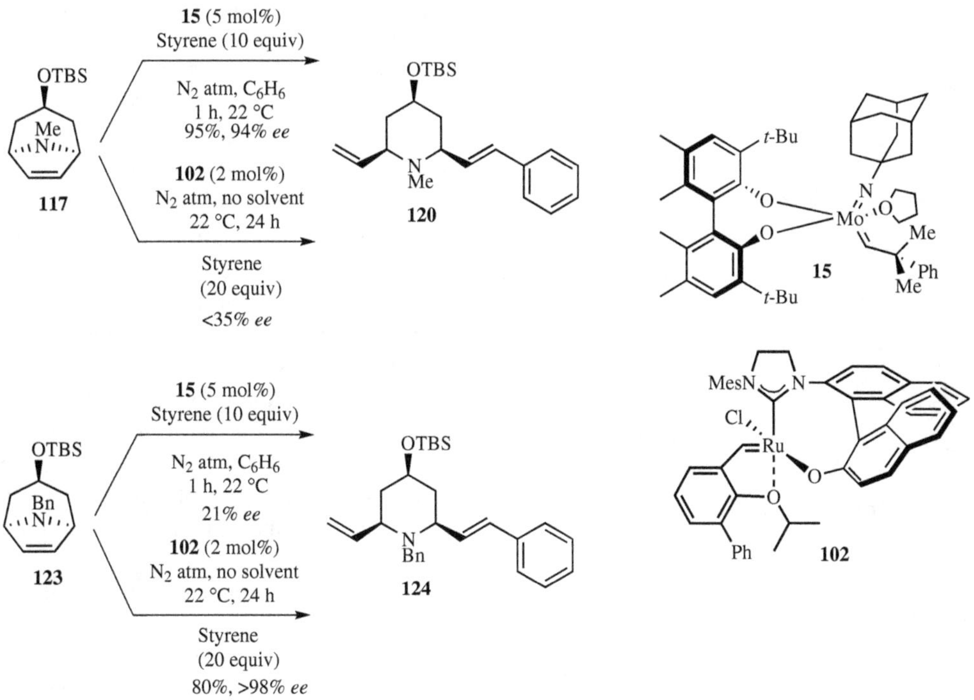

Scheme 6.37 Catalyst complementarity in the tandem ring-opening metathesis/cross metathesis to form piperidines.

125 and styrene as the cross metathesis partner afforded the cyclopentane **130** bearing four contiguous stereocenters and vicinal all-carbon quaternary centers in excellent enantioselectivity (>98% ee) and *E*:*Z* selectivity (23 : 1). The bicycle **131** also underwent AROM/CM with excellent *E*:*Z* selectivity (>30 : 1).

Blechert and coworkers altered their catalyst design that involved the use of NHCs having C_1-symmetric monodentate ligands (Scheme 6.40) (61). The improved variant of catalyst **125** utilized a different chiral NHC unit, which possessed an *N*-aryl group that was fused to the imidazolidinium ring, and both phosphine (**133**) and benzylidene (**134**) versions were prepared. The new chiral ruthenium-based catalysts were evaluated in AROM/CM of *meso*-norbornene derivatives with styrene (Scheme 6.40). Although catalysts **133** and **134** were effective at promoting the tandem AROM/CM of norbornenes **96**, **135**, and **130**, the products were often isolated in high enantiomeric excesses but disappointingly with lower *E*:*Z* selectivities in most cases than were obtained with their previous catalyst **125**.

Catalysts **133** and **134** did allow for an extension of the scope of the cross metathesis partner in the tandem AROM/CM reactions (Scheme 6.41). More electron-rich allyltrimethylsilane could be used as the cross metathesis partner to afford excellent conversions to afford the pyrrolidines **138** and **139**. It should be noted that catalyst **133** required lower temperatures to attain the same levels of enantioselectivity that were obtained with the benzylidene catalyst **134**. Although the conversions of the bicycles were excellent, the resulting *E*:*Z* ratios were poor. The authors did note that both the enantiomeric excesses and *E*:*Z* selectivities were influenced by solvent and temperature.

As mentioned above, AROM/CM reactions have become the method of choice for examining the efficiency and selectivity of new catalyst designs. Another innovative design was reported by Fogg and coworkers (62), who described the incorporation of simple binaphtholate ligands into chiral ruthenium-based catalysts (Scheme 6.42). The incorporation of BINOL motifs resulted in binding of the hydroxyl groups in both the cis and trans positions with respect to the NHC ligand. A pyridine or phosphine ligand occupied the other cis-binding site (in most other catalysts the cis-binding site was occupied by halogen substituents) (see **142** and **143**, Scheme 6.42). Interestingly, the catalysts were isolated as mixtures in which a small amount of a minor product was observed (~10%). In most cases, the minor product was a Ru-based catalyst in which one hydroxyl was still bound to Ru, but the second naphthol unit adopts an η^3-enolate coordination mode. Catalysts **142** and **143** afforded good to excellent conversions of anhydride **140** in AROM/CM reactions, but little to no enantioselectivity was observed. The corresponding PCy_3 η^3-enolate coordinated **144** analog was

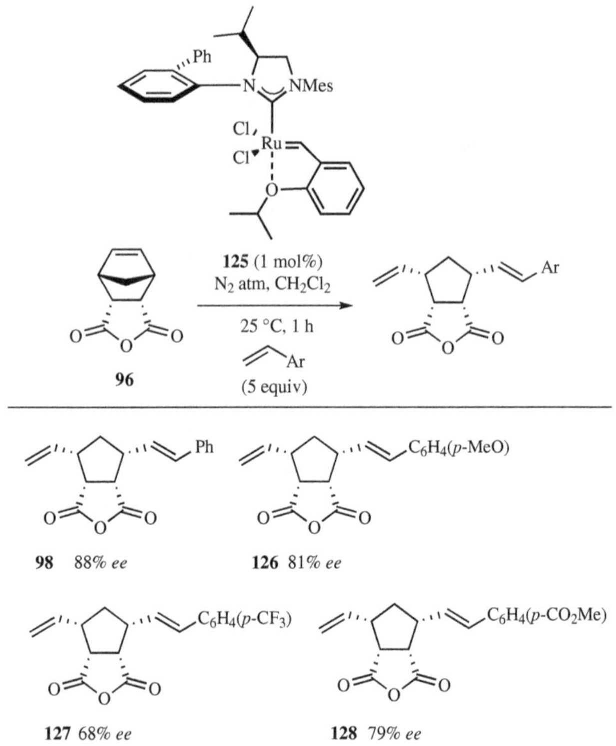

Scheme 6.38 Ruthenium-based catalysts having monodentate NHC ligands in AROCM reactions.

also synthesized and isolated as a pure complex but showed both low reactivity and enantioselectivity in AROM/CM of **140**.

Finally, Schrock and Hoveyda (63) evaluated the utility of their recent *Z*-selective class of stereogenic-at-Mo catalysts in tandem AROM/CM reactions. In 2012, Hoveyda and coworkers (64) prepared a series of chiral-at-Mo catalysts bearing monodentate BINOL-type ligands having different halogens in the 3,3′-position of the bisphenol moiety (Scheme 6.43). The series of Mo catalysts displayed excellent reactivity and allowed the AROM/CM utilizing enol ethers as the cross metathesis partner.

The Mo catalysts **145** through **148** proved to be highly efficient and afforded the products **150** and **151** in typically good yields with very low catalyst loadings (0.07–0.5 mol%). Furthermore, the resulting enol ethers **150** and **151** were obtained in good to excellent enantioselectivity and with a remarkable degree of *Z* selectivity (>98 : 2 in all cases). The efficiency and selectivity of the AROM/CM reactions were influenced by the nature of the cross metathesis partner. The authors suggested that lower conversions could arise from sterically more demanding enol ethers (*p*-methoxyphenylenol ether vs *n*-butylenol ether).

Another interesting observation by the authors concerned the relationship between the nature of the halogen and the efficiency of the catalyst formed *in situ*. Substitution of the fluorides by iodides improved the formation of the *in situ* catalyst 10-fold. Consequently, changing the nature of the halogen translated in most cases into an improvement of conversions, isolated yields and enantioselectivities, and catalyst **147** bearing bromides in the 3,3′-positions was deemed the most efficient for tandem AROM/CM.

Hoveyda and coworkers also examined the effect of the nature of the substituents along the oxabicyclic alkenes upon the efficiency of the AROM/CM process (Scheme 6.44). When the tandem AROM/CM of an exo-substituted oxabicyclic substrate **154** was performed, a large excess of the enol ether (~30 equiv) was required to achieve high enantioselectivity. The authors propose different scenarios

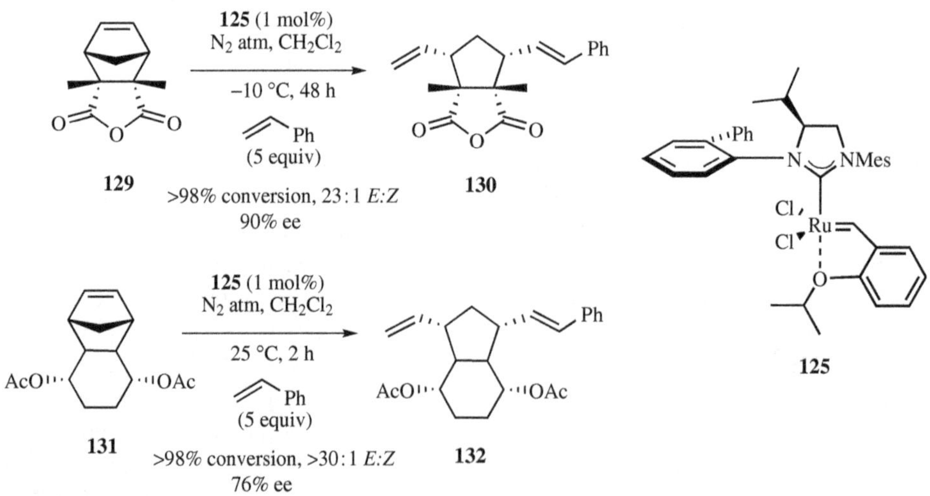

Scheme 6.39 Ruthenium-based catalysts having monodentate NHC ligands in challenging AROM/CM reactions.

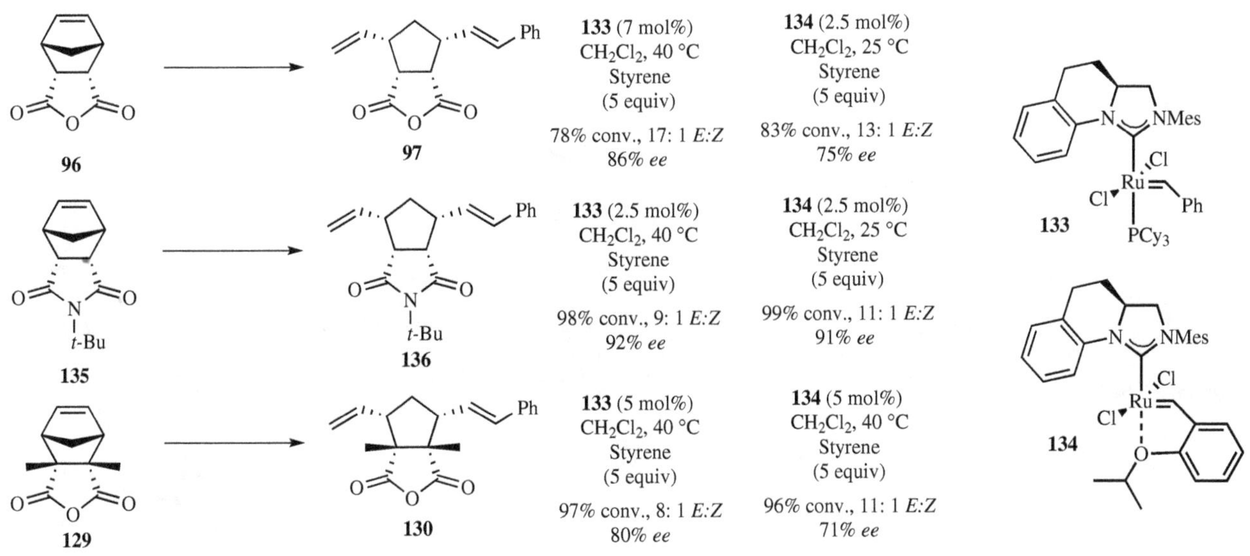

Scheme 6.40 Tandem AROM/CM reactions utilizing ruthenium-based catalysts having polycyclic monodentate NHC ligands.

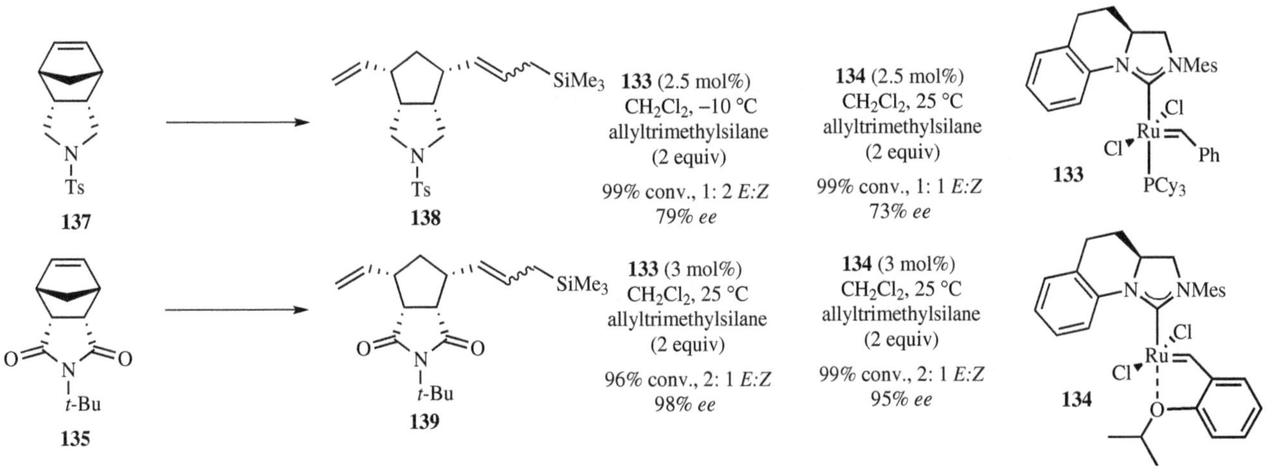

Scheme 6.41 Allyltrimethylsilane as a cross metathesis partner in tandem AROM/CM reactions.

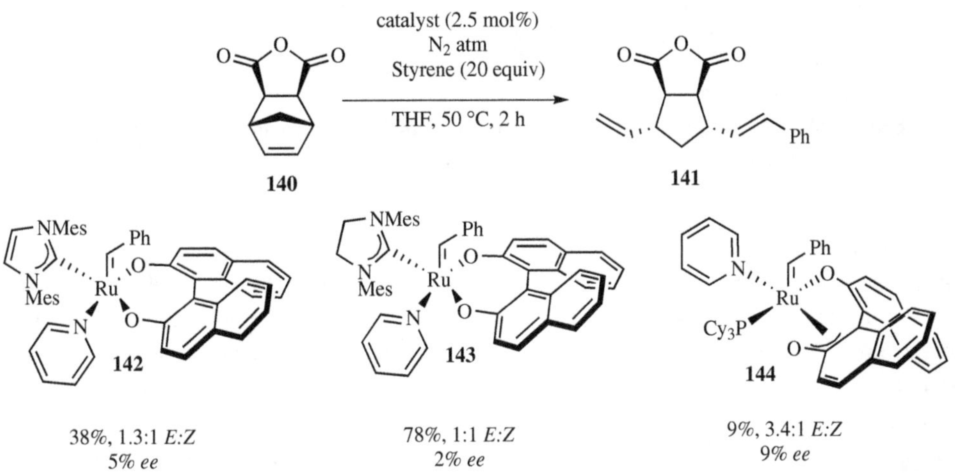

Scheme 6.42 Ruthenium-based catalysts having monodentate NHC ligands in AROCM reactions.

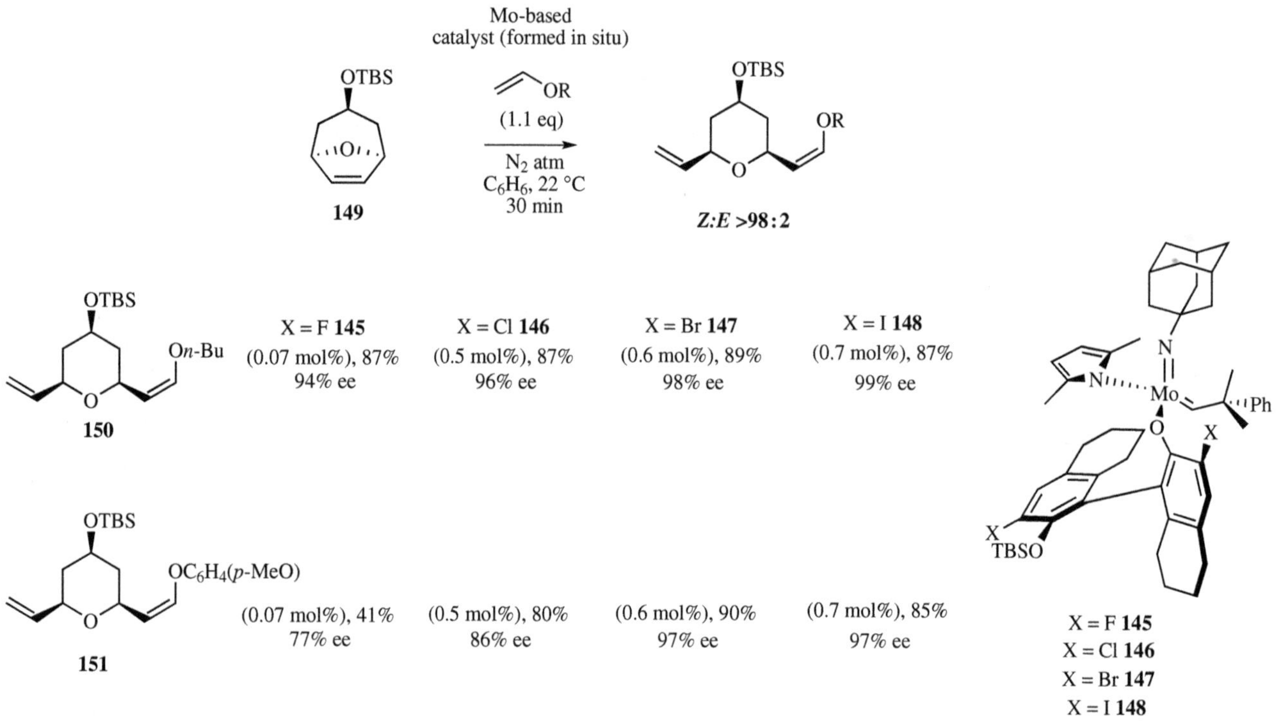

Scheme 6.43 Chiral-at-Mo catalysts in tandem AROM/CM reactions.

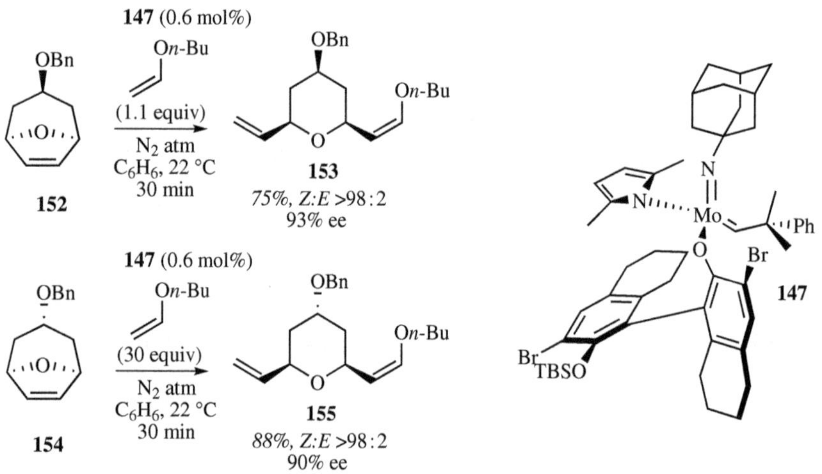

Scheme 6.44 Chiral-at-Mo catalysts in tandem AROM/CM reactions: influence of the nature of the oxabicyclic structure.

that could account for such an observation, including that a large excess of the cross metathesis partner can promote interconversion of the diastereomeric Mo-based complexes through nonproductive olefin metathesis. As such, the more reactive Mo-based complex isomer reacts faster and favors the formation of one enantiomer.

As the Mo-based catalyst **147** was shown to be optimal in the tandem AROM/CM of oxabicycles, it was also investigated in the analogous reactions of different azabicycles, cyclobutenes, and cyclopropenes (Scheme 6.45). The AROM/CM of the azabicycle **117** to afford the piperidine **156** was achieved in excellent yield (90%) but required a large excess of the enol ether cross metathesis partner to achieve high enantioselectivity and *E*:*Z* selectivities. Tandem AROM/CM of azabicycles was found to be less *Z*-selective when compared to the analogous oxabicycle derivatives. In

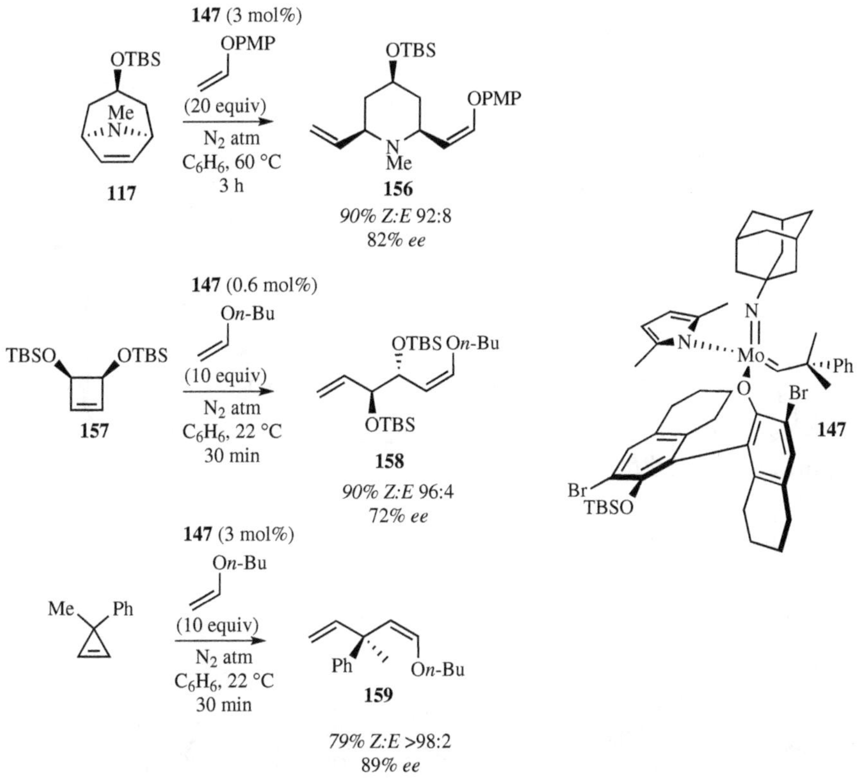

Scheme 6.45 Molybdenum-based catalysts having monodentate NHC ligands in AROCM reactions.

general, azabicycles such as **117** tend to be less reactive, requiring higher catalyst loadings, large excesses of cross metathesis partner, and higher temperatures that can promote alkene isomerization resulting in erosion of stereoselectivity.

Cyclobutenes such as **157** could be ring-opened to afford the acyclic enol ether **158** in good yield (90%), ee (72%), and *E*:*Z* ratio (4 : 96). The authors believed the size of the protecting groups on the hydroxyl moieties were essential in achieving good differentiation between the two faces of the alkene. Cyclopropenes such as **83** could also be reacted with an enol ether to provide the diene **159** possessing an all-carbon quaternary carbon center. Again, the AROM/CM proceeded to afford **159** with good to excellent efficiency (79% yield, 89% ee, *Z*:*E* > 98 : 2).

The chiral-at-Mo catalyst **147** could also be used to promote AROM/CM reactions under solvent-free conditions (Scheme 6.46). When the reaction of oxabicycle **152** with an enol ether under solvent-free conditions was investigated, the desired enol ether pyran **153** was isolated in identical yield when compared to the process conducted in C_6H_6 (74% vs 75% yield, respectively). When the AROM/CM of **152** was conducted neat, the catalyst loading could be reduced, the reaction time was shorter, the ee was higher (98%), and the pyran **153** was isolated in identical *Z*:*E* ratio (>98 : 2).

Given the success of the chiral-at-Mo catalysts to promote AROM/CM with enol ethers, Hoveyda and coworkers revisited the identical process using the stereogenic-at-Ru carbenes previously developed for AROM/CM reactions (65). Surprisingly, excellent *Z* selectivity was observed, and the AROM/CM of alcohol **104** with Ru-based catalyst **81** afforded good yields but with excellent *Z*:*E* selectivity and enantioselectivity using both aryl-substituted vinyl ethers and alkyl-substituted vinyl ethers (Scheme 6.47). The authors commented on the reactivity of the Ru-based catalyst **81**, as it must promote AROM selectively and efficiently but without being active enough to engage in other metathesis events that could erode the catalyst's kinetic selectivity. In general, a variety of enol ethers were tolerated and all provided the products in excellent enantio- and stereoselectivities (up to 98 : 2 *Z*/*E* and >98 : 2 er).

Of note in the tandem AROM/CM using Ru-based catalyst **81** was the use of an oxabicyclic starting material possessing a free hydroxyl group, which supports the contention that chiral Ru-based catalysts tend to be more functional group tolerant than their Mo-based counterpart. As such, Hoveyda and coworkers also attempted AROM/CM of more challenging substrates (Scheme 6.48). The substitution of a vinyl enol ether for a vinyl sulfide was investigated in the AROM/CM of oxabicycle **104**. The desired product **157** was obtained in good enantiomeric excess (92%) and *Z* selectivity

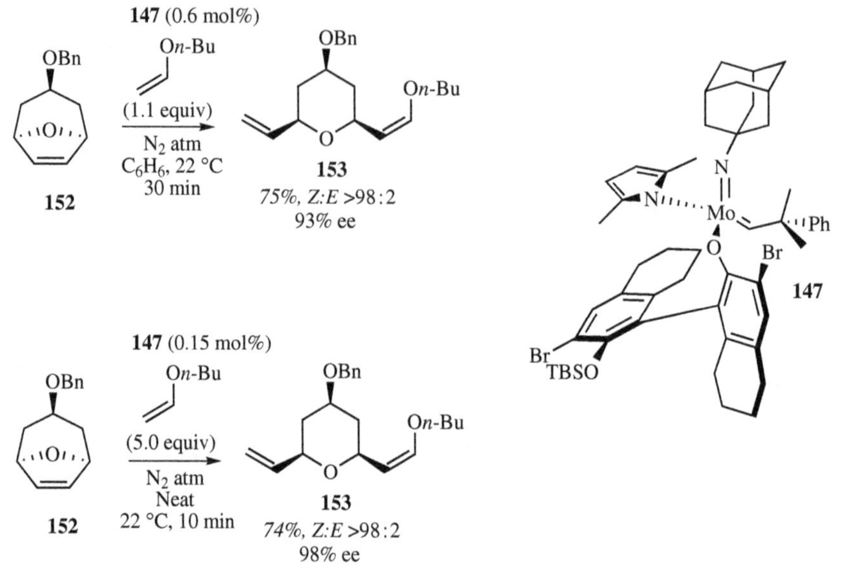

Scheme 6.46 AROM/CM reactions under neat conditions using a chiral-at-Mo catalyst.

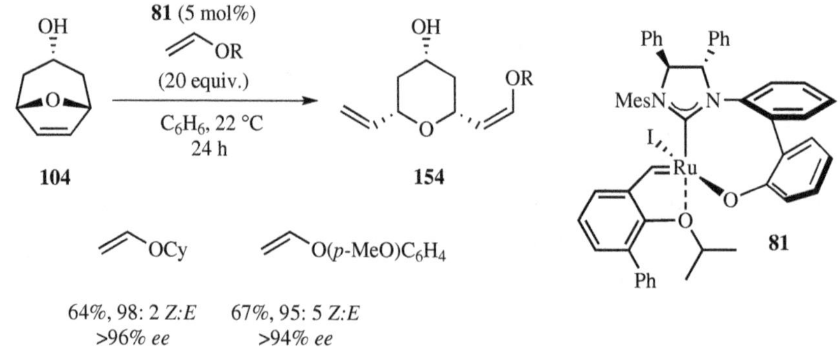

Scheme 6.47 Ruthenium-based catalysts having bidentate NHC ligands in AROCM.

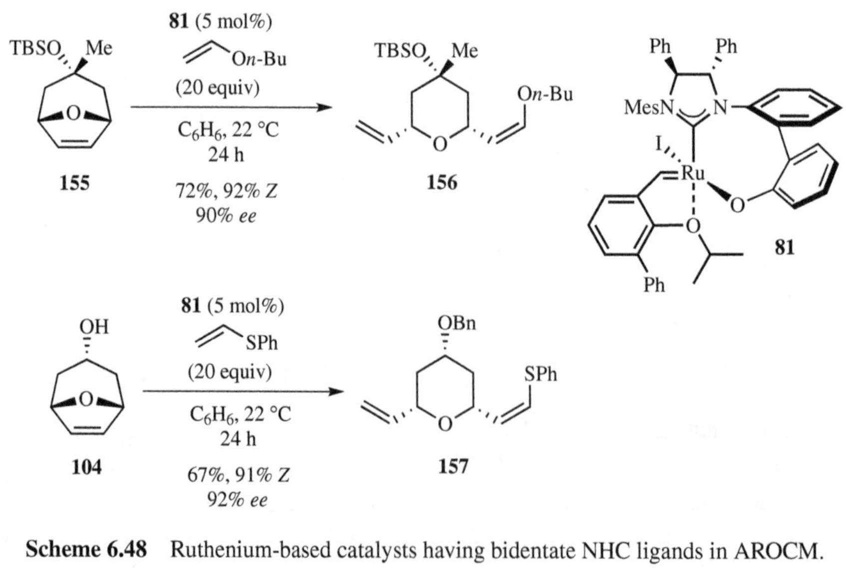

Scheme 6.48 Ruthenium-based catalysts having bidentate NHC ligands in AROCM.

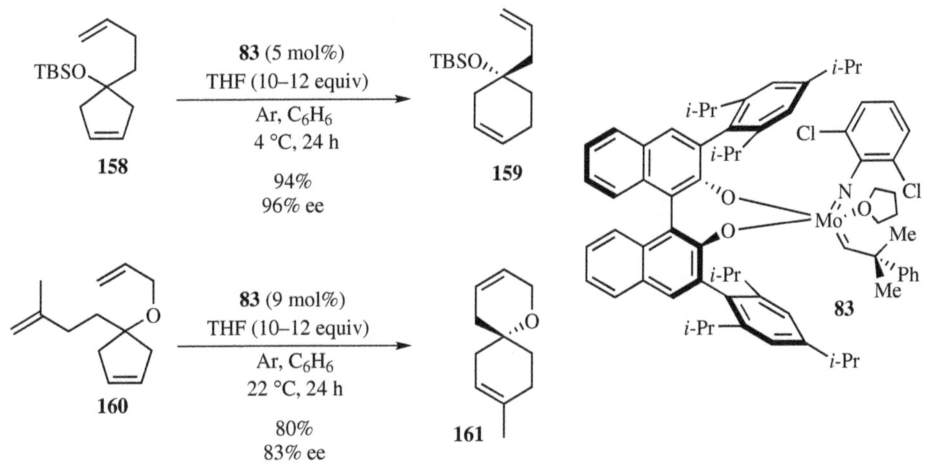

Scheme 6.49 Molybdenum-catalyzed enantioselective rearrangement to form cyclohexenyl tertiary ethers or spirobicyclic ethers (35).

(91%) having the same sense of stereoinduction as was observed for the enol ethers. In addition, the AROM/CM could be promoted on oxabicycles having more sterically encumbered substitution, such as the tertiary carbon center of **155** (Scheme 6.48). The pyran **156** could be isolated in 72% yield with high enantioselectivity and *Z* selectivity.

6.3.3 Ring Opening of Cyclopentenes

The rearrangement of *meso*-cyclopentenes into enantioenriched cyclohexenes affords new cyclic motifs with either a stereodefined quaternary stereocenter or a spirocyclic center. The rearrangements are driven by the formation of the more thermodynamically stable cyclohexene. The silyl ether **158** could undergo rearrangement with Mo-based catalyst **83** to afford the cyclohexene in excellent yield and enantioselectivity (Scheme 6.49). The Mo-based catalyst **83**, although air and moisture sensitive, can be effectively prepared *in situ* from stable and commercially available materials. The rearrangement of cyclopentenes benefitted from a unique solvent effect, as added THF resulted in a significantly higher enantioselectivity (66–68). The ability of the appropriately functionalized cyclopentenes to undergo rearrangement to form spirocycles is particularly noteworthy, as the corresponding heterocycles are not easily prepared by other existing methods. The rearrangement of cyclopentene **160** affords the spirocycle **161** in 80% yield and 83% ee. The olefins in the cyclopentene **160** are differently substituted so as to favor reaction of the catalyst at the terminal olefin and control the site of reaction for the metathesis cascade.

6.3.4 Asymmetric Cross Metathesis (ACM)

ACM of *meso*-dienes is perhaps the most challenging process in asymmetric olefin metathesis. In order to develop a highly efficient ACM process, many different factors must be controlled effectively (Scheme 6.50). In general, the cross metathesis of a terminal olefin **162** and a *meso*-diene **163** would afford the enantioenriched diene **164**. The carbenoid catalyst would have the ability to react with either the terminal olefin or the *meso*-diene. If the catalyst reacts with the terminal olefin to provide species **165**, it could then react with the *meso*-diene to provide the product **164** with a certain level of enantioselectivity (Path A, Scheme 6.50). It is also possible that the metathesis event affords *cis*-**164**, which could have a different level of enantioinduction and complicate product purification. However, if the catalyst reacts with the diene to afford **166**, then the final product would be formed through a cross metathesis with the terminal olefin **162**. Since the nature of the propagating carbene is different, it is possible to get a different level of enantiomeric excess in the final product **164**. Once again, the level of cis- and trans-selectivities and their respective levels of enantiomeric excess would also likely be different. Consequently, control of the nature of the propagating species, its orientation, and enantiotopic olefin selection all must be considered.

To date, only one study has been published on ACM processes by Grubbs and coworkers (69). The olefin cross-coupling partner studied was *cis*-1,4-diacetoxybut-2-ene, often in excess to help control the factors described above, as the reaction of Ru-based catalysts such as **57** occurs faster with the *cis*-olefin than with the corresponding *meso*-diene starting materials (Scheme 6.51). The ACM reaction between the *meso*-diene **167** and the acetate **171** afforded the desired product **168** in only 28% yield and in 44% *ee*. Other *meso*-dienes such as **169** could also participate in the reaction, and the levels of efficiency (48% yield, 37% ee) were similar to what was observed with **168**.

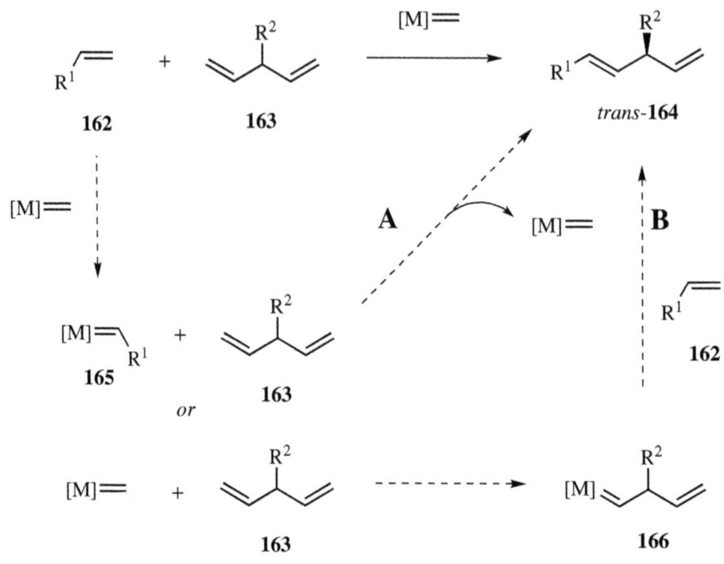

Scheme 6.50 General asymmetric cross metathesis of *meso*-dienes.

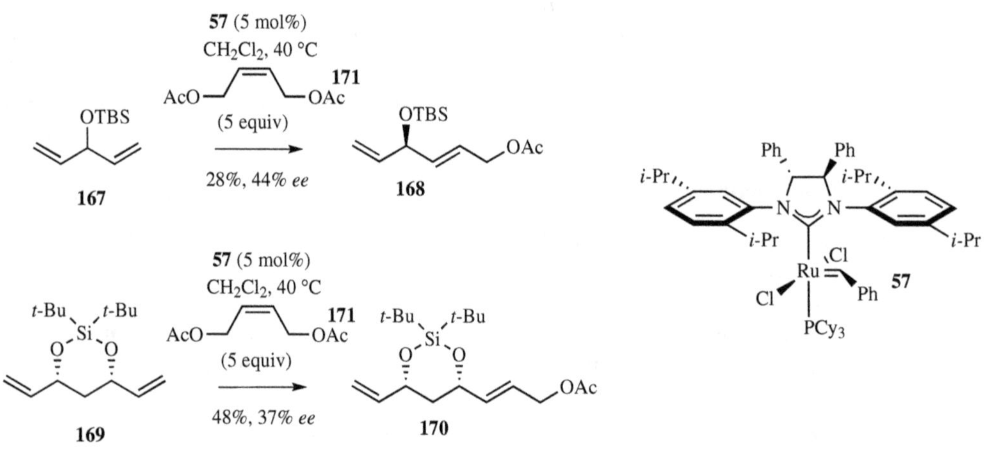

Scheme 6.51 Ruthenium-catalyzed asymmetric cross metathesis of meso-dienes.

General Procedure for ACM (Scheme 6.51) (69)

To a solution of *cis*-1,4-diacetoxy-2-butene (147 μl, 0.93 mmol) in CH_2Cl_2 (570 μl) was added the *meso*-diene **167** (37 mg, 0.186 mmol). The flask was then flushed with argon, and catalyst **57** (0.0093 mmol) was added. The reaction was heated to 40 °C and stirred for 12 h. The reaction was purified by column chromatography (20% Et_2O/hexanes) to afford the diene **168** (14 mg, 28%).

6.3.5 Asymmetric Ene–Yne Metathesis (AEYM)

Asymmetric ring-closing ene–yne metathesis (AEYM) is a relatively unexplored area of enantioselective metathesis. In principle, asymmetric ring-opening reactions, asymmetric cross ene–yne metathesis reactions, and asymmetric ring-closing ene–yne metathesis reactions should all be possible. Surprisingly, only the ring-closing variant has been explored.

In parallel with ARCM reactions, one could imagine both kinetic resolution and desymmetrization processes that could lead to enantioenriched compounds. The only desymmetrization reaction that has been explored involves yne–dienes (Scheme 6.52). In this process, the metathesis catalyst could react with either the alkyne or alkene, although reactions with alkenes tend to be favored (70). Regardless of where the metathesis cycle starts, two products are possible resulting from either an endo- or exo-cyclization mode. With a few rare exceptions, all cyclizations to form small rings catalyzed

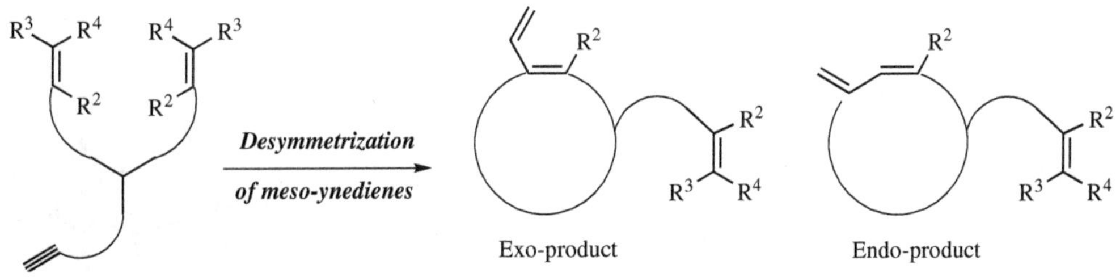

Scheme 6.52 General asymmetric ene–yne metathesis of a *meso*-yne–diene.

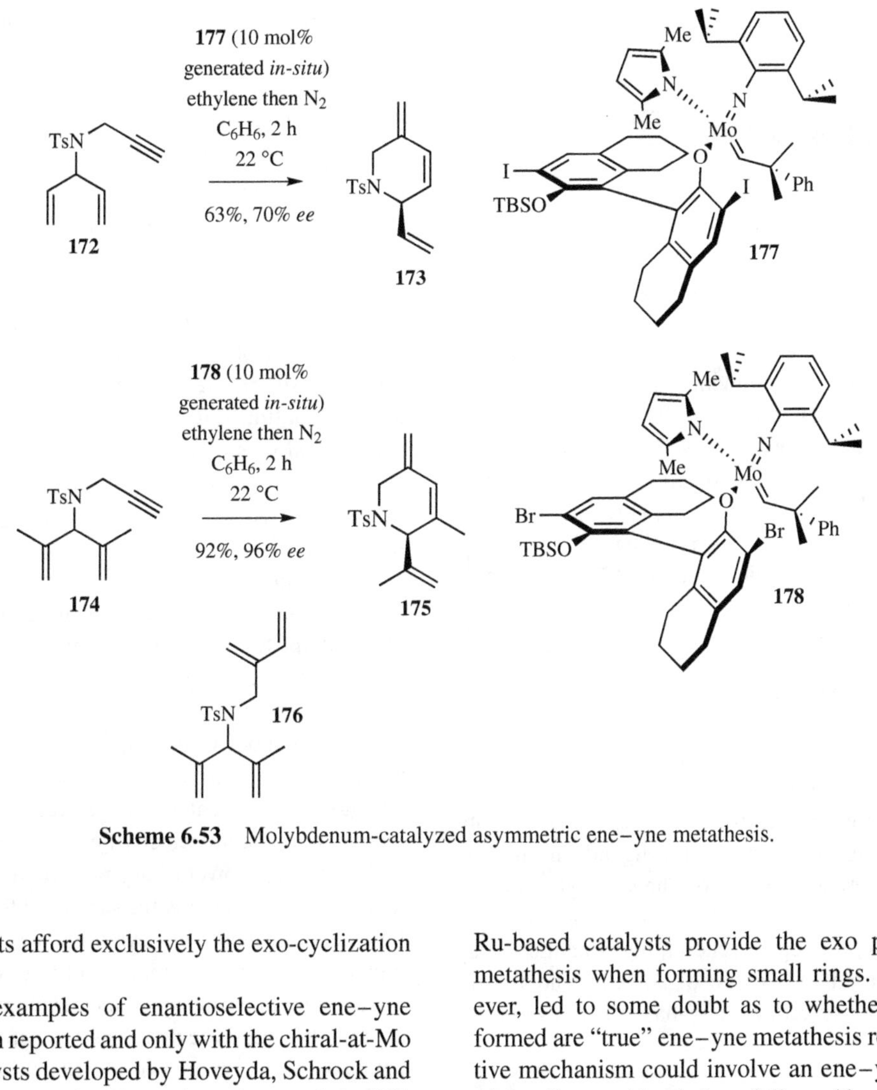

Scheme 6.53 Molybdenum-catalyzed asymmetric ene–yne metathesis.

by Ru-based catalysts afford exclusively the exo-cyclization product (71).

Relatively few examples of enantioselective ene–yne metathesis have been reported and only with the chiral-at-Mo or chiral-at-W catalysts developed by Hoveyda, Schrock and coworkers. The first report by Hoveyda and coworkers (72) describes the desymmetrization of yne–diene substrates using a chiral-at-molybdenum catalyst (Scheme 6.53). The yne–diene **172** could be treated with the catalyst **177** (formed *in situ*) and reacted under an ethylene atmosphere to provide the enantioenriched heterocycle **173** in 63% yield and 70% ee. Interestingly, the desymmetrization ene–yne metathesis process to **173** was completely endo-selective. Traditionally, Ru-based catalysts provide the exo product of ene–yne metathesis when forming small rings. Further study, however, led to some doubt as to whether the endo-products formed are "true" ene–yne metathesis reactions. An alternative mechanism could involve an ene–yne cross metathesis of the alkyne with ethylene followed by a subsequent ARCM reaction. When the yne–diene **174** was reacted with catalyst **178** under an ethylene atmosphere, only the tetraene **176** could be isolated from the reaction mixture. Conversion of **176** to **175** is possible if the ethylene is then removed and replaced with nitrogen. If the identical reaction was performed from the start in the absence of ethylene, no product **175** is ever formed.

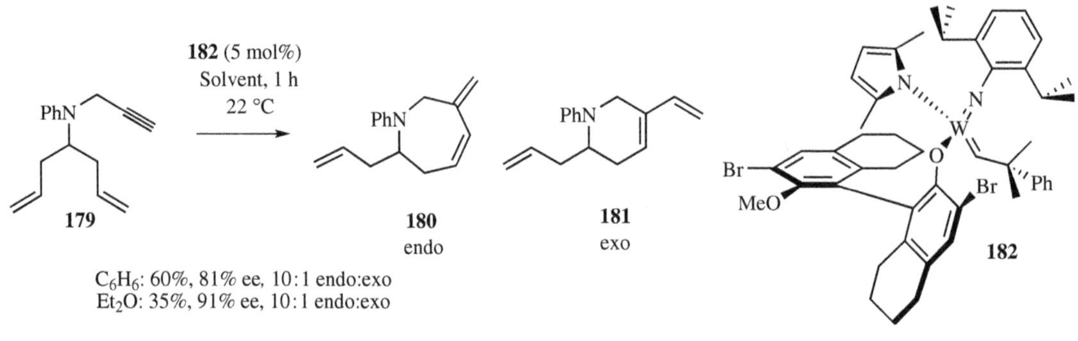

Scheme 6.54 Endo-selective enyne ring-closing metathesis promoted by stereogenic-at-W monopyrrolide complexes.

General Procedure for endo-Selective AEYM (Scheme 6.53) (72)

In an N_2-filled glove box, a solution of molybdenum complex **177** in C_6H_6 (250 μl, 0.02 M, 5.00 μmol) and a magnetic stir bar were added to a 4 ml vial. The vial was capped with a septum and sealed with electrical tape. The vial could then be brought out of the glove box and attached with a balloon of ethylene. A solution of the dienyne **172** (50.0 μmol) in C_6H_6 (500 μl) was added and the resulting mixture was allowed to stir at 22 °C for 2 h. Upon exposure to air, the reaction was quenched and then concentrated in vacuo to afford a brown residue. Purification of the crude mixture by silica gel column chromatography afforded the diene product **173** as a white solid (31.5 μmol, 63%).

Further studies from Hoveyda and coworkers suggest that the analogous chiral-at-W catalysts promote more efficient and endo-selective ene–yne RCM than the corresponding Mo-based variants (Scheme 6.54) (73). Catalysts such as **182** were optimized in an endo-selective ene–yne metathesis of **179**, where the effects of the aryl oxide ligand, the imido ligand, and solvent were all optimized. The *meso*-yne–diene **179** could be cyclized with catalyst **182** to provide a good ratio of the *endo:exo* product. In general, high enantiomeric excesses were observed in a range of solvents and fewer side reactions were observed with W-based catalysts than Mo-based catalysts.

6.4 CONCLUSION

Given the fact that the olefin metathesis reaction forms a carbon–carbon double bond and not a new chiral center per se, it is astounding to witness the level of creativity and imagination that researchers have used to infuse the field of enantioselective metathesis. To date, a variety of enantioselective processes have been developed for every different alternative of the olefin metathesis reaction including ring-opening, ring-closing, and cross metathesis. The complexity and power of asymmetric olefin metathesis is further augmented when the aforementioned processes are combined. The result is the development of enantioselective olefin metathesis reactions, which can furnish numerous types of chiral compounds through multiple sequential bond-forming or -breaking transformations. The kinetic resolution of simple dienes represented the first steps in the field and has continued to remain useful even though the desymmetrization of simple *meso*-triene precursors has begun to dominate the investigation of asymmetric ring-closing reactions. To date, the ring-closing strategies can be exploited to form cyclic structures such as carbo- and heterocycles even with quaternary carbon centers, for use in natural product synthesis, and can also be used to form helical chiral compounds for materials science, or planar chiral metallocenes for use as ligands in asymmetric catalysis. The tandem ring-opening metathesis/cross metathesis cascade has emerged as a powerful transformation that can convert simple small rings into acyclic molecules having multiple contiguous carbon centers and stereodefined olefins in high yields, high enantioselectivities, and high *E*:*Z* selectivities.

Despite the power of asymmetric metathesis, it is reasonable to assume that only the surface of the true potential of the field has scratched. The development of asymmetric ene–yne metathesis has been initiated only in the past 2 years. Without a doubt, one can expect the development of exo-selective asymmetric ene–yne metathesis, most likely using Ru-based catalysts, to be an area of increased interest. The ability to form 1,3-dienes adjacent to a chiral center promises that tandem AEYM/cycloaddition schemes will soon become feasible. Perhaps more important is the field of ACM, which has barely been explored. ACM has the potential to deliver enantioenriched small molecules that can act as integral building blocks in the construction of more complex molecules. The exploration of the ACM process is most likely to experience a surge of interest as the development of catalysts with increased reactivity and *E* or *Z* selectivity continue to expand.

When examining all of the different metathesis processes currently under development, it is easy to imagine further metathesis cascades and reactions that have not been described to date. For example, tandem AEYM/CM cascades which would furnish complex skeletal rearrangement of ene–ynes and then CM of the resulting 1,3-dienes could afford very complex compounds from simple starting materials. The ability to control the *E*:*Z* selectivity of the terminal CM process would result in an even more powerful process. Macrocyclic variants of the desymmetrization of *meso*-trienes or ring rearrangement cascades have yet to appear, despite the intense interest in macrocycles in many pharmaceutical research programs. There are a number of planar or axially chiral motifs that have yet to be targeted by enantioselective processes. Finally, given the ability of some metathesis catalysts to be used for transformations other than olefin metathesis, it is reasonable to assume that asymmetric olefin metathesis catalysts could be harnessed to promote tandem reactions that could end with hydrogenation, oxidation, or atom transfer reactions. The ability to exploit the chiral environment about a metathesis catalyst's metal center suggests that perhaps multiple tandem asymmetric transformations could be possible.

Finally, it should be mentioned that the exploration of asymmetric metathesis has helped fuel innovations in catalyst development in both the Ru and Mo manifolds. The recent development of chiral-at-Mo catalysts was inspired through challenges associated with asymmetric olefin metathesis, but the applications of these catalysts in synthesis and catalysis in general extend far beyond the realms of enantioselective metathesis. In addition, the useful chiral-at-W-based catalysts suggests that perhaps a third general class of catalysts based upon tungsten could soon become commonplace and provide increased complementarity with the more popular Mo- and Ru-based catalysts.

Given that olefin metathesis is considered a "green" technology, many other challenges remain before olefin metathesis becomes an industry standard. Some of these current issues include (i) finding more active catalysts or longer lived catalysts that would require lower catalyst loadings and offer high turnover frequencies and numbers, (ii) the development of new immobilized systems to simplify catalyst handling, recyclability, and purification. When considering the challenges and potential, asymmetric olefin metathesis is primed for even greater levels of innovation and achievement in the future.

REFERENCES

1. Grubbs RH. *Handbook of Metathesis*. Weinheim, Germany: Wiley-VCH; 2003.
2. (a) Schrock RR, Murdzek JS, Bazan GC, Robbins J, DiMare M, O'Regan M. *J Am Chem Soc* 1990;112:3875–3886. (b) Nguyen ST, Johnson LK, Grubbs RH, Ziller JW. *J Am Chem Soc* 1992;114:3974–3975.
3. Schrock RR, Hoveyda AH. *Angew Chem Int Ed* 2003;42: 4592–4633.
4. Hoveyda AH, Schrock RR. *Chem-Eur J* 2001;7:945–950.
5. Hoveyda AH, Zhugralin AR. *Nature* 2007;450:243–251.
6. Hoveyda AH, Gillingham DG, Van Veldhuizen JJ, Kataoka O, Garber SB, Kingsbury JS, Harrity JPA. *Org Biomol Chem* 2004;2:8–23.
7. Totland KM, Boyd TJ, Lavoie GG, Davis WM, Schrock RR. *Macromolecules* 1996;29:6114–6125.
8. Hoveyda AH, Malcolmson SJ, Meek SJ, Zhugralin AR. *Angew Chem Int Ed* 2010;49:34–44.
9. Cossy J, Arseniyadis S, Meyer C. *Metathesis in Natural Product Synthesis*. Weinheim, Germany: Wiley-VCH; 2010.
10. Fujimura O, Grubbs RH. *J Am Chem Soc* 1996;118: 2499–2500.
11. Fujimura O, Grubbs RH. *J Org Chem* 1998;63:824–832.
12. Zhu SS, Cefalo DR, La DS, Jamieson JY, Davis WM, Hoveyda AH, Schrock RR. *J Am Chem Soc* 1999;121:8251–8259.
13. Aeilts SL, Cefalo DR, Bonitatebus PJ Jr, Houser JH, Hoveyda AH, Schrock RR. *Angew Chem Int Ed* 2001;40:1452–1456.
14. Dolman SJ, Sattely ES, Hoveyda AH, Schrock RR. *J Am Chem Soc* 2002;124:6991–6997.
15. La DS, Alexander JB, Cefalo DR, Graf DD, Hoveyda AH, Schrock RR. *J Am Chem Soc* 1998;120:9720–9721.
16. Grandbois A, Collins SK. *Chem Eur J* 2008;14:9323–9329.
17. (a) Wagner G, Herrmann R. In: Togni A, Hayashi T, editors. *Ferrocenes*. Weinheim: Wiley-VCH; 1995; Chapter 4. p 173–175. (b) Hayashi T. In: Togni A, Hayashi T, editors. *Ferrocenes*. Weinheim: Wiley-VCH; 1995Chapter 2. p 105–106. (c) Butsugan Y, Araki S, Watanabe M. In: Togni A, Hayashi T, editors. *Ferrocenes*. Weinheim: Wiley-VCH; 1995Chapter 3. p 143–144. (d) Fu GC. *Acc Chem Res* 2000;33:412–420. (e) Fu GC. *Acc Chem Res* 2004;37:542–547.
18. (a) Patti A, Lambusta D, Piattelli M, Nicolosi G. *Tetrahedron: Asymmetry* 1998;9:3073–3080. (b) Siegel S, Schmalz H-G. *Angew Chem Int Ed Engl* 1997;36:2456–2458. (c) Genet C, Canipa SJ, O'Brien P, Taylor S. *J Am Chem Soc* 2006;128: 9336–9337.
19. Ogasawara M, Watanabe S, Fan L, Nakajima K, Takahashi T. *Organometallics* 2006;25:5201–5203.
20. Ogasawara M, Wu W-Y, Arae S, Watanabe S, Morita T, Takahashi T, Kamikawa K. *Angew Chem Int Ed* 2012;51: 2951–2955.
21. (a) Pape AR, Kaliappan KP, Kündig EP. *Chem Rev* 2000;100:2917–2940. (b) Rose-Munch F, Rose E. In: Astruc D, editor. *Modern Arene Chemistry*. Weinheim: Wiley-VCH; 2002; Chapter 11. p 368–399. (c) Gibson SE, Ibrahim H. *Chem Commun* 2002:2465–2473. (d) Kündig EP, Pache SH. *Sci Synth* 2003;2:155–228. (e) Salzer A. *Coord Chem Rev* 2003;242:59–72.
22. For a representative example see Solladie-Cavallo A, Solladie G, Tsamo E. *J Org Chem* 1979;44:4189–4191.

23. For representative examples see (a) Alexakis A, Mangeney P, Marek I, Rose-Munch F, Rose E, Semra A, Robert F. *J Am Chem Soc* 1992;114:8288–8290. (b) Kondo Y, Green JR, Ho JJ. *Org Chem* 1993;58:6182–6189. (c) Fretzen A, Ripa A, Liu R, Bernardinelli G, Kündig EP. *Chem Eur J* 1998;4:251–259.
24. For representative examples see (a) Kündig EP, Chaudhuri PD, House D, Bernardinelli G. *Angew Chem Int Ed* 2006;45:1092–1095. (b) Murai M, Uenishi J, Uemura M. *Org Lett* 2010;12:4788–4791.
25. Kiely AF, Jernelius JA, Schrock RR, Hoveyda AH. *J Am Chem Soc* 2002;124:2868–2869.
26. Lee A-L, Malcolmson SJ, Puglisi A, Schrock RR, Hoveyda AH. *J Am Chem Soc* 2006;128:5153–5157.
27. (a) Rainier JD, Cox JM, Allwein SP. *Tetrahedron Lett* 2001;42:179–181. (b) Chatterjee AK, Morgan JP, Scholl M, Grubbs RH. *J Am Chem Soc* 2000;122:3783–3784. (c) Arisawa M, Theeraladanon C, Nishida A, Nakagawa M. *Tetrahedron Lett* 2001;42:8029–8033. (d) Okada A, Ohshima T, Shibasaki M. *Tetrahedron Lett* 2001;42: 8023–8027. (e) Sturino CF, Wong JCY. *Tetrahedron Lett* 1998;39:9623–9626. (f) Clark JS, Elustondo F, Trevitt GP, Boyall D, Robertson J, Blake AJ, Wilson C, Stammen B. *Tetrahedron* 2002;58:1973–1982.
28. Heppekausen J, Fürstner A. *Angew Chem Int Ed* 2011;50:7829–7832.
29. (a) Heppekausen J, Stade R, Kondoh A, Seidel G, Goddard R, Fürstner A. *Chem Eur J* 2012;18:10281–10299. (b) Heppekausen J, Stade R, Goddard R, Fürstner A. *J Am Chem Soc* 2010;132:11045–11057.
30. Jernelius JA, Schrock RR, Hoveyda AH. *Tetrahedron* 2004;60:7345–7351.
31. Sattely ES, Cortez GA, Moebius DC, Schrock RR, Hoveyda AH. *J Am Chem Soc* 2005;127:8526–8533.
32. Dolman SJ, Schrock RR, Hoveyda AH. *Org Lett* 2003;5:4899–4902.
33. Malcolmson SJ, Meek SJ, Sattely ES, Schrock RR, Hoveyda AH. *Nature* 2008;456:933–937.
34. Harvey JS, Malcolmson SJ, Dunne KS, Meek SJ, Thompson AL, Schrock RR, Hoveyda AH, Gouverneur V. *Angew Chem Int Ed* 2009;48:762–766.
35. For representative examples see (a) Schuman M, Trevitt M, Redd A, Gouverneur V. *Angew Chem Int Ed* 2000;39:2491–2493. (b) Slinn CA, Redgrave AJ, Hind SL, Edlin C, Nolan SP, Gouverneur V. *Org Biomol Chem* 2003;1:3820–3825. (c) Dunne KS, Lee SE, Gouverneur V. *J Organomet Chem* 2006;691:5246–5259. (d) McReynolds MD, Dougherty JM, Hanson PR. *Chem Rev* 2004;104:2239–2258.
36. (a) Seiders TJ, Ward DW, Grubbs RH. *Org Lett* 2001;3:3225–3228. (b) Funk TW, Berlin JM, Grubbs RH. *J Am Chem Soc* 2006;128:1840–1846.
37. For examples in Cu-NHC complex catalyzed reactions see: (a) Kündig EP, Jia Y, Katayev D, Nakanishi M. *Pure Appl Chem* 2012;84:1741–1748. (b) Wang F, Liu L-j, Wang W, Li S, Shi M. *Coord Chem Rev* 2012;256:804–853.
38. (a) Vehlow K, Maechling S, Blechert S. *Organometallics* 2006;25:25–28. (b) Dinger MB, Nieczypor P, Mol JC. *Organometallics* 2003;22:5291–5296. (c) Ledoux N, Allaert B, Pattyn S, Vander Mierde HV, Vercaemst C, Verpoort F. *Chem Eur J* 2006;12:4654–4661. (d) Bornand M, Chen P. *Angew Chem Int Ed* 2005;44:7909–7911. (e) Adlhart C, Chen P. *Angew Chem Int Ed* 2002;41:4484–4487.
39. (a) Fournier P-A, Collins SK. *Organometallics* 2007;26:2945–2949. (b) Fournier P-A, Savoie J, Stenne B, Bedard M, Grandbois A, Collins SK. *Chem-Eur J* 2008;14: 8690–8695.
40. Savoie J, Stenne B, Collins SK. *Adv Synth Catal* 2009;351: 1826–1832.
41. Stenne B, Timperio J, Savoie J, Dudding T, Collins SK. *Org Lett* 2010;12:2032–2035.
42. Ogasawara M, Watanabe S, Nakajima K, Takahashi T. *J Am Chem Soc* 2010;132:2136–2137.
43. (a) Silverstein RM, Brownlee RG, Bellas TE, Wood DL, Browne LE. *Science* 1968;159:889–891. (b) Mori K. *Tetrahedron* 1989;45:3233–3298.
44. Burke SD, Müller N, Beaudry CM. *Org Lett* 1999;1: 1827–1829.
45. Holub N, Blechert S. *Chem Asian J* 2007;2:1064–1082.
46. (a) Barbe G, Charette AB. *J Am Chem Soc* 2008;130: 13873–13875. (b) Wrobleski A, Sahasrabudhe K, Aube J. *J Am Chem Soc* 2004;126:5475–5481. (c) Pfeiffer MWB, Phillips AJ. *J Am Chem Soc* 2005;127:5334–5335.
47. Giudici RE, Hoveyda AH. *J Am Chem Soc* 2007;129: 3824–3825.
48. Miege F, Meyer C, Cossy J. *Org Lett* 2010;12:248–251.
49. Binder WH, Pulamagatta B, Kir O, Kurzhals S, Barqawi H, Tanner S. *Macromolecules* 2009;42:9457–9466.
50. Cortez GA, Baxter CA, Schrock RR, Hoveyda AH. *Org Lett* 2007;9:2871–2874.
51. La DS, Sattely ES, Ford JG, Schrock RR, Hoveyda AH. *J Am Chem Soc* 2001;123:7767–7778.
52. Van Veldhuizen JJ, Garber SB, Kingsbury JS, Hoveyda AH. *J Am Chem Soc* 2002;124:4954–4955.
53. Van Veldhuizen JJ, Gillingham DG, Garber SB, Kataoka O, Hoveyda AH. *J Am Chem Soc* 2003;125:12502–12508.
54. Gillingham DG, Kataoka O, Garber SB, Hoveyda AH. *J Am Chem Soc* 2004;126:12288–12290.
55. Van Veldhuizen JJ, Campbell JE, Giudici RE, Hoveyda AH. *J Am Chem Soc* 2005;127:6877–6882.
56. Khan RKM, Zhugralin AR, Torker S, O'Brien RV, Lombardi PJ, Hoveyda AH. *J Am Chem Soc* 2012;134:12438–12441.
57. Gillingham DG, Hoveyda AH. *Angew Chem Int Ed* 2007;46:3860–3864.
58. Cortez GA, Schrock RR, Hoveyda AH. *Angew Chem Int Ed* 2007;46:4534–4538.
59. Tiede S, Berger A, Schlesiger D, Rost D, Lühl A, Blechert S. *Angew Chem Int Ed* 2010;49:3972–3975.
60. Kress S, Blechert S. *Chem Soc Rev* 2012;41:4389–4408.
61. Kannenberg A, Rost D, Eibauer S, Tiede S, Blechert S. *Angew Chem Int Ed* 2011;50:3299–3302.
62. Blacquiere JM, McDonald R, Fogg DE. *Angew Chem Int Ed* 2010;49:3807–3810.

63. For additional more information concerning Mo monopyrrolide–monoaryloxide catalysts see: (a) Sattely ES, Meek SJ, Malcolmson SJ, Schrock RR, Hoveyda AH. *J Am Chem Soc* 2009;131:943–953. (b) Jiang AJ, Zhao Y, Schrock RR, Hoveyda AH. *J Am Chem Soc* 2009;131:16630–16631. (c) Marinescu SC, Schrock RR, Müller P, Hoveyda AH. *J Am Chem Soc* 2009;131:10840–10841. (d) Marinescu SC, Levine DS, Zhao Y, Schrock RR, Hoveyda AH. *J Am Chem Soc* 2011;133:11512–11514.
64. Yu M, Ibrahem I, Hasegawa M, Schrock RR, Hoveyda AH. *J Am Chem Soc* 2012;134:2788–2799.
65. Khan RKM, O'Brien RV, Torker S, Li B, Hoveyda AH. *J Am Chem Soc* 2012;134:12774–12779.
66. Weatherhead GS, Ford JG, Alexanian EJ, Schrock RR, Hoveyda AH. *J Am Chem Soc* 2000;122:1828–1829.
67. Cefalo DR, Kiely AF, Wuchrer M, Jamieson JY, Schrock RR, Hoveyda AH. *J Am Chem Soc* 2001;123:3139–3140.
68. Teng X, Cefalo DR, Schrock RR, Hoveyda AH. *J Am Chem Soc* 2002;124:10779–10784.
69. Berlin JM, Goldberg SD, Grubbs RH. *Angew Chem Int Ed* 2006;45:7591–7595.
70. Galan BR, Giessert AJ, Keister JB, Diver ST. *J Am Chem Soc* 2005;127:5762–5763.
71. Dolhem F, Lievre C, Demailly G. *Eur J Org Chem* 2003: 2336–2342.
72. Lee Y-J, Schrock RR, Hoveyda AH. *J Am Chem Soc* 2009; 131:10652–10661.
73. Zhao Y, Hoveyda AH, Schrock RR. *Org Lett* 2011;13: 784–787.

7

OLEFIN METATHESIS POLYMERIZATION

ASTRID-CAROLINE KNALL AND CHRISTIAN SLUGOVC
Institute for Chemistry and Technology of Materials, Graz University of Technology, Graz, Austria

7.1 INTRODUCTION

Since the discovery of olefin metathesis in the 1950s, this fundamentally unique transformation was used for making polymers. Actually, olefin metathesis was discovered while studying the polymerization of olefins, and ring-opening metathesis polymerization (ROMP) was the first olefin metathesis-based transformation described in the literature (1). From this time on until the 1980s of the last century, a wide range of undefined initiator compositions based on metals such as titanium, vanadium, chromium, molybdenum, tungsten, iridium, or ruthenium was used to polymerize strained cyclic olefins, mainly norbornene. Based on the mechanistic understanding published by Hérisson and Chauvin in 1970 (2), the underlying principles of olefin metathesis reaction could be rationalized and initiator research slowly started to focus on the preparation of well-defined transition metal carbene complexes, which nowadays are usually employed for olefin metathesis polymerization. Several publications cover the history of olefin metathesis rendering a repetition here redundant (3). Olefin metathesis polymerization can be subdivided in two important polymerization reaction classes: the chain-growth polymerization technique ROMP and the step-growth condensation polymerization acyclic diene metathesis (ADMET). In ROMP, monocyclic or polycyclic olefins undergo ring opening and thus form a polymer chain, whereas in ADMET, mainly terminal diolefins react upon liberation of a small olefin, in most cases ethylene, to give macromolecules (see Fig. 7.1).

Provided the use of appropriate initiators and monomers, ROMP can be a controlled and living process (4), which enables the precise and fast preparation of diverse polymer architectures with narrow molecular weight distributions. In ROMP, a functional unit is typically located in the side chain of the polymer. ADMET can be used to prepare macromolecules with the functional unit as part of the main chain but those polymers are typically characterized by a broader molecular weight distribution. Both polymerizations are nowadays typically initiated (ROMP) or catalyzed (ADMET) by group VI or group VIII carbene complexes. In particular, ruthenium-based initiators/catalysts are used because they tolerate most functional groups, and exclusion of moisture or air is not necessary. This high functional group tolerance is the most important feature of olefin metathesis polymerization and makes it a versatile tool in modern macromolecular chemistry, aiming at obtaining precision and diversity of macromolecular architectures bearing various functionalities. Nevertheless, the scope of olefin metathesis polymerization is not restricted to tailor-made functional (co)polymers, and there are many examples covering diverse applications such as separation science, construction materials, damping, isolation, and self-healing polymers. This chapter intends to give a flavor of the abilities of olefin metathesis polymerization in macromolecular chemistry.

7.2 ROMP—THE MECHANISM

The basic mechanism of ROMP initiated by a ruthenium-based compound is shown in Figure 7.2. The initiation step comprises the loss of a labile ligand (L_2) and coordination of the monomer followed by the formation of the

Olefin Metathesis: Theory and Practice, First Edition. Edited by Karol Grela.

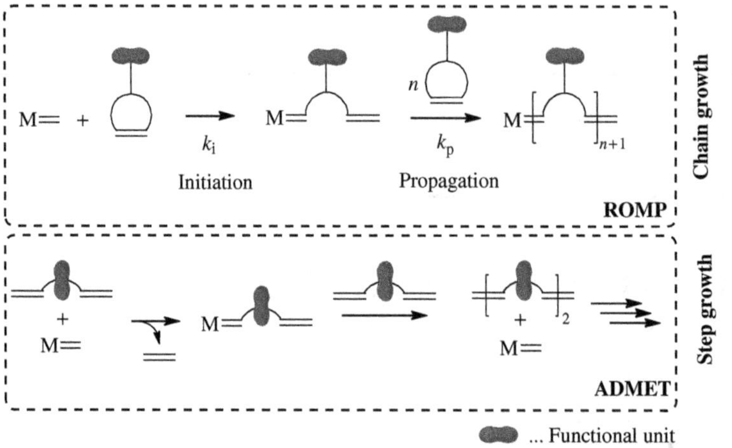

Figure 7.1 Basic description of ROMP and ADMET.

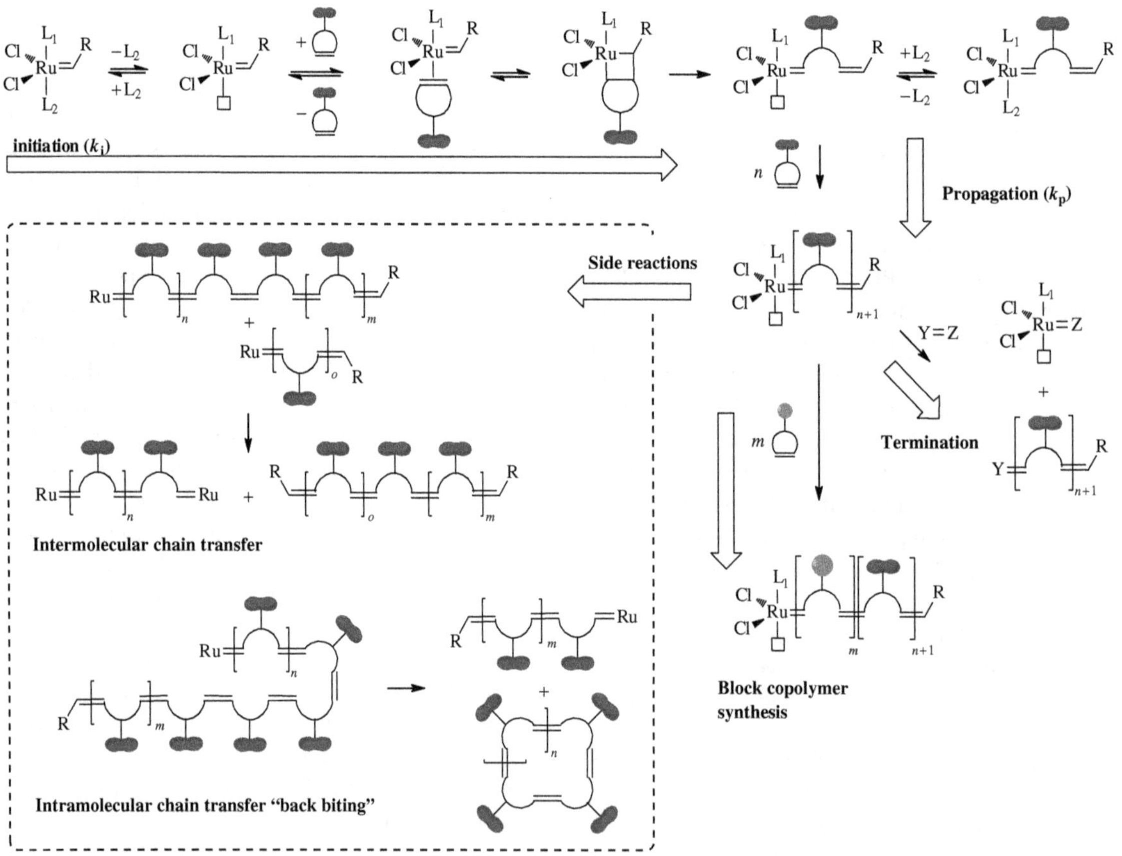

Figure 7.2 Simplified mechanism of ROMP (using a Ru-based initiator).

ruthenacyclopropane derivative and cyclo-reversion leading to the insertion of the first monomer. All reaction steps are reversible except the cyclo-reversion step, which is irreversible when the monomer is a strained cyclic olefin. Thereby, the release of the ring strain is the driving force and the dissociation of L_2 is believed to be the rate-determining step of the initiation reaction in many cases (5). The initiation proceeds with the initiation rate constant (k_i) and is followed by the propagation, which is the successive incorporation of further monomers by the initiated species. Propagation proceeds with the propagation rate constant (k_p) and is determined by the nature of the monomer, the co-ligand L_1, and the presence of donor ligands (L_2) in the reaction mixture that compete with the monomer for the vacant coordination site (□) at the central atom. After consumption of the monomer, an intended termination can be performed by adding a termination agent $Y=Z$ (which is, in most cases, a vinylether derivative yielding the corresponding Fischer carbene ($Z=CHOR$) and a vinyl-terminated polymer chain ($Y=CH_2$)) releasing the now inactivated initiator from the polymer chain. Using suitable termination agents, semi-telechelic macromolecules can be prepared (6). Possible side reactions include unintended termination as well as intermolecular and intramolecular chain transfers. In the latter cases, a macroinitiator undergoes olefin metathesis with a double bond from another macroinitiator's polymer chain or with its own polymeric tail, respectively, leading to broad molecular weight distributions of the resulting macromolecules. Chain transfer reactions become less important with increasing steric shielding of the monomers' double bonds.

To sum up, ideal prerequisites for the preparation of block copolymers (i.e., controlled living polymerization) comprise a k_i distinctly higher than k_p (i.e., all polymer chains are initiated within a short time frame), absence of any side reaction, and a similar or lower k_p of the subsequently added second monomer than the k_p of the monomer used for the first segment. Accordingly, controlled living polymerization is possible with a combination of fast initiating initiators and strained cyclic olefins as monomers which give polymers with no or minimal tendency to chain transfer reactions.

7.3 THE INITIATORS/CATALYSTS

For the majority of metathesis polymerizations, up to now, one of the commercially available initiators/catalysts depicted in Figure 7.3 was used. The Grubbs' first-generation (**G1**) (7) and second-generation initiators (**G2**) (8) as well as Schrock's molybdenum (**S-Mo**) (9) and tungsten alkylidenes (**S-W**) (9) were state of the art until 2002. Then, pyridine-containing species, sometimes termed *Grubbs' third-generation initiators* (**G3**), were discovered to be excellent initiators for ROMP (10,11). From then on, **G3** and its indenylidene analog **M31** (12) constitute the state-of-the-art initiators for controlled living ROMP. All these compounds can also be used for ADMET (13). Here, until now, the best suited commercially available catalyst with high functional group tolerance is the Grubbs' first-generation catalyst (**G1**) because second-generation catalysts tend to give extensive undesired double-bond isomerization as a side reaction (for a deeper discussion see later this chapter) (14).

Schrock's compounds **S-Mo** and **S-W** are highly active and provide, in many cases, access to controlled living polymerization of typical monomers (cf. below). However, their functional group tolerance is rather limited and protic substances or oxygen are not tolerated at all (9,15,16). **G1** provides controlled living polymerization in many cases (7,17), but because of its lower polymerization rate and the somewhat lower functional group tolerance compared to **G2** (18), **G3**, or **M31**, **G1** became less important. **G2**, bearing a *N*-heterocyclic carbene ligand, on the other hand, is much more active than **G1** but does not allow for controlled polymerization (8) because of the distinctly lower initiation rate constant (k_i, cf. Fig. 7.2) compared to the propagation rate constant (k_p) resulting in a low initiation efficacy. This leads to polymers with high molecular mass and broad molecular weight distribution (19). The pyridine-bearing initiators **G3** and **M31** combine the respective advantages of **G1** and **G2** since they exhibit fast propagation and an even faster initiation rate ($k_i \gg k_p$) (11). Accordingly, **G3** and **M31** are suited for the controlled polymerization of many monomers, and furthermore, block copolymers

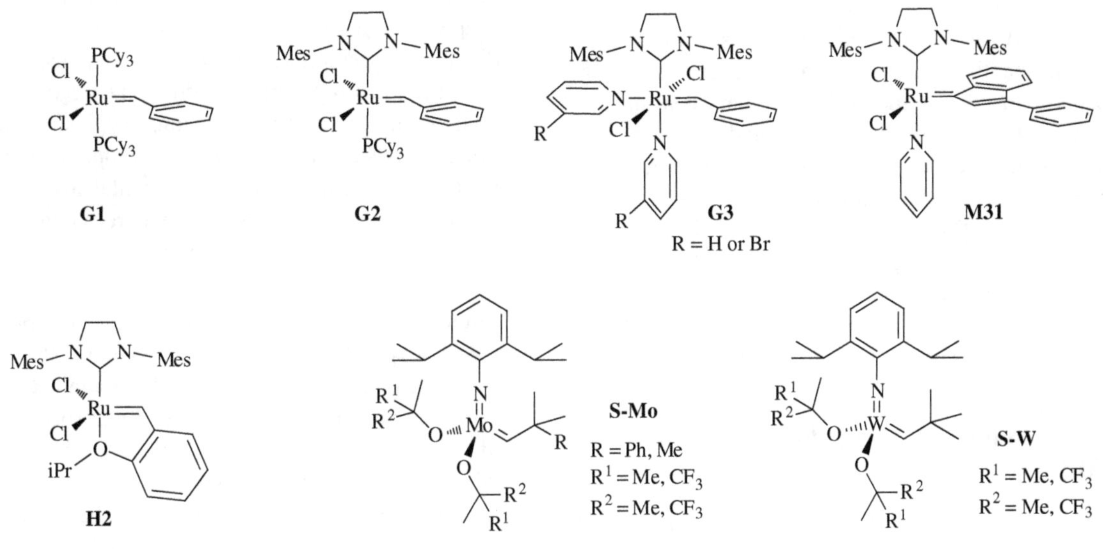

Figure 7.3 Most often used commercial initiators/catalysts.

can be prepared with both initiators (12,20). Moreover, these ruthenium complexes are relatively stable when stored at room temperature under air, which makes their use pretty convenient. The initially introduced variant of **G3** bearing 3-bromopyridines as co-ligands is less stable under these conditions and therefore lost importance during the last years. Besides these initiators/catalysts, a lot of other derivatives have been used for polymerization purposes. In many of these cases, the initiators/catalysts have been designed to meet special requirements, and some examples for this will be given in the forthcoming sections.

7.4 ROMP—THE MONOMERS

The archetypical monomers for ROMP are norbornene derivatives. Their high degree of ring strain (~27 kcal/mol (21) for norbornene) affords rapid polymerization and they are easily accessible via the Diels–Alder reaction of cyclopentadiene (or furan for oxa-norbornene derivatives) and olefins. Some challenges arise due to the fact that endo-substituted norbornenes are polymerized much slower than the corresponding exo-diastereomers (22,23). Accordingly, diastereopure monomers should be used for block copolymer synthesis. Not only the stereochemistry of the substitution of the norbornene derivatives but also the so-called anchor group itself (the group connecting a functional unit to the polymerizable, e.g. norbornene, moiety) is determining the polymerization rate (24). Anchor groups containing donor atoms tend to coordinate in chelating manner to the ruthenium center during propagation (such as L_2 in Fig. 7.2), thus diminishing the propagation rate (25,26). Generally, a functional monomer comprises the polymerizable group (i.e. the strained cycloolefin) and the anchor group (A) which connects a functional unit (F) to the polymerizable group. In many cases, a spacer (S) is used between the anchor group and the functional unit (cf. Fig. 7.4a). A further consequence of substitution is a decreased probability of chain transfer reactions during polymerization. In most works, the monomers bearing the functional unit were prepared, purified, and characterized before the polymerization, and also post-polymerization functionalization approaches for introducing functional units after polymerization have been reported (27–29).

Monomers depicted in Figure 7.4b were used to prepare an amphiphilic triblock random copolymer using **G3** as initiator whose preparation may serve as an example for making block (random) copolymers via ROMP.

General Procedure for Making Block Copolymers via ROMP

Monomer **A** (0.13 mmol, 100 equiv) is dissolved in dry and degassed CH_2Cl_2 (4 ml) under an inert atmosphere of argon. A solution of the initiator **G3** (0.0014 mmol, 1 equiv) in CH_2Cl_2 (0.5 ml) is then added to the reaction mixture. Owing to the living nature of ROMP, the molar ratio of **A** and **G3** determines the length of the first block. Here, it is important that the monomer, the initiator, and the emerging polymer are soluble in the used solvent. Typically, solvents such as CH_2Cl_2, toluene, or tetrahydrofuran (THF) are used, but all solvents that do not coordinate strongly to ruthenium can be used. Furthermore, solvents should not be stabilized or contaminated with olefins. The reaction is stirred under inert atmosphere at room temperature until full consumption of the monomer. The reaction progress should be monitored, for example, by thin layer chromatography (completeness of monomer consumption is proven by the absence of the monomer spot). At this stage, a small part of the reaction mixture can be withdrawn, reacted with ethyl vinyl ether, and then subjected to gel permeation chromatographic (GPC) analysis. Then, monomer **B** (0.27 mmol, 200 equiv) is added and the reaction mixture is stirred at room temperature until full conversion of the second monomer. Again a small part of the reaction mixture can be used to control the block copolymer preparation via GPC. The peak from the first sampling should be absent and a new peak with a shorter elution time should be present. A third segment, in this case, a random segment, is synthesized upon simultaneous addition of monomers **C** (0.0041 mmol, 3 equiv of the platinum dye and 0.023 mmol, 17 equiv of the carbazole derivative, respectively) dissolved in CH_2Cl_2 (0.5 ml) to the reaction mixture. Upon complete incorporation of the monomers, the polymerization reaction is quenched by adding ethyl vinyl ether (50 µl). After stirring for another 15 min, the reaction mixture is concentrated by partially removing the solvent under reduced pressure. Finally, the reaction mixture is added dropwise to a stirred nonsolvent, in this case *n*-pentane (80 ml), to precipitate the polymer. Purification is done by subsequent repeated precipitation (two times) and the final product is dried in vacuum. Typical characterization methods for the final product comprise nuclear magnetic resonance spectroscopy, elemental analysis, infrared spectroscopy, and gel permeation chromatography.

In total, this procedure results in a block copolymer with an $\mathbf{A}_{100}$–$\mathbf{B}_{200}$–$\mathbf{C}_{20}$ composition, whereby the third block is built up of the two, randomly distributed, **C** monomers. The whole procedure is finished in 5 h at room temperature and is a good showcase for a polymer architecture typically achievable by controlled living polymerization. The amphiphilic triblock random copolymer chains self-assemble into nanometer-scaled aggregates when dissolved in a selective solvent. Aggregation goes in hand with energy transfer

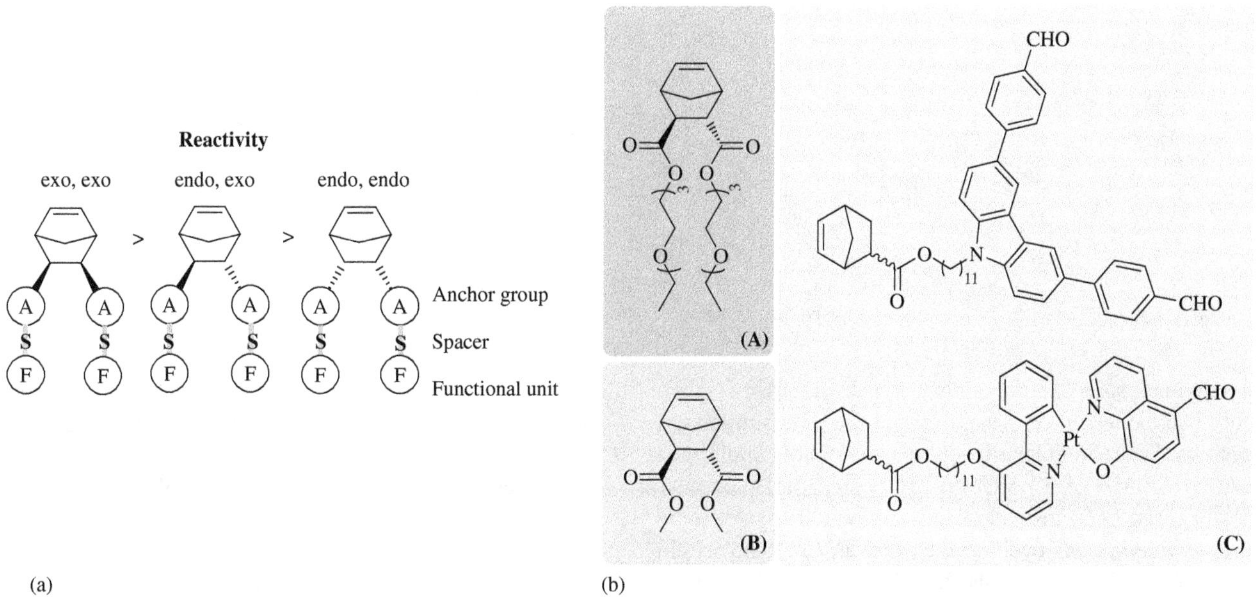

Figure 7.4 (a) Reactivity of disubstituted norbornene derivatives. (b) Exemplary norbornene-based monomers: (A) hydrophilic monomer; (B) hydrophobic monomer; (C) fluorescent donor and a phosphorescent acceptor dye bearing monomers.

from the fluorescent *N*-alkylcarbazole donor to the phosphorescent platinum(hydroxyquinolinolate) emitter. Upon addition of an unselective solvent, the aggregates are swelling and only the emission of the fluorescent dye is observed, which results in the change of the color of the emitted light from red to greenish blue (30).

In addition to norbornenes, cyclobutene and cyclooctene derivatives are other frequently used monomers for ROMP. Cyclobutenes have a higher ring strain than norbornenes (~30 kcal/mol (21)) and are, for this reason, ideally suited for the preparation of bottlebrush copolymers from macromonomers (31). ROMP of 3,4-disubstituted cyclobutenes (such as **1** in Fig. 7.5) yields polymers with a high density of functional groups and with a strictly linear 1,4-linked polybutadiene backbone, not accessible using other polymerization methods (32,33). Moreover, 1-cyclobutene derivatives such as cyclobutene-1-carboxamide (**2**) can be used as monomers (34,35). Interestingly, 1-cyclobutenecarboxylic acid esters (**3**) undergo only the initiation step but propagation does not take place. This reactivity can be exploited for the preparation of perfectly alternating copolymers with cyclohexene derivatives. This process is termed *alternating ring opening metathesis polymerization* (*AROMP*) and works without huge excess of one of the monomers (35,36). AROMP has been used for making cyclic alternating polymers (37) and antimicrobial polycationic polymers (38).

cis-Cyclooctene derivatives have only approximately a third of the ring strain of norbornenes (~7 kcal/mol (21)). Furthermore, cyclooctene-derived polymers are characterized by a relatively broad molecular weight distribution which is due to the relatively high probability of chain transfer reactions. Thus, cyclooctene and its derivatives are mostly used in cases where linear polymer backbones are desired. Recent work with this monomer class demonstrated that 3-alkyl- or aryl-substituted cyclooctenes (**4**) could be polymerized in a regioselective and stereoselective manner (39) and, upon hydrogenation, precision linear low density

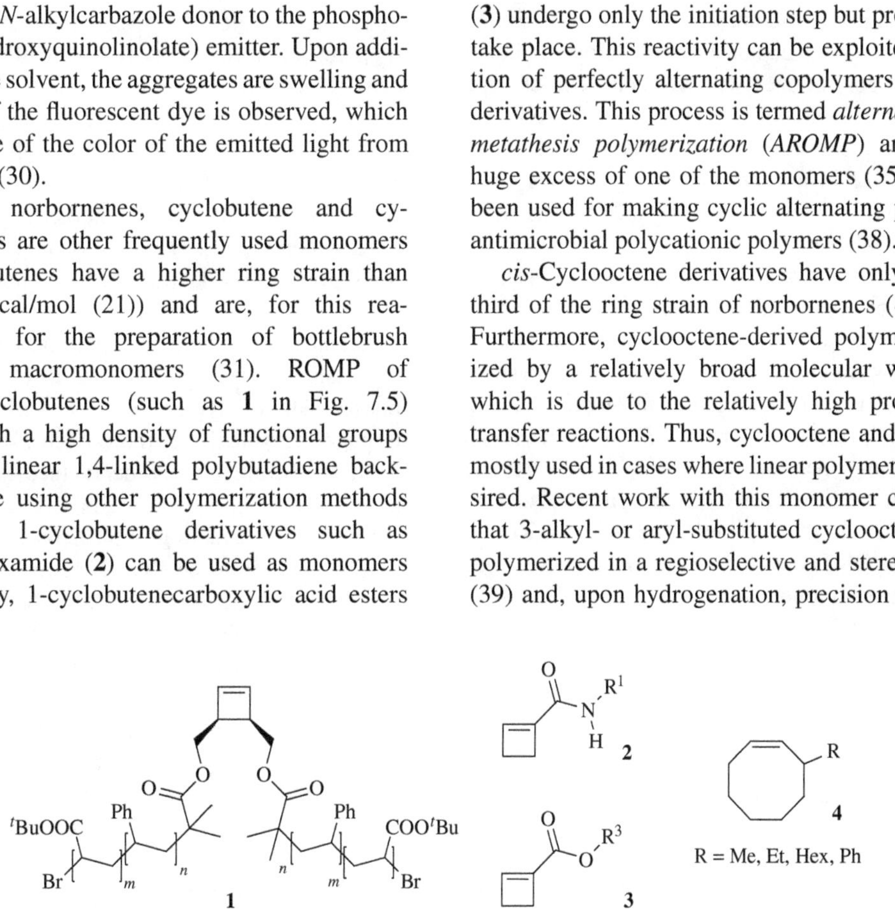

Figure 7.5 Examples for cyclobutene- and cyclooctene-based monomers.

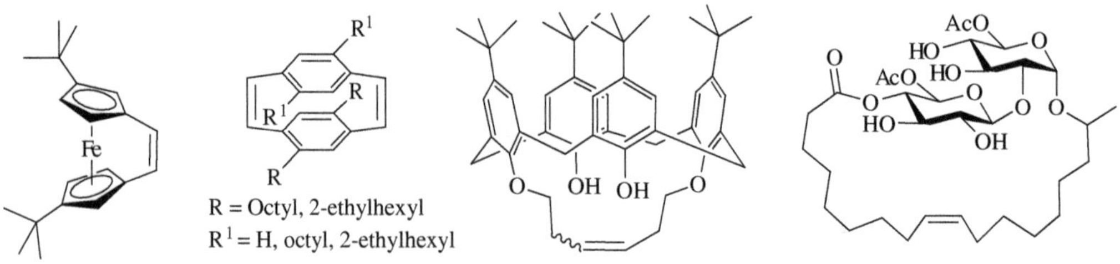

5 6 7 8

Figure 7.6 Unconventional monomers for ROMP.

polyethylene samples containing a substituent on every eighth backbone carbon were obtained (40,41). Similarly, a way toward stereoregular ethylene-vinyl alcohol copolymers made by ROMP of a protected cyclooctenediol monomer was disclosed (42). Instead of low strained *cis*-cyclooctene more strained *trans*-cyclooctene derivatives (ring strain ~17 kcal/mol (21)) can be used, and even controlled living polymerization can be achieved with **G1** upon addition of triphenylphosphine (43). Other low strained cyclic olefins have been studied with respect to their applicability in ROMP (44). Besides hitherto described monomer types, unconventional more complex strained cyclic structures can also be ROMPed. An example is the ansa-(vinylferrocene) **5** (see Fig. 7.6) bearing a *t*-butyl group on every dicyclopentadienyl ligand, which was polymerized with **G2** and Schrock-type initiators leading to soluble polymers of the ferrocenylvinylene type (45). Cyclophanediene-type monomers (**6**) were polymerized with **G2** in a controlled living manner (46), which led to poly(phenylenevinylenes) and, furthermore, block copolymers (47,48).

Ring opening of strained cyclic olefins (over-) compensates the entropic penalty of the polymerization by the release of the strain energy, that is, by the enthalpic contribution, and also virtually nonstrained macrocyclic olefins can be ROMPed. This process is called *entropy-driven ring opening metathesis polymerization* (*ED-ROMP*). Here, the entropy gain on ring opening, which originates from the higher conformational flexibility of the polymer relative to the cyclic monomer, is the driving force (49,50). Recent examples of ED-ROMPed macrocyclic olefins comprise a calix[4]arene-based monomer (**7**, see Fig. 7.6), which was copolymerized with cyclooctene and norbornene using **G2** (51), or the naturally occurring sophorolipid monomer **8** (52).

7.5 ROMP—APPLICATIONS

ROMP is an established and frequently used polymerization technique. Many research papers on ROMP-derived polymers have been published and, a comprehensive overview of all these works is provided by others (53,54). In the following section, three particular and important implementations of ROMP that are substantiated by unique characteristics of this polymerization technique are described.

A recent booming application of ROMP, as mentioned briefly before, is the preparation of bottlebrush copolymers via the macromonomer approach. The high ring strain of typical monomers (norbornenes and cyclobutenes) provides the driving force for obtaining high molecular weight polymers from macromonomers. Furthermore, the high functional group tolerance of ruthenium-based initiators makes ROMP particularly reliable for this application, if not the best suitable polymerization method in this research field. The macromonomer approach enables to prepare bottlebrush copolymers with a maximum of graft density, something which is hard to achieve with other usually used synthetic strategies, such as growing of the grafts from a preformed polymer chain (55,56) or attachment of preformed arms to a preformed polymer (57,58). The ROMP-based macromonomer approach was pioneered by Bowden et al. (59,60) who prepared bottlebrush polymers from poly(L-lactide) bearing monomers initiated with **G2**. Nowadays **G3** is the preferred initiator in this field and macromonomers have been polymerized in a controlled living manner opening the possibility of synthesizing block bottlebrush copolymers (61,62). An impressive example in terms of complexity is the bottlebrush copolymer obtained from the bivalent monomer **9** (see Fig. 7.7) containing a poly(ethylene glycol) arm and a drug (doxorubicin) linked via a photo-cleavable group (63,64). The precision that can be obtained from the ROMP of macromonomers even enabled the realization of photonic crystalline materials on the self-assembly of bottlebrush block copolymer blends (65,66). Several other works on bottlebrush polymers and copolymers and their applications are available and summarized in a recent review (67). Further examples include work on dendronized macromonomers, for example, monomer **10** (68). When polymerizing dendronized macromonomers, a long spacer between the dendronized moiety and the polymerizable group is of particular importance for obtaining a high macromonomer conversion (69).

Another contemporary research field, where ROMP is the key for success, is the preparation of the so-called

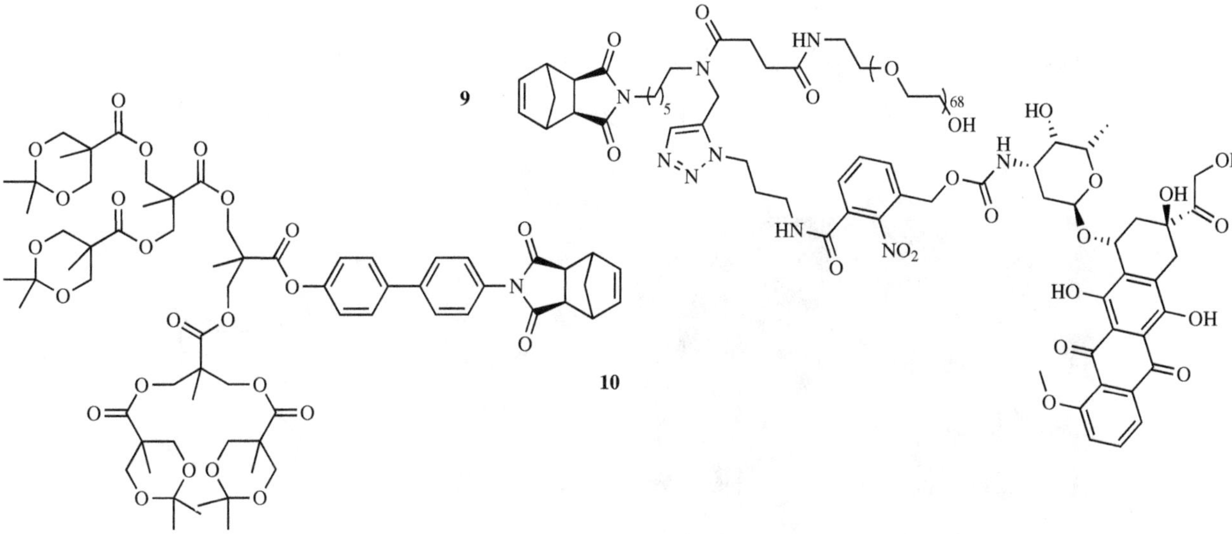

Figure 7.7 Typical macromonomers used for the preparation of bottlebrush copolymers.

ladderphanes. Ladderphanes consist of multiple layers of linkers covalently connected to two (or more) polymer chains (see Fig. 7.8). The polymer chains are made of intentionally designed norbornene derivatives meeting two important prerequisites for ladderphanes, that is, a relatively rigid polymer backbone and spacing between two repeating units in about the distance typical for $\pi-\pi$ stacking. Usually, **G1** is used for the polymerization and with this particular monomer class, trans-isotactic polymer strands are obtained (70). A huge variety of different ladderphanes has been prepared over the last 7 years comprising linkers containing, for example, transition metal complexes (see **11** in Fig. 7.8) or planar aromatic moieties (**12**) (71). Research concerning the replication of polymer strands via ladderphanes is particularly intriguing (72,73). Recently, a comprehensive overview on the work in the field of ladderphanes has been published (74).

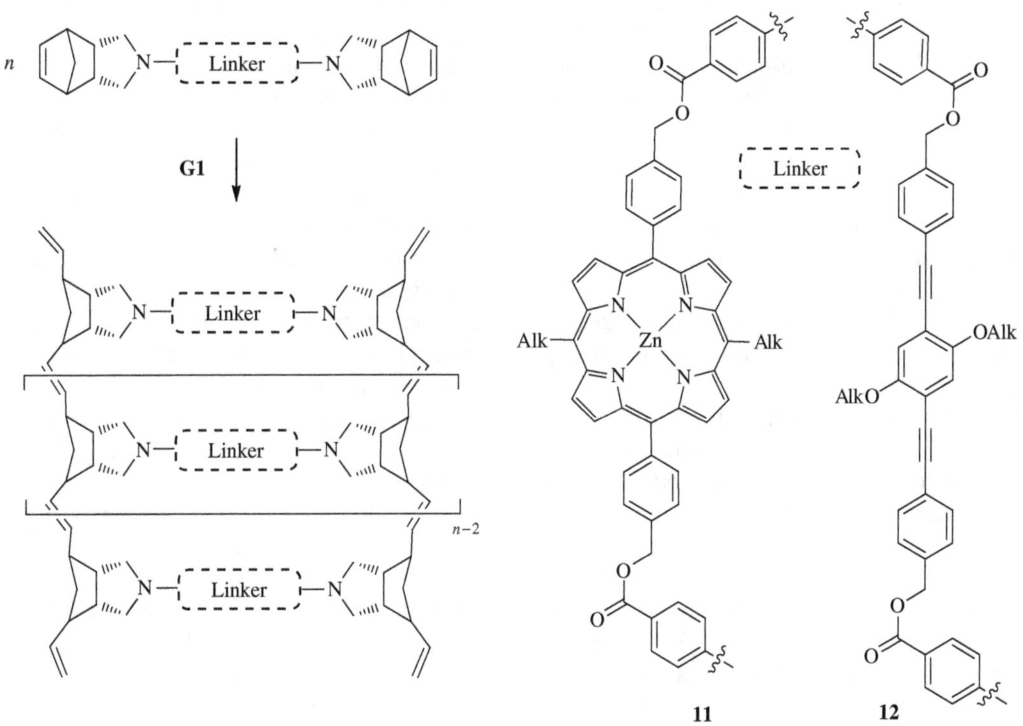

Figure 7.8 ROMP as tool for preparing ladderphanes.

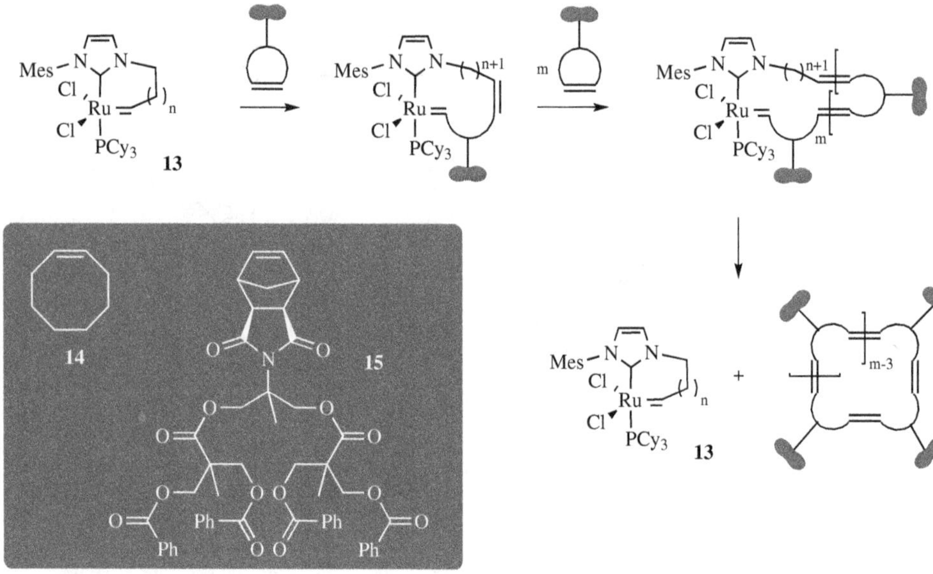

Figure 7.9 Mechanism of REMP and the monomers that were used in REMP.

The preparation of cyclic polymers is a third unique feature offered by ROMP. Compound **13**, initially synthesized in the Fürstner group (75), is able to provide access to cyclic polymers as disclosed by Grubbs et al. (76). Upon monomer addition, this carbene complex produces a polymer chain, which remains attached to the initiator at both ends (cf. Fig. 7.9). This situation eventually leads to intramolecular chain transfer via olefin metathesis (the so-called back-biting) of the growing polymer chain, releasing the cyclic poly(olefin) and the original carbene complex **13**. Therefore, in this particular case, it is better to use the term *catalyst* and not initiator as in other ROMP reactions. This process was termed *ring expansion metathesis polymerization* (*REMP*) (76,77). Making use of this principle, Grubbs et al. (76) prepared cyclic poly(ethylene) from the ROMP of cyclooctene (**14**) and subsequent hydrogenation of the polymer.

Not only simple monomers like cyclooctene, but also functionalized monomers that are not prone to back-biting, for example, dendronized monomer **15**, can also be used in REMP (78), opening new exciting possibilities in the field of macrocyclic polymers (79).

Besides the aforementioned examples that utilized these unique features, many other applications benefit from the high reliability and convenience of ROMP, out of which two showcase applications will be briefly introduced in the following.

The first example is an ROMP-based platform for the synthesis of antibacterial and cell-penetrating polycationic polymers introduced by the group of Tew. The convenient and modular tuning of the overall hydrophobicity and charge density of the water-soluble polymers gives access to high antimicrobial activity and low cytotoxicity of the macromolecules. In this way, synthetic mimics of antimicrobial peptides can be easily obtained (80). Water-insoluble derivatives are contact biocides, which can be used as coatings with nonfouling properties or as additives to equip commodity polymers with biocidic properties (81).

The second example is the (in most cases) covalent functionalization of diverse surfaces. Here, surface-initiated ring opening metathesis polymerization (SI-ROMP) is one of the methods of choice.

In SI-ROMP, a surface is functionalized with olefin-bearing molecules that are in turn used to prepare a surface-bonded initiator upon reaction with a conventional initiator similar to those shown in Figure 7.3. Then, monomers are added and polymer strands grow from the surface. Finally, the macroinitiators are deactivated or used for the preparation of a further segment (see Fig. 7.10). SI-ROMP was used to decorate gold surfaces with alkylated (82) and fluorinated (83) polymer strands. Also, block copolymer synthesis from gold substrates is feasible (84). Silicon substrates (85), silica (86), or ruthenium nanoparticles (87) as well as cellulose fibers (88) or carbon nanotubes (89) can be functionalized with SI-ROMP approaches. Usually norbornene derivatives are used for growing the surface grafts but recently monomers with very low ring strain, for example, cyclopentene, were also used for this purpose. In this case, the monomers were applied from the gas phase and thus chain transfer reactions could be suppressed, which is the key for a successful SI-ROMP (90). Further work of using SI-ROMP includes the preparation of functional supports (91). A comprehensive overview of SI-ROMP and the use of ROMP-derived monoliths for separation purposes is available (54).

A particular ROMP-derived polymer is poly(dicyclopentadiene) (pDCPD). pDCPD is a thermoset material with

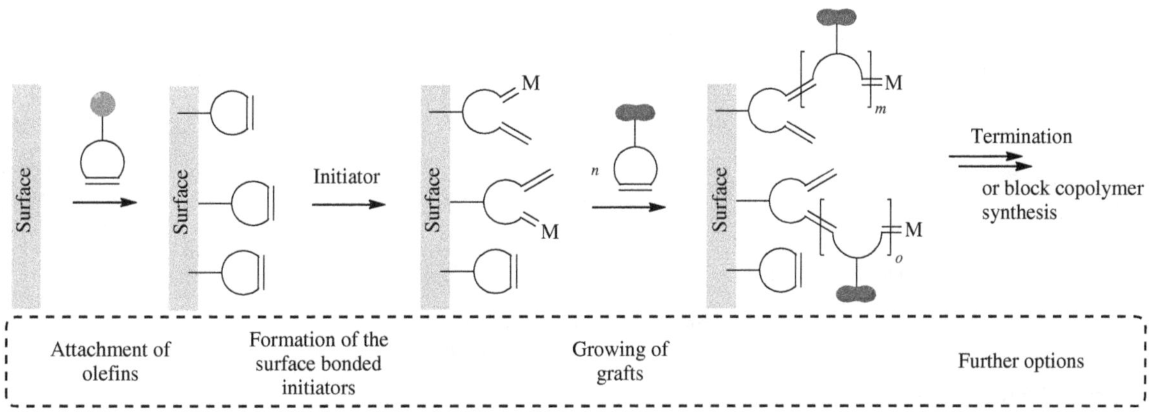

Figure 7.10 Schematic representation of SI-ROMP.

attractive properties such as high toughness, high impact resistance, optical transparency, low price, and high corrosion resistance (92). pDCPD is one of the commercialized ROMP-derived polymers and is applied in, for example, body panels for buses and trucks (cf. Chapter 10). There are many current developments for the use of pDCPD, most of them focusing on substituting epoxy-resins in composite materials. Moreover, even more advanced applications are in the focus of current research. The post-polymerization functionalization of pDCPD by bromination or epoxidation was studied and the assembly of amines on the surface of the functionalized pDCPD was demonstrated (93). These functionalized and also unfunctionalized pDCPD membranes were used for the separation of small molecules (94). Nanostructured pDCPD membranes were made by a templating approach with reactive block copolymers (95) and a route toward macroporous pDCPD beads has been developed (96). High internal phase emulsion templating was used to prepare macroporous pDCPD featuring porosities up to 90% and a throughout open porous structure (97). These monolithic materials combine the favorable mechanical properties of pDCPD and the macroporous structure of polymerized high internal phase emulsions and are the toughest and strongest examples among all macroporous polymeric materials disclosed so far (98). Also, work on pDCPD aerogels was done and materials with a very low density of 0.02–0.05 g/cm^3 have been described (99). Furthermore, pDCPD submicron fibers produced by electrospinning have been prepared (100). All these works have in common that initiators with characteristics different to those favourable in the case of controlled living polymerization are needed. For that reason, initiators that can be triggered by a certain event, for example, heat (101), light (102), mechanical force (103), addition of acids (104), or anions (105) have been developed and research on this topic has been reviewed recently (106).

Even though ROMP is a very reliable tool for obtaining many kinds of polymers and macromolecular materials, there are still some issues that are not resolved in a satisfying manner. Most importantly, control over the microstructure of the polymers or the sequence of monomer incorporation into the polymers is not an easy task with ROMP. As ROMP yields polymers with double bonds in the backbone, control over the stereochemistry of these double bonds is desirable, but standard initiators (cf. Fig. 7.2) exhibit only poor cis/trans selectivity that is also strongly dependent on the structure of the monomer (107). A great leap forward was made by recently disclosed molybdenum-based initiators that allow for cis-selective and, moreover, syndio-selective polymerization of at least some norbornadiene derivatives (108). The scope of molybdenum-based initiators in stereo- and tacticity control has been described in a review article (15). Functional group-tolerant ruthenium-based initiators cannot compete in selectivity with the above-mentioned initiators, but a new initiator design holds promise to improve this particular drawback (109). Obtaining alternating copolymers via ROMP is another target of current research endeavors that already showed some limited success. Careful selection of the monomers (35,36,110) and/or development of designed initiators (111) led to first successful examples in this regard. A more general way to prepare alternating copolymers by olefin metathesis is ring-opening insertion metathesis polymerization (ROIMP) (112). In this case, a ROMPable monomer is homopolymerized, and subsequently, the homopolymer is transformed by cross metathesis with an electron-deficient diolefin present in the reaction mixture (see Fig. 7.11).

Similarly α,ω-dienes and diacrylates can be used as the starting materials, the corresponding reaction was termed *alternating diene metathesis polycondensation* (*ALTMET*) (113). In contrast to the ring-opening polymerization techniques presented in the previous section, ALTMET is governed by a step-growth mechanism similar to ADMET, which will be discussed in detail in the next chapter.

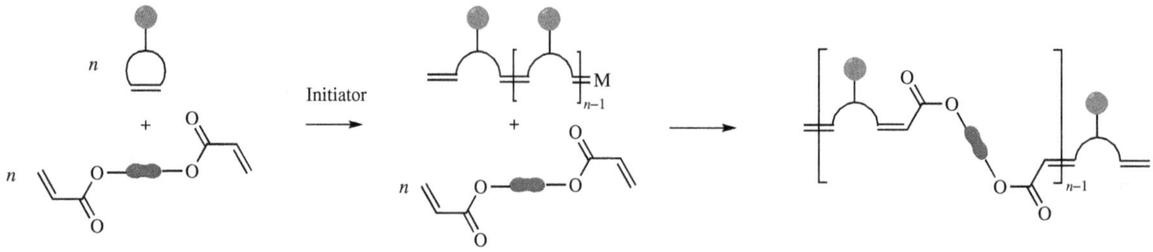

Figure 7.11 Principle of ROIMP.

7.6 ACYCLIC DIENE METATHESIS

ADMET is preferentially performed on α,ω-dienes and produces strictly linear polymers with an unsaturated backbone. The driving force is the release of a small olefin (in most cases, ethylene), which has to be removed from the equilibrium. Typically, neat substrates are transformed and vacuum is applied to facilitate the removal of the small olefin (cf. Fig. 7.12). In the first report of ADMET, a tungsten catalyst was used (114). Nowadays, ADMET is mainly performed with ruthenium-based catalysts because of their high functional group tolerance. However, this comes at the price of (especially severe at high reaction temperatures) double-bond isomerization that leads to less defined polymers (cf. Fig. 7.12) (14). Accordingly, **Mo-S** or **G1** are used as catalysts for ADMET, when double-bond isomerization should be avoided. If the high functional group tolerance of ruthenium is needed and **G1** is not active enough, then second-generation catalysts together with additives such as 1,4-benzoquinone are used, which scavenge the ruthenium hydride species (75, 115) held responsible for the isomerization reactions (116).

By using properly designed α,ω-dienes, ADMET results in a precise spacing between two functional groups (e.g., R in Fig. 7.12 = aliphatic units, OH, OR, COOH, COOR, substituted benzenes, phosphonic acids, oligo(ethylene glycol) grafts, etc.) and has therefore been used in several works, in combination with subsequent hydrogenation, for the preparation of substituted "precision" polyethylenes with particular properties. The work in this field has been comprehensively covered in a recent review (13).

ADMET has been used to prepare a great many polymers for diverse applications (13), of which some examples will be highlighted in the following. ADMET is well suited to prepare main chain liquid crystalline polymers (117). In this material class, the mesogenic group is placed within the main chain of the polymer. A typical example is shown in Figure 7.13. Monomer **16** is a liquid crystalline fluorescent monomer that gives a main chain liquid crystalline polymer via ADMET. The polymer is characterized by an enlarged stability of the mesophase and shows polarized absorption of light when aligned on a proper substrate (118). The preparation of this polymer may serve as a typical example for an ADMET protocol and is presented in the following:

General Procedure for an ADMET Reaction

A Schlenk tube was charged with monomer **16** (120 mg, 100 equiv) and **G2** (1 equiv) dissolved in dichloroethane (10 ml in total) under an inert argon atmosphere. The reaction mixture was stirred at 60 °C and high vacuum (<1 mbar) was employed every 10–15 min for 2 s to remove volatiles (afterward, the tube was flushed with argon again). After 2 h, the remaining solvent was evaporated under reduced pressure and the reaction mixture was stirred for 1 h at 60 °C under vacuum. Then, the reaction mixture was cooled to room temperature, dissolved in CH_2Cl_2 (2 ml), and then, ethyl vinyl ether (100 μl) was added. After further stirring for 30 min at room temperature, the reaction mixture was slowly added to vigorously stirred methanol. The resulting precipitate was sampled and twice reprecipitated using the same solvents. Drying the residue in vacuum gave the desired polymer in 80% yield.

The preparation of (fully conjugated) electroactive polymers, in particular poly(arylene-vinylene)s, is another important application of ADMET. The state of the art in this application including all aspects of olefin metathesis polymerization toward conjugated polymers has been reviewed recently (119). Some monomers for this purpose are depicted in Figure 7.13. The 1,4-divinylbenzene derivative **17** was subjected to ADMET catalyzed by **H2**, and the obtained poly(*p*-phenylenevinylene) derivative was characterized by an all-trans configuration of the double bonds and a high molecular weight of about 100,000 g/mol (120). Monomer **18** was polymerized with the aid of **G2** in a solid-state polymerization setup (121). A copolymerization approach of monomers like **18** with others like **19** is feasible and demonstrates the suitability of ADMET to prepare conjugated copolymers with tunable optoelectronic properties (122).

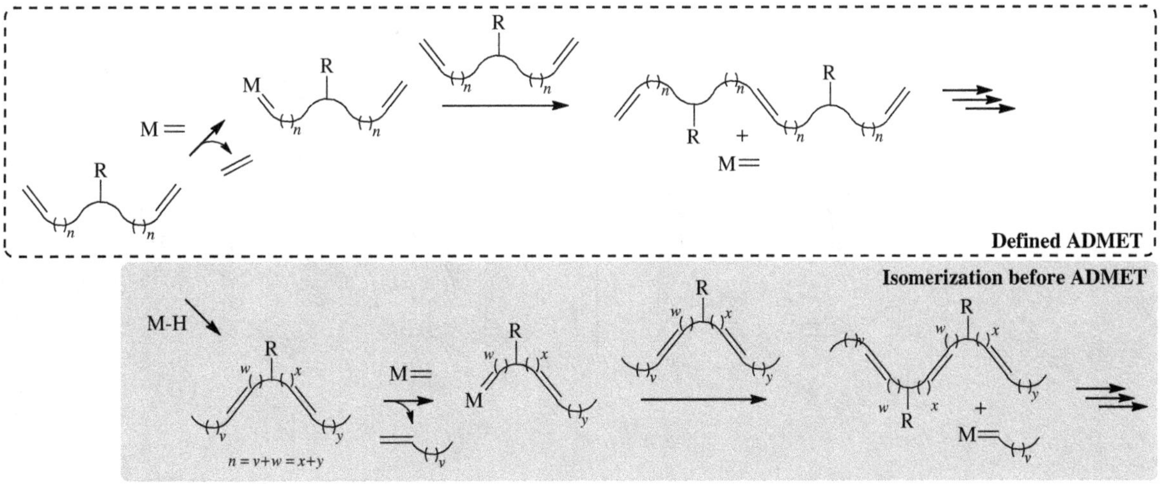

Figure 7.12 ADMET reaction versus ADMET of isomerized olefins.

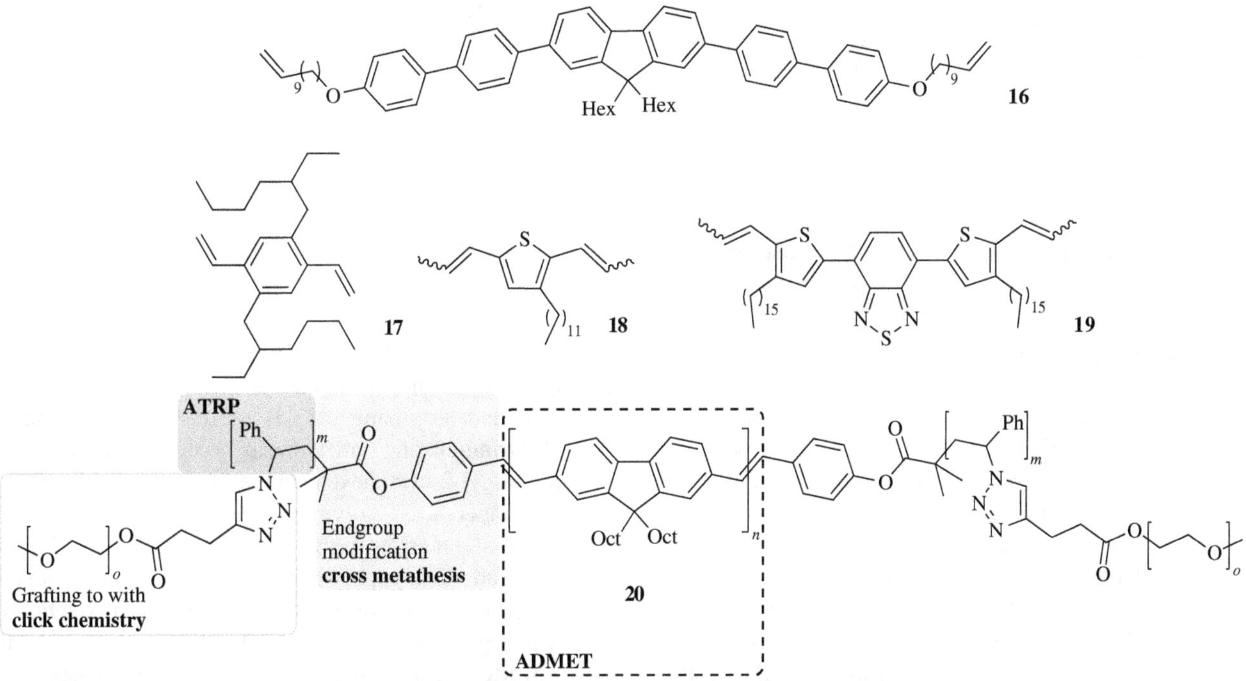

Figure 7.13 Some monomers for ADMET (**16–19**) and a complex polymer architecture obtained from a combination of ADMET, atom transfer radical polymerization (ATRP), and click chemistry (**20**).

As final example, demonstrating the compatibility of ADMET with state-of-the-art polymer chemistry, the preparation of the ABCBA block copolymer **20** is explained in the following. The core unit of this macromolecule was constructed by ADMET chemistry and using a cross metathesis setup, the endgroups of the ADMET-derived polymers were functionalized with a unit capable of initiating atom transfer radical polymerization (ATRP). Now, continuing with nonmetathesis chemistries, a poly(styrene) segment was synthesized and the termini of the two poly(styrene) chains were modified (by conversion of the terminal bromides into azides), which enabled the authors to attach a preformed oligo(ethylene glycol) segment by azide-alkyne click chemistry (123). Another vivid research field is the preparation of polymers from renewable resources, and the scope of ADMET in this context has been reviewed recently (13).

7.7 CYCLOPOLYMERIZATIONS

The polymerization of 1,6-heptadiynes via metathesis yields fully conjugated poly(ene)s (124). As shown in Figure 7.14,

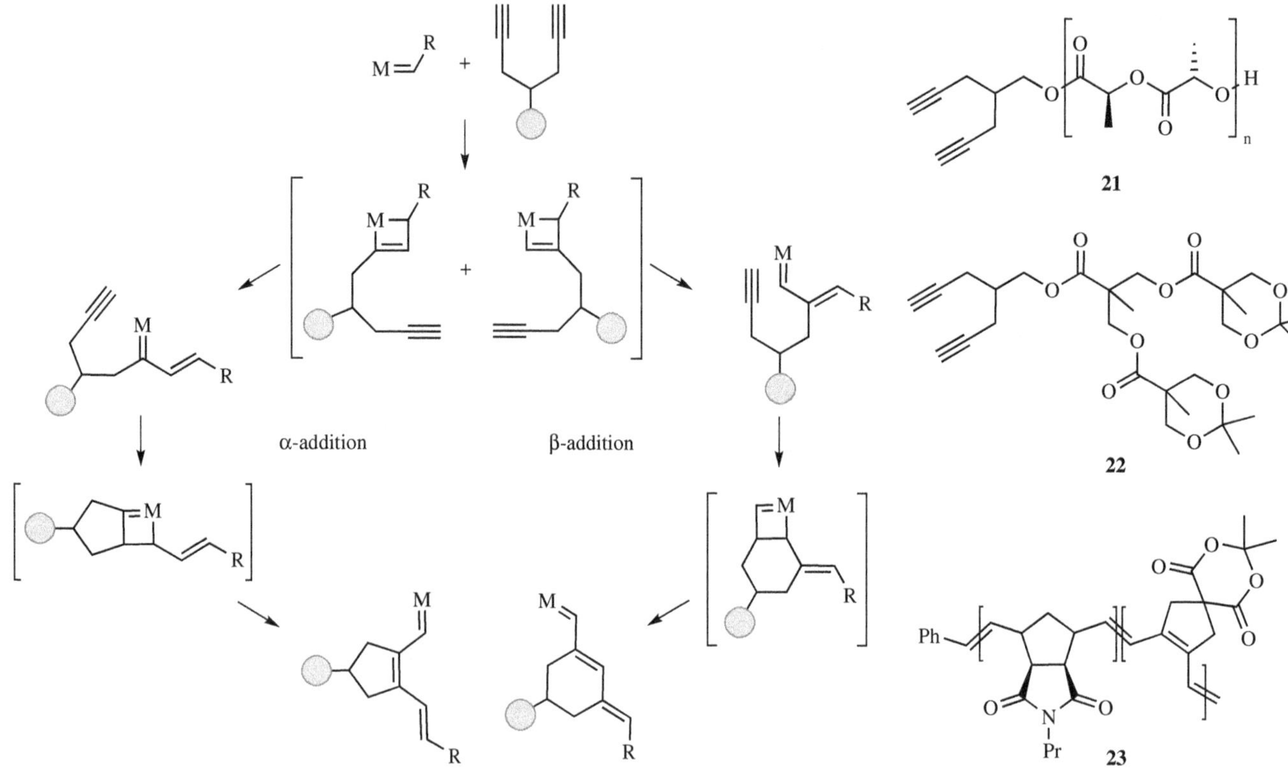

Figure 7.14 Cyclopolymerization of 1,6-heptadiynes.

α-addition and β-addition of the first alkyne group leads to either five-membered or six-membered rings as repeating units. α-Addition-derived polymers are characterized by higher effective conjugation lengths and lower solubility when compared to β-addition-based products (125). Particularly, Schrock-type molybdenum-based initiators and also Grubbs-type initiators are used and offer access to stereo- and regioselective polymerization of 1,6-heptadiynes. In some cases, even controlled living polymerization has been achieved and block copolymers consisting of two different 1,6-heptadiynes were prepared by using a Schrock-type initiator related to **S-Mo** (126). Graft copolymers are accessible via the macromonomer approach, for example, using poly(L-lactide)-substituted 1,6-heptadiynes (**21**, see Fig. 7.14) and using **H2** as the initiator in this case (127). Similarly, dendronized-conjugated polymers were prepared with **G3** as the initiator, which provides controlled living polymerization with many 1,6-heptadiyne derivatives, for example, monomer **22**, and allows to prepare fully conjugated diblock copolymers with narrow molecular weight distributions (128). 1,6-Heptadiynes can also be copolymerized with norbornene derivatives as it was demonstrated by Choi et al. (129), who prepared random and block copolymers (see Fig. 7.14, **23**) with **G3** as initiator.

Similarly, 1,7-octadiynes yield fully conjugated poly(ene)s, and both Schrock-type initiators and Grubbs-type initiators (in particular, **G3**), are suited for polymerizing these monomers (130,131). A somewhat related and surprising case is the extremely fast tandem ROMP/ring-closing metathesis polymerization of an *N*-propargyl-2-cyclohexeneamine derivative with **G3**. At low temperature, polymers with controlled molecular weights and narrow polydispersity indices could be obtained, and block copolymer synthesis with norbornene and 1,6-heptadiyne derivatives was feasible (132). This very recent report clearly demonstrates that new and exciting discoveries in polymer synthesis with metathesis-based techniques are still possible.

7.8 CONCLUSIONS

Olefin metathesis polymerization techniques are a substantial part of the contemporary toolbox of macromolecular chemists. Especially the high functional group tolerance of ruthenium initiated/catalyzed polymerizations is a unique feature, making the use of protecting groups largely redundant. ROMP is one of the best suited methods for preparing tailor-made copolymers of various architectures and is especially useful for preparing graft copolymers by the macromonomer approach. Furthermore, ROMP offers the unique opportunity to prepare cyclic macromolecules in a convenient and easy manner. However, challenges still remain and olefin metathesis polymerization in all its facets will certainly develop in the next decade, which will enable access to new and exciting polymeric materials.

REFERENCES

1. (a) Anderson AW, Merckling NG. US Patent 2721189. 1955. (b) Eleuterio HS. US Patent 3074918. 1963. (c) Tuett WL, Johnson DR, Robinson IM, Montague BA. *J Am Chem Soc* 1960;82:2337–2340.
2. Hérisson J-L, Chauvin Y. *Makromol Chem* 1971;141: 161–176.
3. (a) Black G, Maher D, Risse W. *Living ring-opening olefin metathesis polymerization*. In: *Handbook of Metathesis*. Vol. 3. Weinheim: Wiley-VCH; 2003. p 2–71. (b) Rouhi AM. *Chem Eng News* 2002;80:34–38. (c) Astruc D. *New J Chem* 2005;29:42–56. (d) Grubbs RH. *Living ring-opening olefin metathesis polymerization*. In: *Polymer Science: A Comprehensive Reference*. Vol. 4. Amsterdam: Elsevier; 2012. p 21–29.
4. Szwarc M. *Nature* 1956;178:1168–1169.
5. Sanford MS, Love JA, Grubbs RH. *J Am Chem Soc* 2001;123:6543–6554.
6. Hilf S, Kilbinger AFM. *Nat Chem* 2009;1:537–546.
7. Nguyen ST, Grubbs RH, Ziller HW. *J Am Chem Soc* 1993;115:9858–9859.
8. Bielawski CW, Grubbs RH. *Angew Chem Int Ed* 2000;39:2903–2906.
9. Schrock RR. *Acc Chem Res* 1990;23:158–165.
10. Slugovc C, Demel S, Stelzer F. *Chem Commun* 2002:2572–2573.
11. (a) Love JA, Morgan JP, Trnka TM, Grubbs RH. *Angew Chem Int Ed* 2002;41:4035–4037. (b) Slugovc C, Riegler S, Hayn G, Saf R, Stelzer F. *Macromol Rapid Commun* 2003;24:435–439. (c) Riegler S, Demel S, Trimmel G, Slugovc C, Stelzer F. *J Mol Catal A* 2006;257:53–58.
12. Burtscher D, Lexer C, Mereiter K, Winde R, Karch R, Slugovc C. *J Polym Sci A: Polym Chem* 2008;46:4630–4635.
13. Mutlu H, de Espinosa LM, Meier MAR. *Chem Soc Rev* 2011;40:1404–1445.
14. (a) Lehman SE Jr, Schwendeman JE, O'Donnell PM, Wagener KB. *Inorg Chim Acta* 2003;345:190–198. (b) Fokou PA, Meier MAR. *J Am Chem Soc* 2009;131:1664–1665.
15. Schrock RR. *Dalton Trans* 2011;40:7484–7495.
16. Trnka TM, Grubbs RH. *Acc Chem Res* 2001;34:18–29.
17. Robson DA, Gibson VC, Davies RG, North M. *Macromolecules* 1999;32:6371–6373.
18. Demel S, Riegler S, Wewerka K, Schoefberger W, Slugovc C, Stelzer F. *Inorg Chim Acta* 2003;345:363–366.
19. Slugovc C. *Macromol Rapid Commun* 2004;25:1283–1297.
20. Choi T-L, Grubbs RH. *Angew Chem Int Ed* 2003;42:1743–1746.
21. (a) Schleyer PvR, Williams JE, Blanchard KR. *J Am Chem Soc* 1970;92:2377–2386. (b) Allinger NL, Sprague JT. *J Am Chem Soc* 1972;94:5734–5747.
22. Sutthasupa S, Terada K, Sanda F, Masuda T. *J Polym Sci A: Polym Chem* 2006;44:5337–5343.
23. Sutthasupa S, Sanda F, Masuda T. *Macromol Chem Phys* 2008;209:930–937.
24. Sutthasupa S, Terada K, Sanda F, Masuda T. *Polymer* 2007;48:3026–3032.
25. Slugovc C, Demel S, Riegler S, Hobisch J, Stelzer F. *Macromol Rapid Commun* 2004;25:475–480.
26. Haigh DM, Kenwright AM, Khosravi E. *Macromolecules* 2005;38:7571–7579.
27. Carrillo A, Gujraty KV, Rai PR, Kane RS. *Nanotechnology* 2005;16:416–421.
28. Strong LE, Kiessling LL. *J Am Chem Soc* 1999;121: 6193–6196.
29. Schaefer M, Hanik N, Kilbinger AFM. *Macromolecules* 2012;45:6807–6818.
30. Niedermair F, Sandholzer M, Kremser G, Slugovc C. *Organometallics* 2009;28:2888–2896.
31. Morandi G, Montembault V, Pascual S, Legoupy S, Fontaine L. *Macromolecules* 2006;39:2732–2735.
32. Lapinte V, de Fremont P, Montembault V, Fontaine L. *Macromol Chem Phys* 2004;205:1238–1245.
33. Morandi G, Pascual S, Montembault V, Legoupy S, Delorme N, Fontaine L. *Macromolecules* 2009;42:6927–6931.
34. Lee JC, Parker KA, Sampson NS. *J Am Chem Soc* 2006;128:4578–4579.
35. Song A, Lee JC, Parker KA, Sampson NS. *J Am Chem Soc* 2010;132:10513–10520.
36. Song A, Parker KA, Sampson NS. *J Am Chem Soc* 2009; 131:3444–3445.
37. Song A, Parker KA, Sampson NS. *Org Lett* 2010;12: 3729–3731.
38. Song A, Walker SG, Parker KA, Sampson NS. *ACS Chem Biol* 2011;6:590–599.
39. Martinez H, Miró P, Charbonneau P, Hillmyer MA, Cramer CJ. *ACS Catal* 2012;2:2547–2556.
40. Kobayashi S, Pitet LM, Hillmyer MA. *J Am Chem Soc* 2011;133:5794–5797.
41. Zhang J, Matta ME, Hillmyer MA. *ACS Macro Lett* 2012;1:1383–1387.
42. Scherman OA, Walker R, Grubbs RH. *Macromolecules* 2005;38:9009–9014.
43. Walker R, Conrad RM, Grubbs RH. *Macromolecules* 2009;42:599–605.
44. Hejl A, Scherman OA, Grubbs RH. *Macromolecules* 2005;38:7214–7218.
45. Masson G, Lough AJ, Manners I. *Macromolecules* 2008; 41:539–547.
46. Yu C-Y, Turner ML. *Angew Chem Int Ed* 2006;45: 7797–7800.

47. Yu C-Y, Kingsley JW, Lidzey DG, Turner ML. *Macromol Rapid Commun* 2009;30:1889–1892.
48. Yu C-Y, Horie M, Spring AM, Tremel K, Turner ML. *Macromolecules* 2010;43:222–232.
49. Xue Z, Mayer MF. *Soft Matter* 2009;5:4600–4611.
50. Monfette S, Fogg DE. *Chem Rev* 2009;109:3783–3816.
51. Yang Y, Swager TM. *Macromolecules* 2007;40:7437–7440.
52. Gao W, Hagver R, Shah V, Xie W, Gross RA, Ilker MF, Bell C, Burke KA, Coughlin EB. *Macromolecules* 2007;40:145–147.
53. (a) Leitgeb A, Wappel J, Slugovc C. *Polymer* 2010;51:2927–2946. (b) Sutthasupa S, Shiotsuki M, Sanda F. *Polym J* 2010;42:905–915.
54. (a) Moerdyk JP, Bielawski CW. *Architectures of polymers synthesized using ROMP*. In: *Polymer Science: A Comprehensive Reference*. Vol. 4. Amsterdam: Elsevier; 2012. p 523–550. (b) Buchmeiser MR. *Current and forthcoming applications of ROMP-derived polymers: functional surfaces and supports*. In: *Polymer Science: A Comprehensive Reference*. Vol. 4. Elsevier; 2012. p 597–632. (c) Kiessling LL, Mangold SL. *Current and forthcoming applications of ROMP polymers – biorelated polymers*. In: *Polymer Science: A Comprehensive Reference*. Vol. 4. Elsevier; 2012. p 695–717. (d) Bielawski C, Grubbs RH. *Living ring-opening metathesis polymerization* in *Controlled and Living Polymerizations*, Müller, AHE; Matyjaszewski K. (Eds.) p 297–342, Wiley-VCH, Weinheim, 2009. (e) Bielawski CW, Grubbs RH. *Prog Polym Sci* 2007;32:1–29.
55. Charvet R, Novak BM. *Macromolecules* 2004;37: 8808–8811.
56. Lu H, Wang J, Lin Y, Cheng J. *J Am Chem Soc* 2009; 131:13582–13583.
57. Gibbs JM, Park S-J, Anderson DR, Watson KJ, Mirkin CA, Nguyen ST. *J Am Chem Soc* 2005;127:1170–1178.
58. McLaughlin CK, Hamblin GD, Hänni KD, Conway JW, Nayak MK, Carneiro KMM, Bazzi HS, Sleiman HF. *J Am Chem Soc* 2012;134:4280–4286.
59. Jha S, Dutta S, Bowden NB. *Macromolecules* 2004;37: 4365–4374.
60. Runge MB, Dutta S, Bowden NB. *Macromolecules* 2006;39: 498–508.
61. Xia Y, Kornfield JA, Grubbs RH. *Macromolecules* 2009;42: 3761–3766.
62. Xia Y, Olsen BD, Kornfield JA, Grubbs RH. *J Am Chem Soc* 2009;131:18525–18532.
63. Johnson JA, Lu YY, Burts AO, Xia Y, Durrell AC, Tirrell DA, Grubbs RH. *Macromolecules* 2010;43:10326–10335.
64. Johnson JA, Lu YY, Burts AO, Lim Y-H, Finn MG, Koberstein JT, Turro NJ, Tirrell DA, Grubbs RH. *J Am Chem Soc* 2011;133:559–566.
65. Miyake GM, Weitekamp RA, Piunova VA, Grubbs RH. *J Am Chem Soc* 2012;134:14249–14254.
66. Miyake GM, Piunova VA, Weitekamp RA, Grubbs RH. *Angew Chem Int Ed* 2012;51:11246–11248.
67. Nomura K, Abdellatif MM. *Polymer* 2010;51:1861–1881.
68. Kim KO, Choi T-L. *ACS Macro Lett* 2012;1:445–448.
69. Jung H, Carberry TP, Weck M. *Macromolecules* 2011;44: 9075–9083.
70. Lin W-Y, Wang H-W, Liu Z-C, Xu J, Chen C-W, Yang Y-C, Huang S-L, Yang H-C, Luh T-Y. *Chem Asian J* 2007;2:764–774.
71. Chou C-M, Lee S-L, Chen C-H, Biju AT, Wang H-W, Wu Y-L, Zhang G-F, Yang K-W, Lim T-S, Huang M-J, Tsai P-Y, Lin K-C, Huang S-L, Chen C-H, Luh T-Y. *J Am Chem Soc* 2009;131:12579–12585.
72. Lin N-T, Lin S-Y, Lee S-L, Chen C-H, Hsu C-H, Hwang L-P, Xie Z-Y, Chen C-H, Huang S-L, Luh T-Y. *Angew Chem Int Ed* 2007;46:4481–4485.
73. South CR, Weck M. *Macromolecules* 2007;40:1386–1394.
74. Luh T-Y. *Acc Chem Res* 2013;46:378–389.
75. Fürstner A, Ackermann L, Gabor B, Goddard R, Lehmann CW, Mynott R, Stelzer F, Thiel OR. *Chem Eur J* 2001; 7:3236–3253.
76. Bielawski CW, Benitez D, Grubbs RH. *Science* 2002;297:2041–2044.
77. (a) Bielawski CW, Benitez D, Grubbs RH. *J Am Chem Soc* 2003;125:8424–8425. (b) Boydston AJ, Xia Y, Kornfield JA, Gorodetskaya IA, Grubbs RH. *J Am Chem Soc* 2008;130:12775–12782. (c) Xia Y, Boydston AJ, Yao Y, Kornfield JA, Gorodetskaya IA, Spiess HW, Grubbs RH. *J Am Chem Soc* 2009;131:2670–2677.
78. Boydston AJ, Holcombe TW, Unruh DA, Fréchet JMJ, Grubbs RH. *J Am Chem Soc* 2009;131:5388–5389.
79. (a) Zhang K, Lackey MA, Cui J, Tew GN. *J Am Chem Soc* 2011;133:4140–4148. (b) Zhang K, Tew GN. *ACS Macro Lett* 2012;1:574–579.
80. (a) Tew GN, Lienkamp K. *Chem Eur J* 2009;15: 11784–11800. (b) Som A, Reuter A, Tew GN. *Angew Chem Int Ed* 2012;51:980–983.
81. (a) Kreutzwiesner E, Noormofidi N, Wiesbrock F, Kern W, Rametsteiner K, Stelzer F, Slugovc C. *J Polym Sci A: Polym Chem* 2010;48:4504–4514. (b) Colak S, Tew GN. *Langmuir* 2012;28:666–675.
82. (a) Berron BJ, Graybill EP, Jennings GK. *Langmuir* 2007;23:11651–11655. (b) Guntari SN, Goh TK, Blencowe A, Wong EHH, Caruso F, Qiao GG. *Polym Chem* 2013;4:68–75.
83. (a) Faulkner CJ, Fischer RE, Jennings GK. *Macromolecules* 2010;43:1203–1209. (b) Ye Q, Wang X, Li S, Zhou F. *Macromolecules* 2010;43:5554–5560. (c) Escobar CA, Zulkifli AR, Faulkner CJ, Trzeciak A, Jennings GK. *ACS Appl Mater Interfaces* 2012;4:906–915.
84. Kong B, Lee JK, Choi IS. *Langmuir* 2007;23:6761–6765.
85. (a) Yang L, Lua Y-Y, Tan M, Scherman OA, Grubbs RH, Harb JN, Davis RC, Linford MR. *Chem Mater* 2007;19:1671–1678. (b) Lee W-K, Caster KC, Kim J, Zauscher S. *Small* 2006;2:848–853.
86. Jordi MA, Seery TAP. *J Am Chem Soc* 2005;127:4416–4422.
87. Ren F, Feldman AK, Carnes M, Steigerwald M, Nuckolls C. *Macromolecules* 2007;40:8151–8155.

88. Carlsson L, Malmström E, Carlmark A. *Polym Chem* 2012;3:727–733.
89. (a) Jeong W, Kessler MR. *Chem Mater* 2008;20: 7060–7068. (b) Izuhara D, Swager TM. *Macromolecules* 2009;42:5416–5418.
90. Lerum MFZ, Chen W. *Langmuir* 2011;27:5403–5409.
91. (a) Faisal S, Ullah F, Maity PK, Rolfe A, Samarakoon TB, Porubsky P, Neuenswander B, Lushington GH, Basha FZ, Organ MG, Hanson PR. *ACS Comb Sci* 2012;14:268–272. (b) Long TR, Faisal S, Maity PK, Rolfe A, Kurtz R, Klimberg SV, Najjar U-R, Basha FZ, Hanson PR. *Org Lett* 2011;13:2038–2041.
92. Mol JC. *J Mol Catal A: Chem* 2004;213:39–45.
93. (a) Perring M, Bowden NB. *Langmuir* 2008;24: 10480–10487. (b) Perring M, Long TR, Bowden NB. *J Mater Chem* 2010;20:8679–8685.
94. (a) Long TR, Gupta A, Miller AL II, Rethwisch DG, Bowden NB. *J Mater Chem* 2011;21:14265–14276. (b) Gupta A, Long TR, Rethwisch DG, Bowden NB. *Chem Commun* 2011;47:10236–10238. (c) Gupta A, Bowden NB. *ACS Appl Mater Interfaces* 2013;5:924–933.
95. (a) Amendt MA, Chen L, Hillmyer MA. *Macromolecules* 2010;43:3924–3934. (b) Amendt MA, Roerdink M, Moench S, Phillip WA, Cussler EL, Hillmyer MA. *Aust J Chem* 2011;64:1074–1082.
96. Della Martina A, Graf R, Hilborn JG. *J Appl Polym Sci* 2005;96:407–415.
97. Kovačič S, Krajnc P, Slugovc C. *Chem Commun* 2010;46:7504–7506.
98. (a) Kovačič S, Jeřabek K, Krajnc P, Slugovc C. *Polym Chem* 2012;3:325–328. (b) Kovačič S, Matsko NB, Jeřabek K, Krajnc P, Slugovc C. *J Mater Chem A* 2013;1:487–490. (c) Kovačič S, Kren H, Krajnc P, Koller S, Slugovc C. *Macromol Rapid Commun* 2013;34:581–587.
99. (a) Lee JK, Gould GL. *J Sol–Gel Sci Technol* 2007;44:29–40. (b) Kim SH, Worsley MA, Valdez CA, Shin SJ, Dawedeit C, Braun T, Baumann TF, Letts SA, Kucheyev SO, Wu KJJ, Biener J, Satcher JH Jr, Hamza AV. *RSC Adv* 2012;2:8672–8680. (c) Mohite DP, Mahadik-Khanolkar S, Luo H, Lu H, Sotiriou-Leventis C, Leventis N. *Soft Matter* 2013;9:1516–1530. (d) Mohite DP, Mahadik-Khanolkar S, Luo H, Lu H, Sotiriou-Leventis C, Leventis N. *Soft Matter* 2013;9:1531–1539.
100. Bellan LM, Coates GW, Craighead HG. *Macromol Rapid Commun* 2006;27:511–515.
101. (a) Leitgeb A, Abbas M, Fischer RC, Poater A, Cavallo L, Slugovc C. *Catal Sci Technol* 2012;2:1640–1643. (b) Leitgeb A, Szadkowska A, Michalak M, Barbasiewicz M, Grela K, Slugovc C. *J Polym Sci A: Polym Chem* 2011;49:3448–3454. (c) Szadkowska A, Gstrein X, Burtscher D, Jarzembska K, Woźniak K, Slugovc C, Grela K. *Organometallics* 2010;29:117–124.
102. (a) Keitz BK, Grubbs RH. *J Am Chem Soc* 2009;131:2038–2039. (b) Wang D, Wurst K, Knolle W, Decker U, Prager L, Naumov S, Buchmeiser MR. *Angew Chem Int Ed* 2008;47:3267–3270. (c) Wang D, Wurst K, Buchmeiser MR. *Chem Eur J* 2010;16:12928–12934.
103. Piermattei A, Karthikeyan S, Sijbesma RP. *Nat Chem* 2009;1:133–137.
104. Monsaert S, Lozano Vila A, Drozdzak R, Van Der Voort P, Verpoort F. *Chem Soc Rev* 2009;38:3360–3372.
105. Zirngast M, Pump E, Leitgeb A, Albering JH, Slugovc C. *Chem Commun* 2011;47:2261–2263.
106. Vidavsky Y, Anaby A, Lemcoff NG. *Dalton Trans* 2012;41:32–43.
107. Yang H-C, Huang Y-C, Lan Y-K, Luh T-Y, Zhao Y, Truhlar DG. *Organometallics* 2011;30:4196–4200.
108. (a) Flook MM, Jiang AJ, Schrock RR, Müller P, Hoveyda AH. *J Am Chem Soc* 2009;131:7962–7963. (b) Flook MM, Gerber LCH, Debelouchina GT, Schrock RR. *Macromolecules* 2010;43:7515–7522. (c) Flook MM, Ng VWL, Schrock RR. *J Am Chem Soc* 2011;133:1784–1786.
109. (a) Keitz BK, Fedorov A, Grubbs RH. *J Am Chem Soc* 2012;134:2040–2043. (b) Liu P, Xu X, Dong X, Keitz BK, Herbert MB, Grubbs RH, Houk KN. *J Am Chem Soc* 2012;134:1464–1467.
110. Ilker MF, Coughlin EB. *Macromolecules* 2002;35:54–58.
111. (a) Vehlow K, Wang D, Buchmeiser MR, Blechert S. *Angew Chem Int Ed* 2008;47:2615–2618. (b) Torker S, Müller A, Chen P. *Angew Chem Int Ed* 2010;49:3762–3766.
112. Choi TL, Rutenberg IM, Grubbs RH. *Angew Chem Int Ed* 2002;41:3839–3841.
113. Demel S, Slugovc C, Stelzer F, Fodor-Csorba K, Galli G. *Macromol Rapid Commun* 2003;24:636–641.
114. Wagener KB, Boncella JM, Nel JG. *Macromolecules* 1991;24:2649–2657.
115. Hong SH, Day MW, Grubbs RH. *J Am Chem Soc* 2004;126:7414–7415.
116. (a) Hong SH, Sanders DP, Lee CW, Grubbs RH. *J Am Chem Soc* 2005;127:17160–17161. (b) Mutlu H, Montero de Espinosa L, Türünc O, Meier MAR. *Beilstein J Org Chem* 2010;6:1149–1158.
117. Trimmel G, Riegler S, Fuchs G, Slugovc C, Stelzer F. *Adv Polym Sci* 2005;176:43–87.
118. Sovic T, Kappaun S, Koppitz A, Zojer E, Saf R, Bartl K, Fodor-Csorba K, Vajda A, Diele S, Pelzl G, Slugovc C, Stelzer F. *Macromol Chem Phys* 2007;208:1458–1468.
119. Bunz UHF, Mäker D, Porz M. *Macromol Rapid Commun* 2012;33:886–910.
120. Weychardt H, Plenio H. *Organometallics* 2008;27: 1479–1485.
121. Delgado PA, Liu DY, Kean Z, Wagener KB. *Macromolecules* 2011;44:9529–9532.
122. Speros JC, Paulsen BD, Slowinski BS, Frisbie CD, Hillmyer MA. *ACS Macro Lett* 2012;1:986–990.

123. Abdellatif MM, Nomura K. *ACS Macro Lett* 2012;1: 423–427.

124. Buchmeiser MR. *Adv Polym Sci* 2005;176:89–119.

125. Krause JO, Wang D, Anders U, Weberskirch R, Zarka MT, Nuyken O, Jäger C, Haarer D, Buchmeiser MR. *Macromol Symp* 2004;217:179–190.

126. Sudheendran M, Horecha M, Kiriy A, Gevorgyan SA, Krebs FC, Buchmeiser MR. *Polym Chem* 2013;4:1590–1599.

127. Kang E-H, Lee I-H, Choi T-L. *ACS Macro Lett* 2012;1:1098–1102.

128. Kang E-H, Lee IS, Choi T-L. *J Am Chem Soc* 2011;133:11904–11907.

129. Kim J, Kang E-H, Choi T-L. *ACS Macro Lett* 2012;1: 1090–1093.

130. Naumann S, Unold J, Frey W, Buchmeiser MR. *Macromolecules* 2011;44:8380–8387.

131. Lee IS, Kang E-H, Park H, Choi T-L. *Chem Sci* 2012;3:761–765.

132. Park H, Choi T-L. *J Am Chem Soc* 2012;134:7270–7273.

SECTION 3

APPLICATIONS OF OLEFIN METATHESIS REACTIONS

8

APPLICATIONS IN THE SYNTHESIS OF NATURAL AND BIOLOGICALLY ACTIVE MOLECULES

JANINE COSSY

Laboratoire de Chimie Organique ESPCI ParisTech, Paris Cedex, France

8.1 INTRODUCTION

In the last decades, owing to the synthesis of robust and chemoselective catalysts, olefin metathesis has become a powerful tool for synthetic chemists as this reaction is atom and step economical. Metathesis catalysts have revolutionized the way that chemists were designing their strategies to access complex molecules (1–4).

In this chapter, applications of olefin metathesis to the synthesis of natural and/or biologically active molecules are reported; however, considering the significant volume of work in this area, only selected examples related to ring-closing metathesis (RCM), relay metathesis (RM), tethered metathesis (TM), cross-metathesis (CM), ring-opening metathesis (ROM), and domino/cascade metathesis are described. Different catalysts have been used to perform the metathesis reactions, [Ru]-**0**, [Ru]-**I**, [Ru]-**II**, [Ru]-**III**, [Ru]-**IV**, [Mo]-**I**, and more recently, [Mo]-**II**, [Mo]-**III** and [W]-**I**. We have to point out that the most commonly used catalysts, in the synthesis of complex molecules, are [Ru]-**I**, [Ru]-**II**, [Ru]-**III** as well as [Mo]-**I**; the latter catalyst is more reactive however more sensitive than the ruthenium catalysts (Fig. 8.1). The most common solvents used to realize metathesis are methylene chloride (CH_2Cl_2) and toluene at rt or at reflux.

In considering the synthesis of a complex molecule, the first step is the design of a retrosynthetic scheme, keeping in mind that convergent synthesis is more efficient than linear synthesis. Depending on the functionalities present in the molecule, different disconnections can be envisaged; however, a double bond is an obvious disconnection when a metathesis is planed to access a molecule. On the other hand, as an alkene can be the precursor of a C–C bond, an epoxide, a diol, a hydroxyketone, a diketone, an amino-alcohol, a diamine, etc., the introduction of these functionalities as well as the formation of C–C bonds can be envisaged from an alkene resulting from a metathesis reaction between two alkenes (Fig. 8.2).

The most common metathesis reactions used to build the skeleton of complex molecules are the RCM and the CM. In addition, these reactions can be involved in tandem and/or domino/cascade reactions using an ROM (Fig. 8.3). It is worth noting that the construction of heterocyclic compounds (X = N, O) can be realized by utilizing an RCM applied to 1,ω-dienes **A** as well as to 1,ω-dienes **B** (Fig. 8.3).

8.2 RING-CLOSING METATHESIS (RCM)

Carbocycles and heterocycles of different sizes, from five-membered rings to macrocycles, are present in a great variety of natural and/or biologically active compounds and the construction of their skeleton can be realized by utilizing an RCM.

8.2.1 Small-Sized Rings (Five-, Six-, and Seven-Membered Rings)

Construction of five- to seven-membered rings using an RCM is easy from 1,ω-dienes **A** (X = C, O, N) or **B** (X = O, N). However, when a nitrogen is present in 1,ω-dienes, the protection of the nitrogen by an electron-withdrawing

Olefin Metathesis: Theory and Practice, First Edition. Edited by Karol Grela.

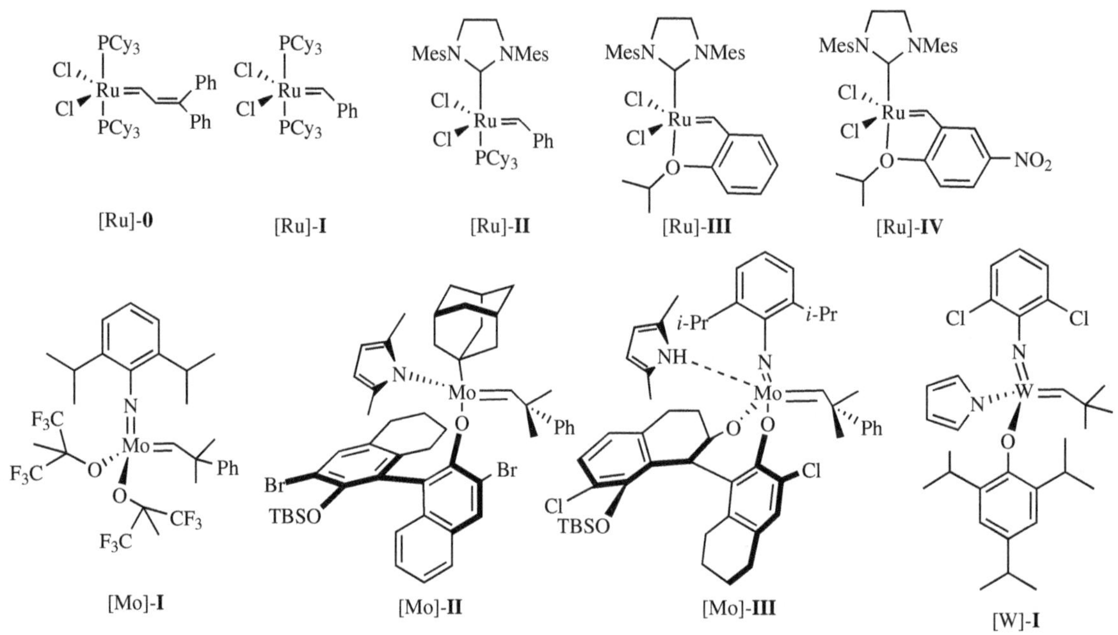

Figure 8.1 Olefin metathesis catalysts.

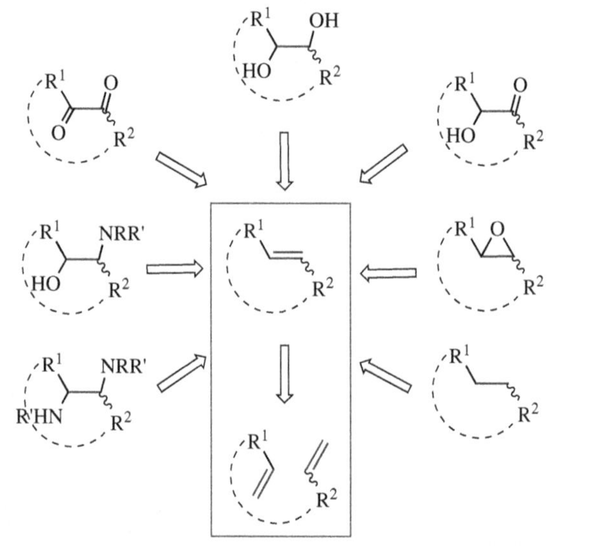

Figure 8.2 Retrosynthetic analysis involving an olefin metathesis step.

group (Ts, Ns, Boc, carbamate, etc.) or its protonation is necessary, otherwise the [Ru]-catalysts and [Mo]-catalysts are rendered inactive. For carbocyclic compounds, nitrogen and oxygenated heterocycles, one example is given for each size of cycle.

8.2.1.1 Carbocyclic Compounds The cyclopentane skeleton of the chinase inhibitor, (−)-allosamizoline, was elaborated by applying an RCM to 1,6-diene **1**. By using [Ru]-**II** (10 mol%) in toluene at 90 °C, cyclopentene **2** was obtained in 88% yield and then transformed in a few steps to (−)-allosamizoline (5) (Fig. 8.4).

Procedure

A solution of [Ru]-**II** (0.1 mmol, 10 mol%) in degassed toluene (10 ml) was added to a solution of dialkene **1** (1.6 mmol) in toluene (10 ml). The reaction mixture was stirred in a sealed flask for 7 h at 90 °C before being evaporated in vacuo. Purification of the residue by chromatography on SiO_2 (EtOAc/petroleum ether = 5 : 2) afforded cyclopentene **2** as a clear oil (88%).

Among antiangiogenic agents, fumagillin and related naturally occurring or semisynthetic analogs have been extensively studied (6). Structure-activity relationship (SAR) studies have shown that some analogs can be 1000 times more active than the parent fumagillin (7). Among the different analogs, fumagillol has been synthetized by utilizing an RCM to construct the cyclohexane ring (8–10). In the first metathesis-based synthesis of fumagillol, the C7–C8 bond of this compound was formed through an RCM applied to **3** by using [Ru]-**I** (10 mol%) in the presence of $Ti(OiPr)_4$ (11) in refluxing CH_2Cl_2. The resulting substituted cyclohexenone **4** was then converted to fumagillol (8). This approach was also used to synthetize other analogs by using [Ru]-**II** and, as this catalyst is more reactive than [Ru]-**I** and $Ti(OiPr)_4$ was not needed (8) (Fig. 8.5).

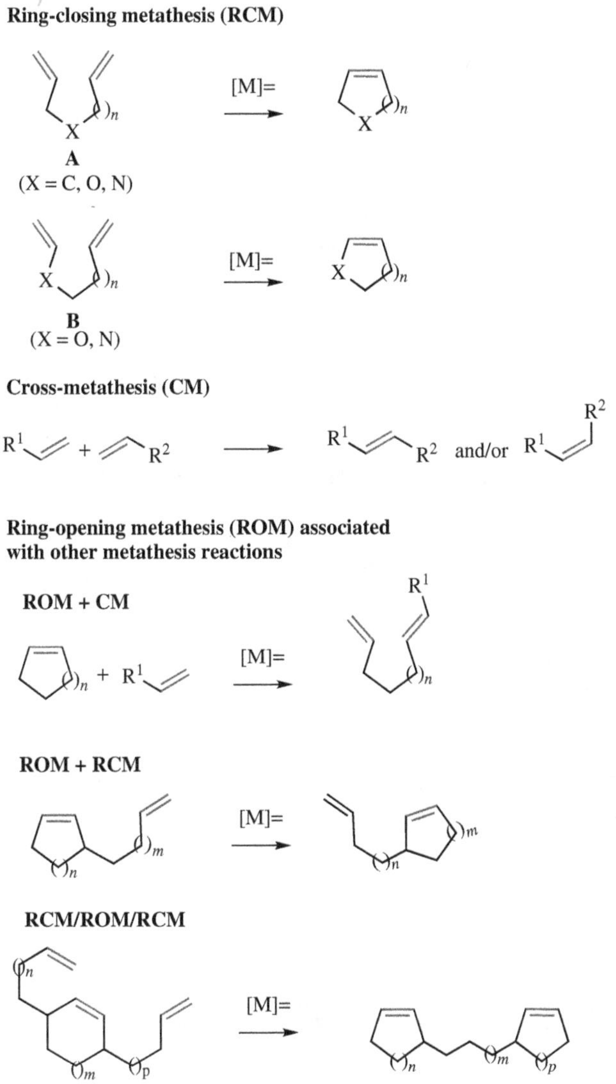

Figure 8.3 Different types of olefin metatheses.

> **Procedure**
>
> $Ti(OiPr)_4$ (12 μl) was added to a solution of **3** (40 mg) in CH_2Cl_2 (1 ml). The mixture was stirred for 30 min at 30 °C and [Ru]-**I** (9 mg, 5 mol%) was added. The mixture was placed in a tightly stoppered flask, heated at 55 °C and stirred for 10 h at which time more [Ru]-**I** (9 mg, 5 mol%) was added. The reaction was allowed to proceed for a further 14 h. After standard workup and preparative TLC on SiO_2 (cyclohexane/AcOEt = 8 : 2), cyclohexenone **4** was isolated as a colorless oil (20 mg, 53%).

Frondosin B was isolated from the marine sponge *Dysidea frondosa* collected from the sea around the Pacific island, Pohnpei, Federated States of Micronesia (12). It was shown that this compound inhibited binding of interleukin-8 (IL-8) to its receptor and also protein kinase C in low micromolar range. The construction of the seven-membered ring of (±)-frondosin was realized from 1,8-diene **5** by utilizing [Ru]-**I** in refluxing benzene (13). Compound **6** was isolated in 88% yield and then transformed in a few steps to (±)-frondosin B (Fig. 8.6).

8.2.1.2 Nitrogen Heterocycles The construction of heterocycles by RCM was first reported by Grubbs and Fu (14) who found that *N,N*-diallyltrifluoroacetamide underwent cyclization in the presence of the Schrock's catalyst, [Mo]-**I**, to give the corresponding pyrrolidine. Since then, many pyrrolidine akaloids have been synthesized by RCM. For example, (+)-broussonetine G (15, 16), initially discovered in the leaves of *Derris eliptica* and then in the branches of the deciduous tree *Broussotia kakinoki*, has been synthesized from *N,N*-diallylamine **8** (R = Bn) that was transformed smoothly to the azabicyclo[3.3.0]octane derivative **9** in 85% yield, using [Ru]-**II** (1.2 mol%, CH_2Cl_2, reflux). It is worth noting that when **7**, possessing a non-protected alcohol (R = H), was treated with [Ru]-**II**, the reaction was troublesome due probably to the complexation of the catalyst by the hydroxy group (Fig. 8.7).

> **Procedure**
>
> To a solution of **8** (23.53 g, 86.1 mmol) in CH_2Cl_2 (250 ml), degassed with argon for 30 min and heated to 40 °C, was added [Ru]-**II** (650 mg, 0.85 mol%) in CH_2Cl_2 (2 ml). After 12 h, a second portion of [Ru]-**II** (250 mg, 0.35 mol%) in CH_2Cl_2 (2 ml) was added to the reaction mixture. The reaction was stirred for an additional 3 h, then cooled to rt and concentrated. The residue was purified by flash chromatography on SiO_2 (Et_2O/petroleum ether = 4 : 1) to yield **9** as a pink solid (color due to ruthenium byproducts). The product was dissolved in CH_2Cl_2 (500 ml) and DMSO (5 ml), and the reaction was stirred for 12 h under N_2 and then purified by flash chromatography on SiO_2 (Et_2O/petroleum ether = 9 : 1) to yield **9** as a white crystalline solid (19.1 g, 91%). Recrystallization from hot Et_2O (300 ml/5 g of product) gave **9** as white prisms (85%).

Among the methods reported to access piperidine alkaloids, RCM-based processes have been used intensively, particularly in the synthesis of (+)-sedamine, a piperidine alkaloid isolated from *Lobelia inflate* used for the treatment of respiratory disorders such as asthma, bronchitis, and pseudomonia. While several racemic syntheses have been reported, only a few enantioselective syntheses were achieved. One enantioselective synthesis was achieved from benzaldehyde by utilizing an RCM. Benzaldehyde was transformed

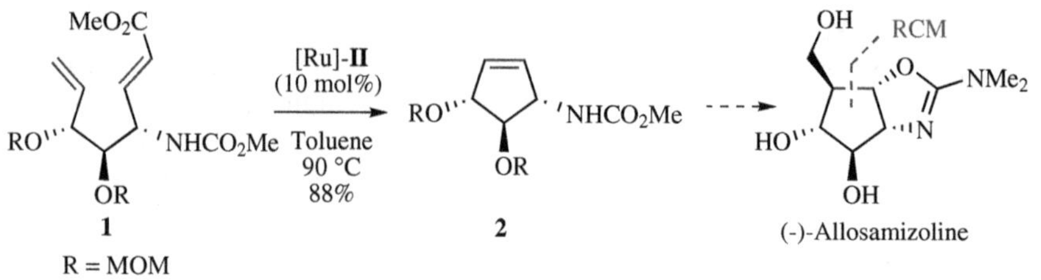

Figure 8.4 Synthesis of (−)-allosamizoline.

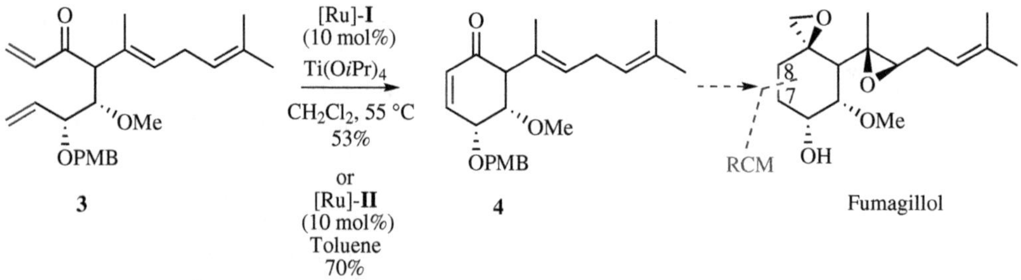

Figure 8.5 Synthesis of fumagillol.

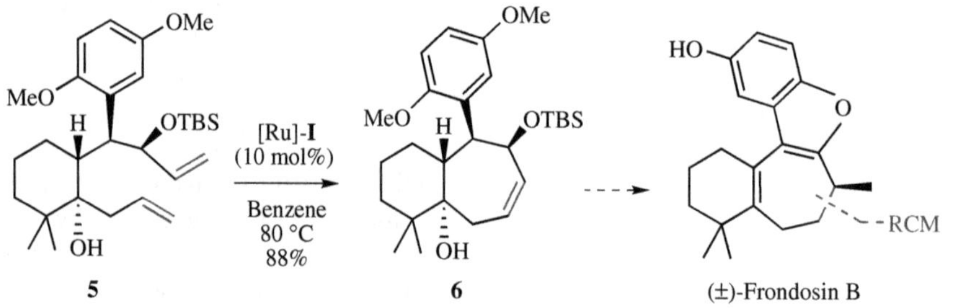

Figure 8.6 Synthesis of (±)-frondosin.

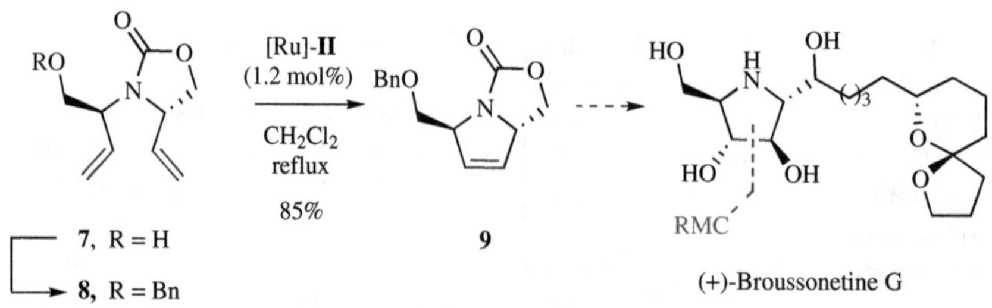

Figure 8.7 Synthesis of (+)-broussonetine G.

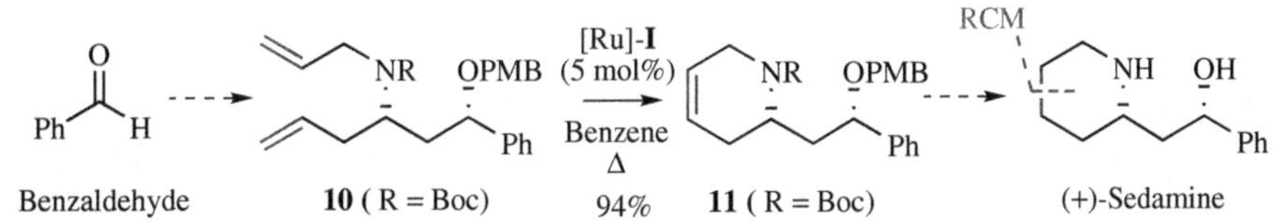

Figure 8.8 Synthesis of (+)-sedamine.

to the optically active ω-unsaturated *N*-allylamine **10**. After conversion of this amine to dihydropiperidine **11** by using [Ru]-**I** (5 mol%) in refluxing benzene, only three steps were necessary to transform **11** to (+)-sedamine (17, 18) (Fig. 8.8).

> **Procedure**
>
> A solution of [Ru]-**I** (0.14 g, 0.2 mmol, 5 mol%) in benzene (3.5 ml) over 4 h at rt was added to a solution of **10** (1.52 g, 3.4 mmol, 1 equiv) in anhydrous benzene (37 ml). The mixture was then concentrated in vacuo. The crude residue was purified by chromatography on SiO_2, (Et_2O/petroleum ether = 1 : 9 to 3 : 7) to afford **11** (1.34 g, 94%).

Balanol, an azepane derivative, is a metabolite produced by *Verticillium belonoides* (19) and *Fusarium merismoides* (20). Owing to its interesting structure and to its high protein kinase C inhibitory activity, balanol has attracted many synthetic chemists. The azepane ring of balanol was constructed by utilizing an RCM (21). The synthesis of this natural product started with the transformation of D-serine to **12** in a few steps. By using [Mo]-**I** catalyst (10 mol%, CH_2Cl_2, reflux), 1,8-diene **12** was transformed to unsaturated azepane **13** in 94% yield. After a few steps, **13** was converted to balanol (Fig. 8.9).

> **Procedure**
>
> A solution of diene **12** (72 mg, 0.193 mmol) and [Mo]-**I** (15 mg, 10 mol%) in carefully degassed CH_2Cl_2 (20 ml) was refluxed for 30 min. Evaporation of the solvent followed by flash chromatography of the crude product (hexane/EtOAc = 20 : 1 to 10 : 1) gives compound **13** as a pale syrup (63 mg, 94%).

*8.2.1.3 **Oxygenated Heterocycles*** Excellent reviews have been published for the construction of polycyclic ethers by RCM (3, 22). Small size cyclic oxygenated heterocycles are present in a number of natural products including lactones, polycyclic ethers (brevetoxin, ciguatoxins, etc.). The most common structure formed by RCM is probably the six-membered ring lactone as illustrated by the synthesis of fostriecin (23). Fostriecin was isolated from *Streptomyces pulveraceus*, and has shown cytotoxic activity against a broad range of cancer cell lines (24). The construction of its unsaturated six-membered ring lactone was achieved using an RCM applied to compound **14**. Treatment of **14** by [Ru]-**I** (20 mol%, CH_2Cl_2, reflux) led to lactone **15** in 82% yield. We have to point out that the RCM is very chemoselective as only the C2–C2′ and C3–C3′ double bonds in **14** were involved in the RCM (25). The chemoselectivity is due to the [Ru]-catalyst, which is more reactive with terminal

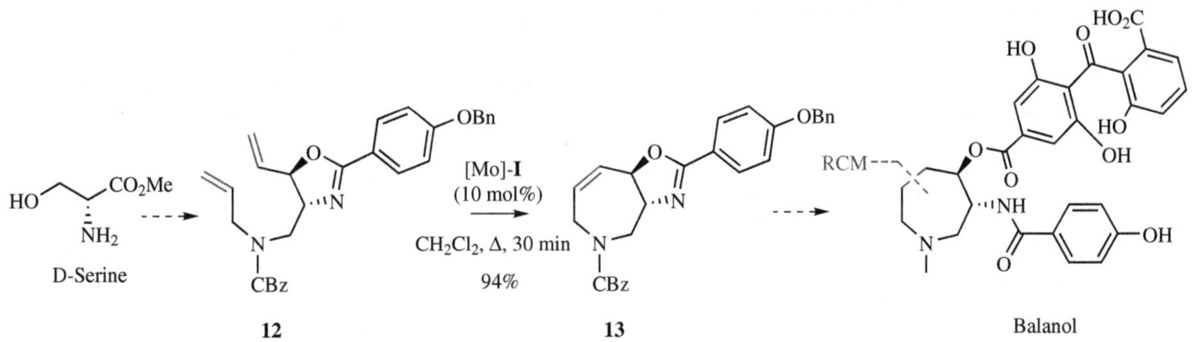

Figure 8.9 Synthesis of balanol.

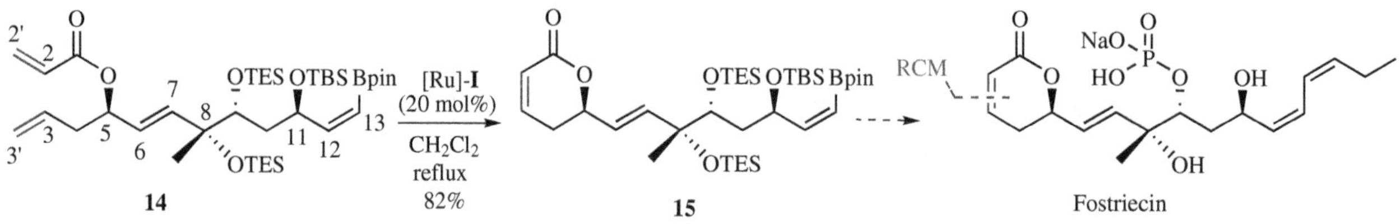

Figure 8.10 Synthesis of fostriecin.

double bonds than with disubstituted double bonds. In addition, the C6–C7 double bond is deactivated by the proximal ester group at C5 and is protected by the silyl ether present at C8; the C12–C13 double bond is a disubstituted double bond and is also protected by a silyl ether at C11 and thus, this double bond is not reactive (Fig. 8.10).

Procedure

[Ru]-**I** (6 mg, 20 mol%) was added to a solution of **14** (25 mg, 0.043 mmol) in CH_2Cl_2 (3.5 ml). The reaction was heated at reflux for 3 h. The solvent was removed under reduced pressure and the residue was purified by flash chromatography on SiO_2 (hexane/EtOAc = 80 : 20) to provide lactone **15** as a colorless oil (20 mg, 82%).

Gambierol is a ladder toxin isolated from the culture cells of *Gambierdiscus toxicus* (26), which has shown toxicity against mice (LD_{50} = 50 μg/kg) with symptoms resembling those of ciguatoxins, so it seems that this compound is involved in ciguatera poisoning (27). The construction of the seven-membered ring (ring H), was realized from a 1,ω-alkenyl enol (1,ω-dienes **B**, Fig. 8.2). The unsaturated ester **16** was transformed to the corresponding enol ether **17** ($TiCl_4$, Zn, $PbCl_2$, CH_2Br_2, TMEDA) and then subjected to [Mo]-**I** or [Ru]-**II** to give the cyclized product **18**. With the more reactive catalyst [Mo]-**I**, **18** was isolated in 88% yield (28). It is worth noting that an RCM was also used to build the D and E rings of gambierol (Fig. 8.11).

Procedure

In a glove box, [Mo]-**I** (0.011 g, 11 mol%) was added to a solution of the acyclic enol ether **17** (0.083 g, 0.122 mmol) in hexanes (24 ml) . The reaction mixture was removed from the glovebox and heated to 50 °C for 16 h in a sealed tube, then cooled to rt and concentrated. Flash chromatography (hexanes/EtOAc = 25 : 1) provided **18** as a colorless oil (69.5 mg, 88%).

8.2.2 Medium-Sized Rings (8-, 9-, and 10-Membered Rings)

In 1995, R. H. Grubbs (29) wrote "… a notable limitation for the RCM was proven to be the synthesis of carbocycles and heterocycles of more than seven atoms …." However, depending on the ring strain or the conformational aspect, these rings can be formed by RCM. In general, high catalyst loading is necessary as well as long reaction time and elevated temperature. To illustrate this purpose, three examples are given.

By imposing a conformational constraint, an eight-membered ring can be formed by RCM. For example, in the synthesis of (±)-mycoepoxydiene, isolated from fermentation of a rare fungus, OS F6661726 (30), an RCM was applied to the dienic tetrahydropyran **19** leading to compound **20** in 83% yield by utilizing [Ru]-**I** (20 mol%) under diluted conditions (31). In addition, benzene was the solvent of choice as in CH_2Cl_2, the decomposition of diene **19** was observed (Fig. 8.12).

Procedure

The reaction was carried out with bubbling N_2. [Ru]-**I** (80 mg, 5 mol%) was added to a stirred solution of **19** (844 mg, 1.94 mmol) in benzene (650 ml). The mixture was refluxed for 2 h and [Ru]-**I** (80 mg, 5 mol%) was added. The mixture was refluxed for 3 h and [Ru]-**I** (80 mg, 5 mol%) was added. The mixture was refluxed for 14 h and [Ru]-**I** (80 mg, 5 mol%) was added. After being stirred for 12 h, the mixture was concentrated in vacuo. The residue was purified by column chromatography on SiO_2 (EtOAc/hexane = 1 : 80) to provide **20** as a colorless oil (653 mg, 83%).

Similarly to the formation of eight-membered rings, the formation of nine-membered rings by RCM is uncertain but depends on the functionality present in the metathesis precursor. One example is related to the synthesis of ophirin B, a cladiellane diterpenoid isolated from the gorgonians *Muricella* sp. and *Astrogorgia* sp., exhibiting significant toxicity

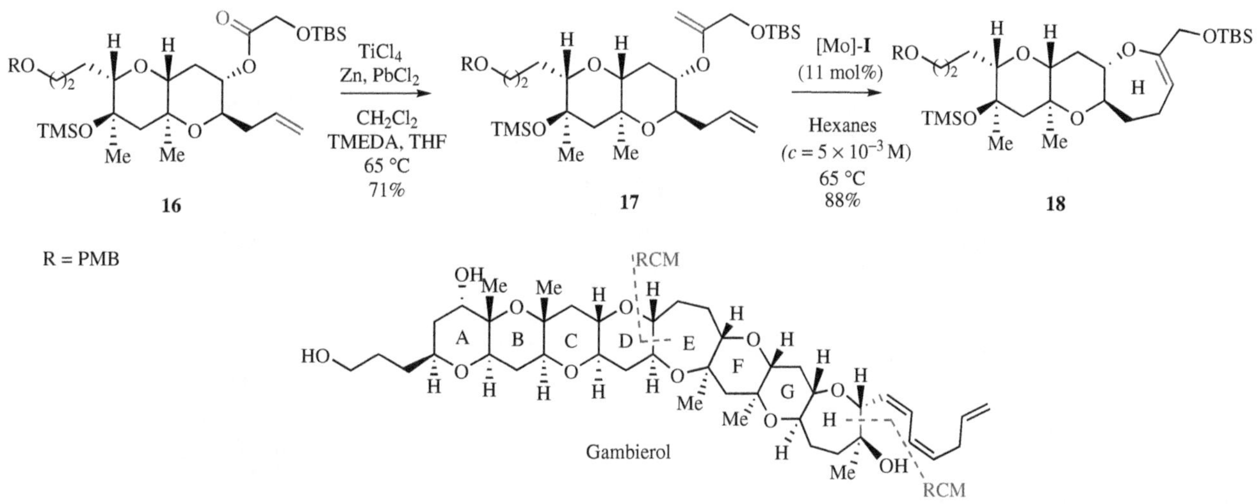

Figure 8.11 Synthesis of gambierol.

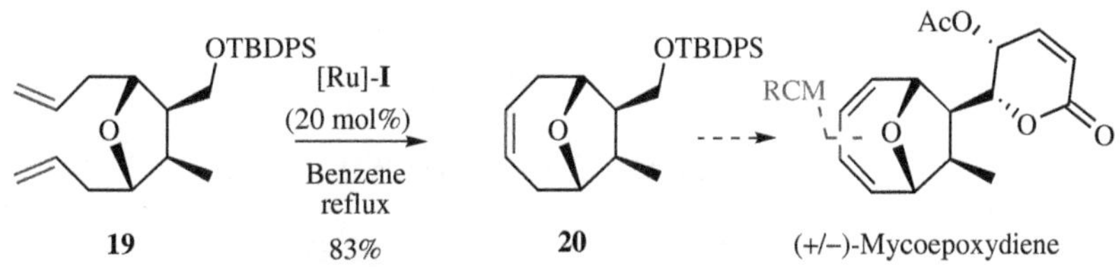

Figure 8.12 Synthesis of (±)-mycoepoxydiene.

and brine-shrimp lethality (32). The synthesis of ophirin B began with diene **21**. When this diene was treated with [Ru]-**I** or [Ru]-**II** at 40 °C in CH_2Cl_2, dimer **22** was exclusively formed, probably due to an unfavorable conformation related to the oxazolidinone that might have been hindering the cyclization. Therefore, the oxazolidinone was reduced and the resulting alcohol **23** was treated with [Ru]-**II** (5 mol%) in refluxing benzene at a concentration of 2×10^{-3} M. Pleasingly, alcohol **23** was transformed to oxonene **24** in 83% yield and only traces of dimer **22** were observed (**24**/**22** = 15 : 1) (33) (Fig. 2.13).

Procedure

Diene **23** (1.0 g, 2.28 mmol) in benzene (1.2 l) was added to a 2 l flask equipped with a reflux condenser. Using a needle, argon was bubbled through the stirring solution for 1.5 h. The needle was removed and [Ru]-**II** (0.194 g, 5 mol%) was added in one portion. The solution was heated at 80 °C for 5 h and then cooled to rt. The solution was stirred vigorously, opened to air overnight, and then concentrated in vacuo. Purification by flash chromatography provided oxocene **24** as a colorless oil (0.832 g, 89%).

To synthesize eremantholide, a compound isolated from *Eremanthus elaeagnus*, a rare woody composite found in Brazil (34) possessing antitumor effects against human KB nasopharyngeal carcinoma (35), epimers **25** and **26** were prepared. By using [Ru]-**III** catalyst, only isomer **26** led to the cyclic compound, **27** (54% yield), which was subsequently transformed to eremantholide, whereas the other isomer, **25**, cross-metathesized (36) (Fig. 2.14).

Procedure

A solution of **26** (20 mg, 0.0512 mmol) and [Ru]-**III** (6.4 mg, 20 mol%) in dry toluene (50 ml) was heated at reflux for 14 h. The solvent was removed in vacuo and the crude residue was purified by multiple-elution preparative TLC (petroleum ether/EtOAc 3 : 1). The desired product **27** was isolated as a colorless oil following flash chromatography on SiO_2 of the concentrated residue using the same eluent (10 mg, 54%).

The first examples of 10-membered ring carbocycle synthesis, utilizing an RCM, were reported in 2001 (37). A formal synthesis of eleutherobin, a potential anticancer agent was reported by applying an RCM to 1,11-diene **28**. Diene

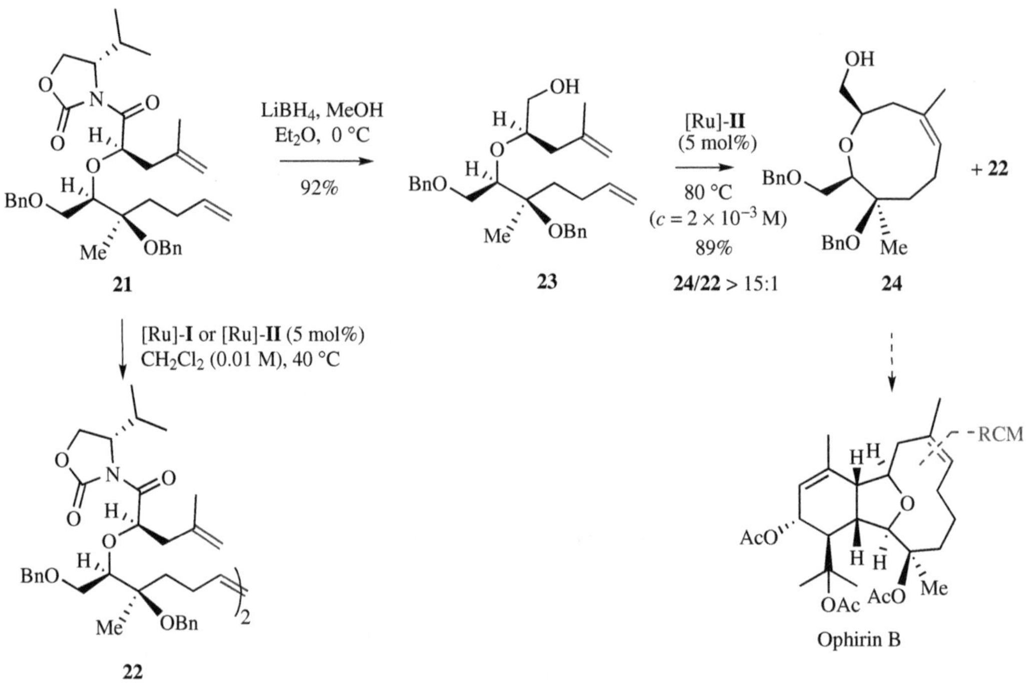

Figure 8.13 Synthesis ophirin B.

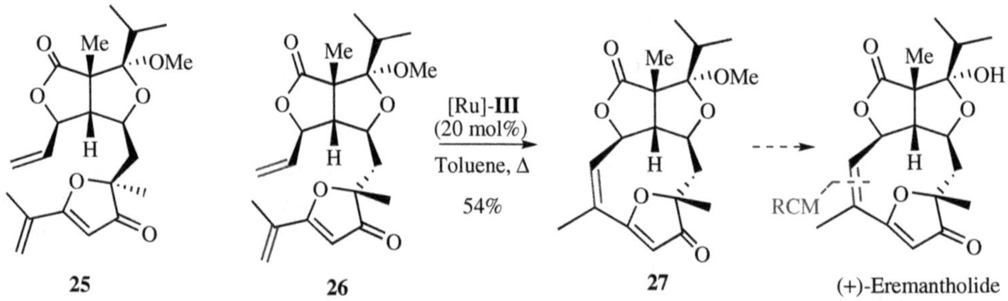

Figure 8.14 Synthesis of eremantholide.

28 was cyclized to the 10-membered carbocycle **29** by using [Ru]-**II** and compound **29** was transformed to eleutherobin via the tricyclic compound **30**. We have to point out that the RCM conditions have to be forced to form the 10-membered ring as 30 mol% of [Ru]-**II** was used and slowly added to **28** in refluxing toluene. Under these conditions, only the *E*-cyclic isomer **29** was obtained (38) (Fig. 2.15).

Procedure

A freshly prepared solution of [Ru]-**II** (16 mg, 30 mol%) in toluene (3.4 ml) was added through a syringe pump over a period of 2.5 h to a heated (110 °C) and stirred solution of compound **28** (42 mg, 0.062 mmol) in toluene (6.8 ml). After 4 h at 110 °C, the reaction mixture was cooled to rt, treated with DMSO (0.67 µl, 0.94 mmol), and stirred for 15 h at rt. Purification of the crude mixture of products by flash chromatography over two consecutive columns (first: petroleum ether/EtOAc = 85 : 15; second: $CH_2Cl_2/iPr_2O = 97:3$) afforded **29** as a white amorphous solid (26 mg, 64%).

8.2.3 Macrocyclic Lactones and Lactams

8.2.3.1 ***Synthesis of Cyclic Alkenes from 1,ω-Dienes*** The use of an RCM to synthesize macrocyclic compounds (superior to 12-membered ring) started in 1996 with the synthesis of exaltolide, a 16-membered ring natural product, which is a musk-odored perfume ingredient (39). The precursor of exaltolide, macrolactone **32**, was obtained by slow addition of [Ru]-**0** (4 mol%) to a solution of 1,ω-diene **31** in CH_2Cl_2. The unsaturated macrolactone **32** was isolated as a mixture of two stereoisomers *E* and *Z* in a ratio of 46/54 in favor of the

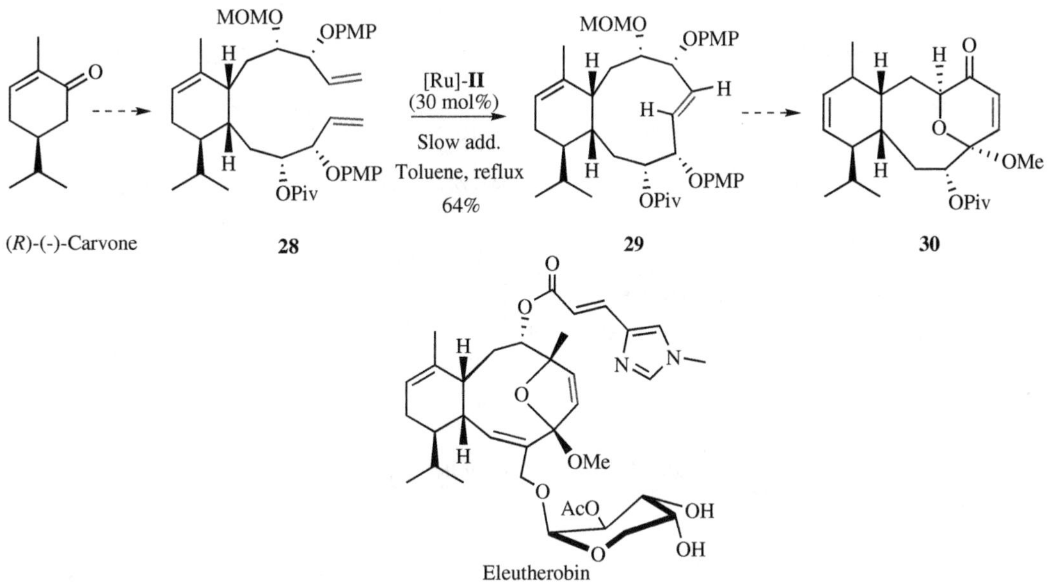

Figure 8.15 Synthesis of a precursor of eleutherobin.

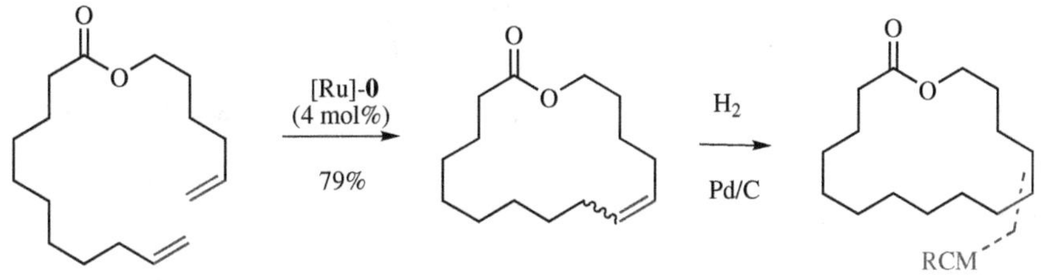

31 **32** ($E/Z = 46{:}54$) Exaltolide

Figure 8.16 Synthesis of exaltolide.

Z-isomer. After hydrogenation, **32** was transformed to exaltolide (Fig. 2.16).

> **Procedure**
>
> A solution of substrate **31** (298 mg, 1.12 mmol) in CH_2Cl_2 (100 ml) and a solution of [Ru]-**0** (50 mg, 4 mol%) in CH_2Cl_2 (100 ml) were simultaneously added dropwise over a period of 24 h to CH_2Cl_2 (50 ml) at rt. After being stirred for another 6 h, the solvent was removed in vacuo and the residue was purified by flash chromatography (*n*-hexane/EtOAc = 100 : 1) to afford **32** as a colorless syrup (219 mg, 79%).

Since 1996, a number of macrocyclic natural products have been synthesized, from 12-membered rings to 72-membered macrocyclic lipids (3). To perform such macrocyclizations, the RCM has to be realized at high dilution (less than 10^{-3} M).

In the synthesis of resorcinylic-type macrolides, the position of the two alkenes of the 1,ω-diene is of importance. For example, in the synthesis of lasiodiplodin, when an RCM was performed on 1,ω-diene **33**, using [Ru]-**0** (6 mol%) in CH_2Cl_2 at rt, the diene was transformed to the 12-membered ring lactone **34** in 94% yield. On the contrary, when the RCM was applied to diene **35**, the cyclized product **36** was not observed owing to the chelating carbonyl group, which attenuates the catalytic activity of the ruthenium catalyst (intermediate **C**) (40, 41) (Fig. 8.17).

> **General Procedure for Macrocyclization**
>
> A solution of [Ru]-**0** (2–6 mol%) predissolved in CH_2Cl_2 (20 ml) was added via cannula to a solution of an acyclic diene (0.360 mmol) in CH_2Cl_2 (80 ml). The solution was stirred at rt or heated at 40 °C for 20 h. The solution was concentrated under reduced pressure and the residue was purified by chromatography on SiO_2.

In the synthesis of epothilone A, an anticancer agent, the position of the alkenes in the precursor of the RCM is also of

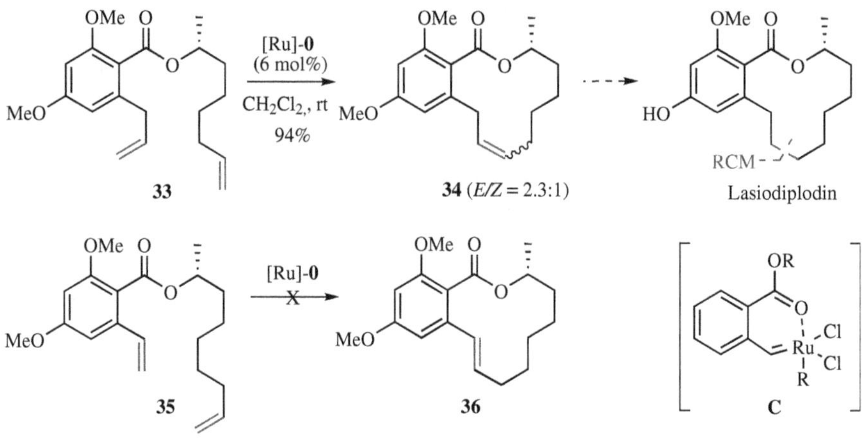

Figure 8.17 Synthesis of lasiodiplodin.

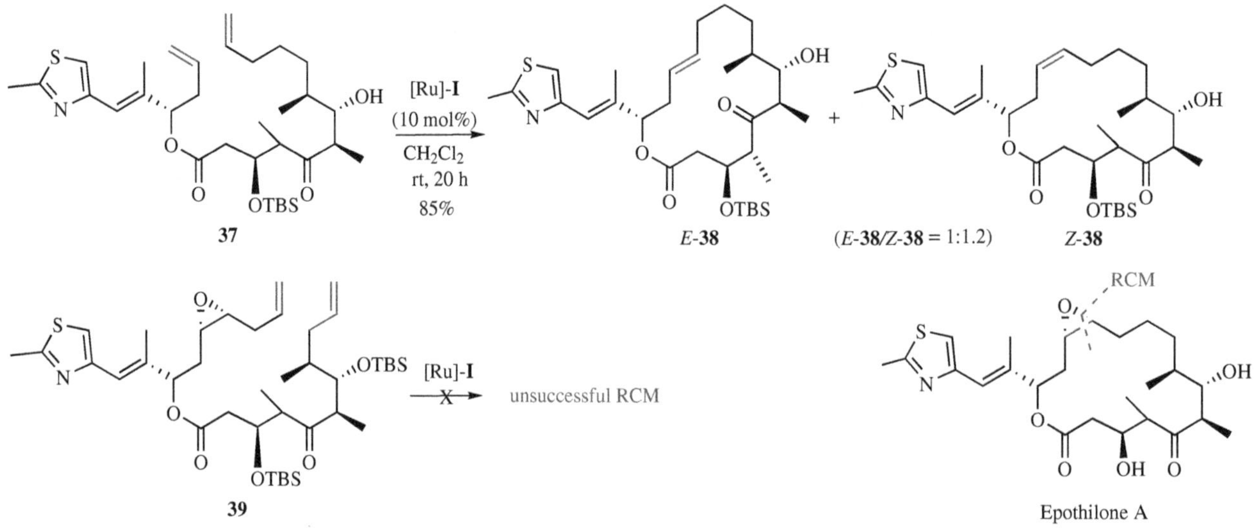

Figure 8.18 Synthesis of epothilone.

importance, as 1,ω-diene **37** was transformed to the desired macrocyclic lactone as a mixture of two stereoisomers *E*-**38** and *Z*-**38** in a ratio of 1/1.2 when [Ru]-**I** (10 mol%) was used (42); but using the same ruthenium catayst, the RCM was unsuccessful when applied to 1,ω-diene **39** (43) (Fig. 8.18).

Procedure

A solution of **37** (0.186 g, 0.3 mmol) in CH_2Cl_2 (200 ml, $c = 0.0015$ M) was treated with [Ru]-**I** (25 mg, 10 mol%) for 20 h at rt to produce *E*-**38** (70 mg, 39%) and *Z*-**38** (83 mg, 46%) after flash chromatography on SiO_2 (EtOAc/hexanes = 7 : 93 to 25 : 75).

Many natural macrocyclic lactones with interesting biological properties possess several alkenes, dienes, and trienes in their structure. If an RCM is planed to synthesize these polyunsaturated macrocyclic lactones, only two specific unsaturations have to react. Fortunately, [Ru]-catalysts are chemoselective as they are more reactive with monosubstituted alkenes than with disubstituted and trisubstituted alkenes. In addition, vinyl halides, alkenes substituted by an electron-withdrawing group, are less reactive than monosubstituted alkenes (terminal double bonds). An RCM was planed to build the C13–C14 bond of palmerolide, a marine natural product isolated from *Synoicum adareanum* and possessing potent and selective activity against the melanoma cancer cell line UACC-62. The highly unsaturated compound **40** was prepared and treated with [Ru]-**II** (10 mol%) in CH_2Cl_2 at rt. The RCM proceeded smoothly to afford the desired macrocyclic derivative **41** as a mixture of two isomers *E* and *Z* in a ratio of 95/5 with a yield of 66% (44), showing the highly synthetic potential of the RCM (Fig. 8.19).

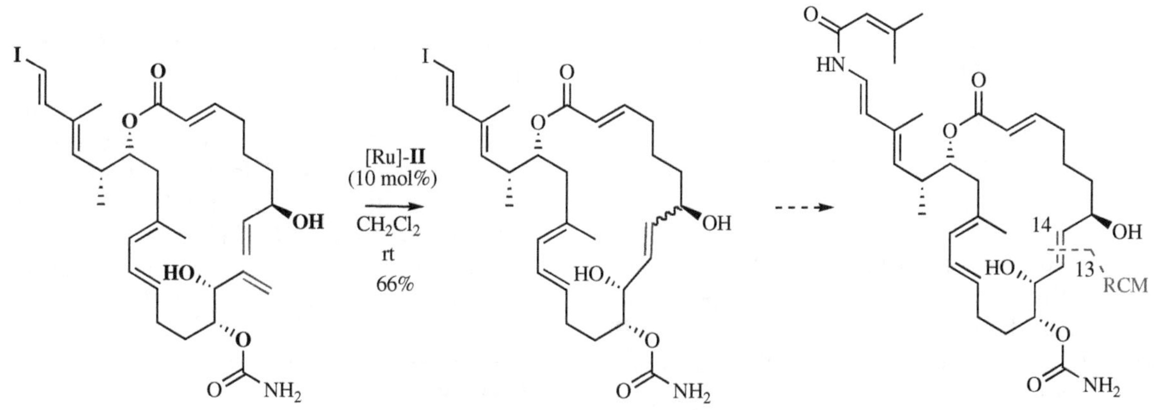

40

41 (*E/Z* = 95:5)

Palmerolide

Figure 8.19 Synthesis of palmerolide.

> **Procedure**
>
> [Ru]-**II** (16 mg, 10 mol%) at rt was added to a solution of diol **40** (61 mg, 0.095 mmol) in CH_2Cl_2 (50 ml). The mixture was stirred for 2 h before it was concentrated in vacuo. Flash chromatography on SiO_2 ($MeOH/EtOAc/Et_3N = 2:100:1$) afforded **41** as a colorless oil (30 mg, 66%).

Macrocyclic *N*-methyl enamides are not common structures but one of these compounds, palmyrolide A, a potential neuroprotective agent, was isolated recently from *Leptotyngbya* cf. sp. and *Oscillatoria* sp. (45). As the synthesis and the RCM of vinyl enamides can be problematic, palmyrolide A was synthesized from *N*-allyl amide **42**. After the cyclization of **42**, using [Ru]-**II** (10 mol%) in dichloroethane at 60 °C, an isomerization of the resulting macrocyclic *N*-allylated tertiary amide **43** was realized with $RuH(PPh_3)_3COCl$ affording the macrocyclic vinylamide that corresponds to palmyrolide A. This compound was isolated in 64% yield when the RCM and the isomerization were realized separately and if a one-pot RCM/isomerization sequence was achieved, the yield was increased to 83% (46). It is worth noting that the C4–C5 bond of palmyrolide A comes from the C5–C6 double bond of **44** that was constructed by RCM (Fig. 8.20).

> **Procedure**
>
> A solution of [Ru]-**II** (4 mg, 5 mol%) in 1,2-dichloroethane (1 ml) was added to a solution of diene **42** (36 mg, 0.1 mmol) in degassed (Ar bubbled through the solvent) 1,2-dichloroethane (30 ml) heated to reflux and the reflux was maintained for 24 h. A second portion of [Ru]-**II** (4 mg, 5 mol%) in 1,2-dichloroethane (1 ml) was added and the reflux was continued for a further 24 h. The reaction mixture was concentrated in vacuo and the residue was purified by flash chromatography on SiO_2 (hexanes/EtOAc = 1 : 1) to give **43** as a dark yellow oil (25 mg, 74%).
>
> A solution of [Ru]-**H** (4 mg, 20 mol%) in toluene (1 ml) was added to a solution of **43** (9.7 mg, 0.029 mmol) in toluene (2 ml) at reflux and the reflux was maintained

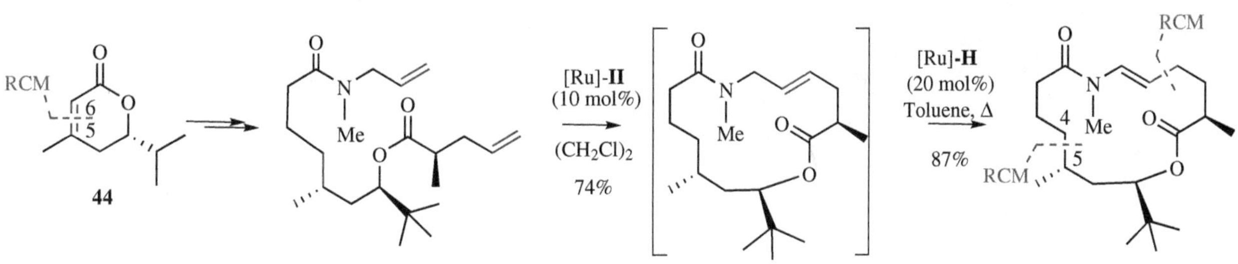

44

42

43

Palmyrolide A

[Ru]-**H** = RuH $(PPh_3)_3COCl$

Figure 8.20 Synthesis of palmyrolide A.

for 24 h. The solvent was concentrated in vacuo and the crude product was purified by flash chromatography on SiO_2 (hexanes/EtOAc = 1 : 1) to give palmyrolide A as a yellow oil (8.5 mg, 87%).

8.2.3.2 Synthesis of 1,3-Dienic Macrocycles from ω-Alkenyl 1,3-Dienes Macrocylic compounds possessing 1,3-dienes can be formed from ω-alkenyl 1,3-dienes **D**. The success of this cyclization depends on the ruthenium catalyst. To form these 1,3-dienic macrocycles **E** from ω-alkenyl 1,3-dienes **D**, the terminal double bond of the 1,3-dienes has to be more reactive than its internal double bond. If the yield in cyclic 1,3-dienes **E** is poor or if the RCM leads to cyclic alkenes of type **F**, the internal double bond of the 1,3 dienes **D** has to be protected by a silyl group at C4; this silyl group will prevent the participation of the internal alkene in the RCM process and, after cleavage of the silyl group, cyclic dienes **E** (R = H) will be obtained. It is worth pointing out that if ω-alkenyl 1,3-dienes **D** are substituted by a methyl group at C3 or C4, the internal double bond is not involved in the RCM process (Fig. 2.21).

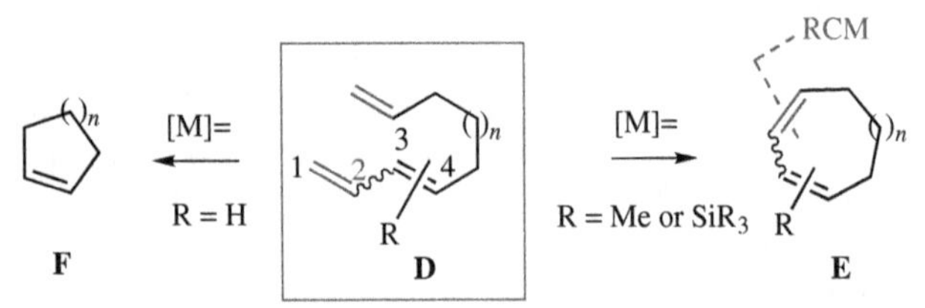

Figure 8.21 General scheme for the formation macrocyclic 1,3-dienes from ω-alkenyl dienes.

The 24-membered ring polyenic compound, iejimalide B, was extracted from the tunicate *Eudistoma* cf. *rigida* collected off Ie island, Okinawa province, Japan and revealed to be highly cytotoxic against the panel of 60 standard human cancer cell lines in the low nanomolar range (47). The macrocycle of iejimalide B was synthesized from **45**. Treatment of **45** with [Ru]-**II** (10 mol%) in toluene at 50 °C afforded **46** in 63% yield and, after treatment of **46** with tetrabutylammonium fluoride, iejimalide B was isolated (99%) (48) (Fig. 8.22).

Procedure

[Ru]-**II** (3.2 mg, 10 mol%) was added to a solution of **45** (32 mg, 0.044 mmol) in toluene (50 ml). The mixture was stirred at 50 °C for 4 h before the reaction was quenched with ethyl vinyl ether (about 0.30 mg). After stirring at rt for 30 min, the mixture was concentrated and the residue was purified by flash chromatography (hexanes/EtOAc = 10 : 1 to 5 : 1 containing 1% of Et_3N), to give the macrocycle **46** as a colorless oil (20 mg, 63%).

8.2.3.3 Synthesis of Trienes from 1,ω-di-1,3-Dienes Macrocyclic trienes can be formed when an RCM is applied to 1,ω-di-1,3-dienes. A new class of ansamycins such as cyclotrienin A bearing a (*E,E,E*)-triene, was isolated from *Streptomyces* sp., RK95-74, and has shown important antitumoral proprieties (49). Cyclotrienin A was synthesized from tetraene **47**. When [Ru]-**II** was used, the insertion of the ruthenium catalyst took place on one of the di-substituted double bonds of one 1,3-diene, affording the 19-membered ring macrolactam **48** (41%). However, when [Ru]-**I** (20 mol%) was utilized in refluxing CH_2Cl_2, the desired 20-membered ring **51** was formed in good yield

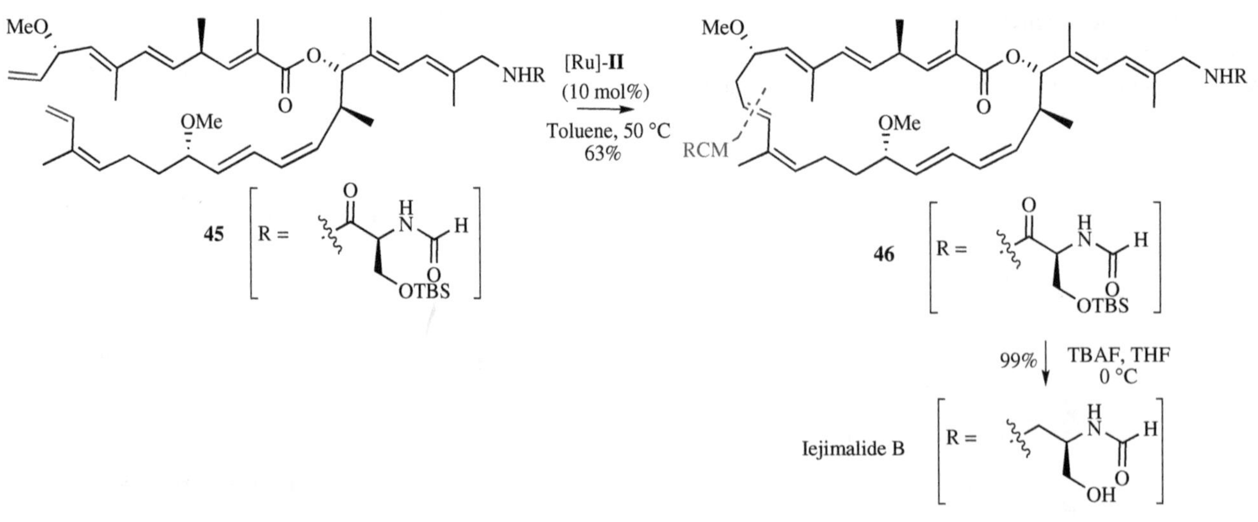

Figure 8.22 Synthesis of iejimalide B.

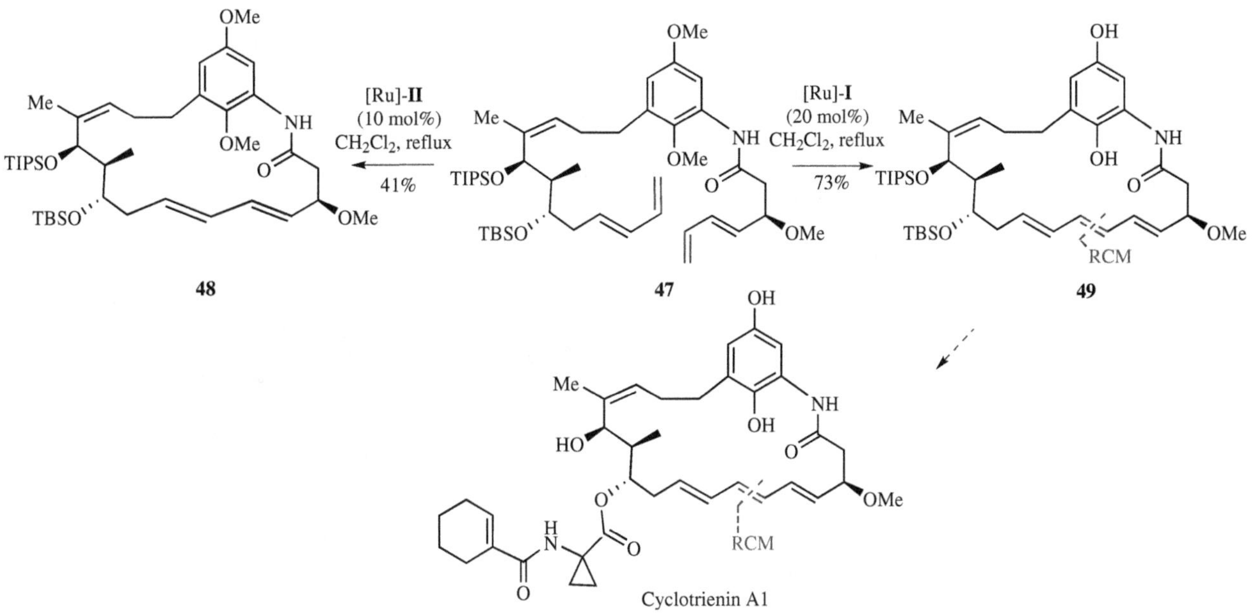

Figure 8.23 Synthesis of cyclotrienin A1.

(73%) and the triene with the (*E,E,E*) configuration was obtained (50) (Fig. 2.23).

> **Procedure**
>
> Synthesis of **48**: [Ru]-**II** (0.5 mg, 10 mol%) was added into a 15 ml pressure tube containing bis-diene **47** (5.0 mg, 6.4 μmol) dissolved in degassed CH_2Cl_2 (3 ml, 0.002 M). The resulting solution was heated to reflux for 2 h, then concentrated and purified by flash chromatography on SiO_2 (hexanes/EtOAc = 98 : 2 to 90 : 10) to afford the cyclic diene as a pale oil (1.9 mg, 41%).
>
> Synthesis of **49**: [Ru]-**I** (7.1 mg, 10 mol%) is added to a solution of bis-diene **47** (68 mg, 87 μmol) in CH_2Cl_2 (43 ml, 0.002 M). The resulting orange-purple solution was heated to reflux for 15 h, another portion of [Ru]-**II** (7.1 mg, 10 mol%) was added, and the reaction was heated to reflux for another 5 h. The reaction mixture was then cooled to rt and air was passed through the solution. After 15 min, the crude mixture was concentrated and purified by flash chromatography on SiO_2 (hexanes/EtOAc = 91 : 7 to 70 : 30) to afford the cyclic triene as a pale yellow foam (48 mg, 73%).

8.2.3.4 E/Z Problem in Macrocyclizations As previously reported in the synthesis of epothilone A, when the RCM of **37** was achieved with [Ru]-**I**, two stereoisomers *E*-**38** and *Z*-**38** were obtained in a ratio 1:1.2 (Fig. 8.18). Recently, it has been reported that the precursor of epothilone C was obtained with a highly *Z*-stereoselectivity by treatment of **50** with [W]-**I** (5 mol%) in toluene at rt, under 0.02 torr. 1,17-Diene **50** was transformed to *Z*-**51** and *E*-**51** in a ratio of 96:4 (51) and the macrocyclic unsaturated lactone *Z*-**51** was then transformed to epothilone C (Fig. 8.24).

> **Procedure**
>
> A 250 ml Schlenk flask, fitted with a connecting adaptor attached to an argon-filled manifold, was flame-dried and charged with diene **50** (0.219 g, 0.298 mmol). After azeotropic distillation with benzene (three times, freeze-pump), the apparatus was charged with [W]-**I** (21.9 mg, 7.5 mol%, weighed in air), evacuated, back-filled with argon, and charged with mesitylene (50 ml). The mixture was exposed to vacuum (0.02 torr) and stirred for 4 h at 22 °C, after which the reaction was quenched by the addition of wet Et_2O (1 ml). Purification by chromatography on SiO_2 (hexanes/Et_2O = 20 : 1) afforded **51** as a mixture of *Z/E* isomers as a white foam (0.72 g, 82%).

An excellent *Z*-stereoselectivity for the alkene present in **53** was also observed when **52** was treated with [W]-**I** (5 mol%) under 0.07 torr, as **53** was produced in a *Z/E* ratio of 97:3 (51). It is worth noting that the *Z/E* selectivity was better with [W]-**I** (*Z/E* = 97/3) than with [Mo]-**II** (*Z/E* = 69 : 31) (Fig. 8.22). Compound **53** was transformed to nakadomarin A (52), a marine alkaloid of the manzamine family isolated from the sponge *Amphimedon* sp., with cytotoxic activity against lymphoma L1210 cells, antimicrobial activity

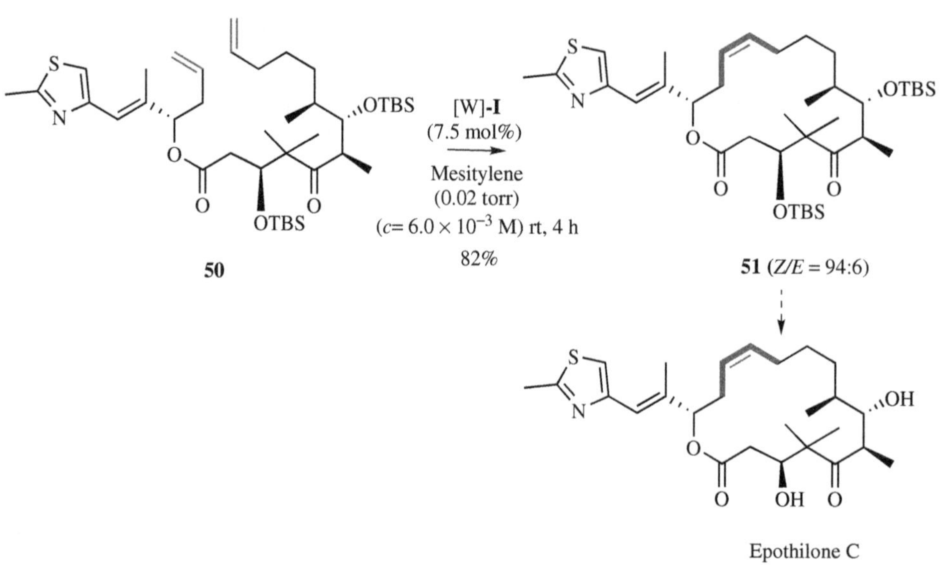

Figure 8.24 Synthesis of epothilone C using [W]-I.

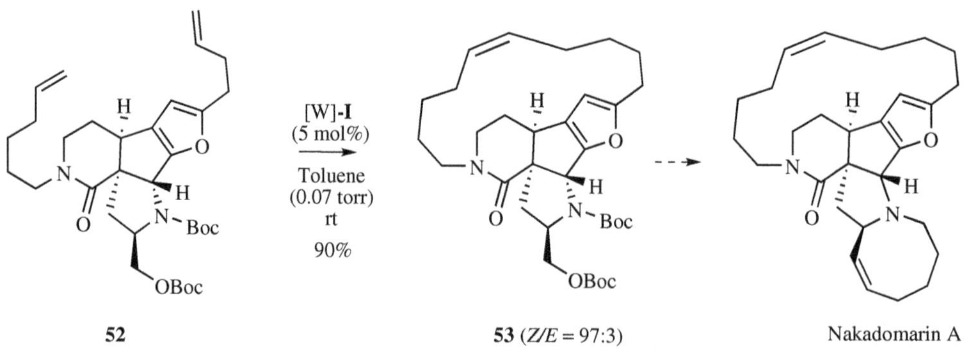

Figure 8.25 Synthesis of nakadomarin A.

against the fungus *Trichophydon mentagrophytes*, and activity against Gram-positive bacterium *Corynebacterium xerosis* (Fig. 8.25).

8.3 RELAY RING-CLOSING METATHESIS (RRCM)

Despite the power of the RCM and the significant advantages made available by modification of external factors (catalyst, use of additive, choice of solvent, temperature), not every 1,ω-diene **G** can be successfully cyclized to cycloalkenes **K**. However, chemists have an alternative when such a limitation is encountered: a relay ring-closing metathesis (RRCM). Altering the parent 1,ω-dienic substrate **G** by installation of an extender arm that includes a tether element in **H** provides the opportunity to introduce the ruthenium catalyst to the more reactive terminal alkene (intermediate **I**). After an RCM, an unsaturated five-membered ring is excluded, leading to intermediate **J** that can induce a cyclization to produce the desired cyclic alkene **K** (Fig. 8.26).

Owing to the difficulties encountered in the formation of the macrocycle of oximidine III (53) from 1,ω-alkenyl diene **54** by treatment with [Ru]-**I** or [Ru]-**II**, the use of the relay-activated substrate **55** in the presence of [Ru]-**III** solved the problem due to the generation of the ruthenium carbene adjacent to the oxirane via the intermediate **56** which led to **57** and then to the cyclized diene **58** (Fig. 8.27). The macrocyclic lactone **58** was then transformed to oximidine III, a natural product isolated from *Pseudomonas* sp. QN05727, an inhibitor of V-ATPase (54).

Procedure

55 (100 mg, 155 μmol) in dry 1,2-dichloroethane (3 ml) was added via a syringe pump over 30 min to a solution of [Ru]-**III** (9.4 mg, 10 mol%) in dry 1,2-dichloroethane (75 ml) at 50 °C. After 2 min, the reaction was quenched with ethyl vinyl ether (2 ml), cooled to rt, and concentrated in vacuo. Purification on SiO_2 (hexanes/EtOAc = 85/1) provided **58** as a colorless oil (59 mg, 71%).

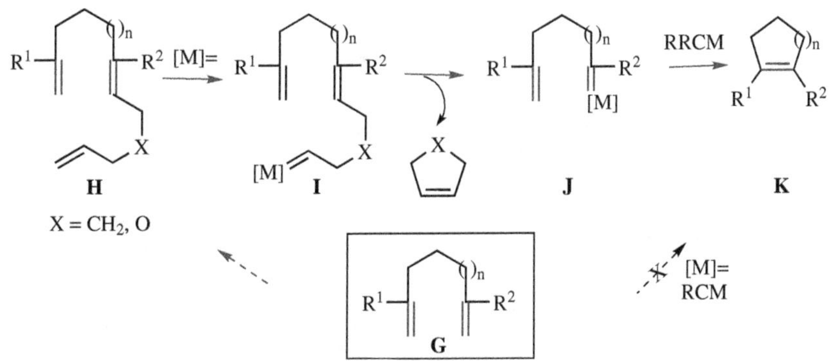

Figure 8.26 General scheme for the RRCM.

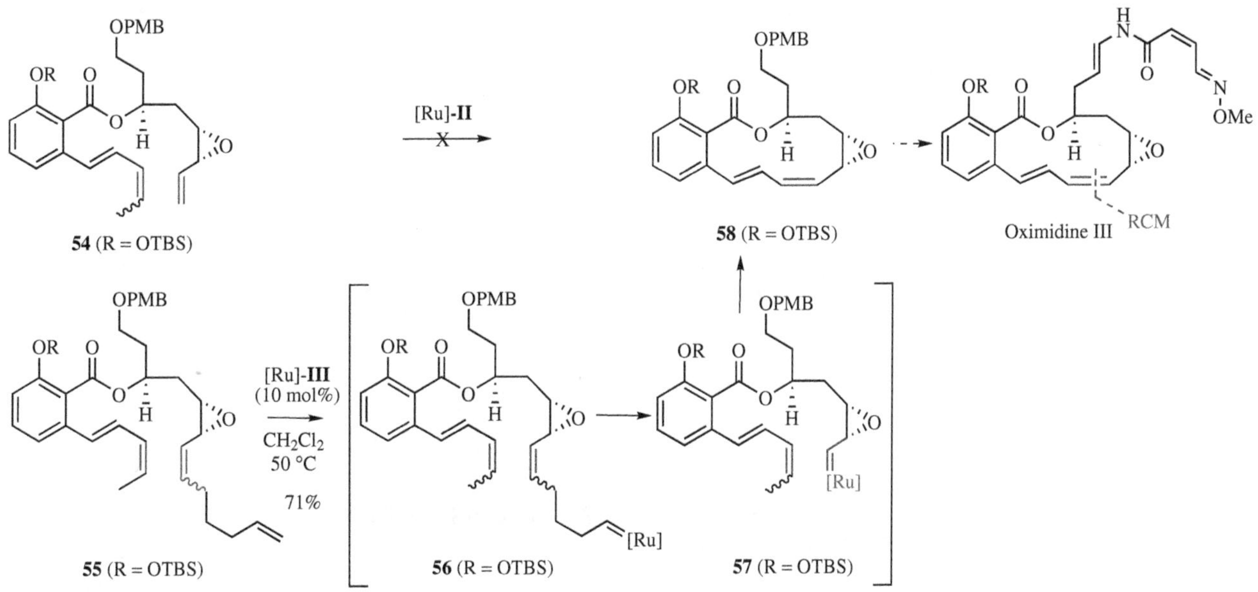

Figure 8.27 Synthesis of oximidine III.

8.4 TEMPORARY TETHER AND RCM (TT-RCM)

Temporary tethered reactions provide an important strategy for target-directed synthesis, as they generally circumvent the problems encountered with entropically unfavorable reactions and they can be useful to couple two alkenes. Among various temporary tethers, silyl ethers have proven to be the most versatile ones. By synthesizing silyl ketal intermediates **L**, an RCM can be performed and, after cleavage of the silyl ethers in **M**, a *Z*-double bond is present in compounds **N**. The Temporary Silicon-Tether (TST)/RCM is the equivalent of a CM between two allylic alcohols however, in contrast to the CM, which produces *E*-alkenes (cf *vide infra*), *Z*-alkenes are obtained by TST/RCM (Fig. 8.28).

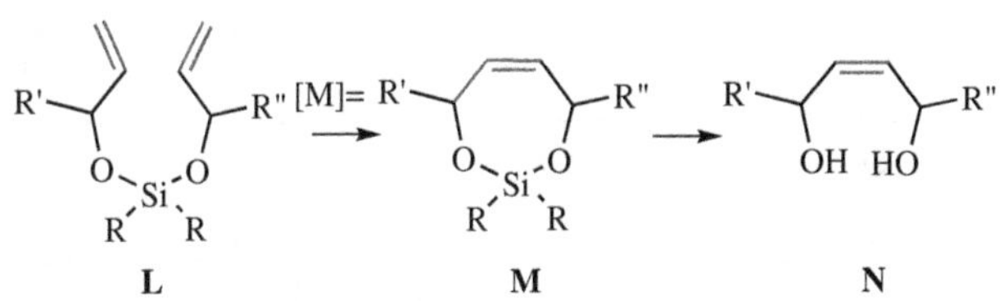

Figure 8.28 General scheme for the temporary silicon tether (TST)/RCM strategy.

The synthetic utility of the TST/RCM reaction was useful in the synthesis of (+)-gigantecin, a compound isolated from the bark of *Goniothalamus giganteus* in Southeast Asia (55) and from the seeds of the Brazilian plant, *Annona coriacea* (56). The synthesis of this compound combines a TST/RCM in conjunction with a CM to facilitate the assembly of the two tetrahydropyranic fragments contained in (+)-gigantecin. The bis-alkoxysilane **61** was obtained by realizing a sequential coupling of the allylic

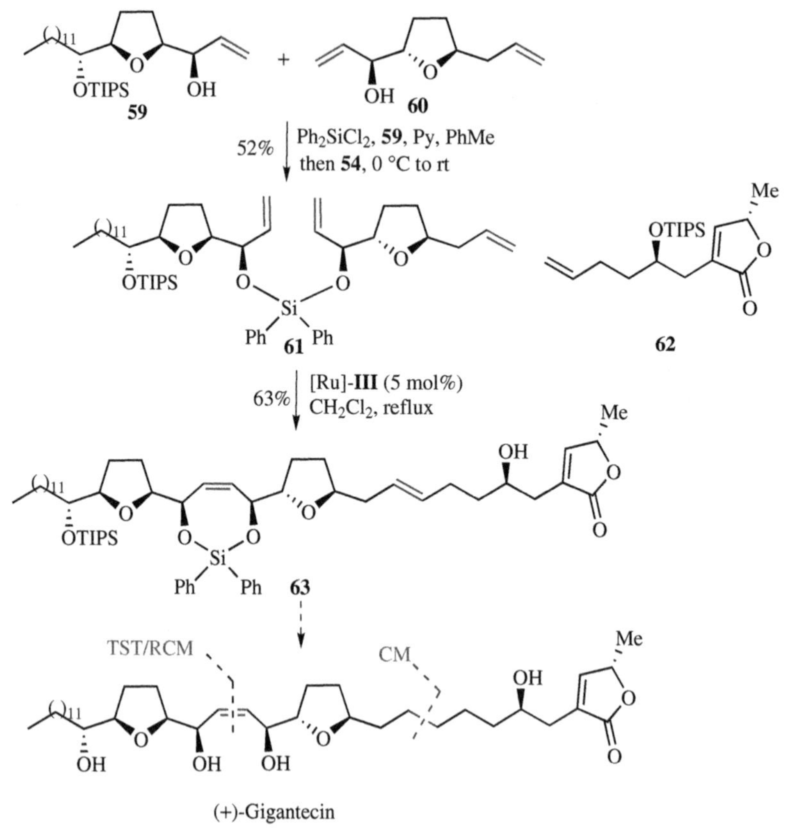

Figure 8.29 Synthesis of (+)-gigantecin.

alcohols **59** and **60** with dichlorodiphenylsilane. Treatment of **61** and **62** (1/4 ratio) with [Ru]-**III** (20 mol%) in refluxing CH_2Cl_2 led to **63** (63% yield), which is the result of a CM/TST/RCM sequence. Compound **63** was then transformed to (+)-gigantecin. We have to point out that the reverse order, for example, a TST/RCM/CM sequence does not produce the desired compound **63** (57) (Fig. 8.29).

Procedure

[Ru]-**III** (4.1 mg, 5 mol%) in CH_2Cl_2 (0.36 ml), under a nitrogen atmosphere, was added via syringe pump over 9 h to silaketal **61** (20 mg, 24 µmol) and butenolide **62** (34 mg, 96 µmol), in CH_2Cl_2 (0.3 ml). The reaction vessel, a septum-capped tube, was kept in a warm bath at 45 °C. At the end of the addition period, little solvent remained. The resultant mixture was allowed to cool to rt and concentrated in vacuo. The crude material was filtered through a plug of SiO_2 (hexanes/EtOAc = 9 : 1). MPLC (hexanes/EtOAc = 50 : 1) afforded the seven-membered cyclic CM–RCM product **63** as a clear colorless oil (17 mg, 63%).

8.5 CROSS-METATHESIS (CM)

The most efficient reaction to couple two alkenes is the CM. CM can be used to functionalize a terminal olefin appending a side chain to the core of complex molecules, or it can be used to couple two alkenes to build the skeleton of a target molecule. The two alkenes can be electron-rich or electron-poor or, one can be electron-rich and the other one electron-poor, in general, one alkene is used in excess. To have a successful CM, the classification of alkenes established by Grubbs et al. (Type I to Type IV alkenes) has to be taken into consideration (58), as alkenes of Type IV are spectators in the CM (1,1-disubstituted alkenes, nitro olefins, quaternary allylic carbon-containing alkenes, protected tertiary allylamines, protected trisubstituted allylic alcohols) (58). When a CM takes place between two alkenes, in most cases, the double bond of the resulting major product is of *E* configuration.

In the synthesis of prosophylline, a prosopis alkaloid isolated from the leaves of *Prosopis africana* and possessing antibiotic and anesthetic properties (59), the CM allowed the construction of the carbon skeleton of the molecule. Thus, when **65** was treated with [Ru]-**II** (10 mol%) in refluxing

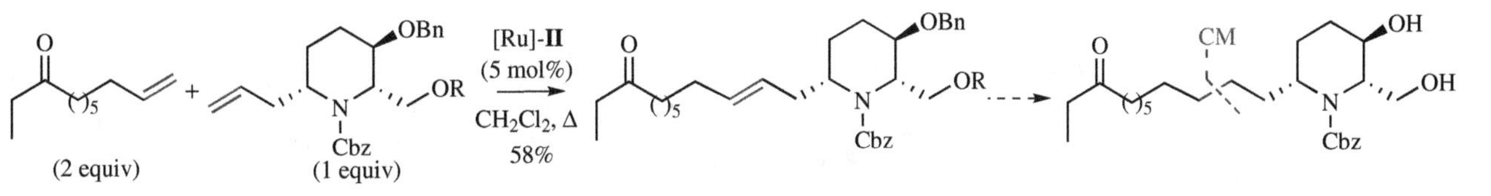

64 65 (R = TBDPS) 66 (R = TBDPS) Prosophylline

Figure 8.30 Synthesis of prosophylline

CH_2Cl_2 in the presence of an excess of alkene **64**, the CM product **66** was isolated in 58% yield. After hydrogenation and deprotection, prosophylline was isolated (54%). As two electron-rich olefins were involved in the CM, the less valuable alkene **64** was used in excess (2 equiv) (60) (Fig. 8.30).

Procedure

A solution of **65** (0.86 mg, 1.36 mmol, 1 equiv) and **64** (0.46 g, 2.72 mmol, 2 equiv) in CH_2Cl_2 (3 ml) was added to a solution of [Ru]-**II** (58 mg, 5 mol%) in CH_2Cl_2 (9 ml). The solution was stirred under argon for 4 h at 50 °C and then concentrated in vacuo. The crude product was purified by flash chromatography on SiO_2 (petroleum ether/Et_2O = 10 : 2 to 10 : 3) to afford **66** as a colorless oil (0.61 g, 58%).

In the synthesis of faranal, the trail pheromone of the common ant *Monomorium pharaonis*, the functionalization of the double bond in **67** was realized with *S*-ethyl thioacrylate **68** using [Ru]-**III** (5 mol%) in refluxing CH_2Cl_2. The unsaturated thioester **69** was isolated in 83% yield, transformed to **70**, and then to faranal (61) (Fig. 8.31).

Procedure

In a dry Schlenk tube, under a N_2 atmosphere, [Ru]-**III** (31.3 mg, 2.5 mol%) was added to a solution of *S*-ethyl thioacrylate **68** (229 ml, 2.0 mmol) and **67** (176 mg, 1.0 mmol) in CH_2Cl_2 (2.5 ml). The resulting green solution was heated for 6 h at reflux. The mixture was allowed to cool, a second portion of the catalyst was added (31.3 mg, 2.5 mol%), and the mixture was heated for another 18 h at reflux. The mixture was concentrated in vacuo and purified by flash chromatography on SiO_2 (Et_2O/pentane = 5 : 95), which afforded an inseparable mixture constituted by **70** and the dimer of **67** in a ratio 20 : 1 as a colorless oil [236 mg (**69** + dimer of **67**), 83% corrected yield for **70**].

It is worth pointing out that the terminal olefin of 1,3-dienes can be involved in a CM with alkenes to produce substituted 1,3-dienes. In the synthesis of meayamycin B, a potential inhibitor of pre-mRNA splicing, the C7–C8 bond present in this molecule was obtained by realizing a CM between unsaturated tetrahydropyran **71** and diene **72**. By using [Ru]-**IV** (7 mol%) in 1,2-dichloroethane at 45 °C, meayamycin B was formed with a yield of 44%. It was noticed that the terminal olefin of the 1,3 diene in **72** was more reactive than the disubstituted olefin and the unsaturated amide (62) (Fig. 8.32).

Procedure

A solution of **72** (18 mg, 99 μmol) was prepared in 1,2-dichloroethane (0.20 ml) at 23 °C under an open atmosphere. A stirred solution of **71** (26 mg, 62 μmol) in 1,2-dichloroethane (0.20 ml) was treated with an aliquot of the solution of **72** (0.10 ml), benzoquinone (1.4 mg, 12 μmol), then [Ru]-**IV** (1.5 mg, 2.2 μmol) at 23 °C under an open atmosphere. The reaction mixture was heated to 45 °C. After 3 h at the same temperature, additional [Ru]-**IV** (1.5 mg, 2.2 μmol) and the remaining of the solution of **72** (0.10 ml) were added. After 11 h, the reaction was concentrated in vacuo. The residue was purified by flash chromatography on SiO_2 (hexanes/EtOAc = 90 : 10 to 80 : 20) to give meayamycin B (10.4 mg, 44% corrected yield).

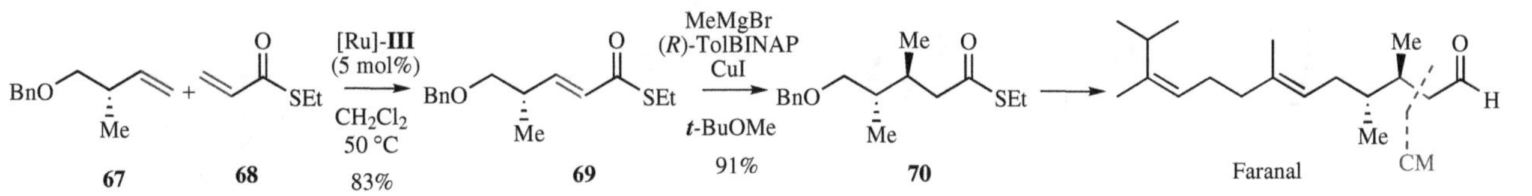

67 68 69 70 Faranal

Figure 8.31 Synthesis of faranal.

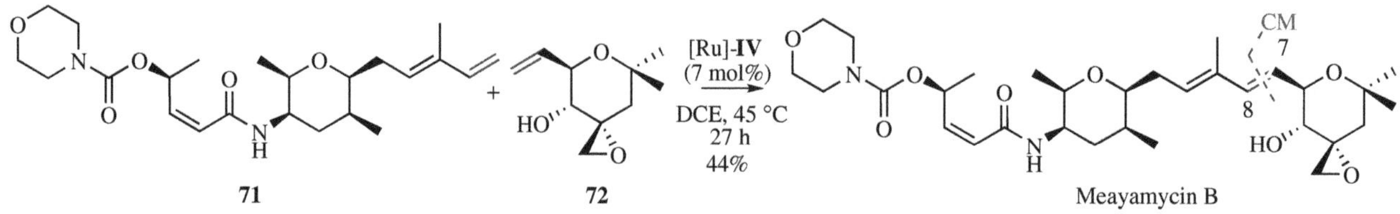

Figure 8.32 Synthesis of meayamycin B.

8.6 RING-OPENING METATHESIS (ROM)

As we have seen previously, the formation of eight-membered rings by RCM is difficult and uncertain. One way to solve the problem is to achieve an ROM of strained rings, such as cyclobutenes and cyclopropenes, and to involve the resulting dienes in thermal processes, rearrangements, or other metathesis reactions (cascade and/or domino processes).

8.6.1 ROM and Sigmatropic Rearrangements

When cyclobutene **73** was treated with [Ru]-**II** under 1 atm of ethylene in refluxing benzene, divinylcyclobutane **74** was formed and a sigmatropic rearrangement took place at 90 °C to produce an eight-membered ring. The synthesis of **75** would have been difficult by using a traditional RCM. The tricyclic compound **75** was then transformed to asteriscanolide (63) (Fig. 8.33).

Procedure

A solution of **73** (50 mg, 263 μmol) in benzene (1 ml) was added over 2 h to a solution of [Ru]-**II** (11 mg, 5 mol%) in freshly distilled benzene (5 ml) under ethylene (1 atm) at rt. The resulting reaction mixture was stirred at 50 °C for an additional 6 h. At this point, **74** and **75** were present in the solution in approximately a ratio of 1 : 2. The dark solution was purged with argon and refluxed (oil bath at 90 °C) for 10 h. After cooling to rt, the dark brown reaction mixture was diluted with pentanes/Et_2O (1 : 1) and filtered through a plug of SiO_2. The yellowish filtrate was concentrated in vacuo and the residue was purified by flash chromatography on SiO_2 (pentanes/Et_2O = 100 : 1) to give **75** as a colorless oil (42 mg, 74%).

8.6.2 ROM and Cascade Reactions

The ROM can be combined with other metathesis reactions such as CM, RCM, or with several metathesis reactions. Depending on the order of the different metathesis reactions, infinitely many strategies can be imagined to have an efficient access to complex molecules. In this paragraph, only three examples will be described, one will be dealing with an ROM and a CM, a second one will describe the combination of an ROM and an RCM, and the third one will be related to an ROM/RCM/CM sequence.

8.6.2.1 ROM and CM An efficient cascade reaction, induced by [Ru]-**II** and involving an ROM of cyclopropene ketal **77** followed by a CM with alkene **76**, led to the protected divinylketone **78**. This sequence was used to synthesize routiennocin, an antibiotic extracted from *Streptomyces routienni* Huang sp. nov. (ATCC 39446) (64). Thus, the metathesis sequence began with the RCM of cyclopropenone acetal **77** with alkene **76** in the presence of [Ru]-**II** (3 mol%). After removal of the acetal *in situ*, under acidic conditions ($HClO_4$), which was required for the second metathesis step to be productive, the resulting enone **78** (50%) was treated with alkene **79** again in the presence of [Ru]-**II** (15 mol%) to produce dienone **80** in 59% yield. After hydrogenation and saponification of the methyl ester, routiennocin was isolated (65) (Fig. 8.34).

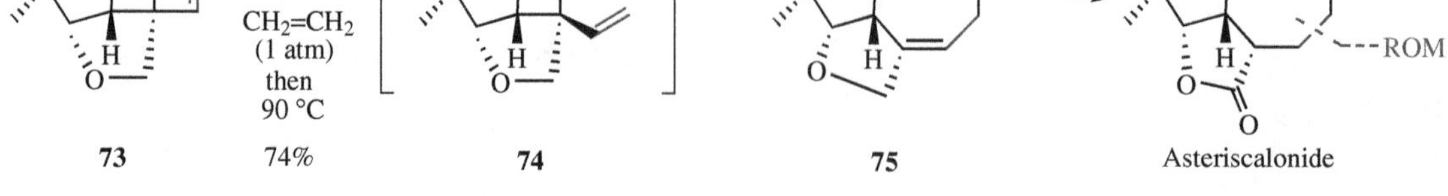

Figure 8.33 Synthesis of asteriscalonide.

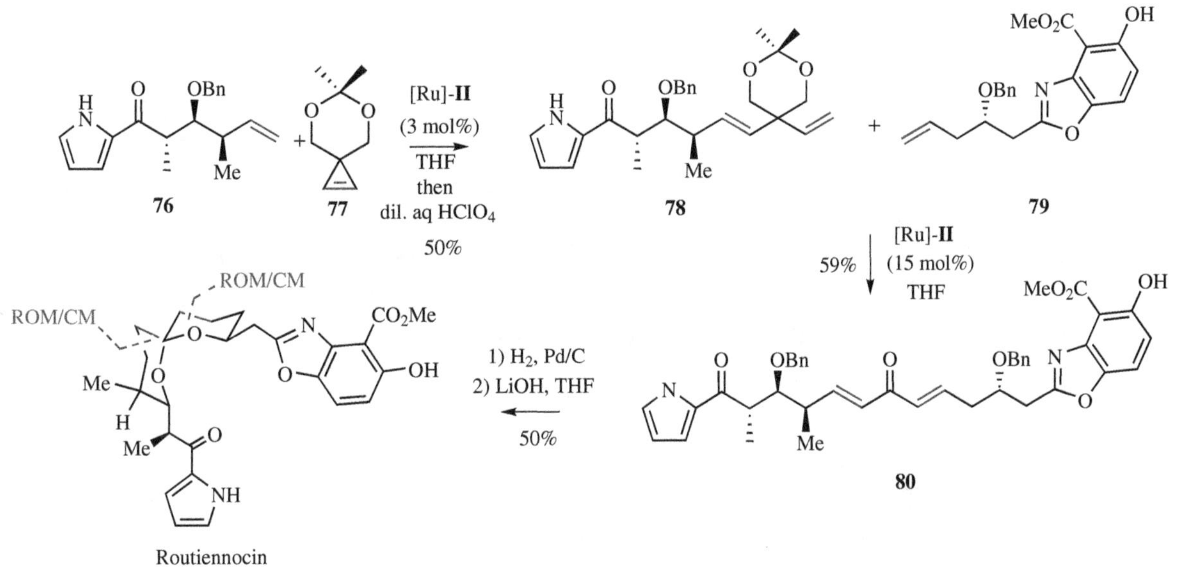

Figure 8.34 Synthesis of routiennocin.

> **Procedure**
>
> Synthesis of **78**: [Ru]-**II** (25 mg, 3 mol%) was dissolved in anhydrous THF (2 ml) and treated dropwise (30 min) with a solution of **76** (90 mg, 0.30 mmol) and cyclopropenone acetal **77** (84 mg, 0. 60 mmol) in THF (3 ml) at rt. After 1.5 h, the reaction mixture was cooled to 0 °C, treated with $HClO_4$ (0.050 ml, 30% solution in H_2O), and stirred for 10 min at 0 °C until TLC indicated complete consumption of the starting material. The reaction mixture was quenched with a saturated $NaHCO_3$ solution and diluted with EtOAc/hexane. The resulting mixture was washed with bromine and dried over $MgSO_4$. The solvent was evaporated and the residue was purified by flash chromatography on SiO_2 (EtOAc/hexane = 1 : 5 to 1 : 3) to afford **78** as a pale brown oil (53 mg, 50%).
>
> Synthesis of **80**: [Ru]-**II** (13 mg, 15 mol%) and **78** (64 mg, 0.18 mmol) were dissolved in anhydrous THF (2 ml) and treated dropwise (30 min) with a solution of **79** (38 mg, 0.10 mmol) in THF (3 ml) at rt. After 6 h, the solvent was removed in vacuo and the residue was purified by flash chromatography on SiO_2 (EtOAc/hexane = 1 : 1) to yield **80** as a brown oil (42 mg, 59%).

8.6.2.2 ***ROM and RCM*** Norbornene derivatives are common substrates for ROM. The metallacarbene formed during the process can be involved, for example, in an RCM. In the synthesis of (±)-*trans*-kumausyne, isolated from a red algae of the genus *Laurencia* in Japan (66), oxanorbornene **81** was treated with [Ru]-**I** (5 mol%) under 1 atm of ethylene (CH_2Cl_2, rt) to furnish **82** resulting from an ROM/RCM cascade. Compound **82** was then transformed to (±)-*trans*-kumausyne (67) (Fig. 8.35).

> **Procedure**
>
> [Ru]-**I** (0.021 g, 2.5 mol%) was added to a stirred solution of **81** (0.15 g, 1 mmol) in CH_2Cl_2 (100 ml). The reaction was stirred for 8 h at rt and a further portion of [Ru]-**I** (0.021 g, 2.5 mol%) was added. After stirring for a further 12 h at rt, the reaction was concentrated and purified by flash chromatography (EtOAc/hexanes = 27 : 75) to yield **82** as an oil (0.124 g, 83%).

8.6.2.3 ***ROM/RCM/CM*** (−)-Lasubine II is a quinolizidine alkaloid, isolated from the leaves of *Lagerstroemia*

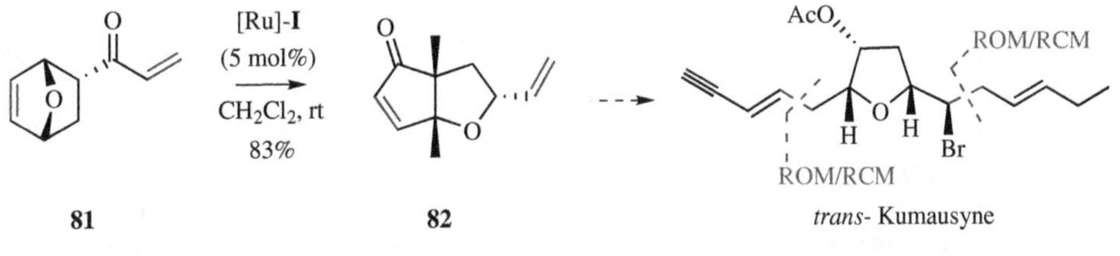

Figure 8.35 Synthesis of (±)-*trans*-kumausyne.

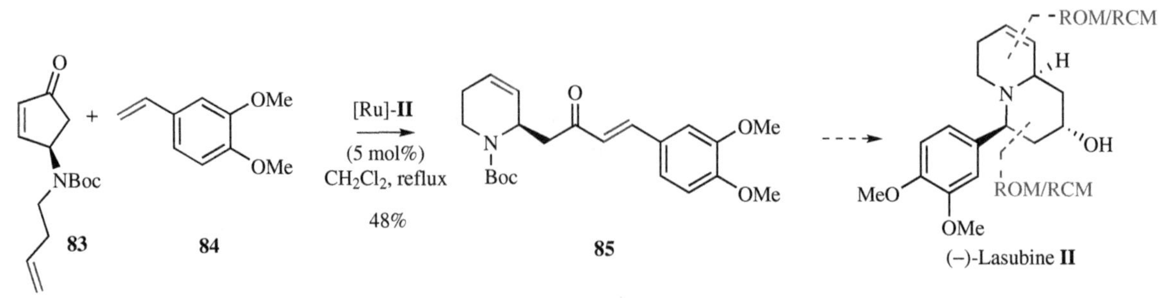

Figure 8.36 Synthesis of (−)-lasubine II.

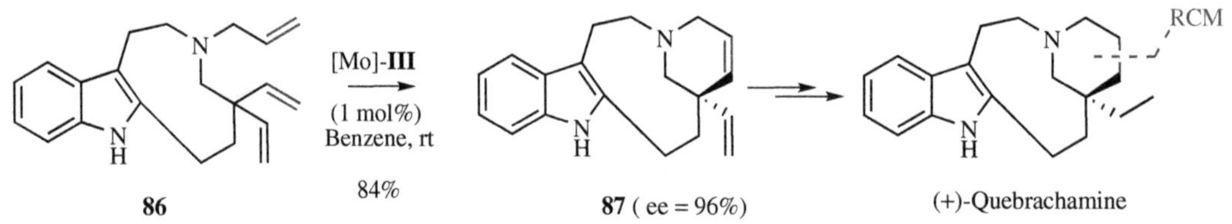

Figure 8.37 Synthesis of (+)-quebrachamine.

subcostata Koehne (68) and an ROM/RCM/CM domino sequence was utilized to achieve its synthesis. The synthesis started with cyclopentenone **83**. When this enone was treated with [Ru]-**I** (5 mol%) in the presence of dimethoxystyrene **84**, only traces of **85** were detected. Fortunately, by using [Ru]-**II** (5 mol%) in refluxing CH_2Cl_2, or [Ru]-**III**, the desired (*E*)-configurated compound **85** was obtained in 48% and 44% yield, respectively (69). The main by-product formed is the result of a CM between **83** and **84**, which demonstrates the main concurrent reaction and the main difficulty of this type of domino reactions (Fig. 8.36).

Procedure

Compounds **83** (1.16 g, 4.62 mmol) and **84** (2.27 g, 13.85 mmol) were refluxed in CH_2Cl_2 (90 ml) under a N_2 atmosphere. [Ru]-**II** (196 mg, 5 mol%) was added and the solution was stirred for further 12 h. The solution was concentrated in vacuo and the residue was purified by flash chromatography (hexane/MTBE = 3 : 1 to 1 : 1) to give **85** as a light yellow oil (859 mg, 48%).

8.7 ASYMMETRIC RCM AND ROM

The desymmetrization of *meso* compounds is an attractive strategy to synthesize optically active molecules. When two enantiotopic sites are available for the metathesis reaction, they can be differentiated in the presence of a chiral metallacarbene to afford optically active compounds.

In the synthesis of (+)-quebrachamine, an *Aspidosperma* alkaloid, the enantioselective desymmetrization of triene **86** was achieved with [Mo]-**III** generated *in situ*, and the RCM product **87** was isolated with a remarkable yield (84%) and enantiomeric excess (ee = 96%). After hydrogenation (PtO_2, H_2 1 atm, EtOH, 22 °C, 1 h), the natural product (+)-quebrachamine was isolated in 97% yield (70) (Fig. 8.37).

Procedure

In a N_2-filled glovebox, a 4 ml vial equipped with a magnetic stirring bar was charged with **86** (10.1 mg, 0.033 µmol) and benzene (50 µl). The solution was treated with the *in situ*-generated chiral complex [Mo]-**III** (16.5 µl, 0.02 M, 1 mol%, final substrate concentration = 0.5 M) and stirred for 1 h at rt. The reaction was then quenched by exposure to air and concentrated in vacuo. After chromatography on SiO_2 (petroleum ether/Et_2O = 8 : 1 washed with 2% v/v concentrated aqueous NH_4OH), **87** was isolated as a colorless oil (7.7 mg, 84%).

The first example of an enantioselective ROM associated with a CM was realized with [Ru]-**V** catalyst to achieve the synthesis of (−)-baconipyrone C, a polyketide natural product isolated from *Siphonaria baconi* collected from itertidal rock platforms near Melbourne in Australia (71). When the oxabicyclic compound **88** was treated at −15 °C for 20 h with [Ru]-**V** (2 mol%) in the presence of styrene **89** (8 equiv)

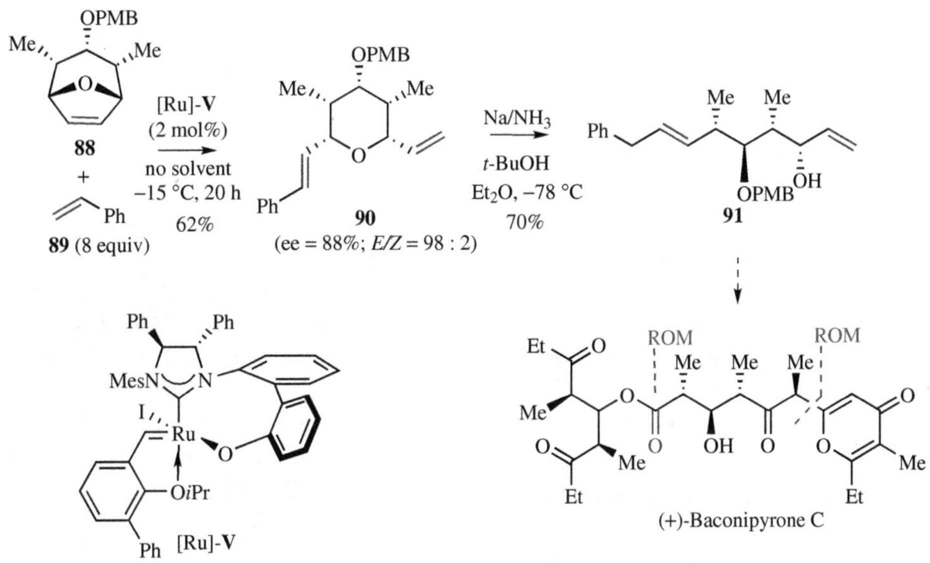

Figure 8.38 Synthesis of (−)-baconipyrone C.

and without any solvent, the optically active tetrahydropyran **90** was formed in 62% yield, with an enantiomeric excess of 88%. Subsequent ring-opening of the tetrahydropyran moiety of **90** under Birch conditions provided alcohol **91**, which was transformed to (−)-baconipyrone C (72) (Fig. 8.38).

Procedure

A solution of catalyst [Ru]-**V** (2 mol%) [prepared *in situ*] in THF (250 μl) was cooled to 0 °C and added through cannula to a solution of **88** (35 mg, 0.13 mmol) in styrene (8 equiv). The THF was evaporated with a nitrogen purge and the mixture was cooled to 15 °C and stirred for 20 h. The resulting mixture was loaded onto a plug of silica gel and eluted with petroleum ether/Et_2O = 19 : 1, to give **90** as a clear colorless oil (30 mg, 62%).

8.8 CONCLUSION

With powerful catalysts and different metathesis reactions in hand, synthetic chemists have considered new disconnections to access natural and or/biologically active compounds. If the reaction conditions need to be improved (solvent, temperature, etc.), the design of new, efficient catalysts is very important. The "perfect" catalyst has to be very efficient, for example, it should have a high turnover; it should be very chemo-, stereo-, and enantioselective, and be commercially available and cheap. In addition, if this catalyst can be transformed *in situ* to another catalyst in order to perform a multistep one-pot synthesis, this catalyst will be of great interest for synthetic chemists to produce complex molecules without any purification. With powerful metathesis catalysts, we can get close to the "ideal synthesis," for example, to achieve a multistep one-pot synthesis of a complex natural and/or biologically active molecule, in quantitative yield. Will the dream become a reality in the near future?

REFERENCES

1. Grubbs RH, editor. *Handbook of Metathesis*. Weinheim: Wiley-VCH; 2003.
2. (a) Some key references on metathesis: Schmalz HG. *Angew Chem Int Ed Engl* 1995;34:1833–1836. (b) Grubbs RH, Miller SJ, Fu GC. *Acc Chem Res* 1995;28:446–452. (c) Schuster M, Blechert S. *Angew Chem Int Ed Engl* 1997;36:2036–2056. (d) Grubbs RH, Chang S. *Tetrahedron* 1998;54:4413–4450. (e) Fürstner A. *Angew Chem Int Ed* 2000;39:3012–3043. (f) Trnka TM, Grubbs RH. *Acc Chem Res* 2001;34:18–29. (g) Grubbs RH. *Tetrahedron* 2004;60:7117–7140. (h) Hoveyda AH, Zhugralin AR. *Nature* 2007;450:243–251.
3. Cossy J, Arseniyadis S, Meyer C. *Metathesis in Natural Product Synthesis*. Weinheim: Wiley-VCH; 2010.
4. Richard J-A, Yen NG S, Chen DY-K. In: Cossy J, Arseniyadis S, editors. *Modern Tools for the Synthesis of Complex Bioactive Molecules*. Weinheim: Wiley-VCH; 2012. p 155–188.
5. Donohoe TJ, Rosa CP. *Org Lett* 2007;9:5509–5511.
6. Liu S, Widom J, Kemp CW, Crews CM, Clardy J. *Science* 1998;282:1324–1327.
7. Han CK, Ahn SK, Choi NS, Hong RK, Moon SK, Chun HS, Lee SJ, Kim JW, Hong CI, Kim D, Yoon JH, No KT. *Bioorg Med Chem Lett* 2000;10:39–43.

8. (a) Boileau JG, van de Weghe P, Eustache J. *Org Lett* 2001;3:2737–2740.(b) Rodeschini V, Boiteau J-G, van de Weghe P, Tarnus C, Eustache J. *J Org Chem* 2004;69:357–373.
9. (a) Picoul W, Urchegui R, Haudrechy A, Langlois Y. *Tetrahedron Lett* 1999;40:4797–4800. (b) Picoul W, Bedel O, Haudrechy A, Langlois Y. *Pure Appl Chem* 2003;75:235–249. (c) Bedel O, Haudrechy A, Langlois Y. *Eur J Org Chem* 2004:3813–3819.
10. Takahashi S, Hishinuma N, Koshino H, Nakata T. *J Org Chem* 2005;70:10162–10165.
11. Fürstner A, Langemann K. *Synthesis* 1997:792–803.
12. Patil AD, Freyer AJ, Killmer L, Offen P, Carte B, Jurewicz AJ, Johnson RK. *Tetrahedron* 1997;53:5047–5060.
13. Mehta G, Likhite NS. *Tetrahedron Lett* 2008;49:7113–7116.
14. Fu GC, Grubbs RH. *J Am Chem Soc* 1992;114:7324–7325.
15. Trost BM, Horne DB, Woltering MJ. *Chem Eur J* 2006;12:6607–6620.
16. Trost BM, Horne DB, Woltering MJ. *Angew Chem Int Ed* 2003;42:5987–5990.
17. Cossy J, Willis C, Bellosta V, BouzBouz S. *Synlett* 2000: 1461–1463.
18. Cossy J, Willis C, Bellosta V, BouzBouz S. *J Org Chem* 2002;67:1982–1992.
19. Kulanthaivel P, Hallock YF, Boros C, Hamilton SM, Janzen WP, Ballas LM, Loomis CR, Jiang JB, Katz B, Steiner JR, Clardy J. *J Am Chem Soc* 1993;115:6452–6453.
20. Ohshima S, Yanagisawa M, Katoh A, Fujii T, Sano T, Matsukuma S, Furummai T, Fujiu M, Watanabe K, Yokose K, Arisawa M, Okuda T. *J Antibiot* 1994;47:639–647.
21. Fürstner A, Thiet OR. *J Org Chem* 2000;65:1738–1742.
22. Clark JS. *Chem Commun* 2006:3571–3581.
23. (a) Tunac JB, Graham BD, Dobson WE. *J Antibiot* 1983; 36:1595–1600.(b) Stampwala SS, Bunge RH, Hurley TR, Wilmer NE, Brankiewick AJ, Steinman CE, Smitka TA, French JC. *J Antibiot* 1983;36:1601–1605. (c) Hakanson GC, French JC. *J Org Chem* 1985;50:462–466.
24. Cossy J, Pradaux F, BouzBouz S. *Org Lett* 2001;3:2233–2236.
25. Gao D, O'Doherty GA. *Org Lett* 2010;12:3752–3755.
26. (a) Satake M, Murata M, Yasumoto T. *J Am Chem Soc* 1993;11:361–362. (b) Morohashi A, Satake M, Yasumoto T. *Tetrahedron Lett* 1998;39:97–100.
27. For biological studies on gambierol see: (a) Fuwa H, Kainuma K, Tachibana K, Tsukano C, Satake M, Sasaki M. *Chem Eur J* 2004;10:4894–4909. (b) Ito E, Suzuki-Toyota F, Tashimori K, Fuwa H, Tachibana K, Sataki M, Sasaki M. *Toxicon* 2003;42:733–740. (c) Fuwa H, Kainuma H, Satake M, Sasaki M. *Bioorg Med Chem Lett* 2003;13:2519–2522.
28. (a) Majumder U, Cox JM, Johnson HWB, Rainier JD. *Chem Eur J* 2006;12:1736–1746. (b) Johnson HWB, Majumder U, Rainier JD. *Chem Eur J* 2006;12:1747–1753. (c) Johnson HWB, Majumder U, Rainier JD. *J Am Chem Soc* 2005;127:848–849.
29. Grubbs RH, Miller SJ, Fu GC. *Acc Chem Res* 1995; 28:446–452.
30. Cai P, McPhail AT, Krainer E, Katz B, Pearce C, Boros C, Caceres B, Smith D, Houck DR. *Tetrahedron Lett* 1999;40:1479–1482.
31. Takao K-I, Watanabe G, Yasui H, Tadano K-I. *Org Lett* 2002;4:2941–2943.
32. Seo Y, Rho J-R, Cho KW, Shin J. *J Nat Prod* 1997;60:171–174.
33. Crimmins MT, Brown BH. *J Am Chem Soc* 2004; 126:10264–10266.
34. Raffauf RF, Huang P-KC, Le Quesne PW, Levery SB, Brennan TF. *J Am Chem Soc* 1975;97:6884–6886.
35. Hladon B, Twardowski T. *Pol J Pharmacol Pharm* 1979; 31:35–43.
36. Li Y, Hale KJ. *Org Lett* 2007;9:1267–1270.
37. Nevalainen M, Koskinen AMP. *Angew Chem Int Ed* 2001;40:4060–4062.
38. (a) Castoldi D, Caggaino L, Bayon P, Costa AM, Capella P, Sharon O, Gennari C. *Tetrahedron* 2005;61:2123–2139. (b) Castoldi D, Caggaino L, Panigada L, Sharon O, Costa AM, Gennari C. *Angew Chem Int Ed* 2005;117:594–597. (c) Castoldi D, Caggaino L, Panigada L, Sharon O, Costa AM, Gennari C. *Chem Eur J* 2006;124:51–62.
39. Fürstner A, Langemann K. *J Org Chem* 1996;61:3942–3943.
40. Fürstner A, Kindler N. *Tetrahedron Lett* 1996;37:7005–7008.
41. Fürstner A, Thiel OR, Lehmann CW. *Organometallics* 2002;21:331–335.
42. (a) Nicolaou KC, He Y, Vourloumis D, Vallberg H, Roschangar F, Sarabia F, Ninkovic S, Yang Z, Trujillo JI. *J Am Chem Soc* 1997;119:7960–7973. (b) Yang Z, He Y, Vourloumis D, Vallberg H, Nicolaou KC. *Angew Chem Int Ed Engl* 1997;36:166–168.
43. Meng D, Bertinato P, Balog A, Su D-S, Kamenecka T, Sorensen EJ, Danishefsky SJ. *J Am Chem Soc* 1997; 119:10073–10092.
44. Nicolaou KC, Sun Y-P, Guduru R, Banerji B, Chen DY-K. *J Am Chem Soc* 2008;130:3633–3644.
45. Pereira AR, Cao Z, Engene N, Soria-Mercado JE, Murray TF, Gerwick WH. *Org Lett* 2010;12:4490–4493.
46. Wadsworth AD, Furkert DP, Sperry J, Brimble M. *Org Lett* 2012;14:5374–5377.
47. (a) Kobayashi J, Cheng T, Ohta H, Nakamura H, Nozoe S, Hirata Y, Ohizumi Y, Sasaki T. *J Org Chem* 1988;53:6147–6150. (b) Kikuchi Y, Ishibashi M, Sasaki T, Kabayashi J. *Tetrahedron Lett* 1991;32:797–798.
48. (a) Fürstner A, Nevado C, Tremblay M, Chevrier C, Teply F, Aissa C, Waser M. *Angew Chem Int Ed* 2006;45:5837–5837. (b) Gagnepain J, Moulin E, Fürstner A. *Chem Eur J* 2011;17:6964–6972.
49. Kakeya H, Zhang HP, Kobinata K, Onose R, Onozawa C, Kudo T, Osada H. *J Antibiot* 1997;50:370–372.
50. Evano G, Schaus JV, Panek JS. *Org Lett* 2004;6:525–528.
51. Hoveyda AH, Schrock RR. *Nature* 2011;479:88–93.
52. (a) Kobayashi J, Watanabe D, Kawasaki N, Tsuda M. *J Org Chem* 1997;62:9236–9239. (b) Kobayashi JI, Tsuda M, Ishibashi M. *Pure Appl Chem* 1999;71:1123–1126.

53. Wang X, Bowman EJ, Bowman BJ, Porco JA. *Angew Chem Int Ed Engl* 2004;43:3601–3605.
54. Hayakawa T, Tomikawa K, Shin-Ya K, Arao N, Nagai K, Susuki K-I. *J Antibiot* 2003;56:899–904. (b) Hayakawa T, Tomikawa K, Shin-Ya K, Arao N, Nagai K, Susuki K-I. *J Antibiot* 2003;56:905–908.
55. (a) Alkofahi A, Rupprecht JK, Liu YM, Chang C-J, Smith DL, McLaughlin JL. *Experentia* 1990;46:539–541. (b) Rupprecht JK, Hui Y-H, McLaughin JL. *J Nat Prod* 1990;53:237–278. (c) Gu Z-M, Fang X-P, Zeng L, Song R, Ng JH, Wood KV, Smith DL, McLaughin JL. *J Org Chem* 1994;59:3472–3479.
56. Yu J-G, Hu XE, Ho DK, Bean MF, Stephens RE, Cassady JM. *J Org Chem* 1994;59:1598–1599.
57. Hoye TR, Eklov BM, Jeon J, Khoroosi M. *Org Lett* 2006;8:3383–3386.
58. Chatterjee AK, Choi T-L, Sanders DP, Grubbs RH. *J Am Chem Soc* 2003;125:11361–11370.
59. (a) Ratle G, Monseur X, Das BC, Yassi J, Khuong-Huu Q, Goutarel R. *Bull Soc Chim Fr* 1966:2945–2947. (b) Khuong-Huu Q, Ratle G, Monseur X, Goutarel R. *Bull Soc Chim Belg* 1972;81:443–458.
60. Cossy J, Willis C, Bellosta V. *Synlett* 2001:1578–1580.
61. Van Zijl AW, Szymanski W, Lopez F, Minnard AJ, Feringa BL. *J Org Chem* 2008;73:6994–7002.
62. (a) Albert BJ, Sivaramakrishnan A, Naka T, Czaicki NL, Koide K. *J Am Chem Soc* 2007;129:2648. (b) Osman S, Albert BJ, Wang Y, Li M, Czaicki NL, Koide K. *Chem Eur J* 2011;17:895. (c) Osman S, Koide K. *Tetrahedron Lett* 2012;53:6637–6640.
63. Limanto J, Snapper M. *J Am Chem Soc* 2000;122:8071–8072.
64. Cullen WD, Cemer WD, Chappel LH, Huang MT, Jefferson MT, Ishiguro M, Maeda H, Nishiyama S, Oscarson JR, Shibakawa R, Tone J. *J Ind Microbiol* 1988;2:349–357.
65. Matsumoto J, Kozmin SA. *Adv Synth Catal* 2008; 350:557–560.
66. Suzuki T, Koizumi K, Suzuki M, Kurosawa E. *Chem Lett* 1983:1643–1644.
67. Chandler CL, Phillips AJ. *Org Lett* 2005;7:3493–3495.
68. Fuji K, Yamada T, Fujita E, Murata H. *Chem Pharm Bull* 1978;26:2515–2521.
69. Zaja M, Blechert S. *Tetrahedron* 2004;60:9629–9634.
70. (a) Malcolmson SJ, Meek SJ, Sattely ES, Schrock RR, Hoveyda AH. *Nature* 2008;456:933–937. (b) Sattely ES, Meek SJ, Malcolmson SJ, Schrock RR, Hoveyda AH. *J Am Chem Soc* 2009;131:943–953.
71. Manker DC, Faulkner DJ, Stout TJ, Clardy J. *J Org Chem* 1989;54:5371–5374.
72. Gillingham DG, Hoveyda AH. *Angew Chem Int Ed* 2007;46:3860–3864.

9

MULTIFOLD RING-CLOSING OLEFIN METATHESES IN SYNTHESES OF ORGANOMETALLIC MOLECULES WITH UNUSUAL CONNECTIVITIES

TOBIAS FIEDLER AND JOHN A. GLADYSZ

Department of Chemistry, Texas A&M University, College Station, TX

9.1 INTRODUCTION

In 2003, one of the authors of this chapter noted that using a transition metal catalyst to carry out a reaction on a transition-metal-containing substrate might at first seem to be a risky proposition (1). However, strategies for directed syntheses of complex organometallic molecules are continually advancing, and nowadays nearly all of the transformations that can be applied to organic compounds see use in organometallic systems in some context.

In this regard, it is now well established that olefin metatheses can be carried out in diverse types of metal coordination spheres (e.g., metals, ligand sets, oxidation states, and charge states). This subject has been more generally reviewed earlier (1). In this chapter, the authors seek to focus on a more challenging subset of processes, namely, those involving *multifold* ring-closing metatheses and the generation of organometallic molecules with unusual connectivities. Transformations that may involve multiple olefin metatheses but do not yield multiple cycles—for example, ring-opening metathesis polymerizations—are excluded. Another exclusion, somewhat regrettable in the authors' opinion, involves metatheses that yield catenanes, knots, or direct precursors thereof. Assemblies with interlocked or threaded rings are derived from topologically more complicated processes than those outlined below, and may receive attention in a follow-up review.

Regardless of the exact scope of this chapter, to the authors' knowledge, the first olefin metatheses in metal coordination spheres were reported by Rudler in 1984 (2,3). These involved olefin-containing tungsten Fischer carbene complexes and the catalyst system $WOCl_4/Ph_2SiH_2$. However, turnover numbers did not exceed four. With regard to multifold ring-closing metatheses in metal coordination spheres, the first example the authors can locate comes from Sauvage and Grubbs (4–6), as depicted in Scheme 9.1 (4). However, this produces a catenane precursor with interlocked rings, **2**. As noted above, these types of cyclizations are not further treated.

The reader is cautioned that this is not intended to be a comprehensive review. Rather, the goal is to illustrate the various types of multifold ring-closing metatheses in a logical order, as opposed to exhaustively cataloging them. Nonetheless, the authors have attempted to make the reference list as complete as possible. All of the reported examples involve ruthenium catalysts, and "Grubbs' catalyst" will be taken to mean his first-generation system. However, Schrock-type catalysts continue to multiply and become increasingly available, and many of these have excellent activities as well as stereoselectivities in ring-closing processes (7).

9.2 STRATEGIC CONSIDERATIONS

A twofold ring-closing olefin metathesis requires four olefinic moieties. In Scheme 9.2, four representative acyclic arrays, designated A–D, are analyzed from the standpoint of possible products that lack interlocked rings. In order

Olefin Metathesis: Theory and Practice, First Edition. Edited by Karol Grela.

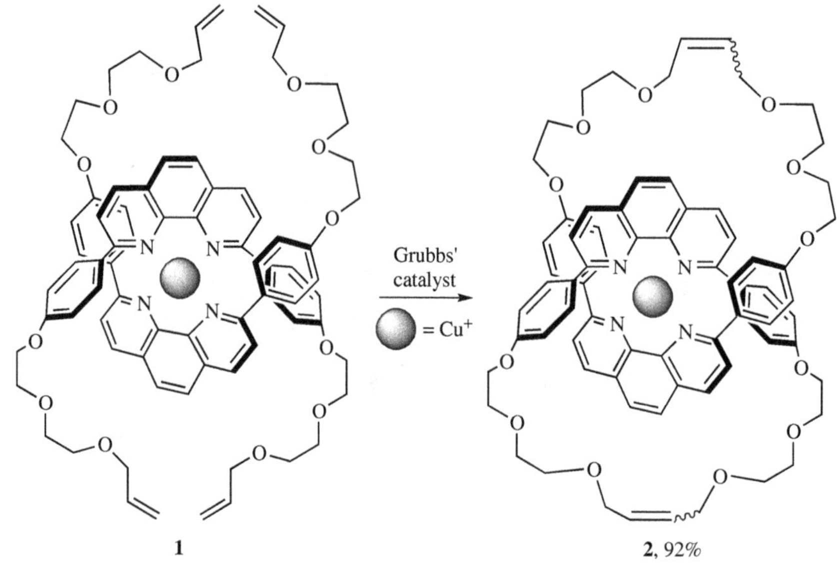

Scheme 9.1

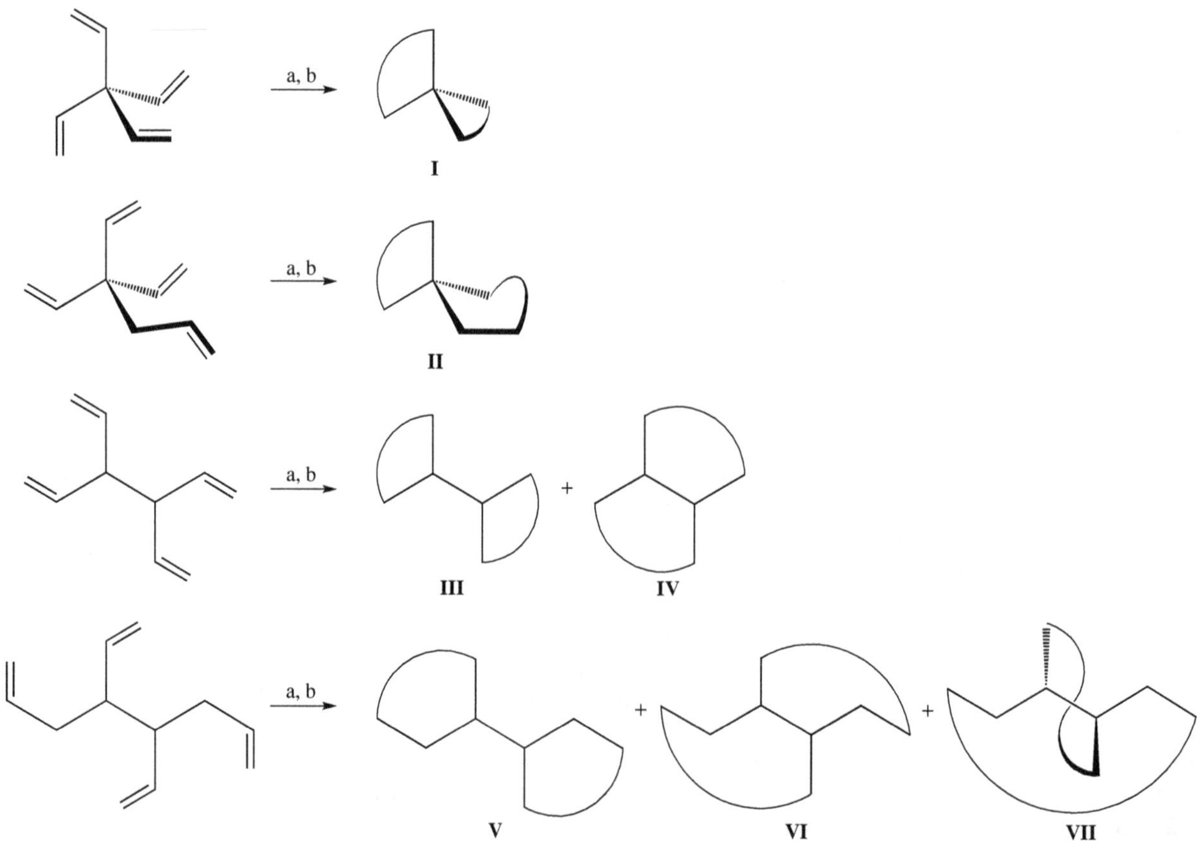

Scheme 9.2 Some types of twofold ring-closing metatheses: (a) C=C metathesis; (b) hydrogenation.

to minimize the number of stereoisomers and facilitate analyses, the C=C linkages have been hydrogenated.

Obviously, a substrate with tetrahedral symmetry such as A can only give rise to one product, **I**. The same holds when the substrate is desymmetrized by altering one of the four olefin-containing moieties, as illustrated by B and **II**. When the tetrahedral center in A is "elongated" to an axis as in C, two distinct cyclization modes are now possible.

The first product (**III**) is derived from metatheses of olefins on the same terminus of the axis or "*intra*ligand" in the case of a central metal. The second product (**IV**) results from metatheses of olefins on opposite termini of the axis or "*inter*ligand" for a suitable substrate. Many complexes described below can be viewed as analogs of C. When the four olefin moieties are arrayed along a chain as in D, three products are possible. These can be viewed as arising from [1,2 + 3,4] (**V**), [1,3 + 2,4] (**VI**), and [1,4 + 2,3] (**VII**) cyclization modes.

Possible cyclization modes become considerably more numerous and complex for substrates with six olefinic moieties. Three representative systems that can give threefold ring-closing metatheses are presented in Scheme 9.3, and obviously many others can be formulated. For the first, an octahedral substrate (E), three products that lack interlocked rings can be defined. These involve metatheses between olefin-containing substituents with cis/cis/cis (**VIII**), cis/cis/trans (**IX**), and trans/trans/trans relationships (**X**), respectively. The last two can also give rise to threaded systems, one of which is illustrated (**XI**).

The next substrate, F, is related to E in the same sense as C and A in Scheme 9.2. Such reactants feature two termini with three olefin-containing groups, and now only two cyclization modes are possible. In the first one depicted (**XII**), all metatheses involve olefinic groups on opposite termini. In the second (**XIII**), one metathesis takes place between groups on opposite termini, and the other two between groups on identical termini.

Substrate G represents an alternative deconstruction of E, resulting in three termini, each with two olefin-containing groups. Now three cyclization modes are possible. In the first one depicted (**XIV**), metatheses are restricted to olefinic groups on the same termini. In the second, all metatheses involve different termini (**XV**). In the third, one metathesis takes place between groups on the same terminus and the other two involve groups on different termini (**XVI**).

As will be seen below, organometallic molecules with even greater numbers of olefins have been subjected to ring-closing metatheses. In addition, the olefins can be arrayed about cyclic core units, as opposed to the acyclic substrates in Schemes 9.2 and 9.3. However, rather than abstractly diagramming even more possibilities, these will be analyzed as encountered.

9.3 EARLY SYSTEMATIC STUDIES WITH PHOSPHINE LIGANDS

Efforts to systematically probe some of the selectivity issues outlined in Schemes 9.2 and 9.3 were undertaken in the authors' group (8,9). One of these involved a series of *trans*-bis(phosphine) platinum complexes **3d,e,f,h** (Scheme 9.4, top; indices **a–h** are coded to the number of methylene groups throughout) (8). The phosphine ligands, $PhP((CH_2)_n(CH{=}CH_2))_2$ ($n = 4, 5, 6, 8$), featured two alkyl substituents terminating in vinyl groups. This allowed for either intraligand or interligand metathesis, akin to the generalized substrate C in Scheme 9.2. As shown in Scheme 9.4, only the latter pathway was observed, giving **4d,e,f,h** in 5–55% yields after workups. An authentic sample of the intraligand metathesis product was independently synthesized in the case of **3f** ($n = 6$). However, it could not be detected in the crude reaction mixture.

When the methylene chains were too short to allow for the formation of *trans*-spanning diphosphine ligands, intraligand metatheses were observed, as shown with **3b** and **4′b** in Scheme 9.4 (middle). Analogous bis(phosphine) complexes in which each phosphorus atom featured three olefin-containing substituents, allowing in principle for threefold intramolecular ring-closing metathesis, were also investigated. These are akin to the generalized substrate F in Scheme 9.3. However, as shown for **5** in Scheme 9.4 (bottom) only oligomeric and polymeric products were detected. As detailed below, when the pentafluorophenyl ligand is replaced by smaller ligands, monometallic products can be isolated.

Another substrate capable of threefold ring-closing metathesis, an octahedral tungsten complex with three facially disposed monophosphine ligands (**6**) (9), is shown in Scheme 9.5. This represents a model for G in Scheme 9.3. Reaction with Grubbs' catalyst gave a complex and almost intractable mixture of all three types of cyclization products, **7–9**, corresponding to **XIV–XVI** (Scheme 9.3). These formed as various C=C and/or P–Ph stereoisomers, as assayed by HPLC and mass spectrometry before hydrogenation. Crystal structures of two isomers of the threefold interligand metathesis product **8** were obtained.

It should be noted in passing that threefold ring-closing metatheses of some organic hexaolefins of the type G give systems analogous to **XV** with high selectivities (no type **XIV** or **XVI** products detected) (10). Highly selective twofold ring-closing metatheses of organic tetraolefins of the type C have also been described (11).

9.4 APPLICATIONS OF TWOFOLD RING-CLOSING OLEFIN METATHESES USING PHOSPHINE LIGANDS

The authors' group has had a particular interest in complexes in which sp carbon or polyynediyl chains bridge two transition metals. For several reasons, it was sought to "insulate" the $-(C{\equiv}C)_n-$ linkages with two flexible sp^3 carbon chains. The diplatinum octatetraynediyl complexes **10f–i** in Scheme 9.6 feature four phosphine ligands with a single olefinic group, two with *trans* relationships on each terminus, and are closely related to the system C in Scheme 9.2. Interestingly,

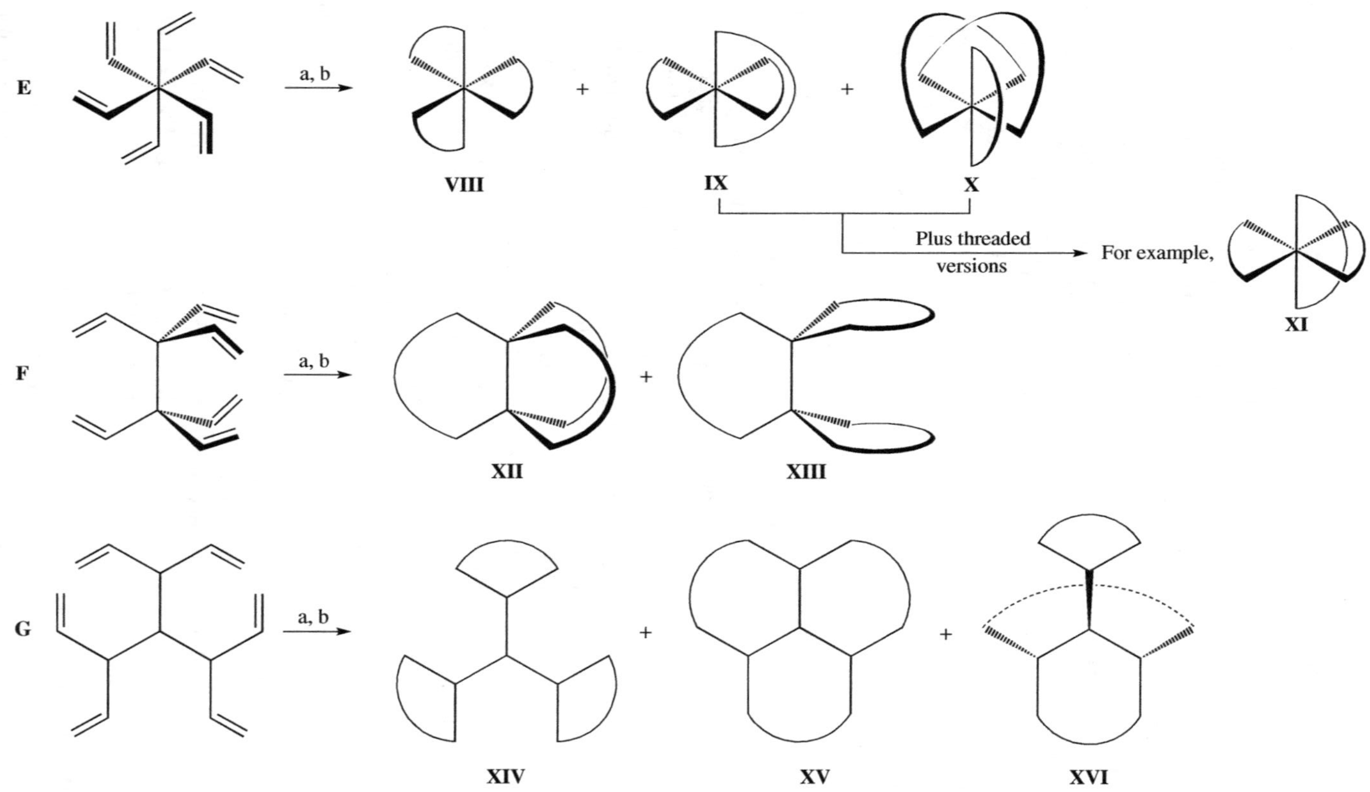

Scheme 9.3 Some types of threefold ring-closing metatheses: (a) C=C metathesis; (b) hydrogenation.

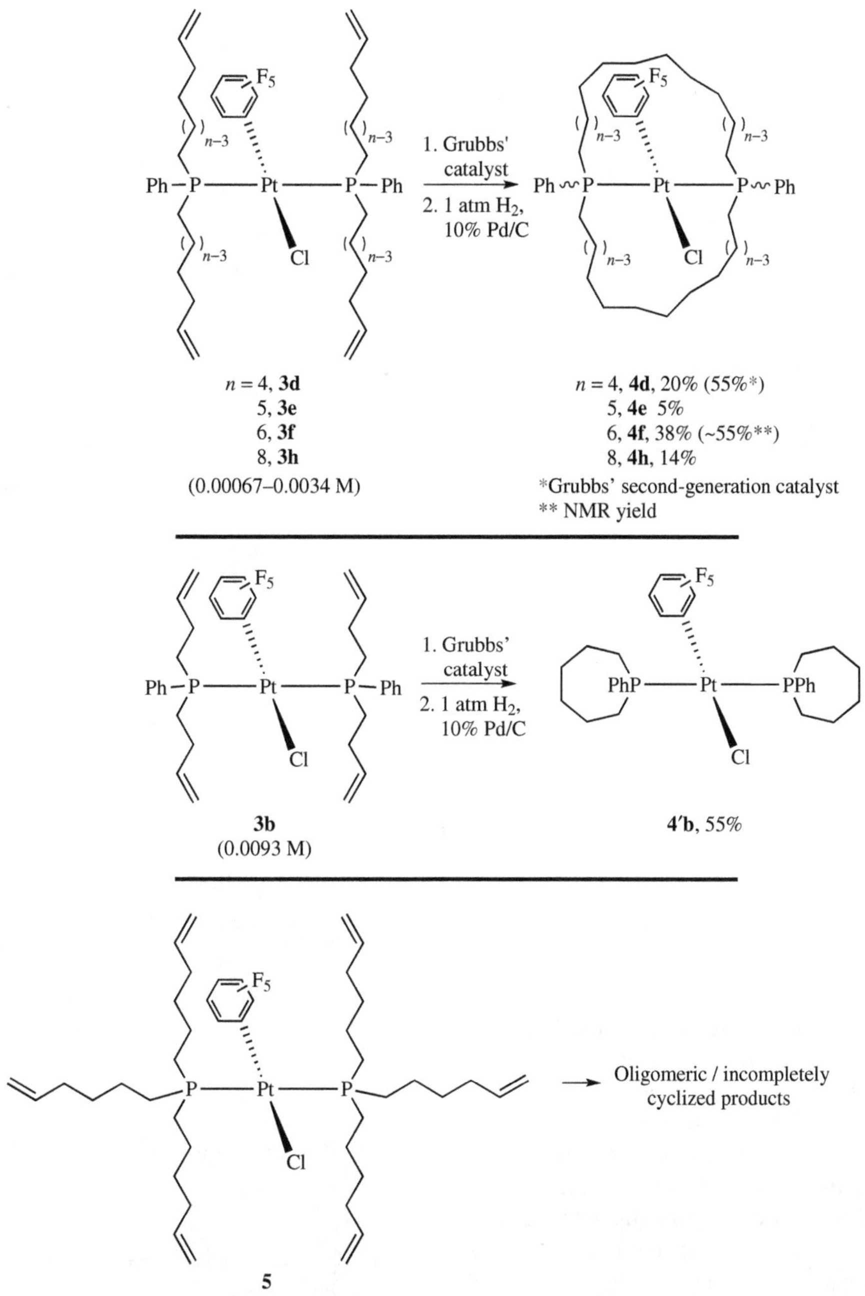

Scheme 9.4

twofold metathesis/hydrogenation sequences yield mainly the target complexes **11f–i**, derived from metatheses between phosphine ligands on opposite platinum termini (see **IV**, Scheme 9.2) (12). In three cases, lesser amounts of the alternative ring-closing products **12**, derived from metatheses between phosphine ligands on identical termini, also formed (see **III**).

Comparable reactions could be realized with diplatinum complexes with longer sp carbon chains. However, as shown in Scheme 9.7, substrate **13**, in which geminal dimethyl groups have been introduced on the sp^3 carbon segments, exhibited an opposite cyclization selectivity (13). Complex **15**, derived from metatheses between phosphine ligands on identical termini, greatly dominated. Possible contributing factors have been analyzed, but remain speculative. As shown in Scheme 9.8, substrate **16**, in which ether oxygen atoms have been inserted into the sp^3 carbon segments, cyclizes with the "normal" selectivity (13).

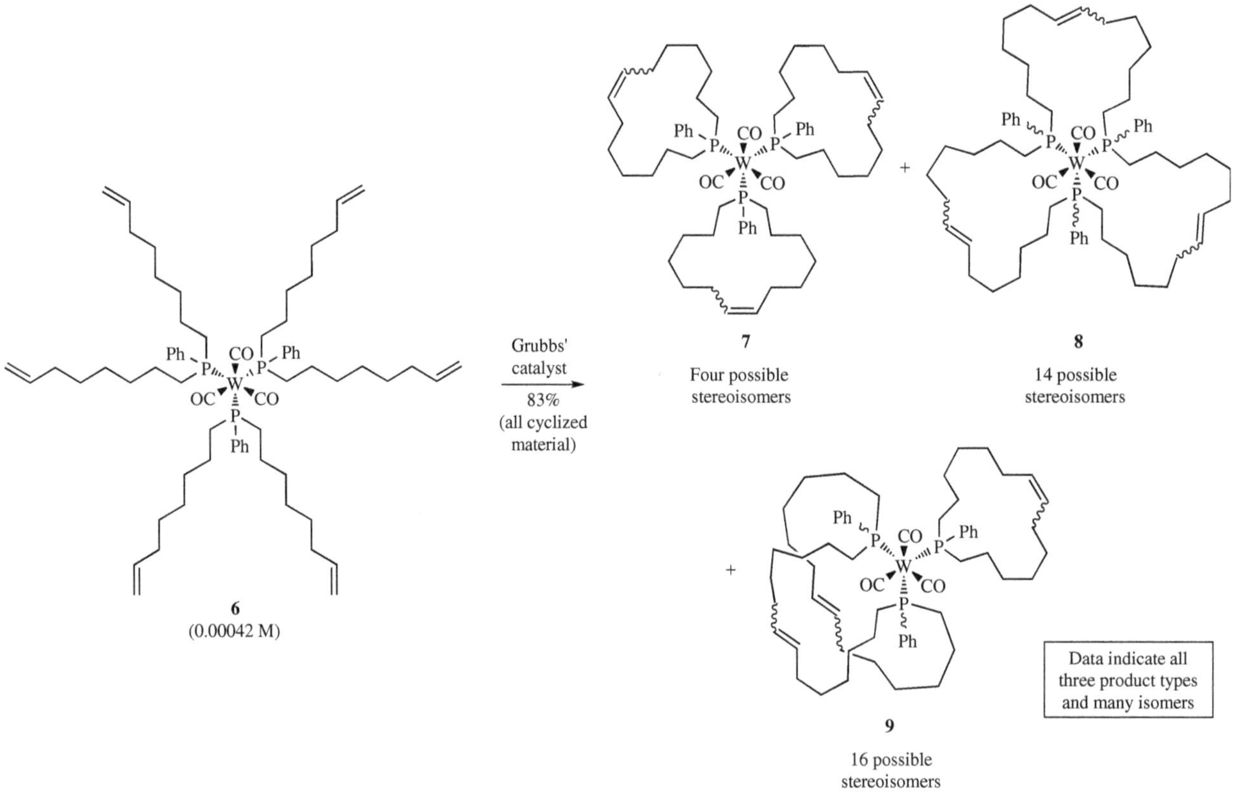

Scheme 9.5

9.5 STUDIES WITH *TRANS* BIS(PYRIDINE) COMPLEXES

An early example of a twofold ring-closing metathesis in a metal coordination sphere is shown in Scheme 9.9 (top) (14). This features a *trans*-bis(pyridine) palladium complex **18** in which the 2,6 positions of the pyridine ligands are substituted with methylene chains that terminate in vinyl groups. As originally reported by Lambert (14,15) and later reinvestigated in the authors' laboratory, only interligand metathesis to give **19** is observed. One trans-spanning linkage extends above the coordination plane, and the other below. Although steric constraints would seem to strongly disfavor intraligand metathesis, this cyclization mode has been observed with similarly substituted monopyridine complexes (16,17). Analogous interligand metathesis products are obtained with the ethereal analogs **20a,b** in Scheme 9.9 (bottom), albeit in lower yields (15).

Monopyridine complexes have also been synthesized in which the 3,5-positions are substituted with methylene chains that terminate in vinyl groups. Here, ready intraligand metatheses have been observed. Accordingly, close relatives of the complexes in Scheme 9.9 with two 3,5-disubstituted pyridines should be capable of either intraligand or interligand metatheses (see **III** and **IV**, Scheme 9.2). As shown by **22d,e,f,h** and **24** in Scheme 9.10, only interligand metatheses to give **23d,e,f,h** and **25** are observed (15).

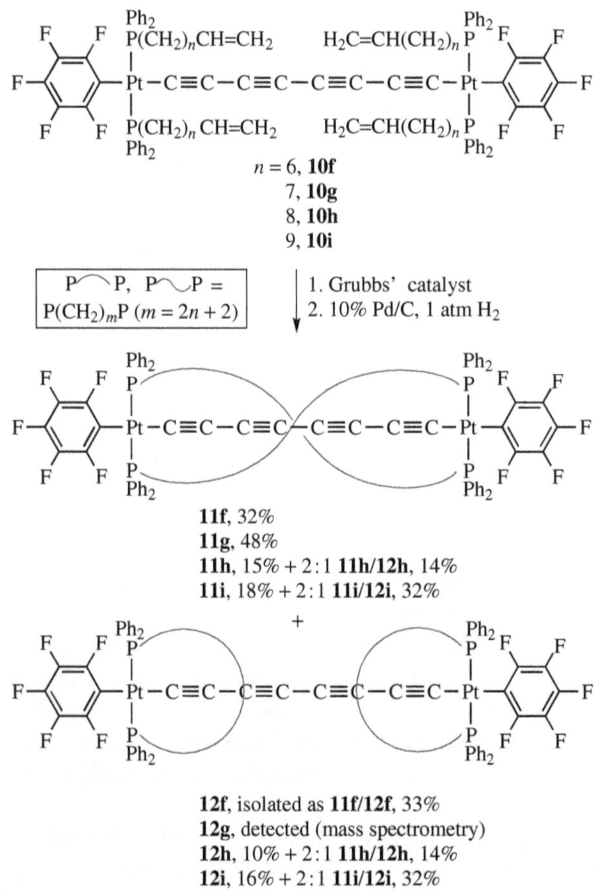

Scheme 9.6

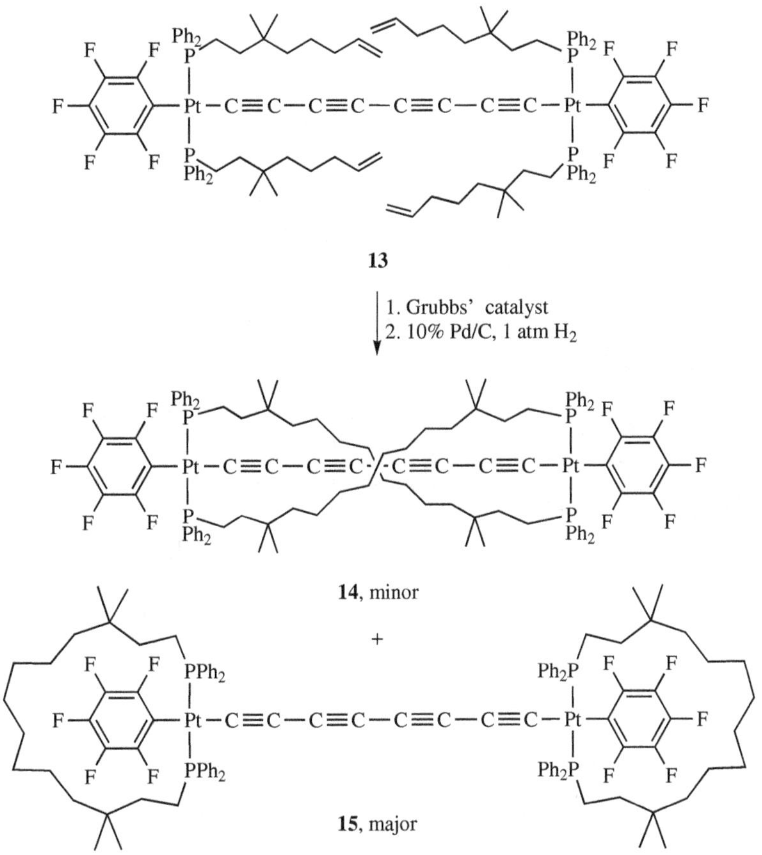

13

1. Grubbs' catalyst
2. 10% Pd/C, 1 atm H_2

14, minor

+

15, major

Total yield **14**, **15**, and oligomers: 79%
15 after chromatography: 14–25%

Scheme 9.7

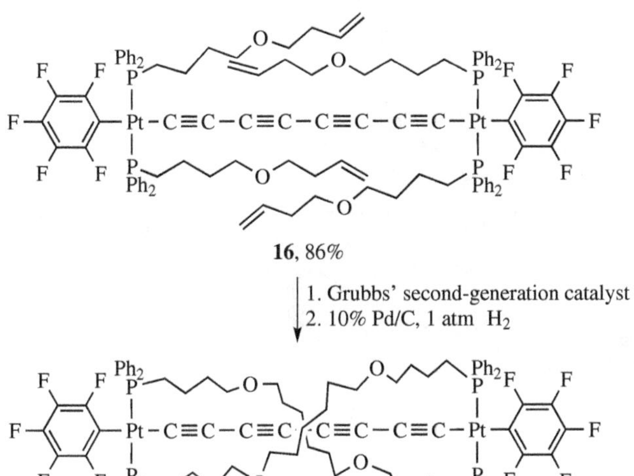

16, 86%

1. Grubbs' second-generation catalyst
2. 10% Pd/C, 1 atm H_2

17, 26%

Scheme 9.8

9.6 STUDIES WITH ARENE AND CYCLOPENTADIENYL LIGANDS

Various polyolefinated arene and cyclopentadienyl complexes have been investigated by Astruc (18) and provide a nice bridge from the twofold ring-closing metatheses that have dominated the previous sections to higher order metatheses. First, as shown in Scheme 9.11 (top), the metathesis of the tetraolefinic iron–arene complex **26** was examined. This substrate features two $H_2C{=}CHCH_2{-}C{-}CH_2CH{=}CH_2$ or "diallylmethane" segments, and gave only the twofold 1,1-ring-closing metathesis product **27** (76%), akin to cyclization mode **III** for C (Scheme 9.2). However, the alternative metathesis product involving allyl groups with 1,3-relationships, which could be viewed as a paddlane, would be highly strained.

Next, Astruc investigated the octaolefinic arene complex **28** shown in Scheme 9.11 (middle), which features four $CH(CH_2CH{=}CH_2)_2$ substituents. This gave the fourfold ring-closing metathesis product **29** in 70% yield, again akin to cyclization mode **III** for C (Scheme 9.2). However, in this case, cyclization between adjacent diallylmethane segments (see **IV**) would not be prohibitively strained. Another series of substrates were synthesized in which the arene ligands featured one to three "triallylmethane" or $C(CH_2CH{=}CH_2)_3$ substituents (**30**; Scheme 9.11, bottom). However, in all cases, initial cyclopentene ring formation was followed by intermolecular metatheses to give diiron and triiron products.

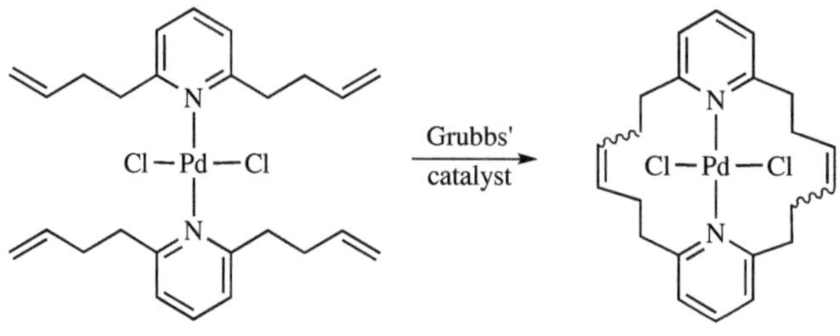

18 **19**, 80%

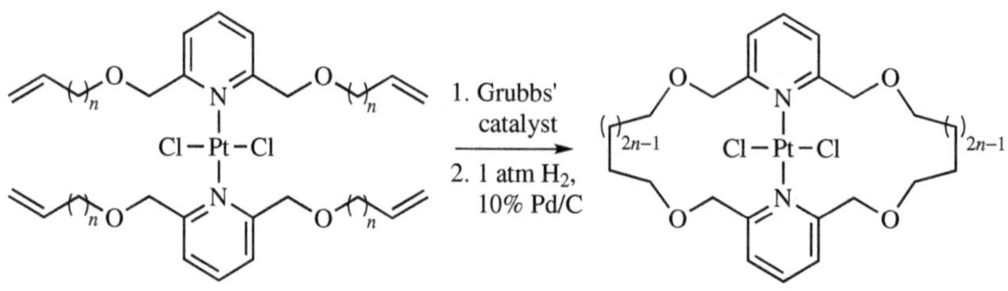

n = 1, **20a**
2, **20b**

n = 1, **21a**, 10%
2, **21b**, 22%

Scheme 9.9

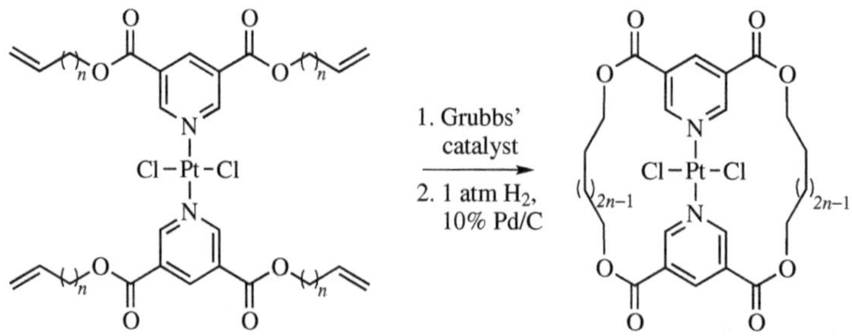

n = 4, **22d**
5, **22e**
6, **22f**
8, **22h**

n = 4, **23d**, 20%
5, **23e**, 45%
6, **23f**, 18%
8, **23h**, 14%

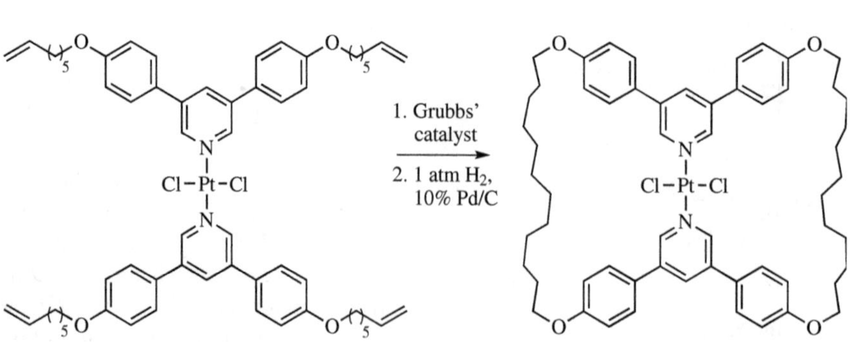

24 **25**, 40%

Scheme 9.10

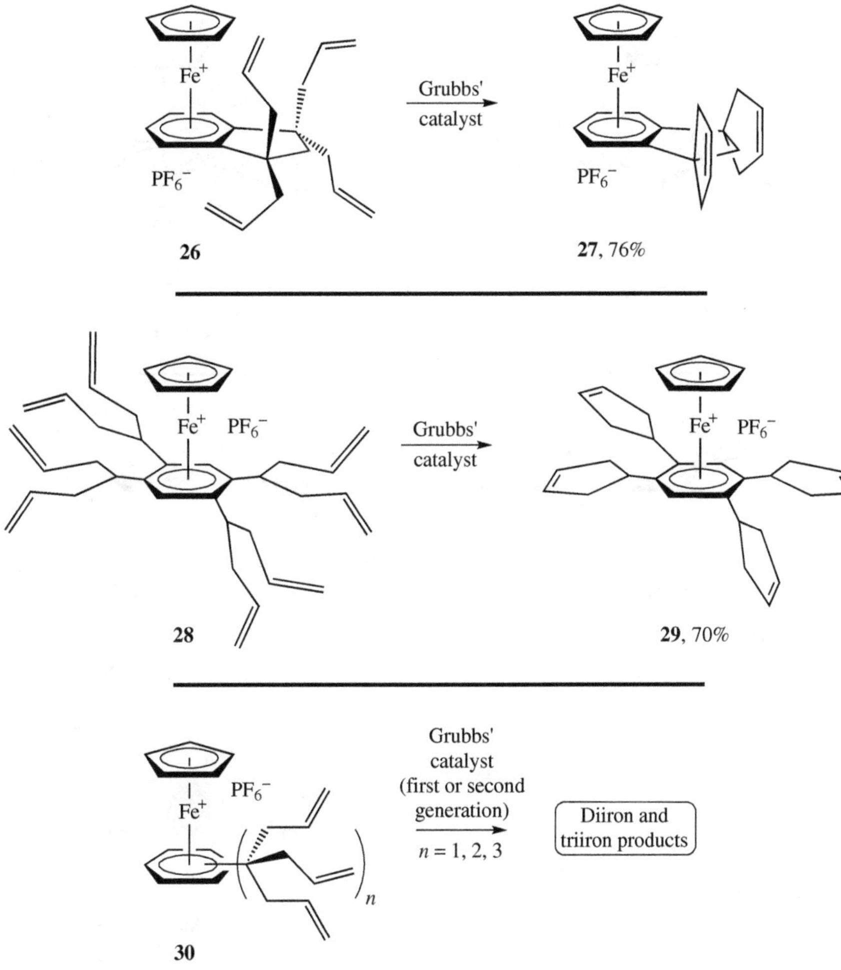

Scheme 9.11

Attention was then turned to the cobaltocene **31** in Scheme 9.12 (top), in which one cyclopentadienyl ligand is completely substituted with five $CH(CH_2CH{=}CH_2)_2$ substituents. This decaolefin underwent a fivefold ring-closing metathesis to give the pentakis(cyclopentene) **32** in 81% yield. As with **28** in Scheme 9.11, cyclization took place within the substituents, as opposed to between them.

Some relevant work from the Buchowicz (19) laboratory is depicted in Scheme 9.12 (bottom). The nickel and iron metallocenes **33a,b** were prepared, in which each cyclopentadienyl ligand features two 1,3-disposed alkyl substituents that terminate in vinyl groups. With the nickelocene, Grubbs' second-generation catalyst afforded a single stereoisomer of a *diansa* product in 50% yield (**34a**; *E*/*E*). With the ferrocene, a mixture of *diansa* stereoisomers was isolated (**34b**), two of which could be crystallographically characterized (*E*/*E* and *E*/*Z*). These results are akin to cyclization mode **IV** of C (Scheme 9.2). Intraligand metathesis products analogous to **III** should be disfavored, as the resulting nine membered rings would include the equivalent of a *trans* C=C linkage, since they span the 1,3 positions of a cyclopentadienyl ligand.

9.7 STUDIES WITH POLYPYRIDINE AND POLY(TERPYRIDINE) COMPLEXES

Another series of higher order ring-closing metatheses involve pyridine adducts of platinum aryl complexes that are arrayed about a benzenoid core, such as the tricationic triplatinum pincer species **35** in Scheme 9.13 (16,20). Van Koten found that this hexaolefin underwent a threefold metathesis to give **36** as a mixture of *E*/*Z* isomers in 67% yield after workup. This corresponds to the cyclization mode **XV** for G in Scheme 9.3. The resulting tripyridine ligand, a 69-membered macrocycle, could be liberated from the triplatinum template upon addition of NaCl.

Kang and Ko (21) later extended this protocol to the hexametallic dodecaolefin **37** shown in Scheme 9.14. This

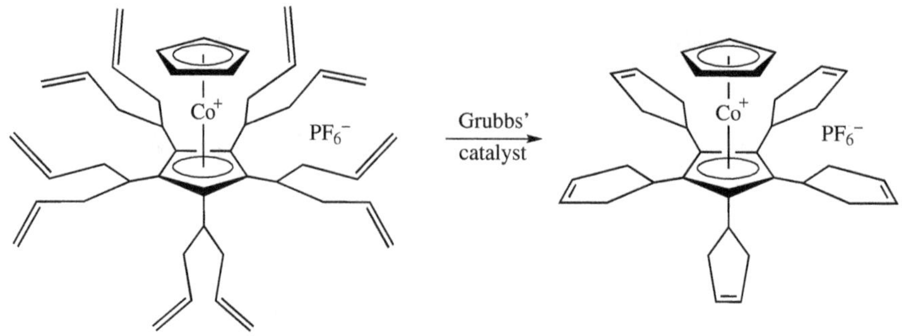

31 **32**, 81%

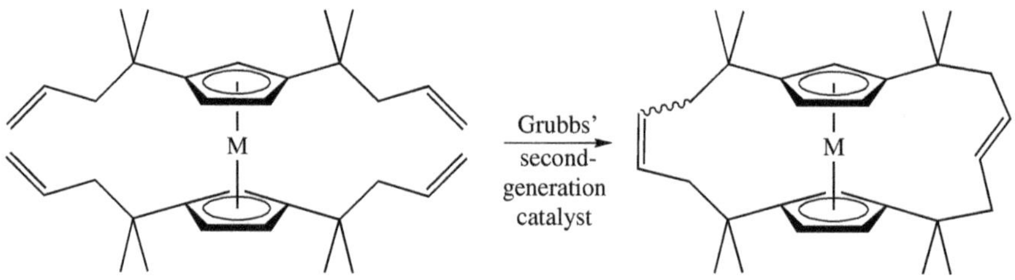

M = Ni, **33a**
Fe, **33b**

M = Ni, **34a**, 50% (*E*/*E*)
Fe, **34b** (crystal structures)

Scheme 9.12

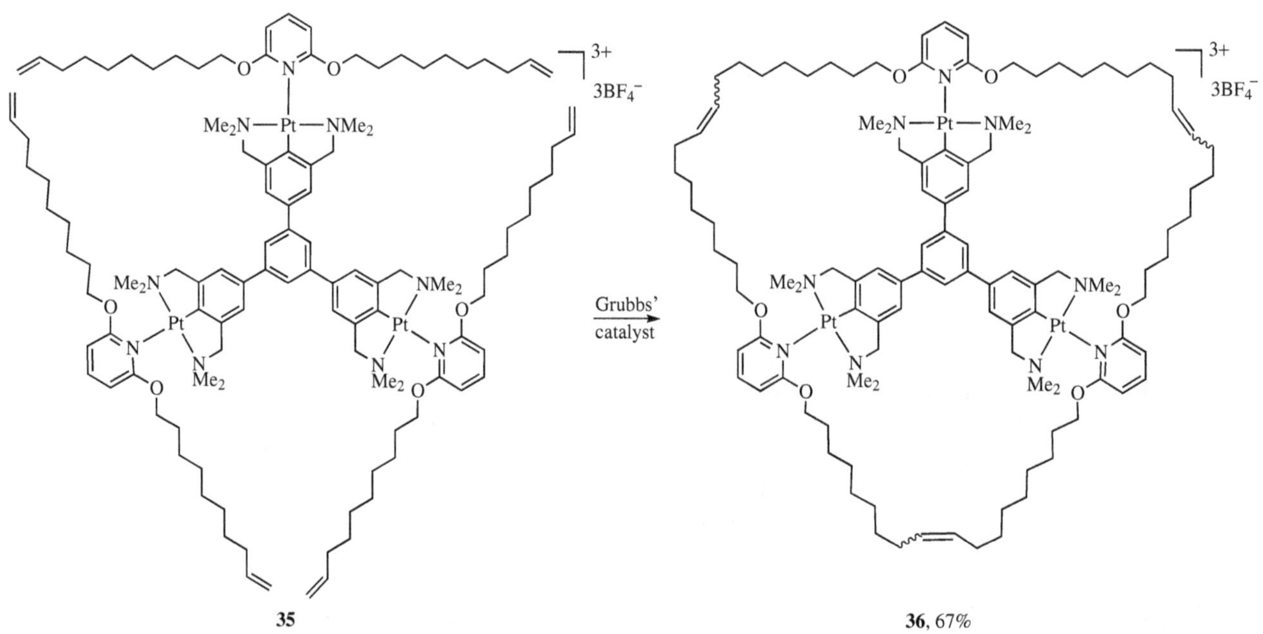

35 **36**, 67%

Scheme 9.13

underwent a related sixfold ring-closing metathesis to give **38** in 80% yield. To the authors' knowledge, this and the following example represent the highest order intramolecular ring-closing metatheses realized to date in metal coordination spheres, excepting three-dimensional capsule compounds (Section 9.9). The resulting hexapyridine ligand, a 90-membered macrocycle, could be liberated from the hexaplatinum template upon addition of NaI.

The largest macrocycle accessed by the preceding strategy is depicted in Scheme 9.15 (22). Newkome first showed that the hexairon dodeca(terpyridine) system **39** could easily be synthesized by self-assembly. Sixfold ring-closing

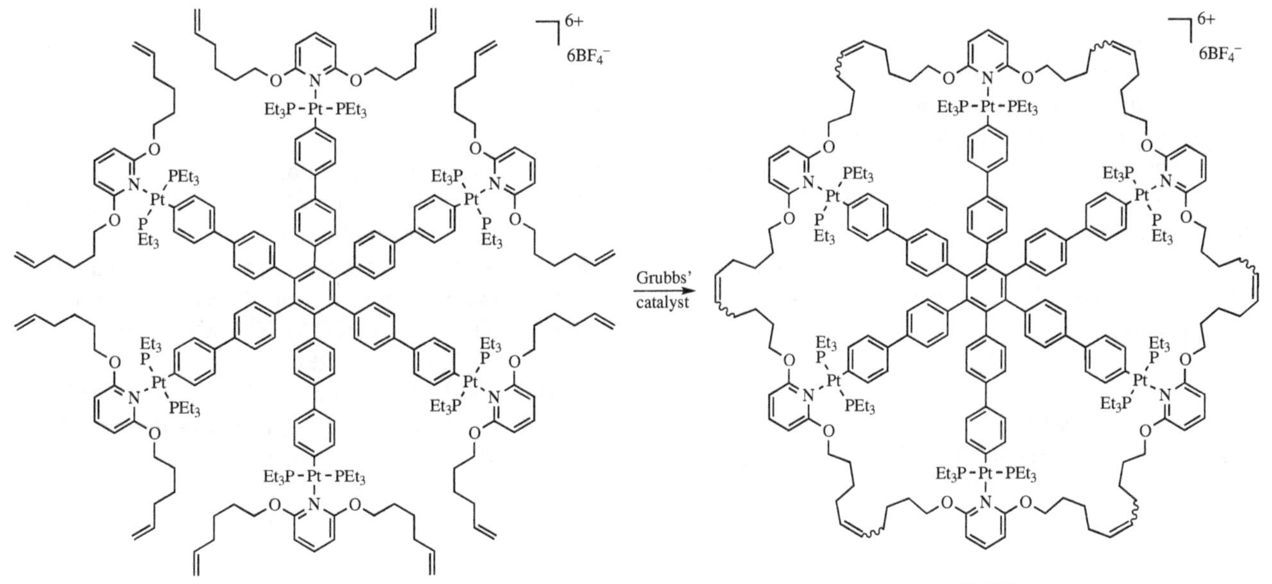

37

38, 80%

Scheme 9.14

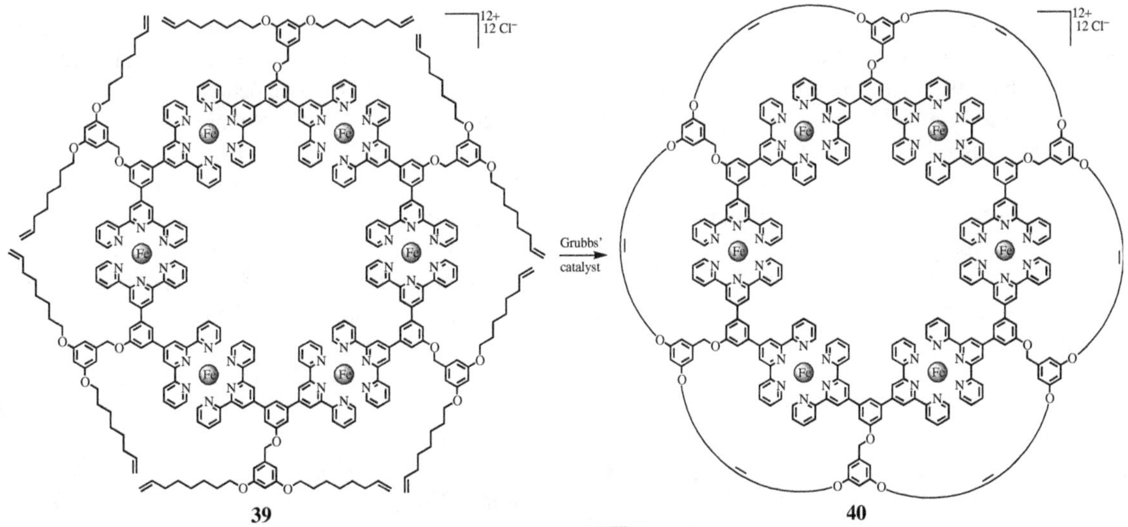

39

40

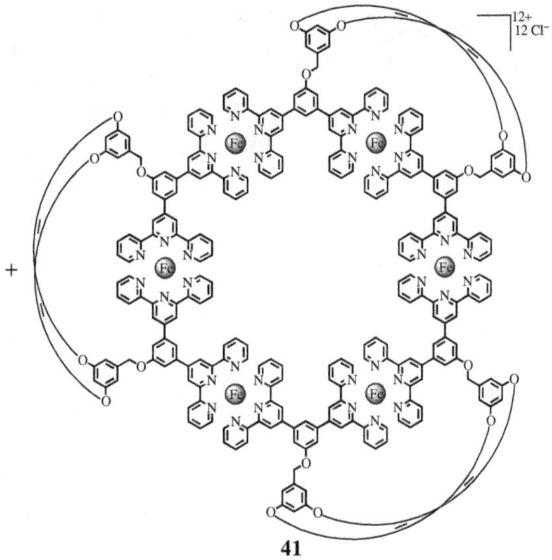

41

Scheme 9.15

metathesis then gave two products: (i) **40**, derived from reactions of the olefins on a given resorcinol moiety with those on two different resorcinol moieties, analogous to the products in Schemes 9.13 and 9.14, and (ii) **41**, derived from reactions of both olefins on a given resorcinol moiety with those on a single other resorcinol moiety. Here, both products represent interligand metatheses. The structures were established by decomplexing the ligands (K_2CO_3/DMF; good overall yields) and, after hydrogenations, characterization of the resulting macrocycles. That derived from **40** features a 114-membered ring.

9.8 GYROSCOPE-LIKE COMPLEXES DERIVED FROM *TRANS*-BIS(PHOSPHINE) COMPLEXES

An extensive series of metal complexes featuring *trans*-phosphine ligands of the formula $P((CH_2)_n(CH{=}CH_2))_3$ have been investigated in the authors' laboratories. As illustrated by F in Scheme 9.3, two modes of threefold ring-closing metatheses are possible (**XII, XIII**). Of these, **XII** results in a class of cagelike complexes that the authors have termed *molecular gyroscopes*. Interestingly, this cyclization mode dominates in all of the coordination geometries examined to date (trigonal bipyramidal, square planar, octahedral), as detailed below. When such metatheses are attempted in the absence of a templating metal, for example, using the phosphine borane $H_3B{\cdot}P((CH_2)_6(CH{=}CH_2))_3$, the yields of cagelike diphosphorus products are miniscule (23).

As shown in Scheme 9.16, the trigonal bipyramidal iron tricarbonyl complexes *trans*-$Fe(CO)_3(P((CH_2)_nCH{=}CH_2)_3)_2$ ($n=4$, 5, 6, 8; **42d,e,f,h**) were treated with

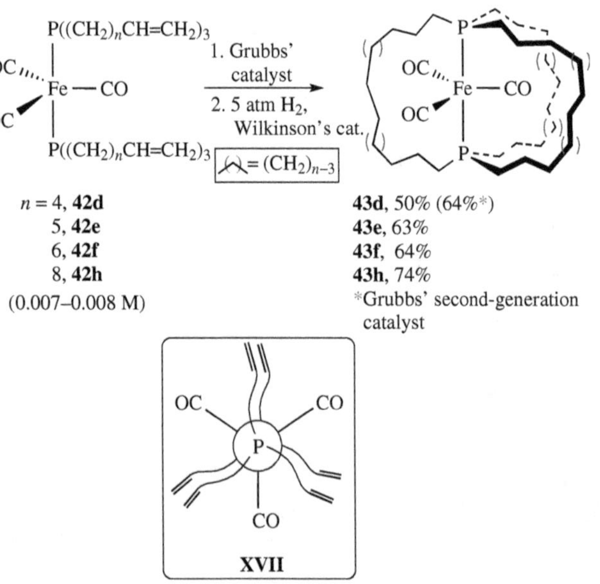

Scheme 9.16

Grubbs' catalyst and then hydrogenated. Workups gave the target gyroscope-like complexes **43d,e,f,h** in 50–74% overall yields (24, Wang, L., unpublished results, Universität Erlangen-Nürnberg.). No products derived from alternative cyclization modes were detected. The smallest macrocycle features thirteen membered rings, and was obtained in somewhat higher yield with Grubbs' second-generation catalyst. This trend has also been found with related complexes with identical ring sizes, but is not observed with the larger cycles.

Parallel sequences have been effected with analogous iron bis(phosphite) complexes (25). However, the products could only be obtained as oils and yields were much lower.

Why do **42d,e,f,h** exhibit such high selectivities for threefold interligand metathesis? Since both the iron fragment $Fe(CO)_3$ and the phosphines possess threefold symmetry axes, the CO ligands and $(CH_2)_nCH{=}CH_2$ substituents can adopt staggered arrangements, as shown in **XVII** in Scheme 9.16 (bottom). As a result, the $(CH_2)_nCH{=}CH_2$ substituents on the *trans*-phosphine ligands become eclipsed, and are directed into the same OC–Fe–CO interstice. Consequently, **42d,e,f,h** can be viewed as preorganized for threefold interligand metatheses.

Analogous metathesis/hydrogenation sequences involving square planar $PdCl_2$, $PtBr_2$, and $PtCl_2$ adducts have been investigated (23,26). The last are representative and are depicted in Scheme 9.17. Square planar complexes are not preorganized for interligand metatheses as proposed for **42** in Scheme 9.16, and no monometallic complexes were detected for substrates with $n=4$ or 5. However, with $n=6$ (**44f**), threefold interligand metathesis dominated to give the gyroscope-like complex **45f**. On smaller scales, **45f** could be reproducibly isolated in 43% yield as the exclusive product. However, on scales larger than 1 g, the yield of **45f** appeared to drop and some of the alternative metathesis product **46f** was isolated. The latter is analogous to product **XIII** from F in Scheme 9.3.

When one more methylene group was added to the phosphine substituents ($n=7$, **44g**), yields of monoplatinum complexes decreased dramatically. Furthermore, both modes of threefold ring-closing metathesis were observed (**XII, XIII**). When another methylene group was added ($n=8$, **44h**), yields of monoplatinum complexes greatly increased. Threefold interligand metathesis to give **45h** dominated (28% after workup), but the yield of the alternative metathesis product **46h** was also significant (23%). Some related organic hexaolefins in which the P–M–P axis is replaced by a phenylene or *p*-CC≡C–C_6H_4–C≡CC moiety also undergo threefold intramolecular ring-closing metatheses (27). So far, only the second cyclization mode (**XIII**) has been observed.

An obvious question is whether comparable olefin metatheses can be effected when such phosphine ligands bear cis relationships. As depicted in Scheme 9.18, the *cis*-dichloroplatinum complexes **47e,f** undergo threefold

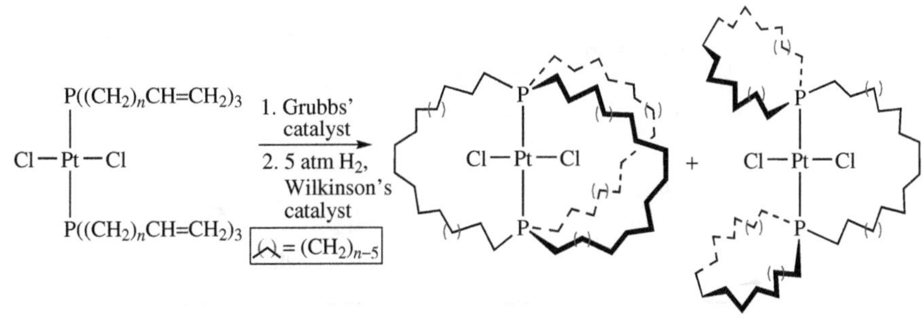

$n = 6$, **44f**
7, **44g**
8, **44h**

45f, 43% (32%)[a]
45g, 4%[b]
45h, 23%

46f, — (5%)[a]
46g, 4%[b]
46h, 23%

[a] Larger scale reaction (see text).
[b] Yields from different experiments.

Scheme 9.17

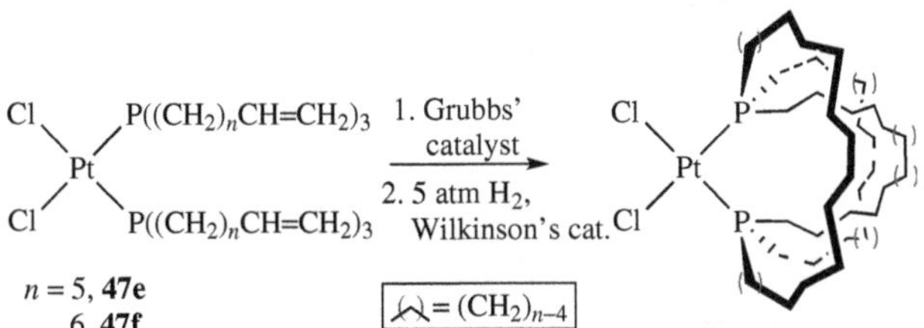

$n = 5$, **47e**
6, **47f**

48e, 6%
48f, 40%

Scheme 9.18

interligand metatheses to give the "parachute" systems **48e,f** (28). No products derived from intraligand metathesis were detected. Although **45f** (Scheme 9.17) and **48f** are isomeric, they exhibit no tendency to interconvert at temperatures of greater than 150 °C.

As shown in Scheme 9.19, square planar rhodium complexes with *trans*-phosphine ligands have also been investigated (29,30). In contrast to the palladium and platinum complexes, these feature two different ligands on the "rotator," chlorine and carbon monoxide. The analog of the platinum complex **44f**, **49f**, underwent a similar threefold interligand metathesis to give **50f** in 40% yield after workup (29). Interestingly, analogous cyclizations could be effected with the triarylphosphine complexes **51e,f** in which the para positions have been substituted with $O(CH_2)_nCH{=}CH_2$ groups (30). The products **52e,f** feature 25- and 27-membered macrocycles. In none of these reactions have alternative cyclization modes been detected.

As illustrated in Scheme 9.20, a related palladium tribenzylphosphine complex, in which an ortho position in each ring has been substituted by a $O(CH_2)_4CH{=}CH_2$ group, has been investigated (**53d**) (31). With this substrate, the only product isolated (**54d**, 86%) was derived by a combination of interligand and intraligand metathesis, analogous to **XIII**. Some related complexes with phosphine ligands with only two benzyl substituents were also investigated. These results, which are more relevant to the types of reactions in Section 3, are depicted in Scheme 9.21.

Analogous metathesis/hydrogenation sequences involving two series of octahedral metal complexes with *trans*-phosphine ligands $P((CH_2)_n(CH{=}CH_2))_3$ have been investigated. It was thought that such substrates, with their more congested equatorial planes consisting of four ligands, would be particularly prone to intermolecular metatheses and oligomerization. However, as shown in Scheme 9.22, the *mer* rhenium tricarbonyl chloride and bromide complexes **60f** and **63f** ($n = 6$) afforded the gyroscope-like complexes **61f** and **64f** in 61% and 37% yields (32). The lower yield of the latter has been ascribed to the increased size of the bromine ligand, a trend also seen with related square planar palladium complexes. With **60f**, a second product, derived from a combination of interligand and intraligand metatheses, was obtained in 6% yield (**62f**; see **XIII**).

As with palladium and platinum analogs, no monometallic products could be detected with substrates with shorter phosphine ligand methylene chains. When an additional methylene group was added to each chain ($n = 7$, **60g** and **63g**), little or no gyroscope-like products could be detected. Rather, complexes analogous to **62f**, namely, **62g** and **65g**, were isolated in 27% and 20% yields (Scheme 9.22). When yet another methylene group was added to each ($n = 8$, **60h** and **63h**), gyroscope-like products again dominated (**61h**, 25%; **64h**, 18%). However, some of the alternative interligand metathesis products were also obtained (**62h**, 11%; **65h**, traces).

As shown in Scheme 9.23, similar results have been obtained with related octahedral osmium dicarbonyl dihalide complexes (33). Although a more extensive data set would

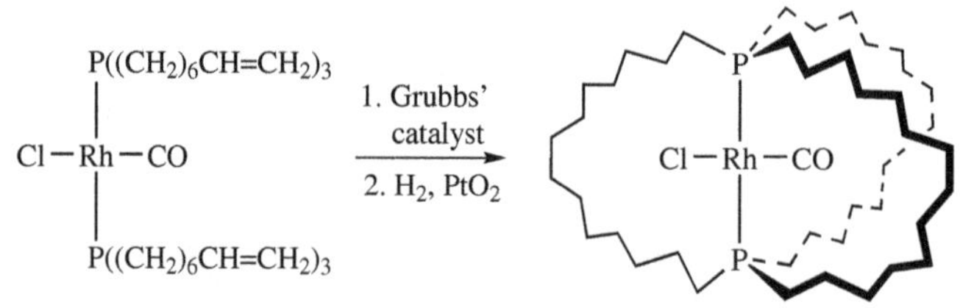

49f **50f**, 40%

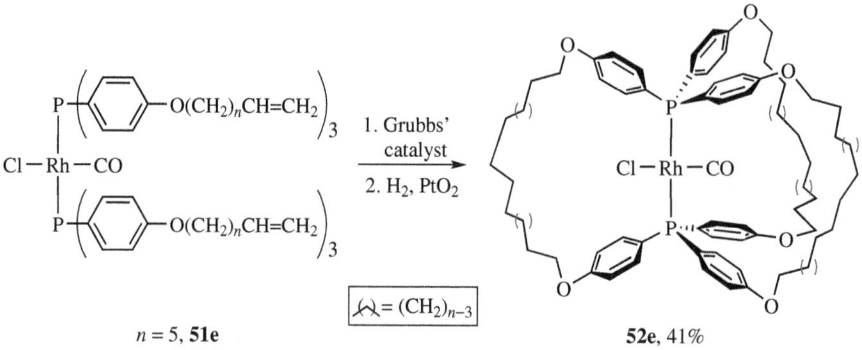

$n = 5$, **51e**
6, **51f**

52e, 41%
52f, 26%

Scheme 9.19

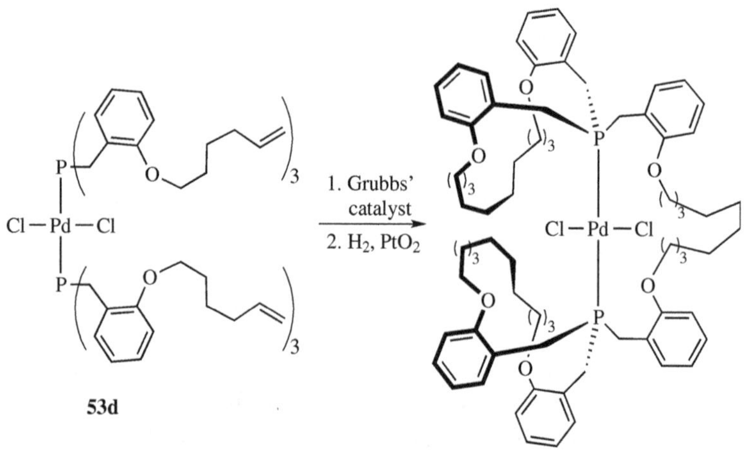

53d **54d**, 86%

Scheme 9.20

be desirable, it does seem that when there are even numbers of methylene groups in every phosphine substituent, there is a marked preference for products of the type **XII** (see F, Scheme 9.3), and when there are odd numbers of methylene groups, there is a marked preference for products of the type **XIII**. Even/odd relationships have been found for a number of physical and chemical properties of polymethylene-containing compounds (34), but the factors underlying the yield trends in Schemes 9.22 and 9.23 remain to be elucidated.

9.9 OTHER RELEVANT MULTIFOLD RING-CLOSING METATHESES

In collecting the literature for this chapter, the authors encountered additional examples of higher order intramolecular ring-closing metatheses in metal coordination spheres that, owing to their overall complexity, seemed beyond the scope of this treatment (35). However, just as the introduction featured a single example of a catenane precursor with interlocked rings, it was thought to conclude the descriptive

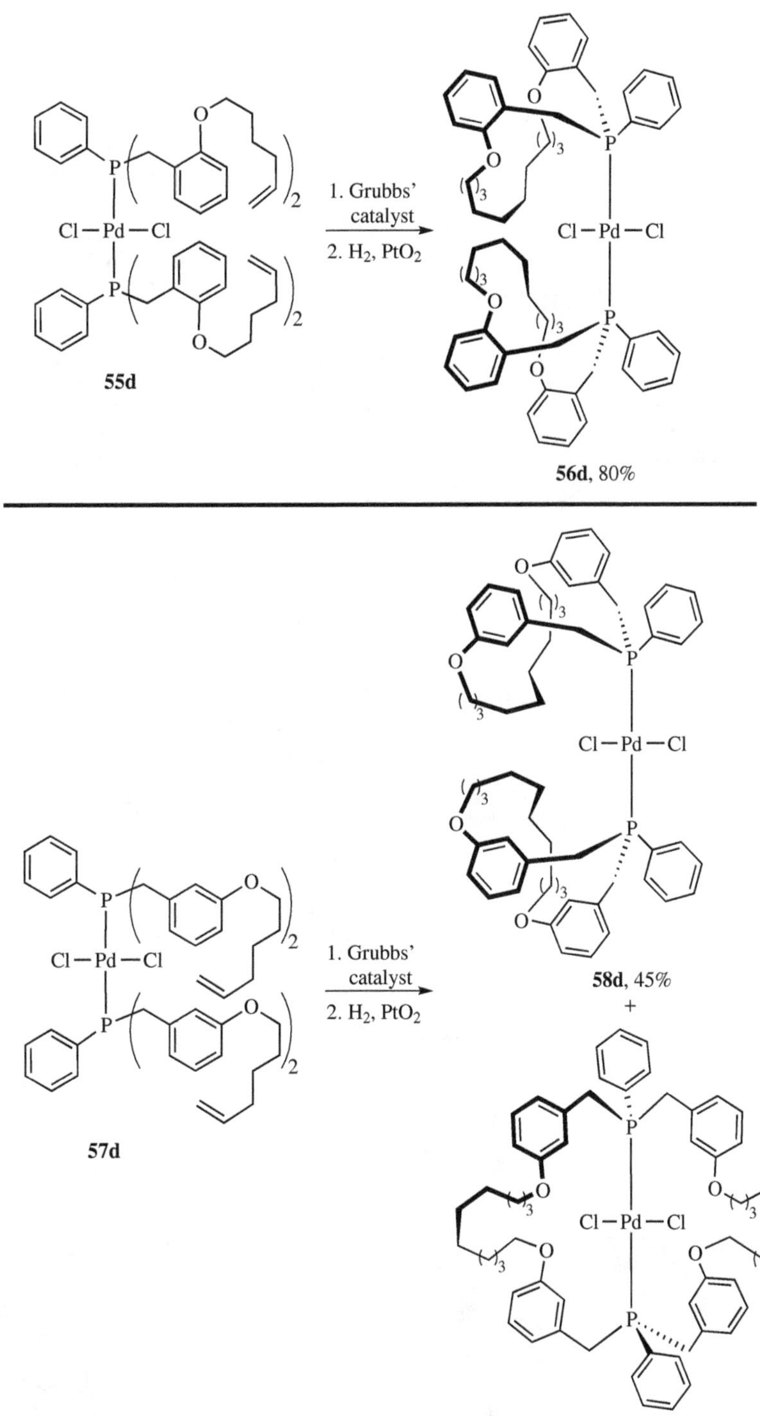

Scheme 9.21

chemistry in this review with an example that illustrates both an emerging frontier for and the immense synthetic power of this methodology.

The octahedral hexapalladium capsule compounds **73** shown in Scheme 9.24 can be synthesized in high yields by self-assembly.[35c] On each trigonal face, three tethers of the formula $O(CH_2)_nCH{=}CH_2$ ($n = 2$–4) were installed. For the cases with $n = 3$ and 4, reactions with Grubbs' catalyst effected 12-fold olefin metatheses to give what might be viewed as "stapled" octahedra, **74**, with each trigonal face tied to all three neighbors. Reactions were complete within 1–2 days, as assayed by mass spectrometry, and

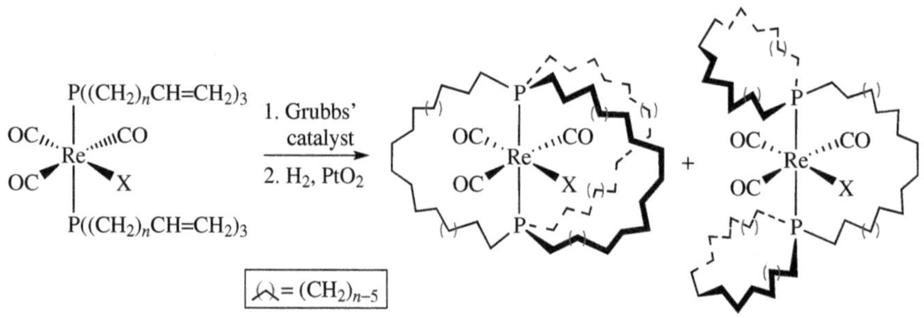

n/X = 6/Cl, **60f**	**61f**, 61%	**62f**, 6%
7/Cl, **60g**	**61g**, traces	**62g**, 27%
8/Cl, **60h**	**61h**, 25%	**62h**, 11%
6/Br, **63f**	**64f**, 37%	**65f**, —
7/Br, **63g**	**64g**, —	**65g**, 20%
8/Br, **63h**	**64h**, 18%	**65h**, traces

Scheme 9.22

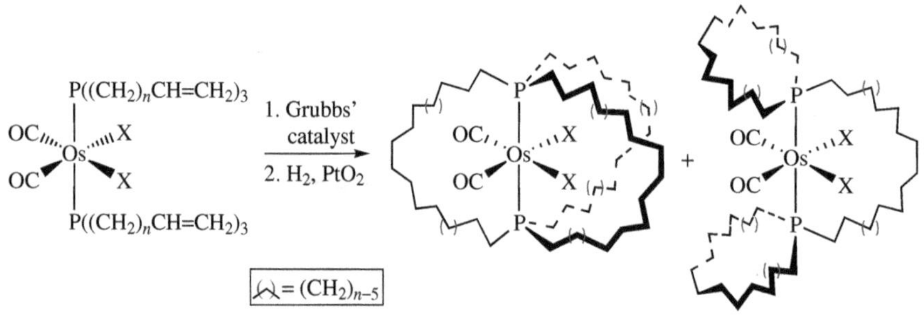

n/X = 6/Cl, **66f**	**67f**, 28%	**68f**, 13%
7/Cl, **66g**	**67g**, 5%	**68g**, 51%
8/Cl, **66h**	**67h**, 23%	**68h**, 12%
6/Br, **69f**	**70f**, 31%	**71f**, 18%
7/Br, **69g**	**70g**, 5%	**71g**, 40%
8/Br, **69h**	**70h**, 27%	**71h**, 24%

Scheme 9.23

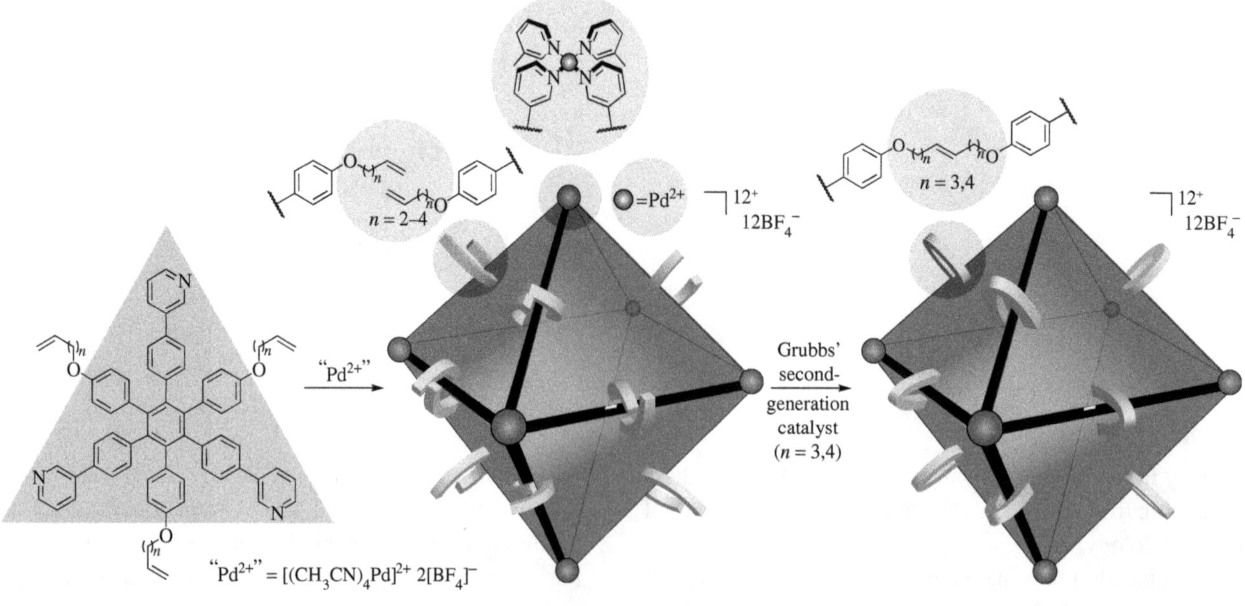

72 **73** $[Pd_6\mathbf{72}_8]^{12+}\ 12[BF_4]^-$ **74**

Scheme 9.24 (*See plate section for color version.*)

mixtures of Z/E C=C isomers were obtained. Similarly to the chemistry in Section 9.7, the metals were removed and the C=C linkages hydrogenated, leaving purely organic capsules containing only covalent bonds. For the case with $n = 3$, the overall yield was an impressive 43%.

9.10 KINETIC AND THERMODYNAMIC CONTROL

Most researchers, including the authors, have made no special attempt to distinguish kinetic versus thermodynamic control in any of the preceding olefin metatheses. Most reactions are believed to be under kinetic control, although equilibrium has been unambiguously reached in certain cases (36). For example, Grubbs' second-generation catalyst is able to further transform the oligomeric or polyplatinum by-products initially generated from **3d** in Scheme 9.4, greatly enhancing the yield of monoplatinum complex **4d**. Prior to hydrogenation, the equilibrium can be reversibly shifted between precursors to **4d** and unsaturated oligomers by increasing or decreasing the concentration.

In any event, the deliberate application of kinetic or thermodynamic metathesis conditions represents an obvious approach to enhancing selectivities for the various cyclization motifs in Schemes 9.2 and 9.3. Furthermore, a variety of unsuccessful multifold ring-closing metatheses have been noted (37), which for obvious reasons were not treated in the preceding narrative. There are probably many more "failures" that have remained unreported. Perhaps some of these sequences would be feasible under carefully optimized conditions.

One approach to probing kinetic versus thermodynamic control in many of the above schemes would involve separation of the two types of metathesis products prior to the customary hydrogenation. Each species would then independently be subjected to further metathesis, ideally achieving the same equilibrium ratio. However, the necessary separations are rendered more challenging by the fact that mixtures of Z/E isomers are commonly obtained. Perhaps newly developed metathesis catalysts with high Z-selectivities (7) could be helpful here.

Importantly, Fogg has emphasized that oligomers can dominate kinetically in certain classes of ring-closing metatheses. However, she has shown how they can be converted to monocyclic products given sufficient time, catalyst, and ethylene coproduct (38). Thus, there remains much potential for optimization in the preceding chemistry; it is only a matter of effort.

9.11 PROSPECTIVE

The preceding sections have established the immense utility of multifold intramolecular ring-closing metatheses in the syntheses of a variety of unusual organometallic complexes. In retrospect, one lesson is that one should never let oneself be "talked out" of trying such speculative sequences. If it is possible for a target structure to be realized from a polyolefin precursor without engendering significant ring strain, the synthesis should by all means be attempted, even if numerous other products are possible.

It is likely that as more experience is gained, additional control over the various types of product distributions (e.g., Schemes 9.2 and 9.3) will be realized. One approach will involve exploiting kinetic and thermodynamic control, which as noted above has frequently been ignored in what are often admittedly "run and gun" studies. Other approaches may involve alkyne metatheses, which do not suffer from complicating Z/E product mixtures and for which increasing numbers of catalysts are becoming available (39). In any event, it is hoped that this review of a rather scattered literature will help stimulate thinking and activity in targeted syntheses of topologically novel organometallic molecules with unusual connectivities.

ACKNOWLEDGMENTS

The authors thank the US National Science Foundation for support (CHE-0719267) and Dr. Giovanni P. Rachiero for helpful comments.

REFERENCES

1. Bauer EB, Gladysz JA. Metal-catalyzed olefin metathesis in metal coordination spheres. In: Grubbs RH, editor. *Handbook of Metathesis*. Weinheim: Wiley/VCH; 2003 Volume 2. pp. 403–431.
2. Toledano CA, Parlier A, Rudler H, Daran J-C, Jeannin Y. *J Chem Soc, Chem Commun* 1984:576–578.
3. Alvarez C, Pacreau A, Parlier A, Rudler H, Daran J-C. *Organometallics* 1987;6:1057–1064.
4. (a) Mohr B, Weck M, Sauvage J-P, Grubbs RH. *Angew Chem Int Ed Engl* 1997;36:1308–1310 *Angew Chem* 1997;109:1365–1367. (b) Weck M, Mohr B, Sauvage J-P, Grubbs RH. *J Org Chem* 1999;64:5463–5471.
5. (a) Dietrich-Buchecker C, Rapenne G, Sauvage J-P. *Chem Commun* 1997:2053–2054. (b) Dietrich-Buchecker C, Sauvage J-P. *Chem Commun* 1999:615–616. (c) Rapenne G, Dietrich-Buchecker C, Sauvage J-P. *J Am Chem Soc* 1999;121:994–1001. (d) Belfrekh N, Dietrich-Buchecker C, Sauvage J-P. *Inorg Chem* 2000;39:5169–5172. (e) Arico F, Mobian P, Kern J-M, Sauvage J-P. *Org Lett* 2003;5:1887–1890. (f) Frey J, Kraus T, Heitz V, Sauvage J-P. *Chem Commun* 2005:5310–5312. (g) Frey J, Kraus T, Heitz V, Sauvage J-P. *Chem Eur J* 2007;13:7584–7594.
6. Loren JC, Gantzel P, Linden A, Siegel JS. *Org Biomol Chem* 2005;3:3105–3116.

7. Yu M, Wang C, Kyle AF, Jakubec P, Dixon DJ, Schrock RR, Hoveyda AH. *Nature* 2011;479:88–93.
8. Shima T, Bauer EB, Hampel F, Gladysz JA. *Dalton Trans* 2004:1012–1028.
9. Ruwwe J, Martín-Alvarez JM, Horn CR, Bauer EB, Szafert S, Lis T, Hampel F, Cagle PC, Gladysz JA. *Chem Eur J* 2001;7:3931–3950.
10. Cao J, Song Z-P, Zhu X-Z, Chen C-F. *Tetrahedon Lett* 2010;51:3112–3115.
11. Representative cases: (a) Camacho DH, Salo EV, Guan Z, Ziller JW. *Organometallics* 2005;24:4933–4939. (b) Simpkins SME, Kariuki BM, Cox LR. *J Organomet Chem* 2006;691:5517–5523.
12. de Quadras L, Bauer EB, Mohr W, Bohling JC, Peters TB, Martín-Alvarez JM, Hampel F, Gladysz JA. *J Am Chem Soc* 2007;129:8296–8309.
13. de Quadras L, Bauer EB, Stahl J, Zhuravlev F, Hampel F, Gladysz JA. *New J Chem* 2007;31:1594–1604.
14. Ng PL, Lambert JN. *Synlett* 1999:1749–1750.
15. Zeits PD, Rachiero GP, Hampel F, Reibenspies JH, Gladysz JA. *Organometallics* 2012;31:2854–2877.
16. (a) Dijkstra HP, Chuchuryukin A, Suijkerbuijk BMJM, van Klink GPM, Mills AM, Spek AL, van Koten G. *Adv Synth Catal* 2002;344:771–780. (b) Chuchuryukin AV, Dijkstra HP, Suijkerbuijk BMJM, Klein Gebbink RJM, van Klink GPM, Mills AM, Spek AL, van Koten G. *Russ J Org Chem* 2003;39:422–429. (c) Chase PA, Lutz M, Spek AL, van Klink GPM, van Koten GJ. *Mol Catal A: Chem* 2006;254:2–19.
17. For related relevant results, including catenane syntheses, see Fuller A-ML, Leigh DA, Lusby PJ, Slawin AMZ, Walker DB. *J Am Chem Soc* 2005;127:12612–12619.
18. (a) Martinez V, Blais J-C, Bravic G, Astruc D. *Angew Chem Int Ed* 2003;42:4366–4369 *Angew Chem* 2003;115:4502–4505; (b) Martinez V, Blais J-C, Bravic G, Astruc D. *Organometallics* 2004;23:861–874.
19. Buchowicz W, Furmańczyk A, Zachara J, Majchrzak M. *Dalton Trans* 2012;41:9269–9271.
20. (a) Chuchuryukin AV, Dijkstra HP, Suijkerbuijk BMJM, Klein Gebbink RJM, van Klink GPM, Mills AM, Spek AL, van Koten G. *Angew Chem Int Ed* 2003;42:228–230 *Angew Chem* 2003;115:238–240. (b) Chuchuryukin AV, Chase PA, Dijkstra HP, Suijkerbuijk BMJM, Mills AM, Spek AL, van Klink GPM, van Koten G. *Adv Synth Catal* 2005;347:447–462.
21. Song KH, Kang SO, Ko J. *Chem Eur J* 2007;13:5129–5134.
22. Wang P, Moorefield CN, Newkome GR. *Angew Chem Int Ed* 2005;44:1679–1683 *Angew Chem* 2005;117:1707–1711.
23. Nawara-Hultzsch, AJ.; Stollenz, M.; Barbasiewicz, M.; Szafert, S.; Lis, T.; Hampel, F.; Bhuvanesh, N.; Gladysz, JA.. *Eur. J. Chem.*2014; 20: in press. DOI: 10.1002/chem.201304419.
24. Shima T, Hampel F, Gladysz JA. *Angew Chem, Int Ed* 2004;43:5537–5540 *Angew Chem* 2004;116:5653–5656.
25. Skopek K, Gladysz JA. *J Organomet Chem* 2008; 693:857–866.
26. Nawara AJ, Shima T, Hampel F, Gladysz JA. *J Am Chem Soc* 2006;128:4962–4963.
27. Nuñez JE, Natarajan A, Khan SI, Garcia-Garibay MA. *Org Lett* 2007;9:3559–3561.
28. Skopek K, Barbasiewicz M, Hampel F, Gladysz JA. *Inorg Chem* 2008;47:3474–3476.
29. Wang L, Shima T, Hampel F, Gladysz JA. *Chem Commun* 2006:4075–4077.
30. Wang L, Hampel F, Gladysz JA. *Angew Chem Int Ed* 2006;45:4372–4375 *Angew Chem* 2006;118:4479–4482.
31. Han J, Deng C, Fang R, Zhao D, Wang L, Gladysz JA. *Organometallics* 2010;29:3231–3234.
32. (a) Hess GD, Hampel F, Gladysz JA. *Organometallics* 2007;26:5129–5131. (b) Hess GD, Doctoral dissertation, Universität Erlangen-Nürnberg; 2010.
33. (a) Fiedler T, Bhuvanesh N, Hampel F, Reibenspies JH, Gladysz JA, manuscript in preparation. (b) Fiedler T, Doctoral dissertation, Universität Erlangen-Nürnberg; 2011.
34. See for example Thalladi VR, Boese R, Weiss H-C. *Angew Chem Int Ed* 2000;39:918–922 *Angew Chem* 2000; 112:942–946.
35. (a) Inomata T, Konishi K. *Chem Commun* 2003:1282–1283. (b) Shoji O, Tanaka H, Kawai T, Kobuke Y. *J Am Chem Soc* 2005;127:8598–8599. (c) Hiraoka S, Yamauchi Y, Arakane R, Shionoya M. *J Am Chem Soc* 2009;131:11646–11647.
36. See Ref. 19 of Ref. 8.
37. Baker MV, Brown DH, Skelton BW, White AH. *Aust J Chem* 2002;55:655–660.
38. (a) Conrad JC, Eelman MD, Duarte Silva JA, Monfette S, Parnas HH, Snelgrove JL, Fogg DE. *J Am Chem Soc* 2007;129:1024–1025. (b) Monfette S, Fogg DE. *Chem Rev* 2009;109:3783–3816.
39. (a) Jyothish K, Zhang W. *Angew Chem Int Ed* 2011;50:8478–8480 *Angew Chem* 2011;123:8628–8630. (b) Haberlag B, Freytag M, Daniliuc CG, Jones PG, Tamm M. *Angew Chem Int Ed* 2012;51:13019–13022 *Angew Chem* 2012;124:13195–13199.

10

INDUSTRIAL APPLICATIONS OF OLEFIN METATHESIS POLYMERIZATION

CHRISTIAN SLUGOVC
Institute for Chemistry and Technology of Materials, Graz University of Technology, Graz, Austria

10.1 INTRODUCTION

Since 1950, research on the preparation of polymers via ring opening metathesis polymerization (ROMP) has been conducted and led to several industrially relevant polymeric materials. Here, only those products which are currently available on the market are discussed. In other overviews of industrially relevant olefin metathesis, more information is also available on products which are no longer commercialized or failed to capture a market. Such papers have been published over the last 20 years and each of them discusses the topic with emphasis on specific facets (1–5). In general, olefin metathesis polymerization-derived materials are niche products with very distinguished properties. Only the thermosetting poly(dicyclopentadiene) (pDCPD) holds promise of becoming a downmarket product, substituting to some extent epoxy-resins (worldwide market for epoxy-resins: US$ 17 billion in 2012).

10.2 POLY(DICYCLOPENTADIENE)

pDCPD is perhaps the most important olefin metathesis polymerization-derived product. The estimated global market for pDCPD is about 25,000 metric tons per year (1) and pDCPD is currently sold under the trade names Telene®, Metton®, Pentam® and Proxima™. DCPD, obtained from the C5-stream in oil refineries, is a potentially cheap starting material. However, high purity grade DCPD is necessary for the production of pDCPD (even traces of conjugated dienes have to be removed) and pDCPD manufacturers had to set up their own facilities to prepare DCPD of sufficient purity.

The business based on ill-defined group VI catalyst systems is nowadays in the hands of Rimtec Corporation (brands: Telene®, Metton®, Pentam®). pDCPD parts are produced by reaction injection molding (RIM) combining two monomer streams in a mold (Fig. 10.1). A typical formulation comprises in one stream the monomers (DCPD and optionally tricyclopentadienes), a precatalyst (e.g., tungsten(VI) hexachloride and tungsten(VI) oxytetrachloride or tetrakis(tridodecylammonium)octa-molybdate), a solubilizer for the precatalysts (e.g., nonylphenol) and optional fillers and additives and in the other stream monomers and the activator for the pre-catalysts (e.g., $EtAlCl_2$ or $EtAlCl_2$, propanol, and $SiCl_4$) as well as fillers and additives (6). The two streams are impingement-mixed in a mixing head located directly before the mold. Upon combination of the two streams and a short induction period (which can be adjusted), a highly exothermic reaction takes place, which warms the molded product to about 200 °C. The molded part is then cooled down (to about 100 °C) and removed from the mold. Typical cycling times are in the range of 2–5 min. A constraint of this technology is the air- and moisture-sensitivity of the pre-catalysts and activators, which excludes a series of potential fillers and additives and demands the establishment of an oxygen- and moisture-free environment during storing and curing of the resins. The basic RIM process outlined above has been modified by introducing a third stream, allowing for the addition of further additives that are otherwise difficult

Olefin Metathesis: Theory and Practice, First Edition. Edited by Karol Grela.

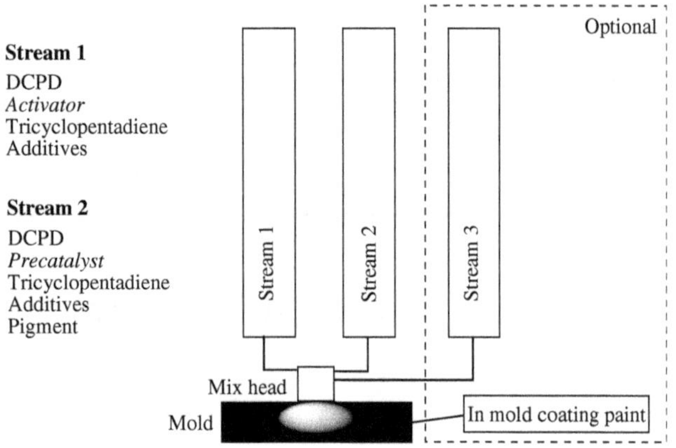

Figure 10.1 Scheme of the reaction injection molding (RIM) for DCPD.

to incorporate in the conventional streams. In this way Telene grades with superior rigidity or flame resistance are available. Another processing variant of the conventional two-stream RIM process comprises a mold-coating step. In this case, the injection of a paint layer in the mold after curing of polymer takes place, which increases the cycle time by up to 2 min but reduces the overall post-molding cost.

Ruthenium initiator-based formulations (Proxima™) offer even more design freedom, as they tolerate a broader range of additives and fillers in addition to air and moisture. Further, molding processes other than RIM, for example, resin transfer molding (RTM) or vacuum-assisted resin transfer molding (VARTM), are feasible with such resins. In general, pDCPD offers several processing advantages in RIM such as low molding pressure, excellent mold release, and no necessity for post-curing steps. Further, because the pDCPD streams are characterized by a distinctly lower viscosity (10–50 cP) compared to, for example, epoxy resins (about 200–500 cP), large and complex parts with variable wall thicknesses can be easily molded (up to 700 kg/part).

pDCPD features very favorable characteristics such as good chemical resistance, high heat distortion temperature, and high toughness and impact resistance. These properties are due to the steric bulk of the bicyclic structure, the unsaturation of the polymer-backbone and the crosslinking. DCPD is self-crosslinking via olefin metathesis and to some degree via olefin addition (7), meaning that no additional crosslinking monomer has to be added to the formulation (Fig. 10.2). Typical pDCPD grades are characterized by a specific gravity of 1.03 g/cm^3, a mold shrinkage of approximately 1% and lower, a heat deformation temperature of more than 100 °C, an elastic modulus of 1.6–1.9 GPa, a flexural modulus of 1.8–1.9 GPa, a compressive strength of 59–62 MPa at yield, a flexure strength at yield of 70–80 MPa, a tensile yield strength of 35–46 MPa, a hardness of 72–84 (Shore D), an elongation at break of 5–70% (typically 5–7%) and a notch impact strength of 31–46 kJ/m^2. pDCPD shows low water absorption (0.09% weight increase after 24 h in water) and is resistant to corrosion, acids, and alkalines, mostly because the surface of pDCPD is oxidizing and this process is self-inhibiting, leading to a "passivation" of the pDCPD surface (8). This process also facilitates painting and bonding; painting is sufficient to prevent uncontrolled oxidation.

The biggest application of pDCPD is as body panels for agriculture and construction equipment and for buses and trucks, competing with traditional glass-reinforced plastic. Reinforced grades (e.g., Telene® 1650 or 1690) are used for these applications. Further applications include areas where the high resistance against acid and base is important, such as compact waste water units or in the chloralkali industry (electrolyser cell covers or butterfly valves). There are many current developments for the use of pDCPD, especially

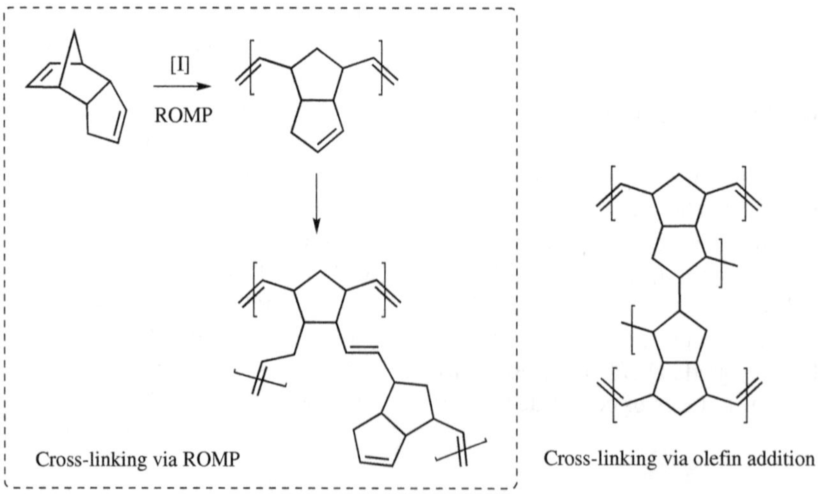

Figure 10.2 ROMP of DCPD.

substituting epoxy-resins in composite materials (e.g., for making rotor blades for wind turbines) and the future will tell if pDCPD will find increasing commercial success.

10.3 POLY(NORBORNENE)

Poly(norbornene), commercialized under the trade name Norsorex®, is known as a very unique rubber. The polymerization process developed in the late 1970s by CdF Chimie of Paris and now followed by Astrotech uses norbornene (bicyclo[2.2.1]-2-heptene) as the monomer and a catalyst system consisting of $RuCl_3$, hydrochloric acid, and butanol and operates in air (2,9). The amorphous poly(norbornene) prepared in this process is characterized by a very high molecular weight of about 3,000,000 g/mol and a high *trans* content (~90%) of the double bonds in the backbone and is manufactured in a powder form of slightly porous particles (Fig. 10.3). The cyclopentane-ring in the main chain gives poly(norbornene) a T_g of 35–37 °C, and thus it is not rubbery at ambient temperatures and requires plasticization to achieve elastomeric behavior. Formulations with up to 300–400 phr plasticizer (mainly mineral oils) are possible, which allows lowering of the softness of vulcanizates far below the limits known for other polymers. The high binding power for large quantities of diverse fillers is another unique feature of this polymer. Norsorex® can be cured by both sulfur and peroxides but requires protection against oxygen, ozone, and UV light. Poly(norbornene) can be blended with many other elastomers, for example, natural rubber or styrene-butadiene rubber, to adjust the properties of the vulcanizates to the needs of the application (10).

Norsorex-based formulations find application in several niche markets. The good damping properties led to the development of diverse Norsorex-based personal protection products such as motorcycle clothing or materials for noise and vibration damping, for example, in railway applications. Another market is tires; here the main advantage is high tensile strength, in combination with a high friction coefficient over a wide temperature range. Finally, Norsorex® is applied for absorbing hydrocarbons such as heavy oils, light oils, and fuels. In this application, Norsorex® can absorb typically 10 times its dry weight in hydrocarbons without losing mechanical properties. Thus the saturated specimen can

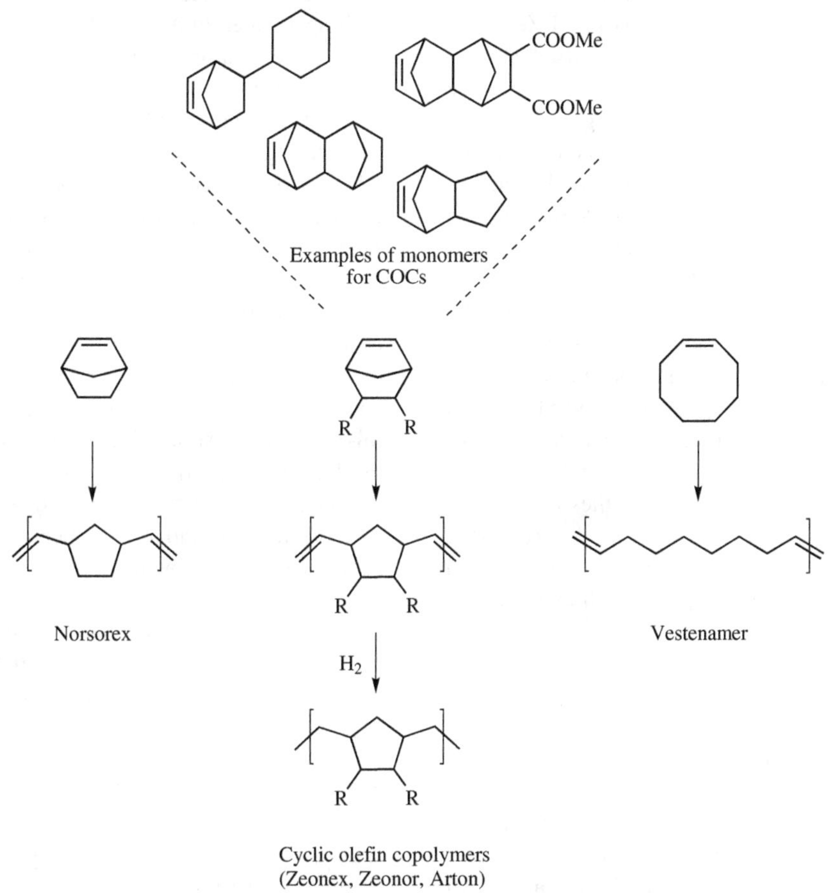

Figure 10.3 Preparation of Norsorex, cyclic olefin copolymers, and Vestenamer.

be mechanically removed from the spill, to be submitted to a controlled combustion.

10.4 CYCLIC OLEFIN COPOLYMERS

Poly(norbornene) exhibits poor thermal and oxidative stability due to the unsaturation of the polymer's backbone. Upon hydrogenation, an opaque crystalline polymer is obtained, which is characterized by a melting temperature of 141.8–143 °C (11,12). Upon changing norbornene to, for example, dicyclopentadiene, 2,3,3a,4,7,7a-hexahydro-1*H*-4,7-methanoindene, 5-cyclohexylbicyclo[2.2.1]hept-2-ene, or 1,2,3,4,4a,5,8,8a-octahydro-1,4:5,8-dimethanonaphthalene amorphous, transparent materials with high glass-transition temperatures are obtained upon hydrogenation of the ROMP derived polymers (Fig. 10.3). The overall production process comprises the synthesis of the monomer, the ROMP step, the hydrogenation step, two steps for the removal of the catalyst and the solvents, and a step for palletizing and packing the resin. The latter step is done under clean room conditions to fulfill the purity requirements of most of the applications of the polymers (13,14). Cyclic olefin polymers of that type are produced at Nippon Zeon and are marketed under the trade names Zeonex® (10 grades available) and Zeonor® (4 grades available), and at Japan Synthetic Rubber Corporation, marketed as Arton® (6 grades available). The thermoplastic materials are characterized by good isolating properties, good precision molding, low specific gravity, low water absorbency, good anti-heat aging, good chemical resistance and, most important, high transparency in the visible range of the electromagnetic spectrum, as well as good transparency in the ultraviolet region, combined with low birefringence and high heat resistance. Accordingly, cyclic olefin copolymers find application in optics, for example, lenses and prisms for cameras, especially for cameras of mobile phones or light guide plates used in liquid crystal displays. Further applications include containers for use in the semiconductor business or for food and medicines, making use of the high moisture barrier, the high purity, and the transparency of the polymers. Zeonex®, Zeonor®, and Arton® compete with poly(methylmethacrylate) (Plexiglas®) or poly(carbonate) and also with cyclic olefin copolymers prepared via addition polymerization, such as Avatrel®, Apel®, or Topas®.

10.5 POLYALKENAMERS

Polyalkenamers are linear polymers obtained from ROMP of cyclic unsaturated monomers such as cyclooctene, 1,5-cyclooctadiene, or 1,5,9-cyclododecatriene. All monomers are prepared by catalytic di- or trimerization of butadiene and subsequent hydrogenation (in case of cyclooctene). Only poly(cyclooctene) is marketed under the trade name Vestenamer® by Evonik (formerly known as *Degussa-Hüls AG*) and is made via WCl_6 catalyzed ROMP of cyclooctene in hexane at elevated temperature (Fig. 10.3). Under these conditions, almost complete conversion of the monomer is reached and a relatively high molecular weight poly(cyclooctene) (molecular weights >500,000 g/mol) is formed. Concomitant secondary metathesis leads to the formation of mainly low molecular weight cyclic poly(cyclooctene) (~25 wt%) resulting in a weight average molecular weight of about 75,000 g/mol and a high polydispersity index of about 5 (data for the commercially available grade Vestenamer 8012). The *cis* to *trans* ratio of the double bonds, which is decisive for the degree of crystallinity of the polymer, can be controlled by the reaction conditions and is typically less than 1. The combination of high content of secondary metathesis products (i.e., cyclic structures), the low content of branched structures, the high polydispersity, and the high trans-double bond content leads to unique elastomeric properties, particularly a low-melt viscosity of the polymers, which are exploited in two commercial grades: Vestenamer 8012 (trans-content: 80%; melting temperature: 54 ± 4 °C; glass-transition temperature: −65 °C; crystallinity at 23 °C: 30%) and Vestenamer 6213 (trans-content: 60%; melting temperature: <36 °C; glass-transition temperature: −75 °C; crystallinity at 23 °C: 10%) (15). Vestenamer 8012 is used as viscosity reducer and compatibilizer in rubber formulations, as modifier for thermoplastic materials, calendaring aid, and asphalt modification. Vestenamer 6213 is optimized for applications at lower temperature. In extrusion, the flowability is improved upon addition of Vestenamer and a higher output of extrudates, exhibiting an improved surface finish, can be obtained. Upon addition of Vestenamer to injection molding formulations, the molding of large parts and precision parts is improved and cycle times can be reduced. In rubber recycling, ground rubber waste can be reprocessed by coating the crumbs with a mineral oil-Vestenamer solution and such treated crumbs can be used as filler in fresh rubber formulations or in asphalt production (16). In the latter application, Vestenamer is used to cross-link ground tire rubber to the asphalt, creating a rubberized matrix in the asphalt which prevents cracking, rutting, and shoving. Moreover, an easier, faster, and more uniform mixing of ground tire rubber, bitumen, and fillers is provided, leading to a lower production cost per kilometer and a longer service life of the road.

10.6 RUBBER MODIFICATION

Lanxess markets a series of hydrogenated acrylonitrile-butadiene rubber (HNBR) under the trade name Therban®. This particular class of HNBRs is prepared by partially degrading the unsaturated acrylonitrile-butadiene rubber (NBR), that is, reducing the molecular weight of the precursor polymer. This step is done by utilizing a second generation ruthenium catalyst. The degraded NBR is subsequently partially or completely hydrogenated, creating a portfolio of different Therban grades. All grades are characterized by a considerably reduced viscosity compared to conventional HNBRs, leading to superior performance in many applications. Important areas of applications include the automotive, mechanical engineering, and oil industries because HNBRs are uniquely suited for applications in aggressive environments (resistance to heat, oil, chemicals, and ozone).

10.7 RENEWABLE FEEDSTOCK COMMERCIALIZATION

Elevance commercializes a wax named NatureWax® which is produced from soybean oil. A second generation ruthenium catalyst is directly added to soybean oil, leading to the oligomerization of the triglyceride. This intermediate is subsequently hydrogenated. NatureWax finds applications in personal care and candle production.

10.8 CONCLUSIONS

Only a handful of products relying on olefin metathesis polymerization are in the market today. For 20 years, overviews like this one have claimed that there are several new applications coming up in the future. Actually, most of them were proved wrong when successful commercialization was considered. The lack of market pull, patent issues, and process challenges might be identified as the main reasons for this situation. Nevertheless, new developments in olefin metathesis polymerization are under permanent observation by the relevant industries and in the next decade new products based on olefin metathesis polymerization are expected to enter the market.

REFERENCES

1. Nickel A, Edgecombe BD. *Polymer Science: A Comprehensive Reference*. Vol. 4. Amsterdam: Elsevier; 2012. p 749–759.
2. Mol JC. *J Mol Catal A: Chem* 2004;213:39–45.
3. Trimmer MS. *Handbook of Metathesis*. Vol. 3. Weinheim: Wiley-VCH; 2003. p 407–414.
4. Dragutan V, Streck R. *Catalytic Polymerization of Cycloolefins*. Amsterdam: Elsevier; 2000.
5. Ivin KJ, Mol JC. *Olefin Metathesis and Metathesis Polymerization*. London: Academic Press; 1997.
6. Breslow DS. *Prog Polym Sci* 1993;18:1141–1195.
7. Davidson TA, Wagener KB, Priddy DB. *Macromolecules* 1996;29:786–788.
8. Mühlebach A, van der Schaaf PA, Hafner A, Setiabudi F. *J Mol Catal A: Chem* 1998;132:181–188.
9. Claverie JP, Soula R. *Prog Polym Sci* 2003;28:619–662.
10. Marbach A, Hupp R. *Rubber World* 1989;6:30.
11. Cataldo F. *Polym Int* 1994;34:49–57.
12. Lee L-BW, Register RA. *Macromolecules* 2005;38: 1216–1222.
13. Yamazaki M. *J Mol Catal A: Chem* 2004;213:81–87.
14. Shin JY, Park JY, Liu X, He J, Kim SC. *Pure Appl Chem* 2005;77:801–814.
15. Schneider WA, Müller MF. *Makromol Chem* 1988;189: 2823–2837.
16. Diedrich KM, Burns BJ. *Polym Recycl* 1999;4:237–246.

11

COMMERCIAL POTENTIAL OF OLEFIN METATHESIS OF RENEWABLE FEEDSTOCKS

ANDREW NICKEL AND RICHARD L. PEDERSON
Materia Inc., Pasadena, CA

11.1 INTRODUCTION AND BACKGROUND

Renewable feedstocks are gaining traction in supplementing and replacing traditional petrochemical products (1,2). Progress has been made in both the production of individual products and in integrated processes with multiple end products from a single renewable feedstock. The latter approach, which mimics the traditional petrochemical paradigm, is commonly referred to as a *biorefinery* approach (3). Olefin metathesis has been studied in both paradigms (4,5). The olefin metathesis reaction, discovered in 1964 (6), has had numerous commercial successes, but the sensitivity of early ill-defined catalysts to polar functionality and impurities in the feed streams limited their use to purified petrochemical feedstocks (7,8). Research into broader substrate scopes accelerated with the development of well-defined homogenous metathesis catalysts by Grubbs, Schrock, and others (9–11). The Grubbs-type catalysts (Fig. 11.1) have been the most widely used olefin metathesis catalysts with renewable feedstocks, mainly due to their high tolerance to air and water and the polar functional groups present in partially refined renewable streams. The Schrock molybdenum systems are more sensitive to moisture and air, which has limited their use in this context. However, very recently, Schrock and Hoveyda have reported a cis-selective tungsten catalyst that is tolerant to moisture and can be handled in air (Fig. 11.2) (12,13).

11.2 SEED OIL SUBSTRATES

The United States Department of Agriculture (USDA) reported that for 2012–2013, the world supply and distribution of major oilseeds was 462 million metric tons (14). The large-volume seed oils include soybean, rapeseed (canola), sunflower, and palm. Of these, palm yields the most oil per hectare: the average yield of palm oil is 3.6 t/ha, rapeseed oil is 0.74 t/ha, sunflower oil is 0.56 t/ha, and soybean oil is 0.4 t/ha (15). Seed oils are attractive renewable feedstocks because they are available on a large scale at low prices and contain multiple functional groups for further modifications. Further, a broad range of chemical products can be obtained from a single feedstock, making natural oils a good fit for a biorefinery (16) (Fig. 11.3).

11.2.1 Synthesis of Methyl 9-Decenoate by Ethenolysis and Alkenolysis

A long-standing goal of olefin metathesis has been to convert seed oils to the truncated, value-added products: methyl 9-decenoate (Me9DA), 1-decene, 1-heptene, and 1,4-pentadiene. In order to reach economic viability on a large scale, the metathesis turnover number (TON) (17) for these products should be greater than 50,000 (18). This section will describe the evolution of producing

Olefin Metathesis: Theory and Practice, First Edition. Edited by Karol Grela.

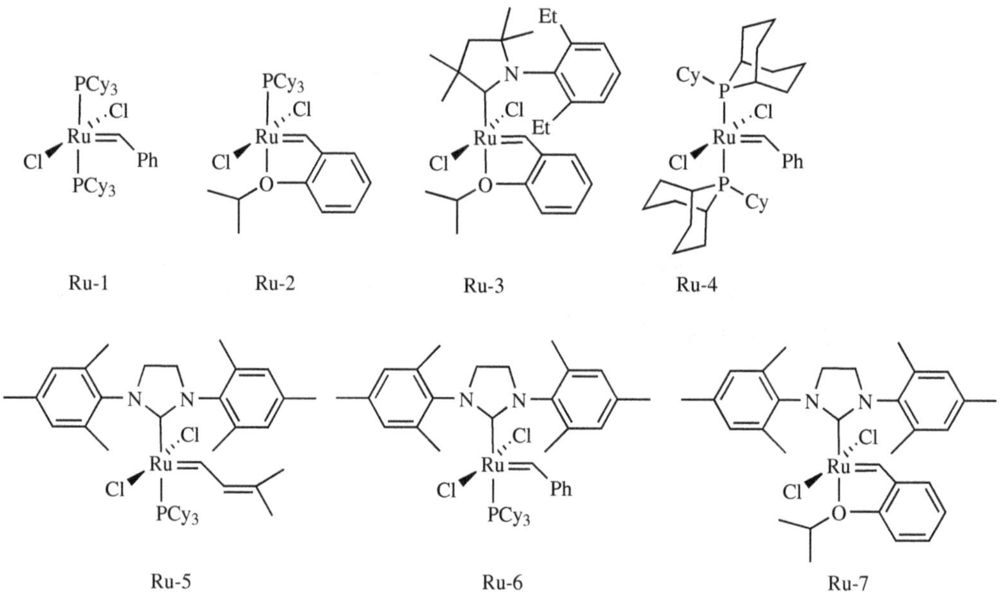

Figure 11.1 Well-defined ruthenium metathesis catalysts discussed in this chapter.

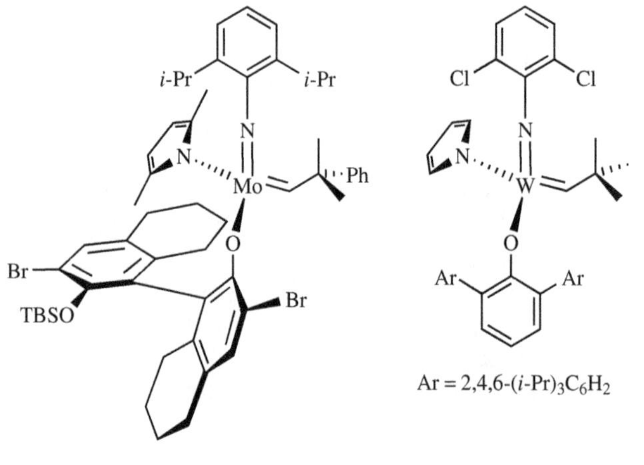

Figure 11.2 Schrock's cis-selective molybdenum and tungsten catalysts.

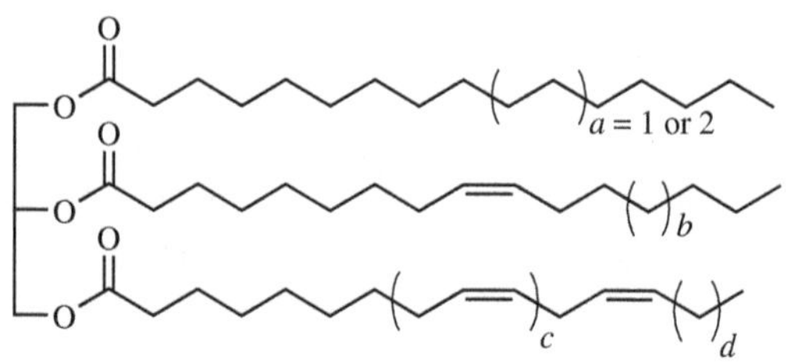

Soybean, canola, and palm oils are triglycerides containing a mixture of oleates, linoleate, linolenate, palmitate, and stearate.

Fatty acid	%Soybean	%Canola	%Palm
Palmitate is when $a = 1$,	11%	4%	35%
Stearate is when $a = 2$,	5%	2%	0%
Oleate is when $b = 3$,	24%	60%	55%
Linoleate is when $c = 1$ and $d = 4$	53%	24%	10%
Linolenate is when $c = 2$ and $d = 1$	7%	10%	0%

Figure 11.3 Common seed oil compositions.

Me9DA from the cross metathesis (CM) of methyl oleate (MO) with ethylene to the discovery of using ethylene surrogates with soybean oil fatty acid methyl esters (FAMEs) to achieve Me9DA with a TON >190,000 (17,19).

11.2.1.1 Ethenolysis The CM of a substrate with ethylene is known as *ethenolysis*. Ethenolysis of Δ^9 FAMEs produces Me9DA, 1-decene, 1-heptene, and 1,4-pentadiene. These products are useful as intermediates for surfactants, polymer additives, surface coatings, and antimicrobials (18,20–22).

Naturally occurring seed oils are mixtures of saturated, mono-, di-, and tri-unsaturated esters, which produce complex product mixtures, and therefore most studies to date have focused on the use of purified MO as a simplified model substrate. Even with the mono-unsaturated MO, there are still several products formed (Scheme 11.1). MO can react with ethylene to give Me9DA and 1-decene ($1C_{10}$), or with itself to form the self-metathesis products 9-octadecene ($9C_{18}$) and 1,18-dimethyl-octadec-9-enedioate (9-ODDE) (23–25).

A number of groups have studied the ethenolysis of MO (18,23,24,26), Grubbs (27), Warwel (20), and Burdett

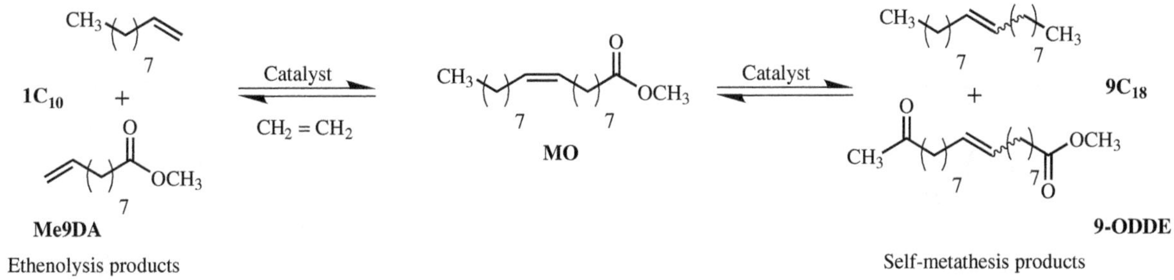

Scheme 11.1 Competing self-metathesis and ethenolysis of MO.

TABLE 11.1 Summary of Ethenolysis of MO[a]

Entry	Catalyst	Loading (ppm)	% Conversion	% Selectivity	% Yield	TON (17)	References
1	Ru-1	100	58%	93%	54%	5,400	(24)
2	Ru-1	10			16%	15,800	(39)
3	Ru-2	100	51%	94%	48%	4,800	(24)
4	Ru-1[b]	10	16%	96%	15%	15,400	(18)
5	Ru-4[c]	30	43%	98%	42%	14,047	(28)
6	Ru-5	100			48%	4,800	(39)
7	Ru-6	100	64%	44%	28%	2,800	(24)
8	Ru-7	100	60%	33%	20%	2,000	(24)
9	Mo-1[d]	2,000	94%	>99%	94%	470	(26)
10	Mo-1[d]	200	58%	>99%	62%	2,900	(26)
11	Mo-1[e]	200	95%	>99%	95%	4,750	(26)
12	W-1[f]	2,000	62%	>99%	58%	310	(26)
13	W-1[d]	2,000	48%	>99%	42%	240	(26)
14	Ru-3	100	73%	73%	53%	5,300	(23)
15	Ru-3	10	42%	83%	35%	35,000	(23)

[a]General conditions: neat; 150 psi of ethylene at 40 °C. Catalyst loadings are defined as the moles of catalyst per million moles of seed oil double bonds (MO has 1 mole of double bonds per mole).
[b]60 psi ethylene at 30 °C.
[c]145 psi ethylene at 60 °C.
[d]59 psi ethylene at rt.
[e]147 psi ethylene at rt.
[f]59 psi ethylene at 50 °C.

and coworkers (18) used the first-generation Grubbs catalyst (Ru-1, Ru-2), while Forman and coworkers (28) used a cyclohexyl-phoban first-generation Grubbs-type catalyst (Ru-4), all with limited success, that is, TON < 16,000 (Table 11.1).

The second-generation Grubbs catalysts (Ru-5, Ru-6, Ru-7) display high activity but poor selectivity and favor the formation of products from the self-metathesis of MO. Recently, Schrock and Hoveyda (26) collaborated to report the ethenolysis of MO with molybdenum (Mo-1) and tungsten (W-1) monoaryloxide-pyrrolide catalysts. These catalysts yielded excellent selectivity (>99%) but only modest activity, with Mo-1 TON of 5000 and W-1 TON of 325. New cyclic alkyl amino carbene (Ru-3) catalysts recently developed through collaboration with the Bertrand and Grubbs groups are interesting in that they have high selectivity and activity which resulted in the highest reported ethenolysis TON of 35,000 (23,24).

The observed ethenolysis TONs in Table 11.1 are considerably lower than metathesis TONs obtained with MO in other processes. Mol and Jackson (29,30) each reported TON values of greater than 440,000 in CM of MO with second-generation Grubbs catalysts where only 1,2-disubstituted olefins were present (Scheme 11.2).

Metathesis of internal olefins in the presence of ruthenium catalysts will produce alkyl-substituted ruthenium carbenes, or alkylidenes, which Grubbs showed to be more stable than the unsubstituted ruthenium carbenes (methylidenes) (31). Grubbs (32) later demonstrated that the ruthenium systems decompose faster in the presence of ethylene. Further, it was proposed that ethenolysis is complicated by the presence of the terminal olefin products, which have been reported to cause catalyst inhibition (18). However, the demonstrated TON of 295,000 using the second-generation Grubbs catalyst Ru-6 reported by Mol (29) for the self-metathesis of 1-octene shows that these species are

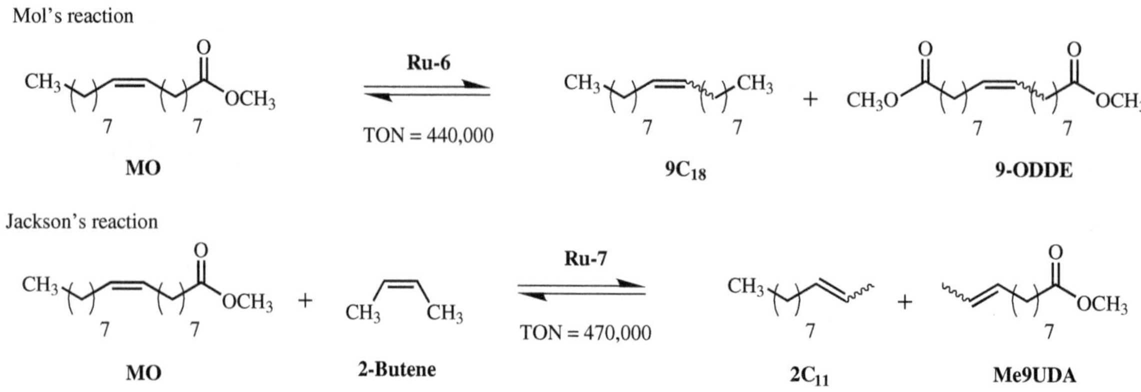

Scheme 11.2 Mol's and Jackson's high turnover number reactions. *Source*: Reproduced from Ref. (19) with permission from Springer.

not inherent catalyst poisons. This raised the question, "Why are ethenolysis TONs so low compared to Mol's and Jackson's TONs?" This question led to the investigation of ethylene surrogates for use in transforming natural oil feedstocks.

11.3 ALKENOLYSIS: ETHYLENE SURROGATES

To help understand the effect of ethylene on catalyst efficiency, the CM of MO was examined using alpha olefins as ethylene surrogates. Materia coined the term *alkenolysis* to describe the CM of an internal olefin with a terminal olefin (other than ethylene) (see Scheme 11.3).

The genesis of the alkenolysis idea was to increase the starting alpha olefin concentration in the metathesis reaction. Ethylene is a non-condensable gas and has poor solubility in seed oils, estimated at 0.1 M at 25 °C (18) which results in a poor driving force for the conversion of starting materials to products (33). Alpha olefins (i.e., propene, 1-butene, 1-hexene, and 1-octene) are soluble in all concentrations. For example, liquid neat 1-butene is 11.3 M and neat MO is 3.0 M; therefore, alkenolysis reactions can be run with [alpha olefin] ≫ [seed oil], unlike ethenolysis reactions. Solvents were avoided in ethenolysis and alkenolysis as they would have reduced our throughput and added more cost to the process.

The first set of experiments compared the metathesis efficiencies of subjecting the first-generation Grubbs catalyst and the second-generation Grubbs catalyst to MO and alpha-olefins (24). Remarkably, comparing Ru-5 in ethenolysis (Table 11.1, entry 7 TON_{Me9DA} = 600) and in octenolysis (Table 11.2, entry 6, TON_{Me9DA} = 23,000) resulted in nearly 40 times more Me9DA produced in the reaction using 1-octene. Similar octenolysis TON_{Me9DA} values were achieved with the same catalyst loadings of other second-generation catalysts Ru-6 and Ru-7 (Table 11.2, entries 1 and 9). Also of note, octenolysis efficiencies were not significantly lowered when soy FAME was used as a substrate in place of MO (Table 11.2, entries 15 and 16) (34). We were also pleasantly surprised that the alkenolysis process worked equally well with canola FAME (Table 11.2, entries 20 through 22). The compatibility with FAMEs is noteworthy since they are commodity feedstocks (biodiesel), and not specialty chemicals like MO (34). First-generation catalysts

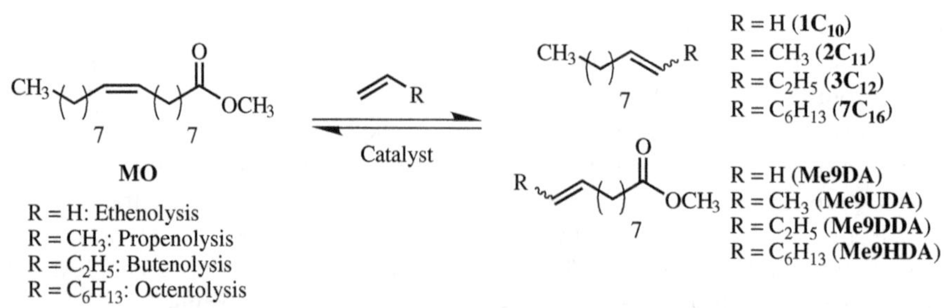

Key: **MO** is methyl oleate, $\mathbf{1C_{10}}$ is 1-decene, $\mathbf{2C_{11}}$ is 2-undecene, $\mathbf{3C_{12}}$ is 3-dodecene, $\mathbf{7C_{16}}$ is 7-hexadecene, **Me9DA** is methyl 9-decenoate, **Me9UDA** is methyl 9-undecenoate, **Me9DDA** is methyl 9-dodecenoate, **Me9HDA** is methyl 9-hexadecenoate

Scheme 11.3 Ethenolysis and alkenolysis of MO.

TABLE 11.2 Summary of Alkenolysis of Seed Oil FAMEs

Entry	Substrate	Olefin[a]	Catalyst (ppm)[b]	% Me9DA	TON_{Me9DA}[c]
1	MO	1-Octene	Ru-7 (10)	20.4%	20,400
2	MO	1-Octene	Ru-7 (5)	16.0%	32,000
3	MO	1-Octene	Ru-7 (5)[d]	12.7%	25,400
4	MO	1-Octene	Ru-7 (5)[e]	28.5%	57,000
5	MO	1-Octene	Ru-7 (5)[f]	29.5%	59,000
6	MO	1-Octene	Ru-5 (10)	23.0%	23,000
7	MO	1-Octene	Ru-5 (5)	18.7%	37,400
8	MO	1-Octene	Ru-5 (5)[g]	18.7%	37,400
9	MO	1-Octene	Ru-6 (10)	18.6%	18,600
10	MO	1-Octene	Ru-6 (5)	18.9%	37,800
11	MO	1-Octene	Ru-6 (5)[g]	20.8%	41,600
12	MO	1-Butene[h]	Ru-6 (5)	18.9%	37,800
13	MO	1-Butene	Ru-5 (5)	18.7%	37,400
14	MO	1-Butene	Ru-7 (5)	16.0%	32,000
15	Soy FAME	1-Octene	Ru-5 (10)	12.9%	12,900
16	Soy FAME	1-Octene	Ru-6 (10)	15.4%	15,400
17	Soy FAME	Propylene[i]	Ru-5 (10)	20.1%	20,134
18	Soy FAME	1-Butene[h]	Ru-2 (100)	7.5%	750
19	Soy FAME	1-Butene	Ru-1 (100)	8.5%	850
20	Canola FAME	Propylene[i]	Ru-5 (5)	16.9%	33,860
21	Canola FAME	1-Butene[g]	Ru-7 (10)	20.0%	20,000
22	Canola FAME	1-Octene	Ru-7 (10)	20.7%	20,700

[a]General conditions: neat; 40 °C; 3 equiv alpha olefin seed oil/double bond, 4 h.
[b]Catalyst loadings are defined as the moles of catalyst per million moles of seed oil double bonds in substrate (methyl oleate has 1 mole of olefins per mole; soy FAME has ~1.5).
[c]TON_{Me9DA} (turnover number based on Me9DA only) = 10,000 × (GC % of Me9DA)/(catalyst loading in mole parts per million).
[d]1 equiv of 1-octene used, 40 °C, 4 h.
[e]10 equiv of 1-octene used, 40 °C, 4 h.
[f]40 h.
[g]40 °C, 20 h.
[h]60 °C, 3–4 h.
[i]130 psi, 60 °C, 4 or 6 h.

produced lower yields of Me9DA even with high 100 ppm catalyst loadings (see Table 11.3, entries 18 and 19).

Catalyst loadings were reduced further by purification of the metathesis feedstocks to remove hydroperoxides. These catalyst poisons arise in aged samples of fatty acid esters, particularly those containing high levels of linoleates and linolenates (i.e., skipped diene- and triene-fatty acids) where the intermediate allylic radical is doubly stabilized (35). When soy FAME was treated with Magnesol® (36) (a magnesium silicate used to regenerate used oil in the food industry), significant improvements in TONs were observed (Table 11.3). Using as little as 1 mol ppm of catalyst Ru-5 per mole of substrate olefin was sufficient for reactivity, resulting in an impressive TON of 190,000 (39).

The improved efficiencies in alkenolysis versus ethenolysis can be rationalized by considering the ruthenium methylidene species (v) (Scheme 11.4, ligands removed for clarity). Ruthenium methylidenes have been shown to decompose more rapidly than substituted alkylidenes (32). While ethenolysis *must* proceed through methylidene intermediate (v), it is possible for alkenolysis to proceed without forming (v) (i.e., (i) → (ii) → (iii) → (iv) → (i)). Methylidene formation is known to be kinetically disfavored when an alternative mode of reactivity is possible (37), and it has been proposed that the benefits of alkenolysis are a

TABLE 11.3 The Effect of Magnesol® Treatment on Propenolysis of Soy FAME[a]

Entry	Pretreatment	Mole Parts Per Million of Ru-5[b]	GC % Me9DA[c]	TON_{Me9DA}[d]
1	None	25	24	9,470
2	None	5	<1	2,140
3	Magnesol®	5	24	48,120
4	Magnesol®	1	19	192,200

[a]General conditions: neat soy FAME; 130 psi of propylene; 60 °C; 4 h.
[b]Mole parts per million of catalyst defined as the number of moles of catalyst per million moles of olefins in soy FAME.
[c]Percentages refer to GC area %. The equilibrium amount of Me9DA under these conditions is approximately 25%.
[d]TON_{Me9DA} (turnover number based on Me9DA only) = 10,000 × (GC % of Me9DA)/(catalyst loading in mole parts per million).

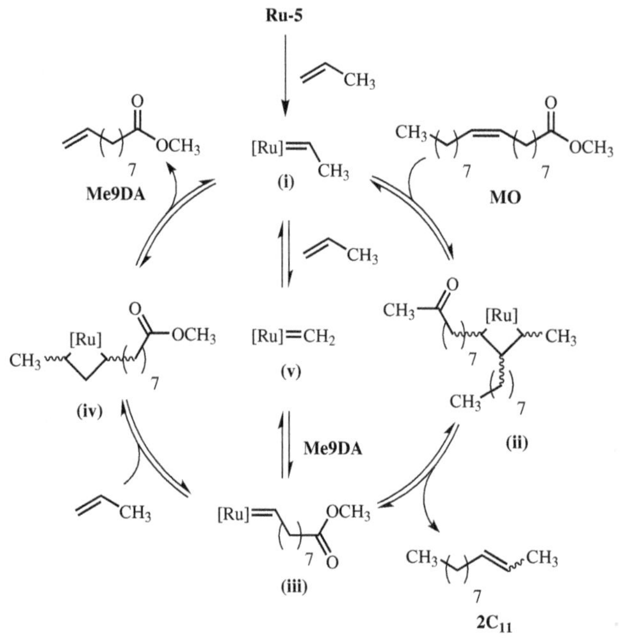

Scheme 11.4 Mechanistic proposal for alkenolysis circumventing ruthenium methylidene (v). *Source*: Reproduced from Ref. (19) with permission from Springer.

result of bypassing this kinetically disfavored and unstable intermediate (19).

The alkenolysis technique has proved to be commercially viable. In June 2010, Elevance Renewable Sciences announced a joint venture with Wilmar International to build a world-scale bio-refinery facility. This commercial-scale manufacturing facility will initially have the capacity to process 180,000 metric tons (~400 million pounds) of seed oils, with room for future expansion. The facility is under construction in Surabaya, Indonesia (16). In May 2011, Elevance Renewable Sciences announced it had completed a 1 million-pound (~500 metric tons) production run to produce specialty chemicals. That production run marked the largest scale in which a Grubbs ruthenium metathesis catalyst has been used (38).

General Procedure for the Ethenolysis of Methyl Oleate

In an argon-filled dry box, a Fisher–Porter vessel equipped with a pressure gauge and dip-tube was charged with MO (12 g MO, 40.5 mmol) and a solution of olefin metathesis catalyst (0.02 M) in dichloromethane was added to obtain the correct catalyst loading. The system was sealed, removed from the dry box, purged three times with ethylene, pressurized to 150 psi, and heated to 40 °C. Samples (0.2 ml) were collected via dip tube at routine intervals and immediately quenched by the addition of a solution of tris(hydroxymethyl)phosphine (1.0 ml, 1.0 M) in isopropanol. The samples were then heated for 1 h at 60 °C, partitioned between a 1 : 1 mixture of distilled water and hexanes (2.0 ml) and the organic phase was separated and analyzed by GC.

General Procedure for the Octenolysis of Methyl Oleate

In an argon-filled dry box, an oven-dried 20 ml scintillation vial equipped with a magnetic stir bar was charged with MO (3.00 g, 10.1 mmol) and 1-octene (3.41 g, 30.4 mmol). The vial was sealed with a cap containing a PTFE septum and heated to 40 °C. Olefin metathesis catalyst (0.02 M) in dichloromethane was added to obtain the correct catalyst loading via a syringe through the PTFE septum while stirring. After 4 h, a solution of tris(hydroxymethyl)phosphine (1.0 ml, 1.0 M) in isopropanol was introduced via syringe, the mixture heated at 60 °C for 1 h, partitioned between a 1 : 1 mixture of distilled water and hexanes (10 ml) and the organic phase was separated and analyzed by gas chromatography.

Magnesol® Treatment of Soy FAME

A three-necked, 500 ml round bottom flask equipped with a magnetic stir bar was filled with 300 g of soy FAME and agitated while sparging with nitrogen. The oil was heated to 80 °C for 45 min before Magnesol® (2.5 wt%) and Celite® (1.5 wt%) were added. Heating was continued, allowing adsorption to take place until optical clarity was achieved (≥1 h). The mixture was cooled to 40 °C and sparged with nitrogen before being filtered through a Buchner funnel fitted with #4 filter paper once and then twice through #2 filter paper. The filtrate was collected in amber glass containers, sparged with nitrogen for 5 min then blanketed with nitrogen for 1 min before being sealed.

General Procedure for the Butenolysis of Soy FAME

Soy FAME (6.8 kg, 22.9 mol, 34.3 mol of double bonds) and Ru-5 metathesis catalyst (0.71 g, 0.859 mmol, 25 mol ppm/double bond) were added to a 20 l Parr reactor and sparged with argon for 1 h. The mixture was heated to 60 °C and 1-butene (24–59 psi) was introduced using a one-way check valve to prevent back flow. After 4 h, a 5 ml sample was collected and immediately quenched by the addition of tris(hydroxymethyl)phosphine (1.0 ml, 1.0 M) in isopropanol. The sample was heated for

1 h at 60 °C, partitioned between a 1 : 1 mixture of distilled water and hexanes (2 ml), and the organic phase was separated and analyzed by gas chromatography, providing the values for % Me9DA shown in Table 11.2. The pressure was subsequently released and vented into a fume hood. When the reactor was at ambient pressure, a solution of tris(hydroxymethyl)phosphine (50 ml,1 M) in isopropanol was added and the reactor was sparged with argon and heated to 60 °C overnight (~18 h). The reactor was cooled to ambient temperature and the contents were transferred to a 12 l flask with bottom drain. The product was washed with 4 l of water and 4 l of brine. The neat organic phase was dried over sodium sulfate, filtered and distilled under reduced pressure to yield Me9DA (1.356 kg), Me9DDA (1.02 kg), 1-decene (459 g), and 3-dodecene (475 g).

11.3.1 Hydrogenated Metathesized Soybean Oil (HMSBO) Wax

Elevance Renewable Sciences, Inc. is commercializing waxes based on hydrogenated metathesized seed oils for use in a broad range of products, under the name NatureWax®. Hydrogenated metathesized soybean oil (HMSBO) can hold roughly double the fragrance of petroleum paraffin waxes in candles (40). This effect is due to the increased polarity of HMSBO relative to paraffin, which results in better suspension and dispersion of the polar fragrance components. HMSBO has also found use in skin care, hair care, and color cosmetics. Dow Corning the company commercializing products made from HMSBO, markets them as natural alternatives to petrolatum-based formulations with additional product benefits (41). HMSBO has found use in a number of other specialty wax applications as well, such as in paintball formulations, where it acts as a homogenizer (42).

Today paraffin is the primary material used in the production of wax-containing products such as candles. Paraffin wax displaced renewable beeswax as the common wax for candles as the petroleum refining industry matured and large amounts of paraffin were produced as a byproduct of fuel refining. As the cost of beeswax increased, due to its relative scarcity and inability to meet the market demand, paraffin wax became the low cost alternative. Recently, as the price of oil has increased and refineries are extracting higher value products from the refining process, the production of low cost paraffin is decreasing.

Partially hydrogenated soybean oil is referred to as *soywax*; it is a renewable and biodegradable alternative to paraffin candle wax but has had limited success in candles due to poor melting and solidification properties. Fully hydrogenated soybean oil has improved melting properties but has a brittle texture that is not acceptable in most candle applications (43). To overcome these shortcomings, researchers at Cargill subjected soybean oil to a self-olefin metathesis reaction prior to hydrogenation. This rearranges the groups around the triglyceride olefins, forming some higher oligomers and excising some alkyl tails. The overall effect after complete hydrogenation is a HMSBO with a broad melting point 10–16 °C lower than the traditional hydrogenated soybean oil product and suitable for most candle applications (43,44). This innovation represents an opportunity to replace petroleum-based paraffin with natural seed oil-based wax (40).

By using deoxygenated soybean oil containing less than 1 ppm hydroperoxides, self-metathesis equilibrium was reached using 0.005 mol% (50 mol ppm) of the second-generation Grubbs catalyst Ru-5 per double bond of seed oil (40,45,46). Subsequent hydrogenation in the same pot produced the HMSBO wax. Fortuitously, the process of hydrogenation over a supported catalyst removed the ruthenium metal from the product, delivering HMSBO with less than 0.1 ppm of ruthenium, following a simple filtration (47).

Synthesis of Hydrogenated Metathesized Soybean Oil (HMSBO) (40,47)

The metathesis of soybean oil has been demonstrated on multi-ton batch scale. In a typical reaction, 18,300 lb of soybean oil (RBD, Cargill) was sparged with nitrogen while warming to 70 °C. The metathesis catalyst, 451 g of catalyst Ru-5 (~50 mol ppm catalyst per mol of soybean oil) was stirred into 4 gallons of soybean oil and this slurry was added to the reactor. After 30 min, the reaction reached equilibrium. The reactor was charged with 12.5 kg of PRICAT 9925 (Johnson Matthey) hydrogenation catalyst and the headspace charged with hydrogen while warming to 120 °C. When the reactor reached 120 °C, hydrogen was introduced at 50 psi and over the next 3 h the exothermic reaction was carefully monitored to keep the temperature below 185 °C. Complete hydrogenation was confirmed by the iodine value being less than 1.

After the hydrogenation was complete, the reactor was cooled to 84–90 °C and the viscous product was filtered through diatomaceous earth to remove the supported nickel hydrogenation catalyst, delivering HMSBO containing less than 70 ppb of residual ruthenium.

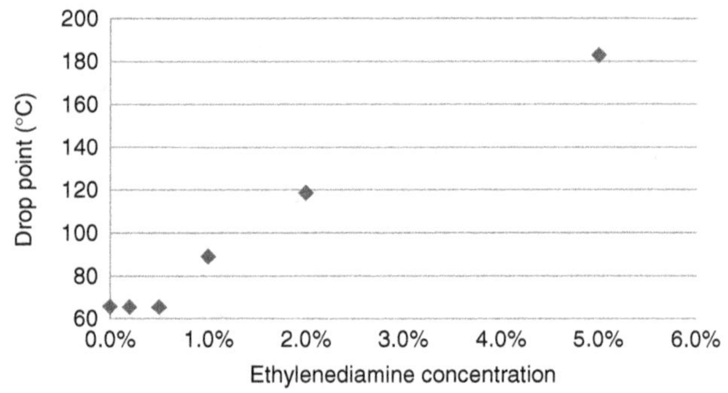

Figure 11.4 Relationship of ethylenediamine concentration on HMSBO wax drop point.

Higher Melting Point HMSBO Compositions

HMSBO has a relatively low melting point range (48–60 °C) which makes it well suited for candle wax and certain cosmetic applications; however a higher melting point is desired for use in cosmetic structuring agents (i.e., lipsticks and sunscreens), emulsifying/thickening agents (i.e., hair pomade, hand lotions), and in hot melt adhesives. These products are traditionally served by microcrystalline polyethylene and Fischer–Tropsch-based waxes (48).

A simple process to increase HMSBO's melting point is to cross link with ethylenediamine. In typical reactions, HMSBO is added along with the appropriate amount of ethylenediamine and 0.5% lithium carbonate. The reaction is degassed with nitrogen and heated to 120–180 °C for several hours. After heating, a vacuum is applied to remove water and any unreacted ethylenediamine. A trend between wax drop point and amount of ethylenediamine used was observed in concentrations from 0.5% to 5.0% (48) (Fig. 11.4).

11.3.2 α,ω-Diacids

For the past 40 years, starting with the pioneering fatty acid metathesis work of Boehlhouwer in 1972 (49), metathesis of fatty acids has been an attractive target for researchers. Significant improvements have been made over the last decade in the synthesis of renewable α,ω-diesters which are useful in coatings, surfactants, and polymers. Direct strategies into α,ω-diesters have included the self-metathesis of fatty acid esters such as MO to provide a statistical mixture of 9-octadecene and 1,18-dimethyl-octadec-9-enedioate, along with starting MO. Self-metathesis of oleic acid (neat) can produce higher yields because the α,ω-diacid crystallizes out of the reaction mixture, shifting the equilibrium toward products. However, the presence of the free carboxylic acid requires the use of higher catalyst loadings. Using 0.1 mol% of the second-generation Grubbs catalyst, 79% conversion of oleic acid is observed, delivering a TON of 790 (0.005 mol% gives 54% conversion and a TON of 10,800) (50). While self-metathesis is the most direct route to the α,ω-diacids, the use of naturally occurring mixtures containing polyunsaturated esters produces a complex mixture of α,ω-diacids products. As using purified MO or oleic acid is not economically viable, new higher yielding routes to α,ω-diesters/acids have been investigated.

Meier's group (51) performed a systematic study of the CM of seed oil substrates and methyl acrylate using ruthenium catalysts. CM of MO and methyl acrylate produced a mixture of 1,11-dimethyl undec-2-enedioate (2-UDDE), methyl 2-undecenoate (Me2UDA), and self-metathesis products 9-ODDE and $9C_{18}$, though increasing the amount of methyl acrylate from 5 to 10 equiv almost completely suppressed formation of the self-metathesis products 9-ODDE and $9C_{18}$ (Scheme 11.5). The first-generation Grubbs catalyst (Ru-1) performed poorly while the second-generation Grubbs (Ru-6) and second-generation Grubbs–Hoveyda (Ru-7) performed well. The CM of MO with 10 equiv of methyl acrylate, run neat, using 0.2 mol% Ru-7 at 50 °C for 18 h resulted in 97% conversion with 92% resulting from CM and only 5% from self-metathesis (51).

Separation of 2-UDDE and Me2UDA can be challenging. Therefore to simplify isolation of the desired α,ω-diesters, the use of terminal olefin containing fatty acid derivatives was examined. Methyl 10-undecenoate, produced from castor oil, smoothly led to high yields of 1,12-dimethyl-2-dodecadienoate. The CM of methyl 10-undecenoate with 10 equiv of methyl acrylate, run neat, using 0.1 mol% Ru-7 at 50 °C for 30 min yielded 99% conversion, 99% CM yields, with no self-metathesis products. Employing 5 equiv of methyl acrylate under the same reaction conditions yielded 99% conversion, with 96% CM yield and only 3% self-metathesis products. The driving force in this reaction was the removal of the ethylene as it was produced. A series of unsaturated α,ω-diesters (1,8-, 1,11-, 1,12-, 1,15-, and 1,20-) were prepared and characterized (51).

Behr's group (52) produced the α,ω-diester, 1-ethyl 12-methyl-dodec-2-enedioate (EMC_{12}), by CM of methyl 10-undecenoate (Me10UDA) with diethyl maleate. Figure 11.5 contains the catalysts used in this work and Scheme 11.6 shows the CM reaction.

Table 11.4 shows selected results from Behr's paper. The first observation is how well the first-generation catalysts (entries 1 and 3) performed, compared to the second-generation catalysts (entries 2, 4, and 6). Increasing the equivalents of diethyl maleate from 2 to 8 had no effect on the yield (entries 5 and 7).

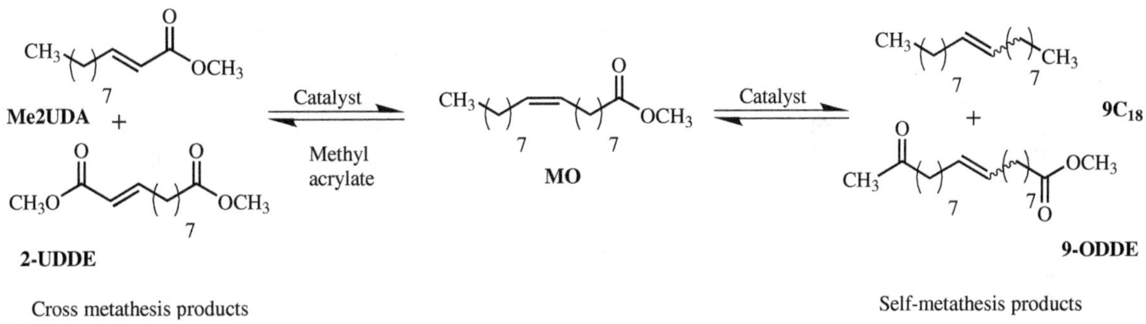

Scheme 11.5 Cross-metathesis of MO with methyl acrylate.

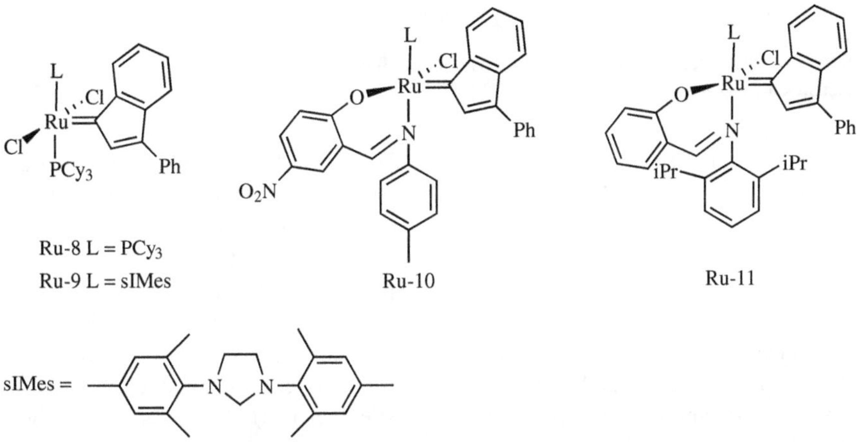

Figure 11.5 Additional well defined metathesis catalysts discussed in Behr's work.

However, increasing the temperature by 10 °C had a significant effect on higher yields of EMC_{12} (entry 8) compared to 50 °C (entry 5). The higher yields of EMC_{12} (entries 8 and 9) are attributed to the higher temperatures facilitating the secondary CM reaction of the by-products ethyl acrylate and the 1,20-dimethyl ester of 10-eicosene (10-ECDE) to produce EMC_{12}.

Several groups have reported the synthesis of ω-functionalized fatty acids using a palladium catalyzed isomerization–methoxycarbonylation process to produce 1,19-dimethyl esters from various seed oils with selectivities of greater than 95% (53–55). Jackson and coworkers (56) reported a clever process to produce 1,12-dimethyl dodecanedioate by a one-pot metathesis–isomerization–methoxycarbonylation–transesterification process from various seed oils (56). This group reported a TON of 470,000 for the 2-butenolysis of MO (refer back to Scheme 11.2), with the key to these high TONs being the use of triply distilled MO (30). Metathesis of MO, sunflower oil, or linseed oil with 10 equiv of 2-butene and 100 ppm of second-generation Grubbs catalyst produced greater than 98% of the 9-undecenoate intermediates. Interestingly, these three seed oils worked equally well, considering linseed oil contains a high percentage of polyunsaturated fatty esters (66% polyunsaturates) compared to high oleic sunflower oil (12% polyunsaturates) and MO (≪1% polyunsaturates). The palladium isomerization–methoxycarbonylation process produced excellent conversions (>98%) and selectivities (>95%) for the synthesis of dimethyl dodecanedioate from

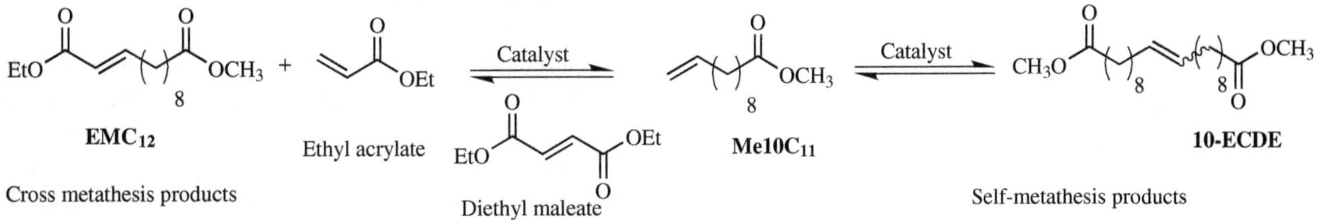

Scheme 11.6 Cross-metathesis of methyl 10-undecenoate with diethyl maleate.

TABLE 11.4 Cross Metathesis of Methyl 10-Undecenoate and Diethyl Maleate[a]

Entry	Catalyst	Temperature (°C)	Equivalents of DEM[b]	% Conversion	% Yield EMC_{12}[c]	% Yield 10-ECDE[c]
1	Ru-1	50	2	83	46	35
2	Ru-6	50	2	84	60	25
3	Ru-8	50	2	84	43	39
4	Ru-9	50	2	98	71	27
5	Ru-10[d]	50	2	98	68	29
6	Ru-11[d]	50	2	88	63	23
7	Ru-10[d]	50	8	98	71	27
8	Ru-10[d]	60	2	99	93	2
9	Ru-10[d]	70	2	99	91	4

[a]4 mol% catalyst, ω(toluene) = 0.5, 0.158 mol methyl 10-undecenoate, 900 rpm.
[b]DEM, diethyl maleate.
[c]EMC_{12}, 1-ethyl 12-methyl-dodec-2-enedioate; 10-ECDE, 1,20-dimethyl ester of 10-eicosene.
[d]100 equiv $PhSiCl_3$ to catalyst added.

the 2-butenolyzed metathesis product from MO, high oleic sunflower oil, and linseed oil.

11.4 TERPENES

Despite the fact that terpenes are a ready and abundant source of renewable oleochemical feedstocks, they have received considerably less attention than natural oils as feedstocks in metathesis reactions. In the 1990s, Nugent and coworkers (57,58) at DuPont turned to ring closing metathesis (RCM) of β-citronellene as a method to prepare the enantioenriched building block 3-methylcyclopentene (Fig. 11.6). The catalyst system they developed for this process was $WOCl_2(OAr)_2$ (Ar = 2,6-dibromophenyl) in combination with tetraethyl lead. This methodology is particularly powerful as both enantiomers of the terpene are readily available. In an early study of the reactivity of the first-generation Grubbs catalyst, Hoye and Zhao (59) performed a series of competition experiments between linalool and linalool analogs. At first glance, linalool would appear to be a challenging substrate for RCM because one of the olefins is trisubstituted and the other is flanked by a fully substituted tetrahedral carbon.

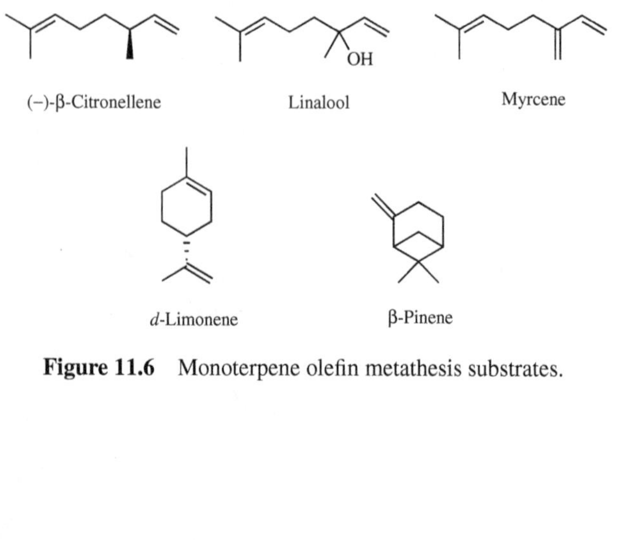

Figure 11.6 Monoterpene olefin metathesis substrates.

However, they found that the allylic alcohol in linalool promotes the RCM reaction, resulting in a fast ring closure. This process was more rapid for linalool than for linalool methyl ether or even citronellene, where the hydroxyl group is replaced by a hydrogen atom.

Harvey and coworkers (60) at the US Navy recently leveraged Hoye's observation in a concise and scalable route to convert linalool to a high density jet and missile fuel (60). They performed the RCM of neat linalool using the second-generation Hoveyda Grubbs catalyst (0.01 mol%) and obtained 44% yield of methylcyclopentenol (Scheme 11.7). This represents a TON of 4400, further underscoring the remarkable reactivity of the linalool system in RCM. The gaseous byproduct isopropylene is a value-added component to renewable fuel production as its hydrogenated oligomers can be blended into jet fuel. The researchers found that $AlPO_4/MgSO_4$-mediated dehydration of methylcyclopentenol led to isomeric mixtures of methylcyclopentadiene in good isolated yield, which could be converted to the known high density fuel RJ-4 (a mixture of hydrogenated methylcyclopentadiene dimers). More recently, other groups have reported the use of these and other monoterpenes to make small molecules using both RCM (61,62) and CM (63,64).

Hoye and Hillmyer (65) at the University of Minnesota collaborated to convert myrcene to 3-methylenecyclopentene and evaluate several polymerization methodologies (Scheme 11.8). The monoterpene myrcene is available in large quantities from the industrial processing of turpentine (66). Myrcene was treated with 0.2 mol% of the second-generation Grubbs catalyst (Ru-6) in decalin at 40 °C to provide 68% conversion of myrcene. Methylenecyclopentene, the low molecular weight diene, was then isolated by distillation along with the co-product isobutylene, which is itself a valuable raw material typically made from petrochemical sources. Akin to butadiene and isoprene, methylenecyclopentene was found to be compatible with

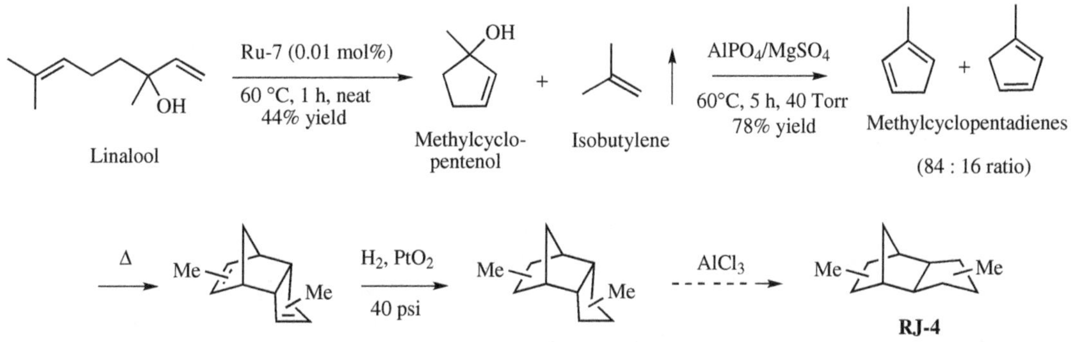

Scheme 11.7 Conversion of linalool to the high density fuel RJ-4.

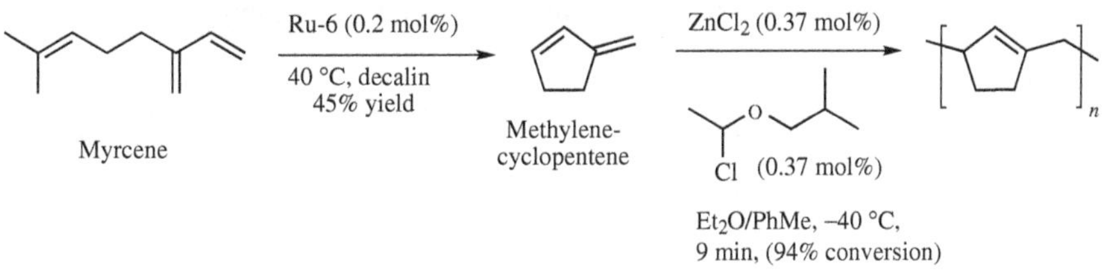

Scheme 11.8 Synthesis and polymerization of 3-methylenecyclopentene.

radical, anionic, and cationic polymerization processes. In particular, the living $ZnCl_2$-initiated cationic polymerization resulted in a polymer with a fully 1,4-polymer microstructure and T_g values between −17 and 11 °C depending on molecular weight. Though the various homopolymers may be of interest, it may also be possible to copolymerize this versatile monomer with butadiene or isoprene to access an even larger portfolio of materials.

11.5 NATURAL RUBBER

Natural rubber is a high molecular weight syndiotactic polyisoprene. Harvested from the latex of the rubber tree, most of the more than 10 million metric tons of production per year come from Southeast Asian plantations. Though there remains a strong demand for natural rubber for use in tires, consumer goods, and industrial products, there is interest in modifying natural rubber to produce grades with modulated properties. Owing to the polyunsaturated nature of natural rubber, any metathesis reactions involving it will be accompanied by intramolecular reactions resulting in chain scission and molecular weight reduction. A number of studies on the modification of natural rubber using metathesis have been reported since the breakthrough report by Alimuniar and Kohjiya that the combination of WCl_6 and $SnMe_4$ will affect this molecular weight reduction (67–71).

Tlenkopatchev and coworkers (72,73) recently used terpenes as chain terminating agents in the metathesis depolymerization of natural rubber (Scheme 11.9) (72,73). β-Pinene and *d*-limonene are excellent chain terminators because they both contain 1,1-disubstituted olefins, which are Type II olefins with respect to the Grubbs second-generation catalyst (74)—they can participate as chain terminators but will be slow to react with themselves to produce tetrasubstituted olefins.

11.6 OTHER RENEWABLE FEEDSTOCK OLEFINS

Cardanol is an abundant but underutilized natural raw material produced as a byproduct of the cashew nut industry. The process of roasting cashew nuts separates the nut from the inedible shells and also causes the shells to secrete an oil known as *cashew nut shell liquid* (*CNSL*), which is collected for various resin applications. When technical grade CNSL is processed by distillation, the resulting oil is comprised predominantly of cardanol, the name generally referring to the mixture of fatty phenols shown in Scheme 11.10 (75). Cardanol is similar to fatty acid esters in that it contains a lipophilic olefin connected to a functional tail. As with fatty acid esters, olefin metathesis offers the possibility to append functionality to the tail of cardanol to make bifunctional molecules. Vasapollo and coworkers (76) have recently published a series of reports exploring this possibility.

Reaction of cardanol with diethyl fumarate or *Z*-3-hexendioic acid in the presence of the Ru-7 delivered (R = CO_2Et and R = CH_2CO_2H), respectively. These molecules have the potential to be used as polymer co-monomers or functional additives. Vasapollo and coworkers (76) have subsequently extended this proof of

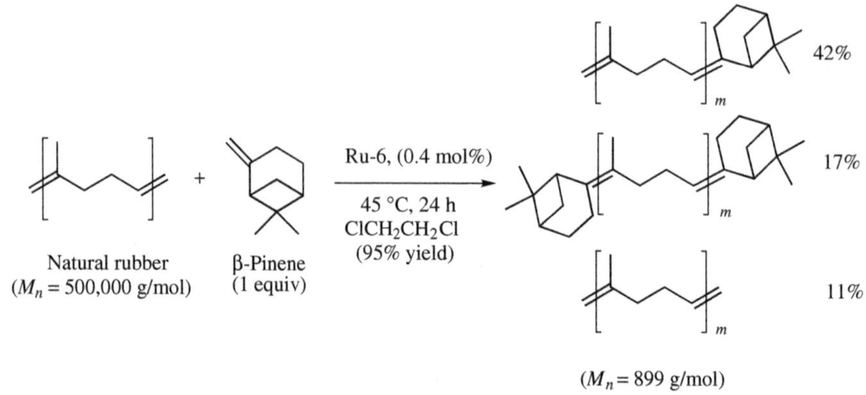

Scheme 11.9 Depolymerization of natural rubber with terpenes.

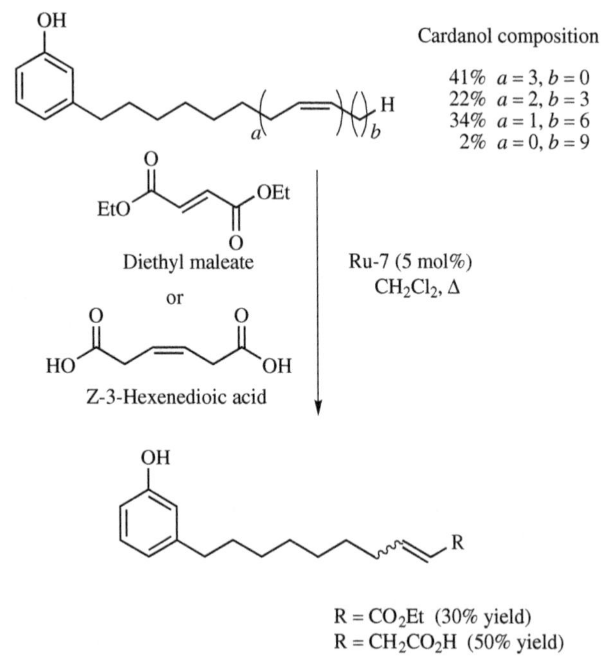

Scheme 11.10 Cardanol CM with fumarate and Z-3-hexenedioic acid.

concept to use cardanol to prepare a number of macrocyclic and dimeric porphyrins.

Though it does not naturally occur directly in significant amounts, renewable ethylene is poised to be the backbone of the renewable polymer and chemical industry. Braskem, the Brazilian chemical giant, has been producing high density polyethylene at its Triunfo with a capacity of 200,000 metric tons per year since 2010. The Braskem process begins with the fermentation of sugar cane to produce ethanol, which is dehydrated to make renewable ethylene and polymerized using standard methodology. In principle, the renewable ethylene could be used as a feedstock for a wide variety of traditional petrochemicals, and Braskem is planning the first step in this direction. They have announced plans for a 30,000–50,000 metric ton per year green polypropylene plant to come online in 2013. The green propylene process will consist of dimerization of green ethylene to butene, then metathesis of the produced 2-butene with ethylene to produce fully renewable propylene (77). This technology has been commercialized elsewhere and is a relative of the earliest commercialized olefin metathesis reaction, the Phillips Triolefin Process (7). The long-term commercial sustainability of this approach to green polymers is yet to be proved, but Braskem is leading the charge.

11.7 CONCLUSION

The replacement of traditional petrochemicals with renewable feedstocks is still in the early stages. This will no doubt remain an important area of research as petroleum reserves continue to be depleted. Though it is not clear today what technologies will ultimately prove to be sustainable, olefin metathesis has demonstrated the potential to make a significant impact on this worthwhile endeavor.

REFERENCES

1. Corma A, Iborra S, Velty A. *Chem Rev* 2007;107:2411.
2. Vennestrøm PNR, Osmundsen CM, Christensen CH, Taarning E. *Angew Chem Int Ed* 2011;50:10502.
3. Clark, J.H.; Deswarte, F.E.I. The biorefinery concept—an integrated approach Introduction to Chemicals for Biomass. Clark, J. and Deswarte, F., John Wiley and Sons 2008. ISBN 978-0-470-05805-3.
4. Mol JC, Buffon R. *J Braz Chem Soc* 1998;9:1.
5. Mol JC. *Green Chem* 2002;4:5.
6. Banks RL, Bailey GC. *Ind Eng Chem Prod Res Dev* 1964:170.
7. Mol JC. *J Mol Catal A: Chem* 2004;213:39.
8. Trimmer MS. In: Grubbs RH, editor. *Handbook of Metathesis: Catalyst Development*. Vol. 3. Weinheim, Germany: Wiley-VCH Verlag GmbH; 2008 Chapter 12.

9. Trnka TM, Grubbs RH. *Acc Chem Res* 2001;34:18.
10. Schrodi Y, Pederson RL. *Aldrichim Acta* 2007;40:46.
11. Schrock RR. *Chem Rev* 2009;109:3211.
12. Hartford B. Z-Selective metathesis of macrocycles. *Chem Eng News* 2011;89(45):11.
13. Yu M, Wang C, Kyle AF, Jakubec P, Dixon DJ, Schrock RR, Hoveyda AH. *Nature* 2011;479:88.
14. USDA. Oilseeds: world markets and trade monthly circular. http://www.fas.usda.gov/oilseeds/circular/Current/. See Table 1: Major Oilseeds: World Supply and Distribution (Commodity View).
15. Biermann U, Bornscheuer U, Meier MAR, Metzger JO, Schafer HJ. *Angew Chem Int Ed* 2011;50:3854.
16. http://www.elevance.com/media/news-releases/elevance-renewable-sciences-announces-joint-venture-with-wilmar-international-to-build-world-scale-biochemical-refinery/.
17. When using MO, well defined products were produced, the TON was defined as (% conversion multiplied by % selectivity) divided by catalyst loading. When SBO or FAMEs, were used, complicated product mixtures were produced. Therefore to compare the alkenolysis data in a meaningful manner, the TON_{Me9DA} was used. TON_{Me9DA} was defined as % GC yield of Me9DA divided by catalyst loading.
18. Burdett KA, Harris LD, Margl P, Maughon BR, Mokhtar-Zadeh T, Saucier PC, Wasserman EP. *Organometallics* 2004;23:2027.
19. Nickel A, Ung T, Mkrtumyan G, Uy J, Lee CH, Stoianova D, Papazian J, Wei W-H, Mallari A, Schrodi Y, Pederson RL. *Top Catal* 2012;55:518–523.
20. Warwel S, Brüse F, Demes C, Kunz M, Rüsch gen Klaas M. *Chemosphere* 2001;43:39.
21. Zullo JL, Anderson JC, Kaido H, Pederson RL, Schrodi Y, Sperber WH, Tupy MJ, Wagener EH. US Patent no. 7,951,232, to Elevance Renewable Sciences, Inc.
22. De Souza ML, Zullo JL, Anderson JC, Pederson RL, Schrodi Y. WO 2008/140469 A2, Priority Date 2006 Oct 13, to Elevance Renewable Sciences, Inc.
23. Anderson DR, Ung T, Mkrtumyan G, Bertrand G, Grubbs RH, Schrodi Y. *Organometallics* 2008;27:563.
24. Schrodi Y, Ung T, Vargas A, Mkrtumyan G, Lee CW, Champagne TM, Pederson RL, Hong SH. *Clean Soil Air Water* 2008;36:669.
25. Thomas RM, Keitz BK, Champagne TM, Grubbs RH. *J Am Chem Soc* 2011;133:7490.
26. Marinescu SC, Schrock RR, Müller P, Hoveyda AH. *J Am Chem Soc* 2009;131:10840.
27. Grubbs RH, Nguyen ST, Johnson LK, Hillmyer MS, Fu GC. WO 9604289. 1996.
28. Forman GS, McConnell AE, Hanton MJ, Slawin AMZ, Tooze RP, van Rensburg WJ, Meyer WH, Dwyer C, Kirk MM, Serfontein DW. *Organometallics* 2004;23:4824.
29. Dinger MB, Mol JC. *Adv Synth Catal* 2002;344:671.
30. Patel J, Mujcinovic S, Jackson WR, Robinson AJ, Serelis AK, Such C. *Green Chem* 2006;8:450.
31. Ulman M, Grubbs RH. *J Org Chem* 1999;64:7202.
32. Hong SH, Wenzel AG, Salguero TT, Day MW, Grubbs RH. *J Am Chem Soc* 2007;129:7961.
33. We made the assumption that the concentration of ethylene in the seed oil was at its saturation concentration throughout the ethenolysis reaction.
34. Soy FAME contains methyl oleate (~24%), methyl linoleate (~53%), methyl linolenate (~7%), and saturated methyl esters (~16%).
35. Lysenko Z, Maughon BR, Bicerano J, Burdett KA, Christenson CP, Cummins CH, Dettloff ML, Schrock AK, Thomas PJ, Varjian RD, White JE, Maher JM. US Patent no. 7,745,652, assigned to Dow Global Technologies, Inc.
36. Magnesol® XL. http://www.dallasgrp.com/web/products.aspx?id=35.
37. Wenzel AG, Blake G, VanderVelde DG, Grubbs RH. *J Am Chem Soc* 2011;133:76429.
38. http://www.elevance.com/media/news-releases/elevance-renewable-sciences-increases-production-of-specialty-chemicals/.
39. Schrodi Y, Pederson RL, Kaido H, Tupy MJ. US Pat. App. 2010/0145086, assigned to Elevance Renewable Sciences, Inc.
40. Murphy TA, Tupy MJ, Abraham TW, Shafer A. US Patent Application 2009/0217568.
41. Product Information Personal Care Dow Corning® HY-3051 Soy Wax Blend. http://www4.dowcorning.com/DataFiles/090007c8803140aa.pdf.
42. Murphy TA, Aufdembrink BA. US Patent Application 2009/0126602.
43. Refvik MD, Larock RC, Tian Q. *J Am Oil Chem Soc* 1999; 76:93.
44. Refvik MD, Larock RC. *J Am Oil Chem Soc* 1999;76:99.
45. Uptain K, Tanger C, Kaido H. US Patent Application 2011/031380.
46. Catalyst loadings are defined as moles of metathesis catalyst to moles of seed oil double bonds. Soybean oil has an average of 4.5 double bonds per triglyceride.
47. Tupy MJ, Amore F, Kaido H, Meng X. US Patent 8,115,021.
48. Mujkic M, Cade D, Lee CW, Starch MS, Swanton BL. WO 2012/006324.
49. van Dam PB, Mittelmeijer MC, Boehlhouwer C. *J Chem Soc Chem Commun* 1972:1221–1222.
50. Ngo HL, Jones K, Foglia TA. *J Am Oil Chem Soc* 2006;83:629.
51. Rybak A, Meier MAR. *Green Chem* 2007;9:1356.
52. Behr A, Gomes JP, Bayrak Z. *Eur J Lipid Sci Technol* 2011; 133:189.
53. Jimenez-Rodriguez C, Eastham GR, Cole-Hamilton DJ. *Inorg Chem Commun* 2005;8:878.
54. (a) Quinzler D, Mecking S. *Angew Chem* 2010;122:4402. (b) Quinzler D, Mecking S. *Angew Chem Int Ed* 2010;49:4306.
55. (a) Cole-Hamilton DJ. *Angew Chem* 2010;122:8744. (b) Cole-Hamilton DJ. *Angew Chem Int Ed* 2010;49:8564.
56. Zhu Y, Patel J, Mujcinovic S, Jackson WR, Robinson AJ. *Green Chem* 2006;8:746.

57. Nugent WA, Feldman J, Calabrese JC. *J Am Chem Soc* 1995; 117:8992.
58. Feldman J, Nugent WA. US Patent 5,516,953. 1996 May 14.
59. Hoye TR, Zhao H. *Org Lett* 1999;1:1123.
60. Meylemans HA, Quintana RL, Goldsmith BR, Harvey BG. *ChemSusChem* 2011;4:465.
61. Conrad JC, Parnas HH, Snelgrove JL, Fogg DE. *J Am Chem Soc* 2005;127:11882.
62. Vieille-Petit L, Clavier H, Linden A, Blumentritt S, Nolan SP, Dorta R. *Organometallics* 2010;29:775.
63. Bilel H, Hamdi N, Zagrouba F, Fischmeister C, Bruneau C. *Green Chem* 2011;13:448.
64. Borré E, Dinh TH, Caijo F, Crévisy C, Mauduit M. *Synthesis* 2011;13:2125.
65. Kobayashi S, Lu C, Hoye TR, Hillmyer MA. *J Am Chem Soc* 2009:7960.
66. Behr A, Johnen L. *ChemSusChem* 2009;2:1072.
67. Alimuniar A, Yarmo MA, Rahman MZA, Kohjiya S, Ikeda Y, Yamashita S. *Polym Bull* 1990;23:119.
68. Solanky SS, Campistron I, Laguerre A, Pilard J-F. *Macromol Chem Phys* 2005;206:1057–1063.
69. Wolf S, Plenio H. *Green Chem* 2011;13:2008.
70. Kongparakul S, Ng FTT, Rempel GL. *Appl Catal A: Gen* 2011; 405:129.
71. Kongparakul S, Ng FTT, Rempel GL. *Top Catal* 2012;55:524.
72. Gutiérrez S, Tlenkopatchev MA. *Polym Bull* 2011; 66:1029–1038.
73. Martínez A, Gutiérrez S, Tlenkopatchev MA. *Molecules* 2012;17:6001–6010.
74. Chatterjee AK, Choi T-L, Sanders DP, Grubbs RH. *J Am Chem Soc* 2003;125:11360.
75. High production volume (HPV) chemical challenge program: exposure and use data for cashew nut shell liquid. Downloaded from the EPA website http://www.epa.gov/hpv/pubs/summaries/casntliq/c13793rr3.pdf on 2013 Jan 12.
76. Vasapollo G, Mele G, Del Sole R. *Molecules* 2011;16:6871.
77. Harracksingh R. Bioplastics near commercialization. *ICIS Chem Bus* 2012:32.

12

CHALLENGES AND OPPORTUNITIES FOR SCALING THE RING-CLOSING METATHESIS REACTION IN THE PHARMACEUTICAL INDUSTRY

KEITH R. FANDRICK, JOLAINE SAVOIE, NATHAN YEE, JINHUA J. SONG, AND CHRIS H. SENANAYAKE
Chemical Development, Boehringer Ingelheim Pharmaceuticals Inc., Ridgefield, CT

12.1 INTRODUCTION

The metathesis reaction has emerged as a fundamental carbon–carbon bond-forming process in organic synthesis (1). There are two major variants of the metathesis reaction: the intermolecular cross-metathesis (CM) reaction and the intramolecular ring-closing metathesis process (RCM, Fig. 12.1). The RCM process is further subdivided between the ene–ene/yne–yne RCM and ene–yne versions (2). Although significant work has been conducted in academia with the RCM reaction, the implementation of this technology in industry is less pronounced. The CM (3) reaction has been utilized in the petrochemical industry (4) since the 1950s using heterogeneous catalysis. Some notable examples where this technology has been utilized on the commercial scale are the Shell higher olefin process (SHOP) (5), the polymer industry (6), and the refinement of plant oils (7). With the advent of the homogeneous metathesis catalyst, the intramolecular version (RCM) has seen significant advancement in academia since the seminal publications of Grubbs and Fu in 1992 (8). Because of the need for increased complexity in active pharmaceutical ingredients (APIs) due to diminishing patent space and requisites for selectivity, this technology has seen increased usage in medicinal chemistry. Owing to the inherent delay in the hand-off of chemical entities from discovery to process development, the initial reports of the use of the RCM in large-scale pharmaceutical settings have begun to emerge.

12.2 CATALYST

There are specific attributes required for metathesis catalysts for large-scale pharmaceutical synthesis. First, the requirement for bioavailability generally imparts highly polar and coordinating functionality to the API. Second, the APIs are becoming more complex due to diminishing intellectual property (IP) space, along with increased requirements for site selectivity. This complexity sometimes results in diverse functional groups within a single structural unit. Third, in order to maintain a robust catalyst system in large-scale plant operations, air and moisture sensitivity is a concern where a single batch failure can cost in excess of several hundred thousand dollars and jeopardize the API delivery. In contrast to typical methodological development where diversity in substrate tolerance is desired, in process development the substrate is fixed and diversity in catalyst structure and reaction parameters is advantageous. Overall, process robustness in catalyst stability and performance is the main requirement for large-scale RCM catalysts.

The API attributes and large-scale constraints significantly limit the usefulness of the Mo-based catalysts. In particular, Schrock's Mo catalyst **I** (Fig. 12.2) (9) has found limited success in large-scale API synthesis due to its air and moisture sensitivity, coupled with its limited compatibility with polar functionality. These limitations are significantly modulated with the Ru based systems (10). In particular, Grubbs's first generation catalyst **II** (11) is tolerant of diverse

Olefin Metathesis: Theory and Practice, First Edition. Edited by Karol Grela.

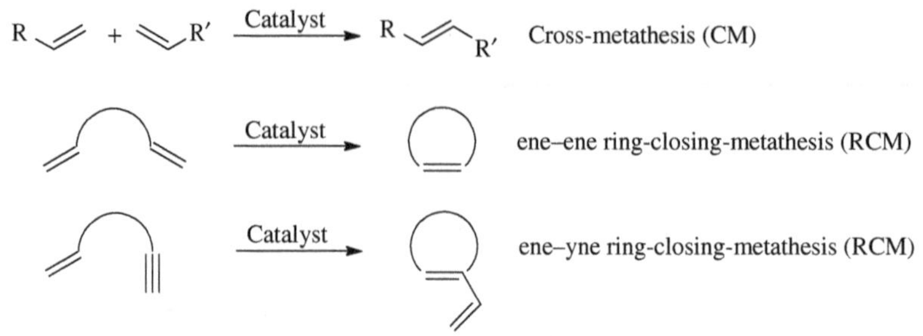

Figure 12.1 Variants of the Metathesis reaction.

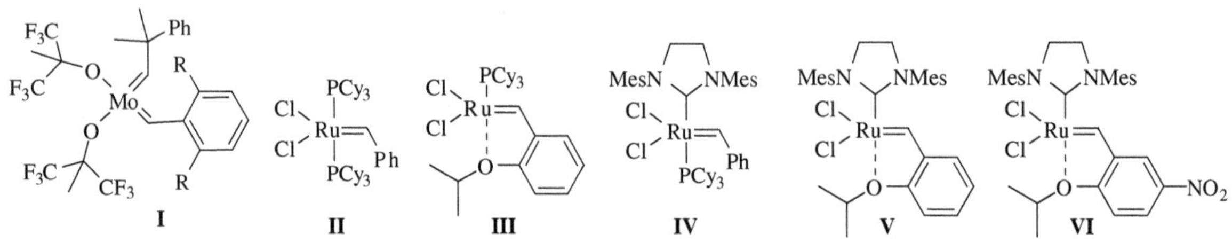

Figure 12.2 Commonly utilized metathesis catalysts.

polar functional groups but the system still suffers from limitations regarding sterics and electron-rich and -deficient olefins. In an effort to increase the stability of the Ru-based metathesis catalysts, Hoveyda and co-workers (12) modified Grubbs' first generation catalyst with a chelated ether linkage (catalyst **III**). This modification not only increased the air and moisture stability of the complex but also offered possibilities for recycling. As chemical entities move through development and onto commercialization, the ability to recycle the catalyst will be an important economic consideration for a commercial process. Grubbs and coworkers (13) subsequently modified the first generation catalyst by replacing the axial phosphine ligand with an *N*-heterocyclic carbene (catalyst **IV**). This catalyst is commonly referred to as the *Grubbs second generation catalyst*. This complex not only shows increased air and moisture stability but significantly higher reactivity. Catalyst loadings as low as 0.05 mol% have been reported with the Grubbs second generation catalyst (13). Hoveyda (14) combined the stability of the chelated ether linkage with the activity increase with the *N*-heterocyclic carbene ligand in catalyst **V**. This catalyst is both highly reactive but also shown to be recoverable. Furthermore, Grela and coworkers (15) were able to increase the activity of the Hoveyda–Grubbs catalyst **V** while retaining the desirable catalyst stability by introducing a strongly electron withdrawing group on the appended isopropoxyphenyl ring (nitro, catalyst **VI**). These catalysts are highly modular, allowing significant structural modification (16,17), to adapt to the structural requirements of the API/intermediate and the constraints of large-scale synthesis. The commercial availability of these catalysts (18) and derivatives allows rapid screening for API development. However, the IP positions of many of these systems impart limitations in both economic considerations (19) as well as supplier dependability.

12.3 LARGE-SCALE CHALLENGES

One of the tools to maintain cost-competitive API production and development within the reality of increasing API complexity is the introduction of new technologies such as transition metal catalysis (20). This technology has demonstrated significant economic and safety advantages for large-scale API production. Two main facets to cost competitiveness is reduction in time to market (rapid API development) and synthetic efficiencies. In addition, RCM offers new opportunities to synthesize complex chemical entities efficiently and this technology has increasingly been employed in medicinal chemistry. As the compounds move through development to large-scale API synthesis, chemical development has started to address some of the unique challenges of RCM and the catalyst thereof.

The two main drivers for large-scale API development can be classified as economic and safe (Fig. 12.3). As is the case for the development of transition metal catalysts on a large scale (21), the factors governing the economics of the catalyst are important. Most of the RCM catalysts are commercially available both on a small scale for screening and bulk quantities for API production (22). However, most of the commonly utilized ruthenium-based catalysts are currently IP protected and thus present strategic business challenges such as single supplier security (23) as well as long-term production arrangements. Furthermore, the complexity of the catalyst, along with the fixed metal cost (24), generally

	Challenge	Optimization Parameter
Economic	• IP status/catalyst cost • Catalyst bulk availability	Low catalyst loading
	• Catalyst robustness	Catalyst design
	• Batch efficiency • Green chemistry	High batch concentration
Safety	• Toxic metal exposure (patient and plant operator) • Carryover metal to noncompatible chemistry	Low catalyst loading Metal remediation
	• Impurity control • Process safety	Process optimziation and design

Figure 12.3 RCM catalyst challenges and optimization parameters.

results in high cost of goods. The most direct means for decreasing the catalyst cost is to focus on lowering the catalyst loading. Generally the RCM catalysts are employed on 2–10 mol% in academia, but for a cost-effective process these loadings are targeted to less than 1 mol%. Furthermore, the stability of the catalyst both in the given process and plant operational handling are important parameters to consider in the development of RCM reaction for production. Although catalyst and/or metal recycling have been developed for production-scale catalysts, this technology has not been explored for the RCM catalysts due to their relatively recent implementation in large-scale API synthesis.

The most important consideration in large-scale development is the safety of the patient and clinical trial volunteers, as well as the plant operators. As with all transition metal catalysts, residual metal content in the final API and intermediates requires control strategies to mitigate the risks associated with these impurities. The 2007 draft European Medicines Agency (EMEA) guidelines for residual metals (25) list ruthenium as class 1B (26) as there is known genetic toxicity of some ruthenium complexes (27). The proposed permitted daily exposure (PDE) of ruthenium is 100 μg/day for oral exposure and 10 μg/day for parenteral exposure. Furthermore, the carryover of the metal to downstream chemistries and intermediates can present significant safety hazards, especially if non-compatible reagents or conditions are employed. Significant resources are dedicated toward process optimizations and analytics for trace metal remediation and control.

The throughput of a given batch is an important target parameter for optimization in large-scale development. A single 1 m^3 batch (1000 l) can cost in excess of \$100,000 and increasing batch efficiency can have a significant effect on the cost competitiveness of a given process. RCM, especially of large ring systems, poses a formidable challenge in this regard as more concentrated reactions are plagued with intermolecular CM and polymerization. Not only can this lead to uncontrolled impurity formation (process robustness) but will also negatively affect all of the economic batch parameters due to the corresponding lower yields.

Transition metal catalysts generally require extensive optimization and development. In contrast to typical methodological development where substrate diversity is desired, the substrate for process development is usually fixed and the focus is the optimization of the reaction parameters and catalysts for one specific or closely related substrate species. This is particularly the case for RCM where a complex fixed target with polar coordinating functionalities is commonly employed. Several major pharmaceutical companies have established dedicated automation/catalysis groups with associated analytical support for expedient catalyst screening and development to shorten the time to market (28). A single day eroded from IP protection on a blockbuster drug (>\$1B in annual sales) costs in excess of \$2.7M. Furthermore, catalyst design within pharmaceutical chemical development has emerged to facilitate this process as well as to discover new catalytic entities that are IP free (29) and designed for the requirements of large-scale API synthesis (30).

The implementation of RCM catalysts on large-scale API synthesis has started to emerge. This review will cover the published examples of the RCM reaction on multi-kilogram scale and the associated optimization and development of this technology with regards to addressing the unique challenges of large-scale RCM catalysts.

12.4 LARGE-SCALE RCM PHARMACEUTICAL SYNTHESIS

Despite a high interest for the RCM reaction as synthetic tools in the pharmaceutical industry, the RCM reaction is not commonly applied to form a bicyclic compound (31). Wallace and colleagues (32) at Merck Research Laboratories reported a selective double RCM reaction of tetraene **1** to afford 1-oxo-7-azaspirodecane **2** (Scheme 12.1), an NK1 receptor antagonist (33). The spiroalkene was previously achieved via a three-step synthesis using reductive Heck couplings. However, the synthesis of the necessary intermediate involves seven steps and requires a low yielding resolution. The initial development targeted a shorter and more economical synthesis of the key spirocyclic compound. Modification of the alkyl chain on the oxygen as well as on the nitrogen was a requisite of the alternate synthesis. The implementation of a diastereoselective double RCM fulfilled these requirements while offering many possibilities of substrate variation.

The RCM *N*-tosylated precursors were readily obtained by the tosylation of the commercially available amino acid ester followed by a cerium mediated double allylation of the carbonyl and allylation of the tertiary alcohol and nitrogen. The RCM reaction was achieved by using a 0.05 M solution of tetraene **5** (Table 12.1) in chloroform with 5–7 mol%

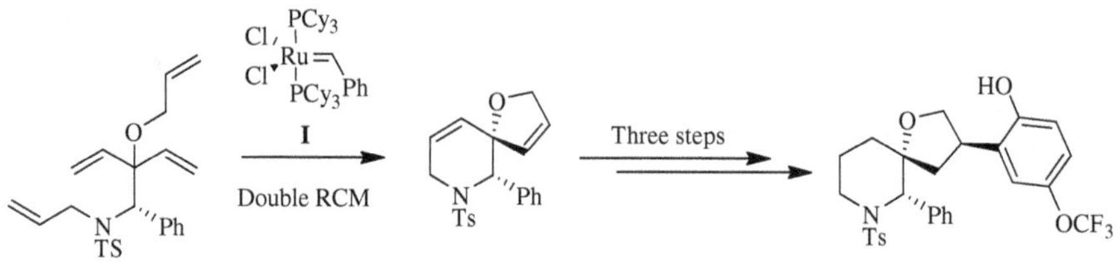

1 2a 3

Scheme 12.1 Double RCM for synthesis of **3**, a NK1 receptor antagonist.

TABLE 12.1 Survey of Substituents in the RCM of Tetraenes 5

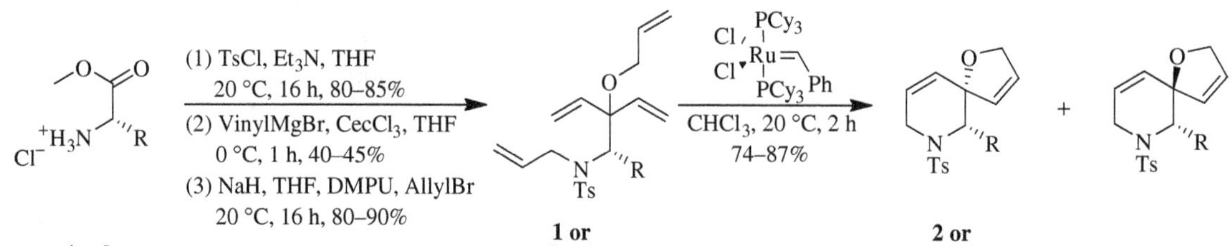

4a–d 1 or 5a–d 2 or 6a–d 7a–d

Tetraene	R	Yield (%)	**6:7**
5a	Me	74	96:4
5b	*i*-Pr	84	96:4
5c	*i*-Bu	76	96:4
5d	Bn	87	96:4
1	Ph	86	85:15

of Grubbs first generation catalyst (**II**). The separable spirocyclic compounds were produced in moderate to good yields (74–86%). Despite the variation of the substituent, the diastereoselectivity remained at approximately 95 : 5 favoring the desired (*5R*, *6S*)-isomer (**5a–d**). However, a lower selectivity was observed with the phenyl substituent (85 : 15). The exact stereochemistry of the spirocyclic products **6** and **7** were confirmed by nuclear overhauser effect (NOE) studies, which also indicated a pseudo axial orientation of the appended phenyl ring.

Molecular modeling studies were performed to explain the variation of selectivity and the source of the stereocontrol. The computational studies (34) indicate no difference in energy between the two diastereoisomers when the substituents are methyl or an isopropyl, thus implying the reaction is under kinetic control. However, when the substituent is a phenyl group, a higher energy difference is observed (~2.6 kcal/mol). The high energy difference between the two diastereomeric products is inconsistent with the observed stereoselectivity and further suggests the reaction is under kinetic control. In order to allow a thermodynamic control, the reaction was performed using a more reactive second generation metathesis catalyst (Scheme 12.2). The survey showed poor selectivity with second generation catalyst **IV** and **V** as compared to the first generation catalyst **II** and **III**.

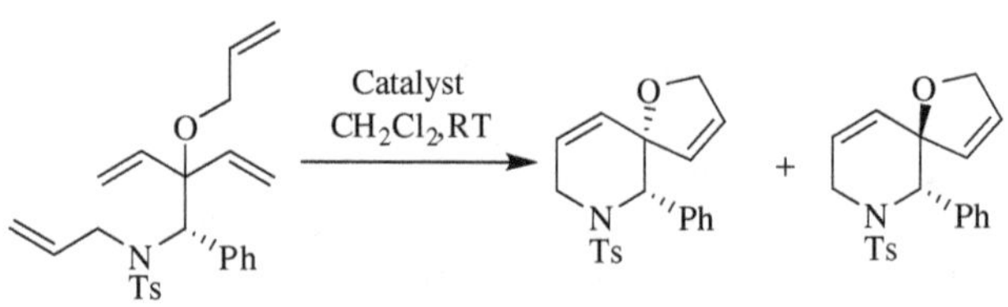

1 2a 2b

Scheme 12.2 Survey of RCM catalyst.

To further study the spirocyclization, an analysis of the corresponding intermediates was undertaken. The first RCM reaction can proceed through either path A, which is the formation of the five-membered ring monocyclic intermediate, or path B, forming the six-membered ring first (Scheme 12.3). Each of those pathways provides two sets of enantiomers. Pathway A is favored approximately 10/1 with no selectivity for **8a** over **8b**, independent of the catalyst. The first generation catalyst favors (90 : 10) **9b** over **9a** when path B occurs. Moreover, no selectivity (1:1) is observed with the second generation catalyst. The expected rate-limiting step in the RCM is usually the intermolecular reaction with the allyl group as the intramolecular cyclization is postulated to be rapid (17). Scientists at Merck proposed that the dominance of path A was derived from the preferred interaction for the catalyst with the oxygen-bound ally group rather than

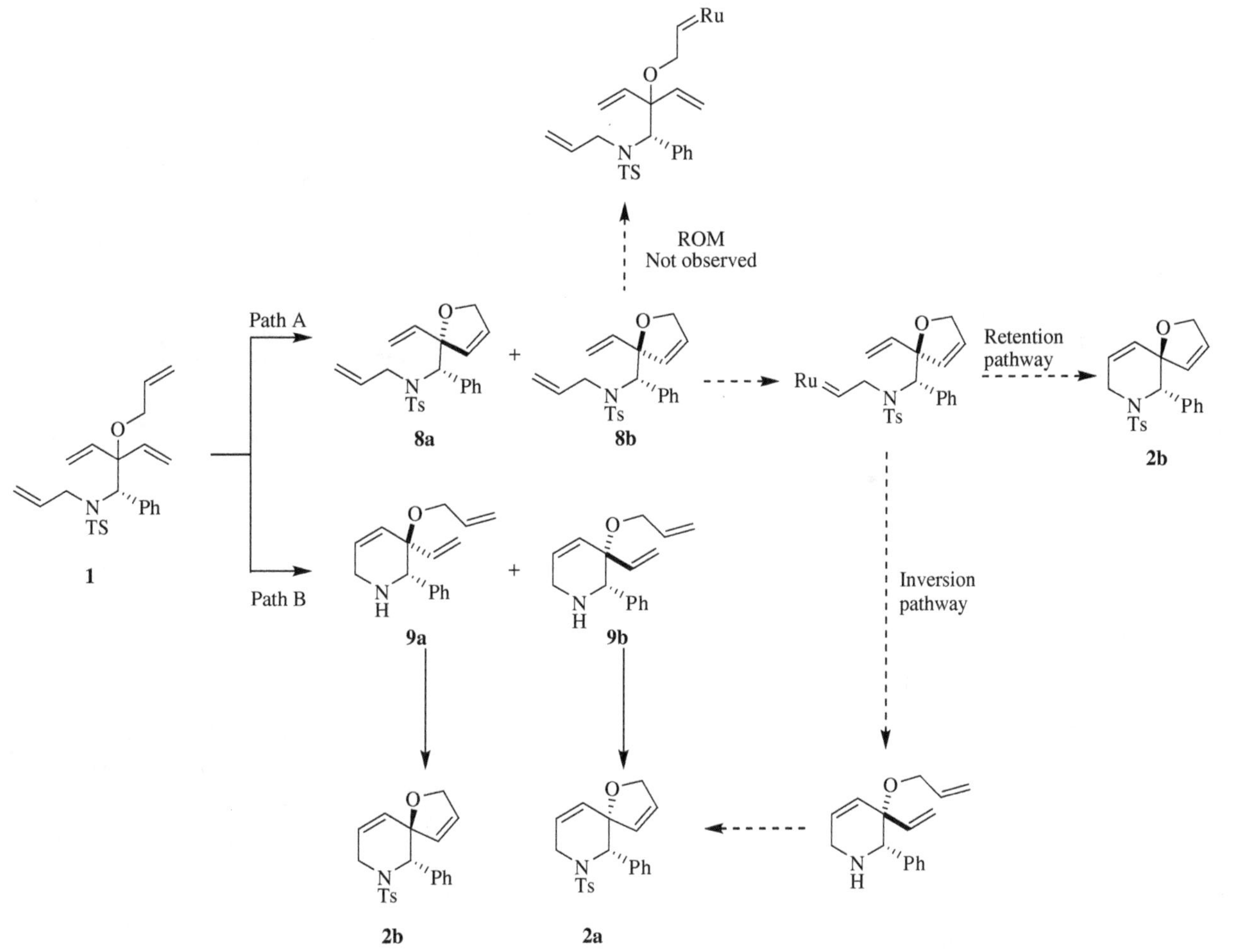

Scheme 12.3 The different possible RCM/ROM pathways.

the kinetic preference of five-membered ring formation. Furthermore, it was proposed that the low selectivity or catalyst dependence may be explained by the rapid imminent formation of the five-membered ring.

The relative ratio of the intermediate **8a/b** and **9a/b** remained the same during the course of the reaction, independent of the catalysts used. Nevertheless, the observed ratio of **8a/b** and **9a/b** is not reflected in the product distribution. It is proposed that the alteration of initial stereochemistry must be the result of the second cyclization. Resubmission of reaction intermediate **8a** and **8b** with all four catalysts gives varied results. When **8a** was submitted with catalyst **I** and **II**, product **2a** was exclusively observed. However, in the presence of the second generation catalyst (**IV** and **V**), 36% of **8b** along with **8a** was observed. The result indicated that an alternate pathway was competing with the cyclization. On the other hand, when **8b** was resubmitted with first generation catalyst **II** or **III**, the inversion pathway increased to 30% compared to 13% with second generation catalyst **IV** or **V**. The inversion pathway was postulated to result from the corresponding ring-closing/ring-opening sequence. Initially, the reaction could start with complexation of the catalyst with the free allyl group on **8b**, which then could direct the cyclization to form **2b**, resulting in retention of stereochemistry. Alternatively, the process could perform a tandem ROM/RCM (ROM, ring-opening metathesis) and permit an inversion of stereochemistry. The ROM of the internal five-membered ring alkene is proposed to be less likely to occur. In contrast, a survey using **9a** and **9b** was also performed with all four catalysts and no inversion was observed, indicating that the formation of the five-membered ring is independent of the catalysts.

Using the optimized RCM conditions, substrate **2a** was isolated in 60% yield and 85 : 15 diastereoselectivity (Scheme 12.4). The double RCM product was submitted to a selective reductive Heck reaction on the diene on the furan ring. Hydrogenation of the olefin followed by tosyl deprotection using sodium naphathalide completed the synthesis of spiropiperidine **3**.

Chemists at Boehringer Ingleheim used an unprecedented macrocyclization RCM to form the core of a 15-membered macrocyclic tripeptide NS3 protease inhibitor ciluprevir

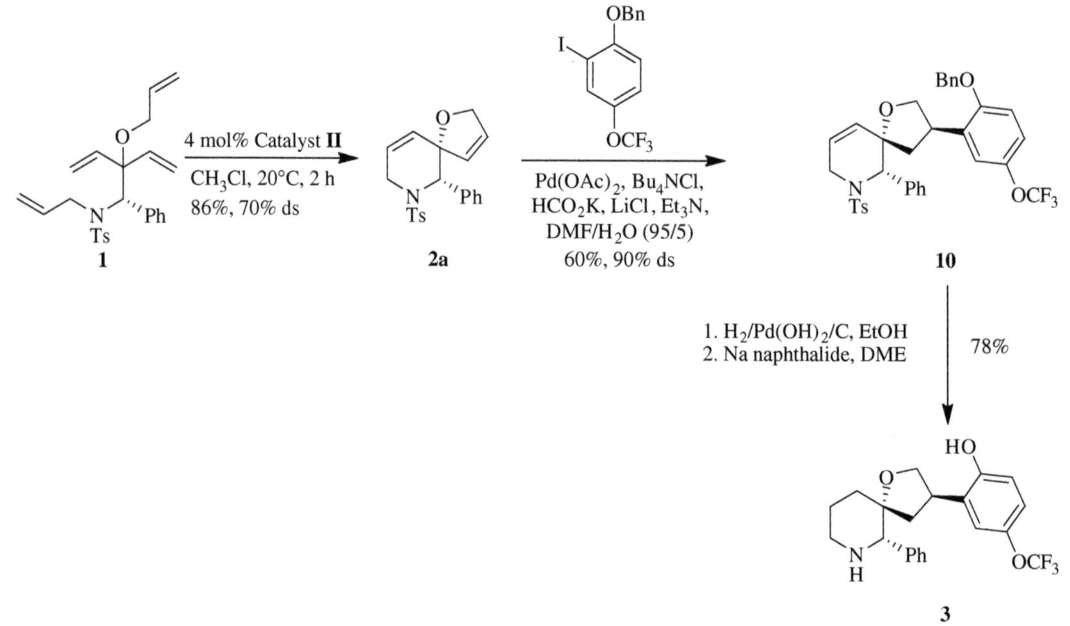

Scheme 12.4 Synthesis of spiropiperidine **3** using optimized RCM conditions.

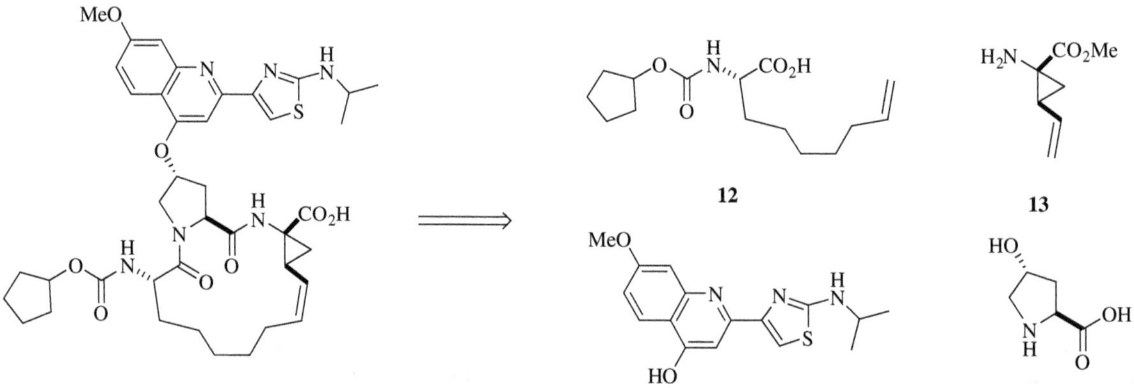

Figure 12.4 Retrosynthetic analysis of BILN 2061 (**11**).

(BIBN 2061, Fig. 12.4), an anti-HCV (hepatitis C virus) drug (35,36). The design of macrocyclic drugs, which are peptidic in nature, is undoubtedly inspired by nature's non-ribosomally formed secondary metabolites, which are often composed of unusual amino acid fragments and are constrained into cyclic peptides with unique bioactive conformations; well-known pharmaceutical examples include cyclosporine, vancomycin, and chloropeptin. However, unlike the production of natural products via fermentation, the chemical manufacturing of a structurally complex macrocyclic peptide, such as ciluprevir, provides a unique synthetic challenge for process chemists to develop a scalable process for commercialization, particularly in developing a practical and efficient chemical process for ciluprevir through RCM (37). After intensive efforts of process research and development, the initial scale-up process for BILN 2061 (11) was identified and focused mainly on the reduction of the raw material cost (38). First, delaying the introduction of the relatively high cost vinylcyclopropane fragment into the synthesis would reduce the raw material cost contribution from this fragment. Second, removal of the Mitsunobu substitution to invert the hydroxy proline stereocenter from the earlier route (8) would eliminate the associated low atom economical reagents from the process. This process was successfully scaled to produce greater than 100 kg of BILN 2061 to support the early preclinical and clinical studies.

Even though the first large-scale process was able to provide sufficient drug substance to support development

TABLE 12.2 Concentration Effects of the RCM on Unprotected and Protected Amide (16a–d)

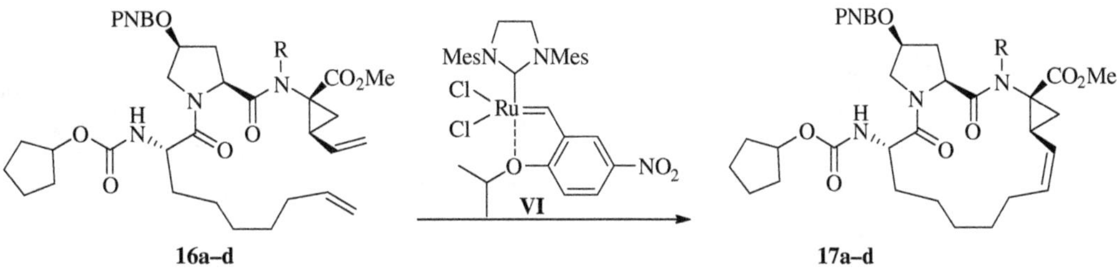

R-Group	Concentration (M)	Yield (%)
H	0.01	85
H	0.10	40
Boc	0.01	97
Boc	0.05	97
Boc	0.10	95
Boc	0.20	93
Ac	0.01	97
Bn	0.01	85

activities, there were still several unmet challenges that required innovative solutions for its further scale-up and commercialization. First, the required high dilution (0.01 M) as well as its high catalyst loading (5 mol%) for the RCM reaction were economically prohibitive on a commercial scale. Second, the higher level of the assembly convergency called for the direct installation of the quinoline heterocycle to the macrocycle through an regioselective nucleophilic aromatic substitution (SNAr) reaction, which avoided the double inversions of the stereocenter at the C-4 position of the hydroxyproline moiety. Therefore, the improvement of the RCM efficiency and its overall assembly convergency would be required for the practical commercialization, which ultimately translated into lower cost of the drug for patients.

A different removable N-protected group was synthesized on the amide moiety and it was proposed that the steric interaction ($A^{1,3}$ strain) may impair the coordinative stabilization of the ester cyclopropyl substituent and therefore favor the catalyst complexation onto the nonenoic acid moiety, which in turn may benefit the RCM. This hypothesis was confirmed by experimental data, as shown in Table 12.2; a higher yield was observed when the Boc-protected amide was used for the RCM reaction. However, a higher concentration (over 0.4 M) of the *N*-Boc amide led to higher dimer formation (39). The acetyl-protected amide provided similar yield. In contrast, when the amide was benzylated, a lower yield was observed, which indicates an electron-withdrawing/or conjugated group as a beneficial element on the RCM reaction. It was postulated that benzyl substitution causes a shift in the initiation to the alkene at the nonenoic acid part rather than the alkene on the cyclopropyl moiety.

The amido-cyclopropane unit within the RCM substrate enforces the double trans-conformation about this segment. This in turn could invoke strain in the corresponding ring system. The beneficial effects of the acylation (or carbonylation) of the cyclopropane-amide on the RCM could result from the electronic electron-withdrawing effect of the substituent, increasing the thermodynamic effective molarity ($EM = k_{intra}/k_{inter}$) (40). Furthermore, the carbonylation of the amide could destabilize the inherent planarity of the segment and in effect improve the EM of the process. The introduction of a carbonyl group at the amine bridge allowed the 50-fold reduction in catalyst (5–0.1 mol%). It was suggested that this improvement was due to the enhancement of the thermodynamic effective molarity of the cyclization and relief of the ring strain originating from the N-substitution.

Computational studies were initiated to understand the strain energy of the macrocycle along with the energetics of the available conformers of the open and cyclized compounds. An initial conformer search was conducted with the Macromodel program to find the lowest energy conformers of the open and cyclized molecules. The lowest energy conformers were then subjected to quantum mechanical (QM) density functional theory (DFT) calculations to obtain more accurate energies. Table 12.3 summarizes the strain energy reduction in the macrocycle due to the Boc substitution using various molecular mechanics (MM) and QM methods. All of the methods studied indicate that the strain energy reduction with the Boc substituent is on the order of approximately 2 kcal/mol, which is in line with the experimental results.

TABLE 12.3 Calculated Reduction of Strain Energy by Boc Substitution on Molecule 16b in Comparison to 17b

Method	OPLS01	MM3	MMFF	DFT/B3LYP
Energy change (kcal/mol)	−3.33	−1.99	−1.10	−2.18

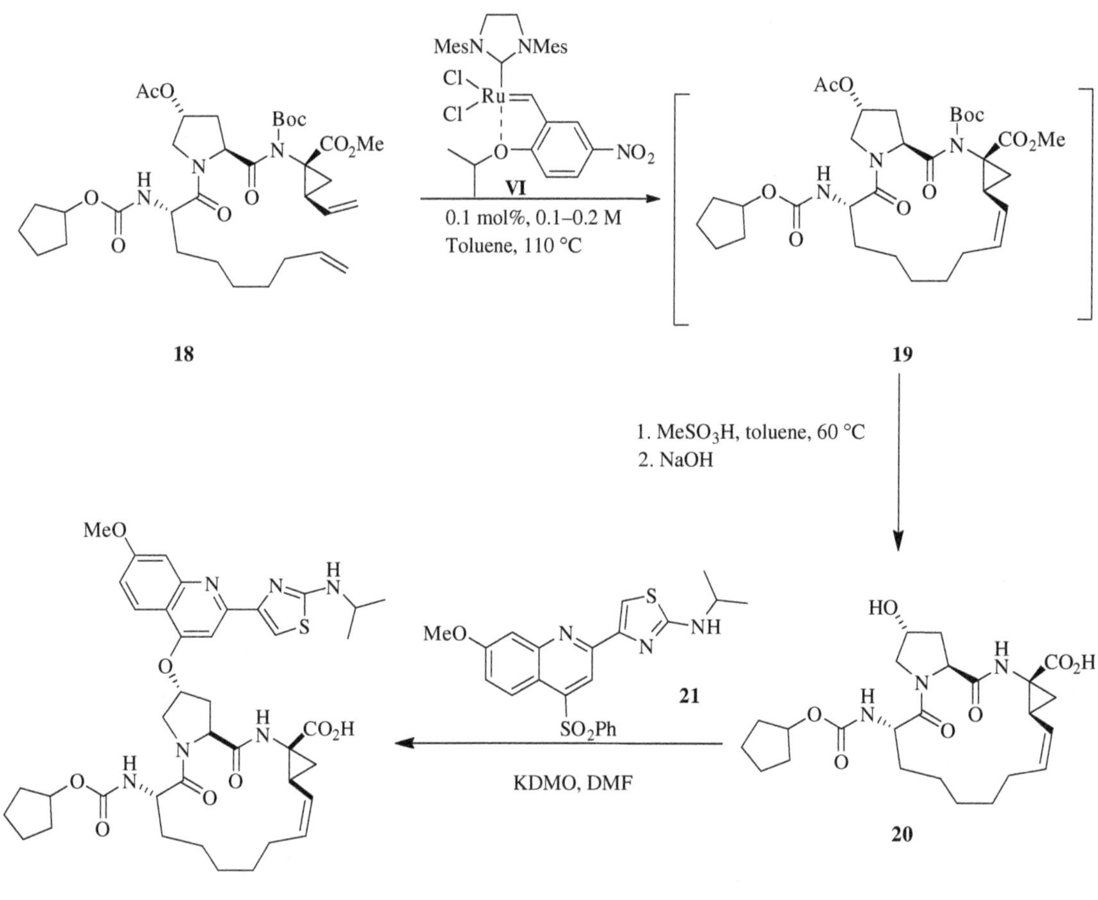

Scheme 12.5 Final scale-up synthesis of BILN 2061 (**11**).

This reduction in strain energy could then be reflected in the improved EM of the process.

The introduction of the removable Boc substituent by Boehringer Ingelheim process chemists was the major breakthrough that allowed the scaling of the macrocyclic RCM (41). Optimization of the RCM precursor (Scheme 12.5) by acylation of the amide NH of the precursor **18** enabled the concentration of the RCM to be dramatically increased to 0.1–0.2 M (10 to 20-folds) from the original 0.01 M, which significantly suppressed the dimerization pathway (<2%). The final process for the RCM involved heating a solution of precursor **18** (0.1 M in toluene) to reflux and a solution of Grela catalyst **VI** (0.1 mol%) was added. The RCM reaction was rapid, with the release of ethylene gas to furnish the desired RCM product **19** in excellent yield. In the same vessel, the crude RCM product **19** was treated with methanesulfonic acid followed by NaOH to sequentially deprotect the Boc group, acetyl group, and ester to provide the hydroxyl acid intermediate **20** which was a highly crystalline compound. The quinolinyl heterocycle was directly attached to the macrocyclic core **20** through an SNAr reaction with quinolinyl sulfone **21** (prepared from the corresponding quinolone) to complete BILN 2061 synthesis (11). While this final process represented the optimal convergent strategy for the assembly of BILN 2061, the most worthy of the accomplishments in this process was Boehringer Ingelheim's innovative solution to their scale-up of the RCM reaction from both academic and industrial perspectives.

Danoprevir (ITMN-191/RG-7227, Fig. 12.5), currently under development by InterMune Inc. and Roche Holding AG, is a promising and potent NS3/4A protease inhibitor for the oral treatment of HCV infection. It was first reported in 2005 (42) and is now in clinical trials. Scientists from InterMune designed a series of HCV protease inhibitors based on the same macrocyclic core developed in Boehringer Ingelheim's first clinical HCV candidate BILN-2061 (ciluprevir) (5) but with a different heterocycle attached to the hydroxyl group of the proline (25) and an alkyl acyl sulfonamide moiety (24) attached to the carboxylic acid (26).

For the key RCM reaction, Roche chemists made significant improvements in the Discovery process to allow for a more economical execution in a pilot plant (43). After intensive screening of different ruthenium catalysts, several penta-coordinated or hexa-coordinated Ru catalysts were

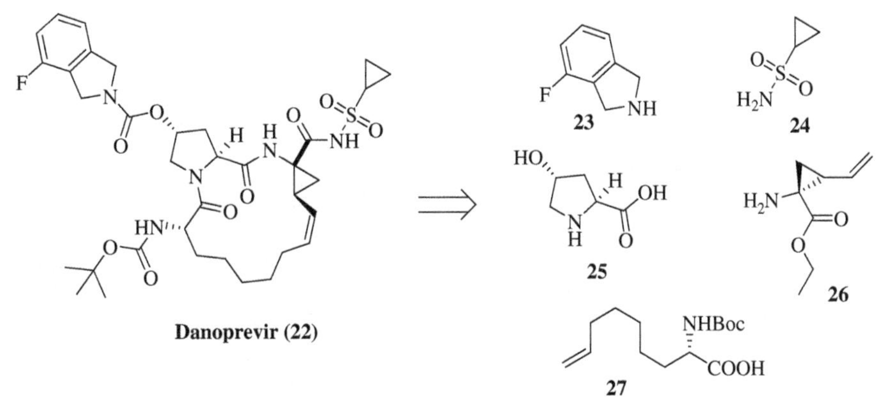

Figure 12.5 Danoprevir and strategic disconnections.

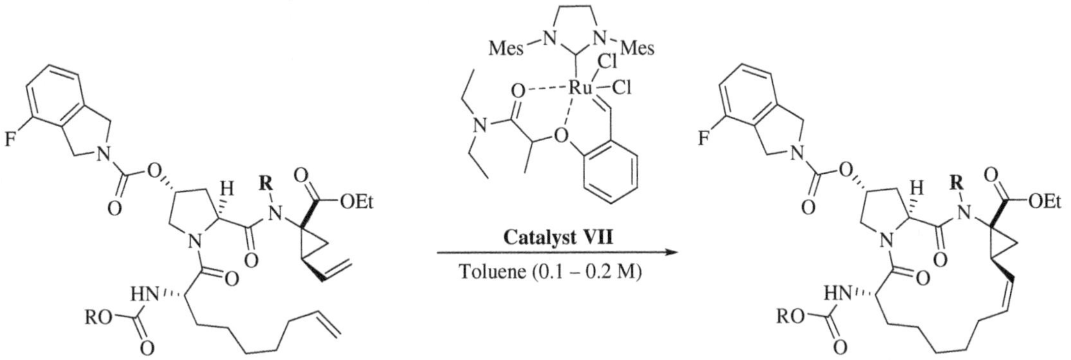

28a–d
a: R=BOC
b: R=Ac
c: R=ArC(O)
d: R=Bz

29a–d
a: R=BOC
b: R=Ac
c: R=ArC(O)
d: R=Bz

Scheme 12.6 Metathesis catalysts identified for the synthesis of Danoprevir.

identified as suitable for this reaction (Scheme 12.6). The optimal catalyst **VII** allowed the catalyst loading to be reduced to 0.1% from the original 25% loading as employed in the Discovery process. The solvent was also changed to toluene from dichloroethane. However, high dilution (~0.01 M) was still required and 10–15% dimer formation was observed. Under these best conditions, the RCM reaction was performed under vacuum (pressure = 0.26 bar) at 60–70 °C in toluene (~0.01 M) using 0.01–0.1% catalyst and an 80–87% yield of the macrocycle **29** was obtained.

Later on, by employing Boehringer Ingelheim's new RCM technology developed for BILN 2061, chemists from both Boehringer Ingelheim (44) and Roche (45) showed that the key RCM reaction of the acylated diene **28** in Danoprevir can be performed at higher concentration (0.10–0.20 M). The large-scale synthesis of Danoprevir was completed by process chemists at Roche after modification of the Discovery synthesis with a focus to facilitate execution in a pilot plant (Scheme 12.7) (46). The low yielding peptide coupling between *N*-Boc vinylcyclopropanecarboxylic acid ethyl ester (26) and *N*-Boc-4-hydroxyproline (25) was replaced with a less expensive amidation procedure. The resulting ester group from the cyclopropane moiety from the RCM product was hydrolyzed and coupled with the cyclopropylsulfonamide fragment **24**. Danoprevir was ultimately isolated as a sodium salt upon treatment with sodium methoxide in methanol and crystallization from ethyl acetate.

Vaniprevir (47) (aka MK7009, Fig. 12.6) was developed by Merck as an HCV NS3/4a protease inhibitor. Some key structural features include the cyclopropyl amino acid and the 20-membered macrocyclic ring. In their quest to develop a practical macrocyclization without using high dilution conditions, Merck scientists explored three strategies for the construction of the macrocyle as indicated by the retrosynthetic disconnections in Figure 12.6.

The first approach was the intramolecular Pd-catalyzed cross couplings (Fig. 12.7) (48). Various reactions were investigated for the macrocyclic alkylation of the aryl bromide, including Heck, Suzuki, and Sonogashira couplings. It was found that in all the cases the macrocyclization gave poor

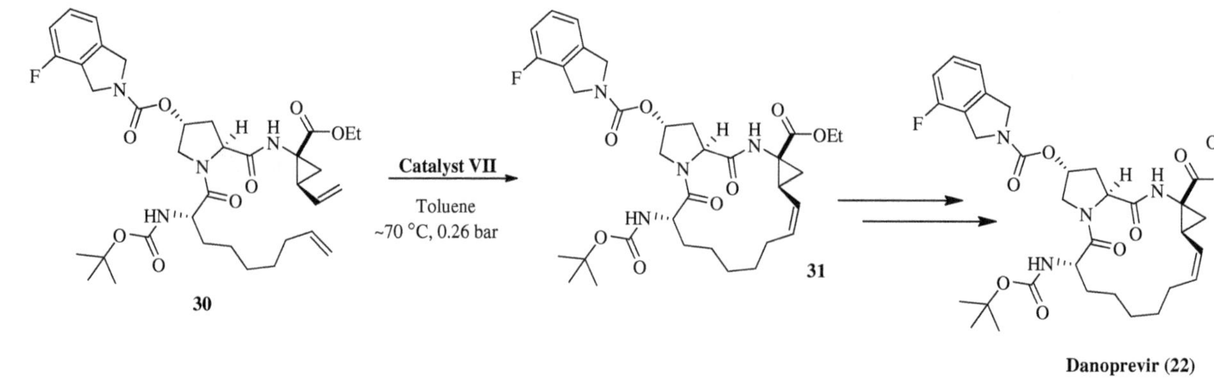

Scheme 12.7 Scale-up of Danoprevir synthesis.

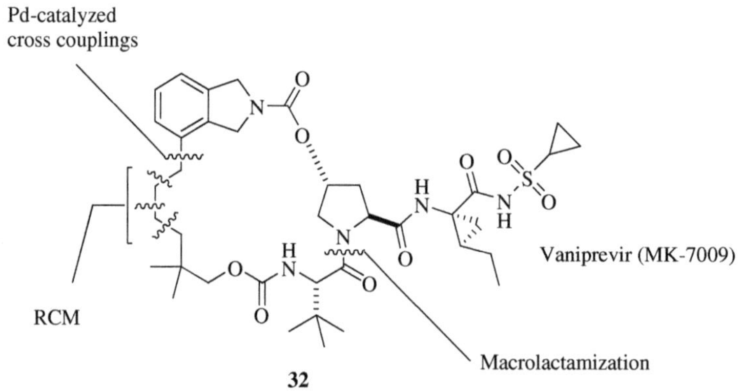

Figure 12.6 Vaniprevir (MK7009).

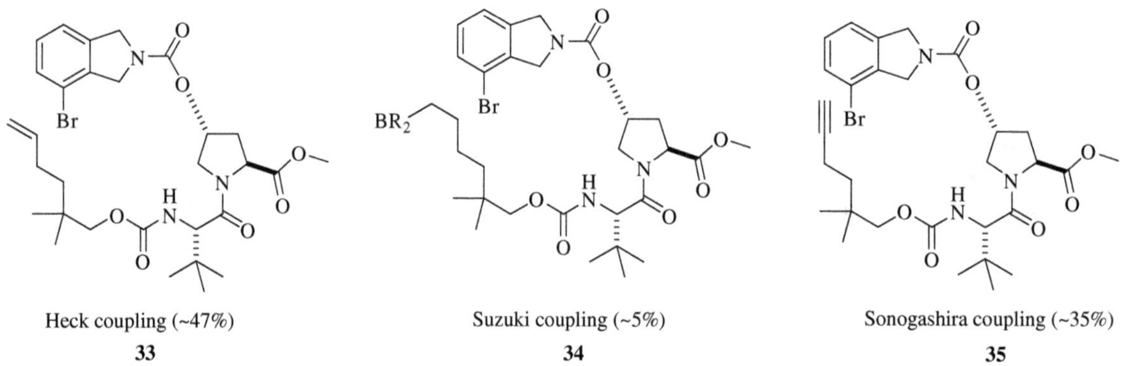

Figure 12.7 Pd-catalyzed macrocyclization.

yields (5–47%), plagued by competing side reactions such as dimerization, oligomerization, and dehydrobromination.

The second approach that was investigated in parallel was macrocyclization via amide bond formation (Scheme 12.8). After having identified HATU (1-[bis (dimethylamino)methylene]-1*H*-1,2,3-triazolo[4,5-*b*] pyridinium 3-oxid hexafluorophosphate) as the best coupling reagent for macrolactamization, Merck process chemists tackled the high dilution issue. As expected, simply increasing the reaction concentration led to decrease in yield as well as formation of various byproducts. The upper acceptable limit for the reaction concentration was 20 g/l (0.035 M) of the amino acid precursor. Since the amide formation is irreversible, the high dilution of the starting amino acid can be mimicked by using a slow addition protocol.

Indeed, as the open-chain precursor was slowly added into a concentrated solution of HATU reagent, the reaction took

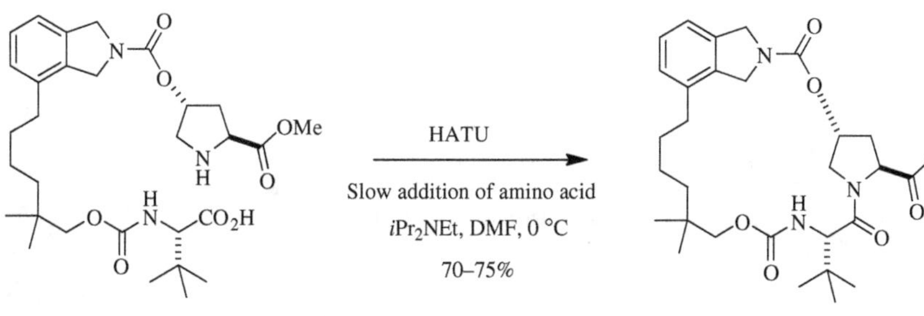

Final substrate concentration = 50 g/l (0.087 M)

36 37

Scheme 12.8 Macrolactamization approach.

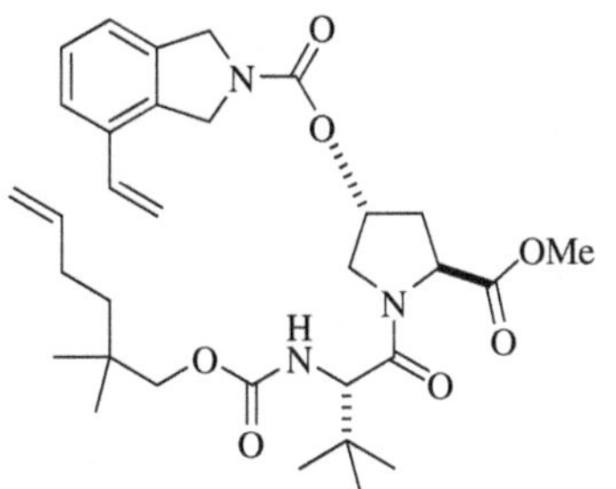

38

Styryl-homoallyl diene (medicinal chemistry diene)

Neolyst 0.3 equiv, <0.003 M (91%)

Zhan 1B 0.1 equiv, <0.003 M (98%)

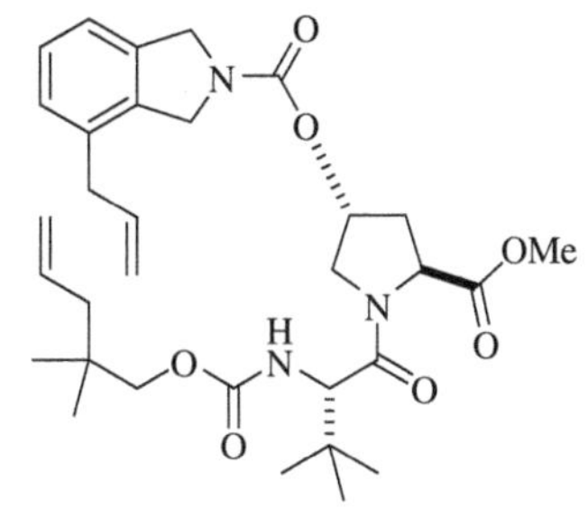

39

Allyl–Allyl diene

Selected for further development

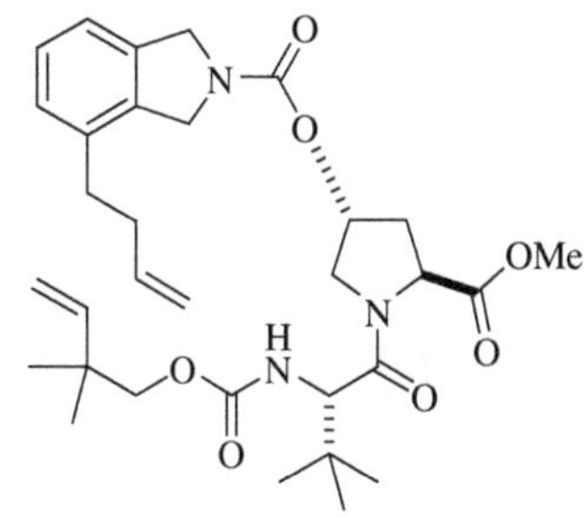

40

Homoallyl-vinyl diene

Higher cost due to vinyl and homoallyl groups

Figure 12.8 Three diene candidates for macrocyclic RCM.

place instantaneously to consume the starting material, thus maintaining an effective low concentration of the amino acid throughout the course of the reaction. It was determined that with the slow addition of amino acid, the reaction can be successfully carried out at a final substrate concentration of 50 g/l (0.087 M) to furnish the desired macrolactam in up to 75% yield. This route involved 10 linear steps (21 total steps) with an overall 30% yield and was used to synthesize the initial large batches of the API.

It turned out that the most efficient way to prepare the macrocycle in MK7009 was the RCM reaction. The initial RCM utilized by the Discovery group at Merck employed the so-called "styryl–homoallyl" diene (Fig. 12.8) as the substrate and Neolyst M1 or Zhan 1B as the catalyst. However, these conditions required high catalyst loadings, high dilutions (<3 mM), and high purity of the starting diene to drive the RCM reaction to completion.

Alternatively, the Merck process chemists examined two other possible disconnections for the construction of the macrocycle via RCM. The "allyl–allyl" diene (Scheme 12.9) was deemed more suitable for further development because of the lower cost of its synthesis.

The initial RCM with the "allyl–allyl" diene using 1 mol% of the Grubbs–Hoveyda second-generation catalyst under high-dilution conditions afforded the desired macrocyclic product in 57% yield after hydrogenation. The use of 5 mol% catalyst increased the yield to 67%. Kinetic data of the reaction provided hints for further optimization. It was observed that the catalyst was very active at the initial phase of the reaction but the conversion rate quickly declined when the reaction progressed further, leading to formation of various byproducts. One hypothesis was that the catalyst decomposition was responsible for this behavior. To circumvent this issue, slow addition was exploited with the hope of minimizing the catalyst deactivation. Indeed, when the 1 mol% catalyst was added over 1 h to a toluene solution of diene (50 ml/g of diene, 0.034 M) at 60 °C, the reaction provided 82% yield (Scheme 12.9).

The current reaction concentration of 50 ml/g was still too low to be practical for large-scale synthesis. Increasing the reaction concentration not only decreased the yield but also led to formation of new byproducts such as the 19-membered macrocycle (Scheme 12.10). It was rationalized that this byproduct was formed through an olefin isomerization-RCM sequence catalyzed by the

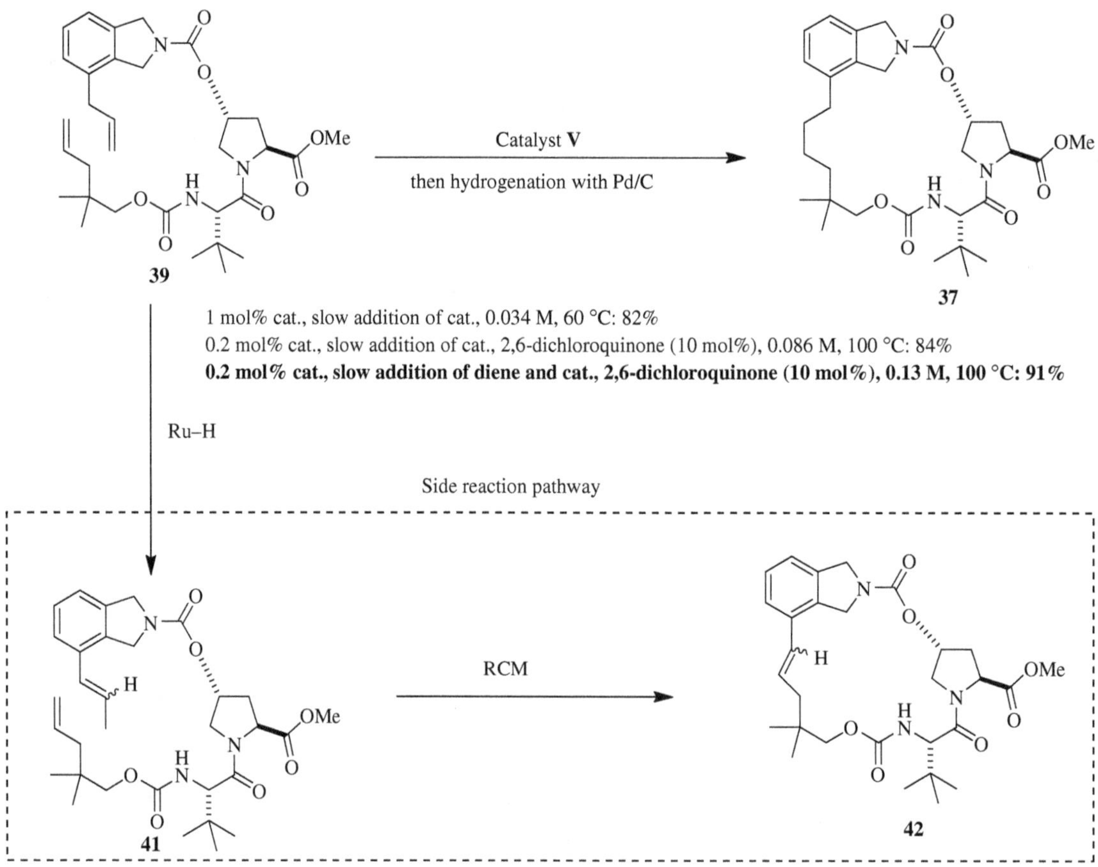

Scheme 12.9 Optimization of RCM with "allyl–allyl" diene.

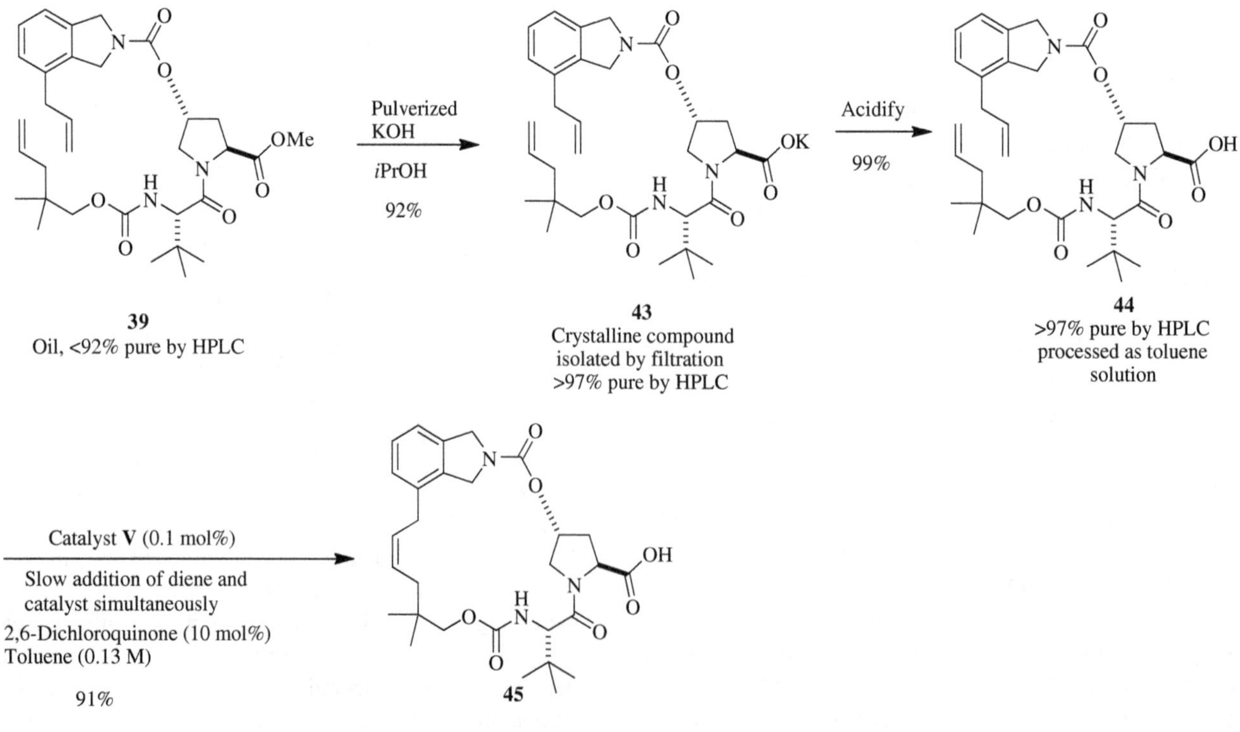

Scheme 12.10 Final process through "allyl–allyl" diene acid.

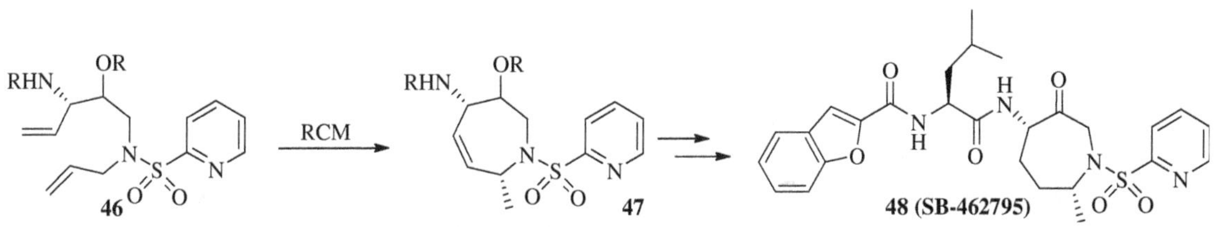

Scheme 12.11 RCM for the construction of the tetrahydroazepine core of SB-462795.

Ru–H complex. The Ru–H complex was a byproduct of catalyst decomposition, which can be inhibited by using 2,6-dichloroquinone. These new conditions allowed the formation of the desired 20-membered macrocycle in 84% yield with only 0.2 mol% catalyst at 20 ml/g (0.086 M) diene concentration. With the purging of ethylene from the reaction mixture, a frequently employed practice to drive thermodynamic equilibrium, the reaction gave consistently approximately 88% yield. To further increase the reaction concentration, the Merck chemists again turned to the slow addition techniques where the catalyst and the diene starting material were concurrently charged into the reaction at a slow rate. This protocol allowed the reaction to be conducted at a final diene concentration of 0.13 M (13.5 ml/g) while maintaining high reaction yield (91%, Scheme 12.10).

It is well known that impurities can negatively impact the RCM reactions, especially at very low catalyst loadings. For large-scale synthesis, it is not practical to purify the starting materials by chromatography. Instead, it is highly preferable to purify through crystallization-based techniques. Unfortunately, the "allyl–allyl" diene is a non-crystalline compound and Merck chemists had to find a work-around for the purification. The corresponding potassium salt of the acid is a crystalline solid (Scheme 12.10). Upon treatment of diene ester with powdered potassium hydroxide in isopropanol, the potassium salt precipitated directly from the reaction mixture and was isolated and purified by simple filtration. Acidification of the potassium salt provided the free acid with enhanced purity (from 92% to >97% by HPLC (high performance liquid chromatography)). The diene acid worked equally well in the RCM as the diene ester to provide the desired macrocycle in 91% yield.

In summary, Merck chemists explored three different approaches for the construction of the 20-membered macrocycle skeleton for MK-7009. Overall, the RCM-based route was most efficient compared to the macrolactamization and Pd-catalyzed cross-coupling approaches. The longest linear sequence of the RCM route was only nine steps. The original challenges of high catalyst loading and low reaction concentration for RCM were overcome through extensive process optimization and development, and with the implementation of the slow addition protocols. In addition, the robustness of the RCM was secured by identification of a highly efficient purification protocol via the crystalline potassium salt of the diene acid.

Wang and coworkers (49) at GlaxoSmithKline (GSK) utilized the RCM for the construction of the tetrahydroazepine core of the cathepsin K inhibitor SB-462795, a potential osteoporosis treatment (Scheme 12.11). The initial development of the RCM process centered around route development with a focus on both cost effectiveness and scalability. In this capacity, it was envisioned that employing the diene **46** without defining the amine stereochemistry would offer economic advantages. The RCM proceeded well (95% yield), employing Grubbs second generation catalyst **III** (50) at relatively high catalyst loading (14 mol%, Scheme 12.12). However, the route was eventually abandoned due to significant purification problems arising from the mixtures of stereoisomers during later purifications, thus highlighting the difficulties of undefined stereoisomer mixtures in crystallization-based purifications (17).

Owing to the purification problems associated with the undefined stereoisomers of diene **46**, Wang and coworkers then focused on the oxazolidinone-dienes **49** and **50** (Table 12.4) for the key RCM reaction. GSK previously had experience with this substrate and RCM for the *des*-C7 methyl analog (51). The RCM reaction of dienes **49** and **50** with Hoveyda's second generation metathesis catalyst **IV** (52) required relatively high catalyst loadings (10 mol%) for complete conversion. Although slightly lower catalyst loading can be achieved with Grubbs catalysts (**II** and **III**) and indenylidene catalyst **VIII** (53), the loadings were still too high to afford a cost-effective process. Furthermore, the residual ruthenium content was found to be difficult to remove before the introduction of reagents in downstream operations that are not compatible with the residual metal (e.g., oxidants and acyl azides). The high catalyst loading was postulated to result from the interference of the chelating oxazolidinone

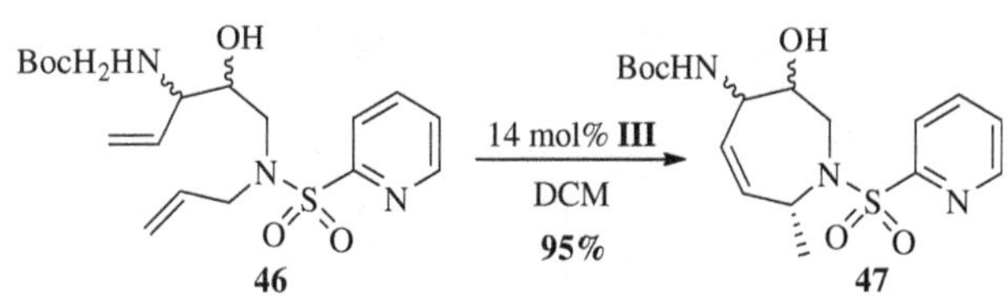

Scheme 12.12

TABLE 12.4 Survey of Metathesis Catalyst for the RCM of Oxazolidinone Dienes 49 and 50

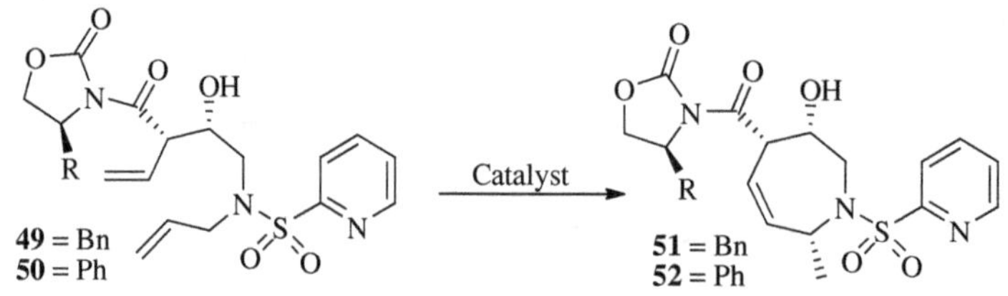

Diene	Catalyst	Conditions	Conversion
49	10 mol% **IV**	Toluene, reflux	100%
50	10 mol% **IV**	Toluene, reflux	100%
49	10 mol% **IV**	Toluene, reflux	60–79%
50	5 mol% **II**	DCM, 40 °C	6%
49	5 mol% **VIII**	DCE, RT to 80 °C	7%
50	5 mol% **III**	DCE, 80 °C	24%

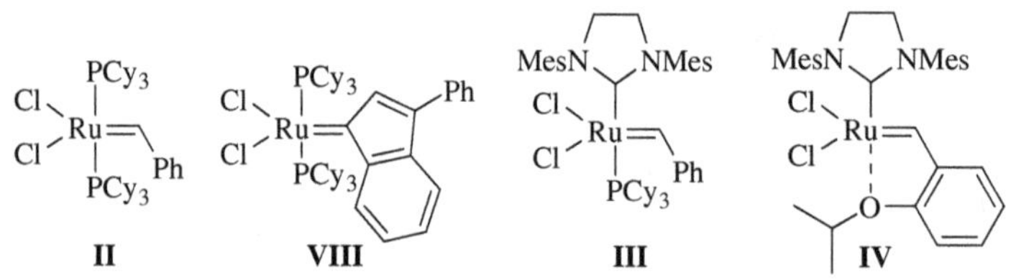

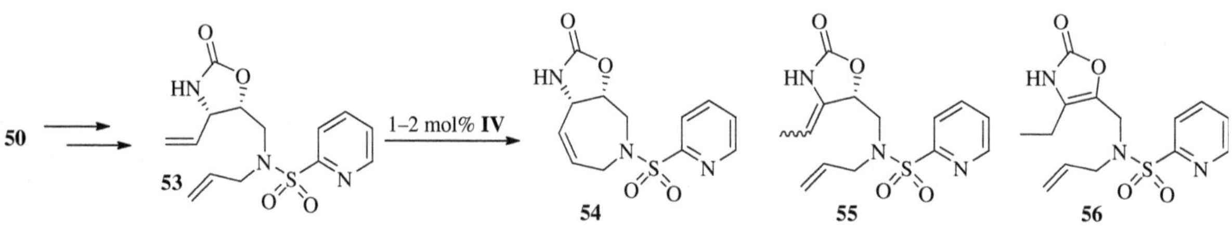

Scheme 12.13

and free adjacent hydroxyl substituent hindering the catalyst turnover. Support for this premise is that the diene with the masked (triethylsilane, TES) hydroxyl group was found to reach complete conversion in the RCM with five times less (2 mol%) of Hoveyda's second generation catalyst **IV**.

In order to avoid the chelation of the hydroxyl-oxazolidinone diene **49/50**, the auxiliary was cleaved and converted to the oxazolidine diene **53** (Scheme 12.13). Initial studies of the RCM with diene **53** showed significant olefin migration in the diene starting material. However, it was found that by employing higher purity diene **53** in the RCM with Hoveyda's second generation catalyst **IV**, the olefin migration can be suppressed. A 90% yield of the desired cyclic product was obtained with 1 mol% of catalyst **IV**. The introduction of a basic cysteine extraction in the processing of the reaction was found to reduce the ruthenium content to parts per million levels (148 ppm).

An alternate substrate that was evaluated for the RCM was the hydroxy-phthalimide diene **57**. Initial results of the RCM with diene **57** gave highly variable results (2–100% conversion) depending on the batch of diene (54). Spiking experiments identified a number of basic impurities that have a negative effect on the RCM (Fig. 12.9). Analysis of the batch histories indicated that 1,8-diazabicyclo[5.4.0]undec-7-ene (DBU) was not observed in any of the batches. A subsequent statistical analysis of the batches indicates that impurities **59** and **61** accounted for the vast majority of variation in batch purity and negative performances of the RCM reaction. A DoE (DoE stands for Design of Experiment) (55) of the four observed impurities in the historical batches was employed to establish specifications for the impurities to ensure high probability of a complete conversion RCM reaction. The specifications were set as follows: NMT (not more than) 0.01 mol% **59**, NMT 0.007 mol% **BTPP**, NMT 0.15 mol% ester **61**, and NMT 0.01 mol% of amine **60**.

With the impurity specifications in place, the reaction parameters of temperature (60–90 °C), batch volume (5–10 volumes), and catalyst loading (0.1–0.3 mol%) were optimized statistically, employing response surface methodology

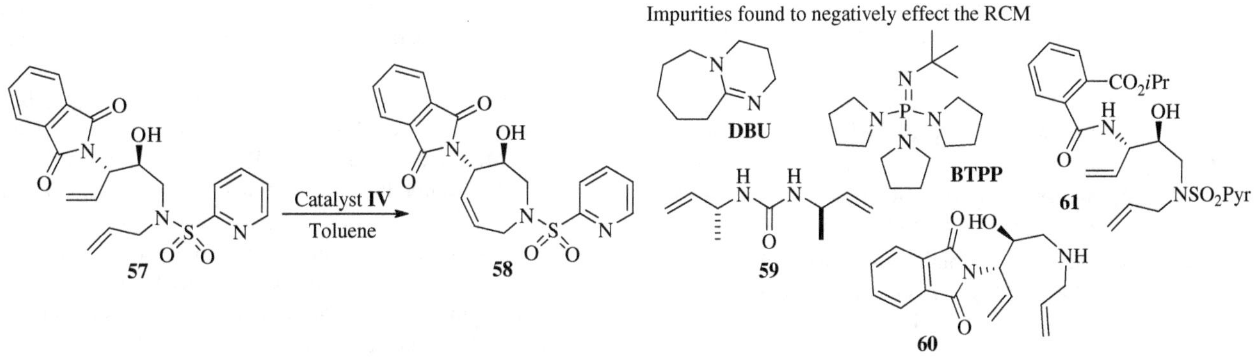

Figure 12.9 RCM of diene **57** and known impurities that negatively affect the RCM.

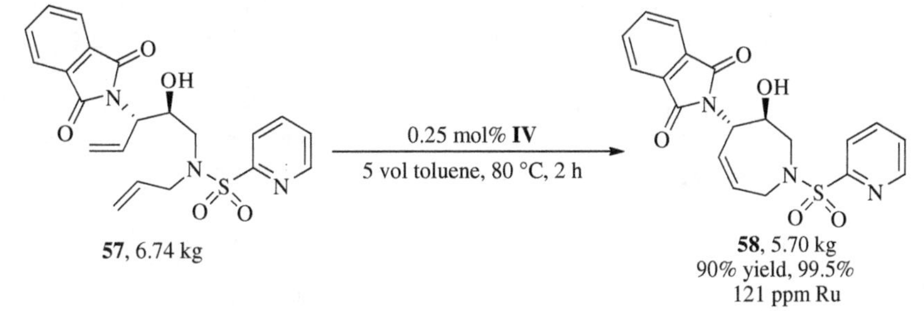

Scheme 12.14

(RSM) (55). From this work, it was found that the batch volume was not statistically relevant for reaction performance, but the RCM conversion was found to depend upon catalyst loading and temperature. The final reaction parameters were set at 80 °C with 0.25 mol% of Hoveyda's second generation catalyst **IV**. Owing to the insoluble nature of the cyclic tetrahydroazepine product **57**, it was found that by simply cooling the batch to 5 °C, the product **57** could be isolated by filtration in high yield and purity. The process was demonstrated on 6.74 kg of diene **57**, and a 90% yield (5.70 kg) of the tetrahydroazepine product was obtained employing 0.25 mol% of Hoveyda's catalyst **IV** (Scheme 12.14). The product was isolated directly by centrifugal filtration and after a toluene wash the material was obtained in 99.5 A% purity and with low ruthenium content (121 ppm). The process development of the RCM for the tetrahydroazepine core of SB-462795 highlights the importance of impurity control, as well as route design and statistical optimizations, to develop a practical and robust RCM large-scale process.

12.5 CONCLUSION AND OUTLOOK

Increasing complexity requirements of drug candidates will necessitate the introduction of new technologies to develop cost-effective processes. In this capacity, the implementation of the RCM reaction fulfills both of these goals. Not only does this technology rapidly increase complexity by the formation of a new ring, it also allows for more direct and efficient synthetic routes. The RCM reaction and associated catalysts have started to emerge in process development and the limited case studies highlight the challenges as well as the solutions in employing this technology on a large scale. These initial successes in scaling the RCM economically and safely on multi-kilogram batches will serve as the drive for future development and implementation of this technology in large-scale API development.

REFERENCES

1. For selected reviews on RCM and associated applications, see: (a) Fürstner A. *Angew Chem Int Ed* 2000;39: 3012–3043. (b) Maier ME. *Angew Chem Int Ed* 2000;39:2073–2077. (c) Schmalz H-G. *Angew Chem Int Ed Engl* 1995;34: 1833–1836. (d) Monfette S, Fogg DE. *Chem Rev* 2009;109: 3783–3816. (e) McReynolds MD, Dougherty JM, Hanson PR. *Chem Rev* 2004;104:2239–2258.
2. For a review on the Ene–Eny metathesis reaction included the RCM variant, see:Poulsen CS, Madsen R. *Synthesis* 2003;1:1–18.
3. For a review on the Cross-Metathesis reaction, see:Connon SJ, Blechert S. *Angew Chem Int Ed* 2003;42:1900–1923.
4. Schuster M, Blechert S. *Angew Chem Int Ed* 1997;36: 2036–2056.
5. Freitas ER, Gum CR. *Chem Eng Prog* 1979;75:73–76.
6. Mol JC. *J Mol Catal A: Chem* 2004;213:39–45.

7. For a review on the Refining of plant oils by olefin metathesis, see:Chikkali S, Mecking S. *Angew Chem Int Ed* 2012;51: 5802–5808.
8. (a) Fu GC, Grubbs RH. *J Am Chem Soc* 1992;114:5426–5427. (b) Fu GC, Nguyen ST, Grubbs RH. *J Am Chem Soc* 1993;115:9856–9857.
9. Schrock RR, Murdzek JS, Bazan GC, Robbins J, DiMare M, O'Regan M. *J Am Chem Soc* 1990;112:3875–3886.
10. For an account on the history of the development of metathesis catalyst and the functional group tolerance of transition metal metathesis catalysts, see:Trnka TM, Grubbs RH. *Acc Chem Res* 2001;34:18–29.
11. (a) Nguyen ST, Johnson LK, Grubbs RH. *J Am Chem Soc* 1992;114:3974–3975. (b) Schwab P, Grubbs RH, Ziller JW. *J Am Chem Soc* 1996;118:100–110. (c) Schwab P, France MB, Ziller JW, Grubbs RH. *Angew Chem Int Ed Engl* 1995;34:2039–2041.
12. Kingsbury JS, Harrity JPA, Bonitatebus PJ Jr, Hoveyda AH. *J Am Chem Soc* 1999;121:791–799.
13. Scholl M, Ding S, Lee CW, Grubbs RH. *Org Lett* 1999;1:953–956.
14. Garber SB, Kingsbury JS, Gray BL, Hoveyda AH. *J Am Chem Soc* 2000;122:8168–8179.
15. (a) Grela K, Harutyunyan S, Michrowska A. *Angew Chem Int Ed* 2002;41:4038–4040. (b) Michrowska A, Bujok R, Harutyunyan S, Sashuk V, Dolgonos G, Grela K. *J Am Chem Soc* 2004;126:9318–9325.
16. For a review of *N*-heterocyclic carbene ligands and the effects of various perturbations, see: Samojlowicz C, Bieniek M, Grela K. *Chem Rev* 2009;109:3708–3742.
17. For selected examples, see: (a) Chatterjee AK, Morgan JP, Scholl M, Grubbs RH. *J Am Chem Soc* 2000;122:3783–3784. (b) Huang J, Stevens ED, Nolan SP, Petersen JL. *J Am Chem Soc* 1999;121:2674–2678.
18. For selected vendors of these catalysts, see: (a) Sigma–Aldrich Corporation, 3050 Spruce St. Louis, Missouri, USA (b) Strem Chemicals, Inc. 7 Mulliken Way, Newburyport, MA, USA.
19. Dependence on an IP related catalyst is strategically not advantageous as risks associated with single supplier and limitations on negotiation leverage are frequently encountered.
20. For a review of asymmetric transition metal catalyst in large scale synthesis, see:Farina V, Reeves JT, Senanayake CH, Song JJ. *Chem Rev* 2006;106:2734–2793.
21. Busacca CA, Fandrick DR, Song JJ, Senanayake CH. Transition metal catalysts in the pharmaceutical industry. In: Crawley ML, Trost BM, editors. *Applications of Transition Metal Catalysis in Drug Discovery and Development an Industrial Perspective*. New Jersery, USA: Wiley, Hoboken; 2012.
22. For selected vendors of the metathesis catalysts, see: (a) Sigma-Aldrich Co. LLC PO Box 14508, St. Louis, MO 63178, USA (b) Strem Chemicals, Inc. 7 Mulliken Way, Newburyport, MA 01950, USA.
23. Single Supplier Security refers to the risk associated with single suppliers of key process components and associated risks of source material interruptions.
24. The cost of the metal component of a given transition metal catalyst is generally fixed and does not decrease upon larger bulk orders.
25. European Medicines Agency (EMEA), Pre-Authorisation Evaluation of Medicines for Human Use, Committee for Human Medicinal Products (CHMP). Draft report: Guideline on the specification limits for residual of metal catalyst. London; 2007 Jan.
26. Class 1 metals, are defined as those metals that pose a significant safety concern and are either known or suspected human carcinogens, or potential agents for significant toxicity.
27. Yasbin RE, Matthews CR, Clarke MJ. *Chem Biol Interact* 1980;31:355–365.
28. Thiel OR. Metathesis reactions. In: Crawley ML, Trost BM, editors. *Applications of Transition Metal Catalysis in Drug Discovery and Development an Industrial Perspective*. New Jersery, USA: Wiley, Hoboken; 2012.
29. IP free with respects to the parent organization.
30. For examples of ligands and catalysts designed in the Chemical Development group at Boehringer Ingelhiem, Pharmaceuticals Inc., see: (a) Busacca CA, Lorenz JC, Grinberg N, Haddad N, Lee H, Li Z, Liang M, Reeves D, Saha A, Varsolona R, Senanayake SH. *Org Lett* 2008;10:341–344. (b) Busacca CA, Grossback D, So RC, O'Brien EM, Spinelli EM. 2003;5:595–598. (c) Tang W, Capacci AG, White A, Ma S, Rodriguez S, Qu B, Savoie J, Patel ND, Wei X, Haddad N, Grinberg N, Yee NK, Krishnamurthy D, Senanayake CH. *Org Lett* 2010;12:1104–1107. (d) Tang W, Patel ND, Xu G, Xu X, Savoie J, Ma S, Hao M-H, Keshipeddy S, Capacci AG, Wei X, Zhang Y, Gao JJ, Li W, Rodriquez S, Lu BZ, Yee NK, Senanayake SH. *Org Lett* 2012;14:2258–2261.
31. (a) Baylon C, Heck MP, Mioskowski C. *J Org Chem* 1999;64:3354. (b) Lautens M, Hughes G. *Angew Chem Int Ed* 1999;38:129–131. (c) Burke SD, Quinn KJ, Chen VJ. *J Org Chem* 1998;63:8628–8627. (d) Bassindale MJ, Hamley P, Leitner A, Harrity JPA. *Tetrahedron Lett* 1999;40:3247–3250. (e) Trevitt M, Gouverneur V. *Tetrahedron Lett* 1999;40:7333–7336. (f) Fu GC, Grubbs RH. *J Am Chem Soc* 1992;114:7324–7325.
32. (a) Wallace DJ, Cowden CJ, Kennedy DJ, Ashwood MS, Cottrell IF, Dolling U-H. *Tetrahedron Lett* 2000;41:2027–2029. (b) Wallace DJ, Goodman JM, Kennedy DJ, Davies AJ, Cowden CJ, Ashwood MS, Cottrell IF, Dolling U-H, Reider PJ. *Org Lett* 2001;3(5):671–674. (c) Wallace DJ. *Tetrahedron Lett* 2005;46:591. (d) Wallace DJ. *J Mol Catal A: Chem* 2006;254:78–84.
33. Baker R, Curtis NR, Elliott JM, Harrison T, Hollingworth GJ, Jackson PS, Kulagowski JJ, Seward EM, Swain CJ, Williams BJ. International Patent WO97/49710. 1997.
34. Mohamadi F, Richards NG, Guida WC, Liskamp R, Lipton M, Caufield C, Chang G, Hendrickson T, Still WC. *J Comput Chem* 1990;11:440–467.
35. Lamarre D, Anderson PC, Bailey M, Beaulieu P, Bolger G, Bonneau P, Boes M, Cameron DR, Cartier M, Cordingley MG, Faucher AM, Goudreau N, Kawai SH, Kukolj G, Lagace L, LaPlante SR, Narjes H, Poupart MA, Rancourt J, Sentjens RE, St. George R, Simoneau B, Steinmann G, Thibeault D,

Tsantrizos YS, Weldon SM, Yong CL, Llinas-Brunet M. *Nature* 2003;426:186–189.

36. Hinrichsen H, Benhamou Y, Wedemeyer H, Reiser M, Sentjens RE, Calleja JL, Forns X, Erhardt A, Croenlein J, Chaves RL, Yong CL, Nehmiz G, Steinmann GG. *Gastroenterology* 2004;127:1347–1355.
37. Grubbs RH. *Handbook of Metathesis*. Weinheim: Wiley-VCH; 2003.
38. (a) Yee NK, Farina V, Houpis IN, Haddad N, Frutos RP, Gallou F, Wang X-j, Wei X, Simpson RD, Feng X, Fuchs V, Xu Y, Tan J, Zhang L, Xu J, Smith-Keenan LL, Vitous J, Ridges MD, Spinelli EM, Johnson M, Donsbach K, Nicola T, Brenner M, Winter E, Kreye P, Samstag W. *J Org Chem* 2006;71:7133–7145. (b) Nicola T, Brenner M, Donsbach K, Kreye P. *Org Process Res Dev* 2005;9:513–515. (c) Dersch W, Samstag W. Process for the preparation of macrocylic compounds by ruthenium complex catalysed metathesis reaction. *Int. Appl.* WO 2009137432. 2004.
39. The main cyclic dimeric product was isolated, and separately converted to **17b** under standard RCM conditions.
40. Mitchell L, Parkinson JA, Percy JM, Singh K. *J Org Chem* 2008;73:2389–2395.
41. (a) Shu C. Ruthenium catalyzed ring-closing metathesis process for the preparation of macrocyclic peptides as antiviral agents. *PCT Int. Appl.* WO 2007030656. 2007. (b) Shu C, Zeng X, Hao M-H, Wei X, Yee NK, Busacca CA, Han Z, Farina V, Senanayake CH. *Org Lett* 2008;10:1303–1306. (c) Farina V, Shu C, Zeng X, Wei X, Han Z, Yee NK, Senanayake CH. *Org Process Res Dev* 2009;13:250–254.
42. (a) Blatt LM, Wenglowsky SM, Andrews SW, Jiang Y, Kennedy AL, Condroski KR, Josey JA, Stengel PJ, Madduru MR, Doherty GA, Woodard BT. Preparation of macrocyclic carboxylic acids and acylsulfonamides as inhibitors of HCV replication. *PCT Int. Appl.* WO 2005037214. 2005. (b) Seiwert SD, Andrews SW, Jiang Y, Serebryany V, Tan H, Kossen K, Rajagopalan PTR, Misialek S, Stevens SK, Stoycheva A, Hong J, Lim S, Qin X, Rieger R, Condroski KR, Zhang H, Do MG, Lemieux C, Hingorani GP, Hartley DP, Josey JA, Pan L, Beigelman L, Blatt LK. *Antimicrob Agents Chemother* 2008;52(12):4432–4441. (c) Rajagopalan R, Misialek S, Stevens SK, Myszka DG, Brandhuber BJ, Ballard JA, Andrews SW, Seiwert SD, Kossen K. *Biochemistry* 2009;48(11):2559–2568.
43. (a) Hildbrand S, Puentener K, Scalone M. Process for the preparation of a macrocycle using ruthenium-catalysed ring closing metathesis. *PCT Int. Appl.* WO 2009080542. 2009 (b) Puentener K, Scalone M. New ruthenium complexes as catalysts for metathesis reactions. *PCT Int. Appl.* WO 2009124853. 2009.
44. Shu C, Senanayake CH, Tan Z, Yee N, Zeng X. Ring-closing metathesis process for the preparation of macrocyclic peptides. *PCT Int. Appl.* WO 2009137432, 2009.
45. Scalone M, Stahr H. Process for the preparation of macrocylic peptides. *PCT Int. Appl.* WO 2010015545. 2010.
46. (a) Seiwert SD, Kossen K, Pan L, Liu J, Buckman BO. Discovery and development of the CHV NS3/4A protease inhibitor danoprevir (ITMN-191/RG7227). 18, 257–271. In: *Antiviral Drugs: From Basic Discovery Through Clinical Trials*. Weinheim: Wiley-VCH; 2011. (b) Gantz F, Stahr H. Process for the synthesis of an HCV protease inhibitor intermediate. *PCT Int. Appl.* WO 2008128921. 2008.
47. Kong J, Chen C, Balsells-Padros J, Cao Y, Dunn RF, Dolman SJ, Janey J, Li H, Zacuto MJ. *J Org Chem* 2012;77:3820 and references cited therein.
48. Song ZJ, Tellers DM, Journet M, Kuethe JT, Lieberman D, Humphrey G, Zhang F, Peng Z, Waters MS, Zewge D, Nolting A, Zhao D, Reamer RA, Dormer PG, Belyk KM, Davies IW, Devine PN, Tschaen DM. *J Org Chem* 2011;76: 7804–7815.
49. Wang H, Matsuhashi H, Doan BD, Goodman SN, Ouyang X, Clark WM Jr, *Tetrahedron* 2009;65:6291–6303.
50. Trnka TM, Grubbs RH. *Acc Chem Res* 2001;34:18–29.
51. Trout REL, Marquis RW. *Tetrahedron Lett* 2005;46:2799–2801.
52. Garber SB, Kingsbury JS, Gray BL, Hoveyda AH. *J Am Chem Soc* 2000;122:8168–8179.
53. Fürstner A, Guth O, Düffels A, Seidel G, Liebl M, Gabor B, Mynott R. *Chem Eur J* 2001;7:4811–4820.
54. Wang H, Goodman SN, Dai Q, Stockdale GW, Clark WM Jr, *Org Process Res Dev* 2008;12:226–234.
55. Anderson MJ, Withcomb PJ. *RSM Simplified: Optimizing Processes Using Response Surface Methods for Design of Experiments*. New York: Productivity; 2005.

PART II

DEVELOPMENT OF THE TOOLS

SECTION 4

DIFFERENT CLASSES OF OLEFIN METATHESIS CATALYSTS

13

WELL-DEFINED OLEFIN METATHESIS CATALYSTS BASED ON METALS OF GROUP 4–7

CEZARY PIETRASZUK

Department of Organometallic Chemistry, Faculty of Chemistry, Adam Mickiewicz University in Poznań, Poznań, Poland

13.1 INTRODUCTION

There are many catalytic systems, hetero- and homogeneous, based on the transition elements Ti, V, Nb, Ta, Cr, Mo, W, Re, Co, Ir, Ru, and Os that are able to initiate olefin metathesis (1,2). Most of them are two- and three-component catalytic systems of undefined structure, developed in the early 1980s. In general, these "classical" systems suffer from sensitivity to air, moisture, and many functionalities, side reactions, and relatively short lifetimes. In such systems, the catalytically active species generated *in situ* are present in very low concentrations and their isolation or even identification is difficult. In contrast to classical systems, well-defined catalysts are (i) essentially identical to (or resemble) the active species in terms of metal oxidation state and ligand coordination sphere; (ii) stable enough to be characterized through spectroscopic means and X-ray analysis; and (iii) permit observation and control of the intermediates, so that upon olefin metathesis they react with olefin to yield a new alkylidene complex derived from that olefin (3). This chapter is concerned with well-defined olefin metathesis catalysts based on metals of group 4–7. Whenever necessary, other catalysts important for the development of olefin metathesis catalysis will be mentioned. The subject is covered by several excellent reviews, which have been referred to in the relevant sections of this overview.

13.2 SURVEY OF OLEFIN METATHESIS CATALYSTS BASED ON TITANIUM, TANTALUM, VANADIUM AND RHENIUM

Early stage development of well-defined group 4–7 metal-based catalytic systems is the subject of a number of excellent reviews (1,2,4–9).

13.2.1 Titanium

In 1979, Tebbe proved that complex (**1**) (presently known as the *Tebbe reagent*) underwent degenerate metathesis of terminal olefins (Eq. 13.1). The reaction has provided the first clear evidence that olefin metathesis is a process in which alkylidene units are exchanged between reaction partners (10).

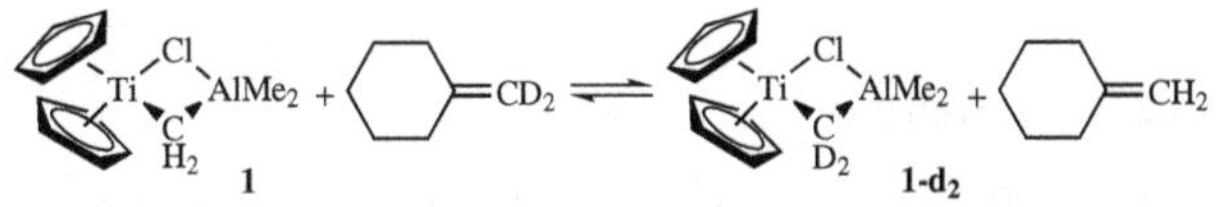

(13.1)

Reaction of complex **1** with 3,3-dimethylbuten-1-ene in the presence of 4-dimethylaminopyridine (DMAP) gives isolable titanacyclobutane **2** (Eq. 13.2) (11).

Olefin Metathesis: Theory and Practice, First Edition. Edited by Karol Grela.

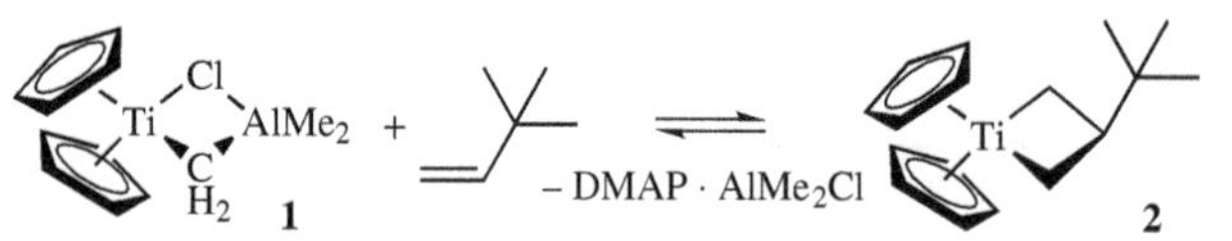

(13.2)

Metallacycles such as **2** were demonstrated to catalyze the exchange of methylene units of terminal olefins (Eq. 13.3) (12).

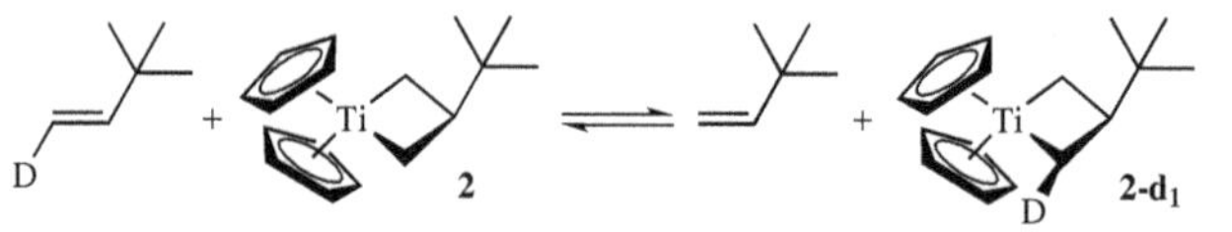

(13.3)

Moreover, Lewis base adducts of Me_2AlCl react with the metallacycles to release the olefin and regenerate the titanium methylene adduct **1** (Tebbe reagent) (Eq. 13.4). For weak bases, the equilibrium of the reaction is shifted to the right.

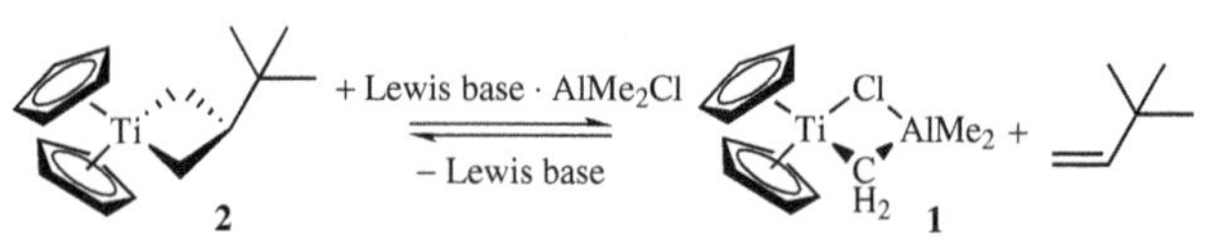

(13.4)

The reaction sequence described above is an elegant demonstration of all relevant steps in the carbene mechanism of olefin metathesis. Titanacyclobutanes obtained by interaction of appropriate alkylidenes, that is, [$Cp_2Ti{=}CH_2$] (**3**) or [$Cp_2Ti{=}CHC(CH_3)_2CH{=}CH_2$] (**4**) with norbornene (NBE) were the first well-defined metathesis catalysts, found to carry out living ring opening metathesis polymerization (ROMP) (13). However, ROMP polymerization with titanacyclobutane catalysts is highly sensitive to moisture. In addition, titanium-based ROMP systems do not tolerate the presence of most functionalities, including aldehyde, ketone, ester, and hydroxyl groups.

13.2.2 Tantalum

[Ta(CH-*t*-Bu)$(OCMe_3)_2Cl(PMe_3)$] (**5**) undergoes metathesis exchange with styrene in the presence of PMe_3 producing the isolable benzylidene complex [Ta(CHPh)$(OCMe_3)_2Cl(PMe_3)_2$] (**6**). Complex **6** has been reported to initiate productive metathesis of *cis*-2-pentene, but the detection of formation of the intermediate ethylidene and propylidene complexes in the system has not been successful. This experiment is quoted as the first example of productive metathesis of a simple olefin initiated by a well-characterized carbene complex (14,15). [Ta(CH-*t*-Bu)$Cl_3(THF)_2$] (THF, tetrahydrofuran) (**7**) (15) and [Ta(CH-*t*-Bu)$(OAr)_3$(THF)] (**8**) (where Ar = 2,6-*i*-$Pr_2C_6H_3$) (16) are short lived catalysts of metathesis of *cis*-2-pentene. [Ta(CH-*t*-Bu)$(OAr)_3$(THF)] (**8**) reacts with NBE to form tantalacyclobutane (**9**) (Eq. 13.5) which initiates ROMP of NBE giving a polymer of narrow molecular weight distribution (PDI = 1.04), when the polymerization is stopped after 75% monomer conversion (17,18).

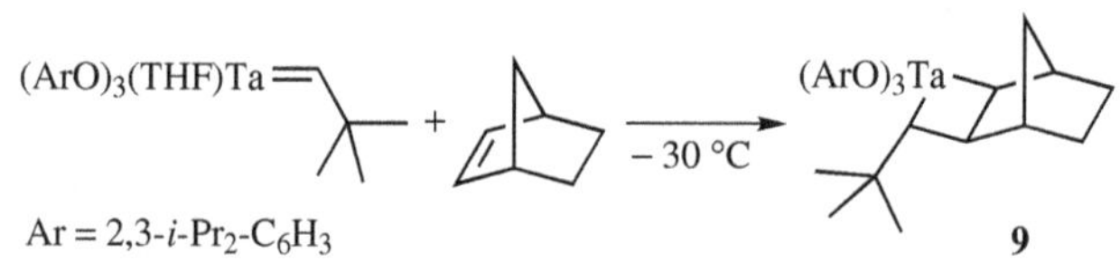

(13.5)

[Ta(CH-*t*-Bu)$(SAr)_3$(THF)] (where Ar = 2,4,6-*i*-$Pr_3C_6H_2$) (**10**) initiates living ROMP of NBE. Polymer chain growth can be terminated by the addition of aldehydes or ketones, which react with the tantalacycle to produce olefin end groups via Wittig-type chemistry (17).

13.2.3 Vanadium

There are no reports on the catalytic activity of the known vanadium(III) and (IV) alkylidene complexes in olefin metathesis. Only [CpV$(CHCMe_3)$(dmpe)] (**11**) has been shown to exhibit, albeit extremely low, activity in promoting ROMP of NBE (19). Vanadium(V) alkylidene complex [CpV(CHPh)(NAr)(PMe_3)] (Ar = 2,6-*i*-$Pr_2C_6H_3$) (**12**) does not react with NBE (20). The chemistry of vanadium(V) alkylidene complexes was developed by Nomura (21,22), who demonstrated the potential of some vanadium(V) alkylidene complexes in promoting ROMP of NBE. Imidoaryloxyvanadium(V) complexes in the presence of aluminum derivatives were found to exhibit exceptionally high catalytic activity for ethylene polymerization. However, the proper choice of a co-catalyst can modify the catalytic system so that [V(NAr′)Cl_2(OAr)]/$AlMe_3$ (where Ar′ = 2,6-$Me_2C_6H_3$; Ar = 2,6-*i*-$Pr_2C_6H_3$) is able to promote ROMP of NBE (23). Dibenzyl complex [V(NAr′)$(CH_2Ph)_2$(OAr)] (where Ar′ = 2,6-$Me_2C_6H_3$, Ar = 2,6-*i*-$Pr_2C_6H_3$) initiated ROMP of NBE at 25 °C, affording a polymer with high molecular weight and uniform molecular weight distribution. Formation of alkylidene species was assumed in the catalytic cycle (23). Catalytic activity in ROMP of NBE was also observed for [V(NAr′)$(CH_2SiMe_3)_2$(N=C-*t*-Bu_2)]. The alkylidene complex [V(NAr′)$(CHSiMe_3)$(N=C-*t*-Bu_2)(PMe_3)] (where Ar′ = 2,6-$Me_2C_6H_3$) (**13**) obtained by reaction of [V(NAr′)$(CH_2SiMe_3)_2$(N=C-*t*-Bu_2)] with an excess of PMe_3 (Fig. 13.6) exhibits remarkable catalytic activity in the ROMP of NBE (24,25). Its activity is low at 25 °C; however, it increases with temperature, so that at 80 °C the observed activity is higher than that of [Ru(CHPh)$(Cl)_2(PCy_3)_2$] (**Ru–I**) under the same conditions, due to the improved thermal stability of the vanadium complex.

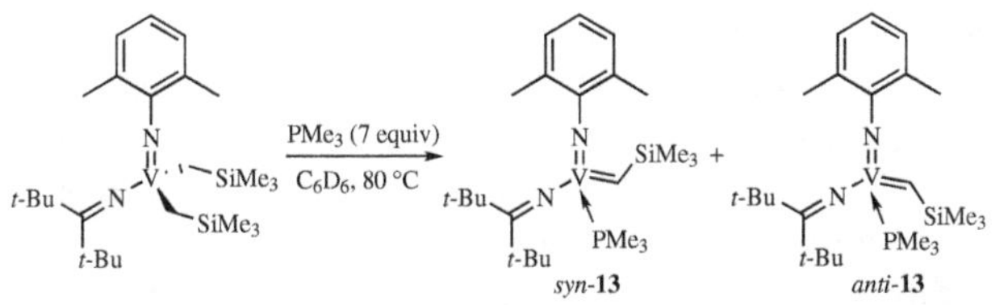

(13.6)

The aryloxide alkylidene analog $[V(NAr')(CHSiMe_3)(OAr)(PMe_3)]$ (**14**) exhibits higher catalytic activity at 25 °C than that of the ketimide analog. It initiates living ROMP of NBE at 25 °C (26). Other vanadium(V)-dialkyl complexes containing an aryloxide ligand like $[V(NAr')(CH_2SiMe_3)_2(OAr)]$ and other complexes containing a (2-anilidomethyl)pyridine ligand show notable catalytic activities for ROMP of NBE in the presence of PMe_3, but attempts at isolating the corresponding vanadium(V)-alkylidene complex were not successful. Treatment of $[V(NAd)(CH_2SiMe_3)_3]$ with NHC at 80 °C in C_6D_6 yielded $[V(NAd)(CHSiMe_3)(CH_2SiMe_3)(NHC)]$ (**15**) (Eq. 13.7) (27). Similarly, the reaction of the arylimido analog with NHC afforded $[V(NAr)(CHSiMe_3)(CH_2SiMe_3)(NHC)]$ (**16**) (Eq. 13.7).

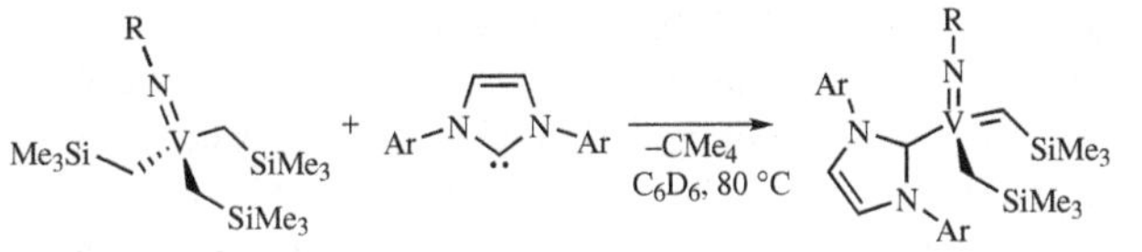

Ar = 2,6-*i*-$Pr_2C_6H_3$

15 R = Adamantyl
16 R = 2,6-$Me_2C_6H_3$

(13.7)

Both vanadium NHC alkylidene complexes were able to initiate ROMP of NBE. However, their activities were lower than those of $[V(CHSiMe_3)(NAr)(N{=}Ct\text{-}Bu_2)(PMe_3)]$.

13.2.4 Rhenium

The metathesis activity of four-coordinate Re catalysts appears to be significantly lower than that of Mo or W catalysts. $[\{Re(CH\text{-}t\text{-}Bu)(C\text{-}t\text{-}Bu)Cl_2\}_n]$ (**17**) is a convenient precursor to a number of bis(alkoxide) complexes of the formula $[Re(CH\text{-}t\text{-}Bu)(C\text{-}t\text{-}Bu)(OR)_2]$ which can exist in the form of *syn-* or *anti*-isomer. $[Re(CH\text{-}t\text{-}Bu)(C\text{-}t\text{-}Bu)\{OCMe(CF_3)_2\}_2]$ (**18**) was shown to be active in olefin metathesis (28). Complex **18** undergoes metathesis exchange with terminal olefins to form observable propagative alkylidene complexes. The reaction with vinylferrocene leads to the formation of an isolable ferrocenylmethylidene complex *anti*-isomer. Metathesis exchange proceeds more rapidly with olefins bearing heteroatoms (O, N, or S) attached directly to the C=C double bond. Complex $[Re(CH\text{-}t\text{-}Bu)(C\text{-}t\text{-}Bu)\{OCMe(CF_3)_2\}_2]$ (**18**) initiates slowly the metathesis of 2-pentene and methyl oleate. The activity of rhenium complexes depends dramatically on the electronic properties of alkoxide ligands. Replacing hexafluoro-*tert*-butoxide by *tert*-butoxide ligands leads to the disappearance of the catalytic activity, so that stable $[Re(CH\text{-}t\text{-}Bu)(C\text{-}t\text{-}Bu)(O\text{-}t\text{-}Bu)_2]$ (**19**) is unable to initiate ROMP of NBE. The neopentylidyne ligand appears to behave solely as an ancillary ligand. However, in the presence of ethylene, the cycloaddition of ethylene to the C≡Re=C unit has been reported (29).

13.3 EARLY TUNGSTEN BASED METATHESIS CATALYSTS

13.3.1 Fischer-Type Complexes

The very first report on metathesis with Fischer-type carbene complexes (although not catalytic and the term metathesis was not mentioned in the paper) was disclosed by Fischer (30). Investigations of the reactions of $[Cr(CO)_5\{=C(OMe)Ph\}]$ with vinyl ethers $H_2C{=}CH(OR)$ led to formation of methoxystyrene $H_2C{=}C(OMe)Ph$, being the metathesis product. Fischer-type metal carbene complexes (**20** and **21**) have been the first carbene complexes able to initiate olefin metathesis (without the need of using any co-catalysts) (31,32).

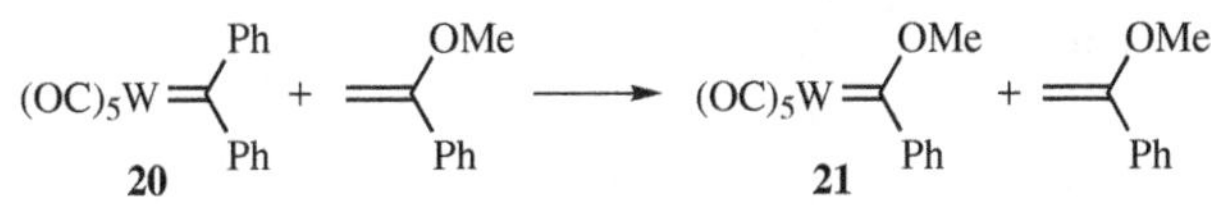

(13.8)

Stoichiometric exchange of alkylidene units for Fischer carbene ligands has been demonstrated by Casey (33) (Eq. 13.8). Diphenylcarbene(pentacarbonyl)tungsten (**20**) synthesized by Casey (34) is able to initiate the metathesis of a variety of alkenes. In the ROMP of cyclooctenes, it permits the formation of a polymer with an exceptionally high *cis*-selectivity (up to 98%) (35). It initiates the ROMP of trisubstituted cycloolefins to give polymers with high regio (head to tail) and stereoselectivity (36–38). Methoxyphenylcarbene(pentacarbonyl)tungsten (**21**) is much less active than **20**. It is able to initiate the ROMP of highly strained cyclobutene to give polybutadiene with a *cis* content up to 90% (39). Moreover, it has been shown to catalyze ring closing metathesis (RCM) of enynes (Eq. 13.9) (38). The reaction was the first example of intramolecular enyne metathesis.

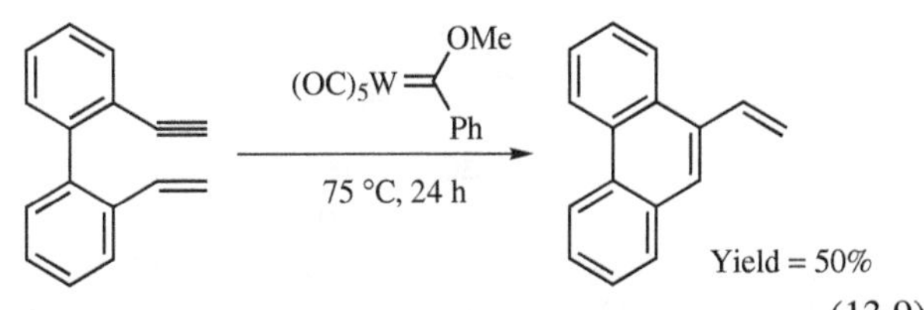

(13.9)

13.3.2 Oxoalkylidene Complexes

$[W(O)(CH\text{-}t\text{-}Bu)Cl_2(PMe_3)_2]$ (**22**) obtained quantitatively in the reaction of oxotungsten tetrakis(*tert*-butanolate) with a tantalum alkylidene complex (Eq. 13.10) (14) was probably the first well-defined tungsten catalyst of olefin metathesis.

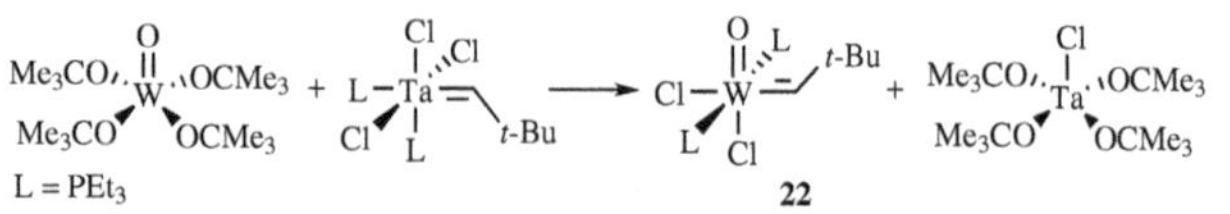

(13.10)

Complex **22** exhibits catalytic activity for the metathesis of terminal and internal olefins. The presence of traces of $AlCl_3$ results in a significant increase in the rate of metathesis (40). For the first time in the reaction of this catalyst with olefins, new propagative alkylidene complexes including methylidene complex $[W(O)(CH_2)Cl_2(PMe_3)_2]$ (**23**) could be observed and identified as the reactive intermediates (14). Complex $[W(O)(CH\text{-}t\text{-}Bu)Cl(PEt_3)_2][AlCl_4]$ (**24**), obtained by addition of 1 equiv of $AlCl_3$ to $[W(O)(CH\text{-}t\text{-}Bu)Cl_2(PMe_3)_2]$ (**25**) slowly metathesizes terminal and internal olefins (41). Isolable, 16-electron complex $[W(O)(CH\text{-}t\text{-}Bu)Cl_2(PEt_3)]$ (**26**) initiates the metathesis of *cis*-2-pentene, without requiring any activator. However, this system is very unstable (40). Aryloxy monophosphine derivatives of the general formula $[W(O)(CH\text{-}t\text{-}Bu)(OAr)_2L]$ (where Ar = 2,6-$Ph_2C_6H_3$; L = PMe_3 (**27**) or PPh_2Me (**28**)) readily catalyse living ROMP of 2,3-dicarbomethoxynorbornadiene and 2,3-bis(trifluoromethyl)norbornadiene leading to polymers of low polydispersity and high *cis* content (>95%) (42). The reaction of $[W(O)Cl_2L_3]$ (where L = $PMePh_2$, $P(OMe)_3$) with 3,3-diphenylcyclopropene in the presence of 2 equiv of $LiOCMe(CF_3)_2$ results in the formation of the phosphane free vinylcarbene complex **29** (Eq. 13.11), that is catalytically active in ROMP of various cycloolefins including cyclooctatetraene, which provides polyacetylene material (43,44).

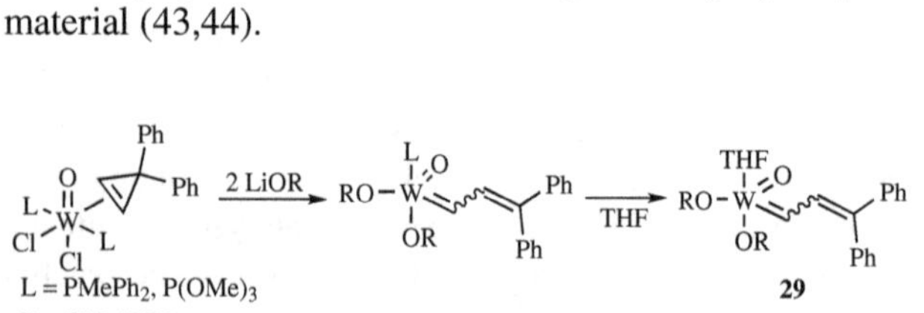

(13.11)

13.3.3 Alkoxide–Alkylidene Complexes

Osborn (45) found that addition of various Lewis acids (e.g., AlX_3; X = Cl, Br) to the oxoneopentyl complex $[W(O)(OCH_2\text{-}t\text{-}Bu)_2(CH_2\text{-}t\text{-}Bu)_2]$ yields metathesis catalysts $[W(OAlX_3)(OCH_2\text{-}t\text{-}Bu)_2(CH_2\text{-}t\text{-}Bu)_2]$. These species were found to transform slowly in solution into the oxo-free alkylidene complexes $[W(CH\text{-}t\text{-}Bu)(OCH_2\text{-}t\text{-}Bu)_2X_2]$ (**30**) (46). Alkylidene complex **30**, when used separately, showed no activity in the metathesis of *cis*-2-pentene. However, the addition of Lewis acids such as $AlBr_3$ or $GaBr_3$ generated highly active species that readily catalyzed the metathesis of *cis*-2-pentene at room temperature. The generation of propagative carbene catalysts can be observed when the reaction is performed at low temperatures (−10 °C) (46). The high catalytic activity of $[W(CH\text{-}t\text{-}Bu)(OCH_2\text{-}t\text{-}Bu)_2Br_2]$ (**31**) in the presence of Lewis acids (e.g., $GaBr_3$) has been shown to be a consequence of the formation of cationic species $[W(CH\text{-}t\text{-}Bu)(OCH_2\text{-}t\text{-}Bu)_2Br]^+[GaBr_4]^-$ (**32**) or $[W(CH\text{-}t\text{-}Bu)(OCH_2\text{-}t\text{-}Bu)_2Br]^+[Ga_2Br_7]^-$ (**33**) (47). Comparing the Lewis acid activated alkoxybromoalkylidene complexes investigated, the highest activity has been observed for $[W(CH\text{-}t\text{-}Bu)(OCH_2\text{-}t\text{-}Bu)_3Br]$ (**34**) (48). The two component systems $[W(CH\text{-}t\text{-}Bu)(OCH_2\text{-}t\text{-}Bu)_2Br]/GaBr_3$ initiated the ROMP of NBE and 2,3-dimethylnorbornene. However, this system was not living and was comprised of several catalytically active components (49–52).

For cyclopentylidene complex **35** (Eq. 13.12), the nature of the interaction with Lewis acids was studied in solution by ^1H NMR spectroscopy and in the solid state (by comparison of the crystal structures of cyclopentylidene dimer with that of its adduct with $GaBr_3$). It has been demonstrated that except for the elongation of one W–Br bond, the Lewis acid does not cause any significant structural changes in complex **35**. The high activity of cyclopentylidene complex **35** in the presence of Lewis acids suggests a formation of cationic species in solution (53). ^1H NMR spectroscopic studies of the interaction of Lewis acid adducts with cycloheptene at 177 K have permitted the observation of a π-olefin complex (Eq. 13.12) (54). It has been probably the only observation of alkylidene/π-olefin intermediate in a metathetically active catalytic system in a high oxidation state.

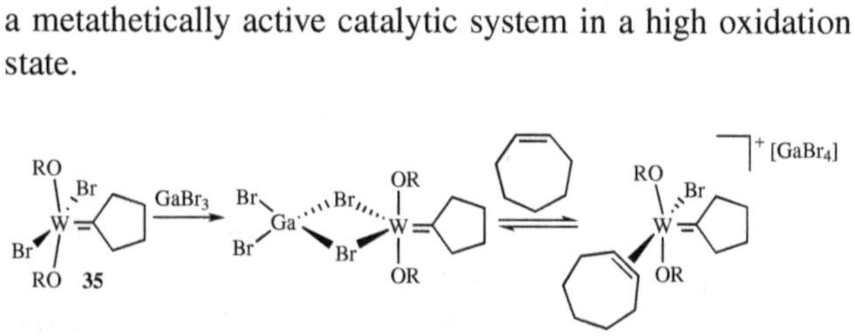

(13.12)

13.3.4 Aryloxide–Alkylidene Complexes

Aryloxide complexes of the type $[W(OAr)_2Cl_4]$ when combined with metal alkyls (e.g., $SnMe_4$) become olefin metathesis catalysts (55). The reaction of $[W(ArO)_2Cl_4]$ (where Ar = 2,6-$R_2C_6H_3$) with 1 and 1.5 equiv of $Mg(CH_2\text{-}t\text{-}Bu)_2$ produced $[W(CH\text{-}t\text{-}Bu)(ArO)_2Cl_2(OR_2)]$ and $[W(CH\text{-}t\text{-}Bu)(ArO)_2(CH_2\text{-}t\text{-}Bu)Cl(OR_2)]$, respectively (56). In the case of the 2,6-diphenylphenoxy ligand, the reaction proceeds with activation of the *ortho* C–H bond of one of the phenyl groups, the formation of a W–C bond and the elimination of CMe_4 (Eq. 13.13)

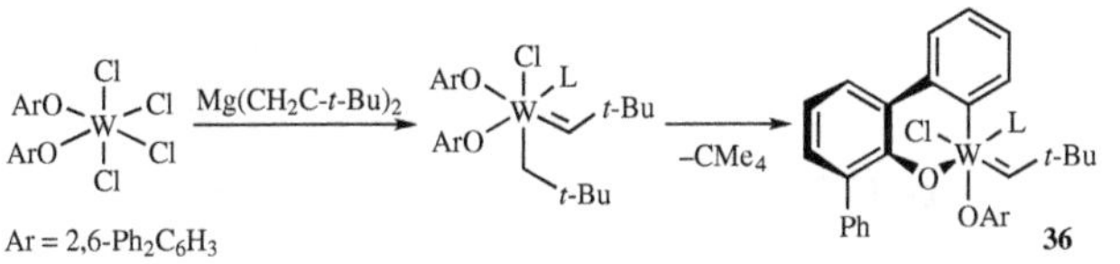

(13.13)

Complex **36** can also be obtained by substitution of two chlorine ligands in [WCl_4(C-*t*-Bu)(dme)] by the 2,6-diphenylphenoxy ones (57). While complexes of the formula [$W(ArO)_2(CH\text{-}t\text{-}Bu)Cl_2(OR_2)$] (**37**) are relatively poor catalysts, the monochloride derivatives [$(ArO)_2W(CH\text{-}t\text{-}Bu)(CH_2\text{-}t\text{-}Bu)Cl(OR_2)$] (**38**) exhibit high activity. The activity of this type of aryloxyalkylidene complexes depends on the electronic properties of the aryloxide ligands. A particularly attractive catalyst has proved to be complex **36**, in whose presence the metathesis of *cis*- and *trans*-2-pentene proceeds stereospecifically (with complete retention of configuration of the starting olefin). Moreover, catalyst **36** initiates the ROMP of 1-methylnorbornene with formation of highly syndiotactic, completely head-to-tail, all-*cis* double bond polymers (57). This catalyst tolerates a wide range of functional groups, that is, alkylsilanes, alkylstannanes, esters, ethers, phosphanes, and thioethers (58). From the point of view of the reaction mechanism, no new alkylidene complexes have been observed in catalytic systems employing complex **36**. Therefore, the species acting in the catalytic cycle have not been identified.

13.4 TUNGSTEN AND MOLYBDENUM IMIDOALKYLIDENE COMPLEXES

13.4.1 Achiral, bis(Alkoxide)Imidoalkylidene Complexes of Tungsten and Molybdenum

13.4.1.1 General Characterization W- and Mo-based complexes of the general formula [M(NR)(CHR′)(OR″)$_2$] constitute a family of well-defined catalysts of olefin metathesis (8,9,59–62). These complexes deserve to be called the first generation W and Mo imidoalkylidene catalysts of olefin metathesis. The first representatives of the family were prepared by Schrock and coworkers in 1986 (63). Bis(alkoxide)imidoalkylidene complexes contain metal (W or Mo) in the high oxidation state and a set of bulky, covalently bound ligands arranged in a pseudotetrahedral geometry. These complexes should be considered 14-electron ones as the imido ligand donates its electron pair to the metal center to form a pseudo triple bond. The oxidation state of the metal in the complex is VI (d^0), while the alkylidene ligand is assumed to act as a dianionic ligand. Owing to their electron deficient character, these complexes are moisture and air sensitive.

The coordination sphere of the first generation of imidoalkylidene complexes (Fig. 13.1) includes the alkylidene, imido, and oxygen-based donor ligands. The stability of the catalyst and its reactivity result from the interplay of steric and electronic effects contributed by each ligand. The metathetically active alkylidene ligand should provide steric hindrance to the complex to protect it from bimolecular decomposition. Complexes with the small methylidene ligand are particularly unstable. The absence of a proton at C_β prevents deactivation via β-elimination. The only successful alkylidene ligands used in this family of complexes are neopentylidene (=$CHCMe_3$) and neophylidene (=$CHCMe_2Ph$). The imido system controls the electron density on the metal (and stabilizes its high oxidation state). Moreover, a sterically demanding aryl group of the imido ligand protects the complex from bimolecular decomposition. The only alkylimido complexes of this class that have been isolated are molybdenum 1-adamantylimido species. The role of the alkoxide ligands is to control the electron density of the metal. Their bulkiness is needed to protect the electron-deficient metal center from bimolecular decomposition. Alkoxides have proved to be the most successful anionic ligands in this group of catalysts in terms of sustaining metathesis activity. Alternative ligands are either not bulky enough to stabilize the electron-deficient metal center or they donate too much electron density to the metal. Complexes containing the relatively small $OCH(CF_3)_2$ or OC_6F_5 ligands can be isolated only as adducts with a suitable base (THF, pyridine). In the presence of imidoalkylidene complexes of Mo and W, as for ruthenium-based catalysts, the olefin metathesis proceeds according to the carbene mechanism (64). However, the initiation of the catalytic cycle differs from that for ruthenium complexes. In tungsten and molybdenum imidoalkylidene catalysts, the ligands attached do not dissociate from the tetrahedral metal center during the catalytic cycle. Therefore, their catalytic performance is

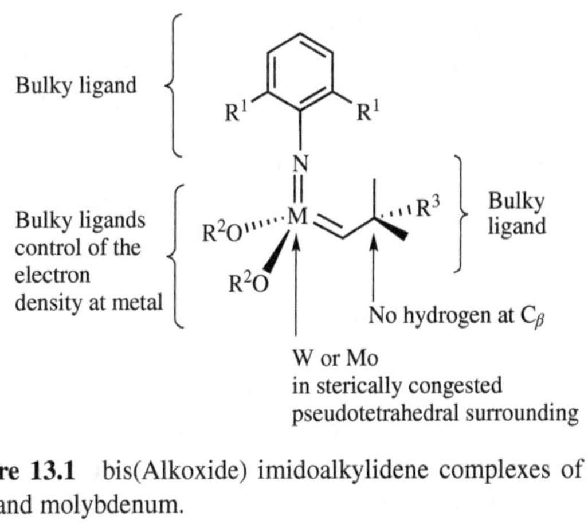

Figure 13.1 bis(Alkoxide) imidoalkylidene complexes of tungsten and molybdenum.

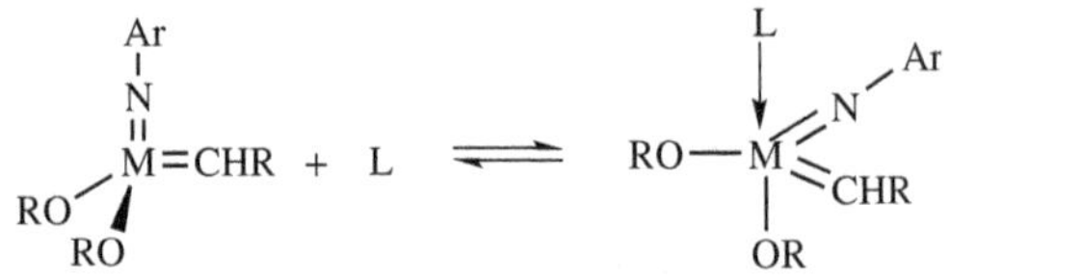

Scheme 13.1 Association of L-type ligand to metal center in W or Mo imidoalkylidene complexes.

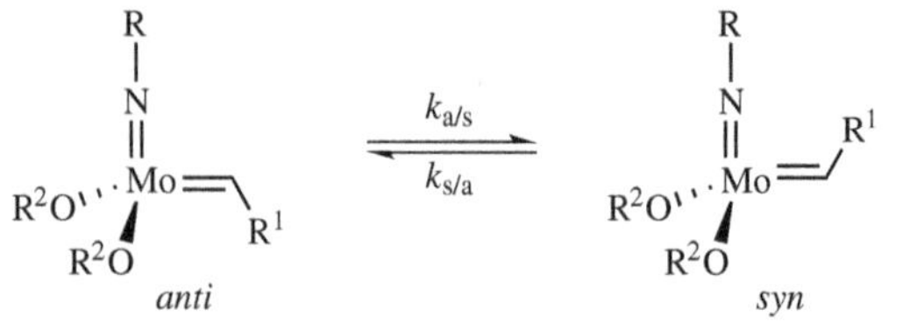

Scheme 13.2 Interconversion of *anti*- and *syn*-rotamers.

influenced by the stereoelectronic properties of each ligand. Hence, a precise control of activity and selectivity of the catalyst is possible by tuning the electronic and steric properties of the ligands. It is generally presumed that the alkylidene complex bearing π-coordinated olefin is a transition state on the pathway to the formation of a metallacyclobutane intermediate. Alkylidene complex bearing associated π-olefin intermediates have not been observed in any imidoalkylidene bis(alkoxide) complexes of Mo and W (54). Therefore, two other electron donors (instead of olefin) have been used as models to study the geometry of the adducts. In all imidoalkylidene complexes, two electron donors (PMe_3, THF, pyridine) have been found to associate with the CNO tetrahedral face (one of the two defined by the imido nitrogen, C_α of alkylidene, and oxygen atom of alkoxide) to afford a trigonal bipyramidal complex in which the new ligand is bound in an axial position (Scheme 13.1).

It has been proposed that the olefin binds to the metal weakly in the same position. The 16-electron five-coordinate adducts bearing two-electron donor ligand other than olefin are inactive in olefin metathesis. It has been assumed that the association of the olefin needs to be preceded by a dissociation of a two electron donor. According to the carbene mechanism, cycloaddition in alkylidene–olefin complex (or transition state) gives metallacyclobutane. Metallacyclobutane complexes of molybdenum and in particular of tungsten are isolable intermediates. Crystallographic studies of tungstacyclobutane complexes indicate that these complexes have either trigonal bipyramidal or square pyramidal geometry (65). Five coordinate metallacycles are relatively stable with respect to bimolecular decomposition. In some systems, metallacyclobutanes tend to decompose to form π-olefin complexes as a consequence of β-hydride elimination. Rearrangement of metallacyclobutanes to olefins in addition to bimolecular coupling are the main decomposition pathways that limit the longevity of a high oxidation state metathesis catalyst in solution (for more comprehensive discussion, see Refs 60 and 61). The problem of catalyst decomposition becomes most important in reaction systems in which ethylene is involved. Ethylene readily reacts with alkylidene complexes with the formation of the methylidene ones. They are believed to be the most reactive (due to the lack of steric protection on the alkylidene ligand) but also the most unstable and tend to decompose by bimolecular coupling and/or by non-metathetic rearrangement of metallacyclobutane to give olefin. An important feature of imidoalkylidene complexes of the general formula $[M(NR)(CHR')(OR'')_2]$ is their occurrence in the form of two stereoisomers: a *syn*-isomer in which the substituent at C_α points toward the imido ligand and an *anti*-isomer bearing the substituent at C_α pointing away from the imido nitrogen. The *syn*-alkylidene complex is energetically more favored, which is explained by an agostic interaction between C_α–H and the metal center. Both isomers can be detected by ^{1}H NMR spectroscopy. Typically, the *anti*-isomer is the minor one and is not always detectable in routine spectra due to its low concentration. *syn*- and *anti*-alkylidene complexes can interconvert (upon thermal or photochemical activation) by rotation about the M=C bond (Scheme 13.2). The rates of interconversion, expressed by $k_{s/a}$ and $k_{a/s}$, depend dramatically on the nature of the alkoxy and the imido ligand and can vary by many orders of magnitude, for example, $k_{s/a}$ for $[Mo(NAr)(CHCMe_2Ph)(O\text{-}t\text{-}Bu)_2]$ (**Mo-I-4a**) is 10^4 times higher than that measured for $[Mo(NAr)(CHCMe_2Ph)\{OC(CF_3)_2Me\}_2]$ (**Mo-I-1**), although both complexes have nearly the same K_{eq} values (K_{eq} = 1200 and 1400, respectively, toluene, 22 °C) (66,67).

The equilibrium constant between *syn*- and *anti*-isomers ($K_{eq} = [syn]/[anti] = k_{a/s}/k_{s/a}$) usually ranges from approximately 20 to around 2000 at 22 °C (66,67). *Syn*- and *anti*-isomers differ in reactivity and catalytic activity. The reactivity differences (determination of which is not possible in many systems) depend dramatically on the nature of the substituent on the alkylidene and the nature of the imido ligand. The lower activity of *syn*-isomers might be due to steric interaction of the alkylidene substituent with the arylimido group upon coordination of the olefin or to lowering of the Lewis acidity of the *syn* form caused by agostic interaction.

The route illustrated in Scheme 13.3 is the preferred method for the synthesis of molybdenum imidoalkylidene complexes. This method can also be used for the synthesis of tungsten analogs. $[Mo(NAr)_2Cl_2(dme)]$ (**I**) can be prepared in a one-step reaction in large quantities from Na_2MoO_4, $ArNH_2$, Me_3SiCl, and triethylamine in dimethoxyethane. Alkylation of bis(imido) derivatives (**I**) with neopentyl or neophyl Grignard reagents gives complexes (**II**), which upon treatment with 3 equiv of triflic acid in dimethoxyethane gives 18-electron bis(triflate) imidoalkylidene complexes (**III**). Formation of the alkylidene

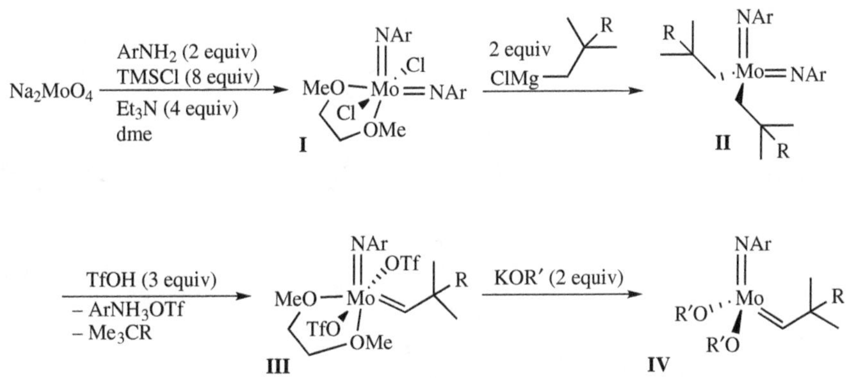

Scheme 13.3 Preferred method for the synthesis of molybdenum and tungsten imidoalkylidene complexes.

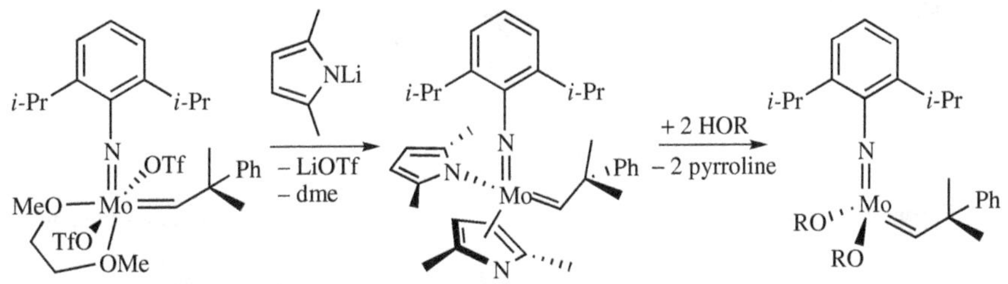

Scheme 13.4 Synthesis of bis(alkoxide)imidoalkylidene complexes.

ligand proceeds via α-H-abstraction. In the presence of various alkoxide or aryloxide salts of lithium, sodium, or potassium, the bis(triflate) complexes (**III**) are converted into bis(alkoxide) complexes IV (Scheme 13.3) (68). This procedure permits synthesis of alkylidene complexes that contain a variety of arylimido ligands as well as adamantylimido ligand.

A triflate derivative (**III**) (Scheme 13.3) (where Ar = 2,6-i-$Pr_2C_6H_3$; R = Ph) is currently commercially available. It readily reacts with lithium pyrrolide to give bis(pyrrolide) complexes (Scheme 13.4) that have been shown to be convenient starting materials for the synthesis of bis- or monoalkoxy derivatives as well as diolate complexes by direct reaction with alcohols (diols) (69).

13.4.1.2 Catalytic Activity Mo-based complexes provide an appreciable level of activity. An important feature of tungsten and molybdenum imidoalkylidene complexes is a distinct substrate/catalyst dependence. Choosing the optimal catalysts for a specific reaction system needs testing of a series of catalysts. Therefore, a large number of complexes has been synthesized and tested as catalysts in metathesis transformations since 1986. The most successful catalysts are presented in Figure 13.2.

Despite the variety of achiral imidoalkylidene complexes described, only one of them (**Mo-I-1** called the *Schrock catalyst*) has won significant interest in the context of use in organic synthesis, most likely because of its commercial availability. In general, catalyst **Mo-I-1** exhibits high catalytic activity in metathetic transformations of a variety of functional olefins, including unsaturated ethers, thioethers, acetals, esters, lactones, lactames, nitriles, phosphanes, substituted silanes, protected amines, amides, and alcohols. Molybdenum catalyst **Mo-I-1** also has been shown to be active in the presence of sterically protected free alcohols and tertiary amines. It can be employed in the presence of a variety of Lewis basic functional groups. However, its usefulness is limited by the sensitivity to oxygen, moisture, and certain polar or protic functional groups, because of the electrophilicity of the high-oxidation-state transition metal. It decomposes in the presence of aldehydes in a Wittig-like reaction. The Mo complex is usually incompatible with carbonyl groups. Application of molybdenum imidoalkylidene complexes in catalysis has been the subject of several reviews (59–61,70–75). Catalyst **Mo-I-1** permits the synthesis of small, medium, and macrocyclic rings via RCM. It efficiently promotes the formation of a variety of trisubstituted as well as five- or six-membered tetrasubstituted (at C=C double bond) rings. The **Mo-I-1** complex is particularly useful for the formation of sterically crowded systems. It efficiently catalyzes RCM involving exocyclic olefins. Besides, **Mo-I-1** has proved active in forming 7 and 8, and less efficient in forming nine-membered carbo- and heterocyclic rings. Its exemplary applications include the formation of an eight-membered unsaturated amide as a step in the synthesis of manzamine A (76), a seven-membered ring Boc-protected

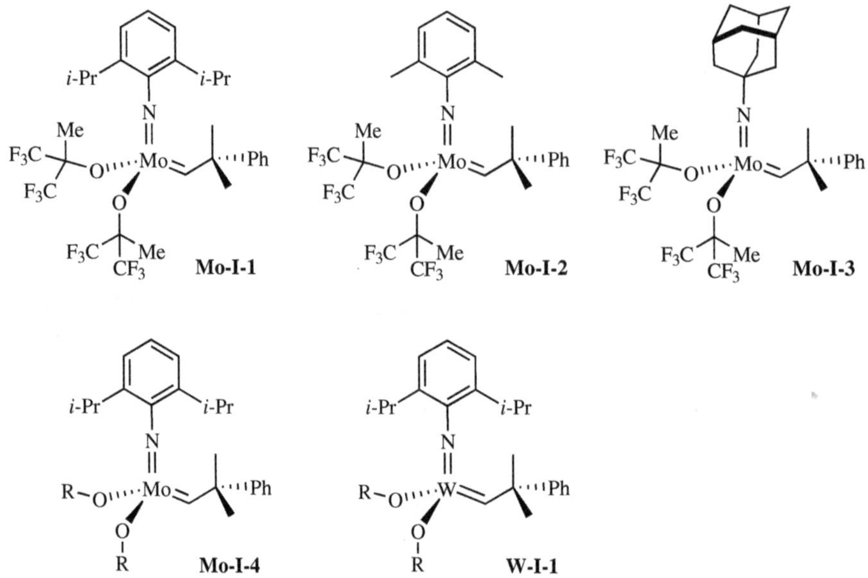

R = *t*-Bu (**a**), CMe_2CF_3 (**b**)   R = *t*-Bu (**a**), CMe_2CF_3 (**b**), $CMe(CF_3)_2$ (**c**)

Figure 13.2 Selected bis(alkoxide)imidoalkylidenes of molybdenum and tungsten.

amine in the total synthesis of (−)-balanol (77), substituted cycloheptenes (78), cyclooctenes (79), seven- and eight-membered ring enol ethers that may be used in the total synthesis of brevetoxin A (80–82), eight- and less efficiently nine-membered ring cyclic ethers (83,84) and S-heterocyclic eight-membered rings (85). RCM seems to be the most straightforward and reliable method for the formation of large ring systems. There are numerous examples of the application of catalyst **Mo-I-1** in metathetic macrocyclization as a step in the total synthesis of natural products or pharmaceuticals. Exemplary applications of catalyst **Mo-I-1** include the formation of a 14-membered lactam ring in the reaction sequence leading to fluvirucin B1 (86), the formation of a 16-membered lactone ring in the total synthesis of (−)-epothilone B (87), and the regioselective formation of a 22-membered ring via a sequential cross-metathesis (CM)/RCM reaction for the synthesis of (−)-cylindrocyclophane A and F, displaying activity against several cancer cell lines (88). **Mo-I-1** exhibits high activity in homometathesis of terminal olefins and some allyl derivatives. Styrene and its derivatives are transformed slowly into homometathesis products but readily react with a number of terminal olefins to give CM products with high selectivities (89). Similarly, acrylonitrile undergoes CM with a number of functional, terminal olefins to give selective products with an untypically high content of *Z*-isomer (90). Allylsilanes undergo selective CM with terminal olefins to give products with preferred *E* selectivity (91). Activity of **Mo-I-1** catalyst has also been reported in the CM of allylstannanes with various terminal olefins (92). In contrast, 1,1-disubstituted olefins have been reported to be spectators in CM catalyzed by **Mo-I-1**.

Tungsten imidoalkylidene complexes have been explored less intensively than Mo catalysts because of their higher sensitivity to a functional group of reagents and higher stability of tungstacyclobutane complexes , especially unsubstituted tungstacyclobutane species (in relation to molybdenum analogs), against loss of olefin. The metathesis activity in the series of complexes **W-I-1a–c** (Fig. 13.2) increases substantially when replacing *tert*-butoxide by $OCMe(CF_3)_2$ ligands, as the transition metal becomes more electrophilic (65). Tungsten imidoalkylidene complex **W-I-1c** (Fig. 13.2) is a very active catalyst for the metathesis of acyclic olefins and metathesizes *cis*-2-pentene at a rate of approximately 1000 turnovers/min at 25 °C, which contrasts with a rate of about 100 turnovers/min for **W-I-1b**. No significant reaction could be detected over a period of 1 day (at 25 °C) when the *tert*-butoxide complex **W-I-1a** was used.

Typical Procedure for RCM with Schrock Catalyst (Mo-I-1) (Eq. 13.14) (93)

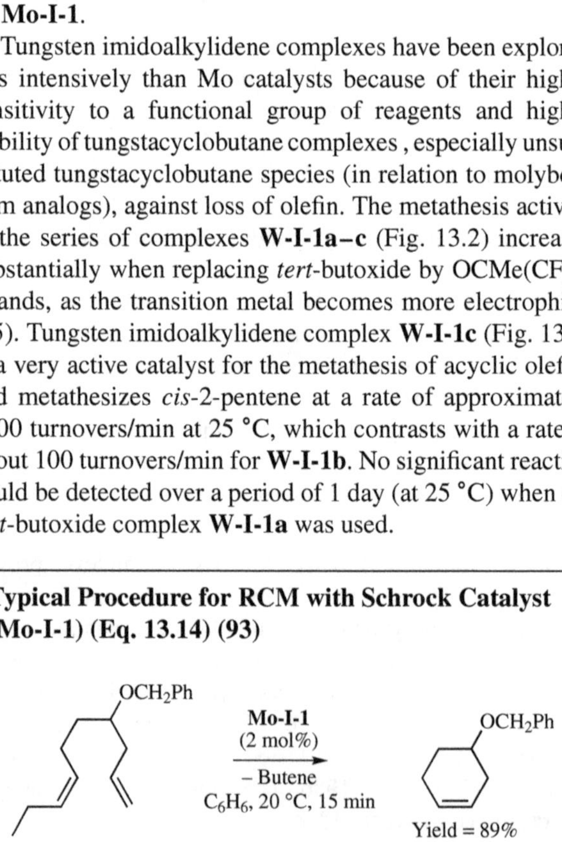

(13.14)

All manipulations were conducted under argon atmosphere. Oven-dried glassware with magnetic stirring was used. Diene (122 mg, 0.50 mmol) was added to a yellow solution of (8 mg, 0.01 mmol) **Mo-I-1** in 17 ml of dry benzene. The reaction mixture was allowed to be stirred at 20 °C for 15 min. The reaction was monitored by thin layer chromatography (TLC) and it was quenched by exposure to air. The content was concentrated and purified by flash chromatography using hexane 100% to 5% EtOAc/hexane as eluent. The procedure permits obtaining a colorless oil (84 mg) identified by ^1H NMR as substituted cyclohexene in a yield of 89%. The reaction can also be performed with CH_2Cl_2 as a solvent.

13.4.2 Chiral, Diolate Imidoalkylidene Complexes of Tungsten and Molybdenum

13.4.2.1 General Characterization Two families of chiral, optically pure tungsten and molybdenum imidoalkylidene metathesis catalysts have been designed for application in asymmetric metathesis. These are "diolate" complexes and complexes stereogenic at metal. Diolate complexes constitute a family of molybdenum and to a less extent tungsten imidoalkylidene complexes, in which two alkoxy ligands are replaced by one chelating, chiral, enantiomerically pure, biphenol, binol, or tetrahydrobinaphthol ligand (8,9,59–62,94). This family of complexes has been termed the second generation of molybdenum and tungsten imidoalkylidene catalysts of olefin metathesis. The first complexes of this family were prepared in 1993 to control the stereochemistry of polymers prepared via ROMP (95). Since then many catalysts containing C_2-symmetric chiral biphenolates or binaphtholates have been prepared and demonstrated to be highly successful catalysts for asymmetric metathesis reactions. Like the first generation catalysts, the diolate family of molybdenum imidoalkylidene complexes also shows a distinct substrate/catalyst dependence. A slight modification in the catalyst's structure can fundamentally influence the outcome of the reaction. Therefore, the identification of systems offering satisfactory reactivity and selectivity needs a large screening of diverse sets of catalysts. The modular structure of diolate complexes permits the synthesis of a range of chiral catalysts, preserving the parent structural motif. The most important representatives of the family of diolate catalysts are illustrated in Figure 13.3. In general, the biphenolates or binaphtholates must contain substituents in the 3 and 3′ positions of the diolate ligand to stabilize the imido alkylidene complexes against bimolecular decomposition and maximize the efficiency of an asymmetric metathesis reaction. Complex **Mo-II-1** is referred to as the *Schrock–Hoveyda catalyst.*

From the point of ease of preparation of optically pure ligands, the synthesis of biphenol and its derivatives is troublesome as it requires a resolution of unavailable racemic material. In contrast to biphenol based ligands, binol ones can be prepared from the commercially available optically pure binaphthol. The beneficial steric properties of the modified biphenol ligands can be at least in part mimicked by using tetrahydrobinaphthol derivatives. This group of ligands can be easily prepared from binaphthol and share structural and catalytic features with both the binol- and biphenol-based systems. The synthesis procedure is analogous to that for bis(alkoxide)derivatives (Scheme 13.3) and involves the reaction of bis(potassium) aryloxide with $[Mo(NAr)(CHCMe_3)(OTf)_2(dme)]$ (95). The improved synthesis employs addition of 1 equiv of biphenol or binaphthol to bis(pyrrolide) precursors, which leads to the rapid and virtually complete formation of diolate complexes (Eq. 13.15) (96). Protocols for the *in situ* generation of these catalysts have been developed (96,97).

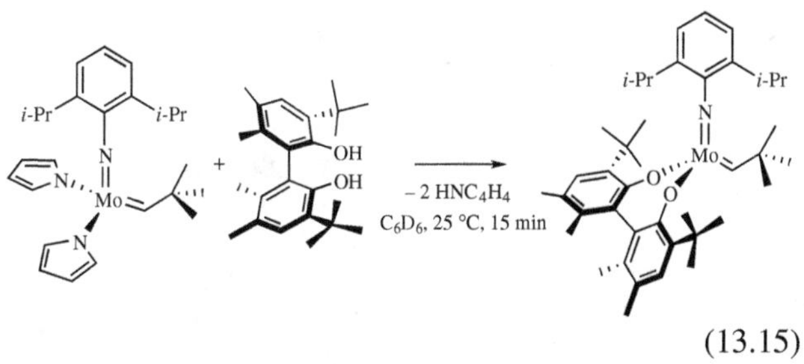

(13.15)

13.4.2.2 Catalytic Activity In the area of olefin metathesis, the term *enantioselective catalysis* refers to three groups of reactions. These are (i) transformations of enantiomerically pure chiral reagents performed in the presence of a non-chiral catalyst, (ii) processes in which a racemic mixture of enantiomers in the presence of a chiral, enantiomerically pure catalyst undergoes a reaction leading to preferred formation of one of the enantiomers (kinetic resolution), and (iii) processes of desymmetrization of prochiral substrates. High asymmetric induction is expected for diolate derivatives of W and Mo imidoalkylidene complexes because the rigid structure and bulkiness of the diolate ligands determine that face of the M=C bond to which the olefin associates. For a review covering the subject of asymmetric olefin metathesis see Refs 94, 98, and 99.

In 1996, Fujimura and Grubbs (100,101) reported the first kinetic resolution of a series of dienes through asymmetric ring closing metathesis (ARCM). The chiral enantiomerically pure molybdenum chelate (Fig. 13.4) used did not permit enantiodifferentiation levels (102) exceeding 3 (typically $k_{rel.}$ <3).

The kinetic resolution of numerous dienes can be carried out efficiently at 22°C in the presence of the catalyst **Mo-II-1** (5 mol%). This catalyst promotes kinetic resolution through RCM of 1,6-dienes with $k_{rel.}$ > 20) (Eq. 13.16) (103) and allyl ethers with $k_{rel.}$ varying from 8 to 23 (104).

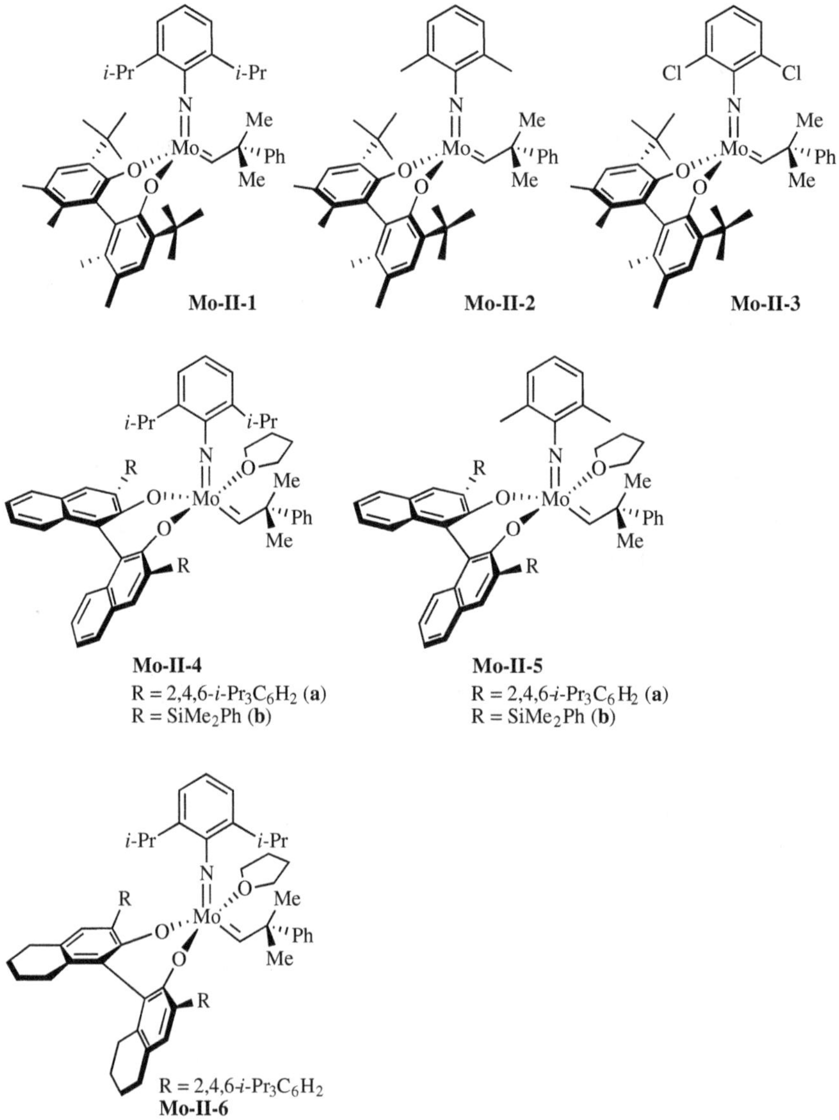

Figure 13.3 Selected diolate imidoalkylidene complexes of molybdenum.

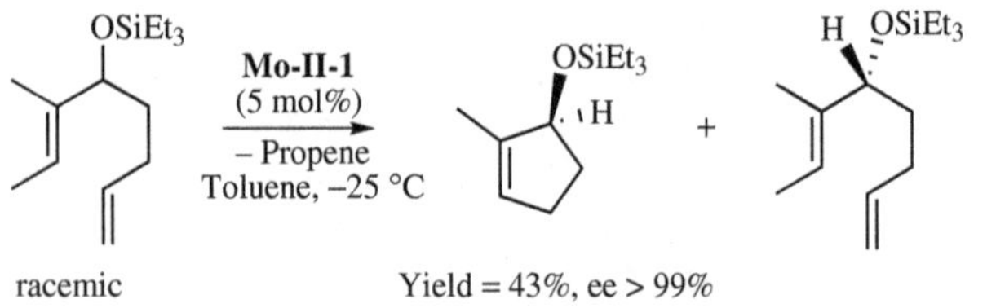

(13.16)

All transformations promoted by Mo diolate complexes can be carried out in toluene/benzene or alkanes with equal efficiency. Importantly, a number of procedures have been proposed to permit kinetic resolutions without the need of a solvent (104–106). Binol-based complex **Mo-II-4a** promotes RCM of 1,7-dienes with a high level of enantiodifferentiation ($k_{rel.}$ = 24 or exceeds 25) (105). The data set presented in Figure 13.5 illustrate a high substrate/catalyst dependence of the diolate system for the kinetic resolution by RCM (99).

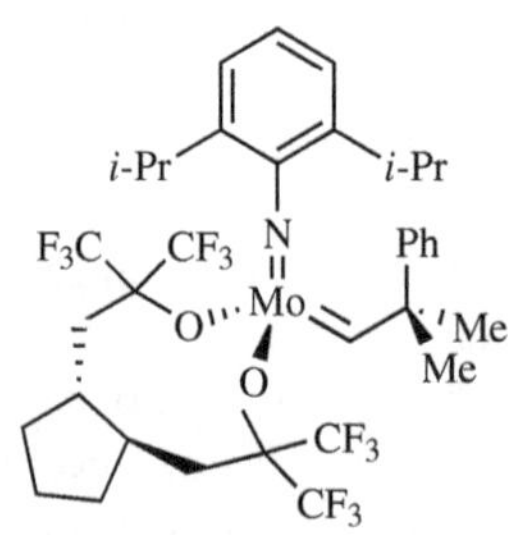

Figure 13.4 First chiral enantiomerically pure molybdenum chelate used for kinetic resolution through ARCM.

A plausible explanation for kinetic resolution through RCM was proposed on the basis of analysis of transition state structures (Fig. 13.6) (98).

The concept of desymmetrization of prochiral/meso compounds via olefin metathesis refers to ARCM of

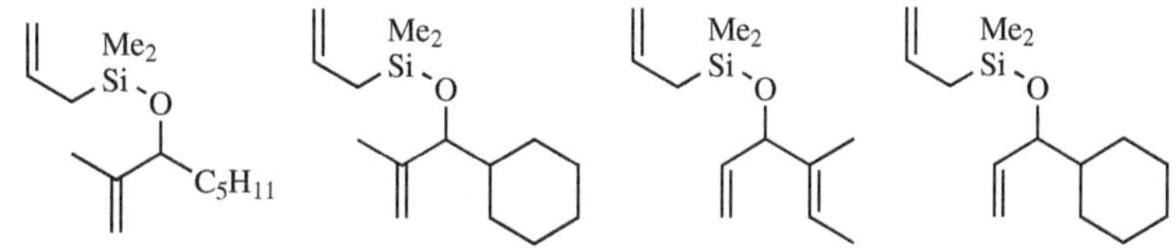

Mo-II-1	$k_{rel} < 5$	$k_{rel} < 5$	$k_{rel} = 21$	$k_{rel} > 25$
Mo-II-2	$k_{rel} = 10$	$k_{rel} = 14$	$k_{rel} < 5$	$k_{rel} < 5$
Mo-II-4	$k_{rel} > 25$	$k_{rel} > 25$	$k_{rel} < 5$	$k_{rel} < 5$

Figure 13.5 Substrate/catalyst dependence in RCM of silyl ethers catalyzed by diolate imidoalkylidene complexes of molybdenum.

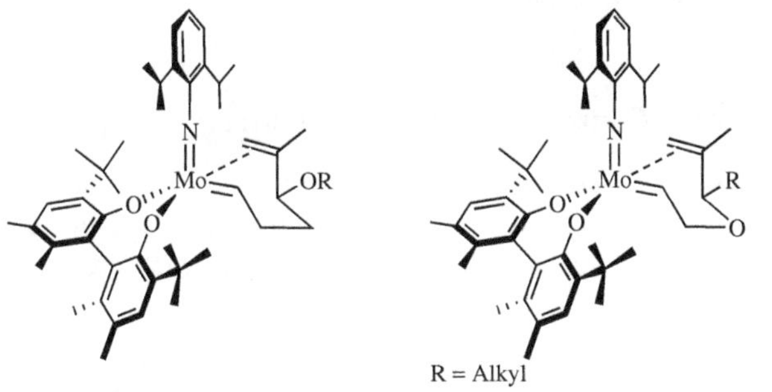

Figure 13.6 Proposed transition state structures responsible for the formation of observed products (98).

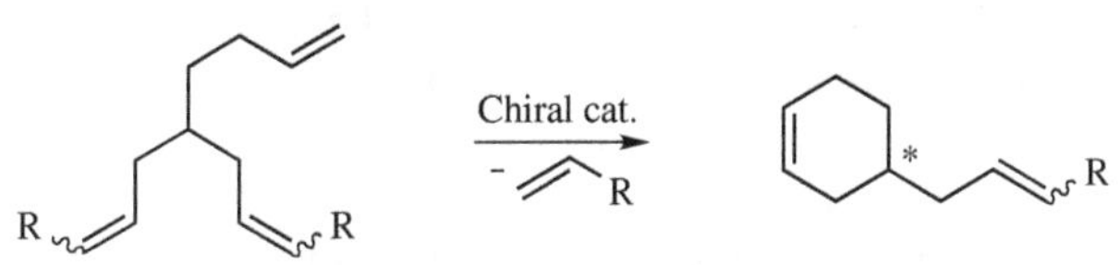

Scheme 13.5 Desymmetrization of trienes through ARCM.

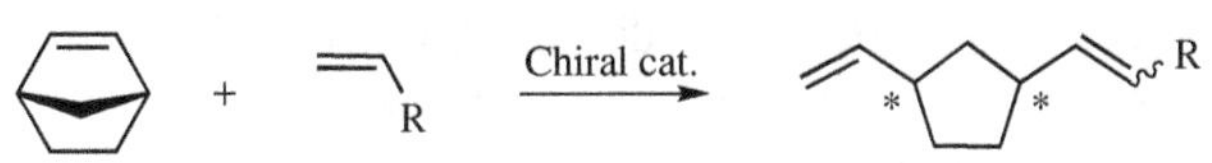

Scheme 13.6 Desymmetrization through AROM/CM.

trienes (Scheme 13.5), asymmetric ring opening metathesis (AROM) followed by CM (Scheme 13.6), and asymmetric CM.

The desymmetrization of prochiral trienes through ARCM is a powerful tool for the enantioselective synthesis of carbo- or heterocycles. Catalyst **Mo-II-1** (Eq. 13.17) promotes desymmetrization of allylic esters to form five-membered rings with enantiomeric excess reaching 99%. It permits the formation of homo- and heterocyclic seven-membered rings (107).

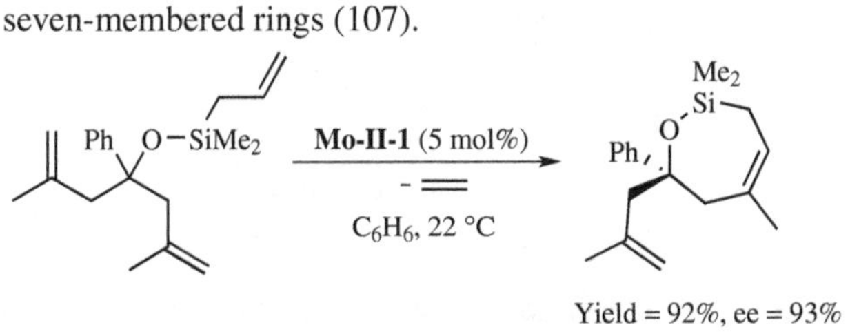

(13.17)

Incorporation of an electron-withdrawing group into the imido ligand (e.g., complex **Mo-II-3**, Fig. 13.3) leads to an increased Lewis acidity of the metal center and can result in a higher catalytic activity. Such complexes have been shown to exhibit increased enantioselectivity toward the other diolate imidoalkylidene complexes (108). Desymmetrization of trienes as a result of ARCM in the presence of **Mo-II-1** has been used in the total synthesis of *endo*-brevicomin (Eq. 13.18) (109).

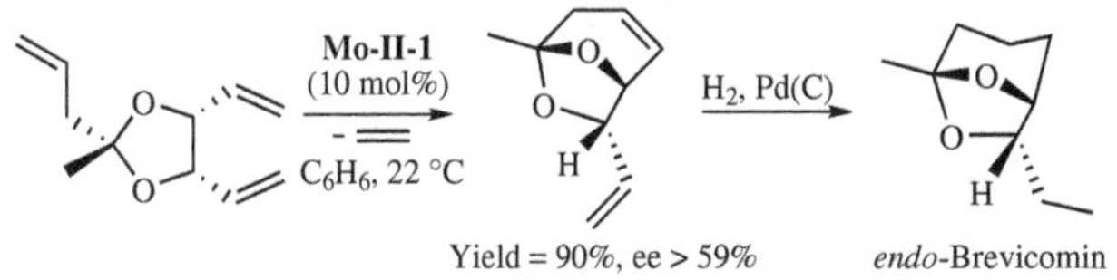

(13.18)

Because of the Lewis basic character of amine groups, their interaction with the catalyst center should be avoided. The synthetic strategies proposed involve the use of sterically hindered amines or reduction of their Lewis basicity through protection with an electron withdrawing group. **Mo-II-1** permits the efficient enantioselective synthesis of six to eight-membered *N*-heterocyclic rings via kinetic resolution of dienes or desymmetrization of *N*-containing trienes (106). Enantioselective synthesis of medium ring unsaturated amines proceeds enantioselectively and can be carried out in the absence of a solvent with a catalyst generated *in situ*. *N*-fused bicyclic amides and amine structures of various sizes (bearing five- to eight-membered rings), as well as a variety of small- and medium-ring cyclic amines (including examples of unprotected secondary amines), can be formed with high enantioselectivity (110). According to the concept of AROM/RCM or AROM/CM (Scheme 13.6), the diastereomeric alkylidene complex formed in the AROM step via interaction of chiral catalyst with prochiral substrate can undergo intramolecular (RCM) or intermolecular (CM) metathesis to form optically enriched compounds.

The selective progress of the reaction requires the use of highly reactive cycloolefins and a sterically hindered terminal alkene. Then a ring opening metathesis (ROM) step precedes the reaction of the catalysts with the terminal double bond. Catalyst **Mo-II-1** promotes AROM/RCM of cyclobutene (Eq. 13.19) and NBE derivatives (Eq. 13.20) to give products with high enantioselectivity. The reactions are the first catalytic AROM reported (111).

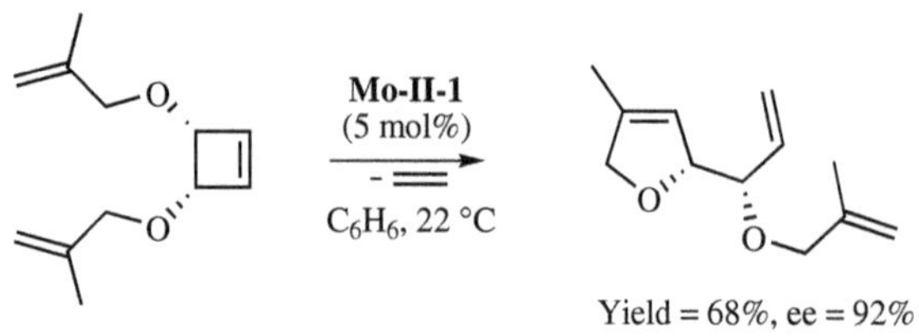

Yield = 68%, ee = 92%

(13.19)

An efficient reaction of NBE derivatives (Eq. 13.20) requires the preliminary *in situ* generation of a more active methylidene complex. For this purpose, diallyl ether has been added to the system. The methylidene complex generated from **Mo-II-1** catalyzes AROM/RCM of *meso*-NBE derivatives to form the *O*-heterocyclic six-membered ring with high enantioselectivity (112).

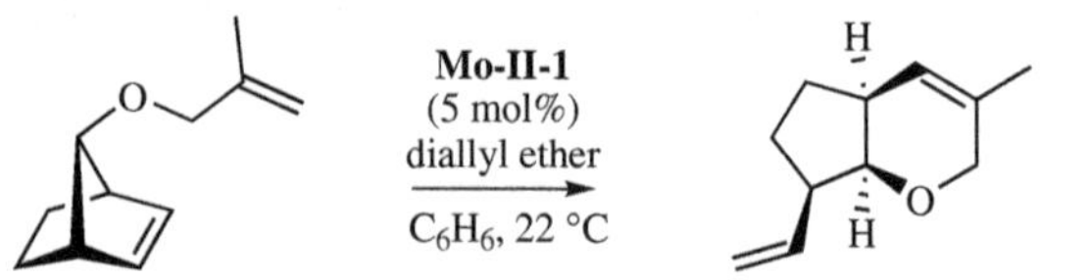

Yield = 54%, ee = 92%

(13.20)

Mo-II-1 was successfully used in AROM/CM of NBE with selected terminal olefins leading to five- and six-membered rings. It promotes AROM/CM of *meso*-NBE derivatives with styrene to give the expected diene with 98% ee and over 98% of *trans*-olefin selectivity (Eq. 13.21) (113). The reaction provides access to highly functionalized cyclopentanes with high optical purity.

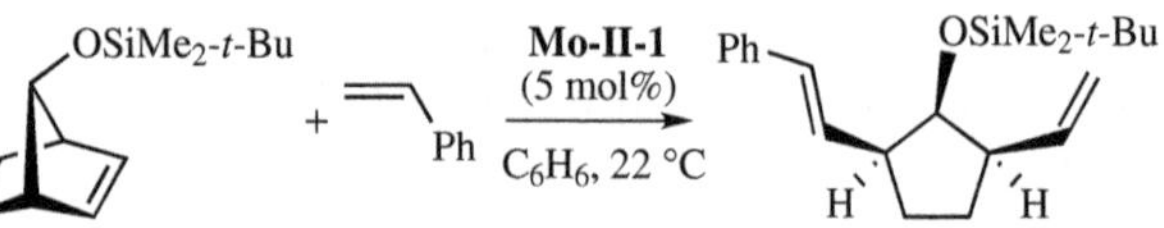

Yield = 57%, ee = 96%
trans content > 98%

(13.21)

Typical Procedure for AROM/CM with the Schrock–Hoveyda Catalyst Mo-II-1 (Eq. 13.22) (113)

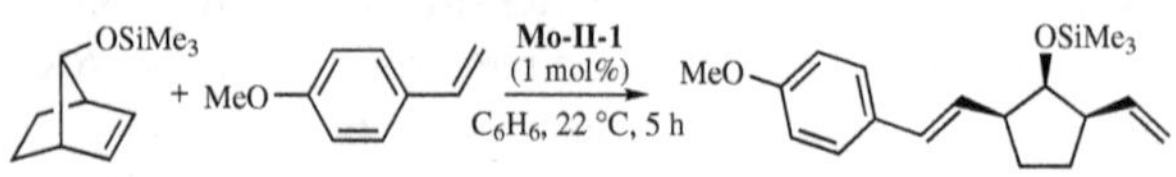

2 equiv Yield = 92%, *E*-isomer content > 98%

(13.22)

In a glove box (163 mg, 0.891 mmol) of 7-trimethylsilyloxynorbornene was dissolved in 4.46 ml of anhydrous benzene. Then 1.78 ml of 1 M solution 4-methoxystyrene and (6.77 mg, 0.009 mmol) of optically active catalyst **Mo-II-1** were added to the vessel. A Teflon cap was secured to the vessel and the reaction was performed at 20 °C for 5 h while stirring. Then the reactor was opened to air and the volatiles were removed in vacuum. The resulting dark brown residue was purified by column chromatography using silica gel as a stationary phase and the mixture of hexane and diethyl ether (volume ratio hexane : ether = 60 : 1) as eluent. The procedure permits obtaining 259 mg of a colorless oil identified by ^1H NMR (92 % yield, *E*-isomer content >98%).

13.4.3 Complexes Stereogenic at Metal

13.4.3.1 General Characterization Synthesis and development of tungsten and molybdenum imidoalkylidene complexes stereogenic at metal led to significant progress in catalysis of olefin metathesis (62,94,97). In these complexes of the general formula $[M(NR^1)(CHR^2)(X)(Y)]$, the stereogenic center at the metal is a consequence of the presence of four distinct, covalently bound ligands arranged in a pseudotetrahedral geometry. In contrast to other generations of imidoalkylidene ligands, X and Y differ in stereoelectronic properties. The "desymmetrization" of the ligand properties has remarkable consequences for the catalytic performance. According to computational studies, high oxidation state complexes bearing a stereogenic metal center and electronically distinct monodentate ligands should be effective promoters of olefin metathesis (114,115). In 2007, Schrock and Hoveyda (116) disclosed the synthesis of monoalkoxide (or monoaryloxide) pyrrolide (MAP) imidoalkylidene complexes, which are the first representatives of this new class of catalysts (Fig. 13.7).

In general, MAP complexes are more efficient olefin metathesis catalysts than other molybdenum-based alkylidene complexes. The increased catalytic activity of MAP complexes was studied by computational methods. The results indicated that for tetrahedral high oxidation state d^0 olefin metathesis catalysts, the potential energy surface in the catalytic cycle is under the influence of two key parameters:

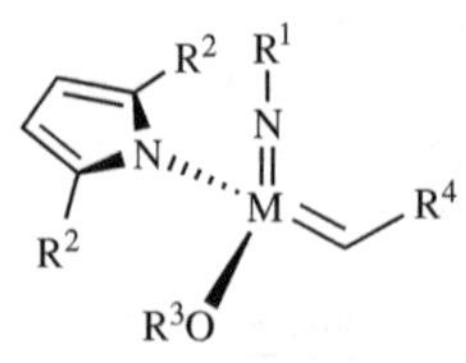

R^1 = Aryl, adamantyl; R^2 = H, Me;
R^3 = aryl, alkyl; R^4 = $CHMe_2Ph$ or *t*-Bu

Figure 13.7 Monoalkoxide pyrrolide (MAP) imidoalkylidene complexes.

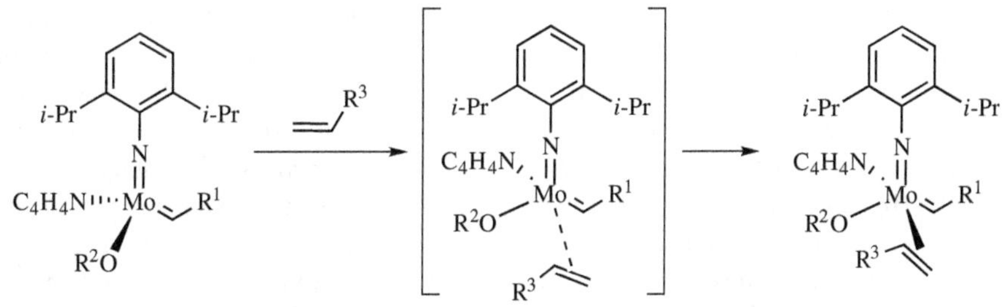

Scheme 13.7 Coordination of an olefin to MAP complex (117).

(i) ease of distortion of the tetrahedral catalyst and (ii) stability of the metallacyclobutane. In a tetrahedral complex, there is no formal empty coordination site and therefore a rearrangement of the structure is necessary to incorporate a fifth ligand (Scheme 13.7) (117). The energy expense required for this structural change is minimized in MAP complexes by the presence of two ligands with different electron-donating ability. The calculations indicated that the coordination of the olefin *trans* to donor ligand is energetically preferred (Scheme 13.7). The coordination of trimethylphosphine (a two-electron donor used to mimic olefin) *trans* to the pyrrolide ligand to form a distorted square pyramidal geometry has been established by crystallographic studies (118).

Metallacyclobutanes with either a trigonal bipyramidal or a square pyramidal geometry were demonstrated to be intermediates in metathesis reactions catalyzed by molybdenum and tungsten imidoalkylidene complexes. Calculations indicated that the most stable metallacyclobutanes are formed in bis(alkoxide) complexes, while the presence of a pyrrolide donor (considered to be a stronger donor than an alkoxide) leads to a destabilization of metallacyclobutanes. This phenomenon partly explains the increased catalytic activity of MAP complexes compared to those of the first and second generation of catalysts. Inversion of configuration at the metal in imidoalkylidene complexes is the point of main concern in the enantioselective metathesis reactions involving MAP catalysts. MAP complexes are configurationally stable in the absence of an olefin. It has been proved that the fast inversion of configuration at the metal center occurs by degenerate metathesis in the presence of ethylene (Scheme 13.8) (119).

Mechanistic considerations for MAP complexes indicate that the olefin attacks the (CNO) face *trans* to pyrrolide ligand. The newly formed olefin (product) leaves the intermediate metallacyclobutane also *trans* to the pyrrolide ligand to give a new alkylidene complex with the opposite configuration at the metal (Scheme 13.9).

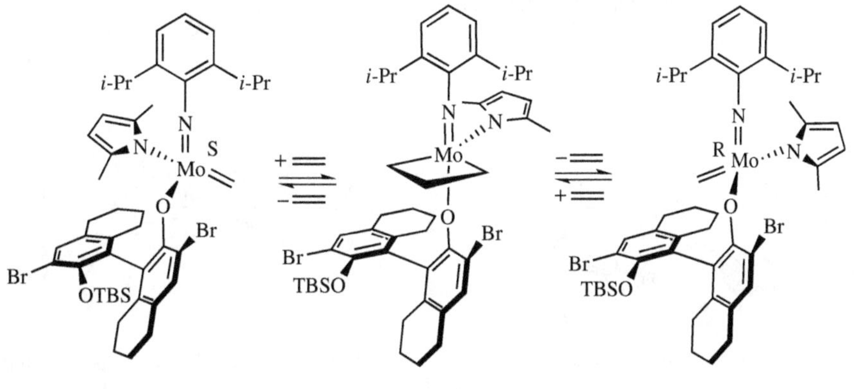

Scheme 13.8 Inversion of configuration at the metal as a consequence of degenerate metathesis with ethylene.

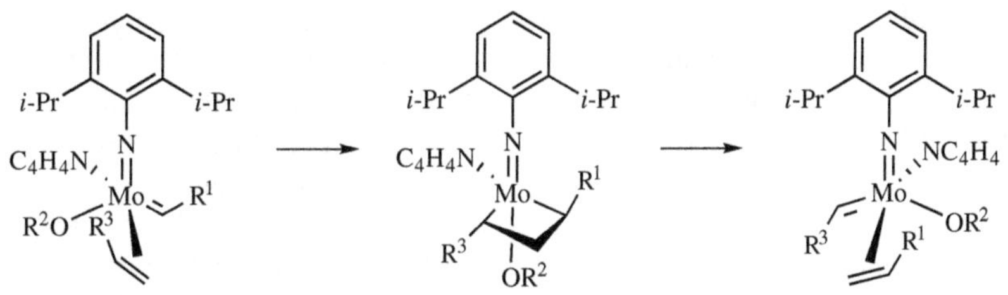

Scheme 13.9 Inversion of configuration at the metal as a consequence of the metathesis step.

As each olefin metathesis catalytic cycle includes two inversion processes, that is, CM leading to a substrate-derived alkylidene complex, and subsequent RCM, ROM, or CM step, a net retention of the configuration at the metal after each metathesis cycle is observed. A substrate-induced inversion of the stereogenic center (by degenerate metathesis) has not been observed (119). Mo-based monoalkoxide monopyrrolide catalysts are easily accessible by direct reaction of bis(pyrrolide) complexes with 1 equiv of alcohol (Eq. 13.23) (69,116).

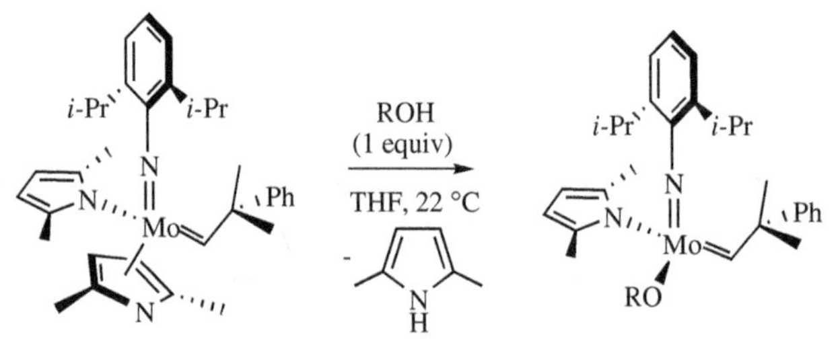

(13.23)

This method has also been successfully used for the preparation of W-based analogs (120). This synthetic approach permits the synthesis of a wide variety of MAP complexes of molybdenum and tungsten. The reaction of bis(pyrrolide) complex with enantiomerically pure alcohols (e.g., Eq. 13.24) results in the formation of a mixture of diastereomeric complexes, which are of particular interest in terms of enantioselective metathesis. From among accessible chiral, bulky alcohols, binaphthol-derivatives have been chosen because of the ease and low cost of synthesis, facility of modification, appropriate steric properties, and electron withdrawing abilities.

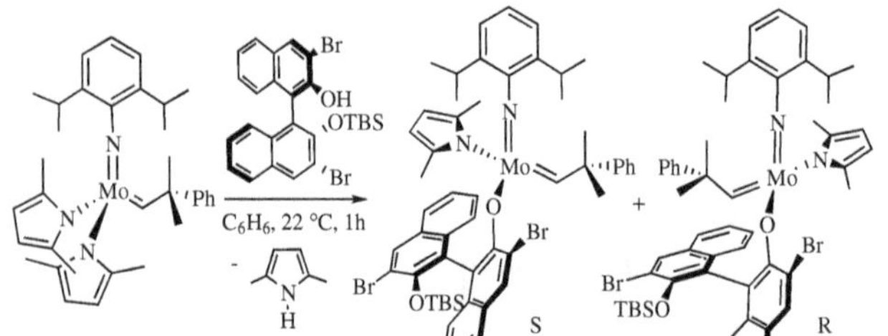

S- and *R*-**Mo-III-5b**, conversion > 98%, dr = 7 : 1 (*S* : *R*)

(13.24)

The synthetic approach described permits the generation of the chiral catalyst *in situ* to promote metathesis and to deliver results that are virtually identical to those found for isolated catalysts (121). The modular structure of MAP complexes and the ease of synthesis permit the formation of a range of catalysts preserving the parent structural motif (Fig. 13.8).

13.4.3.2 Catalytic Activity and Phenomenon of Z-Selectivity The efficient control of the *E*/*Z* selectivity of the products is one of the main challenges in all the investigated fields of olefin metathesis. The only reported examples of *Z*-selective CM (65–90% *Z*) involves substrates with an sp-hybridized substituent (acrylonitrile or enynes) (90,122). The striking feature of appropriately designed MAP catalysts is their extremely high *Z*-selectivity observed in homometathesis, CM, macrocyclic RCM, and ROM/CM. It has been suggested that the *Z*-selectivity of MAP catalysts originates from the trigonal bipyramidal geometry of metallacyclobutane in which the sterically demanding alkoxy ligand and the substantially smaller imido ligand occupy the axial positions (Scheme 13.10).

The free rotation of the alkoxy ligand shields the bottom side of the catalyst, thus forcing the substituent of the incoming olefin to point toward the ligand of the lower steric hindrance. Such a geometry favors the formation of such an all *syn*-metallacyclobutane that produces *Z*-olefins as a result of cycloreversion (Scheme 13.10) (123). A number of MAP complexes of tungsten (Fig. 13.9) and molybdenum (either previously obtained or *in situ* generated) have been tested in homometathesis of 1-hexene, 1-octene (Eq. 13.25), allylboronic acid pinacol ester (Eq. 13.26), and a set of functionalized olefins of the formula $H_2C{=}CHR$ (where R = CH_2Ph, CH_2SiMe_3, $(CH_2)_7CO_2Me$, CH_2OCH_2Ph, CH_2NHTos, CH_2NHPh) and have been shown to be moderately active but highly *Z*-selective.

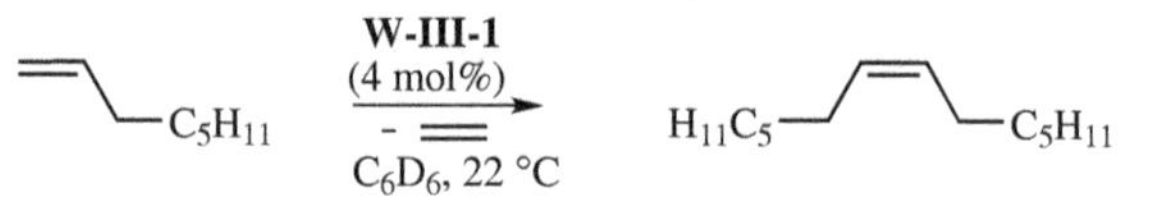

Yield = 33%, *Z* content = 95%

(13.25)

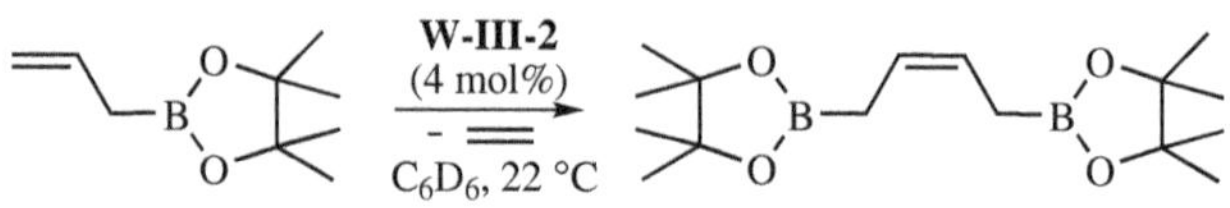

Yield = 70%, *Z* content = 96%

(13.26)

For each of the olefins studied, it has been possible to specify the optimal conditions, thus allowing for *Z*-selectivities of over 90%, often reaching 98% (124).

Particularly high activity and *Z*-selectivity have been observed for the 3,5-dimethylphenylimido tungstacyclobutane complex (**W-III-3**) (Fig. 13.9), which permits the synthesis of *Z*-olefins with selectivities exceeding 99% (125). Mo and W-based MAP complexes, for example, **Mo-III-2b** and **W-III-4** containing a hexaisopropylterphenoxide ligand, have been found to catalyze the homocoupling of simple (*E*)-1,3-dienes to give stereoselectively *E*,*Z*,*E*-trienes (126). MAP complexes have been shown to also exhibit high *Z*-selectivity in the CM of terminal enol ethers and

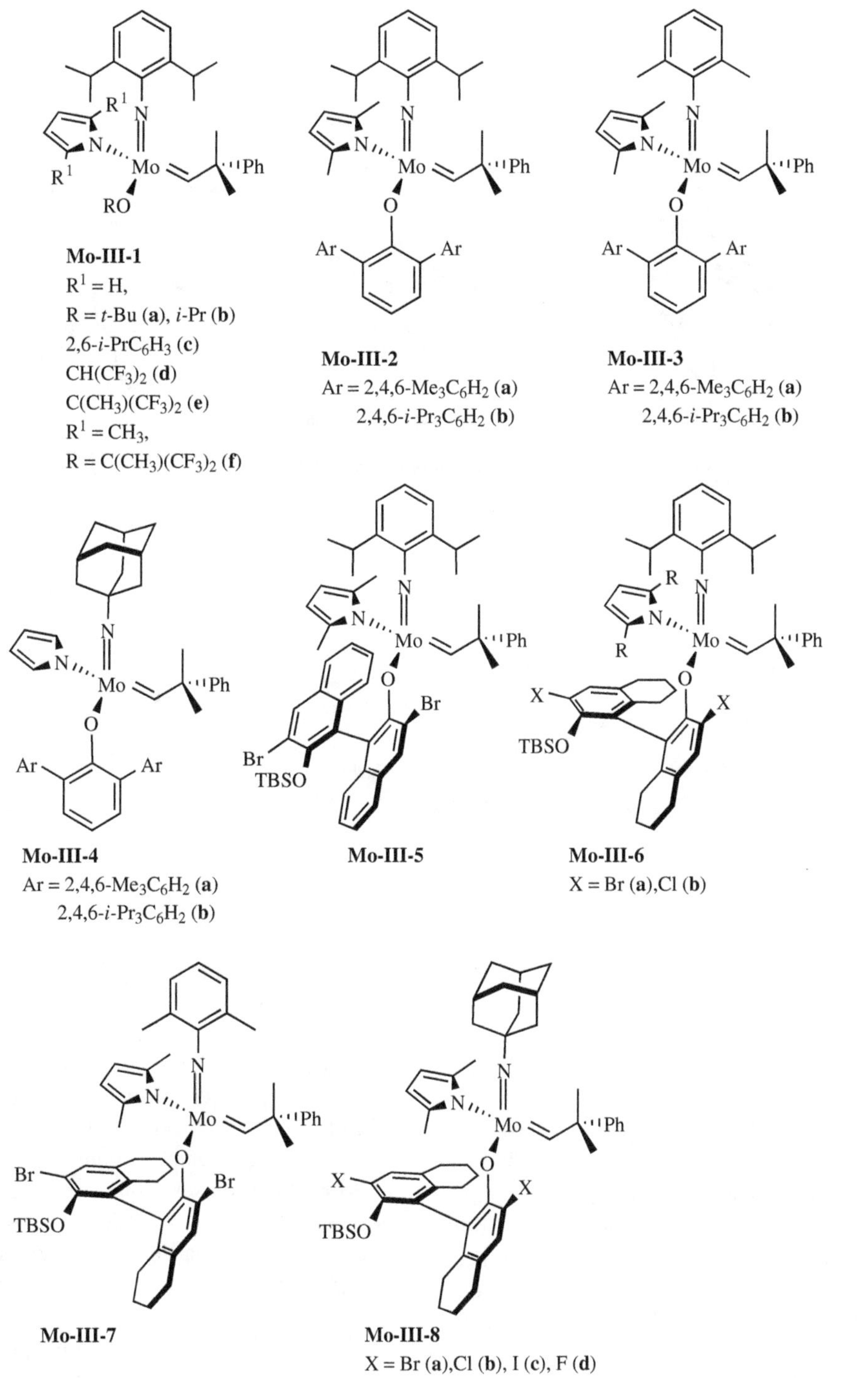

Figure 13.8 Selected MAP imidoalkylidene complexes of molybdenum.

allylamides with terminal olefins (123). High efficiency and *Z*-selectivity was achieved for a wide range of synthetically useful substrates such as vinyl ethers and allyl amides. Vinyl ethers were efficiently transformed in the presence of **Mo-III-7a** to get products with the *Z*-content exceeding 94%. The catalyst of choice for CM of allylamides was **Mo-III-8a** (Fig. 13.8), offering the highest yield, while all MAPs tested exhibited high *Z*-selectivity (in the range of 81–98%) (123). The standard procedure for catalysis with MAP involves the generation of the catalyst *in situ*. *Z*-selective CM was applied for the stereoselective synthesis of anti-tumor agent KRN 7000 (Eq. 13.27).

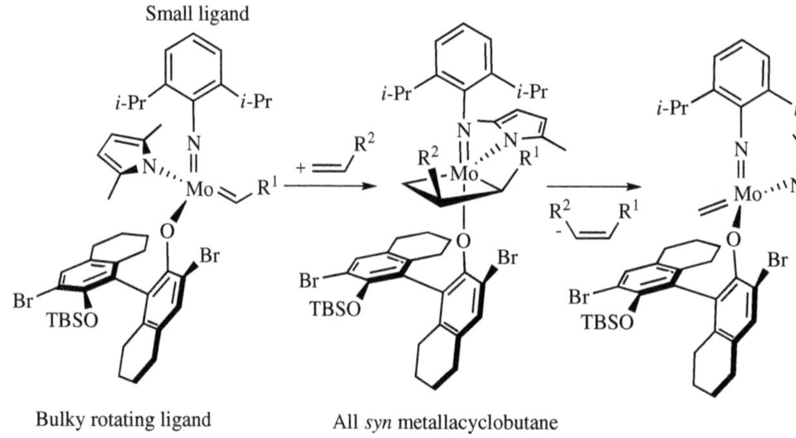

Scheme 13.10 Origin of *Z*-selectivity in CM catalyzed by MAP complexes.

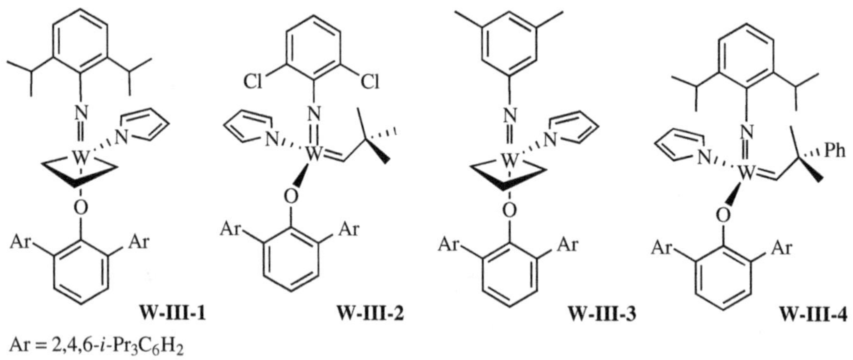

Figure 13.9 Highly *Z*-selective tungsten-based MAP complexes.

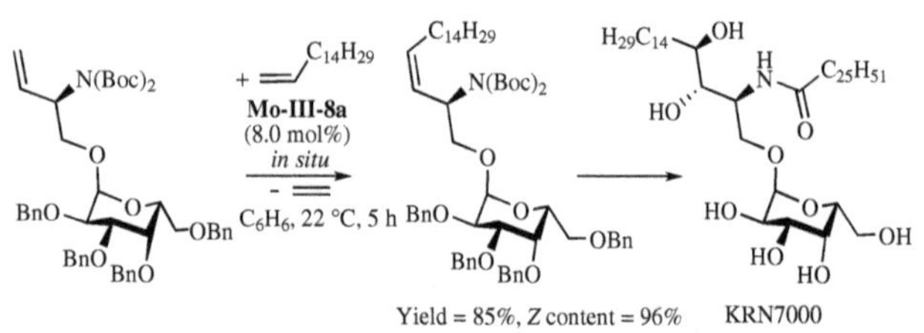

(13.27)

Optimal performance in CM requires the proper choice of the reacting partners as well as a proper reagents ratio to avoid competitive homometathesis. Moreover, it has been established that the presence of ethylene that develops during the process should be minimized. Ethylene competes with cross partner molecules in the reaction with the catalyst and generates the methylidene complex and/or the stable unsubstituted metallacyclobutane. High efficiency without the need for a significant excess of one of the reacting partners has been achieved by conducting the reaction under reduced pressure (123). Complex **Mo-III-6a** exhibits catalytic activity in ethenolysis of routinely purified methyl oleate. Virtually complete conversion has been possible at room temperature under 10 atm of ethylene (127). In the presence of **Mo-III-6a**, a selective ethenolysis of *Z*-isomers of internal olefin was observed (*E*-isomers remaining untouched). This feature of MAP catalysts has been applied in a procedure allowing access to a large variety of stereomerically pure *E*-olefins by selective consumption of *Z*-isomer from the mixture of 1,2-disubstituted olefins (128).

Macrocyclization via RCM is a convenient synthetic tool for the preparation of cyclic structures, particularly applicable in the total synthesis of biologically active molecules (129). Lack of means for controlling the stereoselectivity of the metathetic closure of macrocyclic rings is a serious drawback of this method, limiting its use in total synthesis. It has been found that MAP complexes of Mo and W permit highly *Z*-selective macrocyclization via RCM (Eq. 13.28). The *Z*-selectivity of MAP catalysts in macrocyclization via RCM is likely due to similar principles that result in stereoselective homocoupling and CM reactions.

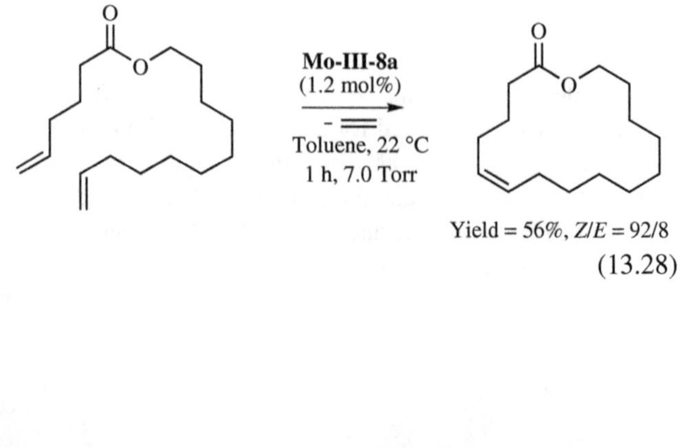

(13.28)

Tungsten imidoalkylidene complex **W-III-2** emerged as a facile and uniquely stereoselective catalyst as has been demonstrated in the example of the RCM step in the total syntheses of epothilone C (Eq. 13.29) and nakadomarin A (Eq. 13.30). The effective procedures developed require the reaction to be performed in vacuum to avoid adventitious isomerization processes leading to reduced *Z*-stereoselectivity (130).

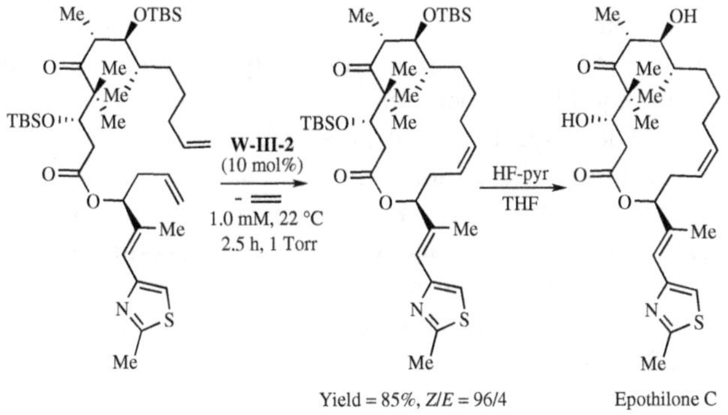

(13.29)

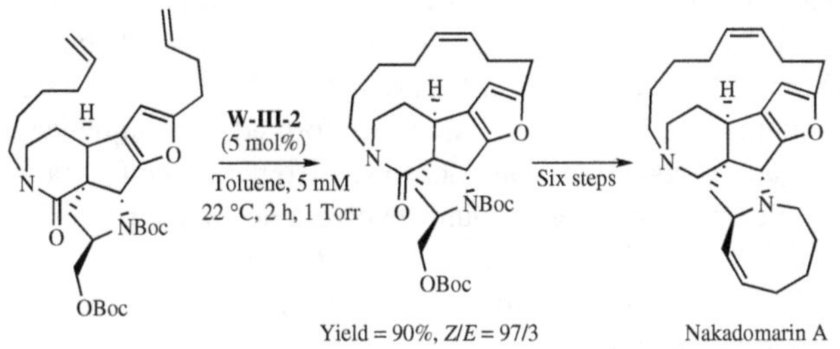

(13.30)

The *Z*-selective macrocyclization catalyzed by monoaryloxypyrrolide complexes offers a valuable synthetic tool, applicable to numerous total syntheses of biologically active molecules. MAP complexes have been shown to be efficient catalysts for desymmetrization of tertiary amine-based prochiral trienes via RCM (117, 121). Among others, the MAP catalyst **Mo-III-6a** efficiently promotes the ARCM of secondary amines that is hardly possible by using other metathesis catalysts (131). Complexes **Mo-III-6** (bearing different halide substituents in positions 3 and 3′ of the aryloxide ligand), unlike diolate complexes or any chiral Ru-based catalyst, efficiently promote the enantioselective RCM of challenging tricyclic triene (Eq. 13.31). The reaction is a metathesis step in a total synthesis of *Aspidosperma* alkaloid quebrachamine (117, 121).

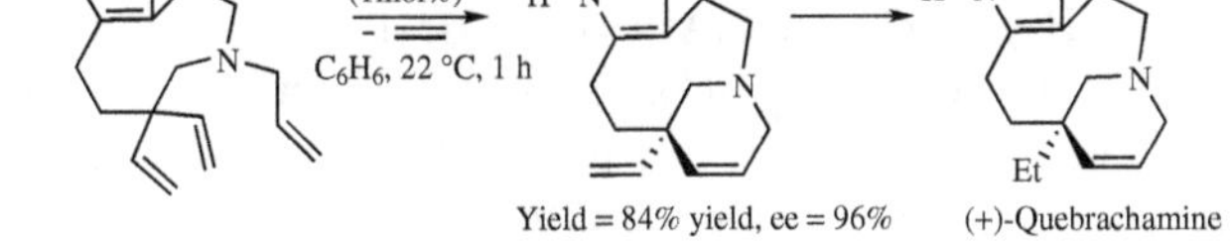

(13.31)

Adamantylimido MAP molybdenum complexes **Mo-III-8** readily catalyze the ROM/CM of strained oxabicyclic alkenes with styrenes to afford trisubstituted pyranes in an efficient way (Eq. 13.32) (132). High enantioselectivity (ee = 86–96%) and *Z/E* ratio (from 80 : 11 to 96 : 4) have been achieved. MAP complexes bearing bulkier imido ligands (e.g., **Mo-III-6a**) do not exhibit catalytic activity in the process.

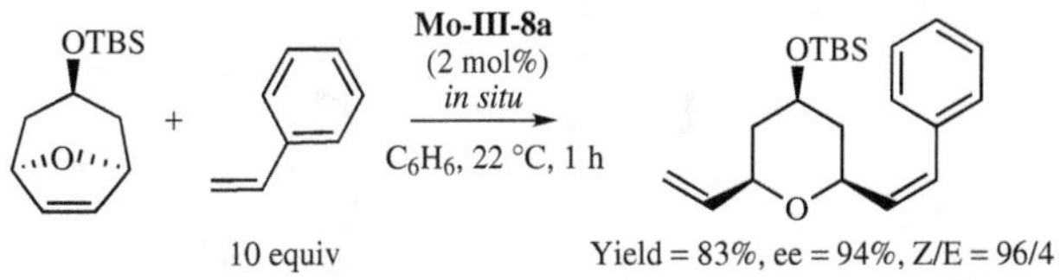

(13.32)

Complexes **Mo-III-1f**, **Mo-III-7a**, and **Mo-III-8** promote the enantioselective ROM/CM of several oxa- and azabicycles, cyclobutenes, and cyclopropenes with aryl or alkyl-substituted vinyl ethers (133). The desired products were formed in up to 90% yield, 99 : 1 enantiomeric ratio, and the *Z*-selectivity exceeding 90%. MAP complexes, for example, **Mo-III-1f**, are also capable of promoting RCM of a variety of enynes (116,134). The reaction leads to the highly selective formation of endo products and offers a convenient method for the synthesis of a number of carbocyclic as well as N- or O-containing heterocyclic *endo*-dienes (e.g., Eq. 13.33). The procedures proposed include the use of MAP catalysts generated *in situ*.

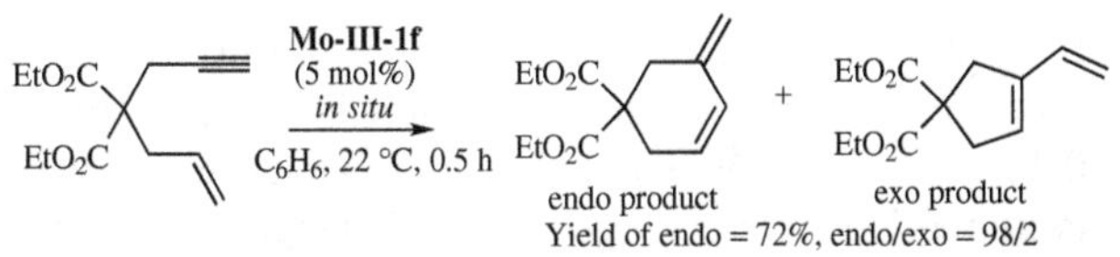

(13.33)

Diastereomeric MAP catalysts **M-III-6a** and **M-III-6c** also proved to be highly enantioselective in the desymmetrization of exemplary enynes (Eq. 13.34).

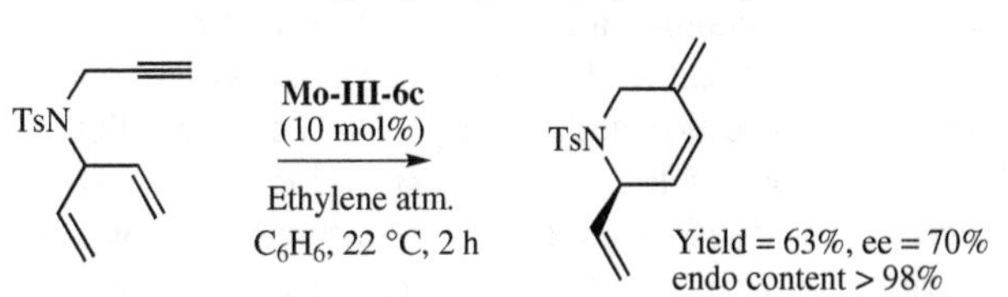

(13.34)

It should be noted that in the presence of other generations of imidoalkylidene catalysts, enynes undergo transformations leading to mixtures of ill-defined, possibly oligomeric, products.

Procedure for *In Situ*-Generation of Complex Mo-III-7 (Eq. 13.35) (123)

In a glove box, a glass vial (volume 4 ml) containing a magnetic stir bar was charged with (20.0 mg, 37.2

μmol) molybdenum bis(pyrrolide) imidoalkylidene complex, (21.1 mg, 37.2 μmol) of alcohol, and (372 μl) of benzene (see Eq. 13.35). The reaction mixture became orange. The vial was closed with a screw cap and the solution was stirred at 22 °C for 1 h.

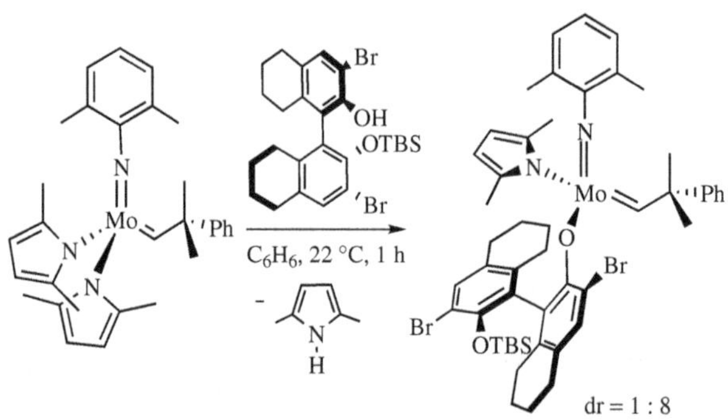

(13.35)

Typical Procedure for Catalytic Z-Selective Enol Ether Cross-Metathesis with MAP Complex Mo-III-7 (Eq. 13.36) (123)

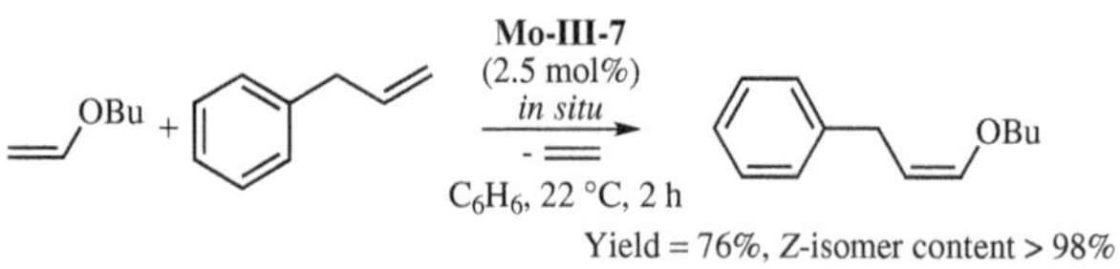

(13.36)

In a glove box, a glass vial (volume 20 ml) containing a magnetic stir bar was charged with (50.0 mg, 0.423 mmol) of allylbenzene. A separate glass vial (volume 2 ml) was charged with (424 mg, 4.23 mmol) butyl vinyl ether and subsequently. (106 μl, 10.6 μmol) of benzene solution of generated *in situ* **Mo-III-7** was added by a syringe (dried at 65 °C). The resulting solution was allowed to stir for 1 min and then transferred by a syringe to the vial containing allylbenzene. The reaction mixture was stirred at 22 °C for 2 h. The vessel was removed from the glove box and about 1 ml of benchtop diethyl ether was added to quench the reaction. The mixture was concentrated *in vacuo*. The resulting brown oil was purified by column chromatography using neutral alumina as a stationary phase and hexane (100%) as eluent. The procedure permits getting 61.3 mg of a colorless oil identified by ^{1}H NMR as (3-butoxyallyl)benzene (76.0% yield, Z-isomer content >98%).

13.4.4 Stereoregular ROMP in the Presence of Imidoalkylidene Complexes of Molybdenum and Tungsten

ROMP initiated by imidoalkylidene complexes of molybdenum and tungsten has been the subject of monographs and numerous review articles (2,74,135–139) and its discussion goes far beyond the volume of this chapter. In this section, only the synthesis of ROMP polymers with a dominating single primary structure is discussed (73,140–142). Well-defined imidoalkylidene catalysts permit the control of the *cis/trans* selectivity and the tacticity of the polymers formed via ROMP. The formation of polymers with a unique structure requires that the polymerization proceeds via a single propagation step. Stereoselectivity control means creating the conditions in which the desired single propagation step is kinetically preferred. Control of the equilibrium between *syn*- and *anti*-isomer of imidoalkylidene, rate of interconversion, and activity of *syn*- and *anti*-isomers toward monomer addition are the key issues to supervise the mode in which monomers are polymerized. In a family of W and Mo-based imidoalkylidene complexes, such a control is possible via the modification of steric and electronic properties of the imido and alkoxy ligands. As demonstrated by Schrock, the *syn* addition of 2,3-bis(trifluoromethyl)norbornene to the *syn*-rotamer gives *cis* double bonds (Scheme 13.11a) while the *syn* addition to rotamer *anti* leads to the formation of *trans*-olefins (Scheme 13.11b) (the rear CNO face of which the olefin attacks the M=C bond was chosen arbitrarily) (66). In accordance with Scheme 13.11, all-*cis* polymers are formed by a sequence of *syn* additions of the monomer to *syn*-isomers. The formation of all-*trans* polymers should involve a sequence of *syn* addition to *anti*-isomer and *syn* to *anti* isomerization, finally followed by *syn* addition of another monomer molecule to the *anti*-rotamer.

The tacticity of ROMP polymers is determined by the fashion in which the monomer approaches the reaction center. If the monomer approaches the M=C bond exclusively via one of the two CNO diastereofaces (A or B, Fig. 13.10), an isotactic polymer is formed. The approach of the monomer to the reaction center via alternating diastereofaces leads to the formation of a syndiotactic polymer.

Polymerization of 2,3-bis(trifluoromethyl)norbornadiene with complex **Mo-I-1** as an initiator performed at room temperature produces a polymer with a high *cis* content (>98%) (143). The formation of high *cis* polymer results from the involvement of the catalyst nearly exclusively in the form of its *syn*-rotamer. (The *anti*-rotamer is rarely available on the time scale of the polymerization). High *cis* content but poor tacticity was also observed for polymers formed via ROMP of 2,3-bis(trifluoromethyl)norbornadiene performed in the presence of $[Mo(NAr)(=CHCMe_2Ph)\{O(CF_3)_2CF_2CF_2CF_3\}_2]$, $[Mo(NAr)(CHCMe_2Ph)\{OC(CF_3)_3\}_2]$, $[Mo(NAr)(CHCMe_2Ph)\{O(CF_3)_2CF_2CF_2CF_3\}_2]$ (Ar = 2,6-$Me_2C_6H_3$), or $[Mo(NAr)(=CHCMe_2Ph)\{(OC(CF_3)_3)_2\}]$ (Ar = 2,6-$Me_2C_6H_3$) (144). 2,3-bis(trifluoromethyl)norbornadiene (giving highly *cis*, poorly tactic polymer in the presence of **Mo-I-1**) affords a highly tactic polymer with 98% *trans* content when the reaction is carried out in the

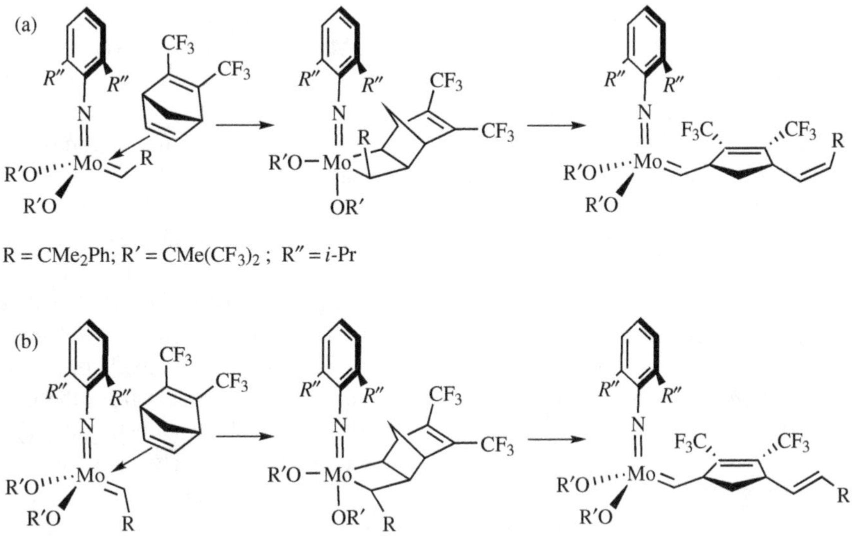

Scheme 13.11 *Syn* addition of a monomer to *syn* (a) and *anti* (b) isomers of a molybdenum imidoalkylidene catalyst. Formation of *cis* and *trans* double bond, respectively.

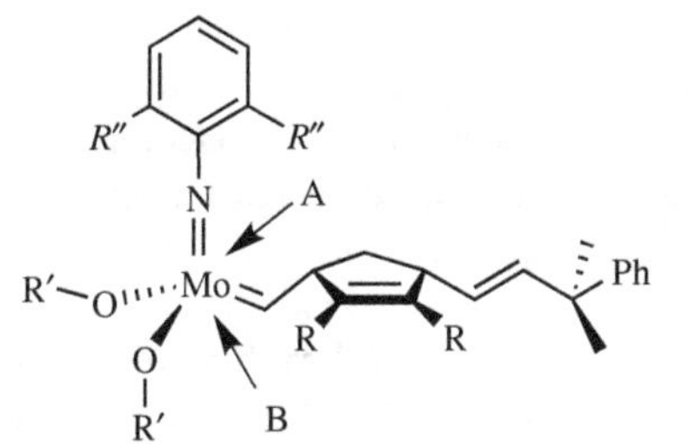

Figure 13.10 Possible approaches of a monomer to alkylidene complex (two CNO diastereofaces).

presence of **Mo-I-4a** (145). The proposed explanation for *trans*-selectivity is the low activity of **Mo-I-4a** catalyst, so that the less active isomer *syn*-**Mo-I-4a** does not react with the monomer. *Syn* addition of the monomer to the more reactive *anti*-**Mo-I-4a** (and subsequent cycloreversion) produces isomer *syn* with a *trans*-arrangement around double bond. Highly *trans* (90–95%) and highly tactic polymers were obtained also via the ROMP of 2,3-di(carbomethoxy)norbornadiene initiated by **Mo-I-4a** at room temperature (146). 1,7,7-Trimethylnorbornene polymerizes slowly in the presence of **Mo-I-1** to give all-*trans* polymers (147). It was proposed that an addition of this monomer to isomer *syn* is impossible due to steric hindrance and the formation of all-*trans* polymers should involve a sequence of *syn* additions of monomer to the *anti*-**Mo-I-1** rotamer with formation of isomer *syn*, interconversion of isomer *syn* to *anti*, and another *syn* addition to the *anti*-rotamer. Such a course of the reaction is in agreement with the kinetic data indicating that the rate of polymerization does not depend on the monomer concentration and is essentially the same as the rate of *syn* to *anti* conversion (147).

Formation of tactic polymers by the first generation of imidoalkylidene complexes is a consequence of the chain end control. The chirality of the β-carbon atom in propagative alkylidene complexes determines the way the monomer approaches the reaction center and thus the tacticity of the polymer formed. Diolate imidoalkylidene complexes (e.g., **Mo-II-5b**) originally synthesized in order to improve the tacticity control in ROMP (95,148), initiate the polymerization of 2,3-bis(trifluoromethyl)norbornadiene to give highly tactic *cis* polymers. The polymerization proceeds via subsequent *syn* additions to the *syn*-isomer of the initiator with the CNO face for monomer attack being determined by the steric properties of binap ligand. Complex **Mo-II-5b** catalyzes the ROMP of 2,3-dicarbomethoxynorbornadiene, 2,3-dicarbomenthoxynorbornadiene, as well as a number of *endo,exo*-2,3-disubstituted NBEs (bearing CO_2Me, CH_2OMe, or CH_3 substituents) to give *cis* polymers with the postulated isotactic structure (148). Although, it is not possible to spectroscopically prove the tacticity of each polymer, it seems reasonable that an appropriate diolate catalyst/monomer combination permits the synthesis of all-*cis*, isotactic polymers through enantiomorphic site control. In *cis*-selective MAP complexes, a combination of a bulky aryloxy ligand and a small aryl- or adamantylimido ligand forces a *syn* addition of the monomer to isomer *syn*, leading to the formation of all-*syn* metallacyclobutane, which in turn rearranges to form a prolonged polymer chain with *cis* geometry around the double bond. For example, catalyst (**Mo-III-4b**) promotes ROMP of cyclooctene and 1,5-cyclooctadiene leading to the formation of polymers with a *cis*-content exceeding 99% (149). The application of stereogenic at metal MAP alkylidene complexes has opened up fascinating opportunities in the synthesis of ROMP polymers via stereogenic metal control. For substituted norbornadienes, MAP catalysts offer a way to syndioselectivity which is proposed to arise as a consequence of a selective *trans* addition of the monomer to the pyrrolide ligand and the inversion of configuration at the metal center in each metathesis step. When chiral 2,3-di(carbomethoxy)norbornadiene undergoes ROMP in the presence of **Mo-III-4b**, as well as in the presence of an analog of **Mo-III-4b** bearing a dimethylpyrrolide ligand, a high *cis*-selectivity is accompanied by the perfect control of tacticity. The reaction yields virtually all-*cis* and all-syndiotactic polymers (Eq. 13.37) (149). Similar results were observed when 2,3-bis(trifluoromethyl)norbornadiene and 2,3-di(carbomenthoxy)norbornadiene were employed as monomers.

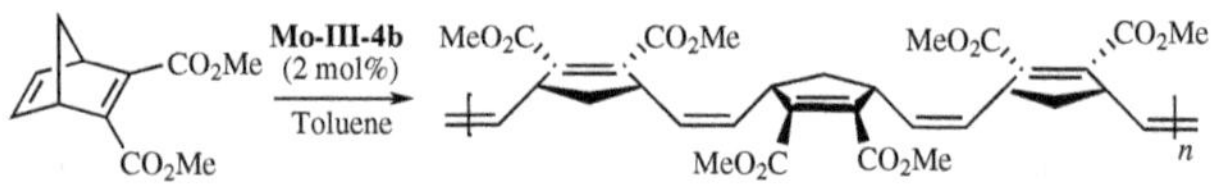

(13.37)

An all-*cis* and highly tactic (most likely syndiotactic) polymer was formed when 3-methyl-3-phenylcyclopropene underwent ROMP in the presence of **Mo-III-4b** (150). A similar polymer microstructure was obtained when 3-methyl-3-phenylcyclopropene underwent ROMP in the presence of $[Mo(NAd)(CHCMe_2Ph)(C_4H_4N)(O\text{-}2,3,5,6\text{-}Ph_4C_6H)]$ at low temperatures (−78°C). For ROMP of a chiral, racemic NBE in the presence of appropriate stereogenic at metal catalysts, the inversion of configuration at metal with each insertion step results in an alternate consumption of enantiomers of the reacting monomer, so that a *cis* syndiotactic structure with an alternating enantiomer sequence in the polymer is formed (Eq. 13.38). Such a scenario was observed for ROMP of *endo,exo*-5,6-dicarbomethoxynorbornene initiated by **Mo-III-4a** (151).

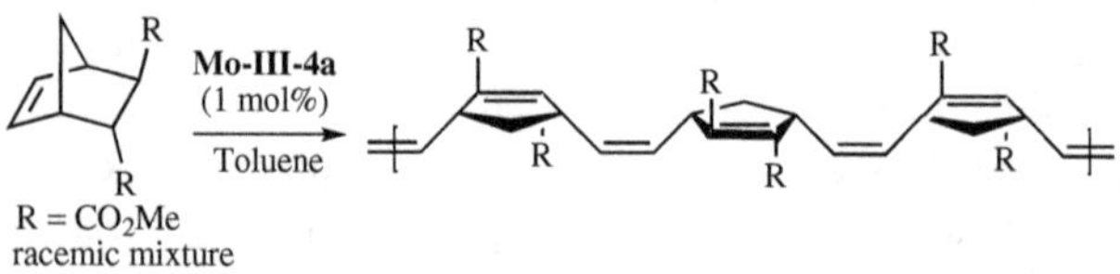

(13.38)

Analogously, racemic 1-methyl-2,3-dicarbomethoxy-7-oxanorbornadiene undergoes ROMP in the presence of **Mo-III-4a** to give a polymer proposed to be *cis* syndiotactic with alternating enantiomer sequence (Eq. 13.39) (151).

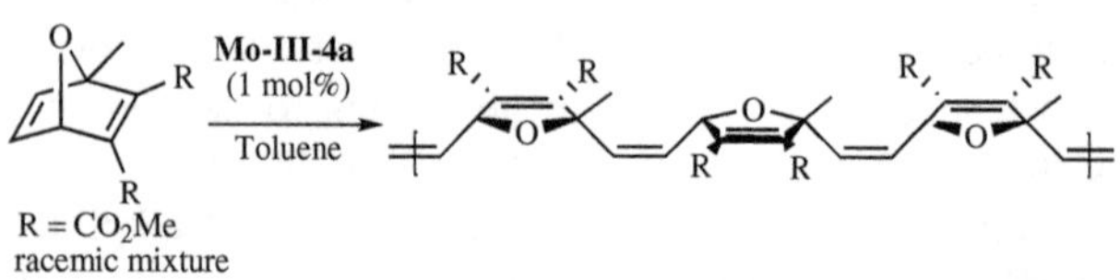

(13.39)

13.5 SUPPORTED IMIDOALKYLIDENE COMPLEXES AS CATALYSTS OF OLEFIN METATHESIS

Heterogenization of metal complexes that are catalysts of certain processes performed in homogeneous systems involves a chemical bonding of a complex, through one or a few ligands, with a polymer or an inorganic support. The aim of complex heterogenization is to obtain a catalyst that would combine the benefits of homo- and heterogenic catalysts, in particular: (i) high activities, selectivities, and reaction rates observed with well-defined homogeneous catalysts, (ii) easy separation of a catalyst from the postreaction mixture, (iii) multiple use of a catalyst, and (iv) products free from inorganic impurities. The immobilization of alkylidene complexes of molybdenum has been described in recent review articles (152–154). In principle, imidoalkylidene catalysts may be connected to the support via an alkoxide (diolate) ligand, imido ligand, or alkylidene ligand. The first well-defined polymer-supported molybdenum imidoalkylidene catalyst was derived from supported chiral biphen-2,2′-diol (Scheme 13.12) (155). When handled under dry and oxygen-free conditions, the catalyst efficiently promotes a range of ARCM and AROM/CM processes and offers similar levels of enantioselectivity and somewhat decreased activity as compared to its homogeneous precursor. The supported system shows comparably low leaching (<5%) resulting in low contamination of the products with Mo. The recycling experiment performed for AROM/CM of 7-benzyloxynorbornene with styrene revealed that the catalyst activity is significantly diminished by the third cycle.

Poly(styrene)-supported molybdenum alkylidene biphenoxide catalysts based on the 2,6-dichlorophenylimido and adamantylimido ligand as well as chiral binaphthyl-based molybdenum alkylidene catalyst containing the 2,6-dichlorophenylimido or adamantylimido ligand were prepared by using similar methodologies (156). Buchmeiser (157) reported synthesis of the alternative biphenoxide catalyst attached to the poly(NBE) support through the biphen ligand. This catalyst exhibits basically similar catalytic performance. Its synthetic procedure involves *in situ* generation of the norbornenyl-substituted catalyst by interaction of functionalized diolate with bis(tosylate)imidoalkylidene precursor followed by ROMP. For all described polymer-supported catalysts, the enantiomeric excess obtained in the series of asymmetric RCM, ROM/CM, and desymmetrization reactions tested was in most cases very similar to the one obtained with the unsupported counterparts. Well-defined silica supported olefin metathesis catalysts can be prepared by grafting the well-defined molecular alkylidene complexes containing a labile ligand (e.g., alkyl, alkoxide, or amido) onto SiO_2 dehydroxylated at 700 °C ($SiO_{2\text{-}(700)}$). The reactive sites of the $SiO_{2\text{-}(700)}$ surface consist mainly of isolated

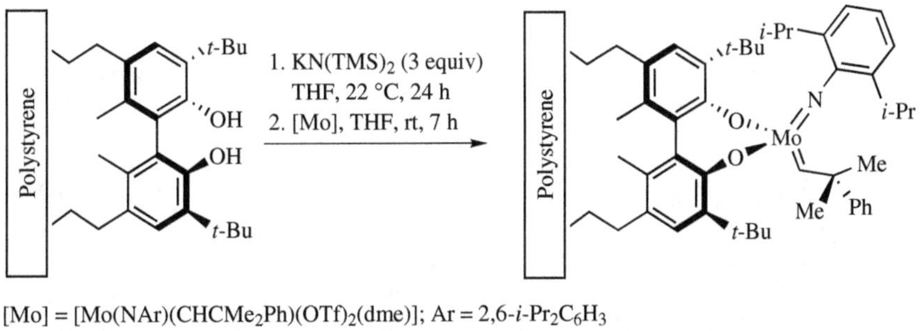

Scheme 13.12 Immobilization of a Mo imidoalkylidene complex on polystyrene bound biphen-2,2′-diol.

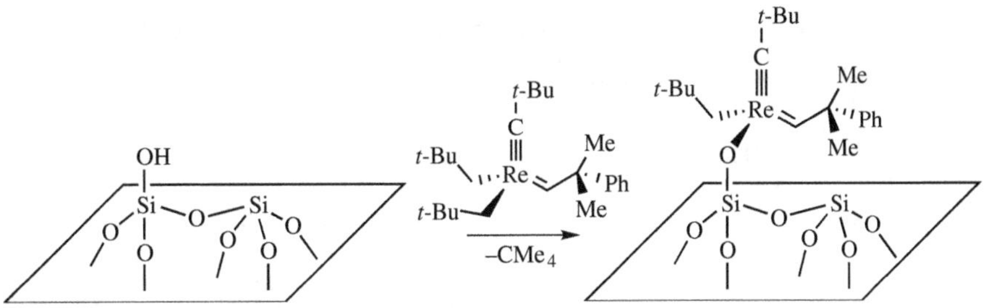

Scheme 13.13 Grafting of [Re(C-*t*-Bu)(CH-*t*-Bu)(CH_2-*t*-Bu)$_2$] onto the surface of $SiO_{2\text{-}(700)}$.

silanol groups. Well-defined silica-grafted complexes are highly active alkene metathesis catalysts. Such supported precursors display higher stability than their molecular equivalents, which is related to site isolation on the silica surface as it prevents bimolecular decomposition. The first fully characterized silica supported system was [(≡SiO)Re(C-*t*-Bu)(CH-*t*-Bu)(CH_2-*t*-Bu)] obtained by grafting of [Re(C-*t*-Bu)(CH-*t*-Bu)(CH_2-*t*-Bu)$_2$] onto the surface silanol groups of $SiO_{2\text{-}(700)}$ (Scheme 13.13) (158).

This system is highly active for metathesis of terminal and internal olefins. It also catalyzes the homometathesis of methyl oleate without the need of a co-catalyst (159). The analogous reactions of the homogeneous precursor [M(NAr)(CH-*t*-Bu)(CH_2-*t*-Bu)$_2$] (Ar = 2,6-*i*-PrC_6H_3) with $SiO_{2\text{-}(700)}$, give surface tungsten (Fig. 13.11a) (160) and molybdenum complexes (Fig. 13.11b) (161). Re, W, and Mo-based catalysts are active in the metathesis of propene and ethyl oleate. Surface catalysts displayed higher activities than those of their homogeneous counterparts. Computational studies of Re-based catalysts have shown that the high activity of this system is associated with the dissymmetry at the metal center, that is, the presence of one strong (alkyl) and one weak σ-donor (siloxy) ligand (114,115). It has been shown that the replacement of the pendant alkyl (CH_2-*t*-Bu) by (i) an amido ligand yields [(≡SiO)Mo(NAr)(CHCMe$_2$Ph)(NR_2)] (where NR_2 = NPh_2 or pyrrolide) (162) or (ii) an alkoxide ligand to get [(≡SiO)Mo(NAr)(CHCMe$_2$Ph)(O-*t*-Bu)] (Fig. 13.11c) (163) is enough to further improve the catalysts. By such replacements, overall higher initial rates, stability, and selectivity can be obtained.

The most efficient d^0 olefin metathesis catalysts developed so far are surface 2,5-dimethylpyrrolide Mo imidoalkylidene complexes (Fig. 13.11d–f). They can be prepared by grafting of bis(dimethylpyrrolide) imidoalkylidene complexes onto $SiO_{2\text{-}(700)}$. Methyl substituents on the pyrrolyl ligand in positions 2 and 5 improve the stability of the system. It has been shown that the activity of the catalyst can be increased by modifying the substituents of the imido ligand. The optimal catalytic performance for homometathesis of ethyl oleate was observed when NAr ligand (where Ar = 2-CF_3-C_6H_4) was employed while the adamantylimido ligand permitted the highest efficiency for propene metathesis (164). For [(≡SiO)(ArN)W(CH-*t*-Bu)(2,5-MeC_4H_2N)], solid state NMR spectroscopy permitted the observation of the propagating alkylidene complex as well as of stable metallacyclobutane intermediates (165).

13.6 MOLYBDENUM-VERSUS RUTHENIUM-BASED CATALYSTS IN OLEFIN METATHESIS

Brilliant achievements in organometallic chemistry have brought about a wide range of diverse, well-defined olefin metathesis catalysts, mostly belonging to the kingdoms of

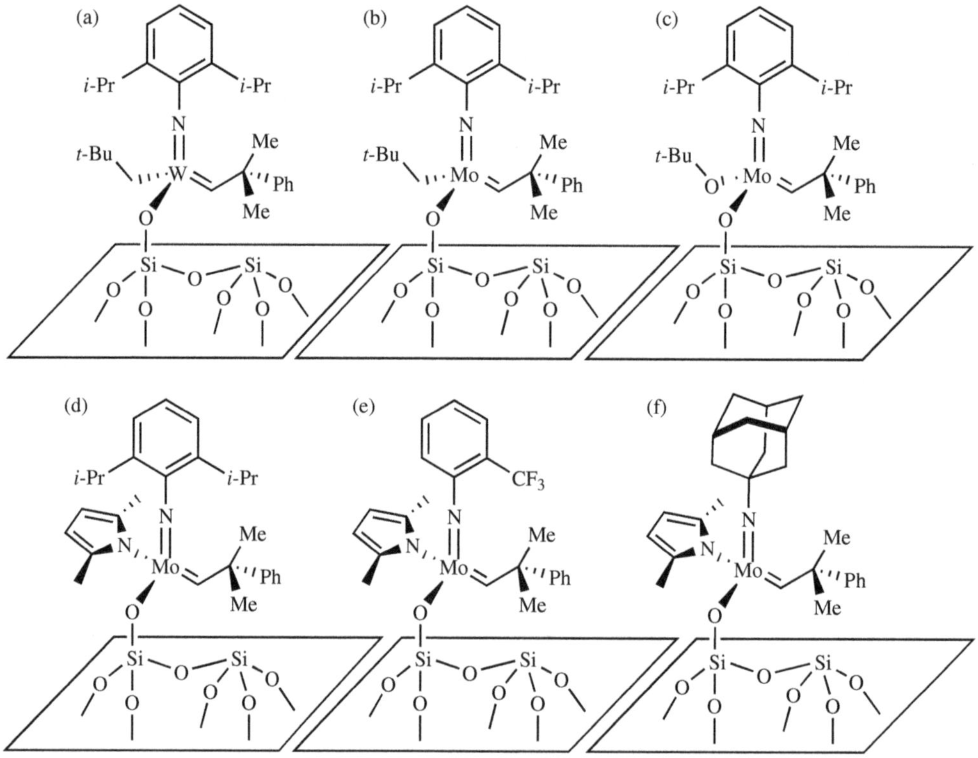

Figure 13.11 (a–f) Well-defined tungsten and molybdenum imidoalkylidene catalysts grafted on $SiO_{2\text{-}(700)}$.

molybdenum and ruthenium alkylidene complexes. These achievements strongly benefit chemists working in the field of organic synthesis and polymer synthesis. From the viewpoint of application, the differences between Mo- and Ru-based catalysts in olefin metathesis concern (i) functional group tolerance, (ii) catalytic activity, and (iii) selectivity. In contrast to ruthenium-based catalysts, molybdenum imidoalkylidene complexes are relatively sensitive to moisture and air, and should be handled under an inert atmosphere and used in anhydrous solvents. They are incompatible with a number of functionalized olefins such as carboxylic acids, aldehydes, most alcohols, and primary amines. On the other hand, molybdenum-based imidoalkylidene complexes tolerate a variety of functionalized olefins including ethers, epoxides, acetals, ketones, esters, secondary and tertiary amines, amides, carbamates, silanes, sulfides, disulfides, and some others (for a summary see Ref. 71). Unlike Ru-based catalysts, Schrock catalyst **Mo-III-1f** remains active in the presence of unprotected phosphanes (166), trimethylvinylsilane and dimethylvinylsilyl ethers. Mo-based MAP complexes exhibit activity in CM with enol ethers as reacting partners (123) while ruthenium catalysts (with a few exceptions) form relatively stable Fischer-type carbene complexes after metathesis exchange with these reagents. Molybdenum imidoalkylidene complexes are relatively stable toward secondary amines and permit the efficient and highly enantioselective synthesis of secondary cyclic amines via ARCM (e.g., 131). Simple generalizations concerning the differences in activity between both groups of metathesis catalysts should be avoided, especially in view of the high substrate/catalyst dependence. In a number of known reaction systems, molybdenum- and ruthenium-based catalysts offer complementary degrees of efficiency and stereoselectivity. A comparison of Ru- and Mo-based chiral olefin metathesis catalysts in asymmetric AROM/CM of oxa- and azabicycles demonstrates some complementarity (167). Recently, the first examples of efficient molybdenum catalyzed RCM of enynes have been reported (116,134). In the presence of the MAP molybdenum imidoalkylidene complexes (e.g., **Mo-III-1f**), the cyclization proceeds with high endo selectivity, while preferential formation of exo products is observed for ruthenium-based catalysts (e.g., the second generation Hoveyda–Grubbs catalyst). Similar complementarity is observed in cyclopolymerization of 1,6-heptadiynes. In the presence of selected ruthenium-based catalysts, reaction leads to the formation of polymers containing five-membered rings (via α-addition) (168,169), while preferential formation of polymers containing either five (170) or six-membered rings (171) was reported for molybdenum-based catalysts.

Generally, the molybdenum catalysts offer the advantage of being able to form rings (also enantioselectively) in sterically more demanding circumstances. A good illustration is the desymmetrization of a challenging prochiral triene via ARCM used in the total synthesis of quebrachamine (Eq. 13.31) (121). From among all available chiral catalysts, the efficient and enantioselective transformation is possible only in the presence of the MAP catalyst **Mo-III-6a**. The level of *Z*-selectivity achieved in homometathesis and CM, ROM/CM, or macrocyclic RCM catalyzed by properly chosen W or Mo-based MAP complexes is at present hardly available with ruthenium catalysts. As far as ROMP is concerned, Mo and W catalysts offer better control of the *cis/trans* ratio and the tacticity of ROMP polymers.

13.7 TOWARD WIDER USE OF MOLYBDENUM-BASED IMIDOALKYLIDENE CATALYSTS

Tungsten and especially molybdenum imidoalkylidene complexes offer enormous potential in olefin metathesis. They permit highly efficient and highly stereoselective syntheses of simple organic molecules and efficient and selective metathesis steps in multistep syntheses of sophisticated natural products, as well as precise control of the microstructure of ROMP polymers. Despite the highly valuable catalytic properties, a widespread use of these catalysts is limited by the common opinion that W- and Mo-based catalysts are difficult to synthesize and are user unfriendly because of their high sensitivity to moisture, oxygen, and a number of functional groups. However, it should be realized that presently, thanks to the commercial availability of the stable precursor $[Mo(NAr)(CHCMe_2Ph)(OTf)_2(dme)]$ (where Ar = 2,6-*i*-$Pr_2C_6H_3$), a variety of alkoxide, aryloxide and diolate catalysts can be easily synthesized. Moreover, useful ligand precursors (both enantiomers of dipotassium salt of octahydro-3,3′-di-*tert*-butyl-binaphthol, enantiomers and racemic mixture of 5,5′,6,6′-tetramethyl-3,3′-di-*tert*-butyl-1,1′-biphenyl-2,2′-diol) as well as the first (**Mo-I-1** and **Mo-I-4a**) and the second generation catalysts (both enantiomers and racemic mixture of **Mo-II-1**) are commercially available. Convenient procedures permitting the synthesis of second generation or MAP catalysts have been proposed. These syntheses are based on the direct reaction of easily accessible bis(pyrrolide) precursor $[Mo(NAr)(CHCMe_2Ph)(C_4R_2H_2N)_2]$ with alcohols (phenols). A number of attempts have been made to make high oxidation state molybdenum catalysts easier to handle. The studies were aimed at the synthesis of stable, air-, and water-sensitive precursors from which, by chemical or thermal activation, active catalysts could be generated *in situ*. A successful example of such an approach has been reported (172). The stable 18-electron, easy-to-handle bipyridine adduct of complex **Mo-II-1** permits the generation of an active catalyst of desymmetrization of trienes by RCM as a result of a chemical or thermal activation. Many procedures of metathetic transformations have been developed, which involve the generation of active second and third generation molybdenum catalysts *in situ* from easy to handle precursors (173,174). The activity and selectivity of the *in situ* generated catalysts are similar to those of the preliminary isolated and purified catalysts.

13.8 CONCLUDING COMMENTS

The development of group 4–7 metal-based catalysts for olefin metathesis has permitted a series of path breaking achievements that have led to the present level of understanding of the process. A great number of diverse catalysts have been synthesized and mechanisms of their operation have been recognized. A continuous progress in the area has been observed since the discovery of metathesis. Important achievements have been made in the study of tungsten and especially molybdenum-based alkylidene complexes. The synthesis and development of stereogenic at metal MAP complexes clearly reveals the potential of molybdenum- and (to a lesser extent) tungsten-based catalysts, regarding a possible increase in activity and stereo- and enantioselectivity. Thanks to the discovery of MAP catalysts, the big challenge of *Z*-selective metathesis has been met. A further rapid development in the area of stereogenic at metal catalysts and in optimization of their catalytic properties by modification of the ligand sphere is to be expected. There is no doubt that in the coming years, new fascinating well-defined catalysts for olefin metathesis will be developed, permiting further improvements in activity, selectivity, stability (longevity), and functional group tolerance.

ACKNOWLEDGMENTS

I wish to express my sincere thanks to Professor Helmut Fischer for reading and commenting the manuscript.

REFERENCES

1. Ivin KJ. *Olefin Metathesis*. Academic Press: San Diego; 1983.
2. Ivin KJ, Mol JC. *Olefin Metathesis and Metathesis Polymerization*. San Diego: Academic Press; 1997.
3. Schrock RR. *J Mol Catal A: Chem* 2004;213:21–30.
4. Grubbs, R. H. *Handbook of Metathesis*; Grubbs, R. H.; Ed.; Wiley-VCH: Weinheim, Germany, 2003, Vol. 1, pp. 4-7.
5. Schrock, R. R. *Handbook of Metathesis*; Grubbs, R. H.; Ed.; Wiley-VCH: Weinheim, Germany, 2003, Vol. 1, pp. 8-32.

6. Coperet, C.; Lefebvre, F.; Basset, J.-M. *Handbook of Metathesis*, Grubbs, R. H.; Ed.; Wiley-VCH: Weinheim, Germany, 2003, Vol. 1, pp. 33-46.
7. Feldman J, Schrock RR. *Prog Inorg Chem* 1991;39:1–74.
8. Schrock RR. *J Chem Soc Dalton Trans* 2001:2541–2550.
9. Schrock RR. *Chem Rev* 2002;102:145–179.
10. Tebbe FN, Parshall GW, Ovenall DW. *J Am Chem Soc* 1979;101:5074–5075.
11. Howard TR, Lee JB, Grubbs RH. *J Am Chem Soc* 1980;102:6876–6878.
12. Straus DA, Grubbs RH. *Organometallics* 1982;1:1658–1661.
13. Gilliom LR, Grubbs RH. *J Am Chem Soc* 1986;108:733–742.
14. Schrock RR, Rocklage SM, Wengrovius JH, Rupprecht G, Fellmann J. *J Mol Catal* 1980;8:73–83.
15. Rocklage SM, Fellmann JD, Rupprecht GA, Messerle LW, Schrock RR. *J Am Chem Soc* 1981;103:1440–1447.
16. Wallace KC, Dewan JC, Schrock RR. *Organometallics* 1986;5:2162–2164.
17. Wallace KC, Liu AH, Dewan JC, Schrock RR. *J Am Chem Soc* 1988;110:4964–4977.
18. Wallace KC, Schrock RR. *Macromolecules* 1987;20: 448–450.
19. Hessen B, Buijink J-KF, Meetsma A, Teuben JH, Helgesson G, Hakansson M, Jagner S, Spek AL. *Organometallics* 1993;12:2268–2276.
20. Buijink J-KF, Teuben JH, Kooijman H, Spek AL. *Organometallics* 1994;13:2922–2924.
21. Nomura K, Zhang W. *Chem Sci* 2010;1:161–173.
22. Nomura K, Zhang W. *Chem Rev* 2011;111:2342–2362.
23. Nomura K, Sagara A, Imanishi Y. *Macromolecules* 2002;35: 1583–1590.
24. Yamada J, Fujiki M, Nomura K. *Organometallics* 2005;24: 2248–2250.
25. Zhang W, Yamada J, Nomura K. *Organometallics* 2008;27: 5353–5360.
26. Nomura K, Onishi Y, Fujiki M, Yamada J. *Organometallics* 2008;27:3818–3824.
27. Zhang W, Nomura K. *Organometallics* 2008;27:6400–6402.
28. Toreki R, Vaughan GA, Schrock RR, Davis WM. *J Am Chem Soc* 1993;115:127–137.
29. Vaughan GA, Toreki R, Schrock RR, Davis WM. *J Am Chem Soc* 1993;115:2980–2981.
30. Fischer EO, Dötz KH. *Chem Ber* 1972;105:3966–3973.
31. Katz, T. J. *Handbook of Metathesis* Grubbs, R. H.; Ed.; Wiley-VCH: Weinheim, Germany, 2003, Vol. 1, pp. 47–60.
32. Katz TJ. *Angew Chem Int Ed* 2005;44:3010–3019.
33. Casey CP, Burkhardt TJ. *J Am Chem Soc* 1974;96: 7808–7809.
34. Casey CP, Burkhardt TJ. *J Am Chem Soc* 1973;95: 5833–5834.
35. Katz TJ, Lee SJ, Acton N. *Tetrahedron Lett* 1976;17: 4247–4250.
36. Katz TJ, McGinnis J, Altus C. *J Am Chem Soc* 1976;98: 606–608.
37. Lee SJ, McGinnis J, Katz TJ. *J Am Chem Soc* 1976;98: 7818–7819.
38. Katz TJ, Sivavec TM. *J Am Chem Soc* 1985;107:737–738.
39. Katz TJ, Acton N. *Tetrahedron Lett* 1976;17:4251–4254.
40. Wengrovius JH, Schrock RR, Churchill MR, Missert JR, Youngs WJ. *J Am Chem Soc* 1980;102:4515–4516.
41. Wengrovius JH, Schrock RR. *Organometallics* 1982;1: 148–155.
42. O'Donoghue MB, Schrock RR, LaPointe AM, Davies WM. *Organometallics* 1996;15:1334–1336.
43. de la Mata FJ, Grubbs RH. *Organometallics* 1996;15: 577–584.
44. Fujimura O, de la Mata FJ, Grubbs RH. *Organometallics* 1996;15:1865–1871.
45. Kress J, Wesolek M, Le Ny J-P, Osborn JA. *J Chem Soc Chem Commun* 1981:1039–1040.
46. Kress J, Wesolek M, Osborn JA. *J Chem Soc Chem Commun* 1982:514–516.
47. Kress J, Osborn J. *J Am Chem Soc* 1983;105:6346–6347.
48. Kress J, Aguero A, Osborn J. *J Mol Catal* 1986;36:1–12.
49. Kress J, Osborn JA, Greene RME, Ivin KJ, Rooney JJ. *J Am Chem Soc* 1987;109:899–901.
50. Kress J, Osborn JA, Amir-Ebrahimi V, Ivin KJ, Rooney JJ. *J Chem Soc* 1988:1164–1166.
51. Kress J, Osborn JA, Ivin KJ. *J Chem Soc* 1989:1234–1236.
52. Kress J, Ivin KJ, Amir-Ebrahimi V, Weber P. *Makromol Chem* 1990;191:2237–2251.
53. Youinou MT, Kress J, Fischer J, Aguero A, Osborn J. *J Am Chem Soc* 1988;110:1488–1493.
54. Kress J, Osborn J. *Angew Chem Int Ed* 1992;31:1585–1587.
55. Quignard F, Leconte M, Basset JM. *J Mol Catal* 1985;28:27–32.
56. Quignard F, Leconte M, Basset J-M. *J Chem Soc Chem Commun* 1985:1816–1817.
57. Couturier JL, Paillet C, Leconte M, Basset J-M, Weiss K. *Angew Chem Int Ed* 1992;31:628–631.
58. Lefebvre F, Leconte M, Pagano S, Mutch A, Basset J-M. *Polyhedron* 1995;14:3209–3226.
59. Schrock RR. *Tetrahedron* 1999;55:8141–8153.
60. Schrock RR, Czekelius C. *Adv Synth Catal* 2007;349:55–77.
61. Schrock RR, Hoveyda AH. *Angew Chem Int Ed* 2003;42:4592–4633.
62. Schrock RR. *Chem Rev* 2009;109:3211–3226.
63. Schaverien CJ, Dewan JC, Schrock RR. *J Am Chem Soc* 1986;108:2771–2773.
64. Herisson JL, Chauvin Y. *Macromol Chem* 1971;141: 161–176.
65. Schrock RR, DePue RT, Feldman J, Schaverien CJ, Dewan JC, Liu AH. *J Am Chem Soc* 1988;110:1423–1435.
66. Oskam JH, Schrock RR. *J Am Chem Soc* 1993;115: 11831–11845.

67. Oskam JH, Schrock RR. *J Am Chem Soc* 1992;114: 7588–7589.
68. Schrock RR, Murdzek JS, Bazan GC, Robbins J, DiMare M, O'Regan M. *J Am Chem Soc* 1990;112:3875–3886.
69. Marinescu SC, Singh R, Hock AS, Wampler KM, Schrock RR, Müller P. *Organometallics* 2008;27:6570–6578.
70. Grubbs RH, Chang S. *Tetrahedron* 1998;54:4413–4450.
71. Armstrong SK. *J Chem Soc Perkin Trans 1* 1988:371–388.
72. Deiters A, Martin SF. *Chem Rev* 2004;104:2199–2238.
73. Schrock RR. *Dalton Trans* 2011;40:7484–7495.
74. Buchmeiser MR. *Chem Rev* 2000;100:1565–1604.
75. Nicolaou KC, Bulger PG, Sarlah D. *Angew Chem Int Ed* 2005;44:4490–4527.
76. Martin SF, Liao Y, Wong Y, Rein T. *Tetrahedron Lett* 1994;35:691–694.
77. Fürstner A, Thiel OR. *J Org Chem* 2000;65:1738–1742.
78. Nakashima K, Inoue K, Sono M, Tori M. *J Org Chem* 2002;67:6034–6040.
79. Fürstner A, Langemann K. *J Org Chem* 1996;61:8746–8749.
80. Clark JS, Kettle JG. *Tetrahedron Lett* 1997;38:123–126.
81. Clark JS, Kettle JG. *Tetrahedron Lett* 1997;38:127–130.
82. Clark JS, Kettle JG. *Tetrahedron* 1999;55:8231–8248.
83. Clark JS, Hamelin O, Hufton R. *Tetrahedron Lett* 1998;39:8321–8324.
84. Clark JS, Hamelin O. *Angew Chem Int Ed* 2000;39:372–374.
85. Barrett AGM, Ahmed M, Baker SP, Baugh SPD, Braddock DC, Procopiou PA, White AJP, Williams DJ. *J Org Chem* 2000;65:3716–3721.
86. Xu Z, Johannes CW, Salman SS, Hoveyda AH. *J Am Chem Soc* 1996;118:10926–10927.
87. May SA, Grieco PA. *Chem Commun* 1998:1597–1598.
88. Smith AB III,, Kozmin SA, Adams CM, Paone DV. *J Am Chem Soc* 2000;122:4984–4985.
89. Crowe WE, Zhang ZJ. *J Am Chem Soc* 1993;115: 10998–10999.
90. Crowe WE, Goldberg DR. *J Am Chem Soc* 1995;117: 5162–5163.
91. Crowe WE, Goldberg DR, Zhang ZJ. *Tetrahedron Lett* 1996;37:2117–2120.
92. Feng J, Schuster M, Blechert S. *Synlett* 1997:129–130.
93. Gregory C, Fu GC, Grubbs RH. *J Am Chem Soc* 1993;115:3800–3801.
94. Kress S, Blechert S. *Chem Soc Rev* 2012;41:4389–4408.
95. McConville DH, Wolf JR, Schrock RR. *J Am Chem Soc* 1993;115:4413–4414.
96. Hock AS, Schrock RR, Hoveyda AH. *J Am Chem Soc* 2006;128:16373–16375.
97. Schrock RR. *Chem N Z* 2011:117–121.
98. Hoveyda, A. H. *Handbook of Metathesis*; Grubbs, R. H.; Ed.; Wiley-VCH: Weinheim, Germany, 2003, Vol. 2, pp. 128-150.
99. Hoveyda AH, Schrock RR. *Chem Eur J* 2001;7:945–950.
100. Fujimura O, Grubbs RH. *J Am Chem Soc* 1996;118: 2499–2500.
101. Fujimura O, Grubbs RH. *J Org Chem* 1998;63:824–832.
102. Kagan HB, Fiaud JC. *Top Stereochem* 1988;53:708–710.
103. Alexander JB, La DS, Cefalo DR, Hoveyda AH, Schrock RR. *J Am Chem Soc* 1998;120:4041–4042.
104. La DS, Alexander JB, Cefalo DR, Graf DD, Hoveyda AH, Schrock RR. *J Am Chem Soc* 1998;120:9720–9721.
105. Zhu SS, Cefalo DR, La DS, Jamieson JY, Davis WM, Hoveyda AH, Schrock RR. *J Am Chem Soc* 1999;121: 8251–8259.
106. Dolman SJ, Sattely ES, Hoveyda AH, Schrock RR. *J Am Chem Soc* 2002;124:6991–6997.
107. Kiely AF, Jernelius JA, Schrock RR, Hoveyda AH. *J Am Chem Soc* 2002;124:2868–2869.
108. Weatherhead GS, Houser JH, Ford GJ, Jamieson JY, Schrock RR, Hoveyda AH. *Tetrahedron Lett* 2000;41:9553–9559.
109. Burke SD, Muller N, Beaudry CM. *Org Lett* 1999;1: 1827–1829.
110. Sattely ES, Cortez GA, Moebius DC, Schrock RR, Hoveyda AH. *J Am Chem Soc* 2005;127:8526–8533.
111. Limanto J, Snapper ML. *J Am Chem Soc* 2000;122: 8071–8072.
112. Weatherhead GS, Ford JG, Alexanian EJ, Schrock RR, Hoveyda AH. *J Am Chem Soc* 2000;122:1828–1829.
113. La DS, Ford JG, Sattely ES, Bonitatebus PJ, Schrock RR, Hoveyda AH. *J Am Chem Soc* 1999;121:11603–11604.
114. Solans-Monfort X, Clot E, Copéret C, Eisenstein O. *J Am Chem Soc* 2007;129:14015–14025.
115. Poater A, Solans-Monfort X, Clot E, Copéret C, Eisenstein O. *J Am Chem Soc* 2009;131:8207–8216.
116. Singh R, Schrock RR, Müller P, Hoveyda AH. *J Am Chem Soc* 2007;129:12654–12655.
117. Malcolmson SJ, Meek SJ, Sattely ES, Schrock RR, Hoveyda AH. *Nature* 2008;456:933–937.
118. Marinescu SC, Schrock RR, Li B, Hoveyda AH. *J Am Chem Soc* 2009;131:58–59.
119. Meek SJ, Malcolmson SJ, Li B, Schrock RR, Hoveyda AH. *J Am Chem Soc* 2009;131:16407–16409.
120. Bailey BC, Schrock RR, Kundu S, Goldman AS, Huang Z, Brookhart M. *Organometallics* 2009;28:355–360.
121. Sattely ES, Meek SJ, Malcolmson SJ, Schrock RR, Hoveyda AH. *J Am Chem Soc* 2009;131:943–953.
122. Hansen EC, Lee D. *Org Lett* 2004;6:2035–2038.
123. Meek SJ, O'Brien RV, Llaveria J, Schrock RR, Hoveyda AH. *Nature* 2011;471:461–466.
124. Jiang AJ, Zhao Y, Schrock RR, Hoveyda AH. *J Am Chem Soc* 2009;131:16630–16631.
125. Marinescu SC, Schrock RR, Müller P, Takase MK, Hoveyda AH. *Organometallics* 2011;30:1780–1782.
126. Townsend EM, Schrock RR, Hoveyda AH. *J Am Chem Soc* 2012;134:11334–11337.
127. Marinescu SC, Schrock RR, Müller P, Hoveyda AH. *J Am Chem Soc* 2009;131:10840–10841.
128. Marinescu SC, Levine DS, Zhao Y, Schrock RR. *J Am Chem Soc* 2011;133:11512–11514.

129. Gradillas A, Pérez-Castells J. *Angew Chem Int Ed* 2006;45:6086–6101.

130. Yu M, Wang C, Kyle AF, Jakubec P, Dixon DJ, Schrock RR, Hoveyda AH. *Nature* 2011;479:88–93.

131. Dolman SJ, Schrock RR, Hoveyda AH. *Org Lett* 2003;5:4899–4902.

132. Ibrahem I, Yu M, Schrock RR, Hoveyda AH. *J Am Chem Soc* 2009;131:3844–3845.

133. Yu M, Ibrahem I, Hasegawa M, Schrock RR, Hoveyda AH. *J Am Chem Soc* 2012;134:2788–2799.

134. Lee Y-J, Schrock RR, Hoveyda AH. *J Am Chem Soc* 2009;131:10652–10661.

135. Grubbs, R. H. *Handbook of Metathesis*; Grubbs, R. H.; Ed.; Wiley-VCH: Weinheim, Germany, 2003, Vol. 3, Chapter 3.2.

136. Bielawski CW, Grubbs RH. *Prog Polym Sci* 2007;32:1–29.

137. Smith D, Pentzer EB, Nguyen ST. *Polym Rev* 2007;47:419–459.

138. Feast WJ, Khosravi E. *New Methods of Polymer Synthesis*, Vol. 2; Ebdon JR, Eastmond GC.; Eds.; Springer Science + Business media: Dordrecht, 1995, pp. 69–116.

139. Buchmeiser MR. *Chem Rev* 2009;109:303–321.

140. Grubbs, R. H. *Handbook of Metathesis*; Grubbs, R. H.; Ed.; Wiley-VCH: Weinheim, Germany, 2003, Vol. 3, pp. 143-179.

141. Ivin KJ, Mol JC. *Olefin Metathesis and Metathesis Polymerization.* San Diego: Academic Press; 1997 Chapters 11 and 13.

142. Breen, D.; Curran, K.; Risse, W. *Stereoselective Polymerization with Single-Site Catalysts*; Baugh, L. S; Canich, J.A.M.; Eds.; CRC Press, 2007, 509-551.

143. Feast WJ, Gibson VC, Marshall EL. *J Chem Soc Chem Commun* 1992:1157–1158.

144. Schrock RR, Lee J-K, O'Dell R, Oskam JH. *Macromolecules* 1995;28:5933–5940.

145. Bazan G, Khosravi E, Schrock RR, Feast WJ, Gibson VC, O'Regan MB, Davis WM. *J Am Chem Soc* 1990;112: 8378–8387.

146. Bazan G, Schrock RR, Khosravi E, Feast WJ, Gibson VC. *Polym Commun* 1989;30:258–260.

147. Feast WJ, Gibson VC, Ivin KJ, Kenwright AM, Khosravi E. *J Chem Soc Chem Commun* 1994:1399–1400.

148. O'Dell R, McConville DH, Hofmeister GE, Schrock RR. *J Am Chem Soc* 1994;116:3414–3423.

149. Flook MM, Jiang AJ, Schrock RR, Müller P, Hoveyda AH. *J Am Chem Soc* 2009;131:7962–7963.

150. Flook MM, Gerber LCH, Debelouchina GT, Schrock RR. *Macromolecules* 2010;43:7515–7522.

151. Flook MM, Ng VWL, Schrock RR. *J Am Chem Soc* 2011;133:1784–1786.

152. Coperet C. *Dalton Trans* 2007:5498–5504.

153. Rendon N, Blanc F, Coperet C. *Coord Chem Rev* 2009;253:2015–2020.

154. Buchmeiser MR. *Chem Rev* 2009;109:303–321.

155. Hultzsch KC, Jernelius JA, Hoveyda AH, Schrock RR. *Angew Chem Int Ed* 2002;41:589–593.

156. Dolman SJ, Hultzsch KC, Pezet F, Teng X, Hoveyda AH, Schrock RR. *J Am Chem Soc* 2004;126:10945–10953.

157. Kroll RM, Schuler N, Lubbad S, Buchmeiser MR. *Chem Commun* 2003:2742–2743.

158. Chabanas, M.; Baudouin, A. ; Copéret, C. ; Basset, J.-M. *J Am Chem Soc* 2001, 123, 2062-2063.

159. Chabanas M, Coperet C, Basset J-M. *Chem Eur J* 2003;9:971–975.

160. Rhers B, Salameh A, Baudouin A, Quadrelli EA, Taoufik M, Copéret C, Lefebvre F, Basset J-M, Solans-Monfort X, Eisenstein O, Lukens WW, Lopez LHP, Sinha A, Schrock RR. *Organometallics* 2006;25:3554–3557.

161. Blanc F, Coperet C, Thivolle-Cazat J, Basset J-M, Lesage A, Emsley L, Sinha A, Schrock RR. *Angew Chem Int Ed* 2006;45:1216–1220.

162. Blanc F, Thivolle-Cazat J, Basset J-M, Coperet C, Hock AS, Tonzetich ZJ, Schrock RR. *J Am Chem Soc* 2007;129:1044–1045.

163. Blanc F, Rendon N, Berthoud R, Basset J-M, Copéret C, Tonzetich ZJ, Schrock RR. *Dalton Trans* 2008:3156–3158.

164. Blanc F, Berthoud R, Salameh A, Basset J-M, Coperet C, Singh R, Schrock RR. *J Am Chem Soc* 2007;129: 8434–8435.

165. Blanc F, Berthoud R, Copéret C, Lesage A, Emsley L, Singh R, Kreickmann T, Schrock RR. *Proc Natl Acad Sci* 2008;105:12123–12127.

166. Slinn CA, Redgrave AJ, Hind SL, Edlin C, Nolan SP, Gouverneur V. *Org Biomol Chem* 2003;1:3820–3825.

167. Cortez GA, Baxter CA, Schrock RR, Hoveyda AH. *Org Lett* 2007;9:2871–2874.

168. Krause JO, Nuyken O, Buchmeiser MR. *Chem Eur J* 2004;10:2029–2035.

169. Kang E-H, Lee IS, Choi T-L. *J Am Chem Soc* 2011;133: 11904–11907.

170. Anders U, Nuyken O, Buchmeiser MR, Wurst K. *Angew Chem Int Ed* 2002;41:4044–4047.

171. Schattenmann FJ, Schrock RR, Davis WM. *J Am Chem Soc* 1996;118:3295–3296.

172. Heppekausen J, Fürstner A. *Angew Chem Int Ed* 2011;50:7829–7832.

173. Aeilts SL, Cefalo DR, Bonitatebus PJ Jr, Houser JH, Hoveyda AH, Schrock RR. *Angew Chem Int Ed* 2001;40: 1452–1456.

174. Teng X, Cefalo DR, Schrock RR, Hoveyda AH. *J Am Chem Soc* 2002;124:10779–10784.

14

RUTHENIUM-BENZYLIDENE OLEFIN METATHESIS CATALYSTS

GEORGIOS C. VOUGIOUKALAKIS
Laboratory of Organic Chemistry, Department of Chemistry, University of Athens, Athens, Greece

14.1 INTRODUCTION TO WELL-DEFINED RUTHENIUM CATALYSTS

The first well-defined ruthenium olefin metathesis catalyst (**1**, Fig. 14.1) was reported in 1992 by Grubbs and coworkers (1–3). The basic structure of the contemporary ruthenium-based metathesis catalysts still resembles that original complex, namely, comprising a ruthenium alkylidene, two halides, and two neutral ligands. However, currently used catalysts (**2–7**, Fig. 14.1) are significantly more reactive and more tolerant toward oxygen, moisture, and functional groups. For example, catalyst **2**, although not as reactive, has much better functional group compatibility than all of the early transition metal olefin metathesis initiators (4,5). In 1999, Grubbs and coworkers substituted one of the tricyclohexylphosphine ligands in **2** with the bulky *N*-heterocyclic carbene (NHC) ligand H_2IMes to obtain **3** (Fig. 14.1), leading to a significant enhancement of catalytic activity, while maintaining the high functional group tolerance and stability of **2** (6). This improvement was attributed to the increased affinity of the NHC-substituted ruthenium center for π-acidic olefins relative to σ-donating phosphines (7). The development of other phosphine-free complexes by the research groups of Hoveyda, Grubbs, and Grela yielded olefin metathesis catalysts **4**, **5** (8), **6** (9,10), and **7** (11) (Fig. 14.1). Compared to their phosphine-containing analogs, precatalysts **5** and **7** demonstrate higher thermal stability, while **6** is among the fastest-initiating ruthenium systems studied thus far (fast initiation is of vital importance in the production of polymers with narrow polydispersity, as well as in the synthesis of block copolymers). Subsequent studies led to the development of enantioselective ruthenium catalysts (12–23), catalysts for applications in aqueous and protic solvent systems (24–32), and ruthenium catalysts that carry out the formation of tetrasubstituted and sterically hindered carbon–carbon double bonds (33,34), as well as *Z*-selective (35–40) and latent metathesis catalysts (41–44).

14.2 RUTHENIUM BENZYLIDENES

14.2.1 First Generation Catalysts

As mentioned earlier, complex **1** (Fig. 14.1) (3), was the first well-defined ruthenium metathesis precatalyst. Although **1** effectively polymerizes norbornene even in the presence of protic/aqueous media, this early complex was metathetically active only with strained and electron rich olefins. Subsequent phosphine exchange studies (45), aimed at improved ruthenium catalysts, led to the conclusion that the catalytic activity of this family of complexes increases in the order $PPh_3 \ll Pi\text{-}Pr_3 < PCy_3$; that is, larger and more basic phosphines afford precatalysts with greater metathesis activity.

However, the turning point to the commercialization of the ruthenium catalysts was the development of a scalable synthetic procedure, based on the reaction of $RuCl_2(PPh_3)_3$ with diazoalkane compounds, leading to substituted ruthenium alkylidenes. This advance, reported by Grubbs and

Olefin Metathesis: Theory and Practice, First Edition. Edited by Karol Grela.

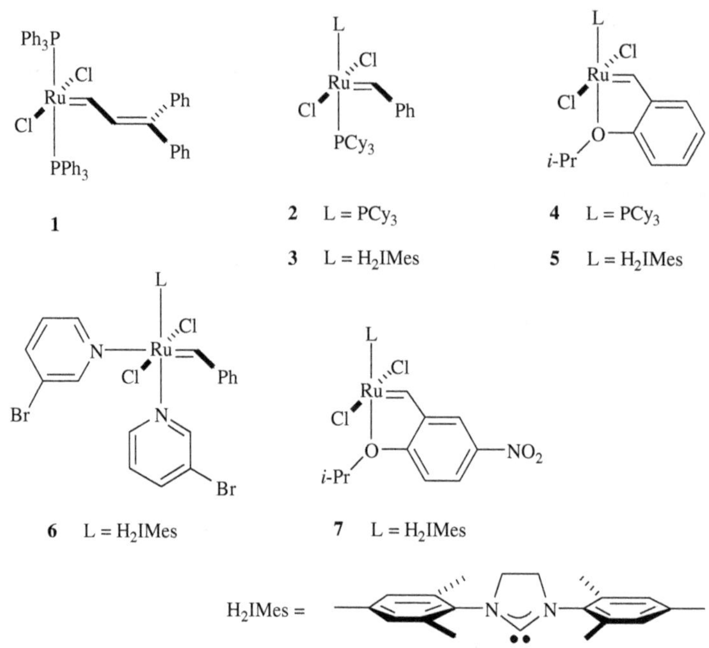

Figure 14.1 Ruthenium olefin metathesis catalysts **1–7**.

coworkers in 1995 (4,5), enabled the widespread use of ruthenium metathesis catalysts that revolutionized the way chemists design and synthesize molecules. Thus, complex **2** (Fig. 14.1), also known as the '*Grubbs first generation catalyst*', is a very efficient olefin metathesis catalyst. More importantly, **2** is stable against oxygen and moisture, but also highly functionally group tolerant. These characteristics make complex **2** an extremely useful catalyst for chemists working in a wide variety of research fields, both in industry and academia (for a selection of catalyst **2** applications, refer to Chapters 8–11). For example, the functional group tolerance of **2** has extended ring opening metathesis polymerization (ROMP) to a more diverse set of monomers. ROMP initiated by **2** are often living polymerizations, leading to nearly monodisperse polymers and offering the possibility to obtain block copolymers (note that the third generation analogs, discussed in Section 14.2.3, are even more competent as regards living ROMP and block copolymers due to their high initiation rates). Besides its impact in polymer synthesis via ROMP, catalyst **2** has changed the way organic chemists approach their synthetic challenges through ring-closing metathesis (RCM) (this is also true for the second generation analogs discussed in Section 14.2.2). Catalyst **2** has thus made it possible to easily and in a straightforward manner form small, medium, and large ring systems, by ring-closing compounds containing diverse functionality and, importantly, usually circumventing the need for time-consuming and tedious protection–deprotection steps (also refer to Chapters 8–11).

14.2.2 Second Generation Catalysts

Carbenes are strong Lewis bases, behave as excellent σ-donors and poor π-acceptors, and afford metal–carbon bonds that are usually less labile than the related metal–phosphine bonds. This decreased lability of carbenes, compared to the phosphine ligands, is one of the reasons for the improved thermal stability of the corresponding organometallic complexes. Furthermore, the electronic properties and the steric environment of heterocyclic carbenes can be easily and, in many cases, rationally, modified by changing the substituents on the carbene framework, thereby fine-tuning the catalytic properties of the resulting organometallic complexes.

The carbenic frameworks used thus far in ruthenium-based metathesis catalysts are illustrated in Figure 14.2 (1). The most successful and well-studied ruthenium catalysts bear imidazol- or imidazolin-2-ylidenes. Triazol-5-ylidenes (46), tetrahydropyrimidin-2-ylidenes (47), and a four-membered ring diaminocarbene (48) have also been utilized, to afford the corresponding complexes. More recently, a series of ruthenium complexes coordinated with cyclic (alkyl)(amino) carbenes (49) and thiazol-2-ylidenes (36) have been synthesized. Following a brief historical review of the development of the "Grubbs second generation catalysts", in this section, we will discuss the most important aspects of the synthesis, structure, and catalytic activity of some representative families of this kind of ruthenium benzylidenes (note that the "second generation Hoveyda–Grubbs catalysts" are being discussed in Chapter 16).

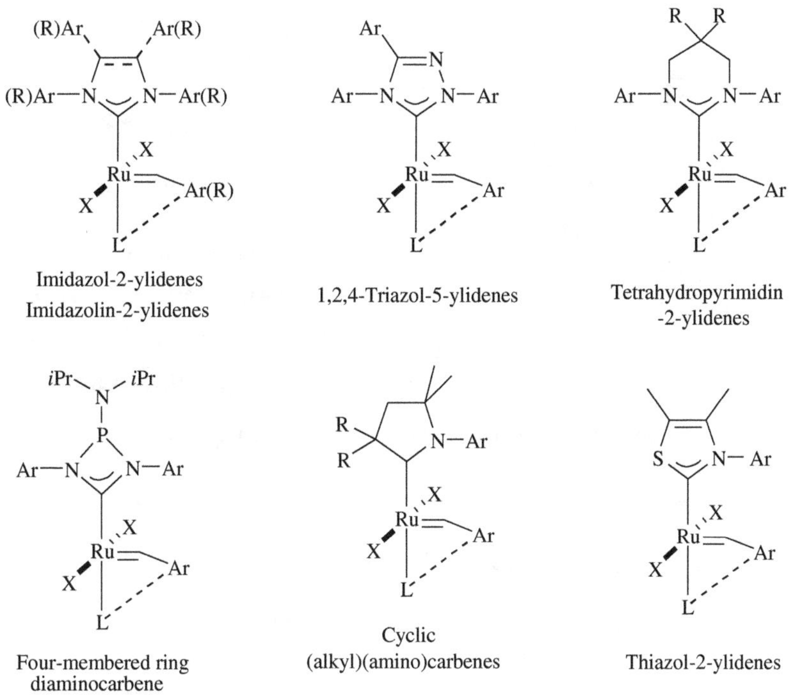

Figure 14.2 Ruthenium heterocyclic carbene-coordinated olefin metathesis catalyst families reported to date.

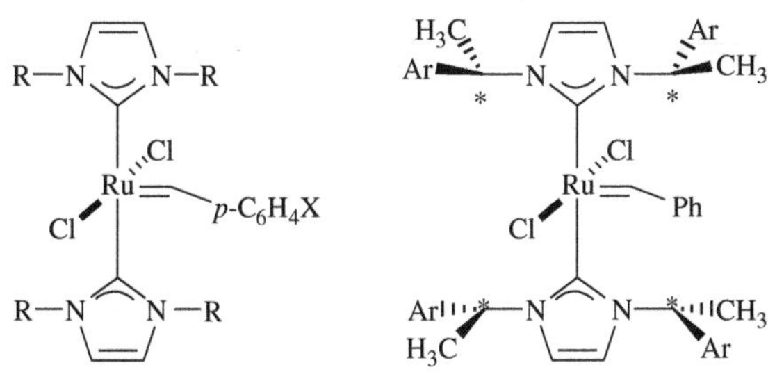

8: R = *i*-Pr, X = H
9: R = Cyclohexyl, X = H
10: R = *i*-Pr, X = Cl

11: Ar = Ph
12: Ar = Naphthyl

Figure 14.3 The first reported ruthenium carbene-coordinated metathesis catalysts.

The first heterocyclic carbene-coordinated ruthenium-based metathesis catalysts (**8–12**, Fig. 14.3), in which both phosphine ligands were replaced by NHCs, were reported in 1998 by Herrmann and coworkers (50). Despite their high stability, these complexes did not show a significant improvement in metathesis activity, mostly due to their slow initiation rates, attributed, in these carbene bis-coordinated complexes, to one of the NHCs (less labile than phosphines) (51,52), which has to dissociate from the metal center for the catalyst to be initiated.

Soon thereafter, Grubbs and Nolan published the synthesis of heteroleptic ruthenium complexes **13a** and **13b** (Fig. 14.4), by combining a non-labile (1,3-dimesityl-imidazol-2-ylidene) with a labile (tricyclohexylphosphine or triphenylphosphine) ligand (53–55). Species **13a** and **13b** exhibited not only higher RCM activity affording even tetra-substituted cycloolefins, at that time out of reach of ruthenium catalysts (53), but also improved thermal stability compared to the parent bis(tricyclohexylphosphine) complex. Almost simultaneously, heteroleptic ruthenium-based metathesis catalysts **14–17** (Fig. 14.4) were also reported by the research groups of Herrmann and Nolan (52,56,57).

These first reports on NHC-coordinated ruthenium catalysts paved the way for one of the major breakthroughs in the field of olefin metathesis, which occurred with the synthesis of "Grubbs second generation catalyst" **3** (Fig. 14.5) (6). The synthesis of carbene-coordinated ruthenium-based complexes **18** and **19** (Fig. 14.5) was also published in that same work. The incentive for the preparation of complexes **3**, **18**, and **19** originated from the expected increased basicity of the corresponding saturated NHCs, compared to their unsaturated counterparts in Figure 14.4. The increased basicity of these saturated NHCs was, in turn, anticipated to translate into an increased activity of the resulting ruthenium complexes (note that more recent studies suggest relatively small differences in the donor capacities between saturated and unsaturated NHCs.). Complexes **3**, **18**, and **19** were synthesized in high yields, starting from "Grubbs first generation catalyst" (**2**), by substituting one of the phosphine ligands with the *in situ* generated free carbenes (6).

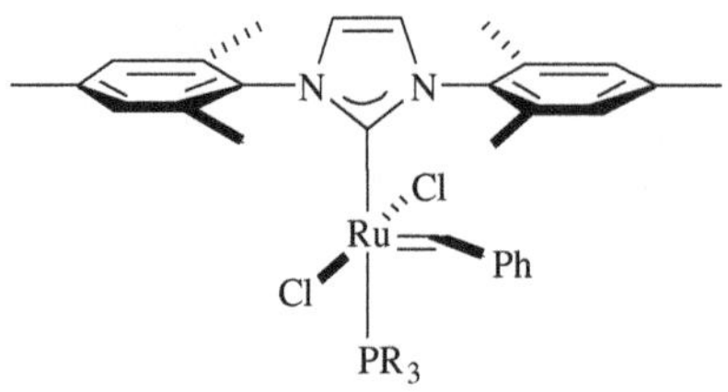

13a : R = Cyclohexyl

13b : R = Ph

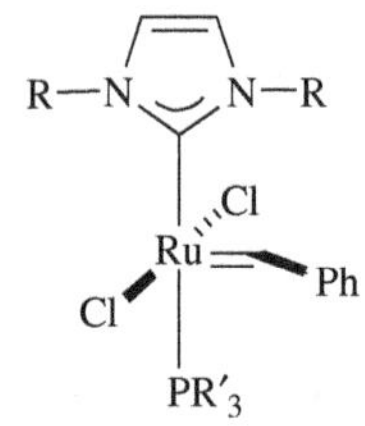

14 : R = *t*-Bu, R′ = Ph

15 : R = Cyclohexyl, R′ = cyclohexyl

16 : R = 4-Me-C_6H_4, R′ = cyclohexyl

17 : R = 4-ClC_6H_4, R′ = cyclohexyl

Figure 14.4 Ruthenium-based metathesis catalysts **13–17**.

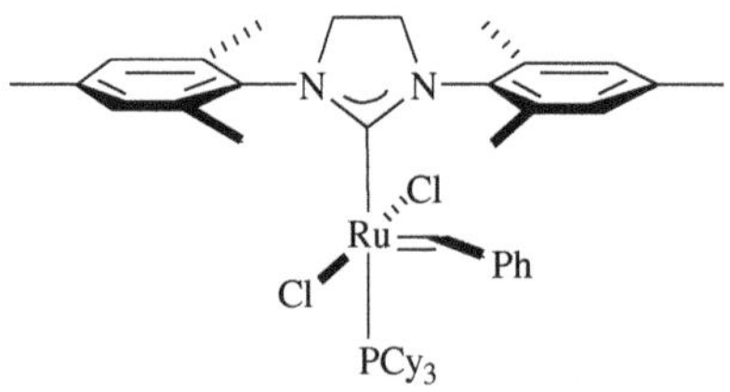

3

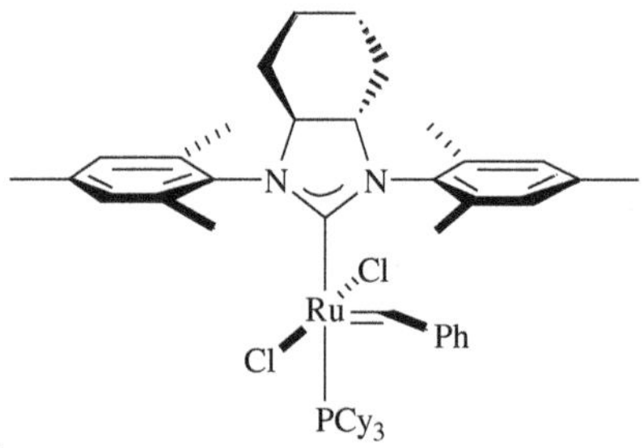

18

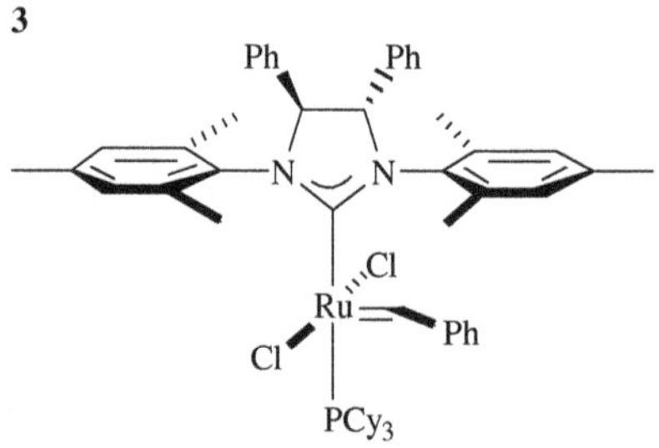

19

Figure 14.5 Ruthenium olefin metathesis catalysts **3**, **18**, and **19**.

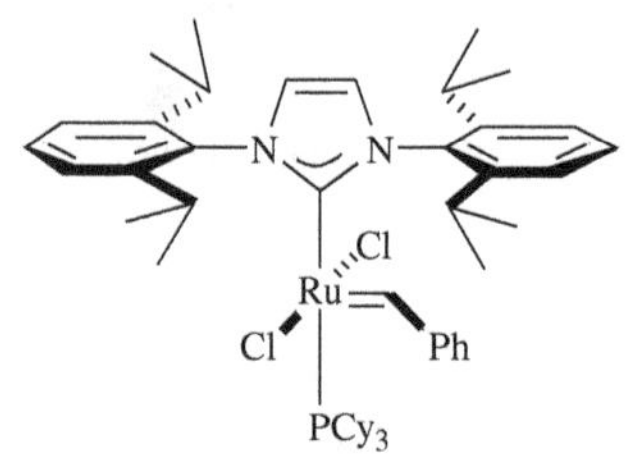

20

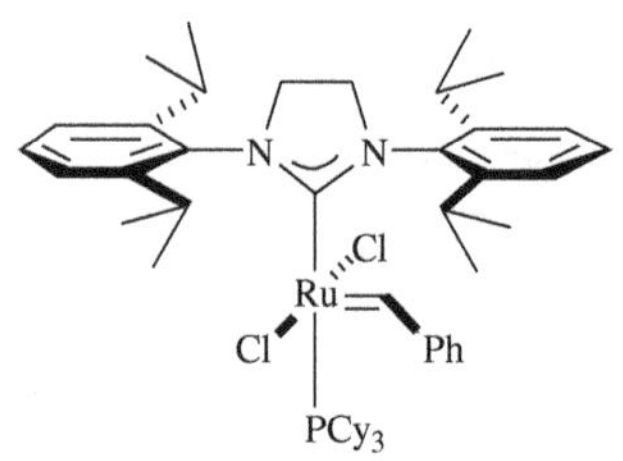

21

Figure 14.6 Ruthenium catalysts **20** and **21** bearing sterically demanding NHCs.

Complex **3** expanded the scope of ruthenium metathesis catalysts significantly, as it was proved to be not only air-, water-, and functional group-tolerant (58), but also highly efficient in the RCM of sterically demanding dienes (6), in the ROMP of low-strain substrates (59), and in the realization of challenging cross-metathesis (CM) reactions (58,60). This kind of reactivity was previously possible only with the more active, though highly air–water-sensitive, early transition-metal catalysts. Moreover, **3** remains efficient at catalyst loadings as low as 0.05 mol% for RCM, and 0.0001 mol% for ROMP reactions. Two alternative synthetic routes for the preparation of complex **3** were published 3 and 4 years later by the groups of Nolan and Grubbs (51,61), respectively.

In 2000, Nolan and coworkers (62) reported the synthesis of complex **20** (Fig. 14.6), bearing two bulky aryl groups (2,6-diisopropylphenyl), in an effort to study the influence of the bulkiness of the NHCs on the catalytic activity of the corresponding complexes. Likewise, the groups of Fürstner (46) and Mol (63) reported the synthesis and the catalytic activity evaluation of complex **21** (Fig. 14.6), the saturated analog of **20**. At ambient temperature, catalyst **21** shows

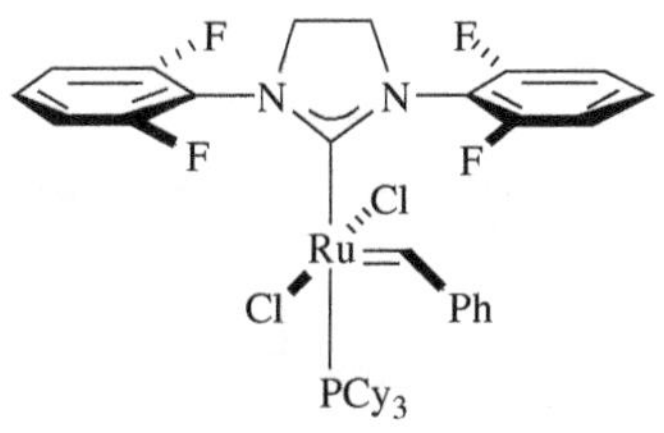

22

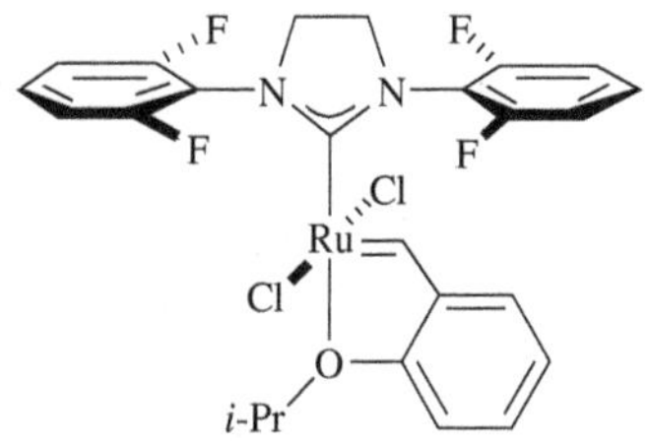

23

Figure 14.7 Fluorinated ruthenium-based catalysts **22** and **23**.

effective turnover numbers six times higher than those of **3** (63), along with very high initiation rates; however, its decomposition rate is also increased (64), especially when utilized in challenging transformations (65).

Complexes **22** and **23** (Fig. 14.7), bearing *o*-fluorinated aryl groups on the NHC ligand, were reported in 2006 by Grubbs and coworkers (66). Phosphine-containing **22** catalyzes the RCM of diethyl diallylmalonate with a significantly increased reaction rate compared to both parent complexes **3** (Fig. 14.5) and **5** (Fig. 14.1), whereas, in the same benchmark reaction, phosphine-free **23** shows a slower rate than both **3** and **5**. This contradictory behavior of **22** and **23** was ascribed to an unprecedented fluorine–ruthenium interaction (Ru–F distance = 3.2 Å) observed in the solid-state structure of **23** via X-ray diffraction.

Despite the development of the highly active and functional group-tolerant catalysts described above, RCM to form tetrasubstituted olefins remained one of the weaknesses of ruthenium-catalyzed metathesis until 2007. In that year, a series of new precatalysts with increased activity for the formation of tetrasubstituted olefins was published by Grubbs and coworkers (33). The design of these catalysts was based on an earlier observation, according to which catalysts with reduced bulk at the ortho position of the two *N*-bound aryl groups of the NHCs exert increased efficiency in the formation of sterically demanding substrates (17,18).

Unfortunately, however, the difficult preparation of the NHCs incorporated in these early catalysts rendered their large scale production uneconomical and imposed significant drawbacks regarding their commercialization. Research in the direction of more easily prepared catalysts eventually led the Grubbs group to develop **24–29** (Fig. 14.8), the synthesis of which can be easily performed on a large scale (34). Catalysts **24–29** proved to be very efficient in the RCM of dimethallylmalonates, a family of sterically demanding benchmark substrates, while *N*-tolyl complexes **24** and **27** were the most successful catalysts in that work. Later on, **27** was also found to exert increased efficiency for the formation of sterically challenging disubstituted olefins by CM (67). On the basis of solution- and solid-state structural data, and in conjunction with a series of theoretical calculations, the outstanding catalytic activity of **24** and **27** toward sterically demanding substrates was proposed to result from a significantly more open steric environment around the ruthenium center (68). This was suggested to originate from the accessibility of conformations in which the *N*-tolyl rings are rotated away from approaching and coordinating olefins.

In another work, based mostly on electrochemical and NMR studies, imidazolylidene- and imidazolinylidene-coordinated complexes **30–47** (Fig. 14.9) were synthesized by Thiele, Plenio, and coworkers (69). The major goal of this study was to investigate the existence

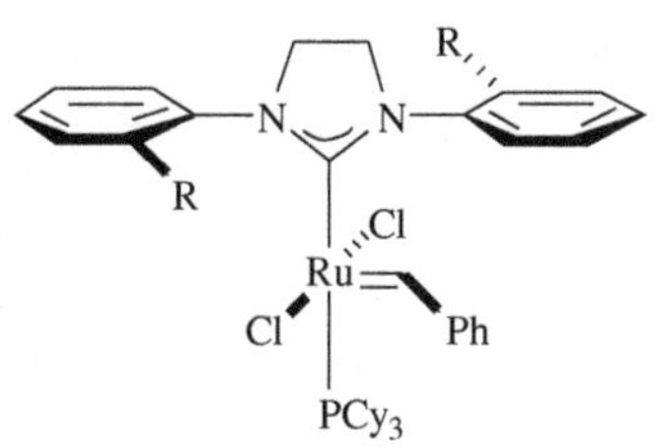

24 (R = Me)
25 (R = Et)
26 (R = *i*-Pr)

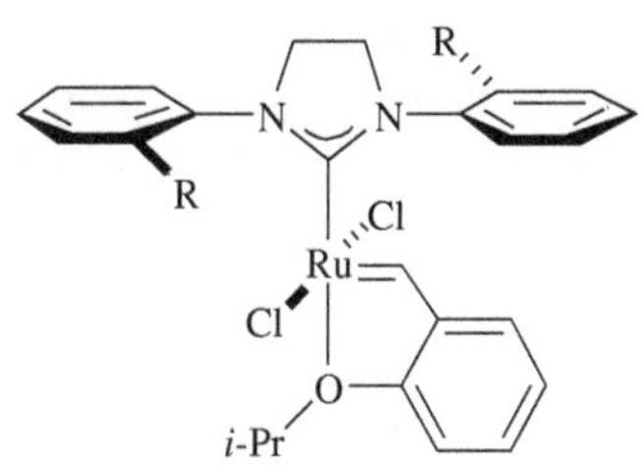

27 (R = Me)
28 (R = Et)
29 (R = *i*-Pr)

Figure 14.8 Ruthenium-based catalysts **24–29** with increased efficiency in the formation of tetrasubstituted olefins via RCM.

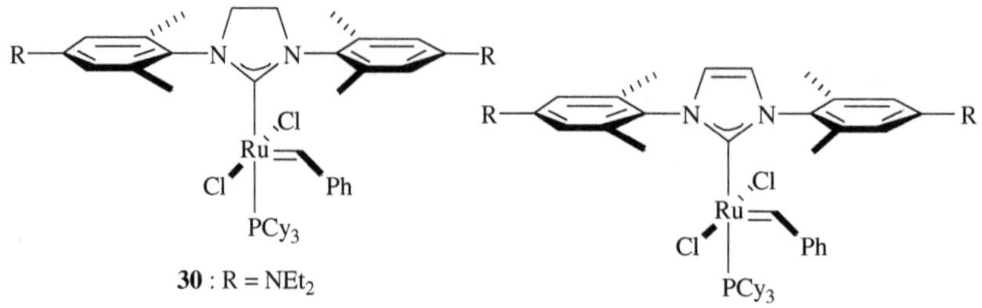

30 : R = NEt_2
32 : R = OMe
34 : R = Me
36 : R = H
38 : R = SMe
40 : R = F
42 : R = Cl
44 : R = Br
46 : R = I

31 : R = NEt_2
33 : R = OMe
35 : R = Me
37 : R = H
39 : R = SMe
41 : R = F
43 : R = Cl
45 : R = Br
47 : R = I

Figure 14.9 Catalysts **30–47** bearing saturated and unsaturated symmetrical NHC ligands.

of intramolecular π–π interactions and whether such interactions influence the electronic density at the ruthenium center, as well as the catalytic ability of the corresponding complexes. The reactivity of some representative complexes in selected RCM and CM reactions was found to systematically depend on the electronic properties of substituents R (Fig. 14.9). Complex **30**, bearing the electron-donating NEt_2 group, was the most catalytically active complex. It was furthermore suggested that the differences in reactivity between saturated and unsaturated NHCs do not originate from different electron densities at the ruthenium center, and that the electron-donating abilities of the saturated and the unsaturated NHCs (bearing the same substituents R) are similar.

On a different note, Grubbs and coworkers (70) recently reported the synthesis and catalytic activity evaluation of ruthenium benzylidene **48** (Fig. 14.10), incorporating both a traditional NHC and a mesoionic carbene. This precatalyst combines the stability and latency of bis-NHC complexes while maintaining low activation temperatures. Thus, it can be very efficiently activated by the addition of a Brønsted acid and then competently catalyze RCM and ROMP reactions. The leaving group that opens up a coordination site for the metathesis cycle to begin is the mesoionic carbene. Moreover, in the RCM of diethyl diallylmalonate, **48** was found to be more efficient than commercially available precatalyst **5** (Fig. 14.1). In that same work, it was found that other conventional bis-NHC complexes may be initiated protonolytically, albeit much less efficiently than **48**.

Although the first ruthenium catalysts having chiral monodentate NHCs (**18** and **19**, Fig. 14.5) were reported in 1999

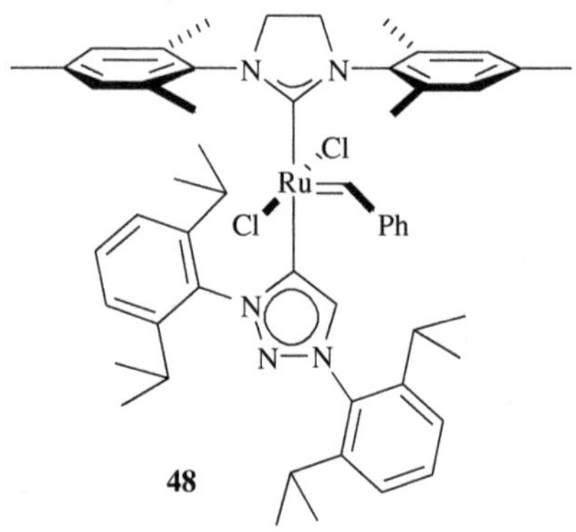

48

Figure 14.10 Ruthenium precatalyst **48** bearing a traditional NHC and a mesoionic carbene.

(6), the first asymmetric metathesis reaction catalyzed by this kind of complex was published by Grubbs and coworkers (12) 2 years later. Besides **18** and **19**, chiral complexes **49–52** (Fig. 14.11) were synthesized and evaluated. In these asymmetric complexes, the chirality is transferred from the four- and five-positions of the NHC imidazolyl ring to the N-bonded aromatic groups, forcing the ortho-substituents of the *N*-aryl rings to reside on the NHC-face opposite to the bulky groups on the backbone (a so-called "gearing" effect). Complexes **49**, **50a**, **51**, and **52a** are air-stable solids easily purified on the bench by column chromatography, whereas **50b**, **50c**, **51b**, and **51c** can be generated *in situ* by the addition of excess LiBr or NaI. Crystallographic evidence of the conformation of these chiral NHCs was obtained by conversion of **50a** to the corresponding bis(pyridine) complex.

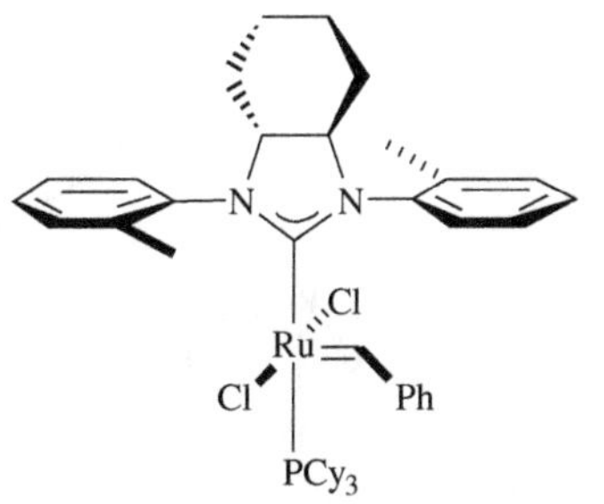

49

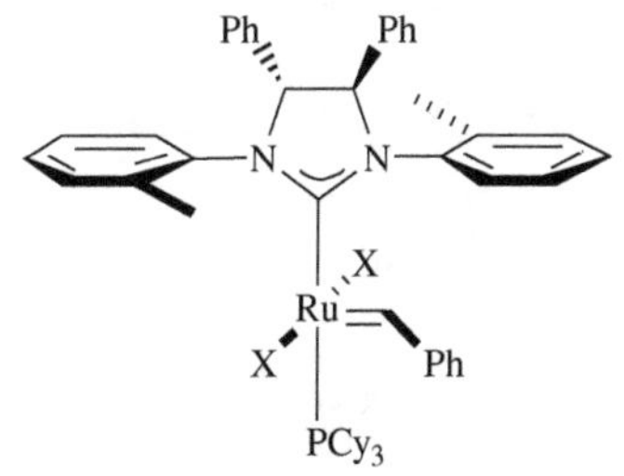

50a–50c

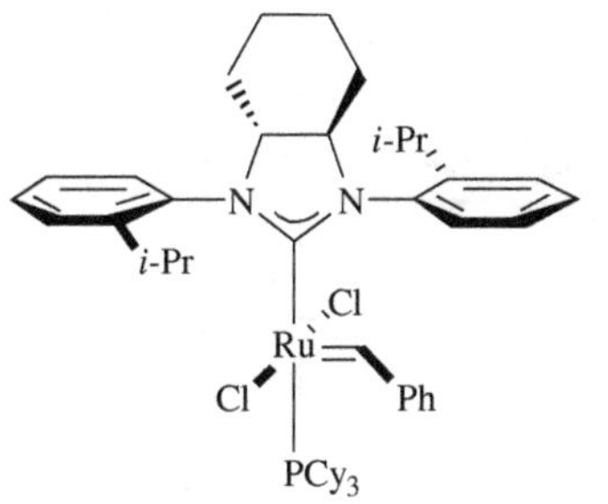

51

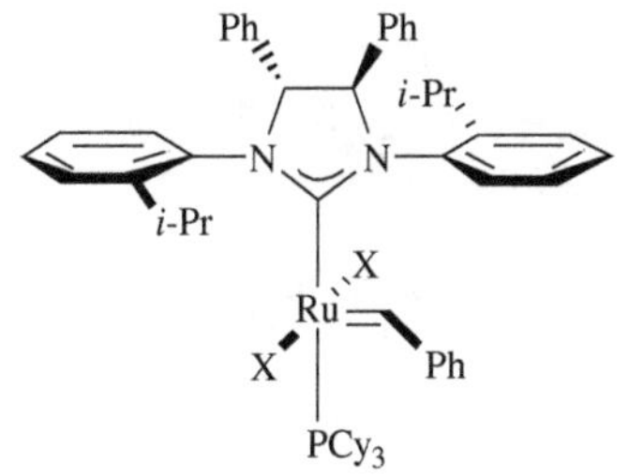

52a–52c

a : X = Cl
b : X = Br
c : X = I

Figure 14.11 Ruthenium complexes **49–52** coordinated with chiral monodentate NHCs.

Complexes **18**, **19**, and **49–52** were evaluated in the enantioselective desymmetrization of achiral trienes (3-allyloxy-2,4-dimethylpenta-1,4-dienes), also known as *asymmetric ring-closing metathesis* (*ARCM*), by Grubbs and coworkers (12). It was found that catalysts encompassing the (1*R*, 2*R*)-1,2-diaminocyclohexane group (**18**, **49**, and **51**) exhibit lower enantioselectivities than those having the (1*R*, 2*R*)-1,2-diphenylethylenediamine moiety (**19**, **50**, and **52**). Replacement of the mesityl substituents (in **18** and **19**) with *o*-methyl-aryl (in **49** and **50**) or *o*-isopropyl-aryl groups (in **51** and **52**) also increases the enantioselectivity. Finally, changing the halide ligands from chlorides to iodides improves the enantioselectivity; however, the conversion to the metathesized products is simultaneously reduced, supposedly due to the lower stability of the diiodide ruthenium intermediates. None of these catalysts showed a significant temperature- or solvent-dependent change in its enantioselectivity. Catalyst **52c** afforded the highest enantiomeric excess measured in that study.

Subsequent studies in the Grubbs group, aimed at the enhancement of the enantioselectivity displayed by **52**, together with the goal of expanding the substrate scope of ARCM, led to the synthesis of chiral complexes **53–56** (Fig. 14.12) (17,18). While **53** and **54** showed enantioselectivities similar to those of catalyst **52**, **55** displayed increased enantioselectivity to the extent that, in a number of substrates, even its dichloride version (**55a**) could be used at very low catalyst loadings (<1 mol%) to afford high enantiomeric excesses and conversions (17). Thus, **52c** and **55a** were utilized to ring-close alkenyl ether- and silyl ether prochiral trienes, affording five- to seven-membered rings; conversions up to greater than 98% and enantiomeric excesses up to 92% were obtained. The influence of solvent and temperature on conversion and enantiomeric excess was also studied.

Grubbs and coworkers also found that precatalysts **52–56** (Figs 14.11 and 14.12) are highly active in asymmetric ring-opening cross-metathesis (AROCM) reactions (18). **54a** was reported to be the most selective of dichloride catalysts **52–56**, whereas the use of diiodide catalyst **54b** slightly improved the enantiomeric excess values (up to 82% in the AROCM of norbornene derivatives with styrene). Nevertheless, since diiodide catalysts are generally less reactive than their dichloride counterparts, the loading of **54b** had to be increased to 3 mol% to achieve activity similar to that observed with 1 mol% loading of **54a**. A ruthenium benzylidene, rather than a ruthenium methylidene as the propagating species, along with a trans coordination pathway, was proposed to be operative in these AROCM reactions. Moreover, in the same work, the first examples of asymmetric cross-metathesis (ACM) reactions, the most challenging of the asymmetric metathesis transformations, were reported (18). The enantiomeric excess values obtained in the ACM reactions

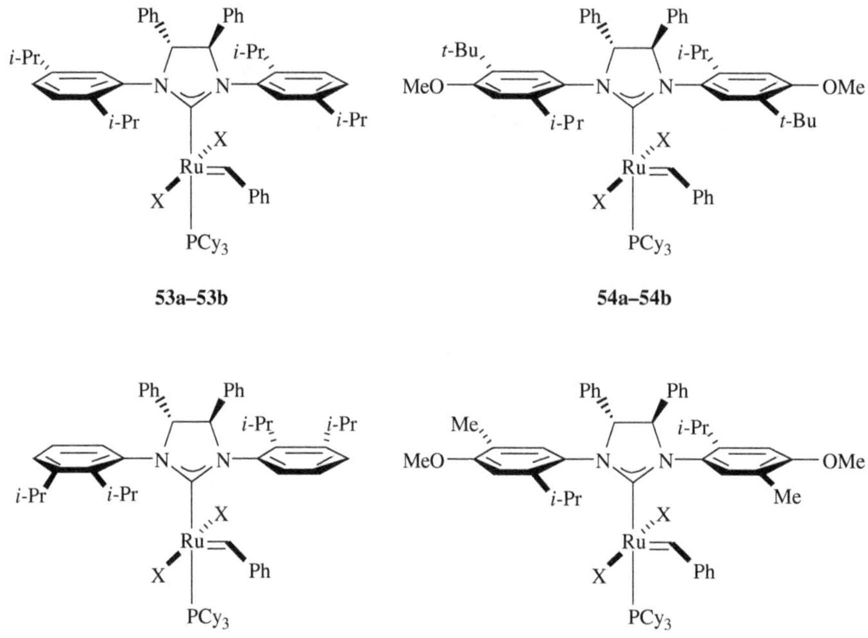

53a–53b 54a–54b

55a–55b a : X = Cl b : X = I 56a–56b

Figure 14.12 Ruthenium-based complexes **53–56** bearing chiral monodentate NHC ligands.

of protected 1,4-pentadien-3-ols, hexa-1,5-diene-3,4-diol, or hepta-1,6-diene-3,5-diol with *cis*-1,4-diacetoxy-2-butene ranged from 37% to 52%.

Collins and coworkers (20,71,72) prepared another family of ruthenium catalysts coordinated with chiral monodentate NHC ligands (**57–60**, Fig. 14.13). Instead of the 1,2-diphenyl enantioinducting backbone of the NHCs in **19**, **50**, and **52–56** (Figs 14.11 and 14.12), complexes **57–60** encompass a bulky 1,2-di-*tert*-butyl unit. Furthermore, the NHCs in **57–60** are unsymmetrical, having one *N*-aryl and one *N*-methyl substituent adjacent to the carbenic center. **57–60** showed high reactivities and high enantiomeric excess values in representative ARCM reactions, without the use of halide additives. It was also proposed that the barriers to the rotation of the NHC ligands in **57–60** play a significant role in determining the reactivity of the corresponding catalysts.

In 2008, the synthesis of complexes **61–64** (Fig. 14.14) was reported by the groups of Buchmeiser, Blechert, and Grisi (73,74). The catalytic activity of **61** was not investigated extensively, as it was essentially prepared as the precursor of the corresponding phosphine-free, pyridine-coordinated complex, utilized in alternating copolymerizations (73). **62–64**, bearing saturated NHCs with two nonaromatic N-substituents, were found to promote RCM, ARCM, CM, and ROMP reactions (74). In these benchmark ROMP and CM reactions, **62–64** showed activities between that of the 'first- and the second-generation Grubbs catalysts' (Fig. 14.1, catalysts **2** and **3**, respectively), with **63** being the most catalytically active complex in the series. In the CM of allyl benzene with *cis*-1,4-diacetoxy-2-butene, **62–64** afforded improved *E/Z* ratios toward *Z* double bond formation. It should be noted that **62**, which does not have chirality in the backbone of the NHC, was completely unable to give enantioinduction in ARCM, whereas, in the same transformation, **64** afforded modest enantioselectivities (33%). This difference in reactivity was a key observation and highlights the significance of chiral substitution on the backbone and the minor role of the chiral N-substituents in chirality induction with this catalyst type.

In 2011, Blechert and coworkers (23) reported the synthesis of a more rigid chiral NHC ligand along with the corresponding tricyclohexylphosphine (**65**, Fig. 14.15) and isopropoxystyrene Ru catalysts. In a series of AROCM of norbornenes with allyltrimethylsilane, **65** afforded enantioselectivities up to 87% (at −10 °C). A clear solvent influence was observed on both the enantiomeric excess- and the *E/Z*-ratios of the products, with CH_2Cl_2 giving the highest enantioselectivities for the reactions studied. As anticipated, the enantiomeric excess increased at low temperatures. Catalyst **65** outperformed the related chiral catalyst **54a** (Fig. 14.12) in terms of enantioselectivity. This was proposed to occur due to a more rigid chiral reaction pocket formed in the case of **65**, in contrast to the one provided in **54a** by its partially rotatable C_2-symmetric NHC.

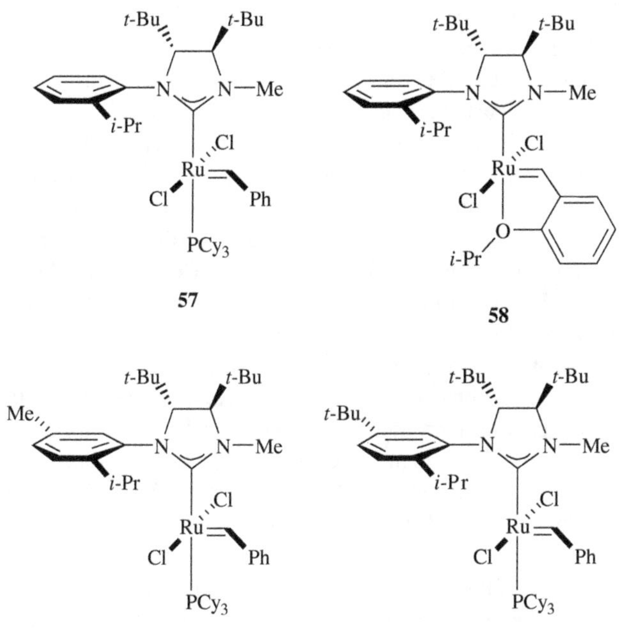

57 58 59 60

Figure 14.13 Ruthenium-based complexes **57–60** bearing chiral monodentate NHC ligands.

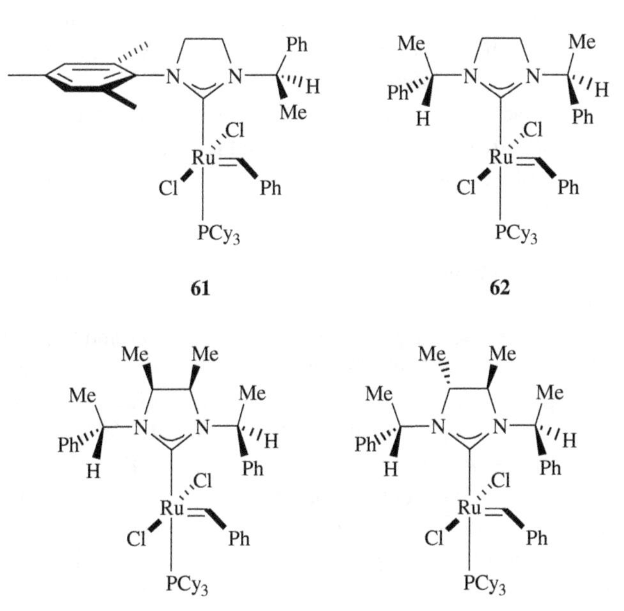

61 62 63 64

Figure 14.14 Complexes **61–64** coordinated with chiral monodentate NHCs.

With regard to phosphine-containing NHC ruthenium catalysts, it should be noted that complexes which are different only in their phosphine ligand provide the same propagating species upon phosphine dissociation (refer to

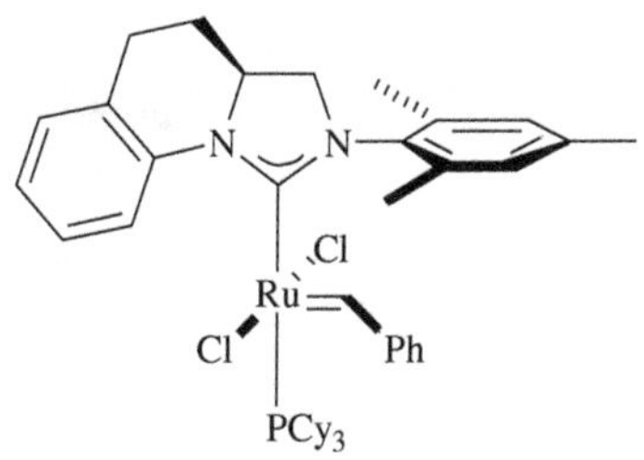

65

Figure 14.15 Enantioselective ruthenium precatalyst **65** bearing a rigid chiral NHC ligand.

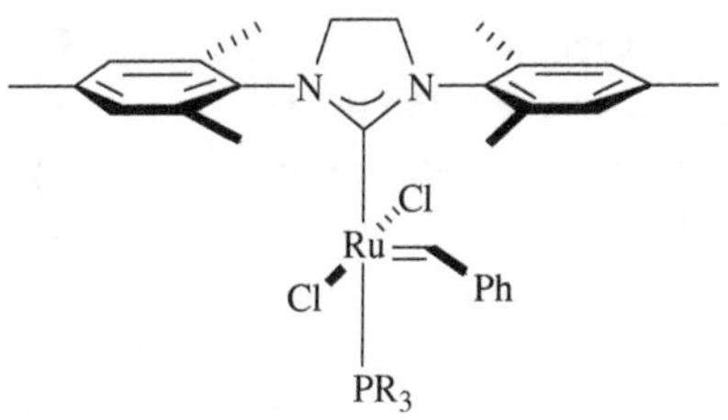

66 : R = Me
67 : R = *n*-Bu
3 : R = Cyclohexyl
68 : R = Benzyl
69 : $PR_3 = P(Ph)_2(OMe)$
70 : R = Ph

71 : $R = 4\text{-}FC_6H_4$
72 : $R = 4\text{-}ClC_6H_4$
73 : $R = 4\text{-}CF_3\text{-}C_6H_4$
74 : $R = 4\text{-}MeC_6H_4$
75 : $R = 4\text{-}MeOC_6H_4$

Figure 14.16 Phosphine-coordinated ruthenium catalysts **3** and **66–75**.

Section 14.5). Consequently, by varying the phosphine, one can manipulate initiation and phosphine rebinding without changing the metathesis ability of the catalyst. Accordingly, phosphine-containing complexes **66–75** (Fig. 14.16) were prepared by the groups of Grubbs and Sponsler (7,9,75,76) for a systematic study of the effect of different phosphine ligands. The data obtained for complexes **70–75** (bearing phosphine ligands with the same cone angle) by either magnetization transfer experiments or by the stoichiometric initiation with ethyl vinyl ether (7,75), revealed the existence of a linear free energy relationship between the phosphine dissociation rate constant and the Hammett constant σ_p (i.e., phosphine σ-donor strength), with the more electron-rich phosphines dissociating at slower rates than electron-poor ones (75). Moreover, arylphosphine dissociation was generally found to be faster than alkylphosphine dissociation. Thus, initiation in H_2IMes phosphine-containing catalysts can be easily adjusted by tuning phosphine electronics. On the other hand, phosphine reassociation showed no direct correlation with phosphine electronics. In addition to electronics, the steric properties of phosphine ligands have a major impact on

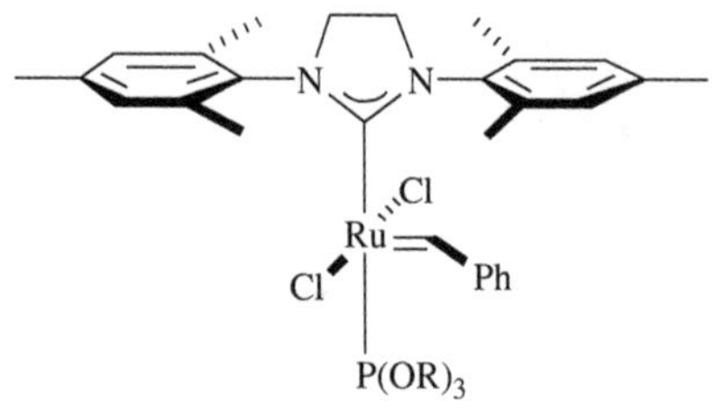

76 : R = Et

77 : R = *i*-Pr

Figure 14.17 Phosphite-containing ruthenium precatalyst **76** and **77**.

phosphine dissociation, and thus complex **66** was essentially metathesis inactive at room temperature (76).

Two phosphite-containing ruthenium benzylidene precatalysts (**76** and **77**, Fig. 14.17) were further reported in 2011 by Cazin and coworkers (77). In the solid state, **76** and **77** present a slightly shorter Ru–P bond compared to the parent complex **3** (Fig. 14.16). This suggests stronger Ru–P bonds in phosphite-containing complexes, obviously due to the π-acidic nature of the phosphites. The catalytic activity of **76** and **77** was evaluated in a variety of RCM and CM reactions at low catalyst loadings. Catalyst **77**, featuring the $P(Oi\text{-}Pr)_3$ ligand, slightly outperformed parent precatalyst **3** in most of these reactions.

14.2.3 Third Generation Catalysts

Further modifications of the ligand environment in heterocyclic carbene-coordinated ruthenium complexes by Grubbs and coworkers led to the synthesis of catalysts **78–81** (Fig. 14.18), bearing one or two pyridine ligands. Because of the lability of pyridine ligands, these complexes can be used as versatile starting materials for the synthesis of other NHC-coordinated ruthenium complexes (9,51,75). Complexes **78**, **80**, and **81** can be easily prepared and purified on a multigram scale, requiring little or no solvent, by simply adding an excess of the appropriate pyridine to **3** (Fig. 14.16) (9,10). As shown in Figure 14.18, pyridine ligands bind in a cis geometry, occupying the coordination sites trans to the benzylidene and the NHC ligands. In the solid-state structure of **78**, the Ru–N bond of the pyridine located trans to the benzylidene is more than 0.15 Å longer than the Ru–N bond of the pyridine positioned trans to the NHC, indicating that the benzylidene exerts a significantly larger trans influence than the NHC (9). **78** reacts instantaneously with PCy_3 to regenerate the parent complex **3** and with NaI to afford mono(pyridine) complex **82** (Fig. 14.18). Mono(pyridine) complex **79** can be prepared from **78** upon loss of one pyridine ligand under vacuum.

Even more importantly, bis(pyridine) ruthenium benzylidenes were proved to be efficient in the challenging CM of acrylonitrile, and are among the fastest-initiating ruthenium systems studied thus far (10). Specifically, the initiation rate in the irreversible reaction of complex **80** with ethyl vinyl ether is at least six orders of magnitude higher than the corresponding initiation rate of second-generation catalyst **3**. This very fast initiation of catalyst **80** has proved to be extremely useful in the production of polymers with very narrow polydispersity and for the synthesis of block copolymers (78).

The procedure for the preparation of complexes **83–89** (Fig. 14.19), reported by the groups of He and Buchmeiser (79–82), is analogous to that of **78** and **80**. The isolation of mono(pyridine)-coordinated **83**, instead of the expected bis(pyridine) complex, was rationalized on the basis of the increased steric bulk and donor ability of the ancillary six-membered NHC ligand, compared to the five-membered NHC in **78** (Fig. 14.18) (82). **83** was used in the ROMP of two enantiomerically pure norbornene derivatives. bis(Pyridine)-coordinated **84** was shown to be highly efficient in the CM of acrylonitrile with various functionalized alkenes (79,80). The activity of **84** decreases in coordinating solvents, whereas utilization of Lewis acids, which prevent the coordination of the cyano functionality with the ruthenium center, improves both the reaction rate and the yield of the CM reactions. In another more detailed study, mono(pyridine) complexes **85–88** were prepared from the corresponding triphenylphosphine-containing NHC-coordinated complex (81). Complex **85** can also be prepared from bis(pyridine)-coordinated **84** under vacuum and, conversely, **85** can be transformed into **84** in the presence of excess pyridine. The metathetic catalytic activity of complexes **84** and **85** was shown to be quite similar, while **87** was the most catalytically active complex studied, namely among complexes **84–89**, in the RCM of diallyl malononitrile and the CM of acrylonitrile with terminal olefins.

In related work by the groups of Buchmeiser and Blechert (73), ruthenium complex **90** (Fig. 14.20) was targeted as a precatalyst with improved initiation efficiency compared to its tricyclohexylphosphine-containing counterpart. Indeed, **90** was shown to be an efficient ROMP initiator in sequence-selective copolymerizations. This alternating ROMP selectivity by **90** was attributed to the steric interaction of the 2-phenethyl substituent of the NHC with the growing polymer chain. Moreover, Schanz and coworkers (32) reported the synthesis of the two pH-responsive ruthenium benzylidene catalysts (**91** and **92**) shown in Figure 14.20. Both catalysts promoted the well-controlled ROMP of two cationic functionalized norbornene monomers in various acidic alcoholic and aqueous media. The lowest polydispersities (1.05), which actually are the lowest polydispersity values reported for ROMP reactions conducted in water-containing media, were accomplished in 2,2,2-trifluoroethanol/0.1 M HCl_{aq} (1/1, v/v). Unfortunately, for slow monomer conversions, noticeable catalyst

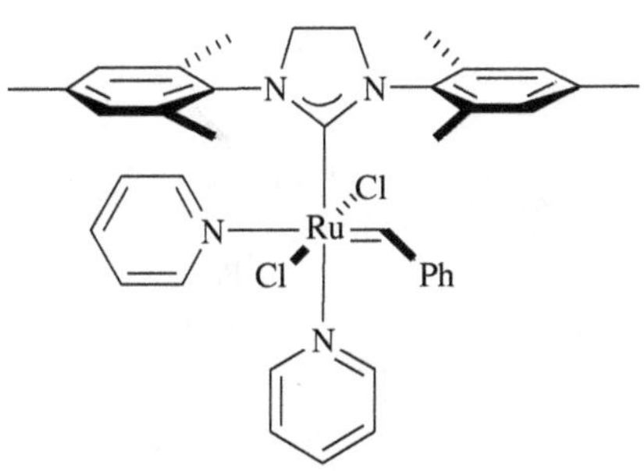

78

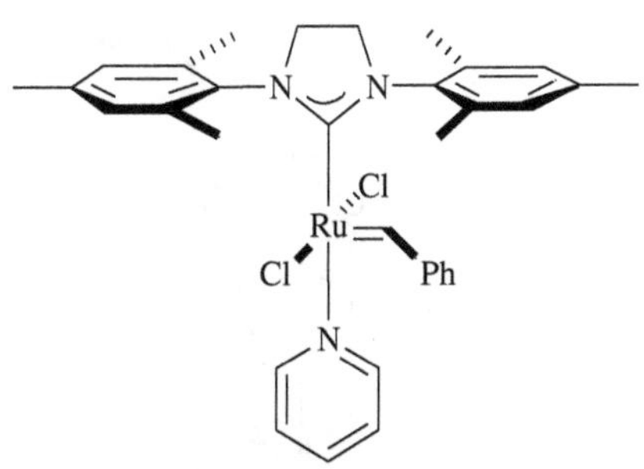

79

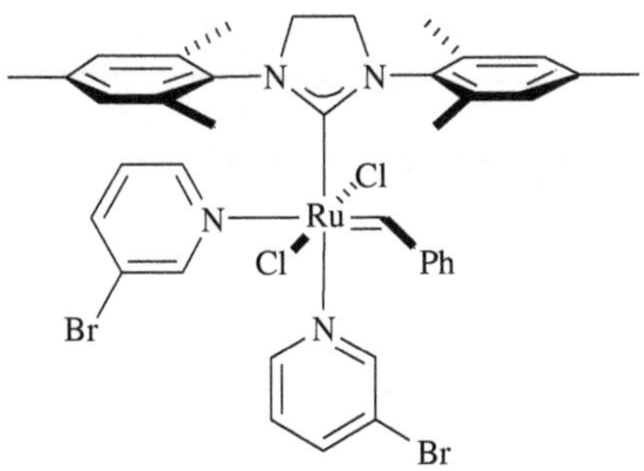

80

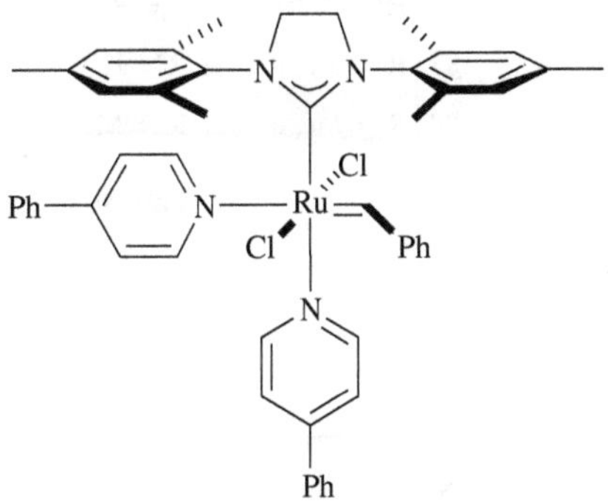

81

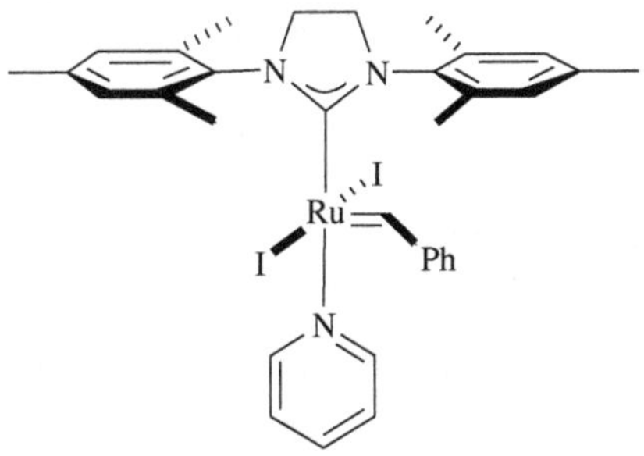

82

Figure 14.18 Pyridine-coordinated H_2IMes complexes **78–82**.

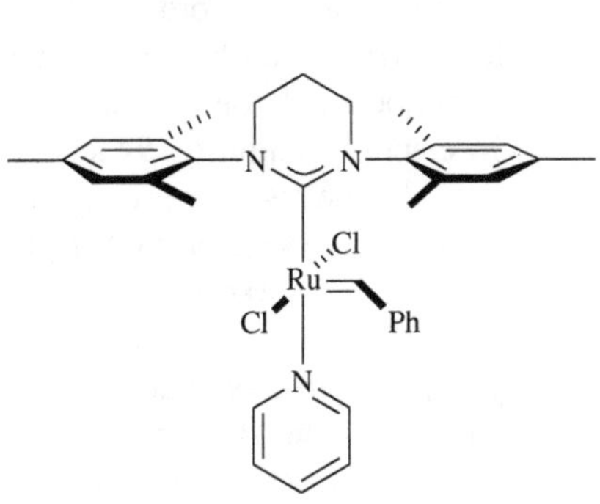

83

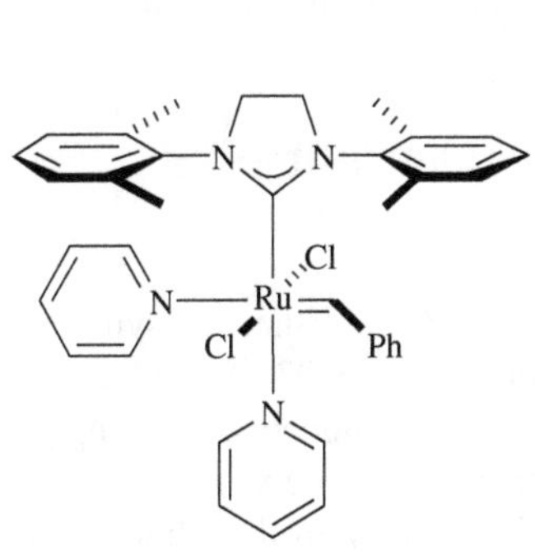

84

Figure 14.19 NHC-coordinated ruthenium complexes **83–89**.

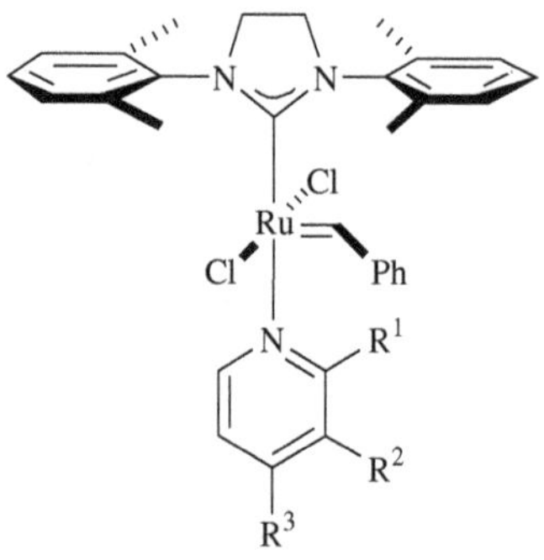

85 : R^1 = H, R^2 = H, R^3 = H

86 : R^1 = H, R^2 = Br, R^3 = H

87 : R^1 = Me, R^2 = H, R^3 = H

88 : R^1 = Me, R^2 = H, R^3 = Me

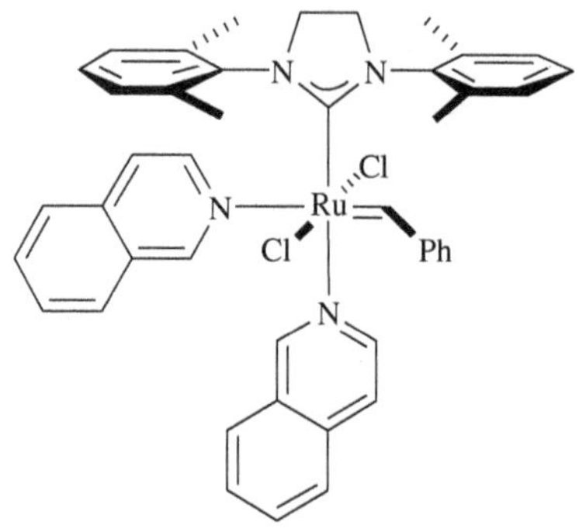

89

Figure 14.19 (*Continued*).

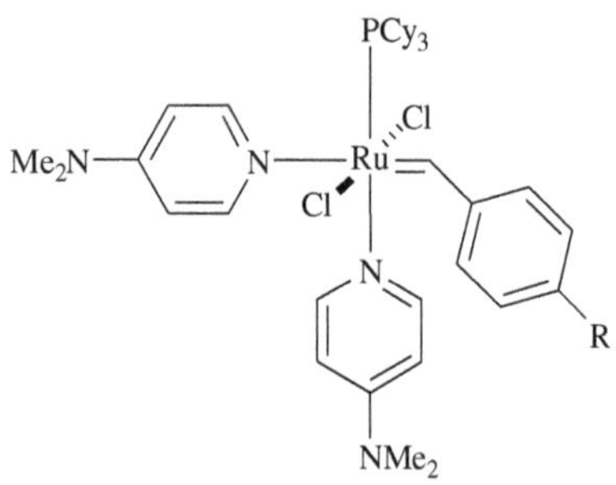

91 : R = CH_2NMe_2

92 : R = NMe_2

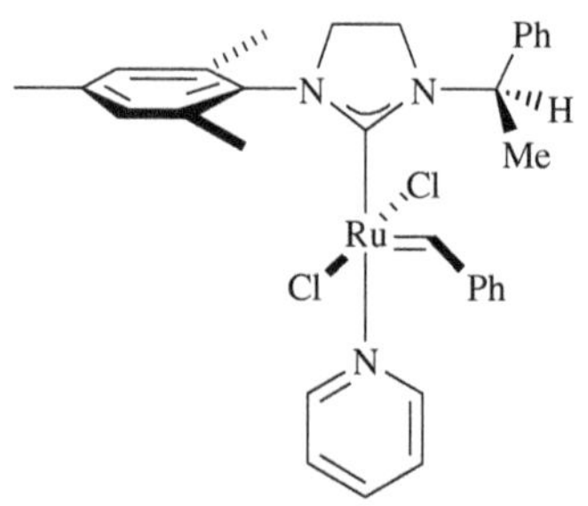

90

Figure 14.20 The sequence-selective copolymerization catalyst **90** and the pH-responsive ruthenium catalysts **91** and **92**.

degradation was also observed. Furthermore, in purely acidic aqueous media, both catalysts show limited solubility, slow overall reaction rates, and significant catalyst decomposition.

14.3 14-ELECTRON PHOSPHONIUM ALKYLIDENES

In 2004, Piers and coworkers (83) published the synthesis of NHC-coordinated ruthenium complex **93** (Fig. 14.21) bearing a 14-electron phosphonium alkylidene. Surprisingly, the four-coordinate complex **93**, which models the presumed active species formed upon dissociation of the labile ligand in NHC-coordinated catalysts, is air- and moisture-stable and, furthermore, highly active in model RCM reactions. More importantly, this system provides rapid metathesis initiation, outperforming even bis(3-bromopyridine) complex **80** (Fig. 14.18). Initiation in these 14-electron phosphonium alkylidenes is more energetically favorable than in the classic five- or six-coordinated systems, as it consists of a low-barrier olefin-binding event without the need for a ligand to dissociate. Soon after the synthesis of **93**, the analogous 14-electron complexes **94** and **95** (Fig. 14.21) were also prepared by the research groups of Grubbs and Piers (84,85).

In a more recent elegant kinetic and thermodynamic study, Piers and coworkers (86) shed some more light on the nature of the complex initiation process in these 14-electron phosphonium alkylidenes. More specifically, they found that the role of the substituents at phosphorous is very important in defining the initiation processes in these systems. Bulky substituents stabilize the 14-electron compounds relative to both decomposition and (deactivating) dimerization pathways; on the other hand, exceedingly bulky substituents increase the barrier for the incoming olefin to coordinate,

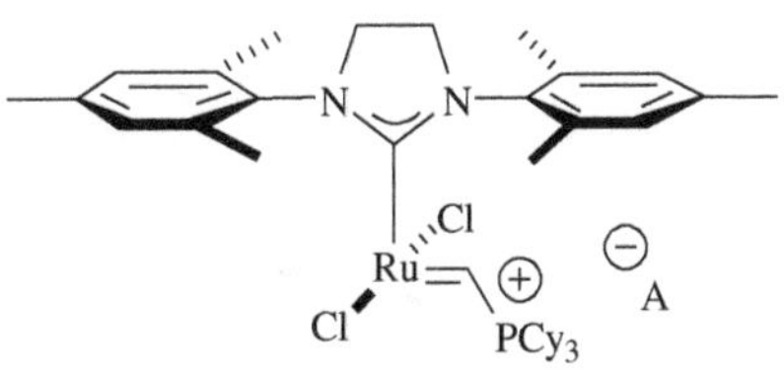

A = OTf, $B(C_6F_5)_4$

93

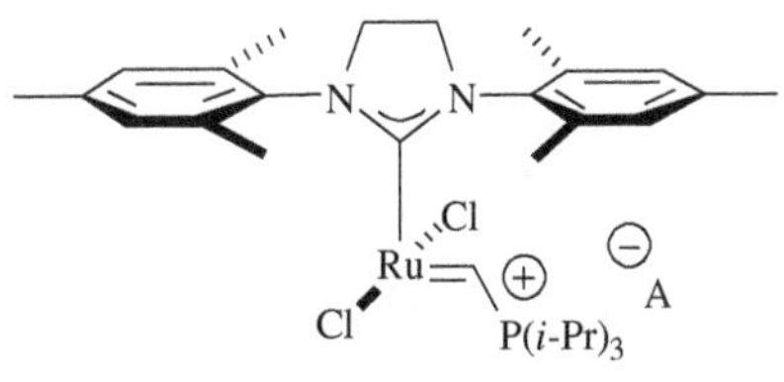

A = OTf, $ClB(C_6F_5)_3$

94

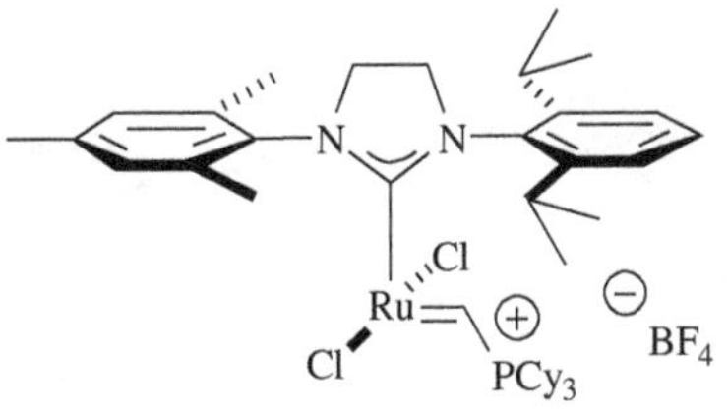

95

Figure 14.21 14-Electron NHC-coordinated ruthenium catalysts **93–95** bearing phosphonium alkylidenes.

slowing down initiation. Although the triisopropyl derivative **94** (Fig. 14.21) was found to strike the balance between steric bulk and activity most successfully, its behavior in solution in the absence of a substrate shows that it decomposes faster than the somewhat more thermally robust tricyclohexyl derivative **93**. Finally, it was also found that the initiation efficiency of these 14-electron phosphonium alkylidenes can be boosted by adding ethylene along with the substrate.

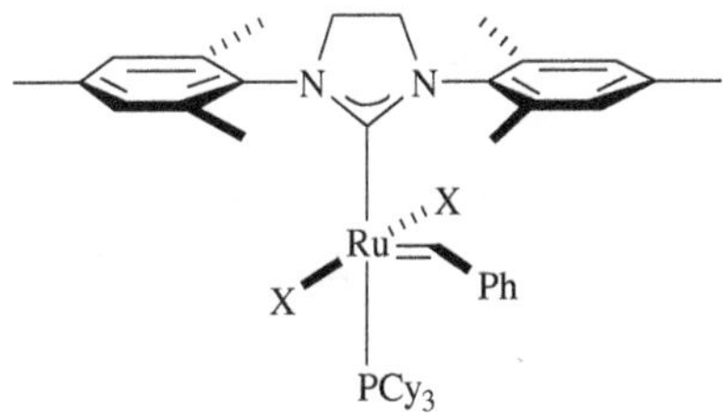

96 : X = Br

97 : X = I

Figure 14.22 Dibromide and diiodide ruthenium catalysts **96** and **97**.

14.4 ANIONIC LIGAND VARIATION

It has been already mentioned that changing the halide ligands from chlorides to iodides often improves the enantioselectivity of chiral ruthenium metathesis catalysts (Section 14.2.2). In the case of chiral monodentate NHCs, the diiodide complexes are prepared *in situ*, by dissolving the dichloride catalyst in tetrahydrofuran (THF) in the presence of NaI (12,17,18), whereas chiral bidentate iodide complexes are stable enough to be isolated and chromatographically purified (15,16). Additionally, halide ligands have been shown to have a significant impact on the initiation rates of second-generation catalysts (7). For example, Grubbs and coworkers (7) showed that dibromide **96** and diiodide **97** (Fig. 14.22) initiate 3 and 250 times faster than the dichloride parent complex **3** (Fig. 14.16), respectively. This initiation rate enhancement was predominantly attributed to the increased steric bulk of bromide or iodide ligands, as cis electronic effects (i.e., between the halide(s) and the phosphine ligand) are generally relatively small in dissociative ligand-substitution reactions. However, despite the increased initiation efficiency of **97**, its olefin metathesis activity is comparable to, or even lower than, that of the parent dichloride complex **3** because of slower turnover rates.

Motivated by the easily tunable steric and electronic properties of aryloxides, and in order to inhibit the formation of chloride-bridged ruthenium species that lead to catalyst decomposition, Fogg and coworkers (87–91) developed "pseudohalide" ruthenium catalysts **98–102** (Fig. 14.23). "Halide-free" **98**, the first NHC-coordinated complex of this type to be reported, proved to be a highly active metathesis catalyst, efficiently carrying out the ring-closing of model α,ω-dienes, even at very low catalyst loadings (87). Equally efficient metathesis catalysts **99** and **100** followed soon thereafter (88,92). Remarkably, **98–100** also showed very high affinity for silica gel, facilitating their efficient removal from the metathesized compound(s). For example, in the RCM of diethyl diallylmalonate using 5 mol% catalyst loading of **98–100**, followed by flash chromatography, ruthenium levels lower than 100 ppm in the product were achieved.

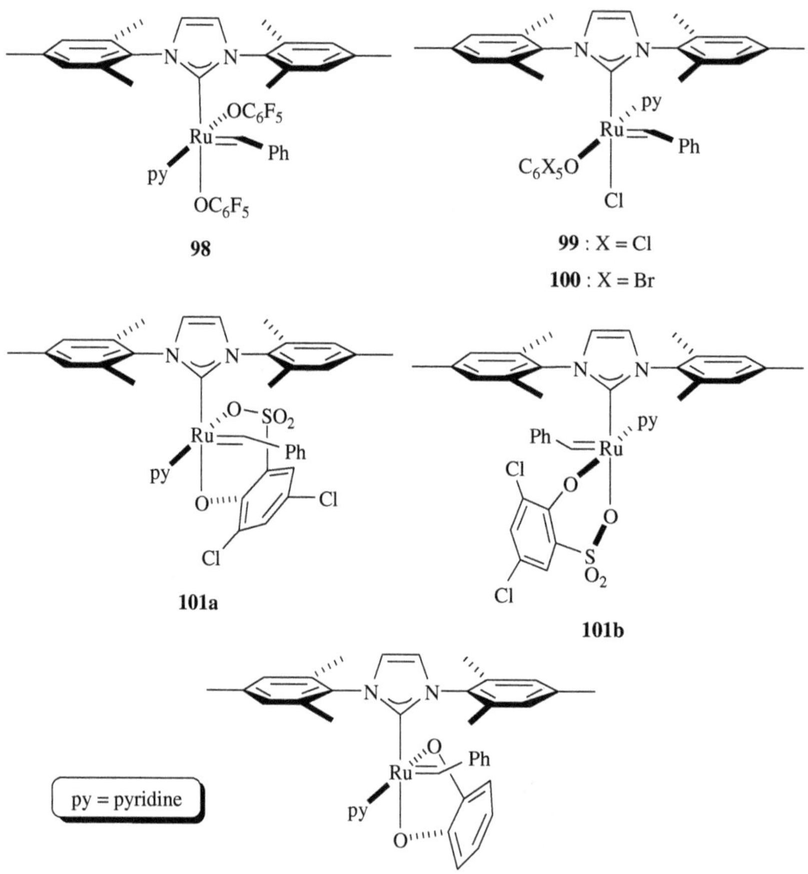

Figure 14.23 Ruthenium-based catalysts **98–102** bearing aryloxide ligands.

This is quite important as difficulties in removing residual ruthenium impose major drawbacks in multistep synthetic procedures that employ metathesis (2).

o-Sulfonato and catecholato aryloxide complexes **101** and **102** were also synthesized and evaluated by Fogg and Monfette (89). As can be seen in Figure 14.23, *o*-sulfonato complex **101** is isolated in the form of two isomers (**101a** and **101b**); while catecholato derivative **102** was shown to be a highly active RCM catalyst, the pyridine ligand in **101** is nonlabile and, consequently, **101** shows modest RCM activity. In a more recent study and in order to evaluate the influence of electronic effects in ruthenium-catalyzed metathesis, Fogg and coworkers (93) synthesized a whole family of ruthenium-catecholate catalysts with variable acidity. According to that work, the metathetical activity drops as the catecholate anionic ligands become more electron deficient. Finally, Fogg and coworkers (94) have also recently reported the synthesis and catalytic activity evaluation of a family of chiral, halide-free binaphtholate catalysts. Unfortunately, the accomplished enantioselectivities by these binol precatalysts in AROCM reactions were very low, suggesting that bulky substituents at the 3,3′-positions of the binaphtholate ligand may be necessary for significant enantioinduction to occur.

Many more types of anionic ligands have been successfully utilized in ruthenium-based metathesis catalysts. These include chelating pyridinyl alcoholates, Schiff base *N,O*-bidentate ligands, and perfluorosulfonates or perfluorocarboxylates (typically connected with the preparation of immobilized catalysts on solid supports). These kinds of ruthenium catalysts are extensively discussed in Chapters 17 and 20 to 24.

14.5 MECHANISM

The generally accepted mechanism for transition-metal-catalyzed olefin metathesis involves olefin binding to a metal alkylidene, metallacyclobutane formation, and subsequent generation of another olefin and metal alkylidene species. However, depending on its metal center and ligand environment, every catalyst has its own mechanism of action (i.e., initiation, olefin binding geometry, intermediates, etc.), the in-depth understanding of which enables the design of more efficient catalysts and metathesis transformations. In the following two sections, we examine the basic mechanistic features of ruthenium catalysts' metathetical action. Given that the related theoretical work is discussed in Chapter 19, only studies related to experimental work are included in this section.

14.5.1 Initiation

The reasons for the increased activity of the "second generation Grubbs catalysts," compared to their first-generation analogs, have been investigated both experimentally and theoretically. Initially, it was assumed that the higher activity of the second-generation catalysts originates from the higher electronic trans-influence of the NHCs, compared to the phosphine ligands, that was in turn expected to lead to a lower barrier to phosphine dissociation (faster initiation). However, a series of mechanistic studies from the Grubbs group (7,95), showed that this is not the case. Thus, whereas both the first-generation (**2**) and second-generation (**3**) tricyclohexylphosphine-containing catalysts were found to initiate via a dissociative mechanism (Scheme 14.1), the initiation rate (k_1) of **2** was measured to be two orders of magnitude higher than that of **3**. Nevertheless, the overall catalytic activity of **3** was found to be about two orders of magnitude higher than that of **2**. To account for these contradictory observations, it was proposed that the partitioning (k_2/k_{-1}) between the coordination of the alkene substrate (k_2) and the phosphine ligand (k_{-1}, return to the resting state of the catalyst), by the corresponding 14-electron ruthenium intermediates (**103**, Scheme 14.1), is about four orders of magnitude greater for **3** relative to **2** (7,95). Therefore, the increased activity of the second-generation catalyst(s) was rationalized on the basis of an increased affinity of the NHC-substituted ruthenium center for π-acidic olefins relative to σ-donating phosphines. It should be noted that improvement of initiation alone does not necessarily lead to a better catalyst, as catalyst efficiency depends on initiation, phosphine rebinding, and reaction of the 14-electron ruthenium intermediate with olefin, as well as the rate of catalyst decomposition.

The catalytic mechanism of the 'Hoveyda–Grubbs precatalysts' is slightly different from that discussed above for the 'Grubbs-type ones'. Initially, 14-electron intermediate **104** is formed through the dissociation of the benzylidene ether chelating group (Scheme 14.2). Coordination of the alkene substrate, followed by metathesis, leads to the formation of the catalytically active species **105** and a molecule of isopropoxystyrene (or a related derivative) (8,35,96,97). Therefore, the initial precatalyst exists in equilibrium with the catalytically active species, and when the olefin is completely consumed, the catalyst may return to its resting state by rebinding the isopropoxystyrene that was eliminated at the first step (8). Also note that both the 'Grubbs-type' (Scheme 14.1) and the 'Hoveyda–Grubbs-type' precatalysts (Scheme 14.2) provide the same propagating species (**105**,

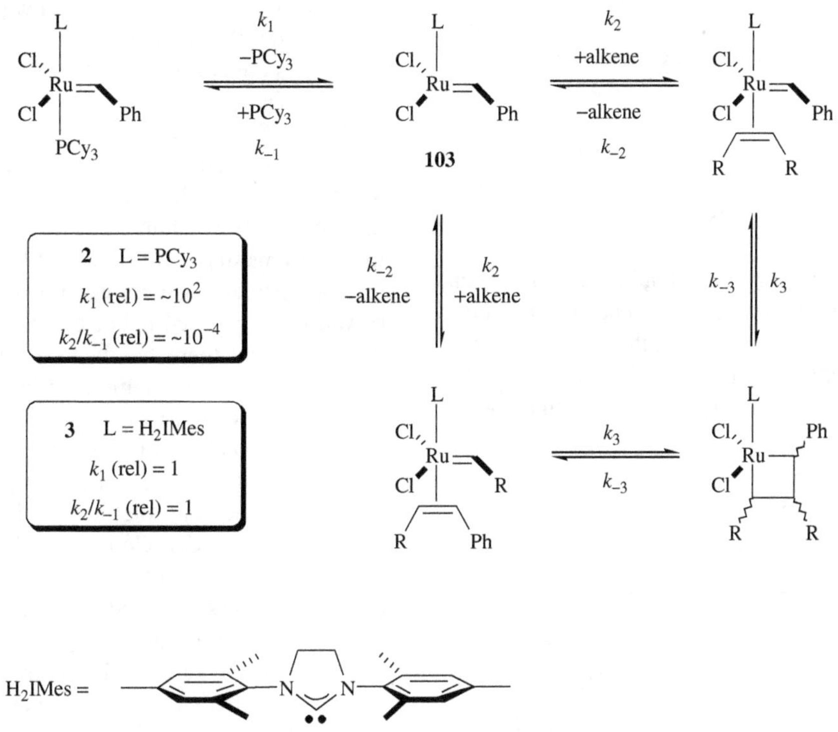

Scheme 14.1 The proposed catalytic mechanism of 'Grubbs first and second generation catalysts'.

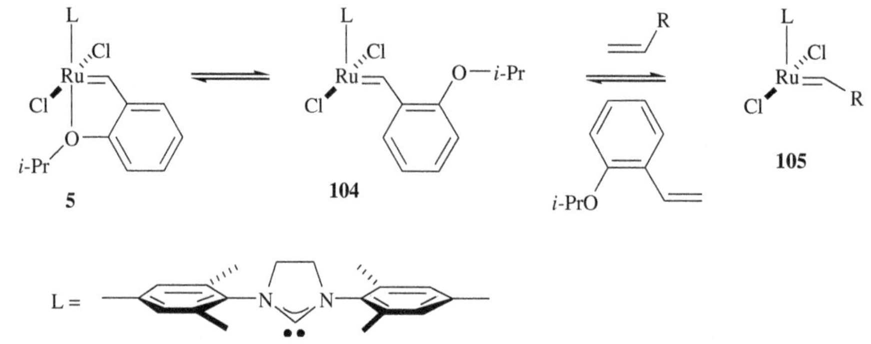

Scheme 14.2 Catalytic mechanism of isopropoxystyrene-coordinated ruthenium catalysts.

Scheme 14.2) after a single turnover. For a more detailed discussion of the mechanism of the chelating benzylidene ether ruthenium complexes ('Hoveyda–Grubbs-type' pre-catalysts) refer to Chapters 16 and 19.

14.5.2 Olefin Binding Geometry

Based on the ligand effects and the stereochemical outcome in a series of enantioselective olefin metathesis studies published by the Grubbs group in 2001, a bottom-face (trans to the NHC) olefin binding pathway (intermediate **106** in Fig. 14.24) was excluded. Among the two side-on (cis to the NHC) olefin binding pathways in Figure 14.24, intermediate **108** was favored, although **107** was not excluded (12). On the contrary, theoretical work published in 2004 by Costabile and Cavallo suggested bottom-face olefin binding (**106**, Fig. 14.24), given that the other two intermediates (**107** and **108**) were calculated to be of remarkably higher energy (98).

Besides being very efficient metathesis catalysts, 14-electron phosphonium alkylidenes (Section 14.3) initiate at very low temperatures. This has proved useful in a series of low-temperature mechanistic studies, carried out by the Grubbs and the Piers groups (84,85,99,100), which resulted in the direct observation of ruthenacyclobutane intermediates relevant to olefin metathesis. Given that ruthenacyclobutanes are known to play a key role in the determination of the regio- and stereochemical outcome of

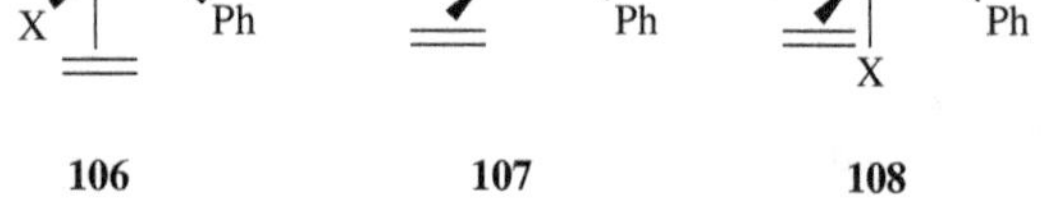

Figure 14.24 Possible geometries of the intermediate olefin complex.

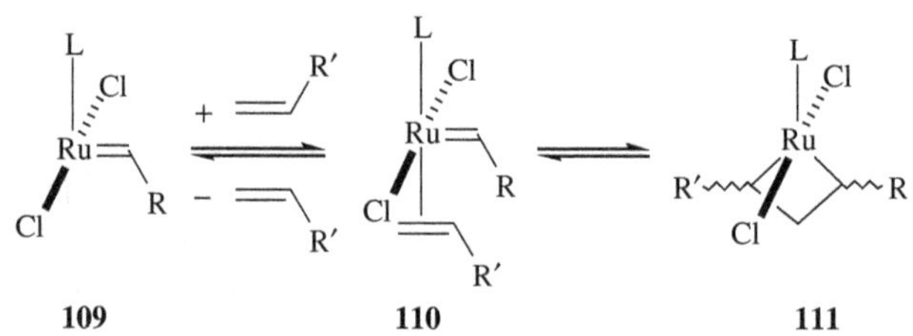

Scheme 14.3 Initial steps proposed for the bottom-face pathway in the ruthenium-catalyzed olefin metathesis mechanism.

metathesis, a better understanding of their geometry is essential to the rational design of diastereo- and enantioselective catalysts. These studies are suggestive of bottom-face olefin coordination and metallacycle generation, that is, trans to the NHC ligand. Spectroscopic data for all reported ruthenacyclobutanes are consistent with a symmetric structure with a flat, kite-shaped four-membered ring (species **111**, Scheme 14.3) (84,85,99,100).

Ruthenacyclobutanes show a dynamic structure, proceeding through a series of nonproductive metallacycle formations/cycloreversions prior to olefin exchange. However, most systems studied to date are relatively simple and, therefore, one should be very careful in generalizing these observations to more complicated ruthenacyclobutane species. Also note that there has been a long-standing debate regarding the site of olefin coordination to the ruthenium catalyst that leads to ruthenacyclobutane formation (side- or bottom-bound). Thus, in 1997, Snapper and coworkers (101) reported the isolation of complex **112** (Scheme 14.4) in which a chelating olefin coordinates trans to the PCy_3 ligand (bottom-face). Complexes **113a** and **113b** (Scheme 14.4) were subsequently isolated by Grubbs and coworkers (102), suggesting a side-bound olefin intermediate. In a similar vein, the reaction between 1,2-divinylbenzene and a series of NHC-coordinated ruthenium complexes led to the formation of two types of side-bound olefin adducts (**114** and **115**, Scheme 14.4) that undergo dynamic interconversion (68,103,104).

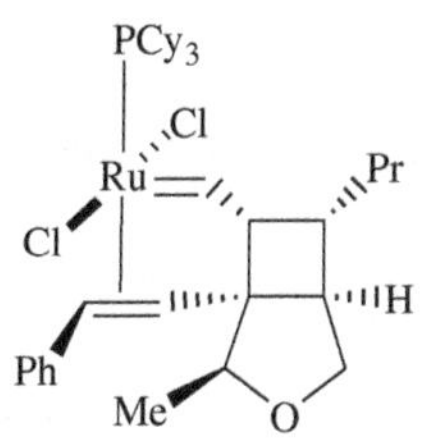

112

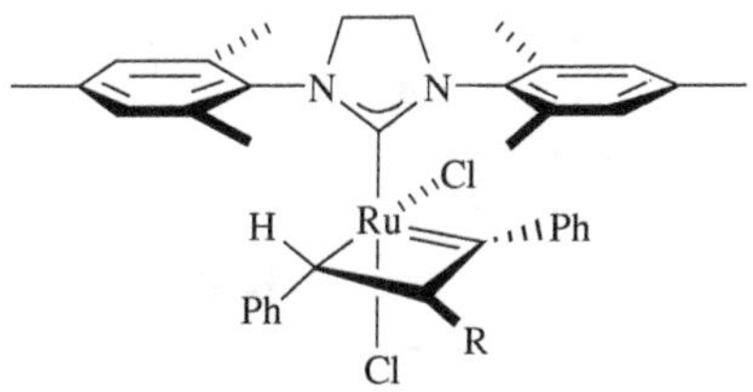

113a : R = Me

113b : R = Ph

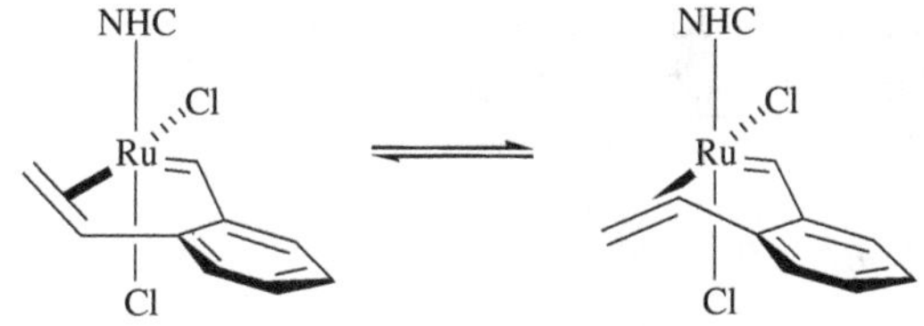

114 115

Scheme 14.4 Side-on and bottom-face bound ruthenium olefinic complexes.

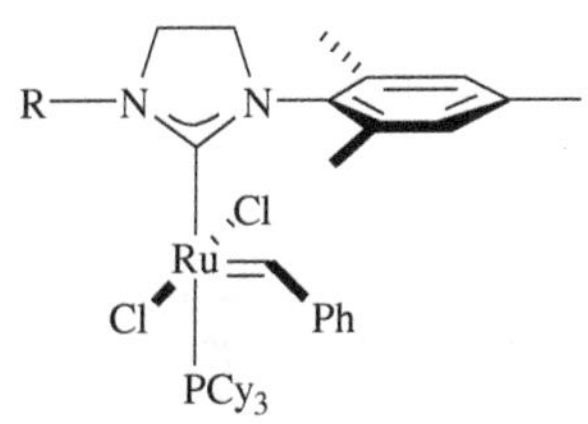

116 : R = Me

117 : R = Et

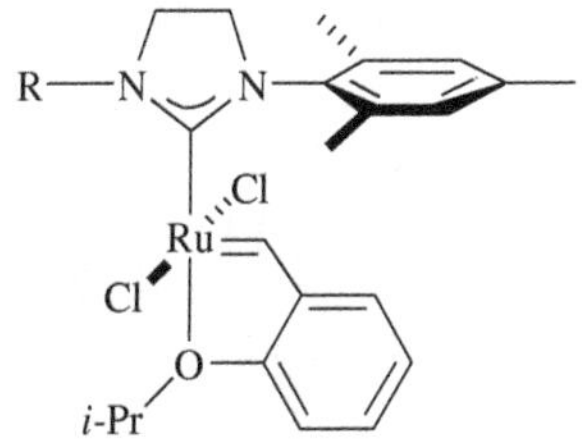

118 : R = Me

119 : R = Et

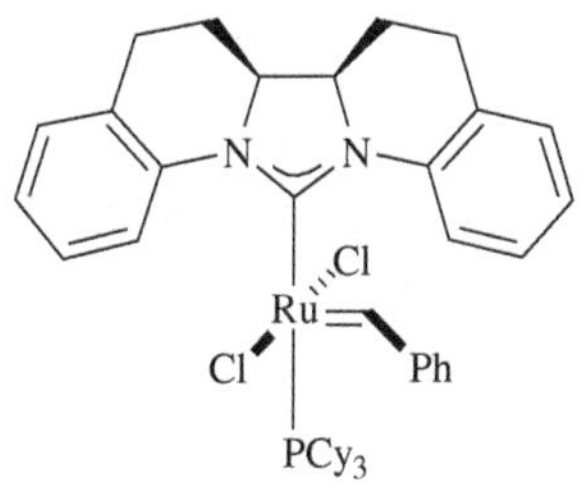

120

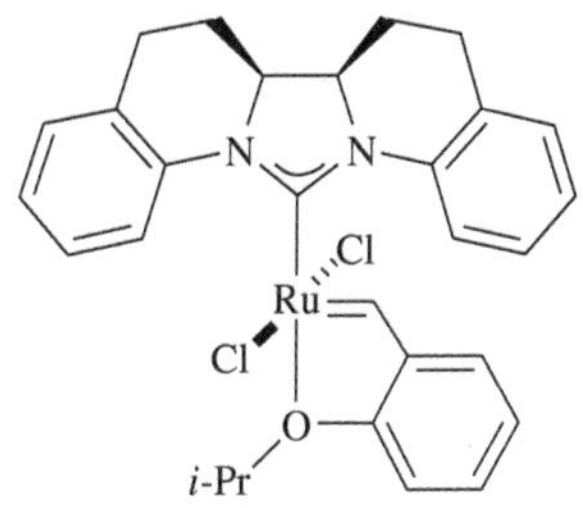

121

Figure 14.25 Ruthenium-based precatalysts **116–121**.

14.6 *E/Z* SELECTIVITY

The majority of olefin metathesis catalysts prefer to afford the thermodynamically favored *E*-olefin. Also, secondary metathesis often promotes isomerization of the product(s) to the favored *E* isomer(s). As a result, the synthesis of *Z*- or *cis*-olefins via metathesis constitutes one of the most persistent problems in the field.

In this context, the design of complexes **116–119** (Fig. 14.25) by Blechert and coworkers (105) was based on the anticipation that the unsymmetrical nature of their NHC ligands might alter the environment of key intermediates in the metathesis pathway, leading to improved *E/Z* selectivity in CM reactions and diastereoselectivity in RCM reactions. In a model RCM reaction, complexes **116–119** showed activities similar to those of their parent bis(mesityl)-substituted complexes **3** and **5** (Fig. 14.1). Moreover, **116** and **118** showed significantly different *E/Z* ratios in selected CM transformations and improved selectivities in a diastereoselective RCM reaction compared to the parent complexes **3** and **5**. In another report from the same group, the synthesis of ruthenium-based catalysts **120** and **121** (Fig. 14.25) was described (106). Complexes **120** and **121** were initially targeted with the aim of increasing the diastereoselectivity of ring-rearrangement metathesis reactions. Although both **120** and **121** were found to be of limited stability in solution, even in the absence of olefin substrates, **120** showed some promising results in diastereoselective ring-rearrangement metathesis reactions, affording improved *E/Z* selectivities.

It was only recently that the research groups of Schrock and Hoveyda (107) reported highly *Z*-selective Mo and W metathesis catalysts (for a detailed discussion, refer to Chapter 13). Soon thereafter, Grubbs and coworkers (38–40,108,109) synthesized and evaluated a series of functional group-tolerant, ruthenium-based catalysts with similar levels of activity and selectivity. The key structural element of this family of catalysts is a five-membered chelating NHC ligand connected with the ruthenium center via the carbenic center and a C–H activated *N*-adamantyl substituent. Given that these ruthenium complexes are all "Hoveyda–Grubbs-type" catalysts, they are discussed in Chapter 16.

14.7 CONCLUSIONS AND PERSPECTIVES

The development of highly active and functional group-tolerant ruthenium-based metathesis catalysts has led to the rapidly growing use of this transformation in many different fields of chemical synthesis. The fact that these catalysts can be handled without rigorous air and moisture exclusion, sometimes even in the open air, made them extremely popular and allowed for the synthesis of new pharmaceutical compounds, polymeric materials, and fine chemicals. On the basis of the high numbers of new ruthenium catalysts that appear in the literature every year, further exciting advancements in the field are soon to come.

ACKNOWLEDGMENTS

I am grateful to Professor Robert H. Grubbs for introducing me to the field of olefin metathesis and for being a wise advisor and mentor. I also thankfully acknowledge funding from a Foundation for Education and European Culture scholarship.

REFERENCES

1. Nguyen ST, Johnson LK, Grubbs RH, Ziller JW. *J Am Chem Soc* 1992;114:3974–3975.
2. (a) Astruc D. *New J Chem* 2005;29:42–56. (b) Schrodi Y, Pederson RL. *Aldrichimica Acta* 2007;40:45–52. (c) Colacino E, Martinez J, Lamaty F. *Coord Chem Rev* 2007;251:726–764. (d) Dragutan V, Dragutan I, Delaude L, Demonceau A. *Coord Chem Rev* 2007;251:765–794. (e) Deshmukh PH, Blechert S. *Dalton Trans* 2007:2479–2491. (f) Samojłowicz C, Bieniek M, Grela K. *Chem Rev* 2009;109:3708–3742. (g) Diesendruck CE, Tzur E, Lemcoff NG. *Eur J Inorg Chem* 2009:4185–4203. (h) Vougioukalakis GC, Grubbs RH. *Chem Rev* 2010;110:1746–1787. (i) Anderson EB, Buchmeiser MR. *Synlett* 2012;23:185–207.
3. Vougioukalakis GC. *Chem Eur J* 2012;18:8868–8880.
4. Schwab P, France MB, Ziller JW, Grubbs RH. *Angew Chem Int Ed Engl* 1995;34:2039–2041.
5. Schwab P, Grubbs RH, Ziller JW. *J Am Chem Soc* 1996;118:100–110.
6. Scholl M, Ding S, Lee CW, Grubbs RH. *Org Lett* 1999;1:953–956.
7. Sanford MS, Love JA, Grubbs RH. *J Am Chem Soc* 2001;123:6543–6554.
8. (a) Kingsbury JS, Harrity JPA, Bonitatebus PJ Jr, Hoveyda AH. *J Am Chem Soc* 1999;121:791–799. (b) Garber SB, Kingsbury JS, Gray BL, Hoveyda AH. *J Am Chem Soc* 2000;122:8168–8179.
9. Sanford MS, Love JA, Grubbs RH. *Organometallics* 2001;20:5314–5318.
10. Love JA, Morgan JP, Trnka TM, Grubbs RH. *Angew Chem Int Ed* 2002;41:4035–4037.
11. (a) Grela K, Harutyunyan S, Michrowska A. *Angew Chem Int Ed* 2002;41:4038–4040. (b) Michrowska A, Bujok R, Harutyunyan S, Sashuk V, Dolgonos G, Grela K. *J Am Chem Soc* 2004;126:9318–9325.
12. Seiders TJ, Ward DW, Grubbs RH. *Org Lett* 2001;3:3225–3228.
13. Van Veldhuizen JJ, Garber SB, Kingsbury JS, Hoveyda AH. *J Am Chem Soc* 2002;124:4954–4955.
14. Van Veldhuizen JJ, Gillingham DG, Garber SB, Kataoka O, Hoveyda AH. *J Am Chem Soc* 2003;125:12502–12508.
15. Gillingham DG, Kataoka O, Garber SB, Hoveyda AH. *J Am Chem Soc* 2004;126:12288–12290.
16. Van Veldhuizen JJ, Campbell JE, Giudici RE, Hoveyda AH. *J Am Chem Soc* 2005;127:6877–6882.
17. Funk TW, Berlin JM, Grubbs RH. *J Am Chem Soc* 2006;128:1840–1846.
18. Berlin JM, Goldberg SD, Grubbs RH. *Angew Chem Int Ed* 2006;45:7591–7595.
19. Giudici RE, Hoveyda AH. *J Am Chem Soc* 2007;129:3824–3825.
20. Fournier P -A, Savoie J, Stenne B, Bedard M, Grandbois A, Collins SK. *Chem Eur J* 2008;14:8690–8695.
21. Savoie J, Stenne B, Collins SK. *Adv Synth Catal* 2009;351:1826–1832.
22. Tiede S, Berger A, Schlesiger D, Rost D, Luhl A, Blechert S. *Angew Chem Int Ed* 2010;49:3972–3975.
23. Kannenberg A, Rost D, Eibauer S, Tiede S, Blechert S. *Angew Chem Int Ed* 2011;50:3299–3302.
24. Lynn DM, Mohr B, Grubbs RH, Henling LM, Day MW. *J Am Chem Soc* 2000;122:6601–6609.
25. Lynn DM, Grubbs RH. *J Am Chem Soc* 2001;123: 3187–3193.
26. Rölle T, Grubbs RH. *Chem Commun* 2002:1070–1071.
27. Gallivan JP, Jordan JP, Grubbs RH. *Tetrahedron Lett* 2005;46:2577–2580.
28. Hong SH, Grubbs RH. *J Am Chem Soc* 2006;128: 3508–3509.
29. Binder JB, Guzei IA, Raines RT. *Adv Synth Catal* 2007; 349:395–404.
30. Gułajski Ł, Śledź P, Lupa A, Grela K. *Green Chem* 2008; 10:271–274.
31. Gawin R, Czarnecka P, Grela K. *Tetrahedron* 2010;66:1051–1056.
32. Dunbar MA, Balof SL, Roberts AN, Valente EJ, Schanz H -J. *Organometallics* 2011;30:199–203.
33. Berlin JM, Campbell K, Ritter T, Funk TW, Chlenov A, Grubbs RH. *Org Lett* 2007;9:1339–1342.
34. Stewart IC, Ung T, Pletnev AA, Berlin JM, Grubbs RH, Schrodi Y. *Org Lett* 2007;9:1589–1592.
35. Vougioukalakis GC, Grubbs RH. *Chem Eur J* 2008; 14:7545–7556.
36. Vougioukalakis GC, Grubbs RH. *J Am Chem Soc* 2008; 130:2234–2245.
37. Teo P, Grubbs RH. *Organometallics* 2010;29:6045–6050.
38. Endo K, Grubbs RH. *J Am Chem Soc* 2011;133:8525–8527.
39. Keitz BK, Endo K, Herbert MB, Grubbs RH. *J Am Chem Soc* 2011;133:9686–9688.
40. Keitz BK, Endo K, Patel PR, Herbert MB, Grubbs RH. *J Am Chem Soc* 2012;134:693–699.
41. Ung T, Hejl A, Grubbs RH, Schrodi Y. *Organometallics* 2004;23:5399–5401.
42. Diesendruck CE, Vidavsky Y, Ben-Asuly A, Lemcoff NG. *J Polym Sci A: Polym Chem* 2009;47:4209–4213.
43. Monsaert S, Ledoux N, Drozdzak R, Verpoort F. *J Polym Sci A: Polym Chem* 2010;48:302–310.
44. Samec JSM, Keitz BK, Grubbs RH. *J Organomet Chem* 2010;695:1831–1837.

45. Nguyen ST, Grubbs RH, Ziller JW. *J Am Chem Soc* 1993;115:9858–9859.
46. Fürstner A, Ackermann L, Gabor B, Goddard R, Lehmann CW, Mynott R, Stelzer F, Thiel OR. *Chem Eur J* 2001;7:3236–3253.
47. Yun J, Marinez ER, Grubbs RH. *Organometallics* 2004;23:4172–4173.
48. Despagnet-Ayoub E, Grubbs RH. *Organometallics* 2005;24:338–340.
49. Anderson DR, Lavallo V, O'Leary DJ, Bertrand G, Grubbs RH. *Angew Chem Int Ed* 2007;46:7262–7265.
50. Weskamp T, Schattenmann WC, Spiegler M, Herrmann WA. *Angew Chem Int Ed* 1998;37:2490–2493.
51. Trnka TM, Morgan JP, Sanford MS, Wilhelm TE, Scholl M, Choi T-L, Ding S, Day MW, Grubbs RH. *J Am Chem Soc* 2003;125:2546–2558.
52. Weskamp T, Kohl FJ, Hieringer W, Gleich D, Herrmann WA. *Angew Chem Int Ed* 1999;38:2416–2419.
53. Scholl M, Trnka TM, Morgan JP, Grubbs RH. *Tetrahedron Lett* 1999;40:2247–2250.
54. Huang J, Stevens ED, Nolan SP, Petersen JL. *J Am Chem Soc* 1999;121:2674–2678.
55. Jafarpour L, Nolan SP. *Organometallics* 2000;19:2055–2057.
56. Weskamp T, Kohl FJ, Herrmann WA. *J Organomet Chem* 1999;582:362–365.
57. Huang J, Schanz H -J, Stevens ED, Nolan SP. *Organometallics* 1999;18:5375–5380.
58. Chatterjee AK, Morgan JP, Scholl M, Grubbs RH. *J Am Chem Soc* 2000;122:3783–3784.
59. Bielawski CW, Grubbs RH. *Angew Chem Int Ed* 2000;39:2903–2906.
60. Chatterjee AK, Grubbs RH. *Org Lett* 1999;1:1751–1753.
61. Jafarpour L, Hillier AC, Nolan SP. *Organometallics* 2002;21:442–444.
62. Jafarpour L, Stevens ED, Nolan SP. *J Organomet Chem* 2000;606:49–54.
63. Dinger MB, Mol JC. *Adv Synth Catal* 2002;344:671–677.
64. Courchay FC, Sworen JC, Wagener KB. *Macromolecules* 2003;36:8231–8239.
65. Ritter T, Hejl A, Wenzel AG, Funk TW, Grubbs RH. *Organometallics* 2006;25:5740–5745.
66. Ritter T, Day MW, Grubbs RH. *J Am Chem Soc* 2006;128:11768–11769.
67. Stewart IC, Douglas CJ, Grubbs RH. *Org Lett* 2008;10:441–444.
68. Stewart IC, Benitez D, O'Leary DJ, Tkatchouk E, Day MW, Goddard WA III, Grubbs RH. *J Am Chem Soc* 2009;131:1931–1938.
69. Leuthäußer S, Schmidts V, Thiele CM, Plenio H. *Chem Eur J* 2008;14:5465–5481.
70. Keitz BK, Bouffard J, Bertrand G, Grubbs RH. *J Am Chem Soc* 2011;133:8498–8501.
71. Fournier P-A, Collins SK. *Organometallics* 2007;26: 2945–2949.
72. Grandbois A, Collins SK. *Chem Eur J* 2008;14:9323–9329.
73. Vehlow K, Wang D, Buchmeiser MR, Blechert S. *Angew Chem Int Ed* 2008;47:2615–2618.
74. Grisi F, Costabile C, Gallo E, Mariconda A, Tedesco C, Longo P. *Organometallics* 2008;27:4649–4656.
75. Love JA, Sanford MS, Day MW, Grubbs RH. *J Am Chem Soc* 2003;125:10103–10109.
76. Bolton SL, Williams JE, Sponsler MB. *Organometallics* 2007;26:2485–2487.
77. Schmid TE, Bantreil X, Citadelle CA, Slawin AMZ, Cazin CSJ. *Chem Commun* 2011;47:7060–7062.
78. Choi T-L, Grubbs RH. *Angew Chem Int Ed* 2003;42:1743–1746.
79. Bai C-X, Lu X-B, He R, Zhang W-Z, Feng X-J. *Org Biomol Chem* 2005;3:4139–4142.
80. Bai C-X, Zhang W-Z, He R, Lu X -B, Zhang Z-Q. *Tetrahedron Lett* 2005;46:7225–7228.
81. Zhang W-Z, He R, Zhang R. *Eur J Inorg Chem* 2007:5345–5352.
82. Wang D, Yang L, Decker U, Findeisen M, Buchmeiser MR. *Macromol Rapid Commun* 2005;26:1757–1762.
83. Romero PE, Piers WE, McDonald R. *Angew Chem Int Ed* 2004;43:6161–6165.
84. Wenzel AG, Grubbs RH. *J Am Chem Soc* 2006;128:16048–16049.
85. Van der Eide EF, Romero PE, Piers WE. *J Am Chem Soc* 2008;130:4485–4491.
86. Leitao EM, Van der Eide EF, Romero PE, Piers WE, McDonald R. *J Am Chem Soc* 2010;132:2784–2794.
87. Conrad JC, Amoroso D, Czechura P, Yap GPA, Fogg DE. *Organometallics* 2003;22:3634–3636.
88. Conrad JC, Parnas HH, Snelgrove JL, Fogg DE. *J Am Chem Soc* 2005;127:11882–11883.
89. Monfette S, Fogg DE. *Organometallics* 2006;25:1940–1944.
90. Conrad JC, Snelgrove JL, Eeelman MD, Hall S, Fogg DE. *J Mol Catal A: Chem* 2006;254:105–110.
91. Conrad JC, Fogg DE. *Curr Org Chem* 2006;10:185–202.
92. Kotyk MW, Gorelsky SI, Conrad JC, Carra C, Fogg DE. *Organometallics* 2009;28:5424–5431.
93. Monfette S, Camm KD, Gorelsky SI, Fogg DE. *Organometallics* 2009;28:944–946.
94. Blacquiere JM, McDonald R, Fogg DE. *Angew Chem Int Ed* 2010;49:3807–3810.
95. Sanford MS, Ulman M, Grubbs RH. *J Am Chem Soc* 2001;123:749–750.
96. Kingsbury JS, Hoveyda AH. *J Am Chem Soc* 2005; 127:4510–4517.
97. Vougioukalakis GC, Grubbs RH. *Organometallics* 2007; 26:2469–2472.
98. Costabile C, Cavallo L. *J Am Chem Soc* 2004; 126:9592–9600.
99. Romero PE, Piers WE. *J Am Chem Soc* 2005;127: 5032–5033.

100. Romero PE, Piers WE. *J Am Chem Soc* 2007;129: 1698–1704.

101. Tallarico JA, Bonitatebus PJ Jr, Snapper ML. *J Am Chem Soc* 1997;119:7157–7158.

102. Trnka TM, Day MW, Grubbs RH. *Organometallics* 2001;20:3845–3847.

103. Anderson DR, Hickstein DD, O'Leary DJ, Grubbs RH. *J Am Chem Soc* 2006;128:8386–8387.

104. Anderson DR, O'Leary DJ, Grubbs RH. *Chem Eur J* 2008;14:7536–7544.

105. Vehlow K, Maechling S, Blechert S. *Organometallics* 2006;25:25–28.

106. Vehlow K, Gessler S, Blechert S. *Angew Chem Int Ed* 2007;46:8082–8085.

107. (a) Marinescu SC, Levine DS, Zhao Y, Schrock RR, Hoveyda AH. *J Am Chem Soc* 2011;133:11512–11514. (b) Marinescu SC, Schrock RR, Müller P, Takase MK, Hoveyda AH. *Organometallics* 2011;30:1780–1782. (c) Meek SJ, O'Brien RV, Llaveria J, Schrock RR, Hoveyda AH. *Nature* 2011;471:461–466.

108. Keitz BK, Fedorov A, Grubbs RH. *J Am Chem Soc* 2012;134:2040–2043.

109. Herbert MB, Lan Y, Keitz BK, Liu P, Endo K, Day MW, Houk KN, Grubbs RH. *J Am Chem Soc* 2012;134:7861–7866.

15

RUTHENIUM-INDENYLIDENE AND OTHER ALKYLIDENE CONTAINING OLEFIN METATHESIS CATALYSTS

César A. Urbina-Blanco, Stefano Guidone, Steven P. Nolan, and Catherine S. J. Cazin
School of Chemistry, University of St. Andrews, St Andrews, UK

One of the ground-breaking developments in olefin metathesis chemistry has been the discovery of well-defined ruthenium-alkylidene complexes (Fig. 15.1). These complexes address the functional group tolerance problems of earlier systems based on molybdenum or tungsten and, in addition, present high stability toward oxygen and water. Although benzylidene complexes are the most commonly used metathesis catalysts, several families of well-defined catalysts have been developed over the past 20 years. In the following sections, the most prominent non-benzylidene families are described.

15.1 Ru-INDENYLIDENE COMPLEXES

The chemistry revolving around ruthenium-indenylidene complexes is one of the fastest growing areas of olefin metathesis, and nowadays these catalysts represent an efficient alternative to the benzylidene congeners. This is because of their straightforward synthesis, enhanced stability under harsh reaction conditions (temperature and functional group tolerance) compared to their benzylidene counterparts, and the commercial availability of the first and later generation examples and easily derivatized Ru precursors.

The history of Ru-indenylidene complexes began when, after reacting 1,1-diphenyl-2-propyn-1-ol with $[RuCl_2(PPh_3)_3]$, Hill (1) reported the isolation of the first coordinatively unsaturated group 8 allenylidene complex **All-1** (Fig. 15.2). However, the NMR spectroscopic data proved complex and several groups hypothesized that the actual structure was different from that proposed by Hill.

It was not until Nolan (2) published the crystal structure of a 1,3-bis(2,6-diisopropylphenyl)imidazol-2-ylidene (IPr) derivative of **Ind-2**, $[RuCl_2(Ind)(IPr)(PCy_3)]$ (**Ind-3**, Ind, 3-phenylindenylid-1-ene) that the real structure of this family of complexes was established. The development of Ru-indenylidene complexes has paralleled the development of ruthenium benzylidene complexes, and the plethora of catalysts developed can be grouped in generations according to structural motifs.

15.1.1 First-Generation Ru-Indenylidene Complexes

The synthesis of **Ind-1** is currently carried out on an industrial scale in a high-yielding and reliable process; however, during the early days of Ru-indenylidene chemistry, the reproducibility of the synthesis was a major issue and, depending on the quality of the starting material $[RuCl_2(PPh_3)_3]$ (**1**) and reaction conditions (solvent, temperature), products of different quality and purity (even composition) were obtained.

Even though the first optimized synthesis for first-generation Ru-indenylidene complexes was reported by Fürstner in 2001 (3), a real answer to the reproducibility issues came several years later when Schanz reported a detailed mechanistic study of the indenylidene rearrangement. Schanz disclosed that the key to obtaining the desired **Ind-1**

Olefin Metathesis: Theory and Practice, First Edition. Edited by Karol Grela.

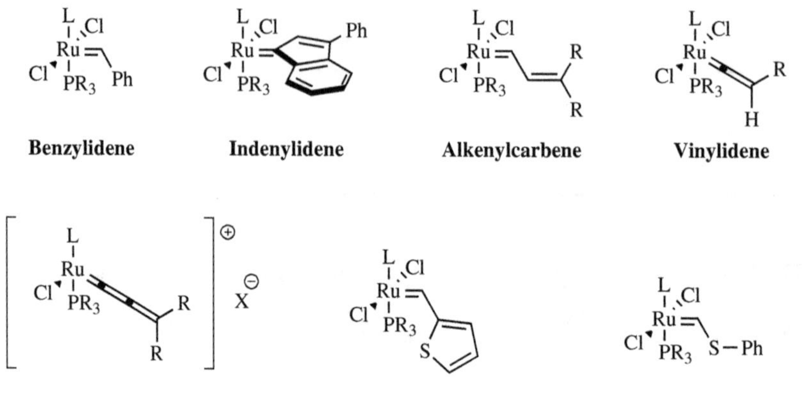

Figure 15.1 Well-defined ruthenium olefin metathesis pre-catalyst families.

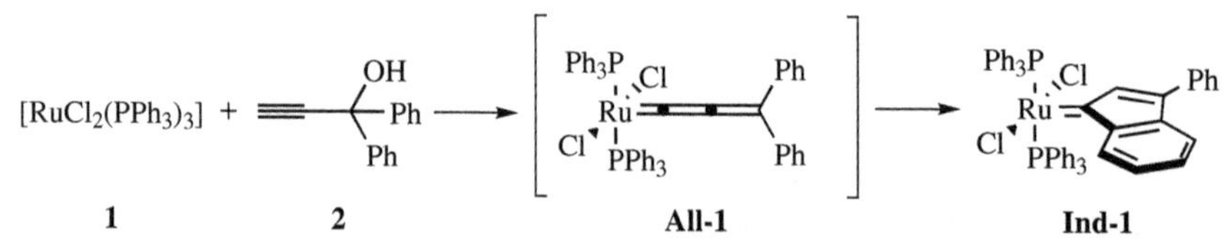

Figure 15.2 Synthesis of **Ind-1**.

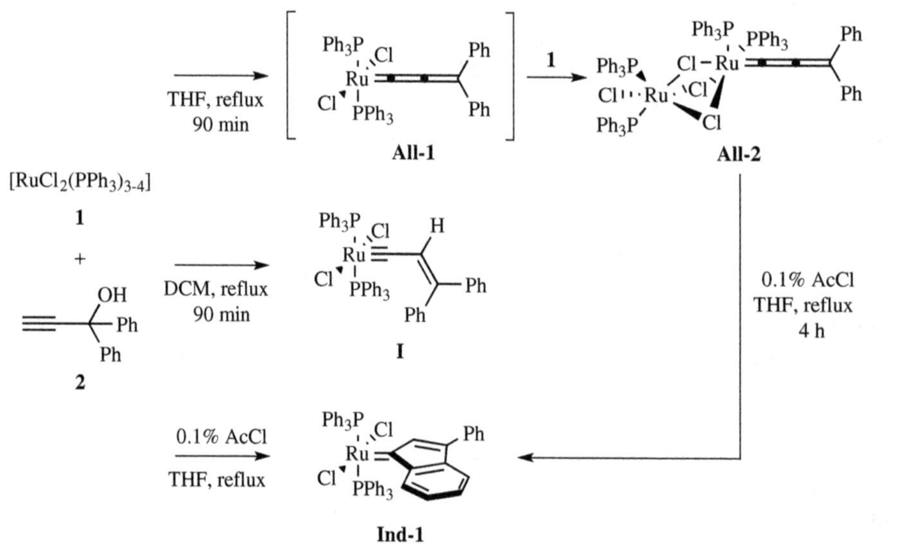

Figure 15.3 Multiple complexes obtained from the reaction of **1** and **2** under different reaction conditions.

lies in the addition of an acid source, such as acetyl chloride. The most common products obtained under different reaction conditions are presented in Figure 15.3.

The unusual rearrangement that takes place in the synthesis of **Ind-1** has been the subject of several studies. The proposed mechanism for the rearrangement that takes place in the formation of the indenylidene moiety is shown in Figure 15.4 (4). The first step is the formation of the allenylidene complex **All-1** that reacts rapidly with catalytic amounts of acid to form intermediate **I**. After reaction of **I** with tetrahydrofuran (THF) to form a cationic carbene species, the α-carbon atom in complex **II** is highly electrophilic. Therefore, this carbon atom is activated toward intramolecular nuclephilic attack by one of the benzene rings

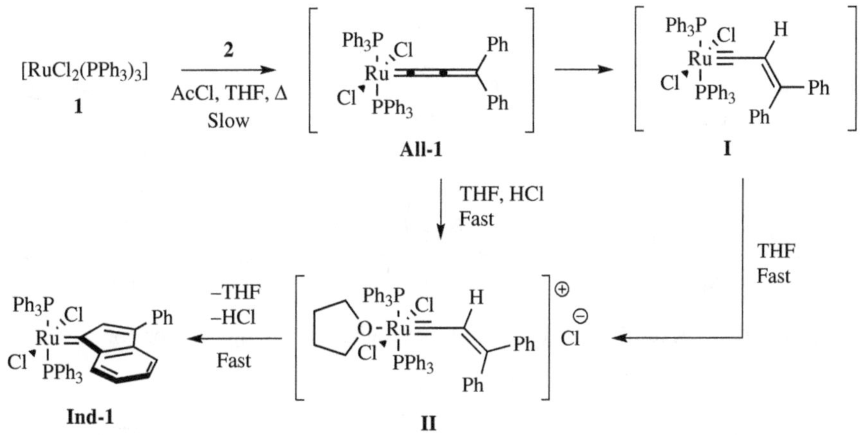

Figure 15.4 Proposed mechanism for the indenylidene formation.

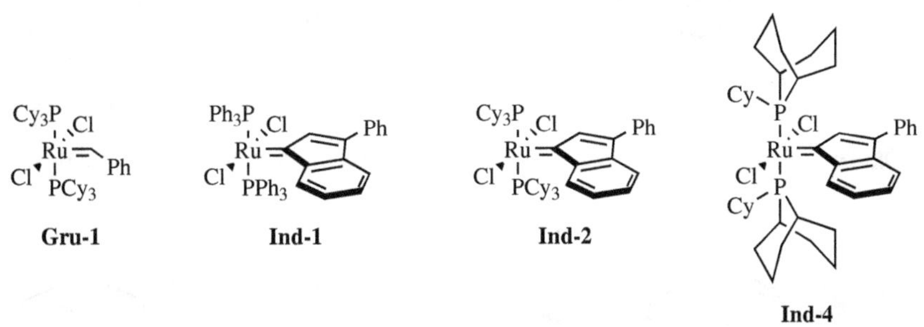

Figure 15.5 Commercially available first-generation complexes.

attached to C_γ to form the 3-phenylindenylidene moiety. Complexes **I** and **II** have been isolated and fully characterized by NMR spectroscopy and by X-ray single crystal diffraction studies (4).

As for the benzylidene first-generation catalyst, more active Ru-indenylidene pre-catalysts can be obtained by substituting triphenylphosphine for more electron-donating phosphines. Only two examples have been reported in the literature: **Ind-2** bearing tricyclohexylphosphine and **Ind-4** featuring cyclohexylphoban, reported by Sasol (5). All three first-generation catalysts are commercially available (Fig. 15.5).

First-generation indenylidene complexes have been evaluated in several metathesis transformations where they have exhibited, in general, similar reactivity to their benzylidene counterparts (6).

15.1.2 Second-Generation Ru-Indenylidene Complexes

Since the report by Nolan of the improved activity and stability of second-generation indenylidene complexes, when compared to their benzylidene counterparts, numerous groups have focused their research efforts on tuning the activity of second-generation catalysts by varying the *N*-heterocyclic carbene (NHC) moiety attached to the metal center (Fig. 15.6).

Second-generation catalysts are easily synthesized by reacting **Ind-1** or **Ind-2** with a free NHC under mild conditions (2,7). The final product is usually separated from the reaction mixture by precipitation with pentane or hexane and washing with similar solvents to remove the free phosphine released during the reaction (Fig. 15.7).

Even though the free carbene route is the most popular synthetic pathway to second-generation catalysts, several alternatives have been reported in the literature in which the free carbene is generated *in situ* by thermal decomposition of a NHC adduct (Fig. 15.8). The use of NHC adducts is more user-friendly than the free carbene protocol, as these can be manipulated under air; however, their use adds a step to the overall synthetic pathway as the NHC adducts are synthesized from the free carbenes themselves.

Verpoort (8) was the first to report the synthesis of second-generation complexes starting from chloroform adducts (**SIMes·HCCl$_3$**), obtaining complexes **Ind-8** and **Ind-9** in very good yields. The scope of this method was further expanded by the synthesis of **SIMe** containing complexes **Ind-10** and **Ind-11** also by Verpoort (9), and more recently by Delaude (10), who obtained a better yield for

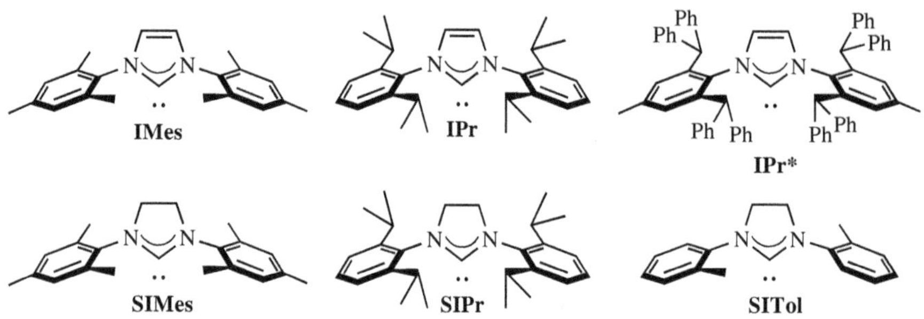

Figure 15.6 A few *N*-heterocyclic carbenes found in second-generation complexes.

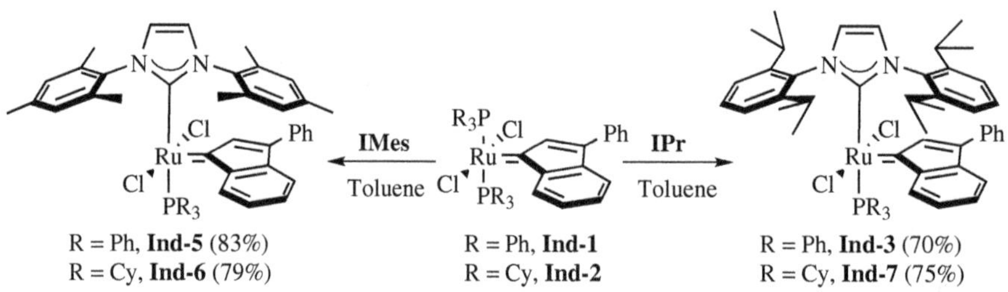

Figure 15.7 Synthesis of second-generation catalysts by reaction with a free carbene.

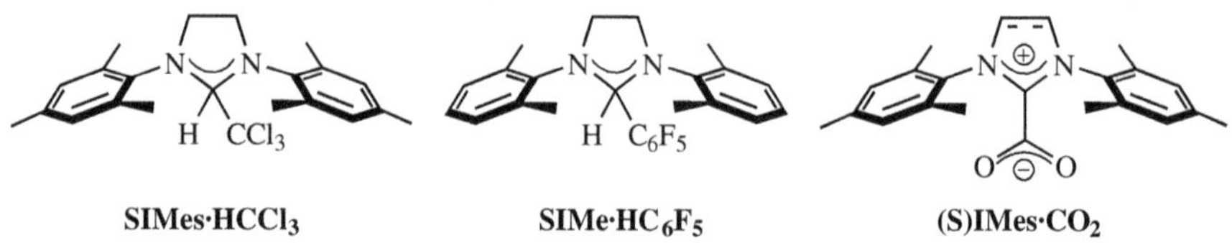

Figure 15.8 NHC adducts employed in the synthesis of second-generation catalysts.

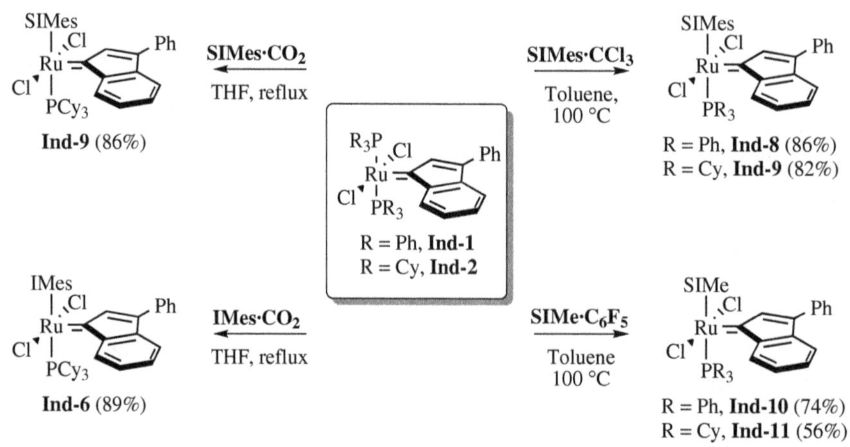

Figure 15.9 Synthesis of second-generation catalysts by reaction with NHC adducts.

the synthesis of **Ind-9** and **Ind-6** by using the corresponding **SIMes**·CO_2 and **IMes**·CO_2 adducts (Fig. 15.9).

The isolation of complexes bearing bulky NHCs such as **SIPr** proved more difficult than that of those bearing its unsaturated analog **IPr** due to the high solubility of the products (11), making it difficult to separate the complex from the released phosphine. In the case of **Ind-13**, analytically pure samples could be obtained only when the crude

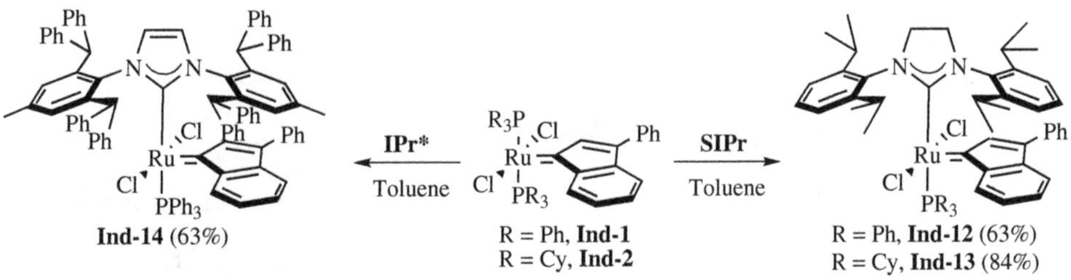

Figure 15.10 Synthesis of second-generation catalysts bearing bulky substituents.

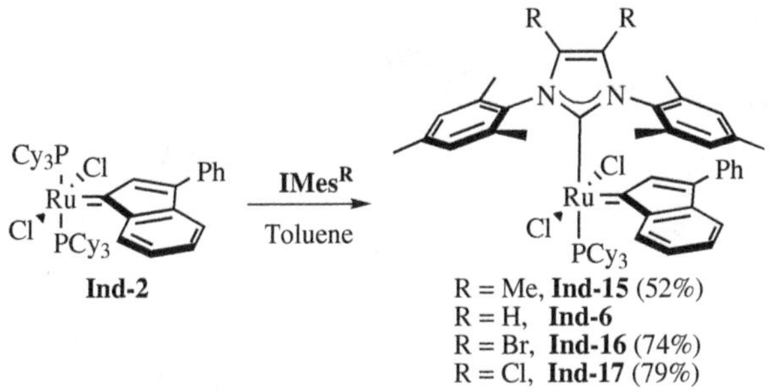

Figure 15.11 Synthesis of second-generation complexes bearing backbone-substituted NHCs.

reaction mixture was subjected to flash chromatography (11b). This problem was not encountered with **Ind-12** (11a), and although this complex is very soluble in most organic solvents, it can be cleanly obtained by washing the crude mixture with small amounts of cold pentane (Fig. 15.10).

Ind-12 and **Ind-13** are highly active pre-catalysts for the synthesis of di- and tri-substituted olefins at room temperature by ring-closing metathesis (RCM), enyne, and cross metathesis; however, they perform poorly in the synthesis of tetra-substituted olefins (11). A further increase in the steric bulk was detrimental to the activity toward the synthesis of trisubstituted olefins, as in the case of **IPr*** derivative **Ind-14** (12). However, the catalyst is most effective for the synthesis of di-substituted double bonds (12).

Overall, pre-catalysts containing bulkier NHCs exhibit higher initiation rates than their counterparts, which also leads to lower thermal stability of the complexes, so they are the best choice when fast initiation and short reaction times are required.

The effect of the electron-donating ability of the NHCs on the activity in RCM, enyne, and cross metathesis has also been studied in Ru-indenylidene complexes (Fig. 15.11) (13). Nolan published a series of complexes bearing IMes ligands featuring substituents in the backbone and concluded that more electron-withdrawing substituents are beneficial for the synthesis of tetra-substituted olefins. This observation was attributed to improved stability of the catalyst under the reaction conditions, derived from a slower initiation rate caused by the less electron-donating nature of the NHC (13).

Second-generation Ru-indenylidene catalysts can also be obtained by the reaction of third-generation catalysts with a phosphine (8,9,14). This synthetic protocol allows access to a wide range of complexes and has been widely used in the literature (7,8,13), especially to study the effect of the electronic properties of the phosphine in second-generation catalysts. Nolan (14) examined the catalytic activity of a series of complexes featuring SIMes and *para*-substituted triphenylphosphines (Fig. 15.12). Complexes were readily synthesized by the reaction of commercially available **Ind-18** with the desired tertiary phosphine, affording the complexes in very good yields (14).

Recently, other P-donor ligands such as phosphites have been studied by Cazin (15). The pyridine adduct **Ind-18** reacted with 1 equiv of triisopropylphosphite leading to the isolation of both the kinetic and the thermodynamic products **Ind-23** and **Ind-24** (the latter is also known as ***cis*-Caz-1**, Fig. 15.13).

Both complexes **Ind-23** and **Ind-24** are very active in olefin metathesis (15); however, their catalytic behaviors differ dramatically. While **Ind-23** is active at room temperature, **Ind-24** exhibits latent reactivity at temperatures below 40 °C. High conversions of several substrates were observed when using **Ind-24** at 80 and 110 °C in toluene at very low catalyst loadings. Indeed, **Ind-24** is one of the state-of-the-art pre-catalysts for the synthesis of tetra-substituted double bonds by RCM (15). The high activity of **Ind-24** comes from its unusual structure; it was proposed that during the course of the reaction the *cis*-species **Ind-24** isomerizes to the *trans*-isomer **Ind-23** and then undergoes olefin metathesis (15), which renders the *cis*-complex a stable reservoir of active species during the reaction.

The high stability of **Ind-24** derived from the *cis* geometry and a phosphite ligand allowed the isolation of its cationic version **Ind-25** (***cis*-Caz-1**$^+$) by reaction of **Ind-24** with silver hexafluoroantimonate (Fig. 15.14). Although, as with its predecessor, it requires thermal activation, **Ind-25** is the first cationic Ru-based complex proven to be highly active in various types of olefin metathesis reactions (16).

Another variation of Ru-indenylidene complexes that results in high activity toward the synthesis of tetra-substituted

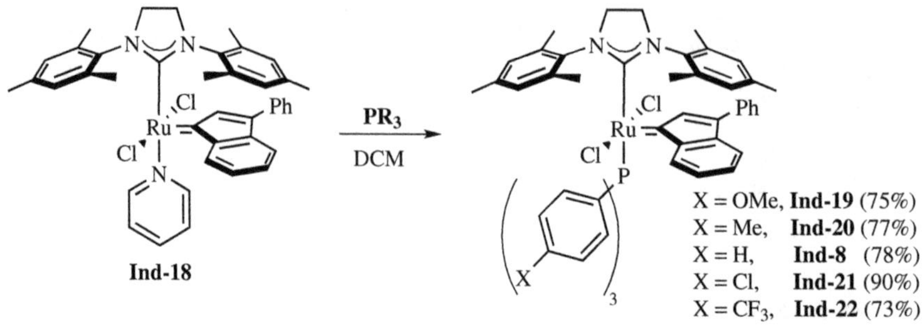

Figure 15.12 Synthesis of second-generation catalysts by reaction of a tertiary phosphine with a third generation catalyst.

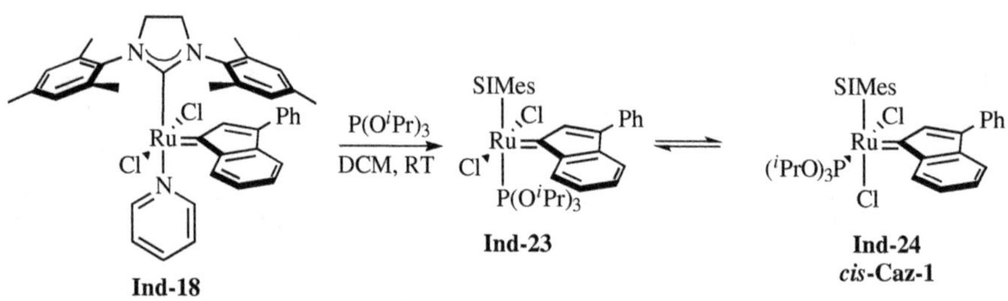

Figure 15.13 Synthesis of phosphite containing pre-catalysts.

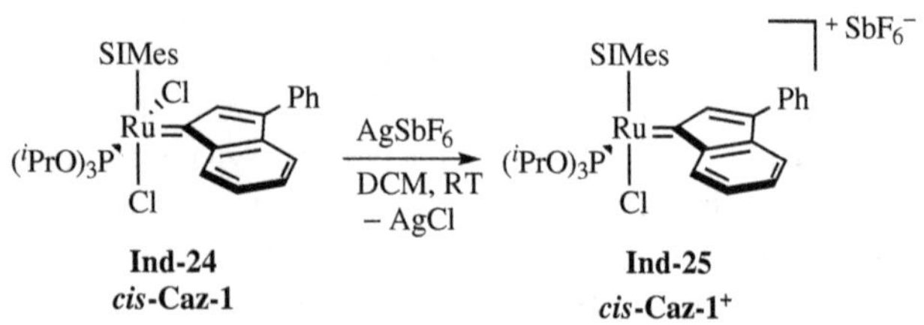

Figure 15.14 Synthesis of **Ind-25**.

olefins was the introduction of bis-NHC complexes. Nolan (17) and Peeck and Plenio (18) simultaneously reported the synthesis of Ru-indenylidene complexes bearing SIMes and an unsaturated NHC. These complexes can be synthesized by reaction of **Ind-18** with either a free carbene, or its silver salt, and are isolated in very good yields (Fig. 15.15). The increased activity was explained by the increased stability of the catalysts at higher temperature, and the concurrent slower initiation rates when compared to their benzylidene analogs (18).

15.1.3 Third-Generation Ru-Indenylidene Complexes

Third-generation Ru-indenylidene complexes are especially useful in ring-opening metathesis polymerization (ROMP) reactions due to complete and efficient initiation, increased propagation rates compared to phosphine-bearing second-generation analogs, and improved stability when compared to their benzylidene counterparts.

In addition, and as described previously, third-generation catalysts are also important synthons that allow access to a wide range of catalysts. Indeed, by facile ligand substitution reactions involving a pyridine displacement, complexes can be accessed which bear two NHCs, less electron-donating phosphines than PCy_3, or chelating carbene ligands (19).

Pyridine-containing catalysts can be synthesized easily by simple ligand exchange between first- or second-generation catalysts and excess pyridine (Fig. 15.16).

Recently, Nolan reported a one-pot procedure for the synthesis of mixed NHC-pyridine complexes starting from **Ind-1** (11a). This new protocol reduces the amount of waste generated as it avoids the use of costly and difficult-to-remove PCy_3 (Fig. 15.17).

15.1.4 Ru-Indenylidene Complexs Bearing Chelating Ligands

With the aim to develop more efficient catalysts having improved thermal stability and/or latent character, several groups have concentrated their efforts toward developing new ruthenium indenylidene complexes bearing chelating ligands. These complexes are especially important in the synthesis of polymers. In some cases, it is highly desirable to be able to mix together the catalysts and the monomers without concomitant polymerization, as this allows for longer handling times of catalyst–monomer mixtures or even their storage over long periods of time.

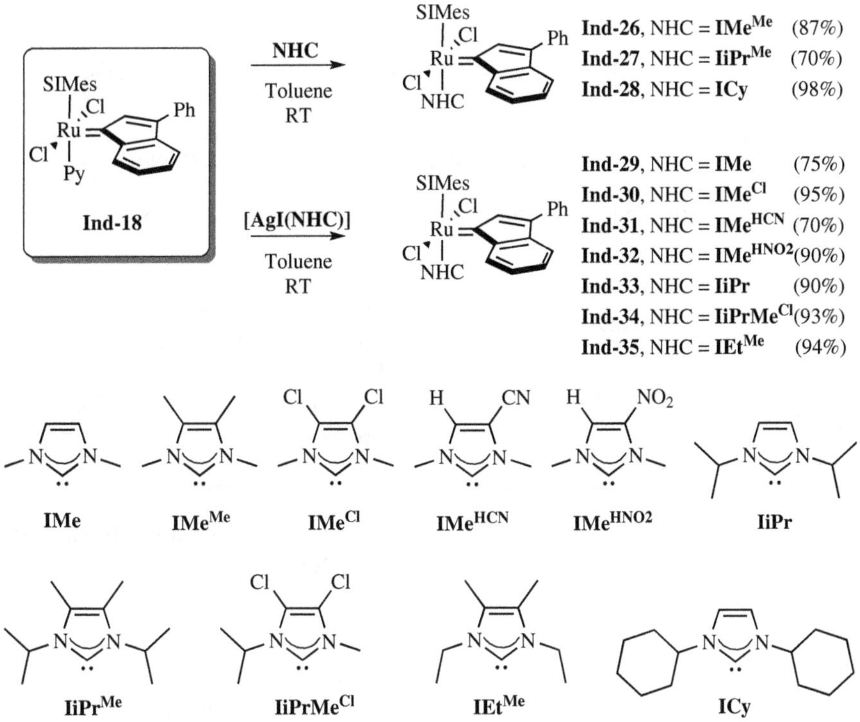

Figure 15.15 Synthesis of bis-NHC Ru-indenylidene complexes.

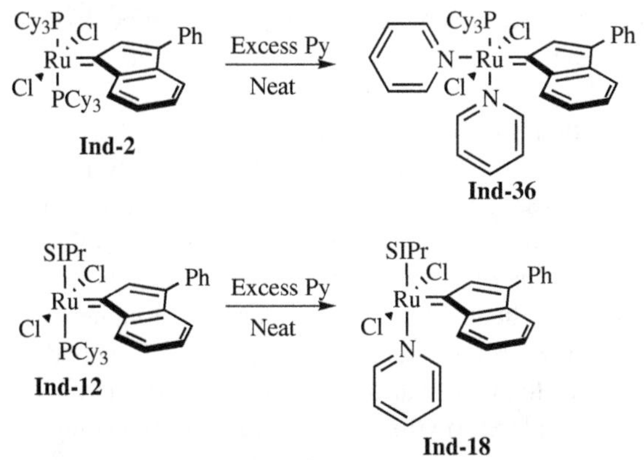

Figure 15.16 Synthesis of third-generation catalysts.

Opstal and Verpoort (20) were the first to report the synthesis of a ruthenium indenylidene complex bearing a chelating ligand **Ind-38** (Fig. 15.18), followed by the report of **Ind-39** and its activity in ROMP of cyclopentene and cyclooctene. These complexes bearing a Schiff base ligand are synthesized by reacting the ligand with thallium ethoxide, affording the corresponding thallium salts, which are then reacted *in situ* with **Ind-2**, to obtain the catalysts in good yields (Fig. 15.18).

Complexes bearing NHC and Schiff bases have also been described in the literature (21). This family of complexes performs very well in cross metathesis transformations when acid activation is used and are commercially available from Umicore (Fig. 15.19) (21b).

Another example of a Ru-indenylidene complex bearing a chelating ligand was reported by Limbach (22), using a pyridine alkoxide ligand. **Ind-42** was synthesized by reacting **Ind-9** with the lithiated pridine alkoxide ligand **5**, formed *in situ* by addition of ^{t}BuLi to a solution of the ligand in THF (Fig. 15.20). The indenylidene moiety in this complex proved important in achieving better conversions as **Ind-42** outperformed all other alkylidene analogs in several metathesis transformations (22). The most interesting feature of **Ind-42** was its increased affinity for silica, rendering the complex easy to separate from products (22).

Another reported variation is the synthesis of a chelating indenylidene moiety. Independently, Schrodi (23), and Fischmeister and Bruneau (24) published the synthesis of a substituted Ru-indenylidene complex bearing an ether functional group strategically placed so that chelation can occur. Schrodi reported the *in situ* synthesis of complex **Ind-43** and its use in catalysis. This complex can be obtained using the two different synthetic procedures described in Figure 15.21. **Ind-43** exhibits catalytic activity similar to well-defined first-generation Hoveyda catalysts.

Fischmeister and Bruneau (24) were able to isolate the chelating indenylidene complex **Ind-45** and characterize it by single crystal X-ray diffraction (24). **Ind-44** showed increased thermal stability compared to Hoveyda-Grubbs type first-generation (**Hov-1**) or **Ind-2** (Fig. 15.22).

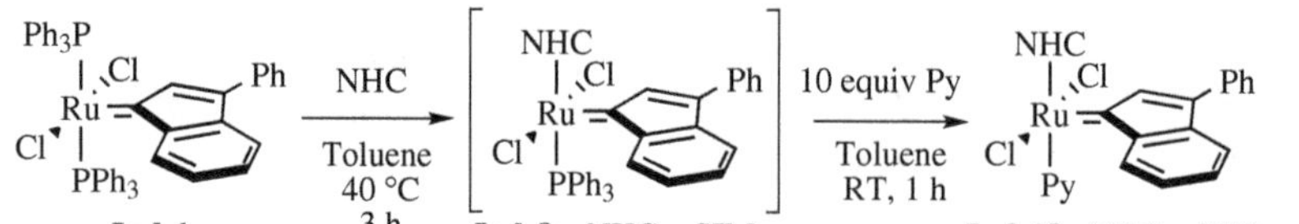

Ind-1 **Ind-8**, NHC = SIMes, **Ind-12**, NHC = SIPr, **Ind-18**, NHC = SIMes, **Ind-37**, NHC = SIPr,

Figure 15.17 One pot synthesis of third generation catalysts.

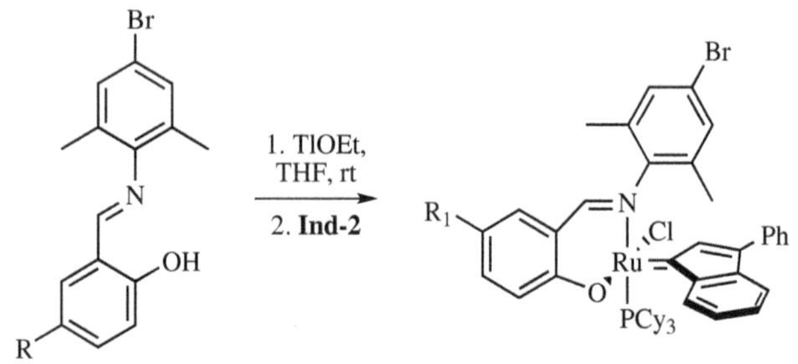

3, R = H
4, R = NO_2

Ind-38, R = H (79%)
Ind-39, R = NO_2 (80%)

Figure 15.18 Synthesis of Schiff base containing Ru-indenylidene complexes.

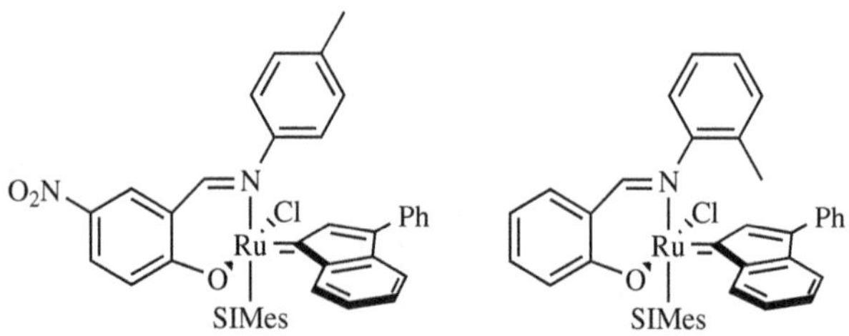

Ind-40 **Ind-41**

Figure 15.19 Commercially available Ru-indenylidene complexes bearing a Schiff base.

Only 20% decomposition was observed in CD_2Cl_2 at room temperature after one month for **Ind-45**, compared to 32% for Hoveyda first-generation under the same conditions. **Ind-2** is completely decomposed after 18 days. Catalytically, **Ind-45** also outperformed **Hov-1** in RCM (24). Other variations of this chelating indenylidene moiety featuring electron-withdrawing groups in the phenyl ring have also been reported, but typically resulted in lower activity than **Ind-45** (25).

15.1.5 The Activation Mechanism of Ru-Indenylidene Complexes

Recently, Nolan and Cavallo (26) reported a study of the activation mechanism of a series of Ru-indenylidene complexes and showed that Ru-indenylidene complexes do not always follow the traditional activation mechanism described for Grubbs-type complexes (Fig. 15.23).

The traditional mechanism for olefin metathesis (using RCM as a specific incarnation of the general reaction), using Ru-indenylidene complexes can be divided into three separate events: initiation, propagation, and termination (Fig. 15.23) (27).

The first step of the most common mechanism is the release of a tertiary phosphine (PR_3) from **Ind-I** to form a 14-electron species (**Ind-II**) that then coordinates the olefin. Formation of a metallacycle (**Ind-IV**) followed by rearrangement of the bonds to release the benzylidene moiety initially attached to the metal center leads to a new carbene (**Ind-V**) (28). Subsequent coordination of the second double bond leads to the formation of the metallacycle (**Ind-VI**) that is rearranged to form the product and the propagating species [Ru(=CH_2)Cl_2L] (**Ind-VII**), which can react with further olefins and proceed along the catalytic cycle, or react with a phosphine and form a resting species (**Ind-VIII**) that does not lead to any further catalytic turnovers.

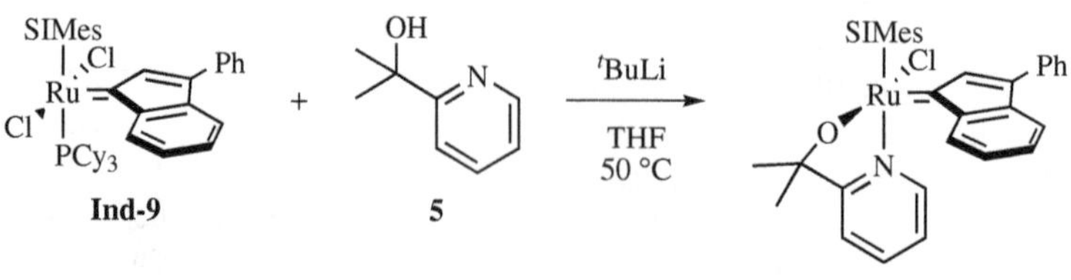

Ind-9 **5** **Ind-42 (75%)**

Figure 15.20 Synthesis of **Ind-42**.

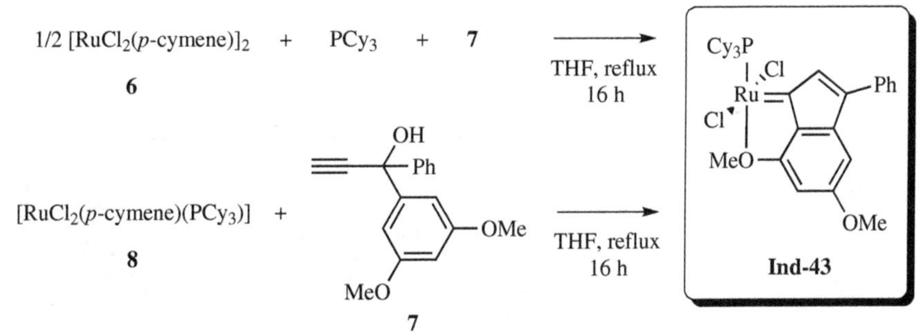

Figure 15.21 Synthesis of **Ind-43**.

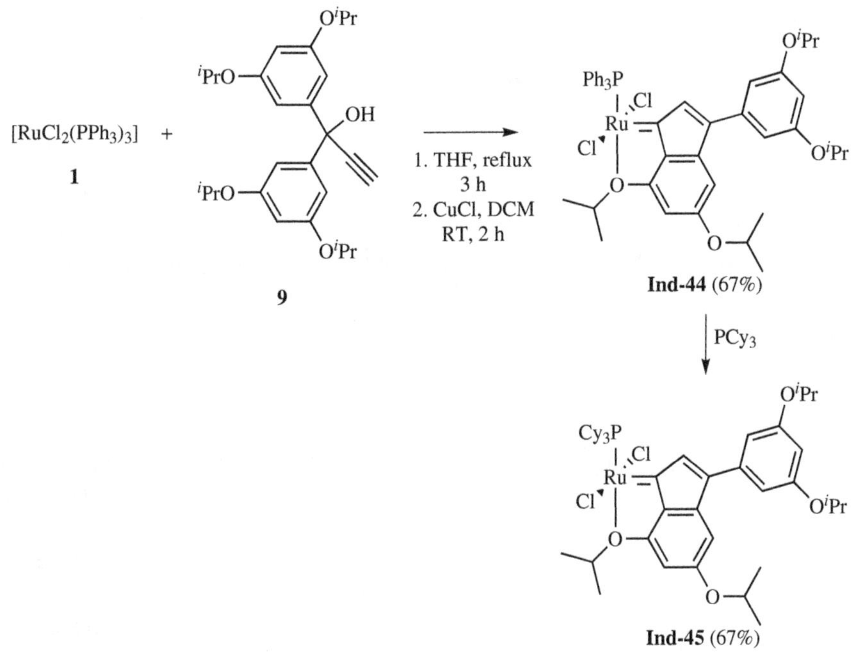

Figure 15.22 Synthesis of well-defined **Ind-44** and **Ind-45**.

Overall, Ru-indenylidene complexes initiate slower than their benzylidene counterparts, which agrees with the experimental finding that indenylidene complexes are more thermally stable than their benzylidene relatives, as catalyst decomposition is directly linked to the amount of catalytically active species present in solution (29).

In addition, Nolan and Cavallo showed that complexes **Ind-8** and **Ind-18-22** bearing *para*-substituted triphenylphosphine do not follow the traditional dissociative initiation mechanism, but an associative/interchange mechanism (Fig. 15.24), and concluded that the preference for Ru-indenylidene complexes to follow a dissociative rather than an associative/interchange mechanism is significantly small energetically. Therefore, small variations in the system, such as substrates and conditions, can shift the balance toward one or the other of the two activation pathways (26).

15.1.6 Arene Indenylidene Complexes

The allenylidene to indenylidene rearrangement has also been observed for η^6-arene ruthenium complexes. Indeed, while studying the acid activation of allenylidene complex **All-2**, Castarlenas and Dixneuf (30) observed that complex **All-2** transformed *in situ* into **Ind-46** (Fig. 15.25). The **Ind-46** prepared *in situ* is very sensitive to temperature as the excess HOTf catalyzes its decomposition. However, the acid-free isolated product is more stable and can catalyze a wide range of reactions. The allenylidene to indenylidene rearrangement can also be catalyzed by thermal activation and has been extensively studied by

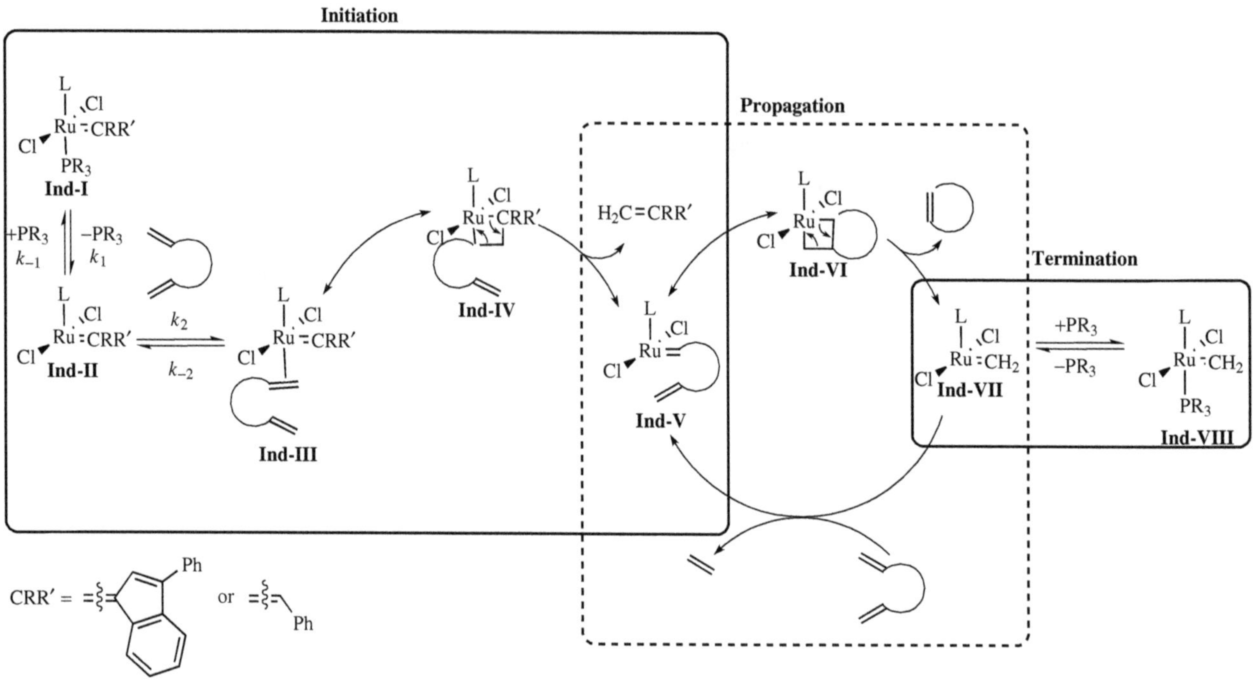

Figure 15.23 Initiation mechanism of Ru-indenylidene complexes.

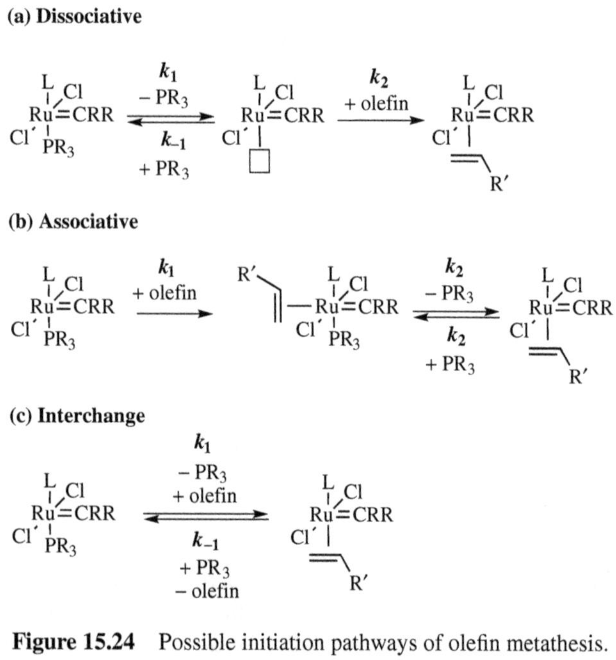

Figure 15.24 Possible initiation pathways of olefin metathesis.

Dixneuf (30,31). In general, the rearrangement proceeds faster in benzene than in dichloromethane, which is explained by the stronger interaction of the counteranion with the cationic complex in the less polar solvent. This depends on the nature of the counteranion and follows the order BF_4^- > $CF_3SO_3^-$≫ PF_6^-> SbF_6^- (31b).

Dixneuf (31b) also studied the effect of the electronic nature of the indenylidene ring (Y in Fig. 15.26) and discovered that electron-donating groups stabilize the allenylidene complex and reduce the rate of the rearrangement dramatically.

The proposed mechanism for the rearrangement (Fig. 15.26) is an intramolecular electrophilic aromatic substitution, which involves a reversible C–C bond-forming step in which the C_α is attacked by the *ortho*-carbon of the phenyl ring, followed by irreversible migration from the benzene to the β-carbon in the cyclopentadiene ring, with the anion acting as a proton shuttle between the two carbons (31b).

Contrary to other indenylidene catalysts, it has been proposed that the mechanism of initiation of **Ind-46** for olefin metathesis follows the pathway described in Figure 15.27. The first step, initiation, occurs by decoordination of the η^6-bound arene, forming the 14-electron species **Ind-46-I** that is stabilized by coordination of the anion. Then, **Ind-46-I** reacts with olefin **A** following the traditional pathway previously described for indenylidene complexes, forming a series of four-membered ring metallacycles leading to the formation of the propagating species **Ind-46-II**, and products **B** and **C**.

Although several pre-catalysts generated *via in situ* formation of arene indenylidene complexes have been reported, these systems exhibit some limitations as they require acid activation or higher temperatures than **Ind-2** to react, and therefore their use is more limited.

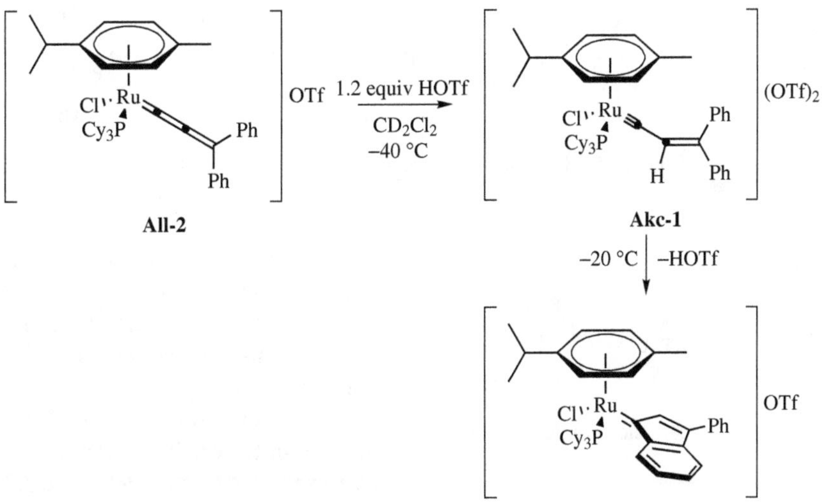

Figure 15.25 Indenylidene rearrangement in Ru-arene complexes.

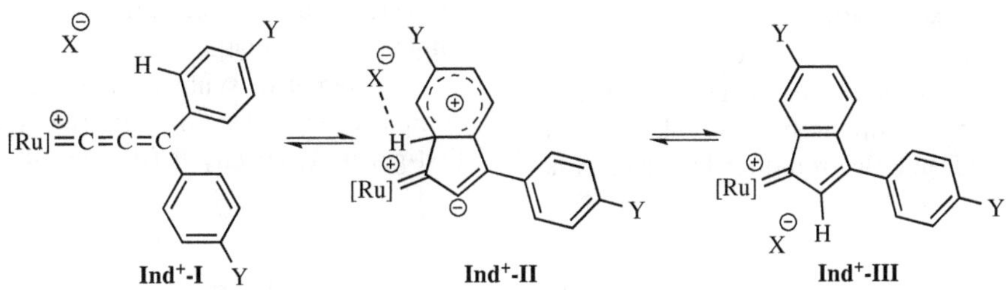

$X^- = BF_4^- > CF_3SO_3^- >> PF_6^- > SbF_6^-$
$Y = F > Cl > H >> OMe$

Figure 15.26 Proposed mechanism for the indenylidene rearrangement of Ru-arene indenylidene complexes (31b).

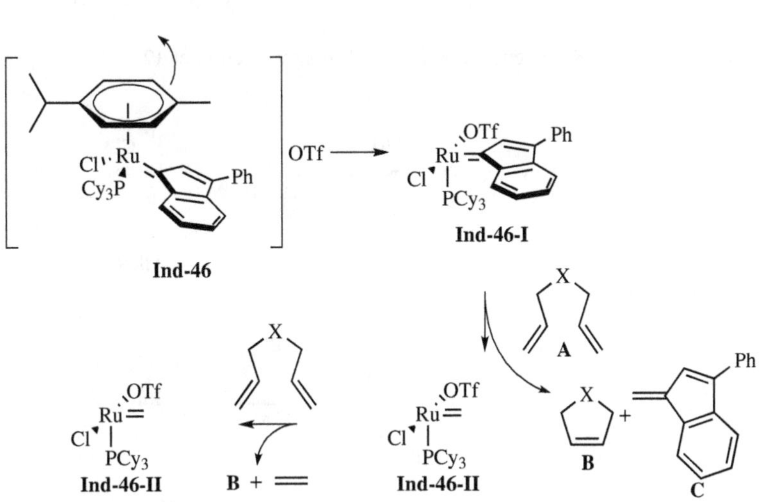

Figure 15.27 Activation mechanism of **Ind-46**.

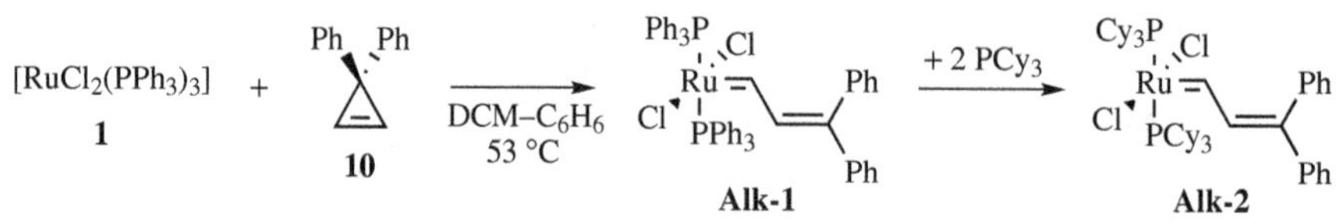

Figure 15.28 Synthesis of Ru-alkenylcarbene complexes.

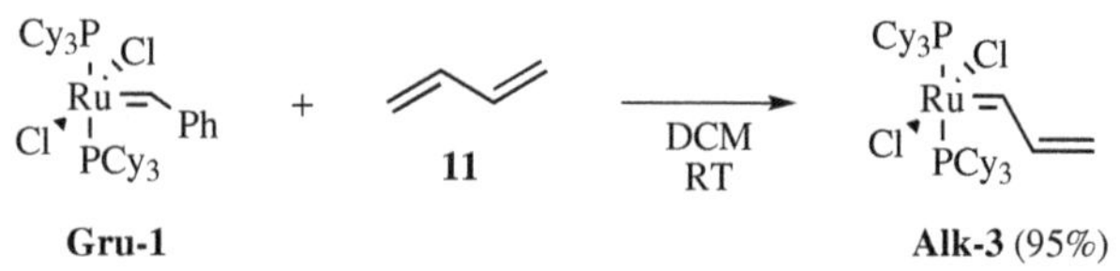

Figure 15.29 Synthesis of Ru-alkenylcarbene complex **Alk-3** by cross metathesis.

15.2 Ru-ALKENYLCARBENE COMPLEXES

Grubbs (32) reported the first well-defined metathesis active ruthenium catalyst: the alkenylcarbene **Alk-1** (Fig. 15.28). However, this complex was able to catalyze only ROMP reactions involving highly strained olefins such as norbornene.

By exchanging the triphenylphosphine in **Alk-1** with a more sterically hindered and electron-donating phosphine such as tricyclohexylphosphine, the activity of the catalyst (**Alk-2**) improved significantly (33). Complex **Alk-2** was able to catalyze ROMP of a large number of olefins, and was also active in RCM, among other metathesis reactions (34).

Complexes **Alk-1** and **Alk-2** represented a major breakthrough in Ru-catalyzed olefin metathesis, as they were the first examples of well-defined catalysts and provided valuable information about architectural scaffolds needed to create ruthenium-based catalysts.

Despite **Alk-2** exhibiting both high metathesis activity and remarkable stability toward various functional groups, the multistep synthesis (and thermal stability of the cyclopropene) leading to the carbene and the low initiation rates limited its use in large-scale reactions (35).

An alternative synthetic pathway to Ru-alkenylcarbene complexes involves cross metathesis of butadiene (**11**) with first-generation catalysts (Fig. 15.29).

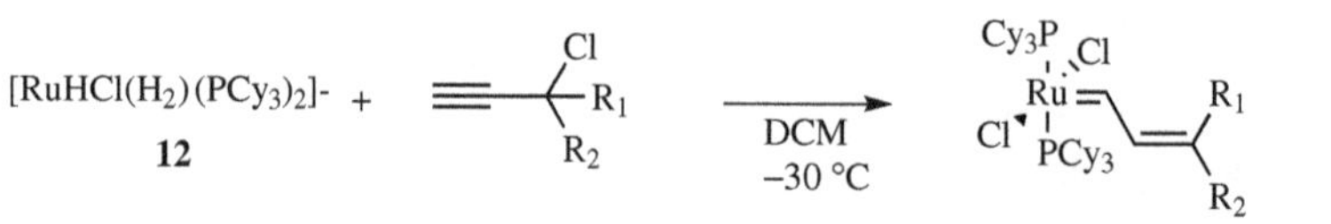

13, $R_1 = R_2 = Me$ | **Alk-4**, $R_1 = R_2 = Me$
14, $R_1 = R_2 = -(CH_2)_5-$ | **Alk-5**, $R_1 = R_2 = -(CH_2)_5-$
15, $R_1 = H$, $R_2 = Ph$ | **Alk-6**, $R_1 = H$, $R_2 = Ph$
16, $R_1 = H$, $R_2 = Me$ | **Alk-7**, $R_1 = H$, $R_2 = Me$
17, $R_1 = R_2 = H$ | **Alk-8**, $R_1 = R_2 = H$

Figure 15.30 Synthesis of Ru-alkenylcarbene complexes by reaction of **12** with propargyl chlorides.

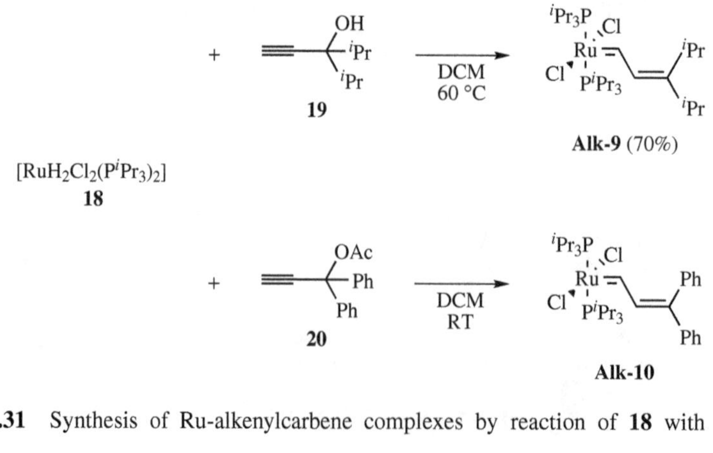

Figure 15.31 Synthesis of Ru-alkenylcarbene complexes by reaction of **18** with propargylic alcohols.

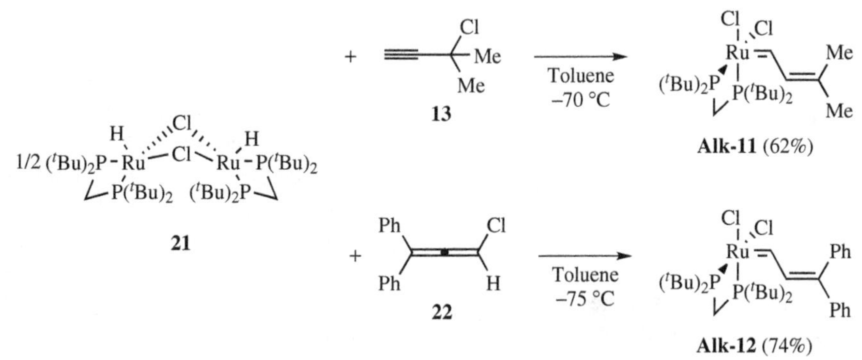

Figure 15.32 Synthesis of **Alk-11** and **Alk-12**.

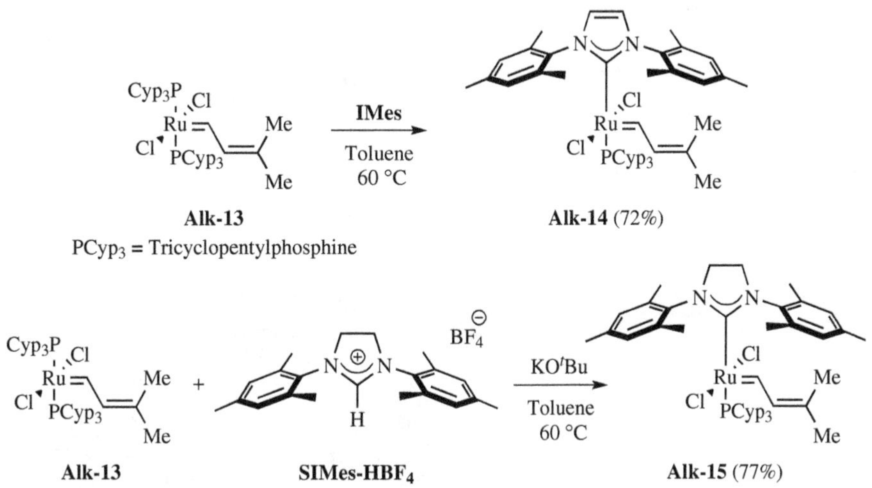

Figure 15.33 Synthesis of second-generation Ru-alkenylcarbene complexes.

A more useful protocol for the synthesis of alkenylcarbene complexes is the reaction of propargyl chlorides with $[RuH(H_2)Cl(PCy_3)_2]$ (**12**) (Fig. 15.30) (36). This synthetic route shows improved yields with sterically demanding R substituents; a Ru(IV) byproduct is observed when the less sterically demanding propargyl chlorides (such as **20**) are employed, in a product:byproduct ratio of up to 0.8 : 1.

Propargylic alcohols have also been used to synthesize Ru-alkenylcarbene complexes. As observed in Figure 15.31, **Alk-9** and **Alk-10** can be easily accessed by reaction of commercially available propargylic alcohols with $[RuH_2Cl_2(P^iPr_3)_2]$ (**18**).

A variation of the previous protocols allowed for the formation of the first *cis* Ru-alkenylcarbene complex (37). By reaction of a series of propargyl chlorides with chloroallenes, Hoffman synthesized a series of Ru-alkenylcarbene complexes bearing a chelating bisphosphine (Fig. 15.32). As for Ru-indenylidene complexes, the *cis* complexes exhibit lower activity than *trans* analogs (37,38).

Like other families of metathesis catalysts, second-generation Ru-alkenylcarbene complexes have also been reported (39). These can be easily accessed by reaction of a first-generation complex with a free carbene (39a). The reaction also proceeds when the free carbene is generated *in situ* (Fig. 15.33) (39b,c).

15.3 Ru-VINYLIDENE COMPLEXES

Ru-vinylidene complexes can be easily accessed by reaction of the appropriate ruthenium source with an alkyne. The first vinylidene complex of the general formula $[RuX_2(=C=CHR)L_2]$ was reported by Wakatsuki *et al.* (40) and was synthesized by treatment of $[RuCl_2(PPh_3)_3]$ (**1**) with 3,3-dimethyl-1-butyne (**23**) (Fig. 15.34).

Later, Werner (41) synthesized **Vin-2** by using complex **18** as the ruthenium source. However, this route also leads to the isolation of **25** as a small impurity (Fig. 15.35).

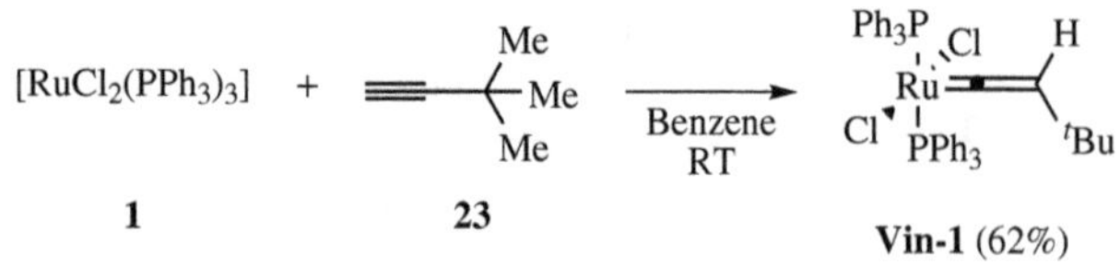

Figure 15.34 Synthesis of the first Ru-vinylidene complex.

Katayama and Ozawa (42) also reported the synthesis of Ru-vinylidene complexes by reaction of alkynes with several ruthenium precursors (Fig. 15.36). The first route requires the *in situ* synthesis of the ruthenium polymer $[RuCl_2(P^iPr_3)_2]_n$ (**27**) which then reacts with alkynes **24** or **23** to yield complexes **Vin-2** and **Vin-3**, respectively, in moderate yields. In comparison, the second route involving the use of $[RuCl_2(p\text{-cymene})]_2$ (**6**) affords better yields and allows access to a wider range of complexes in good to excellent yields. Variations of the first route have been employed to synthesize other vinylidene complexes bearing water-soluble phosphines (43).

Like other families of catalysts, new Ru-vinylidene complexes can be prepared by phosphine exchange. This method was employed by Werner (44) to synthesize a series of Ru-vinylidene complexes bearing chelating bisphosphines (Fig. 15.37). The low yield in the synthesis of **Vin-7** was attributed to the formation of an insoluble complex, possibly a ruthenium polymer (44).

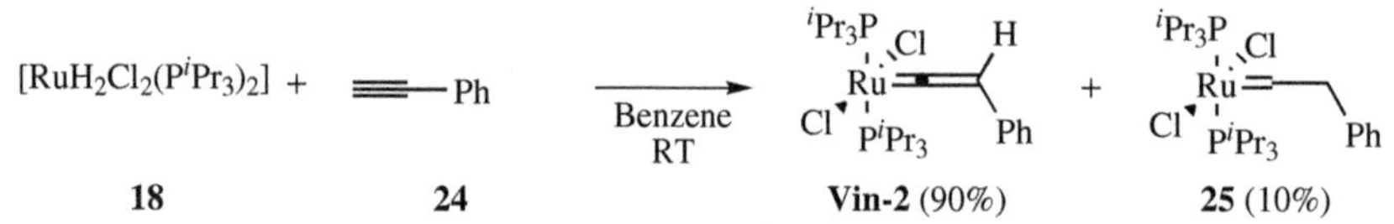

Figure 15.35 Synthesis of **Vin-2**.

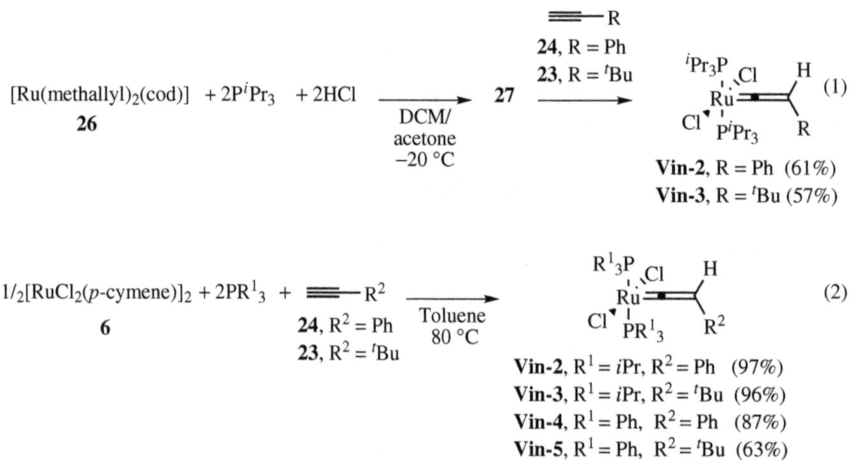

Figure 15.36 Synthesis of Ru-vinylidene complexes using **26** and **6** as the ruthenium source.

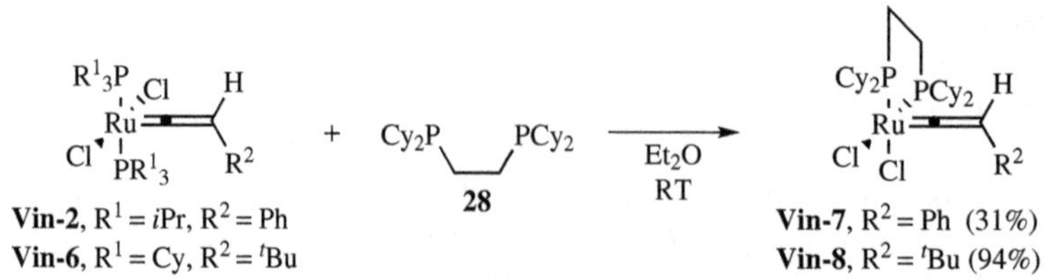

Figure 15.37 Synthesis of Ru-vinylidene complexes bearing a chelating phosphine.

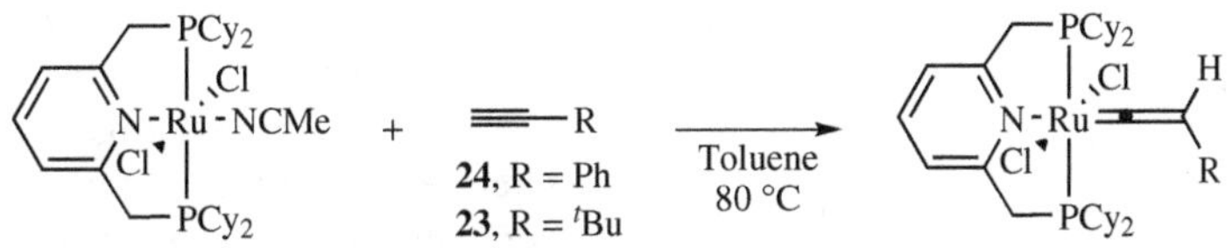

Figure 15.38 Synthesis of Ru-vinylidene complexes bearing a pincer ligand.

Complexes bearing pincer-type tridentate ligands have also been reported (45). In these complexes, the vinylidene moiety is usually introduced in the last step of the synthesis, by reaction of the appropriate ruthenium precursor bearing the chelating ligand with an alkyne, as in the case of **Vin-9** and **Vin-10** (Fig. 15.38).

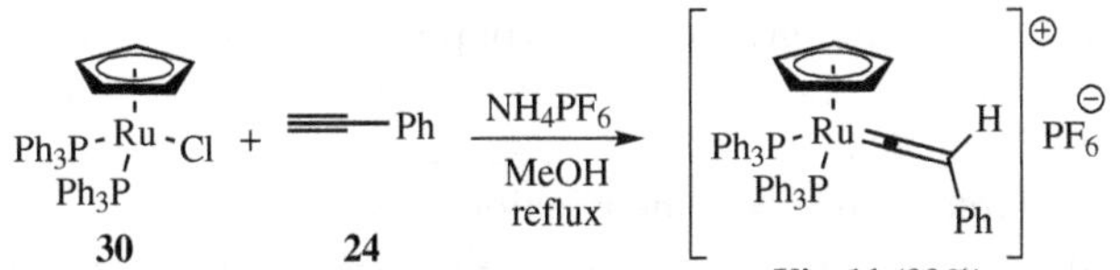

Figure 15.39 Synthesis of the first cationic 18-electron Ru-vinylidene complex **Vin-11**.

Cationic 18-electron Ru-vinylidene complexes have also been described (46). Bruce and Wallis (46a) reported the first complex of this kind, **Vin-11**. It was easily obtained by the reaction of $[RuCpCl(PPh_3)_2]$ (Cp, cyclopentadienyl, **30**) with **24** in the presence of NH_4PF_6 in very good yield (Fig. 15.39). Several other cationic complexes have been synthesized following similar protocols (46b).

The reaction of a nucleophile with a Ru-allenylidene or a Ru-alkenylcarbyne complex also affords Ru-vinylidene complexes in good yields (47). These reactions proceed with a wide range of nucleophiles: protic nucleophiles in the case of reaction with Ru-allenylidenes, and aprotic nucleophiles with Ru-alkenylcarbynes (Fig. 15.40) (47).

The cross-metathesis of a Ru-benzylidene complex with 1,2-propadiene also affords a Ru-vinylidene complex in excellent yield (Fig. 15.41) (48).

Ru-vinylidene complexes bearing NHCs have also been reported in the literature (49). Like other second-generation complexes, they can be accessed easily by reaction of a free carbene prepared *in situ* with the corresponding bisphosphine Ru-vinylidene complex (Fig. 15.42) (49a).

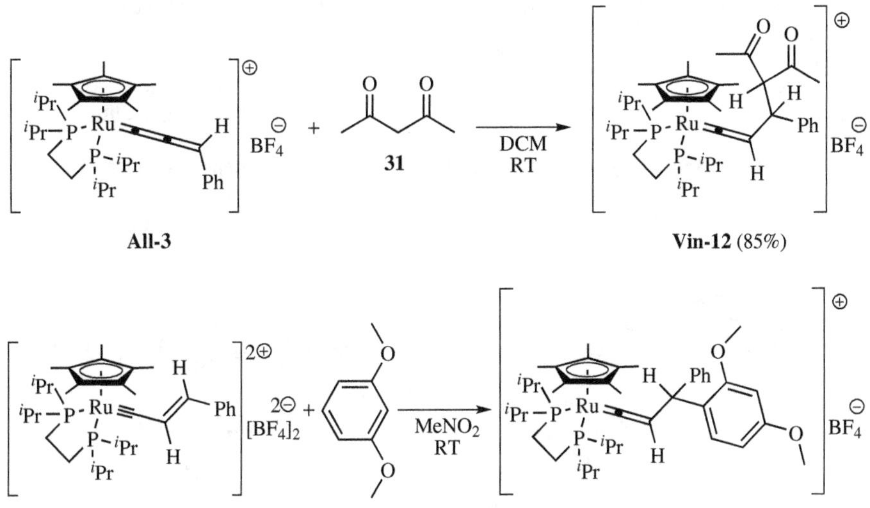

Figure 15.40 Synthesis of Ru-vinylidene complexes by reaction of a nucleophile with a Ru-allenylidene or a Ru-alkenylcarbyne.

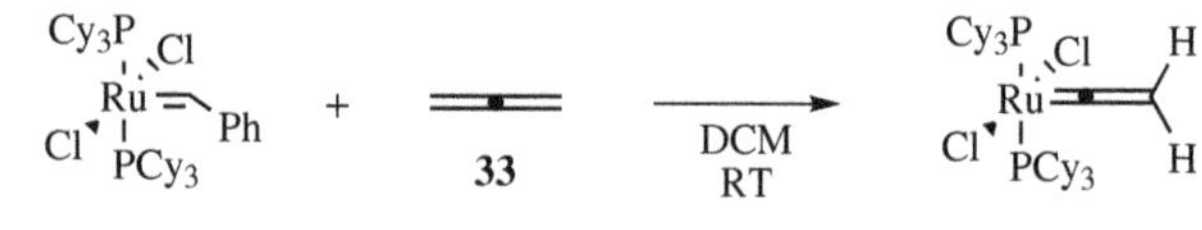

Gru-1 **Vin-14** (98%)

Figure 15.41 Synthesis of a Ru-vinylidene complex by cross metathesis.

Although there are several very efficient synthetic routes to Ru-vinylidene complexes, their activity in olefin metathesis has not been extensively tested. Overall, Ru-vinylidene complexes initiate slower than their benzylidene counterparts and their use in catalysis has been very limited (50).

15.4 Ru-ALLENYLIDENE COMPLEXES

Ru-allenylidene complexes have been extensively studied from a synthetic point of view, as the chemistry related to these complexes dates back to 1982 (51). However, the catalytic activity of these complexes in olefin metathesis has not been studied to the same extent.

Most Ru-allenylidene complexes are prepared following Selegue's protocol (50,51). This method involves the reaction of propargylic alcohols or their derivatives with a suitable 16-electron Ru(II) complex to form the ruthenium–carbon double bond. As illustrated in Figure 15.43, this reaction proceeds through a Ru-vinylidene intermediate, which then dehydrates to form the desired complex **All-3** in good yield (51).

Fürstner and Dixneuf (52) were the first to discover the potential of Ru-allenylidene complexes as olefin metathesis catalysts. They reported the synthesis and catalytic activity of a series of Ru-allenylidene complexes derived from $[RuCl_2(p\text{-cymene})]_2$ (**6**) (Fig. 15.44). **All-4** and related complexes can be easily obtained in a two-step procedure in excellent yields. It is important to mention that this procedure occurs only for sterically demanding phosphines, as with small phosphines MeOH can attack the C_α of **All-4**, resulting in the formation of a catalytically inert Fischer-carbene of the type $[Ru{=}C(OMe){-}CH{=}CPh_2]$ (52b).

Fürstner and Dixneuf also developed an alternative procedure that circumvents the use of protic solvents and results in a more practical and flexible method for the preparation of Ru-allenylidene complexes. As described in Figure 15.44, this protocol involves the reaction of **8** with a silver salt to afford the cationic complexes **35** and **36**, which then react with the propargylic alcohol **2** to yield complexes **All-5** and **All-6**, respectively.

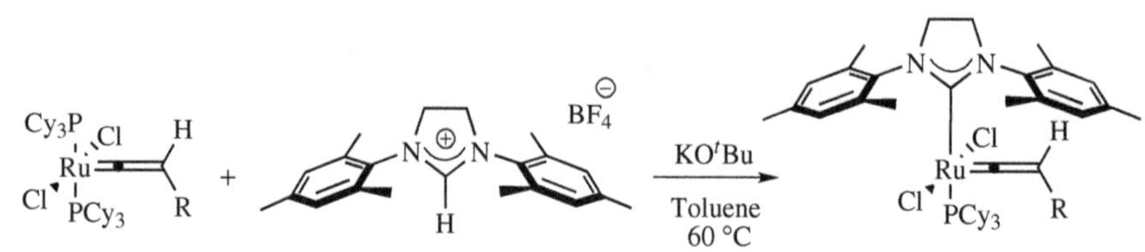

Vin-15, R = Ph
Vin-16, R = $SiMe_3$
Vin-17, R = ^{t}Bu

SIMes·HBF$_4$

Vin-18, R = Ph
Vin-19, R = $SiMe_3$
Vin-20, R = ^{t}Bu

Figure 15.42 Synthesis of second-generation Ru-vinylidene complexes.

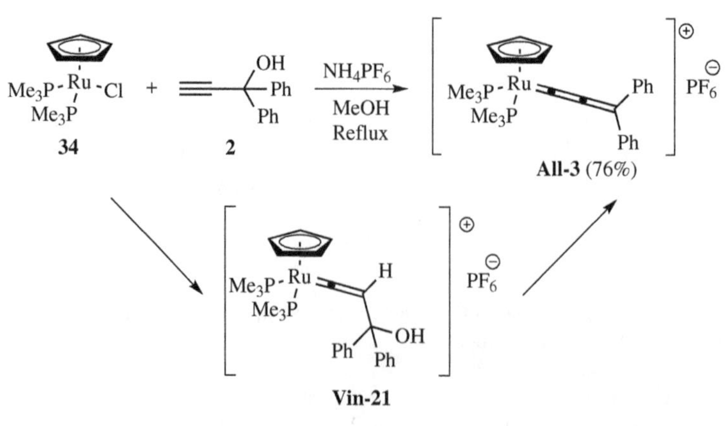

Figure 15.43 Selegue's synthesis of Ru-allenylidene complexes.

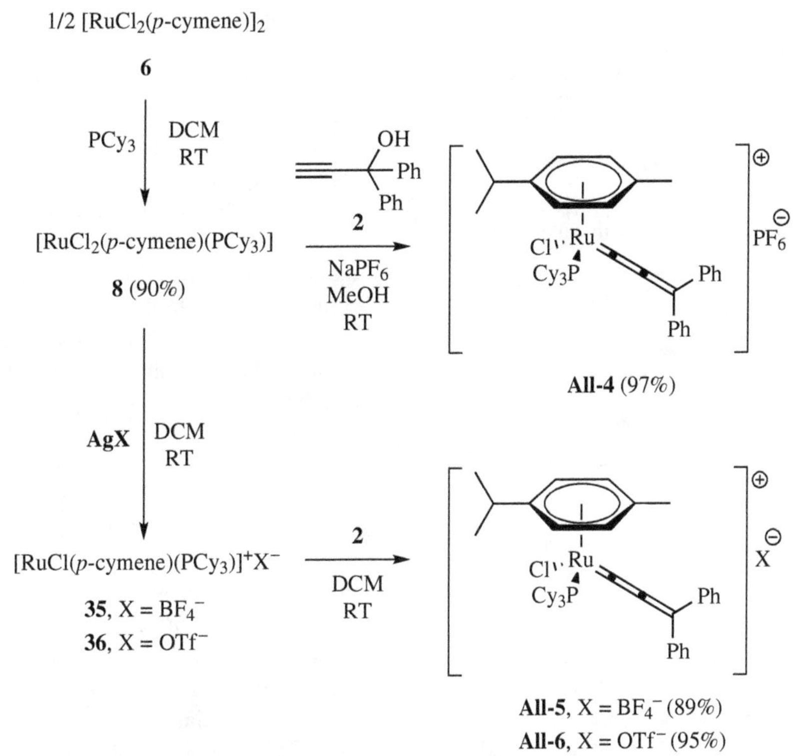

Figure 15.44 Fürstner and Dixneuf synthesis of Ru-allenylidene complexes.

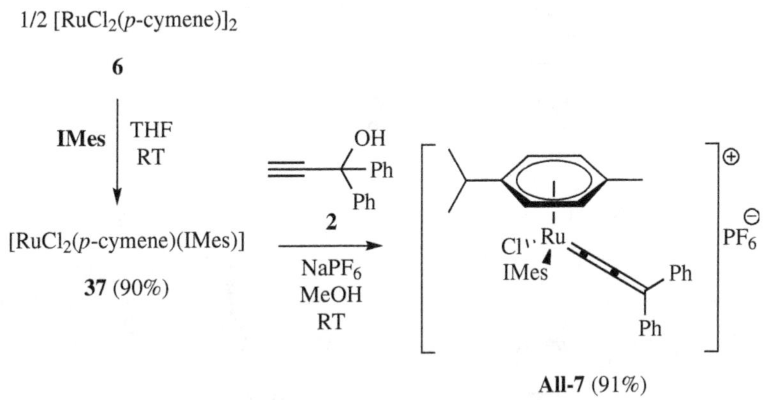

Figure 15.45 Synthesis of NHC-bearing Ru-vinylidene complexes.

Ru-allenylidene complexes bearing NHCs have also been reported (53). The complexes are synthesized using a two-step protocol (Fig. 15.45). First, the free carbene is reacted with **6**, affording complex **37**, which then reacts with the propargylic alcohol **2** to yield complex **All-7** (53).

As discussed earlier (see Section 15.1.6), Ru-allenylidene complexes rearrange *in situ* into their corresponding Ru-indenylidene complexes during the catalytic olefin metathesis reaction and their olefin metathesis activity is strongly related to the reaction rate of the allenylidene to indenylidene rearrangement (31b).

15.5 OTHER Ru-ALKYLIDENE COMPLEXES

In the search for new alkylidene moieties with enhanced stability and activity, several synthetic routes have been explored. Among them, one of the most versatile is the Van der

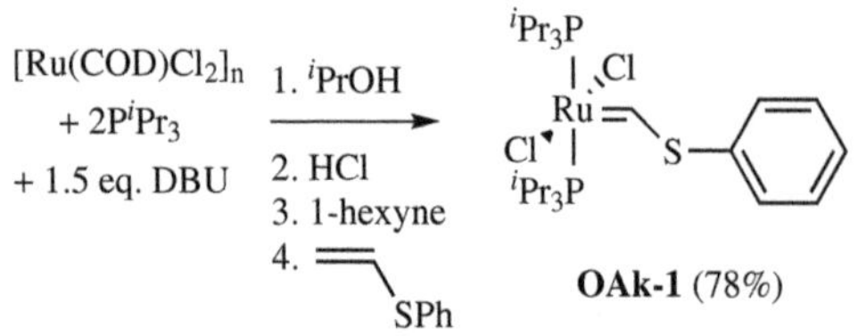

Figure 15.46 Van der Schaaf synthesis of thioalkylidenes.

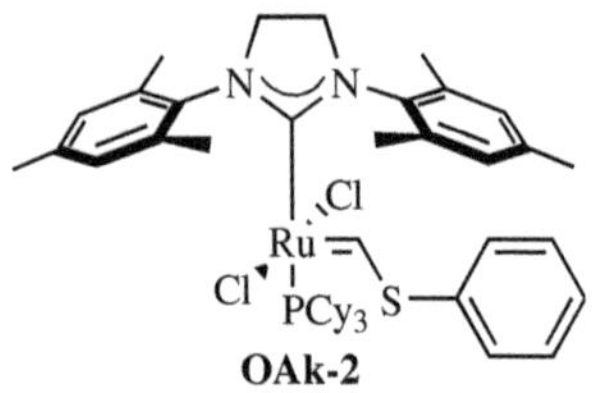

OAk-2

Figure 15.47 Commercially available (phenylthio)methylidene complex **OAk-2**.

Schaaf protocol for the synthesis of sulfur containing (phenylthio)-methylidene **OAk-1** (54). Van der Schaaf reported a one-pot procedure for the synthesis of **OAk-1**, starting from $[RuCL_2(COD)]_n$ (COD, 1,5-cyclooctadiene) as the ruthenium source (Fig. 15.46).

Complex **OAk-2**, an NHC derivative of **OAk-1**, is commercially available and has been reported as a catalyst in a limited number of metathesis transformations (Fig. 15.47) (22,55).

The versatility of the Van der Schaaf protocol was later extended by Kadyrov (56) to the synthesis of a wide range of alkylidenes (Fig. 15.48). The advantages of this protocol are that the alkylidene moiety is synthesized by cross-metathesis at the end of the reaction, which allows for the easy variation of the alkylidene moiety without having to prepare individually tailored starting materials, and the commercial availability of all starting materials which renders the reaction easily scalable.

As observed in Figure 15.48, a wide range of alkylidene moieties can be synthesized by this protocol in moderate to good yields, with groups ranging from simple cyclic olefins to heteroaromatic and aromatic substrates. It is important to highlight that with the exception of **OAk-7** and **OAk-9**, all alkylidenes shown in Figure 15.48 decompose slowly in chlorinated solvents (56).

Further development of complex **OAk-7** led to the isolation of highly active second-generation catalysts that have been commercialized by Evonik under the trade name **catMETium® RF 2–4** (Fig. 15.49). This upcoming family of catalysts performs several types of olefin metathesis

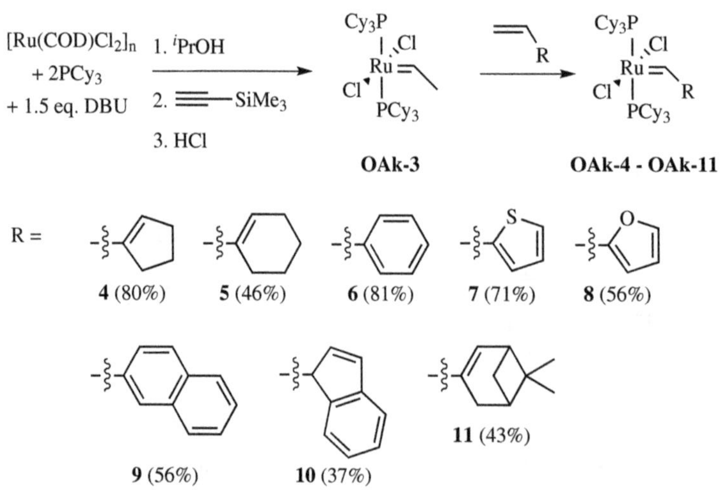

Figure 15.48 Extended protocol for the synthesis of other alkylidene complexes.

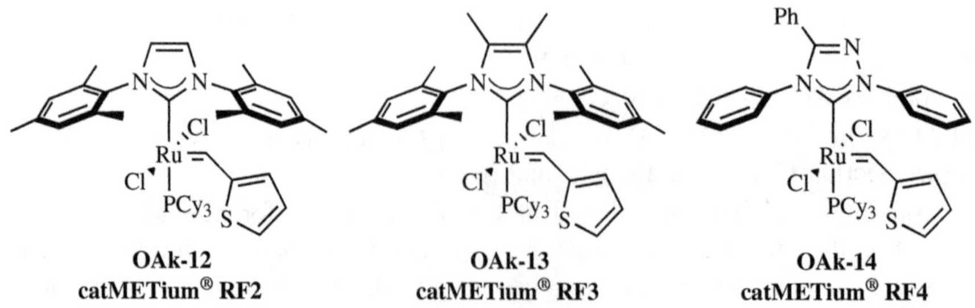

OAk-12 **catMETium® RF2**   **OAk-13** **catMETium® RF3**   **OAk-14** **catMETium® RF4**

Figure 15.49 Commercially available 2-thienylmethylidene complexes.

transformation under mild reaction conditions in very good yields.

15.6 PROTOCOL FOR THE SYNTHESIS OF SELECTED Ru-INDENYLIDENE COMPLEXES

15.6.1 Synthesis of [$RuCl_2$(SIPr)PPh_3(Ind)] (Ind-12)

The following procedure can be adapted to synthesize most second-generation Ru-indenylidene complexes, changing the starting material to **Ind-2** if the corresponding PCy_3 analog is desired.

Inside the glovebox, **Ind-1** (1.00 g, 1.13 mmol) and **SIPr** (528 mg, 1.34 mmol) were charged into a Schlenk flask and dissolved in toluene (3 ml). The reaction was taken out of the glovebox and stirred further at 40 °C for 3 h under Ar. The mixture was then allowed to cool to RT and the solvent removed under vacuum. The reaction crude was washed with cold methanol (2 × 5 ml) and cold hexane (8 × 25 ml) affording **6** (62% yield, 652 mg) as an orange solid. ^{1}H NMR (500 MHz, Tol) δ ppm 7.88 (d, J = 7.7 Hz, 1H), 7.65 (d, J = 7.1 Hz, 2H), 7.40 (dd, J = 7.7, 1.3 Hz, 1H), 7.35 (m, J = 7.4, 7.4, 4.4 Hz, 2H), 7.20–7.30 (m, 9H), 7.09 (dd, J = 7.5, 1.7 Hz, 1H), 6.93–7.02 (m, 11H), 6.89 (m, J = 7.6, 7.6, 1.7 Hz, 7H), 6.79 (dd, J = 7.7 Hz, 1H), 6.77 (s, 1H), 6.68 (d, J = 6.8 Hz, 1H), 6.58–6.63 (m, 1H), 6.57 (d, J = 7.7 Hz, 1H), 4.56–4.66 (m, 1H), 4.07–4.15 (m, 1H), 3.80–3.90 (m, 1H), 3.71–3.79 (m, 1H), 3.57–3.67 (m, 2H), 3.47–3.54 (m, 1H), 3.32–3.41 (m, 1H), 1.77 (d, J = 6.4 Hz, 3H), 1.59 (d, J = 6.6 Hz, 3H), 1.37 (d, J = 6.4 Hz, 3H), 1.26 (d, J = 6.8 Hz, 3H), 1.14 (d, J = 6.8 Hz, 3H), 1.10 (d, J = 6.8 Hz, 3H), 0.85 (d, J = 6.8 Hz, 3H), 0.75 (d, J = 6.6 Hz, 3H). ^{13}C NMR (75.5 MHz, CD_2Cl_2): δ = 301.0 (d, J = 12.52 Hz), 217.4 (d, J = 90.38 Hz), 151.4, 150.9, 148.9, 148.8, 143.5, 143.4, 142.1, 140.2, 138.1, 137.7, 137.2, 135.8, 132.7, 132.2, 131.8, 131.4, 131.0, 130.6, 130.0, 129.8, 129.2, 128.9, 128.1, 126.8, 125.8, 125.2, 125.0, 117.9, 56.5, 56.4, 56.0, 31.1, 30.0, 29.8, 28.5, 28.3, 28.1, 27.9, 25.1, 24.7, 24.3, 23.0. ^{31}P NMR (162 MHz) d = 29.87 ppm. Anal. calcd for $C_{60}H_{63}Cl_2N_2PRu$ C, 70.99; H, 6.26; N, 2.76. Found: C, 70.86; H, 6.35; N, 2.83.

REFERENCES

1. Harlow KJ, Hill AF, Wilton-Ely JDET. *J Chem Soc Dalton Trans* 1999:285–291.
2. Jafarpour L, Schanz H-J, Stevens ED, Nolan SP. *Organometallics* 1999;18:5416–5419.
3. Fürstner A, Guth O, Duffels A, Seidel G, Liebl M, Gabor B, Mynott R. *Chem Eur J* 2001;7:4811–4820.
4. Shaffer EA, Chen C-L, Beatty AM, Valente EJ, Schanz H-J. *J Organomet Chem* 2007;692:5221–5233.
5. Forman GS, Bellabarba RM, Tooze RP, Slawin AMZ, Karch R, Winde R. *J Organomet Chem* 2006;691:5513–5516.
6. Clavier H, Nolan SP. *Chem Eur J* 2007;13:8029–8036.
7. Torborg C, Szczepaniak G, Zielinski A, Malińska M, Woźniak K, Grela K. *Chem Commun* 2013;49:3188–3190.
8. Monsaert S, Drozdzak R, Dragutan V, Dragutan I, Verpoort F. *Eur J Inorg Chem* 2008:432–440.
9. Monsaert S, De Canck E, Drozdzak R, Van Der Voort P, Verpoort F, Martins JC, Hendrickx PMS. *Eur J Org Chem* 2009:655–665.
10. Sauvage X, Demonceau A, Delaude L. *Adv Synth Catal* 2009;351:2031–2038.
11. (a) Urbina-Blanco CA, Manzini S, Gomes JP, Doppiu A, Nolan SP. *Chem Commun* 2011;47:5022–5024. (b) Clavier H, Urbina-Blanco CA, Nolan SP. *Organometallics* 2009;28:2848–2854.
12. Manzini S, Urbina-Blanco CA, Slawin AMZ, Nolan SP. *Organometallics* 2012;31:6514–6517.
13. Urbina-Blanco CA, Bantreil X, Clavier H, Slawin AMZ, Nolan SP. *Beilstein J Org Chem* 2010;6:1120–1126.
14. Broggi J, Urbina-Blanco CA, Clavier H, Leitgeb A, Slugovc C, Slawin AMZ, Nolan SP. *Chem Eur J* 2010;16:9215–9225.
15. (a) Bantreil X, Poater A, Urbina-Blanco CA, Bidal YD, Falivene L, Randall RAM, Cavallo L, Slawin AMZ, Cazin CSJ. *Organometallics* 2012;31:7415–7426. (b) Bantreil X, Schmid TE, Randall RAM, Slawin AMZ, Cazin CSJ. *Chem Commun* 2010;46:7115–7117.
16. Songis O, Slawin AMZ, Cazin CSJ. *Chem Commun* 2012;48:1266–1268.
17. Bantreil X, Randall RAM, Slawin AMZ, Nolan SP. *Organometallics* 2010;29:3007–3011.
18. Peeck LH, Plenio H. *Organometallics* 2010;29:2761–2766.
19. (a) Slugovc C, Perner B, Stelzer F, Mereiter K. *Organometallics* 2004;23:3622–3626. (b) Szadkowska A, Gstrein X, Burtscher D, Jarzembska K, Woźniak K, Slugovc C, Grela K. *Organometallics* 2010;29:117–124.
20. Opstal T, Verpoort F. *Synlett* 2002:935–941.
21. (a) Hendrickx PMS, Drozdzak R, Verpoort F, Martins JC. *Magn Reson Chem* 2010;48:443–449. (b) Behr A, Pérez Gomes J. *Beilstein J Org Chem* 2011;7:1–8.
22. Schachner JA, Cabrera J, Padilla R, Fischer C, Van der Schaaf PA, Pretot R, Rominger F, Limbach M. *ACS Catal* 2011;1:872–876.
23. Jimenez LR, Gallon BJ, Schrodi Y. *Organometallics* 2010;29:3471–3473.
24. Kabro A, Roisnel T, Fischmeister C, Bruneau C. *Chem Eur J* 2010;16:12255–12261.
25. Kabro A, Ghattas G, Roisnel T, Fischmeister C, Bruneau C. *Dalton Trans* 2012;41:3695–3700.
26. Urbina-Blanco CA, Poater A, Lebl T, Manzini S, Slawin AMZ, Cavallo L, Nolan SP. *J Am Chem Soc* 2013;135:7073–7079.
27. (a) Sanford MS, Ulman M, Grubbs RH. *J Am Chem Soc* 2001;123:749–750. (b) Sanford MS, Love JA, Grubbs RH. *J Am Chem Soc* 2001;123:6543–6554. (c) Love JA, Sanford MS, Day MW, Grubbs RH. *J Am Chem Soc* 2003;125:10103–10109.
28. van der Eide EF, Piers WE. *Nat Chem* 2010;2:571–576.
29. Hong SH, Wenzel AG, Salguero TT, Day MW, Grubbs RH. *J Am Chem Soc* 2007;129:7961–7968.
30. Castarlenas R, Dixneuf PH. *Angew Chem Int Ed* 2003;42:4524–4527.

31. (a) Castarlenas R, Vovard C, Fischmeister C, Dixneuf PH. *J Am Chem Soc* 2006;128:4079–4089. (b) Antonucci A, Bassetti M, Bruneau C, Dixneuf PH, Pasquini C. *Organometallics* 2010;29:4524–4531.
32. Nguyen ST, Johnson LK, Grubbs RH, Ziller JW. *J Am Chem Soc* 1992;114:3974–3975.
33. (a) Nguyen ST, Grubbs RH, Ziller JW. *J Am Chem Soc* 1993;115:9858–9859. (b) Fu GC, Nguyen ST, Grubbs RH. *J Am Chem Soc* 1993;115:9856–9857.
34. (a) Miller SJ, Kim S-H, Chen Z-R, Grubbs RH. *J Am Chem Soc* 1995;117:2108–2109. (b) Miller SJ, Grubbs RH. *J Am Chem Soc* 1995;117:5855–5856. (c) Huwe CM, Blechert S. *Tetrahedron Lett* 1995;36:1621–1624. (d) Morken JP, Didiuk MT, Visser MS, Hoveyda AH. *J Am Chem Soc* 1994;116:3123–3124. (e) Kim S-H, Bowden N, Grubbs RH. *J Am Chem Soc* 1994;116:10801–10802. (f) Borer BC, Deerenberg S, Bieräugel H, Pandit UK. *Tetrahedron Lett* 1994;35:3191–3194.
35. Trnka TM, Grubbs RH. *Acc Chem Res* 2001;34:18–29.
36. Wilhelm TE, Belderrain TR, Brown SN, Grubbs RH. *Organometallics* 1997;16:3867–3869.
37. Hansen SM, Rominger F, Metz M, Hofmann P. *Chem Eur J* 1999;5:557–566.
38. Hansen SM, Volland MAO, Rominger F, Eisentrager F, Hofmann P. *Angew Chem Int Ed* 1999;38:1273–1276.
39. (a) Huang J, Schanz H-J, Stevens ED, Nolan SP. *Organometallics* 1999;18:5375–5380. (b) Chatterjee AK, Morgan JP, Scholl M, Grubbs RH. *J Am Chem Soc* 2000;122:3783–3784. (c) Jafarpour L, Nolan SP. *Organometallics* 2000;19:2055–2057.
40. (a) Wakatsuki Y, Yamazaki H, Kumegawa N, Satoh T, Satoh JY. *J Am Chem Soc* 1991;113:9604–9610. (b) Wakatsuki Y, Koga N, Yamazaki H, Morokuma K. *J Am Chem Soc* 1994;116:8105–8111.
41. (a) Grunwald C, Gevert O, Wolf J, Gonzalez-Herrero P, Werner H. *Organometallics* 1996;15:1960–1962. (b) Wolf J, Stüer W, Grünwald C, Gevert O, Laubender M, Werner H. *Eur J Inorg Chem* 1998:1827–1834.
42. Katayama H, Ozawa F. *Organometallics* 1998;17:5190–5196.
43. Saoud M, Romerosa A, Peruzzini M. *Organometallics* 2000;19:4005–4007.
44. Werner H, Jung S, González-Herrero P, Ilg K, Wolf J. *Eur J Inorg Chem* 2001:1957–1961.
45. (a) Beach NJ, Walker JM, Jenkins HA, Spivak GJ. *J Organomet Chem* 2006;691:4147–4152. (b) Del Río I, Van Koten G. *Tetrahedron Lett* 1999;40:1401–1404. (c) Katayama H, Wada C, Taniguchi K, Ozawa F. *Organometallics* 2002;21:3285–3291. (d) Beach NJ, Spivak GJ. *Inorg Chim Acta* 2003;343:244–252.
46. (a) Bruce MI, Wallis RC. *Aust J Chem* 1979;32:1471–1485. (b) Ciardi C, Reginato G, Gonsalvi L, de los Rios I, Romerosa A, Peruzzini M. *Organometallics* 2004;23:2020–2026.
47. (a) Bustelo E, Jiménez-Tenorio M, Puerta MC, Valerga P. *Organometallics* 2006;25:4019–4025. (b) Bustelo E, Jiménez-Tenorio M, Puerta MC, Valerga P. *Organometallics* 2007;26:4300–4309.
48. Schwab P, Grubbs RH, Ziller JW. *J Am Chem Soc* 1996;118:100–110.
49. (a) Opstal T, Verpoort F. *J Mol Catal A: Chem* 2003;200:49–61. (b) Louie J, Grubbs RH. *Angew Chem Int Ed* 2001;40:247–249.
50. Lozano-Vila AM, Monsaert S, Bajek A, Verpoort F. *Chem Rev* 2010;110:4865–4909.
51. Selegue JP. *Organometallics* 1982;1:217–218.
52. (a) Fürstner A, Picquet M, Bruneau C, Dixneuf PH. *Chem Commun* 1998:1315–1316. (b) Fürstner A, Liebl M, Lehmann CW, Picquet M, Kunz R, Bruneau C, Touchard D, Dixneuf PH. *Chem Eur J* 2000;6:1847–1857.
53. Jafarpour L, Huang J, Stevens ED, Nolan SP. *Organometallics* 1999;18:3760–3763.
54. Van der Schaaf PA, Kolly R, Kirner H-J, Rime F, Muhlebach A, Hafner A. *J Organomet Chem* 2000;606:65–74.
55. (a) Wallace DJ. *Adv Synth Catal* 2009;351:2277–2282. (b) Cabrera J, Padilla R, Dehn R, Deuerlein S, Gułajski Ł, Chomiszczak E, Teles JH, Limbach M, Grela K. *Adv Synth Catal* 2012;354:1043–1051.
56. Kadyrov R, Rosiak A. *Beilstein J Org Chem* 2011;7:104–110.

16

HOVEYDA-TYPE OLEFIN METATHESIS COMPLEXES

YAKOV GINZBURG AND NORBERTO GABRIEL LEMCOFF
Department of Chemistry, Ben-Gurion University of the Negev, Be'er-Sheva, Israel

Perhaps the most significant development in the improvement of the properties of the Grubbs catalysts since the introduction of the Arduengo NHC (*N*-heterocyclic carbene) (1,2) was the use of chelating benzylidene ligands. This type of complex was first suggested by Hoveyda in 1997 (3), as a byproduct of a metathesis reaction between a styrenyl ether and a ruthenium benzylidene. This kind of catalysts became the subject of research for chemists around the world as they opened a window to new opportunities in catalyst modification, stability enhancement, and tuning of properties. The first bidentate ruthenium complex, the Hoveyda–Grubbs first generation catalyst (4) (**2**, Fig. 16.1), was the pioneer in this long line of diverse complexes.

Chelation in olefin metathesis catalysts presents not just options for catalyst tuning and modifications but also for compound purification (catalyst and byproducts are easier to separate from the product) and concomitant catalyst recovery and recycling. As is well known, chelating ligands display enhanced binding of metal atoms. In 2000, Hoveyda and Blechert independently introduced to the world the novel phosphine free ruthenium complex which displayed both enhanced activity and stability and later became known as the *second generation Hoveyda–Grubbs catalyst* (5,6). Based on this concept, a plethora of new complexes were obtained that showed significant improvement over previous catalysts (7). In this chapter, we summarize the work in ruthenium-based olefin metathesis catalyst design, using the chelating benzylidene ligands—that is, Hoveyda–Grubbs type catalysts bearing oxygen and other chelating heteroatoms.

16.1 BENZYLIDENE MODIFIED CATALYSTS

The introduction of the Hoveyda–Blechert–Grubbs catalyst challenged the scientific community to explore the possibilities hidden in this type of olefin metathesis catalysis. Some of these works had a vast impact, both academically and industrially, and left a great mark on the development and investigation of olefin metathesis mechanism and activity.

One of the earliest results was produced by Blechert et al. (8,9), detailing the insertion of a phenyl ring into the vacant ortho position of the benzylidene isopropoxy moiety (**5**, Fig. 16.2). Studies showed that increasing the steric bulk in the isopropoxy vicinity caused weakening of the Ru–O bond and an associated substantial increase in the initiation rate. A similar effect was previously noticed by the same group with a 1,1′-bi-2-naphthol (BINOL) benzylidene ligand (**4**, Fig. 16.2) (10). Further investigations of the steric and electronic effects of different electron-withdrawing or -donating substituents showed that electron-withdrawing groups (EWGs) weakened the Ru–O bond and accelerated initiation; however, the synthesis of these complexes was more difficult due to reduced stability (11,12).

Concurrently, Grela et al. presented the illustrious nitro substituted catalyst **8a**, (Fig. 16.3) (13), and a tri-methoxy asarone-based derivative **7** (Fig. 16.3) (14), probing further the influence of EWG and electron-donating groups on the Ru–O bond. Grela (15) later proposed improved syntheses for many Hoveyda–Grubbs type complexes, making them quite accessible to facilitate the study of their properties.

Olefin Metathesis: Theory and Practice, First Edition. Edited by Karol Grela.

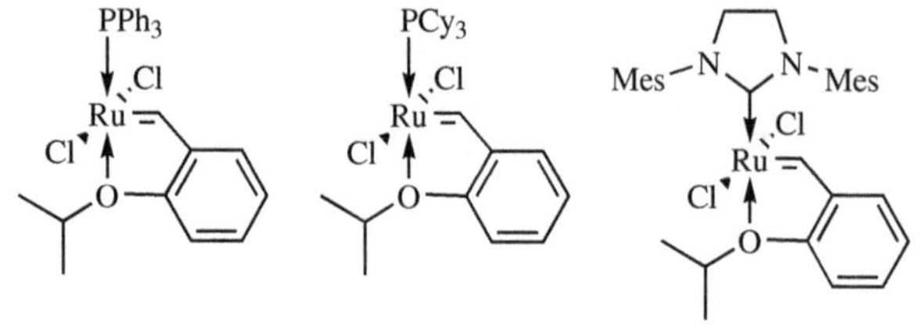

1 2 3

Figure 16.1 First bidentate ruthenium complexes.

The new nitro-substituted catalyst presented extremely fast initiation and high reactivity even in difficult syntheses of trisubstituted olefins and enyne metathesis; however, its sensitivity to air and moisture did not increase. The attempt to combine the steric acceleration with the electronic enhancement (catalysts **8b,c**) resulted in fairly unstable complexes. Also, the addition of the nitro group in the meta or ortho position to the isopropoxy group did not afford more efficient catalyst performance (16).

Grela continued his search for the control of catalyst initiation through electronic factors by attaching functional groups that could be activated *in situ* to afford strongly electron withdrawing elements. Thus, complexes **6i** and **6j** (Fig. 16.2) could be activated by the use of strong Brønsted acids or weak Lewis acids, respectively (17).

In later work (18), other electronic modifications on the chelating oxygen moiety were made, progressively increasing the withdrawing power of the group and diminishing the Ru–O bond strength. As proposed, activity of the complexes increased as the coordinating bond weakened (**9a–c**, Fig. 16.3). Notably, when a $-CF_3$ group was bound directly to the oxygen atom (**9d**, Fig. 16.3), no Ru–O bond could be observed; however, in recent work by Plenio et al. (19), the chelating oxygen was substituted with different aromatic moieties. These new complexes (**10a–d**, Fig. 16.4) showed excellent stability along with full activation in seconds.

In order to further examine the effects of modifications on the chelating moiety, structurally restricted complexes were synthesized where the oxygen was "pulled" away from the ruthenium and the degree of freedom was reduced by a cyclic ether structure. These constraints resulted in a surprising complex structure, a *cis*-dichloro complex (**11**, Fig. 16.5). However, this structure was only seen in solid state, whereas in NMR only the trans complex was observed, implying that crystal packing forces influenced the conformation of the complex. Thus, the activity of this complex was similar to the electronically modified analogs, with stability being preserved. Recent work by Witkowski (20,21) is noteworthy in this context, where both electronic and structural modifications afforded new complexes that, even though they did not present an outstanding performance, may upon further investigation open new possibilities in catalyst tuning and selectivity toward specific reactions (**12**, Fig. 16.5).

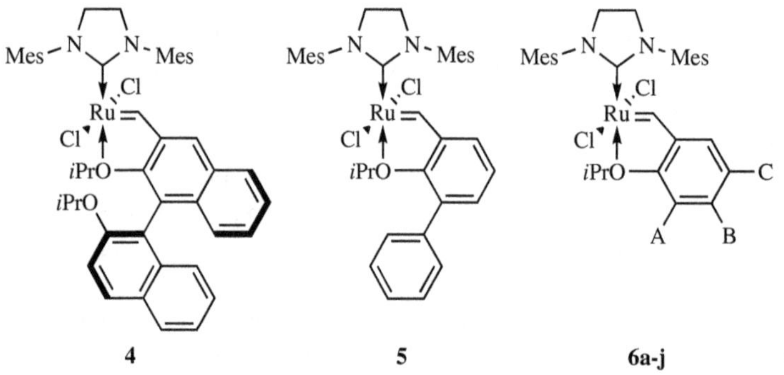

4 5 6a-j

a: A = H, B = H, C = O*i*Pr
b: A = H, B = H, C = Me
c: A = H, B = H, $C = CF_3$
d: A = H, B = H, C = F
e: A = H, B = O*i*Pr, C = H
f: A = H, B = F, C = H
g: A = H, B = CN, C = H
h: A = OMe, B = H, C = H
i: A = H, $B = NH_2$, C = H
j: A = H, B = H, $C = (C(Ph_2)OH)$

Figure 16.2 Blechert and Grela catalysts with modified steric and electronic properties.

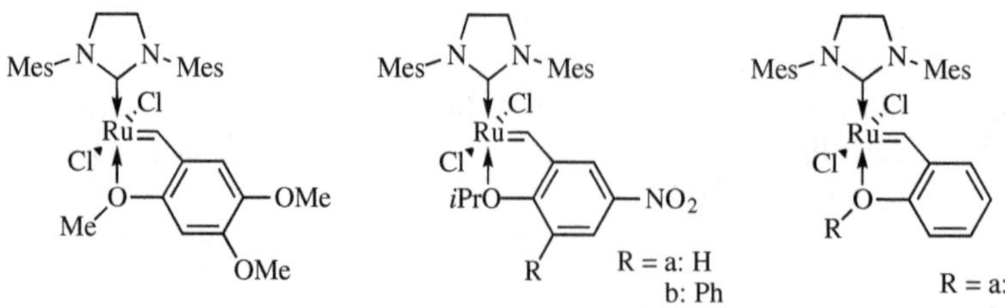

R = a: H
b: Ph
c: OMe

R = a: CH_3
b: Bz
c: 4-Nitro-Bz
d: CF_3 (not formed)

7 8a–c 9a–d

Figure 16.3 Electron withdrawing and donating groups on Hoveyda–Grubbs type catalysts.

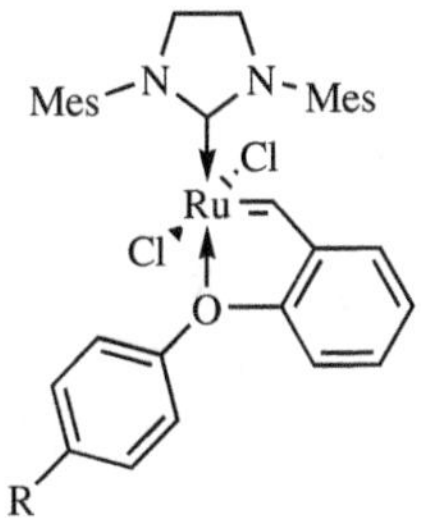

R = a: NMe_2
b: H
c: Cl
d: NO_2

10a–d

Figure 16.4 Plenio's diaryl ether complexes.

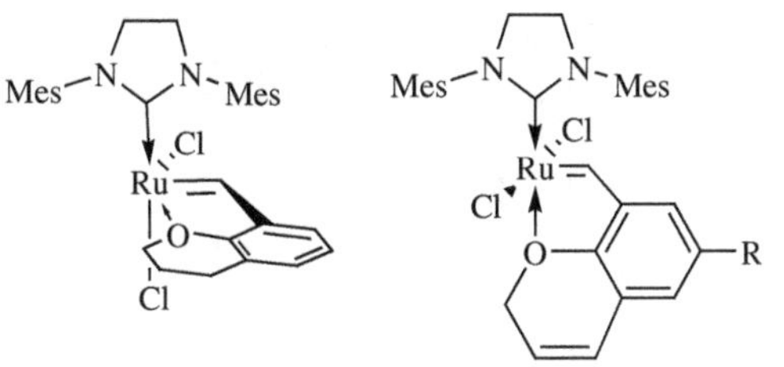

R = a: CH_3
b: NO_2

11 **12a-b**

Figure 16.5 Structural restrictions on Hoveyda type complexes.

An additional aspect that has been used to explicate the properties of oxygen-chelated ruthenium benzylidenes is the degree of pi-electron delocalization in the chelating ring; this topic is discussed in detail in Chapter 18, but must still be mentioned in this chapter due to its importance to the topic (22,23). By adding fused benzene rings, the properties of the ligand were tuned, leading to significant differences in the metathesis activities of the corresponding catalysts, depending on the position of the ring. While the anthracene-like ligand (**14**, Fig. 16.6) promoted high activity, like the parent complex, the phenanthrene-like complex (**13**, Fig. 16.6) displayed no activity toward benchmark ring-closing metathesis (RCM) reactions at normal conditions. The work was further extended to ring opening metathesis polymerization (ROMP) of norbornene-based monomers and once again the anthracene-like complex proved to be the most reactive. The differences in reactivity between the complexes were explained by electronic stabilization in the bent polycyclic-like phenantrene structure versus the linear anthracene shape. This modification exposed the aromatic-mimicking properties of the metallacyclic ring and the fact that the oxygen atom of the isopropoxy group is not just a simple chelating functional group, but participates in a cyclic π-conjugated structure that affects the properties of the catalyst. This work clearly showed that even the slightest changes in the benzylidene ligand may substantially alter the activity and stability of ruthenium catalysts in olefin metathesis reactions and not

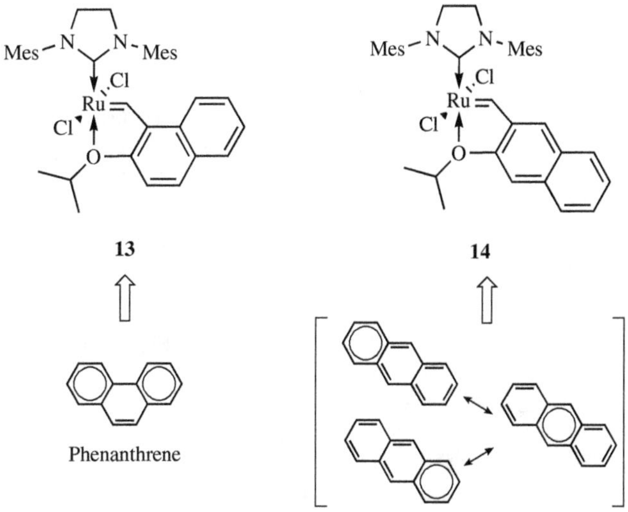

Figure 16.6 Anthracene- and phenanthrene-like structures of catalytic ruthenacycles.

only structural, but also electronic factors must be taken into account, as any physical organic chemist could predict.

Different approaches to ligand modifications were taken in attempts to further enhance the properties of known catalysts, one of which was substituting the isopropoxy ether moiety with an ester group as shown in Figure 16.7. (complexes **15a,b**) (24). These new complexes showed a chelated structure both in solid state and in solution, forming a six-membered ring with the metal. Although the reactivity of these complexes toward benchmark RCM substrates was diminished relatively to the "parent" Grubbs catalysts, the new structure revealed some insight into electronical properties of different ligands (e.g., ether vs ester).

Ester ligands in olefin metathesis were later used in a more creative manner in a joint collaborative work between the Grela, Arlt, and Lavigne groups. The addition of a methyl ester group (25) and other derivatives (26) to replace the isopropyl group next to the chelating oxygen atom led to an additional coordination option that significantly influenced catalyst behavior. These new catalysts presented high stability toward impurities and air on the one hand, but showed reactivity higher than even the Hoveyda–Grubbs catalyst on the other. This was explained by the stabilization of the complexes through the third chelation point, the carbonyl oxygen, which in a solution without the olefinic substrate protects the metal center by ligation to the empty coordination site, creating stable 18-electron complexes **16** and **17** (Fig. 16.7), whereas the high performance was curiously explained by "a more progressive liberation of the active propagating species in the reaction medium".

Recent work by Slugovc et al. revealed two carbonyl chelated benzylidene complexes which differed by their

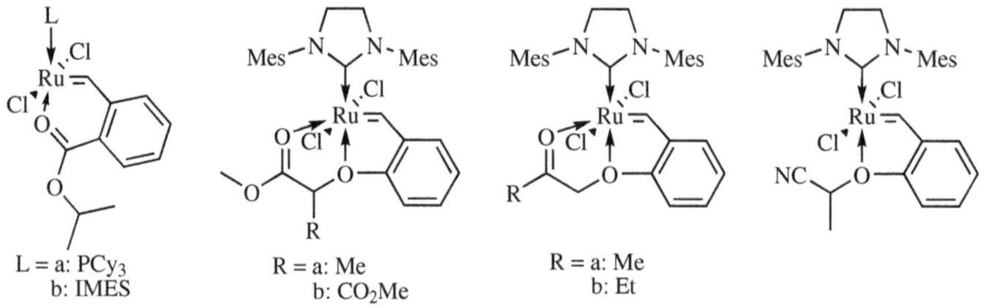

15a,b **16a,b** **17a,b** **18**

Figure 16.7 Carbonyl derived complexes.

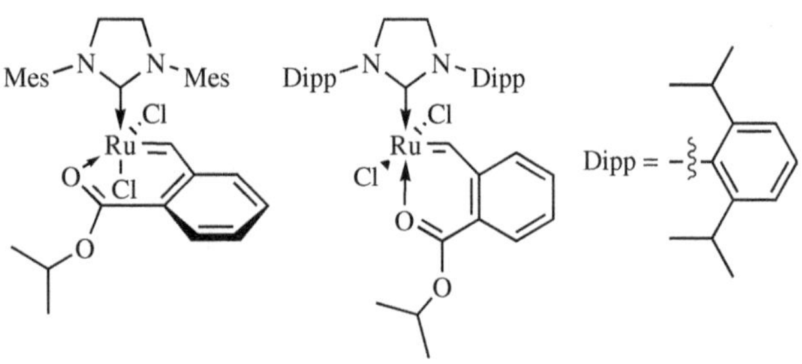

19 **20**

Figure 16.8 *cis-* versus *trans*-carbonyl chelated complexes.

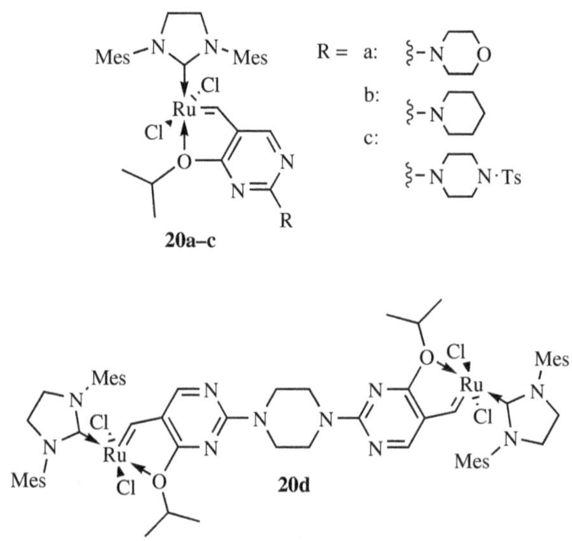

Figure 16.9 Pyrimidine-based complexes.

NHC moieties. Complex **19** (Fig. 16.8) (27) surprisingly afforded a *cis*-dichloro conformation. This complex could be activated by addition of pyridine, which facilitated dissociation of a more labile chloride ligand, trans to the more σ-donating benzylidene carbene. This result was observed for the first time and it suggested the importance of cationic species in cis–trans isomerization. In the second work (28), the mesityl-NHC group was exchanged by a diisopropylphenyl-NHC and only the trans complex was obtained (Fig. 16.8, complex **20**). This was explained by the larger steric bulk of the isopropyl groups toward the ester. Both complexes were submitted to benchmark ROMP and RCM reactions and both showed a latent behavior at ROMP and a similar activity in RCM at elevated temperatures. The difference was found in cross metathesis (CM) reactions, where the latter complex showed much higher activity even upon very low catalyst loadings.

Quite recently, Cao, Chen, and Chen (29) demonstrated a new family of complexes based on pyrimidine derivatives, including a homobimetallic complex (**20a–d**, Fig. 16.9). Different electron-rich substituents were used to increase stability, while retaining excellent activity as well. The bimetallic derivative, **20d**, on the other hand, did not present any unusual activity (see below, complex **45**).

When discussing the properties of olefin metathesis catalysts, one cannot disregard the mechanistic studies that were carried out for the chelated benzylidene complexes. The work by Plenio et al. in 2010 presents an important insight into the initiation mechanism of these complexes (Scheme 16.1). The UV–vis, fluorescent spectroscopic data, and nuclear Overhauser effect spectroscopy (NOESY) NMR analyses, along with other kinetic experiments that were conducted, served to shed light on the subject and present a clearer picture (30,31). The presence of both associative and dissociative mechanisms of initiation were found to be at play, suggesting an interchange mechanism which is governed by the effects of the olefin substrate, the substitutions on the benzylidene ligand and the steric bulk of the ether group, and relative concentrations of the catalyst and substrates. In a parallel work (32), using fluorine-tagged and fluorescent complexes **21a,b**, Plenio showed that no evidence for the previously suggested return-release (boomerang) effect (5,33) could be found and previous reports of catalyst recovery can be explained by recycling of the unreacted catalyst (Scheme 16.2) (34).

An promising work that is worth mentioning in the context of ligand modifications was done independently by Schrodi and Bruneau, who used the indenylidene complexes as a model. In the work of Schrodi (35), a chelated indenylidene

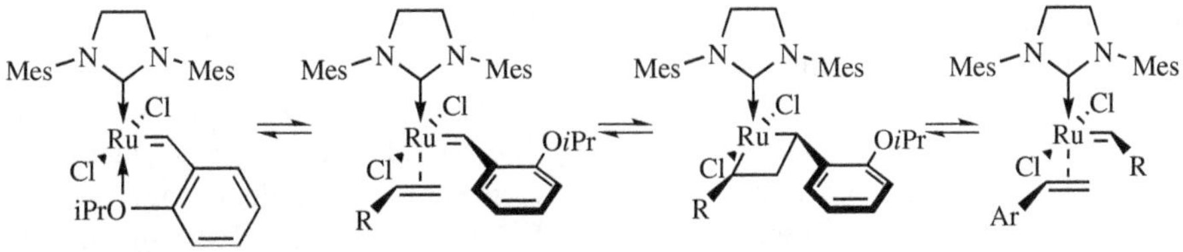

Scheme 16.1 Initiation of olefin metathesis reaction with Hoveyda-type catalysts.

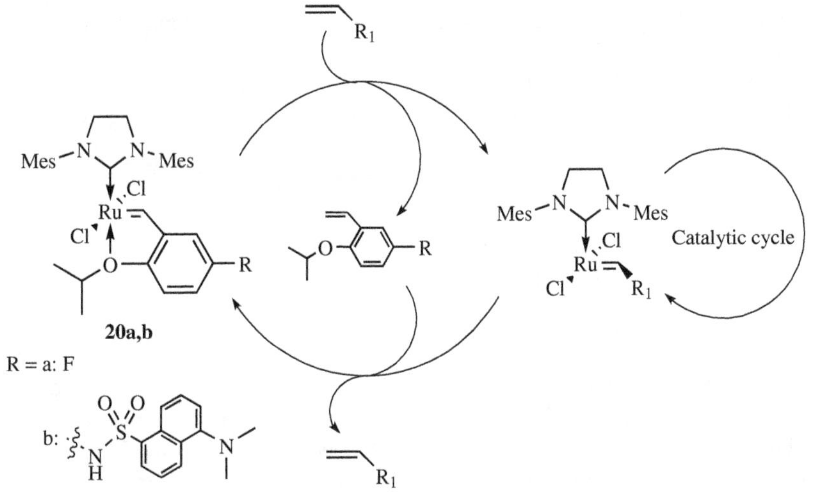

Scheme 16.2 “Boomerang” effect with fluorine- and dansyl-tagged complexes.

complex **22** (Fig. 16.10) was formed *in situ*, but not isolated. Later, an improved method for preparation and purification of this complex was published and the complex was fully investigated for its catalytic properties (36).

Concurrently, Bruneau (37) designed a complex bearing a similar indenylidene ether skeleton but developing the ligand further, by adding isopropoxy groups, to obtain complex **23** (Fig. 16.10) that showed not just high thermal stability but also latency, even for the highly active dicyclopentadiene monomer. Later, this work was expanded to other phenyl-indenylidene ligands, containing also methyl and fluoride substituents (**24a–c**, Fig. 16.10) (38). These complexes, bearing different electronic properties, did not show improved thermal stabilities or activities compared to those of the parent indenylidene or the Hoveyda–Grubbs complexes; however, they revealed a novel family of chelated ligands for olefin metathesis purposes.

16.2 MODIFICATIONS OF THE NHC

Having discussed the diversity of bidentate benzylidene ligands used in oxygen chelated complexes, we now describe complexes based on the ubiquitous isopropoxy-benzylidene

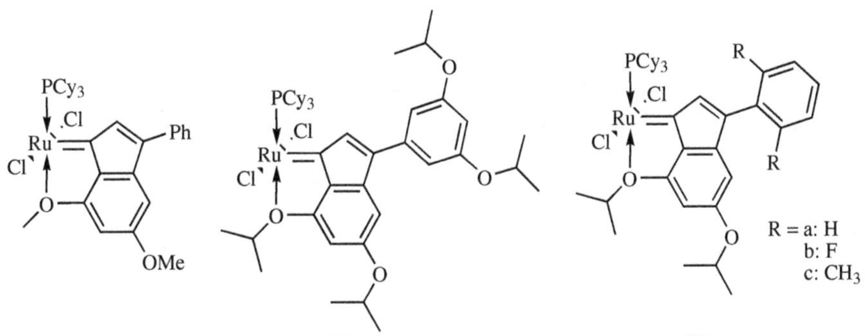

22 **23** **24a–c**

Figure 16.10 Indenylidene chelated complexes of Schrodi and Bruneau.

chelating ligand (derived from 2-isopropoxystyrene) with different NHC groups and their activity in olefin metathesis catalysis.

Throughout the mechanistic cycle of olefin metathesis, the NHC ligand remains bound to the ruthenium center. Thus, modification of the NHC ligand can significantly affect catalytic properties, starting from increased enantio- and stereoselectivity, through enhanced chemo-selectivity for certain substrates, to distinct rate accelerations, improved turn-over numbers (TONs), and stability. Changes on the classical mesitylene substituted Arduengo NHC may be classified into three main types: changing the main imidazoline ring, substitutions on the ring backbone, and functionalization of the aromatic moieties bound to the nitrogen atoms.

16.2.1 Modification of the Imidazoline

In 2005, Despagnet-Ayoub and Grubbs (39) presented a new ruthenium alkylidene complex, bearing a 1,3-diazetidine NHC as shown in Figure 16.11 (complex **25**). Although this new complex did not display unusual properties *per se*, it boasted an unconventional ligand backbone inspiring other modifications, such as Bielawski's complex **26** which has a six-membered *N,N′*-diamidocarbene (DAC) (40). This complex showed some improvement toward RCM of tetrasubstituted olefins. A seven-membered NHC (complex **27**) prepared by Buchmeiser in 2009 (41), initiated ROMP of 1,5-cyclooctadiene (COD) faster even than the Grubbs third generation catalyst.

Other landmark reports on modifications of the NHC ring include modifying not just the ring size but also the atom composition. For example, Vougioukalakis and Grubbs (42) prepared a thiazole-2-yilidene ligand (complex **28**), and Bertrand et al. (43) the 1,2,3-triazol-5-yilidene ligands (complex **29**). When coordinated to the ruthenium alkylidene, both displayed fair activity in benchmark RCM (DEDAM, diethyldiallyl malonate) and ROMP (COD, norbornene) reactions.

To summarize this part, modifications on the NHC ring, whether the ring size or atom composition, turned out to be synthetically quite convoluted and of little recompense, probably the reason for the reduced amount of work published on this specific subject.

16.2.2 Modification of the Backbone

Just as changes in the heterocyclic carbene ring bring about tuning of catalyst properties, so do functional group substitutions on the classical imidazoline framework. As shown upto now, modifications on the NHC ligand can drastically alter olefin metathesis catalysts, causing either beneficial or damaging transformations. This section will deal with variations on the backbone of the NHC ligand, that is, substitutions on the methylenes, without involving

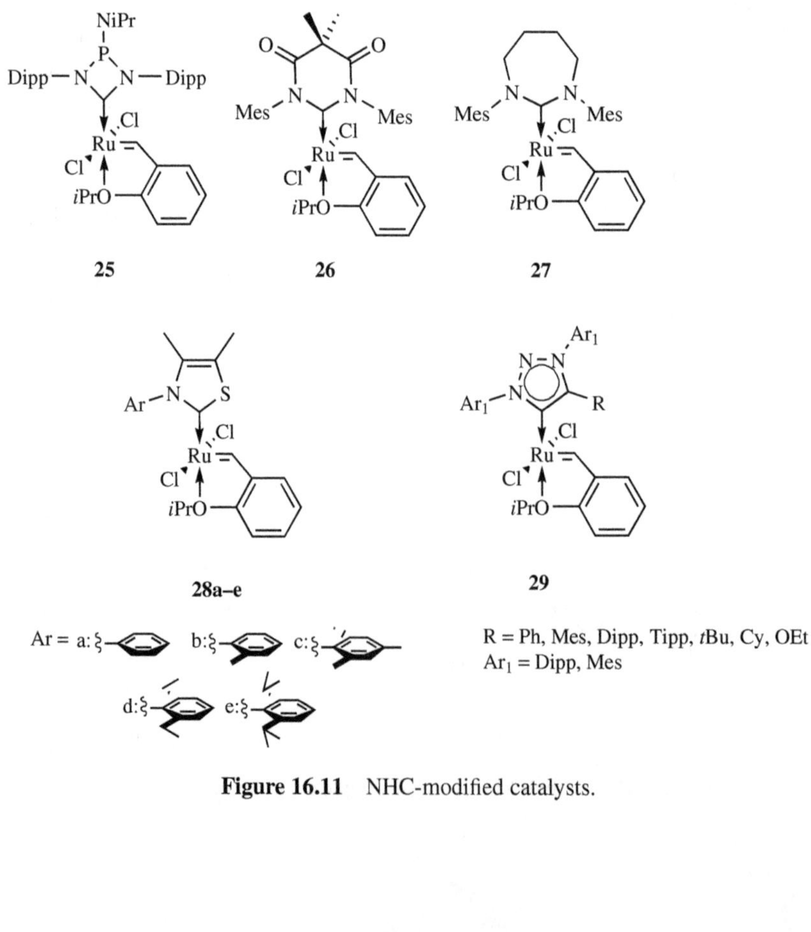

Figure 16.11 NHC-modified catalysts.

amendments on the *N*-mesitylene substituents (dealt with in the ensuing section).

When Grubbs (44) prepared the celebrated second generation catalyst in 1999, it was found that ruthenium catalysts substituted with imidazoline carbenes showed better activity than those substituted with imidazole carbenes (45,46). This was the most simple and earliest modification of the original NHC backbone (the removal of two hydrogens). Within the context of chelated NHC ruthenium complexes, Köhler et al. (47) developed a novel imidazolinium salt which led to the synthesis of a new ruthenium complex, **30a**. When the authors attempted the synthesis of complex **30b**, only the ring-closed ligand could be obtained, proving that this carbon–carbon double bond is active in olefin metathesis. Even though **30a** did not show any increase in catalytic performance over the parent Hoveyda–Grubbs catalyst, the most intriguing feature of this complex was that it contained the aforementioned reactive carbon–carbon double bond, which was untouched by the catalyst in solution (naturally, at much lower concentrations than the substrate) (Fig. 16.12).

Another positive aspect of this complex is that the double bond attached to the NHC could be a good handle for subsequent functionalization of the catalyst, in addition to its asymmetric spatial configuration, leaning as it does toward one side of the complex. Although modifications of the backbone had been reported previously (48) on Grubbs type complexes, this was the first classical bidentate Hoveyda-type complex that presented a modified NHC backbone.

16.2.3 Modification of the Mesitylene Substituents

The aromatic moieties on the NHC ring have a great influence on catalyst activity and stability (49). In all traditional Hoveyda-type catalysts, it was found that a large steric group is needed to maintain stability and prevent decomposition.

Grubbs and Schrodi (50,51) showed that by removing the methyl ortho positioned groups on the mesitylene substituents, it is possible to achieve high efficiency in the formation of hindered tetra-substituted olefins via RCM. However, just replacing the methyl on the ortho position with a hydrogen atom led to faster decomposition of the catalyst due to hydrogen abstraction and ruthenium hydride formation (52). Three different methods were used by Grubbs et al. to prevent phenyl rotation and achieve efficient catalysts to form tetrasubstituted olefins: one was the removal of just one of the methyl groups, leading to complex type **34**; the second method was to add very large substituents on the meta position, complex **33**, while the third approach exchanged the methyl groups with fluorides (complex **31**, Fig. 16.13).

Another approach was also reported by Chung and Grubbs (53) later in 2008. The main idea in this work was to somehow prevent the rotation of a relatively small phenyl aromatic group, to allow synthesis of hindered tetrasubstituted olefins, without the risk of catalyst deactivation by hydrogen abstraction. The addition of large bulky groups on the NHC backbone restricted rotation entirely and a phenyl-NHC containing complex was successfully prepared (Fig. 16.14).

These complexes (**35**, **36**) showed slightly diminished reactivity toward most metathesis benchmark reactions; however, complex **35** presented great reaction rate improvement

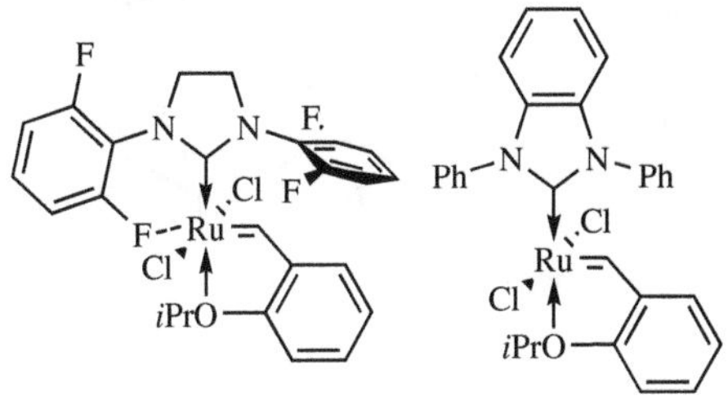

31 **32**

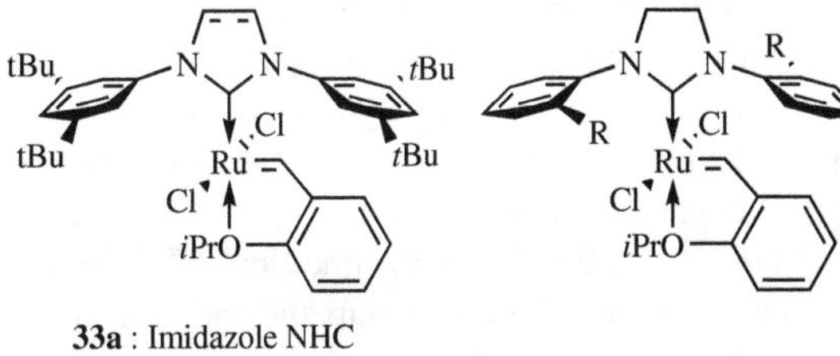

R = a: Me
b: Et
c: *i*Pr

33a : Imidazole NHC
33b : Imidazoline NHC

34a–c

Figure 16.13 Catalysts with reduced aromatic ring size in the ortho position.

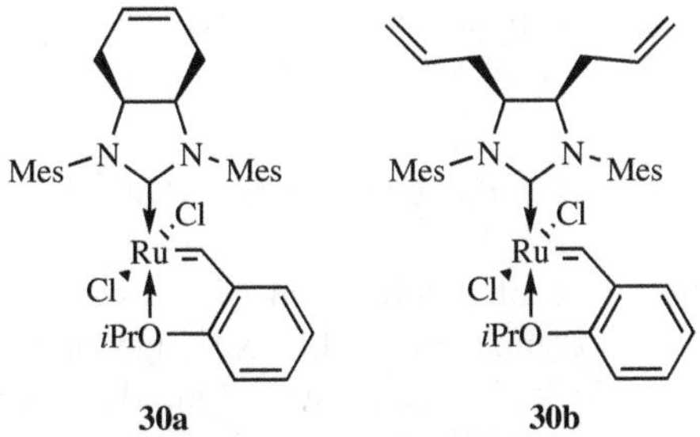

30a **30b**

Figure 16.12 Kohler's backbone-substituted catalysts.

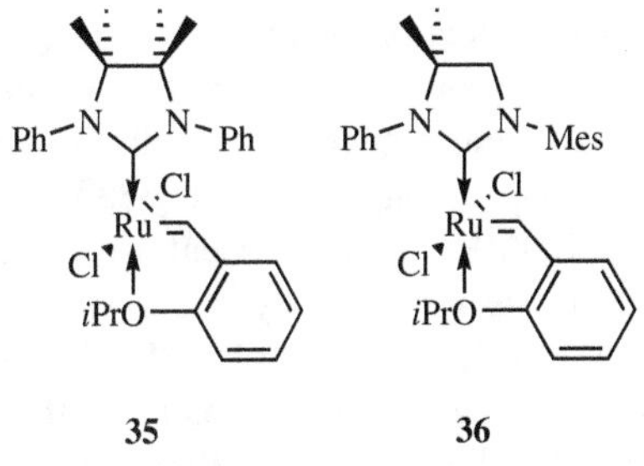

35 **36**

Figure 16.14 Grubbs' phenyl-substituted NHC.

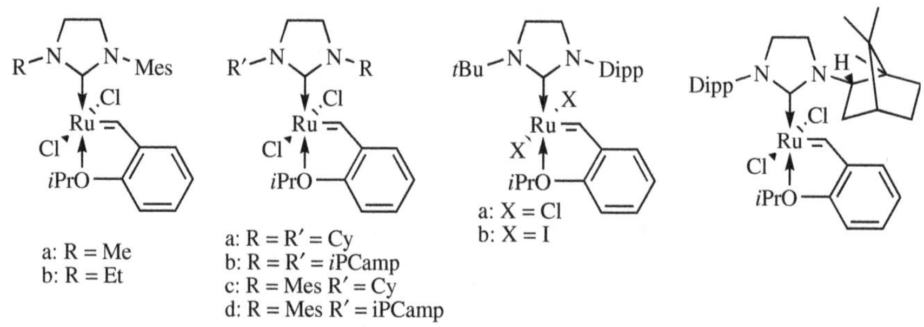

37a,b 38a–d 39a,b 40

Figure 16.15 Catalysts with unsymmetrical or aliphatic NHC groups.

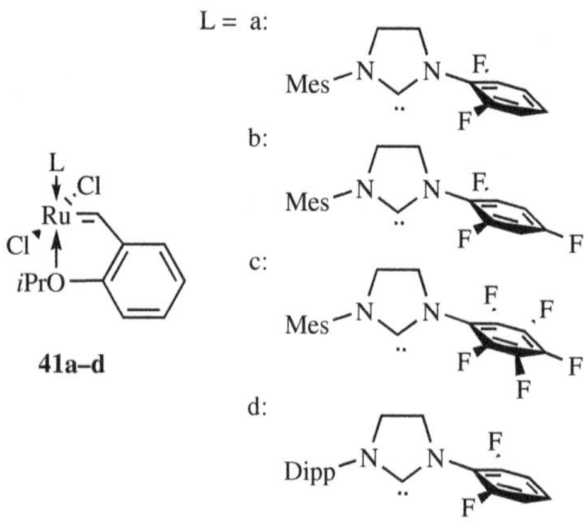

Figure 16.16 Fluorinated unsymmetrical catalysts by Grubbs.

in the RCM of diethyl dimethylallylmalonate (to afford a tetrasubstituted olefin) over the commercial catalysts, proving the efficiency of the backbone substitution.

Until 2006, all works on Hoveyda-type catalysts had only symmetrical NHC groups containing mesityl (Mes) or diisopropylphenyl (Dipp) aromatic moiety. In the work of Blechert (54) in 2006, the first unsymmetrical NHC ligands were reported with an aliphatic group on one side of the NHC ring and an aromatic group on the other. Complexes **37a** and **37b**, although presenting similar reactivity in RCM reactions, provided improved *E*:*Z* ratios in CM reactions and higher selectivity in diastereoselective RCM.

Verpoort (55) later showed that while increase of steric hindrance by introduction of one aliphatic group into the NHC framework as shown in Figure 16.15. (**38c,d**) did not improve the catalytic activity in metathesis reactions, the introduction of two aliphatic groups (**38a,b**) enhanced the reactivity for ROMP. In more recent works, Grubbs presented carefully designed catalysts for specific purposes, such as latency for ROMP (complex **39**) (56) or a plethora of unsymmetrically substituted complexes with guided selectivity toward ethenolysis (complex **40** was the most effective of the 10 complexes synthesized and studied) (57).

When considering modifications on the mesitylene group, naturally heteroatoms come to mind to control the electron-withdrawing or -donating influence of the aromatic ring. The use of heteroatoms on the aromatic groups of NHC was first presented by Grubbs (58,59), by using fluorinated aromatic rings in the catalyst's ligands (see above **31**, Fig. 16.13). This example presented increased efficiency in metathesis due to a unique fluorine–ruthenium interaction, which later led to the development of unsymmetrical fluorinated catalysts **41a–d** (Fig. 16.16), some of which surpassed their commercial predecessors both in activity and *E*:*Z* ratios (60).

The work of Schanz (61,62), where nitrogen atoms were introduced into the aromatic rings, demonstrated pH-controlled catalysts both for acidity-dependent activation and catalyst separation (Scheme 16.3).

Blechert on the other hand demonstrated deactivation of olefin metathesis catalysis through carbene–arene bond by fusing the aromatic ring to the imidazoline framework (**43**, Fig. 16.17) (63) and later expanded this work to obtain *E*-selective catalysis (**44**, Fig. 16.17) (64).

An interesting approach to catalyst modification was put forward by Lemcoff et al. (65), who presented a bimetallic olefin metathesis catalyst **45**, containing two ruthenium atoms (Fig. 16.18). This novel catalyst showed a high selectivity toward "dimer ring-closing metathesis" (DRCM). By using a diene substrate of an appropriate length, that matched the distance between the ruthenium centers, the cyclic dimer was formed with over 80% selectivity over the acyclic diene metathesis (ADMET) and RCM products obtained with the monomeric catalysts.

16.3 ANIONIC LIGAND EXCHANGE

Yet another type of modification of the oxygen chelated Hoveyda-type complexes is the exchange of the anionic ligands. Braddock (66) demonstrated in 2006 the exchange of the anionic chloride ligands to trifluoroacetate groups, but perhaps more importantly he showed that there is a constant

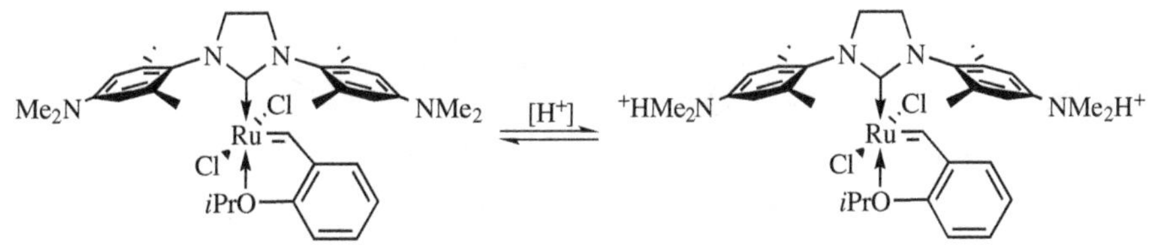

42a 42b

Scheme 16.3 pH-controlled catalyst.

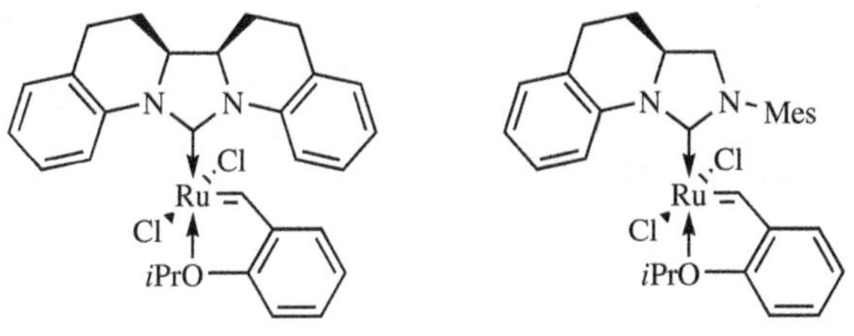

43 44

Figure 16.17 NHC with fused aromatic rings.

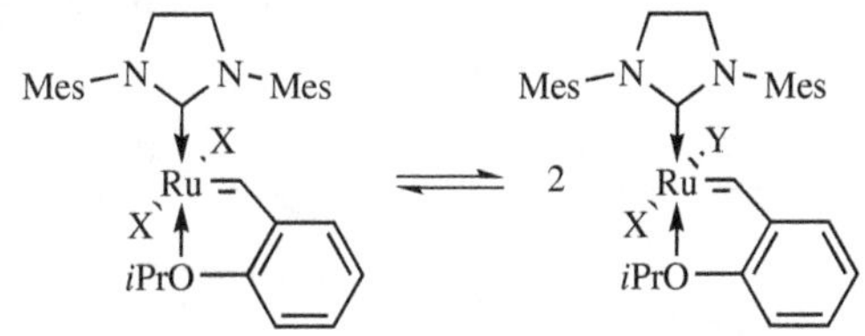

X = a: O_2CCF_3
b: $O_2CC_2F_5$
46a,b

X = O_2CCF_3
Y = $O_2CC_2F_5$
47

Scheme 16.4 Interchange of anionic ligands in equilibrium.

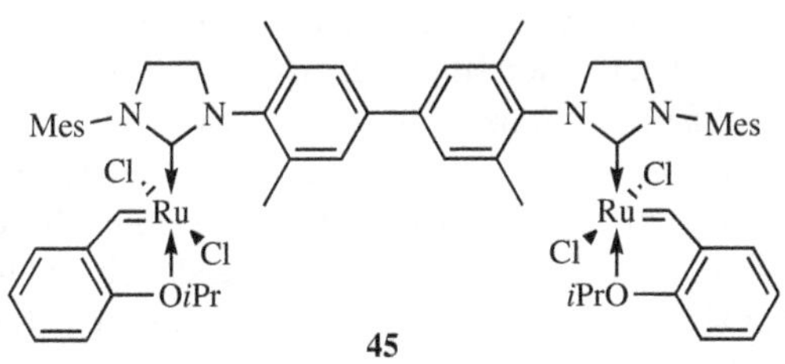

45

Figure 16.18 Bimetallic ruthenium catalyst.

exchange between anionic ligands in solution. This was revealed by mixing two different types of catalysts (**46a** and **46b**) with perfluoroacetate and perfluoropropanoate ligands respectively. ^{1}H-NMR analyses unequivocally indicated that a new complex **47** containing one of each anionic ligand was formed in the statistical equilibrium ratio of 1 : 2 : 1 (**46a** : **47** : **46b**). Naturally, the discovery that anionic ligands freely interchange during the reaction has some serious implications when planning new olefin metathesis catalysts (Scheme 16.4).

In some early studies, Hoveyda (67,68) introduced chelating anionic ligands by using optically pure 2-hydroxy-2′-binaphthylamine for the synthesis of a bidentate complex (complex **48**, Fig. 16.19). Notably, this was the first ruthenium catalyst that was capable of asymmetric olefin metathesis reactions. More recently, Grubbs showed a series of 18-electron oxygen-chelated complexes, **49a–e**, with different anionic ligands and an adamantyl-substituted NHC which also chelates the metal. This type of catalyst demonstrated high selectivity for *Z*-olefin products in CM reactions (69,70).

Gawin and Grela (71) recently reported the synthesis of a novel, tridentate benzylidene ligand, which exchanges one of the chlorides as well when forming the ruthenium alkylidene complex. Substituting the remaining chloride for an iodide gave a remarkably stable complex, which could be activated by the addition of an acid, regenerating an active 16-electron complex. Remarkably, the addition of the iodide "locked" the labile carboxylate ligand and prevented ligand exchange. Complex **50b**, although having an asymmetric center, unfortunately did not display any asymmetric induction toward an appropriate triene substrate (Scheme 16.5).

A new perspective on anion exchange in olefin metathesis catalysts was shown by Slugovc et al. (72). In this work, the authors presented the difference between the two chloride atoms on carbonyl-chelated bidentate *cis*-dihalo complexes **51**. It was found that the halide *trans* to the strong σ-donating NHC group possesses much higher lability than the halide *cis* to the NHC in presence of donor solvents such as water and methanol. It was also suggested that the rapid exchange of this halide goes through a dissociative mechanism, whereas the slow exchange of the *cis* halide, both in dry and wet solvents, was observed and needed a large excess of iodide in the solution and longer time. Thus, iodo-chloro (**51b,d**) and diiodo (**51c**) complexes were obtained by anionic ligand exchange and these complexes were subjected to a benchmark ROMP reaction. Complex **51d** showed much higher activity and this could be explained by its tendency to rearrange to all three possible *trans*-dihalo active species in solution (Fig. 16.20).

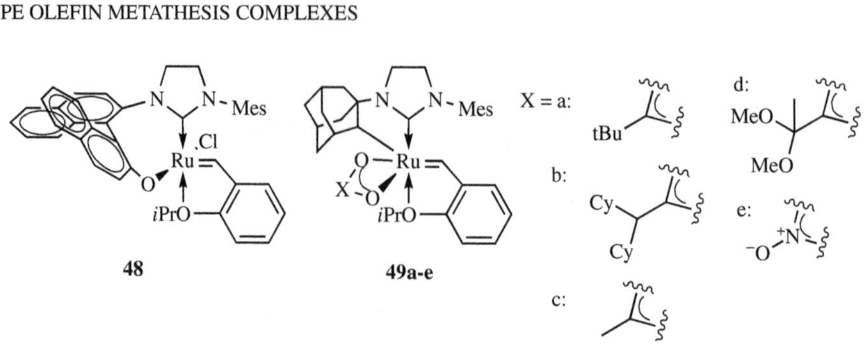

Figure 16.19 Chelating NHC with different anionic ligands.

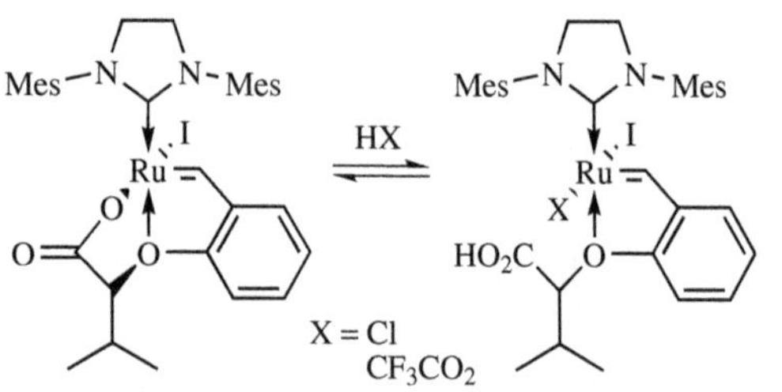

Scheme 16.5 Activation of tridentate complex.

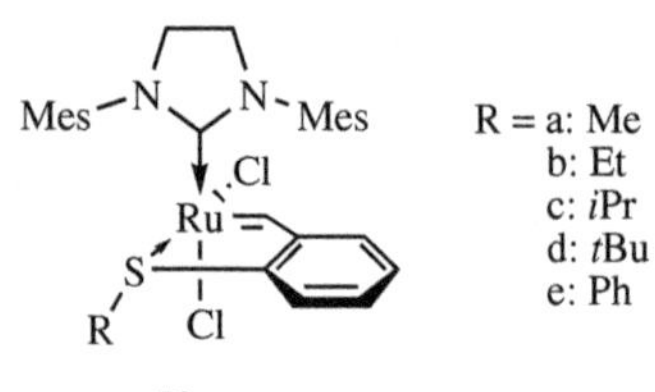

Figure 16.21 Sulfur-chelated catalysts.

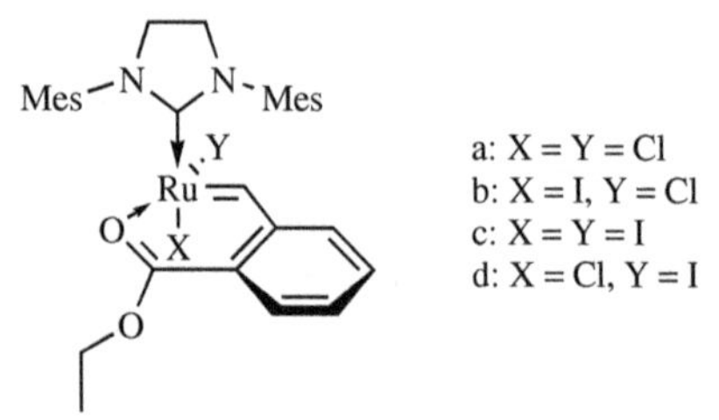

Figure 16.20 *cis*-Dihalo carbonyl-chelated complexes.

16.4 HETEROATOMS IN BENZYLIDENE LIGAND

As presented until now, ruthenium benzylidene chelated complexes have a wide variety of possibilities for tuning and modification. An additional option for catalyst design is presented when considering exchanging the ether oxygen atom for other heteroatoms. The exchange of oxygen with sulfur resulted in a surprising outcome, affording the novel complex **52c** (73), which displayed the unusual *cis*-dichloro conformation (Fig. 16.21). This complex was very stable and was completely inert toward RCM at room temperature; however, when heated to higher temperatures, the complex showed some activity for benchmark RCM and ROMP reactions. Further synthetic work revealed that larger steric groups attached to the sulfur atom accelerated the catalyzed reaction and showed reasonable metathesis activity at lower temperatures (74).

The latent behavior of the complex also led to a different type of polymerization, dubbed "TON controlled polymerization" (75). In contrast to the normal living-type polymerizations obtained with ROMP, the sulfur-chelated initiator afforded polymers that had the same chain length throughout the whole polymerization process. It was concluded that the polymer size was controlled by the TON of the catalyst only because the polymer propagation was much faster than chain initiation, leading to low polydispersities and precise molecular weights, depending only on the reaction conditions and not on the initiator to monomer ratio.

In order to fully understand the mechanism of latency in sulfur and other heteroatom chelated benzylidene complexes, the electronic properties of these catalysts must be understood. In 2009, Lemcoff and Straub (76) predicted by density functional theory (DFT) studies and proved by experimental studies the configuration of novel heteroatom substituted Hoveyda-type complexes. Thus, it was calculated that oxygen- and nitrogen-chelated complexes prefer the trans configuration, while sulfur, selenium, and phosphorus prefer the cis form (Fig. 16.22). Analogs of the unknown

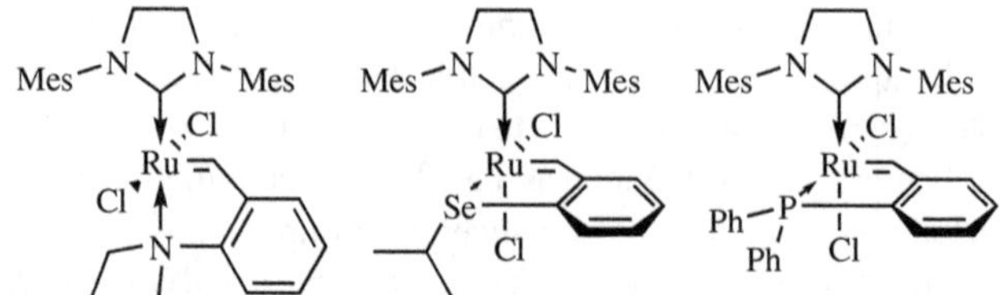

Figure 16.22 Nitrogen, selenium, and phosphorous chelated complexes.

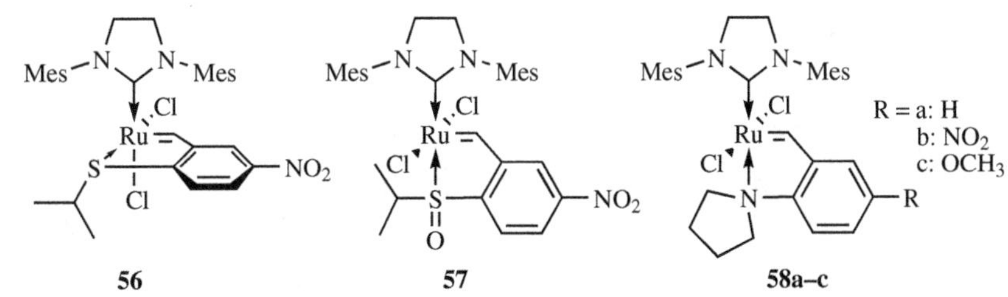

Figure 16.23 Bidentate complexes based on sulfur and nitrogen chelating benzylidenes.

Hoveyda–Grubbs second generation catalyst (chelated sp^3 nitrogen, selenium, and phosphorous) were made to probe the predictions; indeed, they were shown by X-ray crystallography measurements and NMR solution analyses to present the foretold structure. It was proposed that the strong sigma electron donation of chelating atoms, such as sulfur, selenium, and phosphorous, increased the bond strength of the metal and the heteroatom; but the strong trans influence of the NHC ligand repelled the positioning of these heteroatoms in the trans position typically occupied by the chelating oxygen atom. Slugovc et al. (77) later studied the catalytic properties of phosphorous-chelated complex **55**.

In continuation of this work by Grela and Lemcoff in 2010 (78), the authors presented a series of bidentate complexes based on sulfur- and nitrogen-chelating benzylidenes, bearing EWG or electron-donating groups (Fig. 16.23). Spectroscopic and crystallographic data showed that nitrogen-containing complexes could be strongly activated by EWG, whereas the sulfur-chelated complex remained latent even when the aromatic ring was substituted by a *p*-nitro group. As Grela (79) showed, only oxidation of the sulfur increased the activity and also stabilized the complex in the *trans*-dichloro form. The sulfone ligand did not form a chelated complex with this type of ligands. However, later work by Grela disclosed that proper substitution of the sulfur can lead to sulfone and sulfonamide chelated derivatives, introducing a new family of reactive O-chelated ruthenium benzylidenes (80) (complex **59**, Fig. 16.24).

In order to increase the activity of the sulfur-chelated complexes, Lemcoff et al. pursued more efficient ways to activate the latent complexes. It was known that photo-dissociation of a dimethyl sulfoxide (DMSO) molecule from $RuCl_2(DMSO)_4$ can be achieved by UV irradiation (81); thus, Hoveyda-type sulfur chelated complexes were subjected to UV light. Surprisingly, the irradiation led to isomerization of the cis to the trans isomers and olefin metathesis catalytic activity was observed even at room temperature. As shown in Scheme 16.6, the trans could be isomerized back to the cis by heating, giving a switchable system in which activation was initiated by UV light and deactivation by heating (82).

In more recent work (83), Lemcoff et al. showed that upon increase of steric hindrance (phenyl **60**, 2,6-dimethylphenyl **61**, and 2,4,6-triisopropylphenyl **62**) the energy gap between cis and trans isomers is reduced (Fig. 16.25). Also, in polar or halogen-containing solvents, the isomerization from *trans*-dichloro to *cis*-dichloro is faster. Moreover, it was shown that addition of π-coordinating molecules to the solution lowers the barrier for isomerization as well. Structure–activity studies of these complexes suggested that olefin metathesis with chelated complexes happens only through the *trans*-dichloro isomer, present in minute amounts in the equilibrium.

In order to increase the activity gap between the "active" and "resting" isomers of sulfur-chelated catalysts, two new complexes were synthesized (84). Complex **63** contained an electron withdrawing CF_3 group that weakened the ruthenium–sulfur bond, giving a very active catalyst in the *trans*-dichloro configuration, while the *cis*-dichloro remained inactive for most benchmark RCM reactions at room temperature (Fig. 16.26). Activation could be obtained both by irradiation with UV light and by heating. On the other

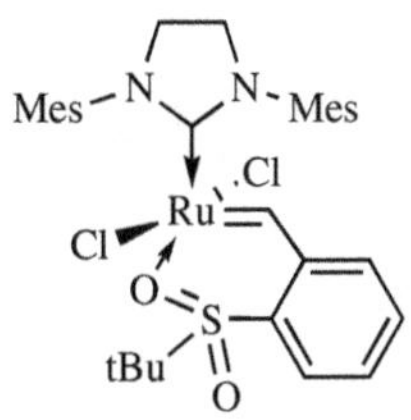

59

Figure 16.24

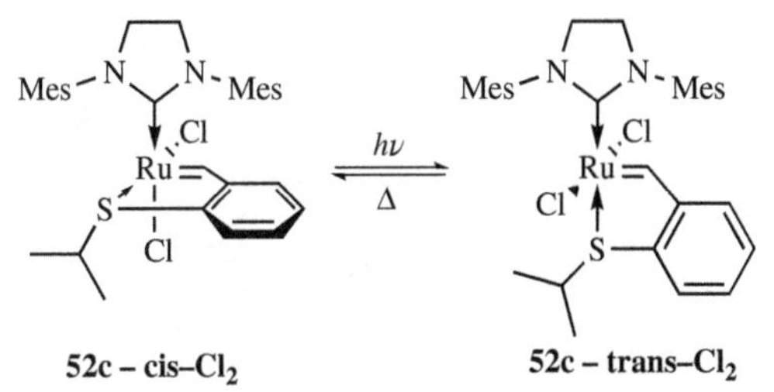

Scheme 16.6 cis–trans isomerization of sulfur-chelated complexes.

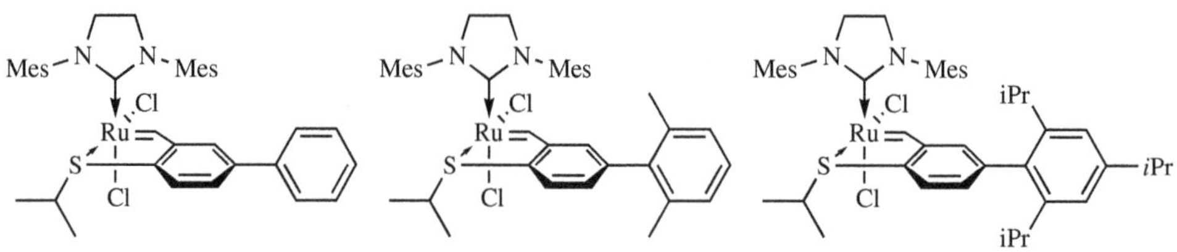

60 61 62

Figure 16.25 Sulfur-chelated catalysts with increasing steric bulk.

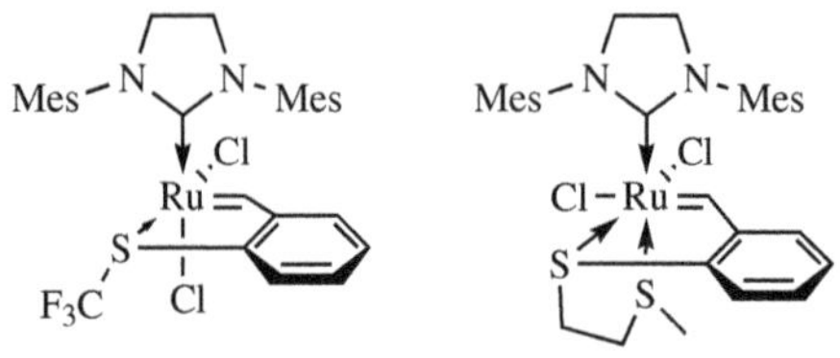

63 64

Figure 16.26 CF_3 substituted and tridentate sulfur-chelated complexes.

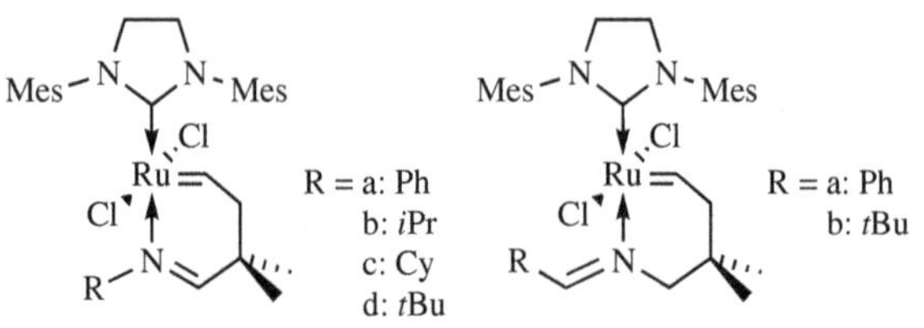

65a–e 66a–b

Figure 16.27 Imine-based alkylidene complexes.

hand, further coordination of the empty site was used to make a tridentate complex **64**, which showed latency even for the highly active dicyclopentadiene (DCPD) monomer. Surprisingly, the chloride atom was shifted to the position trans to the benzylidene carbene, a ligand with very strong trans influence, to prevent coordination of the sulfur in this difficult position.

As discussed above, nitrogen can also be a strong coordinating ligand for ruthenium-based Hoveyda-type complexes. Grubbs demonstrated the importance of the delicate balance between activation and latency, using alkylidene bidentate ligands with an imine (Fig. 16.27). Steric modifications (complexes **65**) along with placement of the imine bond (complexes **66**) were key factors in tuning of the catalytic properties of these complexes. The more bulky substituted **65** catalysts afforded slower initiation rates, probably due to a more difficult approach for the olefin; on the other hand, the complex with the exocyclic imine bond (**66**) presented very fast initiation (85).

In an early work by Slugovc et al. (86), imine-based bidentate complexes containing a five- or six-membered ring (**67** and **68**, respectively) were presented (Fig. 16.28). These complexes revealed slight thermo-switchability and demonstrated different initiation rates for benchmark ROMP reactions, based on the stability of the different rings. In further works by Grela and Slugovc (87,88), quinoline (**69a**) and quinoxaline (**69b**) bidentate complexes were presented, both in cis and in trans configurations. As expected, the trans isomers showed higher activity, due to faster dissociation of the chelating nitrogen.

In more recent work (89), Plenio et al. presented a series of amino-benzylidene-chelated complexes which showed very rapid precatalyst initiation and high stability in air (Fig. 16.29). Moreover, these complexes showed high conversion in various RCM reactions with only 15–150 ppm loadings of catalyst. It was found that complexes with mesityl substituted NHCs (**70a,b**) initiate much faster than those with diisopropylphenyl NHCs (**71**) and the suggested reason was the possibility to have a second rate-limiting step after the first initiation.

Complex **70c**, which was also recently reported by Grela (90), involves a benzylamine group and showed both thermal

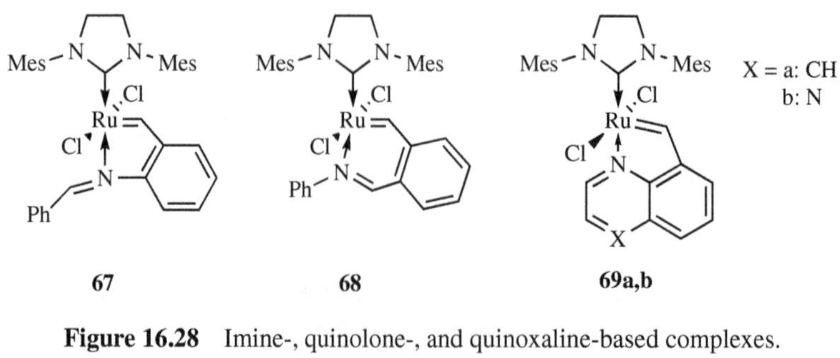

67 68 69a,b

Figure 16.28 Imine-, quinolone-, and quinoxaline-based complexes.

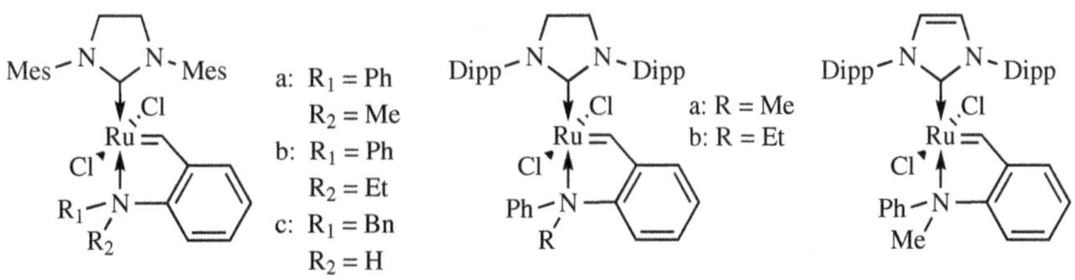

70a–c **71a,b** **72**

Figure 16.29 Amino-benzylidene chelating complexes.

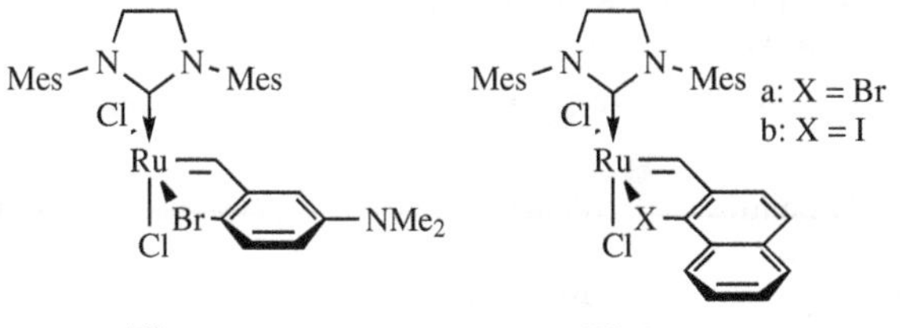

73 **74a,b**

Figure 16.30 Halogen chelated complexes.

activation at and chemoswitchability with trimethylsilyl chloride (TMSCl) as the activating agent. This complex was used in a variety of metathesis reactions and showed fair reactivity for RCM, CM, and ene–yne metathesis.

A new family of fascinating halogen chelated benzylidene complexes (Fig. 16.30) was recently introduced by Barbasiewicz and Grela (91). The authors showed the synthesis of bromo- and iodo-chelated complexes in which the *cis*-dichloro configuration appeared predominant. Dimethyl amino substituent was introduced to stabilise the bromo-chelate (complex **73**, Fig. 16.30), while the unsubstituted bromo-chelate was only observed by 1H-NMR spectroscopy. Increase in stability was also achieved by substituting the benzylidene for a naphthylidene as seen in complexes **74**. These complexes showed fair reactivity for RCM and CM reactions compared to the parent Hoveyda catalyst and were fairly stable in air.

In addition, electron-donating group and EWG were inserted into the iodo-benzylidene moiety (92) (Fig. 16.31).

Unexpectedly, the withdrawing bromo and nitro groups (**75d** and **75e** respectively) reduced the activity of the complexes, whereas electron donating dimethylamino and methoxy groups (**75a** and **75b** respectively) had smaller deactivating effects. The authors proposed two explanations for this phenomenon: one is stabilization of the inactive *cis*-dichloro structure and the other is an alternative activation mechanism in which the halogen–ruthenium bond remains intact in the rate-determining step, resulting in a mechanism with a more associative character.

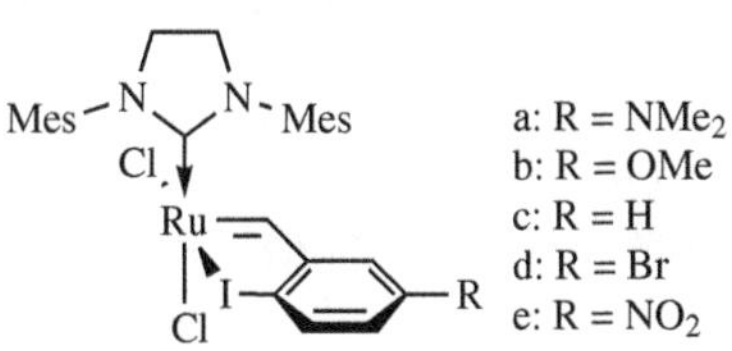

75a–e

Figure 16.31 Halogen chelated complexes with electron donating and withdrawing groups.

REFERENCES

1. Arduengo AJ III, Harlow RL, Kline M. *J Am Chem Soc* 1991;113:361–363.
2. Arduengo AJ III, Kline M, Calabrese JC, Davidson F. *J Am Chem Soc* 1991;113:9704–9705.
3. Harrity JPA, Visser MS, Gleason JD. *J Am Chem Soc* 1997;119:1488–1489.
4. Kingsbury JS, Harrity JPA, Bonitatebus PJ Jr, Hoveyda AH. *J Am Chem Soc* 1999;121:791–799.
5. Garber SB, Kingsbury JS, Gray BL, Hoveyda AH. *J Am Chem Soc* 2000;122:8168–8179.
6. Gessler S, Randl S, Blechert S. *Tetrahedron Lett* 2000;41: 9973–9976.
7. Imhof S, Randl S, Blechert S. *Chem Commun* 2001: 1692–1693.
8. Dunne AM, Mix S, Blechert S. *Tetrahedron Lett* 2003;44: 2733–2736.
9. Wakamatsu H, Blechert S. *Angew Chem Int Ed* 2002;41: 2403–2405.
10. Wakamatsu H, Blechert S. *Angew Chem Int Ed* 2002;41: 794–796.
11. Connon SJ, Rivard M, Zaja M, Blechert S. *Adv Synth Catal* 2003;345:572–575.
12. Zaja M, Connon SJ, Dunne AM, Rivard M, Buschmann N, Jiricek J, Blechert S. *Tetrahedron* 2003;59:6545–6558.
13. Grela K, Harutyunyan S, Michrowska A. *Angew Chem Int Ed* 2002;41:4038–4040.
14. Grela K, Kim M. *Eur J Org Chem* 2003;2003:963–966.
15. Bujok R, Bieniek M, Masnyk M, Michrowska A, Sarosiek A, Stępowska H, Arlt D, Grela K. *J Org Chem* 2004;69:6894–6896.

16. Michrowska A, Bujok R, Harutyunyan S, Sashuk V, Dolgonos G, Grela K. *J Am Chem Soc* 2004;126:9318–9325.
17. Gułajski, Ł; Michrowska, A.; Bujok, R.; Grela, K., *J Mol Catal A: Chem* 2006, 254, 118-123.
18. Barbasiewicz M, Bieniek M, Michrowska A, Szadkowska A, Makal A, Woźniak K, Grela K. *Adv Synth Catal* 2007; 349:193–203.
19. Kos P, Savka R, Plenio H. *Adv Synth Catal* 2013;355:439–447.
20. Hryniewicka A, Morzycki JW, Witkowski S. *J Organomet Chem* 2010;695:1265–1270.
21. Hryniewicka A, Kozłowska A, Witkowski S. *J Organomet Chem* 2012;701:87–92.
22. Barbasiewicz M, Szadkowska A, Makal A, Jarzembska K, Woźniak K, Grela K. *Chem Eur J* 2008;14:9330–9337.
23. Leitgeb A, Szadkowska A, Michalak M, Barbasiewicz M, Grela K, Slugovc C. *J Polym Sci A: Polym Chem* 2011;49:3448–3454.
24. Fürstner A, Thiel OR, Lehmann CW. *Organometallics* 2002;21:331–335.
25. Bieniek M, Bujok R, Cabaj M, Lugan N, Lavigne G, Arlt D, Grela K. *J Am Chem Soc* 2006;128:13652–13653.
26. Bieniek M, Samojłowicz C, Sashuk V, Bujok R, Śledź P, Lugan N, Lavigne G, Arlt D, Grela K. *Organometallics* 2011;30:4144–4158.
27. Zirngast M, Pump E, Leitgeb A, Albering JH, Slugovc C. *Chem Commun* 2011;47:2261–2263.
28. Leitgeb A, Abbas M, Fischer RC, Poater A, Cavallo L, Slugovc C. *Catal Sci Technol* 2012.
29. Wu G -L, Cao S -L, Chen J, Chen Z. *Eur J Org Chem* 2012;2012:6777–6784.
30. Vorfalt T, Wannowius K -J, Plenio H. *Angew Chem Int Ed* 2010;49:5533–5536.
31. Thiel V, Hendann M, Wannowius KJ, Plenio H. *J Am Chem Soc* 2012;134:1104–1114.
32. Vorfalt T, Wannowius KJ, Thiel V, Plenio H. *Chem Eur J* 2010;16:12312–12315.
33. Ahmed M, Barrett AGM, Braddock DC, Cramp SM, Procopiou PA. *Tetrahedron Lett* 1999;40:8657–8662.
34. Rix D, Caïjo F, Laurent I, Gułajski Ł, Grela K, Mauduit M. *Chem Commun* 2007:3771–3773.
35. Jimenez LR, Gallon BJ, Schrodi Y. *Organometallics* 2010;29:3471–3473.
36. Jimenez LR, Tolentino DR, Gallon BJ, Schrodi Y. *Molecules* 2012;17:5675–5689.
37. Kabro A, Roisnel T, Fischmeister C, Bruneau C. *Chem Eur J* 2010;16:12255–12261.
38. Kabro A, Ghattas G, Roisnel T, Fischmeister C, Bruneau C. *Dalton Trans* 2012;41:3695–3700.
39. Despagnet-Ayoub E, Grubbs RH. *Organometallics* 2005;24: 338–340.
40. Moerdyk JP, Bielawski CW. *Organometallics* 2011;30: 2278–2284.
41. Kumar PS, Wurst K, Buchmeiser MR. *Organometallics* 2009;28:1785–1790.
42. Vougioukalakis GC, Grubbs RH. *J Am Chem Soc* 2008;130: 2234–2245.
43. Bouffard J, Keitz BK, Tonner R, Guisado-Barrios G, Frenking G, Grubbs RH, Bertrand G. *Organometallics* 2011;30:2617–2627.
44. Scholl M, Ding S, Lee CW, Grubbs RH. *Org Lett* 1999;1: 953–956.
45. Ackermann L, Fürstner A, Weskamp T, Kohl FJ, Herrmann WA. *Tetrahedron Lett* 1999;40:4787–4790.
46. Scholl M, Trnka TM, Morgan JP, Grubbs RH. *Tetrahedron Lett* 1999;40:2247–2250.
47. Weigl K, Köhler K, Dechert S, Meyer F. *Organometallics* 2005;24:4049–4056.
48. Vougioukalakis GC, Grubbs RH. *Chem Rev* 2010;110: 1746–1787.
49. Samojłowicz C, Bieniek M, Grela K. *Chem Rev* 2009;109: 3708–3742.
50. Berlin JM, Campbell K, Ritter T, Funk TW, Chlenov A, Grubbs RH. *Org Lett* 2007;9:1339–1342.
51. Stewart IC, Ung T, Pletnev AA, Berlin JM, Grubbs RH, Schrodi Y. *Org Lett* 2007;9:1589–1592.
52. Hong SH, Chlenov A, Day MW, Grubbs RH. *Angew Chem Int Ed* 2007;46:5148–5151.
53. Chung CK, Grubbs RH. *Org Lett* 2008;10:2693–2696.
54. Vehlow K, Maechling S, Blechert S. *Organometallics* 2006;25:25–28.
55. Ledoux N, Linden A, Allaert B, Vander Mierde H, Verpoort F. *Adv Synth Catal* 2007;349:1692–1700.
56. Thomas RM, Fedorov A, Keitz BK, Grubbs RH. *Organometallics* 2011;30:6713–6717.
57. Thomas RM, Keitz BK, Champagne TM, Grubbs RH. *J Am Chem Soc* 2011;133:7490–7496.
58. Ritter T, Day MW, Grubbs RH. *J Am Chem Soc* 2006;128: 11768–11769.
59. Vougioukalakis GC, Grubbs RH. *Organometallics* 2007;26: 2469–2472.
60. Vougioukalakis GC, Grubbs RH. *Chem Eur J* 2008;14: 7545–7556.
61. Balof SL, P'Pool SJ, Berger NJ, Valente EJ, Shiller AM, Schanz HJ. *Dalton Trans* 2008:5791–5799.
62. Balof SL, Yu B, Lowe AB, Ling Y, Zhang Y, Schanz HJ. *Eur J Inorg Chem* 2009;2009:1717–1722.
63. Vehlow K, Gessler S, Blechert S. *Angew Chem Int Ed* 2007;46:8082–8085.
64. Kannenberg A, Rost D, Eibauer S, Tiede S, Blechert S. *Angew Chem Int Ed* 2011;50:3299–3302.
65. Tzur E, Ben-Asuly A, Diesendruck CE, Goldberg I, Lemcoff NG. *Angew Chem Int Ed* 2008;47:6422–6425.
66. Tanaka K, Böhm VPW, Chadwick D, Roeper M, Braddock DC. *Organometallics* 2006;25:5696–5698.
67. Van Veldhuizen JJ, Garber SB, Kingsbury JS, Hoveyda AH. *J Am Chem Soc* 2002;124:4954–4955.
68. Van Veldhuizen JJ, Gillingham DG, Garber SB, Kataoka O, Hoveyda AH. *J Am Chem Soc* 2003;125:12502–12508.

69. Endo K, Grubbs RH. *J Am Chem Soc* 2011;133:8525–8527.
70. Keitz BK, Endo K, Patel PR, Herbert MB, Grubbs RH. *J Am Chem Soc* 2011;134:693–699.
71. Gawin R, Grela K. *Eur J Inorg Chem* 2012;2012:1477–1484.
72. Pump E, Fischer RC, Slugovc C. *Organometallics* 2012; 31:6972–6979.
73. Ben-Asuly A, Tzur E, Diesendruck CE, Sigalov M, Goldberg I, Lemcoff NG. *Organometallics* 2008;27:811–813.
74. Kost T, Sigalov M, Goldberg I, Ben-Asuly A, Lemcoff NG. *J Organomet Chem* 2008;693:2200–2203.
75. Diesendruck CE, Vidavsky Y, Ben-Asuly A, Lemcoff NG. *J Polym Sci A: Polym Chem* 2009;47:4209–4213.
76. Diesendruck CE, Tzur E, Ben-Asuly A, Goldberg I, Straub BF, Lemcoff NG. *Inorg Chem* 2009;48:10819–10825.
77. Lexer C, Burtscher D, Perner B, Tzur E, Lemcoff NG, Slugovc C. *J Organomet Chem* 2011;696:2466–2470.
78. Tzur E, Szadkowska A, Ben-Asuly A, Makal A, Goldberg I, Woźniak K, Grela K, Lemcoff NG. *Chem Eur J* 2010;16:8726–8737.
79. Szadkowska A, Makal A, Woźniak K, Kadyrov R, Grela K. *Organometallics* 2009;28:2693–2700.
80. Szadkowska A, Zukowska K, Pazio AE, Wozniak K, Kadyrov R, Grela K. *Organometallics* 2011;30:1130–1138.
81. Alessio E, Mestroni G, Nardin G, Attia WM, Calligaris M, Sava G, Zorzet S. *Inorg Chem* 1988;27:4099–4106.
82. Diesendruck CE, Iliashevsky O, Ben-Asuly A, Goldberg I, Lemcoff NG. *Macromol Symp* 2010;293:33–38.
83. Aharoni A, Vidavsky Y, Diesendruck CE, Ben-Asuly A, Goldberg I, Lemcoff NG. *Organometallics* 2011;30:1607–1615.
84. Ginzburg Y, Anaby A, Vidavsky Y, Diesendruck CE, Ben-Asuly A, Goldberg I, Lemcoff NG. *Organometallics* 2011;30:3430–3437.
85. Hejl A, Day MW, Grubbs RH. *Organometallics* 2006; 25:6149–6154.
86. Slugovc C, Burtscher D, Stelzer F, Mereiter K. *Organometallics* 2005;24:2255–2258.
87. Barbasiewicz M, Szadkowska A, Bujok R, Grela K. *Organometallics* 2006;25:3599–3604.
88. Gstrein X, Burtscher D, Szadkowska A, Barbasiewicz M, Stelzer F, Grela K, Slugovc C. *J Polym Sci A: Polym Chem* 2007;45:3494–3500.
89. Peeck LH, Savka RD, Plenio H. *Chem Eur J* 2012;18: 12845–12853.
90. Żukowska K, Szadkowska A, Pazio AE, Woźniak K, Grela K. *Organometallics* 2011;31:462–469.
91. Barbasiewicz M, Michalak M, Grela K. *Chem Eur J* 2012;18:14237–14241.
92. Barbasiewicz M, Błocki K, Malińska M, Pawłowski R. *Dalton Trans* 2013;42:355–358.

17

SCHIFF BASE CATALYSTS AND OTHER RELATED LATENT SYSTEMS FOR POLYMERIZATION REACTIONS

Yohann Morin and Régis M. Gauvin

Unité de Catalyse et de Chimie du Solide (UMR CNRS 8181), Axe "Catalyse et Chimie Moléculaire", Villeneuve d'Ascq Cedex, France

17.1 INTRODUCTION

The development of ruthenium-based olefin metathesis catalysts owes a lot to combined design and serendipity: Over the years, modification of the ruthenium coordination sphere led to the emergence of several families of catalysts that have their own specificity deriving from the steric, electronic, or peculiar binding properties of the metal-surrounding ligands. Salicylaldimines, most frequently designated as "Schiff base ligands," are one of the main classes of additional ligands to be introduced around the ruthenium center to tune its reactivity (1–3). These ligands are obtained by high-yielding reaction between salicylaldehydes and anilines, and it is obvious that the huge scope of readily available structures offers a unique option to fine-tune the resulting catalysts' steric and electronic properties. The monoanionic phenoxy-imine ($[L,X]^-$ type) ligands comprise two donor atoms having opposite character. Indeed, the hard phenoxide oxygen is known to stabilize high oxidation states of group 8 metals (4), while the softer imine ligand efficiently stabilizes the lower oxidation states of these metals (5).

The phenoxy-imines usually behave as chelating ligands. However, the possibility of having the open (or "dangling") form with the imine not bound onto the metal and the phenate moiety still coordinated was soon postulated (6). Such a dynamic coordination type, known as *hemilability*, implies that a coordination vacancy is temporarily available for substrate binding onto the metal center and thus further completion of a catalytic cycle, which is not operative (or less efficient) in the "closed" (chelating) form of the ligand.

Indeed, the hemilability of this ligand class is an often advocated reason for the robustness and high reactivity of their complexes under precise conditions. Further, the basicity of the imine moiety offers the possibility to trap it in the "open" form with Lewis or Brønsted acids and thus to switch from a chelating to a monodentate ligand. As this creates coordinative unsaturation on the metal center, a sharp increase in reactivity may be obtained via this methodology. Such a behavior is known as *latency*. This is typically evidenced by a catalyst's inertness or very low efficiency toward a given substrate at room temperature, and by its contrasting ability to perform outstandingly when submitted to a stimulus, such as temperature increase or addition of a co-catalyst.

Such features are indeed highly desirable when considering applications in the field of ring-opening metathesis polymerization (ROMP). The production of polymeric materials by ROMP of cyclic olefins is one of the established industrial applications of metathesis (7). Most particularly, this allows converting dicyclopentadiene (DCPD), a side-product from the petrochemical industry, into polydicyclopentadiene (PDCPD), a rigid thermoset with remarkable chemical and physical properties. Due to process constraints, such polymerization is carried out by reaction injection molding (RIM), where the catalyst and the activator have to be stored separately as solutions in the desired monomer. The commonly used catalytic systems are mostly based on group 6 transition metals, combined to alkylaluminum activators. Their replacement by safer, less sensitive, and easy-to-handle ruthenium latent catalysts is a major target. This explains why the Schiff base ligands

Olefin Metathesis: Theory and Practice, First Edition. Edited by Karol Grela.

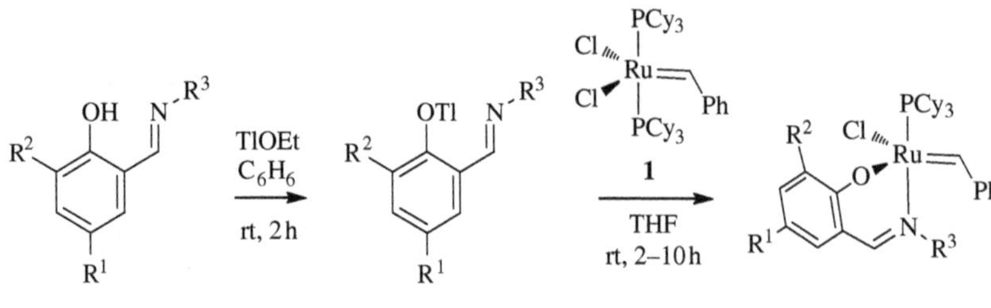

2a–h

2a $R^1 = H, R^2 = H, R^3 = 2,6\text{-}i\text{-}PrC_6H_3$
2b $R^1 = NO_2, R^2 = H, R^3 = 2,6\text{-}i\text{-}PrC_6H_3$
2c $R^1 = NO_2, R^2 = H, R^3 = 2,6\text{-Me-4-MeOC}_6H_2$
2d $R^1 = NO_2, R^2 = H, R^3 = 2,6\text{-Me-4-BrC}_6H_2$
2e $R^1 = NO_2, R^2 = H, R^3 = 2,6\text{-Cl-4-CF}_3C_6H_2$
2f $R^1 = NO_2, R^2 = Me, R^3 = 2,6\text{-}i\text{-}PrC_6H_3$
2g $R^1 = NO_2, R^3 = 2,6\text{-Cl-4-NO}_2C_6H_2$
2h $R^1 = NO_2, R^3 = CH_2$(1-adamantyl)

Scheme 17.1 Synthesis of complexes **2a–h**.

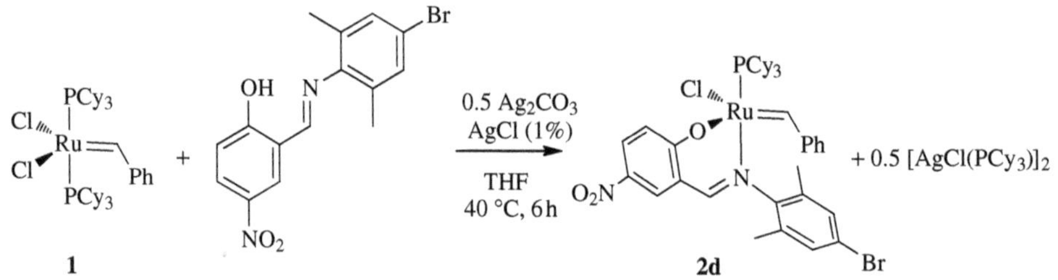

Scheme 17.2 Synthesis of **2d**.

played and still play a key part in the development and successful use of latent ROMP catalysts.

In this chapter, we follow the development of Schiff base catalysts, and get an overview of related systems, namely those featuring chelating ligands with a hemilabile character. Their performances with a focus on ROMP, along with their latency and their mode of activation, are discussed.

17.2 PHENOXY-IMINE $[O,N]^-$ LIGANDS

17.2.1 Phosphine-Containing Catalysts

17.2.1.1 Benzylidene Derivatives The first example of an olefin metathesis catalyst bearing a phenoxy-imine (O–N) was disclosed by Grubbs in 1998 (6). Starting from the first-generation Grubbs catalyst **1**, authors have described the synthesis of complexes **2a–h** (Scheme 17.1). The introduction of the phenoxy-imine was carried out by chloride exchange, using the iminophenoxide thallium salt obtained from reaction between the iminophenol and thallium ethoxide. Use of the toxic thallium derivatives (8) was justified by their higher efficiency compared to other salts, following similar trends observed in the preparation of ruthenium phenoxide complexes (9). Several catalysts have been described, mostly featuring electron-depleted phenoxides combined to 2,6-disubstituted benzymines. Due to high solubility in most organic solvents, isolated yields ranged between 54% and 85%, despite full NMR conversion of the ruthenium starting material.

The yield of these syntheses appeared to greatly depend on the steric properties of the phenoxy-imine considered. It was observed that ligands featuring bulky groups such as *tert*-butyl in the 6-position of the phenoxide moiety did not efficiently react with **1**. The same observation was made for phenoxy-imines bearing triisopropylsilyloxy groups on the 2- and 6- positions of the N-aryl moiety.

In a latter work, the use of thallium salt was bypassed by Bjørsvik and coworkers, who used silver carbonate as a surrogate to thallium ethoxide (10). The carbonate deprotonates the phenol, while the silver cation abstracts the chloride and the tricyclohexylphosphine to form insoluble $[AgCl(PCy_3)]_2$. Interestingly, reaction rates are improved when catalytic amounts of silver chloride are present in the initial reaction mixture. The reported yields of this one-pot procedure are higher in most cases than when using the original Grubbs procedure. However, the efficiency of this method appears to be limited to acidic phenols (namely, bearing a nitro group on the para position), as lower yields were obtained for the simple salicylaldehyde derivative (**2a**).

Procedure for the Synthesis of 2d (Scheme 17.2; 10)

A 50 ml Schlenk flask containing complex **1** (0.300 g, 0.365 mmol), 2-[(4-bromo-2,6-dimethylphenylimino) methyl]-4-nitrophenol (0.127 g, 0.365 mmol), silver(I) carbonate (0.050 g, 0.182 mmol) and silver(I) chloride (0.5 mg, 4 μmol) was evacuated and backfilled with argon. 20 ml dry THF was then added and the reaction mixture was stirred at 40 °C for 6 h, then allowed to cool to room temperature while stirring overnight. Upon cooling at 0 °C, the white precipitate of $[AgCl(PCy_3)]_2$ was separated by filtration. The filtrate was evaporated to dryness under vacuum. The residual solid was recrystallized from pentane at −84 °C using an ethyl acetate/liquid nitrogen slush, to afford catalyst **1d** as a brown powder in a 96% yield. In addition, the $[AgCl(PCy_3)]_2$ side product was isolated with a yield of 82%.

These complexes are stable under air for several days. From the structural point of view, Grubbs and coworkers compared the ^1H NMR chemical shift of the carbenic CH, along with the corresponding $^3J_{P-H}$, for the series of complexes in CD_2Cl_2. Compared to **1** (singlet, 20.1 ppm in CD_2Cl_2), the benzylidene protons in **2a–h** resonate between 18.7 and 19.8 ppm as doublets. Protons of the complexes bearing more electron-withdrawing substituents are shifted further downfield. Interestingly, the proton–phosphorus coupling constant also varied, depending on the nature of the phenoxy-imine ligands. These coupling constants appear more sensitive to the steric bulk than to the electronic properties of the substituents on the Schiff-base ligands (ranging from 2.7 to 13.5 Hz). This pleads for a similar ruthenium coordination sphere, with a slight variation of the relative geometry of each species, depending on the ligands' sterics. The solid-state geometry of **2b** was determined by X-ray diffraction studies (Fig. 17.1). The complex adopts a distorted trigonal bipyramidal structure, similar to **1**, with the

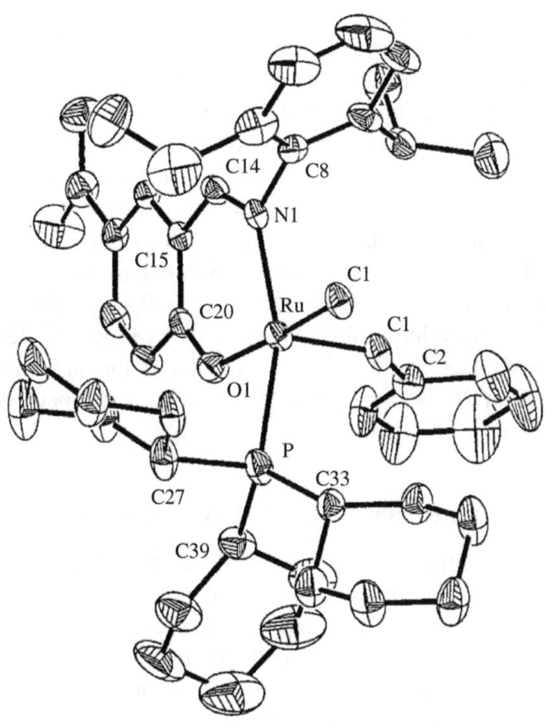

Figure 17.1 ORTEP representation of the molecular structure of **2b**. Reprinted with permission from Ref. 6. Copyright 1998 American Chemical Society.

chlorine, oxygen and carbon atoms in the equatorial positions, and phosphorus and nitrogen atoms in axial positions. The imine is thus located trans to the phosphine: its decoordination would open up a vacant site cis to the carbene.

Ruthenium carbene complexes **2a–h** exhibited catalytic activity in RCM (Ring Closing Metathesis). While these species are generally less reactive at room temperature than the parent complex **1**, their reactivity increases dramatically at higher temperatures. For instance, the RCM of diethyl diallylmalonate ester needs 12 h to proceed to completion at room temperature, using 8 mol% **2g** in dichloromethane, while full conversion is reached in 1 h at 70 °C with 3 mol% loading in benzene. Further, these species are stable and active in polar protic media such as methanol.

A later study on the behavior of such species in ROMP was carried out by Monsaert et al. in 2009. The authors showed that **2d** is inactive toward 1,5-cyclooctadiene (COD) at room temperature (monomer/Ru = 300), but in the presence of 70 equiv of HCl per ruthenium center, it performs better than Grubbs first- and second-generation catalysts (11). However, no significant activity is obtained at the more demanding monomer/catalyst ratio of 3000, which is ascribed to decomposition of the active catalyst.

These species were reacted with chloride abstractors such as silver or silylium salts of weakly coordinating anions, but neither characterization of the resulting species nor their reactivity in metathesis has been described (12). This approach may be of interest, as increased activity in atom-transfer radical polymerization (ATRP) is observed when comparing the activity of neutral catalysts and that of the corresponding cations.

An example of immobilization of a Schiff-base-phosphine ruthenium benzylidene was provided in 2003 (13). Use of a tether on the Schiff base ligand allowed for immobilization on a mesoporous MCM41 support, as demonstrated by a combination of several techniques, including Raman spectroscopy, ICP/MS and solid state NMR. These materials are active in several reactions such as ROMP, RCM, Kharasch addition, ATRP, and vinylation. Regarding ROMP, the heterogeneous systems were successfully applied to a broad range of norbornene derivatives, with a marked solvent dependence (best results in dichloromethane than in toluene). The supported catalysts proved to be recyclable.

17.2.1.2 Vinylidene Derivatives Extension of the Schiff base catalyst examples to vinylidene derivatives was provided by Verpoort and Opstal (14,15). Synthesis proceeds from the reaction of the bisphosphine bischloro derivatives (accessible from the $RuCl_2$(*p*-cymene) dimer, PCy_3, and alkyne (16)) with the thallium salts of the Schiff base ligand (Scheme 17.3). The only reported yields (**3c** and **3d**) are high (81% and 80%, respectively). Regarding the ^1H NMR

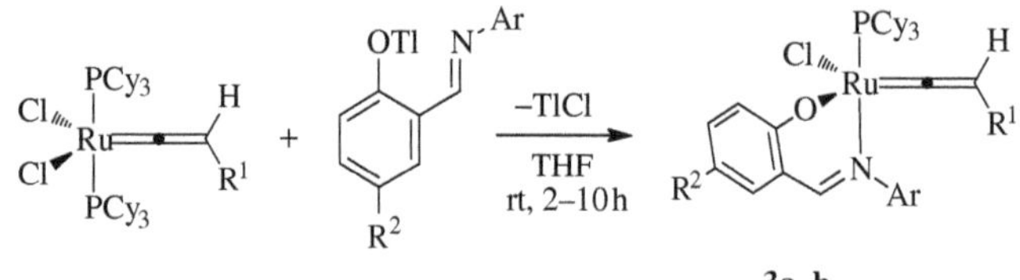

3a–h

Ar = 2,6-Me-4-BrC$_6$H$_2$

3a R^1 = Ph, R^2 = H
3b R^1 = Ph, R^2 = NO$_2$
3c R^1 = SiMe$_3$, R^2 = H
3d R^1 = SiMe$_3$, R^2 = NO$_2$
3e R^1 = *t*-Bu, R^2 = H
3f R^1 = *t*-Bu, R^2 = NO$_2$

Scheme 17.3 Synthesis of complexes **3a–f**.

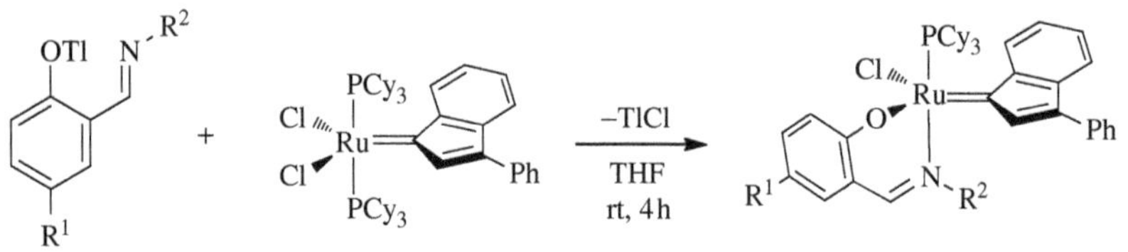

4a–h

4a R^1 = H, R^2 = 2,6-Me-4-BrC$_6$H$_2$
4b R^1 = NO$_2$, R^2 = 2,6-Me-4-BrC$_6$H$_2$
4c R^1 = H, R^2 = 2-MeC$_6$H$_4$
4d R^1 = H, R^2 = 2,4,6-MeC$_6$H$_2$
4e R^1 = NO$_2$, R^2 = 2,4,6-MeC$_6$H$_2$
4f R^1 = NO$_2$, R^2 = 2-ClC$_6$H$_4$
4g R^1 = NO$_2$, R^2 = 2,6-MeC$_6$H$_3$
4h R^1 = NO$_2$, R^2 = 2,,6-*i*PrC$_6$H$_3$

Scheme 17.4 Synthesis of catalysts **4a–h**.

features, the Ru=C=C*H*SiMe$_3$ protons give rise to a characteristic triplet at 3.57 (J = 4.2 Hz) and 3.61 ppm (J = 5 Hz) for **3c** and **3d**, respectively.

Again, the complexes are stable under air both as solids and in solution. The authors claim that this catalyst family presents significant advantages compared to benzylidene derivatives, mostly due to their easier and cheaper synthesis.

ROMP of norbornene is efficiently performed at 80 °C by **3a–f** (monomer/catalyst ratio: 800/1), with a poor initiation efficiency (about 10%). The substrate scope was extended to substituted norbornenes (10 examples), and the system appears to be robust: Sustained activity is observed for ethylidenenorbornene ROMP carried out by consecutive additions over a period of 3 months. The silyl-substituted vinylidenes **3c** and **3d** are the most efficient systems. The best results were obtained with the *para*-nitro derivative **3d**, which indicates an effective electronic effect. The same trend was found for RCM of α,ω-diolefins. Additional reactivity studies have been performed on addition of carboxylic acids on terminal alkynes and terminal alkyne dimerization (14).

17.2.1.3 Phenylindenylidene Derivatives Following the preparation and structure elucidation of the indenylidene analog of **1** (17), its Schiff base derivatives were prepared by Opstal and Verpoort (14,18). The reported yields for **4a** and **4b** were both up to 80%, using the thallium salt of the Schiff base ligand (Scheme 17.4). These two compounds were successfully used in addition to carboxylic acids on terminal alkynes, terminal alkyne dimerization, and ATRP. Short studies on ROMP of cyclooctene and cyclopentene mediated by **4b** revealed good performances. In 17 h at room temperature, cyclooctene is converted to 53% (monomer/catalyst ratio: 10,000/1) with 66% of initiation efficiency, while for the more challenging cyclopentene, a conversion of 41% (monomer/catalyst ratio: 5000/1) is reached, with initiation efficiency of 70%. In both cases, polymers are obtained with molecular weight distribution of about 1.8, as determined by size exclusion chromatography.

Using similar synthetic procedures, the same group later extended the scope of such catalysts (**4c–h**, Scheme 17.4) (19). These were involved in representative cross-metathesis and ring-closing metathesis reactions. Excellent activity of the new complexes was found for the RCM reactions, as well as in cross-metathesis reactions, with reduced isomerization side-reaction compared to the first generation Hoveyda–Grubbs catalyst, in some instances.

Stability tests of compounds **4c–h** in deuterated benzene at 80 °C were conducted by ^{31}P NMR monitoring of the reaction mixture. The Schiff base derivative proved to be more stable than Hoveyda–Grubbs I, which decomposed completely after 3 h. In comparison, Schiff base indenylidene species still had 40–50% of initial catalyst left, with the marked exception of **4h**, for which 86% of the complex remained intact. The half-life of this catalyst, the indenylidene analog of **2b**, is about 6 h at 80 °C. No hydride species was detected by ^{1}H NMR.

17.2.2 Arene- and Diene-Containing Catalysts

In 2001, Verpoort and De Clerq developed phenoxy-imine derivatives of ruthenium arene complexes (**5a–c**) (20,21). These species were accessed from the ruthenium dichloride *para*-cymene dimer and the thallium salt of the selected ligand (Scheme 17.5). No yield was reported.

The potential of species **5a–c** as ROMP catalysts was probed (21). In the presence of 2 equivalents of trimethylsilyldiazomethane (TMSD) per ruthenium center, moderate to good conversions were reached at 85 °C for norbornene and

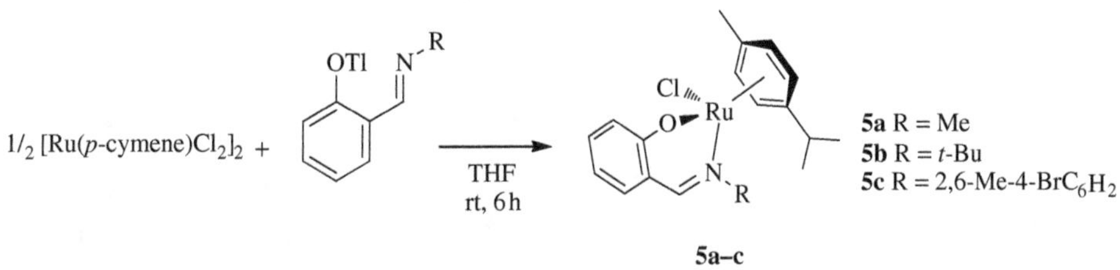

Scheme 17.5 Synthesis of catalysts **5a–c**.

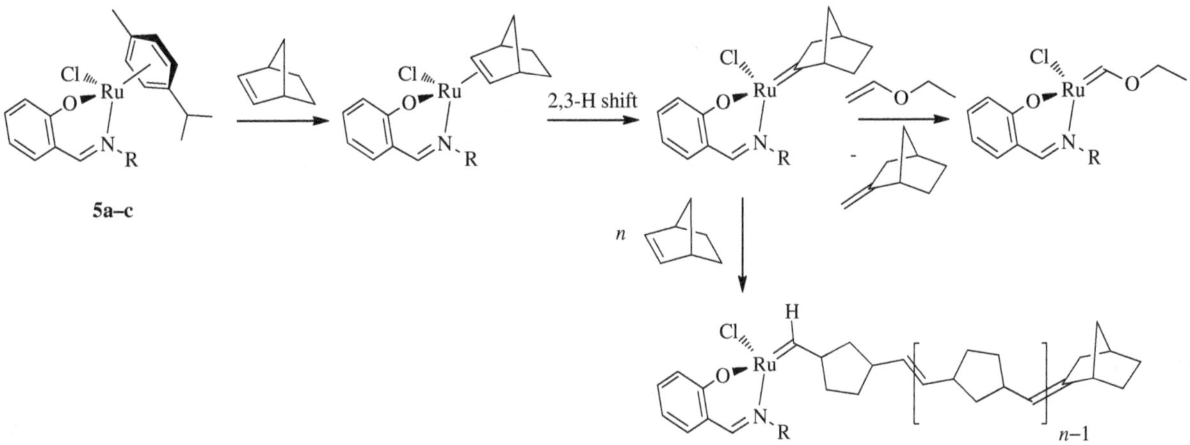

Scheme 17.6 Mechanism of activation of the ROMP of norbornene and trapping of the active species from catalysts **5a–c**.

cyclooctene (respective monomer/catalyst ratio: 800/1 and 200/1), with the *N*-aryl derivative **5c** being the most efficient catalyst. In the absence of TMSD, low conversion of norbornene (6%) was observed for all species. Mechanistic studies allowed reaching a better understanding of these results. Indeed, NMR monitoring of the reaction between **5a–c** and TMSD revealed the presence of a carbene moiety, accounting for 12% of the ruthenium centers, along with free *para*-cymene. Additions of norbornene led to the formation of a new carbenic compound, most probably the propagating species. Heating **5a–c** at 85 °C in the presence of 5 equiv of norbornene also resulted in the formation of the same carbenic species that corresponded to the growing chain, as observed in the case of TMSD-activated systems. When the activation is performed in a stoichiometric ratio, quenching with vinylethylether affords a Fischer-type carbene, along with methylene-norbornane. This is consistent with a carbene formation mechanism that consists of *para*-cymene displacement by norbornene, followed by a 2,3-hydrogen shift, which generates a carbene (Scheme 17.6).

When activated with 2 equiv of TMSD, these complexes catalyze ring-closing metathesis of several substrates, including some sterically hindered dienes, with again **5c** performing the best in the series (20).

The scope of available arene catalysts was expanded by the same authors, who prepared derivatives **5d** and **5e** in good yields. In this case, a bulkier Schiff base ligand was selected, and the electrophilicity of the metal center was tuned by substitution of a chloride by a pentafluorophenyl group, through alkylation of **5d** by the corresponding Grignard reagent (Scheme 17.7) (22). Interestingly, the ligand used here features a 6-*t*Bu group on the phenolic moiety, and contrary to observation of the related complexes **2a–h**, its coordination onto ruthenium proceeds efficiently.

In the absence of TMSD activator, catalyst **5d** is moderately active in ROMP of norbornene (monomer/catalyst: 225/1), achieving 35% of conversion at 85 °C, while **5e** performed better, with 86% conversion. The use of $AlEt_2Cl$ as co-catalyst improves the monomer conversion, allowing high yields (93%) in the case of **5e** with monomer/catalyst ratio of 6000/1. However, initiation efficiency remains low, being estimated to be 1–2% at room temperature, and 2–5% at 85 °C. This was confirmed by NMR mechanistic studies, which provided information in line with what was observed for **5a–c**, as described above.

Along the same lines, the Verpoort group also reported COD and 2,5-norbornadiene (NBD) derivatives of the [Ru(O–N)Cl] core (22). These species can be prepared in moderate yields from the diene ruthenium dichloride precursors and the Schiff base thallium salt (Scheme 17.8).

These compounds catalyze the ROMP of norbornene, with catalytic performances lower than the parent

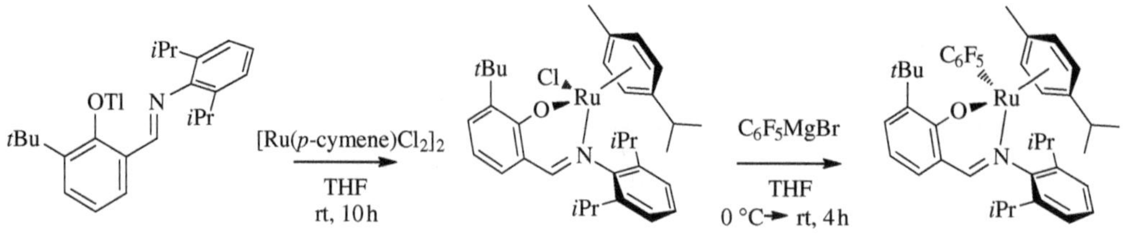

5d **5e**

Scheme 17.7 Synthesis of catalysts **5d** and **5e**.

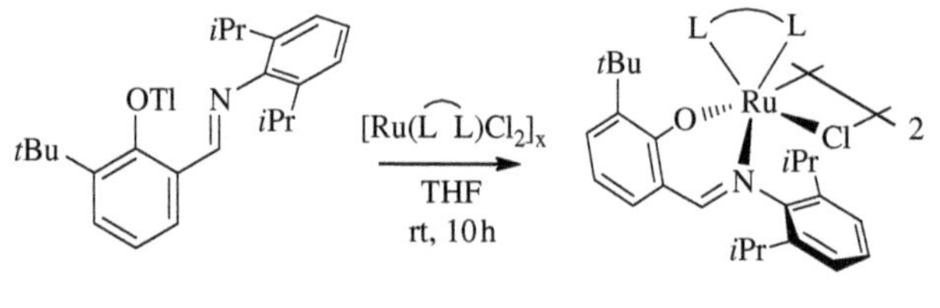

6a: L⁀L = 1,5-cyclooctadiene
6b: L⁀L = 2,5-norbornadiene

Scheme 17.8 Synthesis of arene and olefin derivatives **6a** and **6b**.

para-cymene compounds **5d** and **5e**; similar increase of activity is observed when the $AlEt_2Cl$ co-catalyst (6 equiv per Ru) is used.

Further examples of phosphine and *N*-heterocyclic carbene (NHC)-free Schiff base catalysts were developed by Verpoort and De Clerq, as homobimetallic systems (23). These compounds were prepared in moderate yields (61–73%) from the corresponding derivatives of the **2** family and the dimeric *para*-cymene ruthenium dichloride complex (Scheme 17.9).

The catalytic performances of these bimetallic catalysts in RCM and ROMP were probed and compared to those of the corresponding monometallic phosphine derivatives. The dinuclear catalytic systems display higher activity and improved stability compared to the mononuclear species. Catalysts **7a–f** performed well in RCM of 1,6-heptadiene, 1,7-octadiene, and diallyl ether as well as in ROMP of norbornene and 5-substituted norbornene. However, although for dinuclear complexes, activity in RCM and ROMP follows the same trend (with **4a** being the best catalyst, and **4f** the less active one), mononuclear **2** catalysts do not behave similarly in the two different reactions, as orders of activity are inverted. This was rationalized by the authors as resulting from a dissociative mechanism (from the imine in **2**, and from the $RuCl_2$ (*para*-cymene) fragment for **7**). Further, the dissociation of the nitrogen-bonded arm of the chelated Schiff base ligand was supported by several observations: *inter alia*, addition of the phosphine scavenger CuCl to type **2** catalysts does not result in a significant increase of the activity in ROMP and RCM reactions. Therefore, the observed activity must derive from the imine decoordination.

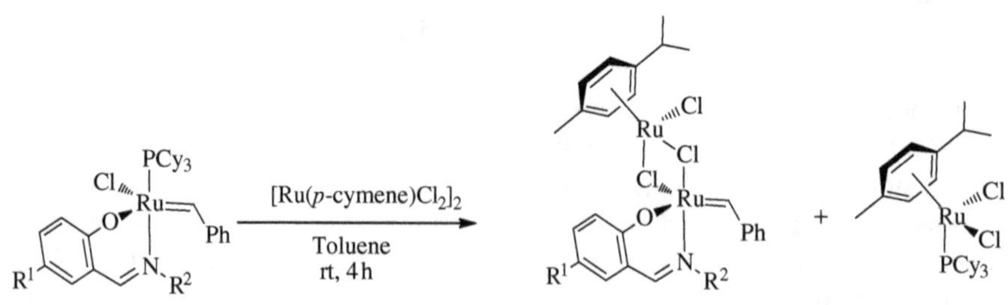

7a-f

2a R^1 = H, R^2 = 2,6-*i*-PrC_6H_3
2b R^1 = NO_2, R^2 = 2,6-*i*-PrC_6H_3
2d R^1 = NO_2, R^2 = 2,6-Me-4-BrC_6H_2
2i R^1 = H, R^2 = 2,6-Me-4-BrC_6H_2
2j R^1 = H, R^2 = Me
2k R^1 = NO_2, R^2 = Me

7a R^1 = H, R^2 = 2,6-*i*-PrC_6H_3
7b R^1 = NO_2, R^2 = 2,6-*i*-PrC_6H_3
7c R^1 = H, R^2 = 2,6-Me-4-BrC_6H_2
7d R^1 = NO_2, R^2 = 2,6-Me-4-BrC_6H_2
7e R^1 = H, R^2 = Me
7f R^1 = NO_2, R^2 = Me

Scheme 17.9 Synthesis of catalysts **7a–f**.

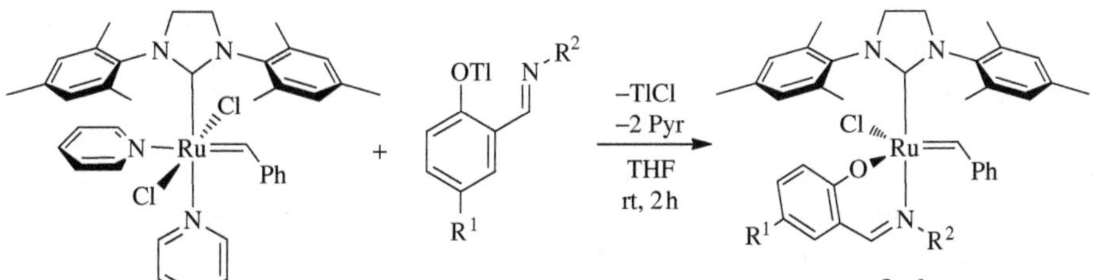

8a–h

8a R^1 = H, R^2 = 2,4,6-MeC_6H_2
8b R^1 = H, R^2 = 2,6-*i*-PrC_6H_3
8c R^1 = NO_2, R^2 = 2,6-*i*-PrC_6H_3
8d R^1 = H, R^2 = 2,6-Me-4-BrC_6H_2
8e R^1 = NO_2, R^2 = 2,6-Me-4-BrC_6H_2
8f R^1 = NMe_3Cl, R^2 = 2,6-Me-4-BrC_6H_2

Scheme 17.10 Synthesis of catalysts **8a–e**.

17.2.3 NHC-Containing Catalysts

17.2.3.1 Benzylidene Derivatives Following the emergence of the use of NHC ligands in the second-generation Grubbs catalysts, the corresponding Schiff base catalysts were first described by Opstal and Verpoort in 2002 (18,24,25). In these reports, the NHC was installed on the metal following exchange of the phosphine ligand in complexes of the **2** family. Reproducibility issues raised by the same group (26) and by others (27) led to the establishment of a new synthetic route, involving reaction of a thallium salt on the NHC-bispyridine benzylidene ruthenium derivative to afford complexes **8a–f** (Scheme 17.10) (28).

Synthesis proceeds with yields ranging from 26% to 84%, with the lowest yields being obtained for **8a** and **8b** (30% and 26%, respectively). This may be ascribed to solubility issues that prevent efficient product separation and purification. The complexes are stable under air, and no sign of decomposition in $CDCl_3$ solution was detected after a month at room temperature.

The NMR chemical shift ranges of the benzylidene proton and carbon are respectively 18.37–18.95 and 296.4–301.4 ppm (CD_2Cl_2), which is comparable to the respective ^{1}H and ^{13}C NMR values of 19.16 and 294.2 ppm (29) for the second-generation Grubbs catalyst [$RuCl_2$($IMesH_2$)(PCy_3)(CHPh)] **9**.

X-ray diffraction studies have been carried out for both **8c** and **8d**, which present overall similar features (26,27). In Figure 17.2 is displayed the solid-state structure of **8d**. The coordination pattern is similar to that of the phosphine analog **2d**, with a distorted square pyramidal structure where the apical position is taken by the benzylidene, and the basal positions by the phenoxy-imine N and O atoms, the NHC and the chloride ligands. The oxygen and the chloride are in mutually trans arrangement.

Binder and coworkers relied on the stabilizing effect of the Schiff base ligand to carry out efficient ring-closing metathesis of both dienes and enynes in methanol and methanol/water mixtures under air (27). Slow initiation of the catalyst appears to protect it from water while providing a steady supply of active species.

Complexes **8a–d** display interesting latent behavior. They are inactive toward 1,4-cyclooctadiene at room temperature, in contrast to what was reported for their indenylidene

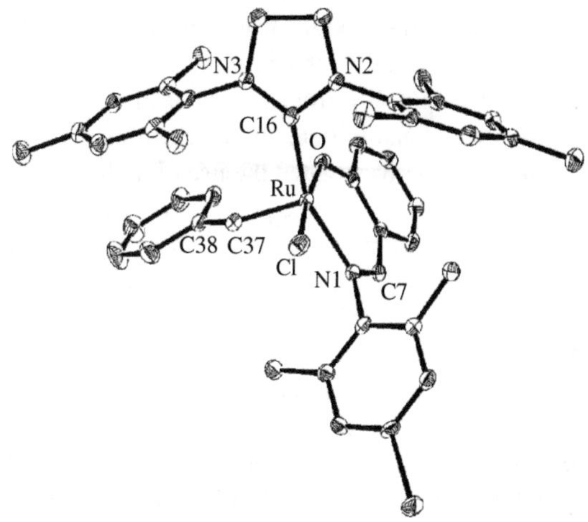

Figure 17.2 ORTEP representation of the molecular structure of **8d**. Reprinted with permission from Ref. 27. Copyright 2007 John Wiley and Sons.

derivatives, but readily mediate ROMP at 90 °C (26). Some trends were observed, such as highest initiation rates for the species bearing the bulky 2,6-*i*-Pr-C_6H_3 Schiff base (ascribed to an eased de-coordination of the imine from the ruthenium center), and the absence of strong effect of a *para*-nitro substituent on the phenolic moiety. In the sole case of **8a**, an induction period was observed, that authors suggest stemmed from slow initiation or a competitive associative mechanism which gained in importance with the decreased steric congestion as in **8a**, compared to the other more bulky members of the series. Compound **8c** does not catalyze the ROMP of DCPD at room temperature (monomer/catalyst ratio: 20,000/1). However, preliminary studies were performed by heating the reaction mixture from room temperature to 150 °C, and an exotherm was detected at 130 °C, in agreement with a thermal activation pathway.

The latent character of these species was also exploited via chemical activation. The Verpoort team showed that catalyst **8e** behaves as a highly controllable latent ROMP catalyst (30,31). Just like its tricyclohexylphosphine counterpart **2d**, it is inactive toward COD at room temperature, but addition of acidic co-catalysts generates very active systems. Several Lewis acids were screened, with best results being obtained with chlorosilanes. More specifically, $HSiCl_3$ gave rise to 630,000 ton, with a monomer/catalyst/$HSiCl_3$ ratio of 3,000,000/1/1000. Activation efficiency was in the order: $HSiCl_3 \approx HSiMeCl_2 > SiCl_4 > HSiMe_2Cl \gg SiMe_2Cl_2$. Other Lewis acids such as BF_3, $AlCl_3$, and CuCl were screened, but did not give rise to efficient catalysis under demanding conditions (i.e., with monomer/catalyst/acid ratio of 30,000/1/70). Attempts at elucidating the nature of the activated species were made through NMR studies. In the case of the $HSiCl_3$/**8e** system (70/1 ratio), no change was detected on the ^1H NMR spectrum. However, the high catalytic activity of the resulting mixture was confirmed. This led the authors to propose that trichlorosilane reversibly forms an adduct with the imine's nitrogen, thus opening a vacant coordination site that can initiate polymerization.

The same group investigated the use of hydrochloric acid as a co-catalyst for **8e**-catalyzed processes (11). Interestingly, HCl triggers high activity in ROMP of COD at room temperature, with monomer/catalyst/acid ratio up to 90,000/1/140 (Fig. 17.3). In this case, full conversion is achieved in 30 min. The monomer/catalyst ratio can be increased to 120,000/1 (210 equiv HCl) with only minor loss in productivity, but activity is suppressed at a value of 300,000/1 (in the presence of 300 equiv HCl). Polymers produced by **8e** have polydispersity indexes ranging from 1.8 to 1.9, with predominantly *trans*-olefin content.

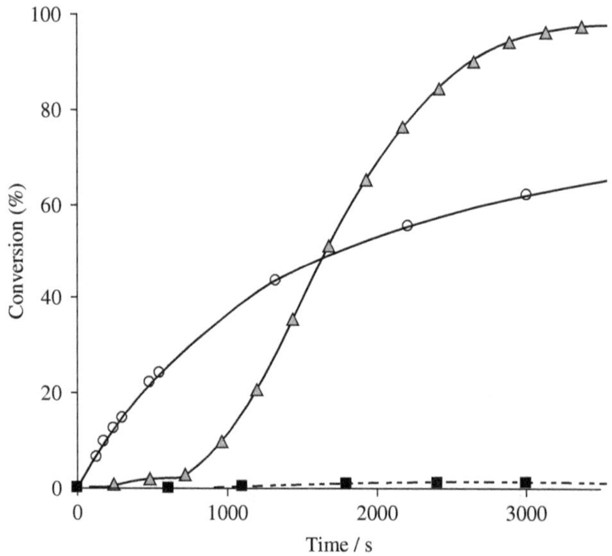

Figure 17.3 Monitoring ROMP of COD by ^1H NMR spectroscopy ($CDCl_3$, 20 °C). Catalysts: –Δ–: **9**; –■–: **2d** + HCl; –o–: **8e** + HCl. Conditions: monomer/acid/catalyst: 1/3000/70/1, catalyst concentration: 0.452 mM. HCl was diluted in COD monomer before addition. Reprinted with permission from Ref. 11. Copyright 2009 Wiley Periodicals, Inc.

The authors also investigated the ROMP of DCPD. Species **8e** is not active at room temperature but high activity is again obtained with excess HCl. Best results are obtained in the 20–40 equiv range, with an optimum for 30 equiv of acid co-catalyst. Thus, curing was obtained with monomer/catalyst ratio up to 60,000. Exothermal temperatures ranged between 160 and 190 °C, depending on the selected ratio of the monomer, acid, and catalyst. In order to further probe the latency of **8e**, a study in time was performed. Formulations of monomer mixture/catalyst (15,000/1 and 30,000/1) demonstrated low increase of viscosity over a time frame of one year, with values still not exceeding the threshold for further implementation in RIM processes. The extended shelf life of the 15,000/1 solution was clearly demonstrated by the fact that only marginal reactivity difference was observed after one year of storage.

In order to address issues about the decrease of HCl in stored solutions, attempts were made at using a combination of protic reagents (alcohol, carboxylic acid) and chloride-containing Lewis acids, each of them stored in a different stream. Upon mixing, HCl was released. The two streams were composed by DCPD solutions of the catalyst and the protic reagent (60 or 90 equiv per Ru center) on one side, and, on the other side, of the Lewis acid (60 or 90 chloride equivalents). The resulting monomer/catalyst ratio is 30,000/1 (i.e., catalyst loading of 333 ppm). Chlorosilanes can be efficiently combined to 1-propanol, with best results obtained with $SiCl_4$ (exotherm maximum of 184 °C reached after 4 min). Highly acidic transition- and main-group metal chlorides such as $TiCl_4$ and $AlCl_3$ proved to be extremely efficient, as exotherm maxima were reached between 1 and 3 min, up to 200 °C ($TiCl_4$) and 170 °C ($AlCl_3$). Variation of the Lewis acid/protic reagent couple allows for fine tuning of the polymerization initiation period. The milder the Lewis acid, the less efficient the polymerization, which indicates that formation of HCl from these reagents is a requisite for efficient catalyst activation.

The authors carried out *in situ* spectroscopic studies to elucidate the activation mechanism (Scheme 17.11). Using ^1H NMR, they demonstrated that in dry solvents, protonation of the imine moiety occurs, resulting in the formation of a 14-electron species with a diagnostic low-field carbenic CH signal. As deduced from reactivity trends in ROMP, this activation is not a quantitative process, as the new species is only present in minor quantities, with unreacted **8e** as the major component. The new complex is stable for hours at room temperature. Upon addition of cyclooctene, a new alkylidene species forms from this postulated unsaturated catalyst, while **8e** remains unchanged. When the same study is performed in non-dried solvents, cleavage of the Ru–O bond and decoordination of the Schiff base ligand is immediately observed, along with the 4-nitrosalicylaldehyde hydrolysis product. The unsaturated bis-chloride species that would thus be formed is not observed, and it is proposed

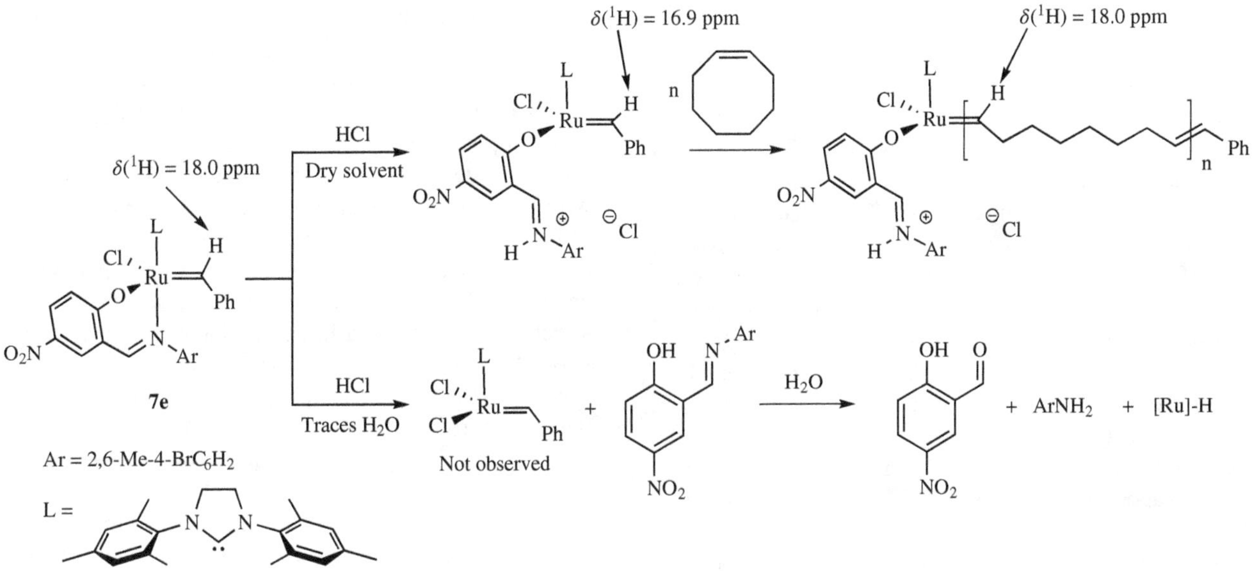

Scheme 17.11 Proposed activation and decomposition mechanisms for **8e** upon HCl activation.

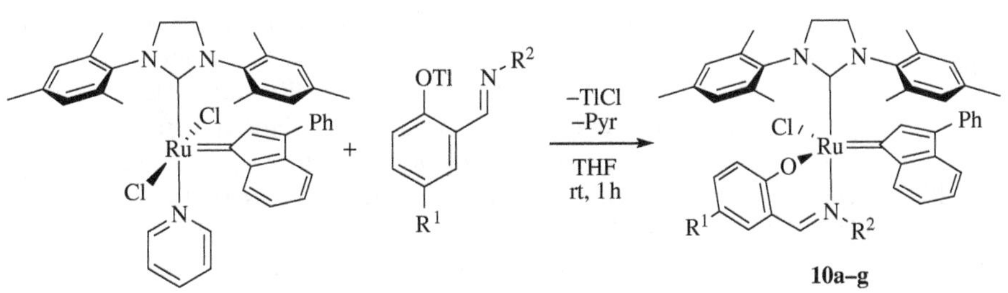

10a $R^1 = NO_2$, $R^2 = 2,6\text{-}MeC_6H_3$
10b $R^1 = OMe$, $R^2 = 2,6\text{-}MeC_6H_3$
10c $R^1 = H$, $R^2 = 2,6\text{-}MeC_6H_3$
10d $R^1 = NO_2$, $R^2 = 2,4,6\text{-}MeC_6H_3$
10e $R^1 = NO_2$, $R^2 = 2,6\text{-}Me, 4\text{-}BrC_6H_2$
10f $R^1 = H$, $R^2 = 2\text{-}MeC_6H_4$
10g $R^1 = NO_2$, $R^2 = 4\text{-}MeC_6H_4$
10h $R^1 = NO_2$, $R^2 = 4\text{-}t\text{BuC}_6H_4$

Scheme 17.12 Synthesis of catalysts **10a–g**.

that it decomposes into (inter alia) hydride derivatives, as deduced from characteristic high-field signals.

*17.2.3.2 **Phenyl-Indenylidene Derivatives*** The preparation of indenylidene Schiff base complexes **10a–c** was reported by the Verpoort group in 2010 (32). Just as for the benzylidene derivatives, the synthetic pathway to these compounds consists of the reaction of the thallium salt with the pyridine adduct of the SIMES-indenylidene-ruthenium dichloride complex, which provides the complexes in very high isolated yields (98%) (Scheme 17.12). These species are very stable in air and heat, tolerating storage for months in solution or as solids in ambient conditions without suffering any degradation. Drozdzak in Telene further expanded the scope of the catalysts by synthesizing compounds **10d–g**, modulating both the electronics of the phenolic moiety and the sterics of the *N*-aryl group (33). Catalysts **10f** and **10g** are commercially available from Umicore, under the respective names of $M4_2$ and $M4_1$.

Detailed structural investigation on compounds **10a–c** was performed by NMR. All the ^1H and ^{13}C NMR signals were unambiguously assigned, thanks mostly to homo- and heteronuclear correlation sequences. These complexes exist in solution as a 55/45 mixture of two rotational isomers, related via a 180° flip along the Ru–C indenylidene bond and separated by a high activation barrier. Other dynamic processes were detected, such as slow rotational exchange around the Ru–C(NHC) bond as well as a fast rotation of the phenyl group located on the indenylidene.

The Telene team probed the use of species **10c–h** as latent ROMP catalysts. Their relative stability in DCPD solution was checked, and only **10g–h** slowly initiated polymerization after standing at room temperature for 7 days, which demonstrates the importance of the *N*-aryl group substitution pattern on the reactivity of these compounds (steric hindrance in *N*-aryl ring increases stability). Activation by *in situ* generated HCl (45 equiv per ruthenium) was applied to these systems. Whereas **10d–e** are inactive, **10c,f–h** are excellent catalysts, with a monomer/catalyst ratio of 15,000/1. Most particularly, **10f** reaches full conversion in about 100 s, with an exotherm maximum of 210 °C. No reactivity difference was observed when the catalysts were stored in DCPD for 7 days prior to acid activation. Thus, ligand design afforded catalysts that combine high stability, robustness against decomposition, and high activity upon chemical activation.

Procedure for the Synthesis of 10a

A solution of thallium ethoxide (0.2 mmol; *Caution: thallium salts and derivatives are toxic and should be handled with appropriate precautions*) in 5 ml of THF was added drop-wise at ambient temperature to a flask containing the Schiff base ligand 2-{[(2,6-dimethylphenyl) imino]methyl}-4-nitrophenol (54 mg, 0.2 mmol) in 10 ml of THF. A yellow solid appeared and the mixture was stirred for 2 h. The thallium salt was isolated in quantitative amount by filtration under argon atmosphere. This compound and [1,3-bis(2,4,6-trimethylphenyl)-2-imidazolidinylidene]dichloro-(3-phenyl-1*H*-inden-1-ylidene)(pyridyl)ruthenium (150 mg, 0.2 mmol) were then stirred in 5 ml THF for 1 h at room temperature. After evaporation under vacuum, the residue was redissolved in toluene and the solution was cooled down to 0 °C, to filter the thallium chloride side-product, which was washed with cold toluene. The filtrate was evaporated to dryness, and the solid obtained was recrystallized from a mixture of dichloromethane and pentane at −70 °C, to afford complex **10a** as a brown-orange solid in a yield of 98% (32).

17.2.3.3 ***Allenylidene Derivatives*** In 2007, Ledoux and coworkers reported the preparation of an allenylidene Schiff base derivative stabilized by a NHC ligand (**11**) (34). They designed a convenient precursor, an allenylidene ruthenium dichloride NHC complex bearing two pyridines as labile ligands, which readily reacts with a thallium salt of a salicylaldimine ligand. Unlike benzylidene complexes **8** and indenylidene derivatives **10**, a mixture of species is obtained, with **11** being the major one (86%), along with two compounds (13% and 1%). Low reaction temperature leads to a decrease of the proportion of **11** to 50%, along with 50% of the formerly second minor isomer and traces of the third compound. Complex **11** was isolated by column chromatography and characterized by NMR and elemental analysis, which confirmed its composition. By analogy, its structure was proposed to be similar with the other Schiff base complexes reported so far. The nature of the other two species remains undetermined and the authors propose that they result from different relative arrangements of the Schiff base and chloride ligands, featuring a cis-configuration of the anionic moieties (Scheme 17.13).

The latency of **11** proved to be inferior to that of the parent benzylidene derivative **8e**, as some activity in ROMP of COD was observed at room temperature. Again, activation by HCl (monomer/catalyst/acid: 3000/1/10) triggered high activity, which however remained inferior to what was recorded for **8e**. Interestingly, the use of only 2 equiv of triflic acid destroyed the catalyst. Switching to $HSiCl_3$ has a beneficial effect, but large excess of the co-catalyst causes catalyst decomposition, again contrasting with **8e**. NMR studies revealed that **11** reacts with excess HCl to afford *in fine* a Schiff base-free alkenylcarbyne complex along with salicylaldimine hydrolysis products (no quantification was given, Scheme 17.14). The carbyne complex was characterized by single crystal X-ray diffraction study.

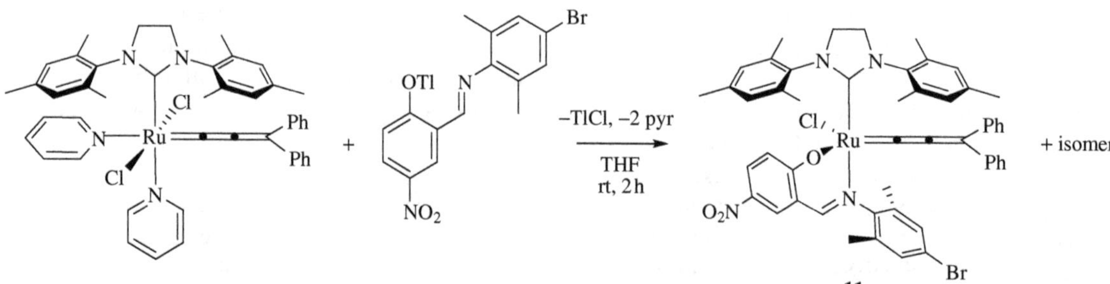

11

Scheme 17.13 Synthesis of complex **11**.

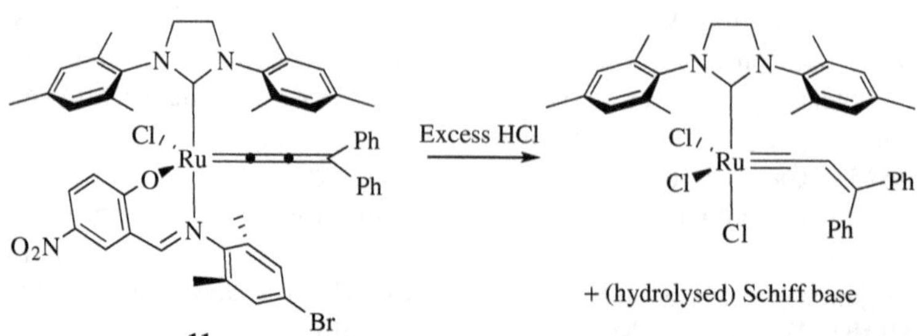

11

Scheme 17.14 Reaction of **11** with excess HCl.

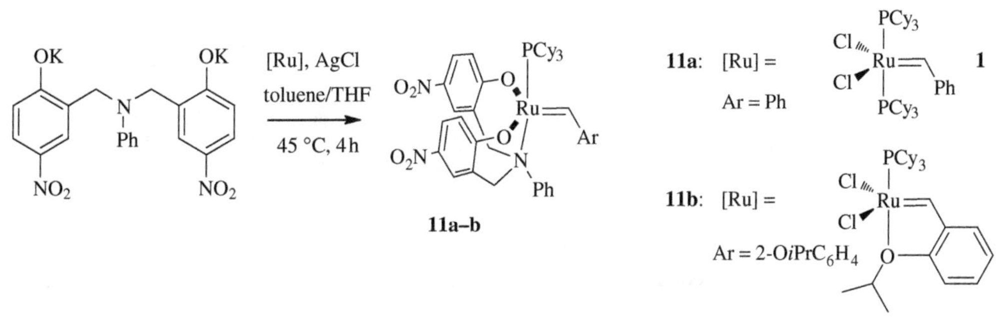

Scheme 17.15 Synthesis of complexes **12a** and **12b**.

From observations in closely related systems, it can be proposed that if the expected activation scheme is operative (protonation of the imine, generation of a coordinatively unsaturated, catalytically active ruthenium species), a second pathway destroys the allenylidene fragment prior to initiation by converting it to the catalytically inert carbyne. On the basis of NMR data, it appears that allenylidene protonation is a fast process. The same study carried out with $HSiCl_3$ indicated that a similar deactivation scheme occurs with this co-catalyst, as formation of a carbyne was also detected. However, no definitive structure can be proposed for the reaction product of **11** with $HSiCl_3$. Reasoning that both HCl and $HSiCl_3$ have similar reactivity toward the allenylidene fragment, $PhSiCl_3$ was tested as co-catalyst. Indeed, good activity in ROMP of COD was reached with monomer/catalyst/acid ratio of 30,000/1/70, as the system allowed full conversion of the monomer in 15 min.

In conclusion, efficient synthetic procedures have been developed to access ruthenium Schiff base derivatives, featuring different spectator ligands (mostly phosphines and NHC) and different metathesis initiating groups (mainly benzylidene and indenylidenes). The most active and robust species are the NHC derivatives of the **8** type: once activated, these compounds feature outstanding catalytic performances. Their latency can be overcome either by thermal or chemical activation, thus achieving very high productivity at catalyst loadings in the parts per million level.

17.3 OTHER EXAMPLES OF CHELATING L_nX SYSTEMS

In parallel to the development of the Schiff base ligand family, the concept of chelating anionic hemilabile ligands has also been addressed with different ligand sets. These are presented in this section, along with their key catalytic performances and discussion on their latency when applicable.

17.3.1 Bisphenoxy-Amine $[O,O,N]^{2-}$ Ligands

Jensen and coworkers described the preparation of (bis) phenoxy-amines, as a further extension of the successful salicylaldimine framework (35). Attempts at generating phenoxy-amine derivatives were unsuccessful, leading in one instance to a phenoxy-imine derivative after amine oxidation. However, the authors managed to isolate bisphenoxy-amine species. Thus, species **12a** and **12b** were prepared from the reaction of a tridentate bis-potassium phenoxy-imine salt with **1** and Hoveyda–Grubbs I catalysts, respectively, in the presence of 1 equiv of silver chloride (Scheme 17.15). The isolated yields are low (21% for **12a** and 34% for **12b**). Complexes **12a** and **12b** are both reasonably stable in air and moisture, and tolerate protic or acidic solvents such as methanol or acetic acid for short periods. However, their thermal stability is rather low: partial decomposition is observed in the solid state after a few days at 4 °C, while compounds are stable for several weeks at −20 °C.

Compound **12a** was characterized by X-ray diffraction. In the solid state, it adopts a trigonal bipyramidal structure featuring a symmetrically coordinated tridentate ligand with the oxygen atoms and the alkylidene in equatorial positions, and the amine and phosphine groups in apical positions. From NMR data, it was concluded that the isopropoxy moiety in **12b** does not bind the ruthenium center.

The constrained geometry of these derivatives featuring such a bis-chelating ligand proved detrimental to catalytic activity. When probed for RCM of diethyl diallylmalonate, they displayed only very low activity. Upon addition of one or more equivalents of a strong Brønsted acid (HCl or H_2SO_4), the activity increases, though the resulting system features a rather short lifetime. Interestingly, **12a** combined with traces of HCl generates Hoveyda–Grubbs I catalyst, which is consistent with decoordination of the tridentate ligand under these conditions.

17.3.2 Alkoxy-Pyridine $[O,N]^-$ Ligands

2-Pyridinylcarbinol is a type of [N,O]-chelating ligand which has been well established in tungsten(VI) catalyst precursors for ROMP reactions (36). Shortly after the discovery of the first Schiff base complexes, Hafner and coworkers from Ciba Specialty Chemicals issued a patent in 1999, describing pyridyl-alkoxide ruthenium carbenes **13a–f** (37). Several

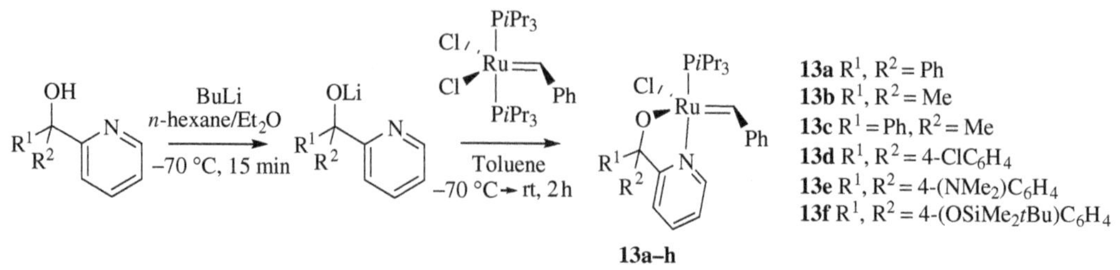

Scheme 17.16 Synthesis of catalysts **13a–f**.

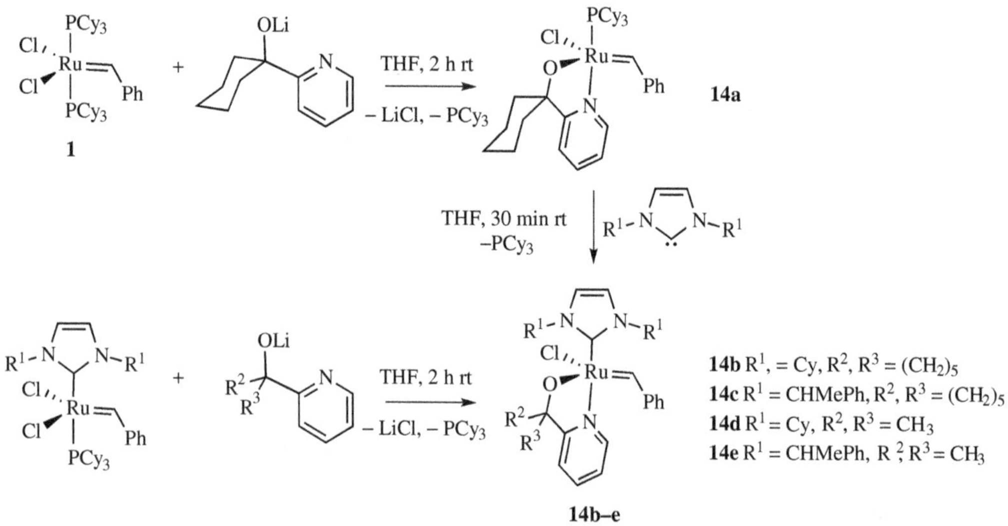

Scheme 17.17 Synthesis of catalysts **14a–e**.

benzylidene derivatives were produced in the example section, from the reaction of the lithium alkoxide with the trisisopropylphosphine derivative of **1**, with yields ranging from 65% to 78% (Scheme 17.16).

The activity of catalyst **13a** in ROMP of DCPD (DCPD/Ru ratio = 140,000/1) was probed by differential scanning calorimetry (DSC), on a temperature range from 0 to 250 °C, with a heating rate of 10 °C/min. An exotherm was observed, with an onset temperature at 106 °C, and a maximum at 130 °C. The resulting material presents a glass transition temperature (T_g) of 155 °C. The initial DCPD/**13a** mixture remains stable at room temperature for more than 24 h but it fully cures within 2 h on heating at 100 °C. After post-curing for one hour at 150 °C, a T_g of 150 °C is measured. This is the first report of thermally activated ROMP of DCPD with a [N,O] ligand-substituted metathesis catalyst. The Ciba team along with Van Koten and coworkers investigated the immobilization of such catalysts on dendrimeric structures, obtaining RCM active catalysts 38). Hermann reported in 2002 on the (unsaturated) NHC derivatives **14b–e**, which were prepared from two synthetic routes, that is, from reaction of the phosphine [N–O] complex **14a** with the free carbenes, or from reaction of the lithium alkoxide with the Grubbs second generation derivatives (Scheme 17.17) (39). Both routes proceed swiftly to the formation of the NHC-pyridyl-alkoxide catalysts with 64–73% yield.

These compounds were probed in the ROMP of cyclooctene and norbornene, with respective monomer/catalyst ratios of 500 and 100. No structure-activity relationship could be drawn from the experiments. While moderate activity is reached at room temperature, strong increase is observed at 60 °C, reaching activity comparable to that of the Grubbs II analog. This is in line with a thermal activation behavior.

A further example was provided in 2006 by Jordaan and Vosloo, who compared the activity of phosphine and (saturated) NHC derivatives of pyridyl-alkoxide catalysts (**14a** and **14f**, respectively) in the self-metathesis of 1-octene (40). The saturated NHC derivative **14f** was prepared from the second generation Grubbs catalyst **9** (Scheme 17.18).

These systems were poorly active at room temperature, but featured improved activity and selectivity at high temperatures compared to the Grubbs catalysts **1** and **9**. Indeed, incorporation of the hemilabile ligand into the ruthenium coordination sphere improved the thermal stability of the catalysts and decreased the extent of secondary metathesis reaction due to isomerization.

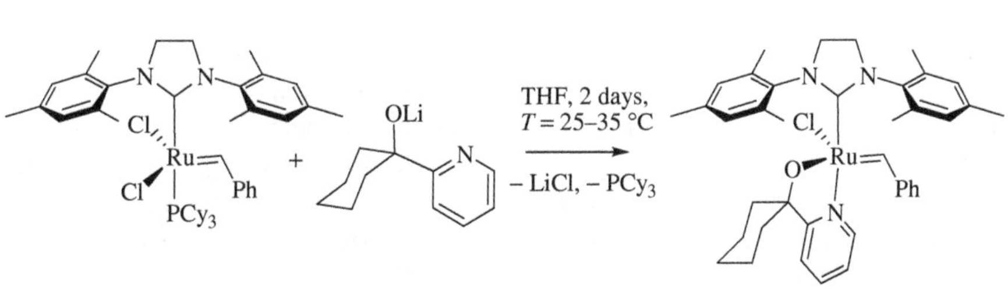

9 **14f**

Scheme 17.18 Synthesis of catalyst **14f**.

NMR investigations on the 1-octene/**14a** and **14f** systems ($CDCl_3$, 50 °C) revealed the presence of different carbenic species, resulting from either vinylidene or heptylidene fragments. While the phosphine derived **14a** ^1H NMR spectra did not provide further information, in the case of the **14f** system, the authors propose that the chelating and dangling systems are both observed, based on the observation of additional Ru=CH signals. Further NMR experiments (through-space correlations and exchange experiments, for instance) would be necessary to confirm this point. These experiments nevertheless indicate that the heptylidene is the catalytically active species that preferentially forms in the 1-octene metathesis reaction with the NHC derivative **14f** (50 °C, $CDCl_3$). In contrast, both the heptylidene and methylidene species form simultaneously during the 1-octene metathesis reaction with **14a**.

17.3.3 Carboxylate-Based Chelating $[O,N]^-$, $[O,P]^-$ and $[O,O]^-$ Ligands

As the carboxylate group is known to strongly bind to ruthenium, several groups investigated ruthenium metathesis catalysts bearing chelating ligands that comprise this moiety along with an N-, P- or O-based additional donor group.

The use of *ortho*-pyridylcarboxylate (picolinate) as a ligand on ruthenium metathesis catalyst was reported by Hahn and coworkers in 2005 (41). The authors reacted 2 equiv of silver pyridine-2-carboxylate with **9**, and obtained complex **15a** in 80% yield (See Scheme 17.19 for its structure). This species proved to be very stable, being amenable to purification by chromatography on silica gel carried out in air using methanol-containing eluent. If **15a** is inactive as such in RCM of diallyl malonate, however, fair activity is triggered by addition of HCl (2 equiv), though the catalytic system does not reach activity of the parent Grubbs complex **9**. However, **15a** enables RCM reactions to take place in methanol with moderate activity. The authors propose that the addition of HCl to **15a** leads to protonation of at least one of the picolinate ligands. Indeed, both picolinic acid and the 2-pyridinecarboxylato, chloride complex resulting from a single Cl/picolinate exchange from **15a** were identified in the reaction mixture after addition of HCl. The synthesis and characterization of this compound and its phosphine counterpart **15b** were revisited by Grubbs in 2010 (Scheme 17.19) (42). Their simplified procedure consists in reacting **1** or **9** with 10 equiv of picolinic acid and 1–1.3 equiv of Ag_2O, which affords the desired complexes in 70–90% yield (Scheme 17.19).

Two isomers (A and B, Scheme 17.19) were obtained for both **15a** and **15b** that could be separated by column chromatography. In the case of **15a**, these were characterized by X-ray diffraction studies, by Grubbs (isomer A) and Hahn (isomer B). Both feature a distorted octahedral geometry, with a pyridine moiety trans to the NHC. In isomer A, the two oxygen atoms occupy mutually trans positions, while in isomer B, they are cis to each other. Exchange between A and B is operative. In the case of **15a**, the thermodynamically more stable isomer is A, while B is more stable for **15b**. Authors track back this observation

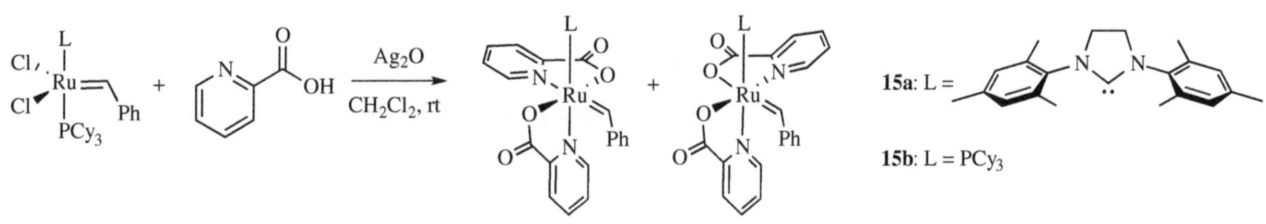

Isomer A **Isomer B**

15a,b

Scheme 17.19 Synthesis of **15a** and **15b**.

to π–π interactions between the NHC mesityl rings with benzylidene and picolinate fragments.

Again, no activity in ROMP of norbornene or RCM of diethyl diallylmalonate was detected after 1 h heating at 80 °C. After addition of HCl (20 equiv relative to Ru), **15a** and **15b** became highly active and initiated ROMP (1% catalyst loading), achieving full conversion in less than 30 min. Species **15a** is inert in DCPD solution (33 ppm Ru loading), allowing efficient mining at room temperature, and upon addition of 20 equiv HCl, polymerization was initiated. The exotherm was reached within 15 s. The resulting polymer features a heat-distortion temperature of 130 °C which is expected for well cured poly-DCPD. Studies were performed to assess the activation process. In the case of diallylmalonate RCM, no rate dependence of HCl is observed for **15b**, contrary to **15a**, which shows faster initiation and increased reaction rates compared to **15b**. Grubbs and coworkers propose three possibilities for activation of **15a–b** by HCl: (1) protonation of pyridinyl moiety leading to a bis-cationic derivative that comprises two bound picolinate ligands; (2) exchange of one bound picolinate by a chloride (as proposed by Hahn) and protonation of the metal-ligated picolinate, and (3) full exchange of picolinates by chlorides (42). To clarify this point, **15a** and **15b** were reacted with β-methyl-2-isopropoxy styrene (Scheme 17.20). Cross-metathesis resulted in the formation of the corresponding Hoveyda–Grubbs derivatives, which demonstrates that the third hypothesis is valid.

He and coworkers described a series of ruthenium catalysts featuring a carboxylate-functionalized phosphine, which were prepared from exchange of both a chloride and a triphenylphosphine using a NHC-PPh_3 benzylidene starting material and the corresponding sodium phosphinocarboxylate ligand (Scheme 17.21) (43). Complexes **16a–c** were isolated in modest to good yields after purification by column chromatography. They can be stored in air without significant decomposition over 4 weeks. Compound **16a** was characterized by X-ray diffraction, and displayed the expected coordination geometry, very close to that of **1**: the phosphine and the NHC ligands were in mutually trans positions, while the carboxylate ligand replaced a chloride.

Complexes **16a–c** were successfully used as catalysts for RCM of diethylmalonate and diallylmalononitrile. The species featuring a five-membered (**16a**) or rigid six-membered (**16c**) chelating ring exhibit enhanced stability and higher catalytic efficiency than the compound bearing a flexible six-membered chelating ligand (**16b**) or the parent PPh_3 derivative of **9**, especially at elevated temperature. This reveals the importance of the hemilability of the chelating ligand that plays a dual, conflicting role as protection of the metal center, and at the same time hinders substrate access to the active species.

Samec and Grubbs investigated amino-acids as another type of chelating N,O ligands, using readily available, chiral L-proline (44). Thus, **1** reacts with 10-fold excess of proline in the presence of copper(I) oxide to afford the product of single chloride substitution, **17a**, in a 40% isolated yield (Scheme 17.22). Only one isomer is obtained.

Compound **17a** efficiently catalyzes RCM of several substrates at room temperature, with rate acceleration in the presence of phosphine scavenger CuCl (at the expense of catalyst stability). However, the initiation mechanism is undetermined. Interestingly, the NHC analog of **17a** (**17b**, vide infra) also sees its activity enhanced by CuCl addition, which raises questions about the fate of the prolinate ligand under these experimental conditions. The electron-withdrawing carboxylate group of proline is crucial for the activity of **17a**

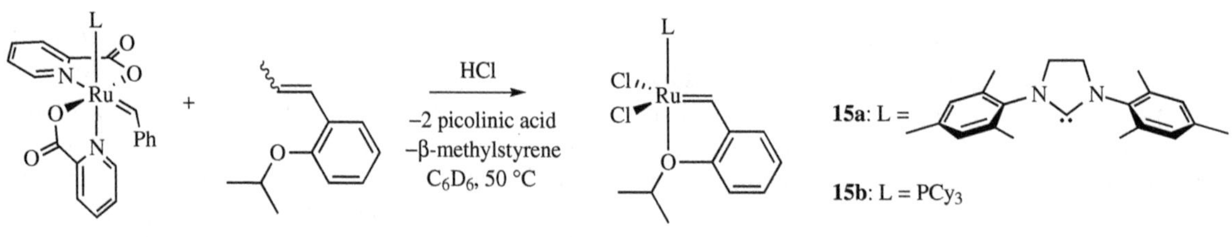

15a–b

Scheme 17.20 Trapping of the active species generated by protonation of **15a** and **15b**.

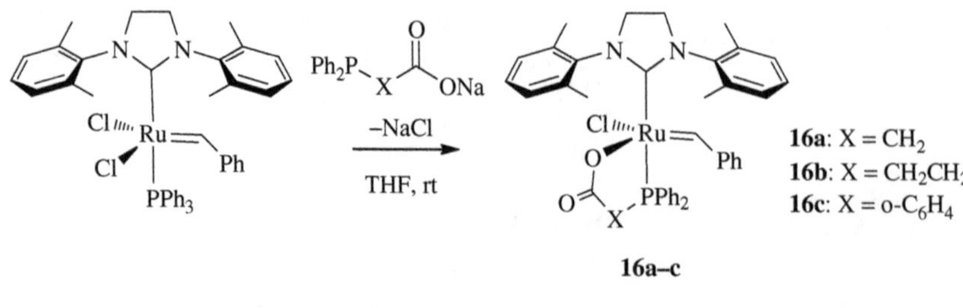

16a–c

Scheme 17.21 Synthesis of catalysts **16a–c**.

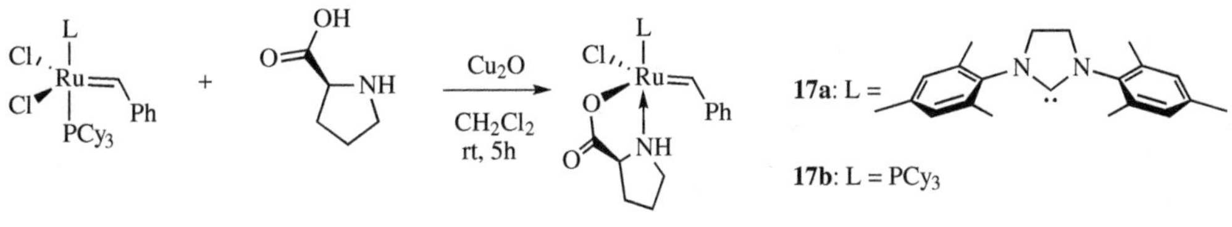

17a–b

Scheme 17.22 Synthesis of **17a** and **17b**.

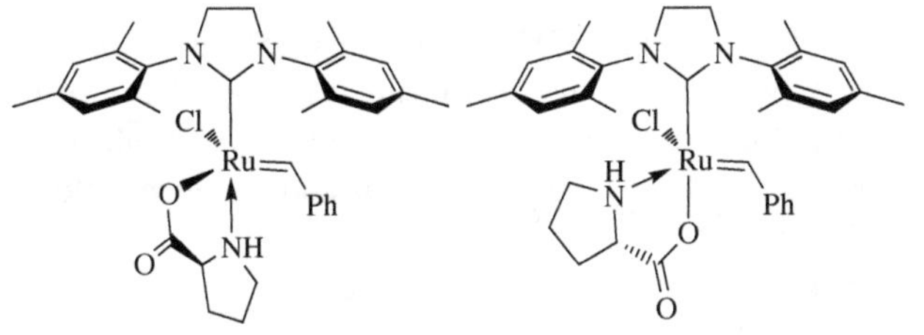

17b-A **17b-B**

Scheme 17.23 Structure of the isomers **17b-A** and **17b-B**.

as it is mentioned that the parent amino-alkoxide ruthenium derivative is considerably less active. Complex **17a** also catalyzes ROMP of COD at 30 °C with low activity.

The NHC derivative **17b** was prepared following the same route with a moderate yield of 54%, as a mixture of isomers (**17b-A** and **17b-B**, Scheme 17.23) (45). Both could be separated by column chromatography. An exchange is operative between the isomers, and the equilibrium ratio at room temperature is 95/5 for **17b-A**/**17b-B**. Diffraction studies on single crystals of **17b-A** were thwarted by poor crystal quality, but showed coordination of the amine trans to the NHC ligand.

Along the same line, tetrahydro-2-furoic acid was reacted with **9** in the presence of Ag_2O to afford complex **18** in an 80% yield, again as an isomer mixture (10/1 ratio), which resisted separation attempts (Scheme 17.24) (45). Their structure should be similar to that of **17b-A** and **17b-B**.

Both **17b** and **18** performed poorly as diethyl diallylmalonate RCM catalysts at room temperature. However, addition of CuCl has a beneficial effect on **17b**'s activity, rendering it efficient toward more demanding hindered RCM substrates. Trapping experiments suggest that activation occurs via coordination of CuCl on the prolinate nitrogen, thus opening a coordination site on the ruthenium center. *In situ* NMR studies exclude the decoordination of the prolinate ligand and its substitution by a chloride.

17.3.4 Tris(pyrazolyl)borate [N,N,N]$^-$ Ligands

A further class of potentially hemilabile, chelating anionic ligand was introduced by Ozawa (46) and by Grubbs (47), who respectively studied ruthenium vinylidene (**19a**) and benzylidene (**19b**) catalysts featuring tris(pyrazolyl)hydridoborate (Tp) ligands (Scheme 17.25). These bulky ligands behave as tridentate anionic donors of six electrons, which occupy three coordination sites. Complex **19a**, originally prepared by Kirchner from [Ru(Tp)$(PPh_3)_2$Cl] (48), is active in ROMP of norbornene, though high temperature (80 °C) and long reaction times (72 h) are required to fully convert 100 equiv of monomer. A more efficient system is formed upon addition of 3 equiv of $BF_3{\cdot}Et_2O$, as full conversion is reached within the same reaction time at a lower temperature (40 °C). On the other hand, the benzylidene derivative **19b**, prepared in good yield from **1** and KTp, does not mediate RCM of diethyl diallylmalonate at 70 °C, even after extended reaction time. Addition of phosphine scavenging agents such as HCl, CuCl, or $AlCl_3$ affords higher catalytic activity, though high catalyst loading (20%) is required. Cationic derivatives were prepared by reaction with silver salts in the presence of Lewis bases (water, pyridine, and acetonitrile),

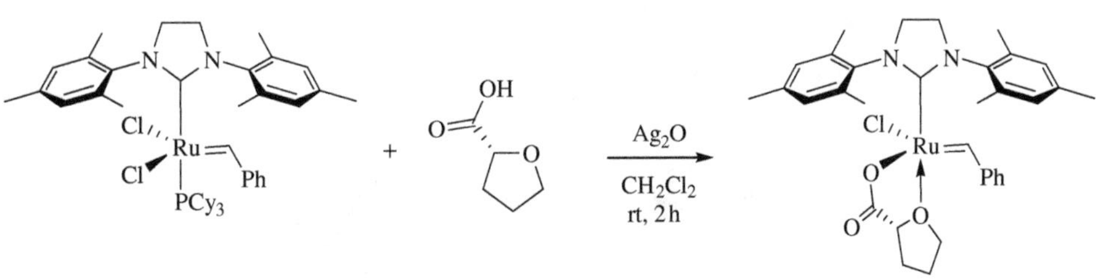

9 **18**

Scheme 17.24 Synthesis of **18**.

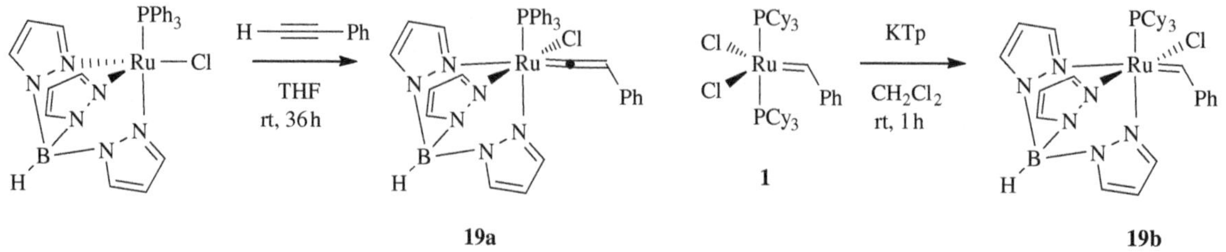

Scheme 17.25 Synthesis of complexes **19a** and **19b**.

but these species were surprisingly inert toward diethyl diallylmalonate.

Further studies were devoted by Slugovc to access other examples of alkylidene Tp derivatives (Scheme 17.26) (49). The reactivity of ruthenium NHC complexes featuring a κ^2-(C,O) alkylidene moiety and a *cis*-dichloro toward the potassium salt KTp configuration was probed. The reactions were not selective and the nature of the products depends strongly on the (C,O) chelate. In the case of an alkylidene-ester, the expected compound **19c** resulting from chloride-tris(pyrazolyl)borate exchange can be isolated, with a low yield of 25%. X-ray diffraction studies indicate a trishapto configuration of the Tp ligand, with a dangling 2-ethylcarboxylatebenzylidene, similar to the bisphenato amine complex **11b**. However, in the case of the complex bearing a chelating alkylidene-aldehyde, decomposition occurred. Single crystal X-ray structure determination of an isolated product (**19d**, isolated yield 13%) revealed the occurrence of a double C–H activation of the *ortho*-methyl substituents of the SIMES ligand, affording a benzylidene type moiety. This also resulted in the release of the 2-formylbenzylidene ligand as *ortho*-methylbenzaldehyde. Further, the Tp's hydride was substituted by a chloride (Scheme 17.26).

Slugovc and coworkers compared the performance of catalysts **19b–d** in ROMP of norbornene-2,3-dicarboxylic acid diethyl ester using DSC (monomer/catalyst = 300/1). High initiation temperatures of the polymerization reaction were observed (**19b**, 138 °C; **19c**, 128 °C; and **19d**, 109 °C).

A study from the Patel group of the introduction of an anionic, bidentate bis(pyrazolyl)borate ligand (κ^2-Bp) in Grubbs first-generation catalyst led to isolation of species **20** that displays high thermal stability in solution and no indication of decomposition over several weeks (Scheme 17.27) (50). Very low activity in RCM of diethyl diallylmalonate was achieved at high catalyst loading (8 mol%), with beneficial effect of CuCl addition.

Overall, bis- and tris(pyrazolyl)borates induce a high degree of catalyst stability and latency toward RCM of

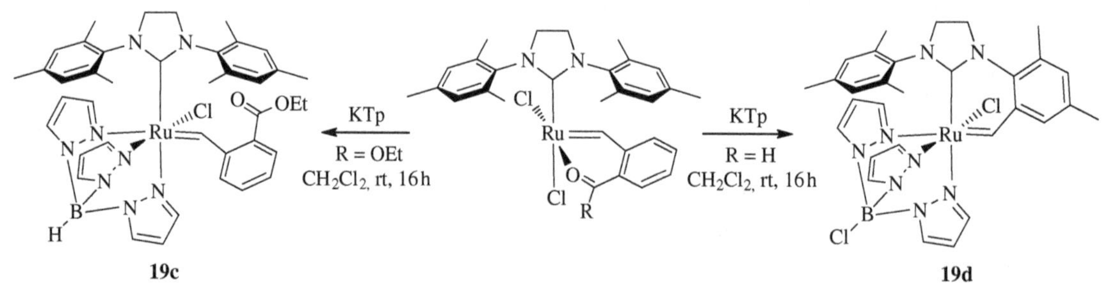

Scheme 17.26 Synthesis of complexes **19c** and **19d**.

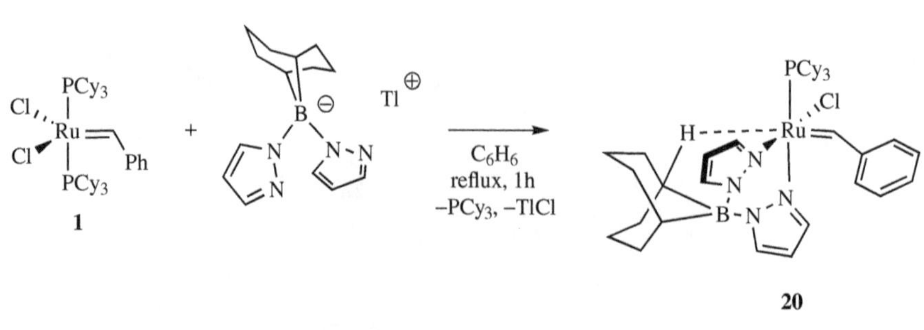

Scheme 17.27 Synthesis of catalyst **20**.

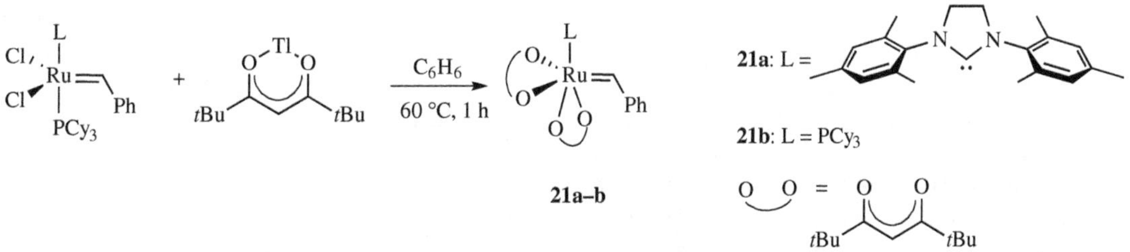

Scheme 17.28 Synthesis of complexes **21a** and **21b**.

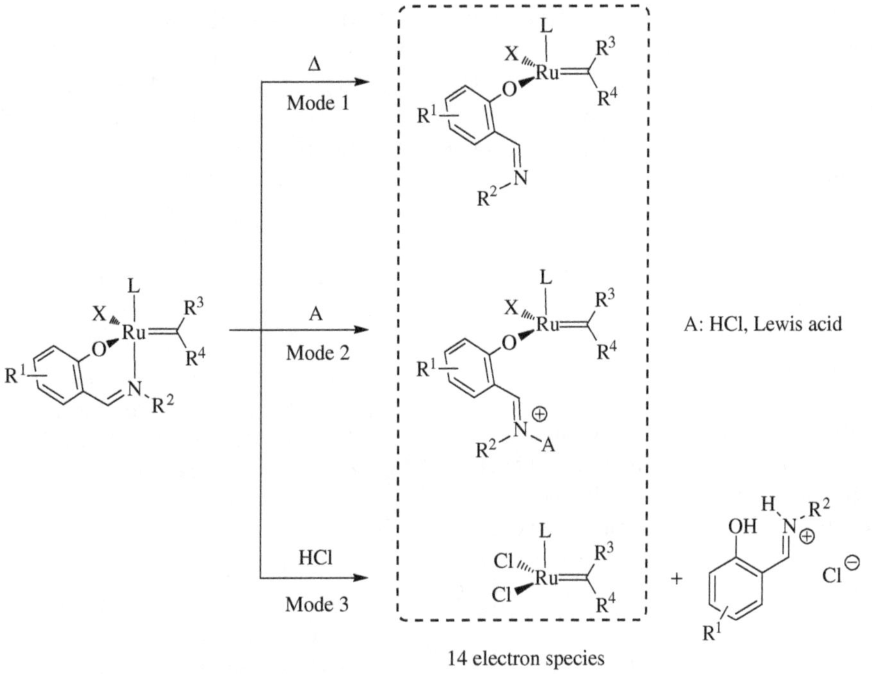

Scheme 17.29 Activation modes of Schiff base catalysts.

dienes and ROMP of strained cyclic olefins. However, thermal activation proved to be difficult, and had to be circumvented by the use of higher catalyst loadings. For these reasons, these types of complexes are unsuitable candidates as potential latent catalysts.

17.3.5 Acetylacetonate [O,O]$^-$ Ligands

A classical type of chloride substituent, namely the acetylacetonate derivatives (acac), was installed by Grubbs and coworkers onto catalysts **1** and **9**, as bidentate κ^2-(O,O) ligands, following reaction with 2 equiv of the thallium salt Tl(acac) (Scheme 17.28). Reaction with the analogs silver salt led to complete ligand exchange, but the resulting products could not be purified.

Complex **21a** proved to be inactive for the ring-opening polymerization of DCPD (in neat monomer) and of 7-oxanorbornene-2,3-dimethoxymethyl (in methanol at room temperature) (51). Catalytic activity is triggered by addition of organic or inorganic acids such as HCl and reactions were completed within minutes, surpassing the activity of the parent Grubbs I complex **1** (no monomer/catalyst ratio is given). More recently, the Grubbs group showed that **21a** and **21b** are readily activated upon irradiation of a catalyst–monomer mixture containing a photoacid generator, and successfully applied these systems in RCM and ROMP (at 5% loading) (52). In this case, trapping experiments with β-methyl-2-isopropoxystyrene resulted in the formation of the Hoveyda–Grubbs I derivative, which indicates substitution of the acetylacetonate groups by chlorides. However, the presence of other active species in the reaction mixture cannot be ruled out.

17.4 CONCLUDING REMARKS

Early reasons for the use of the Schiff base catalyst were related to the broadening of the scope of the metathesis catalysts. Their enhanced stability under aerobic conditions was then often mentioned as a positive factor when justifying their development and use, as such species demonstrated exceptional stability under demanding conditions. From the synthetic point of view, as the sterics and electronics of such salicylaldimine ligands can be most easily tuned, and as high yielding routes to their ruthenium complexes have been designed, this class of catalysts is highly attractive. Indeed, if other types of related (potentially) latent catalysts have been described, they did not comply to as many combined criteria such as high activity, stability, and latency, for instance.

Thus catalyst design has offered latent catalysts able to effect several thousands of turn-over numbers in the industrially relevant ROMP of DCPD, with catalyst loading down to parts per million level. Activation modes of these sleeping beauties can be separated in three modes, which can be operative for the same catalyst depending on the applied or effective activation method (Scheme 17.29):

Mode 1: Thermal activation resulting from the hemilabile character of the ligand, the most simple case, where the latency character depends on the difference in activity at room temperature and at a given higher temperature.

Mode 2: Trapping of the open form by a Brønsted or Lewis acid, leading respectively to protonated imine or imine-adduct formation, thus opening a vacant site on the metal. In such a case, the hemilabile ligand is proposed to stay on the metal. This convenient and efficient way to trigger activity can be easily implemented in large production facilities that rely on RIM processes.

Mode 3: reaction of the phenoxy moiety with HCl to afford a 14 electron species, in analogy with what was initially observed for bisalkoxide derivatives (53). In this case, the Schiff base ligand is cleaved from the ruthenium and the activation generates a bischloride species analogous to active species stemming from catalysts such as **1** or **9**. This has been demonstrated by trapping experiments that resulted in isolation or observation of Hoveyda–Grubbs derivatives. In the absence of systematic checking, it is not always possible to distinguish from the previous activation mode, but the presence of water has been shown to be responsible for such a pathway.

At this stage of the development of catalysts bearing phenoxy-imine ligands, there is still need for deeper understanding of the activation processes and for a rationale for catalyst optimization. The ease of steric and electronic tuning of the Schiff base ligands offers extremely wide possibilities regarding the accessible structures. Thus, lessons from mechanistic and theoretical investigations can definitively be expected to blossom into new and efficient latent catalysts. These should be able to fulfill the requirements for wider use, including industrial applications: high activity, high stability, and extensive shelf-life, while being activated by a simple stimulus. This challenging goal offers great promises for future Schiff-base catalyst development by chemists from both academia and industry.

REFERENCES

1. Drozdzak R, Allaert B, Ledoux N, Dragutan I, Dragutan V, Verpoort F. *Adv Synth Catal* 2005;347:1721–1743.
2. Drozdzak R, Allaert B, Ledoux N, Dragutan I, Dragutan V, Verpoort F. *Coord Chem Rev* 2005;249:3055–3074.
3. Monsaert S, Lozano Vila AM, Drozdzak R, van der Voort P, Verpoort F. *Chem Soc Rev* 2009;38:3360–3372.
4. Bhattacharya S, Boone SR, Fox GA, Pierpont CG. *J Am Chem Soc* 1990;112:1088–1096.
5. Gozwami A, Chakravarty AR, Chakravorty A. *Inorg Chem* 1981;20:2246–2250.
6. Chang S, Leroy J II, Chunming W, Lawrence MH, Grubbs RH. *Organometallics* 1998;17:3460–3465.
7. Mol JC. *J Mol Catal A: Chem* 2004;213:39–45.
8. Galván-Arzate S, Santamaría A. *Toxicol Lett* 1998;99:1–13.
9. Conrad JC, Snelgrove JL, Eeelman MD, Hall S, Fogg DE. *J Mol Catal A: Chem* 2006;254:105–110.
10. Occhipinti G, Jensen VR, Bjørsvik HR. *J Org Chem* 2007; 72:3561–3564.
11. Monsaert S, Ledoux N, Drozdzak R, Verpoort F. *J Polym Sci A: Polym Chem* 2010;48:302–310.
12. De Clerq B, Verpoort F. *Macromolecules* 2002;35:8943–8947.
13. De Clerq B, Lefebvre F, Verpoort F. *Appl Catal A: Gen* 2003;247:345–364.
14. Opstal T, Verpoort F. *Synlett* 2002:935–941.
15. Opstal T, Verpoort F. *J Mol Catal A: Chem* 2003;200:49–61.
16. Katayama H, Ozawa F. *Organometallics* 1998;17:5190–5196.
17. Boeda F, Clavier H, Nolan SP. *Chem Commun* 2008: 2726–2740.
18. Opstal T, Verpoort F. *Angew Chem Int Ed* 2003;42:2876–2879.
19. Lozano Vila AM, Monsaert S, Drozdzak R, Woloviec S, Verpoort F. *Adv Synth Catal* 2009;351:2689–2701.
20. De Clerq B, Verpoort F. *Tetrahedron Lett* 2001;51:8959–8963.
21. De Clerq B, Verpoort F. *J Mol Catal A: Chem* 2002;180:67–76.
22. Opstal T, Couchez K, Verpoort F. *Adv Synth Catal* 2003; 345:393–401.
23. De Clerq B, Verpoort F. *Adv Synth Catal* 2002;344:639–648.
24. De Clerq B, Verpoort F. *Tetrahedron Lett* 2002;43:9101–9104.
25. De Clerq B, Verpoort F. *J Organomet Chem* 2003;672:11–16.
26. Allaert B, Dieltiens B, Ledoux N, Vercaemst C, van der Voort P, Stevens CV, Linden A, Verpoort F. *J Mol Catal A: Chem* 2006;260:221–226.
27. Binder JB, Guzei IA, Raines RT. *Adv Synth Catal* 2007; 349:395–404.
28. Sanford MS, Love JA, Grubbs RH. *Organometallics* 2001; 20:5314–5318.

29. Vehlow K, Maechling S, Blechert S. *Organometallics* 2006;25:25–28.
30. Ledoux N, Allaert B, Schaubroeck D, Monsaert S, Drozdzak R, van der Voort P, Verpoort F. *J Organomet Chem* 2006; 691:5482–5486.
31. Monsaert S, Drozdzak R, Verpoort F. *Chem Today* 2008; 26:93–96.
32. Hendrickx PMS, Drozdzak R, Verpoort F, Martins JC. *Magn Reson Chem* 2010;48:443–449.
33. Drozdzak R, Nishioka N, Recher G, Verpoort F. *Macromol Symp* 2010;293:1–4.
34. Ledoux N, Drozdzak R, Allaert B, Linden A, van der Voord P, Verpoort F. *Dalton Trans* 2007:5201–5210.
35. Occhipinti G, Bjørsvik HR, Törnroos KW, Jensen VR. *Organometallics* 2007;26:5803–5814.
36. van der Schaaf PA, Abbenhuis RATM, van der Noort WPA, de Graaf R, Grove DM, Smeets WJJ, Spek AL, van Koten G. *Organometallics* 1994;13:1433–1444.
37. van der Schaaf P. A., Mühlbach A., Hafner A., Kolly R. Heterocyclyl ligand containing ruthenium and osmium catalysts. WO 99/29701, 1999.
38. Wijkens P, Jastrzebski JTBH, van der Schaaf PA, Kolly R, Hafner A, van Koten G. *Org Lett* 2000;2:1621–1624.
39. Denk K, Fridgen J, Herrmann WA. *Adv Synth Catal* 2002;344:666–670.
40. Jordaan M, Vosloo HCM. *Adv Synth Catal* 2006;349:184–192.
41. Hahn FE, Paas M, Fröhlich R. *J Organomet Chem* 2005;690:5816–5821.
42. Samec JSM, Keitz BK, Grubbs RH. *J Organomet Chem* 2010;695:1831–1837.
43. Zhang W, Liu P, Jin K, He R. *J Mol Catal A: Chem* 2007;275:194–199.
44. Samec JSM, Grubbs RH. *Chem Commun* 2007:2826–2828.
45. Samec JSM, Grubbs RH. *Chem—A Eur J* 2008; 14:2686–2692.
46. Katayama H, Yoshida T, Ozawa F. *J Organomet Chem* 1998;562:203–206.
47. Sanford MS, Henling LM, Grubbs RH. *Organometallics* 1998;17:5384–5389.
48. Slugovc C, Mereiter K, Zobetz E, Schmid R, Kirchner K. *Organometallics* 1996;15:5275–5277.
49. Burtscher D, Perner B, Mereiter K, Slugovc C. *J Organomet Chem* 2006;691:5423–5430.
50. Patel J, Jackson WR, Serelis AK. *Inorg Chim Acta* 2004;357:2374–2378.
51. Lynn DM, Dias EL, Grubbs RH, Mohr B. WO 99/22 865, 1999.
52. Keitz BK, Grubbs RH. *J Am Chem Soc* 2009;131:2038–2039.
53. Sanford MS, Henling LM, Day MW, Grubbs RH. *Angew Chem Int Ed* 2000;39:3451–3453.

SECTION 5

DEVELOPMENT OF CONCEPTS IN OLEFIN METATHESIS CATALYSTS

18

NOVEL CONCEPTS IN CATALYST DESIGN—A CASE STUDY OF DEVELOPMENT OF HOVEYDA-TYPE COMPLEXES

MICHAŁ BARBASIEWICZ
Faculty of Chemistry, Warsaw University, Warsaw, Poland

Understanding of structure–activity relationships plays a fundamental role in the development of catalytic systems. For some of them, the correlations may directly arise from simple mechanistic considerations, while other optimization processes may require more sophisticated concepts being supplemented and corrected by the data obtained in a series of trail-and-error attempts. Sometimes both the situations coexist, for facilitating different aspects of catalyst evolution, delivering both broad rules and detailed knowledge from more advanced studies. For the family of Hoveyda-type complexes, now represented by more than 100 reported structures (1–3), progress in catalyst design is most accurately described by the latter situation. Since the pioneering papers by Hoveyda (4–6), the key structure has continuously been revised and has inspired numerous studies targeting the goal of improved activity, stability, and selectivity, and understanding their mechanism of action at various stages of the catalytic cycle. In this chapter we focus mainly on selected novel concepts, concerning the design of the catalysts, developed recently in our laboratory.

Development of the family of Hoveyda-type metathesis catalysts began in 1998, when the first ruthenium complex with a chelating benzylidene ether ligand was reported (4). In a seminal paper about synthesis of chromene derivatives, the authors described the unexpected quench of catalytic cross-metathesis (CM) reaction of 2-isopropoxystyrene. The chance observation and isolation of phosphane-based complex **1a** in reaction of the substrate with Grubbs complex $(PCy_3)_2Cl_2Ru{=}CHPh$ has sparked a continuing interest in

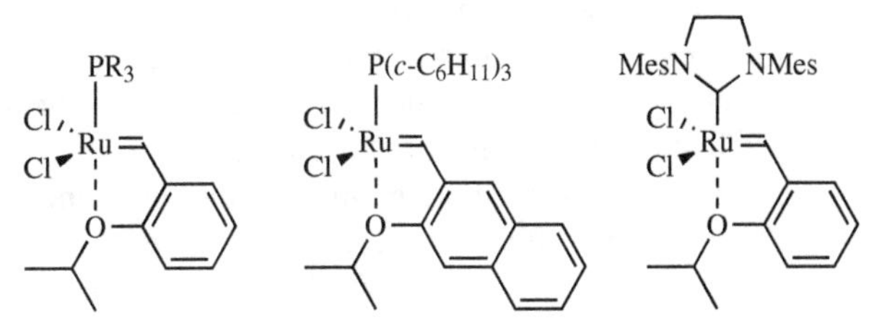

R = c-C_6H_{11}; **1a**
C_6H_5; **1b** — **1c** — **2**

Figure 18.1 Selected early Hoveyda-type complexes with phosphane and NHC ligands (4–6). (Mes—2,4,6-trimethylphenyl).

this class of compounds. In further studies, the complex **1a** was synthesized from intermediate **1b** obtained in a direct reaction of 2-isopropoxyphenyldiazomethane with $RuCl_2(PPh_3)_3$ followed by phosphine exchange (5). In the same report, related complex **1c** was characterized with X-ray structure (Fig. 18.1).

Activity studies on model metathesis substrates revealed that **1a** is a robust catalyst and can be recycled after the reaction in high yield. Particularly, it was suggested that after completion of the catalytic process, the chelating ligand can re-associate with 14ē ruthenium propagating species, in a process called "boomerang" or "release-return" mechanism (7). The assumption was supported by isolation of **1a** when ring-closing metathesis (RCM) reaction of 2-butenyl-(1-phenyl)allyl ether, catalyzed with Grubbs complex, was quenched with a styrene derivative (5).

Olefin Metathesis: Theory and Practice, First Edition. Edited by Karol Grela.

Thereafter, a new complex **2** bearing *N*-heterocyclic carbene (NHC) ligand was described and proved to be superior to the previous first-generation phosphane complexes (**1**) (6). Starting from this discovery, variations of the catalyst's structure concerning NHC, chelate, and anionic ligands are continuously ongoing. Modifications of the chelating arylidene ligands (2,3), which control the initiation process and are responsible for the observed catalytic activity in model metathesis reactions of undemanding substrates, play a prominent role in these efforts.

A common structural motif of the Hoveyda-type complexes consists of benzyl carbene (benzylidene ligand) with the chelating donor atom of oxygen, or other heteroatoms such as nitrogen, sulfur, phosphorus, halogens, etc., present in the adjacent position. In most of the complexes, the coordinating sites are bound with aromatic framework, which enables modulation of properties with substituents. With the aim of improving catalytic activity, complex **2** was modified in numerous variants, and two general modes of activation were described. In the early studies, Grela introduced an electronic effect, which was demonstrated on nitro-substituted complex **3** with superior characteristics (Fig. 18.2) (8,9).

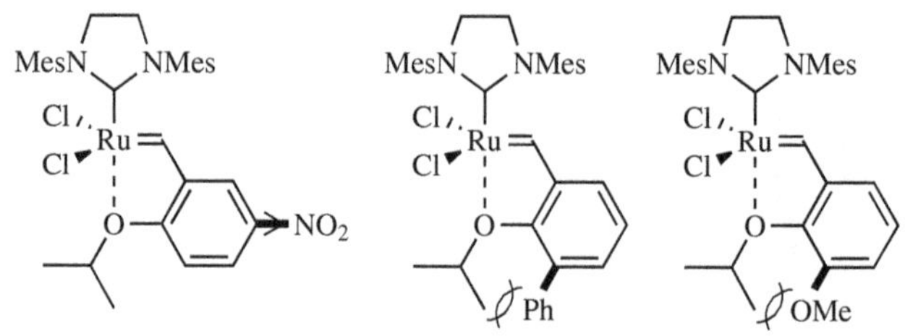

3 **4a** **4b**

Figure 18.2 Modified Hoveyda-type complexes with properties improved by electronic (**3**) (8,9) and steric effect (**4a**,**b**) (13,14).

The presence of the nitro group in *para* position to the chelating ether function decreased its donor properties and facilitated opening of the chelate ring, required for the initiation process. Further studies revealed generality of the electronic effect. In most cases (10) presence of electron-withdrawing groups (EWGs) in benzylidene ring improved catalytic performance, that was attributed to the weakening of Ru···O bond (11). Further support for the hypothesis was obtained in ligand exchange experiments. When the nitro-Hoveyda complex **3** was equilibrated with parent Hoveyda ligand (2-isopropoxystyrene), formation of less active complex **2** was favored (82 : 18) under thermodynamic conditions (12).

Another achievement in the field was the introduction of sterically demanding substituents in *ortho* position to the chelating *i*PrO group. Presence of phenyl (**4a**) (13) or methoxy (**4b**) (14) groups in the benzylidene ring caused out-of-plane distortion of the chelating arm and resulted in increased rate of initiation.

All the studies were carried out on the assumption that opening of the chelate ring is a rate-determining step of the initiation process, and in the end it controls the rate of the entire catalytic reaction. In fact, the initiation process was investigated in detail, and three alternative mechanisms of initiation of **2**, varying with order of Ru· · ·O bond breaking

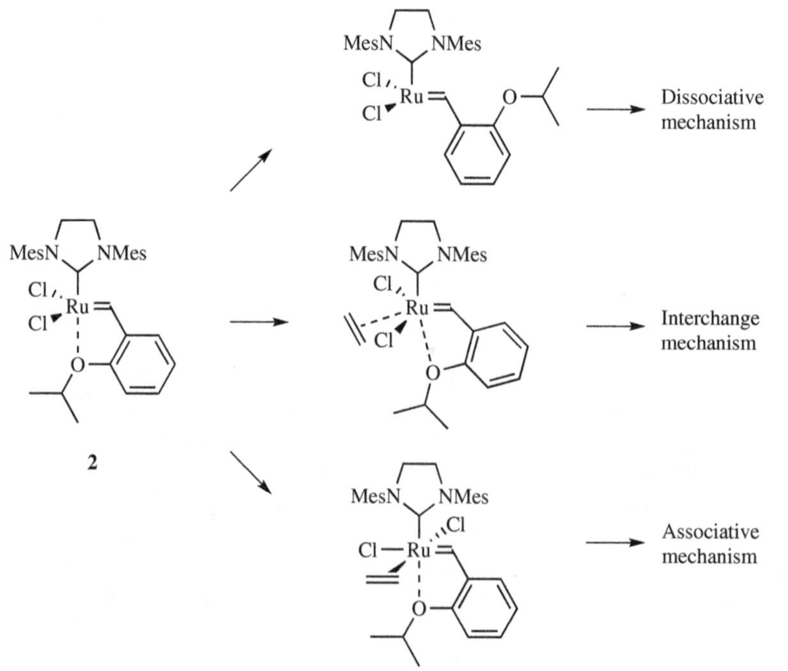

Figure 18.3 Alternative mechanisms of initiation of complex **2**: dissociative, interchange, and associative (15,16).

and olefinic substrate coordination were considered: dissociative, interchange and associative (Fig. 18.3).

Calculations by Vincent et al. inclined toward interchange mechanism, in which incoming olefin is partially bonded to the metal center and Ru···O bond is weakened in the transition state (15). In turn, results of Solans-Monfort (16), supported by experimental results of Plenio (17), revealed that metathesis reactions of sterically demanding substrates (e.g. diethyl diallylmalonate) catalyzed by complexes with *i*PrO chelating group initiate via rate-limiting dissociation of the chelate ring, followed by olefin coordination (dissociative mechanism) (18).

Although weakening of the Ru···O bond was demonstrated to increase the rate of the initiation process, excessive destabilization of the chelate ring may lead to undesired effects. One of the consequences of the proposed mechanisms is possible dissociation of the complex without reaction with substrate, or fast release of active catalytic species at the beginning of metathesis reaction. The two scenarios inevitably result in decomposition of catalysts at low conversions of substrate, limiting overall catalytic performance. The subtle balance between activity and stability of (pre)catalysts was noticed and led to the conclusion that no single catalyst outperforms all others in all cases (19).

In order to improve the catalytic performance with little cost to stability, numerous variants of the benzylidene ether ligands were investigated (Fig. 18.4) (20). The parent *i*PrO substituent present in complex **2** was modified, giving derivatives of dihydrobenzopyran (**5**) (21), benzodioxane (**6**) (22), chromene (**7**) (23), and oxazinone (**8**) (24). In the latter case, synergistic effects between structure of NHC and chelating ligand were demonstrated, suggesting that actually both the fragments control the rate of the initiation process and influence the stability of the system. Also, change of the *i*PrO group into benzyloxy (**9**) (21) and phenoxy substituents (**10**) (25) varied the electronic and steric parameters of the ligand and caused increase in the initiation rate of the catalysts.

An interesting structural feature of the chelating Hoveyda ligand is the presence of π-electron conjugation between the coordinating sites, which enables delocalization of electrons over the chelate ring. The delocalization can influence stability of the complex and therefore its initiation rate, giving a new mode of activity control of the metathesis catalysts. The idea was inspired by a chance observation of intriguing activity differences between naphthalene-based Hoveyda-type complexes (**11a,b**, Fig. 18.5) (26). When the isomeric complexes were tested in model RCM reaction with *N,N*-diallyltosylamine, catalyst **11b** displayed activity similar to **2**, while isomeric **11a** remained inactive over a range of temperatures. Rationalization of the data required consideration of factors that possibly quenched the catalytic activity of **11a**: steric, caused by crowding of the distal aromatic ring of naphthalene ligand with Ru=CH bond and NHC ligand, and electronic, related to the stabilization of the chelate ring. In the report we favored electronic effects, which may be hindered by structures derived from substituted naphthalene molecules (26). Following a well-known property of so-called Fisher-type carbenes (e.g., **12**), which are poorly active in metathesis reactions (27), the structure of the Hoveyda complex (**2**) was considered a vinylogous Fisher-type

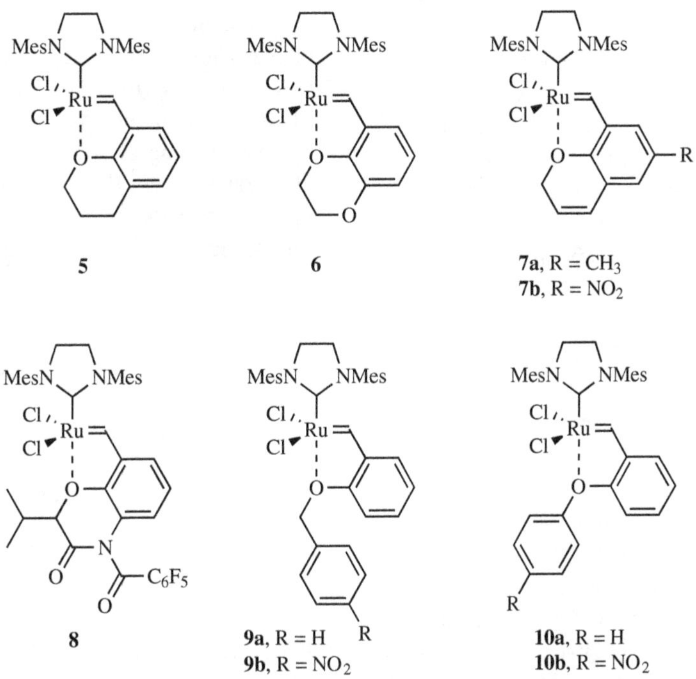

Figure 18.4 Selected Hoveyda-type complexes with modified ether chelating arm (21–25).

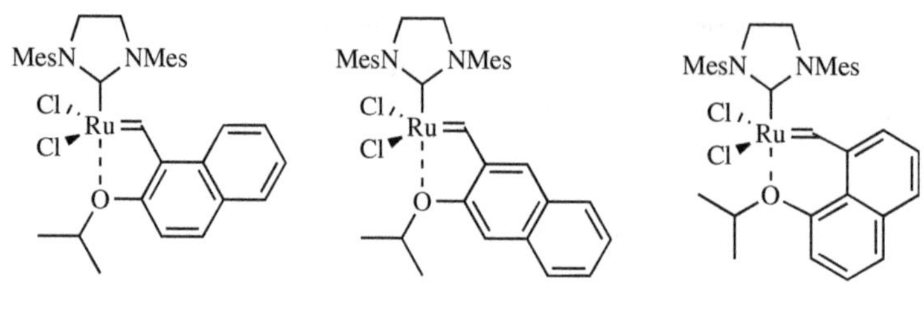

11a, Inactive **11b**, Active **11c**, Very active

Figure 18.5 Naphthalene-based catalysts displaying differences in catalytic activity (26,28).

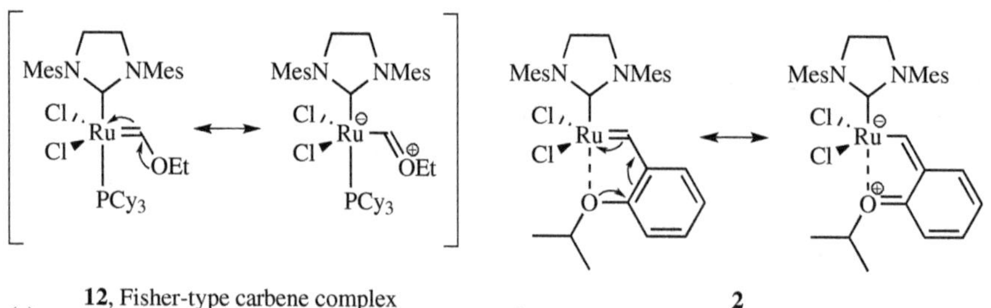

(a) **12**, Fisher-type carbene complex (b) **2**

Figure 18.6 Fisher-type carbene **12** belongs to the class of latent initiators in metathesis reactions (a) (27). Similar resonance structures can be drawn for complex **2** (b) (26).

carbene, in which the electron pair of the coordinating oxygen atom is shared through the conjugated ligand, and increases electron density on the ruthenium metal center (Fig. 18.6). Due to alternation of the naphthalene core with an increased double character of the C1–C2 bond complex with angular tricyclic structure (**11a**) was considered a masked Fisher-type carbene of diminished activity arising from the electronic stabilization. In the latent complex, the naphthalene ligand displays an increased π-electron density on the fragment linking the coordinating sites, and thus electron delocalization in the chelate ring is facilitated, as compared with the more single-bond character of the fragment in **2** and **11b** (C2–C3). Going further, electronic structures of the naphthalene-based complexes were compared with properties of tricyclic aromatic hydrocarbons: phenanthrene and anthracene (considering extension of delocalized structures of ligands with the chelate rings), and referred to as aromatic, on the basis of the empirical Clar rule.

Inspired by the idea, we scrutinized the family of the naphthalene-based complexes and realized the missing isomer is **11c** (Fig. 18.5). By analogy, the complex of *peri* substituted naphthalene ligand with coordinating sites placed at different rings was expected to display a reduced π-electron delocalization, suppressed between *peri*-positions of naphthalene, and facilitate the initiation process. In keeping with the expectations, catalyst **11c** appeared to be an excellent initiator, and despite limited stability (deactivation after ca 75% of conversion of model RCM substrates) at the beginning of the reaction, it initiated even faster than the nitro-substituted complex **3**. The interesting behavior was attributed to distorted molecular structure and suppressed π-electron delocalization within the 6-membered chelate ring (28).

In another study we attempted synthesis of bimetallic systems, in which the benzene ring was bound with two chelates giving tricyclic structures with angular and linear ring junctions (Fig. 18.7) (22). As the initiation process of the homodinuclear catalysts consists of two elementary chelate opening steps, it poses an open question about relative rates of the two processes. Although the idea of cooperativity between catalytic centers was demonstrated on bimetallic metathesis catalysts designed for selective formation of dimeric RCM products (29), initiation behavior, in

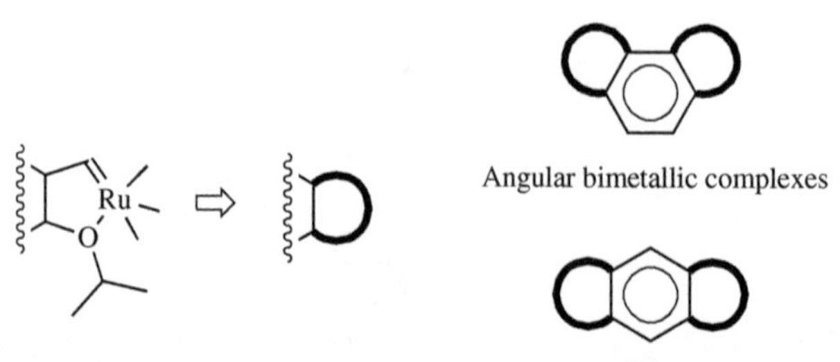

Figure 18.7 A family of bimetallic ruthenium complexes with angular and linear tricyclic structures (22).

which the first step controls the release of the next propagating species from electronic reasons, seemed to be promising. By analogy with properties of complex **11a**, we assumed that bimetallic complexes with angular, phenanthrene-like structures are stabilized with π-delocalization effects, which can affect only the first of the steps. Surprisingly, the angular bimetallic complexes formed unselectively and decomposed on attempts of isolation. ^{1}H NMR studies revealed that the syntheses proceeded unselectively, giving a number of benzylidene-containing products, in contrast to the preparation of monometallic catalysts. Only one homodinuclear ruthenium complex bearing linear tricyclic structure (**13**) was isolated under optimized conditions and characterized with NMR and X-ray techniques. Activity studies revealed that **13** initiates at rate similar to that of the corresponding monometallic complex, suggesting no cooperativity in the initiation process, when two propagating species are consecutively released (Fig. 18.8).

Electronic effects operating in the Hoveyda-type catalysts were also investigated by theoreticians. Solans-Monfort reported that catalytic activity correlates well with $=CH-C_{aryl}$ bond length, rather than the commonly invoked strength of Ru· · ·O*i*Pr bond, and thus (pre)catalyst initiation and recovery are controlled by degree of π-electron delocalization in the chelate ring (30). More detailed studies of electronic properties of chelates **11a,b** were carried out by Sola and Sodupe et al. (31). On the basis of nucleus independent chemical shift (NICS), multicenter index (MCI), and delocalization index (DI) values, they concluded that chelate rings of the complexes display any metalloaromaticity, although there is a certain degree of π-electron delocalization. Although the results definitely made the term “aromatic” inappropriate for the description of the chelate rings, authors did not explain the origins of the observed activity differences between complexes **11a** and **b**. Even more, calculated values of the electronic parameters varied rather slightly for the structures, giving the impression that steric hindrance of the ligands plays an important role in the initiation process, and the electronic effects are in fact smaller than originally suggested.

Another interesting area of development of Hoveyda-type metathesis catalysts was opened by the synthesis of heteroatom analogs of **2** (3). Catalytic activity of the complexes strongly depends on their geometry and type of the coordinating heteroatom, and the two factors are related; heavier analogs of **2** usually prefer a *cis*-Cl_2 geometry over a more typical *trans*-Cl_2 one, observed in structures **1-4** (32). Thus, oxygen and related nitrogen-chelated complexes usually exist as trans isomers, (33) while for sulfur analogs (34,35), both the forms are known, and display an equilibrium with thermodynamic preference toward *cis*-Cl_2 (36). Still heavier selenium and phosphorus chelates are observed exclusively as cis isomers (32). The heavier analogs of the Hoveyda family also display diminished catalytic activity, attributed to strengthening of Ru· · ·Het interaction, when moving down the Periodic Table, and lack influence of the “*trans* effect” caused by the strongly donating NHC ligand, which weakens heteroatom coordination in *trans*-Cl_2 isomers (37). In this respect, recently reported bromine and iodine-chelated complexes were shown to display disparate properties (Fig. 18.9) (38,39). The catalysts combined *cis*-Cl_2 geometry with good application profiles in model RCM, CM, and enyne reactions, in which the activity is comparable to that of the parent complex **2** (38). Besides the unique complex geometry, unusual properties of Ru· · ·X–Ar (X = Br, I) bonds, and the fact that covalently bonded halogen atoms possess any “protective” alkyl groups that hinder approach of substrates to the ruthenium center, make them exceptional members of the Hoveyda family. The unusual complexes were characterized by X-ray studies, and appeared to be stable in solution without the protective atmosphere. Activity of the iodine-chelated complexes varied considerably with temperature, and, what was rather surprising, electronic effects displayed by substituents were reversed (39). Nitro-substituted iodo-chelated complex **15b** was definitely less active than unsubstituted complex **15a**, in contrast to the behavior of ether-chelated Hoveyda complex (**2**), for which introduction of the nitro

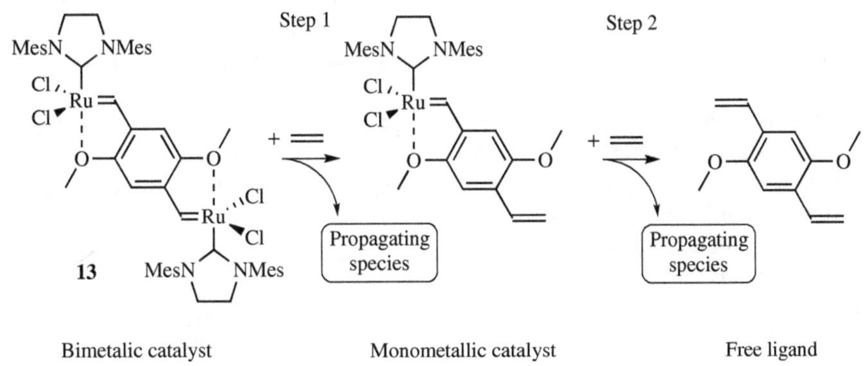

Figure 18.8 Two-step mechanism of initiation of bimetallic complex **13** (22).

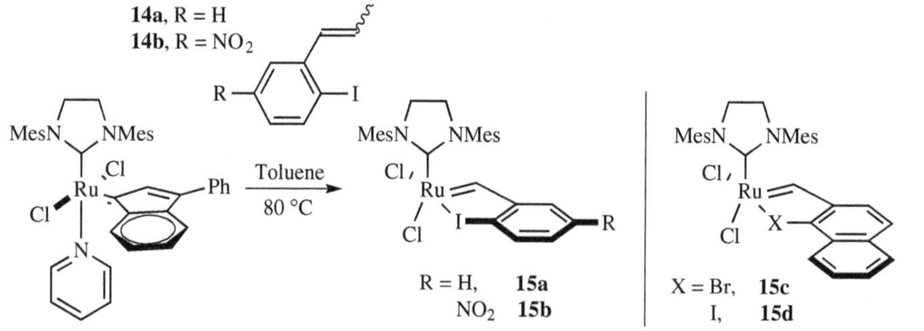

Figure 18.9 Synthesis and structures of selected ruthenium complexes chelated with covalently bonded halogen atoms (**15**) (38,39).

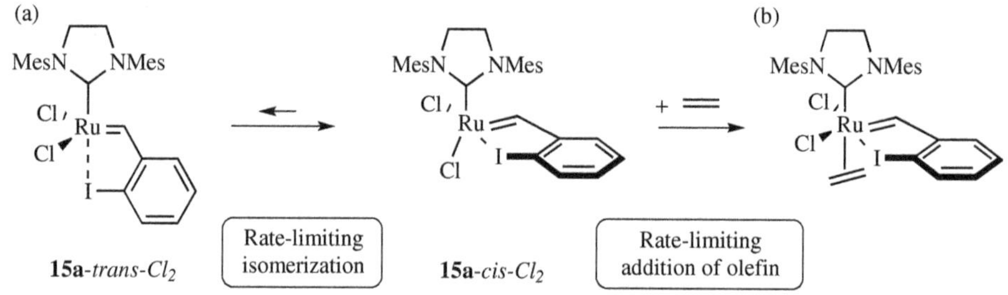

Figure 18.10 Alternative initiation mechanisms of complex **15a** with *cis*-to-*trans*-Cl_2 isomerization (a), or association of olefin (b) as a rate-limiting step (39).

group (**3**) is a well-known mode of activation (8,9). Interestingly, exchange experiments, in which ligands **14a** and **14b** were equilibrated with complexes **15b** and **15a**, respectively, revealed that chelate **15a**, which initiates faster, is more stable than **15b**. The discrepancy between thermodynamic stability of the chelate rings and their kinetic ability to initiation led to the conclusion that the Ru· · ·halogen bond is not broken in the rate-determining step of the initiation process. Alternatively, the experimental data were explained by rate-limiting isomerization of the iodochelates into more active *trans*-Cl_2 forms, or associative mechanism of initiation (cf. Fig. 18.3), in which the highest energy barrier corresponds to olefin coordination (Fig. 18.10) (40).

18.1 CONCLUSIONS

The short story of the development of the Hoveyda-type ruthenium metathesis catalysts is an excellent demonstration of how tangled and unpredictable the optimization process can be. A chance observation by Hoveyda sparked an interest in this class of the ruthenium complexes as excellent metathesis catalysts (4). Later, two broad effects—steric (13,14) and electronic (8,9), which control the initiation process, impeded synthesis of numerous new catalysts with improved application profiles. Further exploration of the subject revealed still other effects related to modifications of the ether chelating arm (21–25), naphthalene ligands (26,28), and alternative modes of initiation of the halogen-chelated catalysts (38–40). Thus, it becomes apparent that the vast area is still surrounded by a veil of mystery and requires further studies to reveal all aspects of the bizarre initiation process of well-defined ruthenium metathesis catalysts.

REFERENCES

1. Vougioukalakis GC, Grubbs RH. *Chem Rev* 2010;110: 1746–1787.
2. Samojłowicz C, Bieniek M, Grela K. *Chem Rev* 2009;109: 3708–3742.
3. Vidavsky Y, Anaby A, Lemcoff NG. *Dalton Trans* 2012; 41:32–43.
4. Harrity JPA, La DS, Cefalo DR, Visser MS, Hoveyda AH. *J Am Chem Soc* 1998;120:2343–2351.
5. Kingsbury JS, Harrity JPA, Bonitatebus PJ Jr, Hoveyda AH. *J Am Chem Soc* 1999;121:791–799.
6. Garber SB, Kingsbury JS, Gray BL, Hoveyda AH. *J Am Chem Soc* 2000;122:8168–8179.
7. (a) Vorfalt T, Wannowius KJ, Thiel V, Plenio H. *Chem Eur J* 2010;16:12312–12315. (b) Ahmed M, Barrett AGM, Braddock DC, Cramp SM, Procopiou PA. *Tetrahedron Lett* 1999;40:8657–8662.

8. Grela K, Harutyunyan S, Michrowska A. *Angew Chem Int Ed* 2002;41:4038–4040.
9. Michrowska A, Bujok R, Harutyunyan S, Sashuk V, Dolgonos G, Grela K. *J Am Chem Soc* 2004;126:9318–9325.
10. Interesting exception in the investigated series is made by halogen atoms. It was reported that presence of chlorine or bromine in *para* position to the chelating *i*PrO group reduces catalytic performance: (a) Ettari R, Micale N. *J Organomet Chem* 2007;692:3574–3576. (b) Grela K, Michrowska A, Bieniek M, Kim M, Klajn R. *Tetrahedron* 2003;59:4525–4531.
11. Zaja M, Connon SJ, Dunne AM, Rivard M, Buschmann N, Jiricek J, Blechert S. *Tetrahedron* 2003;59:6545–6558.
12. Bieniek M, Michrowska A, Gułajski Ł, Grela K. *Organometallics* 2007;26:1096–1099.
13. Wakamatsu H, Blechert S. *Angew Chem Int Ed* 2002;41: 2403–2405.
14. Buschmann N, Wakamatsu H, Blechert S. *Synlett* 2004:667–670.
15. Ashworth IW, Hillier IH, Nelson DJ, Percy JM, Vincent MA. *Chem Commun* 2011;47:5428–5430.
16. Nuñez-Zarur F, Solans-Monfort X, Rodríguez-Santiago L, Sodupe M. *Organometallics* 2012;31:4203–4215.
17. (a) Thiel V, Hendann M, Wannowius K-J, Plenio H. *J Am Chem Soc* 2012;134:1104–1114. (b) Vorfalt T, Wannowius K-J, Plenio H. *Angew Chem Int Ed* 2010;49:5533–5536.
18. For studies of solvent effect on the initiation rate of the catalysts, see: Ashworth IW, Nelson DJ, Percy JM. *Dalton Trans* 2013;42:4110–4113.
19. Bieniek M, Michrowska A, Usanov DL, Grela K. *Chem Eur J* 2008;14:806–818.
20. At Figure 18.4 we omitted complexes, in which the chelating ether function possesses multiple coordinating sites, see for example: (a) Bieniek M, Bujok R, Cabaj M, Noël L, Lavigne G, Arlt D, Grela K. *J Am Chem Soc* 2006;128:13652–13653. (b) Ginzburg Y, Anaby A, Vidavsky Y, Diesendruck CE, Ben-Asuly A, Goldberg I, Lemcoff NG. *Organometallics* 2011;30:3430–3437.
21. Barbasiewicz M, Bieniek M, Michrowska A, Szadkowska A, Makal A, Woźniak K, Grela K. *Adv Synth Catal* 2007; 349:193–203.
22. Grudzień K, Malinska M, Barbasiewicz M. *Organometallics* 2012;31:3636–3646.
23. (a) Hryniewicka A, Morzycki JW, Witkowski S. *J Organomet Chem* 2010;695:1265–1270. (b) Hryniewicka A, Kozłowska A, Witkowski S. *J Organomet Chem* 2012;701:87–92.
24. Nelson DJ, Queval P, Rouen M, Magrez M, Toupet L, Caijo F, Borré E, Laurent I, Crévisy C, Baslé O, Mauduit M, Percy JM. *ACS Catal* 2013;3:259–264.
25. Kos P, Savka R, Plenio H. *Adv Synth Catal* 2013;355:439–447.
26. Barbasiewicz M, Szadkowska A, Makal A, Jarzembska K, Woźniak K, Grela K. *Chem Eur J* 2008;14:9330–9337.
27. Louie J, Grubbs RH. *Organometallics* 2002;21:2153–2164.
28. Barbasiewicz M, Grudzień K, Malinska M. *Organometallics* 2012;31:3171–3177.
29. Tzur E, Ben-Asuly A, Diesendruck CE, Goldberg I, Lemcoff NG. *Angew Chem Int Ed* 2008;47:6422–6425.
30. Solans-Monfort X, Pleixats R, Sodupe M. *Chem Eur J* 2010;16:7331–7343.
31. Nuñez-Zarur F, Poater J, Rodríguez-Santiago L, Solans-Monfort X, Solà M, Sodupe M. *Comput Theor Chem* 2012;996:57–67.
32. Diesendruck CE, Tzur E, Ben-Asuly A, Goldberg I, Straub BF, Lemcoff NG. *Inorg Chem* 2009;48:10819–10825.
33. For examples of X-ray structures of *cis*-Cl_2 oxygen-chelated Hoveyda-type structures, see: Ref. (21), and (a) Leitgeb A, Mereiter K, Slugovc C. *Monatsh Chem* 2012;143:901–908. (b) For an example of X-ray structure of *cis*-Cl_2 nitrogen-chelated ruthenium complex, see: Barbasiewicz M, Szadkowska A, Bujok R, Grela K. *Organometallics* 2006;25:3599–3604.
34. Tzur E, Szadkowska A, Ben-Asuly A, Makal A, Goldberg I, Woźniak K, Grela K, Lemcoff NG. *Chem Eur J* 2010;16:8726–8737.
35. For recent report, see: Żukowska K, Szadkowska A, Trzaskowski B, Pazio A, Pączek Ł. Woźniak, K.; Grela, K. *Organometallics* 2013;32:2192–2198.
36. Aharoni A, Vidavsky Y, Diesendruck CE, Ben-Asuly A, Goldberg I, Lemcoff GN. *Organometallics* 2011;30:1607–1615.
37. Compare: Pump E, Fischer RC, Slugovc C. *Organometallics* 2012;31:6972–6979.
38. Barbasiewicz M, Michalak M, Grela K. *Chem Eur J* 2012;18:14237–14241.
39. Barbasiewicz M, Błocki K, Malińska M, Pawłowski R. *Dalton Trans* 2013;42:355–358.
40. For further studies of the halogen-chelated complexes see: (a) Barbasiewicz M, Malińska M, Błocki K. *J Organomet Chem* 2013;745–746:8–11. (b) Grudzień K, Malinska M, Woźniak K, Barbasiewicz M. *Chem Eur J* 2014;20:2819–2828.

19

THEORETICAL ATTEMPTS: "IN SILICO OLEFIN METATHESIS"—HOW CAN COMPUTERS HELP IN THE UNDERSTANDING OF METATHESIS MECHANISMS AND IN CATALYSTS DEVELOPMENT?

ALBERT POATER

Departament de Química, Institut de Química Computacional, University of Girona, Girona, Catalonia, Spain; Dipartimento di Chimica, Università di Salerno, Fisciano, SA, Italy

LAURA FALIVENE

Dipartimento di Chimica, Università di Salerno, Fisciano, SA, Italy

LUIGI CAVALLO

Dipartimento di Chimica, Università di Salerno, Fisciano, SA, Italy; KAUST Catalyst Research Center, Physical Sciences and Engineering Division, King Abdullah University of Science and Technology, Thuwal, Kingdom of Saudi Arabia

19.1 INTRODUCTION

Although modern olefin metathesis has been pushed by experimental groups, computational groups started to work on this topic quite from the beginning. In the early days, computational insights were basically rationalizing published work, but the growing evidence that computational tools could also accelerate the pace at which new catalysts were discovered, by providing a fertile understanding of the empirical data, allowed computational groups to become highly involved in the direct development of new catalysts. Nowadays, computational insights are pushing forward new advances in the field, rather than being just a simple characterization tool.

The mechanism operative in olefins metathesis is that proposed by Chauvin in 1971 (1), and displayed in Figure 19.1. This mechanism received full experimental support in the late nineties (2), and here we focus on the computational insights (3). The key step is the formation of the metallacycle intermediate, whose opening in the backward or the forward direction connects reactants and products. To date, two families of well-working and well-developed catalysts, based on Mo or Ru as the catalytic metal, have been developed and studied. Their experimental performances have been extensively reviewed (2). Mo-based catalysts are generally considered more active but more delicate, in the sense that they are less tolerant to other functional groups in the substrate. Ru-based catalysts are seen as less active but they tolerate the presence of almost any other functional group in the substrate. Research activity is dedicated to solving the weaknesses of both families of catalysts, and progresses in the field are continuous.

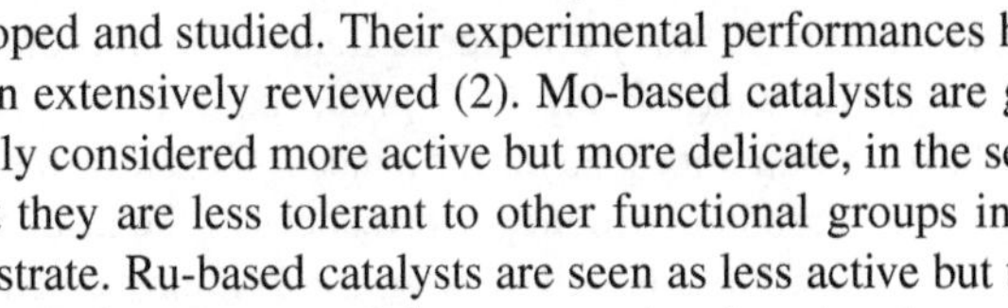

Figure 19.1 The Chauvin mechanism.

Ru-catalyzed metathesis can span a variety of applications, as explained in previous chapters. The most popular catalyst families are shown in Figure 19.2.

Olefin Metathesis: Theory and Practice, First Edition. Edited by Karol Grela.

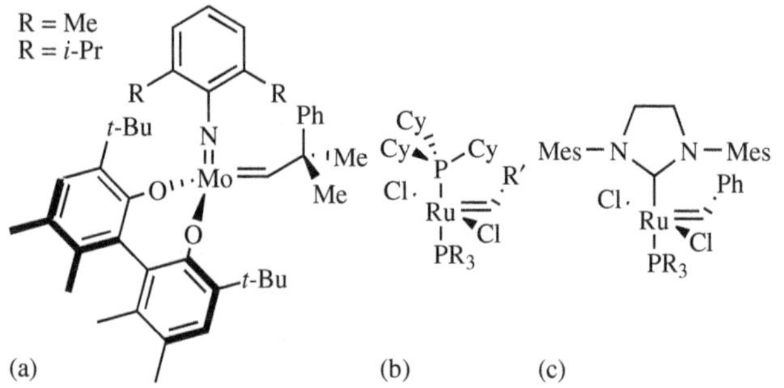

Figure 19.2 Olefin metathesis (pre)catalysts ordered chronologically: (a) Schrock biphenyl catalysts; (b) Grubbs's first generation; and (c) NHC-based second-generation catalysts (Cy = Cyclohexyl; Mes = Mesityl).

Although research interest for the Mo Schrock catalysts is decreasing because of their mature development (2d), there is still interest in their good selectivity and in replacing the metal by rhenium or tungsten (4) as well as in the development of supported catalysts (5). On the contrary, there is still great effort to develop Ru catalysts with better performances or broader scope. Starting from Grubbs's first-generation catalysts with two phosphane ligands, a nonstop race to develop new catalysts started. The first big step forward was to replace one of the phosphane ligands by an *N*-heterocyclic carbene (NHC) ligand (6), followed by modification of the ylidene ligand by new chelating alkylidenes, those based on alkoxy-alkylidene groups among the most effective (3e). The beneficial effect of NHC ligands has been correlated to their higher σ-basicity, which has been suggested to stabilize the catalyst and to improve the key activation step. Indeed, NHCs as ligands have been demonstrated to induce novel or better performances in a broad series of catalysis promoted by transition metals (7). NHCs are rather flexible and versatile ligands whose electronic and steric properties can be tuned to a large extent (8,9). The most relevant NHC ligands used in olefins metathesis, with the commonly used nomenclature, are reported in Figure 19.3. As for the chelating alkoxy-alkylidene ligand, the introduction of an electron-withdrawing NO_2 group para to the chelating alkoxy group improved performances (10). Nevertheless, despite the already excellent performances of the last generation catalysts, there is still large space to improve activity, stability, and selectivity, which justifies the great efforts in the field by both academic and industrial groups.

In the following, we present the main achievements obtained computationally in the field of Ru-catalyzed olefins metathesis, starting with the events associated with activation of the (pre)catalyst. Then we describe events inside the catalytic cycle, and we finish with a description of the mechanisms leading to catalyst deactivation. Since the last improvements in the field focus on catalysts bearing an NHC ligand, we focus this chapter on this catalyst family.

19.2 THE FUNDAMENTAL STEPS OF METAL-CATALYZED OLEFINS METATHESIS

19.2.1 Activation

As observed in previous chapters, the accepted mechanism for both Mo and Ru catalysts can be divided into three separate events: initiation, propagation, and termination (2b). While for Mo catalysts there is a unique and by now well-accepted mechanism (4), for Ru catalysts at least three different possible activation mechanisms are possible, and are normally referred as dissociative, associative, and interchange/associative (Fig. 19.4). In the dissociative pathway, the phosphine is released, forming a 14-electron species that then coordinates to the substrate. In the associative mechanism, the substrate coordinates to the metal to form a six-coordinate Ru-complex, and the phosphine dissociates from this complex. In the interchange/associative mechanism, the substrate binds to the metal center while the originally bound phosphine is released simultaneously.

Once the coordination intermediate **III** is formed, the forthcomimg steps are identical for the three proposed mechanisms. Next, there is the formation of a metallacycle (**IV**) followed by rearrangement of the bonds to release the ylidene moiety initially attached to the metal center, leading to a new alkylidene (**V**), which is able to undergo productive metathesis (2b,11,12).

Initiation is a critical step, and therefore several computational studies have focused on this step (13). Calculations clearly indicated that the associative mechanism is disfavored by large energy barriers and by steric repulsion between the ligands at an overcrowded six-coordinated center (3d,14). Moving on to the dissociative pathway, calculations on second-generation (pre)catalysts confirmed that dissociation of phosphines is more feasible than the dissociation of the NHC ligand, consistent with the experimental evidence that in these systems the phosphine

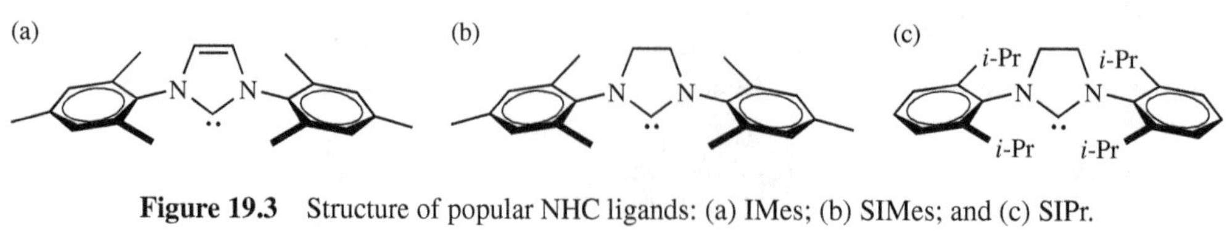

Figure 19.3 Structure of popular NHC ligands: (a) IMes; (b) SIMes; and (c) SIPr.

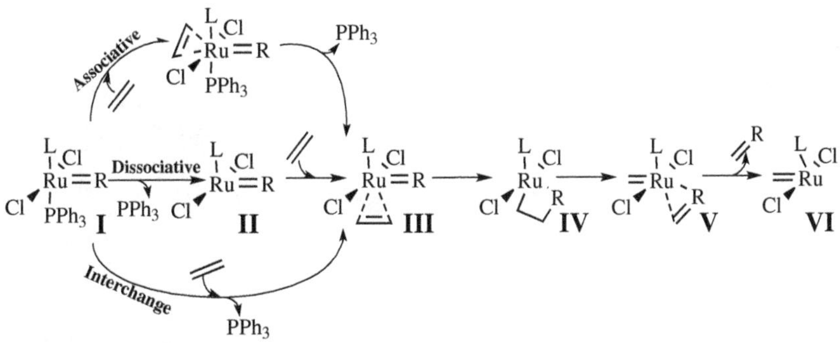

Figure 19.4 Schematic representation of the possible activation mechanisms for Ru-complexes.

rather than the NHC ligand is lost during olefin metathesis (3e,15). Dissociation of a typical NHC, such as the SIMes (N,N′-bis(2,4,6-trimethylphenyl)imidazolin-2-ylidene), costs roughly 5 kcal/mol more than dissociation of a typical phosphine, such as PCy_3 (14). Calculations also showed that the energy required to dissociate the phosphine is reduced in solution, which indicates that care must be taken when quantitatively comparing calculations and experiments (16,17). In this respect, it was clearly demonstrated that inclusion of attractive non-covalent interactions is fundamental for a proper description of the binding of large ligands, such as phosphines (18).

Focusing on (pre)catalysts with an alkoxybenzylidene chelating ligand, and using the unsubstituted parent (pre)catalyst of Figure 19.5 as reference, experiments indicated that an electron-withdrawing nitro group para to the alkoxy position lowers the reaction times and yields full conversion (19), while an electron donor methoxy group in the same position results in longer reaction times (20). This result suggested that electron-withdrawing groups decrease the electron density either of the Ru–alkoxy bond or of the Ru–alkylidene bond, thereby enhancing catalytic activity. Calculations indicated that substitution on the aromatic ring of the chelating alkoxy-arylidene group produces significant variations in the Ru···O and C_a–C_b distances, with an impact on the alkoxy dissociation energy barriers that could be related to the experimentally observed catalytic activity (10,19–21). Calculations showed no direct relationship between the energy barrier and the Ru···O interaction strength; rather, the experimental catalytic efficiencies correlate better with the C_a–C_b bond length, that is, its double bond character, rather than with the Ru···O distance. Correlation between aromaticity in the chelating alkoxy-arylidene group and catalytic activity allowed ascribing the poor performance of (pre)catalysts with the naphthyl bond in the C_c–C_d or C_e–C_f position to an increased conjugation (aromaticity) on the chelating Ru–C_a–C_b–C_g–O ring (19,20). According to Clar's rules, the preserved catalytic activity of the (pre)catalyst with the naphthyl bond in the C_d–C_e position can be related to the reduced conjugation on the chelating ring (22).

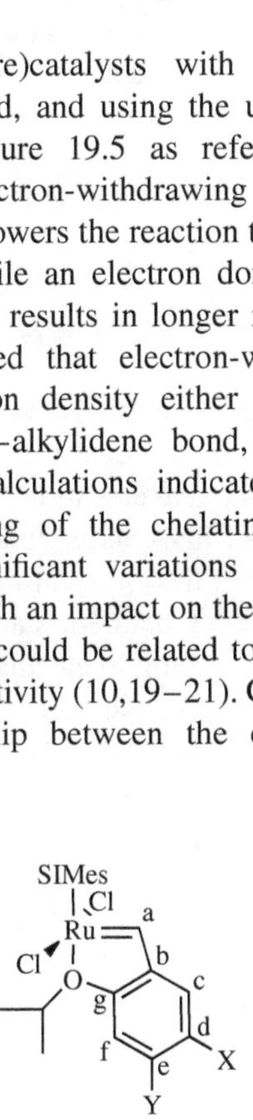

Figure 19.5 Alkoxy-alkylidene (pre)catalysts investigated computationally (X, Y = H, NO_2 and/or OMe).

As for the interchange/associative mechanism, computational studies indicated this as a viable activation mechanism for catalysts having an alkoxy–arylidene moiety (23). This is in agreement with an experimental report indicating that the preference for an associative/interchange or a dissociative initiation mechanism in this family of catalysts is determined by a delicate balance between the electronic and steric configuration of the complex and the olefin studied (24).

19.2.2 Inside the Catalytic Cycle

In the framework that the active species for the most common dissociative mechanism is the 14e Ru-species formed by dissociation of one labile ligand, some early theoretical studies on simplified models confirmed the Chauvin mechanism (3a,25). The intermediates along the reaction pathway, for the popular SIMes-benzylidene catalyst, are reported in Figure 19.6. Starting from the ethene coordination intermediate, with the ethene substrate bound trans to the NHC ligand, the reaction proceeds with the reacting atoms assuming an almost planar four-center geometry at the transition state **II→III**.

This transition state then collapses into the metallacycle intermediate. These conclusions were also supported by the first computational study in which real size catalysts were modeled (3e). This study suggested that only a small energy barrier prevents the coordinated olefin from reacting with the Ru–alkylidene bond, thus precipitating into the

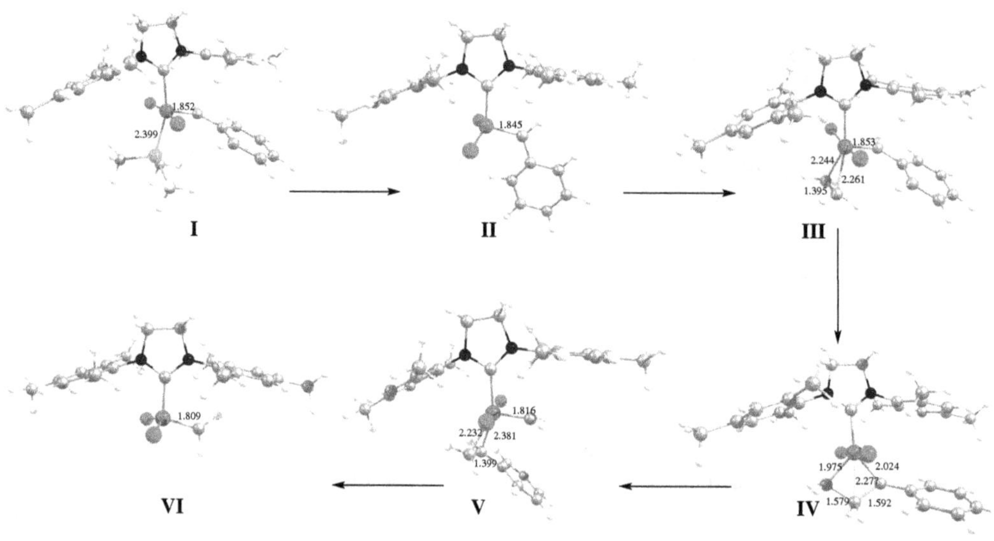

Figure 19.6 DFT geometry of the intermediates of the olefin metathesis reaction of ethene with the SIMes-benzylidene catalyst (distances in Å).

metallacycle, which is the most stable structure along the reaction pathway. Further, it was suggested that steric pressure of the N-substituents on the alkylidene moiety stabilizes the metallacycle, since the formed ruthenacyclobutane develops away from the NHC ligand more than the alkylidene moiety. Focusing on the metallacycle, a synergic experimental and theoretical study showed that the Cα–Cβ bonds of the metallacycle are engaged in an agostic interaction with the Ru center, and that the Cα–Cβ bond within the formed five-membered ring in ring-closing metathesis (RCM) is remarkably longer than the Cα–Cβ bond outside the cycle (Fig. 19.7), which means that the metallacycle intermediate is distorted toward a more stable ring-opening metathesis (ROM) product, rather than toward a less stable RCM product (26).

Coordination of small olefins, such as ethene, presents no or negligible enthalpic barrier, although unfavorable entropic effects could give rise to a free energy barrier (27). An unfavorable enthalpic term could be present in the case of the folding of the substrate prior to an RCM step. Ethene coordination energy in a series of complexes was shown to follow the same trend calculated for phosphines (3e). As

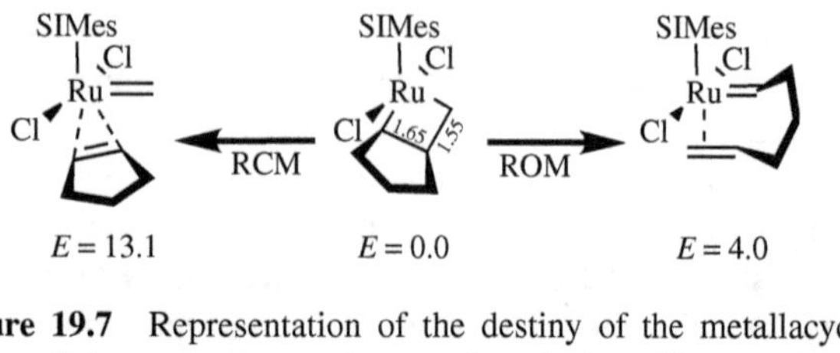

Figure 19.7 Representation of the destiny of the metallacycle intermediate presenting a five-membered ring (distance in Å, energies in kcal/mol).

anticipated, a quite low energy barrier has been calculated for metallacycle formation independently of the system considered (3e,14,21,28). Barriers of this height support the idea proposed by Grubbs that "all of the steps after olefin coordination (particularly metallacyclobutane formation) are fast" (29). Compared to first-generation catalysts, the NHC-based systems have slower initiation rates because of the higher energy required to dissociate the phosphine. Grubbs argued they have a higher activity because olefin coordination is more competitive than rebinding of the phosphine. Nonetheless, the higher activity of the NHC-based systems could be also connected to their substantially lower energy barrier for the metathesis reaction. This could help NHC-based catalysts to perform more metathesis steps than first-generation catalysts, before being trapped back from free phosphine. However, it is worth noting that an alternative explanation for the different reactivity of first- and second-generation catalysts is based on a different flexibility of the Ru-alkylidene bond that can exist in different rotameric states after dissociation of the labile ligand. According to calculations, the 14e species of second-generation catalysts rests in a Ru-alkylidene rotameric state more suitable for an easy coordination of the substrate, thus improving activity (30,31).

Although there is quite a consensus on the above mechanism, there are some modifications which are worth mentioning (32). For example, to rationalize low-temperature NMR experiments on ethene metathesis exchange (33), it has been proposed that a second olefin could coordinate trans to the Ru–alkylidene bond to form an octahedral complex; see Figure 19.8. The second olefins would assist metallacycle opening and dissociation of the formed ethene molecule. It is difficult to extrapolate these results to substrates and temperatures used in productive metathesis.

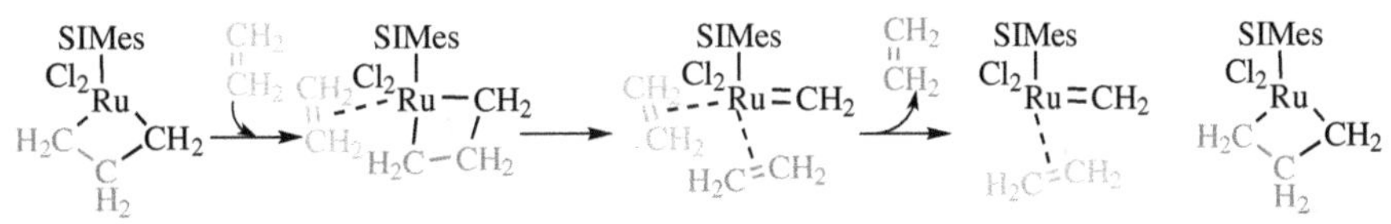

Figure 19.8 Proposed computed mechanism of associative ethene exchange.

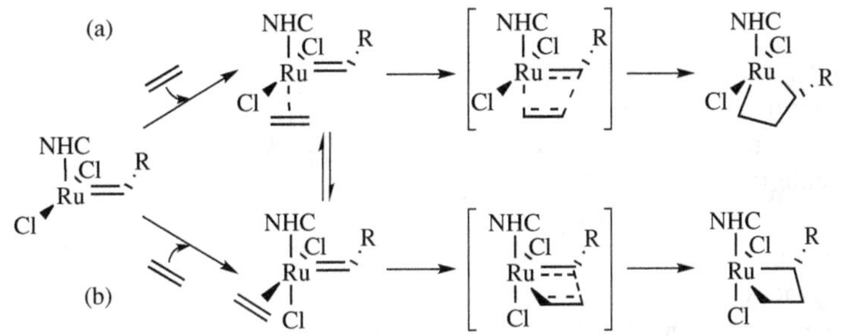

Figure 19.9 Representation of the (a) trans and (b) cis metathesis pathways.

A longstanding question tackled from the experimental and the theoretical side has been whether the exact mechanism in the case of the Ru-based catalysts involves an isomerization from the initial *trans*-Cl to a *cis*-Cl geometry, leading to the so-called cis pathway; see Figure 19.9 (34). A series of computational studies showed that the preferred mechanism with standard bishalide NHC-based catalysts follows the trans pathway (14,28,35). Calculations suggested that the olefin can coordinate both trans and cis to the NHC ligand, and that solvent effects are of paramount relevance to the high stability of complexes with a *cis*-Cl geometry (36). However, the transition state for metallacycle formation with a *cis*-Cl geometry is of much higher energy, and this destabilization was explained by the different orientation of the olefin in the *cis*-Cl geometry on going from the coordination intermediate to the transition state. In the former, the olefin C=C double bond is almost perpendicular to the Ru–alkylidene bond to minimize steric interaction with the NHC ligand. At the *cis*-Cl transition state, instead, the olefin is forced to be parallel to the Ru–alkylidene bond, and this increases the steric interaction between the olefin and the NHC ligand (37). This sort of rule has been recently questioned by theoretical evidence that in the case of Ru-catalysts presenting a chelated NHC ligand, the cis metathesis pathway is preferred and this was used to explain the remarkably good selectivity shown by these catalysts for the formation of Z-olefins (38). Still on the formation of Z-olefins, in this case with classical Ru-catalysts presenting non-chelated NHC ligands, another study indicated that the preferential formation of *E*-olefins, also at low conversions (39), is not determined at metallacycle formation or opening, and that the key step determining the experimentally observed preferential formation of *E*-olefins is, instead, product release (40).

While most of the computational studies have used ethene as representative olefin, a few valuable investigations focused on more complex substrates, such as dienes (21). Calculations showed that also in presence of a more complex substrate all the energy barriers within the metathesis cycle, including the ring-closing step, are rather small and definitely smaller than the energy cost for the initial dissociation of the labile ligand. This suggested that the different activity exhibited by (pre)catalysts bearing different alkoxy groups is related to the ability of the (pre)catalyst to generate the 14e active species, rather than to differences in the catalytic cycle. Still on more complexes substrates, calculations focused on the ring-opening metathesis polymerization (ROMP) of strained norbornene indicated that metathesis of norbornene is clearly exergonic so that the reaction is irreversible (14,16). The termination of norbornene ROMP by ethene or 1,4-dichloro-2-butene (see Fig. 19.10) was also modeled (41).

Another key issue investigated computationally is the competition between the ene and yne functionalities in the case of enynes and dienynes metathesis. Attention was essentially focused on understanding if an ene–then–yne or a yne–then–ene pathway is followed; see Figure 19.11 (42).

In the case of intramolecular metathesis of a model 1,6-enyne (42d), calculations indicated that alkynes coordinate stronger than olefins, but the following metathesis event between the coordinated alkyne and the Ru–alkylidene bond presents quite a high energy barrier. However, un-

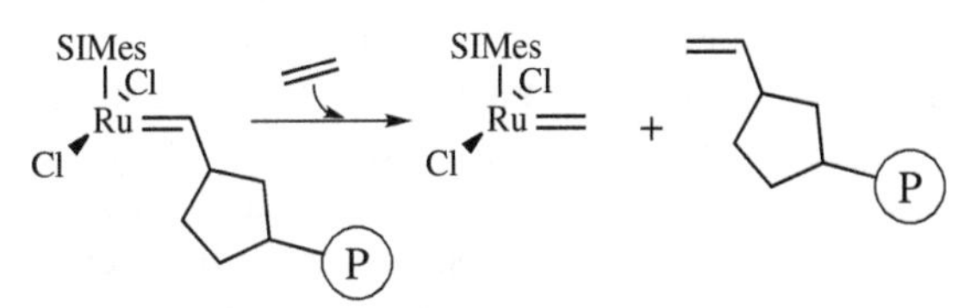

Figure 19.10 Chain termination by ethene in ROMP of norbornene.

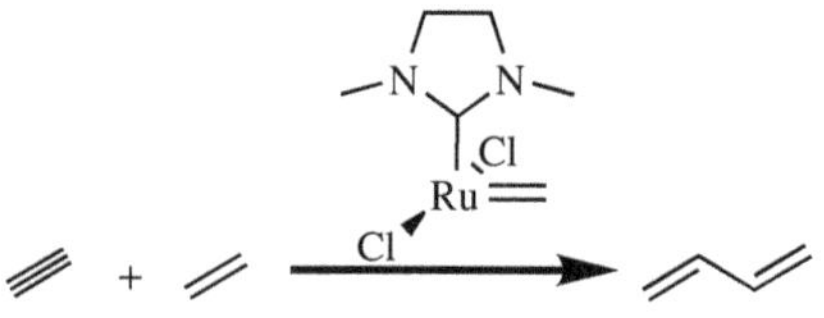

Figure 19.11 Representation of ene + yne metathesis.

like the metathesis of olefins, metathesis of alkynes is irreversible. When more complex substrates, such as tosylamine-containing dienynes, and commonly used catalysts were compared (43), calculations permitted concluding that for first-generation PCy_3 containing catalysts, the ene–then–yne pathway is adopted, whereas for catalysts bearing NHC ligands, both the ene–then–yne and the yne–then–ene pathways are possible, and the steric hindrance of the NHC ligand as well as that around the double and triple C–C bonds appear as a key factor governing the mechanistic pathway adopted (hindrance around the C=C bonds pushes toward the yne–then–ene pathway).

19.2.3 Dynamic Behavior of Ru–NHC Catalysts

Computational studies also indicated that NHC-based Ru-catalysts are highly dynamic structures with practical consequences in catalysis. One example is the isomerization between *cis*-Cl and *trans*-Cl geometries of Ru-precatalysts. *cis*-Cl (pre)catalysts are of practical interest because they can be used as (pre)catalysts whose activation can be controlled via their isomerization to the *trans*-Cl isomer (44). This is particularly relevant in ROMP, since latent catalysts allow for a proper mixing of the monomer and the catalyst before polymerization starts. Calculations indicated that the high stability of *cis*-Cl complexes is essentially due to solvent effects that originates from the strong difference in the dipole moment of the two geometries, which translates into very different solvation energies (36). After this first computational effort, the effect of different σ donors and π acid chelating atoms on the relative stability of the *cis* and *trans*-Cl isomers in a series of structurally related Ru-complexes, such as those in Figure 19.5, was examined (45). Stronger σ-donors were shown to destabilize the *trans*-Cl geometry because of a trans interaction with the strong σ-donor NHC ligand. Coordination of π acid ligands, such as phosphites, destabilizes the *trans*-Cl geometry, or better stabilize the *cis*-Cl geometry, since they prefer to be coordinated trans to a π donor, such as Cl. Conversely, weak σ and π donors, such as ethers, stabilize the *trans*-Cl geometry, since they prefer to coordinate trans to a strong σ donor and π acceptor, such as the NHC ligand (45).

As for the mechanism of isomerization between *cis*-Cl and *trans*-Cl geometries, preliminary calculations indicated the viability of the dissociative pathway shown in Figure 19.12 for a chelating pyridine complex (36).

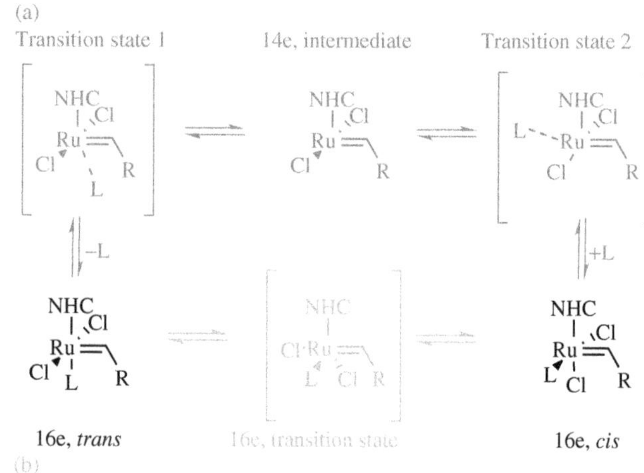

Figure 19.12 Schematic representation of the proposed pathways for the isomerization between the *cis*-Cl and *trans*-Cl geometries. (a) Dissociative pathway and (b) concented.

Extensive analysis of dissociative and concerted pathways suggested a sizeable preference for the concerted pathway in the case of the quinoline complex, whereas the dissociative pathway is only slightly preferred in the case of the pyridine complex (46). As a final remark, recent experimental results indicated that the preference for the *cis* or *trans*-Cl geometry can also be influenced by the nature of the P-ligand. For example, π acid phosphites have been shown to replace effectively phosphines to generate very active catalysts with a peculiar tendency to give *cis*-Cl geometries (47), while the flexibility and the isomerization between different geometries have been also related to the possible dissociation of one of the halide ligands (48).

Still on the flexible behavior of NHC-based Ru-catalysts, those bearing the SIMes NHC ligand are known to exhibit low efficiency in the RCM or CM (cross-metathesis) of sterically encumbered olefins. Reducing the steric bulk of the NHC ligand, such as in the NHC ligand with *o*-tolyl N-substituent of Figure 19.13, resulted in catalysts with improved performance (34a,49). A synergic experimental and theoretical study indicated that the *o*-tolyl-based NHCs can respond actively to the steric requirements of an incoming ligand by adopting conformations that minimize

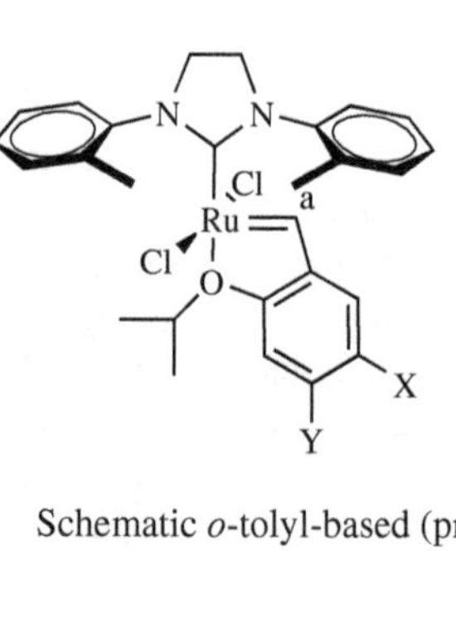

Figure 19.13 Schematic *o*-tolyl-based (pre)catalyst.

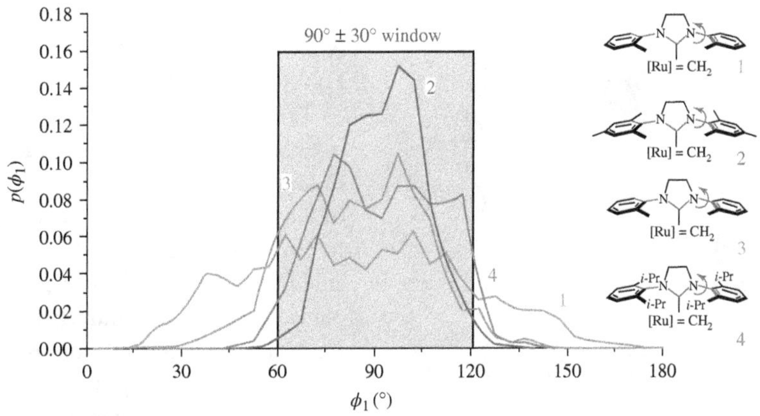

Figure 19.14 Distribution of the ϕ_1 angle in Ru (pre)catalysts. [Ru] stands for the Ru(Cl_2)(PMe_3) moiety. $\phi_1 = \pm 90°$ corresponds to geometries with the aromatic ring of the N substituent perpendicular to the mean plane of the NHC ring, while ϕ_1) 0 or 180° corresponds to geometries with the aromatic ring of the N substituent coplanar with the mean plane of the NHC ring.

their steric bulkiness in the first coordination sphere of the metal (50). Specifically, the *o*-tolyl rings are rotated away from the bound olefin to minimize steric congestion, and the substituted side of the *N*-tolyl rings appears "smaller" than the unsubstituted side, because of rotation of these rings. This result is in agreement with previous computational analysis on asymmetric Ru-catalyzed olefin metathesis (57). A more comprehensive understanding of this point was achieved through *ab initio* molecular dynamics study of a series of 17 NHC-based Ru (pre)catalysts (51). The main results, summarized in Figure 19.14, indicated that the distribution of the ϕ_1 angle around the N-substituent bonds on the Ru–alkylidene side is rather influenced by the nature of the ortho groups. The prototype SIMes system is used as a reference, and its ϕ_1 distribution shows a peak centered at 90°. Conversely, the ϕ_1 distribution of the SIPr (N,N′-bis(2,6-diisopropylphenyl)imidazolin-27-ylidene) based system shows a somewhat broader and jagged peak, while the *N*-phenyl system with no ortho substituent presents a rather broad ϕ_1 distribution with long tails. A quite peaked ϕ_1 distribution was also obtained for the *o*-tolyl system, which means that a single Me group in the ortho position of the N-substituent is enough to reduce flexibility remarkably. In addition, the dynamic analysis indicated that the unsubstituted side of the *o*-tolyl ring is folded toward the Cl–Ru–Cl plane. This result relates with the good catalytic performances of the *o*-tolyl system, since it is less hindered than the SIMes system, but the limited flexibility prevents it from suffering C–H deactivation reactions (52).

The steric pressure introduced by different NHC ligands can be measured by the percentage of the buried volume, $\%V_{Bur}$, defined as the percent of the total volume of the first coordination sphere around the metal occupied by a given ligand; see Figure 19.15 (8b,23,53).

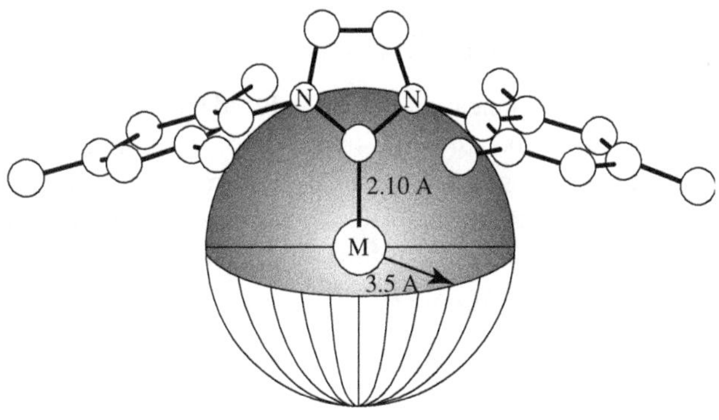

Figure 19.15 Schematic representation of the buried volume.

Comparison of a series of NHC ligands indicated a slightly greater $\%V_{Bur}$ for saturated NHCs than for the unsaturated analogs. For instance, the popular IMes (N,N′-bis(2,4,6-trimethylphenyl)imidazol-2-ylidene) and SIMes NHC ligands display $\%V_{Bur}$ values of 32.7 and 31.6, respectively, and have a bulkiness comparable to that of bulky phosphines, such as PCy_3 (53). The real impact of the NHC ligands in the first coordination sphere of the metal was highlighted by using topographic steric maps, which can be considered chemical analogues of geographical physical maps (51,54). Comparison of the steric maps of (pre)catalysts bearing *o*-tolyl and mesityl N-substituents (see Fig. 19.16) indicated that they shape quite different reactive pockets. A rather flat shape with a constant pressure on the halide–Ru–halide plane in the case of the SIMes system and a steric map of the corresponding complex in the case of the *o*-tolyl-based NHC clearly highlight the higher impact of the unsubstituted side of the ligand in the first coordination sphere of the metal and confirm an overall C_S-symmetric reactive pocket.

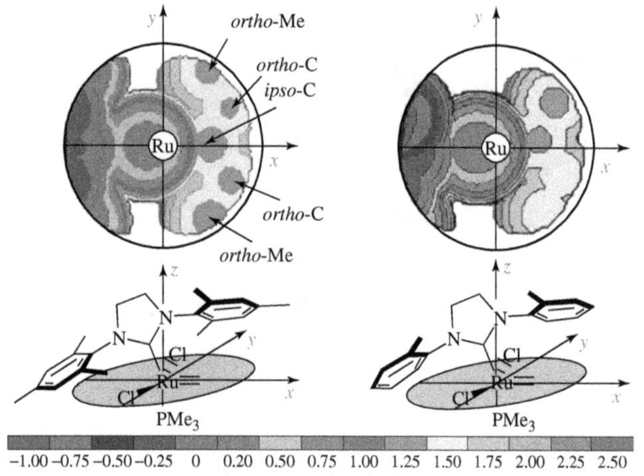

Figure 19.16 Topographic steric map of the NHC ligand of (pre)catalysts bearing a mesityl or a *o*-tolyl N-substituent. The isocontour levels (Å), are also reported. (*See plate section for color version.*)

19.2.4 Asymmetric Metathesis

A challenging application of Ru-catalyzed olefins metathesis is asymmetric metathesis (34b,49,55,56). Several catalysts have been developed to achieve this fundamental transformation that can increase the value of starting achiral reactants by transforming them into chiral products.

Calculations gave a clear mechanistic picture of the origin of stereoselectivity with the prototype catalyst shown in Figure 19.17 (34b,56).

The competing transition states for ring closure, giving rise to the two enantiomeric products, are shown in Figure 19.18. The Ph substituents on the backbone C atoms of the NHC ring impose a chiral folding to the NHC ligand, forcing it to assume a conformation with the unsubstituted side of the *N*-aryl ring pushed down, toward the Cl–Ru–Cl plane. This peculiar folding makes the unsubstituted side of the *N*-aryl ring appear more bulky to the substrate than the *ortho*-isopropyl substituted side, and it is the steric pressure of the unsubstituted side of the N-substituent that generates an energy difference between the two competing transition states, thus promoting stereoselectivity (57). As indicated above, a synergic NMR and DFT (density functional theory) later confirmed that in a strictly related Ru-complex, bearing a NHC ligand with *o*-tolyl N-substituents, the unsubstituted side of the *o*-tolyl ring is folded down toward the Ru–alkylidene moiety, and thus appears bulkier to the substrate in solution (50).

Analysis of the steric map again provided the best evidence of the asymmetry of the chiral pocket shaped by the chiral NHC ligand; see Figure 19.19 (51,54). The steric map is asymmetric both in the tail (left) and head (right)

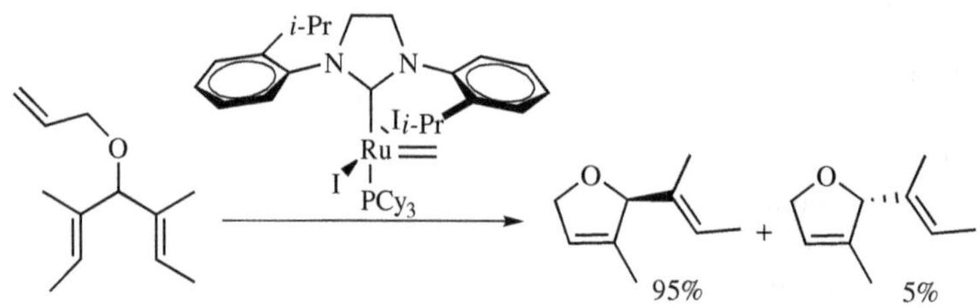

Figure 19.17 Chiral Ru-catalyzed catalyst for desymmetrization of achiral substrates.

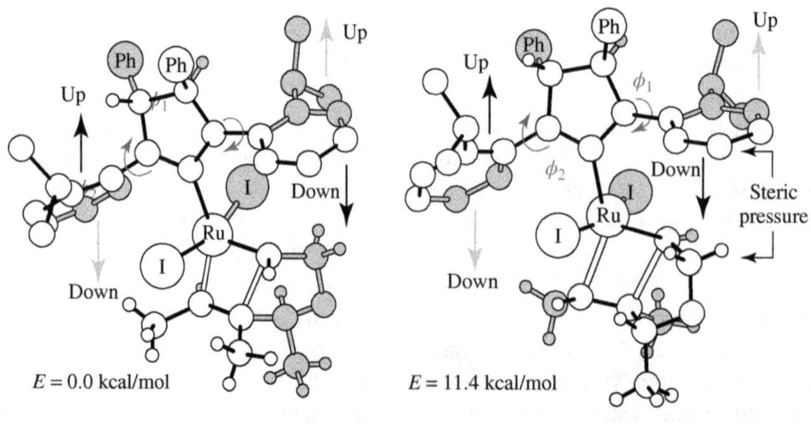

Figure 19.18 Competing transition states for stereoselective RCM.

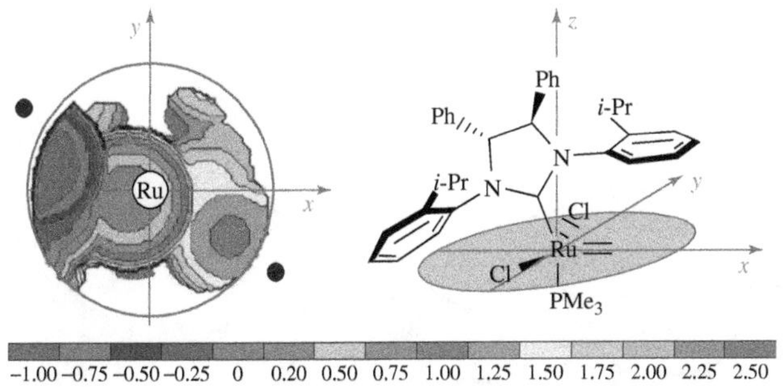

Figure 19.19 Steric map of the chiral NHC ligand is shown. The quadrants occupied by the *ortho-i*-Pr groups are indicated by a black dot. The isocontour levels (Å) are also reported. (*See plate section for color version.*)

quadrants, which confirms that the less encumbered quadrants are those occupied by the *i*-Pr substituted side of the N-substituents (51).

19.2.5 Deactivation

Minimizing decomposition reactions is a key step to designing new and more effective catalysts, which explains the experimental and computational efforts dedicated to understanding the weaknesses of existing catalysts (58). One of the first studies focused on the decomposition of first- and second-generation catalysts in ethene exchange metathesis induced by the substrate (58b). A β−H transfer from the metallacycle intermediate to the Ru center to form a hydride Ru(allyl) species, from which higher olefins are formed, was described as the key step; see Figure 19.20. Calculations indicated that first-generation catalysts are more prone than second-generation catalysts to this deactivation route.

More work was focused on deactivation pathways involving the activation of an *ortho*-C−H bond of the *N*-aryl ring of the NHC ligand of second-generation catalysts (58c–e).

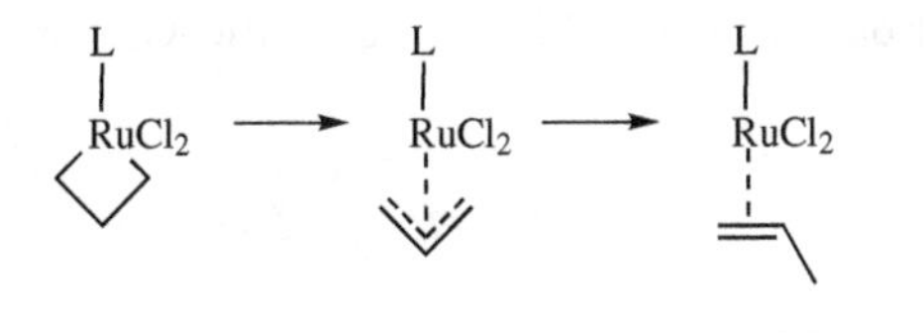

Figure 19.20 Allylic deactivation pathway. L = PR_3 or NHC.

As proposed by Grubbs (52), this deactivation pathway involves a key Ru-hydride intermediate. Calculations were fundamental to connecting correctly the starting complex, the Ru-hydride intermediate, and the decomposition products. As indicated in Figure 19.21, the initial activation of an *ortho*-C−H bond of an *N*-aryl ring transfers the H atom to the benzylidene moiety to form a Ru-benzyl moiety and an ortho-metallated *N*-aryl ring. This step is followed by the formation of the Ru-hydride species with regeneration of the benzylidene group. At this point, the benzylidene group is transferred to the ortho-metallated *N*-aryl ring by reaction with the Ru−C bond. This leads to the first product experimentally observed. Formation of the second product experimentally observed can be explained by the activation of an *ortho*-C−H bond of the other *N*-aryl ring, promoted by free PCy_3. Both the experimental products are stabilized by η^6-coordination of the aromatic ring of the former benzylidene group to the Ru center.

Another computational study clarified the decomposition reaction of the Ru (pre)catalyst promoted by coordination of a CO molecule trans to the Ru−alkylidene bond, investigated experimentally by Diver and coworkers (59); see Figure 19.22. (Pre)catalyst decomposition proceeds through the insertion of the alkylidene moiety into the nearby *N*-aryl ring via a Buchner type ring expansion.

Calculations evidenced continuous energy decay along the reaction pathway, with a cascade of reactions with very small barriers, including the Buchner ring expansion step. This suggested that from a kinetic point of view the system would precipitate into the products with no kinetically stable intermediate, a hypothesis supported by *ab initio* molecular dynamics (58e). The activation role of the CO was related to its ability to weaken the π-back-bonding between the Ru atom and the ylidene ligand, making the latter more electrophilic and disengaging it from the metal center; see Figure 19.23.

Recently, the complex deactivation mechanism of Ru-complexes presenting chelated NHC *N*-aryl rings, which represent an interesting novelty because of their ability to selectively promote the synthesis of *Z*-olefins, was clarified. Since experiments indicated a number of decomposition products, the computational challenge was to properly connect the starting complexes with the experimentally characterized products. The decomposition of C−H acti-

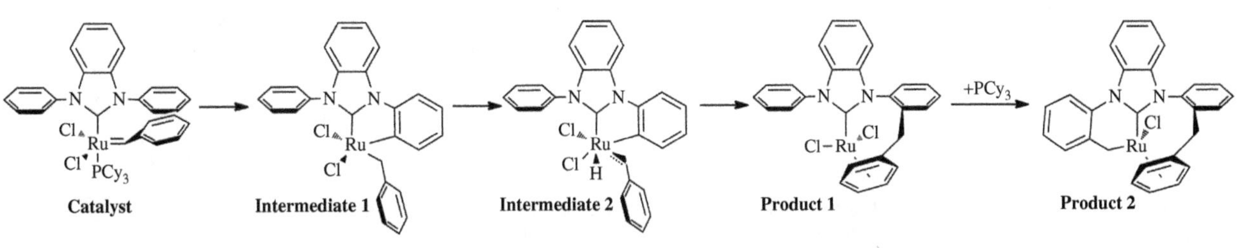

Figure 19.21 Decomposition pathway involving C−H activation of the NHC ligand.

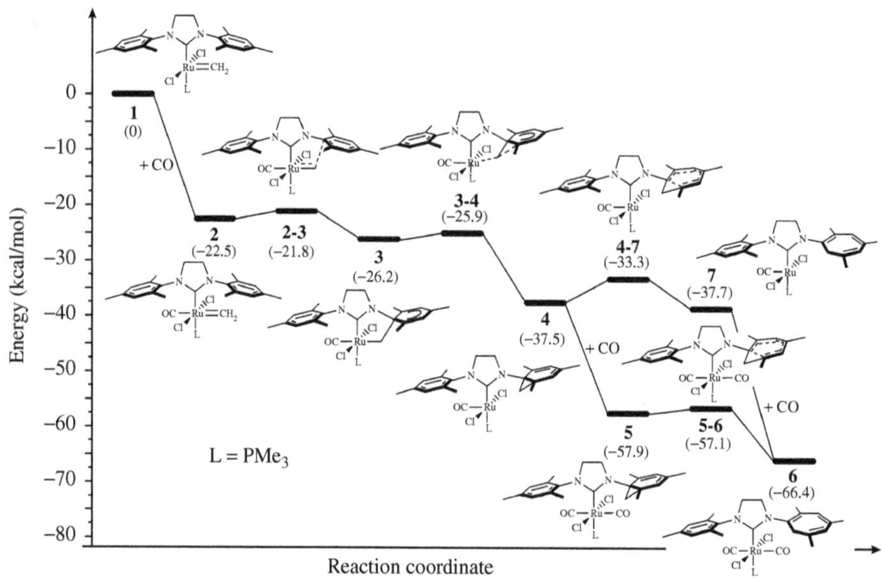

Figure 19.22 Decomposition of Ru (pre)catalysts induced by CO.

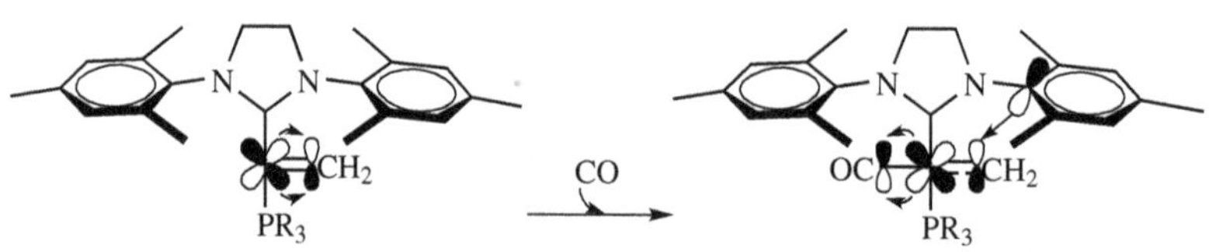

Figure 19.23 Relevant molecular orbitals involved in the CO–Ru–alkylidene interaction.

vated catalysts was suggested to proceed via insertion of the alkylidene into the chelating Ru–C bond to yield a Ru-alkyl intermediate; see Figure 19.24. The presence of a vacant coordination site promotes subsequent hydride elimination from these Ru-alkyl intermediates to generate a Ru-hydrides. The proposed decomposition pathways are markedly distinct from those observed for previous generations of Ru-catalysts, indicating that preventing alkylidene insertion and hydride elimination could be key to developing more robust catalysts (38).

Finally, it is worth noting that decomposition reactions can sometimes open the way to unpredicted results. In one case, the exposure of a classical Ru-precatalyst to alcohols resulted in the transformation of the starting Ru-complex from an olefin metathesis catalyst to an alcohol racemization catalyst; see Figure 19.25.

Calculations were fundamental to indicate that after elimination of HCl, promoted by reaction of a coordinated ROH with one of the Ru–Cl bonds, the key step is the activation of one of the C–H bonds of the Ru–OR ligand by the

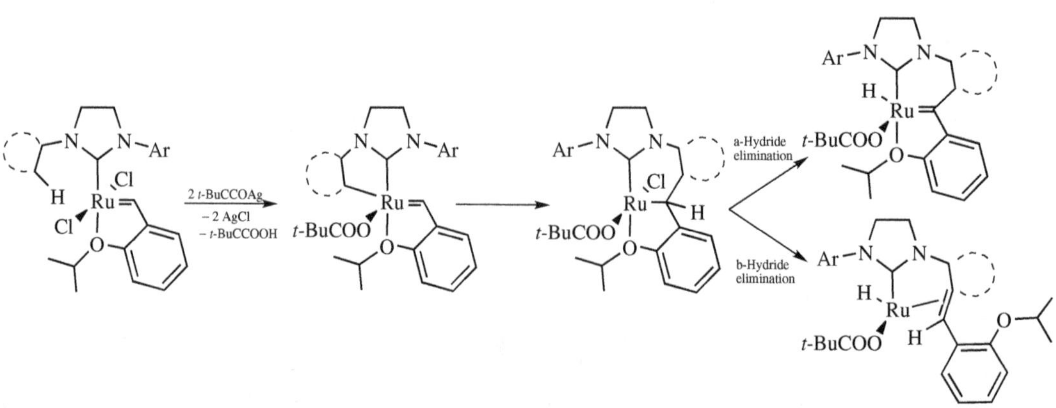

Figure 19.24 Proposed mechanism for decomposition of Ru-chelated catalysts.

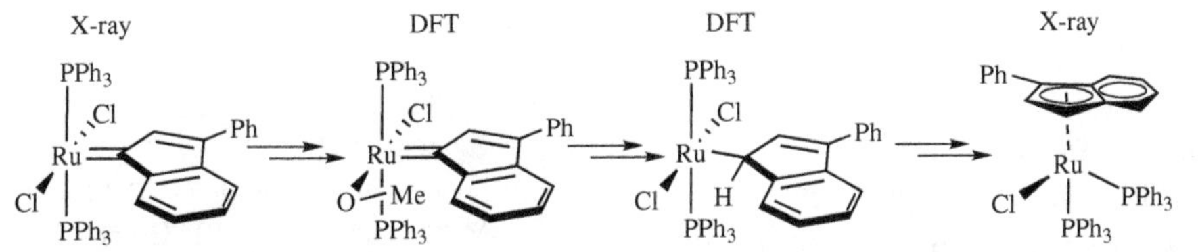

Figure 19.25 Schematic transformation occurring after exposure of the first-generation catalyst to alcohol, with the key DFT intermediates connecting the X-ray characterized reactant and product.

Ru-indenylidene bond, to form a σ-bonded indenyl group. A classical σ to π rearrangement of the indenyl group leads to the experimentally characterized product (60).

REFERENCES

1. Herisson J-L, Chauvin Y. *Makromol Chem* 1971;141: 161–176.
2. (a) Dias EL, Nguyen ST, Grubbs RH. *J Am Chem Soc* 1997;119:3887–3897. (b) Vougioukalakis GC, Grubbs RH. *Chem Rev* 2010;110:1746–1787. (c) Fürstner A. *Angew Chem- Int Ed* 2000;39:3012–3043. (d) Schrock RR, Hoveyda AH. *Angew Chem- Int Ed* 2003;42:4592–4633. (e) Hoveyda AH, Schrock RR. *Chem Eur J* 2001;7:945–950. (f) Trnka TM, Grubbs RH. *Acc Chem Res* 2001;34:18–29. (g) Grubbs RH. *Handbook of Olefin Metathesis*. Vol. 2. Weinheim, Germany: Wiley-VCH; 2003; p 403–431.
3. (a) Aagaard OM, Meier RJ, Buda F. *J Am Chem Soc* 1998;120:7174–7182. (b) Bernardi F, Bottoni A, Miscione GP. *Organometallics* 2000;19:5529–5532. (c) Adlhart C, Chen P. *Angew Chem- Int Ed* 2002;41:4484–4487. (d) Vyboishchikov SF, Bühl M, Thiel W. *Chem Eur J* 2002;8:3962–3975. (e) Cavallo L. *J Am Chem Soc* 2002;124:8965–8973.
4. (a) Solans-Monfort X, Clot E, Copéret C, Eisenstein O. *J Am Chem Soc* 2005;127:14015–14025. (b) Poater A, Solans-Monfort X, Clot E, Copéret C, Eisenstein O. *J Am Chem Soc* 2007;129:8207–8216. (c) Poater A, Solans-Monfort X, Clot E, Copéret C, Eisenstein O. *Dalton Trans* 2006:3077–3087.
5. (a) Solans-Monfort X, Copéret C, Eisenstein O. *J Am Chem Soc* 2010;132:7750–7757. (b) Leduc A-M, Salameh A, Soulivong D, Chabanas M, Basset J-M, Copéret C, Solans-Monfort X, Clot E, Eisenstein O, Böhm VPW, Roper M. *J Am Chem Soc* 2008;130:6288–6297. (c) Blanc F, Basset J-M, Copéret C, Sinha A, Tonzetich ZJ, Schrock RR, Solans-Monfort X, Clot E, Eisenstein O, Lesage A, Emsley L. *J Am Chem Soc* 2008;130:5886–5900.
6. Arduengo AJ III, Harlow RL, Kline M. *J Am Chem Soc* 1991;113:361–363.
7. (a) Marion N, Nolan SP. *Acc Chem Res* 2008;41:1440–1449. (b) Würtz S, Glorius F. *Acc Chem Res* 2008;41:1523–1533. (c) Diez-Gonzalez S, Marion N, Nolan SP. *Chem Rev* 2009; 109:3612–3676.
8. (a) Herrmann WA. *Angew Chem- Int Ed* 2002;41:1290–1309. (b) Cavallo L, Correa A, Costabile C, Jacobsen H. *J Organomet Chem* 2005;690:5407–5413.
9. Kingsbury JS, Harrity JPA, Bonitatebus PJ Jr, Hoveyda AH. *J Am Chem Soc* 1999;121:791–799.
10. Michrowska A, Bujok R, Harutyunyan S, Sashuk V, Dolgonos G, Grela K. *J Am Chem Soc* 2004;126:9318–9325.
11. Van Der Eide EF, Piers WE. *Nature Chem* 2010;2:571–576.
12. Poater A, Ragone F, Correa A, Cavallo L. *Dalton Trans* 2011;40:11066–11069.
13. (a) Nuñez-Zarur F, Solans-Monfort X, Rodríguez-Santiago L, Sodupe M. *Organometallics* 2012;31:4203–4215. (b) Nuñez-Zarur F, Poater J, Rodríguez-Santiago L, Solans-Monfort X, Solà M, Sodupe M. *Comput Theor Chem* 2012;996:57–67.
14. Adlhart C, Chen P. *J Am Chem Soc* 2004;126:3496–3510.
15. Weskamp T, Schattenmann WC, Spiegler M, Herrmann WA. *Angew Chem- Int Ed* 1998;37:2490–2493.
16. Torker S, Merki D, Chen P. *J Am Chem Soc* 2008;130: 4808–4814.
17. Benitez D, Tkatchouk E, Goddard WA III,. *Organometallics* 2009;28:2643–2645.
18. (a) Zhao Y, Truhlar DG. *J Chem Theor Comput* 2009;5: 324–333. (b) Zhao Y, Truhlar DG. *Org Lett* 2007;9: 1967–1970.
19. Grela K, Harutyunyan S, Michrowska A. *Angew Chem- Int Ed* 2002;41:4038–4040.
20. Zaja M, Connon SJ, Dunne AM, Rivard M, Buschmann N, Jiricek J, Blechert S. *Tetrahedron* 2003;59:6545–6558.
21. Solans-Monfort X, Pleixats R, Sodupe M. *Chem Eur J* 2010;16:7331–7343.
22. Barbasiewicz M, Szadkowska A, Makal A, Jarzembska K, Woźniak K, Grela K. *Chem Eur J* 2008;14:9330–9337.
23. Hillier AC, Sommer WJ, Yong BS, Petersen JL, Cavallo L, Nolan SP. *Organometallics* 2003;22:4322–4326.
24. (a) Thiel V, Hendann M, Wannowius K-J, Plenio H. *J Am Chem Soc* 2012;134:1104–1114. (b) Vorfalt T, Wannowius KJ, Thiel V, Plenio H. *Chem Eur J* 2010;16:12312–12315.
25. Adlhart C, Hinderling C, Baumann H, Chen P. *J Am Chem Soc* 2000;122:8204–8214.
26. Rowley CN, Van Der Eide EF, Piers WE, Woo TK. *Organometallics* 2008;27:6043–6045.
27. Woo TK, Blöchl PE, Ziegler T. *J Phys Chem A* 2000; 104:121–129.
28. Benitez D, Tkatchouk E, Goddard WA III. *Chem Commun* 2008:6194–6196.
29. Sanford MS, Love JA, Grubbs RH. *J Am Chem Soc* 2001; 123:6543–6554.

30. Yang H–C, Huang Y–C, Lan Y-K, Luh T-Y, Zhao Y, Truhlar DG. *Organometallics* 2011;30:4196–4200.
31. Straub BF. *Adv Synth Catal* 2007;349:204–214.
32. Webster CE. *J Am Chem Soc* 2007;129:7490–7491.
33. Romero PE, Piers WE. *J Am Chem Soc* 2007;129:1698–1704.
34. (a) Stewart IC, Douglas CJ, Grubbs RH. *Org Lett* 2008; 10:441–444. (b) Seiders TJ, Ward DW, Grubbs RH. *Org Lett* 2001;3:3225–3228.
35. Correa A, Cavallo L. *J Am Chem Soc* 2006;128:13352–13353.
36. Benitez D, Goddard WA III. *J Am Chem Soc* 2005;127: 12218–12219.
37. (a) Tallarico JA, Bonitatebus PJ Jr, Snapper ML. *J Am Chem Soc* 1997;119:7157–7158. (b) Van Der Eide EF, Romero PE, Piers WE. *J Am Chem Soc* 2008;130:4485–4491.
38. Liu P, Xu X, Dong X, Keitz BK, Herbert MB, Grubbs RH, Houk KN. *J Am Chem Soc* 2012;134:1464–1467.
39. (a) Copéret C. *Beilstein J Org Chem* 2011;7:13–21. (b) Anderson DR, Ung T, Mkrtumyan G, Bertrand G, Grubbs RH, Schrodi Y. *Organometallics* 2008;27:563–566.
40. Bahri-Laleh N, Credendino R, Cavallo L. *Beilstein J Org Chem* 2011;7:40–45.
41. Fomine S, Ortega JV, Tlenkopatchev MA. *J Mol Catal A: Chem* 2005;236:156–161.
42. (a) Diver ST. *Coord Chem Rev* 2007;251:671–701. (b) Dieltiens N, Moonen K, Stevens CV. *Chem Eur J* 2007;13:203–214. (c) Lloyd-Jones GC, Margue RG, De Vries JG. *Angew- Chem Int Ed* 2005;44:7442–7447. (d) Lippstreu JJ, Straub BF. *J Am Chem Soc* 2005;127:7444–7457.
43. Clavier H, Correa A, Escudero-Adán EC, Benet-Buchholz J, Cavallo L, Nolan SP. *Chem Eur J* 2009;15:10244–10254.
44. (a) Pruhs S, Lehmann CW, Fürstner A. *Organometallics* 2004;23:280–287. (b) Barbasiewicz M, Szadkowska A, Bujok R, Grela K. *Organometallics* 2006;25:3599–3604. (c) Slugovc C, Perner B, Stelzer F, Mereiter K. *Organometallics* 2004;23:3622–3626. (d) Ung T, Hejl A, Grubbs RH, Schrodi Y. *Organometallics* 2004;23:5399–5401.
45. Diesendruck CE, Tzur E, Ben-Asuly A, Goldberg I, Straub BF, Lemcoff NG. *Inorg Chem* 2009;48:10819–10825.
46. Poater A, Ragone F, Correa A, Szadkowska A, Barbasiewicz M, Grela K, Cavallo L. *Chem Eur J* 2010;16:14354–14364.
47. Bantreil X, Schmid TE, Randall RAM, Slawin AMZ, Cazin CSJ. *Chem Commun* 2010;46:7115–7117.
48. Zirngast M, Pump E, Leitgeb A, Albering JH, Slugovc C. *Chem Commun* 2011;47:2261–2263.
49. Berlin JM, Goldberg SD, Grubbs RH. *Angew Chem- Int Ed* 2006;45:7591–7595.
50. Stewart IC, Benitez D, O'Leary DJ, Tkatchouk E, Day MW, Goddard WA III, Grubbs RH. *J Am Chem Soc* 2009; 131:1931–1938.
51. Ragone F, Poater A, Cavallo L. *J Am Chem Soc* 2010; 132:4249–4258.
52. Hong SH, Chlenov A, Day MW, Grubbs RH. *Angew Chem- Int Ed* 2007;46:5148–5151.
53. Poater A, Cosenza B, Correa A, Giudice S, Ragone F, Scarano V, Cavallo L. *Eur J Inorg Chem* 2009:1759–1766.
54. Poater A, Ragone F, Mariz R, Dorta R, Cavallo L. *Chem Eur J* 2010;16:14348–14353.
55. (a) Giudici RE, Hoveyda AH. *J Am Chem Soc* 2007;129:3824–3825. (b) Fournier P -A, Savoie J, Stenne B, Bédard A, Grandbois A, Collins SK. *Chem Eur J* 2008;14:8690–8695. (c) Grisi F, Mariconda A, Costabile C, Bertolasi V, Longo P. *Organometallics* 2009;28:4988–4995.
56. Funk TW, Berlin JM, Grubbs RH. *J Am Chem Soc* 2006;128:1840–1846.
57. Costabile C, Cavallo L. *J Am Chem Soc* 2004;126:9592–9600.
58. (a) Van Rensburg WJ, Steynberg PJ, Meyer WH, Kirk MM, Forman GS. *J Am Chem Soc* 2004;126:14332–14333. (b) Van Rensburg WJ, Steynberg PJ, Kirk MM, Meyer WH, Forman GS. *J Organomet Chem* 2006;691:5312–5325. (c) Mathew J, Koga N, Suresh CH. *Organometallics* 2008;27:4666–4670. (d) Poater A, Cavallo L. Probing the mechanism of the double C-H (de)activation route of a Ru-based Olefin metathesis catalyst. In: *Green Metathesis Chemistry. NATO Science for Peace and Security Series A-Chemistry and Biology*. Kluwer: Springer; 2010. p 275–280. (e) Poater A, Cavallo L. *Journal of Molecular Catalysis A: Chemistry* 2010;324:75–79. (f) Poater A, Ragone F, Correa A, Cavallo L. *J Am Chem Soc* 2009;131:9000–9006.
59. (a) Galan BR, Gembicky M, Dominiak PM, Keister JB, Diver ST. *J Am Chem Soc* 2005;127:15702–15703. (b) Galan BR, Pitak M, Gembicky M, Keister JB, Diver ST. *J Am Chem Soc* 2009;131:6822–6832.
60. Manzini S, Urbina-Blanco CA, Poater A, Slawin AMZ, Cavallo L, Nolan SP. *Angew Chem- Int Ed* 2012;51:1042–1045.

20

IMMOBILIZATION OF OLEFIN METATHESIS CATALYSTS

MICHAEL R. BUCHMEISER

Lehrstuhl für Makromolekulare Stoffe und Faserchemie, Institut für Polymerchemie, Universität Stuttgart, Stuttgart, Germany; Institut für Textilchemie und Chemiefasern (ITCF) Denkendorf, Denkendorf, Germany

20.1 INTRODUCTION

Metathesis-based coupling reactions are nowadays among the most powerful carbon–carbon bond-forming reactions. With the possibility of preparing cyclic structures including macrocycles with high *cis*- or *trans*-selectivity, metathesis-based reactions, particularly cross-metathesis (CM), ene–yne metathesis and ring-opening cross metathesis (ROCM) certainly outrivaled other potent C–C forming reactions, that is, the Heck reaction in terms of selectivity. So far, however, the high turnover numbers (TONs) of up to 20 million, which are nowadays standard in Heck chemistry, have not yet been accomplished in any metathesis reaction. Instead, typical TONs are rather in the low thousands or, in rare cases, reach few hundred thousands. Consequently, comparably high catalyst loadings, typically in the 0.01–2 mol% range, are required. In homogeneous catalysis, this translates into too high levels of metal and ligand contamination of the products, which turns out to be a real impediment for pharmaceutically relevant products. In order to avoid cost- and time-intense purifications, several approaches to supported metathesis catalysts have emerged over the last 15 years. Besides low contamination of the product with catalyst and its decomposition products, such an approach also entails the advantage of a straightforward and fast separation of the remaining catalyst from the reaction mixture. While supported catalysts can usually be reused to a sometimes quite significant extent, supported metathesis catalysts with their limited lifetime hardly offer this additional benefit. Thus, most reports on repeated use entail loadings of 0.5 mol% or more of catalyst with respect to substrate and are therefore, in view of standard TONs of supported catalysts between $1000 < \text{TON} < 4000$, basically meaningless.

This chapter briefly summarizes the different approaches to supported metathesis catalysts. In doing that, both organic and inorganic supports shall be addressed; however, the main emphasis is placed on the developments made in the last decade. Furthermore, only well-defined metal alkylidenes shall be addressed. Other systems, whether based on oxides, bi- or ternary systems, metal nanoparticles or other progenitors, from which the active species forms *in situ*, will be neglected. Finally, an entirely novel approach to catalyst immobilization, that is, the use of ionic metathesis catalysts working in a supported ionic liquid phase (SILP) and the advantages of such an approach over existing ones is addressed. Recent reviews are recommended for a fully comprehensive reading on that topic (1–6).

20.2 IMMOBILIZATION OF METATHESIS CATALYSTS ON ORGANIC SUPPORTS

20.2.1 Schrock Catalysts

20.2.1.1 Immobilization via the Alkoxides In principle, Schrock catalysts can be immobilized via the alkylidene ligand (i), via substitution of one or two alkoxides (ii) or via the arylimido ligand (iii) (Fig. 20.1). The first polymer-supported version of a Schrock catalyst was reported by Schrock and Hoveyda et al. in 2002 (7). Introducing a chiral biphenyl ligand, which contained two

Olefin Metathesis: Theory and Practice, First Edition. Edited by Karol Grela.

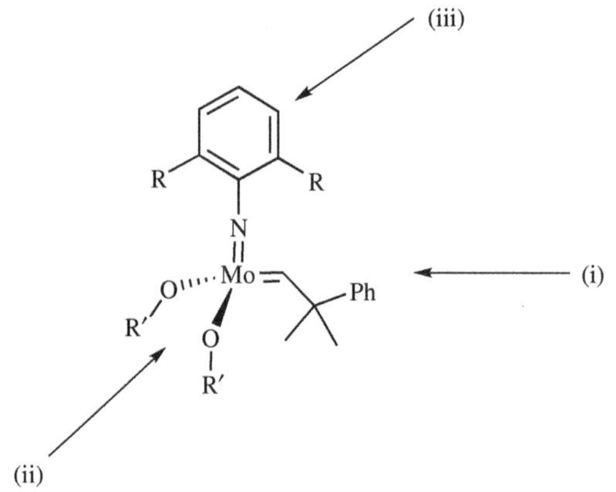

Figure 20.1 Three different possibilities for the immobilization of a Schrock-type catalyst.

styrene moieties, free radical polymerization of the ligand followed by loading with the Mo-bistriflate yielded the desired catalyst (Scheme 20.1). It was successfully used in various enantioselective RCM (ring closing metathesis) and ROCM reactions. In a similar approach, our group used a bis-nornorn-2-ene-substituted binaphthyl ligand. Ring-opening metathesis polymerization (ROMP) of the ligand followed by reaction with the catalyst progenitor yielded the corresponding supported catalyst (Scheme 20.2), which was used in a series of asymmetric RCM and desymmetrization reactions (8). Using the same ligand, a monolith-supported version could be realized by our group (Scheme 20.3). Such monolithic supports (9) consist of a unitary piece of a polymeric material that contains large transport pores, usually in the 5–10 μm regime, while otherwise being nonporous. This allows for running reactions in a continuous manner with high linear velocities while keeping the resulting back pressure well below 200 MPa. In addition, diffusion is restricted to Eddie diffusion since no small or medium-sized pores are present. In turn, this allows for a fast mass transfer between the catalytic sites at the surface of these monoliths and the transport phase containing the educts (10). Consequently, any reduction in diffusivity compared to a homogeneous system is solely related to the tortuosity of the system. Taking advantage of the high porosity of the support (60%), the monolith-supported catalyst was cut in pieces and used in a high-throughput setup were the support simultaneously served as reaction vessel, catalyst support, and filter (11). Following their original concept, Schrock and Hoveyda et al. (12) prepared a series of supported catalysts for use in asymmetric metathesis reactions (Fig. 20.2). Similar to the approach reported by our group, the Schrock and Hoveyda group prepared ROMP-derived supports containing chiral biphenyl-based Schrock catalysts via a one-step synthesis, in which the Schrock catalyst also serves as a catalyst for polymerization (Scheme 20.4) (12).

20.2.1.2 Immobilization via the Arylimido Ligand The immobilization of a Schrock catalyst via the arylimido ligand is highly attractive since such an approach allows for a wider and in fact complementary variability in catalyst structure, since both reactivity and selectivity of (chiral) Schrock-type catalysts are mainly (but not exclusively) (13–15) governed by the nature of the alkoxides, (bi-) phenoxides, and binaphtholates. We reported on a successful realization of that route. A set of 2,6-disubstituted anilines containing an additional ω-halogenoalkyl group in the 4-position was designed. From these anilines, different Schrock-type catalyst progenitors, that is, Mo-bistriflates, of the general formula Mo(NAr′)(CHCMe$_2$Ph)(OTf)$_2$·DME (Ar′=2,5-Me$_2$-4-[(CH$_2$)$_6$-Br]-C$_6$H$_3$; 2,5-*i*Pr$_2$-4-[(CH$_2$)$_6$-Br]-C$_6$H$_3$; 2,5-*i*Pr$_2$-4-[(CH$_2$)$_6$-Cl]-C$_6$H$_3$) (OTf = CF$_3$SO$_3$) have been prepared. The molybdenum triflates were then converted into a set of four different chiral catalysts, that is, Mo(*N*-2,6-Me$_2$-4-(ω-Br-C$_6$H$_{12}$)-C$_6$H$_2$)(CHCMe$_2$Ph)(BIPHEN), Mo(*N*-2,6-Me$_2$-4-(ω-Br-C$_6$H$_{12}$)-C$_6$H$_2$)(CHCMe$_2$Ph)(BIPHEN), Mo(*N*-2,6-Me$_2$-4-(ω-Br-C$_6$H$_{12}$)-C$_6$H$_2$)(CHCMe$_2$Ph)(OCMe(CF$_3$)$_2$), and Mo(*N*-2,6-*i*Pr$_2$-4-(ω-Br-C$_6$

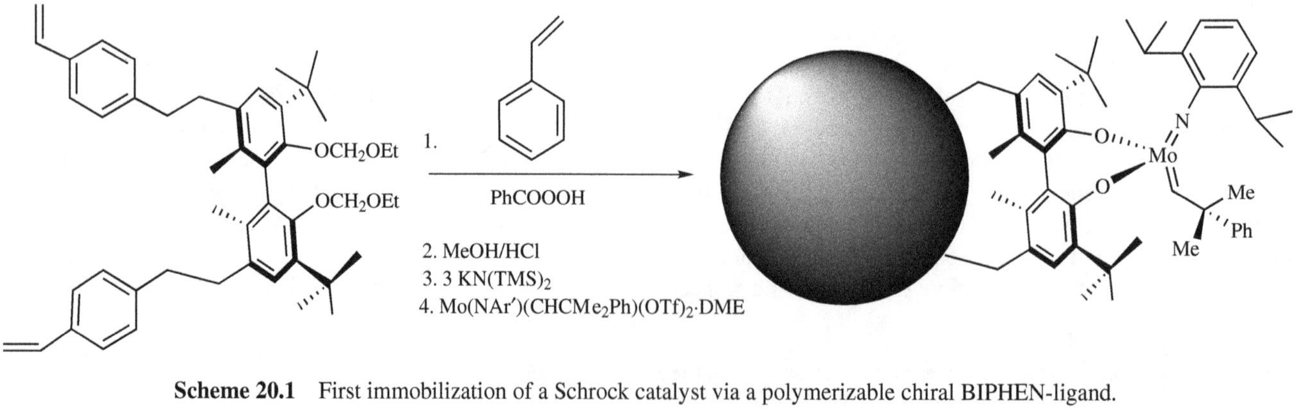

Scheme 20.1 First immobilization of a Schrock catalyst via a polymerizable chiral BIPHEN-ligand.

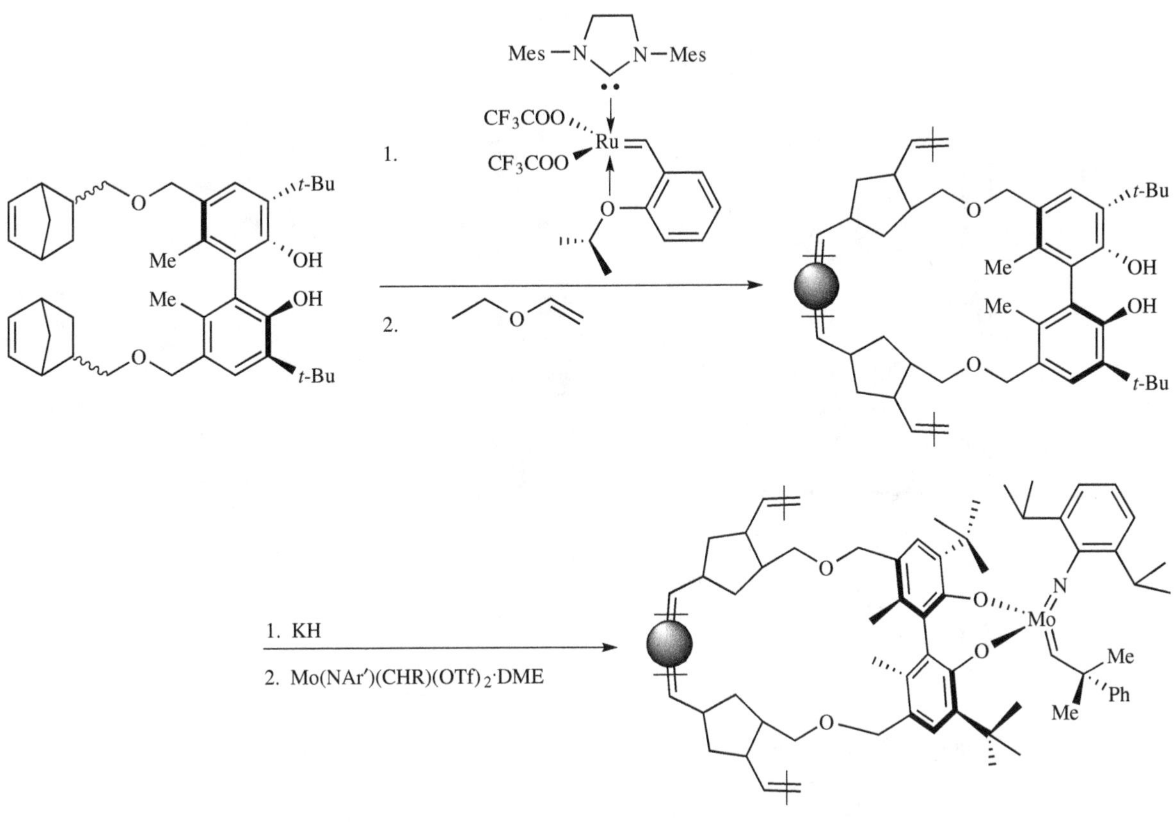

Scheme 20.2 ROMP-based synthesis of a polymeric chiral BIPHEN-ligand and formation of a supported Schrock catalyst.

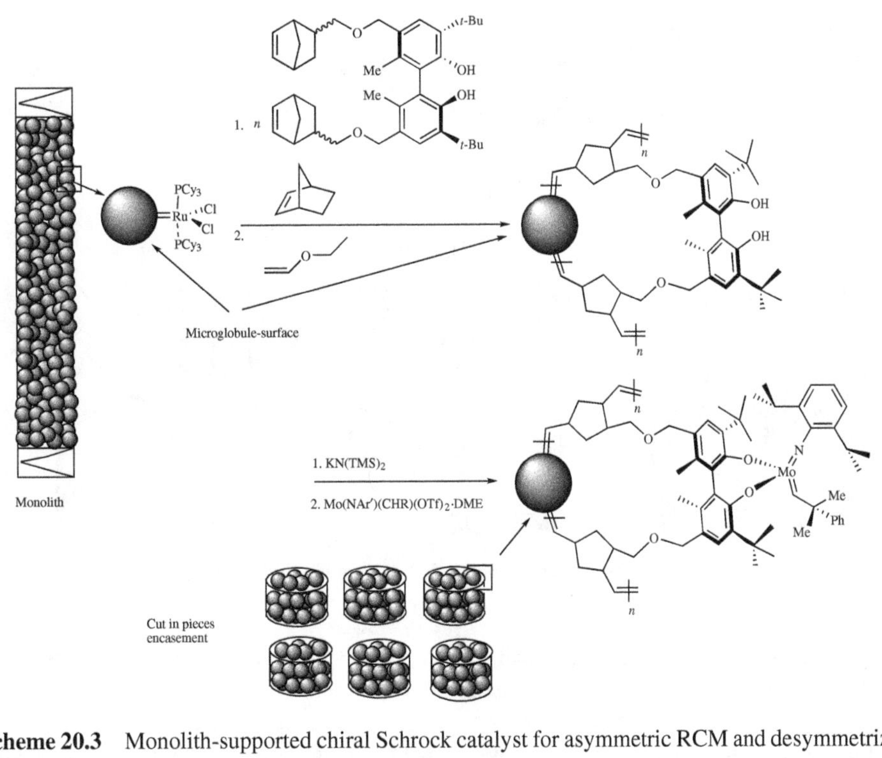

Scheme 20.3 Monolith-supported chiral Schrock catalyst for asymmetric RCM and desymmetrization reactions.

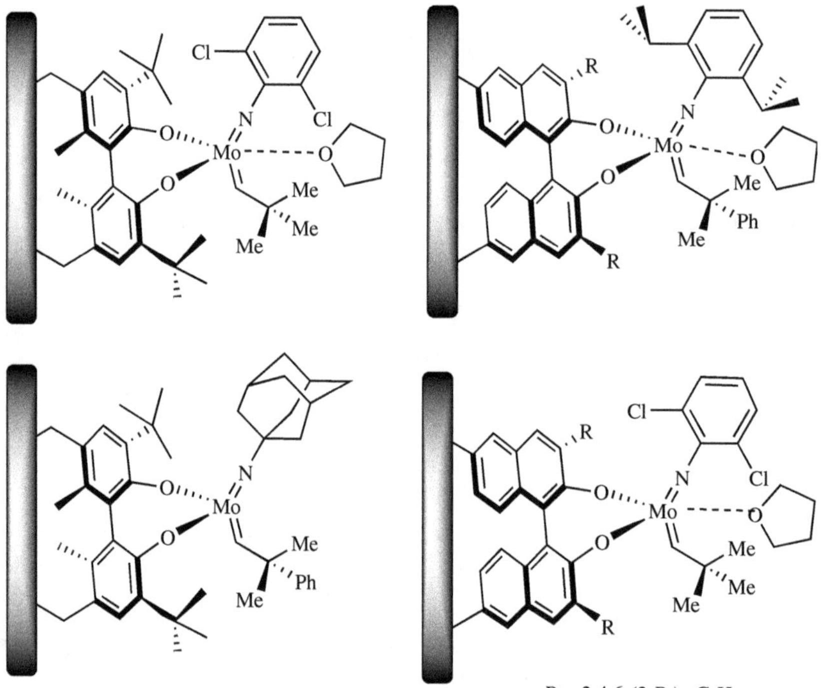

Figure 20.2 Selection of Schrock-type catalysts immobilized on a polymeric support.

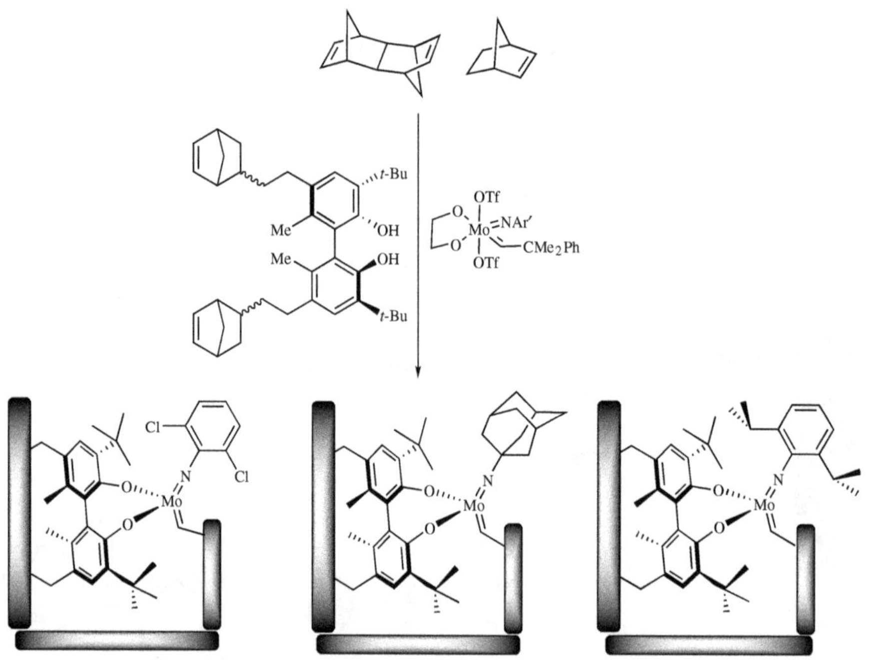

Scheme 20.4 Synthesis of chiral Schrock catalysts via a ROMP-based approach.

H_{12})-C_6H_2)($CHCMe_2Ph$)($OCMe(CF_3)_2)_2$, (BIPHEN=3,3′-di-*tert*-butyl-5,5′,6,6′-tetramethyl-2,2′-biphenolate), which were then, in a single step, subjected to heterogenization using a designed fluorosulfonated polystyrene divinylbenzene (PS-DVB)-based support (Scheme 20.5). With these supported versions of a Schrock catalyst, numerous asymmetric RCM and desymmetrization reactions have been accomplished with values for enantiomeric excess (ee) that were similar to those obtained with the corresponding parent, homogenous catalysts.

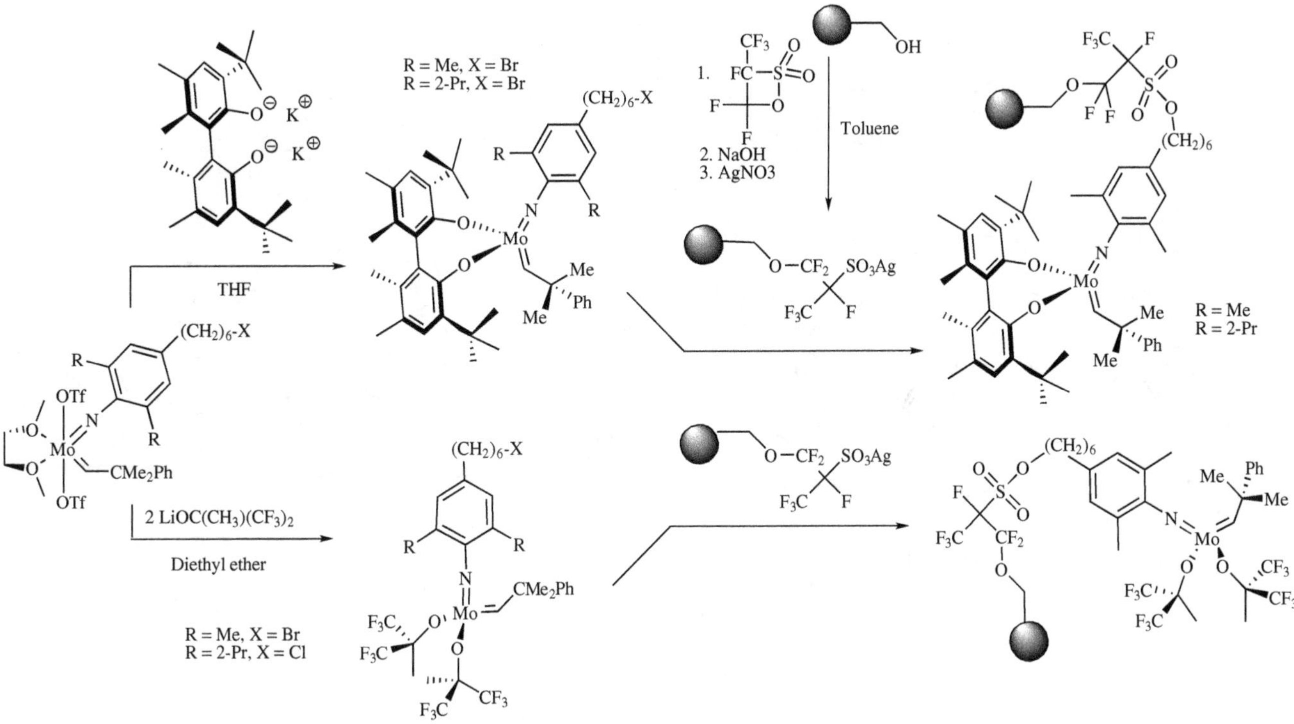

Scheme 20.5 Synthesis of a heterogeneous Schrock catalyst immobilized via the *N*-arylimido-ligand.

20.2.2 Grubbs and Grubbs–Hoveyda Catalysts

Currently, different types of ruthenium alkylidenes are available, usually referred to as first-, second- and third-generation Grubbs (**1–3**) and first- and second-generation Grubbs–Hoveyda catalysts (**4**, **5**, Fig. 20.3). In addition, a huge variety of related catalysts with modifications in the *N*-heterocyclic carbene (NHC) or the alkylidene ligand have been prepared. All these different type of catalysts can be immobilized either (i) via the phosphine or NHC, (ii) the alkylidene or (iii) via exchange of the parent halide (chloride) ligands. In the following, different approaches, which use one of these immobilization strategies, as well as their advantages and disadvantages shall be outlined.

20.2.2.1 Immobilization via the Phosphine, a Tertiary Amine or via the N-Heterocyclic Carbene (NHC) Phosphines, in particular NHCs, coordinate strongly to ruthenium in many of its oxidation states. This makes immobilization of the above-described Ru-alkyidenes via these ligands attractive as the NHCs not only immobilize the parent catalyst but also most of its decomposition products, thereby allowing for very low Ru-contaminations in the products. The first immobilization of a Ru-alkylidene was reported by the Grubbs group, group, which immobilized (Grubb's group is a subject, not Grubbs himself) immobilized $RuCl_2(PPh_3)_2(CHCH{=}CPh_2)$ on phosphine-modified poly (styrene)-*co*-poly(divinylbenzene) (Scheme 20.6) (16). The supported catalyst showed a significantly reduced activity in the metathesis of *cis*-2-pentene as well as in the ROMP of norborn-2-ene. Mainly, the incomplete substitution of phosphine, phosphine chelation effects, and limited diffusion were made responsible for this. No information on the metal content of the products was provided. Gladysz et al. (17) reported on a fluorous version of the second-generation Grubbs–Hoveyda catalyst for use in fluorous biphasic catalysis. For that purpose, fluorinated phosphines have been used. Unprecedentedly, Kirschning et al. (18) reported on the immobilization of both $RuCl_2(PCy_3)_2(CHPh)$ and $RuCl_2(3\text{-Br-py})_2(IMesH_2)(CHPh)$ ($IMesH_2$=1,3-dimesityli midazolin-2-ylidene) on a poly(vinylpyridine)-based resin (Fig. 20.4). However, the non-release mechanism relies on the *selective* dissociation of the second 3-Br-pyridine group. This appears, in view of the similar pK_a values of both pyridine and 3-Br-pyridine, rather unrealistic.

The first report on a Grubbs catalyst supported via the NHC for use in RCM was given by Blechert et al. (Fig. 20.5, top, left) (19). A hydroxymethylated NHC was immobilized to a PS-DVB-based support. In a similar approach, our group immobilized a second-generation Grubbs catalyst to a monolithic support (20–22). In an extension of their work, Blechert et al. (23) also immobilized a Grubbs–Hoveyda catalyst to a Wang resin (Fig. 20.5). Using an unsymmetrically substituted NHC bearing an

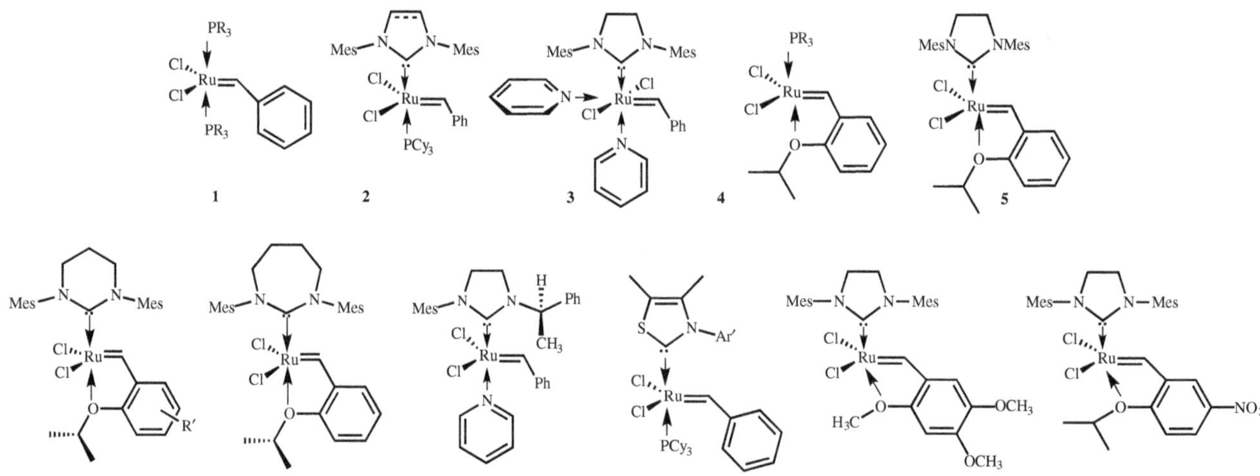

1 2 3 4 5 6 7 8 9 10 11

Figure 20.3 Different types of Ru-alkylidenes. first-, second- and third-generation Grubbs catalysts (**1–3**), first- and second-generation Grubbs–Hoveyda catalyst (**4–5**), Grubbs–Hoveyda-type catalysts based on tetrahydropyrimidin-2-ylides and diazepin-2-ylidenes (**6**, **7**) and on modified benzylidenes (**10**, **11**), Grubbs-type catalysts with unsymmetrically substituted (**8**) and thiazol-2-ylidene based NHCs (**9**).

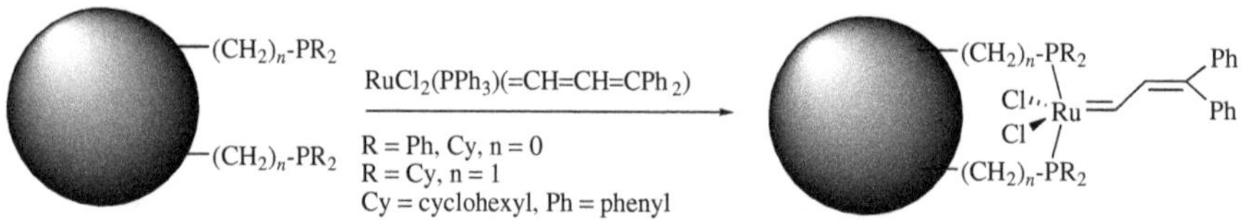

Scheme 20.6 Synthesis of the first immobilized version of Grubbs' catalyst.

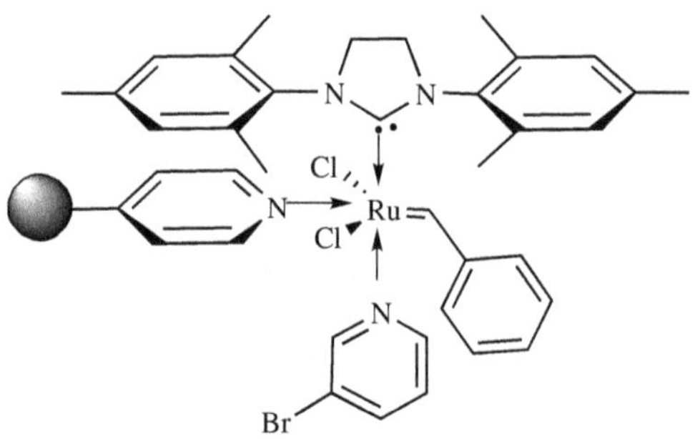

Figure 20.4 A third-generation Grubbs-type catalyst immobilized via the pyridine.

ω-hydroxyhexyl ligand, we could immobilize a Grubbs-type catalyst to a monolithic system (Scheme 20.7). The monolith was cut into pieces 1 cm in height and the devices thus obtained were used in high-throughput experimentation in a series of RCM, ROCM, and ene–yne metathesis reactions (11). Grubbs et al. (24,25) reported on two PEGylated versions of this catalyst for use in RCM and ROMP in methanol or water (Fig. 20.5). Weck et al. reported on ROMP-derived supports containing NHC ligands for use in the immobilization of a second-generation Grubbs catalyst (26) and use in RCM. Very recently, Bergbreiter reported on the use of his poly(isobutylene)-supported version of the second-generation Grubbs–Hoveyda catalyst in ROMP (Fig. 20.6) (27).

20.2.2.2 Immobilization via the Alkylidene Immobilization of *any* metathesis catalyst via the alkylidene ligand strongly relies on a release–catch or "boomerang" mechanism (Scheme 20.8). In 1999, Barrett et al. (28,29) proposed this concept for the first time and prepared the first Grubbs-type catalyst immobilized via the alkylidene ligand by the reaction of vinylated PS-DVB with the first-generation Grubbs catalyst (Fig. 20.7). While the supported catalyst could in fact be reused several times in different RCM reactions, this "recyclability" must rather be attributed to the low substrate loadings than to a real boomerang mechanism. Later, this work was extended to the second-generation Grubbs catalyst. (30–33) With the advent of the second-generation Grubbs–Hoveyda catalyst (34), an increasing number of polymer-supported versions of this catalyst with an alkylidene-based immobilization strategy appeared (Fig. 20.8) (23,35–48). While some arguments that such a mechanism might in fact be active in case of Hoveyda-type ligands have appeared in the literature (49,50), it still remains quite disputable (51). If at all, then batch-type reactions, in which the intermediary formed

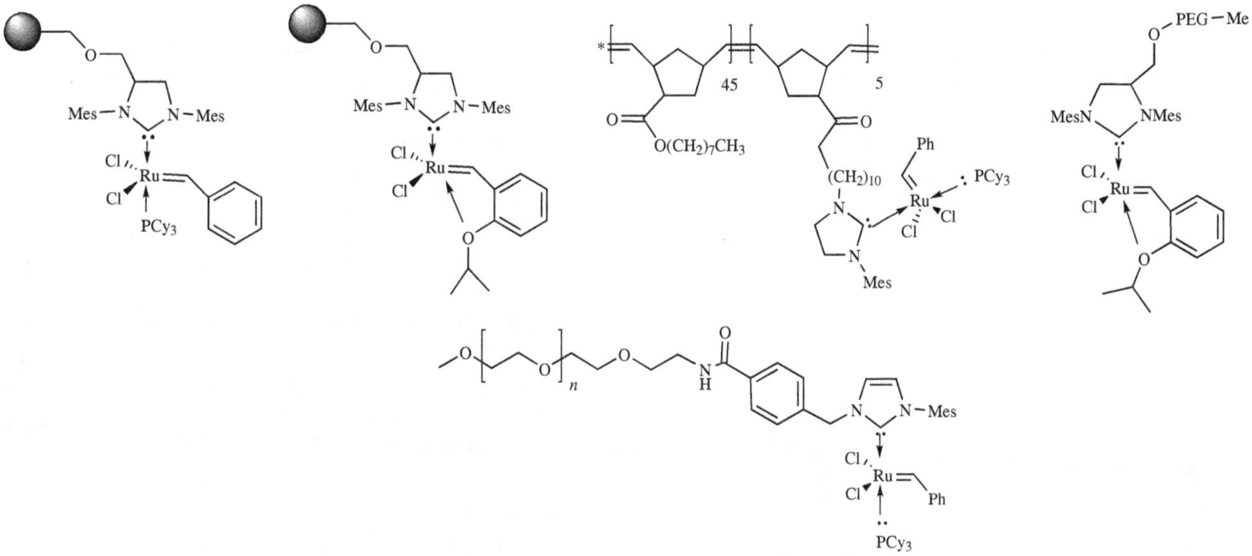

Figure 20.5 Selection of Grubbs-type catalysts immobilized via the *N*-heterocyclic carbene.

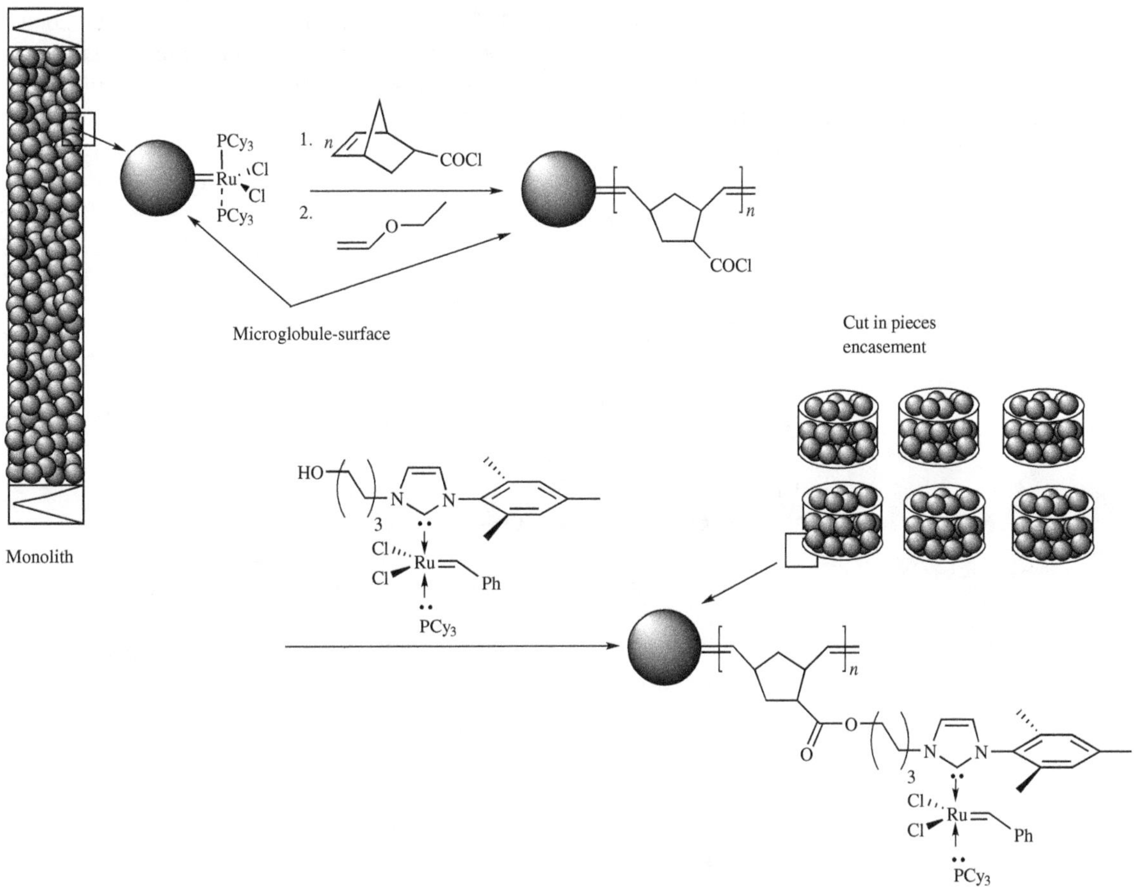

Scheme 20.7 Synthesis of a monolith-supported version of a second-generation Grubbs-catalyst for use in high-throughput reactions.

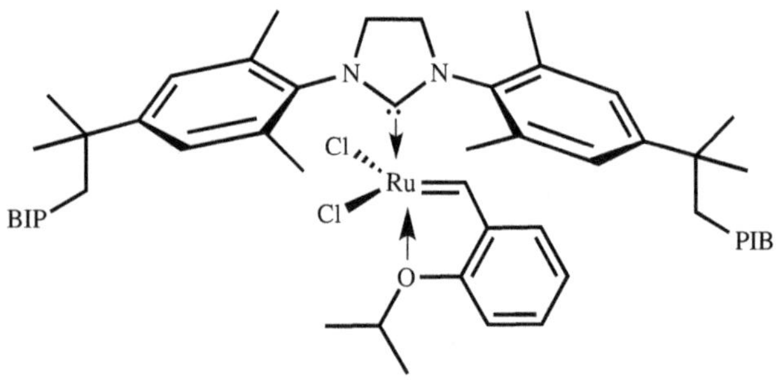

Figure 20.6 Second-generation Grubbs–Hoveyda catalyst with a bis-poly(isobutylene)-substituted NHC.

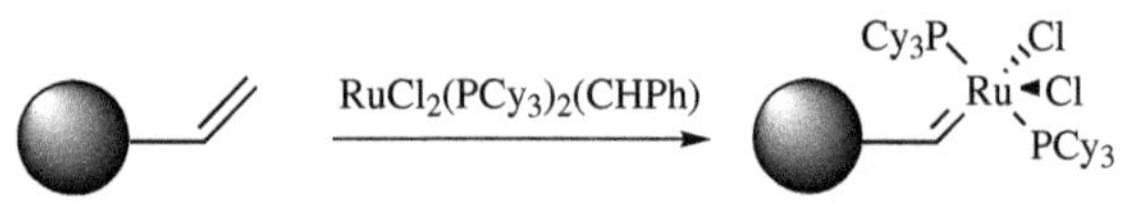

Figure 20.7 "Boomerang"-type catalysts first prepared by Barrett et al.

2-(2-PrO)-styrenes are not removed from the reaction mixture, are required. Consequently, continuous-flow setups, which in fact represent the preferred setup in technical applications, *cannot* be used since the intermediary formed 2-(2-PrO)-styrenes and concomitantly the active species are simply removed from the reaction column. Nonetheless, most probably because of the ease of preparation, most immobilization strategies rely exactly on that type of chemistry, rendering most of the presented systems technically useless. Some typical systems are outlined in Fig. 20.8. Though suffering from the same problems, that is, from incomplete reformation of the initial catalyst, the systems reported by Gladysz et al. address the issue of biphasic fluorous liquid–liquid catalysis. Thus, fluorous polyacrylate labels were used to obtain both first- and second-generation Grubbs–Hoveyda catalysts suitable for running reactions in fluorous solvents (Fig. 20.9) (52).

20.2.2.3 Immobilization via Halogen Metathesis In 2001, Mol (53) reported on the immobilization of the first-generation Grubbs catalyst on a perfluoro carboxylate-modified PS-DVB resin. A rapid decrease in reactivity was observed in the RCM of diethyl diallyl malonate (DEDAM), and a comparably low total TON of about 60 was reported for this reaction. In addition, about 20% of the total amount of ruthenium present leached into the product *after the first run*. Although the data provided by the authors support the proposed structure of the catalyst to a certain extent, the system most probably undergoes rearrangement by ligand scrambling (54–56). In fact, Grubbs-type catalysts containing the ruthenium monochloro-mono(trifluoroacetate) motif are unstable in solution and rearrange into the corresponding ruthenium-dichloro and ruthenium-bis(trifluoroacetate) complexes (Scheme 20.9). A similar rearrangement of the supported catalyst would in fact account for the high

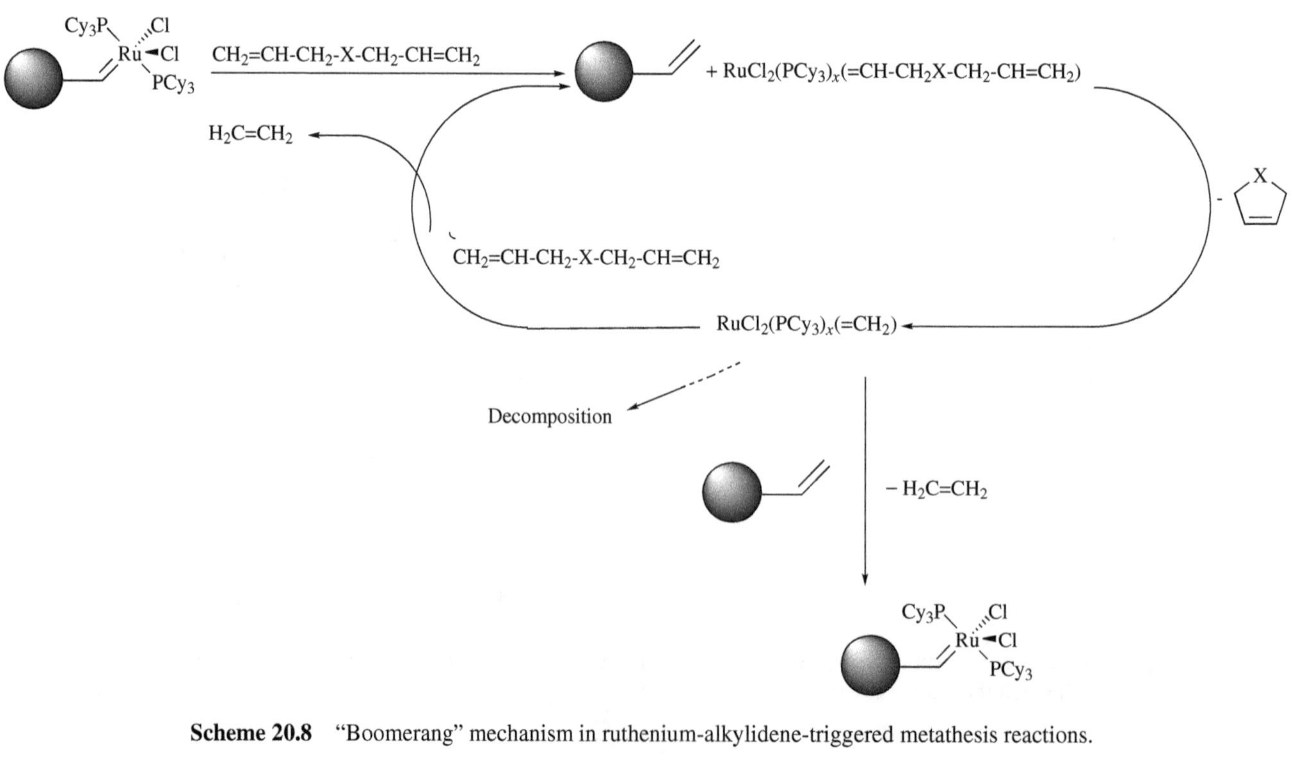

Scheme 20.8 "Boomerang" mechanism in ruthenium-alkylidene-triggered metathesis reactions.

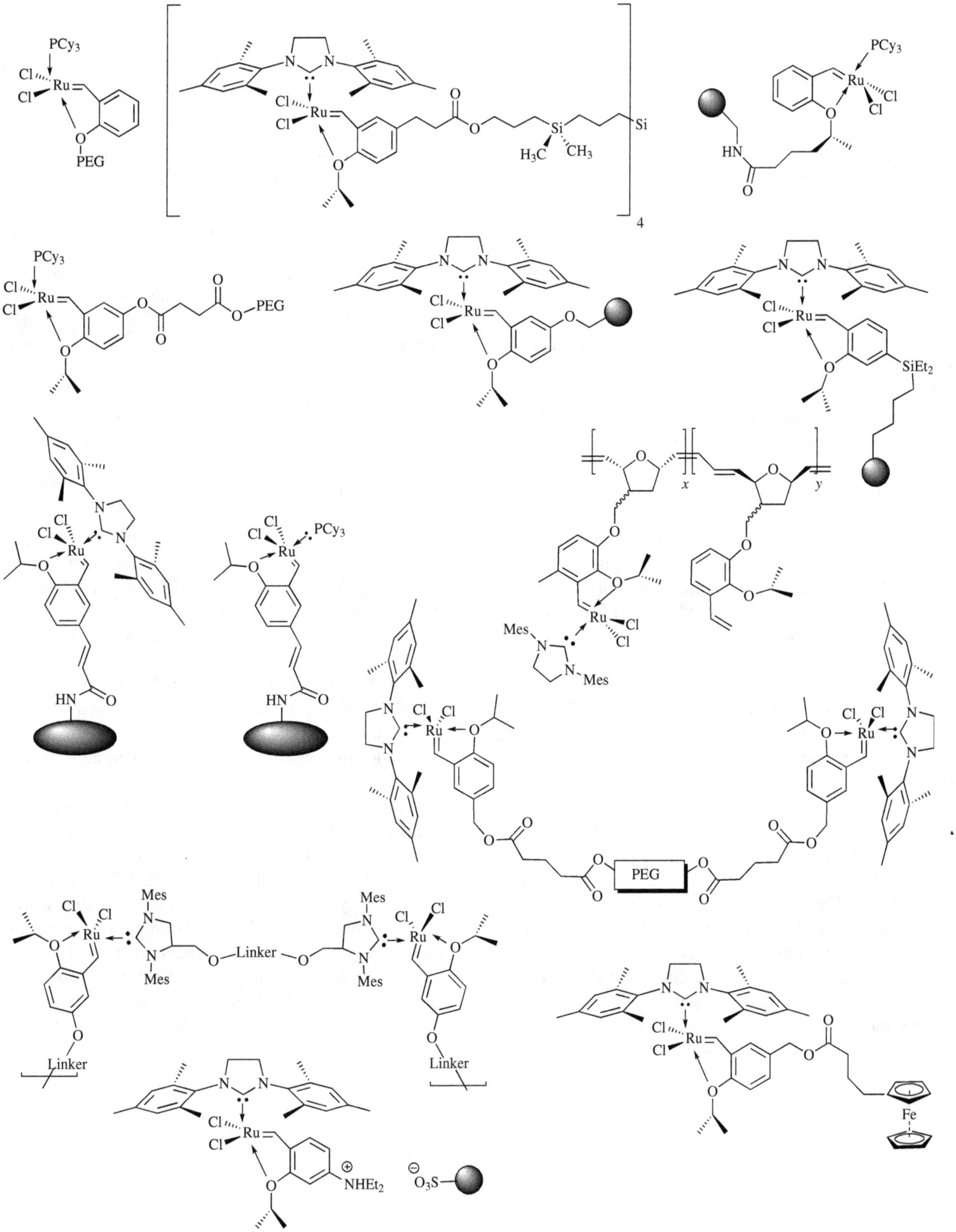

Figure 20.8 Grubbs–Hoveyda catalysts immobilized via the benzylidene ligand.

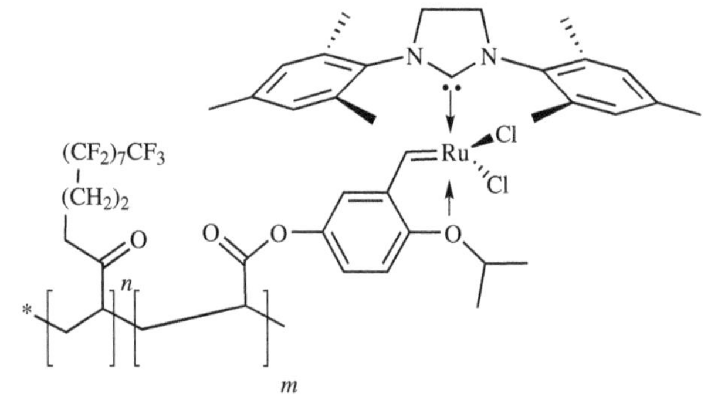

Figure 20.9 Grubbs–Hoveyda catalysts bound to a fluoroacrylate-copolymer via the benzylidene ligand.

ruthenium concentrations in the first batch as it would result in the release of substantial amounts of the initially polymer-bound catalyst into the solution. In an effort to improve the reaction kinetics by minimizing diffusion-related problems (10) and to enhance the stability of carboxylate-based Grubbs–Hoveyda systems, we prepared a series of monolith-supported second-generation Grubbs catalyst containing the saturated $IMesH_2$-ligand (Fig. 20.10) (57). Using an ROMP-based protocol, surface functionalization can be accomplished *in situ* (58,59). In this manner, a series of norborn-2-ene- and 7-oxanorborn-2-ene-based carboxylates and fluorocarboxylates could be attached to the surface of the support and reacted with the corresponding progenitors, that is, with either the second-generation Grubbs or Grubbs–Hoveyda catalyst (57,60–63). Preferably, the second chloride ligand is also substituted by an electron-withdrawing group, that is, by a second fluorinated carboxylate. Notably, homogeneous versions of all catalysts were prepared in order to compare their reactivity (56,60,64). Besides the "standard" IMes- and $IMesH_2$-ligands (IMes = 1,3-dimesitylimidazol-2-ylidene), both tetrahydropyrimidin-2-ylidenes and 1,3-diazepin-2-yidenes were used (65,66). The supported catalysts thus prepared were successfully used in continuous RCM or ene–yne or ROCM reactions, allowing for TONs up to about 4000. Since the Ru-carboxylate bond is comparably strong (pK_a, CF_3COOH = ca 1), metal leaching is very low, that is, in the low parts per million or even in the parts per billion range (67). In fact, this immobilization protocol can be extended to basically any support (68) and it also allows for the synthesis of amphiphilic block copolymers that contain such fluorinated versions of Grubbs–Hoveyda type catalysts (Fig. 20.11) and that can be used in micellar catalysis (69,70).

20.3 IMMOBILIZATION OF METATHESIS CATALYSTS ON INORGANIC SUPPORTS

20.3.1 Schrock Catalysts Based on Mo, W, and Re

As outlined above, the most straightforward way to immobilize Schrock-type catalysts entails the exchange of one or more alkoxide groups by a suitable linker or by the support itself. The first report on an aluminum oxide-supported Schrock catalyst was published in 1998 by Stelzer et al. (71). They reacted Al_2O_3 with hexafluorobisphenol A and Mo $(N\text{-}2,6\text{-}(2\text{-}Pr)_2\text{-}C_6H_3)(CHCMe_3)((OCM(CF_3)_2)_2$. The proposed structure, though with lack of proof, is shown in Fig. 20.12. In fact, for such systems, Buffon et al. (72) later pointed out the possibility of the formation of Mo-amido-(Mo-NHR) groups through 1,2-addition of surface-hydroxyl groups across the Mo=NR bond. With this supported system, polymers with a higher cis content than obtained with $Mo(N\text{-}2,6\text{-}(2\text{-}Pr)_2\text{-}C_6H_3)(CHCMe_3)((OCMe(CF_3)_2)_2$ were obtained. Later, a slightly modified version of this catalyst was used for continuous metathesis reactions (73). Balcar et al. used a similar approach and immobilized $Mo(N\text{-}2,6\text{-}(2\text{-}Pr)_2\text{-}C_6H_3)(CHCMe_2Ph)((OCM(CF_3)_2)_2$ on MCM-41 (Fig. 20.12). This supported catalyst was then used in CM and ROMP.

The art of preparing and, most importantly, thoroughly characterizing well-defined metal alkylidenes immobilized to inorganic surfaces was demonstrated in 1996 by Herrmann et al. and Basset et al. (74). They exposed trisneopentylnitridomolybdenum(VI) to the surface of silica, that is, to the silanol groups. This leads to the 1,2-cross-addition of the Si-OH groups to the Mo-nitride, yielding a monosilyloxytris(neopentyl)imidomolybdenum(VI) species, which rapidly experiences α-H abstraction-triggered elimination of neopentane and formation of the desired Mo-alkylidene species (Fig. 20.13). The supported catalyst was used in the ROMP of norborn-2-ene allowing for turnover frequencies

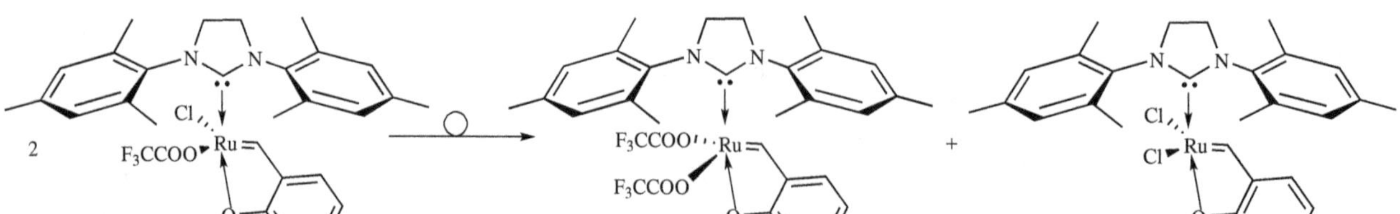

Scheme 20.9 Ligand scrambling in ruthenium alkylidenes containing two different anionic ligands.

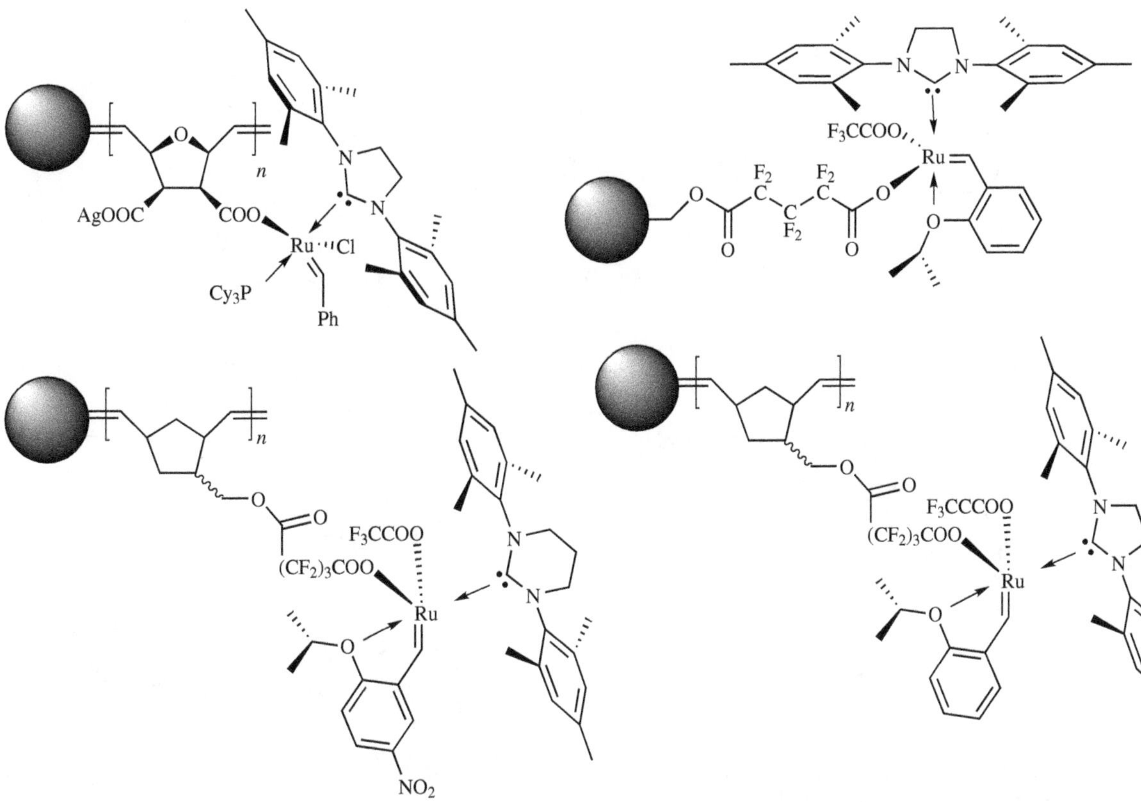

Figure 20.10 Carboxylate- and perfluorocarboxylate-supported versions of second-generation Grubbs and Grubbs–Hoveyda catalysts.

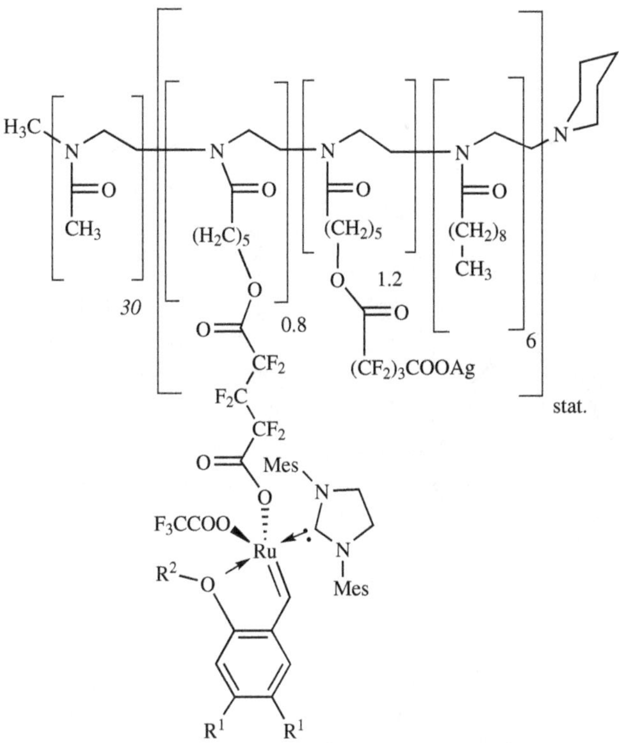

Figure 20.11 Second-generation Grubbs–Hoveyda catalyst bound to an amphiphilic block-copolymer for use in micellar catalysis.

(TOFs) of 0.85 s^{-1}. Further evidence for this chemistry was provided by Copéret et al., who prepared soluble analogs of this catalyst using silsequioxanes (POSS) (75).

In 2001, Copéret and Basset (76) reported on a silica-supported rhenium catalyst (Fig. 20.14) and the system was shown to be active in the self-metathesis of propene. Based on the results obtained with surface-bound Re-alkylidenes, the first well-defined surface-bound Schrock-type catalysts containing one alkyl and one (surface-bound) silyloxy ligand were reported by Copéret, Basset, and Schrock et al. (77). Evidence for the structure of these systems (Fig. 20.15), which were prepared by simply reacting silica with $Mo(N\text{-}2,6\text{-}(2\text{-}Pr)_2\text{-}C_6H_3)(CHCMe_3)(CH_2CMe_3)_2$, was provided by solid-state NMR, FT-IR, and elemental analysis as well as by the preparation of a homogeneous POSS derivative. Initial TOFs (first 5 min) in the self-metathesis of 1-octene and ethyl oleate were 0.06 and 0.04 s^{-1}, respectively. The corresponding tungsten complex was prepared and characterized in a similar manner and showed an initial TOF for propene of 0.14 s^{-1} (78,79). Immobilizing third-generation Schrock catalysts, in which one alkoxide is replaced by an amido ligand such as pyrrolide or diphenylamide (Fig. 20.15), yielded supported systems that allowed for TONs for propene and methyl oleate of 0.6 and 0.15 s^{-1}, respectively. In this context, Schrock, Copéret, and Eisenstein (77,80,81) pointed out the importance of site isolation of the active Mo- and Re-centers by the large

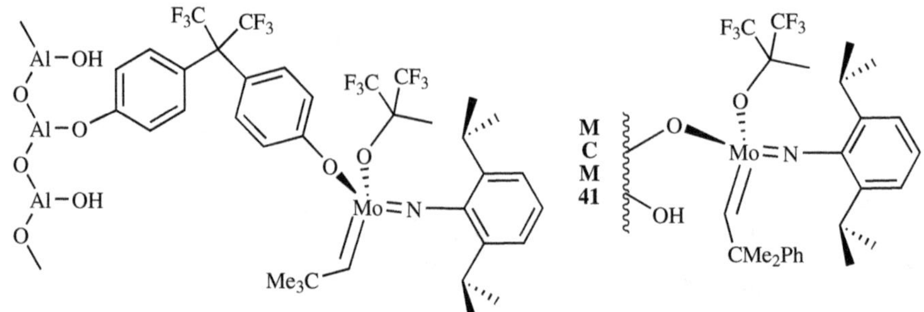

Stelzer et al. Balcar et al.

Figure 20.12 Proposed structure of Schrock catalysts immobilized on Al_2O_3 and MCM41, respectively.

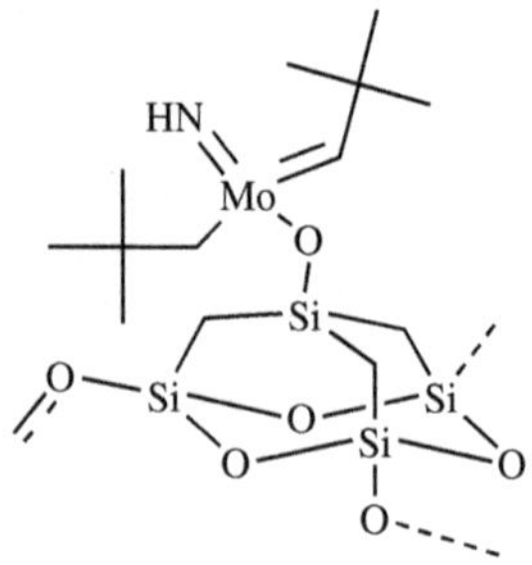

Figure 20.13 A surface-immobilized monosilyloxymolybdenumimidoneopentylneopentylidene metathesis catalyst.

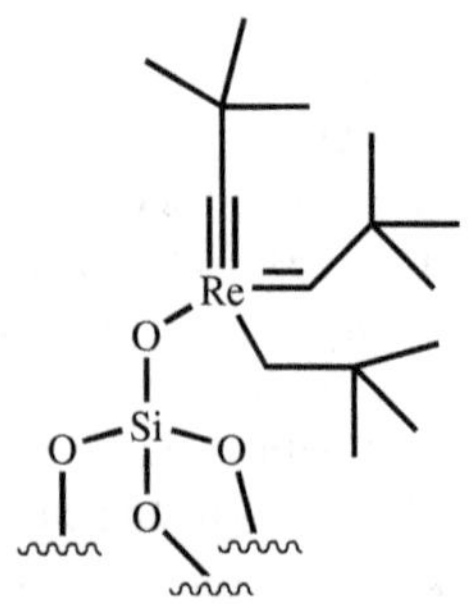

Figure 20.14 Silica-supported rhenium-neopentyl-neopentylidene-neopentylidyne complex.

silyloxy ligand, that is, the support, thereby effectively suppressing dimerization reactions.

Finally, Schrock and Copéret (82) successfully extended the above-described concept to the synthesis of silica-supported monoalkoxy-monosilyloxy-substituted systems (Fig. 20.16). These systems are accessible both from the bis(alkoxide)- and the monoalkoxy-dimethyl pyrrolide-derivatives $Mo(NAr')(CHCMe_2R)(OR')_2$ and $Mo(NAr')(CHCMe_2Ph)((OAr)(Me_2pyr)$. With these systems, TOFs of 0.8 s^{-1} were achieved in propene metathesis. Selectivity on 2-butenes was >99.9%. A similar approach was used for the synthesis of silica-supported molybdenum-

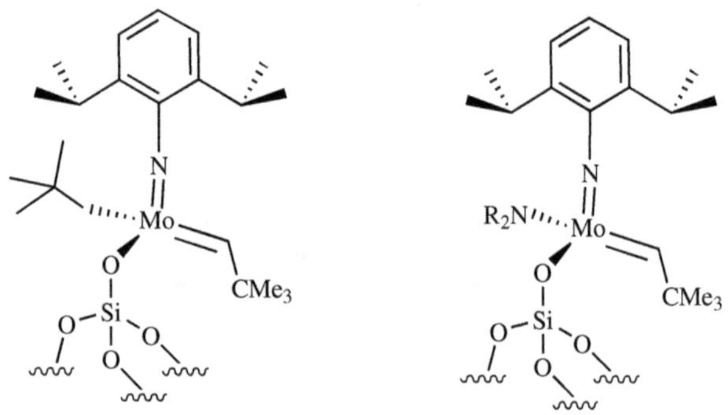

Figure 20.15 Immobilization of a Mo-alkylidene on silica. (a) Mo-imido-alkylosilyloxyalkylidene and (b) Mo-imido-amidosilyloxyalkylidene R_2N = pyrolyl, Ph_2N.

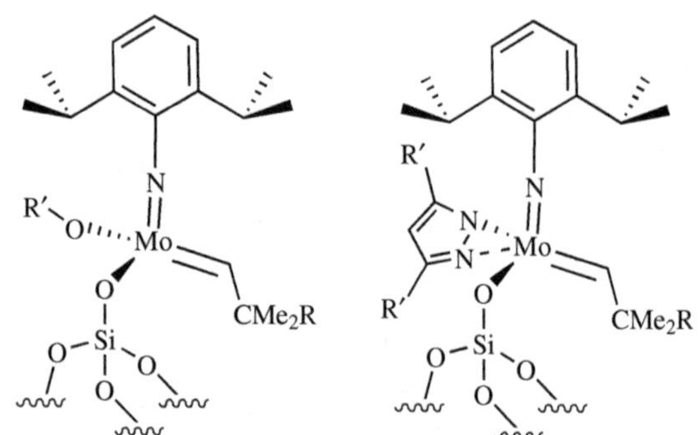

(a) (b)

Figure 20.16 Silica-supported Schrock catalysts bearing one alkoxide and one silyloxy group (a) and one pyrrolide and one silyloxy group (b).

monopyrrolido-monosilyloxy-derivatives (Fig. 20.16) (83); however, no clean formation of this species has been accomplished so far.

In a very recent paper, Basset pointed out the necessity to have tungsten-oxo species such as silica-O-$W(=O)(CH_2CMe_3)_3$, which mimic the industrially used system WO_3/silica, to obtain high and sustained activity in, for

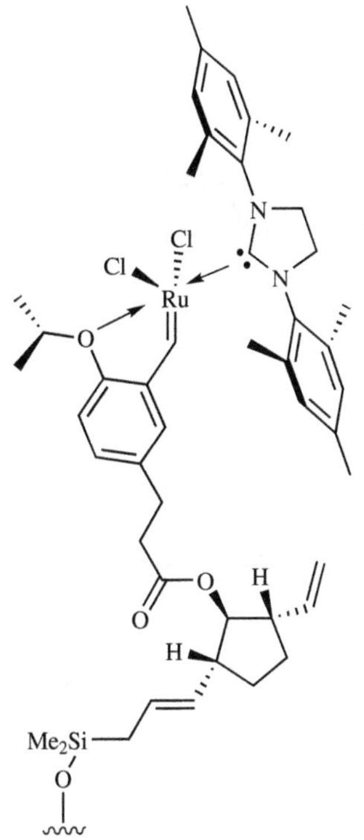

Figure 20.17 Monolithic silica-supported Grubbs–Hoveyda catalyst.

example, propene metathesis (84). However, no alkylidene signals were observed, indicating that only a minor amount of the pre-catalyst converts into the active species under reaction conditions.

20.3.2 Grubbs and Grubbs–Hoveyda Catalysts

20.3.2.1 Immobilization via the Alkylidene Ligand In 2001, Hoveyda et al. (85) reported on the immobilization of well-defined Ru-based metathesis catalyst on a monolithic silica-based support (Fig. 20.17). With this system, a series of high-yield RCM and ROCM reactions were carried out. In selected cases, Ru-contamination in the products was in the 40–60 ppm range; however, for most reactions contaminations in the range of 100–1700 ppm were observed.

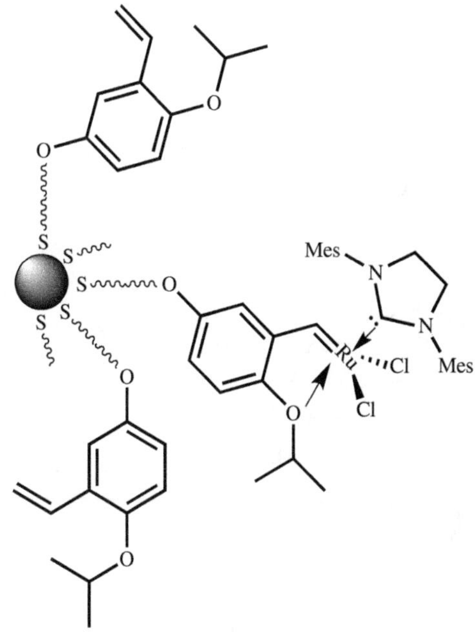

Figure 20.19 Second-generation Grubbs–Hoveyda catalyst supported on a Au-cluster for heterogeneous RCM.

Verpoort et al. reported on a series of MCM-41-supported Grubbs-type catalysts and used them for various ROMP and RCM reactions (Fig. 20.18) (86–88). Lee et al. (89) successfully used gold 'containing decorated with a second-generation Grubbs–Hoveyda catalyst for a series of heterogeneous RCM reactions (Fig. 20.19). An innovative approach to metathesis catalysts immobilized to an inorganic support has been reported by Kirschning and Grela (48). They prepared a dimethylamino-derivative of the second-generation Grubbs–Hoveyda catalyst and immobilized this compound to poly(styrenesulfonic acid) (PS-SO_3H). This polymer-tethered system was then mixed with Raschig rings and converted into an inorganic/organic hybrid material (Scheme 20.10). Used in various RCM reactions, Ru-contaminations were in the range of 21–102 ppm. Rühe et al. (90,91) reported on a second-generation Grubbs–Hoveyda catalyst immobilized to silica according to Scheme 20.11 and pointed out the relevance of the

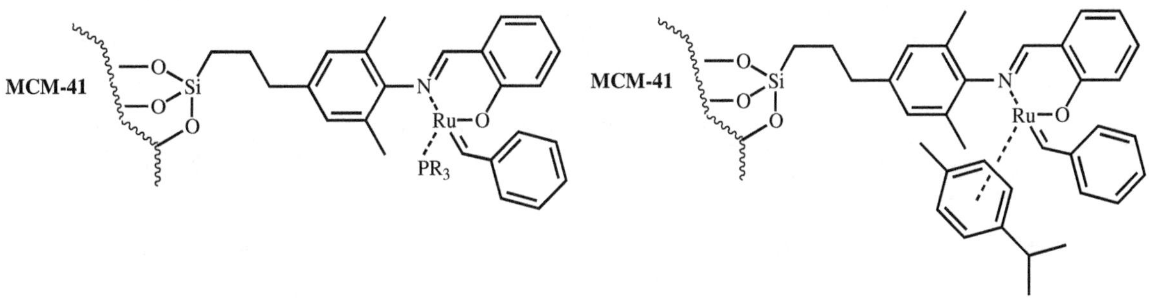

Figure 20.18 MCM-41-supported Schiff-base derived Grubbs-type catalysts.

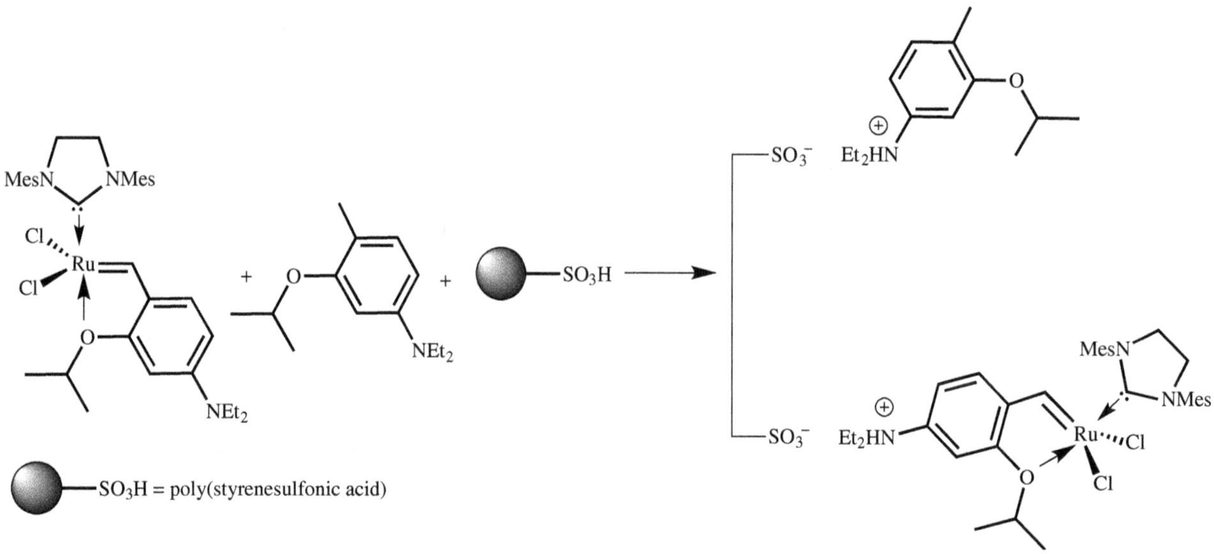

Scheme 20.10 Immobilization of a second-generation Grubbs–Hoveyda-type catalyst on a sulfonated carrier.

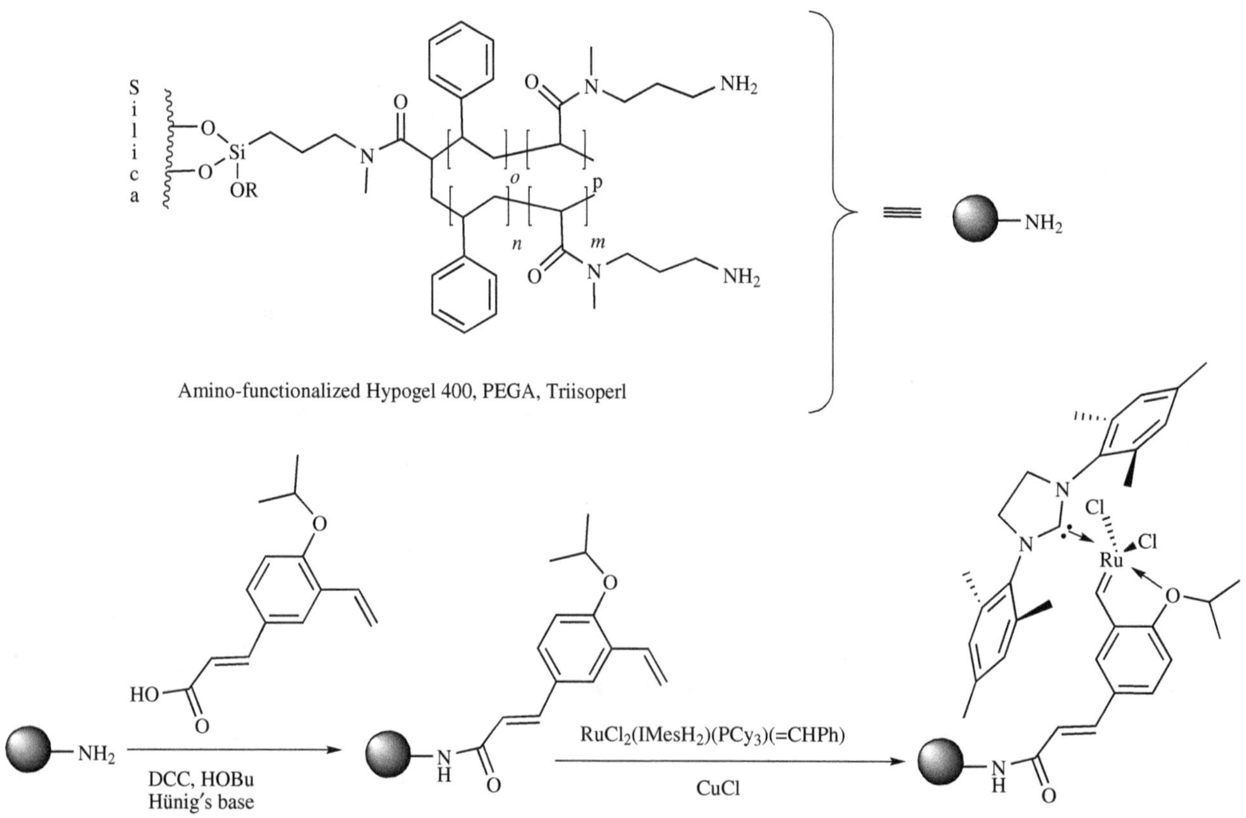

Scheme 20.11 Second-generation Grubbs–Hoveyda catalyst immobilized on a surface-modified silica support.

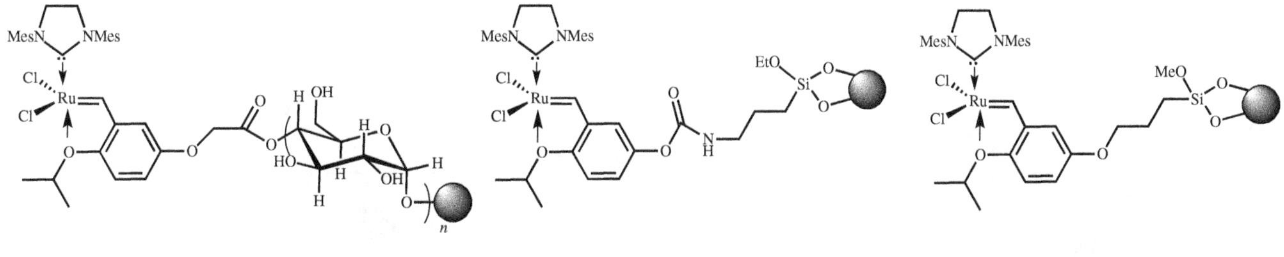

Figure 20.20 Second-generation Grubbs–Hoveyda catalysts immobilized to magnetic nanoparticles, silica, and zirconia.

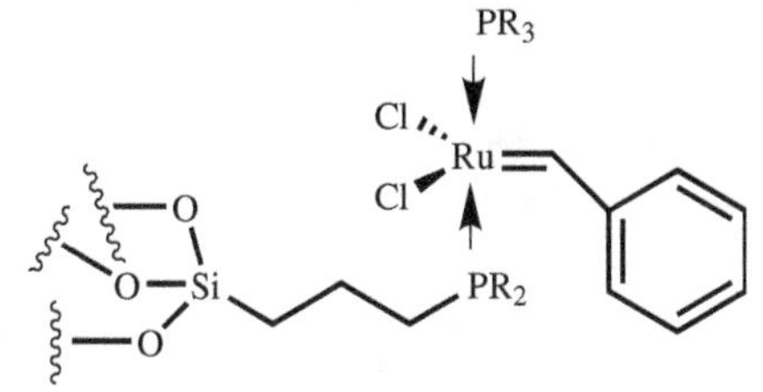

Figure 20.21 First-generation Grubbs catalysts immobilized on phosphonated MCM-41 (R = Ph, cyclohexyl).

silica-support used. In RCM, low Ru-contaminations of the product of about 20 ppm were observed with this system. Using differently substituted 2-(2-propoxystyrenes), a series of groups (92–94) realized silica-, zirconia- and functional nanoparticle-supported Grubbs–Hoveyda type catalysts (Fig. 20.20), which were used in various self-metathesis, CM, and RCM reactions. Ru-leaching into the products was reported to be in the 4 ppm range.

20.3.2.2 Immobilization via the Phosphane or NHC-Ligand Verpoort et al. (95) reported on the immobilization of the first-generation Grubbs catalyst on phosphonated MCM-41 (Fig. 20.21). A series of ROMP and RCM reactions were successfully accomplished. Our group immobilized the second-generation Grubbs catalyst on silica (96). For these purposes, an NHC-ligand with ROMP-active norborn-2-ene groups was subjected to Schrock catalyst triggered ROMP. Termination of the polymerization with triethoxysilylpropylisocyanate yielded the telechelic polymer, which was attached to the silica support. Deprotonation and reaction with the first-generation Grubbs catalysts yielded the desired supported catalyst (Scheme 20.12). The RCM reactions carried out with that support yielded products with Ru-contaminations <5 ppm. Employing amino-functionalized silica, Mingotaud et al. reported on a silica-supported first-generation Grubbs catalyst (Fig. 20.22) and its use in the polymerization of norborn-2-ene. Fürstner et al. chose a classical way of immobilization by using an alkoxysilane-tethered second-generation Grubbs-type catalyst. With this system, a series of RCM reactions were successfully accomplished. Based on that type of chemistry, Grubbs et al. (97) also presented two different catalytic systems (Fig. 20.22), which were reported to allow for RCM reactions with Ru-contaminations in the products <5 ppb. Similarly, Pleixats et al. (98) used a bis(triethoxysilylpropyl)-substituted 1,3-dimesitylimidazo-2-ylidene-based Grubbs-type catalyst for immobilization on silica and used the supported catalysts for various CM, RCM, and ene–yne metathesis reactions (Fig. 20.22). In contrast to other reports, the inorganic carrier was prepared *in situ* via sol–gel synthesis. Finally, Thieuleux et al. (99) proposed another second-generation Grubbs catalyst immobilized on silica (Fig. 20.22). With ethyl oleate, TONS up to 17,000 were reported.

20.3.2.3 Immobilization via Halogen Metathesis In 2003, our group reported on a silica-supported Grubbs–Hoveyda catalyst. The synthesis entailed the reaction of a silica support with norborn-2-enetrichlorosilane, followed by the ROMP-based grafting of 7-oxanorborn-5-ene-2,3-dicarboxylic anhydride. Ring-opening of the anhydride formed the *vic*-dicarboxylate, which was transferred into the corresponding di-silver salt. The latter was then reacted with the second-generation Grubbs–Hoveyda catalyst to yield the desired supported catalyst (Scheme 20.13) (57). In a similar approach, Blechert et al. (100,101) later prepared silica-supported Grubbs–Hoveyda catalysts using trialkoxysilyl-substituted perfluorocarboxylates. With this system, TONs up to 1400 were achieved in RCM.

20.4 NOVEL, ALTERNATIVE APPROACHES

Taking advantage of the high stability and pronounced affinity of second-generation Grubbs–Hoveyda catalysts to silica, a number of groups simply physisorbed this catalyst to silica, MCM-41, and SBA-15 (102–105). Used in various ROMP and CM and RCM reactions, Ru-leaching was strongly dependent on the polarity of the substrates used. Fontaine et al. (105) also outlined the importance of modifying the inner surface of mesoporous SBA-15 in order to increase the longevity of the physisorbed catalyst.

Very recently, SILP technology was successfully applied to metathesis. Wasserscheid et al. (106) presented

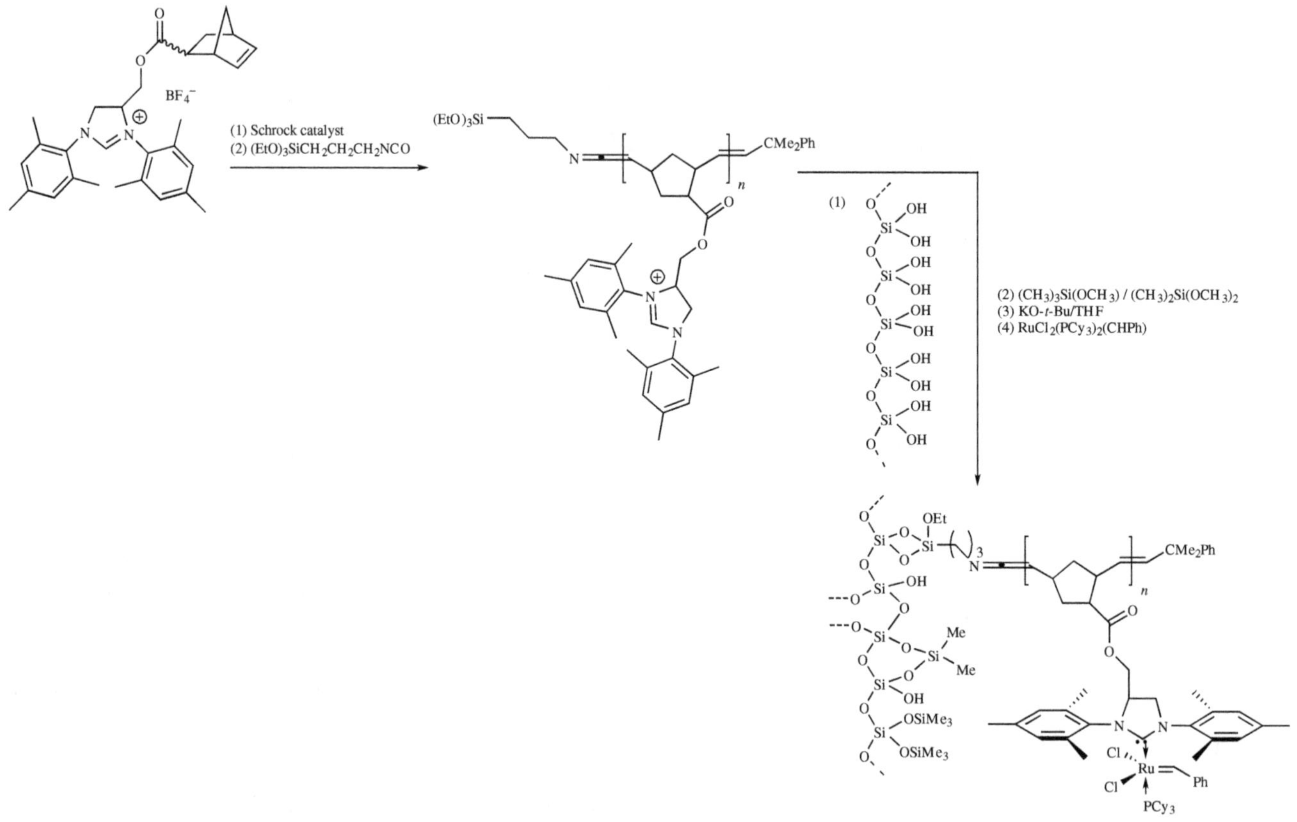

Scheme 20.12 Silica-supported second-generation Grubbs catalyst.

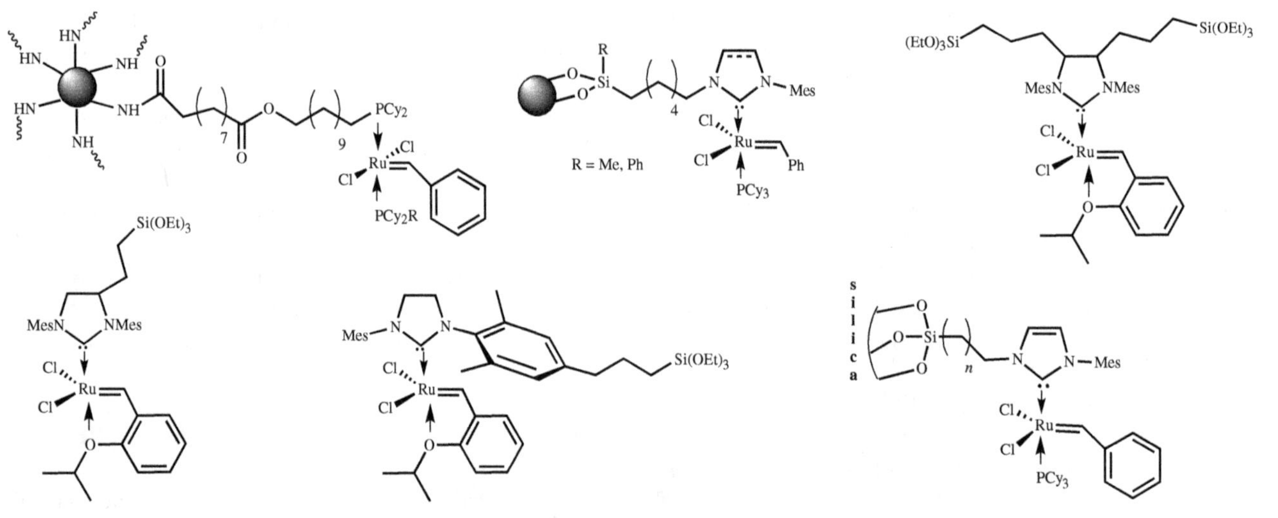

Figure 20.22 Silica-supported first-generation Grubbs catalysts. Top left: immobilization via the phosphine; others: immobilization via the NHC.

an SILP-based system, in which the second-generation Grubbs catalyst was dissolved in an ionic liquid (IL), which was then supported on silica. 1-Olefins such as propene were used as the second gaseous transport phase. This way, a continuous system in which the catalyst remains at a solid phase without being bound to it and the educts are converted to the desired products in the transport phase, was established. Despite its many advantages, the system was restricted to liquid–gas setups because of the good solubility of Grubbs and Grubbs–Hoveyda

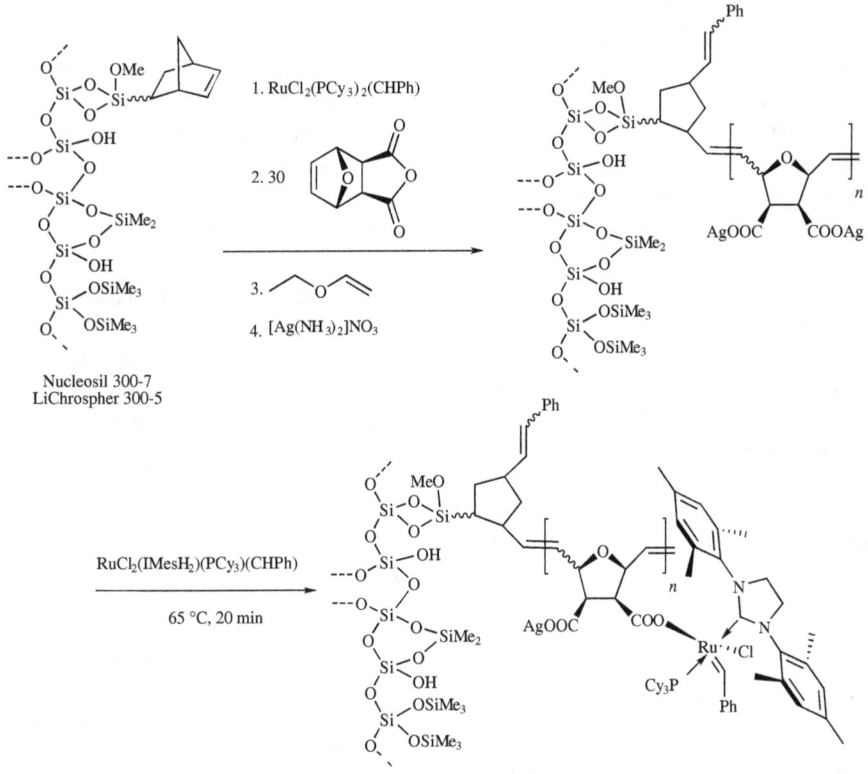

Scheme 20.13 Immobilization of a second-generation Grubbs-catalyst on silica via ROMP of 7-oxanorborn-2-ene-2,3-dicarboxylates followed by chloride metathesis.

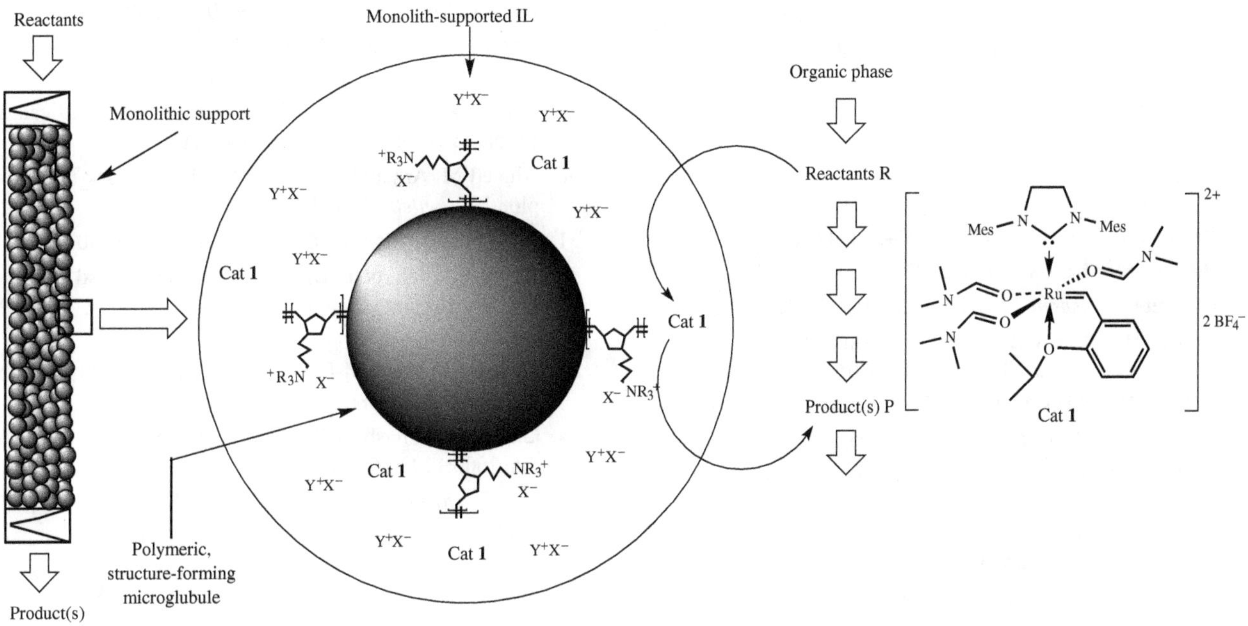

Figure 20.23 Dicationic Ru-alkylidene immobilized on a monolithic support inside an IL and continuous liquid–liquid setup.

catalysts in organic solvents. To overcome this problem, our group presented a novel approach to SILP-immobilized Ru-alkyidenes. A dicationic Ru-catalyst was prepared that selectively dissolved in an IL. This allowed for passing educts in a second organic phase, which is immiscible with the IL, through the monolith (Fig. 20.23). Using this continuous setup, high conversion and excellent selectivity in various CM and RCM reaction were achieved (107).

20.5 SUMMARY/OUTLOOK

With the advent of modern, high-performance metathesis catalysts, an increasing demand for supporting systems evolved. In that context, various supported catalysts, whether based on inorganic or organic supports, have been prepared. However, only a few of them have been really tested under continuous conditions, which certainly presents the preferred industrial setup. While immobilizations via the metal-alkylidene are comparably easy to accomplish, none of these systems allows good catalyst recovery. With Ru-based systems, this is better accomplished by immobilization via the NHC or via chloride metathesis. With Mo-based Schrock catalysts, numerous systems, whether immobilized through the alkoxides or the *N*-arylimido ligand, have been developed and successfully used.

Particularly, with inorganic supports, impressive TONs have already been observed with Mo-, W-, and Re-based catalysts, though in most cases only for simple alkenes. With the comparably low TONs that are observed with metathesis catalysts and functional olefins in general, reloadable supports, for example, as realized via SILP technology, are certainly the way to go.

REFERENCES

1. Buchmeiser MR. *New J Chem* 2004;28:549–557.
2. Buchmeiser MR. *Chem Rev* 2009;109:303–321.
3. Copéret C. *Beilstein J Org Chem* 2011;7:13–21.
4. Clavier H, Grela K, Kirschning A, Mauduit M, Nolan SP. *Angew Chem Int Ed* 2007;46:6786–6801.
5. Sommer WJ, Weck M. *Coord Chem Rev* 2007;251:860–873.
6. Buchmeiser MR. *Catal Today* 2005;105:612–617.
7. Hultzsch KC, Jernelius JA, Hoveyda AH, Schrock RR. *Angew Chem Int Ed* 2002;41:589–593.
8. Kröll RM, Schuler N, Lubbad S, Buchmeiser MR. *Chem Commun* 2003:2742–2743.
9. Buchmeiser MR. *Polymer* 2007;48:2187–2198.
10. Beckert S, Stallmach F, Bandari R, Buchmeiser MR. *Macromolecules* 2010;43:9441–9446.
11. Mayr M, Wang D, Kröll R, Schuler N, Prühs S, Fürstner A, Buchmeiser MR. *Adv Synth Catal* 2005;347:484–492.
12. Dolman SJ, Hultzsch KC, Pezet F, Teng X, Hoveyda AH, Schrock RR. *J Am Chem Soc* 2004;126:10945–10953.
13. Hoveyda AH, Schrock RR. *Chem Eur J* 2001;7:945–950.
14. Schrock RR, Lee J-K, O'Dell R, Oskam JH. *Macromolecules* 1995;28:5933–5940.
15. Buchmeiser MR, Schuler N, Kaltenhauser G, Ongania K-H, Lagoja I, Wurst K, Schottenberger H. *Macromolecules* 1998;31:3175–3183.
16. Nguyen S, Grubbs RH. *J Organomet Chem* 1995;497: 195–200.
17. Corrêa da Costa R, Gladysz JA. *Chem Commun* 2006: 2619–2621.
18. Mennecke K, Grela K, Kunz U, Kirschning A. *Synlett* 2005:2948–2952.
19. Schürer SC, Gessler S, Buschmann N, Blechert S. *Angew Chem Int Ed* 2000;39:3898–3901.
20. Mayr M, Mayr B, Buchmeiser MR. *Angew Chem Int Ed* 2001;40:3839–3842.
21. Mayr M, Mayr B, Buchmeiser MR. *Des Monomers Polym* 2002;5:325–337.
22. Buchmeiser MR. *J Mol Catal A: Chem* 2002;190:145–158.
23. Randl S, Buschmann N, Connon SJ, Blechert S. *Synlett* 2001:1547–1450.
24. Gallivan JP, Jordan JP, Grubbs RH. *Tetrahedron Lett* 2005; 46:2577–2580.
25. Hong SH, Grubbs RH. *J Am Chem Soc* 2006;128: 3508–3509.
26. Sommer WJ, Weck M. *Adv Synth Catal* 2006;348: 2101–2113.
27. Al-Hashimi M, Hongfa C, George B, Bazzi HS, Bergbreiter DE. *J Polym Sci A: Polym Chem* 2012;50:3954–3959.
28. Ahmed M, Barrett AGM, Braddock DC, Cramp SM, Procopiou PA. *Tetrahedron Lett* 1999;40:8657–8662.
29. Barrett, AGM, Braddock, DC, Polymer supported catalysts for olefin metathesis. WO 01/72421 A1.
30. Ahmed M, Arnauld T, Barrett AGM, Braddock DC, Procopiou PA. *Synlett* 2000:1007–1009.
31. Jafarpour L, Nolan SP. *Org Lett* 2000;2:4075–4078.
32. Nolan SP, Jafarpour L, Simply assembled and recyclable polymer-supported olefin metathesis catalysts. US 6.921. 736 B1.
33. Jafarpour L, Heck M-P, Baylon C, Lee H-M, Mioskowski C, Nolan SP. *Organometallics* 2002;21:671–679.
34. Garber SB, Kingsbury JS, Gray BL, Hoveyda AH. *J Am Chem Soc* 2000;122:8168–8179.
35. Yao Q. *Angew Chem Int Ed* 2000;39:3896–3898.
36. Dowden J, Savovic J. *Chem Commun* 2001:37–38.
37. Grela K, Tryznowski M, Bieniek M. *Tetrahedron Lett* 2002;43:9055–9059.
38. Connon SJ, Dunne AM, Blechert S. *Angew Chem Int Ed* 2003;42:1900–1923.
39. Varray S, Lazaro R, Martinez J, Lamaty F. *Organometallics* 2003;22:2426–2435.

40. Blechert S, Connon S, Metathesekatalysatoren, Bayer Chemicals AG. EP 1394190A1.
41. Yao Q, Motta AR. *Tetrahedron Lett* 2004;45:2447–2451.
42. Zarka MT, Nuyken O, Weberskirch R. *Macromol Rapid Commun* 2004;25:858–862.
43. Hongfa C, Tian J, Bazzi HS, Bergbreiter DE. *Org Lett* 2007;9:3259–3261.
44. Koc F, Michalek F, Rumi L, Bannwarth W, Haag R. *Synthesis* 2005;2005:3362–3372.
45. Bannwarth W, Michalek F, Rühe J, Supported transition metal complexes and use thereof in catalysis of olefin metathesis. WO 2007/065907 A1.
46. Chen S-W, Kim JH, Song CE, Lee S-g. *Org Lett* 2007; 9:3845–3848.
47. Lipshutz BH, Ghorai S. *Org Lett* 2009;11:705–708.
48. Michrowska A, Mennecke K, Kunz U, Kirschning A, Grela K. *J Am Chem Soc* 2006;128:13261–13267.
49. Kingsbury JS, Hoveyda AH. *J Am Chem Soc* 2005; 127:4510–4517.
50. Su HL, Hongfa C, Bazzi HS, Bergbreiter DE. *Macromol Symp* 2010;297:25–32.
51. Vorfalt T, Wannowius K-J, Plenio H. *Angew Chem Int Ed* 2009;148:5191–5194.
52. Yao Q, Zhang Y. *J Am Chem Soc* 2004;126:74–75.
53. Nieczypor P, Buchowicz W, Meester WJN, Rutjes FPJT, Mol JC. *Tetrahedron Lett* 2001;42:7103–7105.
54. Tanaka K, Böhm VPW, Chadwick D, Roeper M, Braddock DC. *Organometallics* 2006;25:5696–5698.
55. Braddock DC, Tanaka K, Chadwick D, Böhm VPW, Roeper M. *Tertrahedron Lett* 2007;48:5301–5303.
56. Buchmeiser MR, Ahmad I, Gurram V, Kumar PS. *Macromolecules* 2011;44:4098–4106.
57. Krause JO, Lubbad S, Nuyken O, Buchmeiser MR. *Adv Synth Catal* 2003;345:996–1004.
58. Sinner F, Buchmeiser MR. *Macromolecules* 2000;33: 5777–5786.
59. Sinner F, Buchmeiser MR. *Angew Chem Int Ed* 2000; 39:1433–1436.
60. Krause JO, Wurst K, Nuyken O, Buchmeiser MR. *Chem Eur J* 2004;10:777–784.
61. Krause JO, Mayr M, Lubbad S, Nuyken O, Buchmeiser MR, e-Polymers-conference papers section. 2003, P_007.
62. Krause JO, Lubbad SH, Nuyken O, Buchmeiser MR. *Macromol Rapid Commun* 2003;24:875–878.
63. Krause JO, Lubbad S, Mayr M, Nuyken O, Buchmeiser MR. *Polym Prepr (Am Chem Soc, Div Polym Chem)* 2003;44:790–791.
64. Krause JO, Nuyken O, Buchmeiser MR. *Chem Eur J* 2004; 10:2029–2035.
65. Kumar PS, Wurst K, Buchmeiser MR. *Organometallics* 2009; 28:1785–1790.
66. Yang L, Mayr M, Wurst K, Buchmeiser MR. *Chem Eur J* 2004;10:5761–5770.
67. Yang H, Zhang G, Hong X, Zhu Y. *J Mol Catal A: Chemical* 2004;210:143–148.
68. Halbach TS, Mix S, Fischer D, Maechling S, Krause JO, Sievers C, Blechert S, Nuyken O, Buchmeiser MR. *J Org Chem* 2005;70:4687–4694.
69. Krause JO, Zarka MT, Anders U, Weberskirch R, Nuyken O, Buchmeiser MR. *Angew Chem Int Ed* 2003;42:5965–5969.
70. Kotre T, Zarka MT, Krause JO, Buchmeiser MR, Weberskirch R, Nuyken O. *Macromol Symp* 2004;217:203–214.
71. Preishuber-Pflügl P, Buchacher P, Eder E, Schitter RM, Stelzer F. *J Mol Catal A: Chem* 1998;133:151–158.
72. Wolke SI, Buffon R. *J Mol Catal A: Chem* 2000; 160:181–187.
73. Preishuber-Pflügl P, Eder E, Stelzer F, e-Polymers. 2002, 045.
74. Herrmann WA, Stumpf AW, Priermeier T, Bogdanovic S, Dufaud V, Basset J-M. *Angew Chem Int Ed* 1996; 35:2803–2805.
75. Blanc F, Chabanas M, Copéret C, Fenet B, Herdweck E. *J Organomet Chem* 2005;690:5014–5026.
76. Chabanas M, Baudouin A, Copéret C, Basset J-M. *J Am Chem Soc* 2001;123:2062–2063.
77. Blanc F, Copéret C, Thivolle-Cazat J, Basset J-M, Lesage A, Emsley L, Sinha A, Schrock RR. *Angew Chem Int Ed* 2006;45:1216–1220.
78. Rhers B, Salameh A, Baudouin A, Quadrelli EA, Taoufik M, Copéret C, Lefebvre F, Basset J-M, Solans-Monfort X, Eisenstein O, Lukens WW, Lopez LPH, Sinha A, Schrock RR. *Organometallics* 2006;25:3554–3557.
79. Copéret C. *Dalton Trans* 2007:5498–5504.
80. Solans-Monfort X, Filhol JS, Coperet C, Eisenstein O. *New J Chem* 2006;30:842–850.
81. Sinha A, Lopez PPH, Schrock RR, Hock AS, Müller P. *Organometallics* 2006;25:1412–1423.
82. Rendón N, Berthoud R, Blanc F, Gajan D, Maishal T, Basset J-M, Copéret C, Lesage A, Emsley L, Marinescu SC, Singh S, Schrock RR. *Chem Eur J* 2009;15:5083–5089.
83. Gajan D, Rendón N, Wampler KM, Basset J-M, Copéret C, Lesage A, Emsley L, Schrock RR. *Dalton Trans* 2010;39:8547–8551.
84. Mazoyer E, Merle N, de Mallmann A, Basset JM, Berrier E, Delevoye L, Paul J-F, Nicholas CP, Gauvin RM, Taoufik M. *Chem Commun* 2010;46:8944–8946.
85. Kingsbury JS, Garber SB, Giftos JM, Gray BL, Okamoto MM, Farrer RA, Fourkas JT, Hoveyda AH. *Angew Chem Int Ed* 2001;40:4251–4256.
86. De Clercq B, Lefebvre F, Verpoort F. *New J Chem* 2002; 26:1201–1208.
87. De Clercq B, Lefebvre F, Verpoort F. *Appl Catal A* 2003; 247:345–364.
88. Walter F, De Clerqu B, Metal Complexes Useful in Metathesis and Other Reactions. US 2005/0043541.
89. Lee BS, Namgoong SK, Lee S. *Tetrahedron Lett* 2005; 46:4501–4503.
90. Michalek F, Mädge D, Rühe J, Bannwarth W. *Eur J Org Chem* 2006:577–581.

91. Michalek F, Mädge D, Rühe J, Bannwarth W. *J Organomet Chem* 2006;691:5172–5180.
92. Lim J, Lee SS, Riduan SN, Ying JY. *Adv Synth Catal* 2007;349:1066–1076.
93. Keraani A, Fischmeister C, Renouard T, Le Floch M, Baudry A, Bruneau C, Rabiller-Baudry M. *J Mol Catal A: Chem* 2012;357:73–80.
94. Yinghuai Z, Kuijin L, Huimin N, Chuanzhao L, Stubbs LP, Siong CF, Muihua T, Peng SC. *Adv Synth Catal* 2009;351:2650–2656.
95. Melis K, De Vos D, Jacobs P, Verpoort F. *J Mol Catal A: Chem* 2001;169:47–56.
96. Mayr M, Buchmeiser MR, Wurst K. *Adv Synth Catal* 2002; 344:712–719.
97. Allen DP, Van Wingerden MM, Grubbs RH. *Org Lett* 2009;11:1261–1264.
98. Monge-Marcet A, Pleixats R, Cattoën X, Wong Chi Man M. *J Mol Catal A: Chem* 2012;357:59–66.
99. Karamé I, Boualleg M, Camus J-M, Maishal TK, Alauzun J, Basset J-M, Copéret C, Corriu RJP, Jeanneau E, Mehdi A, Reyé C, Veyre L, Thjieuleux C. *Chem Eur J* 2009;15:11820–11823.
100. Vehlow K, Maechling S, Köhler K, Blechert S. *J Organomet Chem* 2006;691:5267–5277.
101. Vehlow K, Köhler K, Blechert S, Dechert S, Meyer F. *EurJ Inorg Chem* 2005:2727–2736.
102. Van Berlo B, Houthoofd K, Sels BF, Jacobs PA. *Adv Synth Catal* 2008;350:1949–1953.
103. Yang H, Ma Z, Wang Y, Wang Y, Fang L. *Chem Commun* 2010;46:8659–8661.
104. Balcar H, Shinde T, Zilková N, Bastl Z. *Beilstein J Org Chem* 2011;7:22–28.
105. Staub H, Guillet-Nicolas R, Even N, Kayser L, Kleitz F, Fontaine F-G. *Chem Eur J* 2011;17:4254–4265.
106. Scholz J, Loekman S, Szesni N, Hieringer W, Görling A, Haumann M, Wasserscheid P. *Adv Synth Catal* 2011;353:2701–2707.
107. Autenrieth B, Frey W, Buchmeiser MR. *Chem Eur J* 2012; 18:14069–14078.

21

OLEFIN METATHESIS IN WATER AND AQUEOUS MEDIA

BRUCE H. LIPSHUTZ
Department of Chemistry, University of California, Santa Barbara, CA

SUBIR GHORAI
Catalysis R&D, Sigma-Aldrich Chemical Corp., Milwankee, WI

Olefin metathesis has emerged as one of the most widely used transformations in organic synthesis for the formation of carbon–carbon double bonds (1). Like many other organometallic reactions, olefin metathesis is usually carried out under anhydrous conditions and in chlorinated solvents. However, owing to environmental concerns, there is increasing attention being paid to the use of alternative media, and water in particular, as a reaction medium (2).

Grela and coworkers described the performance of unmodified, commercially available olefin metathesis catalysts (Fig. 21.1; **2** and **3**) in water in the absence of surfactants (3). The ultrasonication of water-insoluble substrates and catalysts in water allowed the formation of emulsions in which smooth catalytic ring-closing metathesis (RCM), cross-metathesis (CM), and enyne metathesis took place in excellent yields (Fig. 21.2). Although it is possible to form five- and six-membered rings by using acoustic emulsification, albeit where relatively high catalyst loadings and high temperatures are required, attempts to close larger rings by this method failed (3).

In 2002, Davis and Sinou (4) studied the effect of several phase-transfer catalysts of both ionic (e.g., sodium dodecyl sulfate (SDS), sodium dodecyl sulfonate ($SDSO_3Na$), cetyltrimethylammonium hydrogen sulfate ($CTAHSO_4$), N-hexadecyl-N,N-dimethyl-3-ammonio-1-propanesulfonate (HDAPS), N-dodecyl-N,N-dimethyl-3-ammonio-1-propanesulfonate (DDAPS), and nonionic type surfactants (e.g., Brij 35, Tween 40) on RCM reactions in water. Relatively high percentages (ca 5 wt%) of amphiphile were employed, although in this early study use of the less reactive catalyst **1** precluded formation of tri- and tetrasubstituted olefinic products. The conclusion from this study was that a surfactant may not be essential for RCM reactions in water. On the other hand, Mingotaud and coworkers (5) described the use of micellar catalysis on RCM reactions of **9** in water, using ionic surfactant dodecyltrimethylammonium chloride (DTAC) and catalyst **8**, where an accelerating effect was clearly observed compared to the corresponding homogeneous reaction in CH_2Cl_2 (Fig. 21.3). It was also found that the substrate scope is limited, and the CM reaction was not attempted.

An impressive number of challenging olefin CM reactions in water in the presence of the commercially available catalysts **2** and **5** were reported by Lipshutz and coworkers (6). The key to success was the use of the nonionic amphiphile PTS (Fig. 21.4; **10**, polyoxyethanyl α-tocopheryl sebacate), whereas results using several alternative surfactants (e.g., TPGS, D-alpha tocopheryl polyethylene glycol 1000 succinate) were far less impressive.

A series of difficult CM and ROM–CM reactions involving water-insoluble alkene and electron-deficient substrates in the presence of catalyst **2** (2 mol%) and PTS (2.5 wt%) proceeded at room temperature (RT), in air, with high efficiency and very high selectivity (Fig. 21.5) (6). This influential technology was also used to effect RCM reactions of lipophilic substrates in the presence of catalyst **2** in water, including the formation of five- to seven-membered rings, as well as trisubstituted alkenes (7). The formation of product **13**, which contains a tetrasubstituted C–C double bond, is particularly noteworthy; this reaction is considered to be very challenging under any set of conditions (Fig. 21.5).

Olefin Metathesis: Theory and Practice, First Edition. Edited by Karol Grela.

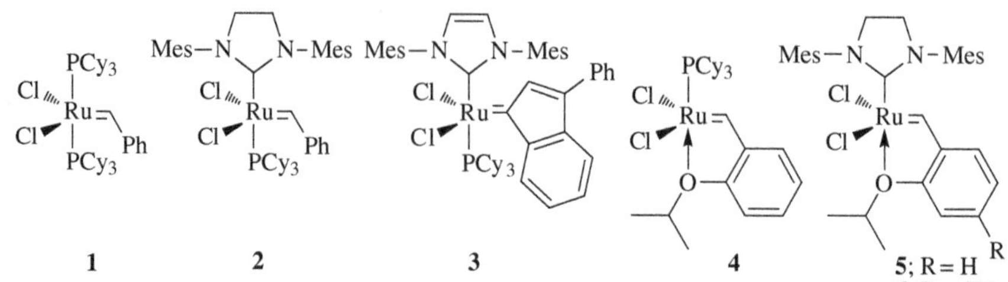

1 2 3 4 5; R = H 6; R = CN 7; R = O^{i}Pr

Figure 21.1 Representative water-insoluble olefin metathesis catalysts.

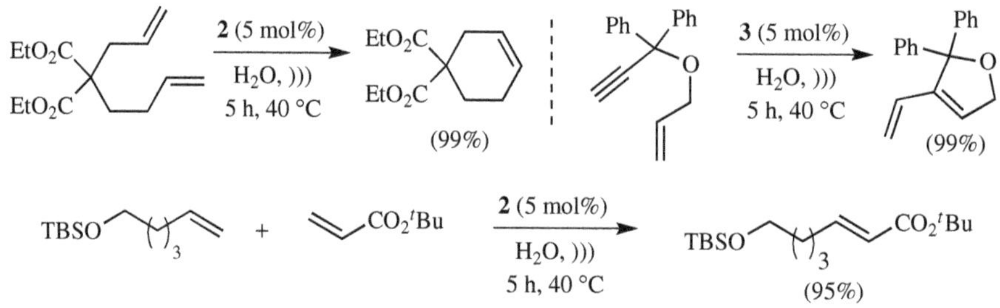

Figure 21.2 Metathesis reactions in ultrasonicated aqueous emulsions. ")))" indicates ultrasound.

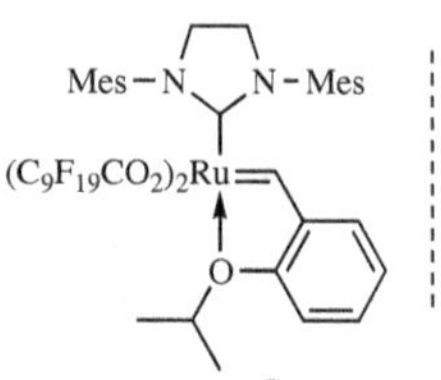

8

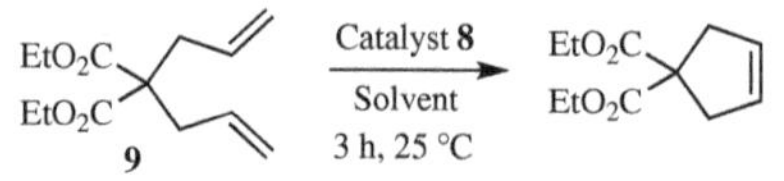

Solvent	8 (mol%)	Conversion (%)
CH_2Cl_2	3	17
H_2O^a	0.5	91

^{a} 0.05 M DTAC.

Figure 21.3 Micellar effect on an RCM reaction using aqueous DTAC.

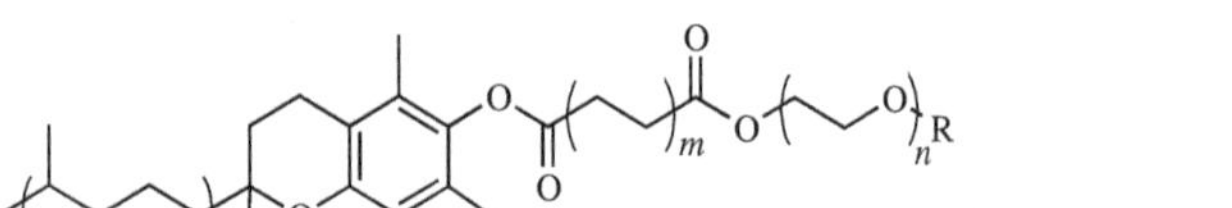

10, PTS; $m = 4$, $n \sim 13$, R = H
11, TPGS-750-M; $m = 1$, $n \sim 19$, R = Me

Figure 21.4 Structure of the nonionic amphiphiles PTS and TPGS-750-M.

Representative CM Reaction in PTS/Water Using Catalyst 2 (Fig. 21.5, Compound 12) (6)

N-Carbobenzyloxy-L-phenylalanine 10-undecenyl ester (229 mg, 0.51 mmol), *tert*-butyl acrylate (202 mg, 1.57 mmol), and catalyst **2** (9.1 mg, 0.0107 mmol) were sequentially added to a Teflon-coated, stir bar containing Biotage 2–5 ml microwave reactor vial at RT and sealed with a septum. An aliquot of 2.5% PTS/H_2O (1.0 ml) was added via syringe, and the resulting emulsion was allowed to stir at RT for 12 h. The homogeneous reaction mixture was then diluted with EtOAc (5 ml) and filtered through a bed of silica gel layered over Celite, and the bed was washed (3 × 10 ml) with EtOAc. The volatiles were removed *in vacuo* to afford the crude material, which was subsequently purified by flash chromatography on silica gel (eluting with 6% EtOAc/hexanes) to yield the product **12** as a colorless oil (269 mg, 96%).

Very recently, Lipshutz and coworkers (8,9) described both the effects of salt (NaCl, CuI) and pH on metathesis reactions in PTS–water mixtures. For example, while racemic citronellol *tert*-butyldimethylsilyl (TBS) ether was smoothly converted to the corresponding enoate **14** over 12 h (84%), the identical reaction run in aqueous PTS containing 3 M NaCl afforded essentially the same yield (86%) but in half the time (8), rationalized due to the complete alteration of the nature of the nanoparticles as verified by cryo-TEM (transmission electron microscopy) analyses (Fig. 21.6). CuI (3 mol%) has also been found to enhance the extent of conversion to product **15** in 80% yield, whereas in the absence of this additive, the desired product was obtained in only 55% yield (Fig. 21.6) (9). CuI has a catalyst stabilizing effect due to iodide replacement of chloride on the metal, while Cu(I) has phosphine-sequestering properties that enhance reactivity of the Grubbs-2 catalyst (**2**). Addition of small amounts of $KHSO_4$ (0.02 M) to the PTS–water mixture had an even greater impact on olefin CM reactions; for example, in the case of product **16**, the acidic conditions

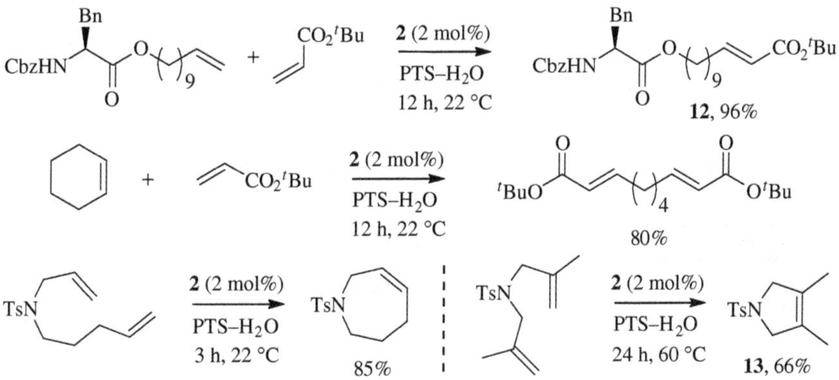

Figure 21.5 Metathesis reactions in the presence of the nonionic amphiphile PTS.

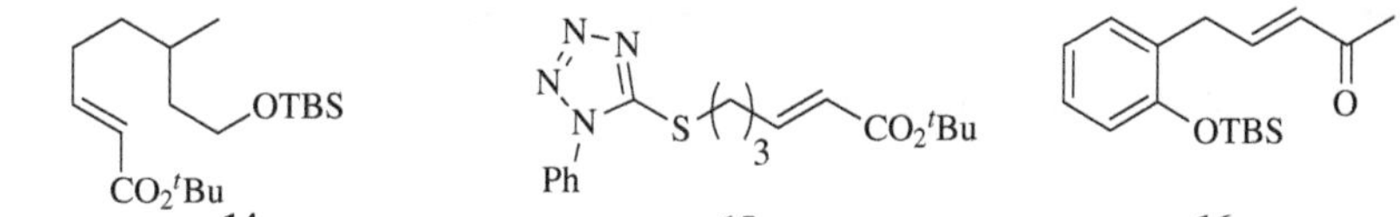

14	**15**	**16**
PTS (12 h): 84%	PTS: 55%	PTS(12 h): 74%
PTS–3 M NaCl (6h): 86%	PTS–CuI: 80%	PTS–$KHSO_4$ (4 h): 91%

Figure 21.6 Effect of salt and pH on olefin CM reactions in aqueous PTS.

(pH 2) drove this typically challenging reaction involving methyl vinyl ketone (MVK) to almost full conversion (8), using the Grubbs-2 catalyst (**2**), which has its phosphine exposed to protons in the water as it undergoes, along with all the other species present in the mixture, exchange between nanomicelles (Fig. 21.6). Very recently, the Lipshutz group (10) introduced a newly engineered, second generation, environmentally benign designer surfactant TPGS-750-M (Fig. 21.4; **11**) that shows promising results that are equal, or superior to, those obtained using PTS.

Grubbs and coworkers (11) explored a number of modified catalysts that impart water solubility, thereby allowing reactions using water as the reaction medium (Fig. 21.7). The reaction of diene **21** was found to proceed as a homogeneous solution in methanol/water with good to excellent conversion using 5 mol% of catalyst **17** (Fig. 21.8) (12). When the catalyst loading of **17** was increased to 10 mol% using water, comparable conversion to that in methanol was achieved. Substrate scope, however, is limited to those partners with water solubility; moreover, catalyst **17** was found to be inactive in CM reactions under these conditions.

In early 2006, Grubbs (13) described an improved, poly(ethylene glycol) (PEG)-tagged Hoveyda-type catalyst, **18** (Fig. 21.7), that is active and stable in water, and showed good activity in ring opening metathesis polymerization (ROMP) and RCM reactions of water-soluble dienes in an aqueous medium. Other water-soluble catalysts containing polar quaternary ammonium groups in the benzylidene fragment were later reported by Grubbs (**19**) (14) and Grela (**20**) (15) (Fig. 21.7). Although **18–20** are very active in

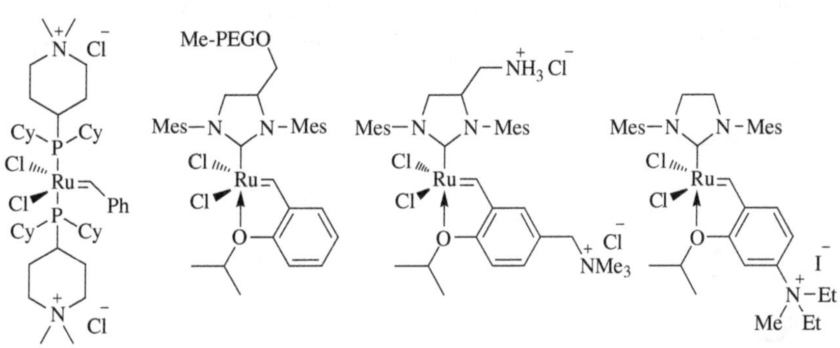

17; Grubbs **18**; Grubbs **19**; Grubbs **20**; Grela

Figure 21.7 Representative water-soluble catalysts for olefin metathesis.

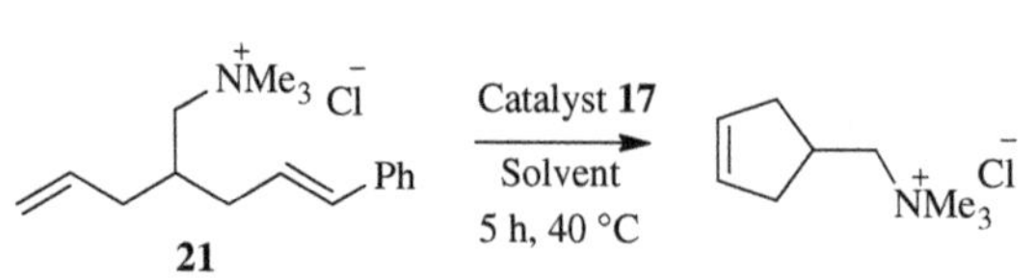

Solvent	17 (mol%)	Conversion (%)
MeOH	5	90
H_2O	5	60
H_2O	10	90

Figure 21.8 CM reaction of an acyclic, charged diene using a water-soluble catalyst.

Substrate	Product	Catalyst (mol%)	*t* (h)	*T* (°C)	Conversion (%)
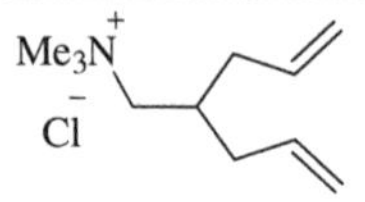	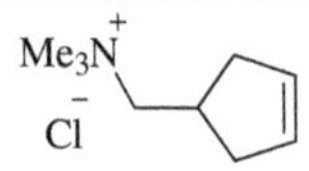	**18** (5)	12	30	>95
		19 (5)	0.5	30	>95
		20 (5)	5	25	99
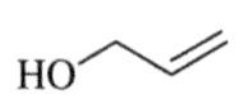	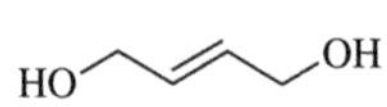	**18** (5)	12	45	>95
		19 (5)	6	45	69 (12)[a]
		20 (2.5)	3.5	25	>99

(a) The yield of the by-product propionaldehyde.

Figure 21.9 Selected metathesis reactions mediated by water-soluble catalysts.

RCM of water-soluble charged dienes as well as CM of allylic alcohols, the undesired isomerization of allyl alcohol to propionaldehyde was observed with catalyst **19**, which was not observed using either **18** or **20** (Fig. 21.9). Catalyst **20** gave low-to-moderate yields for olefin CM reactions with water-insoluble substrates (15).

General Procedure for RCM and CM Reactions with Catalyst 19 (Fig. 21.9) (14)

In an N_2-filled dry-box, **19** (4.8 mg, 6.0 μmol, 0.05 equiv) was weighed into a 1 dram vial. The vial was sealed with a septa-cap and removed from the dry-box. A screwcap NMR tube was also sealed with a septa-cap and removed from the dry-box. The seals of both the vial and the NMR tube were reinforced with Teflon tape. 0.6 ml of a 0.2 M solution of substrate in degassed D_2O was added to the vial, and full dissolution of **19** accelerated with brief (~5–60 s) sonication. (Substrate solution was prepared under inert conditions with degassed D_2O and stored under argon.) The solution was transferred to the NMR tube by gas-tight syringe, and the reaction heated to 30 °C. The reaction was monitored by ^{1}H-NMR spectroscopy.

Very recently, Grela has developed a similar type of quaternary ammonium chloride-tagged catalyst **22**, which is more water soluble than **20** and shows promising activity in the isomerization of (*Z*)-**23** (Fig. 21.10). However, **22** was found to afford lower yields for olefin CM and enyne reactions with water soluble substrates (16).

Lipshutz and Ghorai (17) have recently developed a conceptually new water-soluble metathesis catalyst **24** (Fig. 21.11), where a Ru carbene has been covalently linked to an environmentally benign surfactant PQS (polyethylene glycol–ubiquinol–sebacate). Water-insoluble reactants undergo reaction within its nanometer micelles, formed spontaneously upon dissolution. Using this modified catalyst **24**, RCM reactions were found to give excellent yields at RT with moderate catalyst loading (2 mol%). Unlike the free Grubbs–Hoveyda-1 catalyst (**4**), the PQS derived catalyst, which is engineered to remain in the water, could be recycled without removal from the reaction flask. Ten cycles were demonstrated with minimal loss in reactivity. Elemental analysis of the isolated product revealed that only trace amounts of ruthenium had been lost from the aqueous solution (ca 1.6 ppm) (17). Interestingly, the reaction could be carried out

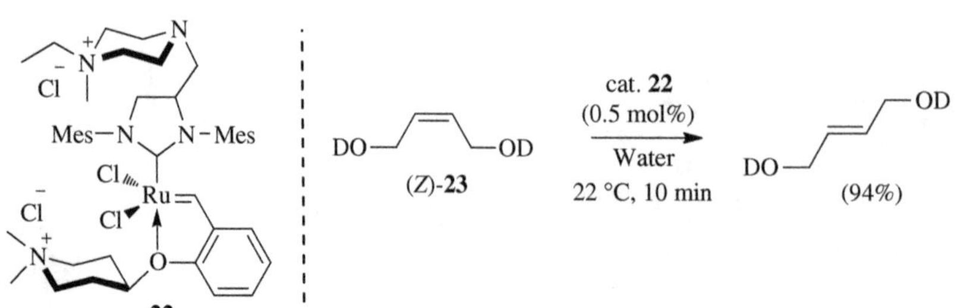

22

Figure 21.10 Isomerization of an olefin using a water-soluble catalyst.

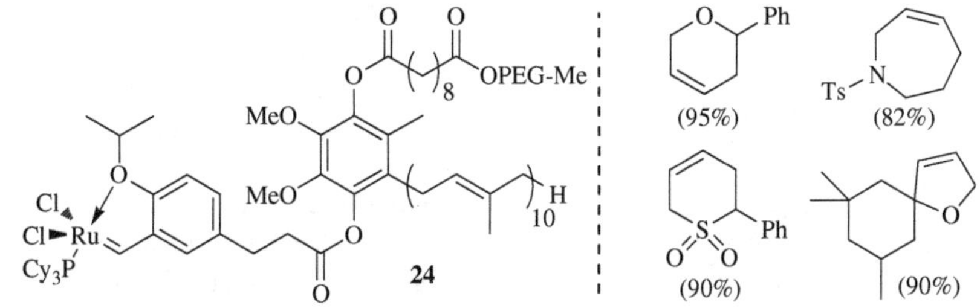

Figure 21.11 PQS Grubbs–Hoveyda-1-mediated RCM reactions in water.

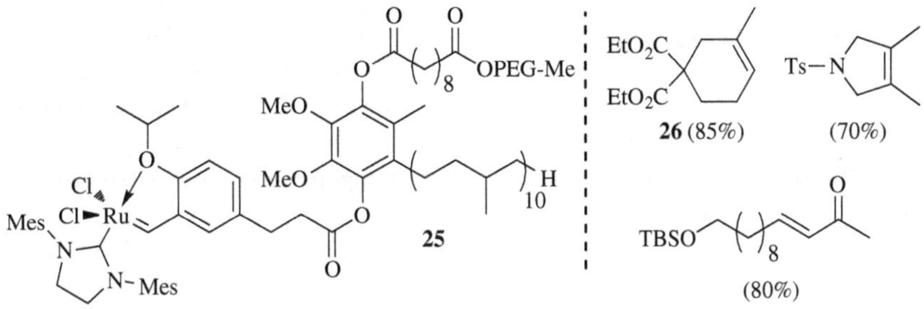

Figure 21.12 PQS Grubbs–Hoveyda-2-mediated RCM and CM reactions in water.

in seawater instead of high performance liquid chromatography (HPLC) grade water, with essentially identical results.

Lipshutz (18) also explored a related water-soluble olefin metathesis catalyst **25**, likewise derived from the important dietary supplement ubiquinol (i.e., the reduced form of coenzyme Q_{10}), in this case incorporating the more reactive Grubbs Hoveyda-2 catalyst, capable of mediating more challenging olefin metathesis reactions (Fig. 21.12). This catalyst system shows very promising results leading to tri- and tetra-substituted products from RCM, as well as CM reactions. Recycling is also possible, again, with minimal loss in catalyst reactivity. One noteworthy advantage of catalyst **25** is that reactions forming trisubstituted cycloolefins can be carried out at RT (18).

Representative RCM Reaction in Water Using Catalyst 25 (Fig. 21.12, Compound 26) (18)

Diethyl 2-(but-3-enyl)-2-(2-methylallyl)malonate (27 mg, 0.10 mmol) and catalyst **25** (7.5 mg, 0.002 mmol) were both added into a Biotage 2–5 ml microwave reactor vial with a Teflon-coated stir bar at RT and sealed with a septum. H_2O (1.0 ml) was added via syringe, and the resulting solution was allowed to stir at RT for 3 h. The homogeneous reaction mixture was then diluted with EtOAc (2 ml), filtered through a bed of silica gel layered over Celite, and the bed further washed (2 × 4 ml) with EtOAc to collect all of the cyclized material. The volatiles were removed *in vacuo* to afford the crude product, which was subsequently purified by flash chromatography using silica gel (2% EtOAc/hexanes) and afforded the product **26** as a colorless oil (20.5 mg, 85%).

Grela and co-workers have also developed a similar surfactant-like metathesis catalyst for both RCM and CM reactions (Fig. 21.13) (19). Best results were obtained from intramolecular processes with catalyst **27**, while CM reactions gave moderate yields and required higher catalyst loadings.

Other than just water, olefin metathesis reactions have been explored using aqueous media (20). The Blechert group (21) compared the effectiveness of two substituted analogues of the Grubbs–Hoveyda catalysts, **2** and **7**, in a mixed solvent system including water (Fig. 21.14). Whereas catalyst **6**, activated by an electron-withdrawing group, was not very active in methanol, the isopropoxy-substituted complex **7** was found to be an excellent catalyst for RCM of alkenes, not only in neat methanol or dimethylformamide (DMF),

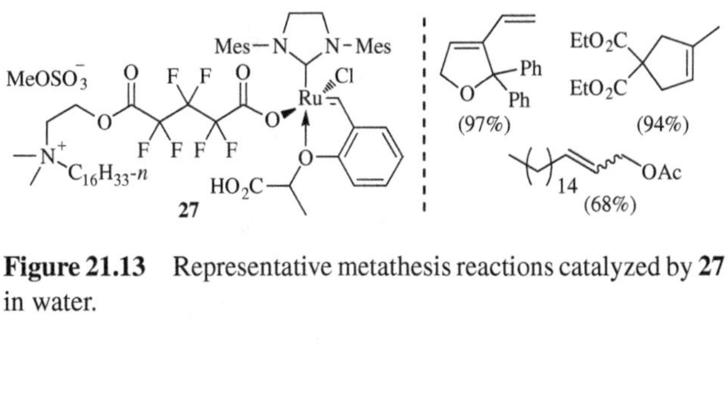

Figure 21.13 Representative metathesis reactions catalyzed by **27** in water.

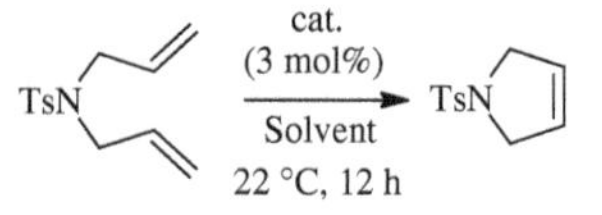

Solvent	Conversion (%)[a]	
	2	7
$MeOH/H_2O$ (3 : 1)	29	87
$MeOH/H_2O$ (1 : 1)[b]	54	90
$MeOH/H_2O$ (1 : 3)[b]	77	94
DMF/H_2O (3 : 1)	91	72
DMF/H_2O (1 : 1)	29	68
DMF/H_2O (1 : 3)[b]	82	94

[a] Determined by ^{1}H NMR spectroscopy.
[b] Substrate not miscible with solvent.

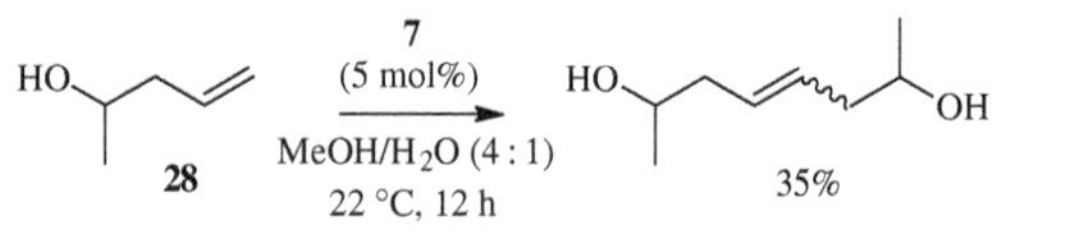

Figure 21.14 Metathesis reactions in aqueous media.

but also in mixtures of these solvents with water. By contrast, low conversion was observed in the homodimerization of the simple homoallyl alcohol **28** with **7** in methanol/water (21).

Raines and coworkers (22) showed catalyst **5** exhibited excellent efficiency in the RCM of a variety of dienes in aqueous dimethoxyethane (DME) and acetone (Fig. 21.15). Interestingly, the use of aqueous 1,4-dioxane and tetrahydrofuran (THF) as reaction media led to much lower conversions. Charged dienes, such as **29**, also underwent cyclization; however, a higher catalyst loading was required for good conversion. Furthermore, the Hoveyda catalyst **5** also promoted the CM reaction of allyl alcohol in aqueous acetone (22).

Recently, the Harding group (23) showed that **5** is an excellent catalyst for reversible aqueous metathesis chemistry between *S*-allylnaphthalimide **30** and β-*O*-allyl glucose in 50% *t*-BuOH/water (Fig. 21.16).

Further, the recent pioneering work of Ward et al. (24) incorporated a biotinylated Hoveyda–Grubbs catalyst (**31**) within avidin protein to produce artificial metalloenzymes that together were found to be highly active in RCM reactions in aqueous dimethyl sulfoxide (DMSO) solution at pH 4 in the presence of $MgCl_2$ (0.5 M) (Fig. 21.17).

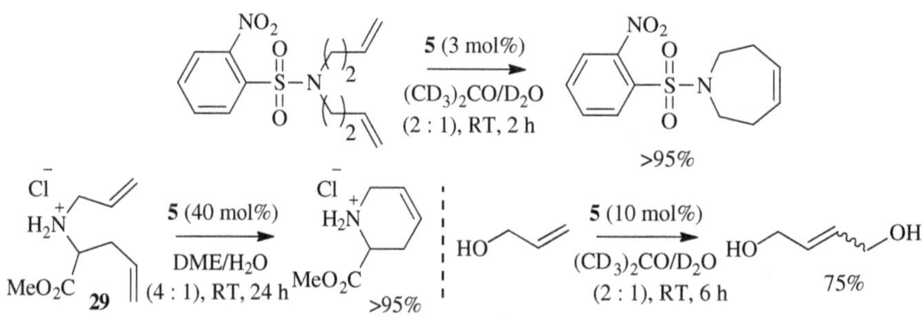

Figure 21.15 RCM and CM reactions in an aqueous medium.

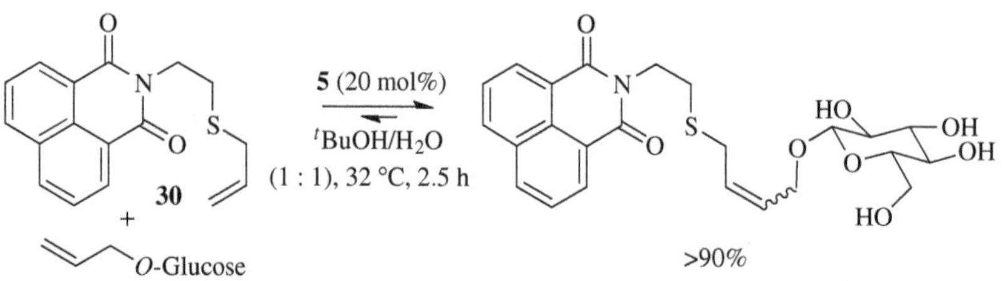

Figure 21.16 A reversible olefin CM reaction in a mixture of *t*-BuOH/water.

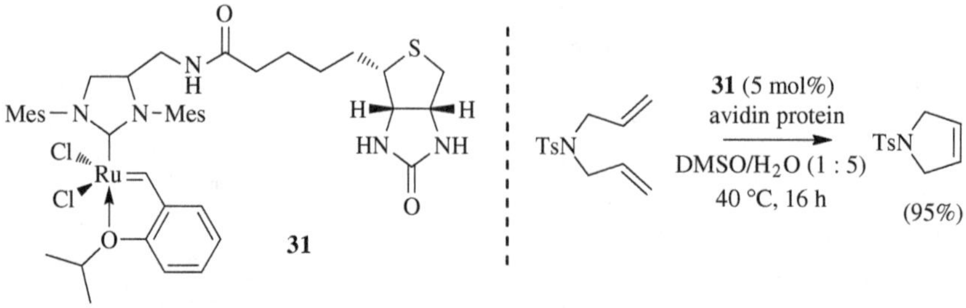

Figure 21.17 RCM reaction using biotin–avidin technology in aqueous DMSO.

In summary, considerable attention is now being directed toward development of metathesis reactions in alternative media, and in water in particular. Such technology avoids the health and environmental issues associated with traditional use of chlorinated solvents. Given the opportunities to adjust various parameters associated with the use of water to synthetic advantage, such as the phase in which the chemistry is performed, as well as the ionic strength and pH, there is every reason to believe that this approach to Nobel Prize-winning olefin metathesis chemistry will continue to evolve.

REFERENCES

1. (a) *Handbook of Metathesis*; Grubbs, R. H., Ed.; Wiley-VCH: Weinheim, 2003 (b) Nicolaou KC, Bulger PG, Sarlah D. *Angew Chem Int Ed* 2005;44:4490–4527. (c) Schrock RR, Hoveyda AH. *Angew Chem Int Ed* 2003;42:4592–4633.
2. (a) *Organic Reactions in Water*; Lindström, U. M., Ed.; Blackwell: Oxford, 2007. (b) Li C-J, Chan T-H. *Comprehensive Organic Reactions in Aqueous Media*. 2nd ed. Hoboken, NJ: Wiley-VCH; 2007. (c) Lipshutz BH, Ghorai S. *Aldrichim Acta* 2008;41:59–72.
3. Gułajski Ł, Śledź P, Lupa A, Grela K. *Green Chem* 2008;10: 271–274.
4. Davis KJ, Sinou D. *J. Mol. Catal. A: Chemical* 2002;177: 173–178.
5. Mingotaud A-F, Krämer M, Mingotaud C. *J. Mol. Catal. A: Chemical* 2007;263:39–47.
6. Lipshutz BH, Aguinaldo GT, Ghorai S, Voigtritter K. *Org Lett* 2008;10:1325–1328.
7. Lipshutz BH, Ghorai S, Aguinaldo GT. *Adv. Synth. Catal.* 2008;350:953–956.
8. Lipshutz BH, Ghorai S, Leong WWY, Taft BR. *J Org Chem* 2011;76:5061–5073.
9. Voigtritter K, Ghorai S, Lipshutz BH. *J Org Chem* 2011;76: 4697–4702.
10. (a) Lipshutz BH, Ghorai S, Abela AR, Moser R, Nishikata T, Duplais C, Krasovskiy A, Gaston RD, Gadwood RC. *J Org Chem* 2011;76:4379–4391. (b) Lipshutz BH, Ghorai S. *Aldrichim Acta* 2012;45:3–16.
11. (a) Lynn DM, Grubbs RH. *J Am Chem Soc* 2001;123: 3187–3193. (b) Mohr B, Lynn DM, Grubbs RH. *Organometallics* 1996;15:4317–4325.
12. Kirkland TA, Lynn DM, Grubbs RH. *J Org Chem* 1998; 63:9904–9909.
13. Hong SH, Grubbs RH. *J Am Chem Soc* 2006;128:3508–3509.
14. Jordan JP, Grubbs RH. *Angew Chem Int Ed* 2007;46: 5152–5155.
15. Gułajski Ł, Michrowska A, Narożnik J, Kaczmarska Z, Rupnicki L, Grela K. *ChemSusChem* 2008;1:103–109.
16. Skowerski K, Szczepaniak G, Wierzbicka C, Gułajski Ł, Bieniek M, Grela K. *Catal. Sci. Technol.* 2012;2:2424–2427.
17. Lipshutz BH, Ghorai S. *Org Lett* 2009;11:705–708.
18. Lipshutz BH, Ghorai S. *Tetrahedron* 2010;66:1057–1063.
19. Gawin R, Czarnecka P, Grela K. *Tetrahedron* 2010;66: 1051–1056.
20. Burtscher D, Grela K. *Angew Chem Int Ed* 2009;48:442–454.
21. Connon SJ, Rivard M, Zaja M, Blechert S. *Adv. Synth. Catal.* 2003;345:572–575.
22. Binder JB, Blank JJ, Raines RT. *Org Lett* 2007;9:4885–4888.
23. Hunter L, Condie GC, Harding MM. *Tetrahedron Lett* 2010;51:5064–5067.
24. (a) Lo C, Ringenberg MR, Gnandt D, Wilson Y, Ward TR. *Chem Commun* 2011;47:12065–12067. (b) Mayer C, Gillingham DG, Ward TR, Hilvert D. *Chem Commun* 2011;47:12068–12070.

22

OLEFIN METATHESIS IN GREEN ORGANIC SOLVENTS AND WITHOUT SOLVENT

CHRISTIAN BRUNEAU AND CÉDRIC FISCHMEISTER

UMR 6226 Institut des Sciences Chimiques de Rennes, Organométalliques: Matériaux et Catalyse, Université de Rennes 1, Rennes Cedex, France

22.1 INTRODUCTION

The necessary consideration of the environmental impact of chemical transformations has led researchers to design new transformations with low(er) environmental impact or to improve and modify existing processes to achieve a similar goal. One of the most easily accessible manifolds to minimize the environmental impact of a chemical transformation concerns the nature of a reaction medium. For this reason, in many domains of chemical synthesis, efforts are being made toward the use of environmentally friendly or greener solvents. In this chapter, we focus on olefin metathesis transformations in green(er) organic solvents, including supercritical carbon dioxide ($scCO_2$) and organic carbonates, in particular dimethyl carbonate (DMC) and poly(ethylene glycol) (PEG). We also review the domain of a solvent very often quoted as the greenest or best solvent in green chemistry textbooks, namely, no solvent or solventless conditions.

22.2 OLEFIN METATHESIS IN SUPERCRITICAL CO_2

Owing to its chemical and physical properties, $scCO_2$ has a high potential as an environmentally benign reaction medium and is a solvent of choice for green chemistry (1). Tuning temperature and pressure of $scCO_2$ makes possible the adjustment of its density and modulation of the interactions with the substrates and the catalysts to reach higher activity and selectivity control (2). The first example reported in the literature was the ring-opening metathesis polymerization (ROMP) of norbornene **1** carried out in the presence of $[Ru(H_2O)_6(Tos)_2]$ as olefin metathesis catalyst precursor at 65 °C under 17–34 MPa of CO_2 either in pure carbon dioxide or in the presence of methanol (3,4) (Scheme 22.1). Under high pressure of pure CO_2, a syndiotactic polymer with high cis stereoselectivity was obtained, whereas in the presence of methanol, an atactic polymer with lower cis/trans ratio was formed.

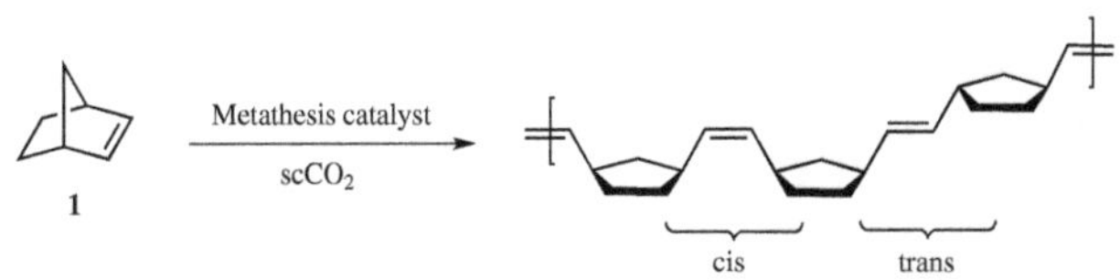

Scheme 22.1 Stereochemical impact on ring opening metathesis polymerization of norbornene in $scCO_2$.

The well-defined ruthenium catalysts $RuCl_2(=CHPh)(PCy_3)_2$ **C1** and $RuCl_2(=CHCH=CPh_2)(PCy_3)_2$ **C2** (Scheme 22.2) provided much higher activities in ROMP of norbornene and cyclooctene than $[Ru(H_2O)_6(Tos)_2]$ (5,6). Operating both in liquid and $scCO_2$ (23 °C < T <56 °C; 56 bar < P < 115 bar) afforded high yields in polymers with cis/trans ratio of about 1 : 3. The molybdenum catalyst **C4** was also very efficient for the polymerization of norbornene in saturated solution of toluene but the cis stereoisomer was the major one.

More recently, it was found that the polarity of the reaction medium had a strong influence on the microstructure of the

Olefin Metathesis: Theory and Practice, First Edition. Edited by Karol Grela.

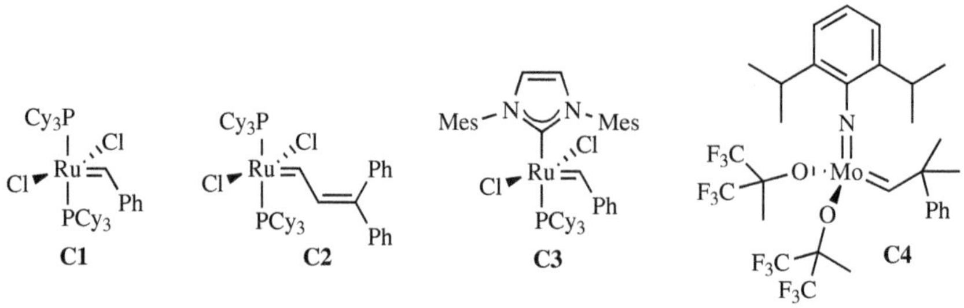

Scheme 22.2 Metathesis catalysts used in $scCO_2$.

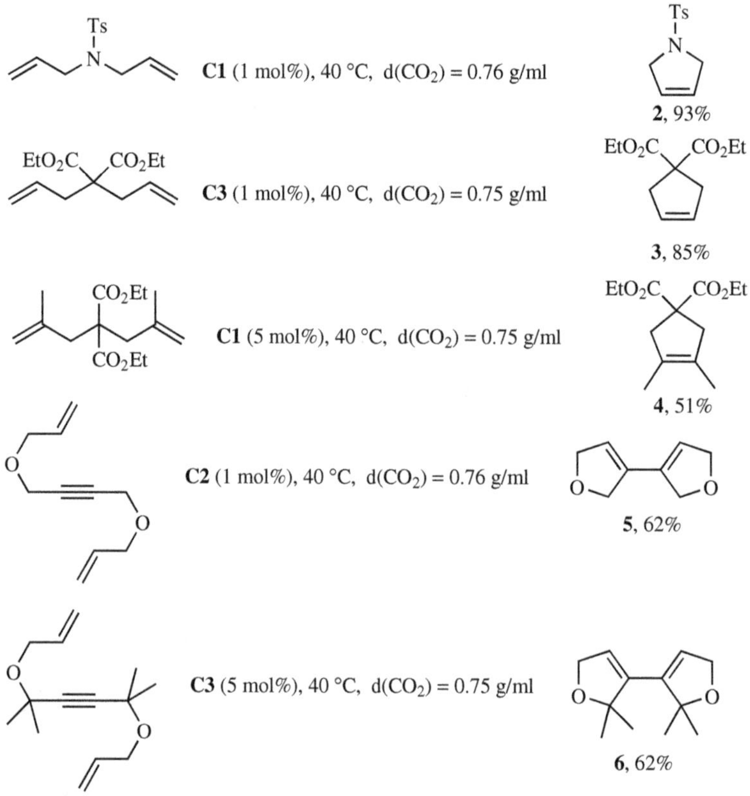

Scheme 22.3 Ring closing metathesis of model compounds in $scCO_2$.

polynorbornenes arising from ROMP, with ruthenium and molybdenum catalysts performing in $scCO_2$ at 70 °C under 206.9 bar. With ruthenium carbene catalysts, the polarity was adjusted by addition of various solvents such as methanol, tetrahydrofuran (THF), toluene, dimethylformamide (DMF), dimethylsulfoxide (DMSO), and the trans/cis ratio of the polynorbornene could be modulated in the range 3.2 : 1–5.5 : 1 (7). It is noteworthy that the nature of the catalyst also played a crucial role, as the trans/cis ratio was reversed when a molybdenum catalyst was used (trans/cis ratio = 1 : 3.9 in pure $scCO_2$). Finally, the CO_2 pressure, the presence of polarity modifiers, and the nature of the metathesis catalyst have a strong influence on the stereochemical outcome of the polymerization of cyclic olefins.

Formation of cyclic products of different ring sizes has been successfully performed via ring closing metathesis (RCM) of dienes in $scCO_2$ with either ruthenium or molybdenum catalysts and the products were easily isolated after simple release of CO_2 (5,6). The dienyne RCM leading to dihydrofurans **5** and **6** was also possible in $scCO_2$ (Scheme 22.3).

Karahanaenone **7** (5,6), a natural olfactory substance featuring a trisubstituted cycloalkene structure and the oxazolidinylpiperidine **8** (8), a precursor of glycosidase inhibitor, were prepared in good yield in $scCO_2$ (Scheme 22.4).

The importance of the density of the supercritical reaction medium has a drastic influence on the fate of the reaction of dienes that are potential precursors of oligomers via

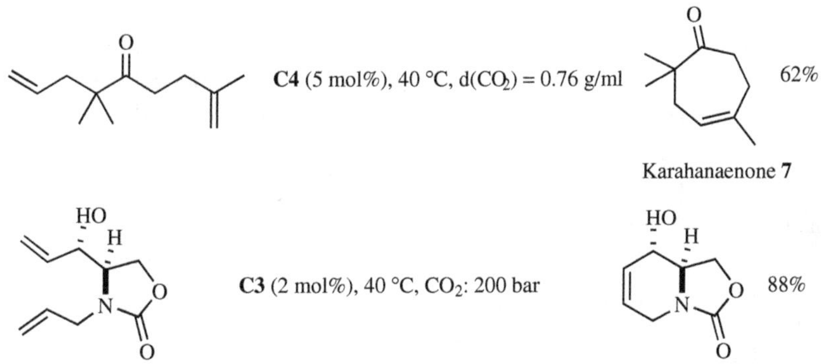

Karahanaenone **7**

Oxazolidinylpiperidine **8**

Scheme 22.4 Preparation of biologically active products by RCM in $scCO_2$.

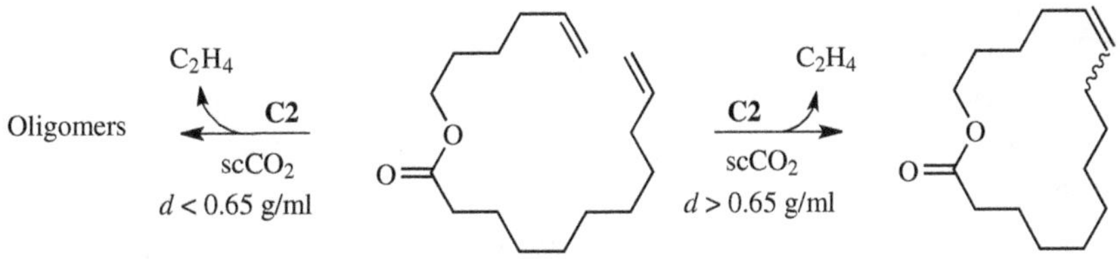

9 **10**

Scheme 22.5 Influence of CO_2 density on reactivity in olefin metathesis.

acyclic diene metathesis (ADMET) or medium size rings via RCM. In conventional solvents, the intramolecular RCM of dienes is favored by high dilution conditions. The high compressibility of supercritical fluids allows easy variations of the reaction medium density and increasing the density leads to a higher number of inert carbon dioxide molecules, which mimics the effect of dilution in classical organic solvents. This has been experimentally observed during the reaction of hex-5-enyl undec-10-enoate **9** at 40 °C in the presence of 1 mol% of the ruthenium catalyst **C2**, which gave mainly oligomers when the CO_2 density was below 0.65 g/ml, and the expected macrocycle **10** when the pressure was increased above this value (88% yield with $d = 0.83$ g/ml) (Scheme 22.5) (5,6,9).

Profit can also be taken from the acido-basic properties of carbon dioxide when the substrate contains an amino group. Indeed, in organic solvent free N–H groups have a tendency to inhibit the activity of olefin metathesis catalysts. A temporary protection of the amino group can be effected in the presence of carbon dioxide, which is known to easily produce ammonium carbamates or carbamic acids. This has been exemplified by the RCM of compound **11**, which gave the corresponding 15-membered ring **12** in 74% yield in the presence of 1 mol% of **C2** at 40 °C in $scCO_2$, whereas no reaction took place in dichloromethane (DCM) (Scheme 22.6) (5,6,9).

The cross metathesis of ethyl oleate **13** with ethylene has also been investigated in $scCO_2$ in the presence of 1 mol% of catalyst **C1** (Scheme 22.7) (10).

It was shown that with this catalyst and under these experimental conditions, very little self-metathesis products were formed and the terminal olefins 1-decene **14** and ethyl 9-decenoate **15** were selectively formed. At 35 °C, the rate and equilibrium conversion of the

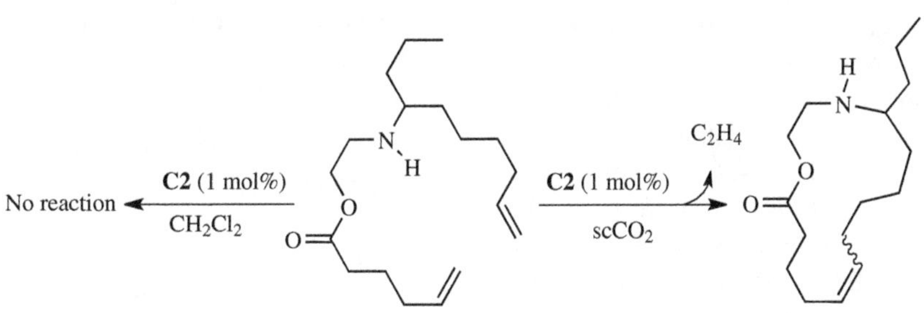

11 **12**, 74%

Scheme 22.6 Ring closing metathesis of amine made possible in $scCO_2$.

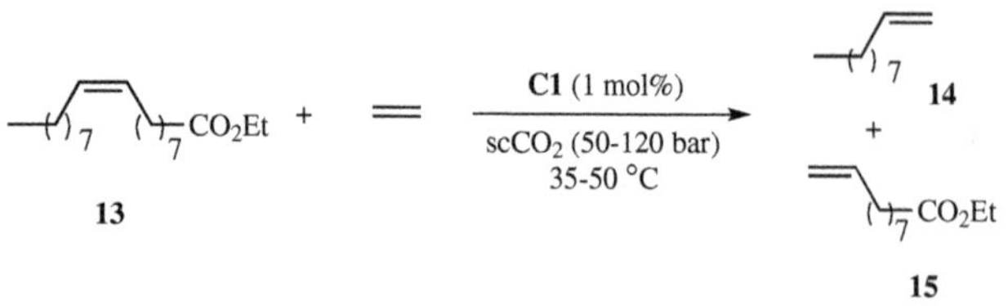

Scheme 22.7 Ethenolysis of ethyl oleate in $scCO_2$.

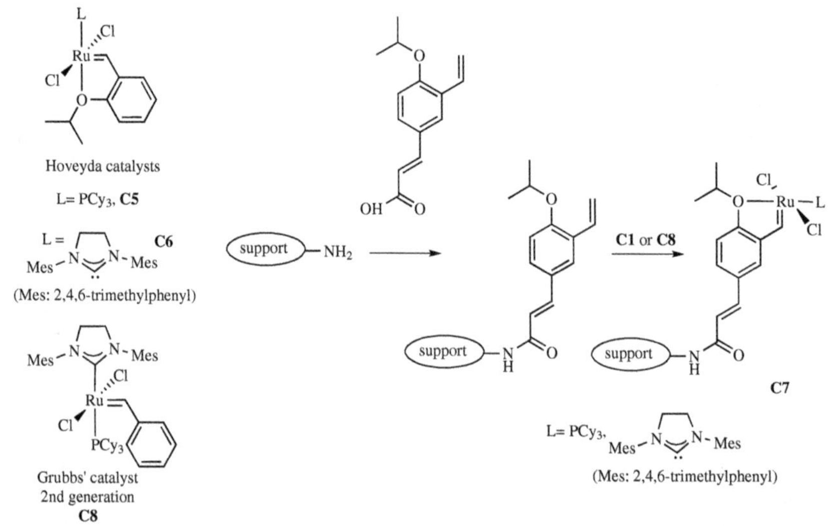

Scheme 22.8 Supported olefin metathesis catalyst.

the self metathesis of terminal aliphatic alkenes such as 1-octene, 1-hexene, and 1-heptene.The reaction were carried out at 35 °C under 90 bar of CO_2 with supported rhenium oxide (Re_2O_7) as catalyst (12,13). Conversions were usually improved in $scCO_2$ as compared to organic solvents such as *n*-heptane or toluene, but it was observed that the nature of reaction were much higher in the presence of compressed CO_2 (50 and 82 bar). Under 82 bar of CO_2 pressure, the solubility of the substrate in the gas phase was low, whereas the solubility of the products was very high, leading to the best efficiency. On the other hand, at 120 bar, the solubilities of the substrate and the products were all high and there was only one gas phase where the catalyst was not well dispersed, leading to slower metathesis transformation.

RCM of model functional dienes producing five-, six-, and seven-membered rings has also been studied with immobilized ruthenium catalysts in $scCO_2$. Owing to their high stability, immobilized versions of Hoveyda catalysts **C5** and **C6** have been prepared by grafting onto various supports leading to supported catalysts **C7** (Scheme 22.8) (11).

In typical conditions, the catalyst loading was 2.5 mol% and the reactions were carried out in $scCO_2$ at 40 °C under a pressure of 140 bar. The performances were lower than with the catalysts **C1** and **C8**, strongly depended on the nature of the support, and did not allow efficient recycling or reuse. The main advantage was the low leaching of ruthenium determined during the RCM of diallyltosylamide. $scCO_2$ has also been used in heterogeneous catalysis for the support had a tremendous influence. Thus γ-aluminoxide support led to improved conversion, whereas no reaction took place when the rhenium oxide catalyst was dispersed over acidic silica.

Finally, it is worth mentioning that $scCO_2$ has been used to remove ruthenium catalyst and its degradation by-products from a crude RCM reaction mixture (14). This easy separation is due to the high solubility of the formed macrocycle in $scCO_2$, the ruthenium species being left in the autoclave. A continuous flow plant was implemented with the Hoveyda first-generation catalyst **C5** (5 mol%) using an autoclave pressurized at 83 bar at 40 °C, which led to 88% of isolated macrocycle with a very low level of residual ruthenium.

22.3 OLEFIN METATHESIS IN ORGANIC CARBONATES

Besides their use as chemical reagents for carbonylation and methylation (15), dialkyl carbonates such as the highly polar propylene carbonate (PC, $\epsilon = 64.8$, $\mu = 4.94$ D) or the less polar DMC ($\epsilon = 3.1$, $\mu = 0.9$ D) and diethyl carbonate (DEC) have recently received attention as environmentally friendly solvents in transition metal catalysis (16). These solvents display several advantages related to safety and toxicity issues, in particular, they are biodegradabale and display a low acute toxicity (15–17). One parameter that is generally neglected with regard to the environmental impact of solvents is their synthesis and the life cycle assessment. In this domain, alkyl carbonate preparation is progressing toward cleaner synthesis. Thus, since the 1980s, several production methods have emerged for the replacement of the phosgene route, including the transition metal catalyzed oxidative carbonylation of methanol and the straightforward dehydrative condensation of alcohols with CO_2 (18). Owing to these improvements, carbonate solvents are attracting increasing interest as solvents in organic synthesis and especially in homogeneous catalysis (16,19). As a result, a pharmaceutical company such as GlaxoSmithKline (GSK) has now included DMC in its solvent selection guide as a greener alternative to conventional chlorinated solvents (20). However, as mentioned earlier, one should keep in mind that carbonate compounds are also reagents in particular for methylation or methoxycarbonylation reactions when reacted at high temperature (>90 °C) in the presence of nucleophiles (15). Therefore, even if they tolerate a broad range of experimental conditions, dialkyl carbonates may not be suitable solvents for any type of reactions.

When considering solvent aspects in olefin metathesis, it is clear that two types of solvents, chlorinated and aromatic benzene derivatives, represent the vast majority of the solvents used (21). DCM is very often used in ruthenium-catalyzed metathesis reactions as it is in general a very good catalyst solvent. However, due to its low boiling point, aromatic solvents such as benzene, toluene, and xylenes are also very often used when higher temperature conditions are required. It is also not rare to find olefin metathesis reactions performed in the high boiling point dichloroethane. All these solvents share a common property which is high toxicity and they are usually "red-listed" in solvent selection guides (20,22). Less hazardous and toxic organic solvents such as ethyl acetate or various ethers have also been reported but they yield in general lower catalytic efficiency (23). DMC was first reported in ruthenium catalyzed olefin metathesis in 2008 (24). It was demonstrated that DMC was suitable for olefin metathesis transformation, leading in some cases to improvement of the catalyst activity. For comparison purposes, the RCM of diethyl diallylmalonate (DEDAM) leading to **3** was performed under standard conditions in DCM and DMC (Scheme 22.9). As depicted in Figure 22.1, the reactions carried out in the presence of catalyst **C6** reached almost full conversion within 20 min whereas 60 min were necessary to reach this conversion in DCM.

DMC was also evaluated in a series of transformations including RCM of sterically hindered dienes, cross-metathesis of *n*-decene with methyl acrylate, and ethenolysis of methyl

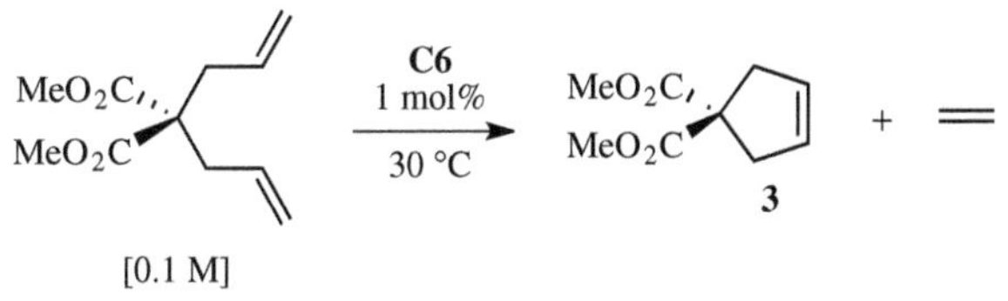

Scheme 22.9 RCM of DEDAM.

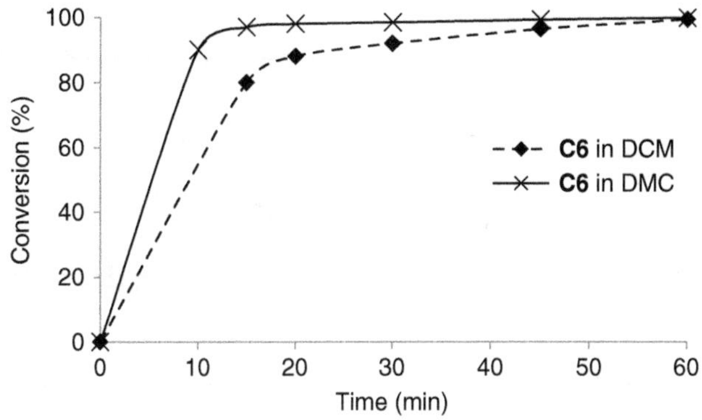

Figure 22.1 Comparative study of the RCM of DEDAM in DMC and DCM.

oleate. In all cases, similar results were obtained in DCM and DMC. Owing to its higher boiling point (b.p. = 90°C), DMC allowed running reactions at high temperature, thus avoiding the use of toluene or dichloroethane when higher temperatures were required. However, it must be noted that reactions performed at refluxing temperature resulted in about 10% double bond migration arising from the formation of ruthenium hydride species. These isomerization species were likely due to the release of methanol into the reaction medium, arising from DMC degradation. This isomerization side-reaction was reduced to trace by simply lowering the reaction temperature to 80 °C without hampering the reaction efficiency. Later, in 2008, DMC was used in membrane nanofiltration of olefin metathesis catalysts (25). This process has recently emerged as a new opportunity for the separation of the catalyst or catalyst residue from the reaction products. This advance was made possible thanks to the development of nanofiltration membranes stable in organic media (26). Thus DMC was evaluated for the nanofiltration of tailor-made olefin metathesis catalyst using a Starmem® 228 polyimide membrane. It was found that this membrane was stable using DMC as solvent thus giving an extra-value to this promising separation process. As mentioned in the very first report, DMC is a suitable solvent for the cross-metathesis transformation of renewable fatty esters. Valuable fine chemicals such as pyrane and lactone derivatives were thus prepared by RCM of pre-functionalized methyl ricinoleate (27). The transformation of such renewable materials was also performed with high efficiency in DMC by means of ene–yne cross-metathesis reactions, leading to conjugated dienes (28). The cross-enyne metathesis of *n*-decene with

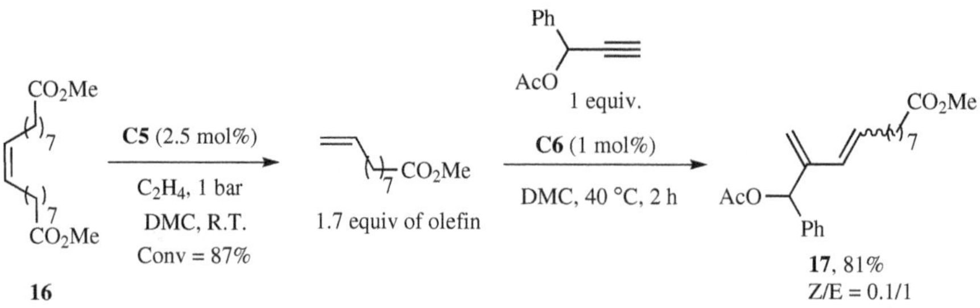

Scheme 22.10 One-pot ethenolysis/ene–yne cross-metathesis in DMC.

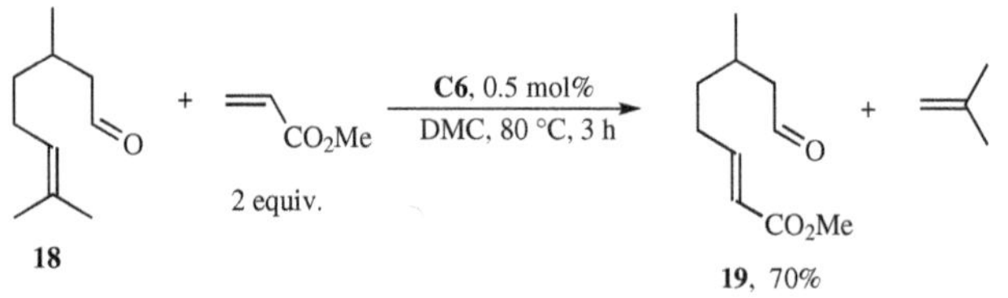

Scheme 22.11 Cross-metathesis of citronellal with methyl acrylate in DMC.

1,4-diacetoxy-2-butyne was used as a test reaction to prove the compatibility of DMC with this type of transformation. Again it was shown that the reaction performed with the same efficiency in DCM, toluene, and DMC. The diester **16** obtained by bio-transformation of oleic acid (29) was transformed by a one-pot sequence ethenolysis/cross-enyne metathesis into the conjugated diene **17** in high yield. Both the ethenolysis and cross-metathesis steps were carried out in DMC (Scheme 22.10). The sustainability of this transformation was further improved employing stoichiometric amounts of alkene and alkyne in DMC (30).

Terpenes are another class of unsaturated natural products. Their transformation by olefin cross-metathesis was carried out with high efficiency in DMC (31). As depicted in Scheme 22.11, the cross-metathesis of citronellal **18** with methyl acrylate led to the corresponding terpenoid **19** in 70% yield. Previously, this compound had been synthesized in 40% yield in DCM (32). Citronellol was similarly transformed in good yields whereas citral led to a moderate yield of 40%. More recently, the same strategy was successfully applied to the transformation of eugenol and eugenol derivatives (33).

22.4 OLEFIN METATHESIS IN NON CONVENTIONAL GREEN SOLVENTS (GLYCEROL, POLY(ETHYLENE GLYCOL), METHYL DECANOATE)

Glycerol, a bio-sourced and biodegradable chemical, has been considered as a potentially green solvent in olefin metathesis reactions (34). It has been applied in RCM of *N,N*-diallyltosylamide and DEDAM under microwave irradiation (Scheme 22.12).

RCM of *N,N*-diallyltosylamide was carried out at 40 °C in the presence of first- and second-generation ruthenium catalysts. Excellent yields were obtained using rather high loading (5 mol%) of second-generation catalysts in short reaction times, except in the case of the Hoveyda catalyst **C6**, which was less efficient. The best result was obtained with Zhan catalyst **C10**, which led to quantitative yield within 15 min. Maintaining a high level of microwave energy by simultaneous cooling did not improve the efficacy of the catalytic systems. Reuse of the catalyst-containing glycerol after extraction of the product by diethyl ether was possible three times with almost full conversion of the diene, but the fourth cycle led to very poor conversion, indicating that progressive catalyst degradation took place in the glycerol medium.

In the case of DEDAM, complete conversions were observed but the isolated yields were modest due to the formation of by-products assumed to result from transesterification reaction of glycerol with the starting diene. This indicates that the use of glycerol as metathesis solvent can be envisioned only from substrates with no reactivity toward alcohol.

PEGs HO-$(CH_2\text{-}CH_2O)_n$-H represent another class of hydroxylated nontoxic, biodegradable solvents suitable for catalysis (35,36). RCM of tosylamide and malonate derivatives leading to various ring sizes were performed at 50–100 °C under microwave irradiation both in solid PEG (PEG-3400) and in *O*-protected MeO-PEG-2000-OMe (37). The low molecular weight liquid PEGs such as PEG-300 did not give satisfactory results because it was difficult to separate the solvent from the reaction products. The involvement of a variety of ruthenium catalysts in the RCM of diallyltosylamide in PEG-3400 revealed that two isomeric *N*-tosylpyrrolines were produced: the expected *N*-tosyl-3-pyrroline RCM product **2** and the *N*-tosyl-2-pyrroline **20** resulting from double-bond migration within the ring (Scheme 22.13). This isomerization is likely due to formation of ruthenium hydride from ruthenium carbene species in the presence of the protic PEG solvent, which are known to catalyze the double bond

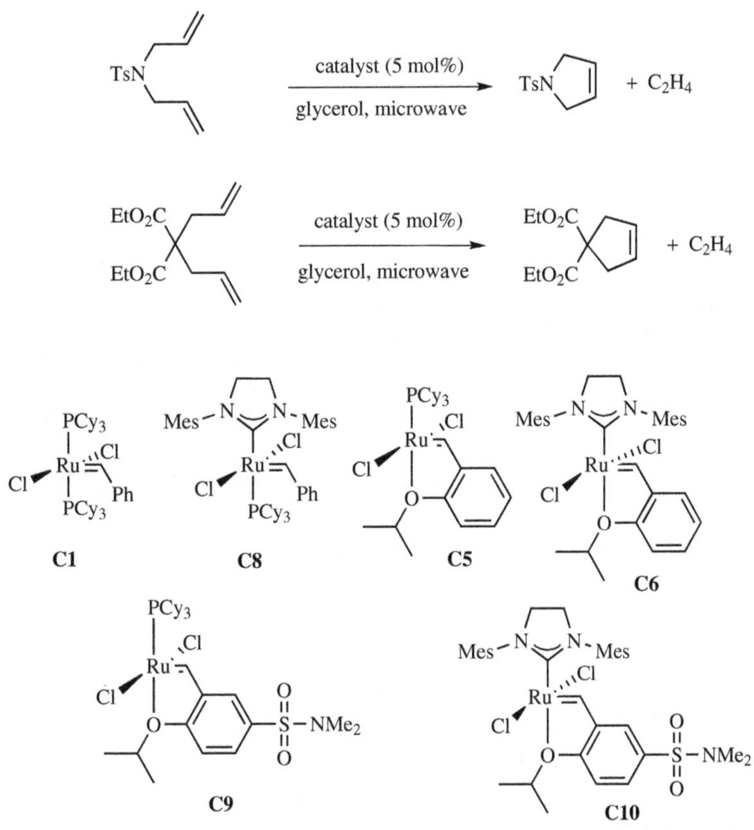

Scheme 22.12 Ring-closing metathesis in glycerol under microwave irradiation.

migration (38). This side reaction was avoided when the OH groups were protected as methyl ethers. Thus, the use of MeO-PEG-2000-OMe as solvent prevented ruthenium hydride formation and as a direct consequence double-bond migration, and the RCM product was obtained in higher yields up to 99% (Table 22.1).

Other diene substrates leading to more or less substituted five- and six-membered tosylamides were also cyclized in good yields in O-protected PEG (the best systems are reported in Scheme 22.14). However, it is noteworthy that with these substrates, the ratio of isomerization products was much lower than from *N,N*-diallyltosylamide, and that good yields in cyclized products **21–23** could be obtained even in unprotected PEG, in particular with **C6** catalyst.

Starting from DEDAM, under similar reaction conditions, the cyclization was more difficult leading to a 68% maximum yield of **3** with the ruthenium indenylidene catalyst **C13**, and the isomerized product was also formed with some catalytic systems. Diethyl allylhomoallylmalonate was fully converted with most catalytic systems in PEG and O-protected PEG but migration of the homoallylic double bond before ring closing took place in some cases leading to a mixture of five- and six-membered rings. In this case again, the use of MeO–PEG–OMe as solvent prevented this migration and yields in six-membered metathesis product located in the range 60–67% were obtained in this solvent in the presence of 5 mol% of catalyst **C13**, **C14**, and **C6**. However, the best yield (90%) was obtained in PEG-3400 with catalyst **C13**, but the five-membered ring was also formed in 10% yield.

Recently, RCM of DEDAM and cross metathesis of allylbenzene with *cis*-1,4-diacetoxy-2-butene were studied, both under solvent-free conditions and in methyl decanoate, a renewable and environmentally benign solvent derived from fatty acid derivatives (39).

For the RCM of DEDAM with 1 mol% of catalyst loading **C6**, **C8**, **C10**, the reaction took place in methyl decanoate but did not compete favorably with DCM, which provided complete conversion when only 66–80% conversion was obtained with methyl decanoate. On the other hand, the reaction performed without solvent gave complete conversion whatever the catalyst used, and even in pure diene, no side product resulting from ADMET were detected. Interestingly, the reaction in bulk could be performed with very low catalyst loading (down to 0.04 mol%) when **C10** was used as catalyst, leading to clean and very reproducible reactions.

In the cross metathesis of allylbenzene with *cis*-1,4-diacetoxy-2-butene, methyl decanoate appeared as a better

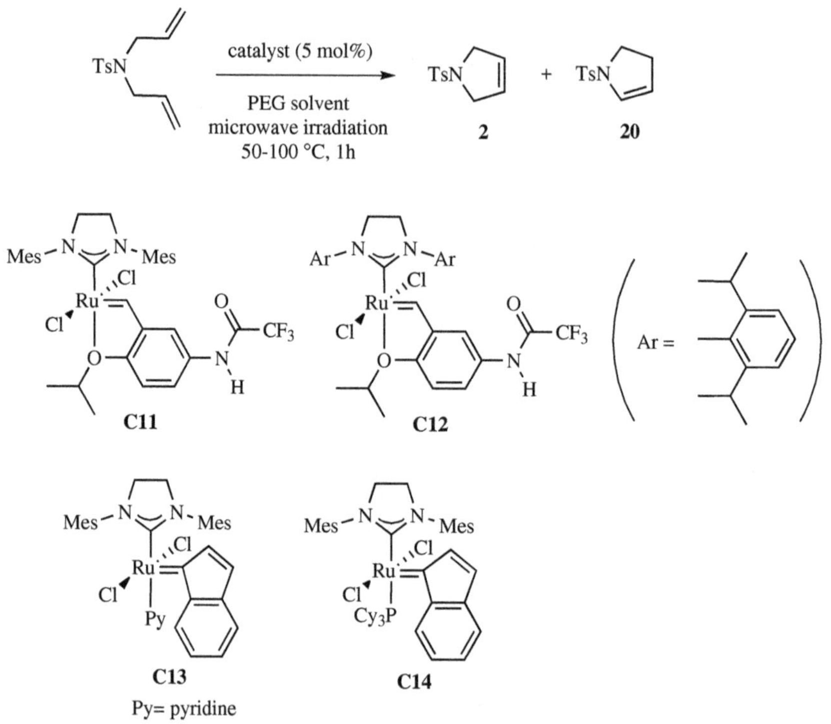

Scheme 22.13 RCM of diallyltosylamide.

TABLE 22.1 RCM of *N,N*-Diallyltosylamide in PEG as Solvent

Entry	Catalyst	Solvent	Yield (**2**) (%)	Yield (**20**) (%)
1	**C1**	PEG-3400	>99	0
2	**C1**	MeO–PEG-2000–OMe	48	0
3	**C6**	PEG-3400	51	42
4	**C6**	MeO–PEG-2000–OMe	92	0
5	**C11**	PEG-3400	35	42
6	**C11**	MeO–PEG-2000–OMe	99	0
7	**C12**	PEG-3400	14	60
8	**C12**	MeO–PEG-2000–OMe	92	0

solvent than DCM in terms of reactivity and selectivity. Indeed, almost no self-metathesis of allylbenzene was observed (<1%) and conversions of 94.8, 91.2, and 89.7 were respectively obtained with 2.5 mol% of catalyst **C8**, **C6**, and **C10**. In this metathesis reaction, the solvent-free conditions were also more appropriate at low catalyst loading and allowed to decrease it to 0.1 mol% while maintaining good efficiency.

22.5 OLEFIN METATHESIS WITHOUT SOLVENT

In 1992, Forbes and Wagener discovered that substituted dienes cyclized by RCM instead of forming the expected ADMET polymers when neat conditions were used (Scheme 22.15). Thus, employing Schrock's catalyst [$Mo(CHCMe_2Ph)(NAr)(OCMe(CF_3)_2)_2$] (Ar = 2,6-diisopropyl-phenyl) **C4**, the diene **24** furnished the cyclized seven-membered cycle **25** due to favorable Thorpe–Ingold effect. However, the longer diene **26** and the unsubstituted diene **27** furnished the ADMET polymer. In addition, it was shown that the nature of the substituent led to ADMET oligomers or mixture of ADMET oligomers and five-membered cycle when shorter dienes were used (40).

In 2003, Vo Thanh and Loupy performed the microwave assisted RCM reaction of a series of dienes under solvent-free conditions. Several five- and six-membered cycles were prepared in high yield using catalyst **C1** and it was demonstrated that the high activity resulted from non-thermal microwave effect (41). More recently, the synthesis of nitrogen containing heterocycles by ruthenium catalyzed RCM using low catalyst loading was studied by

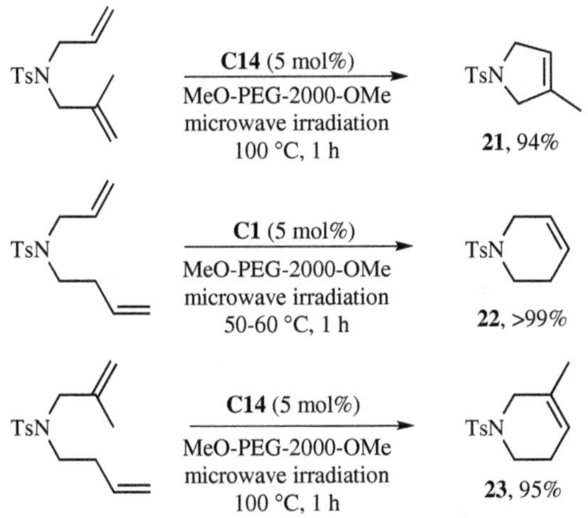

Scheme 22.14 Ring-closing metathesis of *N*-tosylamides in poly(ethylene glycol) solvent under microwave irradiation.

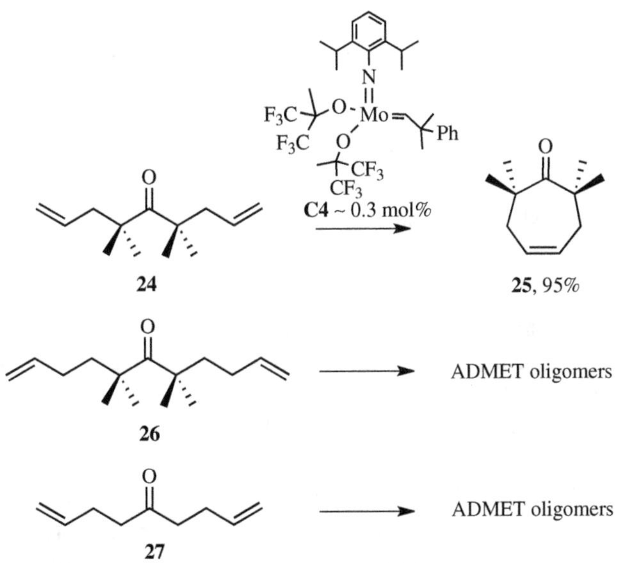

Scheme 22.15 Solvent-free RCM of dienes using olefin metathesis and the Thorpe–Ingold effect.

Grubbs (42). In particular, it was shown that five-membered cycles such as **28** could be prepared under neat conditions, albeit in lower yields than reaction performed in toluene. Larger cycles such as **29** also required more diluted conditions to reach high yields (Table 22.2). Of note, it was also shown that Hoveyda-type catalysts (**C5**, **C6**) performed best under neat conditions. Lower yields obtained with Grubbs type catalysts (**C1**, **C8**) were attributed to potential competitive phosphine-based decomposition pathways. These results may also find rationalization in the recent studies on the activation mechanism of Hoveyda-type catalysts (43).

Renewable unsaturated materials are interesting substrates suitable for transformation by olefin metathesis

TABLE 22.2 RCM of Carbamate Protected Amines[a]

Substrate	Product	Concentration (M)	Yield (%)
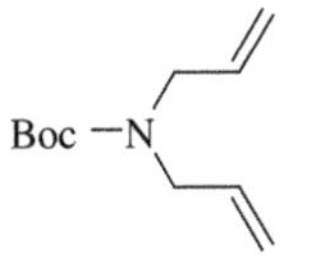	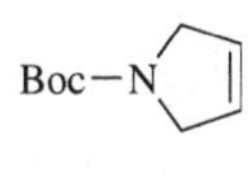 **28**	Neat	87
		1 M	>99%
		0.2 M	92
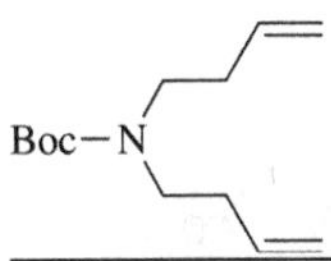	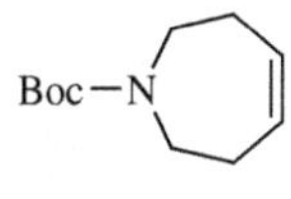 **29**	1 M	46
		0.2 M	82
		0.05 M	90

[a]500 ppm of **C6**, 50 °C, toluene, 8 h.

for the production of raw materials or fine chemicals (44). Two examples of terpenoid transformations by olefin metathesis under neat conditions have been recently reported (Scheme 22.16). Citronellal **30** was transformed by cross-metathesis with methyl methacrylate to furnish compound **31** with preservation of the terpenoic squeleton (31). In the second example, the two double bonds of (−)-citronellene **32** were transformed to furnish the derivative **33** in 32% yield due to competitive RCM reaction (23d).

Fatty acid methyl esters (FAMEs) arising from plant oils have found numerous applications as biofuels and as precursors of raw materials for the chemical industry, in particular as polymer precursors or additives (45). In 2007, Rybak and Meier (46) reported the cross-metathesis of various FAMEs with methyl acrylate under neat conditions for the production of α,ω-dicarboxylic acid precursors of polyesters. For example, methyl 10-undecenoate **34** arising from castor oil (47) was efficiently transformed into the α,ω-diester **35** in high yield with low catalyst loading (Scheme 22.17). In the same manner, cross-metathesis reactions involving protected oleyl alcohol (48) or allyl chloride (49) were also performed under neat conditions.

Although bulk conditions are not very often encountered in organic synthesis, they are more common in polymer synthesis. In 1987, Wagener and coworkers (50,51) reported a study on the ADMET polymerization of dienes, including 1,5-hexadiene under bulk conditions, using $WCl_6/EtAlCl_2$ catalyst system. In a general procedure, the monomer and WCl_6 were mixed at low temperature before addition of the co-catalyst and warming to rt. In some cases, vacuum was used to release the generated gases. In this study, although the ADMET polymers were obtained, side reactions leading to intractable solids occurred. In fact, it was demonstrated with styrene that vinyl addition initiated by the co-catalyst $EtAlCl_2$ occurred. For this reason, the same group turned its attention to metathesis polymerization, employing well defined catalysts that do not require Lewis acid co-catalyst. Thus, polyethers were prepared in high yields from α,ω-diene ethers employing the tungsten based Schrock

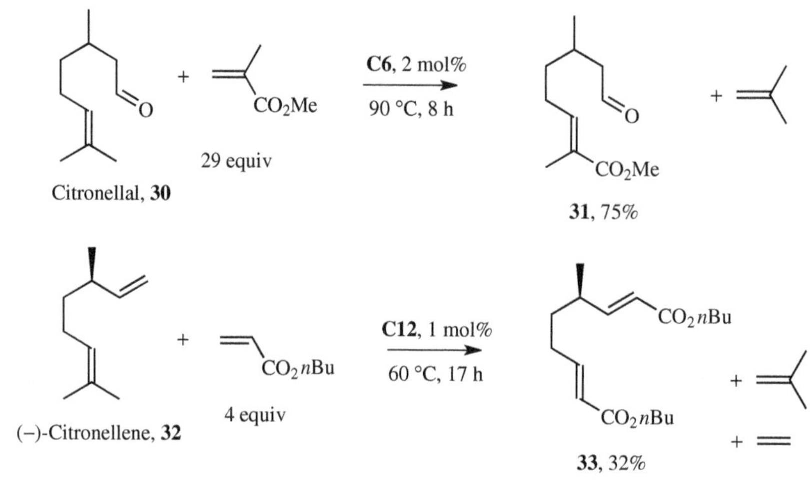

Scheme 22.16 Cross-metathesis transformation of terpenoid derivatives.

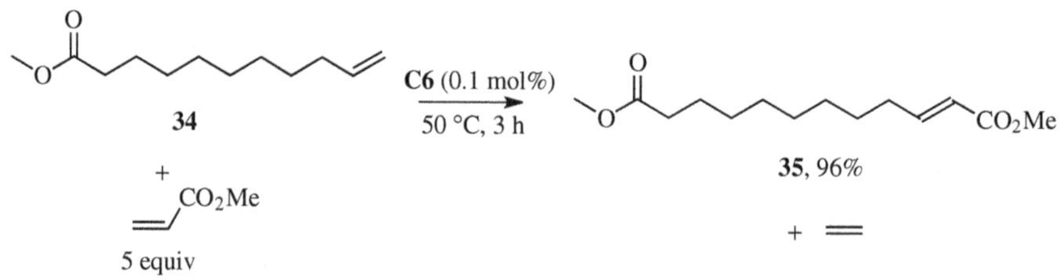

Scheme 22.17 Cross-metathesis of methyl 10-undecenoate.

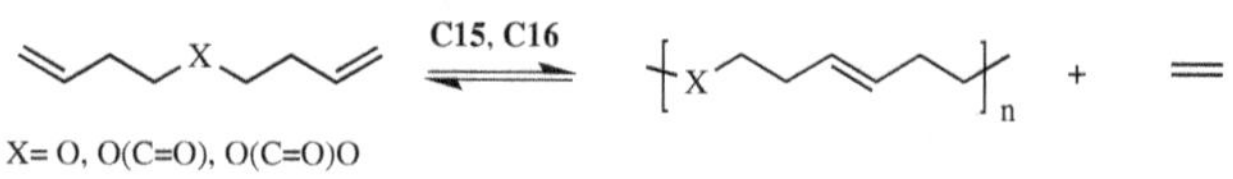

Scheme 22.18 ADMET synthesis of polyethers, -esters, -carbonates.

catalyst W(CH-*t*-Bu)(*N*-2,6-C_6H_3-*i*-Pr_2)-[OCMe(CF_3)$_2$]$_2$ **C15** (52). In the same manner, the first ADMET synthesis of polyester (53) and the synthesis of unsaturated polycarbonates (54) were reported, using the molybdenum version of the tungsten based Schrock catalyst Mo($CHMe_2R$)(*N*-2,6-C_6H_3-*i*Pr_2)-[OCMe(CF_3)$_2$]$_2$ **C16**. In all these cases, the number of methylene spacers between the functional group and the terminal olefin was found to be an important factor in these polymerizations. At least two methylene spacers were necessary to get efficient polymerization reactions (Scheme 22.18).

More recently, ADMET polymerization was used for the synthesis of purely linear polyethylene (55) and precision branched polyethylene (Scheme 22.19) (56). For instance, the ADMET polymerization of 1,9-decadiene followed by hydrogenation of the resulting polyoctenamer led to polyethylene displaying properties (mp, heat of fusion) close to industrial grade high density polyethylene (HDPE). In addition, when a functionalized diene was used as monomer, the resulting polymer contained unambiguously identified and positioned branches. Polymerizations were conducted in bulk, using either Grubbs (first generation) **C1** or Schrock catalyst [Mo(CH*t*Bu)(*N*-2,6-C_6H_3-*i*-Pr_2)(O-*t*-Bu)$_2$] **C17** while the hydrogenations were achieved by heterogenization of the ruthenium metathesis catalyst onto silica or by using a mixture of toluenesulfonylhydrazide and tripropylamine.

ADMET polymerization can also be used for the polymerization of renewable resources, in particular, vegetable oils and FAME derivatives containing unsaturations (45a,57). Thus, Larock reported the polymerization of several unsaturated vegetable oils using **C1** catalyst in the absence of a solvent. Satisfactory yields ranging between 40% and 60% were obtained with a variety of commercially available food-grade vegetable oils such as corn oil, olive oil, soybean oil, safflower oil, and sunflower oil (58). Of note, it was shown that there was no correlation between the number of double bond in the triglycerides and the polymer yields. ADMET polymerization under bulk conditions was used for the synthesis of the bio-sourced polyester **37** (59). The synthesis was accomplished by the initial preparation of the monomer **36** obtained from 10-undecenoic acid and 10-undecenol, both arising from castor oil (Scheme 22.20). **36** was then polymerized by ADMET using second-generation Grubbs and Hoveyda catalysts **C8** and **C6**, respectively. In both cases,

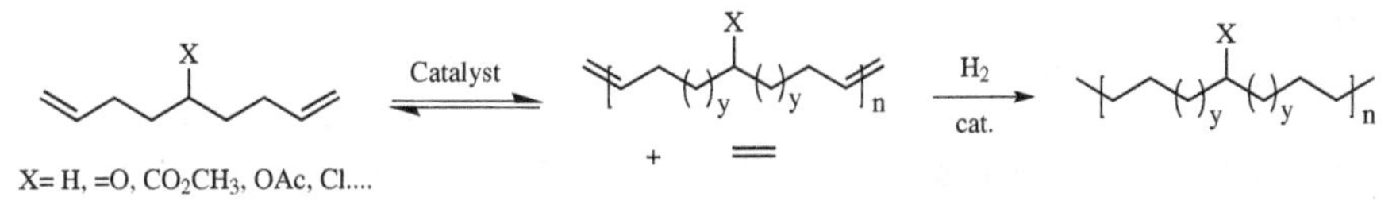

Scheme 22.19 Purely linear and precision branched polyethylene.

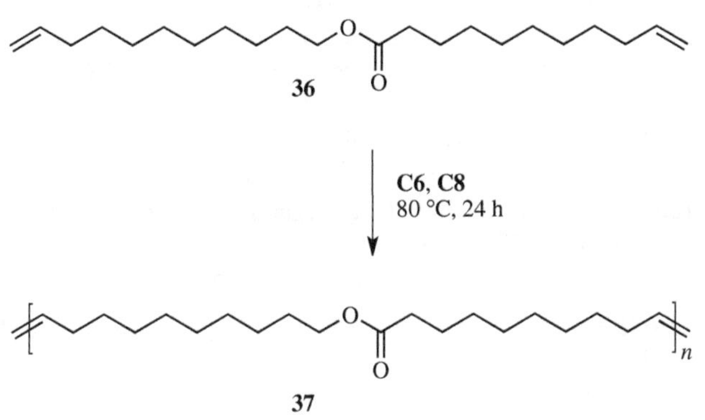

Scheme 22.20 Bio-sourced polyester by ADMET.

polymerization occurred without solvent, with full monomer conversion and with high polymer yields.

Acyclic triene metathesis under bulk conditions was performed to prepare branched macromolecules (60). The triglyceride **38** prepared from glycerol and methyl 10-undecenoate **34** was polymerized in the presence of Hoveyda catalyst **C6** and methyl acrylate as chain stopper. It was found that the amount of methyl acrylate was crucial to control the size of the macromolecule. Hence a large excess of methyl acrylate totally inhibited polymerization, whereas a low amount of chain stopper (1.75 equiv) resulted in the formation of insoluble resins (Scheme 22.21).

Recently, a class of ruthenium catalysts, the so-called latent catalysts, has received much attention due to the need for ROMP catalysts that can be mixed with the monomers without polymerization activity until external stimuli is applied to initiate polymerization. This type of catalyst would allow the preparation of monomer and catalyst mixtures that could be eventually stored and used when required. As in the previous cases, most of the latent catalysts are used under neat conditions in polymerization reactions such as the ROMP of dicyclopentadiene and so far most of the research efforts have been dedicated to the design and synthesis of catalyst architectures (61).

22.6 CONCLUSION

Over the last two decades, the main efforts in olefin metathesis have been dedicated to the design of more efficient, robust, and selective catalysts, as well as the development of new applications. Despite being intrinsically "green" as a catalytic process, efforts are still necessary to attract stronger interest from the industry by making this transformation more sustainable. As presented in this chapter, the use of more

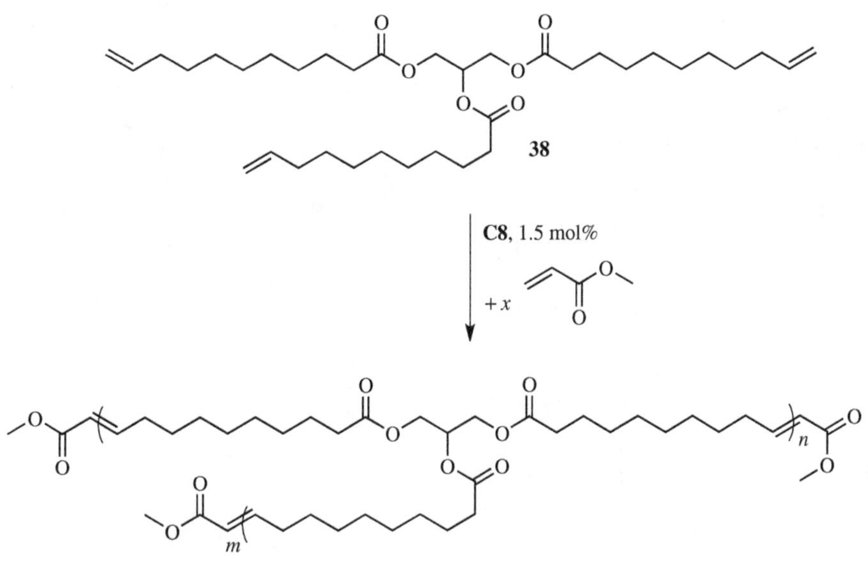

Scheme 22.21 Branched polymers by acyclic triene metathesis polymerization.

environment friendly solvents as reaction media has been under investigation over the last 15 years and it is still a topic of high interest as regulations on volatile organic compound (VOC) emissions are becoming more and more stringent and restrictive. In the examples herein reported, it is important to consider and balance the positive and negative aspects for the selection of an alternative reaction media. If solventless conditions are certainly the greenest ones, they are unfortunately not universal since some transformations do not tolerate high concentration of reagents and also because solvents may be necessary to ensure efficient heat transfers or work-up procedures, for instance. Further when a reagent is used in large excess, as solvent substitute, its recycling should be easy to achieve. In fact, bulk conditions are certainly best suited for olefin metathesis polymerization reactions as it is already the case in many polymerization processes. In the domain of non-conventional solvents, $scCO_2$ has brought very interesting results and offers nice perspectives; however, its large scale development might be hampered by equipment issues. More conventional but greener solvents constitute probably the most appealing solution for academic research as they do not need special equipment or process modifications and are accessible, like any other solvent. However, as presented here, there may be some limitations to take into account. Alcohols such as PEGs or glycerol offer good opportunities but the propensity of olefin metathesis catalysts to generate isomerizing ruthenium hydride species in the presence of alcohols is a serious drawback. In the same manner, dialkyl carbonates are an interesting greener alternative to chlorinated or aromatic solvents as they enable more or less similar reaction performances. However, they also display some limitations as they can be reagents at high temperature if the reaction substrate or the final product contains a nucleophilic site.

REFERENCES

1. Leitner W. *Acc Chem Res* 2002;35:746–756.
2. Leitner W. *Coord Chem Rev* 1996;153:257–284.
3. Mistele CD, Thorp HH, DeSimone JM. *J Macromol Sci Pure Appl Chem* 1996;A33:953–960.
4. Hamilton JG, Rooney JJ, DeSimone JM, Mistele C. *Macromolecules* 1998;31:4387–4389.
5. Fürstner A, Koch D, Langemann K, Leitner W, Six C. *Angew Chem Int Ed* 1997;36:2466–2469.
6. Fürstner A, Ackermann L, Beck K, Hori H, Koch D, Langemann K, Liebl M, Six C, Leitner W. *J Am Chem Soc* 2001;123:9000–9006.
7. Hu X, Blanda MT, Venumbaka SR, Cassidy PE. *Polym Adv Technol* 2005;16:146–149.
8. Martin R, Murruzzu C, Pericas MA, Riera A. *J Org Chem* 2005;70:2325–2328.
9. Leitner W. *C R Acad Sci* 2000;3:595–600.
10. Song J, Hou M, Liu G, Zhang J, Han B, Yang G. *J Phys Chem B* 2009;113:2810–2814.
11. Michalek F, Mädge D, Rühe J, Bannwarth W. *Eur J Org Chem* 2006:577–581.
12. Selva M, Perosa A, Fabris M, Canton P. *Green Chem* 2009;11:229–238.
13. Fabris M, Aquino C, Ward AJ, Perosa A, Maschmeyer T, Selva M. *Top Catal* 2009;52:315–321.
14. Gallou F, Saim S, Koenig KJ, Bochniak D, Horhota ST, Yee NK, Senanayake CH. *Org Process Res Dev* 2006;10:937–940.
15. (a) Tundo P, Selva M. *Acc Chem Res* 2002;35:706–716. (b) Parrish JP, Salvatore RN, Woon Jung K. *Tetrahedron* 2000;56:8207–8237.
16. Schäffner B, Schäffner F, Verevkin SP, Börner A. *Chem Rev* 2010;110:4554–4581.
17. Fischmeister C, Doucet H. *Green Chem* 2011;13:741–753.
18. (a) Sakakura T, Kohno K. *Chem Commun* 2009:1312–1330. For the synthesis of cyclic carbonates from CO_2 and epoxides, see: (b) North M, Pasquale R, Young C. *Green Chem* 2010;12:1514–1539.
19. (a) Behr A, Naendrup F, Obst D. *Adv Synth Catal* 2002;344:1142–1145. (b) Cornely J, Ham LMS, Meade DE, Dragojlovic V. *Green Chem* 2003;5:34–37. (c) Bernini R, Mincione E, Barontini M, Crisante F, Fabrizi G, Gambacorta A. *Tetrahedron* 2007;63:6895–6900. (d) Reetz MT, Lohmer G. *Chem Commun* 1996:1921–1922. (e) Vasapollo G, Mele G, Maffei A, Del Sole R. *Appl Organomet Chem* 2003;17:835–839. (f) Bayardon J, Holz J, Schäffner B, Andrushko V, Verevkin S, Preetz A, Börner A. *Angew Chem Int Ed* 2007;46:5971–5974. (g) Schäffner B, Holz J, Verevkin SP, Börner A. *ChemSusChem* 2008;1:249–253. (h) Torborg C, Huang J, Schulz T, Schäffner B, Zapf A, Spannenberg A, Börner A, Beller M. *Chem Eur J* 2009;15:1329–1336. (i) Schulz T, Torborg C, Schäffner B, Huang J, Zapf A, Kadyrov R, Börner A, Beller M. *Angew Chem Int Ed* 2009;48:918–921. (j) Schäffner B, Andrushko V, Bayardon J, Holz J, Börner A. *Chirality* 2009;21:857–861. (k) Lenden P, Ylioja PM, González-Rodriguez C, Entwistle DA, Willis MC. *Green Chem* 2011;13:1980–1982.
20. Henderson RK, Jiménez-González C, Constable DJC, Alston SR, Inglis GGA, Fisher G, Sherwood J, Binks SP, Curzons AD. *Green Chem* 2011;13:854–862.
21. (a) Fürstner A. *Angew Chem Int Ed Engl* 2000;39:3012–3043. (b) Hoveyda AH, Zhugralin AR. *Nature* 2007;450:243–251. (c) Donohe TJ, Fishlock LP, Procopiou PA. *Chem Eur J* 2008;14:5716–5726. (d) Connon SJ, Blechert S. *Angew Chem Int Ed* 2003;42:1900–1923. (e) Nicolaou KC, Bulger PG, Sarlah D. *Angew Chem Int Ed* 2005;44: 4490–4527.
22. Alfonsi K, Colberg J, Dunn PJ, Fevig T, Jennings S, Johnson TA, Kleine HP, Knight C, Nagy MA, Perry DA, Stefaniak M. *Green Chem* 2008;10:31–36.
23. (a) Buchowicz W, Mol JC. *J Mol Catal A: Chem* 1999;148:97–103. (b) Bai C -X, Lu X -B, He R, Zhang W-Z, Feng X -J. *Org Biomol Chem* 2005;3:4139–4142. (c) Stark A, Ajam M, Green M, Raubenheimer HG, Ranwell A, Ondruschka B. *Adv Synth Catal* 2006;348:1934–1941. (d)

Borré E, Dinh TH, Caijo F, Crévisy C, Mauduit M. *Synthesis* 2011;43:2125–2130. (e) Keitz BK, Endo K, Patel PR, Herbert MB, Grubbs RH. *J Am Chem Soc* 2012;134:693–699.

24. Miao X, Fischmeister C, Bruneau C, Dixneuf PH. *ChemSusChem* 2008;1:813–816.
25. Keraani A, Renouard T, Fischmeister C, Bruneau C, Rabiller-Baudry M. *ChemSusChem* 2008;1:927–933.
26. (a) Silva P, Peeva LG, Livingston AG. Nanofiltration in organic solvents. In: Li NN, Fane AG, Ho WSW, Matsuura T, editors. *Advanced Membrane Technology and Applications*. Hoboken, NJ, USA: John Wiley & Sons, Inc.; 2008. DOI: 10.1002/9780470276280.ch16. (b) Vandezande P, Gevers LEM, Vankelecom IFJ. *Chem Soc Rev* 2008;37:365–405. (c) Volkov AV, Korneeva GA, Tereshchenko GF. *Russ Chem Rev* 2008;77:983–993. (d) Janssen M, Müller C, Vogt D. *Dalton Trans* 2010;39:8403–8411.
27. Dupé A, Achard M, Fischmeister C, Bruneau C. *ChemSusChem* 2012;5:2249–2254.
28. Le Ravalec V, Fischmeister C, Bruneau C. *Adv Synth Catal* 2009;351:1115–1122.
29. Fabritius D, Schäfer H-J, Steinbüchel A. *Appl Microbiol Biotechnol* 1998;50:573–578.
30. Le Ravalec V, Dupé A, Fischmeister C, Bruneau C. *ChemSusChem* 2010;3:1291–1297.
31. Bilel H, Hamdi N, Zagrouba F, Fischmeister C, Bruneau C. *Green Chem* 2011;13:1448–1452.
32. Yoshikai K, Hayama T, Nishimura K, Yamada K-I, Tomioka K. *J Org Chem* 2005;70:681–683.
33. Bilel H, Hamdi N, Zagrouba F, Fischmeister C, Bruneau C. *RSC Adv* 2012;2:9584–9589.
34. Bakhrou N, Lamaty F, Martinez J, Colacino E. *Tetrahedron Lett* 2010;51:3935–3937.
35. Sheldon RA. *Green Chem* 2005;7:267–278.
36. Chen J, Spear SK, Huddleston JG, Rogers RD. *Green Chem* 2005;7:64–82.
37. Bantreil X, Sidi-Ykhlef M, Aringhieri L, Colacino E, Martinez J, Lamaty F. *J Catal* 2012;294:113–118.
38. Schmidt B. *Eur J Org Chem* 2004:1865–1880.
39. Kniese M, Meier MAR. *Green Chem* 2010;12:169–173.
40. Forbes MDE, Patton JT, Myers TL, Maynard HD, Smith DW Jr, Schulz GR, Wagener KB. *J Am Chem Soc* 1992;114:10978–10980.
41. Vo Thanh G, Loupy A. *Tetrahedron Lett* 2003;44:9091–9094.
42. Kuhn KM, Champagne TM, Hyeok Hong S, Wei W-H, Nickel A, Woo Lee C, Virgil SC, Grubbs RH, Pederson RL. *Org Lett* 2010;12:984–987.
43. (a) Thiel V, Hendann M, Wannowius K-J, Plenio H. *J Am Chem Soc* 2012;134:1104–1114. (b) Nunez-Zarur F, Solans-Monfort X, Rodrıguez-Santiago L, Sodupe M. *Organometallics* 2012;31:4203–4215.
44. (a) Corma A, Iborra S, Velty A. *Chem Rev* 2007;107:2411–2502. (b) Marshall A-L, Alaimo PJ. *Chem Eur J* 2010;16:4970–4980.
45. (a) Meier MAR, Metzger JO, Schubert US. *Chem Soc Rev* 2007;36:1788–1802. (b) Biermann U, Bornscheuer U, Meier MAR, Metzger JO, Schafer HJ. *Angew Chem Int Ed* 2011;50:3854–3871. (c) Perego C, Ricci M. *Catal Sci Technol* 2012;2:1776–1786.
46. Rybak A, Meier MAR. *Green Chem* 2007;9:1356–1361.
47. (a) Mutlu H, Meier MAR. *Eur J Lipid Sci Technol* 2010;112:10–30. (b) Van der Steen M, Stevens CV. *ChemSusChem* 2009;2:692–713.
48. Rybak A, Meier MAR. *Green Chem* 2008;10:1099–1104.
49. Jacobs T, Rybak A, Meier MAR. *Appl Catal A: Gen* 2009;353:32–35.
50. For reviews on ADMET polymerization, see: (a) Schwendeman JE, Cameron Church A, Wagener KB. *Adv Synth Catal* 2002;344:597–613. (b) Mutlu H, Montero de Espinosa L, Meier MAR. *Chem Soc Rev* 2011;40:1404–1445.
51. Lindmark-Hamberg M, Wagener KB. *Macromolecules* 1987; 20:2951–2954.
52. Wagener KB, Brzezinska K. *Macromolecules* 1991;24: 5273–5277.
53. Patton JT, Boncella JM, Wagener KB. *Macromolecules* 1992;25:3862–3867.
54. Wagener KB, Patton JT. *Macromolecules* 1993;26:249–253.
55. O'Gara JE, Wagener KB, Hahn SF. *Makromol Chem Rapid Commun* 1993;14:657–662.
56. (a) Smith JA, Brzezinska KR, Valenti DJ, Wagener KB. *Macromolecules* 2000;33:3781–3794. (b) Watson MD, Wagener KB. *Macromolecules* 2000;5733:8963–8970. (c) Watson MD, Wagener KB. *Macromolecules* 2000;33:5411–5417.
57. (a) Meier MAR. *Macromol Chem Phys* 2009;210:1073–1079. (b) Xia Y, Larock RC. *Green Chem* 2010;12:1893–1909.
58. Refvik MD, Larock RC, Tian Q. *J Am Oil Chem Soc* 1999;76:93–98.
59. Rybak A, Meier MAR. *ChemSusChem* 2008;1:542–547.
60. Fokou PA, Meier MAR. *Macromol Rapid Commun* 2008;29:1620–1625.
61. (a) Monsaert S, Lozano Vila A, Drozdzak R, Van Der Voort P, Verpoort F. *Chem Soc Rev* 2009;38:3360–3372. (b) Vidavsky Y, Anaby A, Lemcoff NG. *Dalton Trans* 2012;41:32–43.

23

OLEFIN METATHESIS IN FLUOROUS PHASES AND IN FLUORINATED AROMATIC SOLVENTS

Cezary Samojłowicz, Anna Kajetanowicz, and Karol Grela
Institute of Organic Chemistry, Polish Academy of Sciences, Warsaw, Poland

23.1 OLEFIN METATHESIS IN FLUORINATED ALIPHATIC HYDROCARBONS

Aliphatic fluorinated hydrocarbons (e.g., perfluoro (methylcyclohexane), perfluoro(2-butyltetrahydrofuran)), the so-called fluorous phases, are perfluorinated solvents that, while not being miscible at room temperature with "classical" solvents used in the chemical laboratories (e.g., dichloromethane or toluene), can however form homogeneous mixtures with these "classical" solvents at elevated temperatures (1). This unique observation has been made by Horváth and Rábai in 1994 (2), and since the first publication fluorous chemistry has become a powerful technology.

Well-established first- and second-generation ruthenium (pre)catalysts (3) (Grubbs—**1a–c**, indenylidene—**2a–d**, and Grubbs–Hoveyda—**3a,b** type complexes; Fig. 23.1) perform well in a broad range of solvents and temperatures (4); however, they are not soluble in the fluorous phases, even at high temperatures (5).

In 2003, Yao and Zhang synthesized ruthenium (pre)catalysts anchored to the fluorous polymer **4**, copolyacrylate consisting of fluorous and styrene units in the ratio of 10/1 (6). The authors, using well-established methodology for the synthesis of second generation of Hoveyda–Grubbs type alkylidene (7), obtained (pre)catalysts **5** in conversion greater than 90% (according to ^{1}H NMR, Fig. 23.2) (6).

(Pre)catalyst **5** anchored to the fluorous polymer exhibits high catalytic performance and could be cleanly recovered via extraction by treating the reaction mixture with FC-72 (perfluorohexane), and repeatedly reused. This technique has been successfully applied in ring-closing metathesis (RCM) reaction of *N,N*-diallyltosylamine (and other α,ω-dienes) in the mixture of $PhCF_3/CH_2Cl_2$ using the same catalyst up to 20 cycles. Interestingly, when the authors used FC-72 instead of $PhCF_3$ in the biphasic RCM process, they observed a lower reaction rate. The performance of olefin metathesis (pre)catalysts in $PhCF_3$ and other fluorinated aromatic hydrocarbons is discussed in the next section (6).

Ruthenium complex **5** shows high activity as well as stability and was recovered by FC-72 extraction and reused up to 20 times. The first 15 cycles were performed using 1 mol% of (pre)catalyst **5**; however, cycles 16–20 were performed using the same (pre)catalyst portion with lower loading (0.5 mol%), presenting reluctance for the decomposition of this system (6).

General Procedure for RCM, Product Isolation and Recycling of (Pre)Catalyst 5 by Fluorous Extraction (6)

Freshly prepared (pre)catalyst **5** was placed in a 50-ml flask equipped with a reflux condenser. The flask was

Olefin Metathesis: Theory and Practice, First Edition. Edited by Karol Grela.

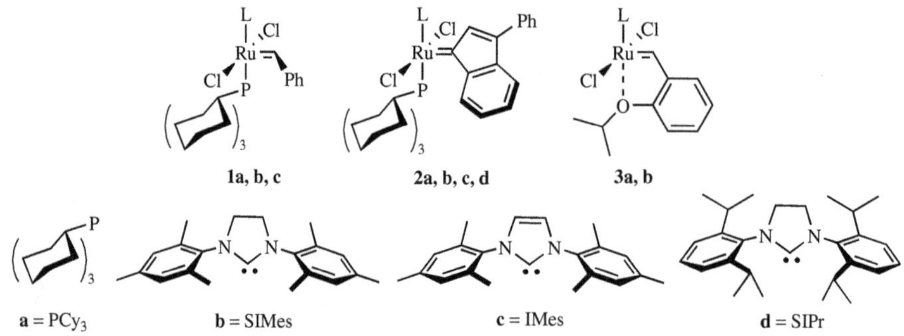

Figure 23.1 Ruthenium-based olefin metathesis (pre)catalysts.

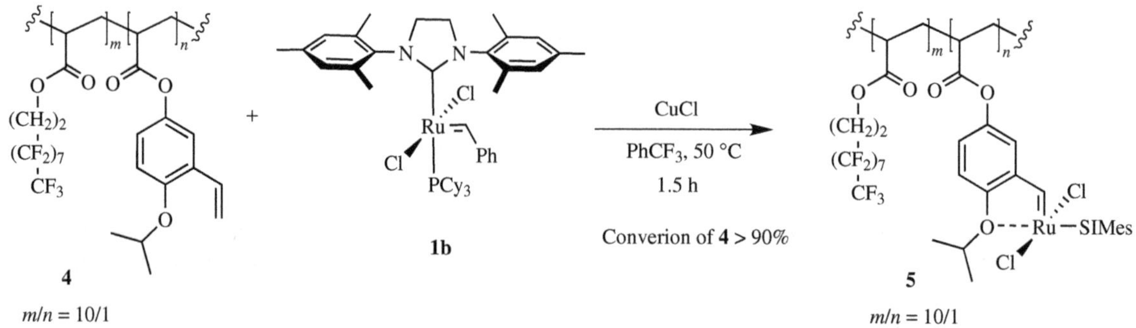

Figure 23.2 Synthesis of polymer-bound ruthenium carbene.

degassed three times by evacuation, filled with argon, and then $PhCF_3$ (0.5 ml) and CH_2Cl_2 (4.5 ml) were added to dissolve the catalyst. Then, *N,N*-diallyltosylamine (126 mg, 0.50 mmol) dissolved in CH_2Cl_2 (5 ml) was added. After heating to reflux (bath temperature 50 °C) for 1 h, the reaction mixture was cooled to room temperature and concentrated to dryness under reduced pressure. The crude reaction mixture was extracted using a mixture of FC-72 (5 ml) and EtOAc (5 ml). The FC-72 layer was separated and the EtOAc layer extracted again with FC-72 (3 ml). The FC-72 extracts were combined and washed with EtOAc (3 ml). After concentration of the combined EtOAc extracts, the crude reaction product was analyzed by ^{1}H NMR, which indicated complete conversion of the starting material and clean, essentially quantitative formation of the RCM product **7**. The FC-72 extracts were concentrated under reduced pressure. After drying under reduced pressure, the recycled (pre)catalyst **5** was used directly in a subsequent RCM reaction. A second run of the RCM was performed in a similar manner as described for the first run using the recycled catalyst. After fluorous extraction of the reaction mixture as described above, ^{1}H NMR analysis of the crude reaction product revealed a complete conversion of the starting material and quantitative formation of cyclic olefin **7**. This reaction was repeated 13 more times (15 times altogether), each time using (pre)catalyst recycled from the previous run (as 1 mol%). To show limits of that system, additional five cycles (cycles 16–20) of experiments were continued using 1.0 mmol of diene **6** in $PhCF_3$ (1 : 19 v/v, 20 ml) with the reaction time being 2 h for each cycle. The results are listed in Figure 23.3.

Promising results obtained by Yao et al. encouraged other researchers to pursue the synthesis of new ruthenium (pre)catalysts containing light fluorous tag (Fig. 23.4), that expected to find new complexes possessing the similar properties compared with complex **5**. However, none of the newly synthesized ruthenium complexes **8–13** have similar affinity to fluorous aliphatic solvents as those of **5**; thus, performing olefin metathesis reactions in fluorous phases was limited (6). The highest partition coefficient at 25 °C between $CF_3C_6F_{11}$ and toluene has been observed for complex **8c** (77.6 : 22.4) (8), which was enough to regenerate ruthenium carbene **8c** after olefin metathesis reaction, but with limited effectiveness. Interestingly, complexes **8a–c** have been used for RCM reactions in CH_2Cl_2/fluorous solvent mixture under biphasic conditions. The latter complexes **8a–c** exhibit acceleration rates comparable to standard conditions (using single solvent). The acceleration rate is related to transfer of the dissociated fluorinated phosphine to fluorous phase (Fig. 23.5), which gives the olefin substrate a better

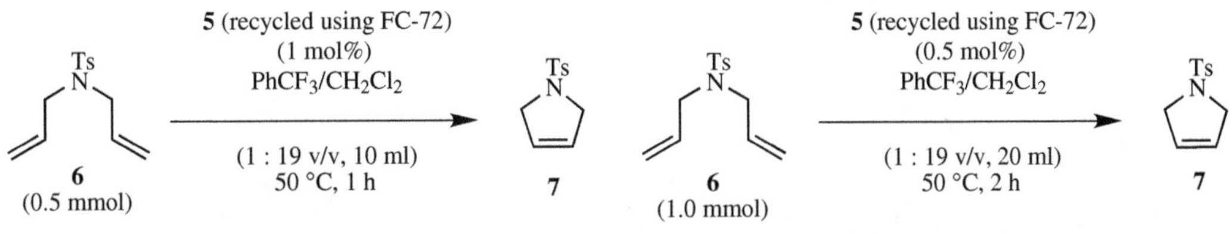

Cycle	1–15
% Conversion	>98

Cycle	16	17	18	19	20
% Conversion	>98	97	97	95	94

Figure 23.3 RCM using (pre)catalysts **5** containing fluorous polymer.

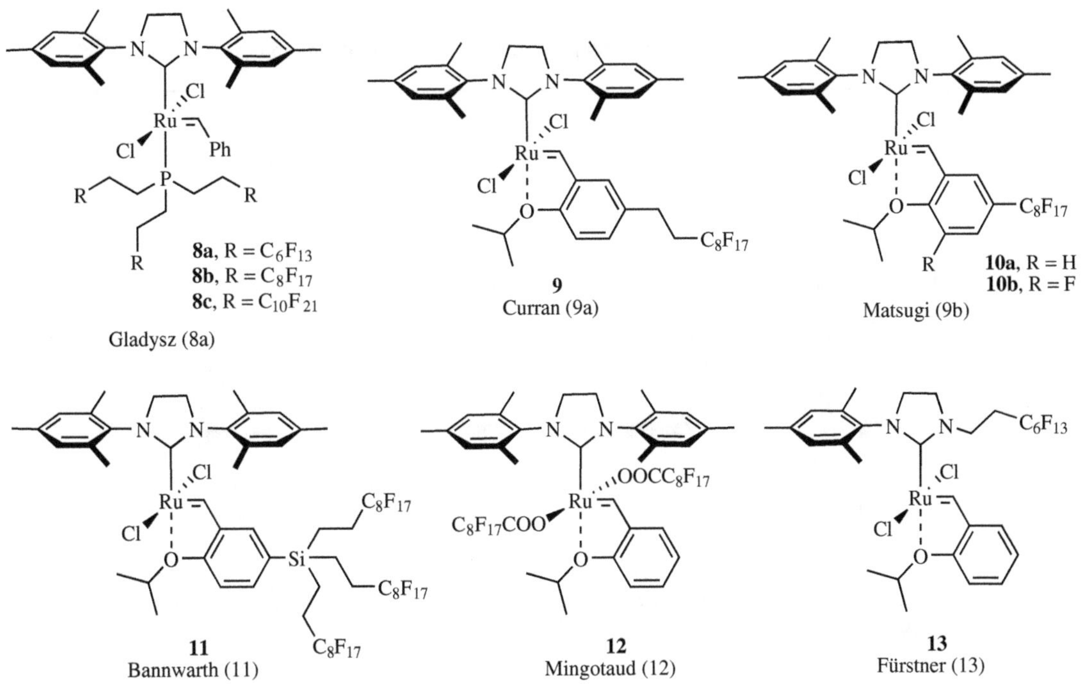

Figure 23.4 Ruthenium (pre)catalysts containing light fluorous tag.

access to the 14-electron ruthenium catalyst (both stayed in dichloromethane phase). As a result, the fluorinated ligand, dissociated from Grubbs-type (pre)catalyst **8**, is removed from the organic reaction phase and cannot compete with the substrate molecules for binding to a reactive metal center (8).

A different approach was applied by Curran and Matsugi who obtained (pre)catalysts **9** and **10a,b**, which can be recycled via fluorous solid phase extraction (F-SPE), utilizing their strong affinity to fluorous silica gel. The fluorous solid phase is typically silica gel with a fluorocarbon bonded phase ($-SiMe_2(CH_2)_2C_8F_{17}$) (9). In this methodology, the reaction mixture was loaded into F-SPE and the olefin metathesis product was separated from the (pre)catalysts via simple solvent switching that allows effective recovery of ruthenium complexes from the reaction mixture and their reuse (10). However, (pre)catalyst **10a** was recovered and reused up to five times, while complex **9** was reused up to seven times (in both cases 5 mol% of ruthenium complex was used) (9) (Fig. 23.6).

General Procedure for RCM Using Fluorous (Pre)Catalysts 10a and F-SPE (8b)

Diethyl diallylmalonate (375 mg, 1.56 mmol) and complex **10a** (81.8 mg, 0.078 mmol, 0.050 equiv) were dissolved in dichloromethane (31 ml, 0.05 M) under a nitrogen atmosphere, and the mixture was stirred for 1 h at room temperature. After removal of the volatile components under reduced pressure, the brownish mixture was submitted for separation by F-SPE. A short column was packed with fluorous silica gel (2.4 g) using

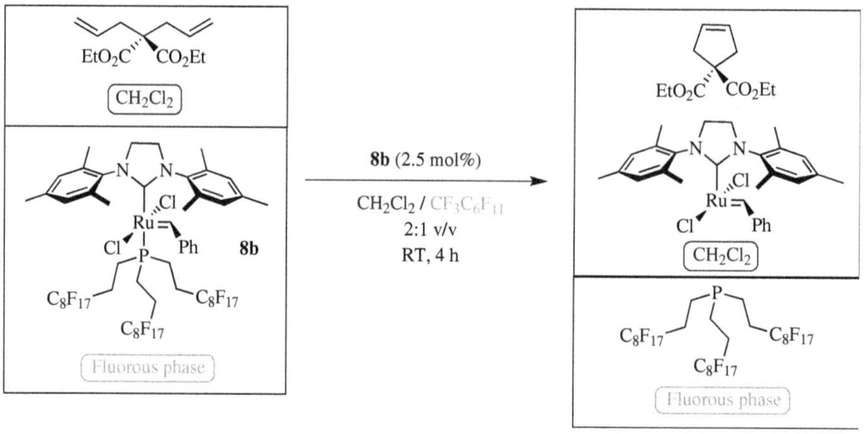

Figure 23.5 Ruthenium complex **8b** presented acceleration rate in biphasic solvents.

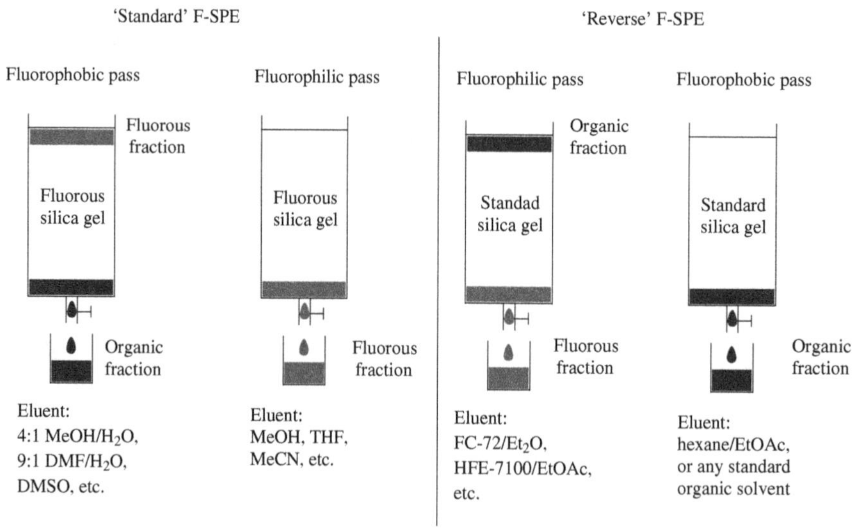

Figure 23.6 A visualization of a "standard" F-SPE and "reverse" F-SPE. The organic fraction is blue and the fluorous fraction is red. (*See plate section for color version.*)

aqueous 80% MeOH as the solvent. The crude reaction mixture was then loaded onto this column and eluted with 9.6 ml of aqueous 80% MeOH (elution of the RCM product) followed by 12 ml of tetrahydrofuran (THF) (elution of ruthenium complex). The evaporation of the aqueous 80% MeOH fraction and the THF fraction under reduced pressure gave the RCM product (diethyl cyclopent-3-ene-1,1-dicarboxylate) in 95% yield (314 mg) and recycled (pre)catalyst **10a** in 97% yield (79.7 mg).

In 2006, Michalek and Bannwarth (11) obtained ruthenium (pre)catalyst **11** which can be noncovalently immobilized on fluorous silica gel and then reversibly released into the reaction mixture. This technique has been designed as an alternative for expensive perfluorinated solvents, because the authors have used smaller amounts of fluorous silica gel as support for the noncovalently immobilized (pre)catalysts **11**. Directly after the reaction, a mixture of methanol and water (4 : 1) was added to the reaction mixture and (pre)catalyst **11** was re-immobilized; thus it can be separated by a simple filtration step and reused for further olefin metathesis reactions. This approach allows for up to seven catalytic cycles. Turnover number (TON) of 294 was reached using 1 mol% of complex **11**.

Mingotaud et al. (12) used complex **12** in ring opening metathesis polymerization performed in micellar solution in water. (Pre)catalyst **12** exhibits low activity in RCM of diethyl diallylmalonate; moreover, no examples of use of **12** in fluorous phases were reported.

Similarly, a unique ruthenium complex **13** bearing fluorous modified unsymmetrical *N*-heterocyclic carbene (NHC) ligand has not been used in fluorinated solvents (13).

23.2 OLEFIN METATHESIS IN FLUORINATED AROMATIC HYDROCARBONS

At the beginning of this section, it is important to point out that fluorinated aromatic hydrocarbons (FAH) cannot be regarded as fluorous solvents (sometimes they are called *faux fluorous*) (1); however, they may exhibit amphiphilic properties. Arene π-electron clouds and sp^2 carbon–fluorine bonds lead to significant intramolecular bond dipole, induced dipole, and quadrupolar interactions with non-fluorous molecules that cause, unlike aliphatic fluorinated hydrocarbons, the full miscibility of fluorinated aromatic hydrocarbons with "classical" organic solvents in a broad range of temperature (1). The early examples of application of hexafluorobenzene and trifluoromethylbenzene in olefin metathesis reactions catalyzed by ruthenium complexes have not proved the positive influence of these solvents on the reaction performance (14). The activating effect related to the use of fluorinated aromatic hydrocarbons in olefin metathesis was claimed for the first time in 2007 by Kadyrov et al. (15). The authors used fluorinated aromatic hydrocarbons in challenging olefin metathesis reactions utilizing commercially available second-generation ruthenium catalysts, which resulted in better conversions as compared to those obtained in "classical" solvents.

In 2008, Blechert et al. (16) reported on the highly efficient formation of tetrasubstituted olefins via RCM using Grubbs-type (pre)catalyst **14** (containing naphthyl substituted NHC) in the presence of hexafluorobenzene. Ruthenium carbenes **14** and **15** used in hexafluorobenzene showed a surprising activating effect, which allowed obtaining a number of challenging products in high yields (Fig. 23.7). Of special note is the RCM reaction of electron-deficient diene **16** catalyzed by 5 mol% of ruthenium complex **14** leading to lactone **17** with an 83% yield (17), when the previously reported reaction of the same substrate catalyzed by 20 mol% of ruthenium alkylidene **1b** gave the desired lactone **17** in only 40% yield (18).

In 2008, Grandbois and Collins (19) published an example of enantioselective synthesis of helicenes in fluorinated aromatic hydrocarbons, which notably extends the methodology of helicene synthesis (20). The asymmetric synthesis of [7]helicene was accomplished with good enantiomeric excess (80%). Three key factors contributed to the success of the kinetic resolution: the use of new ruthenium olefin metathesis (pre)catalysts **20** (bearing C_1-symmetric NHC ligand), addition of vinylcyclohexane to control the nature of the propagating alkylidene, and finally using hexafluorobenzene as a solvent (Fig. 23.8).

In 2008, Grela et al. (21) reported on the utilization of aromatic fluorinated hydrocarbons (especially hexafluorobenzene and octafluorotoluene) as solvents for olefin metathesis reactions catalyzed by commercially available ruthenium (pre)catalysts (**1–3**). This approach makes it possible to obtain substantially higher yields in metathesis of challenging substrates, including natural and biologically active compounds (Fig. 23.9). It was noted that the activating power is directly related to the number of fluorine atoms in the solvent molecule (e.g., $CF_3C_6F_5$ is better

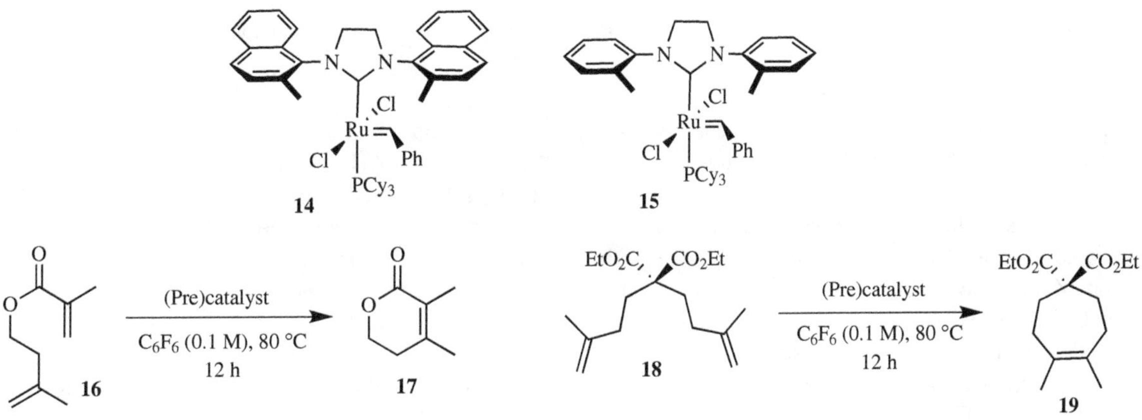

(Pre)catalysts	Conversion (%)
15 (5 mol%)	23
14 (5 mol%)	83
14 (0.25 mol%)	23

(Pre)catalysts	Conversion (%)
15 (3 mol%)	72
14 (3 mol%)	>99

Figure 23.7 Formation of tetrasubstituted C=C in RCM process using perfluorobenzene.

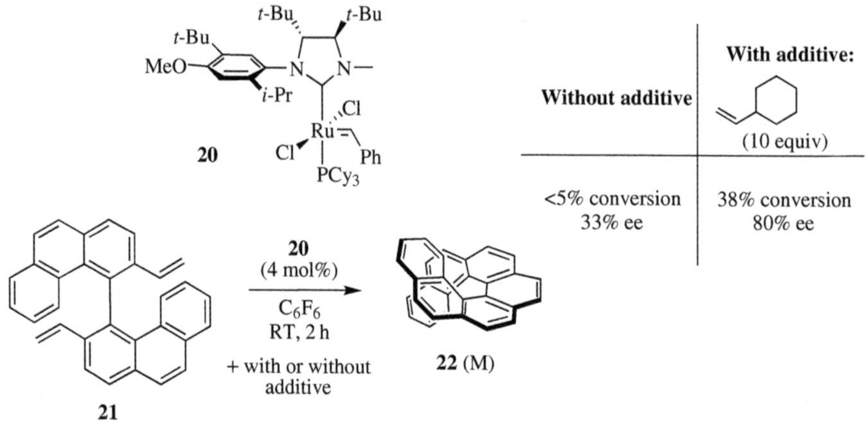

Figure 23.8 Synthesis of [7]helicenes in RCM using perfluorobenzene.

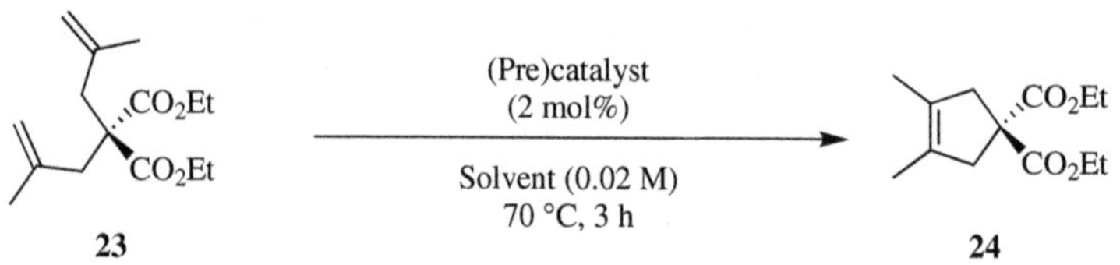

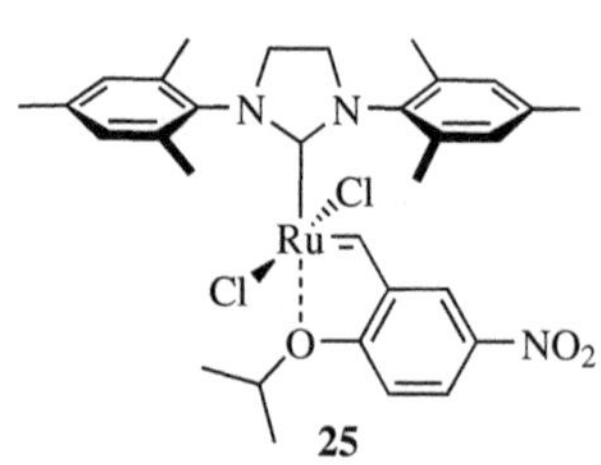

Solvent		**1b**	**1c**	**2b**	**2c**	**3b**	**25**	**26**
		Yield (%)						
"Classical" solvents	$C_2H_4Cl_2$	21	nd	28	38	4	19	19
	$C_6H_5CH_3$	20	nd	27	56	33	29	32
	$C_6H_5CF_3$	36	nd	48	61	13	nd	19
FAH	C_6F_6	61	77	91	74	58	nd	58
	$C_6F_5CF_3$	72	nd	94	85	72	49	61

(Pre)catalyst ← ; Activity ⇓

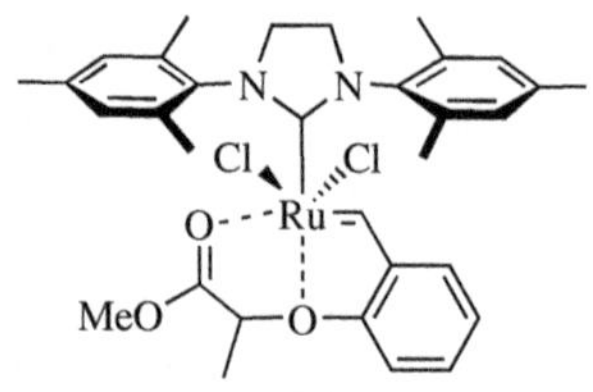

Figure 23.9 Comparative studies using commercially available ruthenium complexes.

than $CF_3C_6H_5$) and that phosphine containing catalysts are usually more activated than phosphine-free Hoveyda-type complexes (22).

Comparing physicochemical properties of the solvent (namely dielectric constant and dynamic viscosity) with the reaction results have not shown any direct correlation between them (23). Nevertheless, other studies have been undertaken to understand the possible role of fluorinated aromatic hydrocarbons: X-ray structural analysis of the second-generation Grubbs and Grubbs–Hoveyda type complexes (**1b** and **3b**) containing perfluorobenzene in the crystal, NMR measurements of ruthenium complexes in solutions containing hexafluorobenzene, and density functional theory (DFT) calculation studies. The initiation step of the ruthenium (pre)catalyst also has been studied in fluorinated aromatic hydrocarbons (24), and it was found that **1b** initiates olefin metathesis slower in FAH than in "classical" solvents (25), contrary to **3b**, which initiates olefin metathesis faster in FAH than in "classical" solvents. In general, it should be noted that interactions between the fluorinated solvent and the second-generation ruthenium complex are unspecific, and it is required to use the FAH in a large excess to observe any doping effect.

In 2011 Grisi et al. (26) reported the use of (pre)catalyst **27** bearing fluorine containing NHC in perfluorobenzene. Authors studied a standard set of dienes (including the challenging diene **23**) for RCM cyclization and they found the activation effect of (pre)catalyst **27** bearing fluorine substituent NHC used in fluorinated aromatic hydrocarbons compared to the same reaction performed in benzene (Fig. 23.10).

In 2011, Grela and Mauduit et al. (27) published an article on using fluorinated aromatic hydrocarbons for reactions performed in a microwave reactor. They observed a synergistic effect in a number of highly demanding olefin

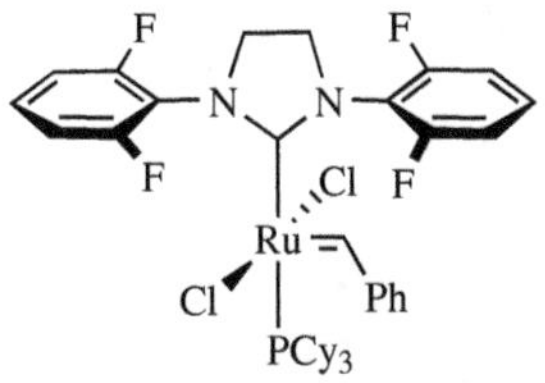

27
Grubbs (26b)
Grisi (26a)

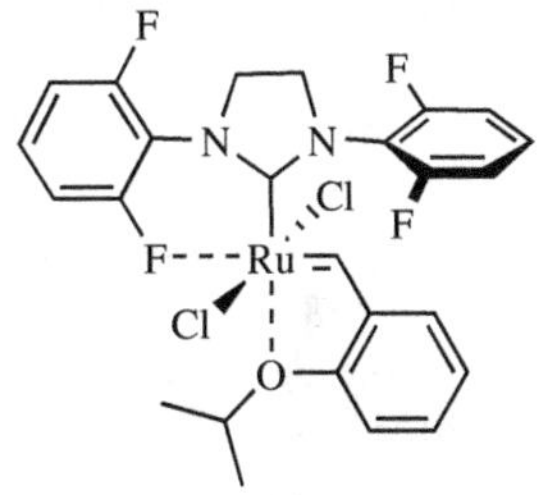

28
Grubbs (26b)

Figure 23.10 (Pre)catalysts bearing fluorine containing NHC.

metathesis reactions (e.g., RCM of **16**, **23**, and **29** as well as enyne metathesis of **31**). Interestingly, cyclization of polar diene **29** bearing chlorosubstituted double bond performed in "classical" condition gave no desired product (lactam **30**) at all. By simple changing toluene into hexafluorobenzene in a conventionally heated reaction (in closed vial), it was possible to obtain the desired product **30** albeit in low yield (29%). Finally, the RCM process of **29** performed in perfluorobenzene heated by microwave irradiation led to the desired product **30** in an acceptable yield of 69%.

In the challenging enyne cycloisomerization process, a similar synergistic effect was also observed. An interesting example was the cyclization of 1,5-enyne leading to cyclobutene derivatives (28). According to previous reports, only a 53% yield of expected product **32** was isolated, using dichloromethane as a solvent. Simply changing the solvent into hexafluorobenzene again allows obtaining the desired product **32** in much higher isolated yield 76% (Fig. 23.11).

General Procedure for the RCM Reactions for Microwave Heated Reactions (27)

Microwave Heating: A microwave vial was filled with diene **29** (0.4 mmol) and solvent (2 ml; hexafluorobenzenene or toluene) followed by addition of the (pre)catalyst **2b** (0.008 mmol). The vial was flushed with argon, sealed with the cap, and then placed in a microwave reactor and heated for the reported time. Progress of the reaction was monitored by thin layer chromatography (TLC) and ^{1}H NMR. Microwave heating was repeated three times until disappearance of the starting material. The content of the vial was then transferred to a round-bottomed flask and the solvent was removed under vacuum. The crude residue was purified by flash column chromatography (*c*-hexane:EtOAc) to obtain the analytically pure product **30**.

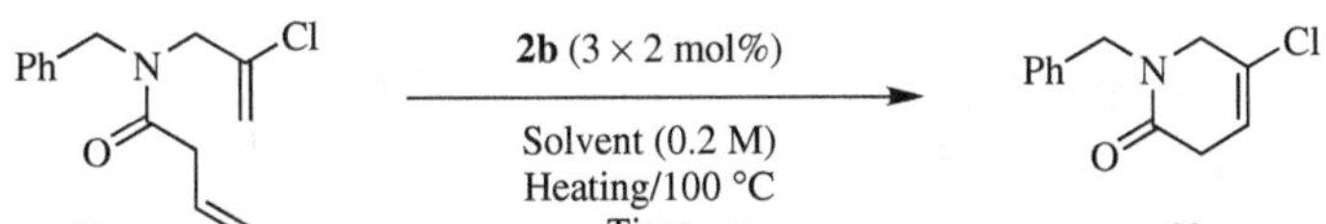

29 30

Entry	Solvent	Conditions/time	Yield (%)
1.	$CD_3C_6D_5$	Conventional/24 h	0
2.	C_6F_6	Conventional/3 × 5 min	29
3.	C_6F_6	Microwave/3 × 5 min	69

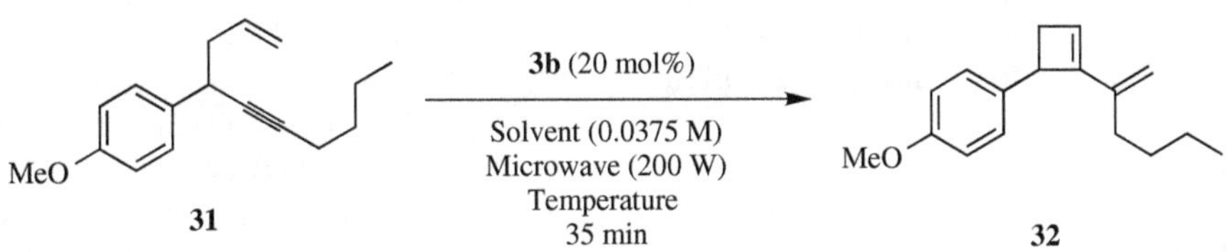

31 32

Entry	Solvent	Temperature	Yield (%)
1.	CH_2Cl_2	70 °C	53%
2.	C_6F_6	100 °C	76%

Figure 23.11 Challenging metathesis reactions utilizing microwave irradiation.

23.3 CONCLUSIONS AND PERSPECTIVE

Aliphatic as well as aromatic fluorinated solvents show a great use in olefin metathesis reactions. Ruthenium (pre)catalysts bearing fluorinated tag in benzylidene, NHC, as well as in anionic ligands part have been successfully obtained (29). Utilization of fluorous phases allow for easy separation of the metathesis product. On the other hand, olefin metathesis reactions performed in fluorinated aromatic hydrocarbons give higher yields of the desired product as compared with reactions conducted in "classical" solvents. Interestingly, positive influence of FAH was also observed in some organocatalytic reactions (30). Also perfluoroalkanes can be used to activate properly designed catalysts (8). In summary, fluorinated solvents can be considered as enabling techniques in modern organic synthesis (31).

ACKNOWLEDGMENTS

CS acknowledges funding from the Iuventus Plus program of the Polish Ministry of Science and Higher Education (Grant No. IP2010 0367 70). KG wishes to thank the Polish Ministry of Science and Higher Education (grant No. NN2044404940).

REFERENCES

1. Gladysz JA, Curran DP, Horvath IT. *Handbook of Fluorous Chemistry*. Weinheim, Germany: Wiley-VCH; 2004.
2. Horváth IT, Rábai J. *Science* 1994;266:72–75.
3. Precisely, the common ruthenium-based olefin metathesis catalysts should be called pre-catalysts: "Compounds that are well-characterized, and that under some conditions will catalyse the metathesis of olefins, but that have not been proven to be essentially identical to the active species for the metathesis reaction, are not well-defined catalysts. They are catalyst precursors, or (pre)catalysts." Schrock RR. *J Mol Catal A* 2004;213:21–30.
4. (a) Grubbs R. H., Eds. *Handbook of Metathesis*, Wiley-VCH, Weinheim, Germany, 2003; Vols. 1–3 (b) Samojłowicz C, Bieniek M, Grela K. *Chem Rev* 2009;109:3708–3742. (c) Vougioukalakis GC, Grubbs RH. *Chem Rev* 2010;110: 1746–1787.
5. (a) Curran DP. *Angew Chem* 1998;110:1230–1255. (b) Curran DP. *Angew Chem Int Ed* 1998;37:1174–1196. (c) Horváth IT. *Acc Chem Res* 1998;31:641–650. (d) Gladysz JA, Curran DP. *Tetrahedron* 2002;58:3823–3825. (e) Curran DP. *Aldrichim Acta* 2006;39:3–9.
6. Yao Q, Zhang Y. *J Am Chem Soc* 2004;126:74–75.
7. Bieniek M, Michrowska A, Gułajski Ł, Grela K. *Organometallics* 2007;26:1096–1099.
8. (a) Corrêa da Costa R, Gladysz JA. *Chem Commun* 2006:2619–2621. (b) Corrêa da Costa R, Gladysz JA. *Adv Synth Catal* 2007;349:243–254. (c) Tuba R, Corrêa da Costa R, Bazzi HS, Gladysz JA. *ACS Catal* 2012;2:155–162. (d) Tuba R, Brothers EN, Reibenspies JH, Bazzi HS, Gladysz JA. *Inorg Chem* 2012;51:9943–9949.
9. (a) Matsugi M, Curran DP. *J Org Chem* 2005;70:1636–1642. (b) Matsugi M, Kobayashi Y, Suzumura N, Tsuchiya Y, Shioiri T. *J Org Chem* 2010;75:7905–7908.
10. Zhang W, Curran DP. *Tetrahedron* 2006;62:11837–11865.
11. Michalek F, Bannwarth W. *Helv Chim Acta* 2006; 89:1030–1037.
12. Mingotaud A-F, Krämer M, Mingotaud C. *J Mol Catal A: Chem* 2007;263:39–47.
13. Fürstner A, Ackermann L, Gabor B, Goddard R, Lehmann CW, Mynott R, Stelzer F, Thiel OR. *Chem Eur J* 2001;7:3236–3253.
14. (a) Imhof S, Randl S, Blechert S. *Chem Commun* 2001:1692–1693. (b) Conrad JC, Amoroso D, Czechura P, Yap GPA, Fogg DE. *Organometallics* 2003;22:3634–3636.
15. Kadyrov R, Bieniek M, Grela K. Verfahren zur Metathese in electroarmen aromatischen Lösungsmitteln. Pat. Appl., DE 102007018148.7. 2007 Apr 16.
16. Rost D, Porta M, Gessler S, Blechert S. *Tetraherdon Lett* 2008;49:5968–5971.
17. Peak S-M. *Molecules* 2012;17:2248–2258.
18. D'Annibale A, Ciaralli L, Bassetti M, Pasquini C. *J Org Chem* 2007;72:6067–6074.
19. Grandbois A, Collins SK. *Chem Eur J* 2008;14:9323–9329.
20. (a) Collins SK, Grandbois A, Vachon MP, Côté J. *Angew Chem* 2006;118:2989–2992. (b) Collins SK, Grandbois A, Vachon MP, Côté J. *Angew Chem Int Ed* 2006;45:2923–2926.
21. Samojłowicz C, Bieniek M, Zarecki A, Kadyrov R, Grela K. *Chem Commun* 2008:6282–6284.
22. (a) Fürstner A, Thiel OR, Ackermann L, Schanz H-J, Nolan SP. *J Org Chem* 2000;65:2204–2207. (b) Ledoux N, Allaert B, Pattyn S, Mierde HV, Vercaemst C, Verpoort F. *Chem Eur J* 2006;12:4654–4661.
23. Samojłowicz C, Bieniek M, Pazio A, Makal A, Woźniak K, Poater A, Cavallo L, Wójcik J, Zdanowski K, Grela K. *Chem Eur J* 2011;17:12981–12993.
24. Ashworth IW, Nelson DJ, Percy JM. *Dalton Trans* 2013;42:4110–4113.
25. (a) Sanford MS, Ulman M, Grubbs RH. *J Am Chem Soc* 2001;123:749–750. (b) Sanford MS, Love JA, Grubbs RH. *J Am Chem Soc* 2001;123:6543–6554.
26. (a) Siano V, D'Auria I, Grisi F, Costabile C, Longo P. *Cent Eur J Chem* 2011;9:605–609. (b) Ritter T, Day MW, Grubbs RH. *J Am Chem Soc* 2006;128:11768–11769. (c) Vougioukalakis GC, Grubbs RH. *Chem Eur J* 2008;14:7545–7556.
27. Samojłowicz C, Borré E, Mauduit M, Grela K. *Adv Synth Catal* 2011;353:1993–2002.

28. (a) Debleds O, Campagne J-M. *J Am Chem Soc* 2008;130:1562–1563. (b) Grela K. *Angew Chem* 2008;120:5586–5589. (c) Grela K. *Angew Chem Int Ed* 2008;47:5504–5507. (d) Graham TJA, Gray EE, Burgess JM, Goess BC. *J Org Chem* 2010;75:226–228.

29. Kvíčala J, Schindler M, Kelbichová V, Babuněk M, Rybáčková M, Kvíčalová M, Cvačka J, Březinová A. *J Fluorine Chem* 2013;53:12–25.

30. (a) Lattanzi A, De Fusco C, Russo A, Poater A, Cavallo L. *Chem Commun* 2012;48:1650–1652. (b) Choudhury AR, Mukherjee S. *Org Biomol Chem* 2012;10:7313–7320.

31. (a) Kirschning A, Solodenko W, Mennecke K. *Chem Eur J* 2006;12:5972–5990. (b) O'Brien M, Denton R, Ley SV. *Synthesis* 2011:1157–1192. (c) Kotha S, Dipak MK. *Tetrahedron* 2012;68:397–421.

24

OLEFIN METATHESIS IN IONIC LIQUIDS

PIERRE QUEVAL, MATHIEU ROUEN, ANNIE-CLAUDE GAUMONT, ISABELLE DEZ, OLIVIER BASLÉ, AND MARC MAUDUIT
Ecole Nationale Supérieure de Chimie de Rennes, Rennes Cedex, France

During the last two decades, metal-catalyzed reactions in nonconventional ionic liquid (IL) solvents has attracted increasing attention (1). Among the numerous valuable transformations evaluated in this dynamic field, olefin metathesis reactions have experienced tremendous improvements, leading to highly efficient, safe, and environmentally respectful processes (2). While the first example of olefin metathesis reaction in IL was described using a tungsten-based complex (3), the increased interest for this transformation has resulted from the development of air, moisture, and functional-group-tolerant ruthenium-based catalysts.

24.1 IONIC LIQUIDS (ILS)

Since their introduction in 1888 (4), ILs have been studied in various research fields such as electrolyte support for battery (5), solvent (1), and task-specific catalyst (6). Many interesting properties have been pointed out such as the very low vapor pressure, the nonflammability, low toxicity and also easy modulation of viscosity by changing the IL counter anion. Actually, the most used room temperature ionic liquids (RTILs) are based on nitrogen aromatics (imidazolium and pyridinium salts) (Scheme 24.1).

Some ILs are immiscible with water and many organic solvents, affording biphasic or triphasic systems, which enable easy separation of reaction products from the IL phase. All these advantages have attracted the attention of chemists and have been considered in the development of industrial applications. As an example, the use of methylimidazole as an acid scavenger led to the formation of IL that can act as a solvent and also as a catalyst. BASF developed the BASIL™ process involving such transformation and an increase in production from 8 kg/m^3/h to 690,000 kg/m^3/h has been observed (7). Another interesting use of IL was in the DIFASOL® process (butene dimerization using nickel catalyst), which involves the use of IL as solvent, allowing the easy separation of the catalyst from the product and permitting batch continuous process (8).

24.2 USE OF NON TAGGED-IL CATALYSTS

Since the development of the olefin metathesis reaction (9), it has been pointed out that ruthenium catalysts are very efficient and have a large functional group tolerance, allowing their use in biologically active product synthesis (10). However, the main drawback of this reaction is the product contamination with residual metal catalysts (11). In order to avoid ruthenium leaching, many approaches have been developed including the use of IL as solvent.

24.2.1 Cationic Ru-Catalysts in RTILs

In the literature, few research groups have focused their attention on cationic ruthenium as an alternative to classical neutral precatalyst (12a–g) (Scheme 24.2).

Even more scarce are examples describing the use of such cationic complexes in IL media. For the first time it was reported by Olivier-Bourbigou and Dixneuf in 2002 (13). Catalyst **4** was easily prepared in a two-step synthesis from the commercially available ruthenium

Olefin Metathesis: Theory and Practice, First Edition. Edited by Karol Grela.

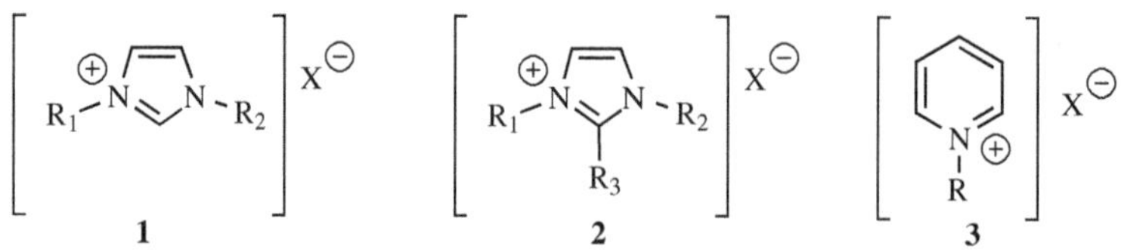

R_1 = Me, R_2 = Et, *n*-Bu, Hex, Oct.
X = BF_4, PF_6, AcO, TfO, NTf_2

Scheme 24.1 Mostly used RTILs.

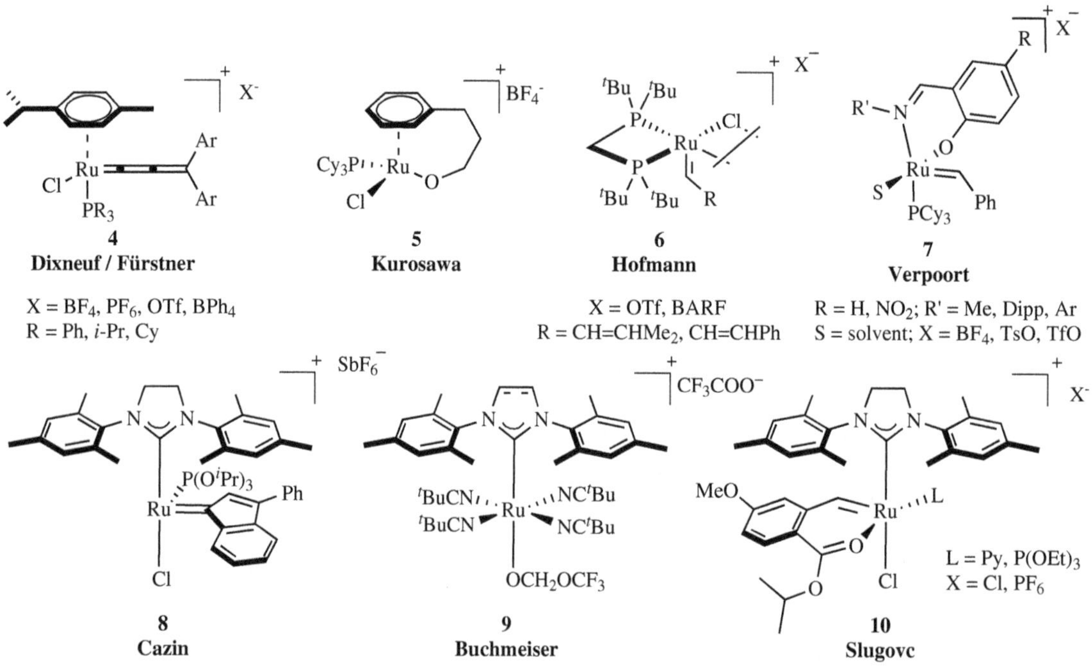

Scheme 24.2 Cationic ruthenium complexes.

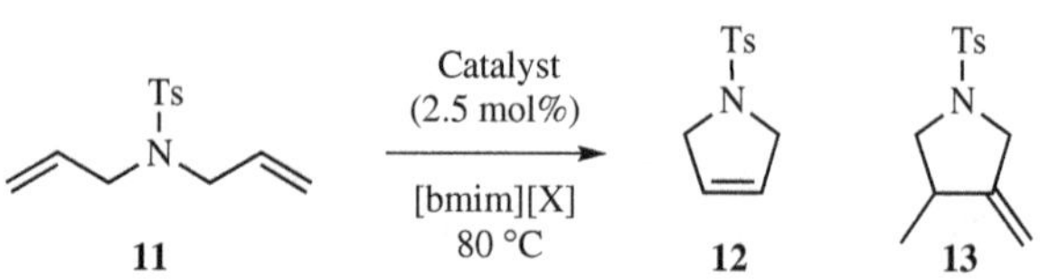

Scheme 24.3 RCM of diallytosylamine **11**.

dimer [{(*p*-cymene)$RuCl_2$)}$_2$]. During this study, the authors pointed out that ring-closing metathesis (RCM) of *N,N*-diallyltosylamine **11** using cationic precatalyst in IL was possible but slower than in toluene phase (Scheme 24.3). Despite this lower reactivity, better selectivity toward **12** formation was noticed, and the reaction strongly depended on the anion nature of the RTIL.

Indeed, when 1-butyl-3-methylimidazolium tetrafluoroborate ([bmim][BF_4]) was used as solvent, lower conversion was observed compared to reactions performed in [bmim][PF_6] or [bmim][OTf]. Moreover, the nature of the precatalyst anion has been modified, and the best results were obtained using the triflate counteranion, allowing complete conversion within 3 h at 80 °C for RCM of **11**. After examination of the scope of a small substrate, the authors attempted to recycle the catalyst immobilized in the ionic phase. Unfortunately, the low stability of the catalyst afforded poor recyclability and no more than two efficient runs (conversion > 86%). Ten years later, Buchmeiser and coworkers reported the formation of a new dicationic ruthenium precatalyst and its evaluation in the supported ionic liquid phase (SILP see Section 24.4.1.2) technology (14). The commercially available Hoveyda–Grubbs second generation **14** was reacted with two equivalents of $AgBF_4$ in DMF, leading to the formation of **15** in a quantitative manner (Scheme 24.4).

Complex **15** demonstrated good catalytic activity at 80 °C with low loading (0.02 mol%) for the metathesis reaction on simple substrates (diethyl diallylmalonate (DEDAM), 1,7-octadiene, methyl oleate (MO) ...) in both classical solvents and alkane/IL biphasic mixtures. Comparison of the cationic complex **15** with the neutral precatalyst **14** evidenced similar reactivity and range of ruthenium leaching, while side product formation were disfavored when

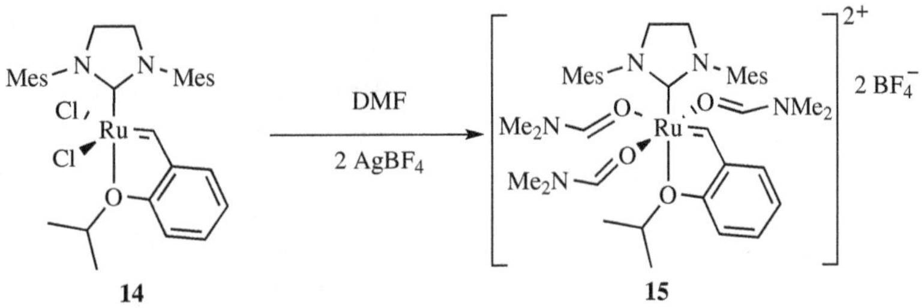

Scheme 24.4 Preparation of catalyst **15** by Buchmeiser.

using **15** (for MO). Interestingly, the reaction with the electron-poor *N,N*-diallyltrifluoroacetamide afforded higher conversion and lower ruthenium contamination with **15** than with **14**.

24.2.2 Neutral Ru-Catalysts in RTILs

Since the development of ruthenium catalysts for olefin metathesis reaction, many research groups have been looking for methods to remove ruthenium residues from the final product. Accordingly, numerous strategies have already been examined for the removal of homogeneous metathesis catalysts, including long purification procedures, use of metal scavengers and sequestering functionalities attached to solid supports (15). In addition, immobilization of homogeneous catalysts without structural modifications has also been considered, especially classical ruthenium catalysts in IL solvents (Scheme 24.5).

In 2001, Buijsman and coworkers reported the ring-closing metathesis reaction of the hydantoin precursor **19** catalyzed by first-generation ruthenium catalyst **16** in various ILs followed by ICP-AES measurement of residual ruthenium contents (Scheme 24.6) (16). High levels of ruthenium contamination (down to 3.2 μg/ml) could be detected in the RCM product, obtained with good conversion (>98%) after extraction with diethylether from [bmim][PF_6]. Optimization of the reaction conditions in [bmim][PF_6] afforded good RCM conversion for various substrates, surpassing in some cases the results obtained in DCM. Moreover, both catalysts **16** and **17** demonstrated recyclability under the optimized conditions for only three cycles. Notably, the NHC-based complex **17** afforded better catalytic activity and generated lower ruthenium contaminations (Table 24.1).

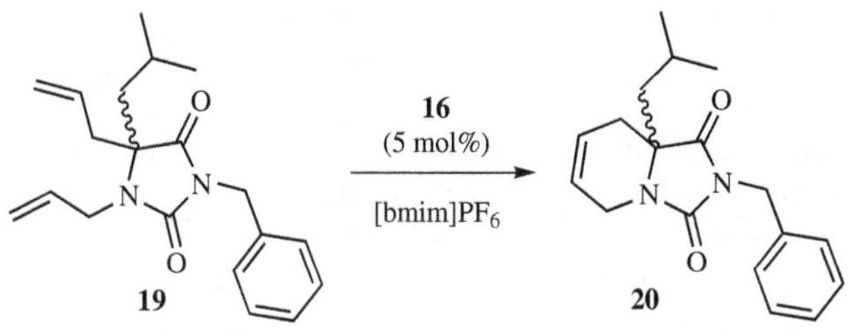

Scheme 24.6 RCM of hydantoin precursor by Buijsman.

TABLE 24.1 Recycling Tests of Ruthenium Catalysts 16 and 17

Entry[a]	Catalyst	Cycle	Conversions (%)[b]	Ru residue (ppm)
1	**16**	1	97	3900
2		2	94	4800
3		3	61	5300
4	**17**	1	95	1600
5		2	88	1600
6		3	74	1300

[a]All reactions were performed at 80 °C, a substrate concentration of 30 mg/ml, and 5 mol% of catalyst for 1 h.
[b]Conversions were measured after extraction with diethyl ether and subsequent HPLC analysis.

In 2002, Kiddle and coworkers (17) described the use of [bmim][BF_4] RTIL as solvent for the conduction of RCM reactions with complexes **16** and **17** under microwave

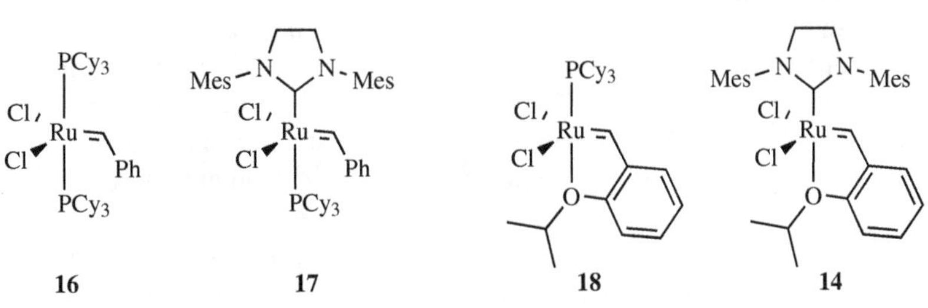

Scheme 24.5 Most used neutral ruthenium catalysts.

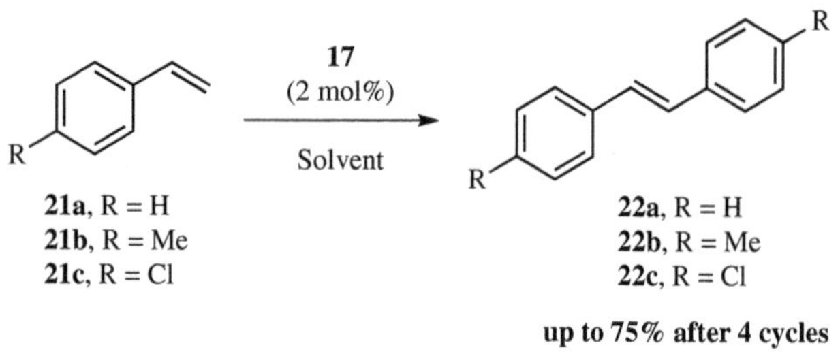

Scheme 24.7 Dimerization of terminal olefin by Tang.

irradiations. Indeed, IL appeared highly sensitive to irradiation allowing rapid increase of the vessel temperature. Moreover, as the irradiation energy is quickly transferred to the reaction, drastic decrease in reaction time could be obtained such as the completion of the RCM reaction within 2 min. During the same year, this methodology involving the use of ILs as solvent for olefin metathesis have been patented by BASF using, for instance, [emim][PF_6] as solvent (18). A few years later, the use of catalyst **17** was reported in the self-dimerization of terminal olefin using [bmim][X] (X = BF_4, PF_6) ILs (Scheme 24.7) (19). Interestingly, catalyst **17** was efficient in the dimerization of various terminal olefins up to four cycles without dramatic loss of activity.

In 2006, Williams and coworkers (20) reported the highly valuable dimerization of 1-octene using RTILs as solvent. The first generation Grubbs catalyst (bearing two phosphine ligands) underwent rapid deactivation in IL and extremely high ruthenium leaching into the organic layer (up to 72%). On the other hand, second-generation Grubbs catalysts demonstrated increased stability and selectivity in ILs. Indeed, the Hoveyda–Grubbs catalyst in the presence of IL, even in small volumes, afforded diminished ruthenium leaching (2%), high conversion (up to 96%), and high selectivity (up to 98%) for the self-metathesis of 1-octene to generate 7-tetradecene in RTILs bearing short alkyl chain. It has been suggested that the nonmiscibility of the primary metathesis product in the IL phase avoided the formation of side products subsequent to double bond isomerization (and metathesis reaction).

In 2009, Marvey and coworkers (21) reported the self-metathesis of MO with Grubbs first- and second-generation catalysts **16** and **17** in RTILs. The reaction in [bmim][X] (X = BF_4, PF_6, and NTf_2) was evaluated and compared to the reaction performed in classical solvents (DCM, DCE, PhMe, and PhCl). Best results were obtained using RTILs as solvent and lead to an increase in conversion and excellent selectivity (>98%). These studies were extended to the ethenolysis of MO and methyl rincenoleate (MR) using Hoveyda-type catalyst in similar media (Scheme 24.8) (22).

Hoveyda-type catalyst afforded better reactivity toward the self-metathesis of such compounds – a reactivity, which was further improved in [bmim][X] (X = BF_4, PF_6, NTf_2) RTILs. Unfortunately, the recycling of the Hoveyda-type catalyst was only efficient for three consecutive runs.

Recently, Rooney and coworkers (23) have classified the reactivity of ruthenium initiators in ring-opening metathesis homo- and copolymerization of norbornene and cyclopentene in RTILs. The neutral first-generation Grubbs catalyst **16** provided better results in pure RTILs whereas cationic initiator **4** worked better in a mixture of toluene and IL. Looking at the best results obtained, **16** showed greater initiation rate than the cationic precatalyst. Moreover, the addition of phenol to the reaction media dramatically increased the activity of **4**, but modification of the cationic precatalyst occurred, leading to a neutral ruthenium complex bearing an indenylidene moiety.

24.3 USE OF TAGGED-IONIC LIQUID CATALYSTS

During the last decade, tremendous improvements have been witnessed in the design of efficient tagged ruthenium catalysts capable of performing metathesis reactions in an IL-based biphasic system while minimizing the contamination of the final product by metal residues. In this context, the covalent incorporation of an ionic-tag motif appeared as one of the most effective strategies to increase the affinity between catalysts and the IL phase. In fact, this immobilization approach afforded stable and reusable catalysts with recent applications in continuous flow processes (24) and membrane reactors (25).

24.3.1 Ionic Liquid-Tag on the Leaving Group

In 2003, the groups of Mauduit (26) and Yao (27) independently reported a new catalyst design considering incorporation of an IL motif attached to the 2-isopropoxybenzylidene ligand (Hoveyda-type leaving group). In both cases, the choice for a catalyst with an imidazolium salt IL-tag (Fig. 24.1) demonstrated high activity and recyclability.

Mauduit, Guillemin and coworkers (26) developed and evaluated the catalyst **25** in the RCM of the model substrate *N,N*-diallyltosylamine **11** in pure [bmim][PF_6]. The catalyst **25** demonstrated very high efficiency over 10 cycles (Table 24.2) while, under the same conditions, a dramatic loss of activity was observed in the second cycle for Grubbs' and Hoveyda's catalysts. The new catalytic system was also applicable to a range of dienic substrates; however, limited recyclability (3–4 cycles) was observed with trisubstituted or oxygen-containing substrates owing to limited stability of **25** under prolonged heating conditions. Therefore, Mauduit and coworkers (28) developed the first second-generation IL-tagged catalyst **27** capable of performing the reaction with *N*-allyl-*N*-(2-methallyl)tosylamine **36** in biphasic [bmim][PF_6]/toluene and to be recycled eight times without loss of catalytic activity. Interestingly, the

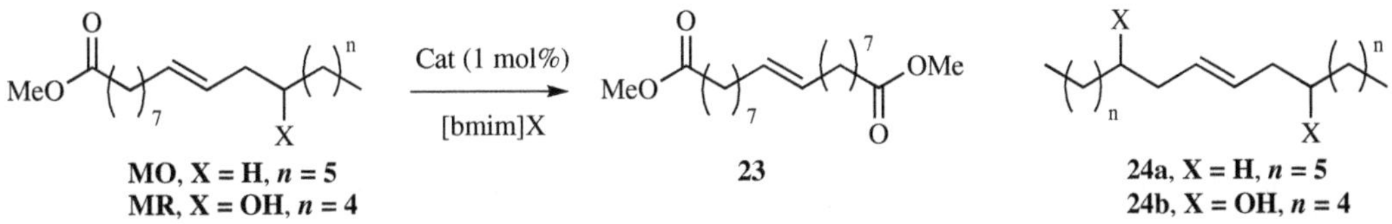

Scheme 24.8 Ethenolysis of MO and MR in RTILs.

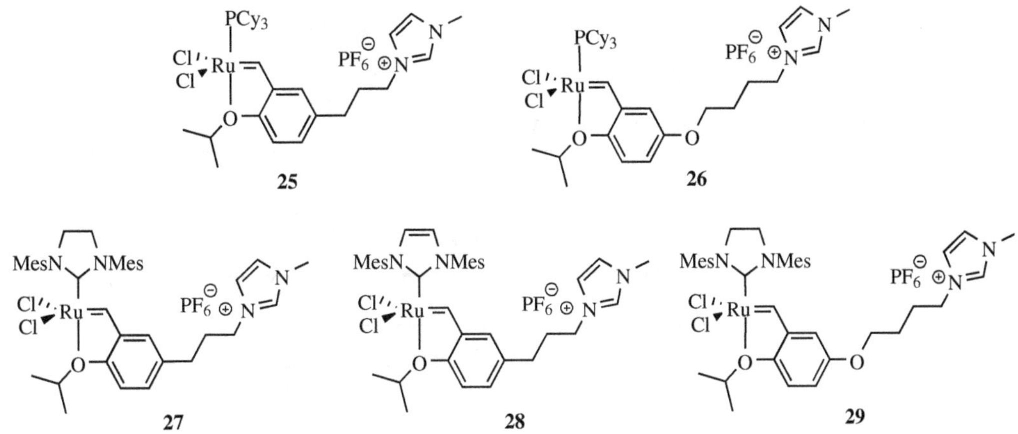

Figure 24.1 First generation (top) and second generation (bottom) Mauduit's catalysts **25, 27, 28** and Yao's catalysts **26, 29**.

TABLE 24.2 Reactivity and Recyclability of Mauduit's Catalyst 25

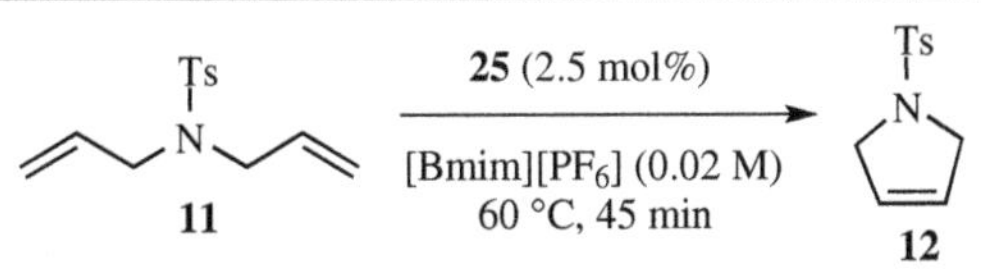

Cycle	1	2	3	4	5	6	7	8	9	10[a]
Conversion (%)	>98	>98	>98	>98	>98	96	92	92	92	95

[a]Performed on *N*-allyl-*N*-(but-3-enyl)tosylamine.

TABLE 24.3 Reactivity and Recyclability of Yao's Catalyst 26

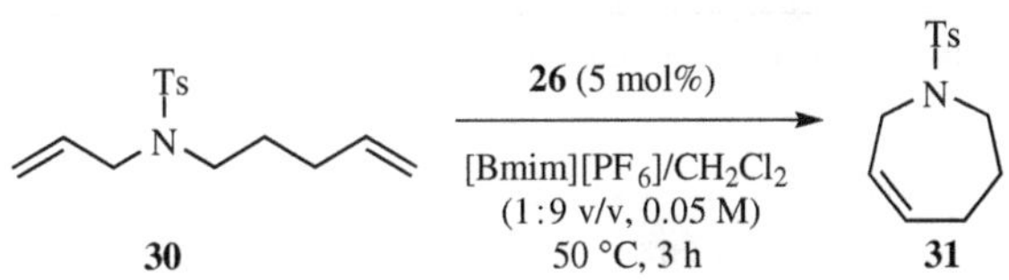

Cycle	1	2	3	4	5	6	7	8	9	10
Conversion (%)	>98	>98	97	96	95	94	92	92	91	90

residual ruthenium contamination ranged from only 1.2 to 22 ppm. Later, Mauduit, Nolan, and coworkers (29) synthesized catalyst **28**, containing an unsaturated NHC ligand, which showed a lower recyclability in cross metathesis than its saturated counterpart **27**.

On the other hand, catalyst **26** developed by Yao and Zhang (27) exhibited very good activity and recyclability in the RCM of the diene **30** in biphasic [bmim][PF_6]/CH_2Cl_2. Indeed, the formation of the seven membered-ring product **31** can be performed 10 times without significant loss of activity of catalyst **26** (Table 24.3). Yao and Zhang also evidenced the efficiency of **26** in RCM with other substrates to synthesize 5-, 6-, or 7-membered ring olefins and enyne metathesis products, regrettably, without information regarding the catalyst's recyclability in the presence of these substrates.

With the objective of increasing catalyst stability, Yao and coworkers (30) prepared in 2005 the second-generation catalyst **29**. Performed on *N*-allyl-*N*-(2-methallyl)tosylamine **36**, catalyst **29** exhibited good recyclability (6 cycles). Catalyst **29** can also be recycled over 17 times when used on the more reactive *N*, *N*-diallyltosylamine substrate **11** and evidenced high reactivity for the RCM of more challenging di-, tri-, and tetrasubstituted diene and enyne substrates although no data about the recycling were reported in those cases.

Further evolutions in catalyst design were reported by Mauduit, Grela, and coworkers based on the ionic tag nature and linker length (Fig. 24.2). Iminium tags (pyridinium (31,32) and imidazolium (26)) were found to possess better recycling properties than quaternary ammonium ones (32) (Table 24.4, entries 1, 4, 5). The authors also evidenced

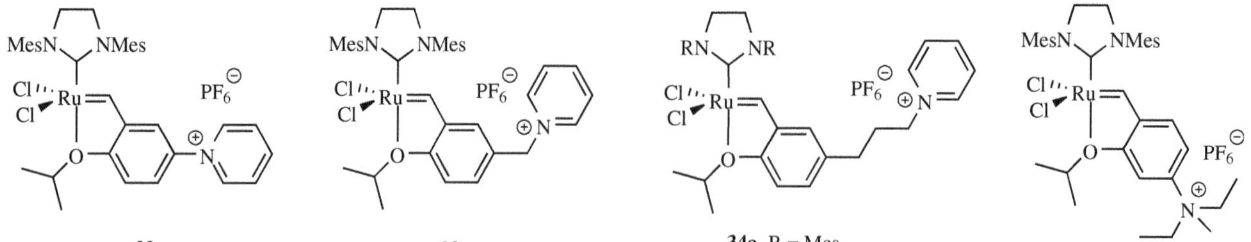

32 33 34a, R = Mes 34b, R = Dipp 35

Figure 24.2 Mauduit and Grela's catalysts.

TABLE 24.4 Recyclability of Mauduit and Grela's Catalysts 32, 33, 34, 35

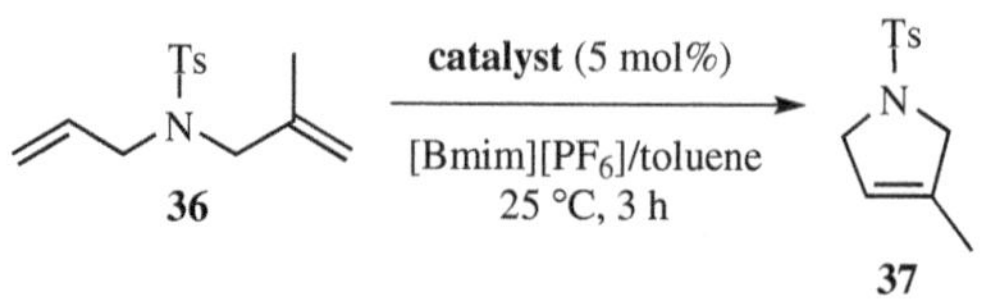

Entry	Catalyst	Cycle (conversion %)														
		1	2	3	4	5	6	7	8	9	10	11	12	13	14	15
1	**25**	>98	>98	>98	>98	>98	>98									
2	**32**	>98	95	65	6											
3	**33**	>98	>98	>98	>98	>98	>98									
4	**34a**	>98	>98	>98	>98	93	91									
5	**34b**[a]	>98	>98	>98	>98	>98	>98	>98	>98	>98	>98	96	95	94	94	93
6	**35**	98	33	2												

[a] 1 mol% catalyst, 60 °C, 30 min.

the necessity to insert at least a methylene spacer between the tag and the benzylidene fragment to allow for optimal catalyst recycling (Fig. 24.2, entries 2–4). In addition, they pointed out that the presence of a SIPr–NHC (saturated 2,6-diisopropylphenyl-*N*-heterocyclic carbene) ligand on catalyst **34b**, instead of SIMes–NHC (saturated mesityl–*N*-heterocyclic carbene) catalyst **34a**, increases the stability of the catalyst and thus permits a higher reusability at higher temperature and with a lower loading (1 mol%), over 15 cycles (Table 24.4, entries 4–5) (33).

In 2007, Dixneuf and coworkers evaluated catalysts **38** and **39** (Fig. 24.3) in the RCM of DEDAM **40** in [bmim][NTf_2]. They deliberately selected experimental conditions to avoid the complete conversion and to evaluate the reactivity and recyclability of catalysts **38** and **39** compared to Grubbs' and Hoveyda's catalysts (34). Compared to catalyst **38** carrying the IL tag on the oxygen coordinated to the ruthenium, catalyst **39** evidenced a better activity for the first two runs but the highly decreasing conversions observed in the following runs indicated a poorer reusability for **39** (Table 24.5). According to the authors, the better activity of **39** is due to both the presence of the *iso*-propoxy group on the ruthenium-coordinated oxygen and the bulkiness generated by a substituent on the *ortho* position from this ether

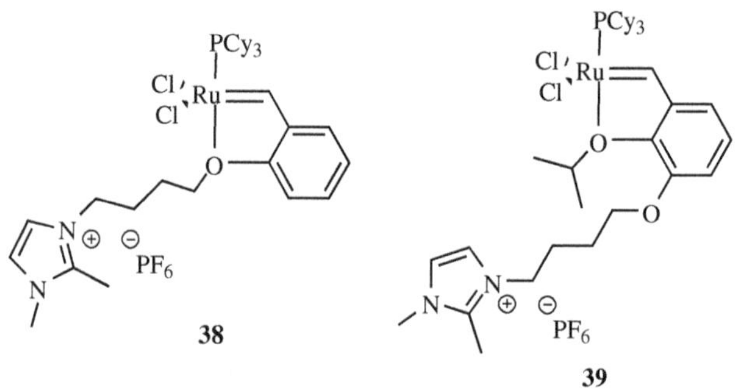

Figure 24.3 Imidazolium-tagged Dixneuf's catalysts.

as they were reported to be observed respectively by Hoveyda and coworkers (35) and by Blechert and coworkers (36) in organic solvents. The authors also explained the better reusability of **38** by the fact that the imidazolium tag was far from the oxygen atom.

The above-mentioned observations were consolidated when using catalysts **38** and **39** in the ethenolysis of methyl-oleate in [bmim][NTf_2]. In fact, **38** displayed a better activity than **39** for the first run but catalyst **38** appeared poorly recyclable also for this transformation (37).

TABLE 24.5 Influence of the tag Position on the Leaving Group

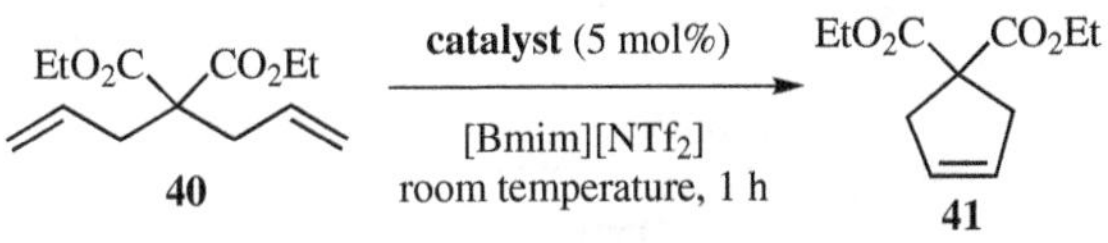

Entry	Catalyst	Cycle (conversion %) 1	2	3	4
1	**38**	24	22	22	16
2	**39**	53	39	23	8

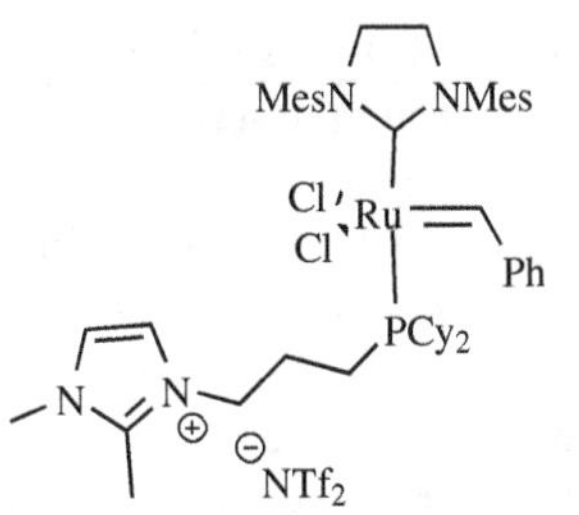

42

Figure 24.4 Consorti and Dupont's catalyst.

The attachment of an ionic phosphine ligand on a second-generation Grubbs-type catalyst was reported in 2008 by Consorti, Dupont, and coworkers (38). A phosphine ligand containing an imidazolium tag was synthesized and incorporated to produce catalyst **42** (Fig. 24.4). The evaluation of **42** in the RCM of the 1,7-octadiene in different ILs/toluene biphasic systems demonstrated a better recyclability in [bmim][PF_6]/toluene as shown by the absence of significant loss of activity after eight cycles. Moreover, the ruthenium contamination measured in the toluene phase after each run was below 2 ppm. Catalyst **42** also efficiently performed RCM on various di- and trisubstituted olefins, excluding tetrasubstituted dienes for which very poor yields were obtained. Consorti, Dupont, and coworkers (39) took advantage of the affinity between their catalyst and the IL media to develop a tandem isomerization–metathesis reaction on linear olefins using two distinct ruthenium catalysts that allowed for the expansion of light olefins into heavier ones.

Others groups have also been interested in the development of imidazolium-tagged catalysts on the leaving group. Wakamatsu and coworkers (40) and Lee and coworkers (41) have independently prepared new catalysts (Fig. 24.5). These catalysts present a quite similar activity for metathesis in ILs as those previously discussed in this chapter, although their reusability was reported to be slightly lower (4–7 cycles when reacted with *N,N*-diallyltosylamine **11**).

To our knowledge only one example of an anionic tagged catalyst used in IL has been reported in the literature. Wang and coworkers (42) have designed and evaluated the original carbollide-tagged ruthenium catalyst **46** in the RCM of various di- and trisubstituted dienes and also in enyne metathesis (Fig. 24.6). Catalyst **46** was found to be very efficient in both organic solvent and the IL [bmim][PF_6]. Interestingly, no significant loss of activity was observed in the RCM of the *N,N*-diallyltosylamine **11** at 40 °C after 10 cycles.

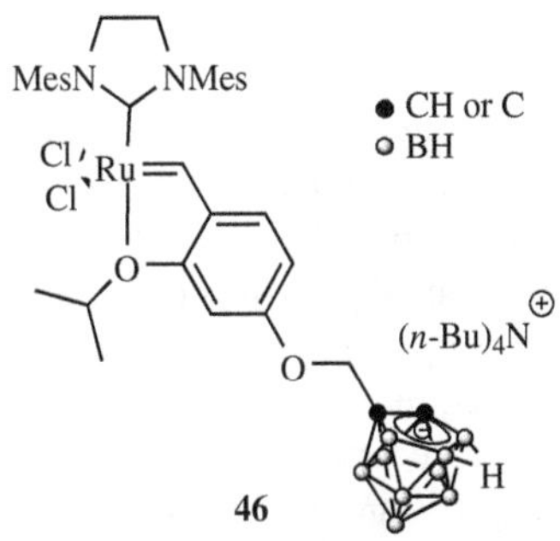

Figure 24.6 Wang's carbollide-tagged catalyst.

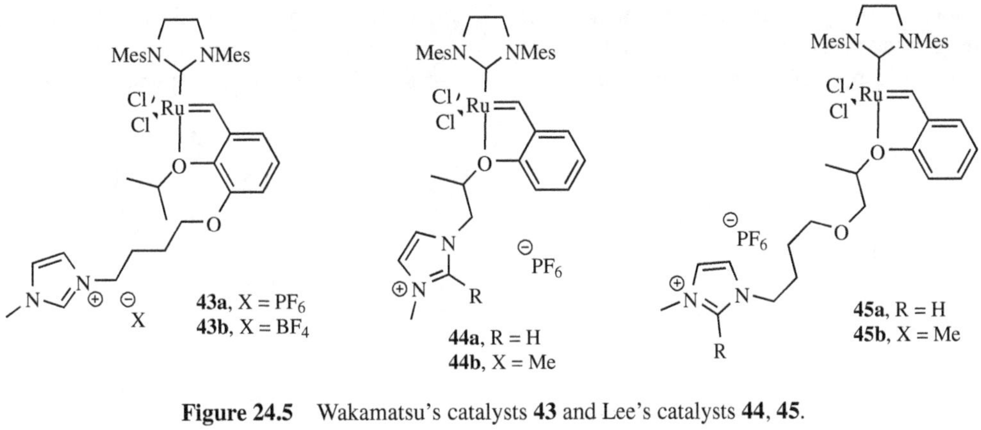

Figure 24.5 Wakamatsu's catalysts **43** and Lee's catalysts **44**, **45**.

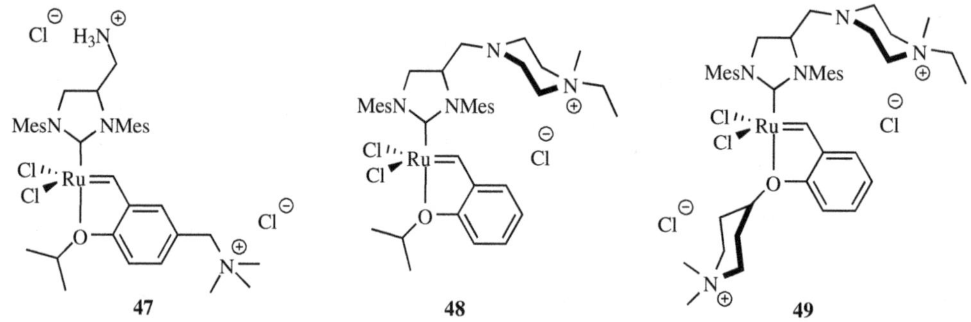

47 48 49

Figure 24.7 Ionic-tagged NHC Grubbs' catalyst **47** and Skowerski and Grela's catalysts **48** and **49**.

24.3.2 Ionic Liquid Tag on the *N*-Heterocyclic Carbene

To the best of our knowledge, no examples referring to the use of ionic tagged NHC ruthenium catalysts in ILs have been reported in the literature. However the catalyst **47** prepared by Grubbs and Jordan (43) and catalysts **48** and **49** developed by Skowerski, Grela and coworkers (44) are worth being mentioned (Fig. 24.7). Although those catalysts have been first and foremost designed to perform metathesis in water, Skowerski, Grela and coworkers mentioned that their immobilizations in IL will be a part of future applications.

24.4 SUPPORTED IONIC LIQUID CATALYSTS

24.4.1 Supported Ionic Liquid Phase (SILP)

Despite the numerous advantages of ILs in the catalytic processes (45), some drawbacks such as the cost of the ionic liquid, the poor recycling of the catalyst, the difficult separation of the catalyst and the product have prevented or limited their use in industry. Indeed, biphasic IL/organic liquid phases generally require significant amounts of IL, which is unattractive from an economical point of view owing to the high cost of many ILs and from an ecological point of view owing to the potent toxicity of many ILs. Furthermore, a considerable number of extractions using an organic solvent is usually required to recover the product from the polar IL. Lastly, a continuous leaching occurred after a number of extractions, leading to the deactivation of the catalytic phase (46). To overcome these problems, an approach using the concept of SILP has recently been proposed (47,48). In these SILP catalysts, a thin film of IL containing the homogeneous catalyst is immobilized at the surface of a solid support. This combination makes easier the separation of the product from the catalyst phase and enables the use of very small amounts of IL. Immobilization of ILs by adsorption or grafting on the support has been used with success in various reactions such as hydroformylation (48,49), Friedel–Crafts acylations (50), Suzuki cross-coupling (51), allylic substitution (52), Heck reaction (53), and hydrogenation (54) using a broad range of solid phases such as silica (47,48,51,55), alumina (46,49a), MFI zeolite (49b), chitosan (52), and carbon nanotubes (56).

Very recently, the SILP technology was applied with success to the ruthenium-catalyzed metathesis reaction. The various Ru-SILP catalysts used were all prepared by physisorption (non-covalent immobilization) of the ruthenium catalyst (neutral or ionically tagged) containing IL phase on a porous support.

24.4.1.1 SILP with Neutral Ru-Catalysts Hagiwara and coworkers first reported on the synthesis and application of SILP metathesis catalysts in 2008 using various inorganic supports to immobilize an IL layer containing Grubbs' I (**17**) and II (**18**) or Hoveyda's catalysts (**14**)(57).

In this work, different SILP catalysts based on complexes **17**, **18** and **14** (Fig. 24.8), different ILs, and various amorphous inorganic solids (silica, alumina powder, hydroxyapatites, molecular sieves, etc.) were prepared. These catalytic materials were applied in slurry phase reactions (solid SILP material dispersed in an organic solvent)

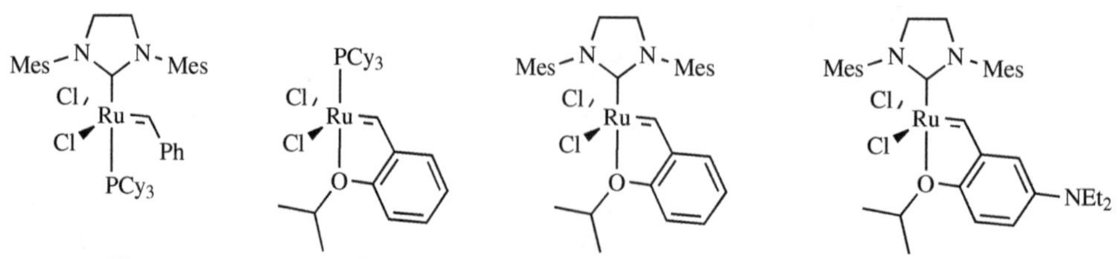

17 18 14 50

Figure 24.8 Most used neutral ruthenium catalysts for SILP.

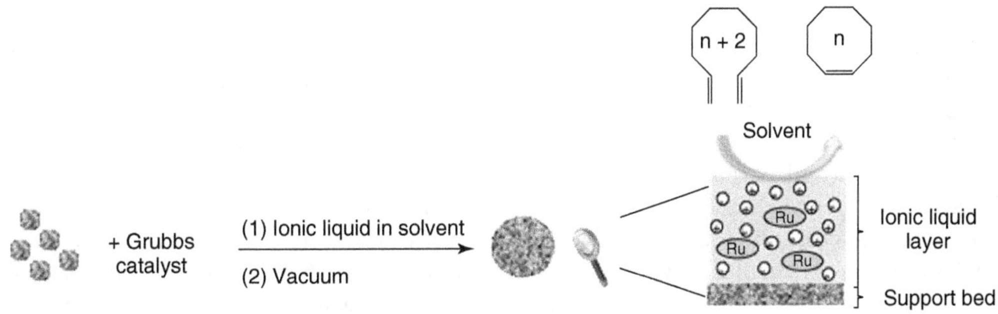

Figure 24.9 The SILP concept in olefin metathesis.

according to Fig. 24.9. The ring-closing metathesis reaction of diallylmalonate was used to tune the parameters of the reaction and define the best catalytic material. The best Ru-supported ionic liquid phase (Ru- (SILP) catalysts was based on an alumina support and [hmim][PF_6] (1-hexyl-3-methylimidazolium hexafluorophosphate) IL enabling a full conversion in benzene in only 1 h with 10 mol% of Grubbs' I catalyst. After reaction, the Ru-SILP was isolated by filtration and could be reused three times before showing clear signs of deactivation. The amount of Grubbs' catalyst could be lowered in this catalytic material (0.5 mol%) but at the expense of reaction time (18 h instead of 1 h for a full conversion).

These satisfactory preliminary results led the authors to use their Ru-SILP material for the efficient synthesis of 13- to 18-membered macrocyclic lactones through ring-closing olefin metathesis of bis-olefinic ester. The results obtained with these Ru-SILP are superior to those under homogeneous conditions. Furthermore, the catalytic materials could be used repeatedly several times (58). In this publication, an additional catalytic material was prepared through immobilization of a solution of complex **50** containing [hmim][PF_6] onto an acidic silica gel. This catalytic material exhibited the highest catalytic activity in the synthesis of the 16-membered lactone compared to the Ru-SILP catalysts based on **17**, **18** or **14**.

The Ru-SILP technology was also applied by the group of Wasserscheid in the gas-phase cross metathesis (CM) of various substrates in order to understand the effect of ethylene on catalyst activity and stability through theoretical (DFT calculations) and experimental studies. To this purpose, various SILP catalysts based on complexes **14**, **51** and **52** (Fig. 24.8 and 24.10), [bmim][X] as ILs (X = BF_4, PF_6, NTf_2, OTf, SbF_6, etc.) and calcinated silica 60 as support were tested in the propene self-metathesis at 40 °C under 1 bar using an average residence time of 32 s in the reactor. The kinetic studies were carried out as continuous gas-phase reactions in classical fixed-bed reactors with direct product analysis via online gas chromatography (59).

Among the different Ru-SILP catalysts tested, those prepared from [bmim][NTf_2] showed a rapid deactivation. This

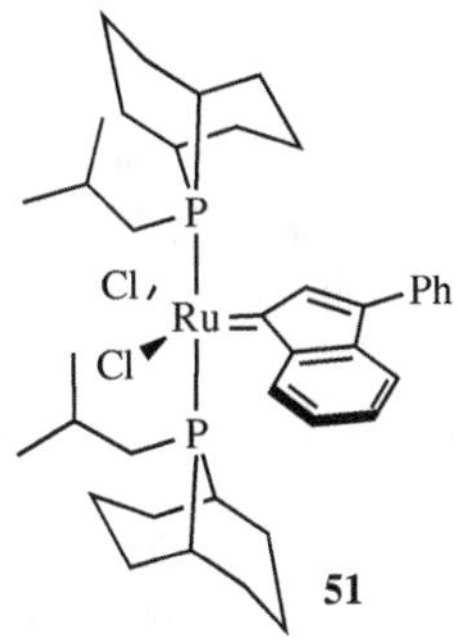

51

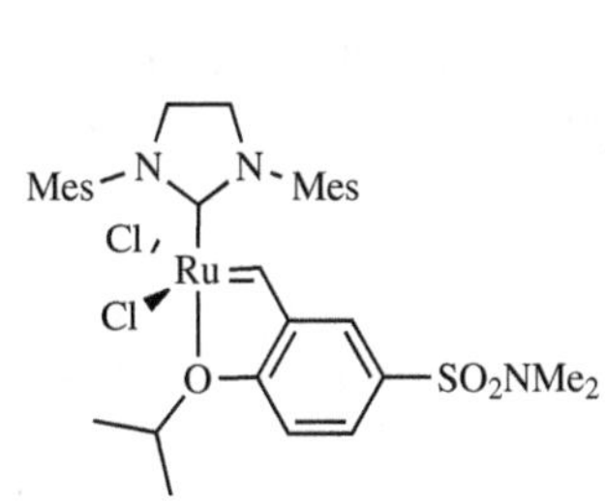

52

Figure 24.10 Neutral catalysts used by Wassercheid in the gas phase CM.

behavior was linked to the high solubility of ethylene in [bmim][NTf_2]. Authors showed through both kinetic studies and DFT calculations that active Grubbs-type metathesis catalysts suffer from temporary inhibition caused by the presence of large amounts of ethylene, because of a reversible shift of active Ru complexes to dormant Ru species during the catalytic cycle. By avoiding the catalytic formation of ethylene, the Ru-SILP catalyst showed constant catalytic activity and selectivity for more than 20 days of continuous gas-phase operation.

24.4.1.2 SILP with Charged Ru-Catalysts In 2010, Bruneau and coworkers (60) reported, for the first time, on the immobilization of an ionically tagged Hoveyda catalyst **53** (Fig. 24.11) in an ionic liquid supported on a solvent-resistant polyimide membrane (Starmem 228).

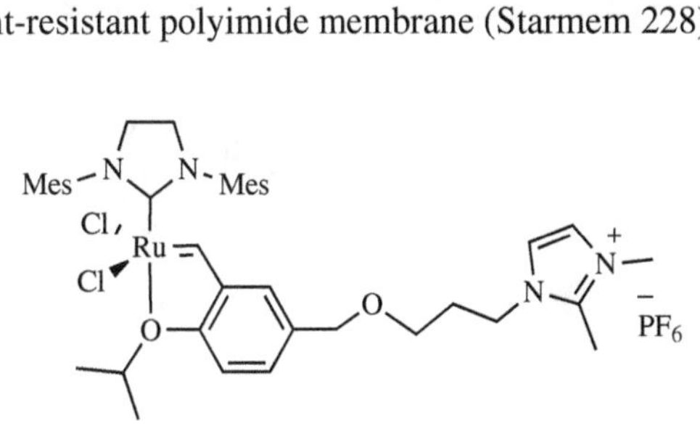

53

Figure 24.11 Ionic-tagged catalyst **53** developed by Bruneau.

The efficiency of this catalytic membrane was demonstrated in the RCM reaction of diallyltosylamine in toluene at 35 °C using a nanofiltration membrane reactor in a discontinuous mode, showing the feasibility of the concept. Although efficient, a deactivation of the catalytic membrane was noticed from the third reuse. One of the explanations proposed by the authors is the leaching of the catalyst due to the partial solubility of the ionic liquid in the solvent used (toluene).

Lastly, in a joint project, the groups of Cole-Hamilton and Mauduit reported the first application of a Ru-SILP catalyst, based on the ionic tagged boomerang Ru catalyst **27**, in a continuous flow olefin metathesis process with supercritical CO_2 (sCO_2) as the flowing medium (SILP-SCF) (24). The choice of sCO_2 as transport vector over systems where liquids flow through SILP catalysts is related to the low solubility of the ionic liquid and the catalyst in sCO_2. It also offers advantages over all gas phase systems using SILP catalysts because the flow rate of less volatile substrates and products can be higher (Fig. 24.12).

The Ru-SILP catalyst was prepared by mixing ionically tagged Ru-precatalyst **27** in [bmim]NTf_2 and silica (Sigma-Aldrich, Merck silica gel 100, grade 10184, 70–230 mesh, 10 nm pore diameter), previously heated to 500 °C to remove surface hydroxides, in dichloromethane as solvent. The catalytic material was obtained as a free-flowing pale green powder after removal of dichloromethane.

The model reaction studied is the RCM of diethyldiallylmalonate. Although a good activity of the catalyst was observed in the first hours, a rapid deactivation of the catalyst was noted. Using 1-octene as substrate in a self-metathesis reaction, it was proposed by the authors that the catalyst deactivation occurred through degradation of ruthenium methylidene, indicating that terminal alkenes were not appropriate substrates in this process. Next, self-metathesis of benchmark internal alkenes was performed. As expected, a high reactivity was observed over at least 10 h with only a slight falloff in activity in time. High total turnover numbers (TONs) > 10,000 are possible over 9 h of reaction. Worthy of note is that products were obtained in high yields and with very low metal contamination. Furthermore, authors studied the metathesis of a compound of industrial importance, methyloleate. Sixty-four percent of conversion was obtained over the first 6 h. From the 7th hour, the conversion dropped to 44% showing some catalyst instability. The cross metathesis of MO and diethyl maleate was carried out successfully in [omim][NTf_2] (1-octyl-3-methylimidazolium) ionic liquid with **27**-Ru-SILP catalyst using batch conditions. However, applied to a flow system, the initially reasonable activity toward the cross metathesis products dropped rapidly, while the self-metathesis of MO was the main reaction.

Lastly, Buchmeiser et al. described the preparation of the new dicationic Ru-alkylidene complex **15** (Scheme 10.4), which was successfully applied in various metathesis reactions in organic solvent and under organic solvent/IL biphasic conditions (see Section 24.2.1) (14). The authors extended its use to continuous metathesis reactions by exploiting the SILP technology. For these purposes, polymeric monoliths were obtained by co-polymerization of norborn-2-ene (NBE) with an NBE-based cross-linker followed by the surface modification of the obtained monolith with an ionic NBE derivative. The macroporous structure of the monoliths makes them ideal supports for continuous metathesis reactions under biphasic conditions. Finally, the support was impregnated by a solution of [bdmim][BF_4] and **15** in CH_2Cl_2, followed by the elimination of the organic solvent under vacuum. The metathesis reaction was performed using the monolith-supported [bdmim][BF_4]/**15** and heptane as mobile phase. The monolith-supported

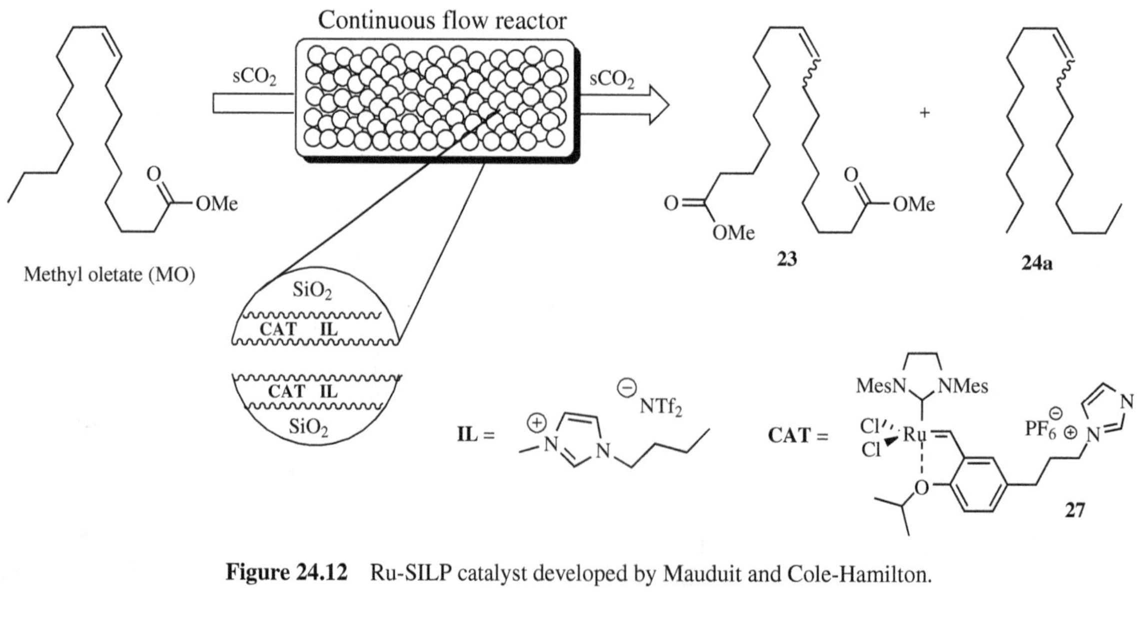

Figure 24.12 Ru-SILP catalyst developed by Mauduit and Cole-Hamilton.

[bdmim][BF_4]/**15** showed high catalytic performances in various RCM reactions better than those obtained with the monolith-supported [bdmim][BF_4]/**14**. Moreover, the metal leaching in the heptane phase was lower by a factor of 4–10 for [bdmim][BF_4]/**15** compared with [bdmim][BF_4]/**14** catalyst. A high TON was also obtained within 45 min at 80 °C in the self-metathesis of the MO which is remarkable with regard to its low solubility in the IL.

24.5 CONCLUSION AND PERSPECTIVES

In summary, the recent years have witnessed tremendous improvements in olefin metathesis in ILs. The early reports involving existing neutral and cationic ruthenium complexes provided limited benefits with poor recyclability and high catalyst leaching. Subsequent evolutions incorporating ionic-tag motifs afforded efficient and highly recyclable catalysts, while minimizing the contamination of the product by metal residues. More recently, the transposition of these homogeneous tagged catalysts to the SILP technology improved the overall catalytic processes with easier separation of the product from the catalytic phase and enabled the use of very small amounts of the costly ionic liquids. Finally, possible future developments may involve more robust ruthenium catalysts containing ionic tags covalently attached to the strongly coordinated NHC-ligand.

REFERENCES

1. For recent reviews on catalysis in IL, see: (a) Hallett JP, Welton T. *Chem Rev* 2011;111:3508–3576. (b) Pârvulescu VI, Hardacre C. *Chem Rev* 2007;107:2615–2665.
2. For a recent review, see: Śledź P, Mauduit M, Grela K. *Chem Soc Rev* 2008;37:2433–2442.
3. (a) Chauvin Y, Olivier-Bourbigou H. *Chemtech* 1995; 25:26–30. (b) Molybdenum-based Shrock catalyst was also evaluated at the early stage of this study: Bayer AG. *Patent Appl.*, EP 1 035 093 A2; 2000.
4. Gabriel S, Weiner J. *Ber Dtsch Chem Ges* 1888;21:2669–2679.
5. Chum HL, Koch VR, Miller LL, Osteryoung RA. *J Am Chem Soc* 1975;97:3264–3265.
6. Lee S-G. *Chem Commun* 2006:1049–1063.
7. Saling P, Maase M, Vagt U. *Handbook of Green Chemistry*. Vol. 6. Wiley; 2010. p 299–314.
8. Olivier-Bourbigou H, Favre F, Forestière A, Hugues F. *Handbook of Green Chemistry*. Vol. 1. Wiley; 2010. p 101–126.
9. For comprehensive reviews on olefin metathesis, see: (a) *Handbook of Metathesis* (Ed.: Grubbs, R. H.), Wiley-VCH, Weinheim, 2003. (b) Vougioukalakis GC, Grubbs RH. *Chem Rev* 2010;110:1746–1787.
10. For a recent review on the use of metathesis in total synthesis, see: (a) Fürstner A. *Chem Commun* 2011;47:6505–6511. For application in biologically active product synthesis, see: (b) Barbaras D, Brozio J, Johannsen I, Allmenedinger T. *Org Process Rev Dev* 2009;13:1068–1079.
11. For a review of different concepts that are used for the synthesis of supported-Ru-based pre-catalysts, see: (a) Clavier H, Grela K, Kirschning A, Mauduit M, Nolan SP. *Angew Chem Int Ed* 2007;46:6786–6801. For an interesting review on the concept of soluble supports, see: (b) Bergbreiter DE, Tian J, Hongfa C. *Chem Rev* 2009;109:530–582.
12. (a) Fürstner A, Leibl M, Lehmann CW, Picquet M, Kunz R, Bruneau C, Touchard D, Dixneuf PH. *Chem Eur J* 2000;6:1847–1857. (b) Miyaki Y, Onishi T, Kurosawa H. *Inorg Chim Acta* 2000;300–302:369–377. (c) Wang D, Wurst K, Knolle W, Decker U, Prager L, Naumov S, Buchmeiser MR. *Angew Chem Int Ed* 2008;47:3267–3270. (d) Volland MAO, Hansen SM, Rominger F, Hofmann P. *Organometallics* 2004;23:800–816. (e) Zirngast M, Pump E, Leitgeb A, Albering JH, Slugovc C. *Chem Commun* 2011;47:2261–2263. (f) Songis O, Slawin AMZ, Cazin CSJ. *Chem Commun* 2012;48:1266–1268. (g) Drozdzak R, Allaert B, Ledoux N, Dragutan I, Dragutan V, Verpoort F. *Coord Chem Rev* 2005;249:3055–3074.
13. Semeril D, Olivier-Bourbigou H, Bruneau C, Dixneuf P. *Chem Commun* 2002:146–147.
14. Autenrieth B, Frey W, Buchmeiser MR. *Chem Eur J* 2012; 18:14069–14078.
15. Vougioukalakis GC. *Chem Eur J* 2012;18:8868–8880.
16. Buijsman RC, Van Vuuren E, Sterrenburg JG. *Org Lett* 2001;3:3785–3787.
17. Mayo KG, Nearhoof EH, Kiddle JJ. *Org Lett* 2002; 4:1567–1570.
18. Guertler C, Jautelat M. EP1035093-A2; 2000.
19. Ding X, Lv X, Hui B, Chen Z, Xiao M, Guo B, Tang W. *Tetrahedron Lett* 2006;47:2921–2924.
20. Bradley D, Williams G, Ajam M, Ranwell A. *Organometallics* 2006;25:3088–3090.
21. Thomas PA, Marvey BB. *Int J Mol Sci* 2009;10:5020–5030.
22. Thomas PA, Marvey BB, Ebenso EE. *Int J Mol Sci* 2011;12:3989–3997.
23. Gallagher MM, Rooney AD, Rooney JJ. *J Mol Catal A Chem* 2009;303:78–83.
24. Duque R, Öschner E, Clavier H, Caijo F, Nolan SP, Mauduit M, Cole-Hamilton DJ. *Green Chem* 2011;13:1187–1195.
25. Keraani A, Rabiller-Baudry M, Fischmeister C, Bruneau C. *Catal Today* 2010;156:268–275.
26. Audic N, Clavier H, Mauduit M, Guillemin J-C. *J Am Chem Soc* 2003;125:9248–9249.
27. Yao Q, Zhang Y. *Angew Chem Int Ed* 2003;42:3395–3398.
28. Clavier H, Audic N, Mauduit M, Guillemin J-C. *Chem Commun* 2004;10:2282–2283.
29. Clavier H, Nolan SP, Mauduit M. *Organometallics* 2008; 27:2287–2292.
30. Yao Q, Sheets M. *J Organomet Chem* 2005;690:3577–3584.
31. Rix D, Clavier H, Coutard Y, Gułajski Ł, Grela K, Mauduit M. *J Organomet Chem* 2006;691:5397–5405.

32. Rix D, Caijo F, Laurent I, Gulajski L, Grela K, Mauduit M. *Chem Commun* 2007:3771–3773.
33. Borré E, Pieck CA, Caijo F, Crévisy C, Mauduit M. *Chim Oggi* 2009;27:74–78.
34. Thurier C, Fischmeister C, Bruneau C, Olivier-Bourbigou H, Dixneuf PH. *J Mol Catal A Chem* 2007;268:127–133.
35. Kingsbury JS, Harrity JPA, Bonitatebus PJ Jr, Hoveyda AH. *J Am Chem Soc* 1999;121:791–799.
36. Wakamatsu H, Blechert S. *Angew Chem Int Ed* 2002; 41:2403–2405.
37. Thurier C, Fischmeister C, Bruneau C, Olivier-Bourbigou H, Dixneuf PH. *Chem Sus Chem* 2008;1:118–122.
38. Consorti CS, Aydos GLP, Ebeling G, Dupont J. *Org Lett* 2008;10:237–240.
39. Consorti CS, Aydos GLP, Dupont J. *Chem Commun* 2010; 46:9058–9060.
40. Wakamatsu H, Saito Y, Masubuchi M, Fujita R. *Synlett* 2008:1805–1808.
41. Chen S-W, Kim JH, Ryu KY, Lee W-W, Hong J, Lee S-G. *Tetrahedron* 2009;65:3397–3403.
42. Liu G, Zhang J, Wu B, Wang J. *Org Lett* 2007;9:4263–4266.
43. Jordan JP, Grubbs RH. *Angew Chem Int Ed* 2007; 46:5152–5155.
44. Skowerski K, Szczepaniak G, Wierzbicka C, Gułajski Ł, Bienek M, Grela K. *Catal Sci Technol* 2012;2:2424–2427.
45. Zhang Q, Zhang S, Deng Y. *Green Chem* 2011;13:2619–2637.
46. Vallette H, Pican S, Boudou C, Levillain J, Plaquevent J-C, Gaumont A-C. *Tetrahedron Lett* 2006;47:5191–5193.
47. (a) Mehnert CP. *Chem Eur J* 2005;11:50–56. (b) Riisager A, Fehrmann R, Haumann M, Wasserscheid P. *Eur J Inorg Chem* 2006:695–706.
48. Riisager A, Fehrmann R, Haumann M, Wasserscheid P. *Top Catal* 2006;40:91–102.
49. Riisager A, Fehrmann R, Flicker S, Van Hal R, Haumann M, Wasserscheid P. *Angew Chem Int Ed* 2005;44:815–819.
50. Valkenberg MH, De Castro C, Holderich WF. *Green Chem* 2002;4:88–93.
51. (a) Jin M-J, Taher A, Kang H-J, Choi M, Ryoo R. *Green Chem* 2009;11:309–313. (b) Hagiwara H, Ko KH, Hoshi T, Suzuki T. *Chem Commun* 2007:2838–2840.
52. (a) Baudoux J, Perrigaud K, Madec P-J, Gaumont A-C, Dez I. *Green Chem* 2007;9:1346–1351. (b) Moucel R, Perrigaud K, Goupil J-M, Madec P-J, Marinel S, Guibal E, Gaumont A-C, Dez I. *Adv Synth Catal* 2010;352:433–439. *Adv Synth Catal* 2010;352:746.
53. Hagiwara H, Sugawara Y, Isobe K, Hoshi T, Suzuki T. *Org Lett* 2004;6:2325–2328.
54. Wolfson A, Vankelecom IFJ, Jacobs PA. *Tetrahedron Lett* 2003;44:1195–1198.
55. Hagiwara H, Sasaki H, Hoshi T, Suzuki T. *Synlett* 2009: 643–647.
56. Rodriguez-Pérez L, Teuma E, Falqui A, Gómez M, Serp P. *Chem Commun* 2008:4201–4203.
57. Hagiwara H, Okunaka N, Hoshi T, Suzuki T. *Synlett* 2008: 1813–1816.
58. Hagiwara H, Nakamura T, Okunaka N, Hoshi T, Suzuki T. *Helv Chim Acta* 2010;93:175–182.
59. Scholz J, Loekman S, Szesni N, Hieringer W, Görling A, Haumann M, Wasserscheid P. *Adv Synth Catal* 2011;353: 2701–2707.
60. Keraani A, Rabiller-Baudry M, Fischmeister C, Bruneau C. *Catal Today* 2010;156:268–275.

25

PURIFICATION STRATEGIES IN OLEFIN METATHESIS

KRZYSZTOF SKOWERSKI AND ŁUKASZ GUŁAJSKI
Apeiron Synthesis S.A., Wrocław, Poland

25.1 THE CORE OF THE PROBLEM

Olefin metathesis has come a long way from the poorly understood process catalyzed by simple metal derivatives that was discovered in 1950s. Development of well-defined molybdenum-based Schrock-type complexes and ruthenium-based Grubbs-type catalysts that can be easily modified by using steric and electronic factors led to the establishment of olefin metathesis as a method of choice in the formation of new C–C double bonds. (1) Nowadays, olefin metathesis is considered to be among the best available techniques (BAT) and therefore its applications in industrially useful processes appear in increasing number. For the synthesis of fine chemicals, ruthenium-based catalysts are considered to be the most suitable mainly because of their high stability, efficiency, and tolerance of functional groups. The usage of olefin metathesis usually allows for shortening of the synthetic pathway, which results in cost reduction and decreases amounts of waste. However, there are some drawbacks related to ruthenium-based catalysts. In addition to their relatively high price and lack of stereoselectivity, the problem associated with impurities containing heavy metal that are present in the product after reaction appears to be the most important. Obviously, removal of ruthenium residue is mandatory in the areas where the level of heavy metal in the final product is regulated by law. This is true not only for pharmaceuticals (e.g., for orally administered drugs, ruthenium content must not exceed 10 ppm) (2) but also for all materials that have direct contact with human body, for example, polymers for biomedical applications (3). A less evident factor that forces chemists to remove traces of ruthenium is related to the side reactions that can be caused, for example, by ruthenium hydrides formed during catalyst decomposition. Such processes can occur during the work-up procedure as well as within the next steps of synthesis and are most likely in the synthesis of active pharmaceutical ingredients (APIs) that usually contain a variety of functional groups. Furthermore, impurities containing ruthenium are strong dyes that cause usually undesired colorization of the product. Finally, traces of ruthenium can potentially have a negative influence on the properties of polymers designed for electronic applications.

25.2 SOLUTION TO THE PROBLEM

A huge potential of olefin metathesis in almost all fields of modern organic and material chemistry encouraged chemists to develop methods for removal of residual ruthenium from crude metathesis products. As a result, several protocols dealing with this issue were established and reviewed recently (4). In this chapter, we describe the most effective methods in view of their possible commercial applications and those originating from the industry. Generally, strategies for ruthenium removal can be divided into two categories, depending on the type of catalyst used. These two approaches are presented in Scheme 25.1. For removal of residual ruthenium from products obtained with the use of classical complexes (Fig. 25.1), some sequestering agents, the so-called "scavengers," are usually applied (Scheme 25.1a). However, the adoption of typical purification methods can also give positive result. Commercial availability of classical catalysts

Olefin Metathesis: Theory and Practice, First Edition. Edited by Karol Grela.

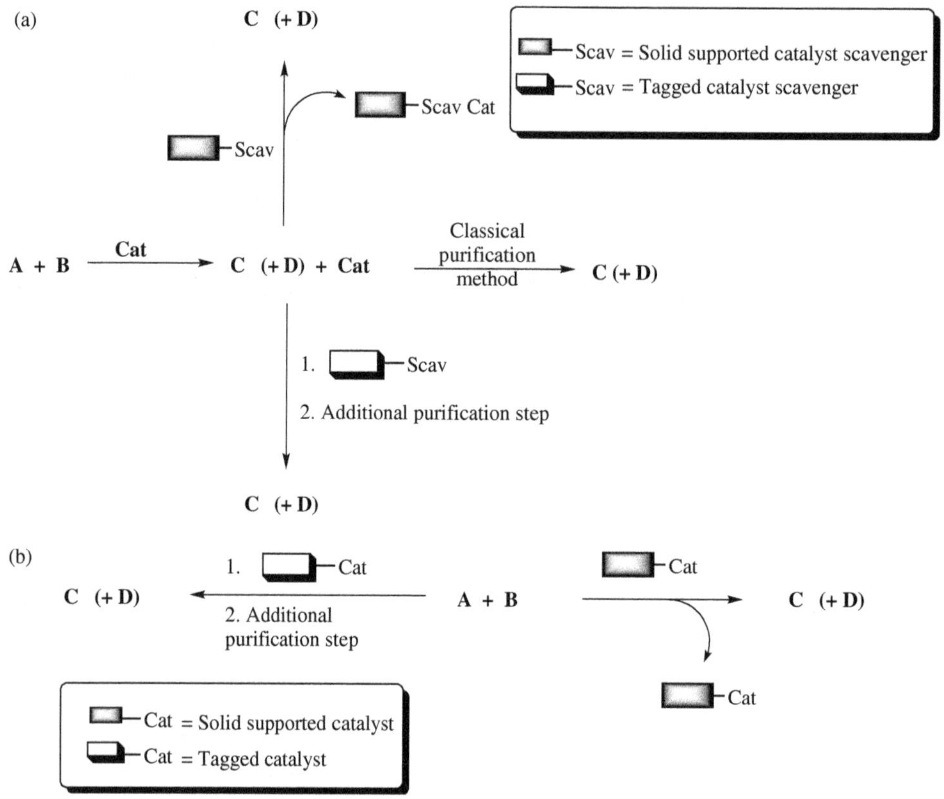

Scheme 25.1 (a,b) Strategies for ruthenium removal from metathesis products.

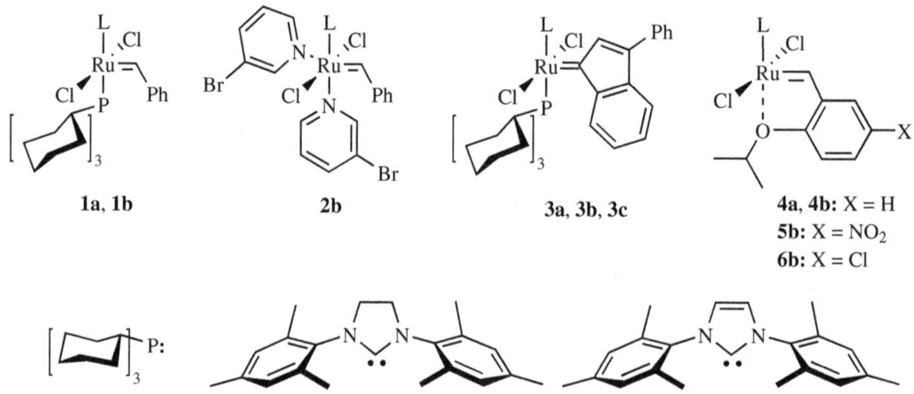

Figure 25.1 Selected classical catalysts for olefin metathesis.

and easy access to several scavengers have to be considered as strengths of this approach. On the other hand, this method is often time- and energy-consuming and requires the use of a large excess of scavengers that can potentially provoke side reactions. The second strategy assumes the use of modified catalysts that can be easily and successfully removed from the post-reaction mixture (Scheme 25.1b). However, modified solid-supported catalysts are not commercially available and usually show inferior efficiency when compared with their homogeneous counterparts. Tagged complexes exhibited better catalytic activity than the solid-supported ones, but it was very recently that the first examples of commercially available tagged catalysts appeared.

Classical unmodified solid-supported catalysts are a very interesting solution that is outside of these two general strategies. Examples of application of such a heterogeneous complex in the synthesis of valuable compounds are given in the last part of this chapter.

25.3 REMOVAL OF RUTHENIUM FROM PRODUCTS OBTAINED WITH CLASSICAL CATALYSTS

25.3.1 Physicochemical Methods

The possibility of residual ruthenium removal after metathesis reactions catalyzed by classical complexes (Fig. 25.1) was extensively studied by Cho and Kim (5). The most efficient method developed by them consists of three purification steps: (i) stirring of the crude reaction mixture with silica gel (SiO_2/**1a** = 10 wt) and filtration through a pad of silica gel, (ii) stirring of the filtrate with activated charcoal (charcoal/crude product = 50 wt), followed by a simple filtration, and (iii) purification by silica gel column chromatography. Using this protocol, the authors were able to reduce the ruthenium content in several ring-closing metathesis (RCM) products to 12–60 ppm as determined by inductively coupled plasma mass spectrometry (ICP-MS). However, this method, owing to the time consumed, is rather too expensive for industrial application, especially because of SiO_2 chromatography. Nevertheless, a compound of pharmaceutical interest was recently also purified by extensive chromatography (6). As a part of the work on BILN 2061 production procedure, Gallou and coworkers from *Boehringer Ingelheim (BI) Pharma* developed a new method for the purification of a macrocyclic product of RCM reaction by the use of supercritical carbon dioxide ($scCO_2$). The story of BILN 2061 is discussed in detail in the last part of this chapter. Recently, van der Gryp et al. (7) have used organic solvent nanofiltration technique for separation and recycling of **1a**, **1b**, **4a**, and **4b**. Filtration of crude reaction mixtures through a commercially available STAR MEM™ membrane reduced the ruthenium content below 9 ppm. The mass and volume of catalysts were sufficient to retain them over the membrane from where they were successfully recovered. Nanofiltration was also successfully used for the separation of mass-extended catalysts (8).

25.3.2 Unsupported Scavengers

It was in 1999 that Grubbs and Maynard (9) reported the first protocol for ruthenium removal after olefin metathesis reaction promoted by **1a**. Utilization of tris(hydroxymethyl) phosphine ($P(CH_2OH)_3$, THP) that coordinates to the metallic center allows for sequestration of ruthenium residues during the work up. Treatment of the crude reaction mixture with a large excess of THP and triethylamine followed by stirring with SiO_2 and filtration resulted in a reduction of ruthenium level to 206 ppm. The use of water extraction gave a worse result (670 ppm). A more effective method that allows for reduction of Ru level to 60 ppm, consisting of the oxidation of ruthenium species and filtration of the reaction mixture through silica gel, was reported in 2000 by Paquette et al. (10). However, the oxidation agent that was used, lead tetra-acetate [$Pb(OAc)_4$], is an environmental pollutant and exhibits serious toxic effects to the human body. A similar method presented recently by Knight and coworkers seems not to be suitable for industrial use, especially in the pharmaceutical sector. In this contribution, the authors oxidize ruthenium species using hydrogen peroxide into a highly insoluble ruthenium dioxide which was separated by filtration of the reaction mixture through SiO_2. Ruthenium level in the products obtained with the use of **1a** and **1b** was reduced only to 3020 and 506 ppm, respectively. Moreover, due to its sensitivity toward the ruthenium dioxide, hydrogen peroxide has to be used in large excess, which can easily induce side reactions. A protocol reported by Georg et al. (11) involves the use of much milder sequestering agents. Precisely, the authors used dimethyl sulfoxide (DMSO) and triphenylphosphine oxide [$P(O)Ph_3$] to remove **1a** or **1b** residues. Reaction mixtures were stirred with 50 M equivalents (relative to catalyst) of either DMSO or $P(O)Ph_3$ overnight and, after concentration, cleaned by chromatographic purification on SiO_2. Ruthenium content was reduced to 240 and 360 ppm, respectively. Later on, Georg and coworkers (12) reported that treatment of the crude reaction mixture with a large excess of SiO_2 (4 g per each 0.01 mmol of catalyst) and 50 equivalents of DMSO (relative to catalyst) gives, after 12 h, products containing as low as 8 ppm of Ru. Diver's research group (13) discovered that metathetic activity of all ruthenium catalysts tested is effectively terminated in 30 min after addition of a small excess of potassium cyanoacetate to the reaction mixture. Subsequent purification of the product on an SiO_2 column reduces the ruthenium level to 120–2200 ppm. Therefore, although being fast and inexpensive, this method cannot be treated as a universal technique for the removal of residual Ru. Nevertheless, potassium cyanoacetate was recently used for purification of ripostatin B precursor after RCM reaction promoted by **5b** (14). Liu and coworkers (15) utilized commercially available di(ethylene glycol) vinyl ether (**7**, Fig. 25.2) and an easily obtainable amine containing derivatives of the vinyl ether (**8**, **9**, Fig. 25.2) for deactivation of several types of commonly used catalysts. Initial screening revealed that **7** is a powerful inhibitor of metathetic activity. Further experiments proved that **7–9** are also very effective sequestering agents. Addition of 4 equiv of **7** (relative to catalyst) to the crude solution of the product obtained with **6b** followed by chromatographic purification reduced ruthenium level to 2 ppm. When 4 equiv of **8** were used, column chromatography was successfully replaced by simple filtration through a silica gel plug (Ru level <43 ppm). Under the same conditions, the reaction of the catalyst with **9** resulted in the formation of Fischer carbene, which was easily washed off with 2 N aqueous HCl. After this simple clean up procedure, the product contained only less than 15 ppm of ruthenium.

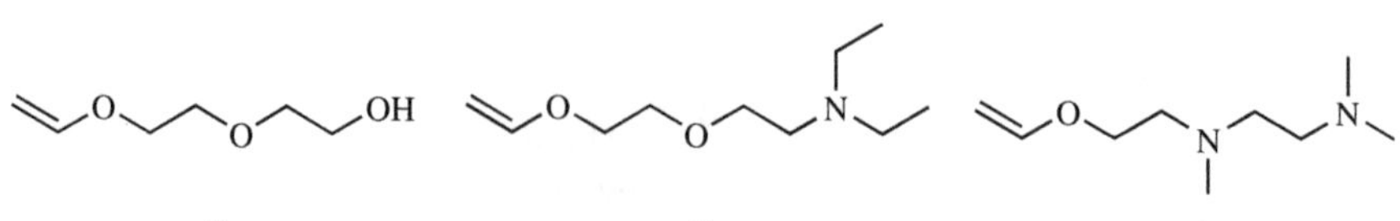

7 8 9

Figure 25.2 Tagged, unsupported scavengers **7–9**.

Among all methods that use unsupported scavengers, the protocol reported by Liu and coworkers seems to be the most suitable for industrial applications. Examples of the practical use of other homogeneous scavengers are presented in the last part of this chapter.

25.3.3 Solid-Supported Scavengers

A resin-bound phosphine **10** (Fig. 25.3) designed for removal of **1a** and its decomposition products was presented by Breinbauer et al. in 2004 (16). With the use of 20 M equivalents of this reagent in combination with charcoal treatment or filtration through SiO_2, the authors obtained products with 1660 and 1120 ppm of Ru, respectively. By using a resin-bound triphenylphosphine oxide **11** (Fig. 25.3), Georg et al. (12) were able to get superior results. When the crude product was refluxed in toluene for 6 h in the presence of 50 M equivalents of **11** and then purified by column chromatography, the Ru level dropped to 344 ppm. Chromatography could be avoided when aside from 50 M equivalents of **11**, SiO_2 (2 g per 0.01 mmol of catalyst) was added. In this case, simple filtration resulted in the isolation of the product contaminated with only 8 ppm of Ru. The same efficiency was observed when the crude product was refluxed in toluene with 100 M equivalents of **11** and 50 M equivalents of DMSO, and further purified by column chromatography. In 2006, Crudden and coworkers (17) presented a series of modified silicates (SBA-15) and SiO_2. Initial screening run with a pure solution of ruthenium chloride hydrate demonstrated that both supports give the best results when modified with aminopropyltriethoxy silane **12** (SBA-15-**12**, SiO_2-**12**). However, these scavengers acted poorly in a more challenging example, that is, the removal of ruthenium residues after RCM reaction promoted by 5 mol% of **1a**. In the best case, two consecutive treatments with SBA-15-**12** (100 mg of scavenger per 0.025 mmol of **1a**) followed by simple filtration reduced the ruthenium level to 760 ppm.

On the basis of the concept previously presented by Liu, in another contribution Slugovc and Kranjec et al. used a vinyl ether derivative to prepare a polymer-based scavenger (18). The obtained material was effective in the removal of **3a** from its solution but efficiency of this scavenger in the purification of crude post-metathesis reaction mixture was not evaluated. An interesting method was published by Macdonald and Veinot (19). In this contribution, the crude metathesis

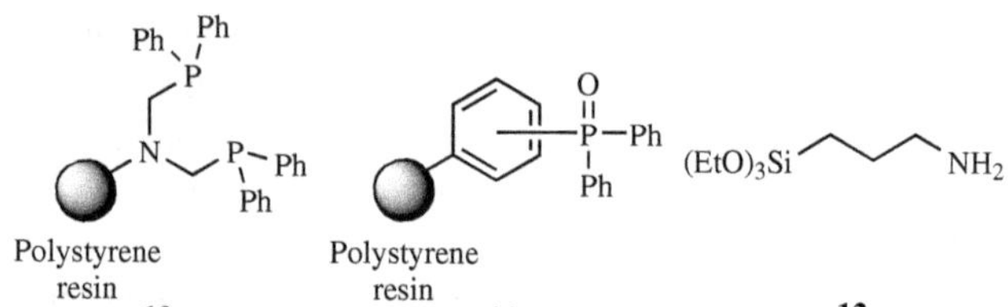

10 11 12

Figure 25.3 Solid-supported scavengers **10–12**.

product was added to a solution of iron-oxide-capped iron nanoparticles (Fe@Fe_xO_y), 3-mercaptopropionic acid, and cysteamine in tetrahydrofuran (THF). The resulting mixture was stirred for 19–20 h and then filtered to provide the product contaminated with 1250 ppm of ruthenium. However interesting, the method is of limited effectiveness and the level of complication and time consumption discourages its use. A valuable research dealing with the removal of ruthenium from product obtained in ring-opening metathesis polymerization (ROMP), including the use of commercially available solid-supported scavengers, is discussed in detail in the last part of this chapter.

25.4 REMOVAL OF RUTHENIUM FROM PRODUCTS OBTAINED WITH MODIFIED CATALYSTS

According to Scheme 25.1b, modified catalysts can be divided into two main groups. These are solid-supported (heterogeneous) catalysts and tagged (homogeneous) catalysts. As shown in Figure 25.4, the concept of modification of catalysts was realized by modification of anionic ligands, alkylidene ligands, as well as neutral ligands (phosphine, pyridine, *N*-heterocyclic carbene (NHC)). Among these possibilities, modification of the NHC ligand seems to be the most reasonable solution as this ligand creates the strongest bond to the ruthenium center and neither dissociates during the metathesis act nor is substituted by other ligands. Solid-supported catalysts usually allow not only for the removal of residual ruthenium by simple filtration or even decantation but also give the opportunity for catalyst recovery and reuse. However, in most cases, these catalysts exhibited lower activity than their homogeneous counterparts. On the other hand, tagged catalysts possess a reactivity similar to unmodified

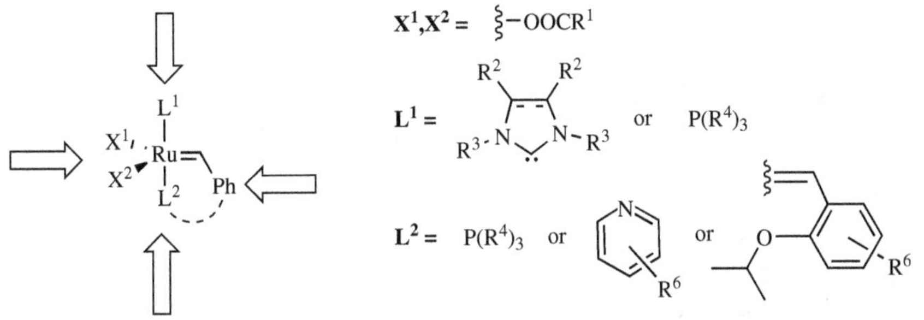

Figure 25.4 Possible sites for olefin metathesis catalysts modifications.

catalysts but require an additional purification technique in order to remove ruthenium residues.

Nevertheless, both classes of catalysts allow for obtaining metathesis products with very low residual ruthenium, usually much lower than obtained after application of sequestering agents. The topic of modified ruthenium catalysts was extensively reviewed (4).

25.4.1 Solid-Supported Catalysts

Several complexes attached to the solid support through the alkylidene ligand can be found in the literature. However, as the alkylidene ligand dissociates during catalytic cycle, the active species containing heavy metal are homogeneous. It was suggested that after completion of the reaction, the original pre-catalyst is reproduced, which was called the *boomerang effect*. Nevertheless, because of the catalyst decomposition, the aforementioned regeneration of pre-catalyst is not quantitative, which is reflected in the ruthenium level in the products that is very often higher than 100 ppm (20).

Rutjes et al. (21) were the first who took advantage of the fact that anionic ligands are more strongly bonded with the ruthenium center than the alkylidene ligand. Catalyst **13a** (Fig. 25.5) was obtained in the reaction of **1a** with a perfluoroglutaric acid-functionalized polystyrene resin. The product of RCM reaction catalyzed by 1 mol% of **13a** contained only 80 ppm of ruthenium after simple filtration of the crude reaction mixture. Later on, a similar concept was realized by the research group of Buchmeiser (22). Complexes **1b**, **4a**, and **4b** were reacted with a perfluoroglutaric-acid-functionalized polystyrene–divinylbenzene resin and subsequently with silver trifluoroacetate to give supported catalysts **14a**, **15a**, and **15b**.

These catalysts gave the products with only 0.083, 0.015, and 0.070 ppm of residual ruthenium, respectively. Interestingly, catalyst **14b**, which was the most active among new complexes, outperformed its homogeneous counterparts with trifluoroacetic anionic ligands. Without any doubts, these results are impressive and the only possible exchange of anionic ligands that could lead to the deheterogenization of catalyst may slightly limit the practical application of these

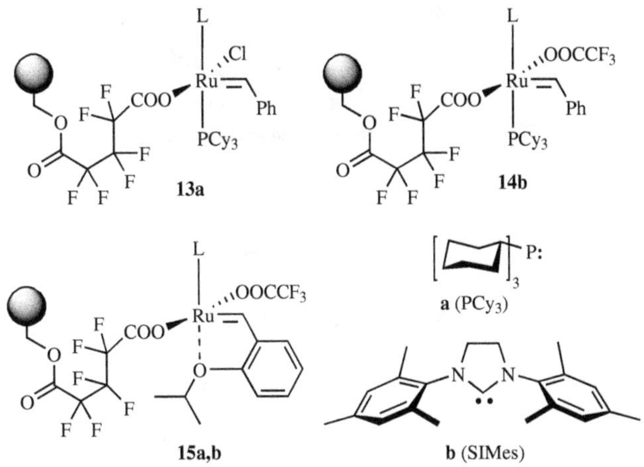

Figure 25.5 Solid-supported catalysts **13–15**.

types of complexes (23). In an earlier work, Buchmeiser and coworkers (24,25) presented catalysts **16** and **17** supported on monolithic materials synthesized by ROMP (Fig. 25.6). These complexes effectively promoted ROMP and RCM reactions in continuous flow mode providing products contaminated with only about 70 ppm of ruthenium (24). More recently, Grubbs et al. (26) developed two heterogeneous complexes **18** and **19** grafted on silica gel.

These catalysts proved to be active in RCM reactions. However, complex **18** turned out to be far more effective and was recycled eight times without a significant loss of activity. After simple filtration of the RCM reaction mixture in which 0.25 mol% of **18** was applied, the ruthenium content was lower than 0.005 ppm.

25.4.2 Tagged, Homogeneous Catalysts

As in the case of solid-supported catalysts, the simplest way to introduce a tag into a catalyst structure is to exchange a classical neutral phosphine or benzylidene ligand with the properly modified ones. One of the most promising approaches is based on the use of ionic liquids (ILs) and catalysts modified with IL tags. A great interest in the use of IL as alternative solvents is related to their high

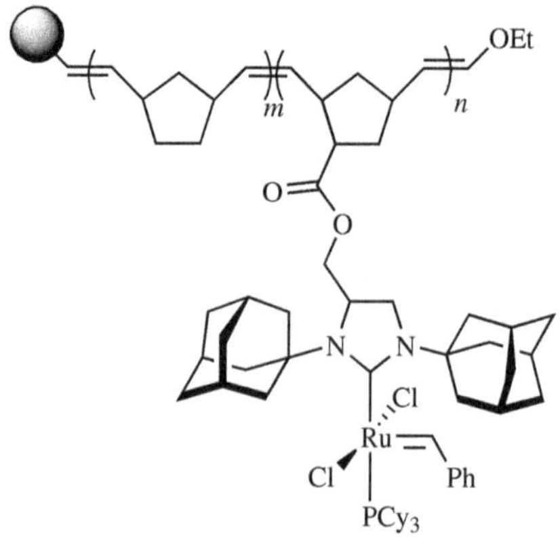

16

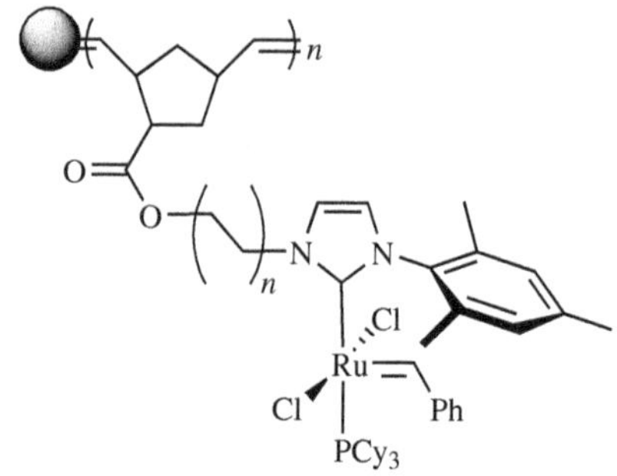

17

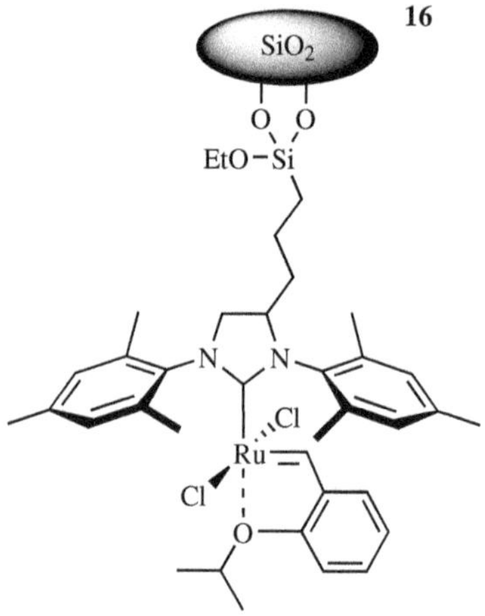

18

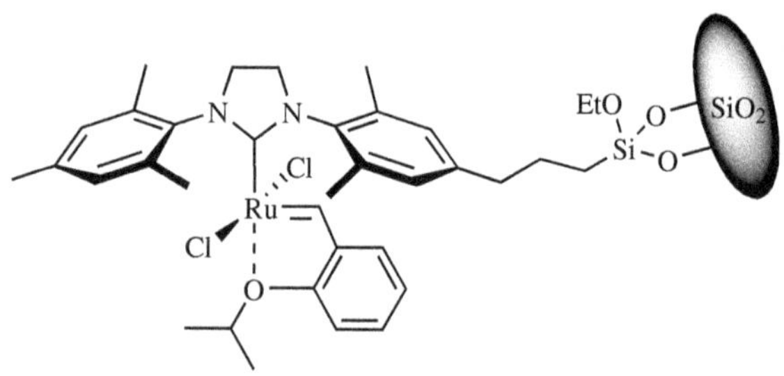

19

Figure 25.6 Solid-supported catalysts **16–19**.

thermal and chemical stability and extremely low vapor pressure. Additionally, chemists exploit the fact that ILs show very low, if any, miscibility with water and organic solvents combined with the ability to solvate a wide range of reactants. IL-tagged olefin metathesis catalysts can be firmly immobilized in IL, which brings the opportunity to carry out biphasic reactions using the IL/organic solvent system. With an appropriate ruthenium complex, such a system allows for an easy catalyst recovery and reuse as well as provides a product with low residual ruthenium. Thus, catalysts **21** and **22a–c** (Fig. 25.7) were synthesized and evaluated in biphasic systems using IL and toluene as the second phase. It was found that electronically activated catalysts **20** (27) and **21** (28) are not suitable for this system since their activity is significantly reduced in the second run. However, both complexes gave metathesis products with relatively low residual ruthenium, 12–68 and 25–173 ppm, respectively, after the filtration of the reaction mixture through a pad of silica gel (29). More stable **22b** (30) and **22c** (31) can be recycled 8 and 17 times, respectively, without a significant loss of activity. Importantly, **22b** provides a product with only 1.2–22 ppm of residual ruthenium, which is a surprisingly good result considering the dissociative nature of benzylidene ligand. In 2006, Grubbs and Hong (32) reported an interesting catalyst **23** (Fig. 25.8) with poly(ethylene glycol) (PEG) attached to the NHC ligand. This complex efficiently promotes different types of olefin metathesis in neat water but is also readily soluble in organic solvents. The authors took advantage of this fact and removed the ruthenium species from an etheral solution of product using extraction with water (33). The ruthenium level in the product after a few extractions was determined to be 41 ppm and could be further reduced below 0.04 ppm by treating this pre-purified product with activated carbon. Very recently, Skowerski et al. (34) reported complexes **24** and **25** with quaternary ammonium groups placed in NHC ligand. Application of these catalysts allows obtaining products with the ruthenium level in the range of 0.004–29 ppm. Direct filtration of crude reaction mixture through a short pad of silica gel and extraction of the reaction mixture with water were almost equally efficient in the removal of residual ruthenium.

As shown above, there are a few examples of very efficient modified catalysts that give virtually Ru-free products and it seems that it is only a matter of time when such complexes

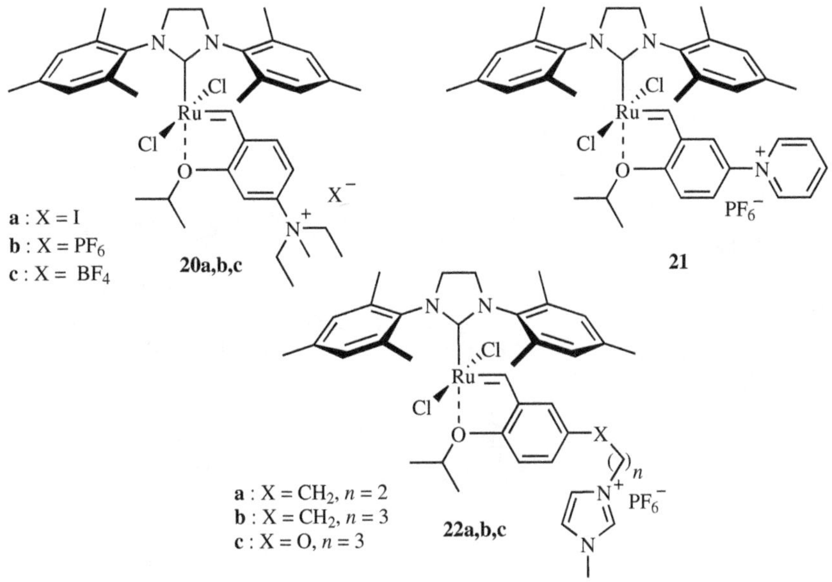

Figure 25.7 Catalysts **20–22** bearing ionic tag in benzylidene ligand.

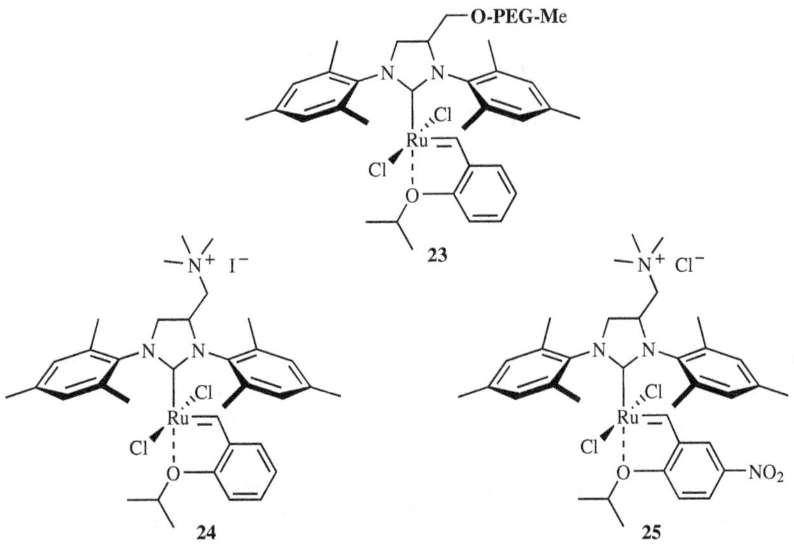

Figure 25.8 Catalysts **23–25** with modified NHC ligand.

will find applications in pharmaceutical or fine chemical synthesis.

25.4.3 Examples from Industry

The very first example of application of metathesis in the synthesis of API was presented in 2005 by chemists from *BI Pharma* (35) who developed a potent hepatitis C virus (HCV) protease inhibitor BILN 2061 (**26**, Scheme 25.2). The scale-up of the RCM reaction promoted by **4a** that provides macrocyclic **27a** was recognized to be a bottleneck in the synthetic route. The RCM was run at 80 °C in high dilution using toluene as a solvent. At laboratory scale, 3 mol% of **4a** added in three portions was sufficient to obtain almost full conversion of substrate. The subsequent work-up consists of several aqueous extractions (water, HCl, $NaHCO_3$) and charcoal treatment of organic phase at 50 °C. Concentration of toluene and precipitation of macrocycle with methyl cyclohexane afforded the product as an amorphous solid that contained from 500 to 1000 ppm of residual ruthenium. It was found that at production scale, a higher amount of **4a** (up to 5.3 mol%) was needed to get full conversion, which could result in even more serious problems with the removal of residual ruthenium. Owing to the difficulties with reducing

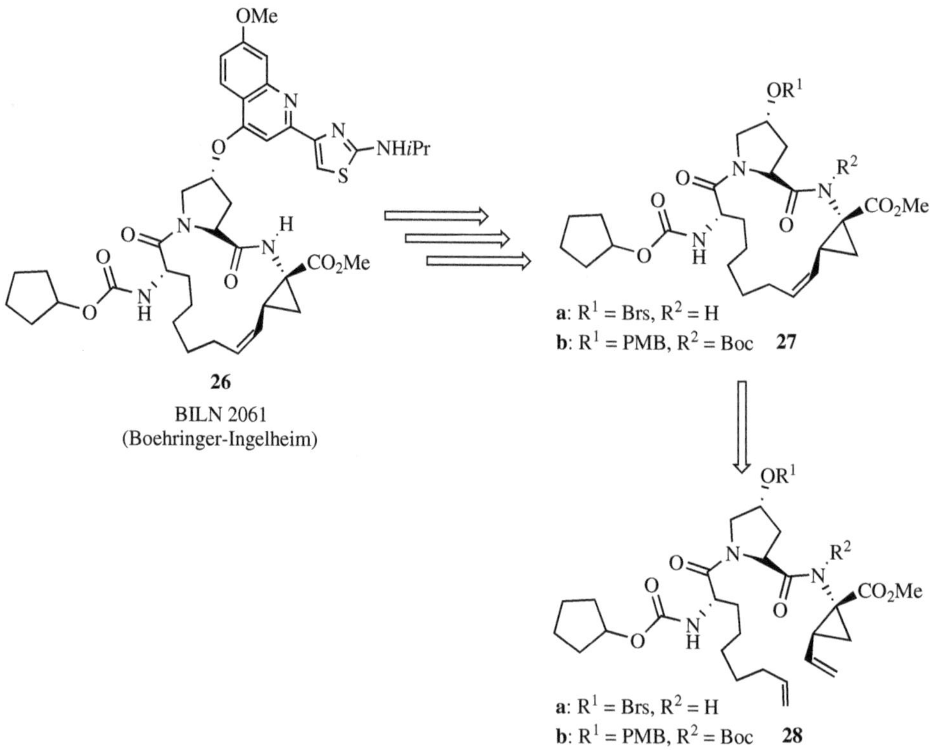

Scheme 25.2 Retrosynthesis of BILN 2061 (Brs, 4-Br$C_6H_4SO_2$; PMB, 4-methoxybenzyl ether).

the heavy metal content in API **26** below 10 ppm, the authors inferred that a different process for the removal of Ru from macrocyclic **27a** is required. Further studies revealed that previously reported methods for Ru scavenging share some disadvantages (36). Besides other issues, high loadings of scavengers (up to 60 equiv of THP), long processing time, and patent infringement encouraged chemists from BI to develop a new, more convenient process for purification of the metathesis product. Researchers turned their attention to the extractive properties of scCO_2 and decided to take advantage of the preferential solubility of macrocyclic **27a** in this medium to separate it from Ru-containing impurities. Preliminary experiments showed that neat **27a** exhibits very poor solubility in scCO_2. However, in the presence of organic solvents, such as dichloromethane or toluene, the solubility dramatically increased.It was found that high dilution favors a good recovery of **27a**. Therefore, subsequent extractions were run with the use of crude reaction mixtures at 0.01 M concentrations. Two extractions made in a batchwise mode proved the high efficiency of this method. At larger scale, 80 g of metathesis product that contained 5 mol% of **4a** was loaded into 1 l autoclave. The temperature and CO_2 pressure were set at 40 °C and 97 bar, respectively. At this condition, the dynamic equilibrium was achieved in 30 min and the product was collected from the CO_2 stream in a cold trap upon depressurization at a flow rate of about 1.5 ml/min. Product recovery reached 90% and Ru level was reduced to 56 ppm. Repetition of the whole procedure with the use of this material resulted in the detection of only 2 ppm of Ru. Finally, the reaction mixture containing 5 mol% of **4a** was purified by extraction in semicontinuous mode. Crude reaction mixture and CO_2 were continuously injected into a 300 ml pressure vessel at a flow rate of 5 and 100 ml/min, respectively. The temperature of 40 °C and pressure of 97 bar were maintained inside the reactor and the product was extracted without a dynamic equilibration. Purification of 250 ml of crude mixture took 50 min and resulted in 88% recovery of **27a** that contained 708 ppm of ruthenium as determined by ICP-MS. It was possible to further reduce the Ru level to 100 ppm by the treatment of the extracted solution with charcoal prior to the evaporation of the solvent. The effectiveness and reproducibility of this process was demonstrated by a second extraction that was run under the same conditions. The other work coming from BI Laboratories, also published in 2006, concerned an efficient large-scale synthesis of BILN 2061 (37). Four commercially available catalysts, namely **1a**, **1b**, **4a**, and **4b**, were tested in RCM of **28a**. Catalyst **1a** was rejected because of the variable amounts of epimerized by-product observed in post reaction mixtures. It was previously found that **4a** gives good yield of RCM product, but it suffers from low reactivity and efficiency, which results in long reaction times (up to 20 h) and low turnover number (TON) (below 30). More active and effective second-generation catalysts **1b** and **4b** were tested to solve these issues. These complexes exhibited significantly higher TON of 174 and 85, respectively. Unfortunately, 8% to 10% of the cyclic dimer was formed, which was difficult to separate from the desired product without the use of chromatographic methods. This was ascribed to the fast reopening of once formed macrocyclic ring and the subsequent dimerization. The RCM product is more kinetically stable in the presence of **4a**. However, it was found that the removal of solvent after large scale RCM reaction promoted by **4a** can lead to the decomposition of the product. It was hypothesized that occasionally, after consumption of substrate, an active methylidene form of catalyst is still present in the reaction mixture, being capable of ring opening product. Extensive studies on catalyst deactivators resulted in the discovery that mercaptonicotinic acid (MNA) effectively inactivates **4a** and additionally, can be extracted with aqueous bicarbonate. Application of this method prevented decomposition of the product during concentration of the reaction mixture and additionally gave an opportunity for removal of some amount of residual ruthenium. Reduction of Ru level in **27** to 159 ppm and appropriate purification of subsequent intermediate and final product (including charcoal treatment and crystallization) afforded **26** with a typical ruthenium content less than 5 ppm. This clearly shows the positives of synthetic path in which metathesis is not the last step. A final approach to the large scale synthesis of BILN 2061 assumed RCM of slightly modified diene but the ruthenium removal aspect was not further discussed in the manuscript. The last reports from the BILN 2061 story appeared in 2008 and 2009 when a second-generation, greener approach to Ru-catalyzed RCM was presented (38). Searching for higher throughput and better overall reaction efficiency, scientists from BI used a more appropriate substrate and a more effective catalyst. The three most important issues related to the first-generation process were high catalyst loading (3–5 mol%), long reaction time, and high dilution (10 mM). Careful evaluation of the conformational effect and catalyst initialization step leads to some important findings. Firstly, it was discovered that carbene transfer occurred mainly at the vinylcyclopropane moiety at which the active Ru species may be stabilized by chelation. Protection of the amide moiety with an easily removable protecting group prevents coordinative stabilization by the ester group, resulting in initiation at the nonenoic acid moiety. Secondly, with the use of new substrate **28b**, formation of the macrocyclic product was more favorable when compared with **28a** because of the lower strain in the product. Consequently, RCM of **28b** promoted by 0.1 mol% of robust and fast initiating Grela's catalyst **5b** could be run at 0.2 M concentration, and provided the cyclic product with 93% yield. Importantly, the amount of MNA used for ruthenium removal was decreased by a factor of 50 when compared with the first-generation process and no charcoal or silica gel filtrations were necessary to reduce the Ru level below 50 ppm. API was obtained with less than 5 ppm of ruthenium without further specific heavy metal removal.

Cathepsin K is a cysteine protease that plays a major role in bone degradation. In 2009, researchers from GlaxoSmithKline laboratories reported on RCM-based synthesis of cathepsin K inhibitor SB-4627695 (**29**, Scheme 25.3) (39). Two approaches to the synthesis of **29** that involved the RCM step were examined. The first route assumed the RCM of diene **32** which, however, proceeds to completion only in refluxing toluene with 10 mol% of **4b**. With 5 mol% loading neither **1a** nor **1b** led to full consumption of the substrate. Addition of $Ti(Oi\text{-}Pr)_4$, with the intention of disrupting the chelation of active form of catalyst by carbonyl moiety and hydroxyl group, caused a significant decomposition of the diene. The detrimental effect of high catalyst loading appeared during the scale-up of this process. Residual ruthenium turned out to be difficult to remove. For example, treatment of crude reaction mixture with previously reported ruthenium scavenger $P(CH_2OH)_4Cl$/NaOH resulted in product decomposition. On the other hand, subsequent reactions involved H_2O_2 for oxazolidinone removal and acyl azide intermediate for Curtis reaction, both being sensitive to heavy metals. For this reason, a new approach to the conversion of aldol adduct into cyclic intermediate was needed. It was assumed that removal of electron-rich oxazolidinone and protection of hydroxyl group allow overcoming the problem with low catalyst turnover. Thus, the hydrochloride of diene **31** was obtained and subjected to the RCM reaction. Only 1 mol% of **4b** was sufficient to complete this reaction in AcOEt at 60 °C. While adsorption on silica gel or alumina removed less than 50% of residual ruthenium, extraction with a basic solution of cysteine or $P(CH_2OH)_4Cl$ reduced Ru level below 20% of the initial amount. Formation of formaldehyde (can be added to carbamate nitrogen) as well as some phosphonium-based by-products (adverse effect in next, hydrogenation step) in the reaction of $P(CH_2OH)_4Cl$ with base favored the cysteine-based work-up. After crystallization, the cyclic carbamate **30** was isolated with 90% yield and contained only a small amount (unfortunately unspecified) of residual ruthenium.

The second route of synthesis of SB-4627695 resulted in the formation of diene **34**, which turned out to be an exceptional substrate for RCM. As low as 0.1–0.2 mol% of **4b** gave very good and reproducible results when diene of high purity was used. Finally, RCM of **34** was accomplished with 0.5 mol% of **4b** in toluene. Favorably, the product had very low solubility in toluene and crystallized from the reaction mixture upon formation. Most of the ruthenium species was removed in this way and the isolated product contained only 359 ppm of residual Ru, which was further reduced by downstream processing. The second route was chosen for large scale synthesis in which more than 200 kg

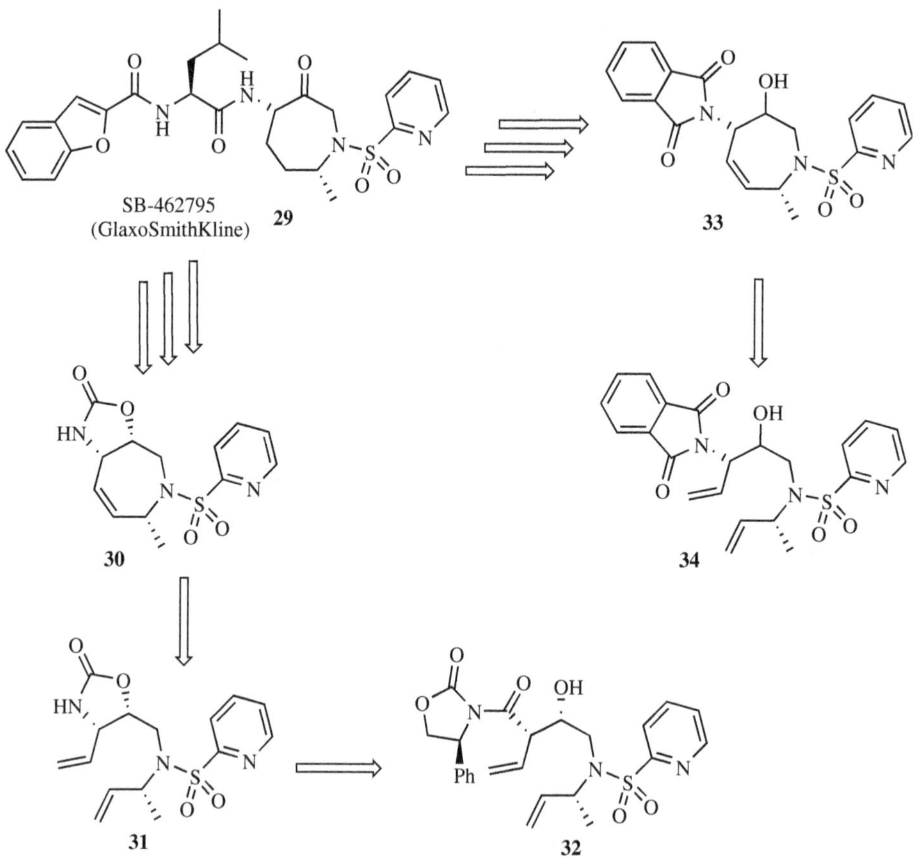

Scheme 25.3 Retrosynthesis of SB-462795.

of SB-4627695 with quality suitable for clinical trials was produced.

ROMP is a powerful tool that allows the preparation of polymers having enormous practical potential. Examples of such polymers were presented by Frerich, Buchmeiser, and coworkers who developed (in cooperation with *NovaTissue GmbH*) porous monolithic materials that were suitable for cell cultivation and tissue engineering. These materials were obtained with the use of **2b** from the mixture of norborn-2-ene and pentaglycerol bis(7-oxanorborn-5-ene-2ylcarboxylate) monoacrylate in varying amounts of a microporogen (toluene) and a macroporogen (2-propanol). The authors used mixture of DMSO, THF, and ethyl vinyl ether (1 : 1 : 1 volume ratio), or alternatively, a mixture of THF and ethyl vinyl ether (1 : 1 volume ratio) to rinse the ruthenium residue (40). Extraction of polymers with one of these mixtures was followed by a methanol wash providing the product with ruthenium level below 0.1 ppm (which is a detection limit of inductively coupled plasma optical emission spectroscopy (ICP-OES)).

More recently, Lambeth et al. (41) from the U.S. Army Research Laboratory published their efforts toward the synthesis of electro-optical (E-O) polymeric material that can be potentially used in photonics. The target copolymer with E-O chromophores in its structure was obtained in ROMP promoted by **2b**. Appropriate tests revealed that E-O activity of this material is far inferior when compared with the best previously recorded results. The authors postulated that the poor E-O properties can be related to the residual ruthenium or oxidative instability of the polymer backbone, which in turn may originate from the presence of ruthenium (42). Thus, model polymers that mimicked the structure of the E-O polymers were prepared with the intention to find an efficient method of removal of residual ruthenium. Some of the previously reported methods, such as precipitation of polymer from methanol (240 ppm Ru), precipitation combined with Soxhlet extraction (110 ppm Ru), treatment with activated charcoal (160 ppm Ru), treatment with THP or DMSO or $Pb(OAc)_4$ followed by SiO_2 filtration and precipitation from MeOH (100, 260, and 100 ppm of Ru, respectively) not only did not remove the ruthenium to the desirable level but also significantly reduced the yield. It was previously suggested that difficulties with the removal of heavy metal from ROMP products can arise from the ability of the olefinic bonds in the polymer backbone to coordinate with ruthenium (42). The experiment in which these double bonds were hydrogenated and the product was precipitated from MeOH supported this hypothesis since the polymer contained only 10 ppm of residual ruthenium. However, the authors noticed that hydrogenation could not be practical in many cases and, therefore, decided to search for a milder method. In subsequent experiments, several commercially available solid-supported scavengers were tested. The purification protocol involved quenching of the reaction, addition of appropriate amount of scavenger, and stirring for 20–24 h at room temperature followed by filtration through Celite, and precipitation of polymer with the use of MeOH. When 50 M equivalents (in relation to **2b**) of silica based scavengers **33–35** or thiol-functionalized polysiloxane beads **36** (Fig. 25.9) were applied, the product was contaminated with 20, more than 10, 20, and 60 ppm of Ru, respectively. Moreover, a filtration-free method was developed with the use of previously reported homogeneous scavengers that are able to increase solubility of the metal-containing spices. Again 50 M equivalents (in relation to **2b**) of the "solubilizing agent" was added to the reaction mixture after its quenching and the resulting mixture was stirred at room temperature for 20–24 h before precipitation of polymer in MeOH. While adoption of MNA and diethylphenylazothioformamide (DEPATF) reduced Ru level to 30 ppm, the THP and *N*-acetyl-L-cysteine proved to be much less effective (110 and 120 ppm, respectively). Finally, the positive effect of the residual ruthenium removal on oxidative stability of the polymer was proved with the use of differential scanning calorimetry (DSC). Reduction of the residual Ru level from 190 ppm (polymer precipitated with MeOH) to 30 ppm (polymer precipitated with MeOH after treatment with DEPATF) resulted in a 21 °C increase in the temperature of oxidation induction. The influence of reduction of ruthenium content on the properties of E-O polymers was not reported yet.

It was shown that weak physical interactions between neutral Hoveyda–Grubbs-type complex and the surface of inorganic materials are sufficient to perform heterogeneous olefin metathesis when appropriate reaction conditions are chosen (43). This was proved by researchers from BASF and Catalysis Research Laboratory (CARLA) who immobilized **4b** via physisorption on silica gel calcined at 550 °C prior to use (44). Preparation of the heterogeneous catalyst involved only dissolution of **4b** in dichloromethane followed by the addition of silica gel and removal of solvent under vacuum. Such obtained material (**4b-SiO$_2$**) was applied in continuous flow ring-opening–ring-closing metathesis (CF RO–RCM) of *cis*-cyclooctene (*cis*-COE). Dimerization of *cis*-COE is one of the shortest routes to macrocyclic skeletons; therefore, it is of high interest to the industry. For CF RO–RCM experiment, the silica material impregnated with ruthenium was charged into a cylindrical steel reactor and 0.7 wt% solution of *cis*-COE in cyclohexane was passed through this reactor at 60 °C. The TON of 4350, 19% of selectivity toward cyclic dimer (2-mer), and 74%

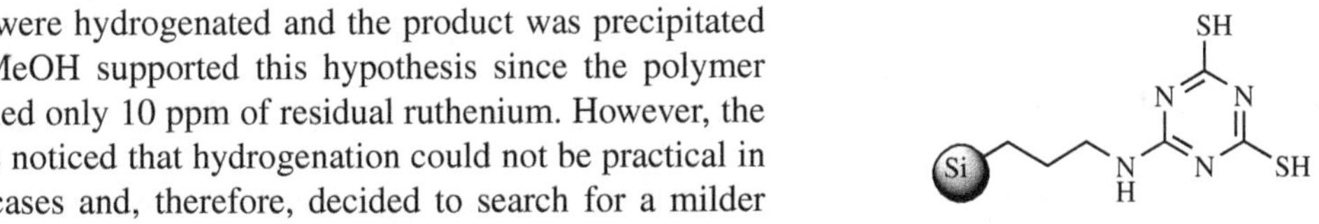

Siliabond® DMT

33

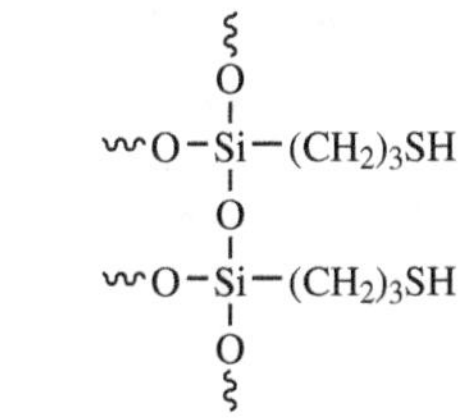

Deloxan® MP

34

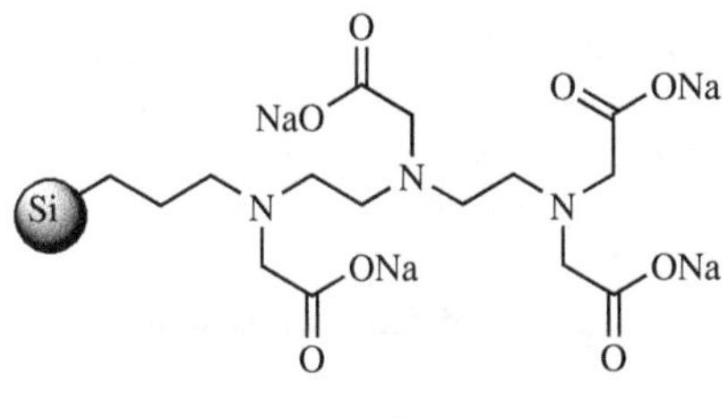

Siliabond® TAAcONa

35

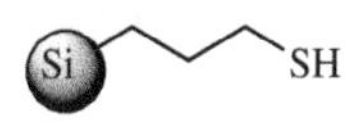

QuadraSil™ MP

36

Figure 25.9 Solid-supported scavengers **33–36**.

of selectivity toward small cyclic oligomers (2- to 5-mer) were observed. This result is far superior when compared to classical, state-of-the-art heterogeneous Re_2O_7 catalyst, both in terms of activity and selectivity (for the latter catalyst TON = 250, selectivity (2-mer) = 6%, selectivity (2- to 5-mer) = 7%). Ruthenium leaching was only 2.99% of the initial amount as determined by a comparison of the amount of ruthenium in the fresh and spent catalysts by ICP-MS. The authors prepared also the set of modified catalysts immobilized on silica in the same manner as **4b**. None of these materials outperformed **4b-SiO$_2$** in RO–RCM of *cis*-COE. However, in more challenging self-metathesis of methyl oleate, silica-supported **37** (**37-SiO$_2$**, Fig. 25.10) performed significantly better than **4b-SiO$_2$** (TON 4950 and 3550, respectively).

For both materials, ruthenium leaching was much more pronounced in this transformation (19% and 12% of initial amount for **37-SiO$_2$** and **4b-SiO$_2$**, respectively), which probably originated from a highly increased polarity of the reaction mixture (50 wt% of methyl oleate in cyclohexane).

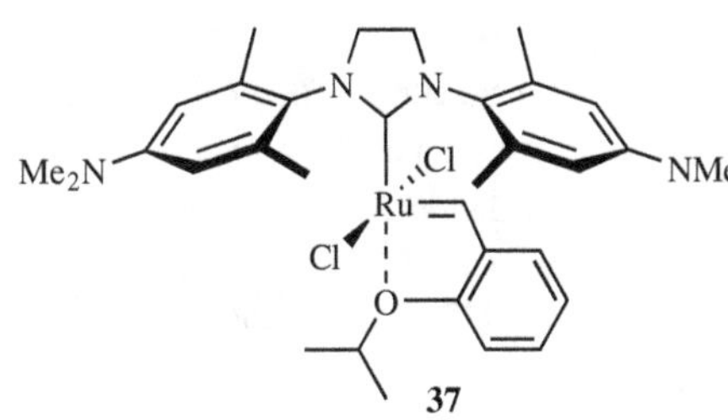

37

Figure 25.10 Ruthenium complex **37** used for preparation of heterogeneous catalyst.

Although there is more than one solution to the problem of heavy metal impurities this issue constitutes a valid research direction. Depending on the character of the purified material different approaches may seem legitimate. Establishment of a universal method working irrespective of both the material and impurity character remains a challenge.

REFERENCES

1. (a) Grubbs RH. *Handbook of Metathesis*. Weinheim, Germany: Wiley-VCH; 2003. (b) M Michalak, Ł Gułajski, K Grela, Alkene metathesis. In *Science of Synthesis: Houben–Weyl Methods of Molecular Transformations*; Vol 47a (Alkenes); Ed.: A de Meijere; Georg Thieme Verlag KG: Stuttgart, Germany. 2010, 327–438.
2. http://www.emea.europa.eu/docs/en_GB/document_library/Scientific_guideline/2009/09/WC500003586.pdf. Accessed 2013 May 6.
3. Madkour AE, Koch AHR, Lienkamp K, Tew GN. *Macromolecules* 2010;43(10):4557–4561.
4. (a) Vougioukalakis GC. *Chem Eur J* 2012;18:8868–8880. (b) Clavier H, Grela K, Kirschning A, Mauduit M, Nolan SP. *Angew Chem Int Ed* 2007;46:6786–6801. (c) Buchmeiser MR. *New J Chem* 2004;28:549–557. (d) Dragutan I, Dragutan V, Delaude L, Demonceau A. *Arkivoc* 2005:206–253. (e) Sommer WJ, Weck M. *Coord Chem Rev* 2007;251:860–873. (f) Copéret C, Basset J-M. *Adv Synth Catal* 2007;349:78–92. (g) Copéret C. *Dalton Trans* 2007:5498–5504. (h) Šebesta R, Kmentová I, Toma S. *Green Chem* 2008;10:484–496. (i) Śledź P, Mauduit M, Grela K. *Chem Soc Rev* 2008;37:2433–2442. (j) Buchmeiser MR. *Chem Rev* 2009;109:303–321. (k) Bergbreiter DE, Tian J, Hongfa C. *Chem Rev* 2009;109:530–582. (l) Copéret C, Beilstein J. *Org Chem* 2011;7:13–21.
5. Cho JH, Kim BM. *Org Lett* 2003;5:531–533.
6. Goldup SM, Pilkington CJ, White AJP, Burton A, Barrett AGMJ. *Org Chem* 2006;71:6185–6191.
7. van der Gryp, P, Barnard, A, Cronje, J-P, de Vlieger, D, Marx, S, Vosloo, HCM. *J Membr Sci* 2010, 353, 70–77.
8. (a) Schoeps D, Buhr K, Dijkstra M, Ebert K, Plenio H. *Chem Eur J* 2009;15:2960–2965. (b) Kajetanowicz A, Czaban J, Krishnan GR, Malińska M, Woźniak K, Siddique H, Peeva LG, Livingston AG, Grela K. *ChemSusChem* 2013;6:182–192.
9. Maynard, HD, Grubbs, RH. *Tetrahedron Lett* 1999, 40, 4137–4140.
10. (a) Paquette LA, Schloss JD, Efremov I, Fabris F, Gallou F, Méndez-Andino J, Yang J. *Org Lett* 2000;2:1259–1261. (b) Méndez-Andino J, Paquette LA. *Org Lett* 2000;2:1263–1265.
11. Ahn YM, Yang K, Georg GI. *Org Lett* 2001;3:1411–1413.
12. Haack K, Ahn YM, Georg GI. *Mol Diversity* 2005;9:301–303.
13. Galan BR, Kalbarczyk KP, Szczepankiewicz S, Keister JB, Diver ST. *Org Lett* 2007;9:1203–1206.
14. Winter P, Hiller W, Christmann M. *Angew Chem Int Ed* 2012; 51:3396–3400.
15. Liu W, Nichols PJ, Smith N. *Tetrahedron Lett* 2009;50: 6103–6105.
16. Westhus M, Gonthier E, Brohm D, Breinbauer R. *Tetrahedron Lett* 2004;45:3141–3142.
17. McEleney K, Allen DP, Holliday AE, Crudden CM. *Org Lett* 2006;8:2663–2666.
18. Pulko I, Sandholzer M, Kolar M, Slugovc C, Krajnc P. *Tetrahedron Lett* 2010;51:5827–5829.
19. Macdonald JE, Veinot JGC. *Langmuir* 2008;24:7169–7177.
20. See for example: (a) Ahmed M, Barrett AMG, Braddock DC, Cramp SM, Procopiou PA. *Tetrahedron Lett* 1999; 40:8657–8662. (b) Jafarpour L, Nolan SP. *Org Lett* 2000; 2:4075–4078. (c) Jafarpour, L, Heck, M-P, Baylon, C, Lee, HM, Mioskowski, C, Nolan, SP. *Organometallics*, 2002, 21, 671–679 (d) Kingsbury JS, Garber SB, Giftos JM, Gray BL, Okamoto MM, Farrer RA, Fourkas JT, Hoveyda AH. *Angew Chem Int Ed* 2001;40:4251–4256. (e) Grela, K, Tryznowski, M ; Bieniek, M. *Tetrahedron Lett*, 2002, 43, 9055–9559 (f) Michrowska A, Mennecke K, Kunz U, Kirsching A, Grela K. *J Am Chem Soc* 2006;128:13261–13267. (g) Borré E, Rouen M, Laurent I, Magrez M, Caijo F, Crévisy C, Solodenko W, Toupet L, Frankfurter R, Vogt C, Kirschning A, Mauduit M. *Chem Eur J* 2012;18:16369–16382.
21. Nieczypor P, Buchowicz W, Meester WJN, Rutjes FPJT. *Tetrahedron Lett* 2001;42:7103–7105.
22. (a) Krause JO, Lubbad SH, Nuyken O, Buchmeiser MR. *Macromol Rapid Commun* 2003;24:875–878. (b) Yang L, Mayr M, Wurst K, Buchmeiser MR. *Chem Eur J* 2004;10: 5761–5770. (c) Krause JO, Nuyken O, Wurst K, Buchmeiser MR. *Chem Eur J* 2004;10:777–784. (d) Halbach TS, Mix S, Fischer D, Maechling S, Krause JO, Sievers C, Blechert S, Nuyken O, Buchmeiser MR. *J Org Chem* 2005;70:4687–4694.
23. Mingotaud A-F, Krämer M, Mingotaud C. *J Mol Catal A* 2007; 263:39–47.
24. Mayr M, Mayr B, Buchmeiser MR. *Angew Chem Int Ed* 2001; 40:3839–3842.
25. Mayr M, Wang D, Kröll R, Schuler N, Prühs S, Fürstner A, Buchmeiser MR. *Adv Synth Catal* 2005;347:484–492.
26. (a) Allen DP, van Wingerden MM, Grubbs RH. *Org Lett* 2009;11:1261–1264. For previously reported similar works see: (b) Prühs S, Lehmann CW, Fürstner A. *Organometallics* 2004;23:280–287. (c) Mayr M, Buchmeiser MR, Wurst K. *Adv Synth Catal* 2002;344:712–719.
27. (a) Michrowska A, Gułajski Ł, Kaczmarska Z, Mennecke K, Kirschning A, Grela K. *Green Chem* 2006;8: 685–688. (b) Michrowska A, Gułajski Ł, Grela K. *Chim Oggi* 2006;24:19–22. (c) Gułajski Ł, Michrowska A, Narożnik J, Kaczmarska Z, Rupnicki L, Grela K. *ChemSusChem* 2008;1:103–109.
28. (a) Rix D, Clavier H, Coutard Y, Gułajski Ł, Grela K, Mauduit M. *J Organomet Chem* 2006;691:5397–5405.For an early work on ionic liquid-tagged ruthenium metathesis catalyst, see: (b) Audic N, Clavier H, Mauduit M, Guillemin J-C. *J Am Chem Soc* 2003;125:9248–9249.
29. Similar efficiency of Ru removal was observed for magnetic separation of magnetic nanoparticles-supported 20: Byrnes MJ, Hilton AM, Woodward CP, Jackson WR, Robinson AJ. *Green Chem* 2012;14:81–84.

30. (a) Clavier H, Audic N, Mauduit M, Guillemin J-C. *Chem Commun* 2004:2282–2283. (b) Clavier H, Audic N, Mauduit M, Guillemin J-C. *J Organomet Chem* 2005;690: 3585–3599.For application of catalyst 23 in continuous flow mode see: (c) Duque R, Öchsner E, Clavier H, Caijo F, Nolan SP, Mauduit M, Cole-Hamilton DJ. *Green Chem* 2011;13:1187–1195.

31. Yao Q, Sheets M. *J Organomet Chem* 2005;690:3577–3584.

32. Hong SH, Grubbs RH. *J Am Chem Soc* 2006;128:3508–3509.

33. Hong SH, Grubbs RH. *Org Lett* 2007;9:1955–1957.

34. Skowerski K, Wierzbicka C, Szczepaniak G, Gułajaski Ł, Bieniek M, Grela K. *Green Chem* 2012;14:3264–3268 These catalysts are commercially available on demand from Apeiron Synthesis S.A. (info@apeiron-synthesis.com).

35. Nicola T, Brenner M, Donsbach K, Kreye P. *Org Process Res Dev* 2005;9:513–515.

36. Gallou F, Saim S, Koenig KJ, Bochniak D, Horhota ST, Yee NK, Senanayake CH. *Org Process Res Dev* 2006;10:937–940.

37. Yee NK, Farina V, Houpis IN, Haddad N, Frutos RP, Gallou F, Wang X-J, Wei X, Simpson RD, Feng X, Fuchs V, Xu Y, Tan J, Zhang L, Xu J, Smith-Keenan LL, Vitous J, Ridges MD, Spinelli EM, Johnson M, Donsbach K, Nicola T, Brenner M, Winter E, Kreye P, Samstag W. *J Org Chem* 2006;71: 7133–7145.

38. (a) Shu C, Zeng X, Hao M-H, Wei X, Yee NK, Busacca CA, Han Z, Farina V, Senanayake CH. *Org Lett* 2008;10: 1303–1306. (b) Farina V, Shu C, Zeng X, Wei X, Han Z, Yee NK, Senanayake CH. *Org Process Res Dev* 2009;13:250–254.

39. (a) Wang H, Matsuhashi H, Doan BD, Goodman SN, Ouyang X, Clark WM Jr, *Tetrahedron* 2009;65:6291–6303. For previous work concerning SB-462795 see: (b) Yamashita DS, Marquis RW, Xie R, Nidamarthy SD, Oh H-J, Jeong JU, Erhard KF, Ward KW, Roethke TJ, Smith BR, Cheng H-Y, Geng X, Lin F, Offen PH, Wang B, Nevins N, Head MS, Haltiwanger RC, Sarjeant AAN, Liable-Sands LM, Zhao B, Smith WW, Janson CA, Gao E, Tomaszek T, McQueney M, James IE, Gress CJ, Zembryki DL, Lark MW, Veber DF. *J Med Chem* 2006;49: 1597–1612.

40. Löber A, Verch A, Schlemmer B, Höfer S, Frerich B, Buchmeiser MR. *Angew Chem Int Ed* 2008;47:9138–9141.

41. (a) Lambeth RH, Pederson SJ, Baranoski M, Rawlett AM. *J Polym Sci A* 2010;48:5752–5757. (b) Lambeth RH, Dougherty JM, Orlicki JA, Rawlett AM, Hoffman RC, Pritchett T, Mott AG. U.S. Army Research Laboratory report. http://www.dtic.mil/dtic/tr/fulltext/u2/a549234.pdf. Accessed 2013 May 13.

42. Mühlebach A, Bernhard P, Bühler N, Karlen T, Ludi A. *J Mol Catal* 1994;90:143–156.

43. (a) Staub H, Guillet-Nicolas R, Even N, Kayser L, Kleitz F, Fontaine F-G. *Chem Eur J* 2011;17:4254–4265. (b) Van Berlo B, Houthoofd K, Sels BF, Jacobs PA. *Adv Synth Catal* 2008;350:1949–1953. (c) Balcar H, Shinde T, Žilková N, Bastl Z. *Beilstein J Org Chem* 2011;7:22–28. (d) Martín-Aranda RM, Čejka J. *Top Catal* 2010;53:141–153. (e) Bek D, Žilková N, Dědeček J, Sedláček J, Balcar H. *Top Catal* 2010;53:200–209. (f) Balcar H, Bek D, Sedláček J, Dědeček J, Bastl Z, Lamač M. *J Mol Catal A: Chem* 2010;332:19–24. (g) Mwangi MT, Runge MB, Bowden NB. *J Am Chem Soc* 2006;128:14434–14435.

44. Cabrera J, Padilla R, Bru M, Lindner R, Kageyama T, Wilckens K, Balof SL, Schanz H-J, Dehn R, Teles JH, Deuerlein S, Müller K, Rominger F, Limbach M. *Chem Eur J* 2012;18: 14717–14724.

PART III

TABLES AND CHARTS

SECTION 6

OLEFIN METATHESIS CATALYSTS—A TABULAR REVIEW

ŁUKASZ WOŹNIAK AND GRZEGORZ ZIELIŃSKI
Institute of Organic Chemistry, Polish Academy of Sciences, Warsaw, Poland

KAROL GRELA
Institute of Organic Chemistry, Polish Academy of Sciences, Warsaw, Poland; Biological and Chemical Research Center, Department of Chemistry, Warsaw University, Warsaw, Poland

TABLE 1 List of Olefin Metathesis Catalysts (Sorted by Dealer)

No.	Structure	MW	CAS No.	Catalyst Trade Name or Code	Available from	References
1	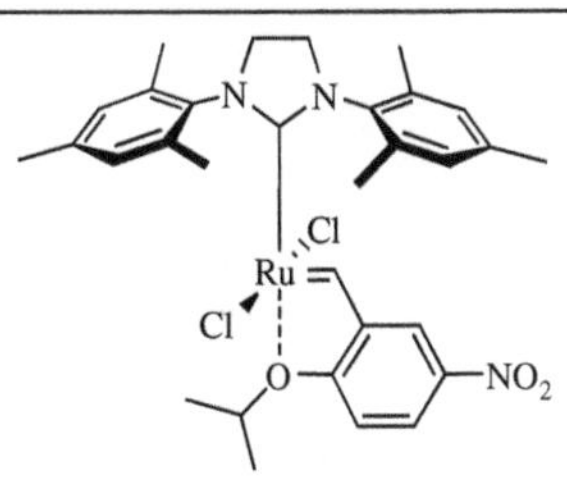	671.63	502964-52-5	nitro-Grela Catalyst AS2032	Apeiron Catalysts	(1)
2	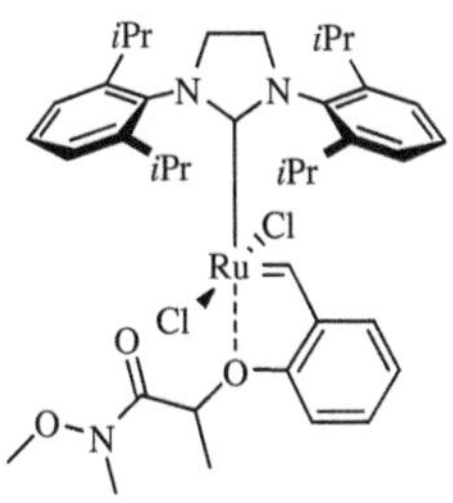	783.84	1448663-06-6	GreenCat AS2034	Apeiron Synthesis	(2)
3	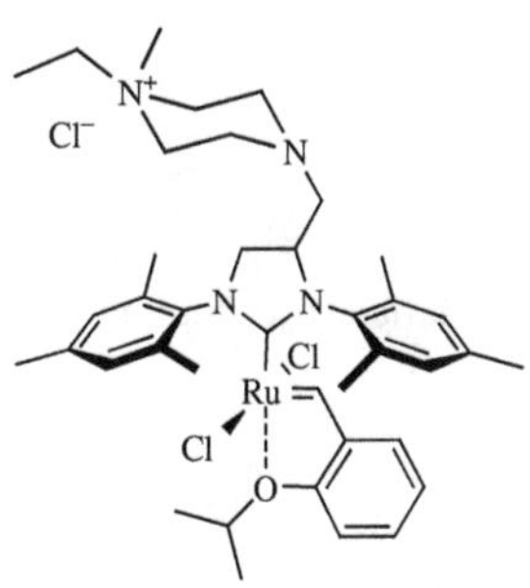	803.31	144707-08-6	AquaCat AS2038	Apeiron Synthesis	(2)

(continued)

TABLE 1 *(Continued)*

No.	Structure	MW	CAS No.	Catalyst Trade Name or Code	Available from	References
4	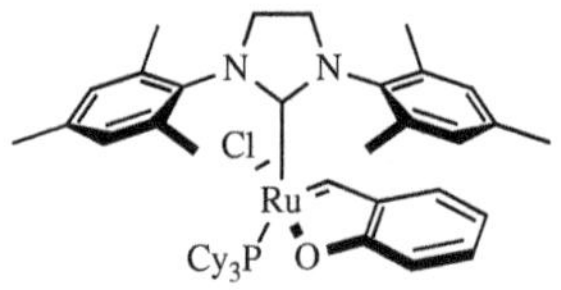	829.53	1407229-58-6	LatMet AS2035	Apeiron Synthesis	(2)
5	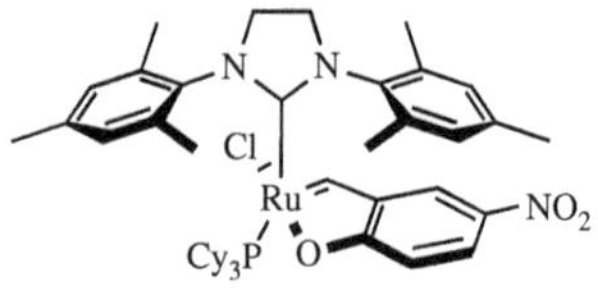	874.52	Not assigned	nitro-LatMet AS2036	Apeiron Synthesis	(2)
6	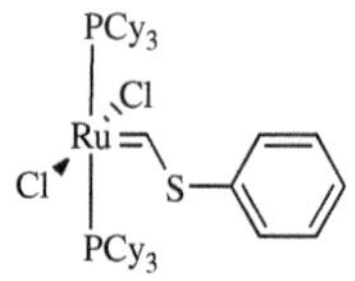	855.02	219770-99-7	44-0073	BASF/Strem	(3,4)
7	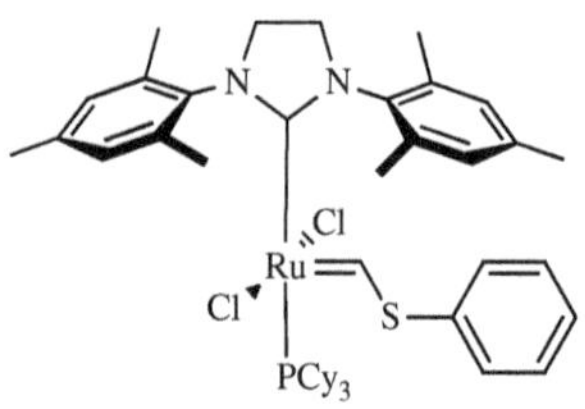	881.04	1155422-69-7	44-7780	BASF/Strem	(3,4)
8	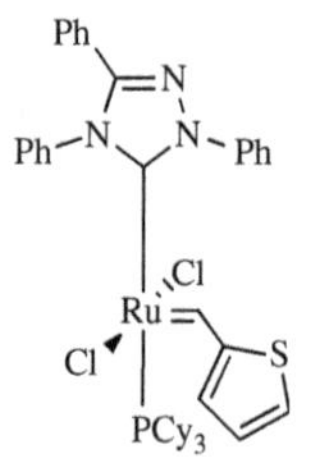	845.91	1190427-51-0	catMETium® RF 4/44-7795	Evonik Industries/Strem	(4,5)
9	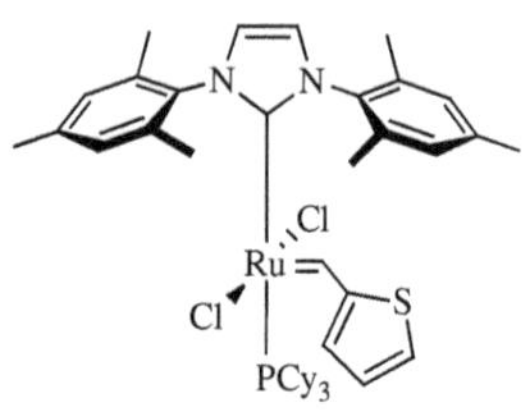	852.98	1190427-49-6	catMETium® RF 2/44-7785	Evonik Industries/Strem	(4,5)
10	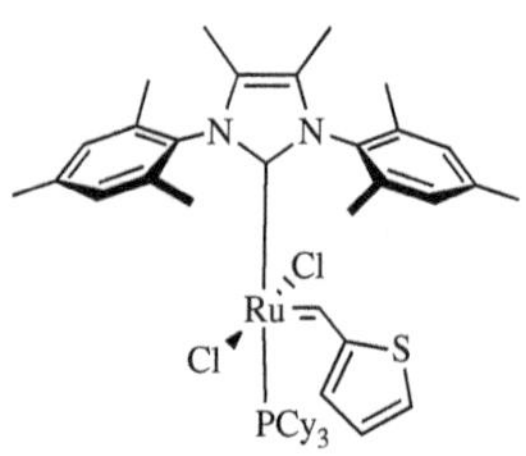	881.04	119427-50-9	catMETium® RF 3/44-7790	Evonik Industries/Strem	(4,5)
11	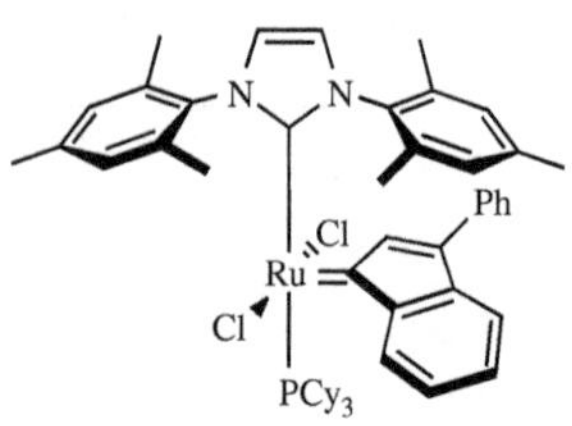	947.07	254972-49-1	catMETium® RF 1/44-7775	Evonik Industries/Strem	(4,5)

TABLE 1 (*Continued*)

No.	Structure	MW	CAS No.	Catalyst Trade Name or Code	Available from	References
12	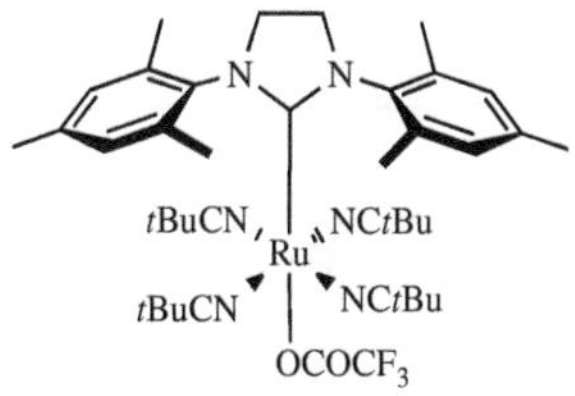	966.08	1034198-65-6	44-7712	ITCF/Strem	(4,6)
13	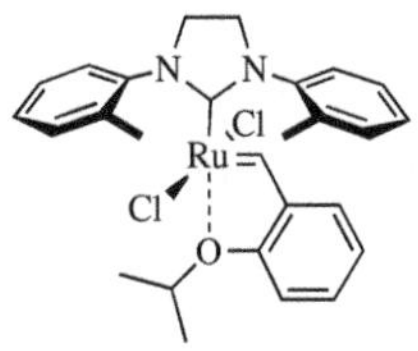	570.52	927429-61-6	C571/682373	Materia/Sigma-Aldrich	(7,8)
14	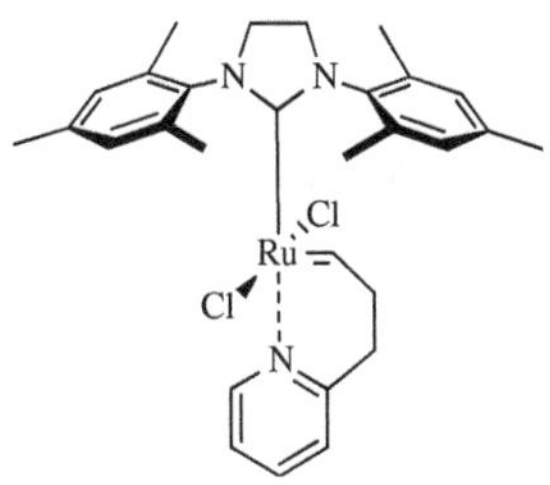	597.58	802912-44-3	C598/682381	Materia/Sigma-Aldrich	(7,8)
15	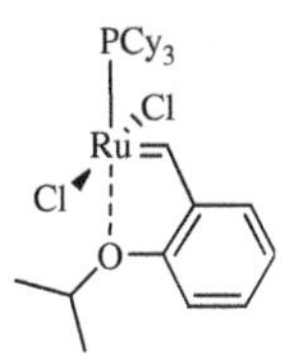	600.61	203714-71-0	C601/577944 Hoveyda–Grubbs Catalyst, first generation	Materia/Sigma-Aldrich	(7,8)
16	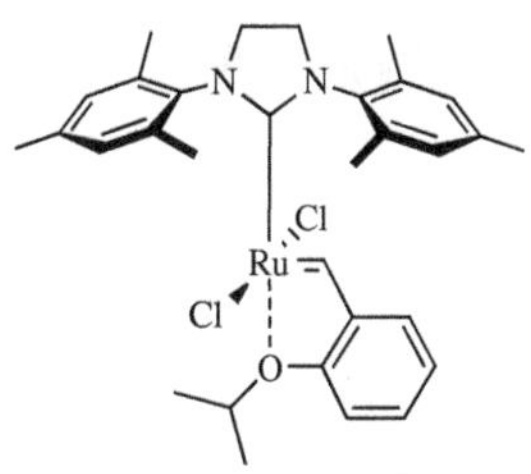	626.62	301224-40-8	C627/569755 Hoveyda–Grubbs Catalyst, second generation	Materia/Sigma-Aldrich	(7,8)
17	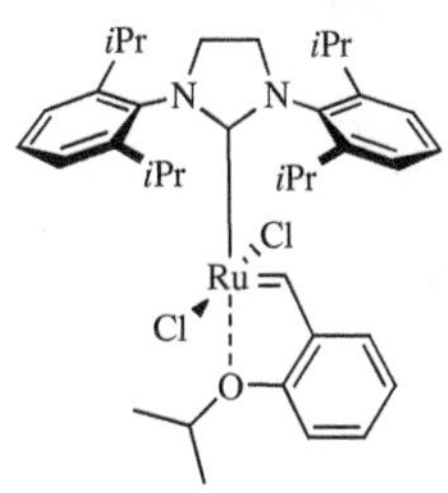	710.78	635679-24-2	C711/729345	Materia/Sigma-Aldrich	(7,8)
18	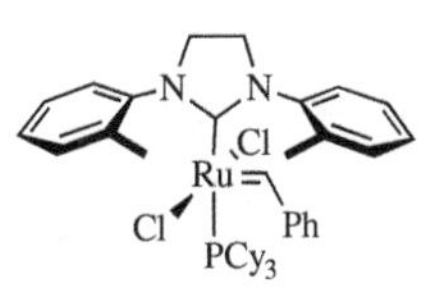	792.87	927429-60-5	C793/682284	Materia/Sigma-Aldrich	(7,8)

(*continued*)

TABLE 1 (*Continued*)

No.	Structure	MW	CAS No.	Catalyst Trade Name or Code	Available from	References
19	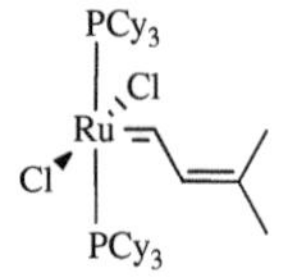	800.95	194659-03-5	C801/578703	Materia/Sigma-Aldrich	(7,8)
20	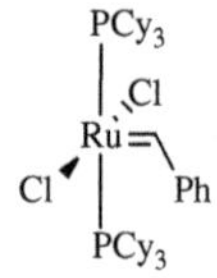	822.96	172222-30-9	C823/579726 Grubbs Catalyst, first generation	Materia/Sigma-Aldrich	(7,8)
21	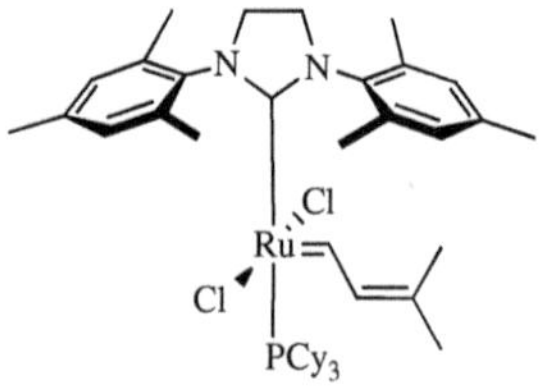	826.97	253688-91-4	C827/682365	Materia/Sigma-Aldrich	(7,8)
22	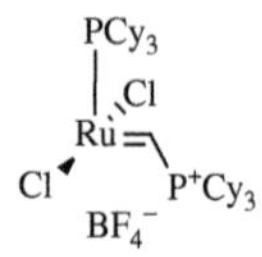	832.66	1020085-61-3	C833/707961 Piers-Grubbs catalyst, first generation	Materia/Sigma-Aldrich	(7,8)
23	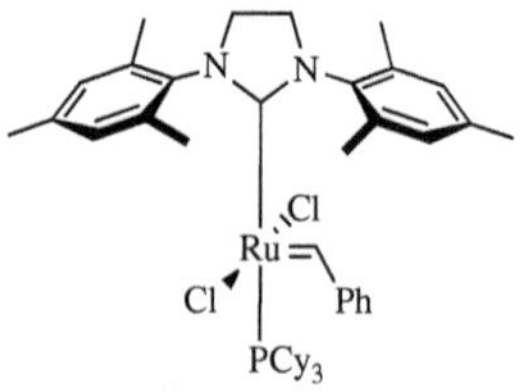	848.97	246047-72-3	C848/569747 Grubbs Catalyst, second generation	Materia/Sigma-Aldrich	(7,8)
24	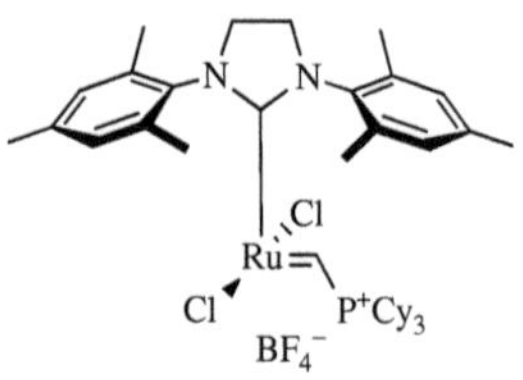	858.67	832146-68-6	C833/707988 Piers-Grubbs catalyst, second generation	Materia/Sigma-Aldrich	(7,8)
25	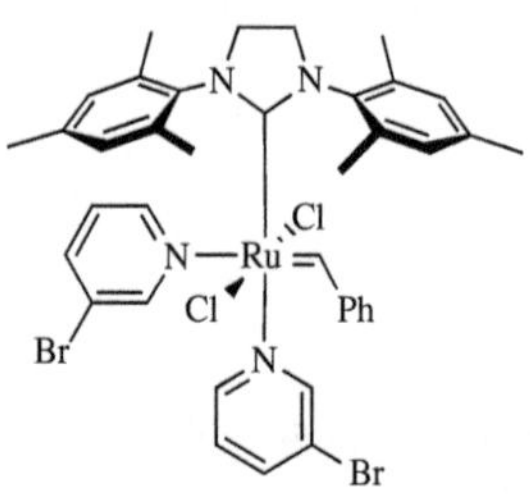	884.54	900169-53-1	C884/682330 Grubbs Catalyst, third generation	Materia/Sigma-Aldrich	(7,8)

TABLE 1 (*Continued*)

No.	Structure	MW	CAS No.	Catalyst Trade Name or Code	Available from	References
26	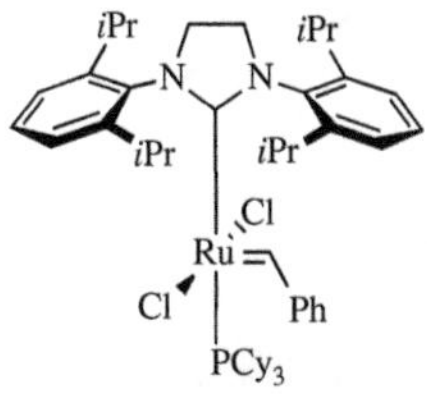	933.13	373640-75-6	C933/729353	Materia/Sigma-Aldrich	(7,8)
27	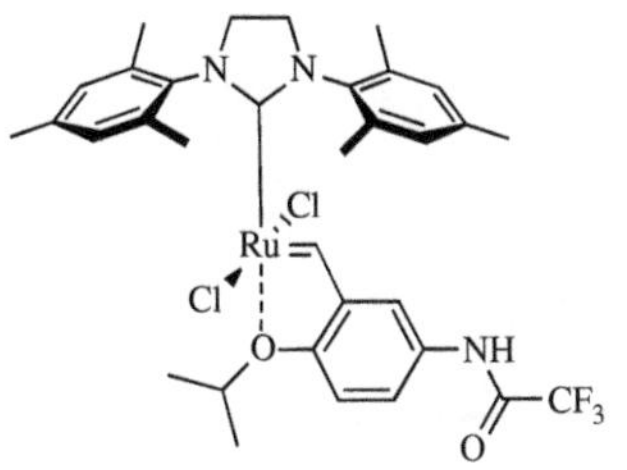	737.64	1025728-56-6	M71-SIMes/ Umicore M71 SIMes	Oméga Cat System/Umicore	(9,10)
28	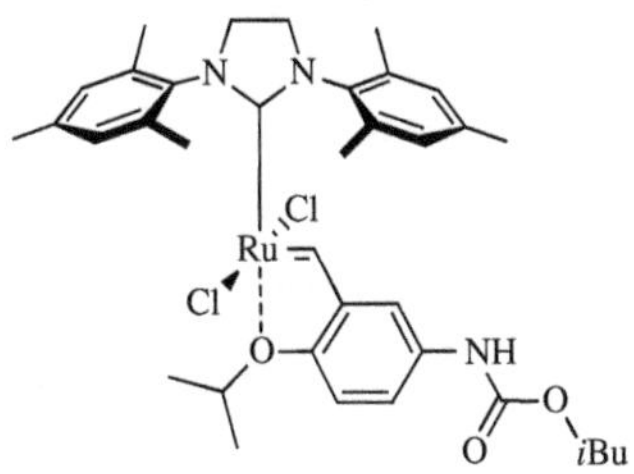	741.75	1025718-57-7	M73-SIMes/ Umicore M73 SIMes	Oméga Cat System/Umicore	(9,10)
29	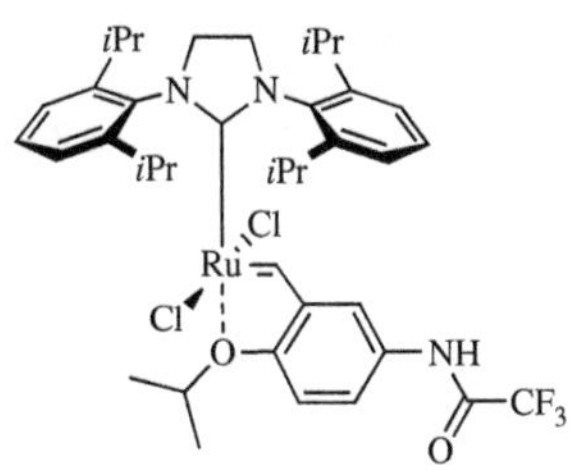	821.80	1212008-99-5	M71-SIPr/44-0055/ Umicore M71 SIPr	Oméga Cat System/Strem/ Umicore	(4,9,10)
30	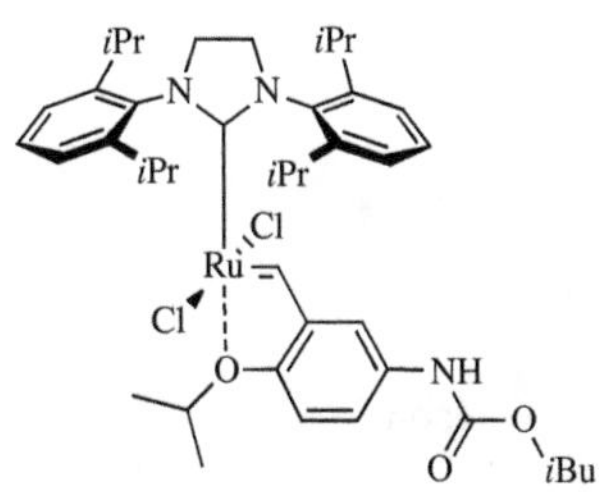	825.91	1212009-05-6	M73-SIPr/ Umicore M73 SIPr	Oméga Cat System/Umicore	(9,10)
31	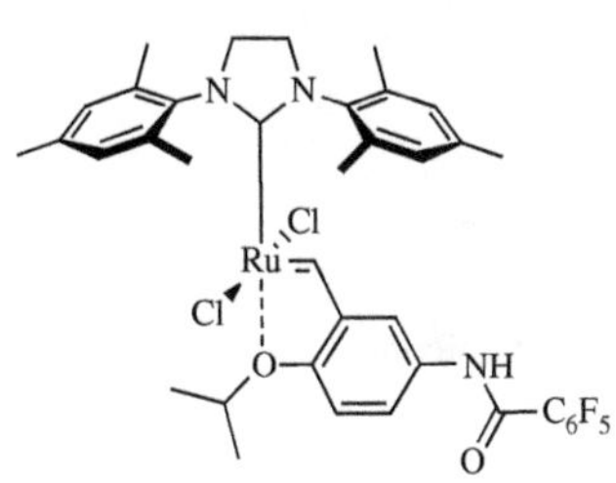	835.70	1030618-02-0	M72-SIMes/ Umicore M72 SIMes	Oméga Cat System/Umicore	(9,10)

(*continued*)

TABLE 1 ***(Continued)***

No.	Structure	MW	CAS No.	Catalyst Trade Name or Code	Available from	References
32	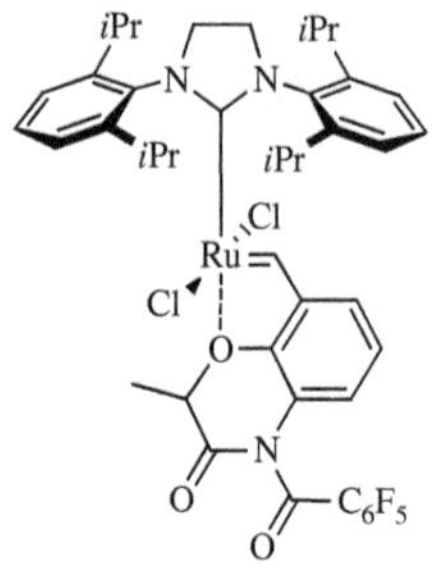	931.82	1357191-65-1	M813	Oméga Cat System	(9)
33	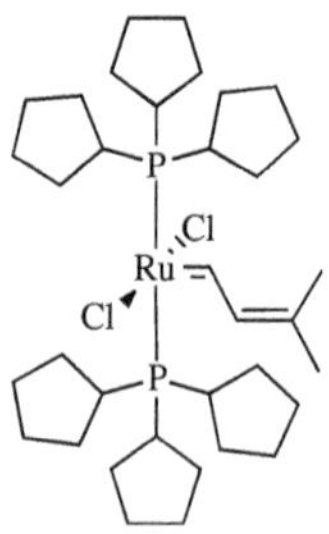	716.79	220883-08-9	578703	Sigma-Aldrich	(8)
34	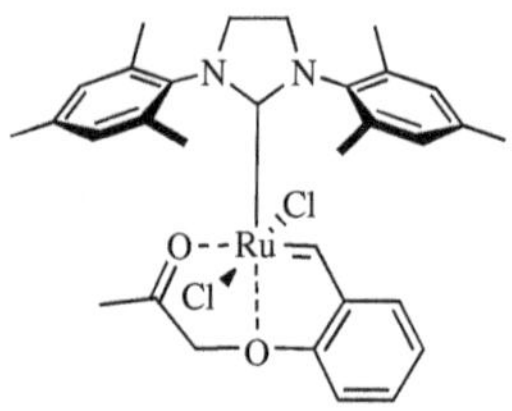	640.12	1014701-61-1	Umicore M52	Umicore	(10)
35	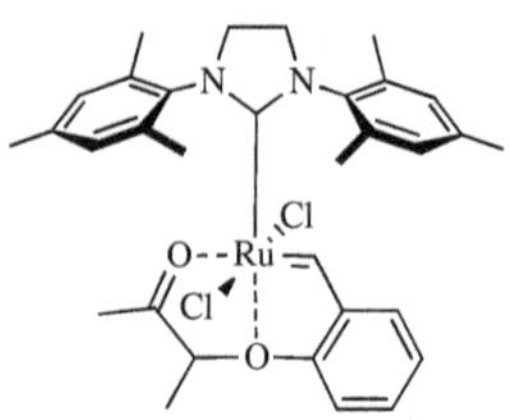	654.63	1031262-71-1	Umicore M51	Umicore	(10)
36	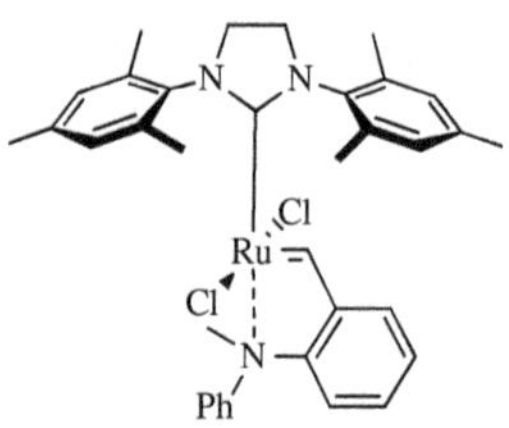	674.16	1416427-09-2	Umicore M61 SIMes	Umicore	(10)
37	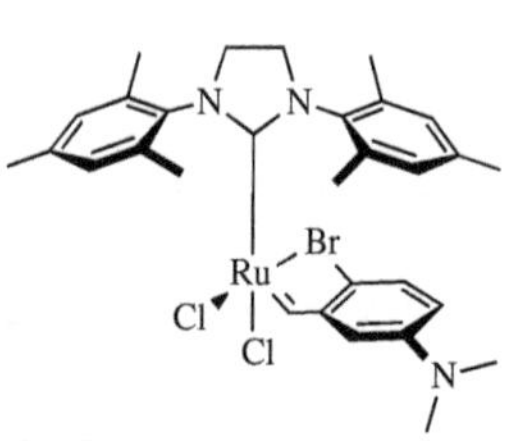	690.06	1415725-68-6	Umicore M93	Umicore	(10)

TABLE 1 (*Continued*)

No.	Structure	MW	CAS No.	Catalyst Trade Name or Code	Available from	References
38	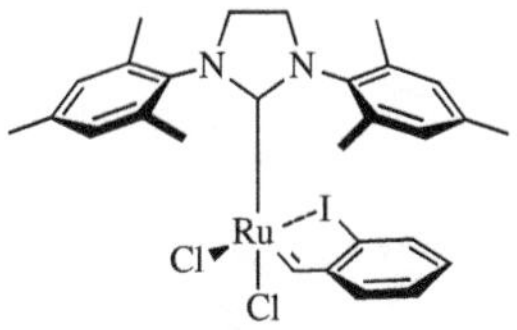	695.00	1415725-62-0	Umicore M91	Umicore	(10)
39	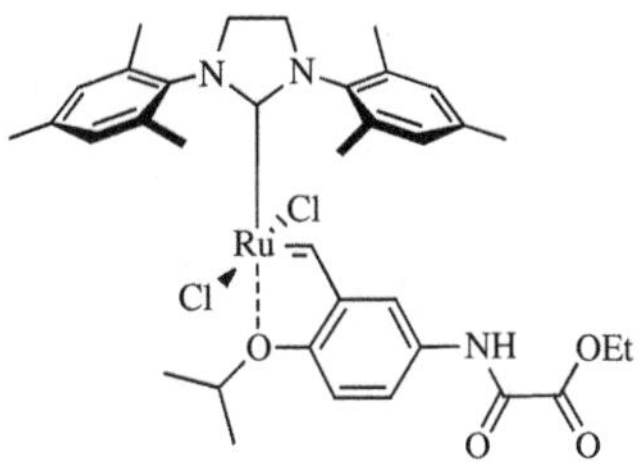	741.84	1030618-11-1	Umicore M74 SIMes	Umicore	(10)
40	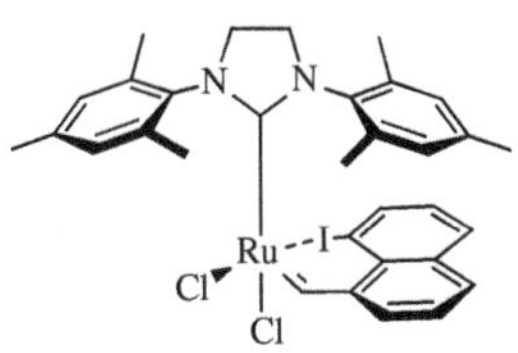	745.02	1417525-73-3	Umicore M92	Umicore	(10)
41	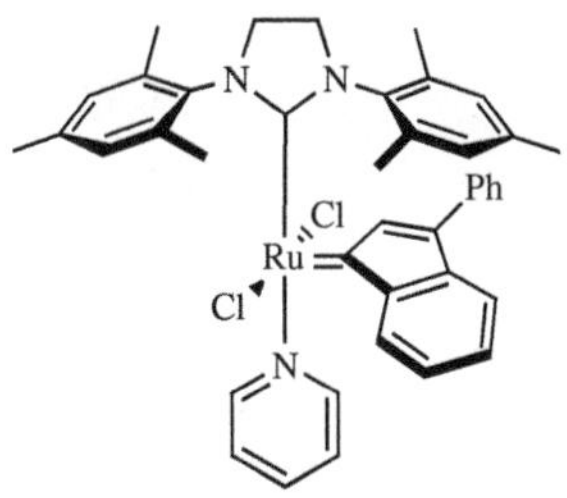	747.76	1031262-76-6	Umicore M31	Umicore	(10)
42	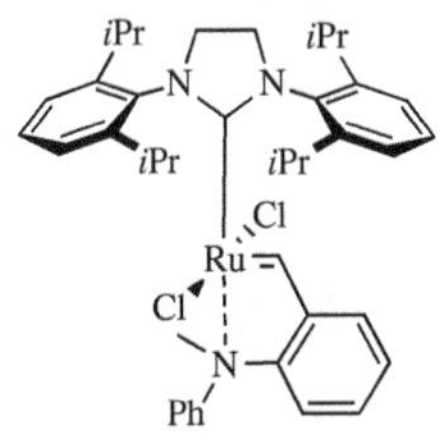	758.26	1416427-12-7	Umicore M61 SIPr	Umicore	(10)
43	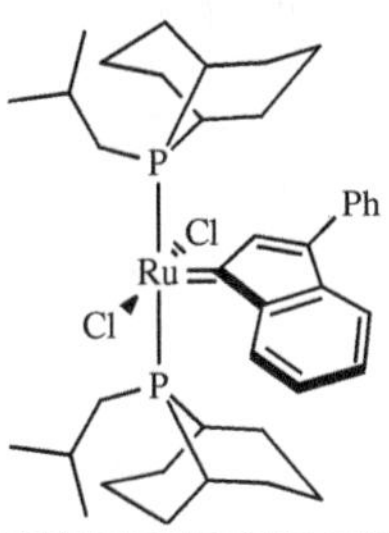	758.78	894423-99-5	Umicore M11/44-7778	Umicore/Strem	(4,10)

(*continued*)

TABLE 1 ***(Continued)***

No.	Structure	MW	CAS No.	Catalyst Trade Name or Code	Available from	References
44	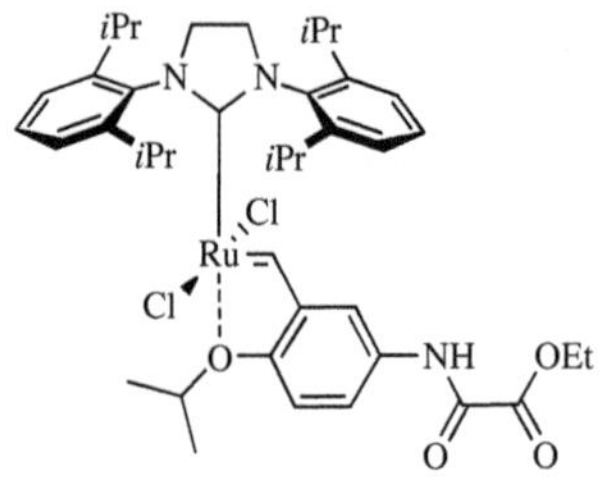	825.88	1212009-08-9	Umicore M74 SIPr	Umicore	(10)
45	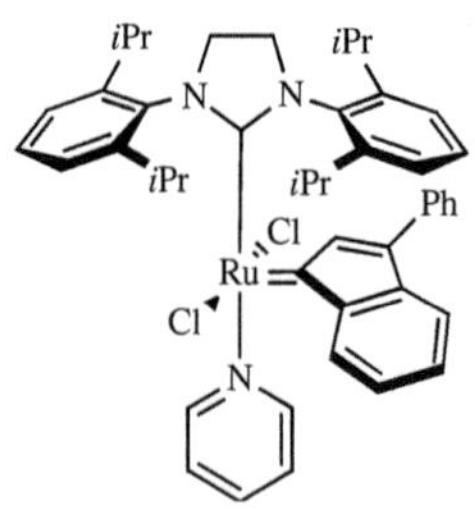	831.93	1304756-39-5	Umicore M32	Umicore	(10)
46	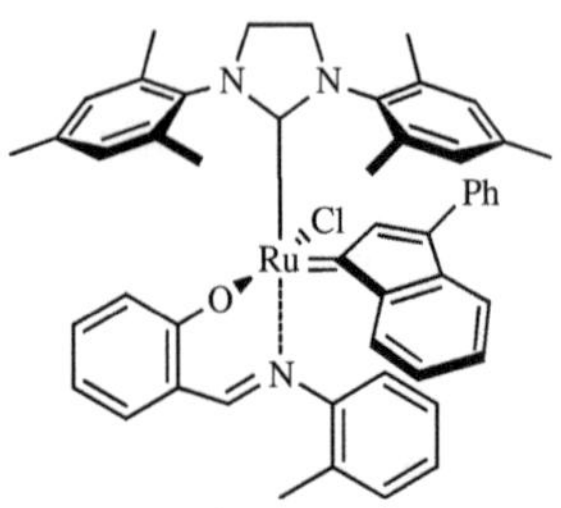	843.46	934538-12-2	Umicore M42/44-0049	Umicore/Strem	(4,10)
47	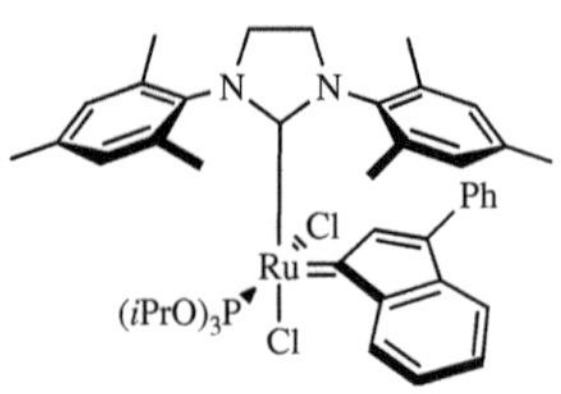	876.89	1255312-90-3	Umicore M22/44-7783	Umicore/Strem	(4,10)
48	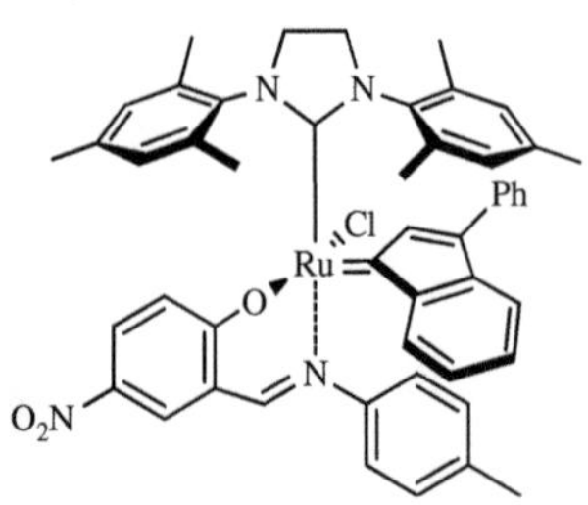	888.47	934538-04-2	Umicore M41/44-0047	Umicore/Strem	(4,10)
49	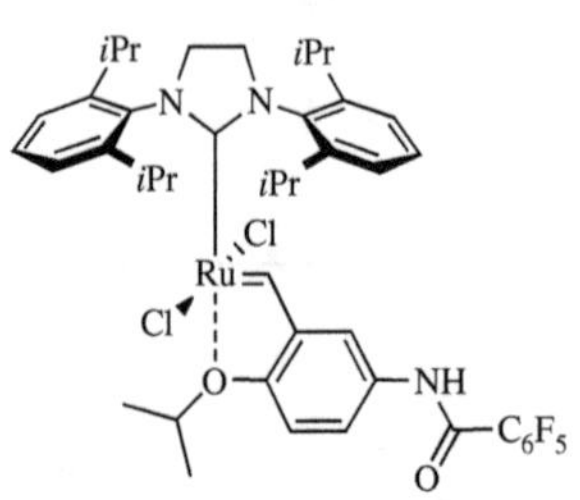	919.87	1354710-90-9	Umicore M72 SIPr	Umicore	(10)

TABLE 1 (*Continued*)

No.	Structure	MW	CAS No.	Catalyst Trade Name or Code	Available from	References
50	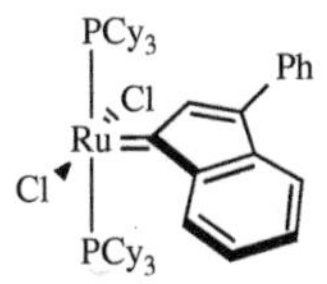	923.03	250220-36-1	Umicore M1/44-0063	Umicore/Strem	(4,10)
51	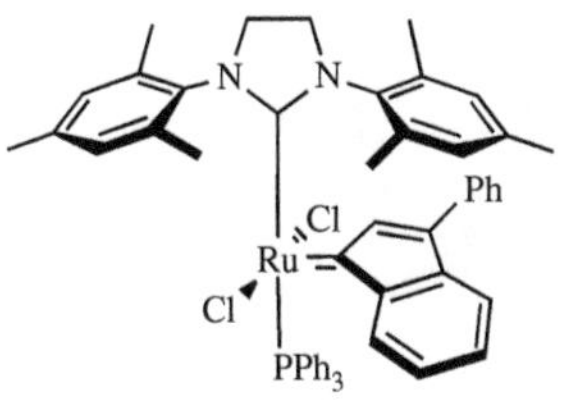	930.96	340810-50-6	Umicore M20	Umicore	(10)
52	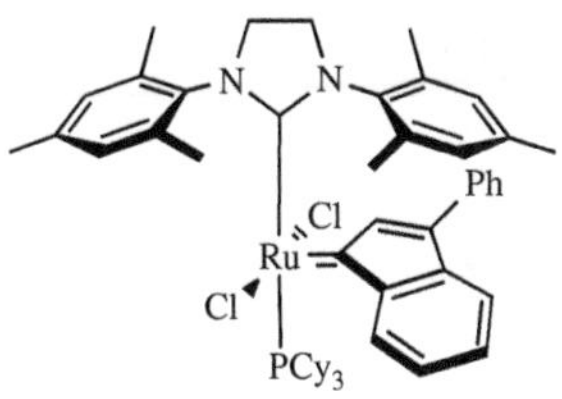	949.09	536724-67-1	Umicore M2/44-7777	Umicore/Strem	(4,10)
53	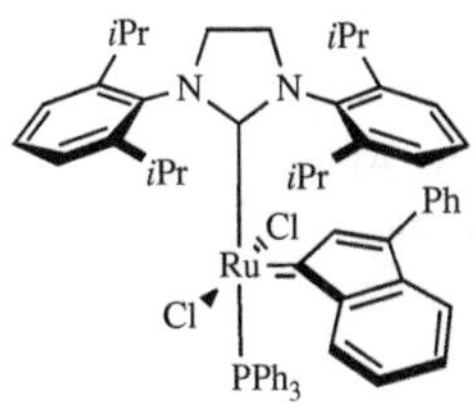	1015.21	1307233-23-3	Umicore M23	Umicore	(10)
54	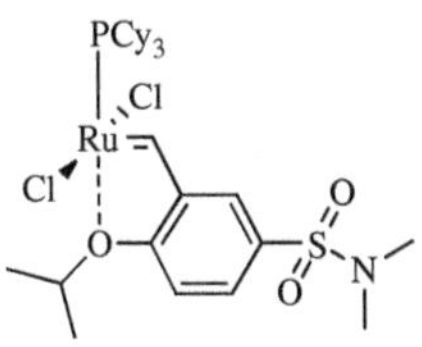	707.74	918871-44-0	RC-304/44-0078	Zannan/Strem	(4,11)
55	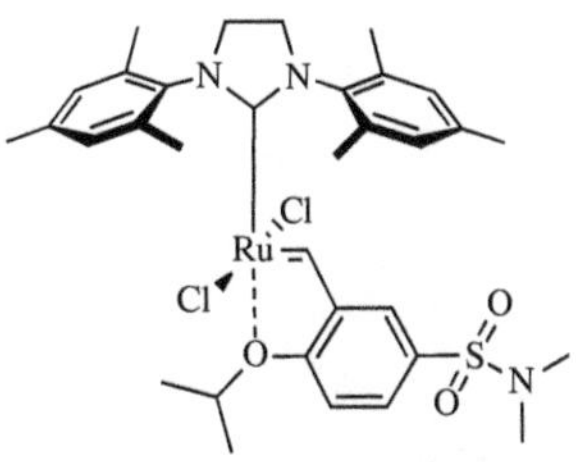	733.75	918870-76-5	RC-303/44-0082	Zannan/Strem	(4,11)
56	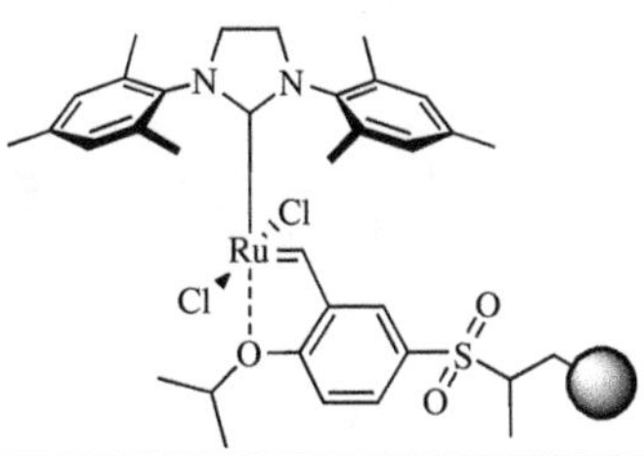	>1000	Not assigned	RC-402/44-0078	Zannan/Strem	(4,11)

(*continued*)

TABLE 1 ***(Continued)***

No.	Structure	MW	CAS No.	Catalyst Trade Name or Code	Available from	References
57	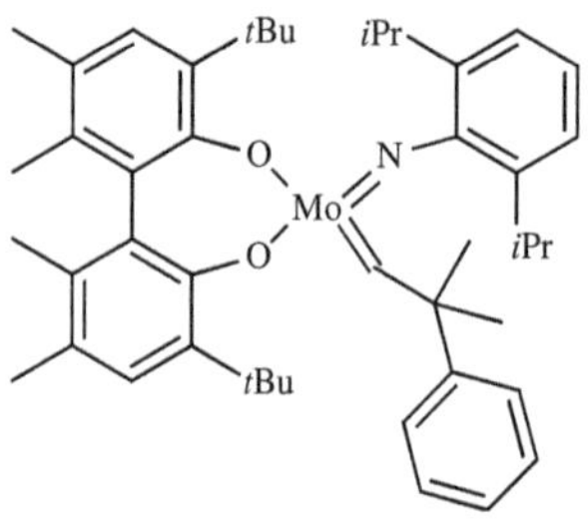	755.92	300344-02-9	42-1212 (rac)-Schrock-Hoveyda Catalyst	Strem	(4)
58	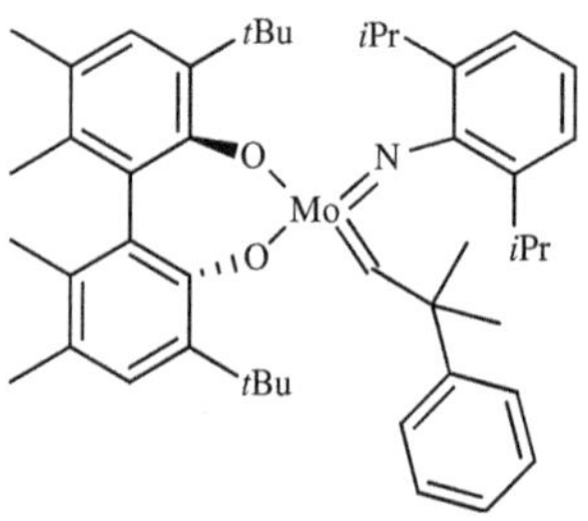	755.92	329735-77-5	42-1213 (R)-Schrock-Hoveyda Catalyst	Strem	(4)
59	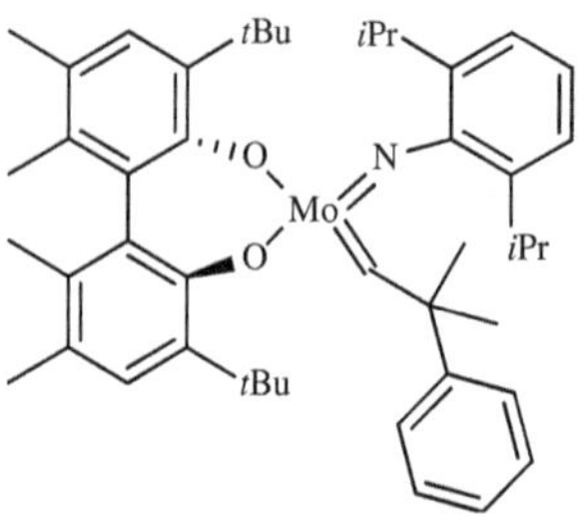	755.92	205815-80-1	73022/42–1214 (S)-Schrock-Hoveyda Catalyst	Sigma-Aldrich/ Strem	(4,8)
60	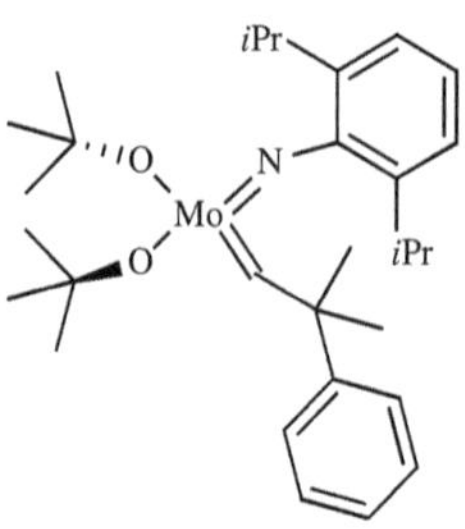	549.65	126949-65-3	42-1200	Strem	(4)
61	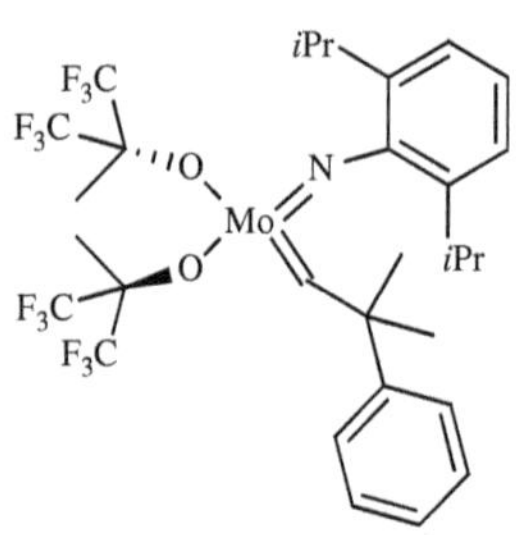	765.53	139220-25-0	42-1205 Schrock Catalyst	Strem	(4)
62	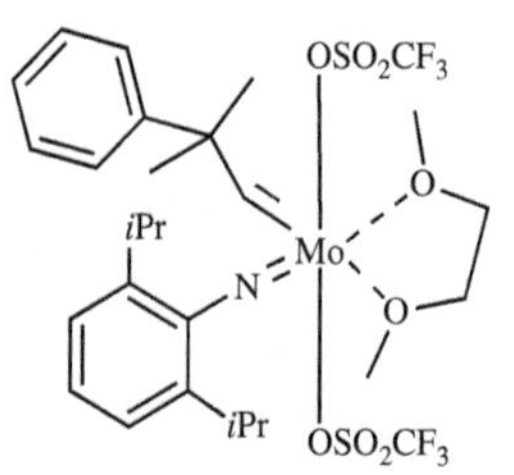	791.68	126949-63-1	42-1210 (precursor for Schrock Catalyst)	Strem	(4)

Recently, also *cis*-selective Grubbs catalyst (CAS Number 1352916-84-7; see structure **49e**, Figure 16.19, Chapter 16) become commercially available from Sigma-Aldrich (catalogue no. 771082).

REFERENCES

1. www.apeiron-catalysts.com. Accessed 2013 Dec 09.
2. www.apeiron-synthesis.com. Accessed 2013 Dec 09.
3. www.basf.com. Accessed 2013 Dec 09.
4. www.strem.com. Accessed 2013 Dec 09.
5. www.evonik.com. Accessed 2013 Dec 09.
6. www.itcf-denkendorf.de. Accessed 2013 Dec 09.
7. www.materia-inc.com. Accessed 2013 Dec 09.
8. www.sigmaaldrich.com. Accessed 2013 Dec 09.
9. www.omcat-system.com. Accessed 2013 Dec 09.
10. www.umicore.com. Accessed 2013 Dec 09.
11. www.zannanpharma.com. Accessed 2013 Dec 09.

OUTLOOK AND PERSPECTIVES

KAROL GRELA
Faculty of Chemistry, University of Warsaw, Warszawa, Poland

In the last decade of the twentieth century, the olefin metathesis reaction gained real significance in advanced organic synthesis and chemical production (1). The development of well-defined catalysts and understanding of the reaction mechanism prompted an extraordinary scientific turnaround, revolutionizing retrosynthetic planning in total syntheses and offering more economical production methods for industrial companies. The importance of "the development of the metathesis method in organic synthesis" has been recognized through the awarding of the Nobel Prize in Chemistry for 2005 jointly to Yves Chauvin, Robert H. Grubbs, and Richard R. Schrock.[i]

Currently, olefin metathesis (and its sister reaction, alkyne metathesis (2), which unfortunately had to be omitted in this book owing to limited space) is one of the most intensively studied transformations in synthetic organic chemistry. Despite its power and applicability, metathesis is still experiencing some limitations, such as troublesome formation of selected functionalized or crowded C–C double bonds, still not perfect stability, and sometimes low activity of existing catalysts, problems with removal of the catalyst after the reaction, and more.

In this book, some solutions to the above problems have been presented. The importance of the proper selection of the catalyst, addition of Lewis acids, effect of microwaves, and/or fluorinated aromatic solvents have been discussed thoroughly in this book. A well-thought-out optimization can sometimes save the whole project, as has been witnessed many times in the literature. Not only the use of sophisticated "enabling techniques" but even a change in the simplest reaction parameters such as temperature and concentration can significantly improve the metathesis outcome. On the other hand, the variety of metathesis catalysts now available on the market makes the proper choice of a catalyst a real problem. The data reported in the literature demonstrate that great care must be taken when choosing an appropriate catalyst for a given metathesis reaction. The book represents an attempt to provide some hints for the selection of the optimal catalysts.

New applications of metathesis methodology are being developed, including sustainable production of valuable chemicals from renewable sources. These advancements create even new challenges, such as conducting metathesis on air with ppm amounts of the catalysts, use of crude fractions of seed oil as a feedstock, etc. Gaining truly fullcontrol on E/Z geometry of the newly formed C–C double bond or preparing of an iron-based catalyst are another "holy gralis" in olefin metathesis. The importance of these challenges is now widely appreciated and more and more discoveries will surely follow.

REFERENCES

1. Grubbs RH, editor. *Handbook of Metathesis*. Vol. 1–3. Weinheim, Germany: Wiley-VCH; 2003.
2. Fürstner A. *Angew Chem Int Ed* 2013;52:2794–2819.

[i]For more details, see http://www.nobelprize.org/nobel_prizes/chemistry/laureates/2005/advanced-chemistryprize2005.pdf

Olefin Metathesis: Theory and Practice, First Edition. Edited by Karol Grela.

INDEX

Olefin Metathesis: Theory and Practice, First Edition. Edited by Karol Grela.

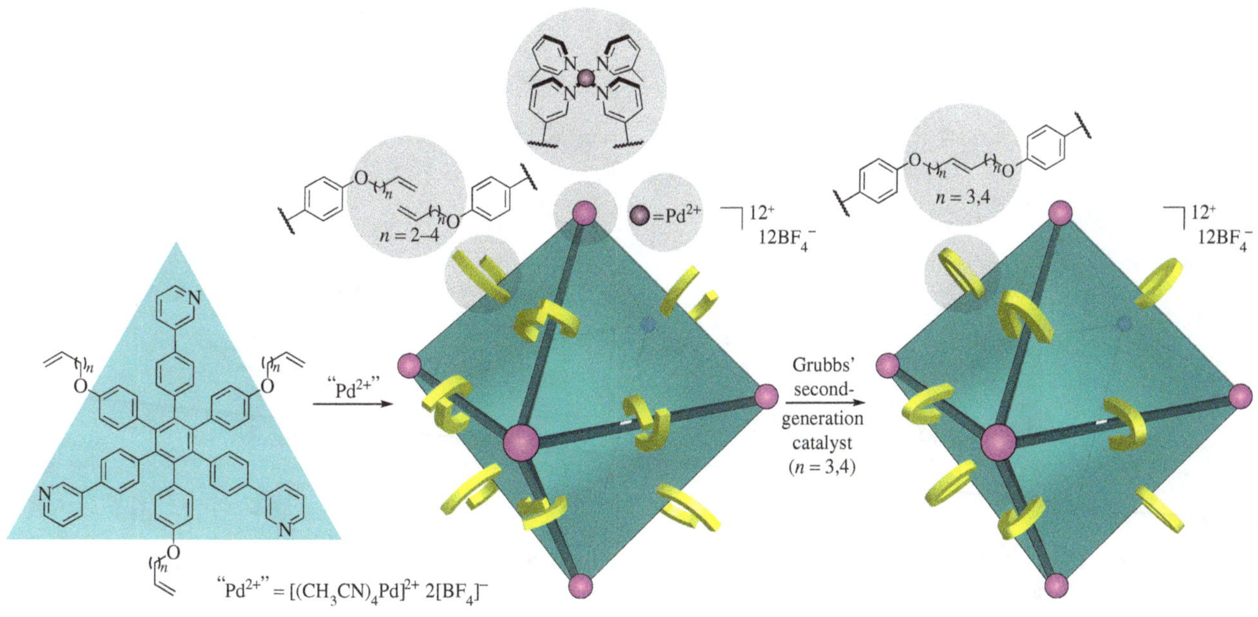

72

73
$[Pd_6\mathbf{72}_8]^{12+}$ $12[BF_4]^-$

74

Scheme 9.24

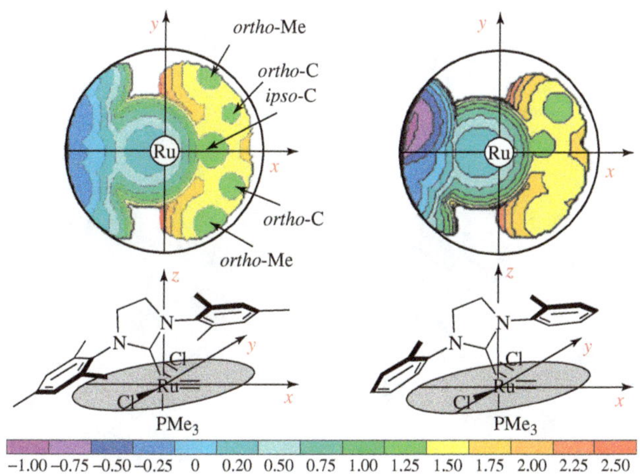

Figure 19.16 Topographic steric map of the NHC ligand of (pre)catalysts bearing a mesityl or a *o*-tolyl N-substituent. The isocontour levels (Å), are also reported.

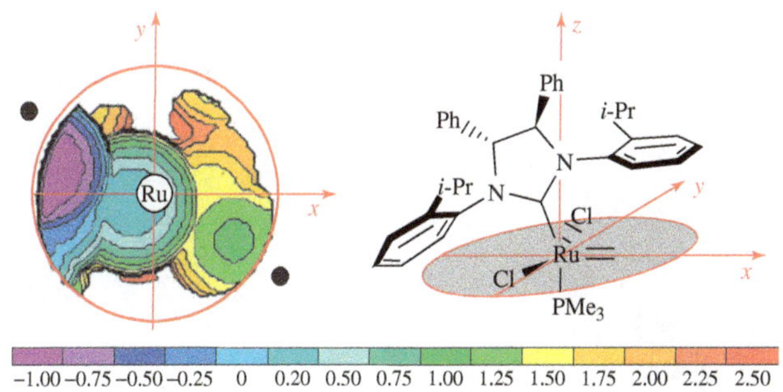

Figure 19.19 Steric map of the chiral NHC ligand is shown. The quadrants occupied by the *ortho-i*-Pr groups are indicated by a black dot. The isocontour levels (Å) are also reported.

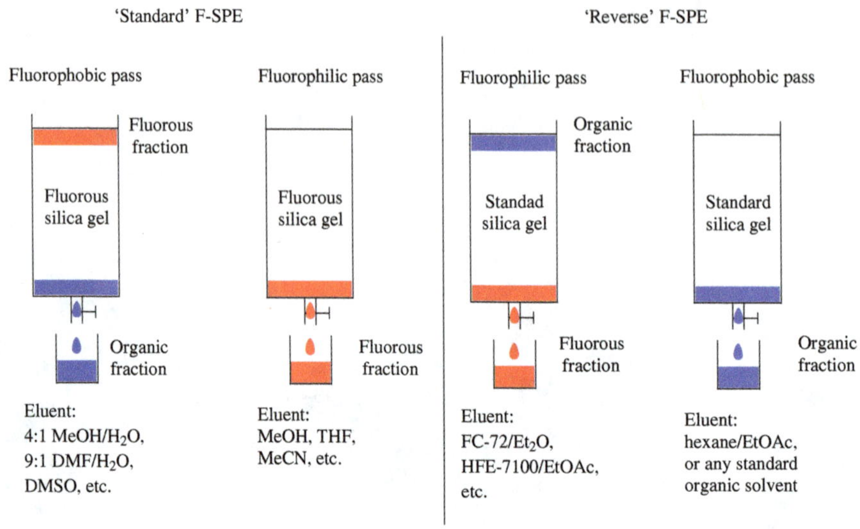

Figure 23.6 A visualization of a "standard" F-SPE and "reverse" F-SPE. The organic fraction is blue and the fluorous fraction is red.

www.ingramcontent.com/pod-product-compliance
Lightning Source LLC
Chambersburg PA
CBHW081608160925
32635CB00005B/362
* 9 7 8 1 1 1 8 2 0 7 9 4 9 *